D0202644

SHEEP & GOAT SCIENCE

(Animal Agriculture Series)

ABOUT THE AUTHOR

Marion Eugene Ensminger completed B.S. and M.S. degrees at the University of Missouri, and the Ph.D. at the University of Minnesota. Dr. Ensminger served, in order, as Manager of the Dixon Springs Agricultural Center (University of Illinois), Simpson, Illinois; and on the staffs of the University of Massachusetts, the University of Minnesota, and Washington State University. Dr. Ensminger also served as Consultant, General Electric Company, Nucleonics Department, and as the first President of the American Society of Agricultural Consultants. Since 1964, Dr. Ensminger has served as President of Agriservices Foundation, Clovis, California, a nonprofit foundation serving world agriculture in the area of World Food, Hunger, and Malnutrition.

Among Dr. Ensminger's many honors and awards are: Distinguished Teacher Award, American Society of Animal Science; the "Ensminger Beef Cattle Research Center" at Washington State University, Pullman, named after him in recognition of his contributions to the University; Faculty-Alumni Award of the University of Missouri; Outstanding Achievement Award of the University of Minnesota; Distinguished Service Award of the American Medical Association (with Mrs. Ensminger); Honorary Professor, Huazhong Agricultural College, Wuhan, China; Doctor of Laws (LL.D.) conferred by the National Agrarian University of Ukraine; and an oil portrait of him was placed in the 300-year-old gallery of the famed Saddle and Sirloin Club, Lexington, Kentucky.

In 1995, Cuba honored Dr. Ensminger by making him an Honorary Member of the Cuban Association of Animal Production; presenting him the 30th anniversary Gold Medal of the Institute of Animal Science, at Havana; making him an Honorary Guest Professor of the Agricultural University (ISCAH), at Havana; and making him an Honorary Guest Professor of the University of Camaguey, Camaguey, Cuba.

In 1995, Dr. Ensminger received the Distinguished Teacher Award, the highest honor of the National Association of Colleges and Teachers of Agriculture (NACTA).

In 1996, Iowa State University awarded Dr. Ensminger the honorary degree, Doctor of Humane Letters for "extraordinary achievements in animal science, education, and international agriculture."

In 1996, Dr. Ensminger was the recipient of the International Animal Agriculture Bouffault Award, Paris, France, and the American Society of Animal Science.

Dr. Ensminger founded the International Ag-Tech Schools, which he directed for more than 50 years. He has directed schools, lectured, and/or conducted seminars in 70 countries. Dr. Ensminger is the author of more than 500 scientific articles, bulletins, and feature articles; and he is the author or co-author of 22 books, which are in several languages and used all over the world. He waives all royalties on the foreign editions of his books in order to help the people. The whole world is Dr. Ensminger's classroom.

SHEEP & GOAT SCIENCE

(Animal Agriculture Series)

by

M. E. Ensminger, B.S., M.A., Ph.D.

SIXTH EDITION

INTERSTATE PUBLISHERS, INC.
Danville, Illinois

Editions:

First	1951
Second	1955
Third	1964
Fourth	1970
Fifth	1986
Sixth	2002

Library of Congress Control Number: 9679330

ISBN 0-8134-3116-6

1 2 3 4 5 6 7 8 9 10   06 05 04 03 02 01

Other books by M. E. Ensminger
available from Interstate Publishers:

Animal Science
Animal Science Digest
Beef Cattle Science (with R. Perry)
Dairy Cattle Science
Feeds & Nutrition (with J. Oldfield & W. Heinemann)
Feeds & Nutrition Digest (with J. Oldfield & W. Heinemann)
Horses and Horsemanship
Poultry Science
Stockman's Handbook, The
Swine Science

Animal Science presents a perspective or panorama of the far-flung livestock industry; whereas each of the other books presents specialized material pertaining to the specific class of farm animals indicated by its title.

Feeds & Nutrition and *Feeds & Nutrition Digest* bring together both the art and the science of livestock feeding, narrow the gap between nutrition research and application, and assure more and better animals in the current era of biotechnology.

The Stockman's Handbook presents the "why" as well as the "how." It contains, under one cover, the pertinent things that a stockman needs to know in the daily operation of a farm or ranch. It covers the broad field of animal agriculture, concisely and completely, and wherever possible in tabular and outline form.

Dedicated

to

Dr. Clair E. Terrill

former National Research Leader for Sheep,

U.S. Department of Agriculture,

whose dedicated efforts gave rise to a

new and modern sheep industry.

PREFACE TO THE SIXTH EDITION

The sheep and goat meat, milk, and fiber industries of the United States have been their own worst enemies. Currently, they are in disarray. In many measurements, the products rank high, yet in actual practice they have failed to measure up.

Since about 1960, predator losses have dominated the sheep industry, and, to a certain extent, the goat industry. United States predator losses increased from 2.3% of all sheep and lambs in 1940 to a peak of 5.8% in 1977, at which point they levelled out.

In 1977, *Sheep Industry News* reported losses of 520,000 sheep and lambs a year to predators. But more than predators are involved! Thus, it is noteworthy that the world's most ambitious predator control project is in Australia, where sheep producers wage constant war against the dingo, the wild dog of down under.

Note: The magnitude of the dingo control fence in Australia may be highlighted by comparing it with the 3,100 miles of The Great Wall of China. Yet, Australia is a great sheep country.

- As the cost of petroleum goes up, the position of wool becomes more favorable. It requires about 17.8 million calories to produce a pound of synthetic fibers compared to 8.3 million calories to produce a pound of wool.

- In 1993, the name of the sheep magazine was changed from *National Wool Grower* to *The National Lamb and Wool Grower*, better to reflect those whom it served.

- In 1995, The National Wool Act was phased out.

- In 1996, the check off (promotion) was voted on, and defeated, twice.

- In 1997, the last issue of *The National Lamb & Wool Grower* rolled off press, thereby ending an 86-year-old tradition.

But take heart! The following facts and figures have evolved in the sheep and goat lexicon:

- "Dyed-in-the-wool" means that its genuine.

- When people have been "fleeced," that means they have been swindled.

- To pull the wool over one's eyes is to fool them.

- Historically, a "spinster" referred to an unmarried woman of the family whose role was to spin wool.

Sheep and goat data are hard to come by. So, the author is especially grateful to the following authorities for their respective contribution to the sixth edition of *Sheep and Goat Science:*

Dr. Larry Deuwer
Dr. Terry Dockerty
Ken Johnson
Dr. Johannes E. Nels
Dr. Clair Terrill
Dr. David L. Thomas
USDA
U.S. Forest Service
U.S. Bureau of Land Management
Dr. Jim Wise, USDA, Standardization Branch
Audrey H. Ensminger
Randall and Susan Rapp

CONTENTS

PART III: GOATS

PART IV: GLOSSARY/FEED COMPOSITION TABLES

PART 1
SHEEP AND GOATS

Because of their useful products, productivity, small size, adaptability, and non-competitiveness with humans for food, sheep and goats were among the earliest animals domesticated.

Chapter 1 traces the origin and domestication of sheep and goats and discusses some of the changes that have occurred primarily in the United States.

Chapter 2 characterizes sheep and their products on a worldwide basis and describes U.S. sheep production, past and present, pointing out problems and potentials of the industry. Additionally, Chapter 2 briefly describes worldwide goat production.

Chapter 3 covers genetics and Chapter 4 covers reproduction. Both of these topics are of primary importance to sheep and goat production.

Chapter 5 covers the fundamentals of sheep and goat nutrition.

Chapter 6 contains valuable information on pasture forages. Both sheep and goats are good foragers.

Chapter 7 includes a discussion on the behavior and environment of sheep and goats.

Chapter 8 reviews the diseases and parasites of sheep and goats, as both sheep and goats share many of the same diseases and parasites.

Chapter 9 contains a discussion of wool—the fiber produced by sheep, and of mohair—the fiber produced by Angora goats.

Chapter 10 details some of the good business practices for sheep and goats that should be applied to any livestock venture.

CHAPTER 1

Beginning with their domestication 7000 B.C., steep topography, much browse, and lack of water and fences have favored sheep rather than cattle. (Photo by David Cornwell. Courtesy, American Sheep Industry Association, Englewood, CO)

HISTORY AND DEVELOPMENT OF THE SHEEP AND GOAT INDUSTRIES

Contents *Page*

Fig. 1-1. He leadeth me beside the still waters. (Psalm 23:2. Photo courtesy, American Colony Photographers)

Fig. 1-2. Sheep in America today—a natural resource of food and fiber. (Courtesy, American Sheep Industry Association, Englewood, CO)

Wild sheep and goats were among the first animals to be domesticated. Archaeological evidence suggests that by 7000 B.C. domestic sheep and goats had become the principal source of meat, wool, tallow, skins, and milk for farmers of southeastern Europe and western Asia.

POSITION OF SHEEP AND GOATS IN THE ZOOLOGICAL SCHEME

Sheep belong to the genus *Ovis*, and goats and their wild relatives make up the genus *Capra.* These two genera of the family Bovidae, *Ovis* and *Capra,* are so closely related that a naturalist never speaks lightly of "separating the sheep from the goats." Goats may be distinguished from sheep, however, by the presence of a beard, by the absence of the foot glands (which sheep have), by the strong smell of the bucks, and by differences in horns and skeleton. Goats are also more intelligent and independent, and they possess greater ability to fight and fend for themselves.

The following outline shows the basic position—the taxonomy—of domesticated sheep and goats in the zoological scheme:

Kingdom *Animalia*: Animals collectively; the Animal Kingdom.

Phylum *Chordata*: One of approximately 21 phyla of the Animal Kingdom, which includes primarily the subphylum *Vertebrata,* the vertebrates.

Class *Mammalia*: Mammals or warm-blooded, hairy animals that produce their young alive and suckle them for a variable period on a secretion from the mammary glands.

Order *Artiodactyla*: Even-toed, hoofed mammals.

Family *Bovidae*: Ruminants having polycotyledonary placenta; hollow, nondeciduous, unbranched horns; and nearly universal presence of a gall bladder.

Genus *Ovis*: The genus consisting of the domestic sheep and the majority of wild sheep. The horns form a lateral spiral.

Species *Ovis aries*: Domestic sheep.

Genus *Capra*: The genus consisting of the domestic and wild goats. Horns of the males are large, bulky, and mostly sabrelike. The males have a beard at the chin.

Species *Capra hircus*: Domestic goats.

Table 1-1 summarizes several basic groups of wild and domestic sheep and goats in the world.

TABLE 1-1
NAMES AND CLASSIFICATIONS OF SHEEP AND GOATS

Common Name	Scientific Name	Distribution	Chromosome Number[1]
			(2n)
Arkhar-argali sheep	*Ovis ammon*	Mountains of Central Asia.	56
Asiatic mouflon	*Ovis orientalis*	Mountainous regions from Asia Minor to southern Iran.	54
Asiatic Urial	*Ovis vignei*	Mountainous regions from northeastern Iran to Afghanistan and northwestern India.	58
Bezoar goat	*Capra aegagrus*	Mountains of Asia Minor and across the Middle East to Sind.	?
Bighorn sheep	*Ovis canadensis*	Rugged mountain slopes of western United States and Canada and northern Mexico.	54
Domestic goat	*Capra hircus*	Worldwide.	60
Domestic sheep	*Ovis aries*	Worldwide.	54
East Caucasian tur	*Capra cylindricornis*	Eastern Caucusus Mountains.	?
European mouflon	*Ovis musimon*	Successfully introduced to many European countries.	54
Ibex	*Capra ibex*	The Alps and the mountains of Spain. In Asia from west of Lake Baikal to Turkestan and Kashmir. Also, the mountains in Arabia and in Ethiopia.	60
Markhor	*Capra falconeri*	Mountains from East Kashmir west to the Hindu Kush and south to Quetta in northern Baluchistan; and also Kazakhstan.	?
Mountain goat	*Oreamnos americanus*	Northern U.S. through Canada to Alaska. The Mountain Goat's ancestors crossed from Asia to North America to become the best mountaineers in the new world.	?

[1]There is a constant number of diploid (2n) chromosomes for each species.

ORIGIN AND DOMESTICATION OF SHEEP

There is more confusion and disagreement about the ancestry and classification of sheep than of any other animal. This difficulty arises from the bewildering number of breeds and the marked changes produced by domestication. There are more than 200 distinct breeds of sheep scattered throughout the world. Although differing widely in body form and wool character, domestic sheep of all breeds are universally timid and defenseless and the least intelligent and least teachable of all the domestic four-footed animals. These traits are plainly the result of selection and are connected with the herding of sheep in large bands, where independence of behavior is a disadvantage. As a result, domestic sheep have become completely dependent on humans. Unlike other farm animals, they are unable to return to a wild life—referred to as becoming *feral*. Though this dependence is a logical final result of domestication, it does appear that the evolution of sheep in this direction may have gone too far, for they are pitifully helpless in emergencies.

Fig. 1-3. Humans have herded sheep and woven wool fibers since the Stone Age; and sheep have played an important part in world history. (A copper engraving from a 17th century edition of Virgil's "Bucolica." Photo courtesy, American Sheep Industry Association, Englewood, CO)

It is certain that domestic sheep came from the wild sheep of Europe and Asia. Although there are considerable differences in appearance between domestic sheep and their wild ancestors, such as tail length, coat color and type, and horn shape, most of these changes occurred very early in the domestication process. The outer coat of wild sheep is stiff and hairy and covers a short woolly undercoat, while in domestic sheep, the hairy outer coat is absent. A statuette of a woolly sheep has been found at Tepe Sarab in Iran, suggesting that selection for woolly sheep occurred as early as 6000 B.C. Features of domestic sheep such as horn shape and hornlessness in ewes, length of tail, and the woolly, white fleece were common in western Asia by 3000 B.C., as evidenced in Mesopotamian art and Babylonian books.

Domestic sheep are thought to descend mainly from the mouflons, *Ovis musimon* and *Ovis orientalis.* The Asiatic urial *(Ovis vignei)* may possibly be an ancestor of domestic sheep, but the difference in chromosome number makes any direct ancestry questionable. Perhaps some modern breeds trace back to other wild stocks, but differing chromosome numbers and geography may limit ancestry (see Table 1-1).

THE MOUFLON

There are two wild stocks of the mouflon: (1) the Asiatic mouflon *(O. orientalis),* a wild sheep still found in Asia Minor and southern Iran; and (2) the European mouflon *(O. musimon),* which is native to Europe and still found in Sardinia and Corsica. These two relatives are closely allied, but the Asiatic mouflon is redder and

Fig. 1-4. The European mouflon *(Ovis musimon)*, one of the ancestors of domestic sheep. Like the wild species from which sheep descended, the mouflon is short-tailed. It appears, therefore, that lengthening of the tail is a characteristic which appeared with domestication. (Courtesy, New York Zoological Society, Bronx, NY)

has a somewhat different twist to the horns. Both of the mouflon stocks are considered as ancestors of domestic sheep. The origin of the European mouflon is, however, unknown and fossil evidence for it is lacking. Possibly, the European mouflon is a relic of the first domestic sheep, and the Asiatic mouflon is the ancestor of all domestic sheep.

Even today relatively unimproved mouflonlike short-tailed domestic sheep exist in different sections of northern Europe. The least modified of these primitive types is the semiferal race of sheep on the uninhabited island of Soay, northwest of Scotland. The only essential difference between the mouflon and the feral Soay sheep is the shorter wool of the latter. The island of Soay is visited once or twice each year by the residents of St. Kilda who hunt down the Soay sheep with dogs and shear them.

Fig. 1-5. These Soay sheep, maintained by the Duke of Bedford on his estate at Woburn, near Bletchley, Buckinghamshire, are believed to be the only pure examples left in Britain. The Soay sheep have lived on the Isle of Soay, off St. Kilda, Hebrides, Scotland, from time immemorial. Their main characteristic is their thick buffalo-type horns. (Courtesy, American Sheep Industry Association, Englewood, CO)

THE ASIATIC URIAL

The Asiatic urial *(O. vignei)*, which is a smaller race of sheep than the mouflon, is native to the mountainous areas stretching from northeastern Iran to Afghanistan and northwestern India. At one time, most familiar breeds of sheep were thought to be descendants of this wild stock.[1] Since domestic sheep possess 54 chromosomes and urials possess 58 chromosomes,

Fig. 1-6. A wild sheep *(Ovis vignei)* native to the province of Punjab in northern India. This is a member of the Asiatic urial. (Courtesy, New York Zoological Society, Bronx, NY)

direct ancestry is unlikely. In north central Iran, however, the urial interbreeds with the Asiatic mouflon, producing a hybrid race of sheep with mixed characteristics.

ORIGIN AND DOMESTICATION OF GOATS

The ancestry of the goat *(Capra hircus)* is far less

Fig. 1-7. Goats are browsers. This Angora goat is browsing as high as it can reach on an oak tree. (Courtesy, Texas A&M University Agricultural Research Station, Sonora, TX)

[1]Different authors have used the same scientific names and the same common names to connote quite different groups of sheep. For example, the name *urial* has been used differently by different authors; thus, sheep that are urials to some scientists are mouflons to others. Another problem stems from the scientific faddism in naming sheep.

confused than that of sheep. It is certain that the bezoar goat *(Capra aegagrus)* was the main progenitor, if not the sole ancestor, of the domestic goat. Goats have not produced nearly so many breeds, nor, except for some of the milk-producing types, such extremely modified breeds. Unlike sheep, goats easily return to a wild state if given the opportunity. In fact, only the domestic cat can equal the goat in returning promptly and successfully to the independent life of a wild creature.

Like sheep, goats were probably among the first animals to be domesticated. Goat remains are found in the archaeological sites in western Asia, such as Jericho, Choga Mami, Djeitun, and Cayonu, which date the domestication of goats around 7000 to 6000 B.C.

Bezoar goats are found in the mountains of Asia Minor and across the Middle East to Sind. Some are found on the Aegean Islands and on Crete, but these goats may constitute relic populations of some very early domestic animals that were transported.

Fig. 1-8. Ancient Babylonian loom in operation. Wool was first used as a clothing material by the Babylonians beginning about 4000 B.C. (Courtesy, The Bettmann Archive, Inc., New York, NY)

WOOL: PRECIOUS FIBER THROUGH THE AGES

The sheep and wool industry began in central Asia 10,000 years ago as early people discovered that sheep could provide two of life's essentials—food and clothing.

Sheep, therefore, may be considered as one of humankind's first helpmates, and the weaving and felting of wool were among the first arts to be developed. In addition to being one of the very first animals to be domesticated, sheep have also been one of humankind's most valuable beasts. Besides providing the wool for cloth, sheep probably gave primitive people skins for raiment and shelter, and meat and milk for food.

The ancient Egyptians, Babylonians, Greeks, and Hebrews did hand spinning and weaving in their homes. The wool industry, like most others, first developed as a household craft, rather than as a primitive factory system.

Sheep raising was known as the earliest pastoral industry, and reference is frequently made to it in Old Testament literature. For example, Abraham, the patriarch of the Old Testament, thrived and prospered through his great flocks and herds. Subjects of the king of Israel were taxed according to the number of their rams. The Bible also refers to Abel as a "keeper of sheep"; and it was shepherds watching over their flocks by night who first saw the star over Bethlehem. Sheep raising was recognized as an early agricultural pursuit, and the early flocks served as a medium of exchange.

Sheep were treated with marked respect in Greece. Individual names were given to them and shepherds would proudly call out their favorites. While they had to be content with loin cloths, flock keepers spread skins over their sheep to protect their fleeces from inclement weather.

When Rome was in its glory, its wealthy and refined citizens boasted of their achievement in producing the finest-quality wool in the world. The Romans established a woolen factory in Winchester,

Fig. 1-9. Ancient Egyptian spinning and weaving. As a household craft, the wool industry had its humble beginning among the Egyptians sometime between 5000 and 4000 B.C. (Courtesy, The Metropolitan Museum of Art, New York, NY)

Fig. 1-10. Photograph of the painting *The Sacrifice of Noah*, by Bernardo Cavallino. This shows a sheep being served up as an offering. In the early days, the contribution of animals extended far beyond their utility value-they were the chief objects of worship and myth and the sacrificial offering of many religious ceremonies. (Courtesy, National Gallery of Art, Kress Collection, Washington, DC)

England, about 50 A.D. Sheep were given extraordinary care, and they were even blanketed so that a luster and gloss might be imparted to the wool. At frequent intervals, the fleece was parted, combed, and moistened with the rarest oils, oftentimes with wine. Surplus stock was usually killed at two years of age, for the Romans believed that the fleece was in its best condition at that period. The distinctive toga, a loose outer garment which was worn by officials of ancient Rome when they appeared in public in time of peace, was made from woolen fabrics.

Domestic sheep are not indigenous to any one country; for they appear to have been cultivated by the earliest peoples in history, and they have gradually spread over the entire face of the globe with the extension of civilization itself. Also, the efforts of flockmasters have been devoted for centuries to the search for methods of improving the quality and increasing the quantity of wool produced.

SHEEP RAISING IN SPAIN

It is said that the merino resulted from a crossing of the Tarentine sheep of Rome with the Laodician sheep of Asia Minor by breeders in the province of Terraconensis in Spain, about the time of Claudius (41 to 54 A.D.). The Merino produced wool of unusually fine fiber, suitable for making such fine, soft fabrics as broadcloth. Before the year 1000 A.D., both Spain and England attached great importance to their flocks; and by the year 1500 A.D., they were recognized as the two greatest sheep-producing countries of the world. Although the Spanish wools were much finer, for several hundred years Spain and England were regarded as competitors on the great wool markets of Flanders.

In Spain, the powerful nobility and clergy engaged in the lucrative sheep industry. In an attempt to produce the finest staple possible, the early Spanish flockmasters drove their sheep from southern to northern pastures in the spring and returned them in the fall. In this manner, it was possible to secure the most favorable grazing and climatic conditions for the flocks. The early laws of the kingdom stipulated that the owners of large flocks should be allowed a path 90 paces wide through all enclosed lands. In the migration process, any animal that failed to keep up with the band was left by the wayside. Presumably, this accounts for the flocking or gregarious instinct of the Merino sheep as well as their hardiness. With the repeal of the migration laws and consequent prohibition of seasonal migration, the Spanish shepherds blanketed their sheep through the colder months—their object being that of keeping equitable temperature, thus producing a more uniform and higher-quality product.

Spain long held a monopoly on merino sheep. It was a criminal offense punishable by death for anyone to send a sheep of this breed out of the country without the king's permission. This monopoly held until early in the 19th century when Spain was invaded by Napoleon, who overthrew the government. In this conquest, a large number of Merino sheep were seized and shipped to other countries, where they added new chapters to the history of sheep raising.

Spain never regained its high position in world sheep production following Napoleon's invasion, but, even today, sheep are the most numerous livestock of the country—exceeding cattle, horses, goats, and pigs in the aggregate. However, coarse-wooled sheep now predominate in Spain, there being three animals of this type to every one Merino.

SHEEP RAISING IN ENGLAND

During the Middle Ages (500-1500 A.D.) in England, sheep were the sheet anchor of farming. Their chief product was not meat, milk, or hides, but wool. Unlike the flocks in Spain, those in England were small; the sheep were not in the hands of a very few powerful owners as they were in Spain, and they were not compelled to travel across the country. The great problem of English sheep farmers, therefore, was to procure sheep that were adapted to their particular locality. This largely accounts for the development of the many types and breeds in that country. The cold winters, the scarcity of winter feed, and the presence of scab and rot made the early sheep raising of England a risky venture. The shepherd's position was highly respected, and lame shepherd's were in great demand

Fig. 1-11. Sheep in the Cotswold Hill district in England, which from earliest times has been a noted sheep center. (Courtesy, *The Field*, 8, Stratton Street, London)

because they were not as likely to overdrive their sheep.

None of the wools from the English breeds was as fine as that of the Merino. Nevertheless, there was a ready market for the English wools because they were more suitable for a variety of uses than those from Spain.

EVENTS GIVING NEW IMPETUS TO SHEEP RAISING IN ENGLAND

The great plague, or Black Death, of 1348–1349 served as a great impetus to the sheep industry of England. Laborers were reduced by the plague and wages rose, forcing the landlords to turn their holdings into pasture. Because of the favorable price for wool and because large numbers of sheep could be herded together with little labor involved, increased attention was given to the sheep industry.

During the reign of Edward III (1327–1377), a grant of special protection was made in favor of all Flemish weavers, dyers, and pullers who would settle in England to follow their trade. Through this grant, England secured artisans skilled in the then most improved methods of cloth making. Later, in the reign of Elizabeth (1557–1603), wool was the chief source of the wealth of traders and of the revenues of the Crown. It even controlled the foreign policy of England.

During the 16th century, some changes in agriculture occurred, even though very slowly. Iron was used more widely in implements, but seeded grasses and roots were still unknown and ewes were still milked. Despite their low quality, in comparison with present standards, sheep were considered to be the most profitable of all English livestock.

The coming of field cultivation of clover and seeded grasses, sometime after 1600, and of roots somewhat later, gave new impetus to all agriculture, including the sheep industry. Winter feed was now assured, and more and better sheep could be maintained.

BAKEWELL'S IMPROVEMENT OF ENGLISH SHEEP

By the time Robert Bakewell of Dishley (1726–1795) entered the livestock-breeding business of England, wool had declined in price until—with the rapidly advancing values of English lands—it alone would no longer justify the keeping of sheep. Bakewell, however, had the foresight to picture the future needs of a growing population in terms of meat, and he set about improving the Leicestershire sheep of the day. He was successful in creating a low-set, blocky, quick-maturing type of animal. He paid little or no attention to fancy points; no animal met with his favor unless it had utility value, as measured by meat production. Bakewell gradually transformed the large, heavy-boned, heavy-framed sheep, that had little or no propensity to fatten quickly, to a short-legged, blocky form with finer bone and quick-fattening propensities. Other breeders followed suit.

EARLY IMPORTATIONS AND IMPROVEMENTS OF SHEEP IN THE UNITED STATES

The Big Horn or Rocky Mountain sheep, prevalent in the Rocky Mountain region from Alaska to California, were native to this continent, but the ancestors of the present-day domestic sheep were imported. Columbus brought sheep and goats to the West Indies on his second voyage in 1493. Cortéz brought Merino sheep into Mexico in 1519. The Spaniards who founded old Santa Fe, New Mexico, were thought to have brought the multi-colored sheep from which the flocks of the Navajo Indians descended. If, as is generally supposed, these early importations were of Spanish Merino extraction, special permission to take them out of Spain must have been granted by the king.

The first sheep of the British breeds to be introduced in this country are said to have been brought into Virginia by the London Company in 1609.[2] Two

[2]In 1607, a shipment of sheep came over on the *Susan Constant*, but these were consumed during the famine of the ensuing winter, thus having no permanent effect.

Fig. 1-12. Hairy, multi-colored, unimproved Navajo sheep on the Navajo Reservation, New Mexico. It is thought that the ancestors of these sheep were brought to America from Spain by the Spaniards who founded old Santa Fe. (Courtesy, Southwestern Range and Sheep Breeding Laboratory, Fort Wingate, NM)

decades later, there were as many as 400 sheep in Charleston, now a part of Boston. These early importations represented very poor specimens of British sheep, the imperfections of which were generally acknowledged. Furthermore, because of the lack of care, inadequate shelter, and promiscuous breeding, the sheep yielded a wool that had lost all pretense to fineness. Predatory animals played havoc with the flocks, and for this reason many of the early sheep were herded on small islands off the coast. Even today, sheep are found on the islands off the coast of Maine and Massachusetts, especially on Martha's Vineyard and Nantucket.

Sheep husbandry was promoted by the colonists primarily in order to furnish wool, rather than to increase the food supply. In addition to the demand for wool in most new countries, the product is well adapted to marketing in an undeveloped area—being (1) light in weight, value considered, and (2) imperishable with respect to time involved in getting it to market.

As early as 1662, there was a woolen mill at Watertown, Massachusetts. At about this time, the exportation of ewes or lambs, except to other colonies, was forbidden. In 1670, in order to encourage sheep growing, Connecticut required every person to labor for one day each year at clearing the underwood to make pasturage. Drastic legislation was passed by all the New England colonies to protect the sheep from dogs. A dog that bit or killed a sheep was often hanged as though it were a human malefactor. The execution was usually carried out in some nearby swamp, whence comes the name *Hang-Dog Swamp*, given to several localities in colonial Massachusetts and Connecticut.

The town common was open to all kinds of livestock, but because sheep were the most defenseless of domestic animals and the hardest to raise, the town regulations concerning their care were numerous. Moreover, each town had shepherds who were considered very important persons.

Owners identified their stock by marks or brands (ear notches or ear holes were the most common means of marking)—a medieval European device revived in the New World. These marks were registered with the town clerk, and sometimes the description was embellished with a crude picture.

In due time, early importations to America and the improvements made by the colonists were to become the sturdy basis for some of the great American flocks. Breeding stock was imported from some foreign countries; and the early flock owners practiced rigid selection in order to improve the quality of wool produced, employing sight and touch, ordinary scales for weighing the fleeces, and in some instances a ruler for measuring fiber length. These empirical methods served well, however, and remarkable improvements in wool were brought about. Fleeces became heavier, and the fibers were longer and more uniform. In 1836, fleeces produced in this country averaged only about 2 lb in weight and were satisfactory only for the coarser woolen fabrics. Half a century later, the average weight had been increased to 5.5 lb. Almost simultaneously, the growing importance of the lamb and mutton trade changed the character of the sheep from one with a small carcass and short, fine fleece to a heavier, larger animal growing a longer-stapled, coarser-grade wool. Thus, a dual-purpose type of animal was developed.

THE MERINO CRAZE

As early as 1660, England had forbidden the export of sheep and wool from any of the American colonies, a prohibition that marked the beginning of the oppressive trade measures that finally helped drive the colonies into rebellion. The British Parliament also forbade the export of blooded English rams to the colonies in order to prevent improvement of the sheep across the water. Despite these restrictions on the part of the home government, the sheep industry rapidly increased in New England. But America's dependence upon English wool was a source of great chagrin.

Napoleon's invasion broke Spain's jealous guardianship of the Merino, a breed monopoly that had prevented their earlier introduction to America. President Jefferson promptly learned of the great liquidation of the Spanish flocks as a result of the invasion, and he at least encouraged their importation, although he did not actually activate the project. James Madison, Jefferson's lifelong friend and successor to the presidency, was also a sheep fancier, who often relaxed from the burden of public affairs on his Virginia farm.

Fig. 1-13. Three Merinos, consisting of two ewes and one ram, were imported to the United States from Spain in 1793, but these animals were butchered. Don Pedro *(above)* was then imported in 1801, by du Pont de Nemours at a cost of $1,000. It is claimed that this ram left a larger impression on the sheep of the United States than any other sheep ever imported. Don Pedro was acclaimed the father of the fine-wool industry in the United States. He weighed 138 lb. After being carefully washed in cold water, his fleece weighed 8.5 lb, and it had a staple length of 1.75 in. (Courtesy, American Sheep Industry Association, Englewood, CO)

Because of the scattering of the native flocks of Spain and the political turmoil that accompanied the fall of Napoleon, the importation of Merinos from Spain was of comparatively short duration.

The American farmer discovered the advantage of wool growing, and the demand for wool was all the heavier because shipments of English wool were stopped by the War of 1812. In 1811, Merino wool sold at 75¢ per pound, and by 1813 it ranged from $2 to $3 per pound. Profits soared, and the demand for Merino sheep became a craze which extended from New England to Ohio. As Spain could no longer supply Merinos and the numbers in the states were inadequate to meet the demands of the boom, American breeders and traders turned to the flocks of Saxony in Germany. Some years before Napoleon's invasion of Spain, the king of Spain had presented the chief ruler of Saxony with a number of Merinos. The pure Saxony Merinos were of high quality, and they thrived most successfully in their adopted land in Germany. Unfortunately, many of the sheep advertised as pure-blooded Saxons and sold to sheep producers in the United States were only grades.

A wave of speculation swept the country. The wildest of claims were fabricated about the yield and quality of the fleeces produced by this marvelous Merino breed. Unfortunately, little attention was given to any characteristics other than the wool itself. In the wild boom, the enthusiasts had failed to foresee that (1) prices of wool were inflated by the temporary halt in British competition, and (2) the constitution of the Merino had dangerously declined through lack of intelligent selection. Finally, when the war with England ended in 1815, British woolen goods began to flow in again and gradually forced down the inflated prices of American wool. Eventually, the Merino bubble burst, and with it went much of the myth that had surrounded the breed. Sheep raising did not again get on its feet until the tariffs of 1828, which helped to protect the U.S. sheep grower and the U.S. woolen manufacturer against outside competition. It is to the everlasting credit of the Merino breed of sheep, however, that they continued to multiply and form the sturdy foundation for much of the sheep industry of the United States.

THE WESTWARD MOVEMENT OF SHEEP

In 1810, the census figures clearly indicated that the northeastern part of the United States—New England and New York—was the sheep-producing center of the nation. At this date, there were an estimated 7 million head of sheep in the United States. By 1840, sheep numbers in the United States had increased to 19 million head. At that time there were no appreciable numbers of sheep in the far West except those owned by the Navajo Indians in northern New Mexico. In fact, the only state west of the Mississippi having sheep in considerable numbers was Missouri. Ten years later, the densest sheep population was centered in the Ohio Valley and Great Lakes region. When this area became somewhat thickly settled and land values rose, many sheep producers, desiring to operate on a large scale, moved farther west where range was cheap and extensive. During this period sheep were maintained primar-

Fig. 1-14. Range band of ewes near Babb, Montana, with Sherburne Peak in the background. (Courtesy, Ernst Peterson, Hamilton, MT)

ily for wool production and the market lamb business, as we know it today, was practically unknown. It was not at all surprising, therefore, to find that with the opening up of cheaper range lands in the West there was also an immediate and marked shift of the sheep population from east to west.

Like other wars, the Civil War caused sharply increased sheep numbers as a result of the demand for and the high price of wool. Following the war, wool prices fell because of lower demand and increased competition from cotton and imported wool. Yet, the westward expansion and the opening up of cheap lands continued.

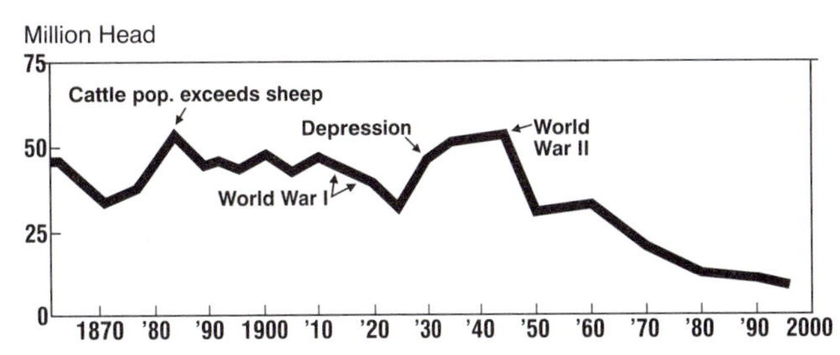

Fig. 1-15. Growth and decline of the U.S. sheep industry. Based on data from the USDA.

GROWTH, DECLINE, AND TRANSITION OF THE U.S. SHEEP INDUSTRY

Fig. 1-15 shows that sheep numbers have fluctuated rather widely from time to time. They peaked in 1884, and they nearly peaked again in 1942, during World War II, when they reached 56 million head. From this near-peak in 1942, the population declined to around 30 million head from 1950 to 1961, when a second decline started that ended in 1979 at about 12.4 million head. A 2% increase in 1980 and 1981 marked the first significant increase in sheep numbers in more than 20 years. On January 1, 1996, there were 8,857,000 breeding (stock) sheep and lambs and market sheep and lambs in the United States.

Several factors contributed to the decline in sheep numbers since 1942, including (1) lower returns and higher risks from sheep than from cattle and some crops in many areas; (2) scarcity and high wages of competent sheep herders; (3) uncertainties in tariff levels and imports, wool incentive payments, and grazing allotments on public domain; (4) more application of science and technology in competing meat and fiber industries; and (5) problems with predator control.

Also, the geography of sheep production shifted. By 1900, the Rocky Mountain region had become the dominant sheep section of the country, but sheep numbers declined more sharply in the range states than in the native states during the 1940s. Since 1950, sheep numbers have declined proportionally in both farmflock and range-band areas. In 1950, the 11 western states and Texas had 68.2% of the nation's stock (breeding) sheep and lambs; in 1996, these same 12 states had 70.1% of the breeding sheep.

CHRONOLOGY OF THE U.S. SHEEP INDUSTRY

A chronology of the U.S. sheep industry is presented in Table 1-2.

TABLE 1-2
CHRONOLOGY OF THE U.S. SHEEP INDUSTRY

Year	Event
1493	Columbus brought sheep to the West Indies on his second voyage.
1519	Cortéz brought Merino sheep into Mexico.
1609	The first sheep of the British breeds were brought to Virginia by the London Company.
1662	Woolen mills were operating in Watertown, Massachusetts. At this time, the export of ewes and lambs, except to other colonies, was forbidden.
1670	In order to encourage sheep and wool production, Connecticut required every person to labor for one day each year at cleaning the underbrush to make sheep pasturage. Also, drastic legislation was passed by all New England colonies to protect sheep from dogs.
1670	During this period, fleeces produced in this country averaged only about 2 lb. 166 years later (1836), they averaged 5.5 lb.
1801	Don Pedro, a Merino ram, was imported from Spain by du Pont de Nemours at a cost of $1,000. He weighed 138 lb and his fleece weighed 8.5 lb.
1860	The Civil War caused sharply increased sheep numbers as a result of the demand, and high price, for wool.
1884	The National Wool Growers Association entered into an agreement with the Axtell Publishing Co. in Pittsburgh to publish a *Wool Growers Quarterly*.
1811	Merino wool sold at 75¢ per pound. Two years later (in 1813), it ranged from $2.00 to $3.00 per pound.

(Continued)

TABLE 1-2 (Continued)

Year	Event
1911	The first issue of the *National Wool Grower* magazine was printed in Gooding, Idaho; and that same year the National Wool Growers Association established a permanent office in Salt Lake City.
1931	The National Wool Growers Association launched its first wool promotion—National Wool Week.
1933	Riboflavin isolated, with milk established as a valuable source of it.
1937	Conrad Elvehjem discovered that niacin (as either nicotinic acid or nicotinic acid amide, isolated from liver) cured blacktongue in dogs. Shortly thereafter, it was established that niacin (as either nicotinic acid or nicotinic acid amide) was essential for humans, monkeys, pigs, chickens, and other species.
1940	The Wool Products Labeling Act became law, stipulating that fabrics must be labeled under one of the following categories: (1) wool, (2) reprocessed wool, or (3) virgin wool.
1943	The Commodity Credit Corporation launched its wool purchase program.
1949	The American Wool Council (created in 1941, lasted only two years) merged its promotional activities with the International Wool Secretariat, forming the Wool Bureau Inc.
1950	The 11 western states and Texas had 68% of the nation's breeding sheep.
1954	USDA discovered parthenogenesis (reproduction without male fertilization in the fowl).
1954	The National Wool Act was passed, providing support prices for wool and mohair.
1977	Gene splicing, also known as recombinant technique, ushered in a new era of genetic engineering, when scientists at the University of California, San Francisco, reported a major breakthrough as a result of altering genes—turning ordinary bacteria into factories capable of producing insulin, a valuable hormone essential to the survival of diabetics. But this was only the beginning! Building upon the present knowledge and understanding of nucleic acids DNA and RNA, *the biotechnology era arrived*.
1986	Animal health and food safety issues loomed large when bovine spongiform encephalopathy (BSE)/transmissible spongiform encephalopathies (TSE), commonly called *mad cow disease*, Crevtzfeldt-Jacob disease (CJD) in humans, broke in Europe, leading to speculation that some people are genetically susceptible to TSEs, which include scrapie in sheep and goats, if they become exposed to the infectious agent.
1989	The announcement of bovine spongiform encephalopathy caused a panic in Great Britain, an export ban on British beef, and cattle prices to plummet. The bottom line for the livestock industries remains that 90% of the U.S. population knows about BSE to one degree or another.
1989	The industry merged the American Sheep Producers Council and the National Wool Growers Association into a new organization named the American Sheep Industry Association.
1990	Scrapie was the hot topic.
1993	The name of the magazine was changed from *National Wool Grower* to *The National Lamb & Wool Grower*, better to reflect those whom they served.
1995	The National Wool Act of 1954 was phased out.
1995	Wool accounted for only 1.8% of U.S. per capita consumption of all fibers.
1996	The Animal Health and Welfare Committee completed the Dairy Sheep Quality Audit—an information database on dairy sheep.
1996	The checkoff was voted on twice. The first vote date was on October 1, 1996; although the referendum passed, the USDA invalidated the vote because of inconsistencies. The second vote date was February 6, 1996, at which time the referendum was defeated.
1996	Estimates put predator kills of sheep and lambs at 520,000 head annually.
1997	In his address at the American Sheep Industry Convention in Nashville, Tennessee, Dr. Hudson Glimp, Range Specialist, University of Nevada, and noted sheep authority, reported that sheep are increasingly being viewed as environmental enhancers.
1997	In February, the last issue of *The National Lamb & Wool Grower* rolled off the press, thus ending an 86-year-old tradition.
1997	The Roslin Institute, Edinburgh, Scotland, announced that they had produced a live lamb (named Dolly) by nuclear transfer from an adult mammary cell. The process is known as *cloning*.

THE U.S. GOAT INDUSTRY

There are 1,900,000 goats in the United States[3], of which about 46% are Angora goats, 33% are dairy goats, and 21% are Spanish or meat goats. Ninety-five percent of the Angoras are located in Texas. New Mexico has the second largest population. California is the most important dairy goat state, and most Spanish or meat-type goats are found in Texas. Dairy goats are found in small herds under intensive management, but Angora and Spanish goats generally are produced under extensive range conditions.

The United States, Turkey, and South Africa contribute about equally to produce 88% of the world's supply of mohair. Angora goats in the United States are sheared twice yearly, yielding a total of 11.7 million lb of mohair in 1994. Significant numbers of the Angora goats in the United States are castrated males that are maintained for fiber production until advancing age reduces hair production and quality, at which time they are salvaged for meat.

Average annual production for U.S. dairy goats with production records is 1,584 lb of milk and 50 lb of fat per goat. Annual world production of milk by dairy goats is less than 440 lb.

[3]From *FAO Production Yearbook*, Vol. 50, 1996, page 194.

Fig. 1-16. Angora goats with guard dogs. (Courtesy, Dr. Maurice Shelton, Texas Agricultural Experiment Station, San Angelo, TX)

Interest in goats is increasing. Participation of U.S. dairy goat breeders in breed association programs has increased significantly in recent years. Official goat shows and animal registrations have increased tenfold, while official testing has increased more than fiftyfold.

The Spanish goats are used mainly for meat production along with brush control. They are usually produced under range conditions where they receive little care. Significant numbers of meat goats are exported to Mexico.

QUESTIONS FOR STUDY AND DISCUSSION

1. Give the taxonomy of domestic sheep or goats.

2. Discuss the ancestry of domestic sheep and goats.

3. There are more than 200 distinct breeds of sheep scattered throughout the world. How do you explain the fact that there are more breeds of sheep than there are of cattle and swine?

4. Apparently the wild stocks of sheep were short-tailed, and the lengthening of the tail followed domestication. How can you explain the selection of long-tailed sheep?

5. Discuss the role of wool in some of the ancient civilizations.

6. Christmas cards and other Biblical scenes depicting shepherds watching over their flocks by night usually show more than one shepherd, because the shepherds brought their flocks together at night. Why did they get together at night?

7. The great plague, or Black Death, of 1348–1349 served as a great impetus to the sheep industry of England because laborers were scarce and the landlords were forced to do less cultivating and more pasturing. Would such circumstances apply in this era of mechanization? Justify your answer.

8. Who was Robert Bakewell?

9. Outline the establishment of sheep in the United States.

10. Did the Merino craze of 1811–1813 have counterparts in cattle and swine breeding? Justify your answer.

11. List reasons back of the growth, decline, and transition of the U.S. sheep industry,

12. Describe the U.S. goat industry.

SELECTED REFERENCES

Title of Publication	Author(s)	Publisher
Agricultural Statistics 1997	Staff	U.S. Department of Agriculture, Washington, DC, 1997
America's Sheep Trails	E. N. Wentworth	Iowa State College Press, Ames, IA, 1945
Domesticated Animals from Early Times	J. Clutton-Brock	British Museum, London, England, and University of Texas Press, Austin, TX, 1981
FAO Production Yearbook, Vol. 50, 1996	Staff	FAO, United Nations, Rome, 1997
Grzimek's Animal Life Encyclopedia, Vol. 13, Mammals, IV	Ed. by B. Grzimek	Van Nostrand Reinhold Company, New York, NY, 1972
History of Livestock Raising in the United States, 1607–1860, Agric. History Series No. 5	J. W. Thompson	U.S. Department of Agriculture, Washington, DC, Nov., 1942
Modern Breeds of Livestock	H. M. Briggs D. M. Briggs	MacMillan Publishing Co., New York, NY, 1980
Our Friendly Animals and Whence They Came	K. P. Schmidt	M. A. Donohue & Co., Chicago, IL, 1938
Productive Sheep Husbandry	W. C. Coffey	J. B. Lippincott Co., Philadelphia, PA, 1937
Sheep and Man-An American Saga	C. A. Kilker C. R. Koch	American Sheep Producers Council, Inc., Denver, CO, 1978
Sheep of the World in Color	K. Ponting	Blandford Press, Ltd., Poole, Dorset, England, 1980
Sheep Production	L. J. Horlacher	McGraw-Hill Book Co., New York, NY, 1927
Sheep Science	W. G. Kammlade, Sr. W. G. Kammlade, Jr.	J. B. Lippincott Co., Philadelphia, PA, 1955
The U.S. Sheep and Goat Industry: Products Opportunities, and Limitations, Report No. 94	Task Force	CAST, Ames, IA, May, 1982
Wild Sheep of the World, The	R. Valdez	Wild Sheep and Goat International, Mesilla, NM, 1982
World Dictionary of Livestock Breeds	I. L. Mason	C.A.B. International, Wallingford, Oxon OX10 8DE, UK, 1988

CHAPTER 2

Sheep production on the White House lawn in 1919. First Lady Edith Wilson (Mrs. Woodrow Wilson) kept the flock to produce wool for charities and to keep the lawn trimmed. (Courtesy, Library of Congress, Washington, DC)

WORLD AND U.S. SHEEP AND GOATS

Contents *Page*

Despite the obscurity surrounding the domestication of sheep and the disagreement about their ancestry and classification, sheep raising followed the conquest and colonization of the Western Hemisphere to Australia, New Zealand, South Africa, and other countries. Today, the sheep industry is worldwide, with numerous breeds providing needed adaptation. Like other industries, sheep raising is affected by wars; national and international policies and politics; supply and demand; wool substitutes; competition for land, labor, and capital; and many other factors.

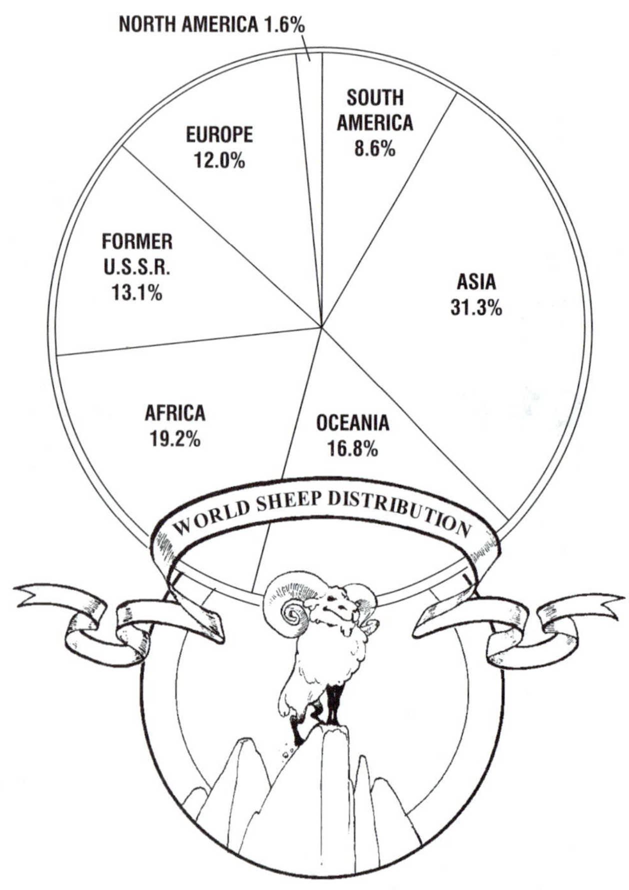

Fig. 2-1. The worldwide distribution of sheep, by major areas. Oceania includes primarily Australia and New Zealand, and North America includes Canada, the United States, and Mexico. *Note:* All figures except U.S.S.R. are for 1994; U.S.S.R. figures are for 1979–81. Because of not having U.S.S.R. sheep numbers for 1994, the percentages in the above illustration total 102.6%. Thus, the percentage figures by areas are approximates. Nevertheless, Fig. 2-1 shows the relative importance of sheep in different areas of the world; it shows that sheep are of minor importance in North America, but of major importance in Asia. (Based on data from the *FAO Production Yearbook*, Food and Agriculture Organization of the United Nations, Rome, Italy, Vol. 48, 1994, pp. 192–194, Table 90)

WORLD DISTRIBUTION OF SHEEP AND GOATS

The United States possesses only 0.95% of the world's sheep, not enough to rank among the 10 leading countries of the world. Fig. 2-1 and Table 2-1 tell the statistical story of the worldwide distribution of sheep. The majority of the world's sheep are concentrated in a relatively small number of countries. The 9 leading countries (see Table 2-1) have 84.4% of the world's sheep. Three countries—China, India, and Australia—account for 63% of the world total. Although the industry is worldwide, it is of greatest importance in those countries which have (1) vast frontierland areas that are sparsely settled, and (2) temperate climates. These conditions prevail in the Southern Hemisphere, and it is there that most of the world's sheep are located. On the other hand, there are areas within countries of the Northern Hemisphere with many great flocks and numerous small ones.

Since World War II, sheep numbers have plummeted in the United States. The less drastic decline in other countries is attributed to sheep raising being an excellent subsistence occupation for people in nations with a high proportion of nonarable land and relatively low living standards.

Favorable prices for wool and lamb and satisfactory grazing and weather conditions usually make for stable sheep numbers in those countries to which they are adapted. Historically, wars have always played a prominent part in stimulating the sheep business.

Generally, the densest Angora goat and meat goat production is found in those areas of the world where grazing is too scanty, rough, or brushy for sheep. Dairy goats have long been a popular milk animal in the old world, where they are often referred to as poor peoples' cows.

Fig. 2-2. Ewes being milked by women of the Bakhtiari tribe, Khuzistan. Even today, sheep are an important source of milk in Asia, southeastern Europe, and the Middle East. (Courtesy, University of Wyoming)

TABLE 2-1
SIZE AND DENSITY OF SHEEP POPULATION OF NINE LEADING SHEEP- AND WOOL-PRODUCING COUNTRIES OF THE WORLD, BY RANK, 2000

Country	Sheep Population[1]	Human Population[2]	Size of Country[2]		Sheep per Capita	Density of Sheep	
			(sq mi)	*(sq km)*		*(sq mi)*	*(sq km)*
China, Peoples Republic	268,143,000	1,203,097,268	3,696,100	*9,572,899*	0.2	72.5	*28.0*
India	180,130,000	936,545,814	1,222,243	*3,165,609*	0.2	147.4	*56.9*
Australia	117,091,000	18,322,231	2,966,200	*7,682,458*	6.4	39.5	*15.2*
New Zealand	46,150,000	3,407,277	104,454	*270,536*	13.5	443.8	*170.9*
Turkey	37,300,000	63,405,526	300,948	*779,455*	0.6	124.0	*47.9*
South Africa	34,910,000	45,095,459	472,281	*1,223,208*	0.8	74.0	*28.5*
United Kingdom	31,080,000	58,295,119	94,251	*244,110*	0.5	330.6	*127.4*
Spain	24,199,000	39,404,348	194,898	*504,786*	0.6	124.7	*48.0*
Russian Federation	18,213,000	149,909,089	6,592,800	*17,075,352*	0.1	2.8	*1.1*
World total	897,310,000	5,734,106,000	55,900,000	*144,781,000*	0.2	16.1	*6.2*

[1]*Agricultural Statistics 2001*, p. VII-27, Table 7-43.
[2]*The World Almanac 1996*, Funk & Wagnalls, Mahwah, NJ.

Fig. 2-3. Goats are browsers. This shows Boer goats being used to control brush on the range. (Courtesy, Dr. J. E. Nel, University of Wyoming, Laramie)

WORLD SHEEP AND GOAT PRODUCTION AND CONSUMPTION

The following statistics attest to the global magnitude of sheep and goat production, and their contribution of food and clothing to people everywhere.[1]

- **Sheep produce 3.4% of the world's annual total meat**—In 1996, 215,169,000 metric tons of all meats were produced globally, including beef and veal, buffalo meat, mutton and lamb, goat meat, pig meat, horse meat, and poultry meat.
- **Sheep produce wool**—In 1996, globally, sheep produced 1,614,000 metric tons of wool (scoured/degreased).
- **World per capita consumption of wool is 0.67 lb.**
- **World per capita consumption of lamb and mutton is 2.7 lb.**
- **World per capita consumption of sheep milk is 3.1 lb.**
- **World per capita consumption of goat meat is 0.001 lb.**
- **World per capita consumption of goat milk is 40.1 lb.**

Note: For comparative purposes, worldwide per capita species milk production data follow: cow milk, 179.6 lb; goat milk, 40.9 lb; buffalo milk, 18.9 lb; and sheep milk, 3.1 lb.

China leads the countries of the world in total meat production, producing about 31% of the total world production; the United States ranks second in total meat production, producing about 16% of the total world production (see Table 2-2).

In lamb, mutton, and goat production, China ranks first and Australia ranks second. The United States ranks seventh in lamb, mutton and goat meat production; but these meats account for only 0.7% of all U.S. meat production (see Table 2-2). Table 2-2 also shows the relative importance of lamb, mutton, and goat meat in the leading meat-producing countries of the world.

[1]All statistics are from *FAO Production Yearbook*, Vol. 50, 1996; and are for the year 1996 (latest year available).

TABLE 2-2
MEAT PRODUCTION IN SPECIFIED COUNTRIES, 1994[1]

Country (by rank of total meat production)	Total Meat Production	Lamb, Mutton, and Goat Meat Production	Lamb, Mutton, and Goat Meat as Percentage of All Meat Production
	- - - (*1,000 metric tons*) - - -		(%)
China, Peoples Rublic	36,957	1,609	4.4
United States	19,361	140	0.7
Brazil	5,775	84[2]	1.5
Russian Federation	5,670	310	5.5
Germany	4,518	41	0.9
France	3,872	154	4.0
Mexico	2,852	142	5.0
Spain	2,825	240	8.5
Australia	2,763	580	21.0
Argentina	2,682	82	3.1
Total meat	120,718	6,311	5.2

[1]From: *Agricultural Statistics 1995–96*, p. VII-48, Table 437. One metric ton = 2,204.6 lb. To change metric ton to short ton (2,000 lb), multiply by 1.102.

[2]*FAO Production Yearbook*, Vol. 48, 1994, p. 202, Table 94.

Among the countries leading in the annual per capita consumption of meat, lamb, and mutton represent a varying percentage of the meat consumed as shown in Table 2-3. Countries with the greatest per capita meat consumption do not necessarily represent those countries with the greatest annual per capita consumption of lamb and mutton.

Fig. 2-4 shows the 10 countries which lead in the annual per capita consumption of lamb, mutton, and goat meat.

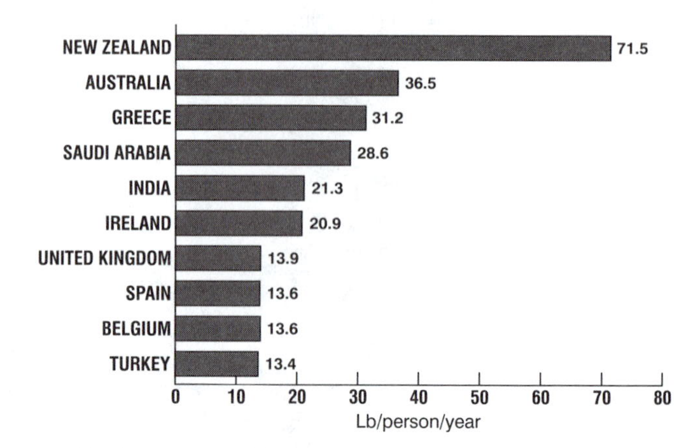

Fig. 2-4. Countries which lead in annual per capita consumption of lamb, mutton, and goat meat, 1996. To convert to kg, divide by 2.2. (Based on data from USDA, Foreign Agricultural Service)

WORLD WOOL PRODUCTION

Table 2-4 reveals that, generally speaking, there is quite a close correlation between sheep numbers and wool production. However, the amount of wool produced per sheep varies widely among countries; with Australia and New Zealand producing more than 8 lb of scoured (degreased) wool per sheep to the world average of 3.2 lb. Australia produces more than 30% of the world's wool.

TABLE 2-3
ANNUAL PER CAPITA MEAT CONSUMPTION IN SPECIFIED COUNTRIES[1]

Country	All Meats		Mutton, Lamb, and Goat Meat		Per Capita Mutton, Lamb, and Goat Meat as Percentage of All Meat Consumption
	(*lb*)	(*kg*)	(*lb*)	(*kg*)	(%)
Czeck Republic	210	*95.5*	Na[2]	*Na*	Na
Austria	170	*77.3*	Na	*Na*	Na
Belgium Luxembourg	164	*74.7*	Na	*Na*	Na
United States	163	*74.2*	1.3	*0.6*	0.8
Australia	159	*72.2*	36.5	*16.6*	23.0
Canada	149	*67.6*	Na	*Na*	Na
France	147	*67.0*	11.4	*5.2*	7.8
Ireland	139	*63.1*	20.9	*9.5*	15.1
New Zealand	138	*62.7*	71.5	*32.5*	52.0
Argentina	137	*62.4*	3.7	*1.7*	2.7
Uruguay	136	*61.7*	Na	*Na*	Na
Germany	136	*61.7*	2.6	*1.1*	1.8

[1]Data for 1996, from Ralph Bean, USDA, Foreign Agricultural Service, Washington, DC.

[2]Na means "not available."

TABLE 2-4
WOOL (CLEAN BASIS) PRODUCTION IN 7 LEADING COUNTRIES OF THE WORLD, BY RANK, 1999–2000[1]

Country	Wool Production	Number of Sheep	Wool (Clean Basis) Produced per Sheep
	(million)		
	(lb)		*(lb)*
Australia	968	115	8.4
New Zealand	417	46	9.1
China	320	134	2.4
Former U.S.S.R.	152	48	3.2
Uruguay	86	14	6.1
Argentina	73	15	4.9
South Africa	75	19	3.9
World total	3,045	1,005	3.0

[1]From: *Cotton and Wool Situation and Outlook*, USDA, ERS, November 2000, p. 61, Table 37.

The Southern Hemisphere countries of Australia, New Zealand, Argentina, Uruguay, and South Africa have been the main surplus-producing and exporting countries (see Table 2-5). As would be expected, with the declining sheep numbers in the United States, the production of wool has been correspondingly lowered in this country. In 1967, the United States ranked sixth in wool production. In 1997, 30 years later, it ranked twenty-first.

Wool production is definitely a frontier industry, thriving in those areas where there is an abundance of cheap range area and where the human population is sparse. On the other hand, wool consumption is greatest in centers of dense population, especially the temperate zones of the Northern Hemisphere. Thus, in 1999–2000, the nine major wool importing nations, by rank, and the millions of pounds that each imported, are presented in Table 2-6.

TABLE 2-5
FIVE MAJOR WOOL EXPORTING COUNTRIES[1]

Country	Clean (Scoured) Wool Exported in 1999–2000
	(million lb)
Australia	867
New Zealand	360
Argentina	43
South Africa	32
Uruguay	20
Total of five countries	1,322

[1]From: *Cotton and Wool Situation and Outlook*, USDA, ERS, November 2000, p. 61, Table 37.

TABLE 2-6
NINE MAJOR WOOL IMPORTING COUNTRIES[1]

Country	Greasy Wool Imported in 1999–2000
	(million lb)
China	342
Italy	270
United Kingdom	147
France	132
Germany	102
Japan	87
Belgium	80
South Korea	52
United States	43
World	1,908

[1]From: *Cotton and Wool Situation and Outlook*, USDA, ERS, November 2000, p. 62, Table 38.

SHEEP RAISING IN AUSTRALIA

Australia is a large country which is best suited to a pastoral type of agriculture and in which sheep seem to have been the animals best adapted to its grazing lands.

Although sheep existed in Australia at an early date, the Merino was first introduced in 1789. Other Merino importations followed, most of which came from Saxony in Germany. The early sheep industry of Australia was financed with English capital, the original aim being to render that country independent of Spanish, German, and other foreign sources of supply. The wool from the Australian Merino flocks met with a ready demand on the part of English manufacturers, thus giving great encouragement to the further expansion of the sheep industry in Australia.

Merino blood still predominates in Australia flocks, although the meat breeds are gradually increasing in favor and greater attention is being given to the meat qualities of the Merino. Most of the wool is marketed in England. Likewise, Australia disposes of most of its surplus mutton by shipping frozen carcasses to England, or by shipping live animals to the Middle East on converted oil tankers.

Most of the flocks of Australia are kept in fenced holdings, rather than being herded as is the most common practice on the western ranges of the United States. The Australian owners prefer ranging on enclosed lands, contending that: (1) the sheep make better use of the range under this system, with the animals scattering out in contrast to each sheep regularly maintaining a fairly definite position in the band as happens in herding; (2) less driving is required, because in herding, the animals must be rounded into camp at

Fig. 2-5. Mustering (called herding in the United States) sheep in Australia. Note that a horse and dogs are used. Australia has wide, flat grasslands. (Courtesy, Australian News and Information Bureau)

night, driven to water, and kept from other bands of sheep; (3) the fences cost less than the added labor in herding; and (4) the fences give considerable protection against predatory animals and help protect the forage from the ravages of rabbits (the fences being rabbitproof). Experienced operators contend that handling sheep in fenced holdings is satisfactory, provided the band can be kept under these conditions the year-round but that it will not work if the animals must be removed from fenced range at intervals and herded, for the band will then be untrained and unmanageable.

SHEEP RAISING IN NEW ZEALAND

Fig. 2-6. Droving Romney Marsh sheep in the Wellington Province, North Island, New Zealand. (Courtesy, Department of Scientific and Industrial Research, Wellington, New Zealand)

New Zealand is a small country, less than twice the area of the state of Illinois. In 1995, however, there were 50 million sheep in New Zealand, or 480 sheep per square mile—the densest sheep population of any country of the world. The sheep in New Zealand run more to the meat type than do the animals in Australia. Practically all the flocks are kept in fenced holdings without herders, in a manner similar to the method followed in Australia. Year-round grazing is available. Cattle and sheep share many areas, to the advantage of each other, with cattle utilizing the coarser vegetation and sheep the finer grasses and legumes. The best lambs in New Zealand are produced by using Southdown rams and Romney ewes.

SHEEP RAISING IN SOUTH AMERICA

There is a considerable sheep industry in Argentina, Chile, Uruguay, Brazil, and Peru.

Without doubt, the finest sheep country in South America, and one of the finest in the world, is in the La Plata River area of Argentina and Uruguay, where sheep compete with cattle for the lush pastures of the Pampas region. Predatory animals are few; winter feeding is seldom necessary; and diseases are rare. In brief, there is probably no other comparable area in the world where the shepherd's life is easier than in this particular territory. Sheep raising in other areas and countries of South America does not compare with the La Plata River section, either in terms of favorable conditions or quality flocks.

In all the South American countries, Merino breeding was used as a foundation and in effecting improvement, but many subsequent importations of the meat breeds have been made, especially the long-wooled breeds. Much of the coarse wool produced in South America is used in the manufacture of rugs.

SHEEP RAISING IN SOUTH AFRICA

The Republic of South Africa's chief claim to fame as a sheep country is in terms of wool production. Merino blood predominates, although a considerable number of representatives of the meat breeds have been introduced in more recent years. The major handicaps to sheep production in South Africa are: (1) prevalence of diseases and parasites, especially sheep scab; (2) unreliable labor; (3) predatory animals, especially jackals; and (4) frequent droughts.

Fig. 2-7. Sheep scene in the Republic of South Africa. Note that the sheep show fine wool breeding. (Courtesy, Embassy of the Republic of South Africa, Washington, DC)

SHEEP AND WOOL PRODUCTION IN THE UNITED STATES

In 2001, there were 6,915,000 sheep and lambs in the United States, with a value of $660,533,000[2]; and in 2000, the wool produced by U.S. sheep had a value of $15,377,000[3].

United States sheep and wool have declined in importance in recent years. In the mid 1990s, U.S. sheep numbers ranked 31st in the world, and U.S. wool production ranked 21st in the world.

In terms of U.S. farm income, cash receipts from farm marketings, sheep and lambs are exceeded by all farm animals and poultry, including cattle and calves, hogs, dairy products, chickens (layers), broilers, and turkeys.

The vast majority of U.S. sheep are still in the western range area, where most of them graze on arid and semi-arid pastures. For the most part, farm flocks utilize untillable areas and waste feeds.

Although the range area still dominates U.S. sheep production, many range operators have either retired or switched to cattle, and few young people have entered the business.

The United States does not produce enough lamb or wool to meet its requirements. It is an importer; importing an amount of lamb and mutton equivalent to more than 15% of production (see Table 2-12), and an amount of wool equivalent to 248% of production for the five-year period 1995–1999 (see Table 2-7).

[2]*Agricultural Statistics 2001*, USDA, p. VII-26, Table 7-41.

[3]*Ibid.*, p. VII-39, Table 7-63.

TABLE 2-7
IMPORTS OF WOOL COMPARED TO U.S. PRODUCTION[1]

Year	Wool Imports	Wool Production	Wool Imports as Percentage of Production
	- - - *(million clean [scoured] wool)* - - -		
	(lb)	*(lb)*	*(%)*
1995	88.8	33.5	265.1
1996	75.4	29.9	252.2
1997	76.4	28.3	270.0
1998	70.5	26.0	271.2
1999	43.0	24.6	174.8
5-yr. avg.	70.8	28.5	248.4

[1]*Agricultural Statistics 2001,* p. VII-35, Table 7-56.

UNITED STATES LAMB AND MUTTON PRODUCTION AND CONSUMPTION

Table 2-8 presents a recent historical account of lamb and mutton production in comparison to beef, veal, and pork production. United States per capita lamb and mutton consumption as a percentage of the per capita consumption of all meats is presented in Table 2-9.

TABLE 2-8
U.S. PRODUCTION OF MEAT[1]

Meat	Year	Production		Production as Percentage of All Meats[2]
		(million)		
		(lb)	*(kg)*	*(%)*
Beef and veal	1975	24,848	*11,271*	67
	1980	22,043	*9,999*	57
	1985	24,242	*10,996*	62
	1990	23,070	*10,405*	60
	1994	24,679	*11,194*	58
Pork (excluding lard)	1975	11,779	*5,343*	32
	1980	16,617	*7,537*	43
	1985	14,805	*6,736*	38
	1990	15,353	*6,964*	40
	1994	17,697	*8,027*	41
Lamb and mutton	1975	411	*186*	1.1
	1980	318	*144*	0.8
	1985	357	*162*	0.9
	1990	362	*164*	0.9
	1994	310	*141*	0.3

[1]*Agricultural Statistics 1990*, USDA, p. 295, Table 452 (1975 and 1980 statistics); *Agricultural Statistics 1995–1996*, USDA, p. VII-47, Table 436 (1985, 1990, and 1994 statistics).

[2]Total production of beef, veal, pork (excluding lard), lamb, and mutton.

TABLE 2-9
ANNUAL U.S. PER CAPITA MEAT CONSUMPTION OF ALL MEATS AND LAMB AND MUTTON, 1991–1999[1]

Year	All Red Meats and Poultry	Lamb and Mutton	Lamb and Mutton as Percentage of All Meats
	(lb)	*(lb)*	*(%)*
1991	170.2	1.0	0.6
1992	174.8	1.0	0.6
1993	174.6	1.0	0.6
1994	178.0	0.9	0.5
1995	178.0	0.9	0.5
1996	176.9	0.8	0.5
1997	175.2	0.8	0.5
1998	180.6	0.9	0.5
1999	186.0	0.9	0.5

[1]*Agricultural Statistics 2001,* p. XIII-5, Table 13-5.

From Tables 2-8 and 2-9, and other information, the following conclusions may be drawn:

1. U.S. lamb and mutton production plummeted in recent years, beef and veal production trended slightly downward, but pork production increased (see Table 2-8).
2. From 1991 to 1999, per capita lamb and mutton consumption as a percentage of all meats declined from 0.6 to 0.5% (see Table 2-9). During this same period, per capita poultry consumption steadily increased.
3. The consumption of fruits, vegetables, and cereal products has increased in recent years.
4. Sheep and goat meat is eaten by Moslems, Jews, and others, who, for religious reasons, do not eat pork.

AREAS OF SHEEP PRODUCTION

Sheep raising in the United States may be divided into two areas; namely, (1) the range-flock states, and (2) the farm-flock sheep states. Each of these areas will be discussed separately.

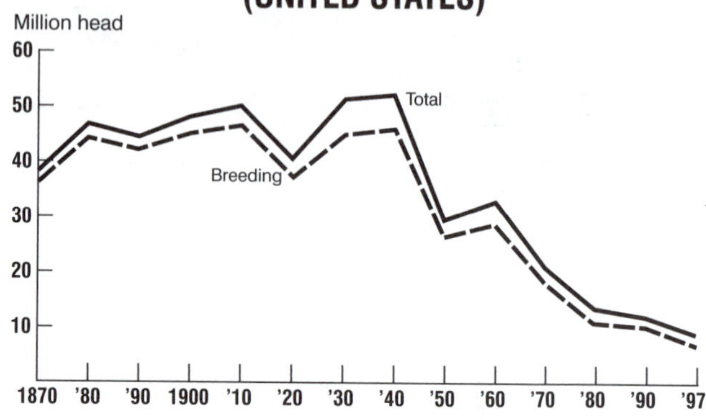

Fig. 2-8. Sheep and lambs in the United States, 1870–1997. (Courtesy, U.S. Department of Agriculture)

RANGE SHEEP STATES

Sheep are produced in every state, but the 17 western range states account for 78.6% of the nation's breeding sheep in 2001.[4] Texas is the leading state, with about 16% of the total stock or breeding sheep. For the most part, the western range sheep is characterized by large bands of from 1,000 to 5,000 ewes. In

[4]*Agricultural Statistics 2001*, USDA, p. VII-29, Table 7-46. The 17 western range states are: Washington, Oregon, California, Idaho, Nevada, Montana, Utah, Colorado, North Dakota, South Dakota, Kansas, Oklahoma, Texas, Arizona, New Mexico, Wyoming, and Nebraska.

TABLE 2-10
BREEDING SHEEP SURVEY PERCENT BY SIZE GROUPS, UNITED STATES 2000–2001[1, 2]

Regions	Operations Having							
	1–99 Head		100–499 Head		500–4,999 Head		5,000+ Head	
	2000	2001	2000	2001	2000	2001	2000	2001
	(%)							
Operations	91.2	90.8	7.2	7.5	1.6	1.6	0.1	0.1
Inventory.	27.9	28.8	22.0	23.8	35.2	33.7	14.8	13.7

[1]*Agricultural Statistics 2001,* p. VII-34, Table 7-55.

[2]Percents reflect distributions of end of year survey.

Fig. 2-9. A U.S. range herd of ewes. (Courtesy, Dr. J. E. Nel, University of Wyoming, Laramie)

general, these bands are run on unenclosed land—extensive operations. As more than half of the ranges of the West are publicly owned and are likely to continue so, fences will not be built simply because (1) it is not prudent to go to so much expense on public lands, and (2) there are game and fish restrictions on fencing public lands that might hinder wildlife migration. On the other hand, scarcity and high cost of labor have resulted in the fencing of a considerable amount of private holdings in the southwestern United States. As most of these ranges are used the year round and the bands are not herded on mountain ranges during the summer months, fenced range has been satisfactory.

Within the western range, there are great variations in topography, rainfall, and vegetation. Thus, in most cases, the sheep production of this area is a migratory type of enterprise, in which deserts, plains, foothills, and mountains may be used during different seasons of the year in such manner as to obtain as nearly year-round foraging as possible. In general, the ranges of the southwestern states do not afford sufficiently good grazing to produce milk-fat market lambs. Wool production with Rambouillet-type sheep, therefore, is of comparatively greater importance in this area than it is in the Northwest. On the better ranges of the Northwest, however, the objective is always that of producing a fat market lamb at weaning time, and feeder lambs result only because the ranges proved inadequate (perhaps through lack of moisture) or because of the usual rejects in culling. Because the vegetation is more abundant in the Northwest, a larger crossbred type of sheep predominates in this area. This type produces a coarser fleece than the Rambouillet-type ewes of the Southwest and yields lambs of more desirable type and quality.

Most of the better range operators provide supplemental feed during a part of the year, especially during the lambing season and during times when the range is covered with snow. Alfalfa from the irrigated areas is the chief hay crop; and corn, oats, barley, and various protein supplements may be used on occasion. In the trade, wools produced in the range states are known as *territory wools*.

Fig. 2-10. White-faced range bands like this one near Kemmerer, Wyoming, produce some of the best wool in the world. (Courtesy, Dr. J. E. Nel, University of Wyoming, Laramie)

FARM-FLOCK SHEEP STATES

This area includes the sheep production throughout the United States, except in the 17 western range states. It embraces the farm flocks, ranging from a few head to a thousand or more, which are kept on the farms of the East, South, and Midwest. In the trade, the wool coming from farm flocks is known as *native wool*.

Fig. 2-11. A Midwestern farm flock. (Courtesy, A. O. Smith, Harvestore Products, Arlington Heights, IL)

Many of the smaller farm flocks of this area are kept primarily as scavengers, and, unfortunately, they are often accorded the neglect of a minor enterprise. In general, the lambs from farm flocks are sold directly for harvest, only a few finding their way to feedlots for further fattening. Farm flocks carry a heavier infestation of parasites than do range bands, and the fleeces lack the care and uniformity accorded to range wool production. Also, some lambs coming from farm flocks are not docked or castrated, a neglect which is a rarity on the western range where sheep production is more of a specialty.

Purebred flocks as well as commercial enterprises characterize the farm sheep flocks. Purebred producers market their surplus stock (1) to other purebred breeders, and (2) as breeding stock for commercial enterprises.

LEADING STATES IN SHEEP PRODUCTION

Some idea of the relative importance of sheep production in the leading states may be obtained through studying Table 2-11.

TABLE 2-11
TEN LEADING STATES IN BREEDING SHEEP, THEIR LAMB CROP, AND WOOL PRODUCTION[1]

State	Breeding Sheep	Lamb Crop	Wool Production
	(thousand)	(thousand)	(thousand)
			(lb)
Texas	950	700	7,506
Wyoming	460	390	4,560
California	380	290	4,000
Utah	360	330	3,060
Montana	340	340	3,315
South Dakota	320	320	2,800
Idaho	245	255	2,190
New Mexico	230	160	2,120
Colorado	210	210	3,310
Iowa	170	210	1,340
U.S. total	5,164	4,733	46,592

[1]From: *Agricultural Statistics 2001*, as follows: The 10 leading U.S. states in breeding sheep numbers in 2000, ranked in descending order, p. VII-29, Table 7-46; lamb crop in thousands of these same 10 states for 2000, p. VII-30, Table 7-47; and wool production in thousands of these same 10 states for 2000, p. VII-38, Table 7-62.

Texas leads substantially in sheep numbers, a position which it also holds in cattle, horse, and goat numbers. Naturally, it also leads in lamb crop and wool production. Nine of the ten leading sheep states are in the western range sheep area of the United States.

LAMB, MUTTON, AND WOOL IMPORTS AND EXPORTS

Sheep producers are prone to ask why the United States buys lamb from abroad. Conversely, consumers sometimes wonder why export any lamb—no matter how insignificant the quantity. There may be some justification for these concerns. In recent years, lamb and mutton imports have been equivalent to about 2.9% of production of all meats (see Table 2-12).

Table 2-12 also reveals that the United States imports more lamb and mutton than it exports, and that lamb imports and exports represent a small portion of all meats imported and exported.

During the period 1996 to 2000, the United States imported an average of 102.2 million lb of lamb and mutton (see Table 2-13). Lamb imports come principally from Australia and New Zealand.

The amount of lamb and mutton imported from abroad depends to a substantial degree on (1) the level of U.S. meat production, (2) consumer buying

TABLE 2-12
U.S. IMPORTS AND EXPORTS OF LAMB AND MUTTON AND OTHER MEATS[1]

Year	Imports: Lamb and Mutton	Imports: All Meats	Imports: Lamb and Mutton as Percentage of All Meats	Exports: Lamb and Mutton	Exports: All Meats	Exports: Lamb and Mutton as Percentage of All Meats
	(million)	(million)		(million)	(million)	
	(lb)	(lb)	(%)	(lb)	(lb)	(%)
1996	72	2,764	2.6	6	2,853	0.2
1997	83	3,061	2.7	5	3,185	0.2
1998	112	3,461	3.2	6	3,407	0.2
1999	113	3,813	3.0	5	3,700	0.1
2000	131	4,174	3.1	6	3,808	0.2
5-yr. avg.	102.2	3,455	2.9	5.6	3,391	0.2

[1]*Agricultural Statistics 2001*, USDA, p. VII-43, Table 7-69. Carcass weight equivalent of all meat.

TABLE 2-13
IMPORTS OF LAMB AND MUTTON COMPARED TO U.S. PRODUCTION[1]

Year	Lamb and Mutton Imports	Lamb and Mutton Production	Lamb and Mutton Imports as Percentage of Production
	---------- (million) ----------		
	(lb)	(lb)	(%)
1996	72	258	27.9
1997	83	253	32.8
1998	112	242	46.2
1999	113	238	47.5
2000	131	225	58.2
5 yr. avg.	102.2	243.2	42.0

[1]From: *Statistical Abstracts of the United States 2001*, p. VII-34, Table 7-53.

power, (3) lamb prices, (4) quotas and tariffs, (5) any restrictions because of disease, and (6) strength of the U.S. dollar. When lamb prices are high, more lamb is imported. Actually, there may be some virtue in judiciously increasing imports of lamb during times of scarcity and high prices, as an alternative to pricing lamb out of the market and to having people get used to going without lamb because of the seasonality in its production. The position of butter versus margarine is proof enough of the price factor, for butter was priced out of the market. Table 2-13 compares lamb and mutton imports to U.S. production in recent years.

Because quotas and tariffs have always been in politics and are subject to change, sheep producers need constantly to increase efficiency of production as a means of meeting foreign competition. The import duties on live goats and sheep, and on goat meat, sheep meat, and lamb are given in Table 2-14.

The amount of lamb exported from this country is dependent upon (1) the volume of meat produced in the United States, (2) the volume of meat produced abroad, and (3) the relative vigor of international trade, especially as affected by buying power and trade restrictions. In 1996–2000, exports of lamb and mutton ranged from 5 to 6 million lb per year (see Table 2-12).

TABLE 2-14
2001 IMPORT DUTIES ON LIVE GOATS AND SHEEP, AND ON GOAT MEAT, SHEEP MEAT, AND LAMB[1]

Item	Unit	Rates of Duty		
		General[2]	Special[3]	Other[4]
		------ ($) ------		
Goats, live	per head	68¢	Free	3.00
Sheep, live	per head	Free		3.00
Goat meat, fresh, chilled, or frozen	per kg	0.7¢	Free	15.4¢
Sheep meat, fresh or chilled	per kg	2.8¢	Free	11¢
Lamb, fresh, chilled, or frozen	per kg	0.7¢	Free	15.4¢

[1]Harmonized Tariff Schedule of the United States (2001). U.S. International Trade Commission.

[2]General or normal trade relations (NTR) rates.

[3]Programs eligible for special tariff treatment..

[4]Afghanistan, Cuba, Laos, North Korea, Vietnam.

Fig. 2-12 shows that, since 1975, United States (1) wool production has fallen with the decline in sheep numbers; (2) wool imports have increased; (3) wool exports, which are small, have increased; (4) wool mill use has held steady; and (5) total wool use has trended upward. Since 1996, wool production has decreased further and imports have remained relatively steady (See Table 2-7).

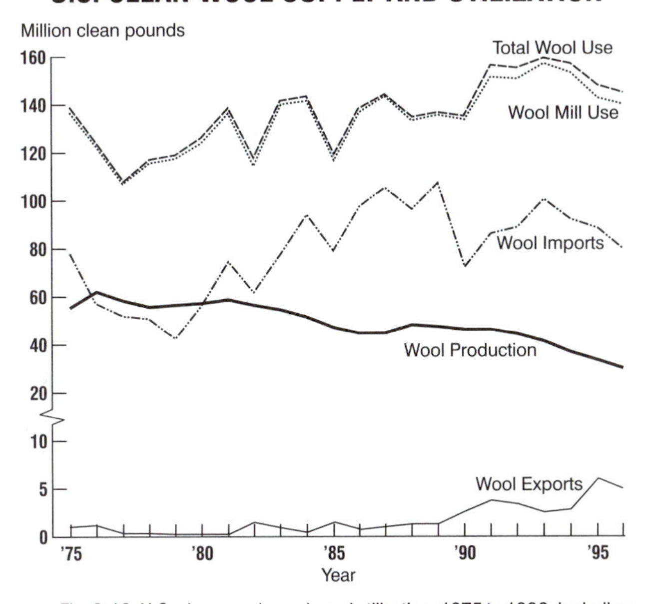

Fig. 2-12. U.S. clean wool supply and utilization, 1975 to 1996; including (1) wool production, (2) wool imports, (3) wool exports, (4) wool mill use, and (5) total wool use. (From: *Cotton and Wool, Situation and Outlook Yearbook*, USDA, ERS, Nov. 1996)

FACTORS FAVORABLE TO SHEEP PRODUCTION

As compared to other classes of livestock, sheep possess the following natural advantages:

1. Compared with beef cows, which may produce 60% of their weight in offspring annually, the ewe can produce 150% or more.

2. For every 22 lb of range forage consumed by ewe-lamb combinations, 1 lb of lamb is produced for human use.

3. Top-quality market lambs can be produced from good grass, other forage, and milk, without grain feeding.

4. They are unexcelled in the utilization of the more arid types of grazing.

5. Sheep are able to convert weeds, brush, grass, and other plants on rangeland and pastureland to useful products, including meat, milk, fiber, leather, and pharmaceuticals. They are excellent scavengers for gleaning fields.

6. Compared to cattle, they produce more liberally in proportion to what they consume.

7. They produce two products, lambs and wool, which are available for market at two different periods of the year; and it seldom happens that both products sell at bottom prices the same year.

8. Their returns come quickly; lambs may be marketed eight months after the ewes are bred.

9. Their habit of bedding down on the highest areas of the field or range leaves the larger part of the droppings at the places where they are most needed. Moreover, the form in which sheep manure is dropped and the way it is tramped into the soil insure a smaller waste than is possible under any other system of stock farming.

10. Their wool clip is easily stored and shipped, thus making wool production ideally suited to a frontier type of agriculture.

11. The energy requirements for sheep production on rangeland are lower than those for other livestock enterprises. Also, the energy requirements for fiber production by sheep are lower than those for synthetic fibers.

12. Sheep are highly adaptable and noncompetitive with humans for food.

13. Special sheep products have evolved in recent years, including (a) the expansion of hand crafting of wool, which has resulted in the development of special types and breeds of sheep to produce a variety of colored wool and coarse wool for hand processing; and (b) the production of high quality cheese from sheep milk.

14. Lower investment costs in breeding animals, along with facilities.

FACTORS UNFAVORABLE TO SHEEP PRODUCTION

There are, however, many factors which are quite unfavorable to sheep production, as many sheep producers will testify to their regret. Some of these are:

1. Wool has always been in politics, and it is apt to so remain. It is rather difficult, therefore, to predict prices over a long period of time.

2. The National Wool Act of 1954, which provided support prices for wool and mohair, was completely phased out in 1995.

3. Consumption of lamb has declined to the point where it is a minor product, with the result that on-foot and retail marketing problems have increased.

4. Sheep are very much subject to attack from numerous predatory animals, including dogs.

5. When disease or injury strikes, sheep have less resistance than other classes of livestock.

6. Herding is not a particularly attractive profession, thus resulting in a scarcity of satisfactory herders.

7. Sheep are quite susceptible to a number of internal or external parasites.

8. The competition from synthetic fibers affects the demand for wool.

9. Harvesting plants for lamb and processing mills for wool have decreased to the point where there is questionable competition in the market.

10. There is a paucity of research and promotional dollars, which will worsen. The industry voted on a referendum on February 6, 1996, but subsequently the U.S. Secretary of Agriculture recalled the vote. The second vote on October 1, 1996, failed. Note: Had the referendum passed, the sheep checkoff called for 1¢/lb on lamb (domestic and imported) and 2¢/lb on domestic raw wool and imported wool tops and textiles, which was projected to generate $13 million annually.

WORLD AND U.S. SHEEP PRESENT STATUS

1. **U.S. sheep population has declined drastically in recent years, whereas world sheep numbers merely trended downward.** In 1979–81, there were 12,670,000 sheep in the United States; in 1994, there were 9,600,000—a decrease of 24% during the 15-year period.[5]

In 1979–81, the world sheep population was 1,088,794,000; in 1994, it was 1,086,661,000—a decrease of only 0.2% during the 15-year period.[6]

2. **Per capita consumption of lamb and mutton has plummeted.** In 1970, it was 3.3 lb; in 1994, it was 1.3 lb—down 60% during the 24-year period (see Table 2-8).

3. **Synthetic fibers have increased.** In 1995, synthetic fibers accounted for 56% of the total per capita fiber consumption (including cotton, wool, synthetics, flax, and silk), whereas wool accounted for only 1.8%, or a mere 1.4 lb, of total per capita fiber consumption. Also, wool remains a minor fiber compared to the per capita consumption of cotton (see Fig. 2-13).

4. **Imports of lamb, mutton, and wool have increased.** Prior to 1957, imports of lamb and mutton represented less than 1% of the nation's production. In

---

[5]*FAO Production Yearbook*, Vol. 48, 1994, p. 102, Table 90.

[6]*Ibid.*

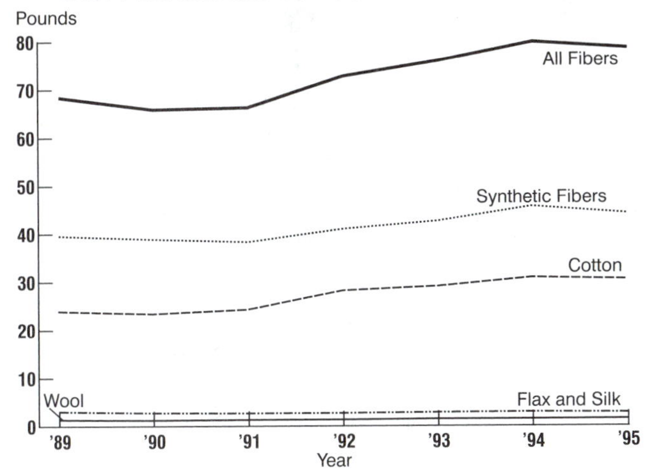

Fig. 2-13. U.S. per capita consumption of fibers totals 78.9 lb, with a percentage breakdown as follows: synthetic fibers, 56%; cotton, 38%; flax and silk, 3.6%; and wool, 1.8%. (Courtesy, USDA; from *Cotton and Wool Situation and Outlook Yearbook*, Nov. 1996, p. 50, Table 31)

the late 1990s, lamb and mutton imports were equivalent to 42.0% of the production (see Table 2-13).

During the five-year period 1995 to 1999, the U.S. imported an annual average of 71 million lb of clean (scoured) wool to augment its production of 29 million lb; wool imports were 248.4% of production (see Table 2-7).

5. **Grazing public lands is becoming more difficult and more costly.** Today, in the 11 western public land states, the federal government owns and administers approximately 320 million acres on which grazing is allowed. At one time or another during the year, domestic sheep and cattle graze about half of these public lands. The number of livestock grazed on the publicly owned rangeland annually accounts for slightly over 50% of the nation's sheep numbers and approximately 4 to 5% of the total cattle numbers. The bulk of federal land is administered by the following six agencies: the Bureau of Land Management, the U.S. Forest Service, the Bureau of Indian Affairs, the Department of Defense, the National Park Service, and the Bureau of Reclamation. A bewildering number of regulations, predators, labor, and higher grazing fees plague both sheep and cattle producers who utilize federal lands.

WORLD AND U.S. SHEEP FUTURE OUTLOOK

The future of the U.S. sheep and wool industry is of concern to all those, including the author, who enjoy (1) really delicious lamb, and (2) a warm woolen blanket on a cold, wintry night.

U.S. sheep are an endangered species, plagued by—

1. Too many breeds, accompanied by lack of uniformity of lamb and wool.
2. Too heavy slaughter weights, too much fat, and too little lean.
3. A paucity of industry-wide promotion dollars, resulting from the twice-failed passage of a checkoff referendum in 1996.
4. Too many diseases and parasites.
5. Herding being an unattractive profession.
6. The uncertain availability of grazing permits on federal lands, greater environmental restrictions, excessive predator losses, scarcity of herders, and higher operating costs of federal grazing lands.
7. Too few markets for live lambs, too few lamb harvests, and too few lamb retailers.
8. Lamb being produced seasonally.
9. Broilers, turkeys, pork, and synthetic fibers being on a roll with these products increasing their share of the consumer's dollar, at the expense of lamb and wool.

The author predicts that all the above will make for fewer and fewer U.S. sheep, along with less and less domestic lamb and wool production.

[*Note:* You haven't lived unless you have eaten New Zealand's famous *Canterbury Lamb*, formerly produced primarily by crossing Southdown rams on Romney ewes and slaughtered at about 65-lb liveweight. M. E. E.]

WORLD GOAT PRODUCTION

Goats are a minor livestock concern in the United States, but in some other areas of the world they are of major importance. Goats provide three products, pri-

Fig. 2-14. Meat (Spanish) goats. (Courtesy, Dr. Maurice Shelton, San Angelo, TX)

marily: milk, meat, and fiber. Table 2-15 lists the major regions of the world, their goat population, meat, and milk production.

■ **Leading goat meat producing countries of the world**—Table 2-16 shows the 10 leading goat meat producing countries of the world, and the total world production. Three countries—China, India, and Pakistan—produce 53% of the goat meat of the world.

Greece, which produced 40 million metric tons of goat meat in 1994, accounts for 44% of the goat meat produced in Europe.[7] Mexico, which produced 39 million metric tons of goat meat in 1994, produces about 80% of the goat meat of North America.

TABLE 2-15
WORLDWIDE GOAT POPULATION, MEAT, AND MILK PRODUCTION, BY MAJOR REGIONS[1]

Region	Number of Goats	Goat Meat Production	Goat Milk Production
	(1,000 hd)	*(1,000 metric tons)*	
Asia	373,005	2,135	6,135
Africa	176,089	662	2,003
South America	22,819	74	159
North America	14,944	49	159
Europe	14,809	90	1,633
Former U.S.S.R.	5,810	1,924	303
Oceania	1,007	22	—
World total	609,488	3,057	10,480

[1]From: *FAO Production Yearbook*, Vol. 48, 1994. Number of goats from pp. 192–194, Table 90; goat meat production from pp. 204–205, Table 95; goat milk production from pp. 218–219, Table 100. All figures except U.S.S.R. are for 1994; U.S.S.R. figures are for 1979–81.

TABLE 2-16
LEADING COUNTRIES IN GOAT MEAT PRODUCTION[1]

Country	Production
	(1,000 metric tons)
China	751
India	470
Pakistan	391
Nigeria	130
Iran	101
Bangladesh	98
Ethiopia	62
Turkey	62
Indonesia	60
Egypt	41
World	3,057

[1]From: *FAO Production Yearbook*, Vol. 48, 1994, pp. 204–205, Table 95. Data for 1994.

[7]The statistics given in the narrative that follows are from *FAO Production Yearbook*, Vol. 48, 1994.

Fig. 2-15. Angora goats and a Border Collie dog. (Courtesy, Mohair Council of America, San Angelo, TX)

Worldwide, the production of goat meat equals 44% of the total production of lamb and mutton, 6.0% of the beef and veal, and 3.9% of the pork.

■ **Leading goat milk producing countries of the world**—Worldwide, the leading goat milk producers are: In Africa, 65% of the goat milk is produced by Algeria, Mali, Somali, and Sudan. In North America, Mexico produces 89% of the goat milk. In South America, Brazil produces 76% of the goat milk. In Asia, Bangladesh, China, India, Indonesia, Iran, Pakistan, and Turkey produce 91% of the goat milk. In Europe, France, Greece, Italy, Spain, and Yugoslavia produce 92% of the goat milk.

■ **Leading mohair producing countries of the world**—At the present time, the goat is bred on a commercial basis for fiber production in five countries: South Africa, United States, Turkey, Argentina, and New Zealand. The mohair production of each of these nations is shown in Table 2-17.

Although goats are rather widely distributed throughout the United States, the production of goats and mohair is of economic importance in only a few states. Table 2-18 summarizes important data relative to the U.S. mohair industry.

TABLE 2-17
LEADING MOHAIR-PRODUCING COUNTRIES[1]

Country	Production	
	(mil. lb)	(mil. kg)
South Africa	12.8	*5.8*
United States	12.0	*5.4*
Turkey	1.3	*0.6*
Argentina	1.1	*0.5*
New Zealand	0.4	*0.2*
Total	27.6	*13.0*

[1]From: Mohair Council, San Angelo, TX. Data for 1995.

TABLE 2-18
MOHAIR PRODUCTION, PRICE AND VALUE BY SELECTED STATES AND 3-STATE TOTAL, 1998–99[1]

State	Goats Clipped		Average Clip per Goat		Production		Price per Pound[2]		Value[3]	
	1998	1999	1998	1999	1998	1999	1998	1999	1998	1999
	(1,000 head)		(lb)		(1,000 lb)		($)		(1,000 $)	
Texas	620	375	7.5	6.8	4,650	2,550	2.59	3.68	12,044	9,384
New Mexico	44	35	5.5	5.5	242	193	1.30	1.85	315	357
Arizona	36	34	4.5	4.5	162	153	1.25	1.80	203	275
Total (3 states)	700	444	7.2	6.5	5,054	2,896	2.49	3.46	12,562	10,016

[1]*Wool and Mohair*, 2000, National Agricultural Statistics, USDA, provided by Mohair Council of America, San Angelo, TX.

[2]Average local market price for mohair sold; does not include incentive payment.

[3]Production multiplied by marketing year average price for individual states; 3-state value is summation of state values.

As noted, Texas is by far the leading state in Angora numbers and mohair production. It accounts for more than 90% of U.S. mohair, with most of the goats located in the Edwards Plateau—a rough and broken area, with much brush and some grass.

U.S. goat numbers declined in the 1960s, primarily in response to low prices for mohair, along with rising labor costs and a resurgence of the predator problems. Other reasons for the drastic reduction in Texas were higher revenues from cattle raising, oil rights, and hunting permits.

Today, the United States is the second largest producer of mohair. In 1999, the goat raisers of this country produced 3 million lb of mohair.

The United States is a major exporter of mohair, the bulk of which goes to Great Britain and the European continent.

As may be noted in Table 2-18, the Angora goats of Texas sheared an average annual clip of 6.8 lb of unscoured fleece per animal in 1999. Purebred herds often clip double this amount. Much of the domestic mohair, especially that produced in the Southwest, is taken off in two clips per year, in the spring and fall, whereas Turkish mohair is usually allowed a full year's growth prior to shearing.

TABLE 2-19
ALL GOATS, TEXAS, JANUARY 1, 1994–97[1]

Year	1,000 Head
1994	1,960
1995	1,850
1996	1,900
1997	1,650

[1]Courtesy, U.S. Department of Agriculture.

TABLE 2-20
ANGORA GOATS, STATES AND TOTAL, JANUARY 1, 1995–97[1]

State	1995	1996	1997
	(1,000 head)		
Arizona	52	50	40
New Mexico	85	85	77
Oklahoma	15	15	10
Texas	1,250	1,250	1,000
Total	1,402	1,400	1,127

[1]Courtesy, U.S. Department of Agriculture.

QUESTIONS FOR STUDY AND DISCUSSION

1. Discuss the factors which account for each of the five leading sheep- and wool-producing countries (see Tables 2-1 and 2-4) holding its respective rank.

2. Discuss the factors which account for each of the five leading goat meat-producing and mohair-producing countries (see Tables 2-2, 2-15, and 2-16) holding its respective rank.

3. Discuss the importance of sheep, lamb/mutton, wool, and sheep milk on a worldwide basis (see Tables 2-1, 2-2, and 2-4).

4. Discuss the importance of goats, goat meat, goat milk, and mohair on a worldwide basis.

5. As a food, what position do lamb, mutton, and goat meat hold around the world? What position do they

hold in the United States, and what factors and changes have contributed to this position?

6. Although Australia ranks third in sheep numbers, what factors cause it to be acclaimed as the world's leading sheep country?

7. Why did sheep numbers increase in many parts of the world after World War II, but, simultaneously, decline in the United States? What is the trend in sheep numbers in the United States today?

8. Assuming that a young person had no "roots" in a particular location, in what area of the United States would you recommend that a sheep enterprise be established? Justify your answer.

9. What is the difference between a farm flock and a range band of sheep?

10. Discuss the factors which account for each of the five leading sheep-producing states (see Table 2-10) holding its respective rank.

11. What is the importance of lamb/mutton, and wool as exports and as imports, and what determines the amount of lamb/mutton, and wool that the United States exports and imports (see Table 2-12)?

12. Looking into the future, do you feel that the factors favorable to sheep production outweigh the unfavorable factors? Justify your answer.

13. Does the future of the sheep industry warrant optimism or pessimism? Justify your answer.

14. Discuss the factors that make for (a) the leading goat meat producing countries of the world (Table 2-16), (b) the leading mohair producing countries of the world (Table 2-17), and (c) the leading goat milk producing countries of the world.

SELECTED REFERENCES

Title of Publication	Author(s)	Publisher
Angora Goat and Mohair Production	M. Shelton	M. Shelton, San Angelo, TX, 1993
Animal Science	M. E. Ensminger	Interstate Publishers, Inc., Danville, IL, 1991
Profitable Sheep	S. B. Collins R. F. Johnson	The Macmillan Company, New York, NY, 1956
Profitable Sheep Farming	M. M. Cooper R. J. Thomas	Farming Press, Ltd., Ipswich, England, 1971
Program for Improving Returns for Domestic Sheep Producers, A		Report prepared for American Sheep Producers Council, Inc., Denver, CO, June, 1962
Sheep Book, The	J. McKinney	John Wiley & Sons, Inc., New York, NY, 1959
Sheep Breeding	Ed. by G. J. Tomes D. E. Robertson R. J. Lightfoot	Butterworth & Co. (Publishers) Ltd., London, England, 1970
Sheep Management and Diseases	H. G. Belschner	Angus and Robertson, Sydney, Australia, and London, England, 1965
SID Sheep Production Handbook	Staff	American Sheep Industry Assn., Englewood, CO, 1997
Sheep Science	W. G. Kammlade, Sr. W. G. Kammlade, Jr.	J. B. Lippincott Co., Philadelphia, PA, 1955
U.S. Sheep and Goat Industry: Products, Opportunities, and Limitations, The, Report No. 94		Council for Agricultural Science and Technology (CAST), Ames, IA, 1982

CHAPTER 3

Chimaera (a sheep X goat hybrid), without pride of ancestry or hope of posterity. Such hybrids are called *chimeras* (after the mythical monster with a lion's head, goat's body, and serpent's tail). (Courtesy, Agricultural Research Council, Cambridge, England)[1]

GENETICS AND SELECTION[2]

Contents *Page*

[1]Scientists at the Institute in Cambridge, England, mingled new embryos from both sheep and goats when each consisted of no more than 4 to 8 cells. Then, these were placed in the wombs of surrogate sheep or goat mothers and allowed to grow to term. Although such experiments will trigger debate, scientists point to practical benefits; for example, it should make it easier to rear embryos of endangered species in the wombs of other species, or even create hybrids as valuable as the indomitable mule.

[2]The author wishes to acknowledge the very considerable assistance of Dr. Clair E. Terrill, U.S. Department of Agriculture, in updating this chapter.

Contents *Page*

While most of the modern advances in genetics are out of their realm, sheep and goat producers must realize the importance of genetics to the profitability of their operations. Sheep and goat producers must select and breed animals whose genetics will (1) increase yearly and lifetime offspring production, (2) increase fiber production, (3) utilize feed efficiently, (4) increase milk and cheese production, and (5) increase adaptability to prevailing environmental conditions on each farm or ranch and to each prevailing ecosystem. To do this, sheep and goat producers need a basic understanding of some of the principles of the sciences of genetics, particularly as applied to sheep and goat production.

In most cases, the principles of genetics discussed herein apply equally to both sheep and goats.

EARLY ANIMAL BREEDERS

The laws of heredity apply to sheep and goat breeding exactly as they do to the breeding of all classes of farm animals. But the breeding of sheep and goats is more flexible than the breeding of cattle or horses because: (1) sheep and goats normally breed at an earlier age, thus making for a shorter interval between generations, and (2) they have more multiple births.

Until very recent times, the general principle that like begets like was the only recognized law of heredity. That the application of this principle over a long period of time has been effective in modifying animal types in the direction of selection is evident from a comparison of present-day types and breeds of sheep and goats.

There can be little doubt that individuals like Bakewell, the English patriarch, and other 18th century breeders had made a tremendous contribution in pointing the way toward livestock improvement before Mendel's laws became known to the world in the early part of the 20th century. Bakewell's use of progeny testing through his ram letting was truly epoch making. He and other pioneers had certain ideals in mind, and according to their standards, they were able to develop some nearly perfect specimens. These men were intensely practical, never overlooking the utility value or the market requirements. No animal met with their favor unless it was earned by meat upon the back, weight and quality of fiber produced, pounds gained for pounds of feed consumed, or some other performance of practical value. Their ultimate goal was that of furnishing better animals for the market and/or lowering the cost of production. It must be just so with the master breeders of the present and future.

Others took up the challenge of animal improvement where Bakewell and his contemporaries left off slowly but surely molding animal types. Thus, during the past 100 years, remarkable progress has been made in breeding better sheep and goats.

MENDEL'S CONTRIBUTION TO GENETICS

Modern genetics was really founded by Gregor Johann Mendel, a cigar-smoking Austrian monk, who conducted breeding experiments with garden peas from 1857 to 1865, during the time of the Civil War in the United States. In his monastery at Brünn (now Brno, in the Czech Republic), Mendel, with a powerful curiosity and a clear mind, revealed some of the basic principles of hereditary transmission. In 1866, he published in the proceedings of a local scientific society a report covering 8 years of his studies, but for 34 years his findings went unheralded and were ignored. Finally, in 1900, 16 years after Mendel's death, three European biologists independently duplicated his findings, and this led to the dusting off of the original paper published by the monk 34 years earlier.

The essence of Mendelism is that inheritance is by particles or units (called genes), that these genes are present in pairs—one member of each pair having come from each parent—and that each gene maintains its identity generation after generation. Thus, Mendel's work with peas laid the basis for two of the general laws of inheritance: (1) the law of segregation, and (2) the independent assortment of genes. Later genetic principles have been added; yet, all the phenomena of inheritance, based upon the reactions of genes, are generally known under the collective term *Mendelism*.

Fig. 3-1. Gregor Johann Mendel (1822–1884), a cigar-smoking Austrian monk, who founded modern genetics through breeding experiments with garden peas. (Courtesy, The Bettmann Archive, Inc., New York, NY)

Thus, modern genetics is really unique in that it was founded by an amateur who was not trained as a geneticist and who did his work merely as a hobby. During the years since the rediscovery of Mendel's principles (in 1900), many additional genetic principles have been added, but the fundamentals as set forth by Mendel have been proved correct in every detail. Inheritance in both plants and animals follows the biological laws discovered by Mendel.

SOME FUNDAMENTALS OF HEREDITY

In the sections that follow, no attempt will be made to cover all of the diverse field of genetics and animal breeding. Rather, the author presents a condensation of a few of the known facts in the field and briefly summarizes their application to sheep and goats.

GENES AND CHROMOSOMES

The bodies of all animals are made up of millions or even billions of tiny cells, microscopic in size. Each cell contains a nucleus in which there are a number of pairs of bundles called chromosomes. Chromosomes carry all the hereditary characteristics of animals. The

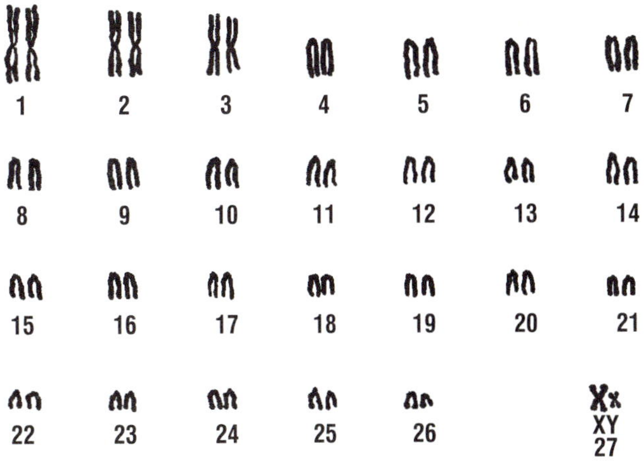

Fig. 3-2. Chromosome pairs of the sheep as they would appear in a photomicrograph made from chromosomes of the metaphase of cell division, when they can be stained and observed. There are 26 pairs of autosomes and 1 pair of sex chromosomes (XY)—all the genetic information necessary to make a male sheep.

central inner portion of each chromosome contains a long, double helical molecule called deoxyribonucleic acid, or DNA for short. The DNA molecule is the genetic material—the genetic code. Genes form a portion of each DNA molecule, each in a fixed or special position, called a *locus*, of each chromosome. Since chromosomes occur in pairs, genes are also in pairs. They are the functional unit of inheritance. The nucleus of each body cell of sheep contains 54[3] chromosomes (sperm and eggs contain half this number), whereas there are perhaps thousands of pairs of genes. These genes determine all the hereditary characteristics of living animals via coded information in their DNA. Thus, inheritance is transmitted by units rather than by the blending of two fluids, as our ancestors thought.

CALLIPYGE GENE[4]

In recent years sheep have appeared in the show rings, club lamb sales, and sire and ewe sales that have exhibited an abnormally large amount of muscling (muscle hypertrophy). These sheep have been successful at selling and placing in these shows and sales with varying degrees of success.

The condition (believed to be genetic in origin) was first identified in a Dorset flock in Oklahoma and has since been identified in several breeds of sheep. The gene was named *callipyge gene*, *i.e.*, nicely proportioned buttocks.

The *callipyge gene* and the increased muscularity which results from its presence has the potential to improve lean yield of lamb carcasses and increase loin eye size and total muscling of the carcass. Poor lean yield, small loin eye and rib crop size have been major impediments to increased lamb consumption. Could this be the answer to these problems faced by the sheep industry?

■ **Two presentations at the University of Wisconsin Sheep Day in 1994**—A very interesting situation is occurring in the U.S. sheep industry. A new gene has emerged which imparts extremely heavy muscling (HM) to leg and loin muscles of animals carrying this gene. It is called the *callipyge gene*, which translates as *thickly proportioned buttocks*. The good news about this gene is that animals receiving it have substantially higher dressing percentages, increased

Fig. 3-3. Four ewe lambs sired by two rams heterozygous for the callipyge gene for heavy muscling. The two ewe lambs on the left are paternal half sisters and the two ewe lambs on the right are paternal half sisters. Within each pair, the ewe lamb on the left received the callipyge gene from her sire and the ewe lamb on the right received the gene for normal muscling. (Courtesy, Sam Jackson, Texas Tech University, Lubbock)

yields of valuable loin and leg muscle, and reduced amounts of carcass fat—exactly what the industry has been seeking. However, the not-so-good news is that scientific investigations are finding that these thicker muscles are significantly less tender. Given these opposing quantitative and qualitative traits in HM animals, what are lamb producers to do at this time?

There are two facts that should be kept in mind. First, the emergence of this gene and the carcass characteristics it produces are very exciting and significant events. The thickness and bulge of the legs and rib-eye sizes of HM lamb carcasses astonish carcass evaluators. Although there may be disadvantages associated with the HM gene, its presence creates potential opportunities to increase dramatically productivity in the sheep industry.

Secondly, information about HM animals is accumulating and there still is a great deal to be learned and confirmed. The evaluation of HM animals may be somewhat different in one or two years than it is right now.

Currently, we know the following about this genetic phenomenon:

1. This gene was first observed in a Dorset ram in a flock in Oklahoma in 1983, which sired lambs with an abnormally large amount of muscling. Although some people have called the condition *double muscling*, the lambs do not have two sets of muscles. Rather, they have a very large expression of muscling in the normal muscle pattern (muscle hypertrophy).

2. The HM condition is passed to offspring through what appears to be a single dominant gene. Rams which carry one dominant gene for HM, when

---

[3]Cattle have 60 chromosomes; swine have 38; horses, 64; and goats, 60. Of course, each sperm and egg has half this number. *Note:* Normal chromosomes occur in pairs in body cells (diploid number).

[4]In the preparation of this article, the senior author drew heavily from *Louisiana Sheep Update*, published by Louisiana State University, Baton Rouge, Louisiana, 1995.

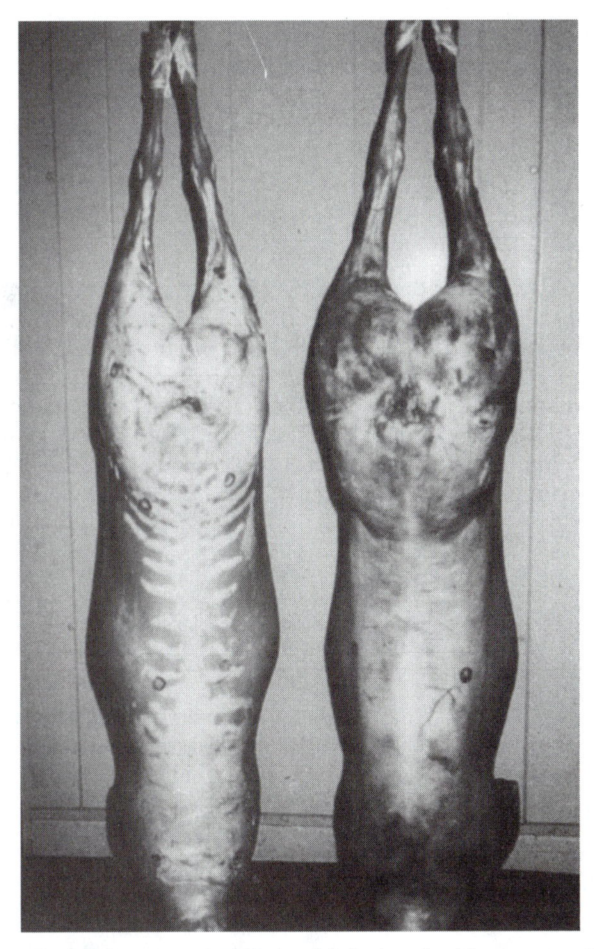

Fig. 3-4. Carcasses of two lambs of similar breeding. The carcass on the right is from a lamb which carried the callipyge gene for heavy muscling. The carcass on the left was from a lamb that carried genes for normal muscling. (Courtesy, Sam Jackson, Texas Tech University, Lubbock)

mated to non-HM ewes, produce about 50% HM offspring (produced equally among ewe and ram lambs). Present information suggests the trait may not be expressed in progeny from HM ewes which complicates the inheritance. Further tests are required to confirm the inheritance pattern of this gene.

3. A study at Texas Tech University on 150 Rambouillet ewes sired by HM rams showed no problem with lambing or lamb survival of HM lambs. The HM condition cannot be detected in lambs until they are about 4 weeks of age. The small amount of information available to determine if HM ewes have any reproductive or lambing problems suggests that they do not.

4. Results by Texas Tech University on 18 ram lambs (9 HM and 9 normal muscled) showed no difference in birthweight or growth rate between HM and normal lambs. HM lambs tend to have a slight advantage in feed efficiency.

5. Carcass evaluation results from the above Texas Tech study showed HM lambs to have higher dressing percents (57.3 vs 53.9%), more carcass lean (63 vs 55%), larger loin eye area (2.8 vs 1.6 sq in.), less carcass fat (15 vs 19%), and less carcass bone (18 vs 22%). The increased muscle mass in the HM lambs occurred almost entirely in loin and leg muscles. Weights of rib and shoulder muscles were about the same between the two groups.

Preliminary reports suggest that the meat from HM lambs is significantly less tender than that of normal lambs. If HM lambs are confirmed to be of inferior tenderness, work must focus on determining if there are genetic strains within the HM population which have acceptable tenderness, or what can be done technologically to improve the tenderness of such lambs.

Meat and animal scientists from agricultural colleges reviewed the current information and offered their perspective: Meat animals and their carcasses exhibit a collection of weaknesses and strengths. The HM condition certainly provides a large advantage in carcass composition. However, the apparent substantial tenderness decrease exhibited by meat from these lambs is a serious liability. At the price consumers pay for lamb products, the eating quality must be very acceptable. Although early information has not yet found reproductive or other physical problems with HM animals, further experience in this area is needed. On balance, the consensus of the animal scientists is that animals or carcasses strongly exhibiting the HM gene deserve to rank near the top.

The take-home message should be that this condition is very exciting and may play a major role in the sheep industry of the future. However, given the current state of information about HM lambs, the industry should go slowly until a clearer picture of their performance and characteristics is available.

At the University Sheep Day in 1994, David L. Thomas, University of Wisconsin Sheep Specialist, reported the following: In 1983, a Dorset ram exhibiting an abnormally large amount of muscling (muscle hypertrophy) was identified in a flock in Oklahoma. It was reported that the muscle hypertrophy of this ram was passed to some of his progeny and descendants, with the trait passed to some of their progeny in later generations. These observations suggested the muscle hypertrophy was of genetic origin capable of being passed from parent to offspring. Interest in this condition has increased in recent years because consumer acceptability of lamb would be improved if the proportion of lean in the lamb carcass was increased and if size of the loin was increased resulting in larger loin and rib chop; assuming there was no decrease in palatability.

The first research projects with sheep exhibiting the muscle hypertrophy condition were started by Sam Jackson, Texas Tech University, Lubbock. Dr. Noelle Muggli-Crockett, Utah State University, Logan, collaborated with Sam Jackson in molecular genetics studies of these sheep. Work continues at both Texas Tech University and Utah State University. Studies on this muscle hypertrophy were also initiated in the fall breeding season of 1992, by Dr. Kreg Leymaster, U.S. Meat Animal Research Center, Clay Center, Nebraska and by Dr. Gary Snowder, U.S. Sheep Experiment Station, Dubois, Idaho. Only preliminary results and no printed information are available from the studies at the two USDA stations.

■ **Understanding the genetics**—While field observations suggest the muscle hypertrophy (MH) is of genetic origin, efficient use of the condition can be realized only if the actual mode of inheritance is understood. The first study to determine the mode of inheritance was reported by Jackson and Green (1993) and Jackson *et al.* (1993a) of Texas Tech University. One hundred fifty Rambouillet ewes of normal muscling were mated to Dorset-cross rams with MH. The Dorset-cross rams were the result of mating rams with MH to ewes of normal muscling. Lambs that resulted from these matings were subjectively classified by three evaluators at weaning (90 to 120 days of age). The proportion of lambs with MH did not differ between male and female lambs. The observed proportion of lambs with MH (48.5%) was not significantly different from the 50% expected if MH is due to an autosomal dominant gene.

Studies conducted at Utah State University to identify the location of the MH gene in the sheep genome have been quite productive. Crockett, *et al.* (1993) named the MH gene the *callipyge*, *i.e.,* nicely proportioned buttocks, *gene*. Two alleles, CLPG and clpg, were proposed with animals with one or two copies of CLPG (CLPG/CLPG or CLPG/clpg) expressing MH and animals with two copies of clpg (clpg/clpg) expressing normal muscling. They (Crockett, *et al.*, 1993) have determined the callipyge gene to be on chromosome 18 (sheep have 27 pairs of chromosomes). While the callipyge gene itself has not been identified, Crockett, *et al.* (1993) have identified two genetic markers on chromosome 18 that are linked to the callipyge gene. The two genetic markers are located on either side of the callipyge gene; each about 20 map units from the gene. Since this chromosome region in humans and cattle is known to contain the gene which encodes for the receptor for insulinlike growth factor 1 (IGF-1), this gene is considered to be the likely candidate for the cause of the MH condition. While the above researchers have suggested that the CLPG is a dominant gene, there is still the possibility that it may act in an additive fashion. Dominance implies that CLPG/CLPG and CLPG/clpg individuals would have the same degree of MH. Individuals of the CLPG/CLPG genotype have not been produced in large numbers because it requires the mating of MH rams with MH ewes. If CLPG/CLPG lambs have greater muscling than CLPG/clpg lambs, then the CLPG gene acts in an additive fashion, *i.e.*, two copies of the CLPG gene result in more muscling than just one copy. Large numbers of animals of all three genotypes (CLPG/CLPG, CLPG/clpg, clpg/clpg) need to be produced before it can be determined if the CLPG gene is dominant or additive.

■ **Growth, feed efficiency and carcass measurements**—Lambs born in the study described above were evaluated for growth, postweaning feed efficiency and carcass traits (Jackson and Green, 1993; Jackson *et al.*, 1993a; Jackson *et al.*, 1993d; Jackson *et al.*, 1993e). Carcass traits were only measured on 18 ram lambs slaughtered at 120 lb liveweight. Results are presented in Table 3-1.

TABLE 3-1
PERFORMANCE OF RAMBOUILLET-CROSS LAMBS WITH AND WITHOUT THE CLPG GENE FOR MUSCLE HYPERTROPHY

Lamb Type	Birth Weight	Weaning Weight	Feed Efficiency Feed/Gain	Dressing Percent	Loin Eye Area
	(lb)	*(lb)*		*(%)*	*(sq in.)*
Ram					
MH (CLPG/clpg)	9.5	44.8	4.9[1]	57.3[1]	2.8[1]
Normal (clpg/clpg)	9.7	45.6	5.4[1]	53.9[1]	1.6[1]
Ewe					
MH (CLPG/clpg)	9.3	43.4	5.7[1]	—	—
Normal (clpg/clpg)	9.1	43.4	6.1[1]	—	—

[1]Significant differences ($P < 0.05$).

There were no significant differences between MH and normal lambs for birth weight or weaning weight. However, feed efficiencies of MH lambs were improved by approximately 8%, *i.e.*, MH lambs required about 8% less feed to produce a pound of liveweight gain than did normal lambs. Of the 18 ram lambs slaughtered, MH lambs had 3.4% higher dressing percentage and 1.2 more square inches (+75%) of loin eye area than normal lambs. There were no differences between MH and normal lambs for carcass fat thickness or for visual estimates of color, texture, or firmness of carcass lean.

■ **Carcass composition**—Carcasses of the 18 ram lambs described in the preceding section were separated into muscle, fat and bone (Jackson *et al.*, 1993b; Jackson *et al.*, 1993c; Jackson *et al.*, 1993d; Jackson *et al.*, 1993e). Weights of 22 individual muscles were recorded (13 leg, 3 loin, and 6 rib shoulder muscles). The dissected muscle and fat were combined and analyzed for protein, moisture, and fat. Results are presented in Table 3-2. Means in Table 3-2 were adjusted for any differences in carcass weight.

TABLE 3-2
CARCASS COMPOSITION OF 18 RAMBOUILLET-CROSS RAM LAMBS WITH AND WITHOUT THE CLPG GENE FOR MUSCLE HYPERTROPHY

Lamb Type	Carcass Composition				Weight of Selected Muscles in—		
	Protein	Water	Fat	Bone	Leg	Loin	Rib-Shoulder
	(%)	(%)	(%)	(%)	(lb)	(lb)	(lb)
MH (CLPG/clpg)	13[1]	50[1]	15[1]	18[1]	6.0[1]	2.6[1]	1.7
Normal (clpg/clpg)	11[1]	44[1]	19[1]	22[1]	4.3[1]	1.8[1]	1.6

[1]Significant differences ($P < 0.001$).

The chemical composition results indicate that MH carcasses had a higher percentage of protein and water and a lower percentage of fat and bone than normal lambs. Since almost all of the protein and water is in the lean, the results in Table 3-2 suggest that carcasses of MH lambs are approximately 63% lean compared with 55% lean for carcasses of normal lambs. This is a very large difference. Furthermore, it appears that the distribution of muscle weight throughout the carcass may be different between MH and normal lambs. Pelvic and loin muscles weighed more in MH than in normal lambs, but weights of rib-shoulder muscles were similar between the two groups. Visual evaluation of live MH lambs also indicates that most of the increased muscularity of MH lambs occurs in the loin and leg areas.

■ **Production of MH lambs**—The CLPG gene would be less valuable to the sheep industry if it resulted in lowered survival. There is no data reported on the survival of MH lambs. This information is needed before production of MH lambs is recommended. Even a small increase in lamb mortality of MH lambs would negate the large advantages they have in carcass composition.

Dystocia may also be a concern with such lambs; however Jackson *et al.,* (1993a) in describing the performance of 200 lambs sired by MH rams and born from 150 normal Rambouillet ewes states: "Concerns about ewes experiencing increased dystocia while having lambs with MH were unfounded. All ewes lambed naturally and no assistance was given to any ewe during the lambing season. It was impossible to determine which lambs expressed MH until they were approximately four weeks of age." Future studies will determine if MH females have any reproductive or lambing problems.

However, even if MH females are undesirable for maternal traits, commercial production systems can have normal ewes and mate them to MH rams with two copies of the CLPG gene. Such a mating system would result in all MH lambs. Of course the seedstock producer of the CLPG/CLPG rams would be producing lambs from MH ewes and would be concerned about any increased reproductive problems with MH females.

■ **Meat palatability**—While increased muscle mass, increased size of the loin muscle and decreased fat content are valuable results of the CLPG gene, its effect on meat palatability needs to be determined. While no results have been published, there has been some concern expressed about tenderness of meat from MH lambs. Subsequently, Texas Tech University, Utah State University, and the U.S. Sheep Experiment Station have presented results showing that grilled loin chops from lambs with the callipyge gene are significantly less palatable, especially in regards to tenderness, than chops from normal lambs. No differences in palatability were observed for oven-roasted legs between the two types of lambs.

■ **Potential**—The CLPG gene and the increased muscularity which results from its presence has the potential to increase lean yield of lamb carcasses to a level that would take several decades to obtain through selection. Poor lean yield and small size of loin and rib chops are major impediments to increased lamb consumption; therefore, the CLPG gene has tremendous economic potential for the sheep industry. The increased dressing percentage alone of MH lambs increases their live value by $3.00 to $5.00/cwt.

However, the sheep industry should not yet make use of the CLPG gene. The effect of the CLPG gene on lamb survival and on meat palatability needs to be determined to make sure that it has little or no detrimental effects on these traits. Research trials to determine these effects are in progress.

DNA (DEOXYRIBONUCLEIC ACID)

In recent years, the molecular basis of heredity has become much better understood. The most important genetic material in the nucleus of the cell is deoxyribonucleic acid (DNA), which serves as the genetic information source. It is composed of nucleotides containing adenine, guanine, cytosine, and thymine. The sequence of these four bases in DNA acts as a code in which messages can be transferred from one cell to another during the process of cell division. The code can be translated by cells (1) to make proteins and enzymes of specific structure that determine the basic morphology and functioning of a cell; (2) to control differentiation, which is the process by which a group of cells become an organ; or (3) to control whether an

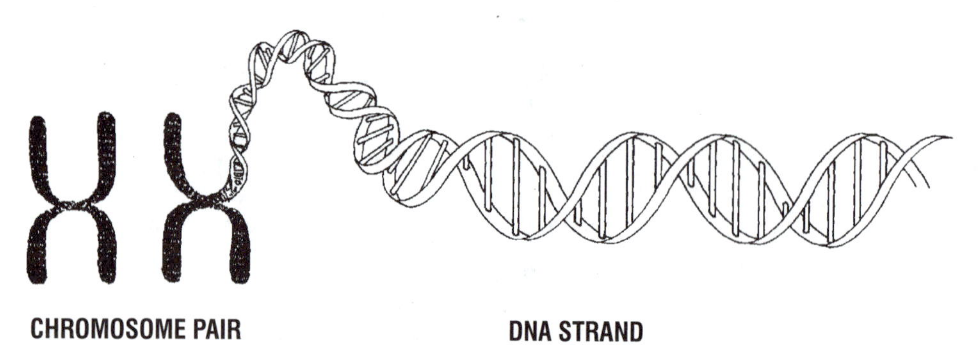

Fig. 3-5. Chromosomes consist of deoxyribonucleic acid (DNA) and a type of protein coiled into a tightly packaged structure. The DNA molecule (uncoiling left to right) has a double helical structure. *Note:* The concept is greatly exaggerated in this diagram.

embryo will become a sheep, a goat, a pig, a chicken, a cow, or a human, etc.

But DNA is far more than a genetic information center, or master molecule. The recent development of the recombinant DNA technique ushered in a new era of genetic engineering—with all its promise and possible hazard.

Recombinant DNA techniques are of enormous help to scientists in mapping the positions of genes and learning their fundamental nature. These may lead to new scientific horizons—of allowing the introduction of new genetic material directly into the cells of an individual to repair specific genetic defects or to transfer genes from one species to another. On the other hand, the opponents of tinkering with DNA raise the specter (1) of re-engineered creatures proving dangerous and ravaging the earth, and (2) of moral responsibility in removing nature's evolutionary barrier between species that do not mate. Nevertheless, molecular biologists are working ceaselessly away in recombinant DNA studies; hence, sheep and goat raisers, along with scientists, should keep abreast of new developments in this exciting field.

The modern breeder knows that the job of transmitting qualities from one generation to the next is performed by the germ cells—a sperm from the male and an ovum, or egg, from the female. All animals, therefore, are the result of the union of two such tiny cells, one from each of its parents. These two germ cells contain all the anatomical, physiological, and psychological characters that the offspring will inherit.

In the body cells of an animal, the chromosomes occur in pairs (diploid numbers), whereas in the formation of the sex cells, the egg (ovum) and the sperm, a reduction division occurs and only one chromosome and one gene of each pair go into a sex cell (haploid number). This means that only half the number of chromosomes and genes present in the body cells of the animal go into each egg and sperm, but each sperm or egg cell has genes for every characteristic of its species. As will be explained later, the particular half that any one germ cell gets is determined by chance. When mating and fertilization occur, the single chromosomes from the germ cell of each parent unite to form new pairs, and the genes are again present in duplicate in the body cells of the embryo.

With all possible combinations in 27 or 30 pairs of chromosomes (the species number in sheep and goats, respectively) and the genes that they bear, any male or female of each species can transmit about 1 billion different samples of its own inheritance; and the combination from both parents makes possible 1 billion times 1 billion genetically different offspring. It is not strange, therefore, that no two animals within a given breed (except identical twins from a single egg split after fertilization, few of which occur in sheep) are exactly alike. Rather, we can marvel that the members of a given breed bear as much resemblance to each other as they do.

Even between such closely related individuals as twins, it is possible that there will be quite wide differences in size, growth rate, temperament, conformation, and in almost every other conceivable character. Admitting that many of these differences may be due to undetected differences in environment, it is still true that in such animals much of the variation is due to hereditary differences. A ram, for example, will sometimes transmit to one offspring much better inheritance than he does to most of his get, simply as the result of chance differences in the genes that go to different sperm at the time of the reduction division. Such differences in inheritance in offspring have been called both the hope and the despair of the livestock breeder.

Generally, there are two genes for each trait. For each locus in one of the members of a chromosome pair there is a corresponding locus in the other member of that chromosome pair. These genes located at corresponding loci in chromosome pairs may be identical in the way they affect a trait, or they may contrast in the way they affect a trait. If the genes are identical in the way they affect a trait, then the individual is said to be *homozygous* at that locus; if they contrast in the way they affect a trait, the individual is said to be *heterozygous* at that locus. It follows that homozygous individuals are genetically pure, since the genes for a trait passed on in the egg or sperm will always be the same. But heterozygous individuals produce sperm or eggs which bear one of two types of genes. Homozygous and heterozygous may also refer to a series of hereditary factors. However, few if any animals are entirely homozygous—producing totally uniform offspring as a

Fig. 3-6. Booroola Merino rams homozygous for the FecB gene for high ovulation rate. (Courtesy, David L. Thomas, University of Wisconsin, Madison)

result of one sperm or egg being just like any other. Rather, animals are heterozygous, and there is often wide variation within the offspring of any given sire or dam. The wise and progressive breeder recognizes this fact, and insists on the production records of all offspring rather than those of a few meritorious individuals.

Variation between the offspring of animals that are not pure or homozygous, to use the technical term, is not be marveled at. Rather, it is to be expected. No one would expect to draw exactly 20 sound apples and 10 rotten ones every time a random sample of 30 is taken from a barrel containing 400 sound ones and 200 rotten ones, although on the average—if enough samples were drawn—one would expect to get about that proportion of each. Individual drawings would of course vary rather widely. The mating of a ewe with a fine show record to a ram that on the average transmits relatively good offspring will not always produce lambs of merit equal to that of their parents. The lambs could be markedly poorer than the parents or, happily, they could in some cases be better than either parent.

Selection and closebreeding are the tools through which sheep and goat producers can obtain animals whose chromosomes and genes contain similar hereditary determiners—animals that are genetically more homozygous.

MUTATIONS

Gene changes are technically known as mutations. *A mutation may be defined as a sudden variation which is later passed on through inheritance and that results from changes in a gene or genes.* Not only are mutations generally harmful, but they are rare. For all practical purposes, therefore, genes can be thought of as unchanged from one generation to the next. The observed differences between animals are usually due to different combinations of genes being present rather than to mutations. Each gene probably changes only about once in each 100,000 to 1,000,000 animals produced.

Once in a great while a mutation occurs in a farm animal, and it produces a visible effect in the animal carrying it. Such animals are sometimes called *sports*. Mutations are occasionally of practical value. The occurrence of the polled characteristic within the Dorset breed of sheep is an example of a mutation or sport of economic importance. Out of this has arisen the Polled Dorset.

Gene changes can be accelerated by exposure to ionizing radiation (X-rays, radium), a wide variety of chemicals (mustard gas, LSD), and ultraviolet light.

SIMPLE GENE INHERITANCE (QUALITATIVE TRAITS)

In the simplest type of inheritance, only one pair of genes is involved. This type of inheritance can be used to demonstrate how genes segregate in the sperm and egg at random. Therefore, the possible gene combinations are governed by the laws of chance (probability) operating in much the same manner as the results obtained from flipping coins. Relatively large numbers are required for certain proportions to be evident. For example, if a penny is flipped often enough, the number of heads and tails will come out about even. However, with the laws of chance in operation, it is possible that out of any 4 tosses, one might get all heads, all tails, or even 3 to 1.

Examples of simple gene inheritance in sheep and goats (sometimes referred to as qualitative traits) include the color of fiber, the color of eyes, the presence or absence of horns, the type of blood, and some lethals.

DOMINANCE AND RECESSIVENESS

When genes at corresponding loci on chromosome pairs are unlike, one of the genes often overpowers the expression of the other. This gene is referred to as the dominant gene, and the gene whose expression is prevented is called the recessive gene. Hence, a sheep's or a goat's breeding performance cannot be recognized by its phenotype (how it looks), which is of great significance in practical breeding.

The polled trait in the Rambouillet and Dorset breeds is apparently dominant, although Rambouillet ewes homozygous for the horned gene have horn knobs and those carrying the polled gene have depressions in the skull in place of horn knobs. Wattles,

Fig. 3-7. Two purebred lambs of the same breed. The black sheep is the result of undesirable recessive genes that were carried by both the sire and the dam. Recessive genes can be passed on for many generations without their presence becoming evident until such time as two animals are mated, both of which carry the same recessive factors. (Courtesy, Washington State University)

which appear in Navajo sheep and in goats, appear to be inherited as a dominant.

Black wool results from a desirable recessive for hand or handicraft processing, but it is an extremely undesirable recessive in white wool used in machine processing because one or a few black fibers in white textiles or textiles to be dyed can be very expensive to remove. If not removed, they leave a blemish. Therefore, black or black spotted sheep should not be kept with white flocks, as contamination of white wool with black fibers is difficult to prevent.

As can be readily understood, dominance often makes the task of identifying and discarding all animals carrying an undesirable recessive factor a difficult one. Recessive genes can be passed on from generation to generation, appearing only when two animals both of which carry the recessive factor happen to mate. Even then, only one out of four offspring produced will, on the average, be homozygous for the recessive factor and show it.

Examples of undesirable recessives in animals are: cryptorchidism, although it is sometimes linked with the desired polled trait; and intersexuality in milk goats. When these conditions appear, one can be very certain that both the sire and the dam contributed equally to the condition and that each of them carries the recessive gene. This fact should be given consideration in the culling program.

Assuming that a pair of genes is responsible for wool color, an example of dominance and recessiveness, and an example of the operation of the law of probability in inheritance, may be illustrated. In Fig. 3-8 the gene for white wool (W) has its full effect regardless of whether it is paired with another just like itself or a recessive gene. When the recessive genes are paired the

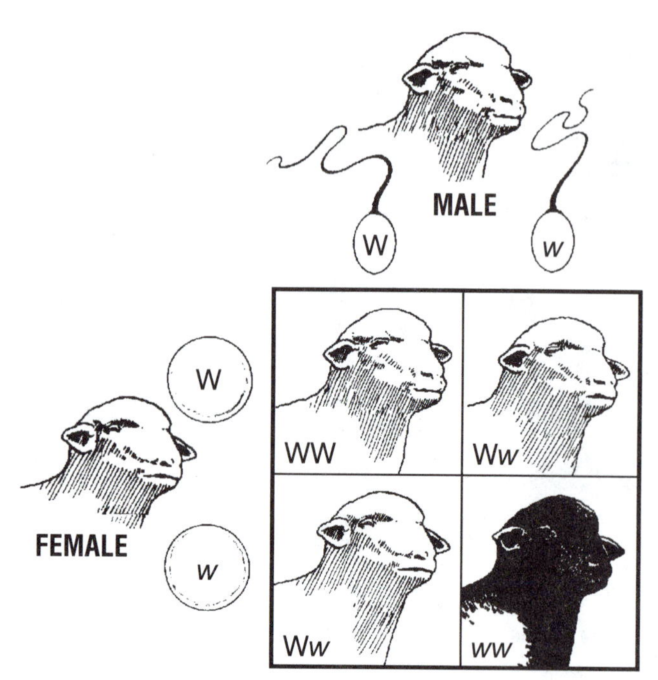

Fig. 3-8. A ram heterozygous for white wool (Ww) is bred to a ewe heterozygous for white wool (Ww). Germ cells (spermatozoa and eggs) with the dominant gene (W) and with the recessive gene are produce in equal proportion. On the average, for a large number of matings, one-fourth of the offspring will have black wool (ww) and one-half of the offspring will be carriers of the recessive gene (w).

animal is black.[5] Animals used as an example in Fig. 3-8 are heterozygous. They have a dominant (W) gene and a recessive (w) gene for wool color. Hence, equal numbers of the sperm produced by the male and the eggs produced by the female will contain the dominant and the recessive genes. When large numbers of matings are involved, phenotypically (how they appear), the results would be as follows: three-fourths of the offspring would be white, and one-fourth would be black. Genotypically, one-fourth would be homozygous dominant for white wool; one-half would be heterozygous; and one-fourth would be homozygous recessive. While one-half of the animals would appear white, they would be carriers of the recessive gene, thereby increasing the chance of black wool.

Assuming that a hereditary defect or abnormality has occurred in a flock and that it is recessive in nature, the breeding program to be followed to prevent or minimize the possibility of its future occurrence will depend somewhat on the type of flock involved—especially on

[5]Recessive black in sheep is due to two pairs of recessive genes for black, either of which when homozygous will result in black. It is postulated that these two pairs of recessive genes were present in the primitive Mouflon sheep, one of the progenitors of modern breeds. (From Warwick, Bruce L., R. O. Berry, and S. P. Davis, *The Journal of Heredity*, Vol. 48, No. 6, p. 255)

whether it is commercial or purebred. In an ordinary commercial flock, the breeder can usually guard against further reappearance of the undesirable recessive simply by using an outcross (unrelated) ram within the same breed or by crossbreeding with a ram from another breed. With this system, the breeder is fully aware of the recessive being present, but has taken action to keep it from showing up.

On the other hand, if such an undesirable recessive appears in a purebred flock, the action should be more drastic. Reputable purebred breeders have an obligation not only to themselves but to their customers among both the purebred and the commercial flocks. Purebred flocks must be purged of undesirable genes and lethals. This can be done by:

1. Eliminating those rams and ewes that are known to have transmitted the undesirable recessive character.
2. Eliminating both the abnormal and the normal offspring produced by these rams and ewes, since approximately half of the normal animals will carry the undesirable character in the recessive condition.
3. In some instances, breeding a prospective stud ram to a number of ewes known to carry the factor for the undesirable recessive, thus making reasonably sure that the new ram is free from the recessive.

Such action in a purebred flock is expensive, and it calls for considerable courage. Yet, it is the only way in which the purebred livestock of the country can be freed from such undesirable genes.

The same steps would be necessary in a herd of purebred goats.

Overall, the phenotypic expression of genes occurs in two ways: additive and nonadditive. Additive expression means that the phenotypic expression of one gene adds to the phenotypic expression of the other gene or genes affecting the expression of a trait. In nonadditive gene action the phenotypic expression of one gene does not add to another. Rather, a particular phenotype is produced.

Unfortunately, sheep and goat genetics are not all a case of dominance and recessiveness. Other types of gene actions occur which influence the appearance and performance of an animal. For example, gene actions can demonstrate lack of dominance, and overdominance.

MULTIPLE GENE INHERITANCE (QUANTITATIVE TRAITS)

Relatively few characters of economic importance in farm animals are inherited in as simple a manner as those listed in the above section. Rather, important characters—such as meat, milk, and fiber production— are due to many genes; thus, they are called multiple-factor characters or multiple-gene (polygenic) characters. Because such characters show all manner of gradation—from high to low performance, for example—they are sometimes referred to as quantitative traits. Still the mechanism of inheritance is the same whether few or multiple genes are involved.

In quantitative inheritance, the extremes (both good and bad) tend to swing back to the average. Thus, the offspring of a grand champion ram (or buck goat) and a grand champion ewe (or doe goat) are not apt to be as good as either parent. Likewise, and happily so, the progeny of two very mediocre (or below-average) parents will likely be superior to either parent.

Estimates of the number of pairs of genes affecting each economically important characteristic vary greatly, but the majority of geneticists agree that for most such characters 10 or more pairs of genes are involved. The growth rate in sheep (or goats), for example, is affected by: (1) the individual animal's appetite or feed consumption; (2) the efficiency of assimilation—that is, the proportion of the feed eaten that is absorbed into the bloodstream; and (3) the use to which the nutrients are put after assimilation—for example, whether they are used for growth or fattening. This example should indicate that such a characteristic as growth rate is controlled by many genes and that it is difficult to determine the mode of inheritance of such characters.

MULTIPLE BIRTHS

From Table 3-3 it is obvious that there are wide specie differences in multiple births—of those listed, ewes and does being the most prolific and beef cattle the least. Of course, it is well-known that sows are

Fig. 3-9. A large percentage of twins is desirable. Apparently, this factor is affected by heredity, environment, and age.

TABLE 3-3
FREQUENCY OF MULTIPLE BIRTHS[1]

	Twins		Triplets		Quadruplets	
	(% of births)	*(ratio)*	*(% of births)*	*(ratio)*	*(% of births)*	*(ratio)*
Sheep and goats[2]	20–60	200–600/1,000	1.0–2.0	10–20/1,000	0.02–0.10	2–10/10,000
Beef cattle	0.5	5/1,000				
Dairy cattle[3]	2.0	20/1,000	0.3	4/14,111	0.01	1/14,111
Horses	1.5	15/1,000				

[1]Importation of Finnsheep, Romanov, East Friesian and probably other highly prolific breeds into the U.S. and the use of modern selection technology will probably bring about a continuous increase of frequency of multiple births.

[2]The average frequency of twinning in the U.S. breeds of goats is about 700 per 1,000 births, or about 70% of the births. Angoras seem to have the lowest twinning rate with only about 150 twins per 1,000 births (15%).

[3]Triplet and quadruplet values are in Brown Swiss cattle.

more prolific than any of these animals, being exceeded only by hens.

Dr. David L. Thomas, University of Wisconsin, Department of Meat and Animal Science, states that "Selection of replacements on basis of the dam's lambing performance can improve prolificacy by annual rates of 1.3 to 2.0 lambs per 100 ewes." Then, he adds: "The estimates of genetic merit for prolificacy generated by The National Sheep Improvement Program should be more accurate than estimates based only on dam's performance and lead to greater genetic improvement."

Increase in use of imported prolific breeds and the use of modern selection technology in both sheep and goats will likely result in increases in percent of lambs and kids weaned in all breeds. This will lead to increases in efficiency of lamb and goat meat production. In time, this will probably aid the survival of family farms and sustainability of food production. Sheep and goats are the only food producing species that can produce high quality food and fiber on forage alone. They are also the only food species which can increase in efficiency and profitability without feeding grain.

SEX DETERMINATION

In a large population, approximately equal numbers of males and females are born in all common species of animals. To be sure, many notable exceptions can be found in individual flocks of sheep (or herds of goats).

Sex is determined by the chromosomal makeup of the individual. One particular pair of the chromosomes is called the sex chromosomes. In farm animals, the female has a pair of similar sex chromosomes, called *X chromosomes*; the male has a pair of unlike sex chromosomes, called *X* and *Y chromosomes*. In the bird, this condition is reversed, the female having the unlike pair and the male having the like pair.

The pairs of sex chromosomes separate when the

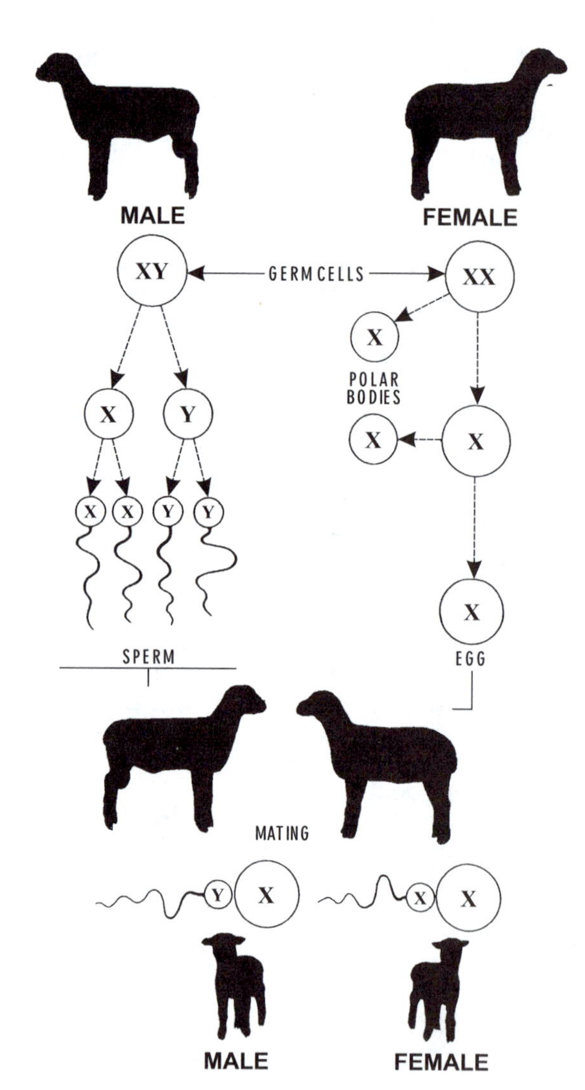

Fig. 3-10. Diagrammatic illustration of the mechanism of sex determination in sheep, showing how sex is determined by the chromosomal makeup of the individual. The ewe has a pair of like sex chromosomes, called X chromosomes; the ram has a pair of unlike sex chromosomes, called X and Y chromosomes. Thus, if an egg and a sperm of like sex chromosomal makeup unite, the offspring will be a ewe. If an egg and a sperm of unlike sex chromosomal makeup unite, the offspring will be a ram.

germ cells are formed. Thus, each of the ova, or eggs, produced by the ewe or doe contains the X chromosome; whereas the sperm of the ram or buck are of two types, 50% containing the X chromosome and the other 50% the Y chromosome. Since, on the average, the eggs and sperm unite at random, 50% of the progeny will contain the chromosomal makeup XX (females) with the other 50% XY (males).[6]

SEX RATIO CONTROL

Through the ages, people have desired to select the sex of their offspring. Rulers were always anxious to have sons. For the producers, controlling the sex of the offspring of farm animals is of great interest because of the economic advantage of being able to produce all males or all females, depending on the need. The normal sex ratio (males to females) is about 50:50, since whether an X- or Y-bearing sperm will fertilize an egg is decided by chance just as flipping a coin for heads or tails.

Many approaches have been tried to alter the sex ratio. Some of these approaches are sophisticated and scientific while others are myths that develop from casual observations.

In 1996, scientists at the U.S. Department of Agriculture Research Center, ARS, Beltsville, Maryland, reported on their development of a sexing technology that skews the sex ratio from the standard 50:50 to 90:10 for either sex. Sperm are sorted by a laser and flow cytometer, based on the fact that X-bearing sperm carry more DNA than Y sperm. Several hundred sheep, cattle, swine, and rabbits of predicted sex have been produced using this technology.[7]

Research and theories on sex preselection will continue because the stakes are high.

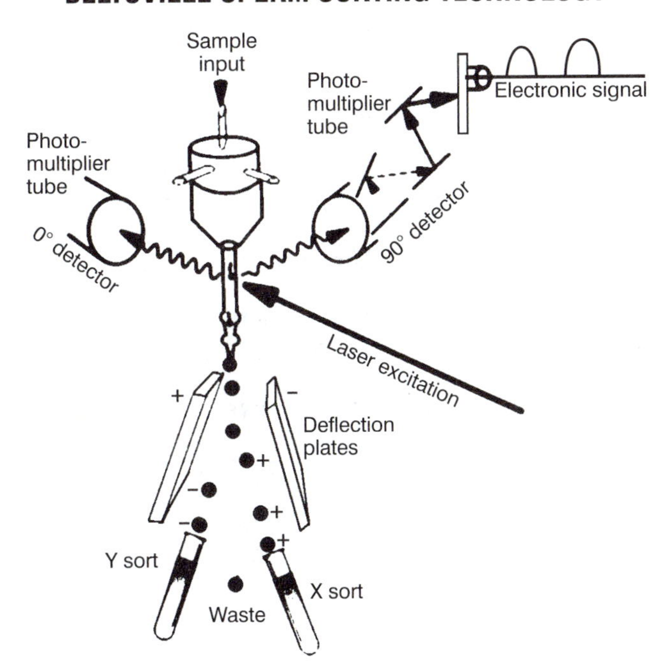

Fig. 3-11. USDA Beltsville sperm sorting technology, using a laser and flow cytometer. (Courtesy, Beltsville Agricultural Research Center, ARS, USDA, Beltsville, MD)

RELATIVE IMPORTANCE OF THE RAM OR BUCK AND THE EWE OR DOE

Since a ram (or a buck goat) can have so many more offspring during a given season or a lifetime than a ewe (or a doe goat), he is, from an hereditary standpoint, a more important individual than any one ewe (or doe) as far as the whole flock is concerned, though both the ram (and buck) and the ewe (and doe) are of equal importance as far as concerns any one offspring. Because of their wider use, therefore, rams and bucks are usually culled more rigidly than ewes and does, and the breeder can well afford to pay more for an outstanding ram or buck than for an equally outstanding ewe or doe.

Experienced sheep producers have long felt that rams often resemble their daughters more closely than their sons, whereas ewes resemble their sons. Some rams and ewes, therefore, enjoy a reputation based almost exclusively on the merit of their sons, whereas others owe their prestige to their daughters. Although this situation is likely to be exaggerated, any such phenomenon that may exist is due to sex-linked inheritance which may be explained as follows: The genes that determine sex are carried on one of the chromosomes. The other genes that are located on the same chromosome will be linked or associated with sex and will be transmitted to the next generation in combination with sex. Thus, because of sex linkage, there are more color-blind men than color-blind women. In poultry breeding, the sex-linked factor is used in a practical way for the purpose of distinguishing the pullets from the cockerels early in life, through the process known as *sexing* the chicks. Thus, when a black cock is crossed with barred hens, all the cocks come barred

---

[6]The scientists' symbols for the male and female, respectively are: ♂ (the sacred shield and spear of Mars, the Roman god of war) and ♀ (the looking glass of Venus, the Roman goddess of love and beauty).

[7]*Ensminger's World Book—state of the world's people, animals, and food*, pp. 94–95, Figs. 280 and 281, published by Agriservices Foundation, Clovis, CA, 1996.

and all the hens come black. It should be emphasized, however, that under most conditions it appears that the influence of the sire and of the dam on any one offspring is about equal. Most breeders, therefore, will do well to seek excellence in both sexes of breeding animals.

PREPOTENCY

Prepotency refers to the ability of the animal, either male or female, to stamp its own characteristics on its offspring. The offspring of a prepotent ram, for example, resemble both their sire and each other more closely than usual. The only conclusive and final test of prepotency consists of the inspection of the get.

From a genetic standpoint, there are two requisites that an animal must possess in order to be prepotent: (1) dominance and (2) homozygosity. Every offspring that receives a dominant gene or genes will show the effect of that gene or genes in the particular character or characters which result therefrom. Moreover, a perfectly homozygous animal would transmit the same kind of genes to all of its offspring. Although entirely homozygous animals probably never exist, a system of inbreeding is the only way to produce animals that are as nearly homozygous as possible.

It is impossible to determine just how important prepotency may be in animal breeding, although many sires of the past have enjoyed a reputation for being extremely prepotent. Perhaps these animals were prepotent, but there is also the possibility that their reputation for producing outstanding animals may have rested upon their being mated to some of the best ewes of the breed.

In summary, if a given ram or buck, or ewe or doe, possesses a great number of genes that are completely dominant for desirable type and performance and if the animal is relatively homozygous, the offspring will closely resemble the parent and resemble each other, or be uniform. Fortunate indeed is the breeder who possesses such an animal.

NICKING

Nicking is the production of offspring of certain matings that are especially outstanding and in general better than their parents. For example, a ewe may produce outstanding lambs to the service of a certain ram, but when she is mated to another ram of apparently equal merit as a sire, the offspring may be disappointing; or sometimes the mating of a rather average ram to an equally average ewe will result in the production of a most outstanding individual from the standpoint of both type and performance.

So-called successful nicking is due to, genetically speaking, the right combination of genes for good characters contributed by each parent, although each of the parents within itself may be lacking in certain genes necessary for excellence. In other words, the animals nicked well because their respective combinations of good genes were such as to complement each other.

The history of animal breeding includes records of several supposedly favorable nicks. Because of the very nature of successful nicks, however, outstanding animals arising therefrom must be carefully scrutinized from a breeding standpoint, because with their heterozygous origin, it is quite unlikely that they will breed true.

HEREDITY AND ENVIRONMENT

A sheep bedded deeply in straw and with a bunk full of feed before it, is undeniably the result of two forces—heredity and environment. If turned to the range, a twin to a pampered sheep would present an entirely different appearance. By the same token, optimum environment could never make a champion out of a sheep with scrub ancestry, but it might well be added that "fat and wool will cover up a multitude of sins."

These are extreme examples, and they may be applied to any class of farm animals; but they do emphasize that any particular animal is the product of heredity and environment. Stated differently, heredity may be thought of as the foundation, and environment as the structure. Heredity has already made its contribution at the time of fertilization, but environment works ceaselessly away until death.

Admittedly, after looking over an animal, a breeder cannot with certainty know whether it is genetically a high or a low producer; and there can be no denying that environment—including feeding, management, and disease—plays a tremendous part in determining the extent to which hereditary differences that are present will be expressed in animals. Recognizing how environment may influence production is important because (1) environmental effects are not passed to the offspring, (2) environmental effects may overshadow heredity effects, (3) the proper environment allows an individual to achieve its genetic potential, and (4) rapid improvements in production can be made by supplying better environmental conditions.

Experimental work has long shown conclusively enough that the vigor and size of animals at birth are dependent upon the environment of the embryo from the minute the ovum or egg is fertilized by the sperm, and some evidence indicates that newborn animals are affected by the environment of the egg and sperm long before fertilization has been accomplished. In other words, perhaps due to storage of substances, the kind and quality of the ration fed to young, growing females may later affect the quality of their progeny. Generally, then, environment may inhibit the full expression of potentialities from a time preceding fertilization until physiological maturity has been attained.

Maximum development of characters of economic importance—growth, body form, wool production, milk production, etc.—cannot be achieved unless there are optimum conditions of nutrition and management. However, the next question is whether a breeding program can make maximum progress under conditions of suboptimal nutrition—such as is often found under some range conditions. One school of thought is that selection for such factors as body form and growth rate in animals can be most effective only under nutritive conditions promoting the near maximum development of those characters of which the animal is capable. The other school of thought is that genetic differences affecting usefulness under suboptimal conditions will be expressed under such suboptimal conditions, and that differences observed under forced conditions may not be correlated with real utility under less favorable conditions. Those favoring the latter thinking argue, therefore, that the production and selection of breeding animals for suboptimal nutritive conditions should be under less favorable conditions and that the animals should not be highly fitted.

In general, the breed or breeds selected for a sheep or goat operation should be selected based on the best available information as to suitability of the breed(s) for the particular purpose, and for adaptability to the environment. Some breeds produce better in certain environments.[8] Moreover, selection and evaluation of breeding animals should be carried on under the same environmental conditions as those under which their offspring will be produced. This is difficult, for purebred males are generally raised under better conditions than the farm or range conditions under which their offspring will be raised.

GENETIC IMPROVEMENT

Sheep and goats are raised primarily for profit, and profit is dependent upon efficiency of production and market value. Fortunately, greater efficiency of production and improved quality of lamb, wool, mohair, and milk can be achieved. In striving for genetic improvement in animals, sheep and goat breeders have two basic tools at their disposal—mating systems and selection.

Dr. Clair E. Terrill, Collaborator, USDA, maintains that the reproductive efficiency of sheep and goats has a higher ceiling than field crops or any other animal species. He contends that the practical ceiling of each birth of sheep or goats can be raised from two to four. Thus, where sheep or goat offspring are born every 7.2 months, this would raise the ceiling to about six young weaned per female per year.

SELECTION

Selection in its simplest form is keeping some animals and culling others, in an effort to improve traits in proportion to their heritability and economic importance.

Early efforts by producers and breeders of sheep and goats to bring about genetic improvement in their animals generally involved importation of breeding animals from abroad, use of purebreds, and emphasis against use of scrub stock. The show-ring was generally used as a measure of quality, and livestock judging was an important part of the training of professional sheep and goat scientists. In early days, wool and mohair were generally the important products of the sheep and goat industries, although surplus animals from breeding and production generally were used for meat.

BASES OF SELECTION

In establishing new flocks or herds or improving old ones, sheep and goat breeders striving for improvement in their animals have (as mentioned in Chapter 12) four bases of selection at their disposal; namely, (1) selection based on type or individuality, (2) selection based on pedigree, (3) selection based on show-ring winnings, and (4) selection based on production testing.

SELECTING SHEEP AND GOATS FOR ECONOMICALLY IMPORTANT TRAITS

Since the amount of selection that can be practiced for any one trait is limited, it is important to emphasize in selection those traits for which the greatest progress can be made and those which are most valuable.

Table 3-4 lists the economically important traits of sheep and gives the approximate heritability of each, while Table 3-5 lists economically important traits of goats and their approximate heritability. That sheep and goats show variations in these characteristics is generally recognized. The problem is to measure these differences from the standpoint of discovering the most desirable genes and then increasing their con-

[8]Hohenboken, W. D., and S. E. Clarke, "Genetic, Environmental and Interaction Effects on Lamb Survival, Cumulative Lamb Production and Longevity of Crossbred Ewes," *Journal of Animal Science*, Vol. 53, No. 4, 1981, pp. 966–976.

TABLE 3-4
ECONOMICALLY IMPORTANT TRAITS IN SHEEP AND THEIR HERITABILITY

Economically Important Characters	Approximate Heritability of Character[1]	Comments
	(%)	
1. Multiple births	15	Where adequate feeds are available, multiple births are desirable because (1) they greatly increase the weight of lambs sold per ewe, and (2) the annual maintenance requirements of ewes are not far different, whether they are producing twins or singles. Australian workers have increased twinning rate in Merino sheep by 2.3% per year by selection. In New Zealand, the Romney has responded to selection for twinning by increasing 1.1% per year.
2. Birth weight of lambs	30	The larger lambs at birth are generally more vigorous and make faster gains.
3. Weaning weight:		Heavy weaning weights are especially important in those areas where cost of production is largely on a per head rather than on a per pound basis, such as the western range.
a. 60 days of age	10	
b. 100 days of age	30	
4. Rate of gain	30	Preweaning rate of gain, or growth rate, is largely a reflection of the milk production of the ewe. It is affected by twinning, sex of lamb, and age of ewe. Postweaning rate of gain is a reflection of inherent growth potential of the individual. It is also positively correlated with mature size. Growth rate is economically important for three reasons: (1) it is highly associated with feed efficiency—rapid growth is efficient growth; (2) rapid growth allows for the sale of a larger amount of product; and (3) it makes for a shorter time in reaching market weight and condition, thus effecting a saving in labor, making for less exposure to risk and disease, and allowing for more rapid turnover in capital. Postweaning growth rate can be measured effectively by average daily gain, either on the farm or in the central station.
5. Type score:		Type can include any or all of the following: (1) characteristics than influence an animal's ability to live and perform in its environment—such as feet and legs, teeth, and udder; (2) traits that indiciate meatiness; and/or (3) breed type. Type is a factor in determining today's market values. Yet, type within itself—unsupported by performance records for other traits—will not likely be sufficient to ensure high selling prices in the future. The determination of optimum type, the evaluation of it, and the use made of the information, should remain the responsibility of the individual breeder.
a. Weaning	10	
b. Yearling	40	
6. Finish or condition at weaning	17	Finish at weaning is largely determined by available feed and is not highly heritable. Yet, it is most important because milk-fat lambs suitable for slaughter at weaning time almost always bring more per pound than thinner lambs that are sold as feeders. For the range area as a whole, about 25% of the lambs lack sufficient finish for slaughter at weaning time.
7. Wrinkles or skin folds:		Sheep with smooth bodies are preferred. Wrinkled sheep are difficult to shear, and lack fiber uniformity.
a. Neck folds (weaning)	39	
b. Body folds (yearling)	40	
8. Face covering	56	Wool-blind ewes do not graze well, require more labor if they are clipped around the eyes, and wean fewer pounds per lamb. At the Western Sheep Breeding Laboratory, ewes with open faces produced 11 lb *(5 kg)* more lamb per ewe bred than those with covered faces.
9. Fleece weight:		Clean fleece weight is most important, especially when wool prices are favorable. However, scoring a whole fleece, or even a sample, requires much time and equipment; hence, it likely can only be justified in the selection of stud rams in purebred flocks of fine- and medium-wool sheep. Since there is a close correlation between clean fleece weight and grease fleece weight, grease fleece weight will suffice under most circumstances.
a. Grease weight	38	
b. Clean weight	40	
10. Staple length:		Fiber length is important because it is a major factor in determining fleece weight and grade.
a. Weaning	39	
b. Yearling	47	
11. Fleece grade	35	The grade of a fleece—which is based primarily on fiber diameter, but with consideration given to length, also—is important because it determines the use and price of wool.
12. Fat thickness over loin eye	23	Fat thickness is a measure of meatiness; excess fat results in an increase in fat trim and a decrease in percent lean cuts.
13. Loin eye area	53	Loin eye area is a good indicator of muscling.
14. Carcass weight/day of age	22	This trait is moderately heritable.
15. Carcass grade	12	High carcass grade is important because it determines eating and selling qualities.
16. Carcass length	31	Long carcasses are usually meaty carcasses.

[1]The rest is due to environment. These heritability figures are averages based on large numbers, so some variation in individual flocks is expected.

TABLE 3-5
HERITABILITY ESTIMATES FOR GOATS[1]

Trait	Approximate Heritability
	(%)
Age at first kidding	55
Multiple births	16
Weaning weight	44
Weight at 7 months	63
Mature body weight	50
Milk yield, per lactation	39
Percent fat content of milk	40
Milking time	62
Mohair production	30
Fiber diameter	16

[1]These heritability figures are averages based on several studies; so, some variation is expected.

centration, and, at the same time, purging the flock of the less desirable traits.

On a practical basis, sheep breeders can and should select for improvement in the following traits:

1. **Lambs marketed per ewe.** Lamb production in the United States, as measured by number of lambs saved per 100 ewes one year old or older on January 1, increased from 85 to 89 lambs in the 1920s to 88 to 98 lambs in the 1950s. In 1995, the percent lamb crop as a percent of ewes one year old and over increased to 106%.[9] Thus, efficiency of sheep production in the United States needs to be improved by increasing the number of lambs marketed per ewe.

Experimental work indicates that ewes having

[9]*Agricultural Statistics, 1995–96*, p. VII-34, Table 416.

SCORE OF 1

SCORE OF 2

SCORE OF 3

SCORE OF 4

Fig. 3-12. Skin folds (wrinkles) can be selected out of a flock. The above photographs show animals that can be used as a guide for scoring skin folds. *Score of 1:* Smooth bodied; neck, brisket, and thighs all free of folds. *Score of 2:* Folds under throat, on neck, slight amount of folds on thigh, shoulder, and side smooth. *Score of 3:* Heavy folds both front and rear, fairly smooth along side. *Score of 4:* Heavy folds front and rear, with folds along side also, nearly entire body covered with folds. (Courtesy, University of Wyoming)

twins wean an average of about 40 lb more lamb per ewe-year than ewes of the same age with singles. Thus, even though multiple births have low heritability, any gain, no matter how small, is worthwhile. Because multiple birth at first lambing is less frequent, young mothers having multiple births are more likely to transmit this trait. To select for multiple births, therefore, the breeder should select as many replacements as possible from among multiple births born to young mothers. Also, crossbreeding with some prolific breeds like Finnsheep can increase rapidly the number of lambs born and the weights of lambs weaned.

2. **Rate of gain.** Preweaning growth is primarily a function of the maternal abilities of the ewe. Rapid gains are associated with efficient gains. So they are desirable. Moreover, faster-growing animals can usually be carried to a heavier weight without becoming excessively fat.

3. **Quality of lamb.** Hand in hand with increasing the percentage lamb crop raised, carcasses without excess fat, with a much larger loin eye, and with a maximum yield of tender, lean meat need to be produced.

4. **Open faces.** Open-faced ewes produce more and heavier lambs, and only slightly less fleece weight; thus, selection should be made accordingly. Since the heritability for face cover is high, selection for open faces should be combined with selection for performance.

5. **Fleece weight and quality.** Fleece weight has leveled out in recent years. In 1985, it was 7.88 lb (grease basis); in 1994, it was 7.73 lb.[10] Lack of fleece weight is attributed to the emphasis on lamb, rather than wool, because of the more favorable price for lamb.

[10] *Agricultural Statistics 1995–96*, p. VII-40, Table 426.

Fig. 3-13. Sheep producers can select for open faces in their flocks. The photographs above show animals that can be used as a guide for scoring face covering. *Score of 1:* Open-faced, clear channel below eyes and no impairment of vision. *Score of 2:* Some wool on face below eyes but not enough to cause blindness. *Score of 3:* Wool covers most of face and even though eyes seem clear, wool blindness is likely. *Score of 4:* Wool blind; wool on face nearly covers eyes and extends down over face to nose. (Courtesy, University of Wyoming)

Fig. 3-14. Neale's device for estimating clean fleece weight by measuring volume of fleece under constant pressure. (Courtesy, New Mexico State University)

Likewise, wool quality—staple length, fleece density, and uniformity of length and fineness—will be accorded more emphasis by producers when, and if, prices paid for wool are more commensurate with quality.

In addition to selecting for the above traits (and traits listed in Table 3-4), producers should always cull-out animals which are unsound, unthrifty, aged (especially shelly animals and gummers), and low producers. This will increase returns and lessen the possibility of transmitting undesirable defects.

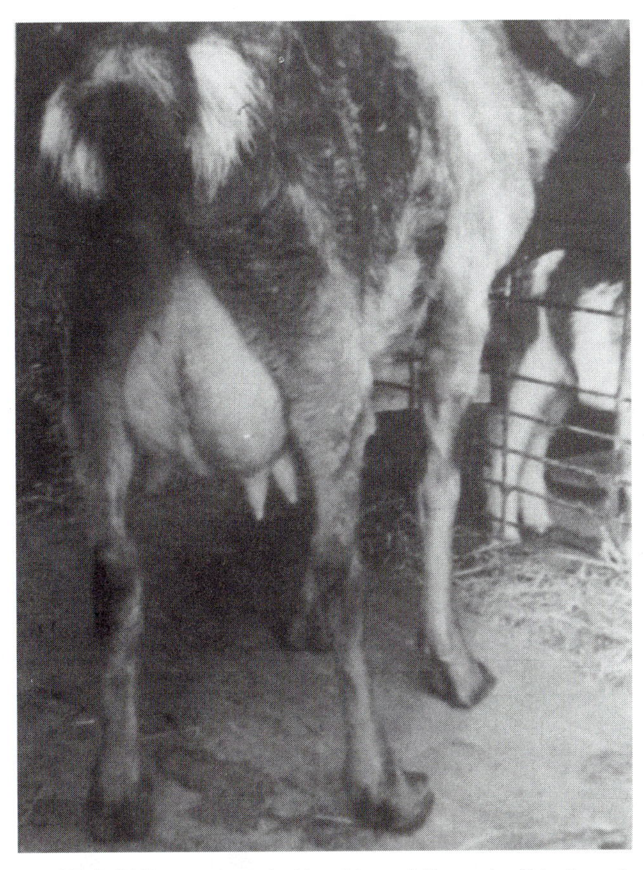

Fig. 3-15. Supernumerary teat in a dairy goat. The mode of inheritance is not clear, but is likely multi-factorial. (Courtesy, Ruth Smelser, Bryan, TX)

LETHALS AND OTHER HEREDITARY DEFECTS

Any hereditary defect will reduce efficiency of sheep production.

The term lethal refers to a genetic factor that causes death of the young, either during prenatal life or at birth. Other defects occur which are not sufficiently severe to cause death but which do impair the usefulness of the affected animals. Some of the lethals and other abnormalities that have been reported in sheep and goats are summarized in Table 3-6. Though strictly nonlethal, other hereditary abnormalities of practical importance in sheep include such things as overshot and undershot jaws. *An overshot jaw is an upper jaw which protrudes beyond the lower one, whereas an undershot jaw is a lower jaw which protrudes beyond the upper one.* Either condition is of practical significance because affected animals cannot graze well. Genetically, these conditions are thought to be due to the interaction of several

Fig. 3-16. Undershot *(left)* and overshot or parrot mouth *(right)* jaws. When undershot, the lower jaw is longer than the upper. When overshot, the lower jaw is shorter than the upper. (Courtesy, U.S. Sheep Experiment Station, Dubois, ID)

TABLE 3-6
LOCI OF SOME LETHAL AND OTHER VISIBLE TRAITS[1]

Locus Name	Gene	Breed	Phenotypic Effect	Allel Name	Symbol	Inheritance
Achondroplastic dwarfism	AchD	Cheviot Merino	Shortening of legs and other dwarfism.	Standard Dwarfed	+ dw	Recessive
Brachygnathia	Br	Australian Merino	Short lower jaw.	Dwe-tel		Recessive
Callipyge	Clpg	Dorset Rambouillet	Heavy muscling.	Muscle hypertrophy	clpg	Dominant
Crytorchidism	Cryp	Merino Rambouillet	Lack of descent of one or both testicles.	Standard Cryptorchidism	+ c	Recessive
Fecundity	Fec^B	Australian Merino Booroola Merino	Increase markedly ovulation rate and fecundity.	Booroola standard	F +	Incomplete dominant
Fecundity Cambridge	FeeC	Cambridge	High range of ovulation rate and fecundity.	Fecund standard	F +	Incomplete dominant
Horn number	HN	Jacob Navajo-Churro	Four horns or more, maybe linked to split eyelid.	Multiple standard	M +	Incomplete dominant
Horns	Ho	Rambouillet Merino	Horn knobs in ewe.	Hornlessness Horns	H h	Recessive
Scurs	Sc	Australian Merino	Aberrant horns, long scurs, short scurs.	Aberrant horns Long scurs Short scurs	Ah ls +	Dominant
Skeletal defect 2, spider syndrome	SD2	Suffolk	Produces larger amount of cartilage which causes deformity of the bones.	Normal Spider syndrome	+ Spi^S	Often lethal

[1]Adapted from *Mendelian Inheritance in Sheep 1996* (MIS 96); editors: J. J. Lauvergene, C. H. S. Dolling, and C. Renieri; publishers COGNOSAG, Clamart (France), University of Camerino, Camerino (Italy).

gene pairs. Affected animals should be culled from the breeding flock.

Many abnormal animals are born on the nation's farms and ranches each year. Unfortunately, purebred breeders, whose chief business is that of selling breeding stock, are likely to "keep mum" about the appearance of any defective animals in their flocks because of the justifiable fear that it may hurt their sales. With commercial producers, however, the appearance of such lethals is simply so much economic loss, with the result that they generally, openly and without embarrassment, admit the presence of the abnormality and seek correction.

Many such abnormalities (commonly known as monstrosities or freaks) are hereditary, being caused by certain bad genes. Moreover, the bulk of such lethals are recessive and may, therefore, remain hidden for many generations. The prevention of such genetic abnormalities requires that the germ plasm be purged of the bad genes. This means that, where recessive lethals are involved, the sheep producer must be aware that (1) both parents carry the gene and (2) a single recessive offspring incriminates the parents. For the total removal of the lethals, test matings and rigid selection must be practiced. The best test mating to use for a given ram consists of mating him to some of his own daughters.

In addition to hereditary abnormalities, there are certain abnormalities that may be due to nutritional deficiencies, or to accidents of development—the latter including those which appear to occur sporadically

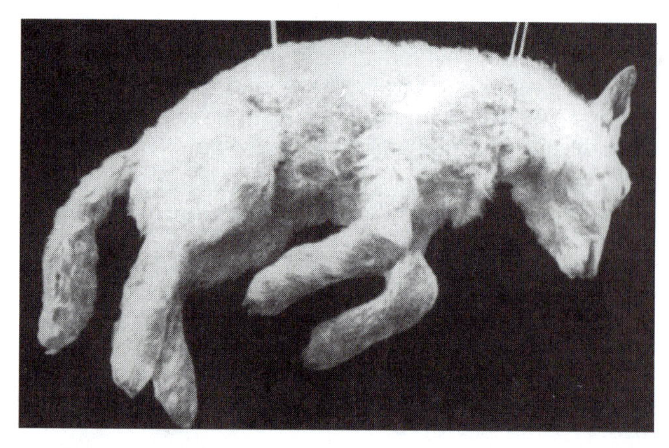

Fig. 3-17. Muscle contracture in the fore limbs. The lamb is suspended by cords to show that the limbs cannot straighten out. (From: *Journal of Genetics*, 21: 57–69, 1929; courtesy, Cornell University)

and for which there is no well-defined reason. When only a few defective individuals occur within a particular flock, it is often impossible to determine whether their occurrence is due to: (1) defective heredity, (2) defective nutrition, (3) viral infection of the dam, or (4) accidents of development. If the same abnormality occurs in any appreciable number of animals, however, it is probably either hereditary or nutritional. In any event, the diagnosis of the condition is not always a simple matter.

The following conditions would tend to indicate a hereditary defect:

1. If the defect had previously been reported as hereditary in the same breed of livestock.
2. If it occurred more frequently within certain families or when there had been inbreeding.
3. If it occurred in more than one season and when different rations had been fed.

The following conditions might be accepted as indications that the abnormality was due to a nutritional deficiency:

1. If previously it had been reliably reported to be due to a nutritional deficiency.
2. If it appeared to be restricted to a certain area.
3. If it occurred when the ration of the mother was known to be deficient.
4. If it disappeared when an improved ration was fed.

If there is suspicion that the ration is defective, it should be improved, not only from the standpoint of preventing such deformities, but from the standpoint of good and efficient management.

If there is good and sufficient evidence that the abnormal condition is hereditary, the steps to be followed in purging the flock of the undesirable gene are identical to those already outlined for ridding the flock of an undesirable recessive factor. An inbreeding program, of course, is the most effective way in which to expose hereditary lethals in order that purging may follow.

APPRAISING CHANGE IN CHARACTERS DUE TO HEREDITY AND ENVIRONMENT

Sheep producers are well aware that there are differences in multiple births, birth weights, weaning weights, body types, fleece weights, milk production, etc. If those animals which excel in the desired traits would, in turn, transmit without loss these same improved qualities to their offspring, progress would be simple and rapid. Unfortunately, this is not the case. Such economically important characters are greatly affected by environment—by feeding, care, management, type of birth (singles vs. twins), age, disease, etc. Thus, only part of the apparent superiority in certain animals is hereditary and can be transmitted to the next generation.

Improvements due to environment are not inherited. This means that if most of the improvement in an economically important character is due to an improved environment, the heritability of that character will be low and little progress can be made through selection. On the other hand, if the character is highly heritable, marked progress can be made through selection. Thus, hair color on the face and legs in sheep—*e.g.*, black, white, or brown—is a highly heritable character, for environment appears to have little or no part in determining it. On the other hand, such a character as finish or condition in sheep is of low heritability because, for the most part, it is affected by environment.

There is need, therefore, to know the approximate amount or percentage of change in each economically important character which is due to heredity and the amount which is due to environment. Tables 3-4 and 3-5 give this information for sheep in terms of the approximate percentage heritability of each of the economically important traits. Similar record forms can be prepared for rams. Likewise, similar record forms can be prepared for buck and doe goats. Actually, heritability estimates for any one trait vary due to differences in breed, accuracy of measurement, importance of environmental effects which cannot be adjusted, and genetic differences in the particular strain or flock involved. Where something is known about these factors, the breeder may be able to judge if estimates from the upper or lower part of the range are more applicable. Where such judgment is not warranted, the midpoint of the range is probably most suitable for use in projecting progress or determining emphasis to place on a given trait in selection. Of course, reliable estimates obtained from the flock in question are most suitable if these are available.

Even though the heritability of many of the economically important characters listed in Tables 3-4 and 3-5 is disappointingly small, it is gratifying to know that the progress from selection is cumulative and permanent.

The following are some of the adjustments that need to be made in order not to confuse nongenetic (environmental) effects with hereditary ones:

1. **Type of birth.** Single lambs generally weigh 6 to 8 lb more at weaning than multiple birth lambs. Also, they gain more slowly between birth and weaning.
2. **Sex.** Males are heavier at birth and gain more from birth to weaning.
3. **Type of rearing.** Lambs born as twins, but reared as singles, gain faster than lambs born and reared as twins, though still slower than lambs born and reared as singles. Management techniques can be employed to increase the performance of multiple birth lambs.
4. **Age of ewe.** When ewes of different ages are compared, several factors should be recognized. Body weight, fleece weight, and lamb production increase in animals up to 3 to 5 years of age, while staple length decreases and fleeces become coarser with age. Lambs from mature ewes wean off 6 to 8 lb heavier than lambs from young ewes.
5. **Weaning age.** If lambs are gaining 0.5 lb daily,

a few days' difference in age can make an appreciable difference in weight.

6. **Year-to-year differences.** Year-to-year differences may be considerable, primarily due to differences in moisture and vegetation.

The above environmental effects may be minimized by selection of lambs of nearly the same age, and within the same year. Also, tables of adjustment factors and formulas are available for adjusting some environmental factors.

■ **Ratios**—To compare animals from different environments, and to determine if an animal is better because of genetics or environment, a breeder often uses ratios. Ratios simply involve the expression of an animal's performance relative to the herd or test group average. They are calculated as follows:

$$\frac{\text{Animal's performance} \times 100}{\text{Average performance of all animals in the group}}$$

For example, a lamb gaining 0.7 lb per day from a group averaging 0.5 lb per day has a daily gain ratio of $(0.7 \times 100) \div 0.5 = 140$. A ratio of 140 implies that the lamb is 40% above the test group average for that trait. Similarly, a ratio of 60 would indicate the lamb performed 40% below the average of contemporary lambs tested. A ratio of 100 indicates the animal is average.

Ratios remove differences in average performance levels among groups (and generally among traits). Thus, they should allow for a more unbiased comparison among individuals that were tested in different groups. This is only true to the extent that average differences among groups are not genetic. Most differences among groups are due to feeding, weather, housing, management, etc. Ratios allow individuals to be compared relative to contemporary groups. They are a useful method for comparing individuals that are not contemporaries, such as those in different tests or herds, or for comparing different traits.

USING FLOCK RECORDS IN SELECTION

On-the-farm records are the first step toward successful production testing and selection. All too often a breeder remembers the good individuals produced by a given animal and forgets those which are mediocre or culls.

A prerequisite for any production data is that each animal be positively identified—by means of earnotches, ear tags, or tattoos. For purebred breeders, who must use a system of animal identification anyway, this does not constitute an additional detail. But the recording of weights and grades does require additional time and labor—an expenditure which is highly worthwhile, however.

In order not to be burdensome, the record forms should be relatively simple. Furthermore, they should be in a form that will permit easy summarization; for example, the record of one ewe should be on one sheet if possible. Figs. 3-18a and 3-18b show an individual ewe or ram record form.

Information on the productivity of close relatives (the sire and the dam and the brothers and the sisters) can supplement that on the animal itself and thus be a distinct aid in selection. This is especially important when traits, such as milk production, lambing rate, and semen production, cannot be measured in both sexes. The production records of more distant relatives are of little significance because, individually, due to the sampling nature of inheritance, they contribute only a few genes to an animal many generations removed.

Flock records have little value unless they are intelligently used in culling operations and in replacement decisions. Also, most producers can and should use production records for estimating the rate of progress and for determining the relative emphasis to place on each character.

Records for goat production are discussed in Chapter 22, Dairy Goats, and Chapter 23, Angora Goats.

Estimating Rate of Progress

For purposes of illustrating how the heritability figures in Table 3-4 may be used in practical breeding operations, the following example is given:

> In a certain flock of sheep, the lambs in a given year average 8 lb at birth. There are available sufficient of the heavier lambs (averaging 10 lb) from which to select replacement breeding stock. What amount of this heavier birth weight (2 lb above the average) is likely to be transmitted to the offspring of these heavier lambs?

Step by step, the answer to this question is secured as follows:

1. $10 - 8 = 2$ lb, the birth weight by which the selected lambs exceed the average from which they arose.
2. By referring to Table 3-4, one finds that the birth weight of lambs is 30% heritable. This means that 30% of the 2 lb can be expected to be due to the superior heredity of the stock saved as breeders, and that the other 70% is due to environment (feed, care, management, etc.).
3. $2.0 \times 30\% = 0.60$ lb, which means that for birth weight the stock saved for the breeding flock is 0.60 lb superior genetically to the stock from which it was selected.

INDIVIDUAL EWE OR RAM RECORD

Breed ____________ Reg. No. ____________ Ear Nick ____________ Tattoo ____________ Birth Date ____________

____________ Type of Birth (Single, Twin, Triplet) ________ Date ____________

Sire ____________ Bred by ____________ Temperament (Gentle, Nervous) ____________

____________ Bought from ____________

____________ Address ____________ Face Covering[6] (as a lamb) ____________

Dam ____________ Date purchased ____________ Face Covering[6] (as a yearling) ____________

____________ Type, Weaned[1] ________ Date ________

Type, Yearling[1] ________ Date ________ Disposed to ____________ Date ________

Back[2] ______ Rump[3] ______ Leg[4] ______ Why Disposed[7] ____________

Defects & Abnormalities[5] ____________

LAMBS (Use one line for each lamb for ewe's offspring; use one line for the average of a ram's progeny for each year.)

Date of Birth	Ear Nick and No.	Vigor at Birth	Type of Birth[8]	Type of Rearing[9]	Sex	Birth Wt.	Defects and Abnormalities[5]	Sire	Milking Ability—Ewe[10]	Weaning Age, Days	Weaning Weight	Weaning Condition[11]	Weaning Type[1]	Disposition[7] or Remarks

[1]Trueness to breed appearance and desired meat conformation: 1 = Excellent, 2 = Good, 3 = Medium, 4 = Fair, 5 = Poor.

[2]Straightness, strength, and spring of rib: 1 = Excellent, 2 = Good, 3 = Medium, 4 = Fair, 5 = Poor.

[3]Width and levelness: 1–2–3–4–5 as above.

[4]Plumpness of thigh: 1–2–3–4–5 as above.

[5]Including overshot or undershot jaw, scurs, black fiber, etc.

[6]1 = Not covered beyond poll; 2 = Covered to eyes; 3 = Covered slightly below eyes, but open-faced; 4 = Covered partially below eyes, but not subject to wool blindness; 5 = Face covered and subject to wool blindness.

[7]Cause of death, reason for disposal, kept for breeding purposes, to whom sold.

[8]S = Single, T = Twin, Tr = Triplet.

[9]S = Single, T = Twin, Tr = Triplet, Gr = Grafted on foster mother and given her number.

[10]Good, medium, poor.

[11]Condition or degree of fatness: 1 = Excellent, 2 = Good, 3 = Medium, 4 = Fair, 5 = Poor.

Fig. 3-18a. Individual ewe or ram record form (see Fig. 3-18b for reverse side of record form).

WEIGHT RECORD OF EWE OR RAM

Date	Age	Weight	Condition[1]	Remarks[2]

REMARKS
(For example: bad udder, poor mother, aborted, veterinary treatment, and nature of ailment.)

Date: Remarks:

FLEECE
(Use one line for each year.)

Length Side[3]	Fineness[4]			Date of Shearing	Days' Growth	Grease Weight	Percent of Yield	Clean Weight	Color of Skin	Purity[5]	Remarks About Fleece
	Shoulders	Side	Thigh								

[1]Condition or degree of fatness: 1 = Excellent, 2 = Good, 3 = Medium, 4 = Fair, 5 = Poor.
[2]Factors affecting weight: *e.g.*, just shorn, soon lamb, etc.
[3]Length of staple, middle at side, to nearest 0.2 cm, just before shearing.
[4]Numerical grade as determined by USDA samples, just before shearing.
[5]Kemp, black fibers, etc.

Fig. 3-18b. Reverse side of individual ewe or ram record form (for front side of record form, see Fig. 3-18a).

4. 8.0 + 0.60 lb birth weight, which is the expected performance of the next generation.

The 8.60 lb birth weight is merely the expected performance. The actual outcome may be altered by environment (feed, care, management, etc.) and by chance. Also, where the heritability of a character is lower, less progress can be made. The latter point explains why the degree to which a trait is heritable has a very definite influence on the effectiveness of mass selection.

Using the heritability figures given in Table 3-4, and assuming certain flock records, one might expect the progress from one generation of selection in a given flock of sheep to appear somewhat as summarized in Table 3-7. Similar tables can be developed for any of the economically important traits listed in Table 3-4, and similar tables can be developed for goats using the heritability values for traits in Table 3-5.

TABLE 3-7
ESTIMATING RATE OF PROGRESS IN SHEEP

Economically Important Characters	Average of Flock	Selected Individuals for Replacements	Average Selection Advantage	Heritability	Expected Performance Next Generation
				(%)	
1. Birth weight of lambs . . (lb)	8	10	2	30	8.6
2. Lambs at weaning time					
a. Weaning weight . . . (lb)	80	90	10	30	83.0
b. Body type at weaning[1]	3	1	2	10	2.8
c. Finish or condition at weaning[2]	3	1	2	17	2.66
3. Fleece					
a. Face covering[3]	3	1	2	56	1.88
b. Yearling grease fleece weight (lb)	9	11	2	38	9.76
c. Staple length at weaning age (cm)	2	3	1	39	2.39

[1]The type grades used herein range from 1 to 5, with No. 1 being the top grade. Any system may be applied.

[2]The finish grades used herein range from 1 to 5, with No. 1 being the top grade. Any system may be applied.

[3]The degree of face covering designates used herein range from 1 to 5, with No. 1 being the most open-faced. Any system may be applied.

Factors Influencing Rate of Progress

Sheep producers need to be informed relative to the factors which influence the rate of progress that can be made through selection. These factors are:

1. **The heritability of the character.** When heritability is high, much of the trait will appear in the next generation, and marked improvement will be evident.

2. **The number of traits for which selection is made at the same time.** If selection is made for four traits, only one-half as much progress would be made in each one as would be achieved if selection were made for only one trait. If selection is made for nine traits, only one-third as much progress will be made as would be achieved if selection were made for only one trait.

This emphasizes the importance of limiting the traits in selection to those having the greatest importance as determined by economic value and heritability. At the same time, it is recognized that it is seldom possible to select for one trait only, and that income is usually dependent upon several traits.

3. **The genotypic and phenotypic correlation between traits.** The effectiveness of selection is lessened by (a) negative correlation between two desirable traits or (b) positive correlation of desirable with undesirable traits. For example, it is reported that the most important factor limiting improvement in the fleeces of Australian Merinos is the negative correlation which exists between fleece weight and crimps per inch. Because of this situation, the rate of improvement in fleece weight is reduced whenever selection pressure is increased for crimps per inch.

4. **The amount of heritable variation measured in such specific units as pounds, inches, numbers, etc.** If the amount of heritable variation—measured in such specific units as pounds, inches, or numbers—is small, the animals selected cannot vary much genetically above the average of the entire flock, and progress will be slow. For example, there is much less spread, in pounds, in the birth weights of lambs than in the 120-day weights (usually there is less than a 4-lb spread in weights at birth, whereas a spread of 30 to 40 lb is common at weaning time). Therefore, more marked progress in selection can be made in the older weights than in birth weights of lambs, when measurements at each stage are in pounds.

5. **The accuracy of records and adherence to an ideal.** A breeder who maintains accurate records and consistently selects toward a certain ideal or goal can make more rapid progress than one whose records are inaccurate and whose ideals change with fads and fancies.

6. **The number of available animals.** The greater the number of animals available from which to select,

the greater the progress that can be made. For maximum progress, enough animals must be born and raised to permit rigid culling. For this reason, more rapid progress can be made with swine than with animals that have only one offspring, and more rapid progress can be made when a flock is being either maintained at the same numbers or reduced than when it is being increased in size.

7. **The age at which selection is made.** Normally, progress is more rapid if selection is practiced at an early age. This is so because more of the productive life is ahead of the animal, and the opportunity for gain is then greatest.

8. **The generation interval.** Generation interval refers to the period of time required for parents to be succeeded by their offspring, from the standpoint of reproduction. The minimum generation interval of sheep is about one year. By way of comparison, the average length of a human generation is 33 years.

Shorter generation lengths will result in greater progress per year, provided the same proportion of animals is retained after selection.

Usually it is possible to reduce the generation intervals of sires, but it is not considered practical to reduce materially the generation intervals of females, Thus, if progress is being made, the best young males should be superior to their sires. Then, the advantage of this superiority can be gained by changing to new generations as quickly as possible. To this end, it is recommended that the breeder change to younger sires whenever their records equal or excel those of the older sires after accounting for age and year differences. In considering this procedure, it is very difficult to adequately compare records made of animals in different years or at different ages.

9. **The caliber of the sires.** Since a much smaller proportion of males than of females is normally saved for replacements, it follows that selection among the males can be more rigorous and that most of the genetic progress in a flock will be made from selection of males. Thus, if 2% of the males and 50% of the females in a given flock become parents, then about 75% of the hereditary gain from selection will result from the selection of males and 25% from the selection of females, provided their generation lengths are equal. If the generation lengths of males are shorter than the generation lengths of females, the proportion of hereditary gain due to the selection of males will be even greater.

Determining Relative Emphasis to Place on Each Trait

A replacement animal seldom excels in all of the economically important characters. The producer must decide how much importance shall be given to each trait. Thus, the sheep producer will have to decide how much emphasis shall be placed on birth weight, weaning weight, body type, rate of gain, weight and quality of fleece, and carcass traits.

Perhaps the relative emphasis to place on each trait should vary according to the circumstances. Under certain conditions, some traits may even be ignored. Among the factors determining the emphasis to place on each character are the following:

1. **The economic importance of the character to the producer.** Table 3-4 lists the economically important characters in sheep, and summarizes (see "Comments" column) their importance to the producer.

By economic importance is meant their dollar-and-cent value. Thus, those traits which have the greatest effect on profits should receive the most attention.

2. **The heritability of the character.** The more highly heritable traits should receive higher priority than those which are less heritable, for more progress can be made thereby.

3. **The genetic correlation between traits.** One trait may be so strongly correlated with another that selection for one automatically selects for the other. For example, rate of gain and economy of gain in meat animals are correlated to the extent that selection for rate of gain tends to select for the most economical gains as well; thus, economy of gain may be largely disregarded if rate of gain is given strong consideration. Conversely, one trait may be negatively correlated with another so that selection for one automatically selects against the other.

4. **The amount of variation in each trait.** Obviously, if all animals were exactly alike in a given trait, there could be no selection for that trait. Likewise, if the amount of variation in a given trait is small, the selected animals would not be very much above the average of the entire flock, and progress would be slow. If variability is high, the selection differential will be high.

5. **The level of performance already attained.** If a flock has reached a satisfactory level of performance for a certain trait, there is not much need for further selection for that trait. Thus, if sheep are almost free of wrinkles, there is little need to select against them; in fact, if sheep are entirely free of wrinkles, it would be impossible to select against them since there would be no wrinkled individuals to cull.

Sufficient selection pressure should be exerted to maintain the flock at the desired level of excellence, as low-producing animals retained in the flock lower the average production.

PREWEANING LAMB PERFORMANCE TEST

The following preweaning lamb performance test

program is recommended for use in all purebred or commercial farm flocks:

Step 1: Record individual data. *Minimum data* should include (a) identification—lamb ear tag number; (b) sire number; (c) dam number; (d) age of dam (in years) at lambing time; (e) birth date of lamb; (f) sex of lamb; (g) type of birth—single, twin, triplet, multiple; and (h) how reared—single, twin, triplet, artificially.

Optional data may include (a) whether or not lamb was creep-fed; (b) slaughter grade; (c) 200-day yearling body weight; and (d) grease fleece weight and staple length to nearest tenth of an inch.

Step 2: Wean and weigh. In advance, decide on weaning age—usually 90, 120, or 140 days. Wean and weigh as near to the intended age as possible.

Step 3: Type score. Even though the heritability of type, or conformation, is low at weaning (10%), it is a factor in determining today's market values; hence, all performance records should be augmented by type scores at weaning time. Also, because of its higher heritability (40%), all animals retained for breeding purposes should be type scored as yearlings.

Type can include any or all of the following: (a) traits that influence an animal's ability to live and perform in its environment—such as feet and legs, teeth, udder, and lethals and sublethals; (b) traits that indicate meatiness; and/or (c) breed type.

Also type score may include face cover score, wrinkle score, and record of the presence or absence of scurs or horns.

The determination of optimum type, the evaluation of it, and the use made of it should remain the responsibility of the individual breeder.

Step 4: Adjust for certain environmental factors. Adjust records for certain environmental factors such as age, sex, type of birth and rearing, and age of dam.

Weaning weights of lambs within a flock may be adjusted to 90, 120, or 140 days of age by finding the weight per day of age, then multiplying by the standardized age desired. Thus, the following formula may be used to provide the estimated 120-day weight of lambs:

$$\text{Adjusted 120-day weight (lb)} = \frac{\text{Actual weaning weight}}{\text{Actual days of age}} \times 120$$

Additional adjustment factors are given in Table 3-8.

To use Table 3-8, multiply the 90-, 120-, or 140-day weight by the appropriate factor. For example, to find the adjusted 120-day weight of a twin-born-and-reared ram lamb from a two-year-old ewe that weighed 90 lb at 110 days of age, make the following calculations:

TABLE 3-8
ADJUSTMENT FACTORS

	Age of Dam		
	1 Year Old	2 Years Old or Over 6 Years Old	3 to 6 Years Old
Ewe Lamb			
Single	1.22	1.09	1.00
Twin—raised as twin	1.33	1.20	1.11
Twin—raised as single	1.28	1.14	1.05
Triplet—raised as triplet	1.46	1.33	1.22
Triplet—raised as twin	1.42	1.28	1.17
Triplet—raised as single	1.36	1.21	1.11
Wether			
Single	1.19	1.06	0.97
Twin—raised as twin	1.30	1.17	1.08
Twin—raised as single	1.25	1.11	1.02
Triplet—raised as triplet	1.43	1.30	1.19
Triplet—raised as twin	1.39	1.25	1.14
Triplet—raised as single	1.33	1.18	1.08
Ram Lamb			
Single	1.11	0.98	0.89
Twin—raised as twin	1.22	1.09	1.00
Twin—raised as single	1.17	1.03	0.94
Triplet—raised as triplet	1.35	1.22	1.11
Triplet—raised as twin	1.31	1.17	1.06
Triplet—raised as single	1.25	1.10	1.00

[1]Scott, G. E., ed., *The Sheepman's Production Handbook*, 2nd ed., Sheep Industry Development Program, Denver, CO, p. 27.

$$\frac{\text{90 lb}}{\text{110 days of age}} = 0.82 \text{ lb} \times 120$$

$$= 98 \text{ lb} \times 1.09 \text{ (adjustment factor)} = 107 \text{ lb}$$

The adjusted 120-day weight of the lamb would be 107 lb.

Step 5: Cull. Cull the lambs that fail to measure up in the preweaning performance test.

NATIONAL SHEEP IMPROVEMENT PROGRAM (NSIP)

The National Sheep Improvement Program (NSIP), P.O. Box 901566, Kansas City, MO 84180-1566, *is a computerized genetic evaluation program.* It calculates estimates of genetic value for every sheep in the flock for several maternal, growth, and wool traits. Flock owners provide the raw data and NSIP calculates the genetic values. NSIP is not a sheep management, ration balancing, or sheep economic program; it is concerned with providing producers with information which will allow them to increase the rate of genetic improvement in their flocks.

■ **What does NSIP measure?**—Producers can choose from several traits, and tailor NSIP to their selection needs. Traits for which estimates of genetic value can be calculated include—

1. **Maternal traits.** Such as, (a) number of lambs born per ewe lambing, and (b) pounds of lamb weaned per ewe exposed.
2. **Growth rate.** Up to three weights of lambs can be recorded for each flock, with weights made at 30, 60, 90, 120, 180, or 360 days of age.
3. **Wool traits.** NSIP can generally evaluate three wool traits: (a) grease fleece weight, (b) staple length, and (c) fiber diameter. Codes for face cover and skin folds may also be recorded.

■ **What does NSIP cost?**—The cost of NSIP is $50.00 per flock per year plus a basic fee of 75¢ per ewe for once-a-year lambing systems or $1.25 per ewe for accelerated lambing systems. This will give the sheep producer three traits: number of lambs born, pounds of lambs weaned, and one growth trait of the producer's choice—plus a ewe productivity index and other information. Additional traits cost 25¢ per ewe. There is no charge for rams.

The annual cost for a 100-ewe once-a-year lambing flock wanting a Flock Genetic Evaluation Summary (FEPDs) on number of lambs born, weight of lambs at 60 days of age per ewe, 60-day weight, 120-day weight, and wool weight is $175.00 ($50.00 flock fee + [100 × 75¢] basic fee + [100 × 50¢] fee for two additional traits). So, the cost is less than $2.00 per ewe per year for the most accurate estimates of genetic value possible for five traits of every ewe, ram, and lamb in the flock.

■ **The goal ahead: to compare FEPDs across flocks**—The ability to compare FEPDs across flocks would allow more rapid genetic improvement of the U.S. sheep population because producers would be able to compare animals from other flocks with their own and select those animals which would improve their flocks. In order to calculate across-flock FEPDs, flocks need to be "genetically tied" through the presence of related individuals in several flocks. In order to obtain across-flock genetic evaluations, more flocks need to be enrolled in NSIP so genetic "connectedness" among flocks can be determined. If adequate genetic ties do not exist, sharing of rams among breeders and use of several sires in many flocks through artificial insemination would create the needed genetic ties.

The U.S. Targee Sheep Association recently completed the first across-flock genetic evaluation of any breed of sheep in the United States. Other breeds have indicated their intent to follow.

PRODUCTION TESTING BY STATE ASSOCIATIONS AND COLLEGES

More states need to initiate production testing programs, and more breeders need to utilize them for flock improvement. Also, production testing programs need to be standardized, or made uniform, from state to state; and where central stations are available and used, sheep producers need to be admonished that on the farm or ranch testing is the eventual goal. In view of this, some of the Land-Grant Universities should offer computerized on-the-farm performance testing. Two of the most used programs are maintained by The University of Wisconsin and The Ohio State University. For a small fee, these programs, using data provided by sheep producers, calculate ewe and lamb indexes based on multiple births, lamb growth, and ewe fleece weights; and provide an annual printout.

For example, the Wisconsin Sheep Improvement Program requires that producers keep the following records on their flocks (see Fig. 3-20).

1. Identification numbers of all ewes in the flock, rams used, and lambs born.
2. Date of birth, sex, and type of birth of each lamb.
3. Age of ewe.
4. Fleece weight of ewe.
5. Breed of ewe.
6. Sire and breed of sire of lamb.
7. Disposition dates and miscellaneous information—death or separation from dam.
8. Birth weight of lamb (recommended).
9. Date ram was first exposed to a ewe (optional).

After submission of this data and computer analysis, each producer will receive: (1) lamb summaries,

Fig. 3-19. Rams going through a 150-day performance test at the University of Wyoming. Selecting rams on the basis of performance testing is winning ground in the sheep industry. (Courtesy, Dr. Johannes E. Nel, University of Wyoming)

WISCONSIN SHEEP IMPROVEMENT PROGRAM

48732701 EDWARD XAVIER AMPLE RT 1 HIGHWAY 8 SHEEP VALLEY WI 53706 PAGE 5

SEASON 1 YEAR

EWE PRODUCTION SUMMARY

EWE				RAM			BIRTH		NUMBER OF LAMBS			AVG. ADJ 90-DAY WWT		AVG. LAMB INDEX		TOTAL WT OF LAMB WEANED	EWE INDEX	
ID	BREED	AGE	FLEECE WT	ID	BREED	DATE IN	DATE	AVG. WT	BRN	RRD	WND	WWT	RATIO	INDEX	RATIO		INDEX	RATIO
75104	TARG	7	10.0	78101	TARG	10- 1	2- 4-82	6.5	3	2	3	102	117	118	118	305	149	122
77111	TARG	5	11.0	78101	TARG	10- 1	1-28-82	8.8	3	3	3	92	105	107	107	275	140	115
77117	TARG	5	11.0	78101	TARG	10- 1	2- 3-82	7.7	3	2	2	91	105	107	107	182	139	114
78111	TARG	4	12.0	78101	TARG	10- 1	1-30-82	3.5	1	0	0	0	0	0	0	0	0	0
78143	TARG	4	8.5	78101	TARG	10- 1	2-11-82	4.5	2	0	0	0	0	0	0	0	0	0
78150	TARG	4	12.0	78101	TARG	10- 1	0- 0- 0	.0	0	0	0	0	0	0	0	0	0	0
78201	TARG	4	9.0	78101	TARG	10- 1	2- 9-82	6.3	2	1	1	70	80	82	82	70	109	89
79101	TARG	3	9.0	78101	TARG	10- 1	1-31-82	9.0	1	1	1	64	74	69	69	64	91	75
79102	TARG	3	8.5	78101	TARG	10- 1	1-24-82	8.5	1	1	1	64	74	69	69	64	90	74
79137	TARG	3	9.0	78101	TARG	10- 1	2- 5-82	7.8	2	2	2	86	99	100	100	172	128	105
80127	TARG	2	9.5	78101	TARG	10- 1	2- 2-82	9.3	2	2	2	79	91	92	92	157	120	98
80134	TARG	2	7.0	78101	TARG	10- 1	2- 1-82	7.3	2	2	2	85	98	98	98	170	120	98
81100	TARG	1	6.5	78101	TARG	10- 1	2- 6-82	8.0	1	1	1	81	93	85	85	81	101	83
81147	TARG	1	10.5	78101	TARG	10- 1	2- 4-82	8.5	2	2	2	86	99	101	101	172	132	108
82147	TARG	1	10.5	78101	TARG	10- 1	2- 4-82	9.5	1	1	1	110	126	127	127	110	142	116
AVERAGE		3.3	9.6					7.5				84	97	96	96	152	122	100
NO. OF EWES		15	15					14	26		21	12	12	12	12	12	12	12

Fig. 3-20. An example of a ewe production summary from the Wisconsin Sheep Improvement Program. (Courtesy, The University of Wisconsin)

(2) sire summaries, (3) an annual ewe production summary, and (4) a flock production summary.

1. **Lamb summaries.** These list each lamb reported and give an adjusted 90-day weight and ratio and a lamb index and index ratio for each lamb. The lambs are ranked from highest to lowest according to the lamb index and ratio.

2. **Sire summaries.** These list all lambs within sex groups by each sire. Basically, the same information is provided in these summaries as in the lamb summaries. The most useful aspect of the sire summaries is that they provide producers with a listing of each ram's progeny, and a summary of how each ram performed in the flock.

3. **Annual ewe production summary.** This summarizes the current productivity of each ewe reported. Each ewe is listed by an identification number. The birth weights, number of lambs dropped, and the number weaned for each ewe are listed. Then the adjusted 90-day weight and ratio, and the average lamb index and ratio of her progeny, are also listed.

4. **Flock production summary.** This provides information useful in evaluating a producer's total breeding program. It summarizes the number of ewes in the flock, number of ewes open, number of ewes lambing, percent open, percent lambing, lambing rate, lambing rate per ewe exposed, percent singles, percent twins, percent triplets or greater, percent lambs born that were weaned, percent lambs weaned per ewe lambing and per ewe exposed, and pounds of lamb weaned per ewe exposed.

Ohio has the only other on-the-farm testing program. It is patterned after the Wisconsin program—the original.

Although it is difficult to measure genetic progress in field situations, because genetic improvement is confounded by nutritional and management improvements, Table 3-9 describes some productivity changes in sheep observed in the Wisconsin Sheep Improvement Program during 31 years.

Most of the state programs are conducted in coop-

TABLE 3-9
CHANGES IN TRAITS MEASURED IN WISCONSIN SHEEP IMPROVEMENT PROGRAM DURING A 31-YEAR PERIOD

Year	Lamb Weight[1]				Twins or Greater	Lamb Weight per Ewe		Wool Weight per Ewe	
	Singles		Twins						
	(lb)	*(kg)*	*(lb)*	*(kg)*	*(%)*	*(lb)*	*(kg)*	*(lb)*	*(kg)*
1950	63	*28.6*	58	*26.3*	52	83	*37.6*	8.6	*3.9*
1960	75	*34.0*	68	*30.8*	65	99	*44.9*	7.9	*3.6*
1970	82	*37.2*	76	*34.5*	64	108	*49.0*	8.4	*3.8*
1980	79	*35.8*	80	*36.3*	69	118	*53.5*	8.8	*4.0*

[1]Since 1971, lamb weights have been adjusted to 90 days of age. Prior to 1971, lamb weights were adjusted to 120 days of age.

eration with state colleges of agriculture. For information relative to these programs, sheep producers should contact their county agents or colleges of agriculture.

PRODUCTION TESTING OF DAIRY GOATS

The National Cooperative Dairy Herd Improvement Program (NCDHIP) was developed to provide information for dairy producers to use in improving the production efficiency of their herds. Although the DHI Program was designed primarily for dairy cows, dairy goat owners may also participate. Records of identification, production, and management enable a dairy goat producer to: (1) cull the least profitable animals; (2) feed for maximum production; (3) make precise management decisions for maximum profit; (4) select superior animals for herd replacements; and (5) breed for improvement for the future.

The DHI is a national cooperative research program between dairy producers, the Cooperative Extension Service, and the U.S. Department of Agriculture.

Individual dairy goat producers benefit by receiving production and management information on their animals. The dairy goat industry benefits when their data are pooled and used in research and genetic evaluation programs.

The DHI Program provides several testing and record-keeping plans. Dairy producers may choose the plan that most appropriately meets their needs. These include *official* and *unofficial* record-keeping plans.

Official records include the DHI and the DHIR type of testing programs. Official records are made under uniform national rules to assure accuracy and reliability, and are used in genetic evaluation, research, and education programs.

Other testing programs not meeting the requirements for official test include the *commercial* and *owner-sampler* types of tests. These records are not used in genetic evaluations or in research programs. They are, however, good records and are used effectively by many dairy producers in herd management.

There are some problems that limit the participation of dairy goats in the DHI Program. These include:

1. The cost of testing goats is high compared to their earning ability.
2. Goat herds tend to have fewer animals than do cow herds, thereby increasing the costs.
3. Goats may be located in areas not served by DHI associations.
4. Goats are seasonal breeders, resulting in a period during the year when all does may be dry at the same time.

Many dairy goat owners have overcome most of these difficulties and are successfully participating in the DHI Program.

Dairy goat owners may obtain credit for their does' production in two ways. They may participate in one-day production tests as specified by their specific breed organization. Arrangements must be made ahead of time with the dairy goat associations and the local Dairy Herd Improvement Association supervisor. The one-day test is not an official DHI program.

Dairy goat owners may also receive credit for their does by testing in the official Dairy Herd Improvement Registry (DHIR) program. Application for participation in the DHIR must be made with the dairy goat association. The program is supervised by the superintendent of official testing of the state university, the county extension personnel, and the DHIA supervisors. The higher costs prevented many dairy goat owners from participating. So, the Group Testing (GT) program was developed. GT is an official DHIR program. It enables dairy goat owners to participate in the DHIR program by allowing them to do part of the work themselves and thereby reduce the costs. Each member of the group is trained as a supervisor to weigh and sample milk of another group member's herd.

In many ways, the GT is more stringent than either DHI or DHIR tests because it has an additional set of GT rules required by the specific breed organization. All GT is conducted under the direct jurisdiction and supervision of the local DHIA supervisor and extension personnel, just as with the other official testing programs.

After DHI supervisors have conducted the butterfat test and obtained basic information from the dairy goat owner on test day, they send the barn sheets to the dairy record-processing center. The data are audited and inspected for completeness and accuracy then entered into a computer. From this data, the goat producer receives lactation reports, herd summaries, monthly goat listings, and management reports—all containing a variety of information designed to help the producer make correct management decisions.

Subsequently, the U.S. Department of Agriculture released the first ever National Dairy Goat Buck Evaluation for milk and fat. This herd sire summary rates dairy goat bucks based on a comparison of the daughters of one, in terms of milk and fat production, with the daughters of another. The data for these buck evaluations were obtained from all of the regional dairy record-processing centers in the NCDHIP, as well as pedigree data from the ADGA. After the available NCDHIP lactation records were edited, 58,562 records, which represented 43,913 does sired by 11,670 bucks in 5 breeds, were selected for analysis.

These buck evaluations represent an important first step toward providing dairy goat breeders with a comprehensive array of genetic information. Thus, sires can be considered on the basis of their production ability as well as type. This will also likely increase the use of artificial insemination in goats as it did in cattle.

MATING SYSTEMS

No one best system of breeding or secret of success exists for any and all conditions. Each breeding program is an individual case, requiring careful study. The choice of the system of breeding should be determined primarily by the size and quality of the flock, by the finances and skill of the operator, and by the ultimate goal ahead.

Goat breeders may also apply these mating systems to their programs.

PUREBREEDING

A purebred animal may be defined as a member of a breed, the animals of which possess a common ancestry and distinctive characteristics; and it is either registered or eligible for registry in that breed. The breed association consists of a group of breeders banded together for the purpose of: (1) recording the lineage of their animals, (2) protecting the purity of the breed, and (3) promoting the interests of the breed.

The term *purebreds* refers to animals whose entire lineage, regardless of the number of generations removed, traces back to the foundation animals accepted by the breed or to animals which have been subsequently approved for infusion. The terms *purebreeding* and *homozygosity* may bear different connotations. Homozygosity refers to the likeness of genes. Yet, there is some interrelationship between purebreds and homozygosity. Because most breeds had a relatively small number of foundation animals, the unavoidable inbreeding and linebreeding during the formative stage resulted in a certain amount of homozygosity. Moreover, through the normal sequence of events, it is estimated that purebreds become more homozygous by from 0.25 to 0.5% per animal generation. Studies have revealed that inbreeding in the Rambouillet and Hampshire breeds occurred at the rate of 0.7 to 0.9% per generation, respectively.

Being a purebred animal does not necessarily guarantee superior type or high productivity. The word *purebred* is not, within itself, magic, nor is it sacred. Many breeders have found to their sorrow that there are purebred scrubs. Yet, on the average, purebred animals are superior to non-purebreds because selection for merit is an important part of the purebred system.

Purebred breeding is a highly specialized type of production. Generally, only the experienced breeder should undertake the production of purebreds with the intention of furnishing foundation or replacement stock to other purebred and commercial breeders. Although there have been many constructive sheep breeders and great progress has been made, only a few achieve sufficient success to classify as master breeders.

INBREEDING

Inbreeding involves the mating of close relatives. If it is intense, it is generally accompanied by loss of hybrid vigor, and is to be avoided. However, close matings to achieve important goals may be quite acceptable, generally for only one generation. Harmful inbreeding can generally be avoided by establishing several sire lines (line breeding) within a closed flock and then mating ewes in rotation to the least closely related sires. Each sire line should be continued indefinitely, within sire line selection, unless a particular sire line becomes definitely inferior to the other sire lines. Such lines might be outcrossed to a highly selected sire from a neighboring flock where the selection goals are similar.

Inbred lines were developed at the U.S. Sheep Experiment Station, Dubois, Idaho, stimulated by the success with hybrid corn. As a result, during and since the 1930s relatively large numbers of inbred lines of sheep have been developed and tested by state and federal agencies. Development of the lines undoubtedly resulted in the redirection or elimination of many undesirable genetic defects of the inbred stock. However, there was generally a depressing affect of inbreeding on reproductive merit and on body and fleece weights.

Closebreeding is the mating of closely related animals, such as sire to daughter, son to dam, and brother to sister.

Fig. 3-21. Purebred Debouillet rams, a fine wool breed. (Courtesy, Dwight McNally, Roswell, NM)

Linebreeding is the mating of animals more distantly related than in closebreeding and in which the matings are usually directed toward keeping the offspring closely related to some highly admired ancestor, such as half-brother to half-sister, female to grandsire, and cousin to cousin.

CLOSEBREEDING

In closebreeding there are a minimum number of different ancestors. In the repeated mating of a brother with his full sister, for example, there are only 2 grandparents instead of 4, only 2 great-grandparents instead of 8, and only 2 different ancestors in each generation farther back—instead of the theoretically possible 16, 32, 64, 128, etc. The most intensive form of inbreeding is self-fertilization. It occurs in some plants, such as wheat and garden peas, and in some lower animals; but domestic animals are not self-fertilized. Closebreeding is rarely practiced among present-day livestock producers though it was common in the foundation animals of most of the breeds.

The **reasons** for practicing closebreeding are:

1. It increases the degree of homozygosity within animals, making the resulting offspring pure or homozygous in a larger proportion of their gene pairs than in the case of linebred or outcross animals. In so doing, the less desirable recessive genes are brought to light so that they can be more readily culled. Thus, closebreeding, together with rigid culling, affords the surest and quickest method of fixing and perpetuating a desirable character or group of characters.
2. If carried on for a period of time, it tends to create lines or strains of animals that are uniform in type and in other characteristics.
3. It keeps the relationship to a desirable ancestor highest.
4. Because of the greater homozygosity, it makes for greater prepotency. That is, selected closebred animals are more homozygous for desirable genes (genes which are often dominant); therefore, they transmit these genes with greater uniformity.
5. Through the production of inbred lines or families by closebreeding and the subsequent crossing of certain of these lines, it affords an alternative approach to sheep improvement. Moreover, the best of the closebred animals are likely to give superior results in outcrosses.
6. Where breeders are in the unique position of having their flocks so far advanced that to go on the outside for seed stock would merely be a step backward, it offers the only sound alternative for maintaining existing quality or making further improvement.

The **precautions** in closebreeding may be summarized as follows:

1. As closebreeding greatly enhances the chances that recessives will appear during the early generations in obtaining homozygosity, it is almost certain to increase the proportion of worthless breeding stock produced. It may include such so-called *degenerates* as reduction in size, fertility, and general vigor. Lethals and other genetic abnormalities often appear with increased frequency in closebred animals.
2. Because of the rigid culling necessary in order to avoid the fixing of undesirable characters, especially in the first generations of a closebreeding program, it is almost imperative that this system of breeding be confined to a relatively large flock and to instances when the owner has sufficient finances to stand the rigid culling that must accompany such a program.
3. It requires skill in making planned matings and rigid selection; thus, it is most successful when applied by master breeders.
4. It is not adapted for use by individuals with average or below-average stock because the very fact that their animals are average or below average means that a goodly share of undesirable genes are present. Closebreeding would merely make the animals more homozygous for the undesirable genes they carry and therefore, worse.

Judging from outward manifestation alone, closebreeding may appear predominantly harmful in its effects—often leading to the production of defective animals lacking in the vitality necessary for successful and profitable production. But this is by no means the whole story. Although closebreeding often leads to the production of animals of low value, the resulting superior animals can confidently be expected to be homozygous for a greater than average number of good genes and thus more valuable for breeding purposes. Figuratively speaking, therefore, closebreeding may be referred to as *trial by fire*, and breeders who practice it can expect to obtain many animals that fail to measure up and that have to be culled. On the other hand, if closebreeding is handled properly, they can also expect to secure animals of exceptional value.

Although closebreeding has been practiced less during the past century than in the formative period of the different pure breeds of sheep, it has real merit when its principles and limitations are fully understood. Perhaps closebreeding had best be confined to use by the skilled master breeder who is in a sufficiently sound financial position to endure rigid and intelligent culling and delayed returns and whose flock is both large and above average in quality.

LINEBREEDING

From a biological standpoint, closebreeding and linebreeding are the same thing, differing merely in intensity. In general, closebreeding has been frowned upon by sheep producers, but linebreeding (the less intensive form) has been looked upon with favor in some quarters.

In a linebreeding program, therefore, the degree of relationship is not closer than half-brother and half-sister or matings more distantly related—cousin matings, grandparent to grand offspring, etc.

Linebreeding may be practiced in order to conserve and perpetuate the good traits of a certain outstanding ram or ewe. Because such descendants are of similar lineage, they have the same general type of germ plasm and therefore exhibit a high degree of uniformity in type and performance.

In a more limited way, a linebreeding program has the same advantages and disadvantages of a closebreeding program. Stated differently, linebreeding offers fewer possibilities for both good and harm than closebreeding. It is a more conservative and safer type of program, offering less probability to either "hit the jackpot" or "sink the ship." It is a middle-of-the-road program that the vast majority of average and small breeders can follow safely to their advantage. Through it, breeders can make reasonable progress without taking any great risks. They can also secure a greater degree of homozygosity of certain desirable genes without running too great a risk of intensifying the undesirable ones.

Usually a linebreeding program is best accomplished through breeding to an outstanding sire rather than to an outstanding dam because of the greater number of offspring of the former. If a sheep breeder owned a great ram—proved great by the production records of a large number of his get—a linebreeding program might be initiated in the following way: Select two of the best sons of the noted ram and mate them to their half-sister, balancing all possible defects in the subsequent matings. The next generation matings might well consist of breeding the daughters of one of the rams to the son of the other, etc. If, in such a program, it seems wise to secure outside blood (genes) to correct a common defect or defects in the flock, this may be done through selecting a few outstanding proved ewes from the outside—animals whose get are strong where the flock may be deficient—and then mating these ewes to one of the linebred rams with the hope of producing a son that may be used in the flock.

Small operators—the owners of a few ewes—can often follow a linebreeding program by breeding their ewes to a ram purchased from a large breeder who follows such a program—thus in effect following the linebreeding program of the larger breeder.

Naturally, a linebreeding program may be accomplished in other ways. Regardless of the actual matings used, the main objective in such a system of breeding is that of rendering the animals homozygous—in desired type and performance—to some great and highly regarded ancestor, while at the same time weeding out homozygous undesirable characteristics. The success of the program is dependent upon having desirable genes with which to start and an intelligent intensification of these good genes.

Some types of flocks should almost never inbreed or linebreed. These include flocks of only average quality.

Owners of grade or commercial flocks run the risk of undesirable results, and, even if successful as commercial breeders, they cannot sell their stock at increased prices for breeding purposes.

With purebred flocks of only average quality, more rapid progress can usually be made by introducing superior outcross sires. Moreover, if the animals are of only average quality, they must have more bad genes than preferred, and these would only be intensified through an inbreeding or linebreeding program.

■ **Inbred lines**—Success with hybrid corn stimulated the interest of sheep geneticists in the possibility of improving sheep through developing and crossing inbred lines. As a result, during and since the 1930s, relatively large numbers of inbred lines of sheep have been developed and tested by state and federal agencies. Although testing of these lines through linecrossing and topcrossing to unrelated stock is still not complete, it seems that there is little evidence to suggest that inbred lines are an effective approach to sheep improvement.

OUTCROSSING

Outcrossing is the mating of animals that are members of the same breed, but which show no relationship close up in the pedigree (for at least the first 4 to 6 generations).

Most of our purebred animals of all classes of livestock are the result of outcrossing. It is a relatively safe system of breeding, for it is unlikely that two such unrelated animals will carry the same undesirable genes and pass them on to their offspring.

Perhaps it may well be that the majority of purebred breeders with average or below-average flocks had best follow an outcrossing program because, in such flocks, the problem is that of retaining a heterozygous type of germ plasm with the hope that genes for undesirable characters will be counteracted by genes for desirable characters. With such average or below-average flocks, a closebreeding program would merely make the animals homozygous for the less desirable characters, the presence of which already makes for their mediocrity.

Judicious and occasional outcrossing may well be an integral part of linebreeding or closebreeding programs. As closely inbred animals become increasingly homozygous with germ plasm for good characters, they may likewise become homozygous for certain undesirable characters even though their general overall type and performance remain well above the breed average. Such defects may best be remedied by the introduction of an outcross through an animal or animals known to be especially strong in the character or characters needing strengthening. This having been accomplished, the wise breeder will return to the original closebreeding or linebreeding program, realizing full well the limitations of an outcrossing program.

GRADING UP

Grading up is that system of breeding in which a purebred sire of a given breed is mated to a native or grade female. Its purpose is to impart quality and to increase performance in the offspring.

Many breeders will continue to produce purebred stock. However, the vast majority of animals in the United States—probably more than 97%—are not eligible for registry. In general, farm animals are sired by purebreds, because of the obvious merit of using well-bred sires. Compared to the breeding of purebreds, such a system requires less outlay of cash, and less experience on the part of the producer. Even with this type of production, grading up of the flock through the use of purebred sires is generally practiced. Thus, one of the principal functions of the purebred breeder is that of serving as a source of seed stock—particularly of sires—for the commercial producer who hopes that concentrated doses of good genes may be secured through the use of purebred sires. As the common stock is improved, this means that still further improvement and homozygosity for good genes are necessary in the purebreds, if they are to bring about further advancement in any grading-up program.

Naturally, the greatest single step toward improved quality and performance occurs in the first cross. The first generation from such a mating results in offspring carrying 50% of the hereditary material of the purebred parent (or 50% of the blood of the purebred parent, as many sheep producers speak of it). The next generation gives offspring carrying 75% of the blood of the purebred breed, and in subsequent generations the proportion of inheritance remaining from the original scrub parent is halved with each cross. Later crosses usually increase quality and performance still more, though in less marked degree. After the third or fourth cross, the offspring compare very favorably with purebred stock in conformation, and only exceptionally good sires can bring about further improvement. This is especially so if the rams used in grading up successive generations are derived from the same strain within a breed.

As evidence that sheep of high merit may be produced through grading up, many present-day champion wethers are sired by purebred rams but from

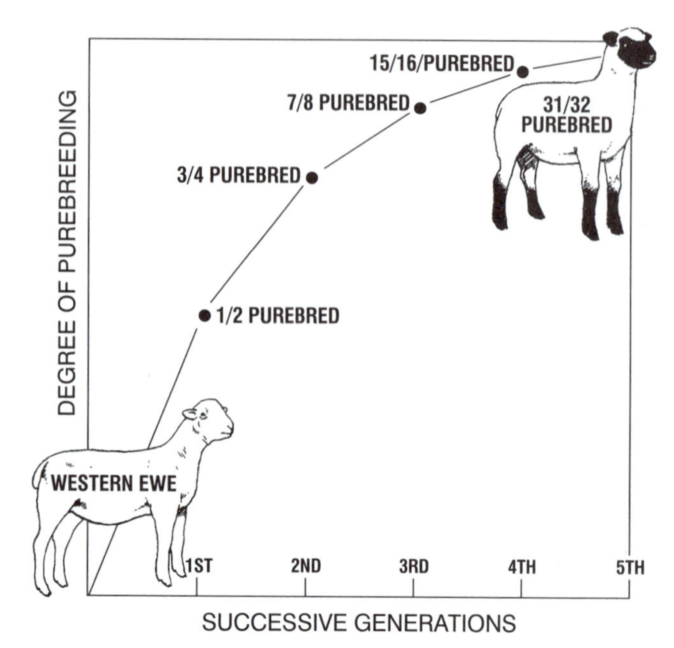

Fig. 3-22. A graph of grading up a western ewe with a purebred ram, showing how the hereditary material changes with each generation.

grade ewes. High-grade animals that are the offspring of several generations of outstanding purebred sires can be, and often are, superior to inferior purebreds.

After some experience, commercial producers who have successfully handled grade animals may add a few purebreds to their flocks and gradually build into the latter, provided that their experience and their market for seed stock justify this type of production.

CROSSBREEDING

Crossbreeding is the mating of animals of different breeds.

No breed is superior in all traits such as prolificacy (multiple births), fast-gaining lambs, carcass grade, and wool production. Rather, there are substantial breed differences in most of these traits. Recognizing this, producers frequently breed wool-type ewes to large, heavily muscled, meat-type rams. This cross takes advantage of the superior wool production and high percentage of lambs weaned on the dam's side, as well as the sire's fast growth rate.

Today, there is great interest in crossbreeding, and increased research is under way on the subject. Crossbreeding is being used by producers to (1) increase productivity over straightbreds because of the resulting hybrid vigor, or heterosis, just as is being done by commercial corn and poultry producers; (2) produce commercial animals with a desired combination of traits not available in any one breed; and (3) produce foundation stock for developing new breeds.

The motivating forces responsible for increased crossbreeding in farm animals are (1) more artificial insemination, thereby simplifying the rotation of sires of different breeds; and (2) the necessity for producers to become more efficient in order to meet their competition, both from within their respective industries and from without.

Crossbreeding will play an increasing role in the production of market animals in the future, because it offers several advantages which are discussed in the following sections.

HYBRID VIGOR OR HETEROSIS

Heterosis, or hybrid vigor, is the name given to the biological phenomenon which causes crossbreds to outproduce the average of their parents. For numerous traits, the performance of the cross is superior to the average of the parental breeds. This phenomenon has been well known for years and has been used in many breeding programs. The production of hybrid seed corn by developing inbred lines and then crossing them is probably the most important attempt to take advantage of hybrid vigor. Today, heterosis is also being used extensively in commercial sheep, swine, layer, and broiler production. An estimated 80% of market lambs are crossbred; 95% of broilers are crosses; and about 90% of the hogs raised for slaughter are crossbred.

The genetic explanation for the hybrid's extra vigor is basically the same, whether it be sheep, cattle, hogs, broilers, hybrid corn, hybrid sorghum, or whatnot. Heterosis is produced by the fact that the dominant gene of a parent is usually more favorable than its recessive partner. When the genetic groups differ in the frequency of genes they have and dominance exists, then heterosis will be produced.

Heterosis is measured by the amount the crossbred offspring exceeds the average of the two parent breeds or inbred lines for a particular trait, using the following formula for any one trait:

$$\frac{\text{Crossbred average} - \text{Purebred average}}{\text{Purebred aver-age}} \times 100 = \text{Percent hybrid vigor}$$

Thus, if the average of the two parent populations for weaning weight of lambs is 56 lb and the average of their crossbred offspring is 64 lb, application of the above formula shows that the amount of heterosis is 8 lb or 14.3%.

Traits high in heritability—like loin eye area and carcass length—respond consistently to selection but show little response to hybrid vigor. Traits low in heritability—like lambing interval and conception rate— usually demonstrate good response to hybrid vigor.

COMPLEMENTARY

Complementary refers to the advantage of one cross over another cross or over a purebred, resulting from the manner in which two or more characters combine or complement each other. It is a matching of breeds so that they compensate each other, the objective being to get the desirable traits of each. Thus, in a crossbreeding program, breeds that complement each other should be selected, thereby maximizing the desirable traits and minimizing the undesirable traits. Complementary in a crossbreeding system is incorporating additive genetic differences between breeds, recognizing that rams and ewes do not contribute equally to the performance of their offspring. For this reason the breeds of sheep are classified as ewe breeds and ram breeds.

- **Ewe breeds**—These are the white-faced breeds which produce fine, medium, or long wool, or crosses among these types. Ewe breeds are selected for adaptability to environmental conditions, reproductive efficiency, wool production, size, milking ability, and longevity. Replacement ewe lambs should be raised from these breed types, or from crossing among these breeds. Ewe breeds include: Corriedale, Debouillet, Delaine-Merino, Finnsheep, Polypay, Rambouillet, Targhee, Columbia, Coopworth, Dorset, Montadale, and Romney.

- **Ram breeds**—These are the meat-type breeds, or crosses of two of these breeds. Ram breeds are raised for production of rams to be crossed on ewe breeds and selected for growth rate and carcass qualities. Ram breeds include: Hampshire, Oxford, Shropshire, Southdown, Suffolk, Dorset, and Texel.

INTRODUCE NEW GENES QUICKLY

Crossbreeding provides a way in which to introduce new and desired genes quickly—at a faster rate than can be achieved by selection within a breed. A practical example of this sort is the introduction of new genes for multiple births by using the prolific Finnsheep or Romanov.

GET HYBRID VIGOR EXPRESSED IN THE FEMALE

Except for a terminal cross, crossbreeding offers an opportunity to have hybrid vigor expressed in breeding females. This is most important in the ewes where it results in increased fertility, survivability of the lambs, milk production, growth rate of lambs, and longevity of the ewes—all factors that mean more profit to the producer.

FACTORS AFFECTING MAGNITUDE OF ADVANTAGES FROM CROSSBREEDING

Many other examples of each of the advantages of crossbreeding could be cited. However, the total magnitude of the advantage of these factors—achieving the 15 to 25% potential immediate increase in yield per female unit through continuous crossbreeding compared to continuous straight breeding—depends upon the following:

1. **Making wide crosses.** The wider the cross, the greater the heterosis.
2. **Selecting breeds that are complementary.** A crossbreeding program should involve breeds that possess the favorable expression of traits desired in the crossbred offspring that will be produced.
3. **Using high-performing stock.** Once a crossbreeding program is initiated, further genetic improvement is primarily dependent upon the use of superior production-tested males.
4. **Following a sound crossbreeding system.** For a continuous high expression of heterosis and maximum output per female, a sound system of crossbreeding must be followed. This should include the use of crossbred females, for research clearly indicates that over one-half the higher profits from a crossbreeding program results therefrom.
5. **Tapping purebreds constantly.** Purebreds must be constantly tapped to renew the vigor of crossbreds. If not, the vigor dissipates.

SYSTEMS OF CROSSBREEDING

Without a planned breeding program, minimum benefits of crossbreeding will be realized. There are numerous different systems of crossbreeding. Among them are the following:

1. **Two-breed cross.** This consists of mating purebred rams to purebred or high-grade ewes of another breed. The two-breed cross is simple, but it does not take advantage of heterosis for maternal traits, since the resulting crossbred ewes are not saved for breeding purposes.
2. **Backcross or crisscross.** This system involves the use of rams of Breed A on ewes of Breed B. Then, the resulting crossbred females are kept for breeding and are mated to nonrelated males of Breed A or B. The advantage of this system is that the ewes and the lambs are crossbreds and will possess hybrid vigor.
3. **Three-breed cross.** This system uses males from each of three breeds in succession on crossbred females. For example, ewes of Breed B are mated to rams of Breed A; selected crossbred ewes (AB) are mated to Breed C rams and all offspring are sold for slaughter. In a rotational three-breed cross, selected crossbred ewes of each generation are bred in rotation to each of the three breeds. Three-breed crosses also take advantage of heterosis in the ewes and the lambs.

CROSSBREEDING SHEEP

Crossbreeding is commonly applied to sheep because of (1) sheep being called upon to produce two products, lamb and wool; (2) the many diverse conditions under which they are produced; and (3) the conviction on the part of many producers that the hybrid vigor of crossbreeding accounts for increased vigor and livability in the lamb crop.

Crossbreeding is extensively followed in commercial sheep production on the western range. The ewe bands are predominantly of Rambouillet extraction; whereas for market-lamb production, Hampshire or Suffolk rams are generally used. Rambouillet ewe bands are desired because of their (1) gregarious or flocking instinct, (2) great hardiness, and (3) superior shearing qualities. On the other hand, lambs of this breeding are not as desirable for market lambs. Thus, meat-type rams are used in order to get large, fast-growing lambs that will attain a good market finish on milk and range vegetation or that can be readily sold to go into feedlots for further finishing. As black-faced crossbred lambs of this type are not suitable as flock replacements, both ewe and wether lambs are marketed. Replacement females are obtained by: (1) outright purchase from a sheep producer who has used white-faced rams (Rambouillet, Columbus, Tracheas, or Panamas) for purposes of raising animals for sale as replacements, (2) using white-faced rams on the band every third year and retaining the ewe lambs (some sheep producers with several bands simply use certain bands for producing lambs for replacement purposes), or (3) using both white-faced and black-faced rams simultaneously on the same ewe band. In the latter type of program, the better white-faced ewe lambs—which are easily recognized as the offspring of the white-faced rams—are selected out for breeding purposes.

Crossbreeding in sheep does make for a considerable problem from the standpoint of producing or purchasing replacement animals. Also, it often makes the ram problem a difficult one. This practice, however, was born of necessity, there being few or no existing breeds or types possessing all the desirable features needed. In recent years, considerable effort has been made toward developing breeds of sheep better adapted to the needs, with the hope of alleviating the necessity of crossbreeding. The Columbia, Targhee, and Panama breeds evolved out of this need.

In addition to the crossbreeding common to the western range, most hothouse lambs are produced through crossbreeding. Usually grade Merino or Dorset ewes are bred to Southdown rams. Ewes of this extraction will breed out of season, and they are excellent

Fig. 3-23. Hothouse lambs produced from a three-breed cross: Dorset X Merino ewes mated to a Southdown ram. (Courtesy, The Pennsylvania State University)

milkers; whereas the Southdown rams impart to their progeny early maturity and meat type. Crossbreeding is popular in Kentucky where crossbred Hampshire-Rambouillet ewes are frequently bred to Southdown rams for the production of grass-finished lambs.

Crossbreeding is extensively followed in commercial sheep production on the western ranges. The ewe bands are predominately of Rambouillet extraction; whereas for market-lamb production, Hampshire or Suffolk rams are generally used. Rambouillet ewe breeds are desired because of their (a) gregarious or flocking instinct, (2) great hardiness, and (3) superior wool qualities. Although present day Rambouillets seem quite good for meat production, meat-type rams may be used in order to get large, fast-growing lambs that will attain an good market finish on milk and range vegetation, or that can be readily sold to go into feedlots for further finishing.

Studies by the U. S. Department of Agriculture[11] demonstrated that (1) fertility, prolificacy, lamb viability, and overall reproductive efficiency was higher in crossbred ewes; (2) crossbred lambs exceeded purebred lambs in birth weight, weaning weight, and average daily gain; and (3) total production as measured by indexes was greater in crossbred ewes. These studies evaluated Hampshire, Targhee, Suffolk, Dorset, and Columbia-Southdale breeds and all possible crosses of these breeds.

In addition to heterosis, crossbreeding enables the commercial sheep raiser to benefit from the complementary of breeds such as long life, flocking instinct, wool production, excellent body conformation, and rapid growth of another breed. Thus, systematic crossbreeding as a mating system for sheep production will continue.

But if crossbred ewes are used and mated to rams of an unrelated breed for maximum commercial production, a minimum of three breeds must be involved. The breeds that are crossed to produce the crossbred ewes (the F_1) that will be retained as the producers in the flock should possess those characteristics which contribute to making highly productive females—such as reproductive efficiency, milk production, maternal instincts, and wool quantity and quality. In turn, these crossbred ewes should be bred to rams of one of the ram breeds. The rams of the ram breed would be mated to the crossbred ewes as a terminal cross, to produce market lambs. An example of a crossbreeding program of this type would be:

1. **Foundation ewe breeds.** These include the Rambouillet, Merino, Columbia, Corriedale, and Targhee. Two of these breeds will be selected and crossed to produce the F_1 females.
2. **F_1 females.** The F_1 females resulting from the above cross will be bred to Suffolk or Hampshire rams as a terminal cross. All lambs will be marketed.

MODERN METHODS FOR SELECTING SHEEP AND MEAT GOATS

Selection of sheep and goats involving progeny tests, calculated indexes and breeding values were all planned and conducted but have generally resulted in very slow progress. Improvement of prolificacy was slow except where prolific exotic breeds were used. Generally too much emphasis was given to individual weaning weights which turned out to be somewhat negatively related to prolificacy. Sometimes these selection programs lengthened the generation intervals which also slowed genetic progress. At best, selection indexes and breeding values are estimates of the real values, while litter weight is identical to real total value of meat. Progeny testing for milk production in dairy cattle has been quite successful but this involves large numbers of progeny tests through use of artificial insemination.

SELECTION FOR LITTER WEIGHT OF DAM

Primary selection for lamb production has been most successful through selection for litter weight from the dam.[12] This measure is available when the lambs

[11]Sidwell, G. M., and L. R. Miller, "Production in Some Pure Breeds of Sheep and Their Crosses," *Journal of Animal Science*, Vol. 32, No. 6, 1971, pp. 1,084-1,098.

[12]Ercanbrack, S. K., and A. D. Knight, 1997, Responses to Various Selection Protocols for Lamb Production in Rambouillet, Targhee, Columbia and Polypay Sheep.

or kids are weaned so that generation lengths can be as short as one year by breeding at about seven months of age. Litter weight gave the most complete overall economic measure of lamb production because it provided a balance between prolificacy, survival from birth to weaning or percent weaned, and weaning weight. Indexes and breeding values tend to give only an estimate of market value.[13]

Selection based on litter weight of the dam puts emphasis directly on total economic value of meat production at weaning age so that breeding at 6 to 7 months of age permits turning male generations every year if breeding to lamb or kid first at about one year of age is practical. This overall measure of economic value of meat production permits more rapid progress in improving efficiency of lamb production in sheep and goats than for any other species or food crop that does not depend on feeding grain, and concentrated feedstuff. Furthermore, increase in efficiency of production or yields can probably continue farther into the future than for most other food crops because yields are measured partly on numbers of offspring rather than entirely on weight. Most field crops have already reached their ceiling or limits in yields while sheep and goats can probably improve for 50 to 100 years before reaching such ceilings. Selection for litter weight is probably the most rapid means of increasing efficiency of food production available to food producers as well as the most effective means of increasing profitability of lamb production and bringing back the family farm with real sustainable production. Unfortunately, it does not increase fleece weights, but they may have reached their practical genetic ceilings. Also, increased lamb production per ewe may have a depressing effect on fleece weights.

SELECTION OF LAMBS JUST AFTER WEANING

Lambs of both sexes should be bred as soon after weaning as feasible not only because this increases total efficiency of production, but because if genetic progress is being made the youngest generation is the most productive on average. The quicker they are brought into production the greater is total efficiency. Supplementary feeding to increase success of early breeding is profitable and often desirable especially under range production. Breeding at 6 to 7 months of age will permit automatic selection pressure for lambing early. Again, adequate feeding for nutritive needs will be well worthwhile.

Female lambs bred at 6 to 7 months of age will increase in lambing at about one year of age as selection for those which succeed continues. In time, only those which succeed in weaning lambs by one year of age will be retained for further breeding. In later years, only those which succeed in weaning at least one lamb from at least one breeding per year should be sold at even three or four years of age because of the shorter generation length.

Weaning litter weight must be adjusted for environmental factors before being used in selection. Records required for each offspring of each dam are sex of lamb, age of animal at weaning in days (based on groups of more than one), and age of dam in years. The mathematical procedure required to make the adjustments is too complicated to present here, so help may be required from a professional animal breeder to complete the calculation of the adjusted litter weights.

SELECTION FOR POST WEANING GAIN AND FEED EFFICIENCY[14]

Ram lambs, above average in litter weights of dam, should be subjected to post weaning gain and feeding tests for feed efficiency and rate of gain as soon as possible after weaning. In fact, ram lambs are weaned at about 80 days at the Dubois Station to permit evaluation at a reasonable age. Use of feeding trials in selection of ram lambs is very important because it permits emphasis in selection on rate of gain and feed efficiency and has a positive effect on carcass merit.

An index for selection on rate of gain and feed efficiency can be calculated by multiplying 3.33 times gain per day plus 0.68 times feed per day plus 0.61 times ending body weight, all measured in kilograms. A suggested ration is 37% barley and 63% alfalfa pellet for the first 10 weeks, and 100% alfalfa pellet for the final 6 weeks. Feeding tests as long as 16 weeks are essential as heritabilities of weight and gains increase with the length of the feeding period from 0.25 to 0.20 in the beginning to 0.75 and 0.71 in the 16th week. This test procedure has proven to be effective.

Lambs selected for efficiency and post weaning gain were about 23% more efficient than the controls, and produced about 23% more profit. Carcasses of the selected lambs were 9.9% heavier, and 10.1% greater total weight of prime cuts, but had 8.6% more bone than control lambs. Selection associated with greater

[13]Terrill, C. E., 1989, Impact of Selection Research on Efficiency of Production on Lambs and Wool. Proceedings, Western Section, American Society of Animal Science, Bozeman, MT, July 12–14.

[14]Ercanbrack, S. T., and A.D. Knight, Selection of Efficiency of Post Weaning Gain in Lambs, U.S. Sheep Experiment Station Progress Report, October, 1988, Dubois, ID.

increase in efficiency had no deleterious effects on critical components of carcass composition and did not increase carcass fat percentage.

BREEDING MORE OFTEN THAN ONCE PER YEAR

Reaching the ultimate in efficient lamb production requires lambing more often than once per year. A common practice is lambing 3 times in 2 years, or every 8 months. A very efficient system developed by Brian Magee, Sheep Farm Manager, at Cornell University involves breeding and lambing every 7 months (the Star System).

It was realized that successful frequent or accelerated lambing system should (1) not violate sheep biology, and (2) fit into a calendar year. It was also apparent that if breeding and lambing could be coincident, overall management would be simplified. The Star System solves the above dilemmas. Pregnancy in the sheep is 146 days, one-half pregnancy (146/2) equals 73 days which is exactly one-fifth of a year. The calendar year contains exactly five half sheep pregnancies as depicted in Fig. 3-24. The annual calendar is depicted in circular fashion and divided into five equal 73-day periods. By connecting the first day of alternate periods, a perfect Star System is formed. By breeding or lambing the first 30 days of a period, the remaining management tasks are immediately apparent. If a ewe lambs in period 1, she is bred in period 2 to lamb in period 4, to be bred in period 5, etc. Each ewe can lamb five times in three years and complete the star. If she has five consecutive lambings with 7.2-month intervals, she is referred to as a *star ewe*. If she has twins at each lambing, she is referred to as an *all-star ewe*. The Star System then: (1) does not violate sheep biology; fits into the calendar year; and (3) simplifies management, especially by making breeding and lambing dates exactly coincident.

The five lambing periods per year then: (1) allows ewes to lamb five times in three years, which equals 1.67 lambings/ewe/year; (2) allows a uniform annual production of market lambs for a uniform supply and cash flow; (3) increases efficiency of utilization of facilities such as lambing areas, etc.; and (4) allows for the identification of ewes that not only will breed out of season but will breed following lambing in any season.

Accelerated systems are not recommended for beginners or for very small farms. It is best to start with a few orphan ewe lambs and to grow from within, with once a year lambing. Thus, any mistakes are made when numbers are small. Start with a job off the farm eventually with a home on the farm and at least 120 acres with enough tillable land to produce about 95 tons of native alfalfa hay per year (in the northern U.S.), and pasture for almost 500 ewes. An additional 100 tons or more of hay in reserve might be desirable if the risk of drought is high.

In the northern U.S. where snow may sometimes be too deep for grazing, a barn of sufficient size to hold 500 ewes and their lambs and a year's supply of hay may be essential. At any one time there might be three groups of ewes and lambs, including (1) breeding and pregnant ewes, (2) lambs and lactating ewes, and (3) growing lambs. The growing lambs are marketed throughout the year. Breeding dates may be varied some to avoid lambing on Easter or Christmas-New Year holidays.

Considerable selection over a number of years may be necessary to adapt ewes to the Star System even though breeds like the Dorset with less seasonal restrictions may be used. It is desirable to maintain breeding under the Star System simultaneously with once a year lambing. Then, as ewe lambs are exposed to the Star System any that fail may be put in the once a year lambing group or culled. Also, ewes which require special attention at lambing should be culled. Ram lambs born from the most difficult breeding period should be favored in breeding.

STAR ACCELERATED LAMBING SYSTEM

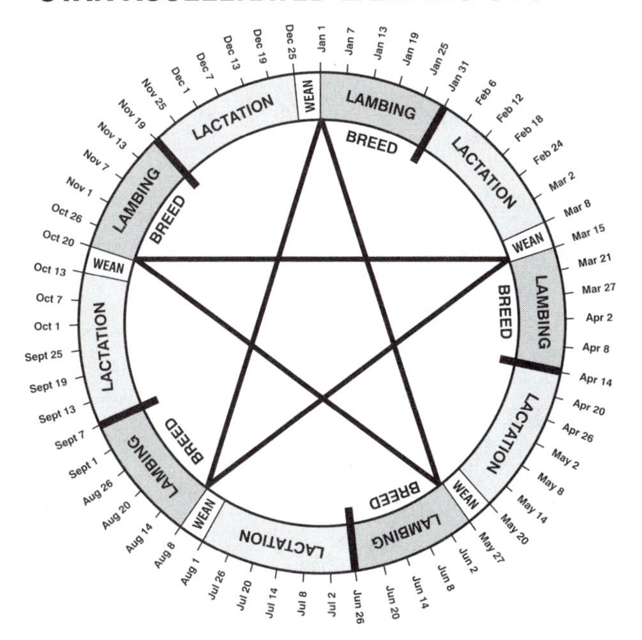

Fig. 3-24. Star Accelerated Lambing System. (Courtesy, Cornell Research Foundation Copyright)

IMPROVEMENT OF HIGH FERTILITY BREEDS

Effective and rapid selection for increased efficiency makes feasible the improvement of high fertility and high vigor breeds like the Romanov so that they can be used with high proportion of high fertility blood from crosses with domestic breeds. The Romanov was developed as a pelt breed and has low quality black and grey wool which is not useful for machine manufacture. The black and grey genes are recessive and can be covered up with dominant white genes from domestic breed crosses. This can be done by private breeders by crossing with appropriate white breeds such as the Rambouillet and then back crossing to the Romanov to retain the prolific and vigorous make up of the Romanov. But many years may be required to eliminate the black and grey genes from the Romanov. This should be done by research stations because they can afford to make the tests for selection of homozygous white animals. Also, they can include selection for efficiency, rate of gain, and carcass merit. New Romanov breeds should be developed from domestic breeds with white fine, medium, and carpet wool, respectively. The Rambouillet or Targhee may be best for fine wool. A number of breeds such as the Columbia or Corriedale would be suitable for medium wool and the Navajo Churro might be best for carpet wool.

SAVING THE FAMILY FARM

A long period of declining per capita production of food such as has occurred in the 1980s and 90s must be reversed if adequate food for the U.S. and the world is to be maintained. Sheep and lambs offer the best hope to reverse this downward trend of production and loss of family farms because they can be produced on the poorest land or any kind of land, can be started with a very low investment, can be increased in numbers and efficiency of production every year almost indefinitely, and can be flexible in date of marketing and in variety of products, such as meat, milk, cheese, pelts, other byproducts, and wool. However, they should be produced only by those who like to work with sheep.

CLONING[15]

Cloning of an animal is the production of an exact genetic copy. In a technical sense, identical twins are clones; they are derived from a single cell, as a result of the embryo splitting early in development to yield what is essentially two carbon copies.

The exciting and much sought technological breakthrough in the cloning of mammals involves the manipulation of embryos.

The dream of cloning is based on the following two pieces of scientific evidence:

1. With few exceptions, all cells in the body of an animal appear to contain the same genetic information. This information is contained in the DNA, a molecule that is located in a sac inside cells called the nucleus. Thus, within an animal, the DNA sequence in the nucleus of a liver cell is identical to that in a skin cell. These cells differ in appearance and function because they make use of different parts of the genetic information, not because the total amount of information differs. Further, all of these cells have the genetic information that was present in the one-cell embryo that developed in the animal. Therefore, if the nucleus of any of these cells were used to replace the genetic information in any one-cell embryo, an exact genetic copy of the animal whose cells donated the nucleus would develop. With such an approach, thousands of cloned copies could be made.

2. The second piece of scientific evidence is that nuclear transplantation experiments have been done successfully with several species of animals, especially frogs and fish, which have the big advantage of their eggs being thousands of times larger than mammalian eggs.

■ **How It Used To Be Done**—Historically, research and development in cloning has passed through the following stages, in order and period of time:

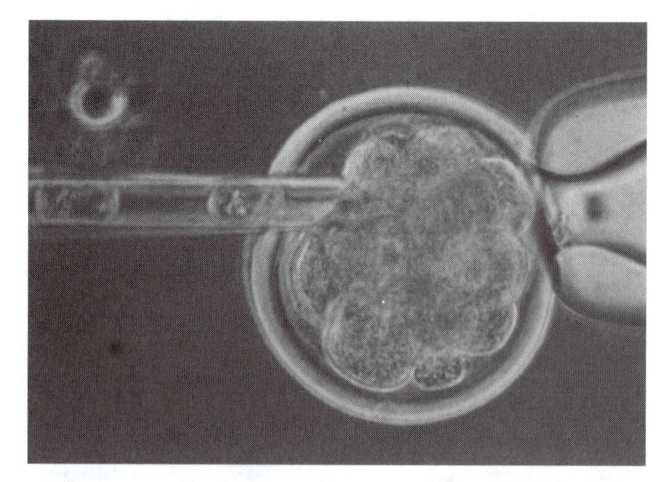

Fig. 3-25. This shows the removal of cells of a valuable embryo to use in cloning, to produce 8 to 20 cloned embryos from this embryo by transfer of each of these cells to an enucleated oocyte. (Courtesy, Dr. Robert Walton, American Breeders Service, a division of W. R. Grace & Co., De Forest, WI)

[15]The review of this section by Dr. Robert Walton, American Breeders Service, De Forest, Wisconsin, is gratefully acknowledged.

1. Identical twin calves were produced by microsurgically splitting embryos, then transferring half embryos to recipients.

2. A bull calf was born as a result of using the laboratory culturing technique (*in vitro*) of maturing an egg, fertilizing the egg *in vitro*, then transferring the fertilized egg to a surrogate mother.

Nuclei have been taken from 16-cell bovine embryos and placed in one-cell bovine eggs whose nuclei had already been removed. The new one-cell embryos were matured and transferred into recipient cows, which subsequently gave birth to cloned female calves.

During the late 1980s, the most advanced cloning procedure consisted of flushing the embryo out of the donor cow at day five of its development (at the 32-cell stage); followed by putting the embryo under a microscope and manually removing one of the cells, then freezing the remaining 31-cell embryo (much like semen is frozen) and putting it away until an order was received. Next, the technician took an unfertilized egg that has been flushed from a low-grade donor cow, removed the nucleus from it, inserted the borrowed cell into the egg, patched up the hole with a short zap of electricity, and the cloned embryo divided day after day and developed into a genetically identical duplicate of the heifer that resulted from the 31-cell embryo.

■ **How New Cloning is Done**—Dr. Ian Wilmut and his colleagues, the Roslin Institute, Scotland, took a short cut! They produced Dolly without going through steps 4, 5, 6, and 7 shown in Fig. 3-27. They selected a mammary cell and produced an exact clone of Dolly's mother without going through either of the above procedures.

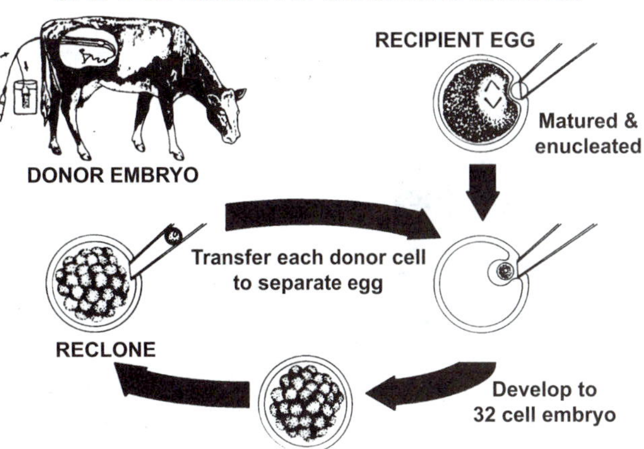

Fig. 3-26. Cattle embryo cloning model. (Prepared by Dr. Robert Walton, American Breeders Service, a division of W. R. Grace & Co., De Forest, WI)

■ **How Future Cloning Will Be Done**—But science continues to take great leaps forward! In 1997, ABS (American Breeders Service) Global, DeForest, Wisconsin, produced a cloned Holstein bull named Gene, using a different technique than the one used to produce Dolly. The company is patenting the process.

Produced without sexual reproduction, a clone is the exact duplicate—the exact same—as another animal.

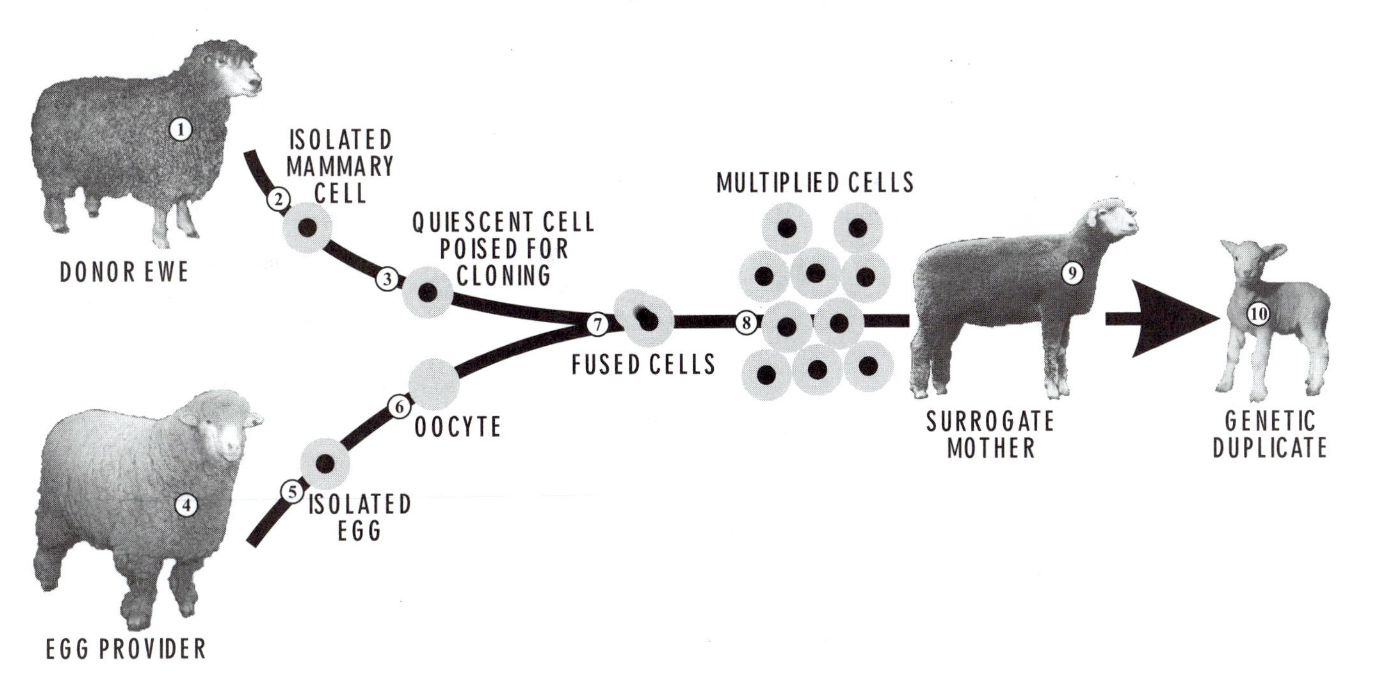

Fig. 3-27. The Roslin Institute, Scotland, produced Dolly without going through steps 4, 5, 6, and 7.

QUESTIONS FOR STUDY AND DISCUSSION

1. What unique circumstances surrounded the founding of genetics by Mendel?

2. Discuss the relationships between chromosomes, genes, and DNA.

3. What is the result of the callipyge gene for a flock of sheep? Is it good or bad? Give the pros and cons.

4. What is the diploid number of chromosomes for sheep and for goats? What is the haploid number?

5. Define homozygous and heterozygous.

6. Give an example of (a) a mutation, (b) dominance, and (c) recessive phenomena in sheep or goats.

7. What should be done about black sheep?

8. What can a sheep producer do to increase multiple births in a flock?

9. Explain how sex is determined. Is it possible to alter the sex ratio? Give the latest in this technology.

10. The "sire is half the herd!" Is this an understatement or an overstatement?

11. Explain why sheep and goats are the result of two forces—heredity and environment.

12. What selection tools are available to sheep and goat herders?

13. Based on (a) heritability, and (b) dollar-and-cent value, what characteristics should receive greatest emphasis in sheep or goat selection?

14. Name and describe six undesirable lethals and other hereditary defects in sheep.

15. When abnormal lambs or kids are born, what conditions tend to indicate each: (a) a hereditary defect, or (b) a nutritional deficiency?

16. Of what importance are flock records?

17. List and discuss some factors influencing the rate of progress that can be made through selection.

18. Describe production testing in goats. What should be the eventual goal in production testing of goats?

19. Outline what trait or traits should be emphasized in selection to make most rapid progress in efficiency and profitability.

20. Compare purebreeding, inbreeding, and crossbreeding. Include the similarities, differences, and relationships.

21. What mating system do you consider to be best adapted to your sheep flock, or to a flock with which you are familiar? Justify your choice.

22. Discuss the concept of selection of litter size in sheep through selection for litter weight of dam.

23. Discuss the concept of selection of lambs of both sexes just 6 to 7 months of age.

24. Explain why ram lambs, above average in litter weight of dam, should be subjected to post weaning gain and feeding tests for feed efficiencies and rate of gain as soon as possible after weaning.

25. Why is breeding more often than once per year important in modern sheep production? How is it accomplished?

26. How may rapid improvement of high fertility breeds be accomplished? Why should this be done by research stations?

27. Why do sheep and lambs offer the best hope of saving the family farm?

28. Discuss the pros and cons of cloning. Is it good or bad?

SELECTED REFERENCES

Title of Publication	Author(s)	Publisher
Can Livestock be Used as a Tool to Balance Wildlife Habitat?	N. V. Reno	U.S. Department of Agriculture Forest Service, 1990
Evolution of Range Ecology Practices and Policy	W. B. Kessler	Rangelands 15(3), 1993
Forest Service Program for Forest and Range — Land Resources	Staff	U.S. Department of Agriculture Forest Service, 1995
Grazing on Public Lands	William A. Laycock	CAST, Iowa State University, Ames, IA, 1996
Grazing Statistical Summary	Staff	U.S. Department of Agriculture Forest Service, 1995
Handbook for Raising Small Numbers of Sheep, A	Staff	University of California, Davis, CA, 1994
Managing Interior Northwest Rangelands	T. M. Quigley H. R. Sanderson A. R. Tie Demann	U.S. Department of Agriculture Forest Service, 1989
Minnesota Sheep Research Report	Staff	University of Minnesota, St. Paul, MN, 1996
Nutrient Requirements of Sheep	Robert M. Jordan, Chairman	National Research Council, National Academic Press, Washington, DC, 1985
Planned Sheep Production, Second Edition	D. Croston G. Pollott	Blackwell Scientific Publications, London, England, 1994
Sharing Common Ground on Western Rangelands	Symposium Report	U.S. Department of Agriculture Forest Service, 1996
Sheep Forage Production Systems	Symposium Report	
Sheep Production Handbook (SID)	Staff	American Sheep Industry Association, Englewood, CO, 1994
Sheep Research Journal: Special Issue	Maurice Shelton, Editor	American Sheep Industry Association (SID), Englewood, CO, 1978
Stockman's Handbook, The, Seventh Edition	M. E. Ensminger	Interstate Publishers, Inc., Danville, IL, 1992

The Jacob Sheep is an ancient breed imported into North America from England. This small, hardy breed is known for its spotted coat and multiple horns in the adult. (Courtesy, Jacob Sheep Breeders Assn.)

CHAPTER 4

Spanish (meat) goats with Angora kids produced by embryo transfer. (Courtesy, M. Shelton, Texas Agricultural Experiment Station, San Angelo, TX)

REPRODUCTION IN SHEEP AND GOATS

Contents *Page*

Contents *Page*

Contents Page

Sheep and goat producers encounter many reproductive problems, a reduction of which calls for a full understanding of reproductive physiology and the application of scientific practices therein. In fact, reproduction is the first and most important requisite of sheep and goat breeding, for if animals fail to reproduce, the breeder is soon out of business.

Many outstanding individuals, and even whole families, are disappointments because they are sterile or they reproduce poorly. The subject of physiology of reproduction is, therefore, of great importance.

Reproduction in the goat is very similar to that in the sheep, but differences between the two do exist. Throughout this chapter the similarities will be mentioned and the differences noted.

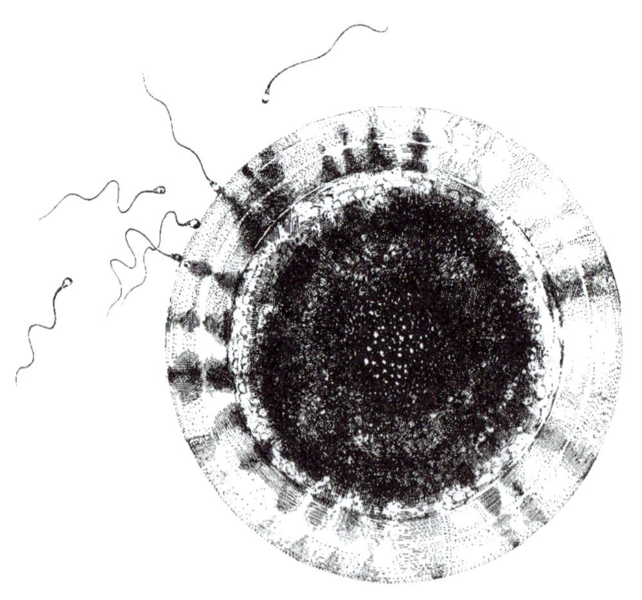

Fig. 4-1. The ultimate objective of sheep or goat breeding is the union of eggs or ova from the females and sperm from the males, which transmits to the offspring all the inheritance from each parent. Each egg from sheep contains 27 half pairs (haploid numbers) of chromosomes as does each sperm; hence, the offspring will have 54 pairs (diploid number) of chromosomes. In goats, both the egg and sperm contain 30 chromosomes. The sperm and egg are drawn to scale.

REPRODUCTIVE ORGANS OF THE RAM AND BUCK

The structure and function of the reproductive organs of the ram and buck are similar.

The male's functions in reproduction are: (1) to produce the male reproductive cells, the *sperm* or *spermatozoa*, and (2) to introduce sperm into the female reproductive tract at the proper time. In order that these functions may be fulfilled, producers should have a clear understanding of the anatomy of the reproductive system of the male and of the functions of each of its parts. Fig. 4-2 shows the reproductive organs of the ram. A description of each part follows:

1. **Scrotum.** This is a diverticulum (outpocketing) of the abdomen, which encloses the testicles. Its chief function is thermoregulatory—to maintain the testicles at temperatures several degrees lower than that of the body proper. In the ram, normal body temperature is about 102°F while that of the testicles averages about 94°F. Thus, sheep herders of long ago learned that it is possible to cause temporary sterility in rams by tying the testicles close to the body wall. Also, among exhibitors the feeling persists that heavy fitting may increase both body and scrotal temperature, and that shearing show rams will lower scrotal temperature and hasten recovery of the reproductive function of the testicles.

2. **Testicles (testes).** The testicles of the mature ram may be 4 in. long and weigh 9 to 10 oz. Their primary function is the production of sperm and the male hormone testosterone. Testosterone is essential for the development and function of male reproductive organs, male characteristics, and sexual drive.

Cryptorchids are males one or both of whose testicles have not descended to the scrotum. The undescended testicle(s) is usually sterile because of the high temperature in the abdomen.

The testicles descend, generally before birth, through the inguinal canal from the pelvic cavity, where accessory organs and glands are located. A weakness of the inguinal canal sometimes allows part of the vis-

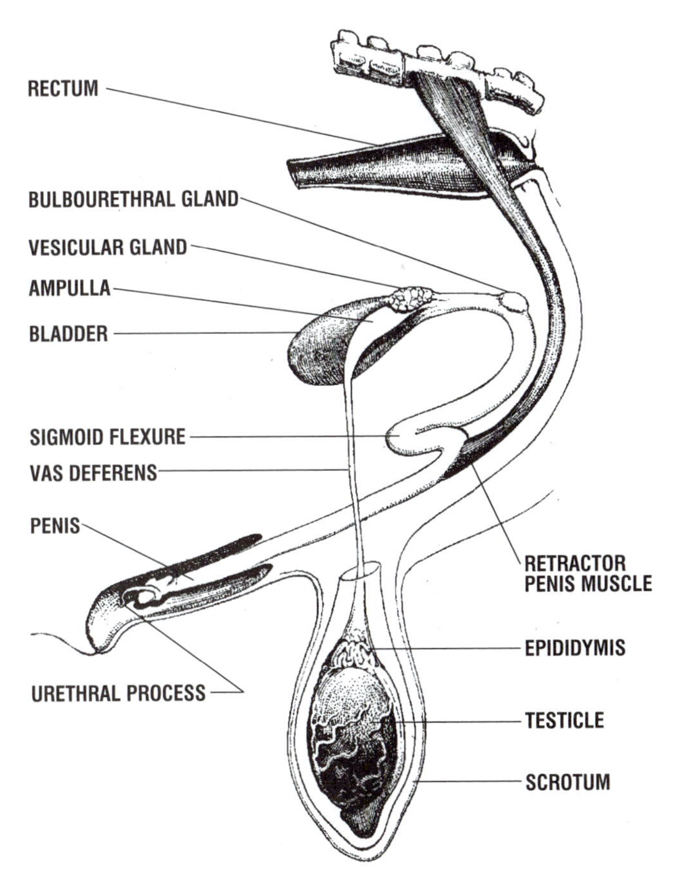

Fig. 4-2. Diagram of the reproductive organs of the ram, showing their location in the body. The testicle, its ducts, the vesicular gland, the bulbourethral gland, and the retractor penis muscle are paired, but, for simplicity, only those on the left side have been drawn.

cera to pass out into the scrotum—a condition called *scrotal hernia*.

Each testicle is covered by a white, tough membrane, known as the *tunica albuginea*.

Removal of the testicles, known as castration, results in the following significant changes in the male: (a) permanent sterility, (b) sexual desire or libido greatly reduced or lacking, (c) the secondary sexual characteristics or masculinity are less prominent, and (d) wethers (castrated male sheep or goats) tend to fatten more readily. This is the reason why wethers differ so much in appearance from rams or bucks.

The following structures within the testicles (see Fig. 4-3) contribute to their function:

a. **Seminiferous tubule.** This is the germinal portion of each testicle, in which are situated the spermatogonia (sperm-producing cells). If laid end to end, it has been estimated that the seminiferous tubules of one testicle of the ram would be 4 mi long! Each gram of testicle is capable of producing about 27 million sperm each day. Once the

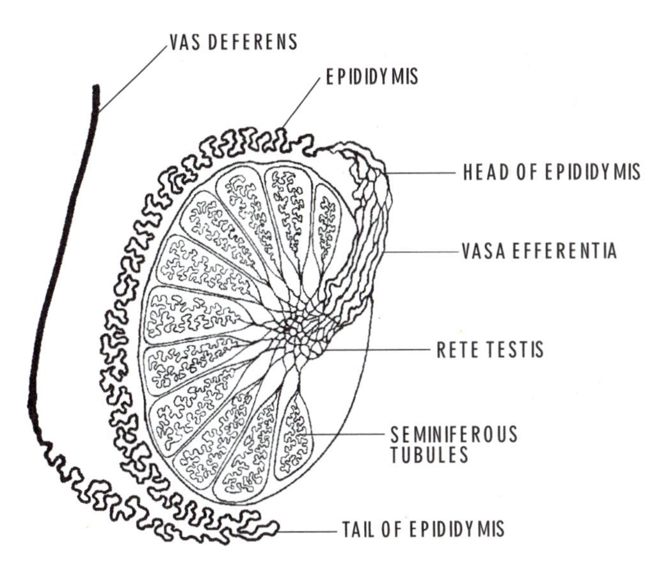

Fig. 4-3. A schematic drawing of the structures within the testicle (sagittal section).

animal reaches sexual maturity, sperm production is a continuous process.

Around and between the seminiferous tubules are the *interstitial (Leydig) cells*, which produce the male sex hormone *testosterone*.

b. **Rete testis.** The rete testis is formed from the union of several seminiferous tubules.

c. ***Vasa efferentia* (efferent ducts).** The efferent ducts carry the sperm cells from the rete testis to the head of the epididymis. Also, it is thought that their secretions are vital to the nutrition and maturing of the sperm cells.

3. **Epididymis.** The efferent ducts of each testicle unite into one duct, thus forming the epididymis. This long and greatly coiled tube consists of three parts:

a. **The head.** Consisting of several tubules which are grouped into lobules.

b. **The body.** The part of the epididymis which passes down along the sides of the testicle.

c. **The tail.** The part located at the bottom of the testicle.

The epididymis has four functions; namely, (a) as a passageway for sperm from the seminiferous tubules, (b) for the storage of sperm, (c) for the maturation of the sperm, and (d) for concentration of sperm.

4. **Vas deferens *(ductus deferens)*.** This slender tube, which is lined with ciliated cells, leads from the tail of the epididymis to the pelvic part of the urethra in the penis—a distance of about 9.5 in. Its primary function is to move sperm into the urethra at the time of ejaculation.

As the vas deferens approaches the urethra, it broadens and thickens, forming the *ampulla*. The ampulla has muscular walls which expel the semen from the vas deferens into the urethra.

The cutting or closing off of the vas deferens, known as *vasectomy*, is the most common operation performed to produce sterility, where sterility without castration is desired. In a vasectomized animal the cells which produce testosterone continue to function, and spermatogenesis continues in the testicles, but sperm are prevented from entering the urethra.

5. **Spermatic cord.** The vas deferens—together with the longitudinal strands of smooth muscle, blood vessels, and nerves, all encased in a fibrous sheath, make up the spermatic cord (two of them) which pass up through an opening in the abdominal wall, the inguinal canal, into the pelvic cavity.

6. **Vesicular glands (seminal vesicles).** These paired, compact, glandular organs with a lobulated surface flank the vas deferens near its point of termination. They are the largest of the accessory glands of reproduction in the male, weighing about 5 g each in the ram, and they are located in the pelvic cavity.

The seminal vesicles secrete a fluid which provides a medium of transport, energy substrates, and buffers for the spermatozoa.

7. **Prostate gland.** This gland, which is diffuse and not easily detected in the ram, is located at the neck of the bladder, surrounding or nearly surrounding the urethra and ventral to the rectum. The prostate gland contributes fluid and salts (inorganic ions) to the semen.

8. **Bulbourethral glands (Cowper's glands).** These two glands, which are deeply embedded in muscular tissue in the ram, are located on either side of the urethra in the pelvic region. They communicate with the urethra by means of a small duct.

Like the other accessory glands, they provide a liquid vehicle for transporting spermatozoa and for flushing the urethra free of urine.

9. **Urethra.** This is a long tube which extends from the bladder to the end of the penis. The vas deferens and vesicular gland open to the urethra close to its point of origin.

The urethra serves for the passage of both urine and semen.

10. **Penis.** This is the male's organ of copulation. Also, it conveys urine to the exterior. It is composed essentially of fibrous tissue. At the times of erection, cavernous spaces in the penis become gorged with blood. Just behind the scrotum it forms an S-shaped curve, known as the sigmoid flexure, which allows for extension of the penis during erection. In the ram a thin, twisted projection, known as the *urethral process*, extends about 1.5 in. beyond the tip of the penis.

In total, the reproductive organs of the ram or buck are designed to produce semen and to convey it to the ewe or doe at the time of mating. The semen consists of two parts; namely, (1) the sperm which are produced by the testicles, and (2) the liquid portion, or seminal plasma, which is secreted by the seminiferous tubules, the epididymis, the vas deferens, the vesicular glands, the prostate, and the bulbourethral glands. Actually, the sperm make up only a small portion of the ejaculate. On the average, at the time of each service, a ram ejaculates about 1 ml of semen, containing 2 to 4 billion sperm. The buck ejaculates 0.5 to 1.5 ml of semen, containing 2 to 6 billion sperm.

REPRODUCTIVE ORGANS OF THE EWE AND DOE

The structure and function of the reproductive organs of the ewe and doe are similar.

The female's functions in reproduction are: (1) to produce the female reproductive cells, the eggs or ova, (2) to develop the new individual, the embryo, in the uterus, (3) to expel the fully developed young at time of birth or parturition, and (4) to produce milk for the nourishment of the young. Actually, the part played by the female in the generative process is much more complicated than that of the male. It is imperative, therefore, that the modern producer have a full understanding of the anatomy of the reproductive organs of the female and the functions of each part. Figs. 4-4 and 4-5 show these organs, and a description of each part follows:

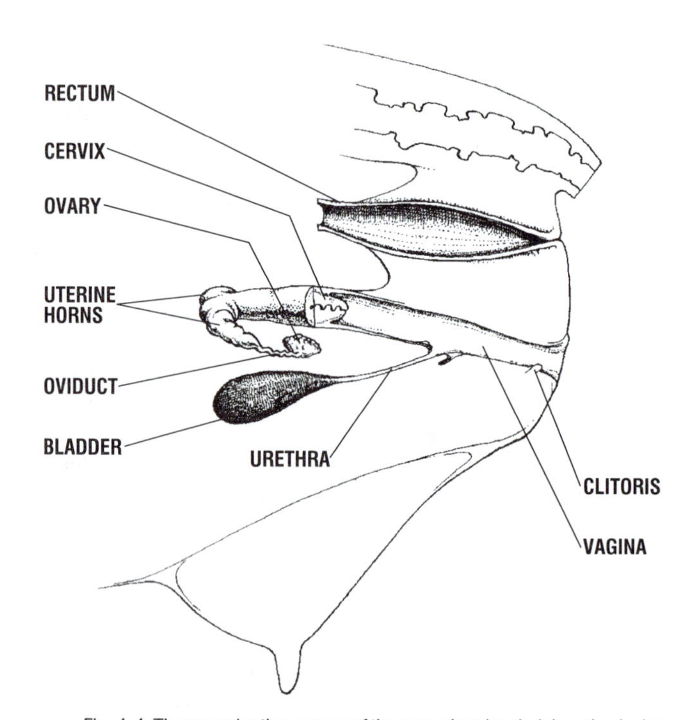

Fig. 4-4. The reproductive organs of the ewe, showing their location in the body.

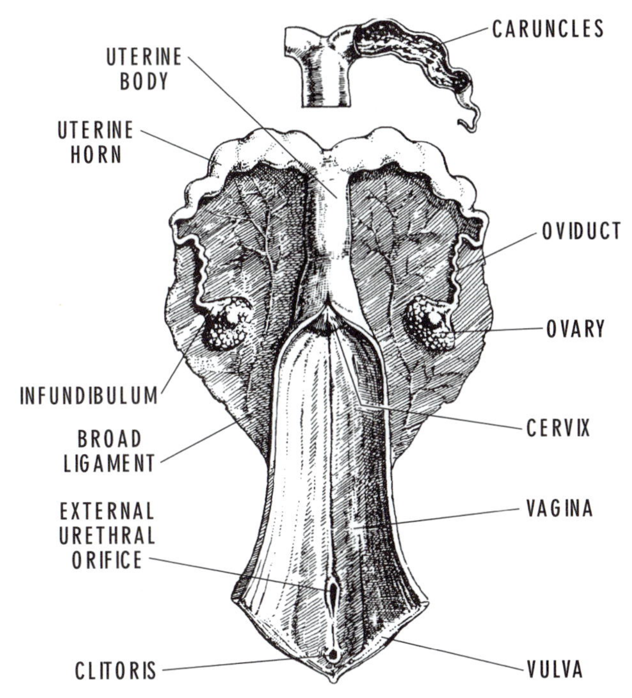

Fig. 4-5. The reproductive organs of the ewe, as viewed from above. The vagina is slit open up to the cervix, and at the top of the diagram a portion of the uterus is duplicated and slit open showing the caruncles.

1. **Ovaries.** The two almond-shaped ovaries of the ewe or doe are supported by a structure called the broad ligament, but lie rather loosely in the abdominal cavity. They average 0.5 to 1.0 in. in length and weigh from 2 to 4 g.

The ovaries have three functions: (a) to produce the female reproductive cells, the *eggs* or *ova*, (b) to secrete the female sex hormones *estrogen* and *progesterone*, and (c) to form the *corpora lutea*, which secrete the hormone progesterone. The ovaries may alternate somewhat irregularly in the performance of these functions.

The ovaries differ from the testicles in that generally only one to three eggs are produced at intervals, toward the end of the heat period or after heat symptoms have passed. Each miniature egg is contained in a follicle, which surrounds the egg with numerous small cells. Large numbers are scattered throughout the ovary. Generally, the follicles remain in an unchanged state until the advent of puberty, at which time some of them begin to enlarge through an increase in the follicular liquid within. Toward the end of heat (estrus), the follicles (which at maturity measure about 0.33 in. in diameter) rupture and discharge the egg. This process is known as *ovulation*. As soon as the eggs are released, the corpora lutea form at the site from which the eggs were released, and begin secreting the hormone progesterone, which (a) acts on the uterus so that it implants and nourishes the embryo, (b) prevents other eggs from maturing and keeps the animal from coming in heat during pregnancy, (c) maintains the animal in a pregnant condition, and (d) assists estrogen and other hormones in the development of the mammary glands. If the eggs are not fertilized, however, the corpus luteum atrophies and allows a new follicle to ripen and another heat period to occur. The corpus luteum may fail to atrophy at the normal time, thus inducing temporary sterility, but this is rare in sheep.

The egg-containing follicles also secrete the female sex hormone estrogen into the blood. Estrogen is necessary for the development of the female reproductive system, for the mating behavior or heat of the female, for the development of the mammary glands, and for the development of the secondary sex characteristics, or femininity, in the ewe and doe.

From the standpoint of the practical sheep or goat breeder, the ovulation of the first mature (graafian) follicle in a ewe lamb or in a doeling generally coincides with puberty; and this marks the beginning of reproduction.

The removal of the ovaries corresponds to castration in the male.

2. **Oviducts (fallopian tubes).** These small, rather tortuous, cilia-lined tubes or ducts lead from the ovaries to the horns of the uterus. They are about 7 in. long and the end of each tube nearest the ovary, called *infundibulum*, flares out like a funnel. The tubes are not attached to the ovaries but lie so close to them that they seldom fail to catch the released eggs.

At ovulation, the egg passes into the infundibulum where, within a few minutes, the ciliary movement within the tube, assisted by the muscular movements of the tube itself, carries it down into the oviduct. If mating has occurred, the union of the sperm and egg usually takes place in the upper third of the oviduct. Then the fertilized egg (embryo) moves into the uterine horn. All this movement from the ovary to the uterine horn takes place in 3 to 4 days.

3. **Uterus.** The uterus is the muscular sac, connecting the oviducts and the vagina, in which the embryo attaches itself and develops until expelled from the body of the female at the time of parturition. The uterus consists of the two horns (cornua), the body, and the neck (or cervix). In the ewe, the horns are about 4 to 5 in. long, the body less than 1 in. long, and the cervix about 1.5 to 4.0 in. long. The horns taper in such manner to their junction with the oviducts that no clear outward distinction between the two exists. In the mature ewe and doe the uterus lies almost entirely within the abdominal cavity.

In the ewe and doe, the fetal membranes that surround the developing embryo form *cotyledons* which fuse with the *caruncles* in the lining of the uterus. This fusion forms the *placentomes*, or functional units of the placenta. There may be 90 to 100 of these units.

4. **Cervix.** Although it is technically part of the uterus, it is often discussed as a distinct organ. The cervix is a thick-walled, inelastic structure. The canal of the cervix, communicating between the body of the uterus and the vagina, is normally constricted, having transverse or spirally interlocking ridges known as *annular rings*. During estrus the cervix relaxes slightly, allowing sperm deposited in the vagina to enter the uterus. Cells within the cervix secrete mucus, and estrogen stimulates the secretion of mucus. During pregnancy, the mucus becomes very thick and occludes the cervical canal. The cervix (1) prevents microbial contamination of the uterus, (2) facilitates sperm transport, and (3) acts as a sperm reservoir. When birth occurs, the cervix must dilate greatly to permit expulsion of the lamb(s) or kid(s).

5. **Vagina.** The vagina admits the penis of the male at the time of service and receives the semen. At the time of birth, it expands and serves as the final passageway for the fetus. In the nonpregnant ewe or doe, the vagina is 3 to 4 in. in length, but it is somewhat longer in the pregnant animal.

6. **Clitoris.** The clitoris is the erectile and sensory organ of the female, which is homologous to the penis in the male. It is less prominent in the ewe than in the mare and cow.

7. **Urethra.** The urine makes its exit through the urethra, which opens into the vaginal vestibule (between the vagina and the vulva).

8. **Vulva (or urogenital sinus).** The vulva is the external opening of both the urinary and genital tracts. It is about 1 in. in length.

The reproductive system of the ewe or doe is regulated by the complex interaction of the nervous and endocrine systems. The functions of the reproductive organs and the occurrence of estrus, conception, pregnancy, parturition, and lactation are all regulated and coordinated by the hormones of the hypothalamus and the pituitary of the brain, the ovarian follicle, the corpus luteum, and the placenta.

NORMAL REPRODUCTIVE CHARACTERISTICS

Perhaps there is as much scientific information about the normal breeding habits of the ewe as there is about any class of farm animals. Even so, not all of the phenomenon is clear, and much work in the field of sheep and goat reproduction remains to be done. Table 4-1 summarizes the normal reproductive characteristics of sheep and goats.

TABLE 4-1
NORMAL REPRODUCTIVE CHARACTERISTICS OF SHEEP AND GOATS

Characteristic	Sheep	Goats
Age at puberty		
Male . . . . (mos)	4–6	4–6
Female[1] . . . . . . (mos)	6–9	5–7
Estrous cycle length . . (days)	14–19	19–22
Length of estrus (heat) . (hrs)	24–36	32–40[2]
Time of ovulation[3] . . . (hrs)	24–30	30–36
Length of gestation . . (days)	140–159	146–151

[1]Fine-wool or Merino-type sheep and Angora goats may be 18 to 20 months of age at first estrus because they often fail to reach puberty during the first breeding season. The Finnsheep and Pygmy goat may reach puberty at 3 to 4 months of age.

[2]Angora goats have an estrus of 22 hrs.

[3]Time from the beginning of estrus.

AGE AT PUBERTY

It is not unusual for ram lambs and buck kids, at 4 to 6 months of age, to have functional sex organs, nor is it unusual to find that some ram lambs have bred a few of the ewes prior to the normal weaning time.

Females, however, are somewhat slower than males in reaching sexual maturity. The age of puberty for females ranges from 5 to 20 months depending on breed, nutrition, and date of birth. Full development of the female may be reached before the onset of estrus or heat, since in some breeds there are long periods (anestrus) when the female organs are not active. Generally, the first estrus in ewe lambs of the meat breeds occurs during the fall of their first year, when they are 8 to 10 months of age. In the more slow-maturing Merino and in the Angora goat, it may be delayed until they are 16 to 20 months of age. The early-maturing Finnsheep reach puberty at 3 to 4 months of age. Selection for early puberty can favorably influence the lifetime reproductive rate.

AGE TO BREED

Most ewes are bred during the first breeding season after they are 1 year of age, producing their first lambs when they are approximately 24 months old.

As a rule, ewe lambs are not used for breeding purposes. Range sheep producers almost never follow the practice, and only comparatively few farm-flock owners breed ewe lambs so that they will drop their first lambs when they are approximately 12 months of age. Experimental work plus practical observations would indicate the following results may be expected from the practice of breeding ewe lambs that are well fed, well grown, and early dropped:

1. Growth of bred lambs is retarded temporarily, but is not stunted permanently.
2. Wool yield is not affected by early breeding.
3. Some ewe lambs do not conceive; birth weights of lambs born to ewe lambs are lighter; and more troubles are encountered at lambing time (more ewes require assistance at lambing and more lambs are disowned).
4. If computed on the basis of total lifetime production up to 5 or 6 years of age, ewes bred as ewe lambs will show more lambs and total pounds of lamb produced per ewe than ewes bred at the normal yearling age.

If ewe lambs are bred to rams of smaller breeds, the smaller progeny will cause less difficulty at lambing time.

Although most dairy goats show estrus their first season, they are usually bred at 9 to 10 months of age or when they have reached 60 to 75% of their mature body weight. This ensures that future productivity is not adversely affected.

HEAT PERIODS (ESTRUS)

The duration of estrus (heat) in ewes ranges from 20 to 42 hours, with an average of 30 hours. Unlike other farm animals, the ewe shows few visible external indications of heat. The acceptance of the ram (or teaser) is the best method of heat detection. Ovulation occurs near the end of the heat period, ranging from about 11 hours before the end to 7 hours after the end of estrus. If the ewe is not bred or if she fails to get in lamb, estrus recurs after an interval of 14 to 20 days, with an average of 16 to 17 days (Table 4-1).

In the doe, the signs of heat are usually easily detected. Does demonstrate frequent nervous bleating, side-to-side tail shaking, a swelling and reddening of the vulva, and a drop in milk production and appetite. Also, does will follow the buck or pace the perimeter of the lot in search of a buck, and in a herd of all does, other does will attempt to mount the doe in heat.

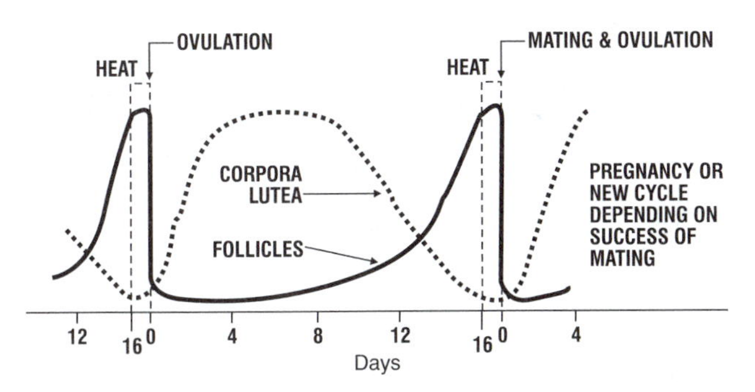

Fig. 4-6. The estrous cycle of the ewe—follicles form, ovulate; corpora lutea form, regress; more follicles form. Unless mating and conception occur, the cycle is normally repeated about every 17 days. Progesterone is secreted by the corpora lutea while the follicles secrete estrogen. In the event of pregnancy, the corpora lutea do not regress but remain functional. Ultimately the control of the estrous cycle resides with the hypothalamus and the pituitary in the brain.

The estrous cycle length of does averages 21 days, and the duration of estrus is about 35 hours for dairy goats (Table 4-1).

MATING

Mating is a brief process in sheep and goats, requiring only seconds, once a ram or buck determines a ewe or doe is in heat. Upon mating, the male deposits 0.3 to 2.0 ml of semen containing 1 to 6 billion sperm.

FERTILIZATION

Fertilization is the union of the male and female germ cells, sperm and ovum. The sperm are deposited in the vagina at the time of service and from there ascend the female reproductive tract. Under favorable conditions, they meet the egg(s) and one of them fertilizes it in the upper part of the oviduct near the ovary. If one egg is produced and fertilized, the ewe or doe will have a single lamb or kid; with two eggs, there may be twins, etc.

In hand mating, or artificial insemination, the breeder's problem is to synchronize insemination with ovulation; to ensure that large numbers of vigorous, fresh sperm will be present in the oviduct at the time of ovulation. This is very difficult for three reasons: (1) There is no reliable way of predicting the length of heat or the time of ovulation since the heat period can vary from 24 to 48 hours and ovulation may occur toward the end of the heat period or several hours afterward; (2) the sperm cells of the male live only 30 to 48 hours in the reproductive tract of the female; and (3) an unfertilized egg will not live over 16 to 24 hours. Like all biological phenomena, there is considerable individual variation in these events.

Clearly, a series of delicate time relationships must be met, and breeding must take place at the right time. For the maximum conception rate, breeding more than once or at the latter part of the heat period is recommended.

Where artificial insemination is used, ewes should be bred twice at the middle or during the second half of estrus, and does should be inseminated 12 hours after the beginning of estrus and then again the next day if they are still in heat.

GESTATION PERIOD

The period of gestation of sheep varies between breeds and between individuals, with the range being from 144 to 152 days and the average being 148 days. Medium-wool breeds, including the down breeds, have short gestation periods of from 144 to 148 days; whereas the fine-wool breeds, such as the Merino and Rambouillet, have long gestation periods ranging from 148 to 152 days. The long-wool breeds, such as the Lincoln and the Romney, have gestation periods intermediate between the medium and fine wools, averaging 146 to 149 days. Individual gestation periods within a breed may vary up to a range of 15 days.

In goats, the length of gestation is fairly constant at 146 to 151 days.

For sheep and goat breeders with breeding dates, parturition dates can be estimated from Table 4-2.

TABLE 4-2
SHEEP AND GOAT GESTATION TABLE BASED ON 148 DAYS

Date Bred	Date Due	Date Bred	Date Due
Jan. 1	May 29	July 5	Nov. 30
Jan. 6	June 3	July 10	Dec. 5
Jan. 11	June 8	July 15	Dec. 10
Jan. 16	June 13	July 20	Dec. 15
Jan. 21	June 18	July 25	Dec. 20
Jan. 26	June 23	July 30	Dec. 25
Jan. 31	June 28	Aug. 4	Dec. 30
Feb. 5	July 3	Aug. 9	Jan. 4
Feb. 10	July 8	Aug. 14	Jan. 9
Feb. 15	July 13	Aug. 19	Jan. 14
Feb. 20	July 18	Aug. 24	Jan. 19
Feb. 25	July 23	Aug. 29	Jan. 24
Mar. 2	July 28	Sept. 3	Jan. 29
Mar. 7	Aug. 2	Sept. 8	Feb. 3
Mar. 12	Aug. 7	Sept. 13	Feb. 8
Mar. 17	Aug. 12	Sept. 18	Feb. 13
Mar. 22	Aug. 17	Sept. 23	Feb. 18
Mar. 27	Aug. 22	Sept. 28	Feb. 23
April 1	Aug. 27	Oct. 3	Feb. 28
April 6	Sept. 1	Oct. 8	Mar. 5
April 11	Sept. 6	Oct. 13	Mar. 10
April 16	Sept. 11	Oct. 18	Mar. 15
April 21	Sept. 16	Oct. 23	Mar. 20
April 26	Sept. 21	Oct. 28	Mar. 25
May 1	Sept. 26	Nov. 2	Mar. 30
May 6	Oct. 1	Nov. 7	April 4
May 11	Oct. 6	Nov. 12	April 9
May 16	Oct. 11	Nov. 17	April 14
May 21	Oct. 16	Nov. 22	April 19
May 26	Oct. 21	Nov. 27	April 24
May 31	Oct. 26	Dec. 2	April 29
June 5	Oct. 31	Dec. 7	May 4
June 10	Nov. 5	Dec. 12	May 9
June 15	Nov. 10	Dec. 17	May 14
June 20	Nov. 15	Dec. 22	May 19
June 25	Nov. 20	Dec. 27	May 24
June 30	Nov. 25		

FERTILITY AND PROLIFICACY

In general, flockmasters have long considered that multiple births are important. This is well illustrated in the following English adage: "Ewes yearly by twinning rich masters do make; the lambs of such twinners for breeders go take." There is now substantial evidence to indicate that this adage was well founded.

Though twinning is inherited, it is not highly so in comparison with certain other traits. Experimental work has shown that the heritability of twinning ranges from 10 to 15%; whereas the heritability of face covering is about 56%.

Prolificacy in sheep is largely determined by the number of eggs liberated by the ovary at the heat period, and by the amount of embryonic mortality. If only one egg is released and fertilized, a single lamb will result unless this egg should divide so that twins are produced. Such division does not seem to occur with much frequency, it undoubtedly being true that most twins and triplets are due to the shedding of a like number of eggs, which are fertilized and complete their development. Although the heritability is low, lambing rates can be increased through selection for multiple births.

Multiple births are desirable; they increase the weight of lamb produced per ewe per unit of time, thereby lowering the relative cost of maintenance and lowering the cost of investment in sheep production.

In general, goats are more prolific than sheep, with two kids usually being born and three being common. The Angora, however, is an exception to this, generally producing only one kid.

Fig. 4-7. Multiple births are desirable. They greatly increase the weight of lambs produced yearly per ewe, and the annual maintenance requirements of ewes are similar whether they produce triplets or singles. (Courtesy, Finnsheep Breeders Assn., Zionsville, IN)

FACTORS AFFECTING REPRODUCTION

Under natural conditions, the fertility and prolificacy of females depends on heredity, age of puberty, age of ewe, light, temperature, association with the male, nutrition, parturition and lactation, diseases and parasites, and fertility of the male.

■ **Heredity**—Some breeds and strains produce a higher percentage of twins and triplets (or more) than do others. Those breeds developed in ecosystems where nutrient resources are limited tend to be less prolific. Those breeds developed in ecosystems and production systems with more plentiful nutrient resources tend to be more prolific.

The Finnsheep, which is widely recognized for its prolificacy, was imported to the United States primarily so that its potential for increasing the productivity of domestic flocks could be assessed. Experiments have shown that Finnsheep can be effectively used in breeding programs to increase productivity. Also, within a flock, selection for twinning can increase productivity. At the University of Wyoming, a flock of 200 white-faced sheep were selected for twin and triplet births. Between 1970 and 1978, the lambing rate increased from 143 to 211%. Dr. David L. Thomas, Department of Meat and Animal Science, University of Wisconsin, reports that selection of replacements on the basis of dam's lambing performance can improve prolificacy by an annual rate of 1.3 to 2.0 lambs per 100 ewes.

Nubian goats frequently produce triplets.

■ **Age of puberty**—Ewes bred to lamb as yearlings demonstrate a greater lifetime production (pounds of lamb) than ewes bred to lamb as two-year-olds. Finnsheep or their crosses breed successfully in their first year. Fine-wool crossbreds, especially Dorset X fine-wool crosses, breed fairly successfully their first year, while slow-maturing breeds, such as the Rambouillet are difficult to breed to lamb at one year.

■ **Age of ewe**—Middle-aged ewes produce a higher percentage lamb crop than young ewes (two years). As long as ewes remain healthy and sound in mouth, body, and udder, it is a good practice to keep them.

■ **Seasonal breeders**—Normally, ewes lamb in the spring when forage is most abundant.

The initiation of the breeding season is primarily controlled by the light-dark ratio. As the days become shorter, the incidence of estrus increases, but because of individual and breed differences, breeding activity occurs all months of the year. In general, however, fertility is the highest during the fall months.

In temperate climates, both male and female goats tend to be seasonal in sexual activity. In northern latitudes sexual activity occurs between September and January.

■ **Temperature**—Temperature demonstrates a marked effect on fertility, embryo survival, and fetal development. A high temperature (90°F or higher), which causes the ewe to be stressed and unable to maintain normal body temperature, results in embryo mortality, fetal death, or the birth of weak, small lambs. Death losses from such lambs is high. Low relative humidity lessens the chance of heat stress when temperatures are high. Increased activity (working animals), excessive feed consumption, and/or excessive body fat may all contribute to the possibility of lowered reproductive efficiency due to heat stress.

Temperature also affects the male reproductive system. High temperatures will cause lower semen quality in rams. During prolonged periods of excessive heat (temperatures of 100°F or more), rams may become sterile. This damage is not permanent, however, and the ram usually becomes sexually sound after 4 to 6 weeks of cooler weather.

■ **Association with the male**—The sight, sound, or smell of the male appears to stimulate ovulation and estrus activity. For this effect, males must be introduced at the critical time of transition between non-breeding and breeding season.

■ **Nutrition**—Proper nutrition is necessary for efficient reproductive performances. Under range conditions, poor nutrition or lack of size and development hinders reproductive efficiency. Size is determined by nutrition and heredity. Generally, larger ewes in a flock have a higher multiple birth rate.

Under good nutrition, twin births may be the rule in dairy goat herds.

■ **Parturition and lactation**—After lambing, the uterus requires 2 to 3 weeks to return to normal, and few ewes will show estrus earlier than 30 days after lambing. Moreover, lactation and/or the nursing reflex depresses the occurrence of estrus activity. Heavy milking does are harder to settle than those producing less milk. Most ewes that lamb in the fall will exhibit a fertile heat 4 to 8 weeks after lambing.

■ **Disease and parasites**—Heavy infestations of internal parasites and certain reproductive diseases lower reproductive efficiency. These are covered in Chapter 8.

■ **Fertility and behavior of the male**—Often infertile or disinterested rams or bucks cause low lambing rates. Factors such as physical condition, heredity, temperature, and disease may affect the performance of a male. Sexual behavior is important and some very aggressive rams will mate with all the ewes in heat, while other less interested rams may mate many times with the same ewe and ignore the others.

One of the most important causes of infertility in dairy goat males is testicular hypoplasia—small, soft testicles not producing sperm. The condition is related to the polled trait.

FLUSHING

Flushing is that practice of feeding thin ewes or does more generously during the period of 2 to 8 weeks immediately prior to breeding. This may be accomplished either by providing more lush pasture or range or by feeding 0.5 to 1.0 lb of grain per head per day.

Although it is not likely that all of the benefits ascribed to flushing will be fully realized under all conditions, the general feeling persists that the practice will result in a 15 to 20% increase in the lamb crop. Flushing may be of little value in farm flocks where the level of nutrition is relatively high during most of the year.

PREPARATION OF EWES AND RAMS FOR MATING

Several practices pertaining to both ewes and rams are important during the breeding season. Discussion of each of these points follows.

TRIMMING AND TAGGING THE EWES

Tagging is the removal of tags, or locks of wool and dirt, about the dock. It is important that this job be done prior to the breeding season in order to prevent the ewes from befouling themselves and to remove obstacles for the service of the ram.

PREPARATION OF THE RAM FOR MATING

As the weather is usually rather warm at the time of the breeding season, shearing the ram just prior to this will make him more active. This is especially true of old show-or-sale rams. Where a ram is not sheared completely, he should at least have the wool clipped from his neck and from his belly in the region of the penis, for this will result in copulation with greater ease. Also, the hoofs need to be properly trimmed prior to the breeding season.

MARKING THE RAM

When a number of rams are turned in with a large band of ewes, it is impossible to detect individual rams that may be failing to settle ewes. Moreover, it is quite likely that a different ram will serve the ewe should there be a recurrence of heat, or perhaps more than one ram may serve the ewe at the time of estrus. When only one ram is being used on a small flock, it is important to know whether the ewes are getting pregnant. Then, too, with a purebred flock, individual breeding records are rather important.

The sheep producer can monitor the ram by using a marking harness (breeding harness), containing a crayon (different colored crayons are available), on the ram, or by smearing the breast of the ram and the area between his fore legs every day or two with colored grease or oil. Then, as the ram serves the ewe, a mark will be left on her rump. Paint or tar should never be used for this purpose.

The color of the crayon should be changed every 17 days (the approximate estrus cycle of the ewe) so that one can determine whether ewes that have been bred are returning in heat. For example, during the first 17-day interval, the color used on the ram can be yellow; for the second 17-day interval, the color can be red; and for the third 17-day interval (if there is still some question about some of the ewes having settled) the color can be black (thus proceeding from light to dark colors).

Table 4-3 lists some coloring agents that producers mix with mineral oil or motor oil and apply to the chest of the ram.

Naturally, if a good percentage of the ewes are found coming in heat for a second time, the ram should be regarded with suspicion, and perhaps another ram should be obtained. Often rams will be sterile or show lower fertility during late summer due to heat stress.

TABLE 4-3
COLORING AGENTS USED FOR IDENTIFYING MATINGS[1]

Agent	Concentration[2]	
	(tsp/pt)	*(ml/liter)*
Cement dust (red or green)	2	21
Lamp black	2	21
Powdered carpenter's chalk[3] (available in four different colors)	2–4	21–42
Rawlins red (iron oxide)	1–2	10–21

[1]Applied to the wool on the chest of the ram every few days.

[2]This is the concentration in 30W motor oil. Powdered carpenter's chalk can also be combined with a small amount of household detergent and suspended in paraffin oil.

[3]Research at the University of Wyoming suggests that carpenter's chalk is less apt to leave wool colored after scouring.

CARE AND MANAGEMENT OF THE RAM

If possible, the ram should be secured considerably in advance of the breeding season. At this time, better rams are available and the prices are more favorable. Then, too, the ram will have an opportunity to become acclimated before being placed in service. In case of show-or-sale rams, it may also be advisable to remove gradually some of their surplus flesh.

Stud rams are usually kept separate from the ewes except during the breeding season. Their quarters need not be elaborate or expensive. Usually, a dry shelter that will provide protection during times of inclement weather is all that is necessary. Plenty of exercise should be provided at all times.

Rams may subsist largely on pasture and dry roughage. If the pasture has been scanty prior to the breeding season, the rams may be fed a little grain, usually not more than 1 lb daily, for conditioning. Rams are usually fed some grain when being fitted for show or sale, but it must be remembered that excess fat may actually be harmful from a breeding standpoint.

AGE AND SERVICE

The number of ewes a ram will serve in a season depends on his age, vigor, and method of handling. Table 4-4 gives pertinent information relative to the use of the ram, including consideration that should be given to age and method of mating.

Vigorous, well-grown, early-maturing lambs may be used on 20 to 25 ewes with no apparent harm. Many good sheep producers believe that there is a definite breed difference in early reproductive capacity. A vigorous ram 1 to 4 years old that is run with the flock during the breeding season is sufficient for 35 to 60 ewes. When the ram is turned in with the flock for only a limited period daily—perhaps an hour in the morning and an hour in the evening—or when a teaser is used for the purpose of locating ewes that are in heat, one mature ram may be sufficient for 50 to 75 ewes. Unless the ram becomes extremely nervous and restless when he is removed, his energy will be conserved if he is kept away from the flock; and he will be available for heavier breeding service. Often he will remain contented if one or two wether lambs or bred ewes are kept with him. With a heavy ram and warm weather, a good plan is to allow him to run with the flock at night and to remove him to separate quarters during the day. Regardless of the system followed, the ewes should be checked daily and accurate records kept of the breeding dates. Though there is considerable variation, most range sheep producers usually plan to have 3 to 4 active mature rams with each 100 ewes.

TABLE 4-4
RAM MATING GUIDE

Age	Number of Matings/Year		Comments
	Hand Mating	Pasture Mating	
Lamb.	20–25	—	Most range operators use 1 ram to 25 to 35 ewes.
Yearling or older . .	50–75	35–60	A ram should remain a vigorous and reliable breeder up to 6 to 8 years of age.

FERTILITY TESTING

Except for an actual breeding test, no one method is capable of determining a ram's fertility with complete accuracy. A semen test can, however, detect rams of low fertility. Semen tests should be conducted on samples collected via natural ejaculation, since an ejaculate of poor quality is more likely when electroejaculation is used to obtain a sample. Samples are evaluated on the basis of pH, motility, concentration, live-dead, and abnormalities.

Semen testing is primarily of value to producers keeping individual progeny records or farm-flock producers using only one or two rams.

Adequate scrotal size (circumference) for their weight may be a valuable tool for determining ram fertility. It is easy to measure and it is highly correlated to testicular development.

CARE AND MANAGEMENT OF THE BUCK

Many dairy goat herds are too small to keep a buck on the premises, so producers use a buck service or artificial insemination. Where bucks are kept, they must be prepared for the breeding season with good nutrition (Chapter 21), parasite control (Chapter 8), foot trimming and a management clip (Chapter 22), and plenty of exercise.

Before bucks are bred, a genital exam should be conducted to examine the testicles for any abnormalities. Each testicle should be plump, firm, and symmetrical. If abnormalities are suspected, the semen should be evaluated. Also, an examination of the prepuce and penis is recommended.

Young bucks should not be used more than once or twice per week their first breeding season. Older bucks, however, are capable of servicing 7 to 15 does per week. Hand breeding is preferable to pasture breeding service, since the frequency of mating can be controlled and precise breeding dates are more easily obtainable.

PREGNANCY TESTING

Barren ewes can no longer return feed costs for their wool production. In an efficient operation they must be identified early for rebreeding or culling. Moreover, identifying pregnant ewe lambs for replacements increases the productivity of the flock. Besides pregnancy testing, detection of multiple fetuses would also be of significance, because feed costs could be lowered on ewes carrying single fetuses.

Does should also be checked for pregnancy and nonpregnant does rebred before the end of the season. It is disappointing and expensive to dry off open does assumed to be pregnant.

Several methods for pregnancy testing sheep have been developed and used experimentally or semipractically in recent years; among them, the following:

1. Direct palpation of the uterus via laparotomy
2. Doppler
3. Hormonal assays
4. Radiographics (X-rays)
5. Rectal-abdominal palpation
6. Ultrasonics
7. Vaginal biopsy

These techniques all demonstrate a high degree of accuracy, with proper equipment and training. Most of these techniques are not adaptable to field conditions and large numbers of animals. Rectal-abdominal palpation is relatively inexpensive and easy to perform in field conditions with more than 90% accuracy after 60 days of gestation.

The rectal-abdominal palpation equipment and supplies required include: a cradle; a hollow plastic rod (palpation rod), $\frac{5}{8}$ in. outside diameter and 21 in. in length with a bullet-shaped tip; a warm, soapy solution of tap water; and a drenching gun to inject the soapy solution into the rectum. Steps to perform the rectal-abdominal palpation include:[1]

1. Fast ewes overnight.
2. Place each ewe on her back in cradle, securing hind legs in a position which keeps the stomach muscles relaxed.
3. Using a drenching gun, gently inject 6 to 8 oz of warm, soapy water into the rectum.
4. Lubricate tip of hollow rod in soapy solution and insert into the rectum, moving the outward end back and forth laterally while exerting a steady but gently forward pressure.
5. After insertion, about 14 to 16 in., press the rod gently but firmly upward (see Fig. 4-8).

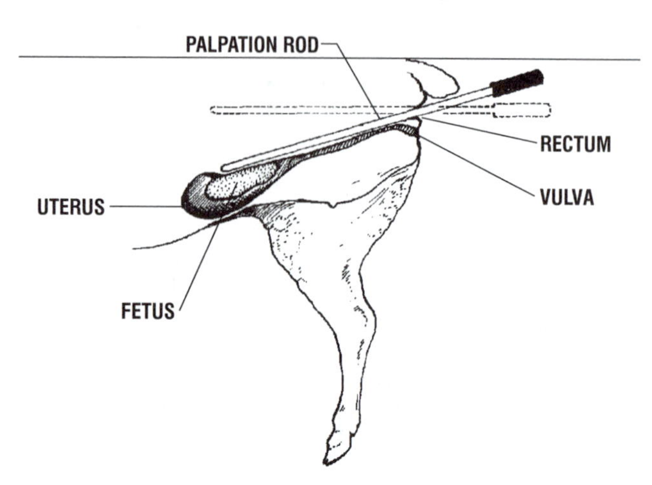

Fig. 4-8. The palpation rod is inserted in the rectum, close to the spine. The objective is to (1) position the rod dorsal to the fetus, and (2) elevate and hold the fetus against the abdominal wall. Then, the free hand is used to feel the fetus through the abdominal wall.

6. Use the free hand to locate the fetus.
7. The few ewes which are of a questionable status should be separated and examined again 1 week later.

With practice, care, and adequate assistance, one person should be able to examine 100 to 120 ewes per hour with more than 97% accuracy at 60 days postmating.[2]

A Doppler instrument may provide the means of detecting pregnancy in ewes. Using a rectal probe, the operator may diagnose pregnancy by detecting the fetal heart pulse. Each examination requires 1 to 2 minutes. At 41 to 60 days postmating, it is about 96% accurate.[3]

Ultrasonic pregnancy detectors are on the market that require less training and less restraint than the Doppler detector. These instruments are placed on the outside of the ewe near the front of the right teat and ultrasound waves are beamed toward the uterus. A colored light and/or a tone indicates pregnancy or lack of pregnancy. This type of ultrasonic detector is 90 to

[1]Hulet, C. V., "A Rectal-Abdominal Palpation Technique for Diagnosing Pregnancy in the Ewe," *Journal of Animal Science*, Vol. 35, No. 4 1972, pp. 814–819. This method was developed through the U.S. Sheep Experiment Station, Dubois, Idaho, from research conducted at the Ruakura Agricultural Research Centre, Hamilton, New Zealand.

[2]Memon, M. A., and R. S. Ott, "Methods of Pregnancy Diagnosis in Sheep and Goats," *Cornell Veterinarian*, Vol. 70, No. 3, 1980.

[3]*Ibid.*

95% accurate when the stage of gestation is 60 to 120 days.

Using sophisticated ultrasonic equipment or radiographic (X-ray) equipment, the operator can detect multiple pregnancies. This method is, however, impractical for field use.

Any of the methods listed may also be applied to does. The cost and/or training, however, may be prohibitive for small herds. Perhaps the most commonly used pregnancy test is the failure to return to estrus. This requires good records and close observation.

CARE OF THE DOE

Care of the pregnant doe and care of the doe at kidding are covered in Chapters 22, 23, and 24. Feeding the pregnant and lactating doe is covered in Chapter 21. Additionally, some of the information contained in the following sections titled "Care of the Pregnant Ewe" and "Care of the Ewe at Lambing Season" may contain information beneficial to the goat breeder as well.

CARE OF THE PREGNANT EWE

The requirements of the pregnant ewe are neither exacting nor difficult to meet. These needs are feed and water, exercise, and shelter.

FEED AND WATER

For successful sheep production, the ewes must be economically and properly fed and watered during the pregnancy period. As these requirements are fully covered in Chapter 14, no further discussion is necessary at this point.

EXERCISE

During periods of inclement weather and when feed is brought into the barn, ewes quite often exercise entirely too little. As a result they become sluggish, and their blood circulation is poor. Scattering a palatable roughage some distance from the shed or driving the ewes at a moderate walk will bring about forced exercise. Above all, overexertion, such as wading through a deep snow or being chased by dogs, should be avoided. With good winter pasture and open weather conditions, no other arrangements for exercise will be necessary.

SHELTER

The shelter should be of such nature as to protect the flock from becoming soaked with rain or wet snow. Dry snow or bitter cold has no harmful effect and, up until lambing time, a shelter open to the south on well-drained ground may be entirely satisfactory.

CARE OF THE EWE AT LAMBING SEASON

Mortality research in the major sheep producing countries of the world show yearly losses of lambs before weaning ranging from 10 to 35%, with an average annual loss of 15 to 20% accepted as normal. Lamb losses before weaning are affected by weather, management practices, nutrition, housing, and genetic factors.[4]

The careful and observant shepherd or herder realizes the importance of having everything in readiness for the lambing season. If pregnant ewes have been properly fed and managed, the next problem is that of saving the newborn animals.

As lambing time approaches, unsheared ewes should be tagged. This consists of shearing the wool from around the udder, flank, vulva, and dock (clipping the wool around the udder allows the lambs easily to find the teats). The ewes should also be placed where they have plenty of room, away from any jamming or crowding. The grain allowance should be materially reduced, but the roughage allowance may be continued, if it is certain that it is of good quality and palatable. Careless feeding at this time is likely to result in milk fever following parturition. If breeding records have not been kept, the signs of approaching parturition—a nervous, uneasy disposition; a sinking in front of the hips; and a fullness of udder—must be relied upon.

LAMBING PEN

Just before lambing, or immediately thereafter, the ewes should be placed in lambing pens made from two hinged hurdles placed together, these pens are usually 4 ft square and are set against the walls of the sheep barn. Use of lambing pens prevents other sheep from trampling on the newborn lambs, eliminates the possibility of the lambs wandering away and becoming chilled, and, through keeping the dams and offspring together, lessens the danger of disowned lambs.

---

[4]Rook, J.S., D.V.M., *National Sheep Reproduction Symposium*, Colorado State University, Ft. Collins, CO, 1989, p. 95.

Lambing pens should be clean, dry, well bedded, and well ventilated and should be located so as to be free from drafts. During extremely cold weather using a heat lamp and/or throwing a blanket over the top of the pen will provide additional warmth for the lambs during the first few hours after their birth.

NORMAL PRESENTATION

A good rule for the shepherd to follow is to be near during parturition but not to disturb a ewe unless she needs help. Normal presentation of the lamb consists of having the forelegs extended with the head lying between them. Common abnormal presentations include: one leg back, both legs back, head back and one or both legs forward, or any combination of these when multiple births are involved and breech.

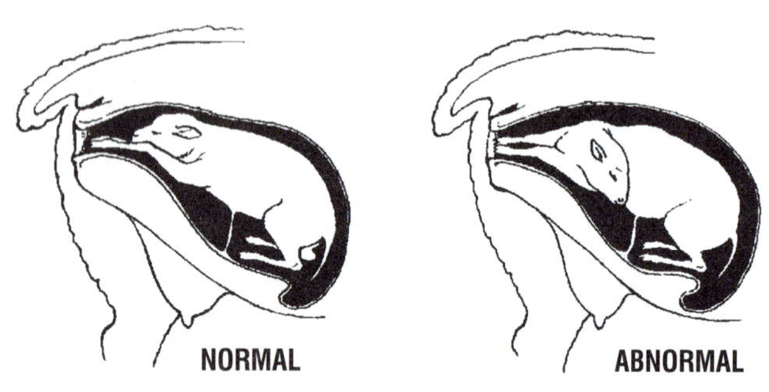

Fig. 4-9. Normal presentation on the left and one type of abnormal presentation on the right.

Even though the lambs are born in clean quarters, tincture of iodine should be applied to their navels soon after birth. This precaution may not be necessary when lambs are dropped on an uncontaminated pasture or range, though many range operators report that they have found it necessary to apply iodine to the navels of lambs born on the range as well as in the shed.

TAKING THE LAMB

If a ewe has labored for some time with little progress or is laboring rather infrequently, it is usually time to give assistance. If the lamb is not in the proper position, such assistance consists of inserting the hand and arm in the vulva and turning the lamb so that the forefeet and head are in position to be delivered first. Pulling the young outward and downward as the ewe strains may then help delivery. Before this is done, however, the fingernails should be trimmed closely, rings and watch removed, and the hands and arms thoroughly washed with mild soap and warm water, disinfected, and then coated with a suitable obstetrical lubricant.

If difficult delivery is due to inadequate dilation of the cervix, dilation can often be achieved with an intramuscular injection of estradiol (estrogen) followed by an intramuscular injection of oxytocin one hour later. If normal lambing fails to occur after three oxytocin injections, a caesarian section should be performed.

CHILLED AND WEAK LAMBS

Lambs arriving during cold weather may become chilled before they have dried. One of the most effective methods of reviving a chilled lamb is to immerse the body, except for the head, in water that is as warm as one's elbow can bear. The lamb should be kept in this for a few minutes and then removed and rubbed vigorously with cloths. It then should be wrapped in an old blanket, a sheepskin, or other heavy material and should be given some warm milk as soon as possible. Another convenient and effective method of drying and warming a chilled lamb consists of putting it in a box containing a light bulb or an electric heater.

When strong, healthy ewes have been properly fed and cared for during pregnancy, there will be a minimum of weak lambs. The shepherd should first make certain that the membrane has been removed from the nostrils and that breathing has started. Blowing into the mouth, lifting the body and dropping it a short distance, working the legs, and pressing the sides are artificial methods of starting breathing that may revive lambs that at first appear lifeless.

After breathing has started and the navel cord has been painted with iodine, an attempt should be made to get the lamb to nurse. Quite often even a very weak lamb will nurse the ewe if it is held to the teat. If it refuses to nurse in this manner, some of the colostrum of the ewe should be milked into a sterilized bottle, and the lamb should be fed a few teaspoons each hour by means of the bottle and nipple, until it gains strength. Weak lambs can also be fed with a stomach tube—a catheter tube 0.25 in. in diameter. With this method, about 2 oz of milk can be administered until the lamb begins to nurse, but one must be certain of the placement of the tube in the stomach and not the lungs. A subcutaneous (under the skin) injection of 25 to 50 ml of a 5% glucose (dextrose) solution may also revive weak lambs.

If the ewe has no milk, an attempt should be made to obtain milk from another ewe that has just lambed, and perhaps in a few hours the normal flow of milk will start. Colostrum is essential to newborn survival, and it must be consumed within 15 hours after the newborn's birth for antibodies in it to be effective. Colostrum can be obtained from ewes, goats, and cows, frozen in ice cube trays, and thawed as needed during the lambing season.

DISOWNED LAMBS

When lambing pens are used, the number of disowned lambs is kept to a minimum. For the most part, disowning of lambs is due to improper feeding during pregnancy or because of a poor milk supply, an inflamed udder, or a maternal instinct that is not sufficiently developed, as is often true in ewes with their first lambs.

For the first few days, a ewe seems to recognize her young by scent or sense of smell. When difficulty is encountered in getting a ewe to own her own lamb or when it is desired to transfer or graft a lamb (as may be necessary with the loss of a single or when there are twins or triplets on an old ewe), deception in the sense of smell is an effective approach. One of the most common practices is to milk some of the ewe's milk on the rump of the lamb and then to smear some of it on the nose of the ewe. Many good shepherds take some of the mucus from the mouth and nose of the newborn lamb and smear it over the nose of the ewe. If these methods fail and the ewe persists in fighting the lamb away, blindfold her so that she cannot see the lamb. As a last resort, and when all other methods have failed, tie a dog in an adjoining pen. Sometimes this will cause latent maternal instincts to rise to a surprising degree.

Occasionally, a ewe will fail to own one of a pair of twin lambs. When this condition exists, about all that can be done is that the shepherd be patient in training the disowned lamb to nurse at the same time as its mate. Both lambs are usually kept from the ewe and turned with her at intervals.

FEED AND WATER AFTER LAMBING

Following parturition, the ewe is in a feverish condition and should be handled carefully. She may be watered immediately after lambing, and at frequent intervals thereafter, but she should never be allowed to gorge. It is also a good plan to take the chill off the water before giving it to her. In general, feeds of a bulky and laxative nature should be provided during the first few days. A mixture of equal parts of oats and wheat bran may be fed in very limited quantities, with all the hay that can be consumed. Heavy grain feeding at this time may cause udder trouble in the ewe and digestive disturbances in the lamb. The feed may be gradually increased until the ewe is on full feed in about a week.

EXAMINATION OF THE UDDER

During the first two days following lambing, the udder should be examined both evening and morning. Sometimes a lamb will nurse one side only. If all the milk is not being taken by the lamb, the udder should be milked out and the ration lessened accordingly. If the udder becomes swollen and feverish, it should be milked out, bathed with warm water, and then dried. Following this, it should be painted with tincture of iodine. This treatment should be repeated once or twice daily, as necessary. Lambs should not be allowed to suckle when their mothers' udders are in such a condition. It is also a good plan to isolate the affected ewes from the rest of the flock. Ewes with unsound udders should be culled.

ARTIFICIAL INSEMINATION (AI) AND EMBRYO TRANSFER (ET)

Artificial insemination is the deposition of spermatozoa in the female reproductive tract by technicians using a pipette rather than by a ram.

In the United States, artificial insemination is more extensively practiced with dairy cattle than with any other class of farm animals.

For some sheep producers, AI provides a tool for meeting the demand for greater efficiency.

Compared to the limited use in sheep, AI is used relatively widely in dairy goats.

For both sheep and goats, some of the **advantages** of artificial insemination are:

1. It increases the use of outstanding sires.
2. It alleviates the bother of keeping a ram or a buck.
3. It makes it possible to overcome certain physical handicaps to mating.
4. It lessens ram and buck costs.
5. It provides the means for rapid improvement in quality of offspring by intensive selection.
6. It helps control diseases.
7. It makes it possible to keep better breeding records.
8. It results in a flock of greater uniformity.
9. It causes the early discovery of nonbreeders.
10. It alleviates distance and time as limiting factors.
11. It may increase profits.

Some of the **limitations** of artificial insemination are:

1. It is difficult to detect ewes or does in estrus, and to time properly the insemination in relation to the time of ovulation.
2. It requires frequent handling of the ewes or

does and dedication of time during the breeding season.

3. It requires extra equipment and skilled technicians.

4. It requires rams or bucks trained and kept for semen collection.

5. It may accentuate the damage done by a genetically inferior ram or buck.

6. It may restrict the ram or buck market.

7. It necessitates more production or progeny testing.

8. It results in a lower conception rate, thus requiring a second, or third, insemination, or natural service.

The procedures in artificial insemination of sheep and goats are similar to those used in cattle and horses. Further, the equipment and supplies used for sheep and goats are similar to those used for cattle, with the exception of a vaginal speculum and light source which aid the placement of the semen in the uterus (Fig 4-10).

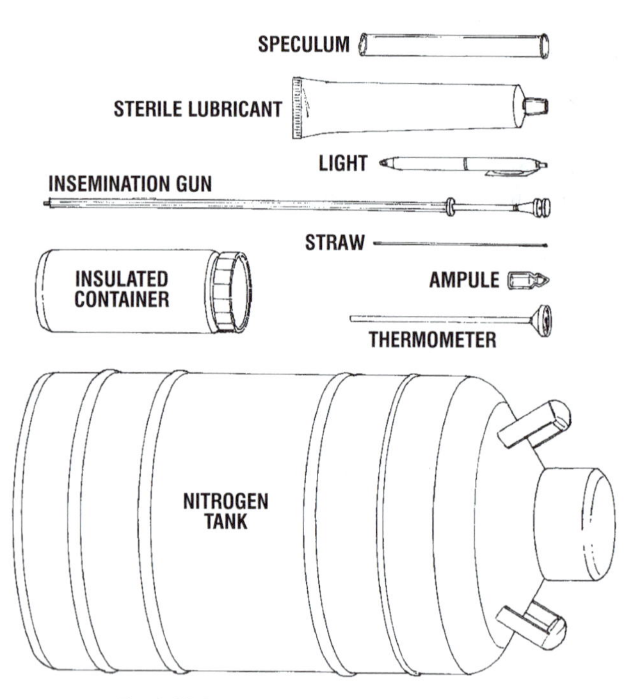

Fig. 4-10. Insemination equipment and supplies.

Briefly, the insemination includes the following steps:

1. Identify females in heat.

2. Properly restrain each ewe or doe. It is desirable to elevate the posterior of the female.

3. Coat speculum with a sterile lubricant and carefully insert into the vagina.

4. Use light source and speculum to locate the cervix.

5. Gently work insemination gun through the cervix.

6. Deposit semen.

7. Withdraw inseminating gun and speculum.

Embryo transfer is another tool that is becoming more readily available to producers. It allows breeders to obtain more progeny from superior males and females within a shorter time span. Numerous embryos that are the result of matings of superior males and superior, superovulated females are surgically removed from the uterus of the superior females and then surgically placed in the uterus of commercial females. For some time to come, however, embryo transfer will probably only be valuable for top purebred breeders because of the cost.

Interestingly, the first embryo transfers in farm animals were performed on sheep and goats, but the commercial interest now is in cattle and horses.

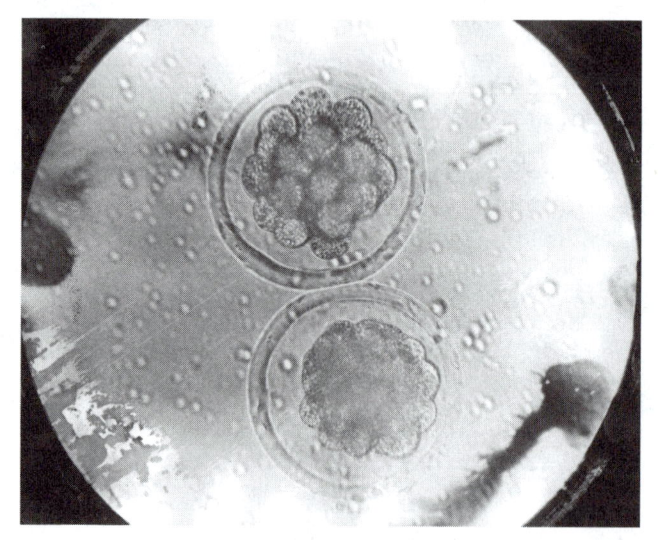

Fig. 4-11. Embryos recovered from a ewe. Numerous embryos flushed from the uterus of a superior ewe, bred to a superior ram, may be placed in the uterus of a commercial ewe where they develop fully. The commercial ewe becomes a surrogate mother. (Courtesy, Mississippi State University)

REGISTRATION OF YOUNG PRODUCED THROUGH ARTIFICIAL INSEMINATION (AI) OR EMBRYO TRANSFER (ET)

Table 4-5 summarizes the rules and regulations of sheep and goat associations that have evolved with rules concerning registering offspring born as a result of artificial insemination (AI) or embryo transfer (ET).

As the AI and ET industry in small ruminants continues to grow, the more populous sheep and goat associations in registration numbers have developed regulations. Many of these associations have rules regulating semen cryopreservation within their respective breeds that affect the AI and ET registration of offspring.

TABLE 4-5
ARTIFICIAL INSEMINATION (AI) AND EMBRYO TRANSFER (ET) RULES OF SHEEP AND GOAT ASSOCIATIONS[1]

Breed	Rules and Regulations
Sheep	
Columbia	Lambs born as a result of AI or ET require a signed breeding certificate.
Coopworth	No reported rules.
Delaine Merino	No reported rules.
Dorset	The ram from whom semen is used (either AI or ET) and the ET donor ewe must be blood typed or DNA tested. Lambs are identified as an AI or ET offspring on the certificate. Contact the Continental Dorset Club.
Finnsheep	Lambs produced by AI, ET, or natural service are so identified on the registration.
Hampshire	The ram from whom semen is used (either AI or ET) must be blood typed or DNA tested. Lambs are identified as AI or ET on the registration certificate.
Katahdin	No reported rules.
Lincoln	No reported rules.
Montadale	Lambs produced by AI or ET must be identified as such. No additional charges.
North Country Cheviot	A signed breeding certificate is required.
Oxford	Owner must identify the offspring as an AI or ET lamb on the registration form. Owner of the ram semen must sign the AI or ET lamb's registration application.
Polypay	No reported rules.
Rambouillet	Lambs produced by AI must be identified on the registry certificate for an additional $2 charge. For frozen semen, the association is to receive the appropriate forms signed by the collector and ram owner.
Romney	Lambs born from AI are identified on the certificate as AI at a $3 fee. Semen from the ram used must be blood typed by an ARBA approved lab ($10 recording fee) and follow association regulations. ET lambs will be identified as Embryo Transplant. The donor ram and ewe must be blood typed by an ARBA approved lab ($10 recording charge each) and proper recording documents ($20 recording charge per embryo) must be signed by the transferring person. AI or ET lambs may be blood typed for compliance.
Shetland	No reported rules.
Shropshire	Lambs resulting from AI or ET procedures shall be designated on the certificate with the initials "AI" or "ET." Proof of such procedure(s) must be furnished to the secretary at the time of registry application.
Southdown	Rams siring AI and ET offspring must have a Notification of Collection completed prior to registration. An Artificial Insemination Form ($2 for each ewe artificially inseminated) must be completed before registration. Embryo Removal/Embryo Transplant forms cost $2 per donor ewe. Ram owners artificially inseminating their own ewes with fresh semen are exempt.
Suffolk	AI or ET lambs born are identified on the certificate as either AI (no charge) or ET ($1 per lamb). The certificate must be signed by the owner of the ram semen. Rams used for either AI or ET must be blood typed or DNA tested and recorded at the national office.
Targhee	A signed breeding certificate is required.
Tunis	No reported rules.
Texel	No reported rules.
Dairy goats	
American Dairy Goat Association	AI sired offspring must have the buck recorded with the ADGA. Does of AI service must be tattooed and the record signed by the inseminator. Embryo transplant kids are identified as ET as the type of birth. NDHIA rules govern the offspring derived from embryo transfer for registration in the ADGA.
American Goat Society	Kids produced by AI results have the following rules: (1) the sire is approved by an official classifier and is on file at the AGS, (2) there is proof of breeding service and identification of the female, (3) reliable and AGS recorded frozen semen is used, and (4) the certificate is completed and sent to AGS at the time of the AI service. Embryo transfer offspring protocol is presently being developed.
Other goat breeds	No reported rules

[1]Compiled by Dr. Dennis D. Gourley, Elite Genetics, Inc., Waukon, IA.

CONTROLLING REPRODUCTION

Considerable research is underway to (1) synchronize heat, (2) increase ovulation rate, and (3) shorten the lambing interval, to boost the efficiency of production from sheep, by producing more lambs. As a result of the success of this research, sheep and goat breeders will need to choose methods which are best suited for their operations.

ESTROUS SYNCHRONIZATION

The capability to synchronize the estrus of a flock of sheep or of a herd of goats offers several benefits: (1) efficient use of rams or bucks, (2) uniform lamb or kid crop, (3) reduced labor cost, (4) intensive management potential, (5) shorter lambing season, and (6) earlier lambing. Procedures for using progesterone (a natural hormone) or a synthetic progestogen[5] are quite standard. The progestogen can be administered by (1) injection, (2) feed, (3) vaginal sponge or pessary, or (4) implant.

A common procedure for synchronizing estrus in cycling ewes is the administration of the progestogen for 12 to 14 days. Upon withdrawal, a majority of the ewes demonstrate estrus within 2 to 3 days. In goats, the progestogen is administered for 18 to 21 days.

Prostaglandin (prostaglandin $F_2\alpha$) can also be used for estrous synchronization during the breeding season. It may be used alone or in combination with a progestogen, but it is only effective when ewes or does are normally cycling.

Successful out-of-season breeding and synchronization can be accomplished by the administration of a progestogen and then the injection of PMSG (pregnant mare serum gonadotropin) which stimulates follicle growth and ovulation.

While the technology is available to control reproductive cycles in ewes and does, its use is curtailed in the United States due to the limited availability of the hormones which require federal clearance.

INCREASED OVULATION RATE AND LAMB CROP

Increased ovulation means increased lambing rate, and a larger lamb crop marketed means more profit. For example, records show that, on the average, range ewes that drop and rear twin lambs wean about 40 lb more lamb than ewes that have single lambs. Because most of the costs of sheep production are about the same regardless of the size of the lamb crop, multiple births are important.

The United States lamb crop as a percentage of ewes one year old and over was 106% in 1995.[6] Lamb crop goals of 125 to 200% are reasonable, depending on the production system.

Ovulation rate is affected by season (highest in the fall), by level of feeding (flushing), and by breeding. Some breeds are more prolific than others. Also fertility, prolificacy, and lamb livability are increased by crossbreeding.

Finnsheep and the Romanov breed seldom have fewer than twins or triplets, and mature ewes of these breeds may have more. Crossbreeding with such prolific breeds can very quickly lead to high twinning rates. Also, as much as a 2% per generation (approximately 1% per year) increase in lambing rate can be realized when multiple birth replacements are selected.

SHORTENED LAMBING INTERVAL AND BREEDING TWICE A YEAR

Most ewes freeload half the year. They are bred in the fall, have a 5-month pregnancy period, and suckle lambs for another 4 to 5 months. They spend a whole year to produce one lamb crop. This is inefficient, and costly, too.

Using breeds of sheep with long breeding seasons (Dorsets, Rambouillets, and Merinos), along with proper management, can achieve lambing intervals of approximately 8 months—without the use of hormones. This means that three lamb crops can be produced in 2 years (one crop each 8 months). It is possible, however, from both time and physiological standpoints, for a ewe to produce two lamb crops in 12 months. Research has shown that through the use of hormones, estrus and ovulation can be induced early enough so that ewes can be rebred once or twice after lambing, and yet produce another lamb crop in the 6-month period following lambing. Currently, the most hopeful efforts to achieve twice-a-year lambing (and three times in 2 years) seem to be selection for ewes with a longer natural breeding season, early postpartum fertility, and a shorter gestation.

Shortening the lambing interval will make for lower costs of production. Also, the lessening of seasonal restrictions will (1) contribute to a more even supply of lamb throughout the year, and (2) allow for greater flexibility in production.

---

[5]Progestogens are female sex hormones including those produced naturally by the female (progesterone) and others which are produced synthetically. Synthetic progestogens include: medroxyprogesterone (MAP), megestrol acetate (MGA), flurogestone acetate (FGA), and chlormadinone acetate (CAP).

[6]*Agricultural Statistics 1995–96*, p. VII-34, Table 416.

CONTROLLING REPRODUCTION WITH LIGHT

Modification of the length of exposure to light can be used to induce out-of-season breeding in sheep and goats. Artificially reducing the amount of light per day to which the animals are exposed may induce changes in the nervous and endocrine systems which initiate out-of-season cycles. Under most production systems, it is difficult to provide a practical way of placing animals under decreasing light.

QUESTIONS FOR STUDY AND DISCUSSION

1. Diagram and label the reproductive organs of the ram or buck, and briefly describe the function of each organ.

2. Diagram and label the reproductive organs of the ewe or doe, and briefly describe the function of each organ.

3. Discuss each of the following reproductive characteristics of sheep or goats: (1) age of puberty, (2) age to breed, (3) heat periods, (4) mating, (5) fertilization, and (6) gestation period.

4. Describe the estrous cycle of the ewe.

5. Multiple births are valuable to sheep producers. What determines a ewe's ability to produce two or more lambs?

6. What factors may influence fertility and prolificacy in sheep and goats?

7. Discuss the preparation of the ewe and the ram for the breeding season.

8. Discuss the preparation of the buck (goat) for the breeding season.

9. How often should a mature ram be used when hand mating is practiced? What guidelines should you follow if pasture mating is practiced?

10. In your opinion, is pregnancy diagnosis a valuable asset to a sheep or goat operation? Defend your answer.

11. What method of pregnancy diagnosis would you choose? List the methods available.

12. Outline some practices that you feel would ensure the survival of the maximum number of lambs at lambing and the first few days afterwards.

13. What advantages could accrue from the practical and extensive use of artificial insemination in sheep and goats?

14. Describe the use of an artificial insemination program that would take advantage of superior on-farm rams.

15. Discuss some of the potential means of controlling reproduction in sheep and goats.

SELECTED REFERENCES

Title of Publication	Author(s)	Publisher
Angora Goat and Mohair Production	M. Shelton	M. Shelton, San Angelo, TX, 1993
Animal Reproduction: Principles and Practices	A. M. Sorenson, Jr.	McGraw-Hill Book Co., New York, NY, 1979
Applied Animal Reproduction	J. W. Fuquay H. J. Bearden	Reston Publishing Co., Inc., Reston, VA, 1980
Artificial Insemination and Genetic Improvement of Dairy Goats	H. A. Herman	American Supply House, Columbia, MO, 1982
Genetics of Livestock Improvement	J. F. Lasley	Prentice-Hall, Inc. Englewood Cliffs, NJ, 1978
National Sheep Reproductive Symposium	Staff	Colorado State University, Ft. Collins, CO, 1989
Out of Season Breeding Symposium	Staff	Iowa State University, Ames, IA, 1992
Planned Sheep Production, Second Edition	D. Croston G. Pollott	Blackwell Scientific Publications, London, 1994
Practical Lambing and Lamb Care, Second Edition	F. A. Eales J. Small	Longman Scientific & Technical, Essex, England, 1995
Reproduction in Domestic Animals	Ed. by H. H. Cole P. T. Cupps	Academic Press, Inc., New York, NY, 1977
Reproduction in Farm Animals	Ed. by E. S. E. Hafez	Lea & Febiger, Philadelphia, PA, 1980
Sheep Breeding	Ed. by G. L. Tomes D. E. Robertson R. J. Lightfoot	Butterworth (Publishers), Inc., Woburn, MA, 1979
Sheepman's Production Handbook, The	Staff	American Sheep Industry Assn., Englewood, CO, 1992
Stockman's Handbook, The, Seventh Edition	M. E. Ensminger	Interstate Publishers, Inc., Danville, IL, 1992

CHAPTER 5

Lavoisier and his beauteous wife. (Courtesy, The Rockefeller Institute)

FUNDAMENTALS OF SHEEP AND GOAT NUTRITION

Contents	Page

Both sheep and goats inherit certain genetic possibilities, but how well these potentialities develop depends upon the environment to which they are subjected; and the most important influence in the environment is nutrition. In turn, all feed comes directly or indirectly from plants which have their tops in the sun and their roots in the soil. This is the nutrition cycle as a whole—from the sun and soil, through the plant, thence to the animal, and back to the soil again.

Therefore, nutrition is more than just feeding. Correctly speaking, *nutrition is the science of the interaction of a nutrient with some part of a living organism.* It begins with a knowledge of the fertility of the soil and the composition of plants; and it includes the ingestion of feed, the liberation of energy, the elimination of wastes, and all the syntheses essential for maintenance, growth, reproduction, lactation, fattening (fitting), and wool and mohair.

A good understanding of nutrition is important because animals and people are dependent upon food nutrients for the processes of life.

PERSPECTIVE OF NUTRITION

Of the basic needs of animals and people, none is more important than food. The ingestive behavior in mammals begins at birth—with suckling. The primary purpose of keeping sheep and goats is to transform feed into meat, milk, wool, and mohair. But the conversion of feed to these uses must be done efficiently and economically. To do this, the principles of nutrition must be applied; and they must be augmented by superior breeding, good health, and competent management.

Like other sciences, nutrition does not stand alone. It draws heavily on the basic findings of chemistry, biochemistry, physics, microbiology, physiology, medicine, genetics, mathematics, endocrinology, and, most recently, animal behavior and cellular biology. In turn, it also contributes richly to each of these fields.

BODY COMPOSITION OF SHEEP AND GOATS

Table 5-1 shows that there is a wide range in the body composition of sheep according to age and nutritional state (degree of fatness). Similar trends apply to goats, also.

TABLE 5-1
BODY COMPOSITION OF SHEEP[1]

Age or Status	Weight		Water	Fat	Protein	Ash
	(lb)	*(kg)*	*(%)*	*(%)*	*(%)*	*(%)*
Lamb, newborn . .	9	*4.1*	72.8	2.0	20.2	5.0
Lamb, feeder . . .	65	*29.5*	63.9	17.0	15.7	3.4
Lamb, fat	100	*45.4*	53.2	29.0	15.0	2.8
Lamb, very fat . . .	125	*56.8*	39.0	44.0	14.4	2.6

[1]Prepared by the author from numerous sources. Body composition is on an ingesta-free (empty) basis.

Based on this table, together with other studies, the following conclusions may be reached:

1. **Water.** On a percentage basis, the water content shows a marked decrease with advancing age, maturity, and fatness. In sheep, the water content from conception to market weight and finish changes as follows: embryo soon after conception, 95%; newborn lamb, 73%; feeder lamb, 64%; fat lamb, 39 to 53%.

2. **Fat.** The percentage of fat normally increases with growth and fattening. In sheep, the body of a newborn lamb contains only about 2% fat, whereas the body of a very fat lamb may run as high as 44% fat.

3. **Fat and water.** As the percentage of fat increases, the percentage of water decreases.

4. **Protein.** The percentage of protein remains rather constant during growth, but decreases as the animal fattens.

On the average, there are 3 to 4 lb of water per 1 lb of protein in the body.

5. **Ash.** The percentage of ash shows the least change. However, it decreases as animals fatten because fat tissue contains less mineral than lean tissue.

6. **Composition of gain.** The data presented in Table 5-1 clearly indicates that gain in weight may not provide an accurate measure of the actual gain in energy of the animal, because it tells nothing about the composition of gain. This is important because efficiency of feed utilization (pounds feed per pound body gain) is greatly influenced by the amount of fat produced.

Also, the chemical composition of the body varies widely between organs and tissues and is more or less localized according to function. Thus, water is an essential of every part of the body, but the percentage composition varies greatly in different body parts; blood plasma contains 90 to 92% water; muscle, 72 to 78%; bone, 45%; and the enamel of the teeth, only 5%. Proteins are the principal constituents, other than water, of muscles, tendons, and connective tissues. Most of the fat is localized under the skin, near the kidneys, and around the intestines. But it is also present in the muscles (known as marbling in a carcass), bones, and elsewhere.

A very small amount of carbohydrates (mostly glucose and glycogen) is present in the bodies of animals and found principally in their livers, muscles, and blood. Although these carbohydrates are very important in animal nutrition, they account for less than 1% of the body composition. The carbohydrate content is one of the fundamental differences between the composition of plants and of animals. In animals, the walls of the body cells are made chiefly of protein, whereas in plants they are composed of cellulose and other carbohydrates. Also, in plants most of the reserve food is stored as starch, another carbohydrate, whereas in animals nearly all the reserve is stored in the form of fat.

DIGESTIVE SYSTEM OF SHEEP AND GOATS

The digestive tract (or gastrointestinal tract) can be considered a continuous hollow tube—open at both ends—with the body built around it. It's a factory assembly line in reverse. Instead of building something, it takes things apart. The digestive tract of sheep and goats includes five main parts: mouth, esophagus, stomach, small intestine, and large intestine.

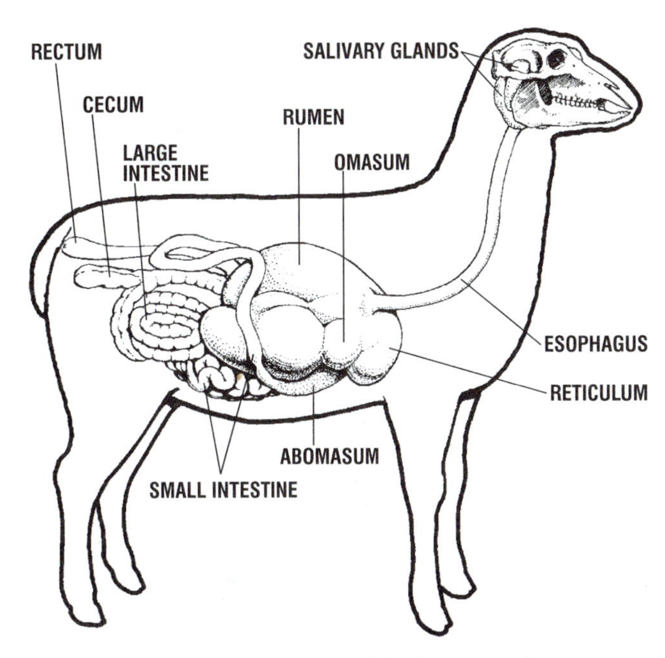

Fig. 5-1. The digestive system of the sheep—a ruminant.

Sheep, goats, and cattle are ruminants. They differ from nonruminant animals (pigs and people) in the following important ways:

1. **Mouth.** Ruminants have no upper incisor or canine teeth. Thus, they depend on the upper dental pad and lower incisors, along with the lips and tongue, for the prehension of feed.

2. **Four stomach compartments.** Ruminants possess four stomach compartments—rumen, reticulum, omasum, and abomasum (true stomach)—whereas monogastrics (nonruminants) have one. Such a digestive system makes for two primary nutritional differences between ruminants and simple-stomached animals:

   a. **More space.** They have the necessary space for processing large quantities of bulky forages to provide their nutrients. For example, the capacity of the digestive tract of sheep is about 12 gal., while the capacity of the human's digestive tract is only about 2 gal.

   b. **More microorganisms.** The rumen provides a highly desirable environment for the enormous population of microorganisms. The number of rumen bacteria varies according to the nature of the diet, feeding regimen, time of sampling after feeding, specie differences, individual animal differences, season, availability of green feed, and the presence or absence of ciliate protozoa.

   Rumen microorganisms serve two important functions:

(1) They make it possible for ruminants to utilize roughage—to digest the fiber therein. They break down the cellulose and pentosans of feeds into usable organic acids, chiefly acetic, propionic, and butyric acid—commonly called the volatile fatty acids (VFA). These VFAs are largely absorbed through the rumen wall and provide the ruminant 60 to 80% of its energy needs. Microbial digestion is of great practical importance in the nutrition of ruminants; it is the fundamental reason why they can be maintained chiefly on roughages.

(2) In exchange for their rumen-housing privileges, the microbes synthesize nutrients for their host, in a true type of symbiotic relationship. Rumen microbes synthesize, or manufacture, all the B complex vitamins and all the essential amino acids. The latter can even be made from nonprotein nitrogen compounds (NPN), such as urea or ammoniated products, or from proteins that are deficient in one or more of the amino acids. Finally, the microorganisms give their lives to their host in payment for food and shelter, being digested farther along in the gastrointestinal tract.

3. **Rumination.** This activity, or phenomenon, which is peculiar to ruminants, is of great practical significance. During rumination, the animal regurgitates and rechews a soft mass of coarse feed particles, called a bolus. Each bolus is chewed for about a minute, then, it is swallowed again. Ruminants may spend 8 hours or more per day in rumination, the amount of time varying according to the nature of the diet. Coarse, fibrous diets result in more time ruminating. Rechewing does not improve digestibility. Rather, rumination has an important bearing on the amount of feed the animal can eat and utilize. Feed particle size must be reduced to allow passage of the material from the rumen. Because high-quality forages contain less fiber than low-quality forages, they require much less rechewing and pass out of the rumen at a faster rate; hence, they allow the sheep or goat to eat more.

4. **Eructation (belching of gas).** Substantially more gas is produced in digestion by ruminants than by simple-stomached animals. The microbial fermentation in the rumen results in the production of large amounts of gases (primarily carbon dioxide and methane) which must be eliminated; otherwise, bloat results. Normally, these gases are expelled quite freely by eructation (belching) and, to a lesser extent, by absorption into the blood draining from the rumen, from which they are eliminated through exhaled air from the lungs.

5. **Stomach of the newborn.** When a lamb or kid is born, the rumen is small and the fourth stomach is by far the largest of the compartments. Thus, digestion in the young lamb or kid is more like that of a simple-stomached animal than that of a ruminant. The milk which the lamb or kid normally consumes bypasses the first two compartments by way of the esophageal groove and goes almost directly to the fourth stomach in which the rennin and other compounds for the digestion of milk are produced. If the lamb or kid gulps too rapidly, or gorges itself, the milk may go into the rumen where it is not digested properly and may cause upsets of the lamb's or kid's digestive system. As the lamb or kid nibbles at hay, small amounts of material get into the rumen. When certain bacteria become established, the rumen develops and the lamb or kid gradually becomes a full-fledged ruminant.

CLASSIFICATION OF NUTRIENTS

Sheep and goats do not utilize feeds as such. Rather, they use those portions of feeds called *nutrients* that are released by digestion, then absorbed into the body fluids and tissues.

Nutrients are those substances, usually obtained from feeds, which can be used by the animal when made available in a suitable form to its cells, organs, and tissues. They include carbohydrates, fats, proteins, minerals, vitamins, and water. (More specifically, the term *nutrients* refers to the more than 40 nutrient chemicals, including amino acids, minerals, and vitamins.) Energy is frequently listed with nutrients, since it results from the metabolism of carbohydrates, proteins, and fats in the body.

Knowledge of the basic functions of the nutrients in the animal body, and of the interrelationships between various nutrients and other metabolites within the cells of the animal, is necessary before one can make practical scientific use of the principles of nutrition.

FUNCTIONS OF NUTRIENTS

Of the feed consumed, a portion is digested and absorbed for use by the animal. The remaining undigested portion is excreted and constitutes the major portion of the feces. Nutrients from the digested feed are used for a number of different body processes, the exact usage varying with the class, age, and productivity of the animal. All animals use a portion of their absorbed nutrients to carry on essential functions, such as body metabolism and maintenance of body temperature and the replacement and repair of body cells and tissues. These uses of nutrients are referred to as *maintenance*. That portion of digested feed used for growth, fattening, or the production of milk, and wool and mohair, is known as *production requirements*. Another portion of the nutrients is used for the development of the fetus and is referred to as *reproduction requirements*.

Based on the quantity of nutrients needed daily for different purposes, nutrient demands may be classed as high, low, variable, or intermediate.

Requirements for milk production are considered *high-demand uses*, whereas wool is a *low-demand use*. The last stages of pregnancy have *variable requirements*. Growth and fattening may be classed as intermediate in nutrient demands. Each of these needs will be discussed in more detail.

MAINTENANCE

Sheep and goats, unlike machines, are never idle. They use nutrients to keep their bodies functioning every hour of every day, even when they are not being used for production.

Maintenance requirements may be defined as the combination of nutrients which are needed by the animal to keep its body functioning without any gain or loss in body weight or any productive activity. Although these requirements are relatively simple, they are essential for life itself. Mature sheep and goats must have (1) heat to maintain body temperature, (2) sufficient energy to keep vital body processes functional, (3) energy for minimal movement, and (4) the necessary nutrients to repair damaged cells and tissues and to replace those which have become nonfunctional. Thus, energy is the primary nutritive need for maintenance. Even though the quantity of other nutrients required for maintenance is relatively small, it is necessary to have a balance of the essential proteins, minerals, and vitamins.

No matter how quietly a sheep or goat may be lying in a pen or pasture, it requires a certain amount of fuel and other nutrients. The least amount on which it can exist is called its *basal maintenance requirement.* With the exception of horses, most animals require about 9% more fuel (calories) when standing than when lying, and even more when they walk or run. This explains why it is desirable, for economic reasons, that finishing animals eat, then lie down as much as possible.

There are only a few times in the normal life of a sheep or goat when only the maintenance requirement needs to be met. Such a status is closely approached by mature males not in service; and by mature, dry, nonpregnant females. Nevertheless, maintenance is the standard bench mark or reference point for evaluating nutritional needs.

Even though maintenance requirement might be considered an expression of the nonproduction needs of an animal, there are many factors which affect the amount of nutrients necessary for this vital function; among them, (1) exercise, (2) weather, (3) stress, (4) health, (5) body size, (6) temperament, (7) individual variation, (8) level of production, and (9) lactation. The first four are *external factors*. They are subject to control to some degree through management and facilities. The others are *internal factors*. They are part of the animal itself. Both external and internal factors influence requirements according to their intensity. For example, the colder or hotter it gets from the most comfortable (optimum) temperature, the greater will be the maintenance requirements.

GROWTH

Growth may be defined as the increase in size of bones, muscles, internal organs, and other parts of the body. It is the normal process before birth and after birth until the animal reaches its full mature size. Growth is influenced primarily by nutrient intake. The nutritive requirements become increasingly acute when young animals are under forced production, such as when ewes are bred to lamb as yearlings.

Growth is the very foundation of sheep and goat production. Young sheep will not make the most economical finishing gains unless they have been raised to be thrifty and vigorous. Likewise, breeding ewes or does may have their reproductive ability seriously impaired if they have been improperly grown.

Knowledge of normal growth and development is useful for a variety of purposes. From a nutritional standpoint, growth curves, as shown in Fig. 5-2, are

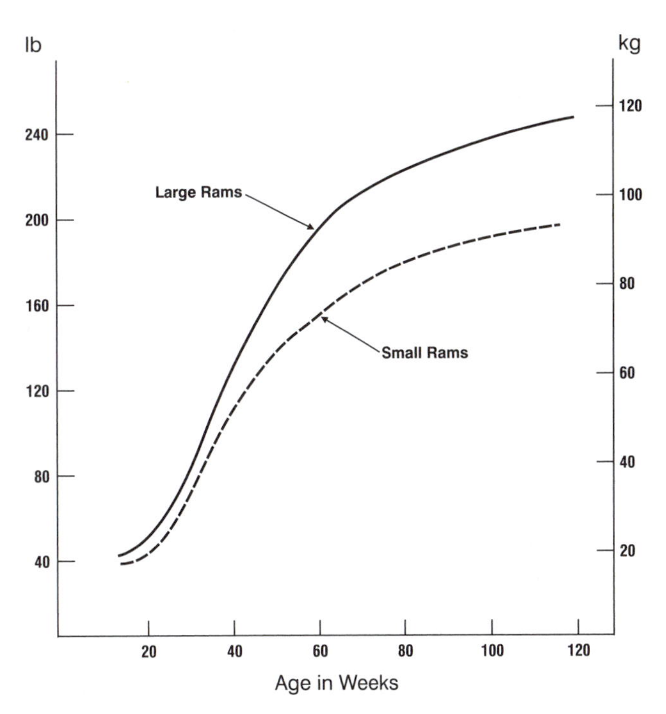

Fig. 5-2. Typical fleece-free, liveweight growth curves of rams.

used primarily as standards against which to gauge the adequacy of nutrient allowances. In fact, such curves are often the entire basis for the allowances set down in dietary and feeding standards. Also, they provide a basis for comparisons of breeding groups and serve as a reference point from which to establish breeding and management objectives. Economically, growth is important, for young gains are cheap gains. This is generally so because, in comparison with older animals, young animals (1) consume more feed according to size, (2) use a smaller proportion of their feed for maintenance, and (3) form relatively more muscle tissue which has a lower caloric value than fat.

REPRODUCTION

Being born and born alive are the first and most important requisites of sheep and goat production, for if sheep and goats fail to reproduce, the producer is soon out of business. A "mating of the gods," involving the greatest genes in the world, is of no value unless these genes result in (1) the successful joining of the sperm and egg, and (2) the birth of live offspring. Still, mortality research in the major sheep producing countries of the world shows yearly losses of lambs before weaning ranging from 10 to 35%, with an average annual loss of 15 to 20% accepted as normal. Moreover, 5% of all ewes bred fail to lamb. Certainly, there are many causes of reproductive failure, but scientists agree that nutritional inadequacies play a major role.

As with all mammalian species, most of the growth of the lamb or kid fetus occurs during the last third of pregnancy. Additionally, females must store body reserves during pregnancy, for the demands for milk production are generally greater than can be supplied by the ration fed during early lactation. Hence, the nutrient requirements are very critical during this period, especially for young, pregnant females.

Also, the ration exerts a powerful effect on sperm production and semen quality. Too fat a condition can lead to temporary or permanent sterility. Moreover, there is abundant evidence that greater fertility of herd sires exists under conditions where a well-balanced ration is provided.

LACTATION

Simply stated, milk production is a byproduct of the reproductive process.

The lactation requirements of females of all mammalian species for moderate to heavy milk production are much more rigorous than the maintenance or pregnancy requirements. For example, it is estimated that at the peak of lactation the net energy requirements of ewes suckling twins are about three times the maintenance requirements. In fact, the nutritive needs for milk production are exceeded only by sustained heavy muscular exercise—like racing a horse. Fortunately, females can store up body reserves of certain nutrients before and during pregnancy, to be drawn upon following parturition. Of course, if there has not been proper body storage, something must give—and that something will be the mother, for nature ordained that growth of the fetus, and the lactation that follows, takes priority over the maternal requirements. Hence, when there is a nutrient deficiency, the female's body will be deprived, or even stunted if she is young, before the developing fetus or milk production will be materially affected.

FINISHING (FITTING)

Finishing is what the name implies—the laying on of fat, especially in the tissues of the abdominal cavity and in the connective tissues just under the skin and between the muscles. It is the normal feeding practice followed prior to slaughter, for the purpose of improving the flavor, tenderness, and quality of meat, better to meet consumer demands. Generally, the higher the degree of finish, the higher the dressing percentage and the lower the protein (red meat) content. Also, it takes more nutrients to produce a pound of fat than a pound of lean; hence, excess finish is wasteful and undesirable.

Fattening is usually achieved through the use of high-energy feeds, feeds high in carbohydrates and fats—a liberal allowance of grains. However, due to world food shortages, the long-time trend is to incorporate more roughages in finishing rations for ruminants. Such rations are lower in net energy and produce smaller gains than high-concentrate rations, but they may make for more net returns when feed grains are scarce and high in price.

The objective of livestock producers is to finish animals to the degree of fleshing and carcass weight desired by consumers, at a maximum of profit for their efforts.

Fitting is the conditioning of animals, usually for show or sale, through careful feeding, grooming, and exercising, to enhance their bloom and attractiveness. Fitting animals for show or sale involves the application of similar principles and practices to those followed in fattening (finishing) livestock for market. Animals intended for show or sale should be fed so as to achieve a certain amount of finish or bloom, but they should not be too fat. In general, most fitting rations are similar to the rations used in commercial fattening operations for animals of like species and comparable ages, except that they are usually higher in protein content; experienced herders feel that they get more bloom by using high-protein rations.

WOOL AND MOHAIR

Wool and mohair are high-protein products. They are especially rich in the sulfur-containing amino acid cystine. This requirement is usually amply met by the cystine of feeds or by methionine—another amino acid which is also rather widely distributed in feeds, as well as derived from rumen synthesis.

NUTRIENTS

Nutrients are utilized in one of two metabolic processes: (1) anabolism, or (2) catabolism. *Anabolism is the process by which nutrient molecules are used as building blocks for the synthesis of complex molecules.* Anabolic reactions are endergonic. That is, they require the input of energy into the system. *Catabolism is the oxidation of nutrient, liberating energy (exergonic reaction) which is used to fulfill the body's immediate demands.*

In the sections that follow, nutrients are grouped into one of five categories: energy, proteins, minerals, vitamins, or water.

ENERGY

Energy is required for practically all life processes—the action of the heart, maintenance of blood pressure and muscle tone, transmission of nerve impulses, ion transport across membranes, reabsorption in the kidneys, synthesis of protein and fat, secretion of milk, and production of wool and mohair.

Lack of energy—hunger—is probably the most common nutritional deficiency of sheep and goats. It may result from lack of feed or from the consumption of poor-quality feed.

Inadequate amounts of feed may result from overgrazing, drought, snow covering the feed, or from a low dry matter content of lush, washy feeds. Also, poorly digested low-quality forage leads to reduced feed intake.

The energy needs of sheep and goats are largely met through the consumption and digestion of forages—pasture, hay and silage. Grains, such as corn, barley, milo, wheat, and oats, are used to raise the energy level of the ration during periods when supplementation is necessary. In general, sheep and goats subsist on an even higher proportion of forages to concentrates than do beef cattle, and this applies to finishing lambs. The bacterial action in the paunch converts forages into suitable sources of energy.

In addition to size, age, pregnancy, lactation, and growth, and their relationship to nutrients such as protein, which must be supplied in adequate amounts, the following factors can affect energy requirements:

1. **Environment.** The energy requirements increase as temperature, humidity, and wind depart from the comfort zone.
2. **Shearing.** The energy requirements may increase at shearing in cold weather due to decreased insulation.
3. **Stress.** Stress of any kind increases energy requirements.
4. **Last six weeks of gestation.** Ewes and does need more energy during the last six weeks of gestation to meet increased requirements for fetal growth and the development of the potential for high milk production. Too much energy during gestation may lead to fattening and birth difficulties; too little energy may result in low birth weights, weak young, and pregnancy disease in ewes.
5. **Lactation.** The lactation requirements are higher than the maintenance or gestation requirements. It is estimated that at the peak of lactation the net energy requirements of ewes suckling twins are 1.7 to 1.9 greater than their maintenance requirements. Also, ewes nursing twin lambs produce 20 to 40% more milk than ewes nursing singles, and milk production during the third and fourth months of lactation is approximately one-half of the production during the first two months.
6. **Lambs for breeding.** The energy requirements of lambs are affected by the following:
    a. **Mature size of breed.** Larger breeds grow more rapidly than smaller breeds and have a higher energy requirement.
    b. **Sex.** Ram lambs gain more rapidly and have higher feed requirements than ewe lambs. Also, intact (uncastrated) male lambs use feed more efficiently for body weight gains than ewe lambs, because of the higher protein and water and lower fat content of the increased body weight.
    c. **Ewe lambs bred to lamb as yearlings.** Ewe lambs bred to lamb as yearlings should be fed more liberally throughout—prior to breeding, during pregnancy, and during lactation.
7. **Finishing lambs.** Care should be exercised in starting finishing lambs on high-energy rations. They should be shifted gradually from roughage rations to more concentrated rations in order to avoid digestive upsets.
8. **Early weaned lambs and kids.** Early weaned lambs and kids (weaned at 5 to 8 weeks of age) lack the rumen development and capacity to utilize bulky

feeds. Hence, they should be fed palatable, high-energy, adequate protein rations.

9. **Controlling energy intake.** Limiting the amount of feed offered, adding fiber or bulk to the ration, feeding every other day, or limiting the time of eating can control energy intake.

■ **Symptoms of energy deficiency**—An energy deficiency is characterized by slowing and cessation of growth, loss of weight, reduced fertility or reproductive failure, lowered milk production and shortened lactation period, reduced quantity and quality of wool or mohair (including breaks in the fiber), lowered resistance to infection with internal parasites, and increased mortality.

A ration must contain carbohydrates, fats, and proteins. Although each of these has specific functions in maintaining a normal body, all of them can be used to provide energy. From the standpoint of supplying the normal energy needs of animals, however, carbohydrates are by far the most important, more of them being consumed than any other compound. Fats are next in importance for energy purposes. Carbohydrates are usually more abundant and cheaper, and most of them are very easily digested, absorbed, and transformed into body fat. Also, carbohydrate feeds may be more easily stored than fats in warm weather and for longer periods of time.

Energy for goats is further discussed in Chapter 21, Feeding Goats.

CARBOHYDRATES

Carbohydrates are organic compounds composed of carbon, hydrogen, and oxygen. This group includes sugars, starches, cellulose, gums, and related substances. They are formed in the plant by photosynthesis as follows: $6CO_2 + 6H_2O$ + energy from sun = $C_6H_{12}O_6$ (glucose) + $6O_2$. On the average, the carbohydrates comprise about three-fourths of all the dry matter in plants, the chief source of animal feed. They form the woody framework of plants as well as the chief reserve food stored in seeds, roots, and tubers. When consumed by animals, carbohydrates are used as a source of heat and energy; and any excess of them is stored in the body as fat.

No appreciable amount of carbohydrates is found in the animal body at any one time, the blood supply being held rather constant at about 0.05 to 0.1% for most animals. However, this small quantity of glucose in the blood, which is constantly replenished when the glycogen of the liver is changed back to glucose, serves as the chief source of fuel with which to maintain the body temperature and to furnish the energy needed for all body processes. The storage of glycogen (so-called animal starch) in the liver amounts to 3 to 7% of the weight of that organ.

FATS

Lipids (fat and fatlike substances), like carbohydrates, contain the three elements—carbon, hydrogen, and oxygen. As livestock feeds, fats function much like carbohydrates in that they serve as a source of heat and energy and for the formation of fat. Because of the larger proportion of carbon and hydrogen, however, fats liberate more heat than carbohydrates when digested, furnishing on oxidation approximately 2.25 times as much heat or energy per pound as do the carbohydrates. A smaller quantity of fat is required, therefore, to serve the same function.

A small amount of fat in the ration is desirable, as fats are the carriers of the fat-soluble vitamins A, D, and E.

MEASURING AND EXPRESSING ENERGY VALUE OF FEEDSTUFFS

One nutrient cannot be considered as more important than another, because all nutrients must be present in adequate amounts if efficient production is to be maintained. Yet, historically, feedstuffs have been compared or evaluated primarily on their ability to supply energy to animals. This is understandable because (1) energy is required in larger amounts than any other nutrient, (2) energy is most often the limiting factor in livestock production, and (3) energy is the major cost associated with feeding animals.

Our understanding of energy metabolism has increased through the years. With this added knowledge, changes have come in both the methods used to measure, and the terms used to express, the energy value of feeds.

Some pertinent energy definitions and energy-term conversions follow:

■ **Calorie (cal)**—The amount of energy as heat required to raise the temperature of 1 g of water 1°C (precisely from 14.5°C to 15.5°C). It is equivalent to 4.184 joules.

■ **Kilocalorie (kcal)**—The amount of energy as heat required to raise the temperature of 1 kg of water 1°C (from 14.5°C to 15.5°C). Equivalent to 1,000 calories.

■ **Megacalorie (Mcal)**—Equivalent to 1,000 kilocalories or 1,000,000 calories. Also, referred to as a *therm*, but the term *megacalorie* is preferred.

■ **British thermal unit (Btu)**—The amount of energy as heat required to raise 1 lb of water 1°F; equivalent to 252 calories. This term is seldom used in animal nutrition.

■ **Joule**—A proposed international unit (4.184J = 1 calorie) for expressing mechanical, chemical, or electrical energy, as well as heat.

■ **Total digestible nutrients (TDN)**—This method of measuring energy expresses the sum of the digestible protein, fiber, nitrogen-free extract, and fat × 2.25. One lb of TDN = 2 Mcal or 2,000 kcal. However, the roughage component in a ration affects its energy value. Thus, when converting all-roughage rations from TDN to calories, some scientists figure that 1 lb of TDN = 1,500 kcal, instead of 2,000.

■ **Hay equivalent (HE)**—This is the energy equivalent of 1 ton of hay which, on the average, contains 800 Mcal of net energy. With an Animal Unit Month (AUM) being equivalent to 320 Mcal of net energy, 2.5 AUM are required to furnish the same amount of energy as 1 ton of hay.

Through various digestive and metabolic processes, much of the energy in feed is dissipated as it passes through the sheep's or goat's digestive system. About 60% of the total combustible energy in grain and about 80% of the total combustible energy in roughage is lost as feces, urine, gases, and heat. These losses are illustrated in Fig. 5-3.

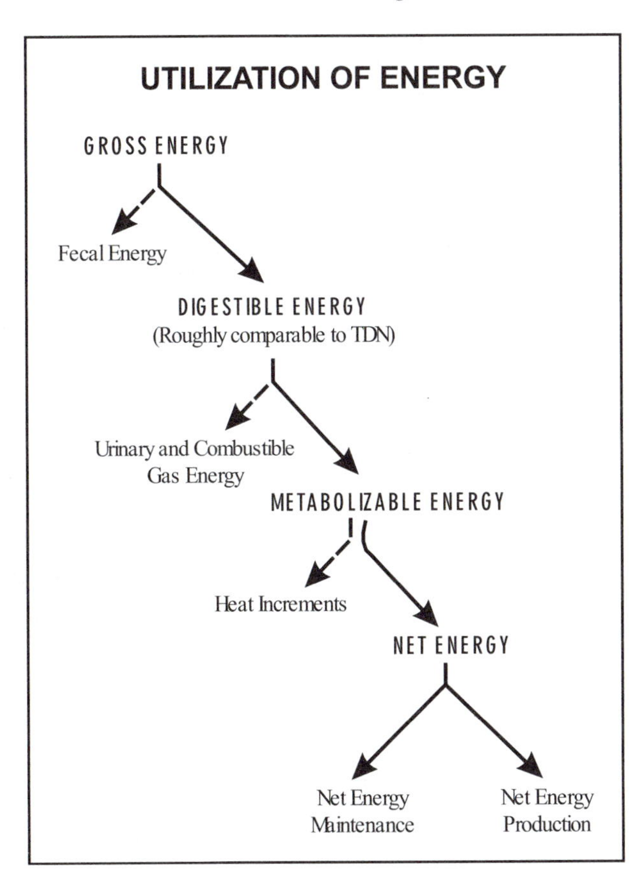

Fig. 5-3. Utilization of energy. Digestible energy is roughly comparable to total digestible nutrients.

As shown in Fig. 5-3, energy losses occur in the digestion and metabolism of feed. Measures that are used to express animal requirements and the energy content of feeds differ primarily in the digestive and metabolic losses that are included in their determination. Thus, the following terms are used to express the energy value of feeds:

■ **Gross energy (GE)**—Gross energy represents the total combustible energy in a feedstuff. It does not differ greatly between feeds, except for those high in fat.

■ **Digestible energy (DE)**—Digestible energy is that portion of the GE in a feed that is not excreted in the feces.

■ **Metabolizable energy (ME)**—Metabolizable energy represents that portion of the GE that is not lost as feces, urine, and gases. Although ME more accurately describes the useful energy in the feed than does GE or DE, it does not take into account the energy lost as heat.

■ **Net energy (NE)**—Net energy represents the energy fraction in a feed that is left after the fecal, urinary, gas, and heat losses are deducted from the GE. The net energy, because of its greater accuracy, is being used increasingly in ration formulations, especially in computerized formulations for large operations.

Although net energy is a more precise measure of the real value of the feed than other energy values. it is much more difficult to determine.

PROTEINS

Proteins are complex organic compounds made up chiefly of amino acids, which are present in characteristic proportions for each specific protein. This nutrient always contains carbon, hydrogen, oxygen, and nitrogen; and, in addition, it usually contains sulfur and frequently phosphorus. Proteins are essential in all plant and animal life as components of the active protoplasm of each living cell.

Crude protein refers to all the nitrogenous compounds in a feed. It is determined by finding the nitrogen content and multiplying the result by 6.25. The nitrogen content of protein averages about 16% (100 ÷ 6 = 6.25).

In plants, the protein is largely concentrated in the actively growing portions, especially the leaves and seeds. Using energy from the sun, plants also have the ability to synthesize their own proteins from such relatively simple soil and air compounds as carbon dioxide, water, nitrates, and sulfates. Thus, plants, together with some bacteria which are able to synthesize these products, are the original sources of all proteins.

In animals, proteins are much more widely distributed than in plants. Thus, the proteins of the animal body are primary constituents of many structural and protective tissues—such as bones, ligaments, wool and mohair, hoofs, skin, and the soft tissues which include the organs and muscles.

Sheep and goats need protein, as do other classes of animals, for maintenance, growth, reproduction, and finishing. Additionally, sheep and goats need protein for the production of wool and mohair—protein products. Wool and mohair are especially rich in the sulfur-containing amino acids, cystine and methionine, which are derived from rumen synthesis. Methionine is usually the most limiting amino acid for wool production.

Green pastures and legume hays (alfalfa, clover, soybeans, lespedeza, etc.) are excellent and practical sources of proteins for sheep and goats in most areas. However, when the ranges are bleached and dry for an extended period, or when legume hays cannot be produced for winter feeding, it may be desirable to provide sheep and goats with such protein-rich supplements as soybean meal, cottonseed meal, linseed meal, peanut meal, sunflower meal, or a commercial protein supplement, at the rate of about 0.25 to 0.33 lb per head per day.

The protein requirements of sheep and goats are affected by growth, pregnancy, lactation, mature size, weight for age, body condition, rate of gain, and protein-energy ratio. The protein requirements of ewes nursing lambs and lactating does are much like those of lactating cows, though correspondingly less because of their smaller body size and lower milk production.

Additional facts pertinent to the protein requirements and utilization by sheep and goats are:

1. **Condition and rate of gain.** Ewes and does beginning pregnancy in a very thin condition have higher protein requirements than ewes in good condition. Also, the protein requirements of finishing lambs increase with rate of gain.

2. **Protein-energy ratio.** A ratio of about 20 g of digestible protein per Mcal of DE has been shown to be adequate for mature, nonlactating ewes and does. Higher levels of digestible protein are required per Mcal of DE for lactation, growth, and finishing. Also, the higher the energy concentrate of the ration of finishing lambs, the greater the protein requirement.

3. **Quality of protein.** Quality and form of protein fed to sheep and goats are more important than formerly thought, because the protein produced by ruminal synthesis does not supply all the amino acids in quality or quantity needed for maximum production. The sulfur-containing amino acid methionine is the first limiting amino acid in microbial protein for both wool and mohair growth and weight gain, followed by lysine, and threonine. But the cystine of feed can replace methionine.

4. **Nonprotein nitrogen (NPN).** Urea or other nonprotein nitrogen can be used to provide all the supplemental nitrogen that may be needed in high-energy, grain-based rations, provided the diets are properly formulated and fed continuously. Among the factors which should be observed for the optimum utilization of urea are:

a. Provide a readily available energy source, such as molasses or grain.

b. Supply adequate and balanced levels of minerals and other nutrients.

c. Achieve a nitrogen-sulfur ratio not wider than 10:1.

d. Where the addition of nonprotein nitrogen to high-roughage rations is planned, either (1) provide a supplement containing a readily available source of energy (molasses and/or grain) and fortified with minerals or vitamins, fed at frequent intervals, or (2) use a slow-released nonprotein nitrogen product, such as Biuret or Starea.

e. Accustom animals gradually to nonprotein nitrogen-containing feeds, which may take as long as 3 to 5 weeks before maximum use of the nitrogen is obtained.

f. Limit urea to not more than 1.0% of the dry matter in the ration (or 33% of the total nitrogen in the ration, or 3% of the concentrate portion of the ration).

g. Reduce the hazard of nonprotein nitrogen toxicity by (1) preventing animals from consuming large amounts of NPN from an empty start and in a short time, (2) gradually accustoming animals to it, and feeding regularly, (3) providing available energy in the ration, and (4) using a slow-release product.

■ **Symptoms of protein deficiency**—A protein deficiency is characterized by reduced appetite, lowered feed intake, and poor feed efficiency. In turn, this makes for poor growth, poor muscular development, weight loss, reduced reproductive efficiency, lowered milk production, and reduced wool or mohair production. Under extreme conditions, there are severe digestive disturbances, nutritional anemia, and edema.

MINERALS

Minerals are inorganic elements, frequently found as salts with either inorganic elements or organic compounds.

At least 16 minerals are known to be essential for sheep and goats; these are: calcium, chlorine, cobalt, copper, fluorine, iodine, iron, magnesium, manganese, molybdenum, phosphorus, potassium, selenium, sodium, sulfur, and zinc. Other minerals that are essential for other species may eventually prove to be essential for sheep and goats.

Those minerals that are needed in relatively large amounts are referred to as *macrominerals*. Others that are needed in very small amounts, often only identified when highly purified experimental diets are fed, are referred to as *trace minerals*, or *microminerals*. These terms do not imply any lesser role for the trace minerals; rather, they represent quantity designations based on the amounts needed by animals. According to this reasoning, the minerals for sheep and goats can be divided into the two groups presented in Tables 5-2 and 5-3.

The general functions of these minerals are as follows:

1. Give rigidity and strength to the skeletal structure.
2. Serve as constituents of the organic compounds, such as protein and lipid, which make up the muscles, organs, blood cells, and other soft tissues of the body.
3. Activate enzyme systems.

TABLE 5-2
MACROMINERAL REQUIREMENTS OF SHEEP (PERCENTAGE OF RATION)[1]

Nutrient	Requirement	
	As-fed[2]	Moisture-free
	(%)	(%)
Sodium	0.08–0.16	0.09–0.18
Chlorine	—	—
Calcium	0.18–0.74	0.20–0.82
Phosphorus	0.14–0.34	0.16–0.38
Magnesium	0.11–0.16	0.12–0.18
Potassium	0.45–0.72	0.50–0.80
Sulfur	0.13–0.23	0.14–0.26

[1]Adapted by the author from *Nutrient Requirements of Sheep*, sixth revised edition, NRC-National Academy of Sciences.

[2]As-fed was calculated using 90% dry matter (moisture-free).

TABLE 5-3
MICROMINERAL REQUIREMENTS OF SHEEP AND MAXIMUM TOLERABLE LEVELS (PPM OR MG/KG OF RATION)[1]

Nutrient	Requirement		Maximum Tolerable Level	
	As-fed[2]	Moisture-free	As-fed	Moisture-free
	(ppm or mg/kg)	*(ppm or mg/kg)*	*(ppm or mg/kg)*	*(ppm or mg/kg)*
Cobalt	0.09–0.18	0.1–0.2	9	10
Copper	6–10	7–11[3]	23	25[4]
Fluorine	—	—	54–135	60–150
Iodine	0.09–0.72	0.10–0.80[5]	45	50
Iron	27–45	30–50	450	500
Manganese	18–36	20–40	900	1,000
Molybdenum	0.45	0.5	9	10[4]
Selenium	0.09–0.18	0.1–0.2	1.8	2
Zinc	18–30	20–33	675	750

[1]Adapted by the author from *Nutrient Requirements of Sheep*, sixth revised edition, NRC-National Academy of Sciences.

[2]As-fed was calculated using 90% dry matter (moisture-free).

[3]Requirement when dietary Mo concentrations are <1 mg/kg DM.

[4]Lower levels may be toxic under some circumstances.

[5]High level for pregnancy and lactation in rations not containing goitrogens; should be increased if rations contain goitrogens.

4. Control fluid balance—osmotic pressure and excretion.
5. Regulate acid-base balance.
6. Exert characteristic effects on the irritability of muscles and nerves.
7. Interact with certain vitamins.

In addition to the general functions in which several minerals may be involved, each essential mineral has one or more specific roles.

SHEEP AND GOAT MINERAL CHARTS

Table 5-4, Sheep Mineral Chart, gives, in summary form, the following pertinent information relative to each mineral listed: (1) conditions usually prevailing where deficiencies are reported, (2) function, (3) deficiency symptoms, (4) nutrient requirements, (5) recommended allowances, and (6) practical sources, while Table 5-5, Goat Mineral Chart, provides similar information on minerals for goats. Also, Table 5-4 groups minerals as (1) major or macrominerals, and (2) trace or microminerals. Further elucidation on certain minerals is contained in the accompanying narrative. Fluorine is discussed because of its toxicity to sheep.

TABLE
SHEEP MINERAL

Minerals Which May Be Deficient Under Normal Conditions	Conditions Usually Prevailing Where Deficiencies Are Reported	Function of Mineral	Deficiency Symptoms; Toxicity
Major or Macrominerals:			
Salt (sodium and chlorine—NaCl)	Negligence; for salt is inexpensive.	Sodium and chlorine are known to have regulatory functions in the body. They maintain osmotic pressure in cells, regulate the acid-base balance, and control water metabolism in tissues.	**Deficiency symptoms**—A deficiency of salt may result in an abnormal appetite, with the sheep trying to satisfy their craving by licking dirt, or eating toxic amounts of poisonous plants; decreased feed consumption; and decreased efficiency in the utilization of nutrients. ***Toxicity**—The maximum tolerable level of salt is 9.0% of the ration.
Calcium (Ca)	Lack of vitamin D. When finishing lambs are fed heavily on concentrates and limited quantities of legume roughage. When the feed consists largely of dried mature grasses or corn silage. Calcium-deficient areas (where pasture and range forages are deficient in Ca) are Florida, Louisiana, Nebraska, and Virginia. Chronic internal parasite infections. Where there is magnesium deficiency.	Essential for development and maintenance of normal bones and teeth. Important in blood coagulation and lactation. Enables heart, nerves, and muscles to function. Regulates permeability of tissue cells. Affects availability of phosphorous and zinc.	**Deficiency symptoms**—Subnormal development of bone; rickets in young animals, and osteomalacia in adults. A high incidence of urinary calculi when there is a low calcium:high phosphorus ratio. To lessen the incidence of urinary calculi, the Ca:P ratio should be about 2:1. ***Toxicity**—If there is adequate phosphorus, sheep can tolerate a calcium-to-phosphorus ratio of 7:1 and as much as 2% calcium in the ration.
Phosphorus (P)	Lack of vitamin D. When sheep subsist for long periods on mature forages (such as dry range or grass or cereal hays). When the ration consists of a high proportion of beet byproducts. When sheep subsist on pastures in phosphorus-deficient areas. Chronic internal parasite infections.	Essential for sound bones and teeth, and for the assimilation of carbohydrates and fats. A vital ingredient of the proteins in all body cells. Necessary for enzyme activation. Acts as a buffer in blood and tissue. Occupies a key position in biologic oxidation and reactions requiring energy.	**Deficiency symptoms**—Slow growth, depraved appetite, unthrifty appearance, listlessness, low level of phosphorus in the blood (less than 4 mg/100 ml of plasma), and development of knock-knees. *Caution:* A high level of phosphorus in the blood is not always an indication of adequacy in the diet; it may result from loss of weight. ***Toxicity**—Phosphorus at levels of 2 to 3 times the requirement can cause increased bone resorption in mature sheep.
Magnesium (Mg)	Tetany most frequently occurs in nursing ewes shortly after they are turned to pasture in the spring (grass tetany), when the magnesium requirements for lactation are high and grass is low in magnesium.	It is a constituent of bone. Also, it is necessary for many enzyme systems and for proper functioning of the nervous system. Closely associated with the metabolism of calcium and phosphorus.	**Deficiency symptoms**—Hypomagnesmic tetany, a hyperirritability of the neuromuscular system. Sometimes this condition is accompanied by hypocalcemia. Acute tetany may occur as a result of insufficient dietary magnesium or inability to mobilize skeletal magnesium. ***Toxicity**—Oral administration of 0.8% magnesium in the ration will produce toxicosis.

5-4
CHART

Mineral Requirements[1]				
Minerals/ Animal/Day	**Mineral Content of Ration, in % or ppm**	**Recommended Allowances[1]**	**Practical Sources of the Mineral**	**Comments**
*Lambs in drylot consume about 9 g of salt daily. Mature sheep in drylot may consume more.	*As-fed*[2] *M-F* *Salt for growing lambs, %: 0.38 0.42 *Na requirement of sheep, %: 0.08–0.16 0.09–0.18 (See Table 5-2.)	*Salt for mature sheep: 0.5% of the complete feed, or 1.0% to the concentrate portion. *Range operators commonly provide 0.5–0.75 lb *(0.25–0.33 kg)* salt/ewe/month. Mature sheep in drylot may consume more.	Free access to salt. Loose salt, rather than block salt, should be provided, for the reason that sheep bite at salt blocks, rather than lick, with the result that their teeth may be broken. In iodine-deficient areas, stabilized iodized salt should always be provided.	Sheep consume about 5 times more salt/100 lb body weight than cattle, which is attributed to their high forage consumption. Sheep can consume high quantities of salt without apparent harm provided water is freely available. In alkaline areas, the water may contain enough salt to meet the requirements, and supplemental salt may not be needed.
Variable, according to class, age, and weight of sheep (see Table 14-1).	*0.18–0.74% *0.20–0.82% (See Tables 14-2 and 5-2.)	Self-feed suitable mineral, or add calcium to the ration as required to bring level of total ration slightly above requirements.	Ground limestone, or oystershell flour. Where both calcium and phosphorus are needed, use bone meal, dicalcium phosphate, or defluorinated phosphate.	Most pasture and range forage contains adequate amounts of calcium. Forage containing from 0.24–0.32% calcium is considered adequate. Calcium requirements are usually met when sheep receive at least one-third of a legume forage. *Blood calcium levels below 9 mg/100 ml of plasma suggest chronic low calcium intake.
Variable, according to class, age, and weight of sheep (see Table 14-1).	*0.14–0.34% *0.16–0.38% (See Tables 14-2 and 5-2.)	Self-feed suitable mineral, or add phosphorus to the ration as required to bring level of total ration slightly above requirements.	Monosodium phosphate or diammonium phosphate. Where both calcium and phosphorus are needed, use bone meal, dicalcium phosphate, or defluorinated phosphate.	The proper calcium-phosphorus ratio should be maintained. Forage containing below 0.16% phosphorus is usually considered deficient for ewes during gestation, and 0.20% borderline during lactation. *A phosphorus deficiency may be manifested when the blood phosphorus level falls below 4 mg/100 ml of plasma.
	*0.11, 0.14, and 0.16% for growing lambs, ewes in late pregnancy, and ewes in early lactation, respectively. | *0.12, 0.15, and 0.18% for growing lambs, ewes in late pregnancy, and ewes in early lactation, respectively. *Where ewes in early lactation are grazing forage with high nitrogen and potassium content, the minimum level of magnesium in the ration is 0.2%.		Plant protein supplements are excellent sources of Mg. Likewise, by-product feedstuffs derived from plants tend to be good sources. The common magnesium supplements are magnesium carbonate, magnesium oxide, and magnesium sulfate.	*Blood serum normally contains about 2.5 mg/100 ml.

(Continued)

TABLE 5-4

Minerals Which May Be Deficient Under Normal Conditions	Conditions Usually Prevailing Where Deficiencies Are Reported	Function of Mineral	Deficiency Symptoms; Toxicity
Major or Macrominerals (Continued):			
Potassium (K)	When finishing lambs are fed high-concentrate and urea rations and limited amounts of dry roughage. When sheep are grazing mature range forage during winter or drought periods. The potassium level of such forage may decrease to less than 0.2%.	It affects osmotic pressure and acid-base balance within the cell. It also aids in activating several enzyme systems involved in energy transfer and utilization, protein synthesis, and carbohydrate metabolism.	**Deficiency symptoms**—Poor appetite and feed conversion, progressive stiffness from front to rear, and dry wool. ***Toxicity**—The maximum tolerable level of potassium for sheep is about 3% of the ration DM.
Sulfur (S)	When finishing lambs are fed high-concentrate and urea rations and limited amounts of roughage.	Functions in synthesis of sulfur-containing amino acids (methionine and cystine) in the rumen and various compounds of the body. Wool is high in sulfur; hence, sulfur is closely related to wool production.	**Deficiency symptoms**—Loss of appetite, reduced weight gains and feed efficiency, and reduced wool growth. Also, excessive salivation, lacrimation, and shedding of wool. ***Toxicity**—It appears that 0.4% is the maximum tolerable level of dietary sulfur as sodium sulfate.
Trace or Microminerals:			
Cobalt (Co)	Cobalt-deficient areas or soils in the U.S. and Canada. The most severely deficient U.S. areas include portions of New England and the lower Atlantic Coastal Plain. Moderately deficient areas include the rest of New England, northern New York, northern Michigan, and parts of the Central Plains.	Promote synthesis of vitamin B-12 in the rumen.	**Deficiency symptoms**—Cobalt deficiency signs are actually signs of vitamin B-12 deficiency. They are: lack of appetite, lack of thrift, severe emaciation, weakness, anemia, decreased fertility, and decreased milk and wool production. ***Toxicity**—Approximately 204.5 mg/100 lb live weight.
Copper (Cu)	In copper-deficient areas (soils), as in Florida and in the coastal plains region of the Southeast. Also, in several of the western states, there are areas where an excess of molybdenum induces copper deficiency.	Anemia is associated with copper deficiency. Animals suffering from inadequate copper intake appear unable to absorb iron at a normal rate, and a deficiency in hemoglobin synthesis results. Steely wool and depigmentation of black sheep.	**Deficiency symptoms**—Signs in suckling lambs include "swayback," muscular incoordination, partial paralysis of the hindquarters, and degeneration of the myelin sheath of the nerve fibers. Lambs may be born weak and may die because of their inability to nurse. Sheep suffering from a copper deficiency may produce "steely" or "stringy" wool, which is lacking in crimp, tensile strength, affinity for dyes, and elasticity. Depigmentation of the wool of black sheep has been noted as a sign of severe deficiency. ***Toxicity**—23 ppm As-fed or 25 ppm M-F (see Table 5-3), but Mo level of the ration is a factor.
Fluorine (F)	*Conditions which may result in flourine toxicity:* High fluorine in the water supply. Use of rock phosphate that contains 3–4% fluorine.		Fluorine deficiency not reported. Rather, the hazard is fluorine toxicity. ***Toxicity**—Acute toxicity can occur at 200 ppm.

(Continued)

Mineral Requirements[1]					
Minerals/ Animal/Day	Mineral Content of Ration, in % or ppm		Recommended Allowances[1]	Practical Sources of the Mineral	Comments
	As-fed[2]	*M-F*			
	*0.45% for growth of lambs. 0.63–0.72 % for lactation and stress.	*0.5% for growth of lambs. 0.7–0.8% for lactation and stress.	0.7 to 1.0% of total air-dry ration.	Roughages usually contain adequate potassium, with the possible exception of nonlegume silage. Potassium chloride and potassium sulfate are the supplements of choice.	The feeding of potassium chloride appears to reduce the incidence of urinary calculi in feedlot lambs. This is especially true with high-milo rations.
	*Mature ewes: 0.13–0.16% 0.14–0.18% *Young lambs: 0.16–0.23% 0.18–0.26%		*It is recommended that a dietary nitrogen-sulfur ratio of 10:1 be maintained.	Sulfate sulfur, elemental sulfur, or sulfur-containing proteins or amino acids. Inorganic compounds are generally more convenient and economical for supplemental feeding.	*Practically all common feedstuffs contain more than 0.1% sulfur. However, mature grass and grass hays are sometimes low in sulfur. Where forages are low in sulfur or high in urea, increased weight gains and wool growth can be obtained by feeding sulfur.
	*0.09–0.18 ppm However, young, rapidly growing lambs may have a slightly higher requirement.	*0.1–0.2 ppm	*Feed cobalt at the rate of 1.4 g/100 lb *(2.5 g/100 kg)* of salt as cobalt chloride or cobalt sulfate.	A cobalt mineral mixture. Other effective methods of providing cobalt are (1) to add cobalt to the soil, or (2) to place cobalt pellets into the rumen.	Several good commercial cobalt-containing minerals are on the market in either block or loose form. Cobalt is much more effective when given by mouth than when given intravenously.
	*6.3–20.7 ppm The Cu requirement varies with (1) the Mo content of the feed, and (2) the growth, pregnancy, lactation, and breed involved. (See the narrative section on "Copper" for details. Also, see Table 5-3.)	*7–23 ppm	*Add copper sulfate to the salt at rate of 0.5%.	Salt containing 0.5% of copper sulfate.	Copper deficiencies may exist alone or along with deficiencies of cobalt and iron. An interesting interrelation exists between copper, molybdenum and sulfur. An excess of molybdenum causes a pathological condition which can be cured only administering copper. Stores of copper in the liver, kidney, heart, lungs, pancreas, and spleen serve as a reserve for as long as 4 to 6 months when animals are grazing copper-deficient forage. Sheep are much more susceptible to copper toxicity than cattle. As much as 25 mg of copper in the daily ration of sheep is considered toxic; and about 9 mg/day is considered the safe tolerance level. Copper toxicity may result from feeding poultry wastes or mineral supplements designed for other species.
	*Breeding sheep should not be fed a ration containing more than 55 ppm (As-fed) or 60 ppm of flourine on a moisture-free basis. *Finishing lambs can tolerate up to 135 ppm (As-fed) or 150 ppm of fluorine in the ration on a moisture-free basis.				Symptoms of fluorine toxicity are loss of appetite; the normal ivory color of bones changes to chalky white; bones thicken, and the teeth, especially the incisors, may become pitted and eroded to such an extent that the nerves are exposed.

(Continued)

TABLE 5-4

Minerals Which May Be Deficient Under Normal Conditions	Conditions Usually Prevailing Where Deficiencies Are Reported	Function of Mineral	Deficiency Symptoms; Toxicity
Trace or microminerals (Continued):			
Iodine (I)	Iodine-deficient areas or soils (in northwestern U.S. and the Great Lakes and Rocky Mountain regions) where iodized salt is not fed. Feeds from iodine-deficient areas.	Formation of thyroxin, a hormone of the thyroid gland.	**Deficiency symptoms**—Lambs born with goiter; usually stillborn or die soon after birth. Usually, such lambs have very little wool. In mature sheep an iodine deficiency may result in reduced wool yield and reduced rate of conception. ***Toxicity**—Maximum tolerable level for sheep is 45 ppm As-fed basis or 50 ppm M-F (see Table 5-3). However, much higher tolerable levels have been reported.
Iron (Fe)	Iron-deficiency anemia sometimes occurs in lambs raised on slotted floors. Loss of blood from parasite infestation can produce a secondary iron-deficiency anemia.	Hemoglobin formation.	**Deficiency symptoms**—Anemia, poor growth, lethargy, increased respiration rate, decreased resistance to infection, and in severe cases high mortality. ***Toxicity**—Signs of chronic toxicity are reduction in feed intake, growth rate, and feed efficiency. In acute toxicosis, animals exhibit loss of appetite, scanty urination, diarrhea, below normal temperature, shock, acidosis, and death.
Manganese (Mn)	Lambs on a purified diet containing less than 1 ppm of manganese over a 5-month period. High calcium and iron may increase manganese requirements.	Skeletal development and reproduction.	**Deficiency symptoms**—Bone abnormalities, lack of coordination in newborn lambs, impaired growth, and depressed or disturbed reproduction. ***Toxicity**—It appears that 1,000 ppm of dietary Mn is the maximum tolerable level.
Molybdenum (Mo)	The major concern about molybdenum is that in excess it may induce a copper deficiency. Excess molybdenum in the soil such as is found in areas of California, Nevada, and England.	It is believed that molybdenum binds and inactivates the copper in the intestine.	**Deficiency symptoms**—A low intake of molybdenum causes excess copper to accumulate in tissues, especially the liver, even when the copper intake is moderate, thus producing fatal jaundice (easily detected in the eyes). This disease can be prevented by increasing the molybdenum intake. ***Toxicity**—High levels of molybdenum (10 to 20 ppm in forage plants) will induce copper deficiency characterized by stringy wool, lack of pigmentation in black wool, anemia, bone disorders, and infertility. Also, sheep start to scour after being turned to high molybdenum pastures (5 to 20 ppm M-F basis). The scouring can be controlled by increasing the copper level in the diet to 5 ppm.
Selenium (Se)	Areas where selenium content of crops is below 0.1 ppm, such as northwestern, northeastern, and southeastern U.S. Parts of South Dakota, Wyoming, and Utah produce forage containing excess selenium which causes toxicity in farm animals.	Component of the enzyme glutathione peroxidase, the metabolic role of which is to protect against oxidation of polyunsaturated fatty acids and resultant tissue damage. Interrelation with vitamin E—they spare each other.	**Deficiency symptoms**—The most commonly noticed lesion from a deficiency of selenium is white muscle disease, which affects lambs 0 to 8 weeks of age, along with reduced growth of lambs. Additional signs of inadequate selenium are unthriftiness, infertility, early embryonic death, and periodontal disease. ***Toxicity**—Chronic selenium toxicity occurs when sheep consume feeds containing more than 3 ppm of selenium on a dry basis over a prolonged period. Toxicity signs include loss of wool, soreness and sloughing of the hooves, and marked reduction in reproductive performance.

(Continued)

Mineral Requirements[1]				
Minerals/ Animal/Day	**Mineral Content of Ration, in % or ppm**	**Recommended Allowances[1]**	**Practical Sources of the Mineral**	**Comments**
	As-fed[2] *M-F* *0.09–0.72 ppm *0.1–0.8 ppm The higher levels are indicated for pregnancy and lactation. When goitrogens are present increase the iodine.	*Free access to stabilized iodized salt containing 0.0078% iodine.	Stabilized iodized salt containing 0.0078% iodine. Calcium iodate.	Do not use iodized salt in a mixture with a concentrate to limit feed intake, as the animals may consume an excessive amount of iodine.
	*27–45 ppm *30–50 ppm	*Intramuscular injections of iron-dextran; 2 injections, 150 mg of iron in each, given 2 to 3 weeks apart.	Ferrous gluconate, ferrous succinate, or ferrous sulfate given orally. Iron dextran injection.	A primary iron deficiency in grazing sheep is very unlikely.
	*18–36 ppm *20–40 ppm	18 ppm in As-fed ration, or 20 ppm in M-F ration.	Manganese gluconate.	
	*The minimum dietary requirement of molybdenum is not known. *The Food and Drug Administration does not recognize molybdenum as safe; hence, the law prohibits adding it to feed for sheep.	The two contrasting situations—(1) high molybdenum and copper deficiency, or (2) low molybdenum and excess copper accumulation—make it very difficult to define nutrient requirements of molybdenum and copper.		A high-molybdenum intake induces a copper deficiency even when the copper content of pasture is quite high; the scouring effect can be prevented by providing an increased copper intake. Sheep are less affected than cattle by highmolybdenum intakes. *In treating copper toxicity, both molybdenum and sulfate should be administered. Drench each lamb with 100 mg of ammonium molybdate and 1 g of sodium sulfate in 20 ml of water.
	*0.09–0.18 ppm 0.1–0.2 ppm	Selenium as either sodium selenite or sodium selenate at the rate of 0.3 ppm of complete feed. Selenium added to salt-mineral mixture fed free-choice at rate of 90 ppm. Selenium in the limit feeding (feed supplements and salt-mineral mixtures) consumption rate for sheep of 0.7 mg per head per day.		Plants grown on the same seleniferous soils vary greatly in their uptake of selenium, with a range of 1,000 ppm to only 10–25 ppm. The most practical way to prevent livestock losses from selenium poisoning is to manage the grazing so that animals alternate between high-selenium and low-selenium areas. Selenium is a cumulative poison, but mild chronic signs can be overcome readily by feeding low-selenium forage.

(Continued)

TABLE 5-4

Minerals Which May Be Deficient Under Normal Conditions	Conditions Usually Prevailing Where Deficiencies Are Reported	Function of Mineral	Deficiency Symptoms; Toxicity
Trace or microminerals (Continued):			
Zinc (Zn)	Diets high in calcium adversely affecting zinc utilization.		**Deficiency symptoms**—Loss of appetite, reduction in growth rate, excessive salivation, parakeratosis, wool loss, and delayed wound healing. Ram lambs show reduced testicular development and defective spermatogenesis. In females, all phases of the reproductive process from estrus to parturition and lactation may be adversely affected. ***Toxicity**—There is a wide margin of safety between zinc requirements and zinc toxicity. However, 0.1% zinc in the diet reduced feed consumption and gain in lambs; and 0.075% induced severe copper deficiency in pregnant ewes and caused a high incidence of abortions and still births.

[1]As used herein, the distinction between "mineral requirements" and "recommended allowances" is as follows: In mineral requirements, no margins of safety are included intentionally; whereas in recommended allowances, margins of safety are provided to compensate for variations in feed composition, environment, and possible losses during storage or processing.

Where preceded by an asterisk (*), the requirements, recommended allowances, and other facts presented herein were taken from *Nutrient Requirements of Sheep*, 6th rev. ed., NRC-National Academy of Sciences.

[2]Estimated 90% dry matter.

TABLE
GOAT MINERAL

Minerals Which May Be Deficient Under Normal Conditions	Conditions Usually Prevailing Where Deficiencies Are Reported	Function of Mineral	Deficiency Symptoms; Toxicity
Major or Macrominerals:			
Salt (NaCl)	Negligence, for salt is inexpensive. Lactating does may require additional salt as milk contains high amounts of sodium.	Sodium chloride helps maintain osmotic pressure in body cells, upon which depends the transfer of nutrients to the cells, the removal of waste materials, and the maintenance of water balance among the tissues. Also, sodium is important in making bile, which aids in the digestion of fats and carbohydrates; and chlorine is required for the formation of hydrochloric acid in the gastric juice so vital to protein digestion. It is noteworthy that when salt is omitted, sodium expresses its deficiency first.	**Deficiency symptoms**—Loss of appetite, depraved appetite and consumption of soil and debris, emaciation, decline in milk production, a general rough appearance with poor coat and lusterless eyes. Acute deficiency symptoms include shivering, weakness, cardiac disturbances, and ultimately death. **Toxicity**—The maximum tolerable level of salt for sheep is 9.0%. For goats, a similar level of salt will likely be toxic.
Calcium (Ca)	Goats in heavy lactation. Lack of vitamin D. Calcium-deficient areas (where pasture and range forages are deficient in calcium) are Florida, Louisiana, Nebraska, Virginia, and West Virginia. Feeds that contain primarily cereal grains.	Essential for the development and maintenance of good strong bones and teeth; maintains the contractability, rhythm, and tonicity of the heart muscles; antagonizes the action of the sodium and potassium on the heart; is required for normal coagulation of the blood; is necessary for proper nerve irritability; and appears to be essential for selective cellular permeability.	**Deficiency symptoms**—In young kids, retarded growth and abnormal bone development. Also, a deficiency of calcium may cause rickets in young animals and osteomalacia in young adults. In lactating does, depressed milk yields and fragile bones. Milk fever can occur when calcium levels in the blood drop. **Toxicity**—If there is adequate phosphorus, sheep can tolerate a calcium:phosphorus ratio of 7:1 and as much as 2% calcium in the diet. It is postulated that goats can tolerate a similar level of calcium.

(Continued)

Mineral Requirements[1]				
Minerals/ Animal/Day	Mineral Content of Ration, in % or ppm	Recommended Allowances[1]	Practical Sources of the Mineral	Comments
	As-fed[2] *M-F* *Growth: 18 ppm 20 ppm *Reproduction: 30 ppm 33 ppm			

Note: Mineral recommendations for all classes and ages of sheep, especially those fed unmixed rations or on pasture, are—

1. *When sheep are on liberal grain feeding*—Provide free access to a two-compartment mineral box, with (a) trace mineralized salt in one side, and (b) in the other side, a mixture of one-third trace mineralized salt (salt included for purposes of palatability), one-third defluorinated phosphate or steamed bone meal, and one-third ground limestone or oystershell flour.
2. *When sheep are primarily on roughage (pasture, hay, and/or silage)*—Provide free access to a two-compartment mineral box, with (a) trace mineralized salt in one side, and (b) in the other side, a mixture of one-third trace mineralized salt and two-thirds defluorinated phosphate or steamed bone meal.

Additionally, in those areas where cobalt and/or copper deficiencies exist in the soil (and plants), and cobalt and/or copper sulfate to either the salt or salt-phosphorus mixture in the proportions indicated. If desired, the mineral supplement may be incorporated in the ration in keeping with the recommended allowances given in this table.

5-5
CHART[1]

Nutrient Requirements[2]	Recommended Allowances[2]	Practical Sources of the Mineral	Comments
	Salt should be provided free-choice or as a component of the ration. In a complete feed, 0.5 to 1.0% salt is recommended, with proportionately higher levels in supplements.	Iodized salt in iodine-deficient areas. Can be offered free-choice or incorporated into the ration. In alkaline areas, water may contain enough salt to meet the requirements.	In range areas, salt may be added to feed to limit feed intake. If self-feeders are located near water, the level of salt in the ration should be high (25–40%). If self-feeders are some distance from water, the level of salt in the ration should be reduced. In arid regions, the salt content of some water sources can reduce intake of water and feed.
Variable according to age, sex, and class (see Tables 21-1, 21-2, and 21-3).	Because milk is high in calcium, lactating does need rations with high calcium levels. In % of ration: 0.78 M-F 0.70 A-F	Ground limestone, steamed bone meal, dicalcium phosphate, and oyster shell.	The recommended ratio of calcium to phosphorus ranges from 2:1 to 4:1. If the ratio falls below 2:1, urinary calculi may develop in males. Under grazing conditions, calcium is seldom a problem with either Angora or meat-type goats.

(Continued)

TABLE 5-5

Minerals Which May Be Deficient Under Normal Conditions	Conditions Usually Prevailing Where Deficiencies Are Reported	Function of Mineral	Deficiency Symptoms; Toxicity
Major or Macrominerals (Continued):			
Phosphorus (P)	When goats subsist on pastures in phosphorus-deficient areas. When goats subsist for long periods on mature, dry forages. Lack of vitamin D.	Essential for sound bones and teeth, and for the assimilation of carbohydrate and fats. A vital ingredient of the proteins in all body cells. Necessary for enzyme activation. Acts as a buffer in blood and tissue. Occupies a key position in biologic oxidation and reactions requiring energy.	**Deficiency symptoms**—Slowed growth, depraved appetite (chewing bones, wood, hair), unthrifty appearance, rickets in young animals, osteomalacia in mature animals, and depressed milk yields in lactating does. **Toxicity**—There is no known phosphorus toxicity in goats. However, excess phosphorus consumption may decrease the absorption of calcium. Also, when phosphorus is high in relation to calcium, urinary calculi may be formed.
Magnesium (Mg)	Animals grazing lush green grass or winter cereal pastures fertilized with nitrogen and potassium.	Required for many enzyme systems and for proper functioning of the nervous system. Also, closely associated with the metabolism of calcium and phosphorus.	**Deficiency symptoms**—Loss of appetite, excitability, and calcification of soft tissues. The most noted problem associated with low magnesium is grass tetany. **Toxicity**—Magnesium toxicity of goats has not been reported under practical conditions.
Potassium (K)	When goats are grazing mature range forage during winter or drought periods. High concentrate rations.	It (1) affects osmotic pressure and acid-base balance within the cells, and (2) aids in activating several enzyme systems involved in energy transfer and utilization, protein synthesis, and carbohydrate metabolism.	**Deficiency symptoms**—Marginal deficiencies result in reduced feed intake, retarded growth, and reduced milk production. Severe deficiencies cause emaciation and poor muscular tone. **Toxicity**—The maximum tolerable level of potassium for sheep is about 3% of the ration DM. It is postulated that the toxicity level of goats is similar.
Sulfur (S)	Possibly with liberal intake of tannic acid-containing plants. This is of concern with range goats, which liberally graze and browse such plants.	Essential for synthesis of the sulfur amino acids (cystine and methionine). Sulfur is particularly high in goat hair.	**Deficiency symptoms**—Depressed appetite, loss of weight, poor growth, excessive salivation, tearing, loss of mohair, depressed milk yields. **Toxicity**—Elemental sulfur is practically devoid of toxicity.
Trace or Microminerals:			
Cobalt (Co)	In cobalt deficient areas when the cobalt level in the feed drops to 0.04 to 0.07 ppm or lower.	The only function of cobalt is that of being an integral part of vitamin B-12.	**Deficiency symptoms**—The deficiency symptoms are actually vitamin B-12 deficiencies. They are loss of appetite, emaciation, weakness, anemia, and decreased production. **Toxicity**—In sheep, about 204.5 mg cobalt/100 lb live weight is toxic. Likely, the same applies to goats.
Copper (Cu) and Molybdenum (Mo)	Copper and molybdenum are interrelated in animal metabolism; hence, they should be considered together. The most common problem occurs when a normal or low level of copper is accompanied by a high level of molybdenum, resulting in copper being excreted and producing a copper deficiency. This condition can be corrected by adding copper.	Copper and iron are mutually involved in the formation of hemoglobin—the red pigment which carries oxygen.	Few studies on copper and molybdenum have been conducted with goats. It appears that sheep are sensitive to copper toxicity and resistant to molybdenosis, but it is not known whether this is also the case with goats.

(Continued)

Nutrient Requirements	Recommended Allowances[1]	Practical Sources of the Mineral	Comments
Variable according to age, sex, and class (see Tables 21-1, 21-2, and 21-3).	Can be offered free-choice or incorporated into the ration. In % of ration: 0.45 M-F 0.40 A-F	Cereal grains. Defluorinated phosphate, dicalcium phosphate, steamed bone meal, monosodium phosphate.	Phosphorus is the mineral most likely to be deficient in range forages. It is, therefore, recommended that it be supplied in range supplements. *The calcium-to-phosphorus ratio should not drop below 1.2:1.
	In % of ration: 0.25 M-F 0.22 A-F	Plant protein supplements and plant by-product feeds are excellent sources of magnesium. The common magnesium supplements are magnesium carbonate, magnesium oxide, and magnesium sulfate.	*Goats have a marginal ability to compensate for low dietary magnesium by reducing the rate of excretion.
*In growing sheep, the potassium requirement is 0.5% of the ration. In lactating dairy cattle, the requirement is 0.8% of the complete ration. These levels are also postulated as the requirements of growing and lactating goats, respectively.	In % of ration: 1.0 M-F 0.9 A-F	Roughage-based rations. Common potassium supplements are potassium chloride, potassium bicarbonate, and potassium sulfate.	
	In % of ration: 0.20 M-F 0.18 A-F A sulfur-to-nitrogen ratio of 1:10 is recommended.	Sulfates, such as sodium sulfate and ammonium sulfate, are the most available forms of sulfur for ration formulation.	Because of mohair production, Angora goats may have an elevated sulfur requirement.
*A level of 0.1 ppm in the M-F ration.	In % of ration: 0.1 to 0.2 ppm M-F 0.09 to 0.18 ppm A-F	*Cobalt sulfate or cobalt chloride added at the rate of 5.45 g per 100 lb *(12 g per 100 kg)* of salt.	
	Add copper sulfate to the salt at the rate of 0.5% Copper in total ration: 5.0 ppm M-F 4.5 ppm A-F	Salt containing 0.5% copper sulfate.	

(Continued)

TABLE 5-5

Minerals Which May Be Deficient Under Normal Conditions	Conditions Usually Prevailing Where Deficiencies Are Reported	Function of Mineral	Deficiency Symptoms; Toxicity
Trace or Microminerals (Continued):			
Fluorine (F)		Necessary for sound bones and teeth.	**Deficiency symptoms**—Fluorine deficiency appears to be rare. Rather, the hazard is fluorine toxicity. **Toxicity**—With sheep, fluorine toxicity occurs at levels above 200 ppm. So, it is postulated that the toxicity level for goats is similar.
Iodine (I)	Iodine-deficient areas or soils (in northwestern U.S., and the Great Lakes and Rocky Mountain regions), unless iodized salt is fed.	Formation of thyroxin, a hormone of the thyroid gland.	**Deficiency symptoms**—Enlarged thyroid gland, a condition called goiter. Kids born weak or dead. **Toxicity**—The maximum tolerable level for sheep is 45 ppm A-F or 50 ppm M-F. It is postulated that the toxicity level for goats is similar.
Iron (Fe)	Iron deficiency may occur in young goat kids because of their minimal body stores at birth and the low iron content of milk.	As a component of blood hemoglobin required for oxygen transport. Iron is also required for some enzyme systems.	**Deficiency symptoms**—Anemia, poor growth, lethargy, increased respiration rate, decreased resistance to infection, and in severe cases high mortality. **Toxicity**—Free iron ions are very toxic, causing loss of appetite, diarrhea, below normal temperature, shock, acidosis, and death.
Manganese (Mn)	High calcium and iron may increase manganese requirements.	Skeletal development and reproduction.	**Deficiency symptoms**—Reluctance to walk, deformity of the forelegs, delayed estrus, more inseminations per conception, more abortions, and 20% reduction in birth weights. **Toxicity**—1,000 ppm appears to be the maximum tolerance level for sheep; so, it is postulated that the toxicity level for goats is similar.
Selenium (Se)			**Deficiency symptoms**—White muscle disease in young kids from birth to a few months of age, which may take one of two forms: (1) sudden unexplained death, or (2) muscular paralysis, particularly of the hind limbs, or stiffness and inability to rise. **Toxicity**—All livestock species, including goats, are susceptible to selenium toxicity. Selenium toxicity in sheep occurs from prolonged consumption of plants containing over 3 ppm Se. It is postulated that the toxicity level for goats is about the same as for sheep.
Zinc (Zn)	Rations excessively high in calcium adversely affect zinc utilization.	Needed for normal skin, bones, and hair. A component of several enzyme systems involved in digestion and respiration.	**Deficiency symptoms**—Reduced feed intake, weight loss, parakeratosis, stiffness of joints, excessive salivation, swelling of the feet and horny overgrowth, small testicles, and low libido. **Toxicity**—Levels of 1,000 ppm may be toxic.

[1]Where preceded by an asterisk (*), the requirements, recommended allowances, and other facts presented herein were adapted from *Nutrient Requirements of Goats*, No. 15, NRC-National Academy of Sciences.

[2]As used herein, the distinction between "nutrient requirements" and "recommended allowances" is as follows: In nutrient requirements, no margins of safety are included intentionally; whereas in recommended allowances, margins of safety are provided to compensate for variations in feed composition, environment, and possible losses during storage or processing.

(Continued)

Nutrient Requirements	Recommended Allowances[1]	Practical Sources of the Mineral	Comments
Iodine in the ration: A-F, 0.09–0.72 ppm; M-F, 0.1–0.8 ppm. The higher levels are indicated for pregnancy and lactation.	Free access to stabilized iodized salt containing 0.0078% iodine. In total ration: 0.5 ppm M-F 0.45 ppm A-F	Iodized salt.	Iodized salt should not be used as a feed-limiter because it could lead to excessive intakes of iodine.
*0.03% ferrous iron in the ration.	*Iron-dextran (150 mg) may be injected in kids at 2 to 3 week intervals if iron deficiencies are observed. In total ration: 50 ppm M-F 45 ppm A-F	Iron-dextran is recommended as an injection; and ferrous sulfate and ferric citrate are recommended for incorporating in rations.	Iron deficiency seldom occurs in mature grazing goats.
	In total ration: 40 ppm M-F 36 ppm A-F	Manganese gluconate.	
	In total ration: 0.15 ppm M-F 0.13 ppm A-F		
*Direct and indirect evidence indicates minimum requirement of 10 ppm in the ration.	In total ration: 50 ppm M-F 45 ppm A-F	Zinc carbonate. Zinc sulfate.	

MAJOR OR MACROMINERALS

The major or macrominerals involved in sheep and goat nutrition are salt (sodium chloride), calcium, phosphorus, magnesium, potassium, and sulfur.

SALT (NaCl)

Sheep and goats are particularly fond of salt and consume considerably more of it per 100 lb body weight than do cattle.

The total salt requirement of growing lambs approximates 0.40% of the dry matter of the ration. The sodium requirements are given in Table 5-2, but the chlorine requirements are unknown.

Range operators commonly provide 0.5 to 0.75 lb salt per ewe per month. Mature sheep in drylot may consume more. Finishing lambs consume about 0.6 lb per head per month. Loose salt, rather than block salt, should be provided, for the reason that sheep bite at salt blocks, rather than lick, with the result that their teeth may be broken. In iodine-deficient areas, stabilized iodized salt should always be provided.

When salt is added to mixed feeds for sheep or goats, it is customary to add 0.5% to the complete ration or 1.0% to the concentrate portion. In the alkaline districts of the West, the water may contain enough salt to meet the requirements, and supplemental salt may not be necessary.

Salt may be used to limit feed intake, provided adequate water is available. Such mixtures range from 10 to 50% salt, depending on the consumption desired. Trace mineralized salt should not be used to govern feed consumption.

■ **Symptoms of salt deficiency**—A deficiency of salt may result in an abnormal appetite, with the sheep or goats trying to satisfy their craving by licking dirt, or eating toxic amounts of poisonous plants; decreased feed consumption; and decreased efficiency in the utilization of nutrients.

CALCIUM (Ca) AND PHOSPHORUS (P)

Calcium and phosphorus utilization depends on the presence of vitamin D and magnesium.

Most pasture and range forage contains adequate amounts of calcium, although calcium supplementation of pastures is required in Florida, Louisiana, Nebraska, Virginia, and West Virginia.

Legumes are an excellent source of calcium; hence, the requirements for this mineral are usually met when the winter forage for breeding ewes or the roughage for finishing lambs consists of one-third or more of a good-quality legume hay. Corn silage is a poor source of calcium. Finishing lamb rations based on low-quality roughage, or high in concentrates, may require calcium supplementation.

Mature pasture and range forage in North America is almost always deficient in phosphorus. Because of the high-phosphorus content of cereal grains, finishing lambs usually obtain an adequate allowance of this mineral, unless a high proportion of beet byproducts or other low-phosphorus feeds are fed.

Under grazing conditions, calcium is seldom a problem with either Angora or meat-type goats, but it can be very important for high-producing dairy goats. Low-calcium diets lead to reduced milk production. Appropriate calcium levels in the diet are also important in the prevention of parturient paresis (milk fever) in dairy goats.

A phosphorus deficiency in grazing goats is more likely than a calcium deficiency. It might be encountered with any type of goat grazing on phosphorus-deficient forages.

Chronic internal parasite infections can cause a deficiency in calcium and phosphorus. Magnesium deficiency interferes with calcium absorption. Low levels of dietary phosphorus decrease the rate of calcium absorption. High levels of aluminum and iron will increase the need for phosphorus.

■ **Symptoms of calcium and phosphorus defi-**

Fig. 5-4. Lamb fed a ration deficient in phosphorus. Note the knock-kneed conformation. (Courtesy, University of Idaho, Moscow)

ciency—Rations that are decidedly lacking in calcium or phosphorus cause rickets in young animals and osteomalacia in adults.

Signs of calcium deficiency due to low intake of calcium develop slowly because the body draws on the store of calcium in the bones until it is greatly reduced. In extreme cases, which may occur in lambs on high-grain diets, low levels of calcium may result in tetany. Blood levels of calcium below 9 mg per 100 ml of serum indicate a calcium deficiency (hypocalcemia).

Phosphorus deficiency symptoms are depraved appetite, slow growth, unthrifty appearance, listlessness, knock-knees, and low level of phosphorus in the blood (less than 4 mg/100 ml of plasma).

MAGNESIUM (Mg)

Magnesium is a constituent of bone. Also, it is necessary for many enzyme systems and for proper functioning of the nervous system. A deficiency of magnesium may result in grass tetany, particularly in animals grazing on lush green grass or winter wheat fertilized with nitrogen and potassium.

The requirement for magnesium on a moisture-free (M-F) basis is 0.12, 0.15, and 0.18% for growing lambs, ewes in late pregnancy, and ewes in early lactation, respectively. Where ewes in early lactation are grazing forage with high nitrogen and potassium content, the minimum level of magnesium in the ration is 0.2%. It is postulated that the magnesium requirements of goats at similar age and productive status are about the same as for sheep.

The use of intraruminal magnesium alloy pellets (bullets) weighing 30 g has effectively prevented hypomagnesemic tetany in lactating ewes.

■ **Symptoms of magnesium deficiency**—Acute tetany, characterized by stiff legs and head retraction, may occur as a result of insufficient dietary magnesium.

POTASSIUM (K)

Potassium is the third most abundant mineral in the body, accounting for approximately 0.3% of the body dry matter. It is primarily present in intracellular fluids (in skin and muscle), where it affects osmotic pressure and acid-base balance within the cells. It also aids in activating several enzyme systems involved in energy transfer and utilization, protein synthesis, and carbohydrate metabolism.

The potassium requirement for growth in lambs and kids appears to be about 0.5% (M-F basis) of the ration. During periods of stress and during lactation, 0.7 to 0.8% may be required.

The possibility of potassium deficiency is very slight under most feeding conditions, because most forages contain adequate potassium. Nevertheless, attention must be given to the potassium supply when lambs are on high-grain rations and when sheep are grazing mature range forage during winter or drought periods. Potassium levels in mature, weathered range forage may decrease to less than 0.2%.

■ **Symptoms of potassium deficiency**—A deficiency of potassium results in decreased feed intake and decreased weight gains. Listlessness, stiffness, impaired response to sudden disturbances, convulsions, and death have also been reported.

SULFUR (S)

Sulfur is used in the synthesis of the sulfur-containing amino acids, cystine and methionine, in the rumen and in various compounds of the body. Also, wool and mohair are high in sulfur; hence, this element is closely related to wool and mohair production.

It is recommended that a dietary nitrogen-sulfur ratio of 10:1 be maintained. In percentage of dry matter, the sulfur requirements are as follows: mature ewes, 0.14 to 0.18%; young lambs, 0.18 to 0.26%.

Most feedstuffs contain more than 0.1% sulfur. However, mature grass and grass hays are sometimes low in sulfur and may not furnish enough of this element for optimum performance.

Where forages are low in sulfur, or where high-urea rations are fed, weight gains and growth of wool and mohair can be increased by feeding a sulfur supplement, such as sulfate sulfur, elemental sulfur, or sulfur-containing proteins or amino acids.

■ **Symptoms of sulfur deficiency**—Loss of appetite, reduced weight gains and feed conversion efficiency, and reduced wool and mohair growth. Also, excessive salivation and lacrimation, and shedding of wool and mohair.

TRACE OR MICROMINERALS

The trace or microminerals involved in sheep and goat nutrition are cobalt, copper, fluorine (because of toxicity to sheep), iodine, iron, manganese, molybdenum, selenium, and zinc.

COBALT (Co)

Cobalt is essential for the synthesis of vitamin B-12 in the rumen. Indicators of the cobalt status of sheep and goats are the levels of vitamin B-12 in the rumen contents, in the blood and liver, and in the feces.

Cobalt should be ingested frequently, preferably daily. This may be accomplished by adding cobalt to the salt, adding cobalt to the soil, placing cobalt pellets into the rumen, or by daily doses of cobalt.

The recommended amount of cobalt in the ration DM is 0.1 to 0.2 ppm. However, young, rapidly growing lambs and kids may have a slightly higher requirement.

Cobalt-deficient areas have been widely reported in the United States and Canada. In known deficient areas, it is recommended that cobalt be added to the salt at the rate of 1.4 g per 100 lb of salt as cobalt chloride or cobalt sulfate.

■ **Symptoms of cobalt deficiency**—Affected sheep and goats show loss of appetite, lack of thrift, severe emaciation, weakness, anemia, decreased fertility, and decreased milk production.

COPPER (Cu)

A copper deficiency may exist alone or in combination with deficiencies of other trace minerals. In practice, copper deficiency is frequently induced by excess molybdenum in forages.

Copper is found in adequate amounts in most feeds throughout the United States. The NRC copper requirements vary depending on the molybdenum content of the feed as follows:

	Recommended Cu Allowance	
	As-Fed	Moisture-Free
	(ppm)	*(ppm)*
Mo contents of diet ppm < 1.0:		
Growth	8.9–9.0	8–10
Pregnancy	8.1–9.9	9–11
Lactation	6.3–7.2	7–8
Mo content of diet ppm > 3.0:		
Growth	15.3–18.9	17–21
Pregnancy	17.1–20.7	19–23
Lactation	12.6–15.3	14–17

Merino sheep are less efficient in absorbing copper from feedstuffs than British breeds; so, they need an additional 1 to 2 ppm in their ration.

Copper-deficient areas have been reported in Florida, in the coastal plains region of the Southeast, and in Nevada, Oregon, and other western states. In such areas, it is recommended that copper sulfate be added to the salt at the rate of 0.5%. Copper is stored in the body; reserves may last as long as 4 to 6 months when animals are grazing copper-deficient forage.

Fig. 5-5. Cobalt deficiency. *Top:* Lamb fed cobalt-deficient ration containing 0.05 ppm cobalt. Lamb weighed 48.5 lb at the end of the experimental period. *Bottom:* Control lamb that received the same ration as the lamb in the top picture plus a daily allowance of 0.1 mg of cobalt as the sulfate. This lamb weighed 92.5 lb at the end of the experiment. (Courtesy S. E. Smith, Cornell University, Ithaca, NY)

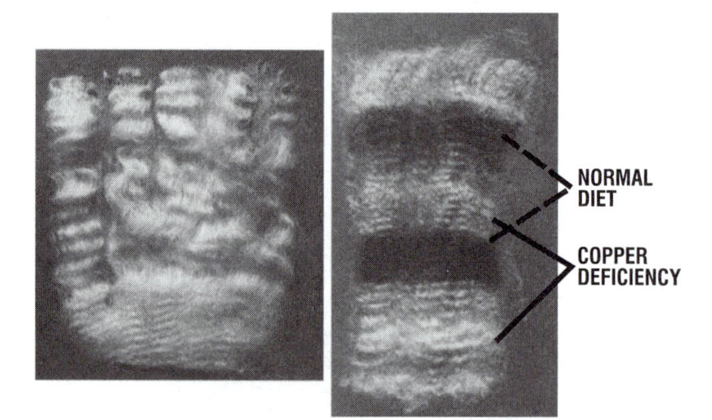

Fig. 5-6. Two samples of Australian wool, both of which show what may happen when sheep are on a copper-deficient ration. *Left:* The outer (bottom) two-thirds of this sample was produced by a sheep on a copper-deficient ration, resulting in hairlike or "steely" wool. Then, copper was added to the sheep's ration, and normal, well-crimped wool was produced. *Right:* Wool sample from a normally black sheep. The white bands appeared at intervals when copper was deficient in the ration, because copper is essential for melanin or pigment production. Where such deficiencies occur under natural conditions, it is recognized that copper deficiencies result in the production of wool of lowered elasticity, tensile strength, and affinity for dyes.

■ **Symptoms of copper deficiency**—Lambs and kids may be born weak and may die because of their inability to nurse. Suckling lambs and kids show muscular incoordination and partial paralysis of the hindquarters.

Copper-deficient sheep produce "steely" wool, lacking in crimp, tensile strength, affinity for dyes, and elasticity. With a severe deficiency, the wool of black sheep is depigmented.

■ **Copper toxicity**—Sheep are extremely intolerant to copper excess. Copper toxicity—characterized by hemolysis (dissolution of red corpuscles with liberation of their hemoglobin), jaundice (easily detected in the eyes), hemoglobinuria, and very dark-colored liver and kidneys— may result when sheep are fed rations high in copper and low in molybdenum. Copper toxicity may be prevented by lowering the copper level in the ration (normal is 8 to 11 ppm), or by a high-zinc ration (100 ppm on a dry matter basis). The recommended treatments for copper toxicity are (1) administering molybdenum and sulfate, or (2) drenching each lamb daily with 100 mg of ammonium molybdate and 1 g of sodium sulfate in 20 ml of water, for about 3 weeks. The Food and Drug Administration does not recognize molybdenum as safe; hence, it is not legal to add it to the feed of sheep.

FLUORINE (F)

Fluorine is a cumulative poison. It occurs in some parts of the world as a result of consuming water high in fluoride or using rock phosphate that contains fluorine in amounts sufficient to be toxic—3 to 4%. Finishing lambs can tolerate up to 150 ppm of fluorine in the ration on a dry matter basis. Acute toxicity occurs at 200 ppm.

■ **Symptoms of fluorine toxicity**—Affected animals exhibit loss of appetite and weight, change in bone color from ivory to chalky white, thickened bones, and pitted and eroded teeth—especially the incisors.

IODINE (I)

Iodine is necessary for the formation of thyroxin, the iodine-containing hormones of the thyroid gland. Northwestern United States and the Great Lakes region are well-known iodine-deficient areas, but other deficient areas are widely scattered throughout the United States.

The iodine requirement is 0.09 to 0.72 ppm in the ration on an as-fed basis, or 0.1 to 0.8 ppm moisture-free basis, with the higher levels indicated for pregnancy and lactation. When goitrogens such as kale or rape are fed, the dietary iodine should be increased.

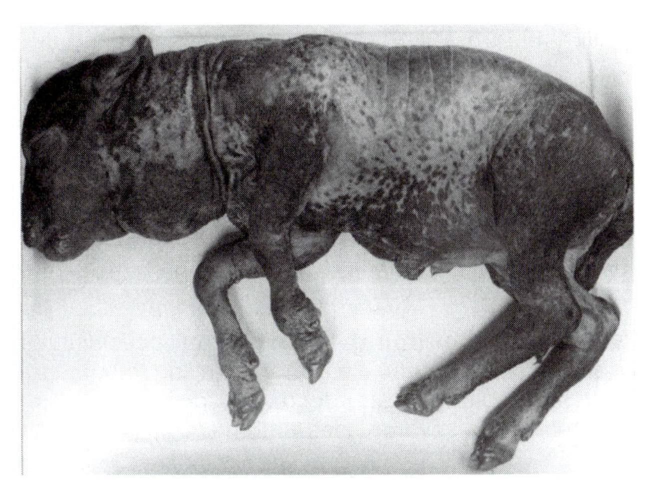

Fig. 5-7. Goiter. Woolless, goitered (big-necked) lamb stillborn due to iodine deficiency. (Courtesy, Montana State University, Bozeman)

Lamb losses can be prevented by feeding gestating ewes iodized salt containing 0.0078% iodine. The iodine in the salt should be stabilized to prevent losses from exposure to sunlight and moisture.

Iodized salt should not be used in a feed mixture to govern feed intake, as the animals may consume too much iodine.

■ **Symptoms of iodine deficiency**—Lambs or kids born with goiter (an enlarged thyroid gland) is the most common deficiency symptom. If the condition is not too advanced, afflicted lambs or kids may survive. Other signs are lambs born without wool; kids born without mohair; and with young born weak or dead. An iodine deficiency in mature sheep and goats may result in reduced fiber yield and conception.

Fig. 5-8. Pygmy goat kid with goiter—an abnormal enlargement of the thyroid gland due to an iodine deficiency. (Courtesy, Christine S. F. Williams, Michigan State University)

IRON (Fe)

Iron-deficiency anemia sometimes occurs in lambs and kids (1) raised in confinement, on slotted floors, or (2) as a result of loss of blood from parasite infestation. It can be prevented (1) by giving lambs or kids two intramuscular injections of iron-dextran, each of 150 mg of iron, 2 to 3 weeks apart, or (2) by allowing free access to a commercial iron compound in the creep area.

■ **Symptoms of iron deficiency**—Iron deficiency is characterized by anemia, poor growth, lethary, increased respiration rate, decreased resistance to infection, and in severe cases high mortality.

MANGANESE (Mn)

Although the exact manganese requirement of sheep and goats is unknown, it appears that 18 ppm in as-fed rations, or 20 ppm in moisture-free rations, will be adequate. It is needed for skeletal development and reproduction, as it is for various other species. A deficiency of manganese was produced in early weaned lambs fed for a five-month period on a ration containing less than 1 ppm of manganese. When a ration containing 8 ppm of manganese was fed to two-year-old ewes for a five-month period prior to breeding and throughout gestation, more services per conception (2.5 vs 1.5) were required for ewes fed a ration containing 60 ppm manganese.

The manganese content of wool or mohair appears to be a good indicator of the manganese status of sheep and goats, respectively.

■ **Symptoms of manganese deficiency**—A manganese deficiency results in impaired growth, skeletal abnormalities and lack of coordination of the newborn, and depressed or disturbed reproductive function.

MOLYBDENUM (Mo)

A molybdenum deficiency in sheep, unrelated to copper, has not been reported. The major concern about molybdenum is its interaction with copper.

A high-molybdenum intake can induce copper deficiency in sheep and goats even when the copper content of pasture is quite high, the effect of which can be prevented by an increased copper intake. On the other hand, when pastures provide a low intake of molybdenum, excess copper may accumulate in the body and result in a fatal jaundice, which can be prevented by increasing the dietary molybdenum.

■ **Molybdenum toxicity**—Molybdenum toxicity (excess molybdenum) has been reported in California, Nevada, Oregon, and England, where it causes a scouring disease. Sheep start to scour within a few days after being turned to high-molybdenum pasture (5 to 20 ppm on a dry matter basis; or on forage containing as little as 1 or 2 ppm molybdenum provided the dietary copper level is low and the sulfate level is high); the feces become soft, the fleece becomes stained, and the animals lose weight rapidly. Molybdenum toxicity can be controlled by increasing the copper level in the ration by 5 ppm.

SELENIUM (Se)

A minimum level of 0.1 ppm of selenium in feeds is considered adequate for preventing a deficiency in sheep and goats. There are extensive areas in the northwestern, northeastern, and southeastern United States where the selenium content of crops is below this level. On the other hand, parts of South Dakota, Wyoming, and Utah produce forage so high in selenium that it causes selenium toxicity.

White muscle disease (stiff lamb disease) in lambs, the main manifestation of selenium deficiency, can be prevented by (1) adding 0.3 ppm of selenium to the complete feed of ewes during gestation through weaning, or (2) adding 90 ppm of selenium to a salt-mineral mixture fed free-choice. Kids, bucks, and does may also benefit from an injection of selenium-tocopherol when indicated.

■ **Symptoms of selenium deficiency**—Selenium deficiency has serious effects on lamb and kid production. The main signs of such deficiency are white muscle disease, sometimes called *stiff lamb disease*, and reduced growth. If the muscle lesions are in the heart, lambs may die suddenly if subjected to exercise. Additional signs of inadequate selenium are unthriftiness, infertility, early embryonic death, and periodontal disease.

Fig. 5-9. Selenium toxicity in ewe. Note loose wool, typical of sheep afflicted with chronic selenium poisoning. (Courtesy, University of Wyoming)

■ **Selenium toxicity**—Chronic selenium toxicity occurs when sheep consume plants containing more than 3 ppm selenium over a prolonged period. The extent to which plants take up selenium varies greatly by species. The most practical way in which to prevent livestock losses in high-selenium areas is to rotate the grazing between high-selenium and low-selenium areas. Although selenium is a cumulative poison, mild chronic signs can be readily overcome by feeding selenium-low forage. Also, small amounts of arsanilic acid are effective in reducing the toxicity of selenium.

ZINC (Zn)

Zinc is essential for sheep and goats. An as-fed dietary level of 18 ppm of zinc appears to be adequate for growth, but a level of 30 ppm is necessary for normal reproduction.

■ **Symptoms of zinc deficiency**—Deficiency signs include impaired growth, excessive salivation, parakeratosis, wool loss, and delayed wound healing. Ram lambs show reduced testicular development and defective spermatogenesis. In females, all phases of the reproductive process from estrus to parturition and lactation may be adversely affected.

Fig. 5-10. Zinc deficiency. *Left:* Zinc-deficient lamb after 15 weeks on the basal ration (2.7 ppm of zinc). *Right:* Lamb of the same age, which was fed the basal ration plus 100 ppm of zinc. Note the stunted growth, loss of wool, and dermatosis of skin on the lower part of the legs of the zinc-deficient lamb. (Courtesy, Purdue University, West Lafayette, IN)

VITAMINS

Vitamins are complex organic compounds that are required in minute amounts by one or more animal species for normal growth, production, reproduction, and/or health.

Many phenomena of vitamin nutrition are related to solubility. Vitamins are soluble in either fat or water. Based on this, vitamins are grouped as follows:

Fat-soluble Vitamins	**Water-soluble Vitamins**
A (carotene)	Biotin
D	Choline
E	Folacin (folic acid)
K	Niacin (nicotinic acid; nicotinamide)
	Pantothenic acid (vitamin B-3)
	Riboflavin (vitamin B-2)
	Thiamin (vitamin B-1)
	Vitamin B-6 (pyridoxine; pyridoxal; pyridoxamine)
	Vitamin B-12 (cobalamins)
	Vitamin C (ascorbic acid)

Mature sheep and goats require the fat-soluble vitamins A, D, E, and K, but they do not need added sources of the B vitamins, since the latter are synthesized in adequate amounts by rumen microorganisms.

Normal sheep and goat rations are adequate in all the fat-soluble vitamins with the exception of the low-carotene and/or vitamin A content of dry winter ranges. However, animals can build up liver stores of vitamin A adequate to maintain production for 3 to 4 months. Vitamin D deficiency is not a problem unless sheep and goats are maintained on vitamin D-deficient diets in environments devoid of sunlight. In some coastal areas of the United States, the latter condition may prevail.

SHEEP AND GOAT VITAMIN CHARTS

Table 5-6, Sheep Vitamin Chart, summarizes the vitamin needs of sheep, and Table 5-7, Goat Vitamin Chart, summarizes the vitamin needs of goats. Further information pertaining to certain vitamins is contained in the accompanying narrative.

FAT-SOLUBLE VITAMINS

The fat-soluble vitamins, which may be stored in varying quantities in the body, of importance in sheep and goat nutrition are vitamin A, vitamin D, vitamin E, and vitamin K.

VITAMIN A

Dietary vitamin A or its precursor, carotene, is necessary for maintaining normal epithelial tissues.

Sheep and goats do not convert carotene to vitamin A as efficiently as laboratory rats. Thus, for sheep and goats, it is suggested that 1 mg of feed carotenes be considered as the equivalent of 400 to 500 IU of vitamin A. (For the rat, 1 mg of beta-carotene = 1,667 IU of vitamin A.)

TABLE
SHEEP VITAMIN

Vitamin Which May Be Deficient Under Normal Conditions	Conditions Usually Prevailing Where Deficiencies Are Reported	Function of Vitamin	Deficiency Symptoms/Toxicity
Fat-Soluble Vitamins:			
A	Vitamin A deficiencies may occur when (1) extended drought results in dry, bleached pastures; (2) winter feeding on bleached hays (especially overripe cereal hays or straws) with little or no green hay or silage; (3) drylot finishing on rations with little or no green forage or yellow corn, especially for feeding periods longer than 2–3 months; and (4) there is high nitrate intake, in either water or feed.	Necessary for maintaining normal epithelial tissue.	**Deficiency symptoms**—Keratinization of the respiratory, alimentary, reproductive, urinary, and ocular epithelia; lowered resistance to infection; abnormal development of bone; birth of lambs that are weak, malformed, or dead; and night blindness. **Toxicity**—Changes in bone composition.
D	When ration consists predominantly of dehydrated hays, green feeds and seeds and their byproducts. Prolonged cloudy weather or when kept inside, especially in fast-growing young lambs. When the vitamin D in feed is lost by oxidation.	Prevention of rickets in young lambs and osteomalacia in older sheep.	**Deficiency symptoms**—Rickets in young lambs and osteomalacia in older sheep. Congenital malformations in newborn lambs from extreme deficiencies. **Toxicity**—Abnormal deposition of calcium in soft tissues and brittle bones subject to deformation and fractures.
E	When lambs are making rapid growth, although this isn't always the case. When old hay is fed, as oxidation destroys vitamin E.	Prevention of white muscle disease (stiff lamb disease). As an antioxidant. Closely associated with selenium in metabolism.	**Deficiency symptoms**—Stiff lamb disease, or white muscle disease; characterized by a stiff, stilted way of moving and a "roached" back. Sometimes a paralysis of hind legs.
K	Vitamin K deficiency may occur when the dicumarol content of hay is excessively high, as when moldy sweet clover hay is fed.	Vitamin K_1 or K_2 is necessary in the blood clotting mechanism.	
Water-Soluble Vitamins:	B-vitamin deficiencies may be evident in poorly fed and unhealthy animals.		

[1]As used herein, the distinction between "vitamin requirements" and "recommended allowances" is as follows: In vitamin requirements, no margins of safety are included intentionally; whereas in recommended allowances, margins of safety are provided to compensate for variations in feed composition, environment, and possible losses during storage or processing.

Where preceded by an asterisk (*), the vitamin requirements, recommended allowances, and other facts presented herein were taken from *Nutrient Requirements of Sheep*, 6th rev. ed., NRC-National Academy of Sciences.

5-6
CHART

Vitamin Requirements[1]		Recommended Allowances[1]	Practical Sources of the Vitamin	Comments
Vitamins/Animal/Day	Vitamin Content of Ration			
Variable, according to class, age, and weight of sheep, as indicate\d by the following: Categories — Mcg of B-carotene/day /lb body wt. / *kg body wt.* All categories . . . . 31 / *69* Late gestation and in lactation . . . 57 / *125* First 6–8 weeks of lactation of ewes suckling twins . . . . . 67 / *147* Categories — IU of Vitamin A/day /lb body wt. / *kg body wt.* All categories . . . . 21 / *47* Late gestation. . . . . 39 / *85* First 6–8 weeks of lactation of ewes suckling twins . . . . . 45 / *100* (See Table 14-1.)	Variable according to class, age, and weight of sheep (see Table 14-2).	Sheep that are deficient in vitamin A and weigh 70 lb *(32 kg)* or more should receive 100,000 IU of vitamin A by injection, and their rations should be adjusted to provide recommended levels of vitamin A or carotene. Ewes deficient in vitamin A should be given vitamin A either orally or by injection prior to breeding.	Stabilized vitamin A. Green pasture. Grass or legume silages. Yellow corn. Green hay not over 1 year old. *The vitamin A value of carotene from two common feeds is as follows: IU/mg Dehydrated alfalfa meal . . 254–520 Silage, corn . . . . . . 436 Where sheep are grazing forage low in carotene for extended periods, vitamin A deficiency can be prevented by (1) intra-muscular injection of vitamin A, or (2) the addition of vitamin A to the ration as a pasture supplement or part of a salt mixture.	*Sheep do not convert carotene to vitamin A as efficiently as rats. For sheep, 1 mg of feed carotene is equivalent to 400–500 IU of vitamin A. *It requires 200 days to deplete entirely liver storage of ewe lambs previously grazing on green feed. Because of this storage, animals that normally graze on green forage during the growing season are able to do reasonably well on a low-carotene ration of dry feed for periods of 4–6 months.
*For all sheep except early-weaned lambs: 250 IU/100 lb body weight. *For early-weaned lambs: 300 IU/100 lb body weight.	Variable, according to class, age, and weight of sheep.	Breeding sheep, 500 to 800 IU/head/day. Feeder lambs, 500 IU/head/day.	Exposure to sunlight, through irradiation. Sun-cured hays. Irradiated yeast. Vitamin D_2 or vitamin D_3, which sheep use equally well.	Newborn lambs are provided with enough vitamin D from their dams to prevent rickets for 4 to 6 weeks if the ewes have adequate storage. Sheep with white skin and short wool receive more vitamin D activity from irradiation by sunlight than do animals with dark skin or long wool.
Categories — *IU Vitamin E A-F basis /lb / *kg* Lambs under 44 lb *(20 kg)* live weight . . . . 8 / *18* Lambs over 44 lb *(20 kg)* live weight and pregnant ewes . . . 6 / *13.5* Categories — *IU Vitamin E M-F basis /lb / *kg* Lambs under 44 lb *(20 kg)* live weight . . . . 9 / *20* Lambs over 44 lb *(20 kg)* live weight and pregnant ewes . . . 7 / *15*		Meet the requirements given in the column headed "Vitamin Content of Ration."	Wheat germ meal, dehydrated alfalfa, some green feeds, and vegetable fats are good sources of vitamin E. Alpha-tocopherol (either in dl or d forms).	Experiments have failed to relate vitamin E deficiency with reproductive failure in sheep. The need for vitamin E in the ration of young nursing lambs is related to the selenium level in the ration. Selenium has a sparing effect on the vitamin E requirement; the higher the selenium level in the diet, the lower the vitamin E requirement, and vice versa.
		Vitamin K_2 is normally synthesized in large amounts in the rumen; no need for dietary supplementation has been established.	Green leafy materials of any kind, fresh or dry, are good sources of K_1.	
		*The B vitamins are not required in the diet of sheep with functioning rumens, because the microorganisms synthesize these vitamins in adequate amounts.		Addition of B vitamins has not been shown to be beneficial to mature sheep. However, young lambs (to about 2 months of age) with undeveloped rumens have been shown to have a dietary need for vitamin B-12, thiamin, pyridoxine, riboflavin, niacin, folic acid, and possibly some of the other B vitamins, since they will not be receiving these in the milk from their dams. Cobalt is necessary for the synthesis of vitamin B-12 in the rumen. *No supplementary dietary need for vitamin C has been shown.

TABLE
GOAT VITAMIN

Vitamin Which May Be Deficient Under Normal Conditions	Conditions Usually Prevalent Where Deficiencies Are Reported	Function of Vitamin	Deficiency Symptoms
Fat-Soluble Vitamins:			
A	During extended dry periods when the supply of green forage is limited.	Required for normal vision. Aids in reproduction and lactation. Needed for maintaining normal epithelial tissue. Aids in resistance to infection.	Keratinization of the epithelia of the respiratory, alimentary, reproductive, and urinary tracts, and of the eye. Multiple infections, poor bone development, birth of abnormal offspring, and vision impairment, including night blindness.
D	Young goats kept in confinement where they have little or no access to sunlight.	Absorption of calcium and phosphorus.	Bone abnormalities, including rickets. Depressed growth.
E	Abnormally high levels of nitrates may produce vitamin E deficiencies. Where soils are very low in selenium.	Serves as a physiological antioxidant. In dairy goats, the vitamin E transferred to the milk is important because of the antioxidant properties that aid in milk storage.	Evidence of spontaneous vitamin E deficiency signs in goats is lacking. The probability of lowered productivity in goats as a result of vitamin E deficiency is remote.
K	Vitamin K deficiency may occur when the dicoumarol content of hay is excessively high, as when moldy sweet clover hay is fed.	Vitamin K or K_2 is necessary in the blood clotting mechanism.	
Water-Soluble Vitamins:			
B vitamins, Vitamin C	B vitamin deficiencies may be evident in poorly fed and unhealthy animals. B-12 may be deficient if cobalt is absent or at extremely low levels, as cobalt is required for the synthesis of vitamin B-12.	B-1 participates as a coenzyme in the utilization of carbohydrates.	

[1]As used herein, the distinction between "nutrient requirements" and "recommended allowances" is as follows: In nutrient requirements, no margins of safety are included intentionally; whereas in recommended allowances, margins of safety are provided to compensate for variations in feed composition, environment, and possible losses during storage or processing.

Vitamin A is fat-soluble and is stored in the body. It takes about 200 days to deplete entirely the liver storage of vitamin A of lambs previously grazing on green feed. This explains why animals which graze green forage during the normal growing season are able to do reasonably well on a low-carotene ration for periods of 4 to 6 months.

Both vitamin A and carotene may be destroyed by oxidation. Sun-cured hay is usually lower in carotene than dehydrated hay. Feedstuffs that contain little green or yellow material or that have been badly weathered, heated, or stored for long periods are low in carotene. Stabilized vitamin A, which is resistant to oxidation, may be added to rations low in carotene.

Sheep that are deficient in vitamin A and weigh 70 lb or more should receive 100,000 IU of vitamin A by injection, followed by an adjustment of their rations to provide recommended levels of vitamin A or carotene. Ewes deficient in vitamin A should be given vitamin A either orally or by injection prior to breeding.

■ **Symptoms of vitamin A deficiency**—Vitamin A-deficient sheep and goats develop keratinization of the respiratory, alimentary, reproductive, urinary, and ocular epithelia, accompanied by lowered resistance to infections. Also, a deficiency of vitamin A interferes with normal development of bone; may result in the birth of weak, malformed, or dead lambs and kids; and causes night blindness, the appearance or nonappearance of which is the most common means of determining the vitamin A status of animals.

5-7
CHART

Nutrient Requirements[1]	Recommended Allowances	Practical Sources	Comments
Variable according to size, sex, age, and class (see Tables 21-1, 21-2, and 21-3).	The recommended allowances should provide margins of safety over and above the requirements. So, add 10 to 20% to the requirements given in Tables 21-1, 21-2, and 21-3.	Synthetic vitamin A. Injectable vitamin A. Yellow corn. Green forages.	Young animals, which have not built up vitamin A reserves, are more susceptible to a vitamin A deficiency than are mature animals. Goats that have had access to green feed can store sufficient vitamin A in the liver and fat to last for 3 months on a low carotene ration without showing signs of vitamin A deficiency.
Variable according to size, sex, age, and class (see Tables 21-1, 21-2, and 21-3).	Add 10 to 20% to the requirements given in Tables 21-1, 21-2, and 21-3 to provide a margin of safety.	Sunlight action on ergosterol, a plant sterol, and on 7-dehydrocholesterol, a sterol of animal origin. Sun-cured hays. Irradiated yeast. Vitamin D_2 or vitamin D_3, which goats use equally well.	Vitamin D should be of little concern when goats are maintained on pasture or range.
		Alpha-tocopherol, added to the diet or injected intramuscularly. Grains are generally high in vitamin E.	Most goat rations contain adequate amounts of vitamin E. Hence, there is little need for vitamin E supplementation.
		Green leafy materials of any kind, fresh or dry, are good sources of K_1. Vitamin K_2 is normally synthesized in large amounts in the rumen; no need for dietary supplementation has been established.	
		Only vitamin B-12 (cobalamin) is likely to be deficient in goats with functioning rumens, because the microorganisms synthesize these vitamins in adequate amounts. Adequate vitamin C is synthesized in body tissues to satisfy requirements.	The B vitamins should be included in the diets of very young kids, animals with poorly functioning rumens, sick animals, and those with radically changed diets.

VITAMIN D

Vitamin D is required, in addition to calcium and phosphorus, for the prevention of rickets in young lambs and kids and osteomalacia in older sheep and goats. The addition of vitamin D to lamb rations low in this nutrient has resulted in increased growth even when there were no signs of rickets.

Since vitamin D is fat-soluble and stored, it is less important in mature animals, except during pregnancy, when the demands are greater. If pregnant ewes have adequate storage of vitamin D, they provide their newborn lambs with sufficient amounts to prevent rickets for 4 to 6 weeks.

Sheep and goats exposed to sunlight obtain some vitamin D through irradiation, which may be sufficient to meet their requirements. Animals with white skins and/or short fleeces receive more vitamin D activity through irradiation than do animals with black skins and/or long fleeces.

Sheep and goats on pasture seldom need additional vitamin D. But the question of adequacy arises during extended cloudy periods or when sheep and goats are kept indoors. Under these circumstances, it is especially important that adequate vitamin D be included in the ration of fast-gaining lambs and kids.

Early weaned lambs require 300 IU of vitamin D per 100 lb body weight, daily. All other sheep require 250 IU per 100 lb body weight, daily. Sheep and goats use either D_2 or D_3 equally well.

Sun-cured hays are fairly good sources of vitamin D. Dehydrated hays, green feeds, seeds, and byproducts of seeds are poor sources. Vitamin D is oxidized, but with greater difficulty than vitamin A. However,

Fig. 5-11. Bilateral bent leg in yearling Rambouillet ram due to a deficiency of vitamin D. (Courtesy, Utah Agricultural Experiment Station)

when mixed with minerals, especially calcium carbonate, its stability is poor.

■ **Symptoms of vitamin D deficiency**—A deficiency of vitamin D results in rickets in young lambs and kids and osteomalacia in older animals. Extreme deficiencies may cause congenital malformations in newborn lambs and kids. *CAUTION:* Vitamin D deficiency should not be confused with "spider leg syndrome," a genetic abnormality found in some sheep.

VITAMIN E

Vitamin E is essential for all sheep and goats, especially for young lambs and kids. Unlike vitamin A, it is not stored in the body in appreciable quantities.

Vitamin E functions as an important biological antioxidant and helps to prevent white muscle disease (stiff lamb disease) in lambs, through its association with dietary selenium in metabolism. Some signs of deficiency, such as white muscle disease or nutritional muscular dystrophy, may respond to either selenium or vitamin E, or may require both. Although vitamin E is a dietary requirement for young nursing lambs and kids, experiments have failed to relate a deficiency of it to reproductive failure in sheep and goats.

Stiff-lamb disease in nursing lambs is characterized by a stiffness (especially in the rear quarters), tucked-up rear flanks, and arched back. On autopsy, the disease is shown as white striations in the heart and other muscles, characterized by bilateral lesions. Affected lambs often die of pneumonia and starvation.

The need for vitamin E in the diet of young nursing lambs and kids is related to the selenium level in the diet. Selenium has a sparing effect on the vitamin E requirement; the higher the selenium level in the diet, the lower the vitamin E requirement, and vice versa. White muscle disease in lambs is prevented by adding alpha-tocopherol and selenium to the diet. The suggested dietary levels of vitamin E are as follows: Lambs under 44 lb in weight should receive 8 IU/lb of as-fed ration; lambs over 44 lb in weight and pregnant ewes should receive 6 IU/lb of as-fed ration. (The IU is defined as 1 mg of dl-alpha-tocopherol acetate; 1 mg dl-alpha-tocopherol has the biological potency of 1.5 IU of vitamin E activity.) The above recommendations assume that dietary selenium levels are <0.05 ppm.

Values for the vitamin E requirements of sheep are presented in Tables 14-1, 14-2, and 5-6. The values presented in Table 14-1 were calculated from values per kilogram of dry feed consumed given in Table 14-2. Table 5-8 presents daily vitamin E requirements for lambs and the suggested amounts of alpha-tocopherol acetate to add to rations to provide 100% of the requirements.

Based on the average alpha-tocopherol content of feedstuffs generally used in lamb growing-finishing rations (corn, soybean meal, and alfalfa hay), the typical ration may contain less than 6.8 mg of alpha-tocopherol/lb, which could result in inadequate intake of vitamin E. In addition, preintestinal destruction of vi-

Fig. 5-12. Lamb afflicted with stiff lamb disease, which in this case was associated with a lack of vitamin E in the ration. (Courtesy, Cornell University, Ithaca, NY)

TABLE 5-8
VITAMIN E REQUIREMENTS OF GROWING-FINISHING LAMBS
AND SUGGESTED LEVELS OF FEED FORTIFICATION TO PROVIDE 100% OF REQUIREMENTS[1]

Body Weight		Alpha-Tocopherol Acetate			Feed Intake Per Lamb		Amount of Vitamin E Added to Concentrate			Amount of Vitamin E Added to Protein Supplement[2]		
(lb)	(kg)	(mg/lamb/day)[3]	(mg/lb/ration)	(mg/kg/ration)	(lb)	(kg)	(mg/lb)	(mg/kg)	(mg/ton)	(mg/lb)	(mg/kg)	(mg/ton)
22	10	5.0	44	20	0.50	0.23	9.1	20	18,200	133	60	120,000
44	20	10.0	44	20	1.00	0.45	9.1	20	18,200	60	133	120,000
66	30	15.0	33	15	2.10	0.96	6.8	15	13,600	45	100	90,000
88	40	20.0	33	15	2.86	1.30	6.8	15	13,600	45	100	90,000
110	50	25.0	33	15	3.50	1.60	6.8	15	13,600	45	100	90,000

[1]Adapted by the author from *Nutrient Requirements of Sheep*, sixth revised edition, NRC-National Academy of Sciences.

[2]Assumes the concentrate diet contains 15% protein supplement.

[3]Rounded values based on approximate diet intake containing recommended vitamin E level.

tamin E of an orally administered dose increases from 8 to 42% as the corn content of the ration increases from 20 to 80%. So, many sheep rations heretofore believed to be adequate in vitamin E may be inadequate, thus explaining the sporadic outbreaks of white muscle disease in areas considered adequate in selenium.

VITAMIN K (K_1 AND K_2)

These are fat-soluble vitamins, one or the other of which is necessary in the blood-clotting mechanism. All green, leafy materials, fresh or dry, are good sources of vitamin K_1. Vitamin K_2 is usually synthesized in large amounts in the rumen, so no need for dietary supplementation with it has been established.

WATER-SOLUBLE VITAMINS

The water-soluble vitamins, which are not stored, include members of the vitamin B complex and vitamin C.

VITAMIN B COMPLEX

Digestion in the young lamb or kid is more like that of a nonruminant animal than that of a ruminant. Thus, up to about two months of age, early-weaned lambs and kids have a dietary need for biotin, choline, folacin, niacin, pantothenic acid, riboflavin, thiamin, vitamin B-6 (pyridoxine), and vitamin B-12, which are usually supplied in the ewe's or doe's milk.

Normally, the B vitamins are not required in the ration of sheep with functioning rumens (older than about two months of age), because the microorganisms synthesize them in adequate amounts. It is noteworthy, however, that polio-encephalomalacia (PEM), a noninfectious disease of sheep, the highest incidence of which occurs in lambs, responds to a parenteral injection of 0.5 g of thiamin hydrochloride, repeated at two-day intervals if necessary. Apparently, PEM is caused by an anti-thiamin enzyme in the rumen.

It is necessary that cobalt be present for the synthesis of vitamin B-12 in the rumen.

VITAMIN C

Vitamin C is synthesized rapidly enough by the tissues to meet the animal's needs; hence, it is not a required dietary constituent for sheep and goats.

WATER

Sheep and goats get water by drinking, and from snow, dew, and feed. The amount of water that sheep or goats voluntarily consume is affected by temperature, rainfall, covering of snow and dew, age, breed, stage of production, number of lambs or kids carried or level of milk production, wool or hair covering, rate of respiration, frequency of watering, kind and amount of feed, and exercise. On the average, mature animals consume approximately 1 gal. of water per day, whereas feeder lambs require about half this amount. However, when foraging on grasses and other feeds of high-moisture content, sheep may go for weeks without drinking water. This condition often prevails on desert ranges in the early spring and on many of the mountain ranges during the summer months.

During cold weather, water should be heated in order to keep it from freezing and to insure adequate intake.

Chapter 21, Feeding Goats, further covers the water requirements of goats.

FEEDS FOR SHEEP

Sheep are adapted to consume a great variety of feeds. Most of the common feeds used by them are of plant origin and bulky in nature. The feeding of concentrates is usually limited to finishing lambs and to the breeding flock at such special periods as the lambing season or just before and during the breeding season. Roughages constitute about 90% of the ration of the vast majority of sheep during most of the year. Most lambs are marketed as milk-fat lambs directly off pastures or ranges without having had any grain.

Fig. 5-13. Ewes and lambs on grass. Note shepherd with crook (staff).

PASTURE

No other class of farm animals is so well adapted to the utilization of maximum quantities of pasture as sheep. Although cattle compete with sheep for many of the same grazing areas and are also ruminants, sheep are unique in that the vast majority of the young are marketed as milk-fat animals directly off pastures. Also, in their grazing habits sheep differ from cattle in that (1) they show a decided preference for short, fine forages, and (2) they have the gregarious or flocking instinct.

Although there are great differences in plants, sheep are able to utilize the various grasses, legumes, weeds, forbs (broad-leafed herbaceous plants commonly called weeds by sheep producers), and browse (broad-leafed woody plants, shrubs, bushes, or small trees) that grow on millions of acres of cultivated and uncultivated land in this and other countries. This characteristic, plus the imperishable nature of wool from the standpoint of storage and transportation, has made sheep raising a frontier industry throughout the world.

Sheep are adapted to the grazing of both fenced and unfenced holdings. In this country, most farm flocks and a limited number of range bands of the Southwest are confined to fenced areas with no herders being necessary. On the other hand, most of the range pastures of the western United States, Spain, the Balkans, and African, and Asian countries are utilized by migratory bands under the supervision of herders.

Regardless of the location of the area, year-round grazing is desired. In order to obtain succulent and palatable pastures, the range bands of the West frequently migrate to different altitudes at different seasons of the year. In the mountainous sections, the summer range is usually at high altitudes, the spring-fall grazing is at intermediate altitudes, and much of the winter range is on the desert or lowland areas. On some ranges, such traveling is not possible, the ranges being used on a year-long basis.

During the winter months and following periods of extended drought, pastures usually become leached and bleached. Although these forages can be and are used by sheep, in comparison with green growing grasses, they are lacking in nutrients, being especially low in protein and carotene (provitamin A) content. When providing supplemental feeds, producers should consider these facts.

Abundant and succulent pastures are ideal for stimulating milk production in ewes. Moreover, pastures of this type are desirable from the standpoint of the limited digestive capacity of the young lambs. Accordingly, the degree of finish carried by lambs at market time is an accurate reflection of the amount and quality of the forage available on the pasture or range.

Sheep may successfully utilize either permanent or cultivated pastures. In the range area, the vast ma-

Fig. 5-14. A band of sheep south of Salt Lake City in Scull Valley, the winter desert range. (Courtesy, Union Pacific Railroad Co., Omaha NE)

Fig. 5-15. Ewes and lambs on a western pasture. (Photo by J. C. Allan and Son, West Lafayette, IN)

jority of pastures are of a permanent type, whereas both types of pastures are used by native sheep.

Chapter 6, Pasture Forages, provides additional information on types of pastures and the management of pastures.

HAY AND OTHER DRY ROUGHAGES

Inclement weather, extreme droughts, overstocked pastures and ranges, and leached and bleached pastures make providing dry roughages necessary. Sheep are fond of good roughage and make good use of it. In general, however, they cannot effectively use as much coarse roughage as cattle.

Fig. 5-16. Feeding hay to a wintering flock of ewes. (Photo by J. C. Allen and Son, West Lafayette, IN)

Hays are the standard winter feed when the sheep cannot be out on the pasture or range or when the condition of the pastures is such as to require supplemental feeding. The choicest hay for sheep is a legume which has been produced on fertile soil, cut at the proper stage, and well cured. Such hay is palatable and rich in protein (and the quality of protein is good), calcium and vitamins A and D. If legume hay cannot be secured, a high-quality grass-legume mixed hay will be entirely satisfactory, and much superior to a straight grass hay. Sheep may do very well for a considerable period of time when fed no feed other than a good-quality legume hay, salt, and water.

Although legume hays are preferable for sheep, nonlegume hays are fed extensively in the sheep-raising and lamb-feeding areas. If straight grass hays must be fed, they should be cut at an early stage of maturity. Even then, they will be lower in protein, calcium, and vitamins than the legumes; hence, for best results, protein and mineral supplements should be provided.

Bright, early-cut corn, sorghum fodder, or stover and early-cut, green cereal hays and straws of many kinds are fed to sheep in different areas. The feeding value of these coarser roughages varies considerably according to the stage of maturity at which they are cut, the amount of leaves, and the green coloration. Although these forages are not satisfactory as the sole roughage, especially during the latter part of gestation and in the suckling period, they may be successfully used when mixed with liberal quantities of good-quality legume hays. Where nonlegume roughages are fed, special attention should be given to providing a suitable protein concentrate and minerals, especially calcium.

SILAGES AND ROOT CROPS

Silage consists of green, succulent plants which are usually harvested at an early stage of maturity, chopped, compressed tightly into a silo, and allowed to ferment. During the fermentation, a part of the sugars in the plants are broken down, with the formation of organic acids—such as lactic, acetic, and butyric acids—and the release of carbon dioxide gas. The resulting product more closely approaches green pastures in nutritive value than dry roughages of comparable quality, being higher in carotene content and perhaps superior in other factors.

Silage for sheep may be made from a great variety of plants, including corn, sorghums, cereal grains, legumes, grasses, cannery refuse, pea vines, potatoes, beets, beet tops, sunflowers, and other materials. When properly preserved and fed, silages made from any of these materials are quite satisfactory.

Most practical sheep producers prefer to limit the silage allowance to about 4 to 6 lb per head per day,

Fig. 5-17. Ewes eating stage from a fenceline bunk. Note gravel back of fence where sheep are standing. (Courtesy, *The Sheepman's Production Handbook,* Denver, CO)

with the balance of the roughage ration consisting of hay. If a nonlegume silage is used, it is important that the hay be a legume. If for some unfortunate reason a legume hay cannot be provided when a nonlegume silage is fed, it is very important that a suitable protein concentrate and minerals be provided.

Roots include all plants whose roots, tubers, bulbs, or other underground vegetative parts are used for feed. The important root crops for sheep are: mangels (stock beets), rutabagas (swedes), turnips, and carrots. These feeds are very succulent in nature, containing from 85 to 90% or more of water. They are highly relished by sheep and have a peculiarly beneficial effect upon the digestion and general thrift of animals. The only objection to their general use is the cost and difficulty of growing, harvesting, and storing them. For the latter reasons, roots are raised mainly by producers of purebred animals, and producers in Europe.

Fig. 5-18. Ewes utilizing a root crop in New Zealand. For the most part, roots for sheep are stored and used as a winter feed in the United States. (Courtesy, Department of Scientific and Industrial Research, Wellington, New Zealand)

Roots are generally fed sliced, though some are fed whole. Where the animal's teeth are good, practical sheep producers feel that little is gained by slicing roots. In this country, roots are usually limited to 5 to 6 lb per head daily, but in other countries up to 12 to 14 lb are often allowed daily per ewe.

QUALITY OF FORAGE

The quality of forage—pasture, hay or other dry roughage, or silage—greatly affects its consumption. Forage of high quality is more digestible and passes through the digestive tract more rapidly than forage of low quality; hence, sheep will consume more of it.

By feeding forage harvested before the protein content has decreased and before lignification of the fiber content has increased, producers can usually obtain the most favorable nutritional response. Loss of leaves, weather damage, fermentation, and leaching losses reduce the value of harvested forage.

When it is necessary to feed sheep low-protein forage, adding a suitable protein supplement fortified with needed minerals and vitamins can increase intake and utilization.

CONCENTRATES

Concentrates include those feeds which are low in fiber and high in nutritive value. For purposes of convenience, concentrates are often further classified as (1) carbonaceous feeds, and (2) nitrogenous feeds.

Ordinarily, sheep are fed few concentrates, except immediately before and after lambing, when ewes and rams are being conditioned for breeding, or when lambs are being finished. During these periods, the most frequently used concentrates consist of the common farm grains—oats, corn, barley, wheat, rye, and the grain sorghums. Numerous by-product feeds are also utilized for sheep, including those from the flour- and corn-milling industries, beet byproducts, and oil meals or cakes made from soybeans, cottonseed, and flaxseed.

FEEDS FOR GOATS

Goats are, by nature, browsers. They are continually traveling about in search of feed; and they can effectively utilize feeds that are normally refused by other livestock.

Fig. 5-19. Angora goats are browsers! Note the brush in this pasture. (Courtesy, Mohair Council of America, San Angelo, TX)

PASTURE AND RANGE FORAGES

For using unimproved pasture, goats are without a peer. Mohair and meat-type goats are used extensively to graze unimproved pastures and range areas where vegetation is generally of low quality. Since goats are good browsers, they can be used effectively to control brush and undergrowth. Numerous types of shrubs and woody plants can be utilized as feed for goats with varying degrees of success.

Forages can provide the vast bulk of the nutrients required for maintenance. Thus, the goat rancher should have a good knowledge of the feeding value of the forages available and should supplement them when necessary. Generally, range forages are very low in phosphorus and salt, and often marginal in levels of vitamin A, calcium, and trace minerals. As forages mature, their nutrient value and digestibility decline.

Good-quality pasture and a supply of minerals are all that is required to feed goats at maintenance levels. For the lactating doe, pasture can replace up to one-half of the concentrate in the ration. When pastures are short or when winter limits the availability of good, fresh grass, it is advisable to provide some supplemental feed. This may consist of whole corn, range cubes, or a salt-feed mixture to limit feed intake.

Improved pasture, properly managed and grazed, is a necessity for does and yearlings. Some of the grasses and legumes that can be effectively used in pasture management for goats are alfalfa, alfalfa-brome mix, clover, clover and grass, timothy, bluegrass, Sudangrass, and millet grasses.

Temporary pasture can provide excellent forage. Rye, wheat, and barley are excellent for early-spring or late-fall grazing. Rape, or a combination of rape and oats, has been used with considerable success.

When stresses, such as the effects of weather, are minimized, production becomes more efficient. If goats are placed on pasture or range, the producer must ensure that adequate shade and water are available at all times. Since goats are ruminants, care should be exercised when fresh, lush legume pastures are first used, as bloat problems could result.

Good-quality pastures are essential for high-producing animals. When proper supplementation of grains and minerals is used, these pastures can provide high-quality feed that is cheap and easy to maintain.

Chapter 6, Pasture Forages, provides additional information on the types of pastures and the management of pastures.

HAY AND DRY ROUGHAGES

With the exception of pasture and range feeds, hay and dry roughages make the most economical feeds for goats. A good-quality legume hay or a mixed legume and grass hay provides an excellent source of highly digestible nutrients. Mixed hays should be at least 50% legume, especially if hay is to be the primary source of feed. Grass hays require supplementation with concentrate. Since the cost of concentrates is generally high, the feeding of grass hay as a sole source of roughage is generally uneconomical. Except for dairy goats in lactation, it is not necessary to provide large amounts of concentrates to goats, especially if they are on maintenance or low-production levels.

Hays with the highest nutritive values are those that have tender stems and are leafy. For this reason, hay from second cuttings is generally better utilized than from first cuttings. Palatability of coarse, stemmy hay can be improved by crushing. The stage at which hay is cut has a direct influence on its feeding value. As the grass or legume matures, there is a steady decrease in crude protein content along with an increase in crude fiber content.

The following kinds of hays are most commonly fed to goats: alfalfa, alsike clover, red clover, ladino clover, soybean hay, vetch, birdsfoot trefoil, and mixed legume and grass hays.

SILAGES, HAYLAGES, AND ROOT CROPS

Silages and haylages have never been used extensively as feeds for goats. This is due to management problems more than it is due to any limitations in the nutritional or feeding value of the crops. The only time silage would be practical in goat feeding is when the animals could be confined in a reasonably small area close to a silo.

A small herd of goats would have to be associated with a cattle operation to have enough volume of silage to justify using it as a practical feed. A trench silo 6 ft wide, by 5 ft high, by 50 ft long would provide 1.5 to 2.0 lb of silage daily for 300 goats for 2½ to 3 months. It is necessary to remove at least 3 in. of silage from the exposed surfaces daily during warm weather to keep the silage from spoiling.

Because silage contains only about 30 to 35% dry matter, 2.5 to 3.0 lb is needed to replace 1.0 lb of hay. Normally, silage should be limited to the replacement of only one-third of the hay—about 1.5 to 2.0 lb of silage daily for a mature goat. A young goat should not be fed silage until its rumen is functional; otherwise, digestive disturbances and scouring may result. Even mature goats should be allowed a period of adjustment when silage is incorporated in the ration. Gradual increases in the amount of silage will prevent any digestive disturbances.

Silage may be produced from grain or a hay crop such as grass and/or legumes. If the silage is to be consumed readily by goats, it must be free from moldy or spoiled spots and have a good, clean silage odor. Silage should be fed fresh daily and in amounts that will be consumed within 3 to 4 hours after feeding. In milking operations, no silage should be available for a period of at least 3 hours before milking time; otherwise, odors and flavors from the silage will be present in the milk. It is a good practice to feed silage to dairy goats immediately following milking, rather than before milking. This will ensure that no off-flavors will be imparted to the milk. Silage should not be permitted to accumulate in the manger from one day to the next as it will spoil and become unsuitable as a feed, in addition to lowering palatability of other feed placed in the same manger.

Only about 2 lb of haylage is needed to replace 1 lb of dry hay. However, haylage requires more care in storage to prevent spoilage and the loss of nutrients than does silage. Airtight silos are often used to store haylage. One of the most common problems with haylage is heat during storage which can turn the feed brown, destroy many of the nutrients, and substantially reduce the digestibility of protein.

Goats are quite fond of root crops and garden products; and these types of feeds can be effectively incorporated into the ration for a change of routine. Carrots, beets, turnips, and cabbage are especially relished by goats. These types of feeds are high in moisture and should be fed in the same manner as silage. Several of these feeds, such as turnips, can create off-flavors in milk if fed too close to milking time. A general rule of thumb is to avoid feeding ingredients which may impart flavors to milk within 3 hours of milking time.

CONCENTRATES

The concentrates used in goat rations can be classified as either energy feeds or protein feeds.

ENERGY FEEDS

Cereal grains are excellent sources of energy. Corn, oats, barley, milo, and wheat are frequently used in goat rations. The amount of cereal in the ration should be determined by the production demands. A dry doe requires little or no supplementation, while a doe at the peak of lactation requires substantial amounts of energy.

Molasses, an excellent energy source, is commonly used to reduce the dustiness of feed and to increase palatability. If too much molasses is included in the ration, the feed becomes sticky and lumpy; so, it is usually limited to 5 to 10% of the mixture.

PROTEIN FEEDS

A wide variety of protein supplements can be used in rations for goats. As is the case with feeding other species of livestock, oil meals are used extensively. Cottonseed meal is probably the most widely used source of protein for goats, but other meals can be and are used, depending on their respective prices and availability. Among the alternative sources are soybean meal, copra meal, peanut meal, sunflower meal, safflower meal, corn gluten feed, brewers' dried grains, and distillers' dried grains.

Urea and other nonprotein nitrogen (NPN) sources are often used in rations for Angora goats, but several precautions should be taken when they are incorporated into the ration. Urea can constitute up to 1% by weight of the total concentrate mix or 33% of the protein in the total ration. Rations containing NPN should be introduced gradually into the feeding scheme of goats, as a certain period of adaptation is required by the microorganisms of the rumen. In addition to urea, other NPN sources are ammoniated cottonseed meal, ammoniated rice hulls, ammoniated citrus pulp, and ammoniated beet pulp. NPN should not be included in rations for lactating dairy goats due to toxicity problems.

QUESTIONS FOR STUDY AND DISCUSSION

1. Why is knowledge of sheep and goat nutrition and feeding so important?

2. Define *nutrition*.

3. Why is knowledge of body composition important to the nutritionist? What is the effect of age and the degree of fatness on body composition?

4. Compare the digestive tract of sheep to that of humans or pigs.

5. Classify feed nutrients. What is the significance of such a classification?

6. Discuss the nutrient needs for each of the following body functions:
   a. Maintenance
   b. Growth
   c. Reproduction
   d. Lactation
   e. Finishing (fitting)
   f. Wool and mohair

7. List and describe at least six factors that may influence the energy requirements of sheep and goats.

8. Carbohydrates and fats can serve as energy sources in the diet. What are the differences between these two energy sources?

9. What is a calorie and how is it determined?

10. How are the total digestible nutrients (TDN) of a feed calculated? Compare the TDN of a feed to the Mcal content of a feed.

11. What is the difference between the gross energy of a feed and the net energy of a feed?

12. Why are proteins important in sheep and goat nutrition?

13. When sheep are fed urea or other nonprotein nitrogen, how is it utilized by their bodies?

14. Distinguish between macrominerals and microminerals. List the macrominerals and microminerals essential to sheep and goats.

15. Describe the deficiency symptoms in sheep and goats of each of the following minerals, then tell how you could distinguish between them: calcium, phosphorus, salt (sodium chloride), magnesium, potassium, sulfur, cobalt, copper, iodine, selenium, and zinc.

16. What are the manifestations of toxicity in sheep and goats of each of the following: copper, fluorine, molybdenum, and selenium?

17. Describe the deficiency symptoms in sheep and goats of each of the following vitamins, then tell how you could distinguish between them: vitamin A, vitamin D, vitamin E, and vitamin K.

18. Most lambs are marketed as milk-finished lambs directly off pastures or ranges without having had any grain. What are the advantages of this, from an economic standpoint? Can this practice be applied to cattle or swine?

19. How do sheep differ from cattle in their grazing habits?

20. Compare the feeds for sheep to the feeds for goats.

21. Why are silages and haylages not routinely fed to goats?

22. What precautions should be taken when silage and root crops are fed to dairy goats?

SELECTED REFERENCES

Title of Publication	Author(s)	Publisher
Angora Goat and Mohair Production	M. Shelton	M. Shelton, San Angelo, TX, 1993
Animal Science, Ninth Edition	M. E. Ensminger	Interstate Publishers, Inc., Danville, IL, 1991
Bioenergetics and Growth	S. Brody	Reinhold Publishing Co., New York, NY, 1945
Body Composition in Animals and Man	National Research Council	National Academy of Sciences, Washington, DC, 1968
Effect of Environment on Nutrient Requirements of Domestic Animals	National Research Council	National Academy Press, Washington, DC, 1981
Feeds & Nutrition, Second Edition	M. E. Ensminger J. E. Oldfield W. W. Heinemann	Ensminger Publishing Company, Clovis, CA, 1990
Feeds & Nutrition Digest, Second Edition	M. E. Ensminger J. E. Oldfield W. W. Heinemann	Ensminger Publishing Company, Clovis, CA, 1990
Fire of Life, The	M. Kleiber	John Wiley & Sons, Inc., New York, NY, 1961
Foods & Nutrition Encyclopedia, Second Edition	A. H. Ensminger M. E. Ensminger J. K. Konlande J. R. K. Robson	CRC Press, Boca Raton, FL, 1994
Fundamentals of Nutrition	L. E. Lloyd B. E. McDonald E. W. Crampton	W. H. Freeman & Co., San Francisco, CA, 1978
Nutrient Requirements of Goats	National Research Council	National Academy of Sciences, Washington, DC, 1984
Nutrient Requirements of Sheep	National Research Council	National Academy of Sciences, Washington, DC, 1985
Planned Sheep Production, Second Edition	D. Croston G. Pollott	Blackwell Scientific Publications, London, England, 1994
Profitable Sheep	S. B. Collins	The Macmillan Company, New York, NY, 1956
Sheep Book, The	J. McKinney	John Wiley & Sons, Inc., New York, NY, 1959
Sheepman's Production Handbook, The	Staff	American Sheep Industry Association, Englewood, CO, 1992
Sheep Science	W. G. Kammlade, Sr. W. G. Kammlade, Jr.	J. B. Lippincott Co., Philadelphia, PA, 1955

CHAPTER 6

Ninety-four percent of the total feed supply of U.S. sheep is derived from forage, most of which is pasture. (Courtesy, USDA)

PASTURE FORAGES

Contents *Page*

Note: Also see Chapter 16, Range Sheep Management, including sections on Range Grazing Systems.

Pasture is a tremendous forage resource, capable of intercepting and storing large amounts of solar energy and consequently supporting high levels of sheep production if managed properly. Also, and most important, no method of harvesting has been devised which is as cheap as that which can be accomplished by grazing animals.

As the ever-increasing human population of the world consumes a higher proportion of grains and seeds directly, there will be increased reliance on grass for meat, milk, wool and mohair production. Sheep and goats are completely recyclable, producing a new crop each year and perpetuating themselves through their offspring, without the need of fossil fuel energy.

Grassland agriculture, better than any other type of agriculture, will continue in the face of economic and social changes to conserve the land and ensure a food supply of the desired quantity, variety, and quality. At its best, it calls for an interdisciplinary approach—for knowledge and application of soil, plant, and animal sciences. This joint focus characterizes the great livestock areas of the world.

CLASSES OF PASTURE

A pasture is an area of land on which there is a growth of forage that animals may graze. Broadly speaking, all U.S. pastures may be classified as either (1) tame (seeded) pastures, or (2) native pastures. Although no sharp line of demarcation exists between the two groups, tame pastures include those which receive more than approximately 20 in. of rainfall annually or those which are irrigated. They are the seeded (cultivated) pastures of the Corn Belt, the South, the East, and the irrigated areas, and the smaller, scattered moderate to high rainfall areas, throughout the West. Six grass species—orchardgrass, reed canarygrass, tall fescue, smooth bromegrass, ryegrasses, and Bermudagrass—account for the major portion of tame forage grass production in the United States.

The native pastures include those range pastures which receive less than 20 in. of rainfall annually. Their vegetative cover, known as native plants, consists of adapted plants developed by natural selection that have existed in the area for many years, and that were not introduced by humans.

In addition to the tame (seeded) pastures, and the native pastures, there is considerable acreage of nonirrigated hill pastures on the west coast of Washington, Oregon, and California which do not fit under either of these traditional classifications. The annual rainfall of the area varies from 12 to 14 in. in California to more than 100 in. in coastal Oregon and Washington. More and more of these brushy hills are being developed into productive tame (seeded) pastures. The higher rainfall areas are being seeded to subterranean clover and perennial ryegrass, with tall fescue included where beef cattle are more important than sheep. In the drier areas, ryegrass, Hardinggrass, and some annual grasses are being seeded. There are an estimated 25 million acres of this brushy hill land, once classified as range, which have the potential for development as improved or tame pasture.

Pastures may be further classified as:

1. **Permanent pastures.** Those which, with proper care, last for many years. They are most commonly found on land that cannot be used profitably for cultivated crops, mainly because of topography, moisture, or fertility. The vast majority of the farms of the United States have one or more permanent pastures, and most range areas come under this classification.
2. **Semipermanent or rotation pastures.** Those that are used as a part of the established crop rotation. These are seeded pastures that are generally used for 2 to 7 years before plowing.
3. **Temporary and supplemental pastures.** Those that are used for a short period; and they are usually annuals, such as Sudangrass, sorghum, millet, rye, barley, wheat, oats, ryegrass, arrowleaf clover, crimson clover, ball clover, rape, kale, and turnips. They are generally seeded for the purpose of providing supplemental grazing during the season when the permanent or rotation pastures are relatively unproductive.

Pasture plants are classed as (1) grasses, (2) legumes, (3) browse, or (4) forbs.

1. **Grasses.** *Botanically, any of the plants of the family Gramineae.* In grassland agriculture, grasses refer to the forage species of Gramineae when grown alone or with legumes.
2. **Legumes.** *Plants, such as alfalfa and the clo-*

Fig. 6-1. Ewes in Illinois grazing turnips less than 60 days after planting. (Courtesy, David L. Thomas, University of Wisconsin, Madison, WI)

Fig. 6-2. Goats have unique preferences for shrubs and tree leaves, whether deciduous or evergreen. (Courtesy, George F. W. Haenlein, University of Delaware)

vers, that obtain nitrogen through bacteria living in their roots.

3. **Browse.** *The edible parts of woody vegetation, such as leaves, stems, and twigs from bushes.*

4. **Forbs.** *Nongrasslike range herbs which animals eat (forbs are generally called weeds by western producers).*

Pastures vary greatly in quality, depending on type (variety), soil, growing conditions, and stage of maturity. Mature grasses, especially those that are leached and bleached, are low in palatability, digestibility, protein, carotene, and some of the minerals. Grasses are usually adequate in calcium, magnesium, and potassium, but they are apt to be borderline or deficient in phosphorus, and they may be low in some of the trace minerals.

Grazing is the process by which animals harvest their own feed (grasses, legumes, browse, and/or forbs) from its growing condition. It is the oldest form of harvesting feed, and in many respects it is still the most reliable and productive method when properly used.

But grass—the nation's largest crop—should no longer be taken for granted. Seeding new and better varieties of grasses and legumes, fertilizing, and wise management, including using scientifically controlled grazing, avoiding overgrazing by both domestic livestock and wild animals, and using supplemental feeding, can improve most grazing areas.

Again and again, scientists and practical farmers and ranchers have demonstrated that the following desired goals in pasture production are well within the realm of possibility:

- To produce higher yields of palatable and nutritious forage.
- To extend the grazing season from as early in the spring to as late in the fall as possible.
- To provide a fairly uniform supply of feed throughout the entire season.

No one plant embodies all the desirable characteristics necessary to meet the desired goals of pasture production. None of them will grow year round, or during extremely cold or dry weather. Each of them has a period of peak growth which must be conserved for periods of little growth. Consequently, the progressive producer will find it desirable (1) to grow more than one species, and (2) to plan pastures for each season of the year. In general, a combination of permanent, rotation, and temporary pastures—accompanied by scientific management—will best achieve these ends.

PASTURES FOR SHEEP AND GOATS

The economic importance of pastures for sheep and goats continues to be demonstrated in many experiments and on thousands of livestock farms and ranches.

Sheep and goats are able to utilize the various grasses, legumes, weeds, herbs, and browse that grow on millions of acres of cultivated and uncultivated lands in this and other countries. This characteristic, plus the gregarious or flocking instinct of sheep, has made sheep raising a frontier industry throughout the world. Furthermore, 94% of the total feed supply of all U.S. sheep and goats is derived from forage; for the most part, this means pasture. No other class of farm animals is so well-adapted to the utilization of maximum quantities of pasture as sheep. They are unique

Fig. 6-3. Sheep on pasture in Vermont. (Courtesy, American Sheep Industry Association, Englewood, CO)

in that the vast majority of their young are marketed as milk-fed animals directly off grass.

ADAPTED AND/OR COMMON GRASSES AND LEGUMES OF THE UNITED STATES

The specific grass or grass-legume mixture will vary from area to area, according to differences in soil, temperature, and rainfall. Fig. 6-4 shows the 10 generally recognized U.S. pasture areas; and Table 6-1 lists the best adapted and/or most common grasses and legumes for each of these areas.

In using Fig. 6-4 and Table 6-1, one must recognize that many species of forages have wide geographic adaptation, but varieties often have rather specific adaptation. Thus, alfalfa, for example, is represented by many varieties which give this species adaptation to nearly all states. Variety then, within species, makes many forages adapted to widely varying climate and geographic areas. County agricultural agents and state agricultural colleges can furnish recommendations for the areas that they serve.

For more specific and individual farm recommendations, farmers and ranchers are urged to seek the advice of local authorities or to write to their state agricultural college.

The following points are pertinent to the recommendations given in Table 6-1:

1. **Fertilizer rates.** Because of the high price of fertilizer, along with concern relative to possible pollution of groundwater, fertilizer rates should be based on soil test values. Although the practice of soil testing is increasing, from authoritative sources the author has determined that of the nation's currently chemically fertilized pastures (not including lime), only an estimated 15 to 20% of the applications are made on the basis of soil tests. *Soil tests are urged.*

After legumes have been lost from a grass-legume sward, it is recommended that they be re-established in the grass sod. If the latter is not feasible, an annual nitrogen application at the rate of 60 to 100 lb of actual nitrogen per acre per year should be applied in split applications or increments.

2. **Varieties.** The best guide for the varietal selection of grasses and legumes from the numerous varieties available is the use of certified seed of an adapted variety.

Fig. 6-4. The 10 generally recognized U.S. pasture areas.

TABLE 6-1
ADAPTED GRASSES AND LEGUMES (INCLUDING BROWSE AND FORBS) FOR SHEEP AND GOAT PASTURES, BY 10 GEOGRAPHICAL AREAS OF THE UNITED STATES (SEE FIG. 6-4 FOR GEOGRAPHICAL AREAS)[1]

	Areas of the United States									
	1	2	3	4	5	6	7	8	9	10
Grasses, shrubs, forbs:										
Bahiagrass			X	X						
Bermudagrass		X	X	X		X		X		X
Bluegrass, big							X		X	
Bluegrass, Kentucky	X	X	X		X		X		X	
Bluestem, big	X	X	X	X	X	X				
Bluestem, Caucasian		X	X			X				
Bluestem, little	X	X	X	X	X	X				
Bluestem, sand	X	X			X	X				
Bristlegrass, plains						X		X		
Bromegrass, meadow					X		X	X	X	
Bromegrass, smooth	X	X			X		X	X	X	X
Buckwheat (wild)								X		
Buffalograss					X	X				
Buffelgrass						X				
Canarygrass, reed	X	X					X		X	
Cottontop, Arizona								X		
Curly mesquite						X		X		
Dallisgrass			X	X						
Digitgrass, pangola			X	X						
Dropseed, sand						X		X		
Fescue, tall	X	X	X				X	X	X	X
Foxtail, creeping							X		X	
Galleta						X		X		
Gamagrass, eastern	X	X	X	X	X	X				
Grama, black								X		
Grama, blue					X	X	X	X		
Grama, sideoats	X	X	X	X	X	X	X	X		
Hardinggrass						X			X	X
Indiangrass	X	X			X	X				
Indianwheat								X		
Johnsongrass			X	X		X				
Kleingrass						X				
Koleagrass, Perla									X	X
Limpograss				X						
Lovegrass, Lehmann							X	X		
Lovegrass, sand	X	X			X	X				
Lovegrass, weeping			X			X				
Maidencane				X						
Millet	X	X	X	X		X				
Muhly, spike							X	X		
Needle-and-thread	X				X					
Needlegrass, green	X				X					
Oatgrass, tall									X	
Oats	X	X	X	X	X	X		X	X	X
Orchardgrass	X	X	X	X	X		X	X	X	X
Paragrass				X						
Pearlmillet		X	X	X		X				
Redtop	X						X		X	
Rescuegrass			X	X					X	
Rhodesgrass			X	X						
Ricegrass, Indian							X	X		
Rye	X	X	X	X	X	X		X	X	X
Ryegrass, annual		X	X	X		X			X	
Ryegrass, perennial	X	X	X						X	
Sacaton, alkali						X	X	X		

	Areas of the United States									
	1	2	3	4	5	6	7	8	9	10
Sage, pitchers	X	X	X		X	X				
Saltbrush, fourwing					X	X	X	X		
Sorghum-Sudan hybrids	X	X	X	X	X	X	X			
Stargrass			X							
Sudangrass	X	X	X	X	X	X	X	X	X	X
Sunflower, Maximilian	X	X	X		X	X				
Switchgrass	X	X	X		X	X				
Three-awn				X	X	X	X	X		
Timothy	X	X					X		X	
Tobosa grass						X				
Wheat	X	X	X	X	X	X	X	X	X	X
Wheatgrass, bluebunch					X		X		X	
Wheatgrass, crested					X		X			
Wheatgrass, intermediate	X				X		X		X	X
Wheatgrass, pubescent					X	X	X			X
Wheatgrass, tall	X				X	X	X	X	X	
Wheatgrass, western	X	X			X	X	X		X	
Wild-rye, basin					X		X			
Wild-rye, Canada	X	X	X		X	X	X			
Wild-rye, Russian					X		X			
Winterfat (white sage)								X		
Wintergrass, Texas						X				
Legumes:										
Alfalfa (lucerne)	X	X	X	X	X	X	X	X	X	X
Alyceclover			X	X						
Black medic (yellow trefoil)			X			X		X		
Bur-clover		X						X		X
Clover, alsike	X	X			X		X	X	X	
Clover, arrowleaf		X	X							
Clover, crimson		X	X							
Clover, Hubam (white sweet clover)	X	X						X	X	
Clover, Kura	X	X			X		X			X
Clover, Ladino	X	X	X	X			X	X	X	X
Clover, prairie						X		X		
Clover, red	X	X	X	X			X	X	X	
Clover, strawberry					X		X	X		X
Clover, subterranean			X	X				X	X	X
Clover, white	X	X	X	X			X	X	X	X
Cowpeas			X	X						
Crown vetch	X	X								
Flat pea			X	X			X			
Hairy indigo				X						
Lespedeza (annual)		X	X	X						
Lespedeza (perenial, sericea)		X	X	X						
Milk vetch, cicer	X				X		X			
Peas, field									X	
Pea shrub								X		
Prairie clover, purple	X	X	X		X	X	X			
Ratany								X		
Soybeans	X	X	X	X			X			
Sweet clover, white	X	X						X	X	
Sweet clover, yellow	X	X			X	X	X	X	X	
Trefoil, birdsfoot	X	X	X				X		X	X
Velvet bean			X	X						
Vetch		X	X	X	X	X		X	X	

[1]For each of the recognized pasture areas, see Fig. 6-4.

SEEDING AND MANAGEMENT OF PASTURES

This section, and the subsections under it, have reference to those pastures that receive above approximately 20 in. of rainfall annually or that are irrigated. These include the pastures of the Corn Belt, the South, the East, and the irrigated valleys and smaller, scattered moderate-to-high-rainfall areas throughout the West.

Fig. 6-5. Sheep on pasture in the state of Washington. (Courtesy, American Sheep Industry Association, Englewood, CO)

ESTABLISHING A NEW PASTURE

The following practices are usually adhered to in the successful establishment of new pastures in the sub-humid, humid, and irrigated areas:

1. Adapted varieties and suitable mixtures are selected (see Table 6-1).
2. The soil is tested and fertilized.
3. High-quality seed is purchased.
4. Scarified legume seed is used to assure quicker and more uniform germination.
5. Legume seed is inoculated with the proper bacteria.
6. A good seedbed is prepared.
7. The seeding operation is timed and carried out as determined by the area and by the species or mixture used.
8. A companion or nurse crop may or may not be included.

IMPROVING OR RENOVATING AN OLD PASTURE

In altogether too many cases, old permanent pastures are merely gymnasiums for livestock. Generally, this condition exists because the least productive areas are used for pastures, and because little attention is given to fertility and pasture management.

Permanent pastures in sub-humid, humid, and irrigated areas that are run down may be brought back into production by either (1) reseeding without growing a crop in the interim, or (2) fertilizing, overseeding, and managing.

FACTORS AFFECTING VALUE OF PASTURE

Many factors affect the value of pasture, including (1) soil and fertilizer, (2) plant species, (3) stage of maturity, (4) rate of growth and season of year, and (5) grazing.

■ **Soil and fertilizer**—Soil and fertilizer affect the growth and composition of pasture crops. Many experiments have been conducted to determine the effect of soil fertility and fertilizer application on pasture. Some of the benefits that generally accrue from pasture fertilization include: (1) increased yields, (2) increased proportion of legumes, (3) extended grazing season, (4) increased protein and palatability, and (5) increased calcium and phosphorus.

■ **Plant species**—Plant species affect the feeding value of pasture. Generally, legumes contain a higher percentage of protein and calcium than nonlegumes. Also, there are marked differences between various kinds of pasture plants in the rate of changes as growth advances. For example, bromegrass retains its palatability and nutritive value over a longer period than most grasses. By contrast, reed canarygrass is readily eaten when young, but becomes woody, high in alkaloids, and unpalatable with maturity. Most legumes retain their palatability and nutritive value as they mature better than most grasses. An exception to the latter rule is lespedeza sericea, which becomes bitter and distasteful with maturity due to the accumulation of tannin in the plants. However, plant breeders have developed sericea that is low in tannin, thereby overcoming this problem to a considerable extent.

■ **Stage of maturity**—Many producers are not aware of the great differences in nutritive value between young, immature pasture and the same plants when they are mature or even at the usual hay stage. These wide differences are shown in Chapter 27, Feed Composition Tables, and in Table 6-2.

Table 6-2 shows that the stage of maturity affects pasture composition as follows:

1. Protein decreases with maturity.
2. Fiber increases with maturity.

TABLE 6-2
SEASONAL VARIATION OF SOME NATIVE RANGE GRASSES
(Dry Matter Basis)[1]

Month	Crude Protein	Crude Fiber	Phos-phorus	Carotene	Units Vitamin A[2]
	(%)	(%)	(%)	(mg/lb)	(IU/lb)
April.	11.1	24.0	0.149	118.0	47,200
May.	9.7	27.9	0.149	121.0	48,400
June	7.3	32.4	0.098	166.0	66,400
July.	6.3	33.9	0.085	96.0	38,400
August.	5.2	36.7	0.083	49.0	19,600
September . . .	4.2	37.2	0.059	22.0	8,800
October	3.2	38.0	0.053	5.0	2,000
November . . .	2.7	39.8	0.039	4.0	1,600
December . . .	2.5	40.4	0.034	0.5	200
January	2.5	40.8	0.038	0	0
February	2.5	41.2	0.045	0	0
March	2.3	40.6	0.030	0	0

[1]Composition of native range grasses near Stillwater, Oklahoma based on chemical analyses made by Oklahoma State University.

[2]One mg of beta-carotene is equal to 400 IU of vitamin A for sheep and goats

3. Phosphorus decreases with maturity.
4. Vitamin value decreases with maturity.

■ **Rate of growth and season of year**—Rapidly growing grass is usually rich in protein and in other nutrients on a dry basis. It is important, therefore, that pasture plants be properly fertilized and managed so that they will keep growing and will not head out.

Grass is usually higher in protein and other nutrients early in the spring than later in the season. If plant growth is sharply checked in the summer—due to drought, hot weather, and/or lack of available plant food—the protein content and the digestibility will be lower than those of grass at the same stage of maturity earlier in the season.

If pasture resumes growth after the fall rains come, it may be nearly as high in protein and other nutrients as spring growth.

Permanent pastures in the warmer parts of the United States—in the South—generally have a much lower percentage of protein than do grasses at the same stage of maturity in the cooler climates of the North.

■ **Grazing**—When pastures are grazed closely throughout the season, the total yield of dry matter is usually 30 to 50% less than when they are allowed to grow to the normal hay stage. This is due to the smaller leaf surface and lowered photosynthesis. This explains why rotational, strip, and green chop grazing usually yield more than close continuous grazing.

The effect of frequent grazing will depend on the kinds of plants. The yield of tall-growing plants—such as timothy, orchardgrass, alfalfa, and the erect clovers—is reduced much more than that of low-growing spreading plants, such as bluegrass, Bermudagrass, and white clover.

In contrast to the lowering of the yield of dry matter, frequent grazing usually results in greater total production of protein for the season than when the crop is cut for hay. Also, because immature plants are lower in fiber and more digestible than mature plants, the yield of total digestible nutrients is not reduced as much by frequent grazing as the yield of dry matter—dry matter production is lowered by 30 to 50%, whereas digestibility is lowered only by 25 to 40%.

Plenty of available forage results in selective grazing—with the animals picking and choosing the leaves and finer parts of stems, which are more tender and more nutritious, and rejecting the coarser, stemmy parts. Thus, the portion consumed under such circumstances may differ appreciably from the chemical composition of the entire plant.

PASTURE MANAGEMENT

Many good pastures have been established only to be lost through careless management. Good pasture management in the sub-humid, humid, and irrigated areas involves the following practices:

1. **Controlled grazing.** Nothing contributes more to good pasture management than controlled grazing. At its best, it embraces: (a) protecting first-year seedings; (b) shifting the location of the salt, shade, and water to promote uniform grazing; (c) deferring spring grazing to give plants a needed start, (d) avoiding close late-fall grazing, (e) avoiding overgrazing, and (f) avoiding undergrazing.
2. **Clipping pastures and controlling weeds.** Pastures should be clipped at such intervals as necessary to control weeds (and brush) and to get rid of uneaten clumps and other unpalatable coarse growth left after incomplete grazing. Good grazing management will reduce the amount of clipping needed. Pastures that are grazed continuously may be clipped at or just preceding the usual haymaking time; rotated pastures may be clipped at the close of the grazing period. Weeds and brush may also be controlled by chemicals and burning.
3. **Topdressing.** In most areas it is desirable and profitable to topdress pastures with fertilizer annually, and, at less frequent intervals, with reinforced manure and lime. Such treatments should be based on soil tests, and are usually applied in the spring or fall.
4. **Scattering droppings.** The droppings should be scattered at the end of each grazing season in order to prevent animals from leaving ungrazed clumps and to distribute the droppings over a larger area. This can best be done by a brush harrow or a chain harrow.

5. **Grazing by more than one kind of animal.** Grazing by two or more species of animals makes for more uniform pasture utilization and fewer weeds and parasites, provided the area is not overstocked. Different kinds of livestock have different habits of grazing; they show preference for different plants and graze to different heights. For example, sheep consume shorter and finer forages and more forbs than cattle.

6. **Irrigating where practical and feasible.** Where irrigation is practical and feasible, it alleviates the necessity of depending on natural precipitation.

EXTENDING THE GRAZING SEASON

In the South and in Hawaii, year-round grazing is a reality on many successful farms. By carefully selecting the proper combination of crops, farmers and ranchers in other areas can make pastures for each month of the year a reality.

In addition to lengthening the grazing season through the selection of species, farmers and ranchers may extend it by (1) obtaining earlier spring pastures, (2) saving fall growth for winter grazing, and (3) using crop residues.

SEEDED PASTURE GRAZING SYSTEMS

Several systems of grazing management have been successfully applied to pastures. Generally speaking, the more intensive the system of management on such pasture, the higher the yield of forage and of livestock products.

It is noteworthy that pasture grazing systems have been changed/adapted by both researchers and farmers; and with such changes/adaptations, they have been given different names. Nevertheless, under whatever name, the basic type of rotation grazing, intensive grazing, creep grazing, strip grazing, and green chop are covered in the sections that follow.

(Also, see the section on Range Grazing Systems in Chapter 16, Range Sheep Management. The principles involved in grazing seeded pastures and western ranges are similar. However, the application differs because western range pastures are generally much larger and have lower rainfall.)

CONTINUOUS GRAZING

The name identifies the practice. *Continuous grazing is the uninterrupted grazing of a specific pasture by livestock throughout the year or grazing season.* It can be successful provided moderate stocking is practiced, with some adjustment in animal numbers to reduce the severity of under or overgrazing.

Fig. 6-6. Goats on continuous grazing. (Courtesy, Ralston Purina Company, St. Louis, MO)

The **advantages** of continuous grazing as compared to rotational grazing are (1) lower costs for fencing and watering facilities, (2) fewer management decisions when animals are not moved from pasture to pasture, and (3) often slightly better individual animal performance when younger animals are grazed.

The **limitations** of continuous grazing are (1) animal numbers are seldom flexible; (2) pastures must be stocked lighter than desired when forage growth is maximal to avoid overgrazing during periods of minimal forage growth; (3) animals selectively graze some species in preference to others and return to graze the regrowth of the same plant, thus selectively reducing plant vigor; and (4) livestock often show preference to grazing certain portions of pastures, resulting in uneven fertilization.

Because animal numbers under continuous grazing are seldom flexible, excess forage can best be harvested for silage or hay, thereby providing a practical means of balancing available forage and animal numbers.

ROTATION GRAZING

Rotation grazing is that system in which two or more pastures are grazed and rested in a planned sequence. In this system, pastures are divided into two or more pastures, with the objective to develop a grazing program where major forage species are harvested and then provided a period of rest enabling the plants to regrow and remain thrifty and vigorous.

Rotation grazing involves the concept of time as a management variable for either the grazing period or the regrowth interval of each pasture. Duration of grazing and rest generally are governed by herbage growth rate, which depends primarily on the time of year, moisture, fertility, and species.

If, during the grazing season, an average regrowth period of 30 days and a grazing period of 5 days is required, it is apparent that each 35-acre area must be subdivided into seven areas of 5 acres each. Generally, each area is grazed intensively for 5 to 7 days and at intervals of 3 to 5 weeks.

The **advantages** of a rotation grazing system are:

1. It permits the farmer to match grazing more adequately to the growth habit of the forage species, condition of the pasture, and animal needs than does continuous grazing.
2. It improves stand persistence and production. Plants are given recovery periods during the growing season for more or less unhampered development of tillers and leaves. This is essential to replenish root reserves. This system of grazing enables the tall-growing legumes and grasses to survive.
3. It increases carrying capacity. Greater amounts of feed nutrients can be removed in the form of herbage with reduced losses due to trampling, fouling, and herbage death and decay.
4. It encourages equalization of grazing. It helps prevent overgrazing and undergrazing, and results in maintaining a better balance of the legumes and grasses. Also, both the palatable and the inferior species are grazed more nearly the same.
5. It often provides more nutritious herbage since the herbage is at the most ideal pasture stage. It will be high in protein and low in crude fiber.
6. It helps prevent the grasses from heading out. This is done by concentrating grazing animals or by mowing when animals are shifted to new pastures. This allows new growth to come back uniformly and keeps it more palatable.
7. It helps control livestock parasites. Life cycles of worms can be broken by proper planning of grazing and rest periods.
8. It makes it more convenient to harvest surplus forages as hay or silage.

The **limitations** of rotation grazing are:

1. It requires a higher input of capital and management than continuous grazing.
2. There is a continuous day-to-day decline in the quality of the available forage, especially on the more intensive systems. At first turn-on, animals have access to leafy, high-quality forage, but the quality of the forage gets poorer and poorer during the grazing period.

INTENSIVE GRAZING

Several ingenious intensive grazing systems have evolved. All of them are designed to provide and harvest the maximum of high quality forage; to utilize the highest quality pastures for the highest producing animals; and to increase profits.

These systems were first described and used in Europe by Voisin and have since been installed to some extent throughout the United States and much of the world. Many designs are utilized for intensive short duration grazing systems. Basically, they fall into either a rectangular design or a circular or wagon wheel design.

■ **Conventional (rectangular) systems**—The rectangular system generally is a series of small pastures, usually of equal size or production, that are fenced in a grid arrangement. Water and salt may be located in each of the pastures or a single source may be used with cattle gaining access by a lane or alley.

■ **Cell or wagon wheel system**—This system was introduced and popularized in the United States by Alan Savory. The pasture fences radiate out from a central hub, giving it the appearance of a wheel, thus the name. Water, salt and mineral, and working pens are usually located at the hub. As livestock return to the hub daily, they can easily be moved from one pasture to the next.

■ **First and second grazers**—This short duration grazing system involves two herds: first grazers and second grazers. It calls for using the best quality pastures for ewes with multiple lambs and does with kids. The groups may consist of any animal species or class. For example, the first grazers may consist of lactating does, and the second grazers may be dry does.

The chief **advantage** of the system of first and second grazers is the enhanced productivity of the first grazers. The main **limitations** are the necessity of maintaining (1) two groups of animals of different productivity levels, and (2) balanced stocking rates and pasture sizes.

■ **Intensive early season stocking**—This grazing system calls for heavy stocking (perhaps twice the average summer carrying capacity of the pasture) in the spring and early summer, when the pasture is of highest quality and the most productive.

The **advantages** of this system are (1) more pounds of product per acre, (2) lower interest charges, because of owning the animals for a shorter period of time, and (3) higher net returns. The main **limitation** is the lack of flexibility relative to removal of half the animals; they must either be sold or moved into the feedlot as scheduled—in mid-summer.

CREEP GRAZING

Creep grazing is a system of grazing nursing young on a high-quality pasture(s) (a grass-legume mixture, all legumes, or high-quality annuals) separated from their dams. This system may be used for ewes and lambs, or for goats and kids. Creep grazing may be accomplished as follows:

1. Allowing young to forward graze ahead of their dams, then following with their dams later. This is similar in principle to "first and second grazers," with the young having access to the choicest most succulent pasture(s) without competition from their dams. It may be accomplished by confining the dams and young in one field for a period, but allowing the young to enter the next choice-quality pasture(s) through a creep opening large enough for the young but small enough to keep the dams out. The dams are kept on each pasture in the rotation until forage is utilized to the desired level, then moved to the new pasture. This same pattern is contained through all pastures in rotation.
2. Keeping the dams and young on a base pasture, and providing an additional creep pasture for the young. The dams and young are kept on a base pasture (or pastures if rotational). In addition the young are given access to a high quality creep pasture(s) through a creep opening.

Limited studies in creep grazing indicate that as much as one-half pound extra production may be obtained by creep grazing.

STRIP GRAZING

In this system, animals are allowed access to a strip which may be large enough for several days of grazing or small enough for one-half to one day of grazing. Heavy stocking rates of upwards to 50 animal units per acre are used by fencing each strip with movable electric fences both in front and behind the grazing livestock. The **advantages** claimed for this method are:

1. Increased utilization of herbage, with wastage reduced to 10 to 20%.
2. Increased meat and milk yields per acre up to 25%.
3. Improved stability of meat and milk yield because the nutritive value of the pasturage consumed is quite constant.
4. Improved utilization of the available forage. Less herbage is soiled by dung, urine, and treading. Under strip grazing, animals are quieter and settle down quickly for steady grazing rather than roaming about and tramping forage.
5. Increased animal units maintained on a given area, although individual animal productivity may not be increased.

IRRIGATED PASTURES

Well managed irrigated pastures can enhance the flexibility and add to the stability of livestock forage programs. Throughout the western United States, irrigated pastures have been used successfully to improve carrying capacities, reduce feed costs, lengthen the grazing season, improve gains and milk production, and improve breeding efficiency.

Irrigated pastures provide forage of high quality at a relatively low cost, often on land unsuitable for other crops. Both perennial and annual irrigated pastures are important feed crops.

Successful pasture irrigation involves special decision making relative to (1) irrigation—the method, frequency, and amount of irrigating, and the removal of excess water; and (2) the kind and amount of fertilizer.

■ **Method of water application**—Two basic methods of irrigating pastures are practiced: (1) flood, and (2) sprinkler. The choice of the method for any given pasture should be determined by soil type, topography, water supply, and funds available for irrigation development.

The efficiency of flood irrigation can be improved with the use of borders. The border-flood method is adapted where a large head of water is available and the land is level or requires only minor movement. When properly used, there is no runoff and efficiency of water utilization is high.

Sprinkler irrigation may be preferable to the flood method (1) on land that is not level enough for surface irrigation or where the cost of leveling would be prohibitive; (2) on soils of variable texture where the amount and frequency of application can be adjusted to the water-holding capacity of the soil; or (3) where water cost is high or the supply of water is limited. Sprinklers are on the increase throughout the United States, with laborsaving, center-pivot and wheel move systems making hand-move systems obsolete.

■ **Frequency and amount of irrigation**—It is recommended that in many parts of the United States, especially the west, irrigation water must be applied when it is available, not necessarily when it is desired, and often it may not be available for part of the season. Such restriction may severely limit pasture production. However, when possible and practical, water should be applied: (1) at a rate and frequency to maintain good soil moisture throughout the root profile; (2) immediately following grazing where grazing is rotated; (3) according to the consumptive use rate of the major species;

and (4) relative to the content of soluble salt, as soils high in salt may be flushed while irrigation water high in salts should be used sparingly.

Many pasture plants, especially the clovers, are shallow-rooted and require more frequent and lighter irrigations than deep-rooted plants. The Washington Station reports that the highest yields per acre from an orchardgrass-ladino clover pasture can be obtained with a summer irrigation frequency of 7 to 11 days, rather than at less frequent intervals, and that more frequent irrigations also give the highest proportion of clover. In important irrigated areas, county extension agents and district conservationists are usually knowledgeable relative to the proper time to irrigate specific crops in the particular area; hence, they should be consulted when developing a schedule.

■ **Excess water**—Excess water in irrigated pastures is caused by either (1) overirrigation and the inability of the excess water to drain from the soil, or (2) subsurface drainage from adjacent and higher land. Allowing excess water to stand on pasture can drown desirable plants, with the resulting area growing up in weeds. Also, standing water is a breeding ground for insects. Surface drains are necessary to remove excess irrigation. Drainage is particularly important wherever there is danger of salt accumulation. Deep drainage ditches spaced at proper intervals help to remove these excess salts and control the level of the water table. In some areas, drainage ditches must be augmented by tile drainage in order to keep the salt content below levels that are harmful to plants.

■ **Plant species**—Selection of species for establishment of irrigated pastures should be dependent on (1) adaptation to the general area, (2) water availability, (3) soils, (4) salinity problems, and (5) forage needs.

■ **Fertilization**—Irrigated pastures require high soil fertility to be productive. The kind and amount of fertilizer should be determined by the level of the productivity desired, and the role of the legumes in the mixture. Production levels of irrigated pastures are increased more by N fertilizer than by other fertilizer elements, with responses also obtained from P and K where soils are deficient in these elements. Nitrogen stimulates grass growth, whereas P increases the legume component. Nitrogen can be supplied by either fertilizer or inoculated legumes. When legumes are a major component of pasture, economic returns from applied N, measured in increased animal production, may not be obtained.

■ **Grazing**—Although continuous grazing has been used effectively in some locations, the potential benefits from irrigated pastures in the West are of such magnitude that some form or rotation grazing should be employed.

QUESTIONS FOR STUDY AND DISCUSSION

1. What are the primary differences between (a) tame (seeded) pastures, and (b) native pastures?

2. What are the primary differences between (a) permanent pastures, (b) semipermanent or rotation pastures, and (c) temporary and supplemental pastures?

3. Define each of the following terms: (a) grasses, (b) legumes, (c) browse, and (d) forbs.

4. Discuss the economic importance of pastures for sheep and goats.

5. How would you go about determining what grass and/or legume to seed on a particular farm or ranch?

6. How may a sheep or goat producer use Table 6-1, Adapted Grasses and Legumes?

7. Outline practices for the successful establishment of a new pasture.

8. Why should pasture fertilizer rates be based on soil tests?

9. How can an old pasture be renovated?

10. Discuss how each of the following factors affects the value of pasture (a) soil and fertilizer, (b) plant species, (c) stage of maturity, (d) rate of growth and season of year, and (e) grazing.

11. Discuss how the stage of maturity affects the nutritive value of pasture.

12. Outline practices for good pasture management.

13. List four ways of extending the grazing season.

14. Discuss and compare each of the following seed pasture grazing systems: (a) continuous grazing, (b) rotation grazing, and (c) intensive grazing, (d) creep grazing, and (e) strip grazing.

15. Is there a need and a place for more irrigated pastures in the United States? Justify your answer.

SELECTED REFERENCES

Title of Publication	Author(s)	Publisher
Crop Production	R. J. Delorit L. J. Greub H. L. Ahlgren	Prentice-Hall, Inc., Englewood Cliffs, NJ, 1984
Feeds & Nutrition, Second Edition	M. E. Ensminger J. E. Oldfield W. W. Heinemann	The Ensminger Publishing Company, Clovis, CA, 1990
Feeds & Nutrition Digest	M. E. Ensminger J. E. Oldfield W. W. Heinemann	The Ensminger Publishing Company, Clovis, CA, 1990
Forage and Pasture Crops	W. A. Wheeler	D. Van Nostrand Company, Inc., New York, NY, 1950
Forages, The Science of Grassland Agriculture	M. E. Heath R. F. Barnes D. S. Metcalfe	Iowa State University Press, Ames, IA, 1985
Grass: The Yearbook of Agriculture, 1948		U. S. Department of Agriculture, Washington, DC, 1948
Manual of the Grasses of the United States	A. S. Hitchcock rev. by A. Chase	U. S. Government Printing Office, Washington, DC, 1950
Pasture Book, The	W. R. Thompson	W. R. Thompson, State College, MI, 1950
Pasture and Range Plants	Phillips Petroleum Company	Phillips Petroleum Company, Bartlesville, OK, 1963
Producing Farm Crops	L. V. Boone	Interstate Publishers, Inc., Danville, IL, 1991
Sheep Research Journal	Staff	American Sheep Industry Assn., Englewood, CO, 1994
Stockman's Handbook, The, Seventh Edition	M. E. Ensminger	Interstate Publishers, Inc., Danville, IL, 1992

CHAPTER 7

Two Oberhasli kids at play. (Courtesy, Dorothea M. Custer, President, Oberhasli Breeders of America, Harvard, IL)

SHEEP AND GOAT BEHAVIOR AND ENVIRONMENT

Contents *Page*

Contents Page

Successful sheep and goat producers are students of animal behavior. For example, they recognize when ewes and does are in heat (estrus), or when parturition is imminent, and they know the bleat of a lamb or a kid in trouble. Furthermore, a producer should be able to walk through the flock or herd and spot sick animals by their behavior, thereby beginning early treatment or taking steps to correct some environmental problem.

This chapter presents some of the principles and applications of behavior. Those who have grown up around farm animals and have dealt with them in practical ways have already accumulated a substantial workaday knowledge of animal behavior. Those with urban backgrounds need to familiarize themselves with the behavior of animals. To all, the principles and applications of animal behavior depend on understanding, which is the intent of this chapter.

ANIMAL BEHAVIOR

Animal behavior is the reaction of animals to certain stimuli, or the manner in which they react to their environment. The individual and comparative study of animal behavior is known as *ethology*. Through the years, behavior has received less attention than the quantity and quality of the meat, milk, eggs, power, and fiber produced by animals. But modern breeding, feeding, and management have brought renewed interest in behavior, especially as a factor in obtaining maximum production and efficiency. With the restriction, or confinement, of herds or flocks, many abnormal behaviors evolved to plague those who raise them, including cannibalism, loss of appetite, stereotyped movements, poor parental care, overaggressiveness, dullness, degenerate sexual behavior, and a host of other behavioral disorders. Confinement not only has limited space but has interfered with the habitat and social organization to which, through thousands of years of evolution, the species became adapted and best suited. This is due to a genetic time lag. People altered the environment faster than they altered the genetic makeup of animals.

HOW SHEEP AND GOATS BEHAVE

Animals behave differently, according to species. Also, some behavioral systems or patterns are better developed in certain species than in others. Moreover, ingestive and sexual behavioral systems have been most extensively studied because of their importance commercially. Nevertheless, most animals exhibit the following nine general functions or behavioral systems, each of which will be discussed:

1. Agonistic behavior (combat)
2. Allelomimetic behavior
3. Care-giving and care-seeking (mother-young) behavior
4. Eliminative behavior
5. Gregarious behavior
6. Ingestive behavior (eating and drinking)
7. Investigative behavior
8. Sexual behavior
9. Shelter-seeking behavior

AGONISTIC BEHAVIOR (COMBAT)

This type of behavior includes fighting and flight, and other related reactions associated with conflict. Among all species of farm animals, males are more likely to fight than females. Nevertheless, females may exhibit fighting behavior under certain circumstances. Castrated males are usually quite passive, indicating

Fig. 7-1. Agonistic behavior in rams.

hormones—especially testosterone—are involved in this type of behavior. Rams fight by backing off and charging at each other headlong. The fight generally continues until one ram gives up, usually after both combatants have bloody noses.

When fighting, goats will frequently rear up on their hind legs, come down, striking their opponents head-to-head. Sheep almost never exhibit this rearing behavior.

ALLELOMIMETIC BEHAVIOR

Allelomimetic behavior is mutual mimicking behavior. Thus, when one member of a group does something, another tends to do the same thing, and because others are doing it, the original individual continues.

CARE-GIVING AND CARE-SEEKING (MOTHER-YOUNG) BEHAVIOR

After parturition, ewes and does lick their newborn, removing moisture and placental membranes. The mother-young bond in sheep and goats is very strong. Ewes and does become attached to their offspring, and vice versa. Although ewes are normally timid and easily frightened, they will defend their young even if the attacker is formidable.

Immediately after giving birth, ewes and does will readily accept alien newborns; thereafter, aliens are vigorously rejected by butting. It is their sense of smell that ewes and does employ to discriminate between their own and alien young.

Abnormal maternal behavior includes: (1) deserting newborn, (2) moving when young attempt to suckle, and (3) butting their own offspring.

Newborn kids are hiders the first few days after birth. Under pasture or range conditions, they will hide when the does are not present. On the other hand, newborn lambs are followers. They follow the ewes everywhere.

Fig. 7-2. Mother-young behavior. (Courtesy, American Polypay Sheep Assn., Sidney, MT)

ELIMINATIVE BEHAVIOR

Sheep deposit their feces in a random fashion. While defecating, sheep often wiggle their tails. To urinate, females assume a squatting position and males may arch their backs and bend their legs.

GREGARIOUS BEHAVIOR

Gregarious behavior refers to the flocking or herding instinct. It is particularly strong in sheep. Moreover, it is more evident in some breeds than in others. The Merinos, and animals carrying Merino breeding, are noted for their flocking instinct. This makes it possible to herd them on the range. Goats are not very gregarious.

The gregarious instinct of sheep diminishes to some extent when they are placed within fenced holdings, instead of herded. As a result, those who handle western range bands do not try to switch back and forth from fenced range to herding, because the bands will become unmanageable from the standpoint of herding once they have been in a fenced holding for an extended period of time.

Packers use the gregarious instinct of sheep by having an old goat, appropriately called a *Judas*, lead sheep to slaughter. A well-trained Judas will lead group after group of sheep to slaughter all day long.

INGESTIVE BEHAVIOR (EATING AND DRINKING)

This type of behavior is characteristic of all mammals of all species and all ages. Animals cannot live without feed and water. Moreover, for high production, animals must have aggressive eating habits. They must consume large quantities of feed.

The first ingestive behavior common to all young mammals is suckling. Within 1 to 2 hours after birth, lambs find the udders and begin to suckle. Initially, lambs are allowed to suckle at any time for as long as they wish, but after 1 to 2 weeks, ewes begin to restrict the frequency and duration of suckling periods. They will generally do this by walking away from them. While suckling, lambs characteristically nudge the udders with their noses and wiggle their tails from side to side.

The suckling behavior of newborn kids is similar to that of lambs.

Sheep graze very much like cattle, but the cleft upper lip allows them to graze vegetation closer to the ground. The lips, the lower incisor teeth, and the dental pad are the primary structures used to grasp food. The tongue does not protrude during grazing as in cattle.

Fig. 7-3. Sheep demonstrating ingestive behavior—drinking. (Courtesy, American Sheep Industry Assn., Englewood, CO)

Leaves and stems are severed by the lower incisors against the dental pad as the sheep jerks its head slightly forward and upward. During a day, sheep will graze 4 to 7 times, amounting to 9 to 11 hours of grazing, covering 1 to 8 miles.

Under grazing conditions, the biggest difference between sheep and goats is the amount of browsing done by goats. They have unique preferences for shrubs and tree leaves. Depending on the feed conditions, goats will spend 6 to 9 hours daily grazing and browsing.

Rumination is the act of chewing the cud. It involves (1) regurgitation of ingesta from the reticulorumen, (2) swallowing of regurgitated liquids, (3) remastication of the solids accompanied by reinsalivation, and (4) reswallowing of the bolus. Sheep will experience about 15 rumination periods per day. These may last anywhere from 1 to 120 minutes, but the total time spent ruminating is 8 to 10 hours per day. Goats ruminate mainly during the night, and each cycle of rumination lasts about 1 minute.

When grazing, large flocks (bands) of sheep generally split into subgroups and occupy separate areas. Different breeds vary in their tendency to move or flock together, with the gregarious trait being the strongest in the Merino and Rambouillet breeds.

The social structure of sheep is dependent upon visual contact and a flocking tendency. Also, group size is important. It has been shown that Merino sheep kept in pairs (just two sheep) gained less liveweight and produced less wool than their counterparts in groups ranging from 4 to 30 animals. The pairs spent less time grazing and more time walking along the fence line trying to keep contact with the flock in the next pasture; thus, pairing produced a level of stress which affected production. The need for visual contact during grazing may well account for this stress. It should be noted, however, that preferred group size and overall flock dispersion are dependent on breed, age, stability of flock membership, and vegetation.

Studies have not shown any correlation between the rank in which sheep reach feed troughs and their competitive ability at the troughs. However, certain individuals are constantly among the first few sheep to reach the troughs. Thus, it appears that going to feed is initiated by a few sheep, then others follow. Aggression is rare during normal grazing and social hierarchies.

■ **Social organization among goats**—A herd of goats forms smaller groups than sheep, usually built up from the extended family group. When competing for feed, goats may rear up and head clash with a downward stroke of the head.

The principles and practices of good goat herding are very similar to those with sheep, with one exception: Rarely do sheep herders work ahead of a range band. However, it is common practice for goat herders to work in front, turning the lead goats back to avoid unnecessary travel.

INVESTIGATIVE BEHAVIOR

All animals are curious and have a tendency to explore their environment. They investigate strange quar-

Fig. 7-4. Angora goats exhibiting investigative behavior. (Courtesy, *Sheep and Goat Raiser*, San Angelo, TX)

ters and objects, approaching objects in a heads-up, ears-forward, eyes-fixed manner. Sheep, however, are much more timid than goats or cattle; thus, they will usually turn and run if an object moves or if something frightens them.

Goats easily find escape routes in pens. They are easily frightened, but learn quickly.

SEXUAL BEHAVIOR

Sexual behavior involves courtship and mating. It is largely controlled by hormones. A ewe in heat (estrus) will seek out a ram and closely associate with him, rub her neck and body against him, and stand for mating. Sexual behavior in the ram includes: following the female, rubbing against the side of the female, pawing the female, biting the wool of the female, nosing the genital region and sniffing female urine then extending the neck with upcurled lips (Flehmen), running the tongue in and out, and raising and lowering one front leg in a stiff-legged striking motion. Actual mating requires only a few seconds.

Sexual behavior in goats is similar to that in sheep, with possibly two notable exceptions: (1) Does will mount each other occasionally, whereas ewes will not; and (2) aroused bucks urinate upon their front legs, briskets, and beards.

SHELTER-SEEKING BEHAVIOR

All species of animals seek shelter—protection from the sun, wind, rain and snow, insects, and predators.

Sheep seek shelter by moving into barns or under trees, by huddling together to keep off flies, by crowding together in extremely cold weather, and by pawing the ground and lying down. Like cattle, during a severe storm they turn their rear ends towards the wind.

When there is no shelter, there is danger of sheep massing together and smothering during a very severe storm.

SOCIAL RELATIONSHIPS

Social behavior may be defined as any behavior caused by or affecting another animal, usually of the same species, but also, in some cases, of another species.

Social organization may be defined as an aggregation of individuals into fairly well-integrated and self-consistent group in which the unity is based upon the interdependence of the separate organisms and upon their responses to one another.

The social structure and infrastructure in herds and flocks are of great practical importance. Some of the ideas on peck order (bunt, or hooking order) have had to undergo changes as a result of increased understanding of the social organization within the flock or herd.

SOCIAL ORDER (DOMINANCE)

Within most groups of farm animals of the same species, there is a well-organized social rank—dominance hierarchy—which is just as real as any social register or blue book of people. Animals observe this order in their relationships just as carefully as protocol demands that it be observed at a state dinner.

With ewes, however, dominance does not seem to be as important or as obvious as with the other farm animals. Unfamiliar ewes wandering into a group are generally subjected to a sniffing investigation but active butting is rare. Even when only small amounts of feed are available, the competition involves primarily pushing and shoving toward the feed rather than active butting.

Dominance is most likely to be observed in rams during mating, in competition for supplementary feed in a restricted space, or at a watering trough.

In flocks of mixed ages and/or mixed breeds, the dominance observed at the feed trough is that the poor competitors—the youngest and the oldest—cease trying, and some breeds demonstrate less competitive spirit.

Dominance and subordination are not inherited as such, for these relations develop due to several factors: (1) age, (2) early experience, (3) size, (4) aggressiveness or timidity, and (5) sex. Of course, social rank becomes important when a group of animals is fed in confinement and doubly important if limited feeding is practiced, since dominant animals crowd subordinate ones away from the feed bunk.

In goat herds, the hierarchy depends on (1) body weight, (2) horns, and (3) age, but in herds where there are many polled animals of the same age, the order is less firmly established. Nevertheless, when goats are offered a palatable new feed and when feeder space is limited, a few goats may try to break the established order, resulting in fights and reduced feed intake.

LEADER-FOLLOWER

Leader-follower relationships are particularly strong in sheep, for lambs follow their mothers from birth. In a naturally formed flock of sheep, the oldest ewes lead, followed immediately by their young lambs. Each is followed less closely by her descendants, with the females followed by their own lambs. Thus, the leader in the flock is usually the oldest ewe with the largest number of descendants. This type of leadership

Fig. 7-5. In single file, leaders lead to the feed while others wait to follow. (Photo by J. C. Allen and Son, West Lafayette, In)

is broken up in flocks where unrelated animals are brought together.

Leadership in goats does not seem to be as established as that in sheep. Some leadership order may be noted, but it is less likely to be clearly related to age.

INTERSPECIES RELATIONSHIPS

Social relationships are normally formed between members of the same species. However, they can be developed between two different species. In domestication this tendency is important (1) because it permits several species to be kept together in the same pasture or corral, and (2) because of the close relationship that exists between people and animals. Such interspecies relationships can be produced artificially, generally by taking advantage of the maternal instinct of females and using them as foster mothers. For example. goats can be used to raise orphaned lambs.

Fig. 7-6. A goat being used as a surrogate mother for an orphan lamb. (Courtesy, Christine S. F. Williams, Michigan State University, East Lansing, MI)

PEOPLE-ANIMAL RELATIONSHIPS

People need pets and pets need people! Both groups desire to love and be loved. Orphaned lambs raised by children contribute richly to the happiness and well being of their caretakers through the human-animal bond.

Fig. 7-7. People need pets and pets need people. This boy is bottle feeding an orphaned lamb. (Courtesy, American Sheep Industry Assn., Englewood, CO)

HOW SHEEP AND GOATS COMMUNICATE

Communication involves a signal by one sheep which upon being received by another sheep influences its behavior. Communication between sheep may be via sound, smell, touch, and/or vision.

■ **Sound**—Typically, if a ewe is separated from her lamb, both animals "baa" or "bleat" until they are reunited. Also, a sheep will "baa" or "bleat" when separated from the flock. During the breeding season, rams and bucks will produce courting grunts as they approach females, and ewes and does demonstrate an increase in nonspecific bleats during courting. Ewes frequently produce a low "m-m-m" when caring for their newborn lambs. A doe can identify her kid(s) by its voice after it is about four days old.

■ **Smell**—Initially, it is primarily the sense of smell that ewes and does employ to find their young. Rams use smell to locate females that are in heat. The buck odor elicits estrous behavior in does.

■ **Touch**—As part of the courting behavior, ewes will rub against rams; and rams will rub against ewes, paw them, bite their wool, and nose them. While suckling, lambs nudge the udders with their noses. Goats will also show nuzzling behavior during courtship.

■ **Vision**—Sheep use vision to maintain contact with other members of the flock. As a flock grazes, each individual throws up its head at intervals, presumably to respond to the position of other members. Vision is also important for mother-offspring recognition in both goats and sheep.

NORMAL ANIMAL BEHAVIOR

The sheep and the goat producer need to be familiar with the behavioral norms of animals in order to detect and treat abnormal situations—especially illness. Many sicknesses are first suspected because of some change in behavior—loss of appetite (anorexia); listlessness; labored breathing; straining to urinate or defecate; unusual posture; reluctant or unusual movement; persistent rubbing or licking, and altered social behavior, such as one animal leaving the flock or herd and going off by itself—these are among the useful diagnostic tools.

Some of the signs of good health are:

1. Contentment.
2. Alertness.
3. Eating with relish.
4. Dense, clean, bright fleece or hair, and pliable, elastic skin.
5. Bright eyes and pink membranes.
6. Normal feces and urine.
7. Normal temperature (102.3°F for sheep and 103.8°F for goats).
8. Normal heart rate (70 to 80 beats per minute for both sheep and goats).
9. Normal respiration rate (12 to 20 breaths per minute for both sheep and coats).

Vision and sleep are also important aspects of normal sheep and goat behavior:

■ **Vision**—The eyes of sheep and goats, like many animals, are on the side of the head. This gives them an orbital or panoramic view—to the front, to the side, and to the back—virtually at the same time. Sometimes this type of vision is referred to as rounded or globular. The field of vision directly in front of a sheep or goat is binocular, but on the sides and toward the back it is monocular. This type of vision leads sheep and goats to different interpretations of their environment than the binocular type vision of humans.

The orbital vision of sheep and goats explains why they will go through a curved chute more easily than a straight one.

Blind sheep at times exhibit rather bizarre behavior such as panicking when approached and running headlong into obstacles.

■ **Sleep**—The normal sleeping posture of sheep is on the stomach but tilted to the side with one front leg folded under the body and the other extended forward. Usually the head is turned to one side and the eyes are closed. Although sheep and goats are usually inactive about half of the day, as with cattle there is considerable debate as to whether they actually sleep. Certainly, sheep and goats do not enter the state of deep sleep that exists in swine, horses, dogs, and cats.

ABNORMAL ANIMAL BEHAVIOR

Abnormal behavior of domestic animals is not fully understood. Like human behavior disorders, more study is needed. However, studies of captured wild animals have demonstrated that when the amount and quality, including variability, of the surroundings of an animal are reduced, there is an increased probability that abnormal behavior will develop. Also, it is recognized that confinement of animals creates a lack of space which often leads to unfavorable changes in habitat and social interactions for which the species have become adapted and best suited to over thousands of years of evolution. Abnormal behavior may take many forms—ingestive, eliminative, sexual, maternal, agonistic and/or investigative. Abnormal sexual behavior is particularly distressing because the whole of production depends on the animal's ability to reproduce.

Sheep in confinement may develop a wool-eating habit. They do not inflict great harm, for they only take small nibbles of wool from each other. The cause is unknown, but it does seem to be associated with an unnatural environment, lack of comfort, and/or boredom.

APPLIED ANIMAL BEHAVIOR

The presentation to this point in this chapter has been for the purpose of reader understanding. However, knowledge and understanding must be put into practice to be of value. Sheep and goat producers must make practical applications of animal behavior.

BREEDING FOR ADAPTATION

The wide variety of livestock in different parts of the world reflects a continuous process of natural and artificial selection which has resulted in the survival of animals well adapted to climate and other environmental factors. Changes in the physical structure of species is dependent upon (1) the ability of animals to mutate and/or respond to selection pressure (natural or artificial), and (2) the effect of environmental pressure on the animals, which results in survival of the fittest. Among the examples of adaptation to environment are haired sheep (devoid of wool) in desert areas, and fat-tailed sheep in arid zones. Such adaptations relate to survival of the animals, but they do not necessarily entail maximum productivity of food for people. It is understandable, therefore, why there have been many attempts to introduce improved livestock into countries in which the productivity of native stock is low. But there are many problems in breed replacement; thus, a large number of experimental introductions of new breeds have not been successful. Disease problems, poor resistance to temperature extremes, and limited feed supplies contribute to the failure of attempts to improve the output of native stock by replacing them with improved imported breeds. To succeed, breed replacement or crossbreeding systems for improved production with improved breeds must accompany improvement of the nutritional, parasitological, disease, and husbandry conditions of the area.

When choosing replacements to improve future generations, the producer should select from among animals kept under an environment similar to that under which it is expected that their offspring will perform. Moreover, these animals should be those that demonstrate high productivity in their environment. Therefore, the producer should select and breed sheep and goats that adapt quickly to their environment—animals that not only survive, but thrive, under the conditions imposed upon them. When properly combined, heredity and environment complement each other, but when one or the other is disregarded, they may oppose each other. (For further details on heredity and environment. see Chanter 3.)

QUICK ADAPTATION—EARLY TRAINING

Early training and experience are extremely important. In general, young animals learn more quickly and easily than adults do; hence, advance preparation for adult life will pay handsome dividends. The optimum time for such training varies according to species. Furthermore, stress can be reduced or avoided entirely if animals proceed through a graduated sequence of events leading to an otherwise noxious experience.

MANURE ELIMINATION

Body waste is a major concern; although unavoidable, it is expensive and time-consuming to handle, and it may create a major pollution problem. But manure handling can be facilitated by an understanding and application of eliminative behavior.

The habit of sheep bedding down on the highest areas of the field or range leaves the larger part of the droppings at the places where they are most needed. Moreover, the form in which sheep manure is dropped and the way it is tramped into the soil insure a smaller waste than any other system of stock production.

COMPANIONSHIP

The best-known animal companionship pertains to high-strung racehorses and stallions, in which all sorts of companions are used—a goat, a sheep, a chicken, a duck, or a pony. Such companions are commonly referred to as *mascots.*

The expression "to get his goat" was born of the common custom of having goats for mascots. Back in the days when skullduggery was as important as form in winning races, the employees of one stable sometimes plotted to kidnap the goat mascot of a rival's horse. By "getting the goat" of a favorite, they cleaned up by betting against a horse that was odds-on to win, but too upset to run at its best.

ANIMAL ENVIRONMENT

Environment may be defined as all the conditions, circumstances, and influences surrounding and affecting the growth, development, and production of animals. The most important influences in the environment are the feed and quarters (space and shelter).

The branch of science concerned with the relation of living things to their environment and to each other is known as ecology.

Producers were little concerned with the effect of environment on animals as long as they grazed on pastures or ranges. But rising feed, land, and labor costs, along with the concentration of animals into smaller spaces, changed all this.

In sheep, goats, and other animals, environmental control involves space requirements, light, air temperature, relative humidity, air velocity, wet bedding, ammonia buildup, dust, odors, and manure disposal. Control or modification of these factors offers possibilities for improving animal performance. Although there is still much to be learned about environmental control, the gap between awareness and application is becoming smaller.

In the present era, pollution control is the first and most important requisite in locating a new livestock establishment, or in continuing an old one. The location should be such as to avoid (1) complaints from neighbors about odors, insects, and dust; and (2) pollution of surface and underground water. Without knowledge of animal behavior, and without pollution control, no amount of capital, native intelligence, and sweat will make for a successful livestock enterprise.

In summary, selection provides a long-term answer to behavioral problems induced by the environment. Animals need to be bred and selected for adaptation to their environments.

Feed and nutrition, weather, health, and stress are environmental factors of special importance, and they should be included in any discussion of animal behavior.

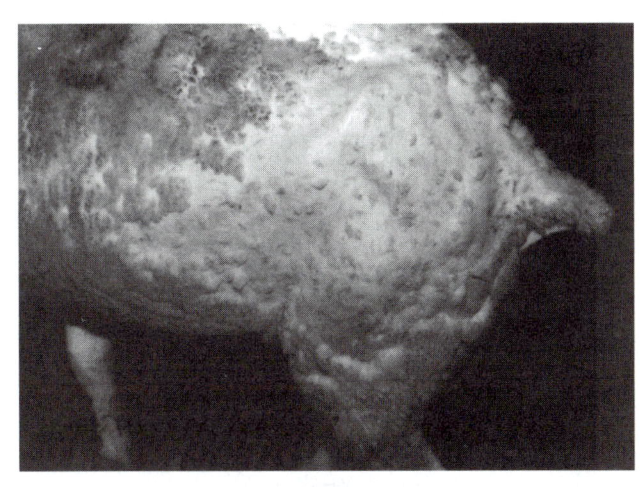

Fig. 7-8. Rear thigh of a ewe in confinement on slatted flooring that has been denuded of wool by other ewes picking wool from here. This caused either by boredom or by too little fiber in the diet of the sheep. (Courtesy, Randy Gottfredson and David L. Thomas, University of Wisconsin, Madison)

HOW ENVIRONMENT AFFECTS ANIMALS

No matter how good the genetics, a good environment is essential to obtain high production. Producers need to concern themselves more with the natural habitat of animals. Thus, heredity in sheep and goats has already made its contribution at the time of fertilization, but environment works ceaselessly away until death. Among the environmental factors affecting sheep and goats are: feed, water, weather, facilities, health, and stress.

FEED/ENVIRONMENTAL INTERACTIONS

Sheep and goats may be affected by (1) too little feed, (2) rations that are too low in one or more nutrients, (3) an imbalance between certain nutrients, or (4) the physical form of the ration—for example, it may be ground too finely.

Forced production and the feeding of forages and grains which are often produced on leached and depleted soils have created many problems in nutrition. These conditions have been further aggravated through the increased confinement of animals, many animals being confined to lots all or a large part of the year. Under these unnatural conditions, nutritional diseases and ailments have become increasingly common.

Also, nutritional reproductive failures plague livestock operations. Generally, energy is more important than protein in reproduction. The level and kind of feed before and after parturition will determine how many females will show heat—and conceive. After giving birth, their feed requirements increase tremendously because of milk production; hence, ewes and does suckling young need approximately 50% greater feed allowance than during the pregnancy period. Otherwise, they will suffer a serious loss in weight.

The following additional feed-environmental factors are pertinent:

1. **Regularity of feeding.** Animals are creatures of habit, hence, they should be fed at regular times each day.
2. **Underfeeding.** Too little feed results in slow and stunted growth of young stock; in loss of weight, poor condition, and excessive fatigue of mature animals; and in poor reproduction, failure of some females to show heat, more services per conception, lowered young crop, and light birth weights.
3. **Deficiency of nutrient(s).** A deficiency of any essential nutrient required by a lactating female will lower milk production and feed efficiency, rather than make for significant changes in the composition of milk.
4. **Overfeeding.** Too much feed is wasteful. In addition, it creates a health hazard; there is usually lowered reproduction in breeding animals, and there is a higher incidence of digestive disturbances. Animals that suffer from mild digestive disturbances are commonly referred to as being *off-feed*.

WATER/ENVIRONMENTAL INTERACTIONS

Animals can survive for a longer period without feed than without water. Water is one of the largest constituents in the animal body, ranging from 40% in very fat, mature animals to 80% in newborn animals. Deficits or excesses of more than a few percent of the total body water are incompatible with health, and large deficits of about 20% of the body weight lead to death.

The total water requirement of animals varies primarily with the weather (temperature and humidity); feed (kind and amount); the species, age, and weight of animal; and the physiological state. The need for water increases with increased intakes of protein and salt, and with increased milk production of lactating animals. Water quality is also important, especially with respect to the content of salts and toxic compounds.

It is generally recognized that animals consume more water in summer than in winter.

The water content of feeds ranges from about 10% in air-dry feeds to more than 80% in fresh, green forage. Feeds containing more than 20% water are known as *wet feeds*. The water content of feeds is especially important for animals which do not have ready access to drinking water. Also, the water on the surface of plants, such as dew, may serve as an important source for sheep and goats on arid ranges, but this supply is rarely sufficient to meet their needs.

Under range conditions, the frequency of watering of sheep and goats is one time per day, although when grazing desert ranges in early spring, they may go for weeks without drinking water. Goats approach camels in their water requirements.

Under practical conditions, the frequency of watering sheep and goats is best determined by the animals, by allowing them access to clean, fresh water at all times.

WEATHER/ENVIRONMENTAL INTERACTIONS

Because of their fleeces, sheep and goats are more tolerant of environmental extremes than swine, cattle, and poultry. Nevertheless, thermal stress affects their maintenance requirement, voluntary feed intake, average daily gain, and feed required per unit of gain.

The comfort zone of sheep and goats varies according to age; for newborn lambs and kids it is 75 to 80°F, for mature sheep and goats it is 45 to 75°F.

Weather may be modified by shelters. Winter shelters and summer shades almost always improve production and feed efficiency. The additional cost, however, incurred by shelter has frequently exceeded the benefits gained by improved performance, particularly in those areas with less severe weather and climate.

■ **Confinement sheep production**—Some producers are using confinement structures and high-density production. With this, building designs, temperature, relative humidity, air velocity, dust, light, odors, and space requirements are critical.

Confinement buildings are costly, but when properly designed and constructed, they create the ultimate in animal health, comfort, and efficiency of feed utilization. Also, they lend themselves to automation, which results in a savings of labor, and they effect a savings in land cost because they minimize space requirements. The decision on whether or not confinement and environmental control can be justified should be determined by economics. Manure disposal and pollution control are also considerations.

Before a confinement system can be designed for animals, it is important to know their (1) heat production, (2) vapor production, and (3) space requirements. This information is presented in Chapter 17, Buildings and Equipment for Sheep.

Dairy goats may be managed under confinement. Information about their requirements is presented in Chapter 25, Buildings and Equipment for Goats.

ADAPTATION, ACCLIMATION, ACCLIMATIZATION, AND HABITUATION OF SPECIES/BREEDS TO THE ENVIRONMENT

Every discipline has developed its own vocabulary. The study of adaptation/environment is no exception. So, the following definitions are pertinent to a discussion of this subject:

Adaptation refers to the adjustment of animals to changes in their environment.

Acclimation refers to the short-term (over days or weeks) response of animals to their immediate environment.

Acclimatization refers to evolutionary changes of a species to a changed environment which may be passed on to succeeding generations.

Habituation is the act or process of making animals familiar with, or accustomed to, a new environment through use or experience.

Species differences in response to environmental factors result primarily from the kind of thermoregulatory mechanism provided by nature, such as type of coat (wool and mohair of sheep and goats).

ENVIRONMENTALLY CONTROLLED BUILDINGS

Environmentally controlled buildings are costly to construct, but they make for the in ultimate comfort, health, and efficiency of feed utilization.

In hot climates, increased use is being made of shades for the purpose of enhancing animal comfort and minimizing maintenance requirements.

Environmentally controlled buildings are rather common in poultry and swine housing, and on the increase for dairy cattle, dairy goats, and milking sheep.

LIGHTING

The number of hours of light in the day affects the initiation of the normal breeding season of ewes and does, both of which are seasonal breeders. It is noteworthy, too, that the reproductive function in poultry and migratory fowl is regulated by the length of daylight.

The ratio of hours of daylight to darkness throughout the year acts on nerves in the region of the pituitary gland, and stimulates or inhibits the release of the follicle-stimulating hormone (FSH). Lengthening the daylight hours activates the pituitary, and causes it to release increasing amounts of the FSH which stimulates ovarian function. Thus, sometime after the daylight period increases, the estrous cycle begins in ewes and does.

Artificial lighting will accomplish the same thing as daylight; hence, it may be used to alter the estrous cycle in both ewes and does.

Normally, ewes and does come in heat during the late summer or early fall, though there is both an area and a breed difference. The breeding season is usually restricted to about four months.

Ewes and does generally begin cycling when the number of daylight hours drops below 14. This is the reason that most breeds of sheep and goats come into heat during the fall months. To initiate estrus, however, it appears that the shorter days must be preceded by longer days.

FACILITIES/ENVIRONMENTAL INTERACTIONS

Optimum facility environments can provide the means for animals to express their full genetic potential or production, but they cannot compensate for poor management, health problems, or improper rations.

Research has shown that animals are more productive and feed-efficient when raised in an ideal environment. The primary reason for having facilities, therefore, is to modify the environment. Proper barns and other shelters, shades, sprinklers, insulation, ventilation, heating, air conditioning, and lighting can be used to approach the desired environment. Also, increasing attention needs to be given to other stress sources such as space requirements, and the grouping of animals as affected by class, age, size, and sex.

The principal scientific and practical criteria for decision making relative to the facilities for animals in modern, intensive operations is the productivity and cost of production of animals , which can be achieved only by healthy animals under minimal stress. So, the investment in environmental control facilities is usually balanced against the expected increased returns.

Temperature, humidity, ventilation, and space recommendations for sheep are given in Chapter 17, and for goats they are given in Chapter 25.

In recent years, there has been a trend to modify the environmental control facilities as much as possible; among such modifications designed for maximum animal comfort and efficiency of production of different animal species are fans, floors, lights, shades, sprinklers/sprayers/foggers, ventilation, wallows, and windbreaks.

HEALTH/ENVIRONMENTAL INTERACTIONS

Health is the state of complete well-being, and not merely the absence of disease.

Environment embraces the forces and conditions, both physical and biological, that (1) surround animals, and (2) interact with heredity to determine behavior, growth, and development.

Disease is defined as any departure from the state of health.

Parasites are organisms living in, on, or at the expense of another living organism.

Feed, air quality, lighting, noise, other animals, and weather are among the many factors that constitute an animal's environment. Extremes or alterations in the environment may subject an animal to stress; and stress may affect health and lead to more diseases and parasites.

Diseases and parasites (external and internal) are ever-present animal environmental factors. Death takes a tremendous toll. Even greater economic losses result from retarded growth and poor feed efficiency, carcass condemnations, decreases in meat and wool quality and milk production, and higher labor and drug costs.

Any departure from the signs of good health constitutes a warning of trouble. Most sicknesses are ushered in by one or more signs of poor health—by indicators that tell expert caretakers that all is not well—that tell them that their animals will go off-feed tomorrow, and that prompt them to do something about it today.

Among the signs of animal ill health are: lack of appetite—the animal does not eat or graze normally; listlessness; droopy ears; sunken eyes; humped-up appearance; abnormal feces—either very hard or watery feces suggests an upset in the water balance or some intestinal disturbance following infection; abnormal urine, repeated attempts to urinate without success, or off-colored urine should be cause for suspicion; abnormal discharges from the nose, mouth, and eyes, or a swelling under the jaw; unusual posture—such as standing with the head down; extreme nervousness; persistent rubbing; dull hair coat; loose wool; or areas where wool is rubbed off, and dry, scurfy, hidebound skin; pale, red, or purple mucous membranes lining the eyes and gums; reluctance to move or unusual movements; higher than normal temperature; labored breathing—increased rate and depth; altered social behavior such as leaving the flock or herd and going off alone; and sudden drop in production—weight gains, milk, wool, or mohair.

STRESS/ENVIRONMENTAL REACTIONS

Stress is any physical or psychological tension or strain. Many kinds of stress affect animals. Among the external forces which may stress animals are weaning, hunger, thirst, poor sanitation, disease, parasites, surgical operations, injury, excitement, presence of strangers, fatigue, number of animals together, space, weather, shearing, trailing long distances, changing corral and corral mates, previous training, previous nutrition, and management.

Many sheep and goat diseases appear as a result of stress. Moreover, sheep and goats already weakened by poor nutrition or internal parasites may be particularly affected by stress.

Animals can be prepared in such a manner as to reduce stress. For example, feeder lambs should be preconditioned by being started on small amounts of grain 2 to 3 weeks before shipment, being drenched for internal parasites, and being vaccinated for type D enterotoxemia at least 2 weeks before shipment. Also, lambs should be well rested and dry before being loaded. Such treatments minimize the stress of working, loading, transporting, and placing in a feedlot.

In the life of an animal, some stresses are normal, and they may even be beneficial—they can stimulate favorable action on the part of an individual. Thus, we need to differentiate between stress and distress. Distress—not being able to adapt—is responsible for harmful effects. The trick is to manage stress so that it doesn't become distress and cause damage and to recognize the warning signals of distress.

The principal criteria used to evaluate, or measure, the well-being or stress of people are: increased blood pressure, increased muscle tension, body temperature, rapid heart rate, rapid breathing, and altered endocrine gland function. In the whole scheme, the nervous system and endocrine system are intimately involved in the response to stress and the effects of stress.

The principal criteria used to evaluate, or measure, the well-being or stress of animals are: growth rate or production, efficiency of feed use, efficiency of reproduction, body temperature, pulse rate, breathing rate, mortality, and morbidity. Other signs of animal well-being, any departure from which constitutes a warning signal, are: contentment, alertness, eating with relish (and cudding by ruminants), sleek coat and pliable and elastic skin, bright eyes and pink eye membranes, and normal feces and urine.

Stress is unavoidable. Wild animals were often subjected to great stress; there were no caretakers to modify their weather, often their range was overgrazed, and sometimes malnutrition, predators, diseases, and parasites took a tremendous toll.

Domestic animals are subjected to different stresses than their wild ancestors, especially to more restricted areas and greater animal density. However, in order to be profitable, their stresses must be minimal.

PREDATOR CONTROL

Since about 1960, predator losses have dominated the sheep industry, and to a certain extent the goat industry. Many sheep producers attribute the decline in sheep numbers chiefly to predator losses resulting from the restriction and banning of the use of toxins to control predators, since toxins offer the only effective and affordable reduction of predator losses. U.S. predator losses increased from 2.3% of all sheep and lambs in 1940 to a peak of 5.8% in 1977, at which point they leveled off.[1]

The sheep industry loses 520,000 sheep and lambs a year to predators, according to American Sheep Industry Chairman of the Legislative Action Council, Frank Moore.[2]

Currently interspecies relationship is being used to protect sheep and goats from predators. By raising puppies, young llamas, or young donkeys with sheep, at maturity they become their protectors (guards).

(Also see earlier section headed "Interspecies Relationships.")

The world's most ambitious sheep predator project is in Australia, where sheep producers wage con-

Fig. 7-9. Lambs with 1080 toxic collars to help control coyote predation. (Courtesy, Dr. Johannes Nel, University of Wyoming)

---

[1]Statistics on predator losses provided by Clair E. Terrill, Ph.D., Collaborator, U.S. Department of Agriculture.

[2]Moore, Frank, American Sheep Industry Fighting for ADC Funding, *Sheep Industry News*, June 1997, Vol. 1, Issue 3.

Fig. 7-10. Llamas with sheep for predator control. Sheep producers have tried many approaches to control coyotes, most without success. (Courtesy, University of Wyoming)

Fig. 7-11. Guard dog protecting Angora goats on Ebeling ranch near Marble Falls, Texas. (Courtesy, *Ranch and Rural Living Magazine*, San Angelo, TX)

stant war against the dingo, the wild dog of down under. It consists of a wire mesh fence stretching 3,307 miles[3] across the interior, barring marauding dingos from southeastern sheep lands. Yet, in a land of flash floods and hard charging kangaroos, holes in the fence happen. Thus, the fence has to be patrolled constantly.

[3]The Great Wall of China extends from east to west across China, a distance of 3,100 miles, according to *China—the impossible dream*, by M. E. And Audrey Ensminger, published by the Ensmingers in 1973.

POLLUTION OF THE ENVIRONMENT

Pollution remains an issue of the decades. It matters little whether pollution is due to agriculture or factories. Everything that defiles, desecrates, or makes impure or unclean streams or atmosphere must be controlled.

Everyone must ever be mindful that life, beauty, wealth, and progress depend upon how wisely we use nature's gifts—the soil, the water, the air, the minerals, and the plant and animal life.

In agriculture, particular attention needs to be given to any pollution that may be caused by manure, fertilizers, insecticides and pesticides, herbicides, growth promotants, dust, muddy lots, and stray voltage.

In recent years, there has been a worldwide awakening to the problem of pollution of the environment (air, water, and soil) and its effect on human health and other forms of life. Much of this concern stemmed from the amount of manure produced by the sudden increase of animals in confinement. Certainly, there have been abuses of the environment—not limited to agriculture only. There is no argument that such neglect should be rectified in a sound, orderly manner, but it should be done with a minimum disruption of the economy and lowering of the standard of living.

MANURE

When sheep and goats are held in confinement, the manure pollution problem, suspicioned or real, will persist.

Of course, there is no one best manure management system for all situations. But, one way or another, science and technology must evolve with ways of disposing of manure, and this must be accomplished without polluting streams or the atmosphere or being offensive to the neighbors.

If not managed properly, animals may produce the following pollutants in troublesome quantities: manure, gases/odors, dust, and flies/other insects. Also, they may pollute water supplies.

When manure and urine are stored and undergo anaerobic digestion, dangerous and disagreeable gases are produced. The ones of primary concern are: hydrogen sulfide, ammonia, carbon dioxide, and methane.

PESTS AND PESTICIDES

Although science and technology have been the great multipliers in increasing our food supply, potential food supplies are still destroyed by the ravages of pests.

A pesticide is a substance that is used to control pests. Pesticides are an integral part of modern agricultural production and contribute greatly to the quality of food, clothing, and forest products we enjoy. Also, they protect our health from disease and vermin. Pesticides have been condemned, however, for polluting the environment, and in some cases for posing human health hazards. Unfortunately, opinions relative to pesticides tend to become polarized. A report by the National Research Council summarized the situation as follows:

> Users of pesticides fear that they will be regulated to the point where pests cannot be effectively controlled, with the concomitant losses of food while opponents of the use of pesticides fear that people are being poisoned and that irreversible damage is being done to the environment.

No pest control system is perfect; and new pests keep evolving. So, research and development on a wide variety of fronts should be continued. We need to develop safer and more effective pesticides, both chemical and nonchemical. In the meantime, there is need for prudence and patience.

POLLUTION LAWS AND REGULATIONS

Invoking an old law (the Refuse Act of 1899, which gave the Corps of Engineers control over runoff or seepage into any stream which flows into navigable waters), the U.S. Environmental Protection Agency (EPA) launched a program to control water pollution by requiring that all cattle feedlots which had 1,000 head or more the previous year must apply for a permit by July 1, 1971. The states followed suit; although differing their regulations, all of them increased legal pressures for clean water and air. Then followed the Federal Water Pollution Control Act Amendments, enacted by Congress in 1971, charging the EPA with developing a broad national program to eliminate water pollution.

Owners/operators of animal feeding facilities with more than 1,000 animal units must apply. Animal units are computed as follows: multiply number of slaughter and feeder cattle by 1.0; multiply number of mature dairy cattle by 1.4; multiply number of swine weighing over 55 lb by 0.4; multiply number of sheep by 0.1; and multiply number of horses by 2.0. (See Table 7-1, footnote 1, for what constitutes 1,000 animal units.)

TABLE 7-1
SUMMARY OF REGULATIONS

Feedlots with 1,000 or More Animal Units[1]	Feedlots with Less than 1,000 but with 300 or More Animal Units[2]	Feedlots with Less than 300 Animal Units
Permit required for all feedlots with discharges[3] of pollutants.	Permit required if feedlot— 1. Discharges[3] pollutants through an unnatural conveyance, or 2. Discharges[3] pollutants into waters passing through or coming into direct contact with animals in the confined area. Feedlots subject to case-by-case designation requiring an individual permit only after on-site inspection and notice to the owner or operator.	No permit required unless— 1. Feedlot discharges pollutants through an unnatural conveyance, or 2. Feedlot discharges pollutants into waters passing through or coming into direct contact with the animals in the confined area, and 3. After on-site inspection, written notice is transmitted to the owner or operator.

[1]More than 1,000 feeder or slaughter cattle, 700 mature dairy cows (milked or dry), 2,500 swine weighing over 55 lb *(24.9 kg)*, 500 horses, 10,000 sheep or lambs, 55,000 turkeys, 100,000 laying hens or broilers with continuous overflow watering, 30,000 laying hens or broilers with liquid manure handling, 5,000 ducks; or any combination of these animals adding up to 1,000 animal units.

[2]More than 300 feeder or slaughter cattle, 200 mature dairy cows (milked or dry), 750 swine weighing over 55 lb *(24.9 kg)*, 150 horses, 3,000 sheep, 16,500 turkeys, 30,000 laying hens or broilers with continuous overflow watering, 9,000 laying hens or broilers with liquid manure handling, 1,500 ducks; or any combination of these animals adding up to 300 animal units.

[3]Feedlot not subject to requirement to obtain permit if discharge occurs only in the event of a 25-year, 24-hour storm event.

POLLUTION POTENTIAL FROM GRAZING FEDERAL LANDS

Little pollution potential exists from pasture systems with low animal densities or numbers, or where pastures are rotated. So, except for high-density pasture systems involving a number of animals, pollution is no problem. Nevertheless, those interested in the grazing of public lands should be well informed.

Grazing influences the environment on federal lands. Under poor range management, the environment is affected adversely; under good range management, such as exists on most ranges today, grazing actually improves the environment.

Eating of plant materials by animals is a natural process in earthly and aquatic systems. Thus, the coming of the white colonists to what is now the United States, along with the introduction of domestic animals, did not constitute an entirely new component in the environment. Rather, domestic animals replaced, or added to, the wild animals that were already there.

Mistakes in grazing practices have occurred in the United States in the past, the most significant of which was the exploitative grazing practiced between 1865

THE WHEEL OF ECOLOGY

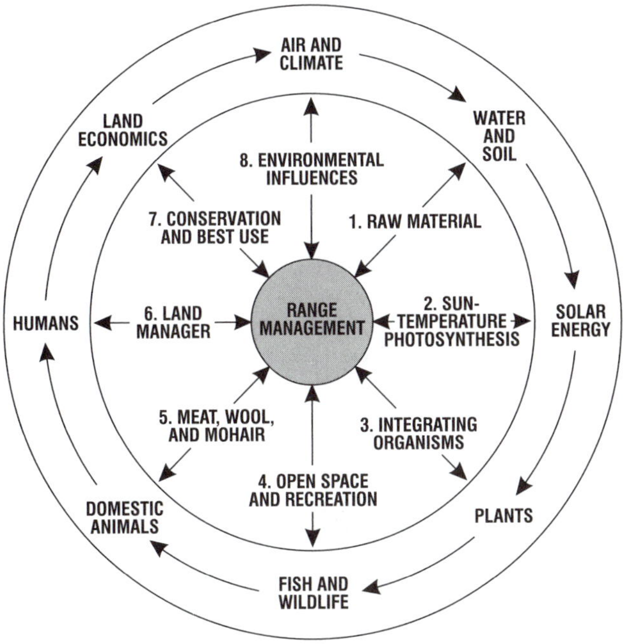

Fig. 7-12. The wheel of ecology. Ranchers share today's increasing national concern for the quality of our environment, and, through scientific range management, they are doing much to improve it.

and the 1930s. The effects were almost catastrophic. Nevertheless, they were not the result of grazing ranges that had never been grazed before. Rather, they resulted from several decades of grazing the western ranges with too many animals for too long, and often at the wrong season of the year. Most range livestock operators of that period were not aware of the benefits that could accrue to them from improved range management.

Scientific management of rangeland began at the turn of the century. Range managers and livestock operators found that controlling grazing improved both range conditions and livestock production. Development of this new concept marked the beginning of the end of the exploitative period of grazing and the introduction of managed grazing on the western ranges.

The environmental effects of grazing depend upon the kind of range, the intensity of grazing, and the kind of management employed to control livestock on the range. it is generally recognized that unregulated heavy grazing results in loss of desirable forage plants, increased runoff and erosion, and other indications of range deterioration. On the other hand, planned seasonal grazing and controlled animal distribution foster rapid vegetational growth. Most grazing experiments have shown that ranges may be improved more rapidly under proper grazing management than with no grazing at all.

There is no evidence that well-managed grazing of domestic livestock is incompatible with a high-quality environment. But there is ample evidence that managed grazing by livestock enhances certain uses and that poor management detracts from them. Properly managed grazing is a reasonable and beneficial use of the range.

Ecologists tell us that good range management will support more wildlife than the wilderness. This explains why big game numbers on federal lands have increased during recent years, and why wildlife production is an increasingly important use of rangelands.

Indeed, ranges actually improve while being properly utilized by domestic livestock. The benefits which accrue to the range include increased vegetation cover, improved plant species composition, improved soil fertility and soil structure, and greater yield of high-quality water. When sheep, goats, and cattle go, rank underbrush takes over, and fire becomes a real hazard.

Both upland birds and big game animals are benefited by grazing that promotes good cover for mating sites and enhanced food supply and other habitat requirements.

On ranges with mixed types of vegetation, herbaceous species increase and browse species decline when grazed only by game. The converse is true when cattle graze the land. The combined grazing by two groups of animals maintains a better balance of browse species—preferred by game animals, and of herbaceous species, preferred by sheep, goats, and cattle.

Heavy livestock grazing is beneficial to irrigated pastures used by geese and other migratory waterfowl. Unless the vegetation is closely cropped, these areas are unattractive to the birds.

Thus, livestock grazing of the public lands is contributing to improved wildlife habitat conditions and increased numbers of game animals. Range development programs, particularly livestock water developments, have made more public land usable by game animals and is partly responsible for the vast increase in game numbers over the years.

On many grass-shrub ranges, livestock grazing reduces the danger of fire by preventing buildup of dry grass, which is highly flammable.

Grazing systems and manipulation of vegetation can create contrast in vegetation color and pattern, thereby improving the aesthetic value of the landscape. Also, the livestock industry is traditional to the West; hence, a well-managed range with its cattle herd and roundup, or with its sheep camp, has recreational values. Indeed, cattle and sheep on the landscape are pleasing to tourists who come to view the "old west."

Ranges properly grazed by hoofed animals produce safe water. Counts of fecal coliform organisms, as indicators of water pollution by warm-blooded animals, relate more closely to the quantity of the fecal material than to the kind of animal. Investigations have shown that the count of harmful bacteria in streams is no greater in areas grazed by livestock than in areas grazed by wild animals alone, and that modern livestock grazing has little effect upon the chemical and physical quality of the water.

It is noteworthy, too, that few western ranges are ever in a stable, natural condition, whether or not they are grazed by domestic animals. Rather, most of them are in a stage of vegetational development following disturbances by such phenomena as drought, flood, avalanche, frost, or fire. Also, cyclical phenomena, such as large numbers of deer, rodent epidemics, or insect plagues, temporarily change the natural ecosystems. Thus, an absolutely stable rangeland is seldom attained or maintained.

Significantly, the greatest diversity of animal and plant species and the highest rates of reproduction occur when the landscape supports many stages of ecosystem development. Fire, grazing, and drought stimulate plants and animals to new growth. Each stage of vegetational development is more productive of certain animal species than of others.

Finally, in an era of food shortages, the contribution of properly managed federal lands in terms of food and fiber production needs to be recognized. More and more grains will be used for direct human consumption. As a result, there will be an increased reliance on ranges for meat, wool, and mohair production. It just makes sense to preserve all the natural food and fiber that we can. Remember that petroleum is not needed to make wool. Remember, too, that sheep, goats, and cattle are completely recyclable. It takes thousand of years to create coal, oil, and natural gas; and when they're gone, they're gone forever. But animals produce a new crop each year and perpetuate themselves through their offspring.

Fig. 7-13. Sheep grazing on the Shoshone National Forest in Wyoming. (Courtesy, U.S. Forest Service)

Approximately 261 million acres of federal land are administered for livestock grazing. In 1980, lands in the 11 western states administered by the Bureau of Land Management and the U.S. Forest Service provided grazing all or part of the year for 5,981,980 head of all classes of livestock—cattle, horses, sheep, and goats.

Both livestock producers and environmentalists need to recognize (1) that forage is a renewable natural resource, which regrows each year and is wasted unless it is utilized annually; (2) that grazing on federal rangelands helps to keep the natural environmental systems active and productive; (3) that we cannot allow overgrazing by domestic livestock, bison, deer, or wild horses; and (4) that grazing must be scientifically controlled and responsive to the needs of all users.

Indeed, it may be said that the influence of people on, and the use of, the environment will determine how well we live—and how long we live.

SUSTAINABLE AGRICULTURE

Fig. 7-14. Sustainable agriculture involves such age old, and 21st century new, practices as terracing. (Courtesy, USDA, Soil Conservation Service)

Endangered species—and more! Today, it is endangered planet, endangered people and animals, and endangered agriculture. Among the deluge of warnings of environmental catastrophes are:

- Pollution-caused warming of the atmosphere, known as the *greenhouse effect*, threatening weather changes that could render large areas of the planet unproductive and uninhabitable.

- Toxic and radioactive wastes and dumped garbage that could poison drinking water and despoil the land.

■ Chemical pollution that is depleting the atmosphere's protective ozone layer.

■ Slashing and burning of tropical rain forests, driving thousands of species to extinction, increasing the amount of carbon dioxide in the atmosphere, and contributing to the greenhouse effect that warms the earth.

Awareness of these and other similar environmental catastrophes are being observed too little and too late. Remember that it took nature thousands of years to form the rain forest, but it took a mere 25 years for people to destroy much of it. And when a rain forest is gone, it is gone forever!

Although less dramatic, the Amazon rain forest story has been, or is being, repeated all over the world in the form of the greenhouse effect, toxicities, polluted streams, and/or other harbingers of threats to our environment. Too long we have managed our nonrenewable resources like there is no tomorrow! Now, the situation is being righted. Worldwide, environmental quality and economic efficiency are in vogue. In the United States, this movement is called *sustainable agriculture.*

Sustainable agriculture is often described as farming that is ecologically sound and economically viable. It may be high or low input, large scale or small scale, a single crop or diversified farm, and use either organic or conventional inputs and practices. Obviously, the actual practices will differ from farm to farm. A definition follows.

Sustainable agriculture is farming with reduced off-farm purchased inputs of pesticides, herbicides and fertilizers, along with reduced negative impact on natural resources and improved environmental quality and economic efficiency, while producing and distributing abundant, nutritious, affordable, high-quality foods and fibers for American and world markets.

The development of improved crops, cropping systems, irrigation, farm management, and marketing will be needed to make farms more profitable and sustainable. Typically, such farms will rely more on biological resources and management than on nonrenewable inputs of energy and chemicals. The foundation of a sustainable farm system is a comprehensive under-

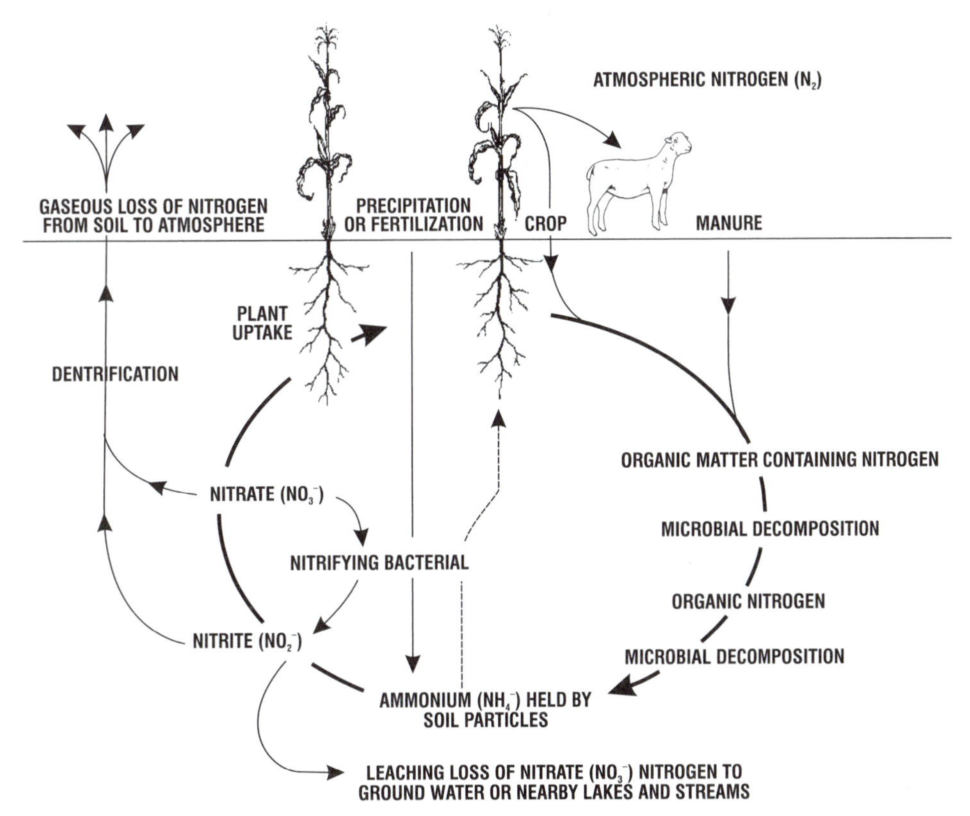

Fig. 7-15. At its best, a sustainable agriculture enhances the *cycle of nature*; manure is applied to soils and decomposed by microbes; complex protein in manure is broken down to release nitrogen as ammonia (NH_4); aerobic microbes convert ammonium nitrogen to nitrite (NO_2), thence to nitrate (NO_3) nitrogen; nitrate is (1) taken up by plants and built back into protein compounds; (2) leached downward when the soil is saturated—contaminating surface and groundwater if excessive nitrogen has been applied to the soil; and/or (3) released into the atmosphere when soils are wet for extended periods of time and the absence of air causes anaerobic microbes to convert the nitrate nitrogen to gaseous form.

standing of the land, the farm resources and operations, and potential short- and long-term markets.

Many of the practices advocated under sustainable agriculture are not new; they involve such timeless agricultural practices as soil erosion control, the protection of groundwater, the use of legumes as a source of nitrogen, biological insect and weed control, and the use of pastures as a primary feed source.

ANIMAL WELFARE/ ANIMAL RIGHTS

In recent years, the behavior and environment of animals in confinement have come under increased scrutiny of animal welfare/animal rights groups all over the world. For example, in 1987 Sweden passed legislation designed (1) to phase out layer cages as soon as a viable alternative can be found; (2) to discontinue the use of sow stalls and farrowing crates; (3) to provide more space and straw bedding for slaughter hogs; and (4) to forbid the use of genetic engineering, growth hormones, and other drugs for farm animals except for veterinary therapy. Also, the law provides for fining and imprisoning violators.

The United Kingdom Farm Animal Welfare Council deems *Five Freedoms* as essential guarantees of animal welfare in any system of livestock husbandry; namely, freedom from: (1) hunger and thirst; (2) discomfort; (3) pain, injury, and disease; (4) fear and distress; and freedom to (5) display normal behavior.[4]

Animal welfarists see many modern practices as unnatural, and not conducive to the welfare of animals. In general, they construe animal welfare as the well-being, health, and happiness of animals; and they believe that certain intensive production systems are cruel and should be outlawed. The animal rightists go further; they maintain that humans are animals, too, and that all animals should be accorded the same moral protection. They contend that animals have essential physical and behavioral requirements, which, if denied, lead to privation, stress, and suffering; and they conclude that all animals have the right to live.

Livestock producers know that the abuse of animals in intensive/confinement systems leads to lowered production and income—a case in which decency and profits are on the same side of the ledger. They recognize that husbandry that reduces labor and housing costs often results in physical and social conditions that increase animal problems. Nevertheless, means of reducing behavioral and environmental stress are needed so that decreased labor and housing costs are not offset by losses in productivity. The welfarist/rightists counter with the claim that the evaluation of animal welfare must be based on more than productivity; they believe that there should be behavioral, physiological, and environmental evidence of well being, too. And so the arguments go!

But wild animals are often more severely stressed than domesticated animals. They didn't have caretakers to store feed for winter or to irrigate during droughts; to provide protection against storms, extreme temperatures, and predators; and to control diseases and parasites. Often survival was grim business. In America, the entire horse population died out during the Pleistocene Epoch.

To all animal caretakers, the principles and application of animal behavior and environment depend on understanding; and on recognizing that they should provide as comfortable an environment as feasible for their animals, for both humanitarian and economic reasons. This requires that attention be paid to environmental factors that influence the behavioral welfare of their animals as well as their physical comfort, with emphasis on the two most important influences of all in animal behavior and environment—feed and confinement.

Animal welfare issues tend to increase with urbanization. Moreover, fewer and fewer urbanites have farm backgrounds. As a result, the animal welfare gap between town and country widens. Also, both the news media and the legislators are increasingly from urban centers. It follows that the urban views that are propounded will have greater and greater impact in the years ahead.

---

[4]*Livestock Environment IV*, Fourth International Symposium, University of Warwick, Coventry, England, July 6–9, 1993, Supplement, American Society of Agricultural Engineers, pp. 1255–1266.

QUESTIONS FOR STUDY AND DISCUSSION

1. Why is there so much interest in the subject of animal behavior?

2. Define *animal behavior* and *ethology*.

3. Why has there been a genetic time lag, with the environment altered faster than the makeup of animals?

4. List the nine general functions or behavioral systems that sheep and goats exhibit.

5. Discuss each of the following behavioral systems as it pertains to sheep and goats:
   a. Gregarious behavior
   b. Care-giving and care-seeking (mother-young) behavior
   c. Ingestive behavior (eating and drinking)
   d. Sexual behavior
   e. Shelter-seeking behavior

6. How important is the dominance hierarchy in sheep and goats?

7. Explain leader-follower behavior.

8. Discuss how sheep and goats communicate with each other.

9. Those who care for sheep and goats need to be familiar with behavioral norms in order to detect and treat abnormal situations—especially illness. Describe a normal sheep or goat.

10. Discuss (a) breeding for adaptation, and (b) quick adaptation—early training.

11. How may environment affect sheep and goats?

12. Discuss how each of the following environmental factors affects sheep and goats:
    a. Feed/environmental interactions
    b. Weather/environmental interactions
    c. Health/environmental interactions
    d. Predator control

13. Would you change the current pollution laws and regulations? If so, how?

14. Discuss the pollution potential from grazing federal lands.

15. Define sustainable agriculture.

16. How would you bring about more sustainable agriculture?

17. Discuss the position of the animal welfarists and the animal rightists.

Sheep in Wyoming. (Courtesy, American Sheep Industry Assn., Englewood, CO)

SELECTED REFERENCES

Title of Publication	Author(s)	Publisher
Angora Goat and Mohair Production	M. Shelton	M. Shelton, San Angelo, TX, 1993
Animal Agriculture: The Biology of Domestic Animals and Their Use by Man	Ed. by H. H. Cole M. Ronning	W. H. Freeman & Co., San Francisco, CA, 1974
Animal Behavior	J. P. Scott	The University of Chicago Press, Chicago, IL, 1958
Animal Behavior	V. G. Dethier E. Stellar	Prentice-Hall, Inc., Englewood Cliffs, NJ, 1970
Animal Behavior in Laboratory and Field	Ed. by A. W. Stokes	W. H. Freeman & Co., San Francisco, CA, 1968
Animal Behavior: A Synthesis of Ethology and Comparative Psychology	R. A. Hinde	McGraw-Hill Book Co., New York, NY, 1970
Animal Science, Ninth Edition	M. E. Ensminger	Interstate Publishers, Inc., Danville IL, 1991
Behavior of Domestic Animals, The	Ed. by E. S. E. Hafez	The Williams & Wilkins Co., Baltimore, MD, 1975
Development and Evolution of Behavior	Ed. by L. R. Aronson, *et al.*	W. H. Freeman & Co., San Francisco, CA, 1970
Ethology, The Biology of Behavior	I. Eibl-Eibesfeldt	Holt, Rinehart and Winston, New York, NY, 1975
Ethology of Free-ranging Domestic Animals	G. W. Arnold M. L. Dudzinski	Elsevier Scientific Publishing Company, New York, NY, 1978
Kinships of Animals and Man	A. H. Morgan	McGraw-Hill Book Co., New York, NY, 1955
Livestock Behavior	R. Kilgour C. Dalton	Westview Press, Boulder, CO 1983
Livestock Environment IV	Ed. by E. Collins C. Boon	American Society of Agriculture Engineers, St. Joseph, MO, 1993
Principles of Animal Behavior	W. N. Tavolga	Harper & Row, New York, NY, 1969
Principles of Animal Environment	M. L. Esmay	Avi Publishing Co., Westport, CT, 1969
Readings in Animal Behavior	Ed. by T. E. McGill	Holt, Rinehart and Winston, New York, NY, 1965
Scientific Aspects of the Welfare of Food Animals		Council for Agricultural Science and Technology (CAST), Ames, IA, 1981
Social Hierarchy and Dominance	Ed. by M. W. Schein	Dowden, Hutchinson & Ross, Inc., Stroudsburg, PA., 1975
Social Structure in Farm Animals	G. J. Syme L. A. Syme	Elsevier Scientific Publishing Company, New York, NY, 1978
Stockman's Handbook, The, Seventh Edition	M. E. Ensminger	Interstate Publishers, Inc., Danville, IL, 1992

CHAPTER 8

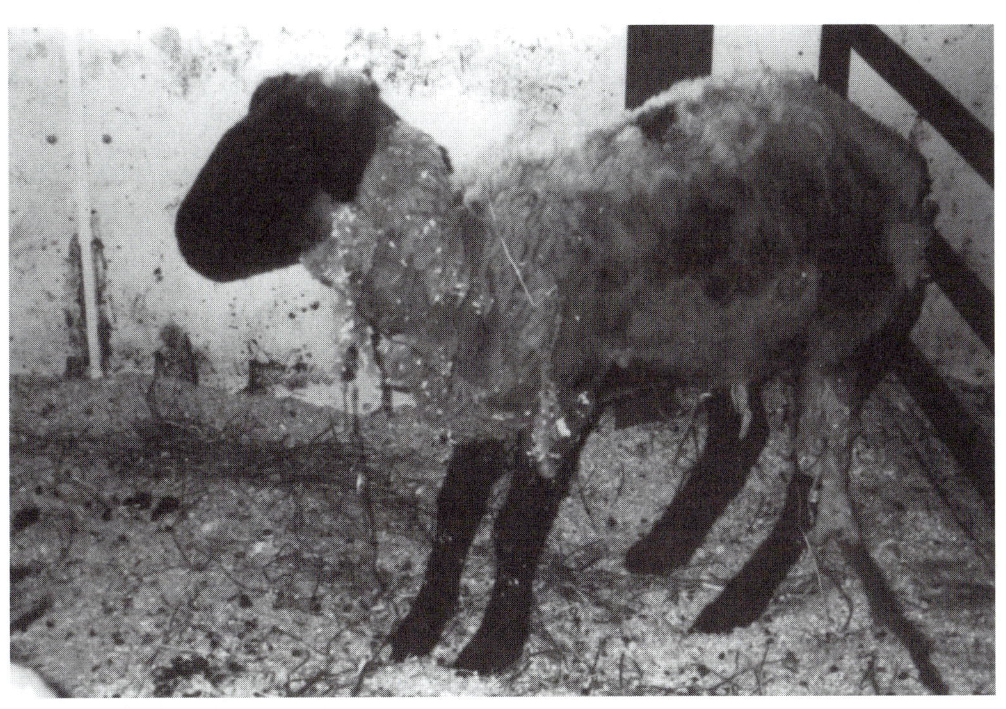

Scrapie. Affected animals rub themselves against fences and other objects to relieve the intense itching, and in doing so scrape off the wool; hence the name *scrapie*. (Courtesy, Dr. John R. Gorham, College of Veterinary Medicine, Washington State University, Pullman)

SHEEP AND GOAT HEALTH, DISEASE PREVENTION, AND PARASITE CONTROL[1, 2]

by

John R. Gorham, D.V.M., Ph.D.
Professor of Veterinary Microbiology and Pathology
USDA Agricultural Research Service
Animal Disease Research Unit
College of Veterinary Medicine
Washington State University
Pullman, Washington

William J. Foreyt, Ph.D.
Professor of Veterinary Microbiology and Pathology
USDA Agricultural Research Service
Animal Disease Research Unit
College of Veterinary Medicine
Washington State University
Pullman, Washington

and

M. E. Ensminger, Ph.D.

[1]Nutrition related disorders are covered in chapter 14, Feeding Sheep, of this book.

[2]The material presented in this chapter is based on factual information. However, when the instructions and precautions given herein are in disagreement with those of competent local authorities and/or reputable manufacturers, always follow the latter.

Contents *Page*

Hopefully, the material presented in this chapter will be of benefit to the producer, the student, and the veterinarian. It is based on the premises that: (1) Sheep and goat diseases and parasites can be largely prevented through good management, proper feeding, and strict sanitation; and (2) once disease troubles are encountered, the enlightened sheep or goat producer will more readily call a veterinarian in order that the condition may be brought under control with a minimum of loss.

Producers should be well-informed relative to the relationship of sheep and goat diseases and parasites to other classes of animals and to human health, because many of them are transmissible between species. It is noteworthy, for example, that about 90 different types of infectious and parasitic diseases can be spread from animals to human beings. Accordingly, the means by which sheep and goat diseases and parasites are transmitted to other classes of livestock and to humans will be stressed in the discussion which follows.

NORMAL TEMPERATURE, PULSE RATE, AND BREATHING RATE OF FARM ANIMALS

Table 8-1 gives the normal temperature, pulse rate, and breathing rate of farm animals. In general, any marked and persistent deviations from these normals may be looked upon as a sign of animal ill health.

Every livestock producer should have an animal thermometer. Also, at the end opposite of the bulb, an animal thermometer has an eye to which a 12-in. length of string with a clip is tied. The temperature is measured by inserting the thermometer full length into the rectum, where it should be left a minimum of three minutes. The clip on the string is attached to the wool or mohair. preventing the loss of the thermometer.

TABLE 8-1
NORMAL TEMPERATURE, PULSE RATE, AND BREATHING RATE OF FARM ANIMALS[1]

Animal	Normal Rectal Temperature		Normal Pulse Rate	Normal Respiration Rate
	Average	Range		
	(°F)	(°F)	(rate/min.)	(rate/min.)
Sheep	102.3	100.9–103.8	70–80	12–20
Goats	103.8	101.7–105.3	70–80	12–20
Horses	100.5	99.0–100.8	32–44	8–16
Cattle	101.5	100.4–102.8	60–70	10–30
Swine	102.6	102.0–103.6	60–80	8–13
Poultry	106.0	105.0–107.0	200–400	15–36

[1]To convert degrees Fahrenheit (F) to degrees centigrade (C), subtract 32, then multiply by 5/9.

In general, infectious diseases are ushered in with a rise in body temperature, but body temperature is affected by stable or outside (ambient) temperature, exercise, excitement, age, feed, etc. It is lower in cold weather, in older animals, and at night.

The pulse rate indicates the rapidity of the heart action. The pulse of sheep is taken on the inside of the thigh where the femoral artery comes in close proximity to the skin. It should be pointed out that the younger, the smaller, and the more nervous the animal, the higher the pulse rate. Also, the pulse rate increases with exercise, excitement, digestion, and high outside temperature.

Placing a hand on the flank, observing the rise and fall of the flanks, or, in the winter, watching the breath condensate in coming from the nostrils can determine the breathing (respiration) rate. Rapid breathing due to recent exercise, excitement, hot weather or poorly ventilated buildings should not be confused with disease. Respiration is accelerated in pain and in febrile conditions.

A PROGRAM OF SHEEP HEALTH, DISEASE PREVENTION, AND PARASITE CONTROL

Among practical sheep raisers, the importance of disease prevention and parasite control is well-known. The following program of sheep health, disease prevention, and parasite control is presented so that sheep producers will use it (1) as a yardstick with which to compare their existing programs, and (2) as a guidepost so that they and their local veterinarians, and other advisors, may develop similar and specific programs for their own enterprises.

I. General Sheep Program

1. Limit visitors, and have them disinfect or change their shoes. Salespersons, rendering truck personnel, shearers, sheep buyers, neighbors, and other visitors can unknowingly carry bacteria or other infectious agents on their bodies, clothing, and footwear. For this reason, (a) unnecessary visitors should be kept out of the sheep premises; and (b) before entering the premises, visitors should be required either to step into a shallow metal vat containing a foam rubber mat immersed in a disinfectant solution, or to put on clean rubber boots or plastic shoe covers provided for them.

2. Clean and disinfect the sheep barn (or sheep quarters) thoroughly, at least once a year. First, remove all manure. Then, either (a) scrub down the pens, walls, floors, and feeders with a lye solution (1 lb of lye to 15 to 20 gal. of water), or (b) apply a high pressur-

ized water sprayer and steam cleaner. Following cleaning and disinfecting, it is highly desirable that the barn be opened for air circulation and left empty for 3 to 6 weeks.

3. Sprinkle a light application of superphosphate (0-46-0) on fresh bedding; to keep the moisture down, to reduce ammonia fumes, and to aid in foot rot control.

4. Avoid overcrowding; provide 12 to 15 sq ft of floor space for each mature ewe of the larger breeds. Provide proper ventilation and keep the quarters clean and dry at all times.

5. Keep the feed and water troughs clean.

6. Force the flock to take plenty of exercise.

7. Control foot rot. Trim the feet of all ewes at shearing time and examine for evidence of foot rot. In problem flocks and on problem farms, control foot rot by a combination of intelligent husbandry, foot trimming, foot bathing, and vaccination.

8. In known bluetongue areas, vaccinate all sheep prior to the insect season (vaccinate at shearing time each year), observing the following precautions: (a) Do not vaccinate ewes during their first 2 or 3 months of pregnancy, for it will likely result in abnormal lambs; and (b) delay vaccinating lambs that are suckling immune dams until the lambs are 3½ months of age, because the ewes' milk will impair the active immunity produced by the vaccine.

9. Rotate the flock on clean, well-drained pastures at about two-week intervals.

10. Eliminate stray dogs, examine all dogs that must come in contact with sheep because the cyst stage of tapeworms may occur in sheep, and then administer the proper worm treatment, when necessary.

11. Give any animal showing evidence of infestation with internal or external parasites prompt and modern treatment, provided that a known and satisfactory treatment exists. Preferably, and when feasible, the parasite should be eliminated by breaking its life cycle before it enters the sheep.

12. Control most internal parasites through the following program:

a. Two weeks before turning to pasture, treat each adult sheep with the wormer of choice.

b. During the pasture season, worm sheep at such intervals as necessary.

c. At the end of the pasture season, again individually treat with the wormer of choice. Use different wormers in rotation, to lessen the hazard of a strain of stomach worm becoming resistant to a wormer.

13. Spray all sheep, including the lambs, with an approved insecticide for tick and wool maggot (blowfly) control, in from 10 to 30 days after shearing and again in July and August. An exception should be made in the case of very thin ewes or young lambs for which spraying should either be delayed or be of lowered concentration.

14. Tag all ewes prior to the breeding season.

15. Keep commercial feeder lambs in isolated areas away from breeding animals. When such lambs are placed in the feedlot, vaccinate them against enterotoxemia (overeating disease). If sore mouth (contagious ecthyma) has been present on the premises, vaccinate against it, also.

16. In the case of sick or dead sheep, promptly consult the veterinarian so that proper control measures can be instituted if necessary

II. Lambing Time

1. Before lambing, trim all tags from the udder and dock. If a warm barn is available, shear the ewes before lambing. Shorn ewes will take better care of their lambs because they will seek shelter, rather than stand outside in the snow or in a chilling rain. Also, more shorn ewes can be housed in the same space, and lambs can nurse shorn ewes better than ewes with heavy fleeces.

2. Clean and disinfect the lambing quarters thoroughly, well in advance of the lambing season.

3. Keep the holding pens as clean and as dry as possible, by applying lime as indicated and by changing the bedding frequently.

4. Treat the navel of the newborn lamb with 7% tincture of iodine.

5. In the spring, vaccinate lambs against overeating disease; and if sore mouth has been present on the premises, vaccinate against it, also.

III. New Stock

1. Avoid the inshipment of sheep from areas known to be infected with such contagious diseases as bluetongue and scrapie.

2. Do not make any additions to the ewe flock during the last three months of pregnancy.

3. Isolate, for a period of 30 days, all sheep, including show sheep, which are brought from the outside to be added to the flock. Also, take the following precautions on such new animals:

a. On arrival, treat for parasites, both external and internal.

b. Vaccinate for overeating disease; and if sore mouth has been present on the premises, vaccinate against its also.

A PROGRAM OF GOAT HEALTH, DISEASE PREVENTION, AND PARASITE CONTROL

The following recommendations should be considered as a guide or a foundation which goat producers, in cooperation with their local veterinarians, can enlarge and build upon to meet their specific needs. Some recommendations for a program of goat health, disease prevention, and parasite control include:

1. Clean all quarters thoroughly at least once a year, and clean and disinfect kidding pens and housing for kids between each generation.
2. Keep housing dry, well-ventilated, and draft-free. Check for moisture condensation in the winter.
3. Raise kids in separate age groups, preferably in movable pens that can be placed on new ground with each group of kids.
4. Keep kids separate from adult stock until they are at least 6 months of age or until they have been bred at least 2 months.
5. Avoid overcrowding. Provide proper amount of floor and lot space for the different age groups.
6. Provide feeders and waterers designed to prevent fecal contamination. Water should be warmed in the winter.
7. Be certain animals are able to get plenty of exercise.
8. Isolate new animals and animals returning from shows for a period of 30 days before adding to the herd.
9. Routinely trim hoofs and check for foot rot. Treat if necessary.
10. Periodically check for mastitis and practice regular teat dipping.
11. Dip the navel cord of each newborn kid in tincture of iodine.
12. Follow a program of internal and external parasite control, dictated by local conditions. Promptly treat any animals showing signs of internal or external parasite infestations
13. Vaccinate for enterotoxemia and tetanus when kids are 3 to 4 weeks of age, and repeat in 2 weeks. Administer booster 2 to 3 weeks before kidding. Vaccinate bucks whenever does are vaccinated.
14. Vaccinate for sore mouth, if goats are to be shown.
15. In abscess-problem herds, institute vaccination program against *Corynebacterium ovis* or *C. pyogenes*. Vaccinate at 1 and 2 months of age, then possibly at 4- to 6-month intervals.
16. Administer selenium and alpha-tocopherol (a) to does 60 days before kidding and possibly again 15 days before kidding, and (b) to bucks at the same time does receive their dose, then again several weeks before the breeding season.
17. Have the veterinarian conduct postmortem examination of dead goats.

By referring to the program suggested for sheep, which preceded this section, the reader may glean additional ideas for a program of goat health, disease prevention, and parasite control.

DISEASES OF SHEEP AND GOATS

Sheep and goat producers, and students, should be well informed relative to the most important diseases so that they can do a better job of preventing and controlling them.

Disease prevention in sheep and goats requires providing adequate feed of a proper ration, providing plenty of clean, fresh water, avoiding overcrowding, forcing the flock to take plenty of exercise, and providing clean, dry, well-ventilated quarters. A thorough knowledge of these preventive measures is a requisite of successful caretakers. They should also have sufficient knowledge of the common diseases so that they can (1) recognize a hazard once it is encountered, and (2) promptly call a veterinarian and carefully carry out treatments and preventive measures.

Only the most common diseases affecting sheep and goats are covered herein.

ANTHRAX (SPLENIC FEVER, CHARBON)

Anthrax is an acute infectious disease affecting all warm-blooded animals and humans. It usually occurs as scattered outbreaks or cases, but hundreds of animals may be involved. Certain sections are known as anthrax districts because of the repeated appearance of the disease. Mature animals on summer pasture and grazing animals are particularly subject to anthrax, especially when pasturing closely following a drought or on land that has been recently flooded. In the United States, most human infections of anthrax result from handling diseased or dead animals on the farm or from handling hides, hair, wool, and mohair in factories.

Historically, anthrax is of great importance. It is one of the first scourges to be described in ancient and Biblical literature; it marks the beginning of modern bacteriology, being described by Koch in 1877; and it is the first disease in which immunization was effected by means of an attenuated culture, Pasteur having immunized animals against anthrax in 1881.

SYMPTOMS OR SIGNS[3]

In sheep and goats, anthrax usually appears as a very rapidly fatal disease, with death occurring within one hour or less. Grinding of the teeth, pounding of the heart, hard breathing, and collapse are generally

[3]Currently, many veterinarians prefer the word *signs* rather than *symptoms*, but throughout this chapter the author accedes to the more commonly accepted terminology among sheep producers and includes the word *symptoms*.

noted. The carcass usually exhibits dark blood about the nose and the anus.

CAUSE, PREVENTION, AND TREATMENT

The disease is identified by a microscopic examination of the blood or the lymph nodes in which will be found the typical large, rod-shaped organisms causing anthrax, *Bacillus anthracis*. These organisms can survive for years in a spore stage, resisting most destructive agents. As a result, they may remain in the soil for extremely long periods.

This disease is one that can largely be prevented by immunization. In the so-called anthrax regions, vaccination should be performed annually well in advance of the time when the disease normally makes its appearance. At least nine types of biologics (serums, bacterins, and vaccines) are now available for use in anthrax vaccination. The choice of the biologic is dependent upon the local situation and should be left to the local veterinarian or state livestock sanitary officials. Also, there should be adequate fly control during the insect season. Flocks or herds that are infected should be quarantined. The farmer or rancher should never open the carcass of a dead animal suspected of having died from anthrax; instead the veterinarian should be summoned at the first sign of an outbreak.

When the presence of anthrax is suspected or proved, all carcasses and contaminated material should be soaked with oil and completely burned or covered with lime and deeply buried, preferably on the spot. Such carcasses should not be taken to the rendering plant. These precautions are important because the spores, which are long lived, can be spread by dogs, coyotes, buzzards, and other flesh eaters and by flies and other insects.

When an outbreak of anthrax is discovered, all sick animals should be isolated promptly and treated. All exposed healthy animals should be vaccinated; pastures should be rotated; the premises should be quarantined; and a rigid program of sanitation should be initiated. These control measures should be carried out under the supervision of state or federal regulatory officials.

Early treatment with massive doses of penicillin may be effective.

BIGHEAD (SWELLHEAD, PHOTOSENSITIZATION)

This is a condition in which animals (mostly sheep and cattle, although it occurs in goats, horses, and swine) become hypersensitive to sunlight following ingestion of certain substances, usually plants, containing sensitizing agents that render the tissue cells abnormally sensitive to light. It is worldwide in distribution. Photosensitization occurs in all breeds, ages, and sexes of sheep, especially in nonpigmented or lightly-pigmented animals.

SYMPTOMS OR SIGNS

The skin (not protected by wool) of white or light-colored sheep is affected. The first indication is a swelling of the ears, eyelids, and lips, accompanied by intense itching. In severe cases, the skin under the ears becomes distended with clear fluid, which finally oozes out and adheres to the surface. Later, the swellings decrease in size, and brownish scabs form over the surface. These symptoms do not appear in black-faced sheep or in sheep kept out of the sunlight.

CAUSE, PREVENTION, AND TREATMENT

The cause of bighead may vary from area to area. Several plants produce the condition, including little leaf horsebrush *(Tetradymia glabrata)*, spineless horsebrush *(Tetradymia canescens)*, agave *(Agave lecheguilla)*, sachuiste *(Nolina texana)*, smartweed *(Polygonum persicaria)*, St. Johns wort *(Hypericum perforatum)*, blossoming buckwheat, and wet alsike clover.

Keeping sheep away from areas where causative plants are located, or allowing supplemental feeding when trailing animals through such plant-infested areas may prevent the disease.

No antidotes have been discovered. Keeping the animals in the shade during the day and letting them graze at night is helpful. Soothing and protective preparations may be applied to affected skin to allay itching and control infection; and the veterinarian may inject antihistamines.

BLACK DISEASE

Black disease affects sheep throughout the United States, wherever fluke-bearing snails abound.

SYMPTOMS OR SIGNS

Black disease usually affects adult sheep (lambs and yearlings are rarely affected) that are on pastures harboring fluke-bearing snails, and strikes during the summer and fall. There is generally rapid death with no distinct symptoms; affected animals lag behind the flock and die within 1 or 2 hours.

CAUSE, PREVENTION, AND TREATMENT

The disease is caused by a spore-bearing anaerobe, *Clostridium novyi*. Immature liver flukes *(Fasciola hepatica)*, which are present in the bile ducts, provide the site for *C. novyi* action.

Recommended preventive measures include (1) destroying fluke-bearing snails (as described elsewhere in this chapter), and (2) vaccinating sheep with a toxoid annually before going to snail-infested pastures (a sheep vaccinated two successive years is very resistant).

Once black disease has become established, there is no effective treatment.

BLACKLEG (BLACKQUARTER, QUARTER ILL, EMPHYSEMATOUS GANGRENE)

This disease is called blackleg because this describes the appearance of infected limb muscles. Blackleg is an acute, infectious, but noncontagious, disease.

SYMPTOMS OR SIGNS

In sheep, the disease is manifested by swellings on any part of the body, usually—though not always—in the region of a recent wound (from shearing, castrating, docking, bruising from fighting, or parturition). The disease may occur in ewes as a result of infection of wounds of the genital tract produced at lambing time. Cases in rams have been observed in which the lesions appeared about the head as the result of the infection of wounds acquired through butting. Pressure over the swellings usually causes a cracking sound due to movement of the entrapped gas formed by the bacteria. The swellings are usually hot and painful. The temperature may run up to 107°F, and there is depression and loss of appetite. Finally, the animals go down and usually die in 24 to 48 hours. Recoveries are rare.

CAUSE, PREVENTION, AND TREATMENT

The disease is caused by *Clostridium chauvoei*.

Vaccination is effective in preventing the disease. Where previous experience has shown that the soil is heavily seeded with the spores, it may be wise to vaccinate 2 to 4 weeks before shearing, castrating, or docking. Proper carcass disposal of animals killed by the disease will prevent reseeding the soil with spores. Also, instruments used in the common management operations on sheep—castrating, docking, marking, etc.—should always be sterilized before use and at frequent intervals during use.

Antibiotics may be used, but usually the disease progresses too rapidly, once symptoms have developed, for any treatment to be effective.

BLUETONGUE

Bluetongue—a disease of sheep, though goats and cattle are sometimes mildly affected—has been known in South Africa since 1876. Since 1949, it has been identified in Cyprus, Palestine (in 1948, Palestine was incorporated into Israel, Jordan, Egypt, and Syria), Turkey, Portugal, Spain, Pakistan, and the United States. It was probably present in Texas, where it was known as *sore muzzle*, as early as 1948. But it was first definitely diagnosed in the United States in 1953. In the United States, it is most common in the Southwest.

In South Africa, mortality rates usually do not exceed 10%, but have run as high as 70%. In the United States, the mortality and morbidity rates vary from 10 to 80%. A severe economic cost results from the loss of condition and loss of wool.

SYMPTOMS OR SIGNS

The symptoms of bluetongue are: a blue tongue, high temperature (104 to 107°F), depression and loss of appetite, rapid and extreme loss of weight, reddened mucous membrane of the mouth which turns purplish or blue, frothing of saliva, formation of lip ulcers, offensive odor, discharge from the eyes and nose, weakness, appearance of a red band at the top of the animal's hoofs, lameness (in extreme cases the hoofs may slough off), and loss of wool. Rams are more susceptible than ewes.

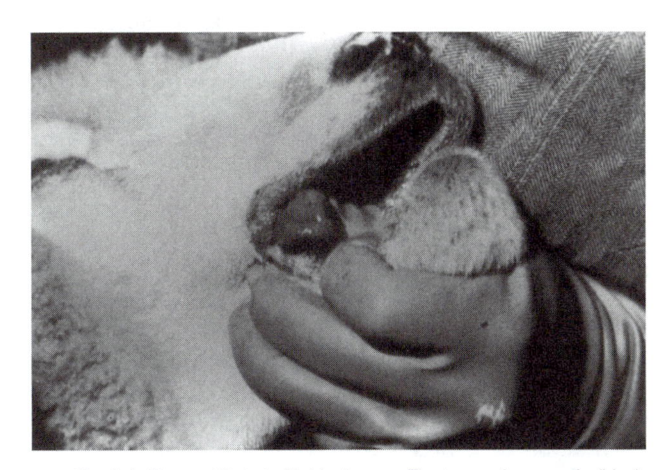

Fig. 8-1. Sheep afflicted with bluetongue. The tongue is cyanotic (blue). (Courtesy, Dr. John Gorham, D.V.M., Ph.D., Professor of Microbiology and Pathology, Washington State University, Pullman)

Bluetongue becomes prevalent in the summer and stops abruptly after the first hard frost, since the disease is carried by small insects known as biting midges, gnats, or "no-see-ums."

When the disease first appeared in the United States, it was erroneously diagnosed as sore mouth, which it somewhat resembles. Other conditions with which it may be confused are bighead, and lip and leg ulceration.

CAUSE, PREVENTION, AND TREATMENT

Bluetongue is an infectious but noncontagious viral disease, transmitted by and multiplied in the insect *Culicoides variipennis*.

Banning the inshipments of sheep and cattle from infected areas will help control the disease, but the flies that transmit the disease are carried about by automobiles, trucks, and planes. Also, these insects breed in mud along the edges of slow-moving streams or where water tanks overflow, so breeding sites should be sprayed and/or eliminated. Night housing of sheep on high land beyond the flying range of the insect vector prevents exposure. Vaccination, by injection of a modified live virus-type vaccine, is recommended in bluetongue areas. All ewes and rams should be vaccinated at shearing time each year, and all replacement lambs should be vaccinated at about 3½ months of age. Do not vaccinate ewes during their first two months of pregnancy, otherwise a large number of "crazy" lambs may be born. Goats are not routinely vaccinated.

To date, treatments have been of little value. Good nursing will save some affected animals. Secondary complications may be treated by the veterinarian with appropriate antibiotics and sulfonamides. It may also be necessary to treat or prevent secondary screwworm infections of the lesions in the Southwest.

Recovered sheep may be immune for life against exposure to the strain of virus that infected them, but not to other strains.

Suspected cases of bluetongue should be diagnosed by the local veterinarian, and then reported to state or federal officials.

BRUCELLOSIS

Brucellosis, which occurs throughout the world, is more common in goats than in sheep.

Brucellosis is an insidious (hidden) disease in which the lesions frequently are not evident. Although the medical term *brucellosis* is used in a collective way to designate the disease caused by the three different but closely related *Brucella* organisms, the species names further differentiate the bacteria as: (1) *Brucella abortus*, (2) *B. suis*, and (3) *B. melitensis*.

Brucellosis control and eradication in farm animals is important for two reasons: (1) the danger of human infection, and (2) the economic loss.

SYMPTOMS OR SIGNS

Unfortunately, the symptoms of brucellosis are often rather indefinite. While the act of abortion is the most readily observed symptom in goats, not all animals that abort are affected with brucellosis and not all animals affected with brucellosis will necessarily abort. On the other hand, every case of abortion should be regarded with suspicion until it is proved noninfectious.

In the laboratory, *B. melitensis* can be cultured and identified from blood, vaginal discharges, milk, lymph nodes, placentas, and aborted fetuses. Agglutination tests may be applied to blood serum and milk for diagnosis.

CAUSE, PREVENTION, AND TREATMENT

In sheep and goats, the disease is caused by *Brucella melitensis*, and is known as Malta fever or abortion.

If the disease appears in sheep or goats, appropriate preventive methods for the prevailing circumstances should be taken. These may include (1) eradication by identifying and slaughtering all infected or exposed sheep or goats; (2) cleaning, disinfecting, and resting infected premises; (3) quarantine of infected area; and (4) immunizing susceptible sheep or goats. All dairy goats should be tested by the state veterinarian.

To date, there is no known medicinal agent that is effective in the treatment of brucellosis in any class of farm animals. Thus, farmers and ranchers should not waste valuable time and money on so-called cures that are advocated by fraudulent operators.

CAPRINE ARTHRITIS AND ENCEPHALITIS (CAE)

Caprine arthritis and encephalitis (CAE) is a newly discovered disease of goats in the United States. In kids, it causes paralysis, and in adults it causes arthritis. The late stages of CAE are similar to rheumatoid arthritis in humans. Between 80 and 90% of the domestic goats in the United States are infected by the virus responsible for CAE.

SYMPTOMS OR SIGNS

In any herd, the expression of the disease ranges from 0 to 25% and in those animals that do show clinical symptoms, the rate of progression and the severity of the disease varies markedly. The expression in kids may vary from a barely noticeable unsteadiness of gait to a rapid fatal paralysis. Diligent nursing may maintain paralyzed goats for months, but the damage is permanent. In mature goats, the joints (front knees, hocks, and stifle joints) become swollen or disfigured, with a loss in body weight and a drop in production. The severity of the arthritis form varies from intermittent lameness or stiffness for years to complete debilitation.

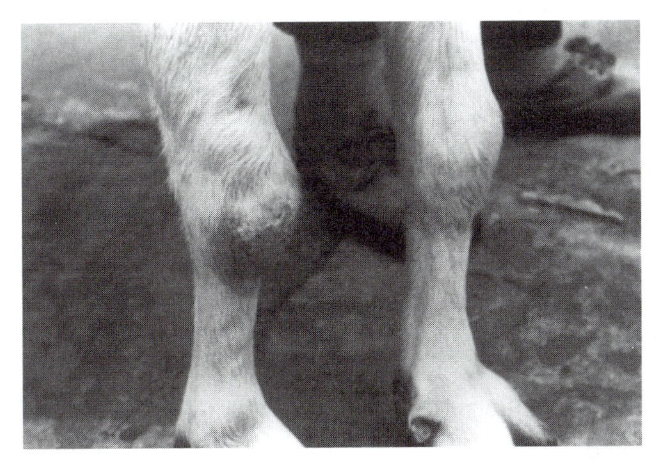

Fig. 8-2. Greatly enlarged knee joints in a long-standing case of caprine arthritis and encephalitis. (Courtesy, Dr. John Gorham, D.V.M., Ph.D., Professor of Microbiology and Pathology, Washington State University, Pullman)

CAUSE, PREVENTION, AND TREATMENT

CAE is caused by a retrovirus, a slow virus that produces disease after a long incubation period, persisting throughout an animal's life. The virus is transmitted from the doe to the kid(s) through the colostrum and milk. Does not showing any symptoms of CAE may carry the virus and transmit it.

Attempts to create a vaccine thus far have not been successful. Currently, the best method of prevention seems to be the establishment of CAE virus-free herds. This involves the routine testing of all animals in a herd with the Agar Gel Immunodiffusion Test (AGID) for antibody against CAE. Animals testing positive are culled. However, if AGID-positive does are kept in the herd, then steps to establish a CAE-free herd begin with the kids. All births are observed and kids are removed from the does immediately at birth. Kids are then fed colostrum only from does identified as negative to the AGID test or heat-treated 132.8°F for 1 hour. After receiving colostrum, kids can be raised on pasteurized goat's milk or milk replacer. Another approach is to raise colostrum-deprived kids, but this requires know-how and experience.

At present, there is no effective treatment available. Aspirin or phenylbutazone gives temporary relief from the arthritis pain. Good nutrition delays wasting, and avoidance of cold eases joint stiffness and pain.

CASEOUS LYMPHADENITIS (LUNGERS, PSEUDOTUBERCULOSIS)

This is an infectious, contagious, chronic disease which occurs predominantly in range bands of sheep, but rarely in farm flocks. Caseous lymphadenitis is a major disease of goats in the United States. The infectious agent gains entrance to the body through docking, castrating, and shearing wounds, or wounds received in pens or corrals in poor repair.

SYMPTOMS OR SIGNS

Affected animals do not recover. Most of the losses occur in sheep four years old or older. Economic losses also result from unthriftiness of infected sheep and goats, and condemnation of infected carcasses.

The first noticeable symptom is rapid breathing after exercise. Breathing becomes progressively more difficult, and respiration more rapid. In the final stage, affected animals show heavy flank breathing even when at rest. Early in the course of this disease, animals cough occasionally; later, there are often prolonged, spasmodic spells of coughing. There is no rise in body temperature unless an acute secondary lung infection sets in, and there is no loss of appetite. Affected sheep lose flesh, and become extremely emaciated and weak. Eventually, death occurs. The course of the disease varies from months to years.

CAUSE, PREVENTION, AND TREATMENT

The cause of caseous lymphadenitis is the bacillus *Corynebacterium pseudotuberculosis* (or *C. ovis*). Upon entry into the body, the bacteria move to the lymph nodes where they grow, multiply, and spread, eventually entering the lungs.

Animals showing symptoms should be removed and slaughtered. In buying stock—particularly old ewes or does—the producer should be careful not to introduce caseous lymphadenitis. In badly infected flocks, (1) lambs and replacement ewes should be shorn first, thereby alleviating the possibility of spreading the disease through shearing wounds; (2) shears should be disinfected between sheep; (3) shearing wounds should be treated with tincture of iodine; (4) freshly shorn sheep should be moved to a clean area as soon as possible; and (5) docking and castrating should be performed in a clean area.

In goats, prevention is primarily that of reducing the likelihood of wounds and not allowing the introduction of infected animals into the herd.

There is no specific treatment. Vaccines have been tried, but found wanting. Although life can be prolonged considerably by good nursing, the desirability of doing so is questionable.

Treatment in goats usually consists of surgical drainage and local disinfectant treatment of the abscess, or complete surgical removal of the abscess intact. Extreme care must be exercised to ensure that the contents of the abscess or drainage from the abscess does not contaminate the environment where goats are kept.

Fig. 8-3. Sheep with circling disease. Note pushing against the wall. Circling disease affects the nervous system. Affected animals show depression, staggering, circling, and strange awkward movements. (Courtesy, Department of Veterinary Pathology and Hygiene, College of Veterinary Medicine, University of Illinois)

CIRCLING DISEASE (ENCEPHALITIS, LISTERIOSIS)

Circling disease is an infectious but noncontagious disease affecting sheep and goats, and other animals and humans. It occurs most often in winter and spring, with a mortality approaching 100%.

Animals of all ages are susceptible. Also, well-cared-for animals are just as prone to infection as those that are accorded neglect. Confined sheep demonstrate a higher incidence than range flocks.

SYMPTOMS OR SIGNS

When the flock of sheep (or herd of goats) is moved, affected animals are sluggish and trail the flock. Since the bacteria attack the brain, nervous signs are evident; *i.e.*, incoordination, depression, circling, and strange awkward movements are noted. Ear drooping is also a common sign. The animals may be seen holding a mouthful of hay for hours. There may be inflammation around the eyes, and abortion may occur. The course of the disease is very short, with paralysis and then death the usual termination. Positive diagnosis can be made only by laboratory examination of the brain.

CAUSE, PREVENTION, AND TREATMENT

Circling disease results from the invasion of the central nervous system by bacteria called *Listeria monocytogenes*. The method of transmission to susceptible sheep or goats is through (1) contaminated feed, especially silage, (2) contaminated water, and (3) secretions and excretions from infected animals. Outbreaks often occur after a stress such as inclement weather, shipping, or crowding.

No commercial vaccine is available in the United States. Early treatment with high doses of penicillin may be effective.

The following program may aid in preventing the disease

1. Do not store silage in a silo that is in poor repair.
2. Do not feed silage from the top layer of an upright silo.
3. Never feed moldy or spoiled silage.
4. Provide clean, dry quarters during inclement weather.
5. Provide clean drinking water.
6. Control parasites.
7. Avoid stress.
8. Choose replacement animals from listeria-free flocks (or herds).

Fig. 8-4. Goat convulsing with listeriosis—generally fatal. (Courtesy, Christine S. F. Williams, Michigan State University)

9. Isolate animals with circling disease. Move unaffected animals to clean premises.

Silage samples can be submitted to a diagnostic laboratory to determine if *Listeria* are present.

Affected sheep or goats seldom respond to treatment.

ENTEROTOXEMIA (APOPLEXY, OVER-EATING DISEASE, PULPY KIDNEY)

Enterotoxemia is an acute, non-contagious disease of sheep and goats of all ages, affecting animals in a high state of nutrition—on a lush feed of grain, milk, or grass. Finishing lambs on high grain rations are most frequently attacked, though young lambs suckling ewes are sometimes affected. Before the development of a commercial vaccine, enterotoxemia rated as the number one cause of economic losses among feedlot and pasture-fattening lambs.

SYMPTOMS OR SIGNS

Affected animals go off-feed and display general sluggishness and diarrhea. Sometimes animals throw their heads back while they walk in a circle. Other animals may be seen staggering blindly about or having convulsions. Usually the course of the disease is acute, with affected animals dying within a few hours. In other cases, the animals may die a few at a time without showing much in the way of symptoms. The death rate is a minimum of 1%, with an average of 3 to 4% in unvaccinated feedlot lambs. In explosive outbreaks, losses range from 10 to 40%.

Enterotoxemia usually affects the biggest and the fastest-growing lambs.

CAUSE, PREVENTION, AND TREATMENT

The cause is a toxin produced by one type of the anaerobic bacterium *Clostridium perfringens*, Type D. This organism is found in the soil and in the digestive tract of nearly every warm-blooded animal.

Several preventive measures can be taken:

1. Make a gradual change from range to feedlot conditions. Precondition lambs by placing them on hay and concentrates before weaning.
2. Vaccinate lambs with either a bacterin or a toxoid soon after their arrival in the feedlot, provided they are in good condition and not wet. Allow at least 10 days after vaccination for immunity to develop.
3. Sometimes revaccination with the bacterin or toxoid (a booster shot) is required 2 to 4 weeks following the first vaccination.

Vaccinating the pregnant ewes may prevent losses of young lambs during their first 6 weeks of life. Ewes that have not been vaccinated previously should be vaccinated twice 2 to 4 weeks apart, with the second vaccination being given 2 to 4 weeks prior to lambing. Thereafter, an annual booster shot should be given 2 to 4 weeks prior to lambing. Vaccinating the ewes prior to lambing ensures that the lambs will receive colostral protection for 2 to 3 weeks. Then, the lambs should be vaccinated at 4 to 6 weeks of age.

The feeling persists among some feeders that a higher than usual incidence of enterotoxemia occurs in feedlot lambs on high-silage rations.

Control of explosive outbreaks late in the feedlot period consists of the following:

1. Reducing the concentrate allowance by 50% for one week or longer.
2. Marketing all lambs carrying adequate condition for slaughter.
3. Vaccinating the remaining lambs with bacterin or toxoid and gradually returning to full feed.
4. Considering the use of Type D antitoxin (which is expensive) to stop the losses. It will confer temporary immunity (2 to 3 weeks), following which vaccinating with bacterin or toxoid may establish a long-lasting immunity.

Once the disease has developed, treatment seems to be of little avail. With those animals that do survive, great difficulty is encountered in getting them back on feed.

Due to the extreme susceptibility of dairy goats to enterotoxemia, sound feeding management along with an immunization program must be used effectively to prevent the occurrence of this disease. In goats, excess feeding, sudden access to palatable feed, changes in feed, or feeding following an unusual period of fasting should be avoided. Goats should be routinely vaccinated with *Clostridium perfringens* toxoid, Types C and D, on the following schedule:

1. Initially, all goats should receive two doses of the toxoid at 2- to 4-week intervals.
2. Does should receive annual boosters about 1 month before kidding.
3. All kids should be vaccinated at 3 to 4 weeks of age, followed by a second dose of toxoid 2 weeks later.
4. All goats in the herd should be vaccinated twice a year—when does are in late pregnancy, and when kids are 4 to 5 months old.

Goats affected with enterotoxemia can be treated with large doses of *C. perfringens* B-C-D antitoxin.

ENTROPION (INVERTED EYELIDS)

In newborn lambs and kids, one or both eyelids are sometimes turned in (inverted)—a condition known as *entropion*. Most frequently, it is the lower lid. The eyelashes irritate the eye, causing blindness if the condition is not corrected or treated.

SYMPTOMS OR SIGNS

The eyes become infected, followed by a discharge and cloudiness, and, if untreated, the animal becomes blind.

CAUSE, PREVENTION, AND TREATMENT

Entropion is a heritable defect; hence, afflicted animals should be marked when treated and not retained for breeding purposes.

The caretaker should examine the eyelids of each newborn lamb or kid. If they are turned in, corrective treatment may be necessary. Unless the condition is serious, a daily application of an eye ointment until the eyelids turn out will usually suffice. Otherwise, minor surgery may be necessary. This consists of either (1) picking up a fold of skin sufficient to draw the edge of the eyelid to a normal position and cutting off the fold so that the eyelid will be drawn from the eye while healing occurs; or (2) applying wound clips (suture clips) to the center of the lower eyelid to hold the fold of the skin in place. In a few days, the clips will probably drop off, but sufficient irritation will develop to hold the edge of the lid in a normal position.

ENZOOTIC ABORTION (CHLAMYDIA)

Enzootic abortion in sheep and goats is a worldwide, infectious disease which manifests by abortion and, to a lesser extent, by stillbirth or premature parturition. In the United States, EAE is more prevalent among sheep of the western states of the intermountain area. In some areas, EAE often causes severe economic losses.

SYMPTOMS OR SIGNS

Abortion in ewes occurs late in pregnancy, or lambs are full-term stillborn or weak and die soon after birth. Ewes may appear somewhat sick and depressed before and after they abort. Often the placental membranes are retained. The placental membranes and aborted fetuses are essential for proper laboratory diagnosis.

Unlike sheep, where abortion generally occurs the last 2 to 4 weeks of pregnancy, abortion in goats occurs at any stage of pregnanacy.

CAUSE, PREVENTION, AND TREATMENT

EAE is caused by a strain of *Chlamydia psittaci* similar to, and possibly identical to, the one causing bovine abortion.

Some prevention is possible by segregation of aborting animals since the infection is primarily transmitted at lambing time. Also, a vaccine may be available that can be given in combination with the vaccine for vibriosis at the time of breeding.

During an outbreak, antibiotics may help, especially when administered to infected newborn lambs and to ewes that carried dead fetuses for some time. Also, secondary bacterial infections resulting from retained placentas necessitate antibiotic treatment.

FOOT-AND-MOUTH DISEASE

This is a highly contagious disease of cloven-footed animals (mainly sheep, goats, swine, and cattle) characterized by the appearance of blisters (vesicles) in the mouth (and in the snout in the case of hogs), on the skin between and around the claws of the hoofs, and on the teats and udders. Fever, diminished rumination, and reduced appetite are other signs of the disease.

Humans are mildly susceptible but very rarely infected, whereas horses are immune.

The disease is not present in the United States, but there were at least nine outbreaks (some authorities claim 10) in this country between 1870 and 1929, each of which was stamped out by the prompt slaughter of every affected and exposed animal. No U.S. outbreak has occurred since 1929, but the disease is greatly feared. Preventing the introduction of the disease into the United States, or, in the case of actual outbreaks, eradicating it, has required drastic measures.

SYMPTOMS OR SIGNS

The disease is characterized by the formation of blisters and by a moderate fever 3 to 6 days following exposure. These blisters are found on the mucous membranes of the tongue, lips, palate, cheeks, on the skin around the claws of the feet, and on the teats and udders. In sheep the blisters are found most commonly about the feet.

The presence of these vesicles, especially in the mouth of cattle, stimulates a profuse flow of saliva that hangs from the lips in strings. Complicating or secondary factors are infected feet, caked udders, abortion, and great loss of weight. The mortality of adult animals is not ordinarily high, but the usefulness and productivity of affected animals is likely to be greatly damaged, thus causing great economic loss.

CAUSE, PREVENTION, AND TREATMENT

The infective agent of this disease is a small virus of which there are at least seven strains; Infection with one strain does not protect against the other strains. The virus is present in the fluid and coverings of the blisters, in the blood and meat, and in the milk, saliva, urine, and other secretions of the infected animals.

Except for the nine outbreaks mentioned, the disease has been kept out of the United States by extreme precautions, such as quarantine at ports of entry and assistance with eradication in neighboring countries when introduction has appeared imminent.

Two methods have been applied in control: slaughter and quarantine. If the existence of the disease is confirmed by diagnosis, the area is immediately placed under strict quarantine; infected and exposed animals are slaughtered and buried, with owners being paid indemnities based on their appraised value. Everything is cleaned and thoroughly disinfected.

Fortunately, the foot-and-mouth disease virus is quickly destroyed by a solution of the cheap and common chemical sodium hydroxide (lye). Because quick control action is necessary, state or federal authorities should be notified the very moment the presence of the disease is suspected.

No effective treatment is known.

Vaccines have not been used in the outbreaks in the United States because they have not been regarded as favorable to rapid, complete eradication of the infection.

FOOT ROT (FOUL FOOT)

Foot rot is a serious infectious disease affecting sheep and goats in practically all countries where they are raised. It is especially troublesome in wet, muddy areas.

SYMPTOMS OR SIGNS

Although foot rot is not highly fatal, it is of considerable economic importance. Affected ewes become unthrifty and produce less wool and milk; lambs become stunted, and rams may be rendered useless at breeding time. It is not uncommon for sheep producers to dispose of entire flocks because of this disease.

Usually the first indication of the presence of foot rot is severe lameness in one or more sheep, though not all lameness is caused by foot rot. In the early stages of the disease, the skin in the cleft between the toes and the soft tissues on the inside surface of the heels is slightly swollen, reddened, and moist. Within 3 or 4 days, this skin becomes grayish-yellow or dead-looking. If not checked at this stage, the disease spreads under the hoof, causing a separation of the hoof from the underlying soft tissue. A grayish-yellow pus oozes from the affected parts.

Foot rot is accompanied by a characteristic foul odor which attracts flies in warm weather. Frequently, the affected feet become infested with maggots.

Foot rot in sheep rarely involves the coronary band or extends above the top of the hoof. Since the tissue from which the hoof is formed is not affected, the hoof continues to grow, resulting in a misshapen hoof.

A sheep may have foot rot in more than one foot at the same time. If both front feet are affected, the animal may kneel when grazing. In cases where more than two feet are involved, the sheep will spend most of the time lying down, and may even refuse to stand.

CAUSE, PREVENTION, AND TREATMENT

Foot rot is a contagious, infectious disease caused by the organism *Bacterioides nodosus* in conjunction with *Fusobacterium necrophorum*. Possibly other organisms such as *Spirocheata penortha* and *Corynebacterium pyogenes* may be involved. Since the disease can be spread only by infected sheep or goats, early diagnosis followed by proper treatment will prevent spread of the disease in a flock or herd and will result in considerable saving in time and expense. Walking sheep or goats over contaminated pastures and areas where infected sheep or goats have been is the principal means of spreading the disease, though the incidence is influenced by the weather and temperature.

To treat infected animals, place them in a clean, dry pen and treat as follows:

1. Examine every foot of every animal. Trim each foot showing infection, removing enough of the horn of the hoof thoroughly to expose all diseased tissue.
2. Walk all animals through a suitable disinfectant solution and move to clean ground. Visibly affected animals should be kept standing in the solution 5 to 10 minutes. The two most widely used disinfectants are (1) formaldehyde, 10%, and (2) copper sulfate, 20%.[4] Repeat foot bath at weekly intervals until foot rot disappears. Then continue at two-week intervals for another

[4]A 10% formaldehyde solution may be made by mixing 1 gal. of 38% formaldehyde and 9 gal. of water, and a 20% copper sulfate solution may be made by mixing 1.66 lb of copper sulfate per gallon of water. Since copper sulfate is corrosive for most metals, it should be prepared in earthenware or wooden containers.

two months. Two weeks after initial antiseptic treatment, examine feet of each animal a second time, to detect and trim infections overlooked the first time or developed subsequently.

3. After trimming, and treatment in a foot bath, place animals in a clean, dry pasture or lot. One that has not been used for 30 days would be considered clean. An animal with foot rot may spread infection up to three years, but contaminated land loses its ability to infect within three weeks.

Sulfonamide or antibiotic therapy may accompany trimming and foot baths with good results.

Prevention of foot rot includes draining muddy pastures and segregating new animals. Foundation and replacement animals should be purchased from known clean sources. If animals are from a questionable or unknown source, pass through a public market, or are transported by a public conveyance, their hoofs should be trimmed on arrival, and then they should be walked through a foot bath and isolated for one month. Also, land previously pastured by sheep should be allowed to remain idle for four weeks before other sheep are turned on it. Cross infections of foot rot between cattle and sheep do not occur, but cross infections between sheep and goats do occur.

A commercially prepared vaccine using the pili (hair-like projections) on the surface of *Bacterioides nodosus* is reported to be effective in preventing footrot. Two injections of the vaccine spaced 30 days apart followed by a yearly booster are required. But vaccination alone will not control the disease in sheep or goats.

HEMORRHAGIC ENTEROTOXEMIA (BLOODY SCOURS)

Hemorrhagic enterotoxemia is an acute, highly fatal disease which affects lambs within the first few weeks of life, causing a bloody (hemorrhagic) infection of the small intestine. The disease is often caused by a change in feed such as beginning creep feeding or a sudden increase in milk as when a twin dies.

SYMPTOMS OR SIGNS

Hemorrhagic enterotoxemia is characterized by the sudden death of seemingly vigorous lambs without an obvious cause. In some instances, lambs become dull, refuse to eat, and may shiver, bleat, and show signs of colic. Also, straining, black or bloody diarrhea, and fever may be observed. An examination of the internal organs after death reveals a spotted or hemorrhagic condition of the small intestines and stomach.

CAUSE, PREVENTION, AND TREATMENT

This highly fatal disease is caused by *Clostridium perfringens*, Type C—anaerobic bacteria living in the soil, manure, and digestive tract of some sheep. These organisms are extremely resistant to heat and drying, and they endure in the environment when pens are unused. It is the beta toxin produced by these rapidly multiplying bacteria that causes the damage.

Immunization and sanitation are the best preventive measures. Pregnant ewes should be vaccinated twice, 2 to 4 weeks apart, with the second vaccination occurring 2 to 4 weeks before lambing. Another preventive measure is to administer *C. perfringens*, Type C, antitoxin to newborn lambs. Moving lambs and ewes from infected facilities to a grazing range or open pasture avoids exposing lambs to large numbers of *C. perfringens*, Type C.

The antitoxin can also be used to treat lambs showing signs of hemorrhagic enterotoxemia. But treatment is usually unrewarding as the course of the disease is too rapid.

JOHNE'S DISEASE (PARATUBERCULOSIS)

This is a chronic, contagious infection of sheep and goats. It is usually fatal. Economic losses result from the prolonged unthriftiness and eventual death of animals.

SYMPTOMS OR SIGNS

The incubation period for Johne's disease ranges from months to years. Affected sheep and goats slowly lose weight, but body temperature and appetite remain normal. Intermittent or continuous diarrhea occurs, though some animals may exhibit soft, rather than fluid, feces.

The disease in sheep and goats is similar, but scouring is less marked than in afflicted cattle, and thickening of the intestines is less obvious.

CAUSE, PREVENTION, AND TREATMENT

Johne's disease is caused by the bacillus *Mycobacterium paratuberculosis*, which is nearly always introduced into a clean flock by an infected animal. Therefore, keeping the flock away from infected animals is an effective prevention method. However, infected animals are difficult to identify, even though Johnin allergic tests, complement fixation tests, and immunofluorescence tests are valuable for identifying infected herds. Control and prevention can be aided by (1) isolating young stock from mature animals, (2) purchasing new or replacement animals from disease-free flocks, and (3) providing good sanitation. Vaccines are not effective.

No satisfactory treatment is known.

MASTITIS (BLUEBAG, GARGET)

This is a disease of ewes and does. To most sheep producers, it is known as *bluebag*; technically, it is called *mastitis*, which simply means inflammation of the mammary gland or udder. In some range bands, up to 5% of the ewes are affected, and there are some deaths.

SYMPTOMS OR SIGNS

Affected ewes are easily detected because they tend to separate themselves from the flock or band. Generally, only one side of the udder is affected. The udder becomes hard, reddened, and swollen. To avoid the pain, the ewe will often limp to prevent the rear leg from hitting the udder. Affected does may walk straddle-legged or hold up a rear leg. The disease is usually accompanied by a rise in temperature, 104 to 106°F. Milk from the affected gland may contain flecks or clots of solid material, or it may be watery.

In a matter of a day or two, the affected udder becomes cold and turns blue (gangrene); hence, the common name *bluebag*. When this happens, death may follow; or if the ewe recovers, a portion of the udder may slough off. Death occurs in about 25% of the cases.

CAUSE, PREVENTION, AND TREATMENT

Mastitis may be caused by a variety of bacteria, but the organisms most commonly involved are *Staphylococcus aureus, Escherichia coli, Pasteurella haemolytica*, and *Corynebacterium pyogenes*. The disease is spread by the lamb of an affected ewe attempting to nurse other ewes or through milk excreted on the bedding area. Injury to the teat or udder, either by the lamb or by some mechanical means (such as wire) may play a role. Thus, when detected and where possible, such injuries should be treated. In dairy goats, faulty vacuum of the milking machine or poor milking technique also play a role.

Good management is the most effective preventive measure. Affected ewes should be isolated immediately, and either new bedding should be provided for the rest of the ewes or they should be bedded down on clean ground. Affected ewes should be marked for culling, and each year, all ewes should be examined before breeding for evidence of mastitis. Any ewes with hard, fibrous udders should be culled.

In dairy goats, sanitation is very important in the prevention and management of mastitis. Washing the teats, dipping the teats, and keeping milking equipment clean are important sanitary measures. The use of a strip cup or the California Mastitis Test (CMT) facilitates the detection of mastitis for early treatment.

A veterinarian should be consulted relative to treatment. Antibiotics and sulfas are the treatments of choice. If treatment is started at the first sign of udder inflammation, the udder may return to normal milk production. A ewe in which the affected gland is not capable of producing milk is usually sold for slaughter. The surgical removal of the gangrenous half of the udder is not successful.

OVINE PROGRESSIVE PNEUMONIA (OPP)

Ovine progressive pneumonia (OPP) is a chronic viral disease of sheep which develops slowly after a long incubation period. Subclinical infection, which may be present in 90% of the flock, does not appear to have an adverse economic effect on ewe wool or lamb production.

SYMPTOMS OR SIGNS

Ovine progressive pneumonia is characterized by progressive weight loss, difficulty in breathing, and development of lameness, paralysis, and mastitis. Appetite appears normal. Sheep may remain infected for life. Some infected sheep never show signs of pneumonia.

Laboratory assistance is required in arriving at a correct diagnosis. Histopathologic examination of lung, mammary gland, brain, and synovial tissue, and serologic testing, must be used to confirm a diagnosis of OPP.

CAUSE, PREVENTION, AND TREATMENT

Ovine progressive pneumonia is caused by a retrovirus.

Methods of control include: (1) removing lambs from seronegative ewes at birth and isolating them from infected sheep, and (2) testing all sheep over one year of age and removing the positive reactors from the flock. However, these methods are time consuming and expensive and should not be considered until the disease has been diagnosed in the flock and its economic impact has been defined.

Treatment of animals with OPP is not effective.

PNEUMONIA

Pneumonia is one of the most common sheep and goat diseases. Healthy animals may carry the bacteria, viruses, or lungworm parasites that cause pneumonia. However, the disease usually develops due to stress from chilling, poor feeding, parasitism, or prolonged exposure, along with the presence of these infectious agents.

SYMPTOMS OR SIGNS

The disease is characterized by fever, labored breathing, and refusal to eat. Later, the animal becomes depressed and may have a discharge from the eyes and nose. Sometimes older sheep and goats die without showing any symptoms.

CAUSE, PREVENTION, AND TREATMENT

The causes of pneumonia are numerous, including many microorganisms, a number of viruses, and inhalation of fluids. Exposure, dampness, fatigue, and chilling are important in predisposing sheep and goats to pneumonia. Thus, it is important to prevent these conditions. To this end, (1) warm, dry, sanitary lambing and kidding pens should be provided; (2) young lambs and kids should be protected from wind and rain; (3) parasites should be controlled; (4) animals should not be dipped, sprayed, or sheared in cold weather unless they are housed in a warm building; (5) housing should be properly ventilated; and (6) animals should be fed and managed properly. In general, the best prevention is the provision of hygienic surroundings and the practice of sound husbandry.

Prompt action is necessary for effective treatment. Sick animals should be segregated and in some cases culled from the flock. Sulfonamides and broad-spectrum antibiotics are effective if given early, except when pneumonia is due to viral infection. Also, pneumonia caused by lungworms will not respond to these treatments.

POLIOENCEPHALOMALACIA (PEM, POLIO, CEREBROCORTICAL NECROSIS)

Polioencephalomalacia is a noninfectious disease of sheep and goats. The disease occurs in all breeds, sexes, and ages of sheep, but the incidence is highest in feedlot lambs 5 to 8 months of age. Outbreaks of the disease may develop in farm flocks of sheep on pasture, especially in lambs following changes from overgrazed to lush pasture. Also, PEM occurs in both feedlot and pasture goats.

SYMPTOMS OR SIGNS

Feedlot sheep with PEM show two degrees of severity—acute and subacute. In the acute form, sheep may be found dead or prostrated.

In subacute PEM, the sheep become blind, uncoordinated, and weak. In the early stages of the disease, some affected animals become isolated from the flock, others are unable to keep up with the flock; when disturbed, they may develop muscle contractions, tremors, and fall down.

The morbidity ranges from a few cases up to 10% of the flock, and 50% of affected animals may die. The course of the disease varies from 2 to 6 days.

CAUSE, PREVENTION, AND TREATMENT

Polioencephalomalacia appears to be due to a thiamin deficiency. Some forages, such as bracken fern or kochia *(Kochia scoparia)*, are high in the enzyme *thiaminase*, which ties up or destroys the thiamin produced in the rumen, thus causing a deficiency. Also, certain bacteria which produce a large quantity of thiaminase may become numerous in the rumen and cause a thiamin deficiency.

Treatment consists of parenteral administration of 0.5 g thiamin hydrochloride, repeated at two-day intervals if necessary. Good nursing will help; affected sheep should be isolated, helped to a standing position, and given feed and water. Rapidity of recovery relates directly to the speed of disease recognition and institution of thiamin treatment.

POLYARTHRITIS (STIFF LAMB DISEASE)

Known as *Chlamydial polyarthritis*, polyarthritis is an acute, contagious, but nonfatal disease of nursing lambs, recently weaned lambs, and feedlot lambs. Economic losses result from reduction in body weight, prolonged fattening time, reduced feed efficiency, extra labor, and cost of treatment. The disease is widespread, occurring in farm flocks, range bands, and feedlots.

SYMPTOMS OR SIGNS

Polyarthritis is characterized mainly by varying degrees of stiffness, lameness, depressed appetite, and conjunctivitis. Affected sheep are depressed, reluctant to move, and often hesitant to stand or bear weight on one or more legs. Following forced exercise, however, they may "warm out" of their stiffness and lameness. Rectal temperatures may range from 102.3 to 107°F.

CAUSE, PREVENTION, AND TREATMENT

The disease is caused by *Chlamydia psittaci*, which can be isolated from the affected joints, and which is excreted in the feces, urine, and conjunctival exudates of affected animals.

Affected animals should be isolated so they can be treated and so other lambs will not be exposed. If treatment is begun early, penicillin, tetracyclines, or tylosin appear to be beneficial. Advanced cases, however, may not respond to this treatment. Daily feeding of 150 to 200 mg of chlortetracycline reduces the incidence of polyarthritis when there is an outbreak in feedlots.

At present, there is no effective means of preventing polyarthritis.

RABIES (HYDROPHOBIA, MADNESS)

Rabies (hydrophobia or madness) is an acute, infectious disease of all warm-blooded animals and humans. This disease is one that is far too prevalent, and, if present knowledge were applied, it could be controlled and even eradicated.

When a human being is bitten by a dog that is suspected of being rabid, the first impulse is to kill the dog immediately. This is a mistake. Instead, it is important to confine the animal under the observation of a veterinarian until the disease, if it is present, has a chance to develop and run its course. If no recognizable symptoms appear in the animal within a period of two weeks after it inflicted the bite, it is safe to assume that there was no rabies at the time. Death occurs within a few days after the symptoms appear, and the dog's brain can then be examined for specific evidence of rabies.

With this procedure, unless the bite is in the region of the neck or head, there will usually be ample time in which to administer treatment to exposed human beings.

As the virus has been found in the saliva of dogs at least five days before the appearance of the clinically recognizable symptoms, the bite of a dog should always be considered potentially dangerous until proved otherwise. In any event, when a human being is bitten by a dog, it is recommended that a physician be consulted immediately.

SYMPTOMS OR SIGNS

Less than 10% of the rabies cases appear in sheep, goats, cattle, horses, and swine. Affected sheep or goats may charge the caretaker, and/or each other. They bleat continually, and show sexual interest. Paralysis occurs just before death.

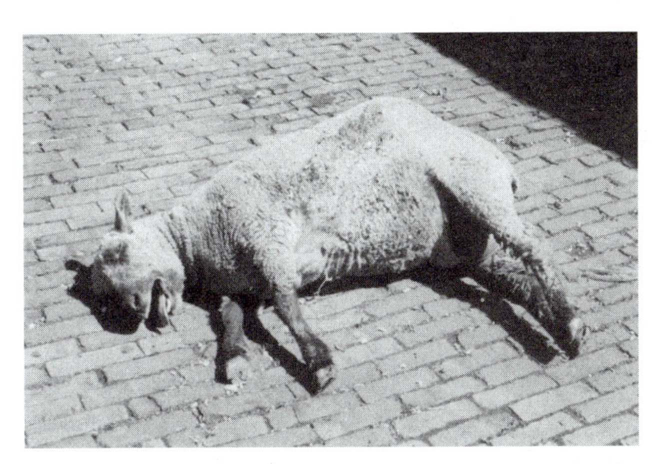

Fig. 8-5. Sheep with rabies. Note the paralytic condition and extreme misery. (Courtesy, Department of Veterinary Pathology and Hygiene, College of Veterinary Medicine, University of Illinois)

CAUSE, PREVENTION, AND TREATMENT

Rabies is caused by a virus which is usually carried into a bite wound by the infected saliva of a rabid animal. The malady is generally transmitted to farm animals by dogs and certain wild animals. such as the fox and the skunk.

Rabies can best be prevented by attacking it at its chief source, the dog. With the advent of an improved vaccine for dogs, it should be a requirement that all dogs be immunized. This should be supplemented by regulations governing the licensing, quarantine, and transportation of dogs. For understandable reasons, the control of rabies in wild animals is extremely difficult.

In some areas, the most practical protection of farm animals is vaccination with a vaccine that is suitable for sheep and other large animals.

Once the disease has been introduced into a farm animal, little can be done. Theoretically, the treatment could be administered immediately after the bite by a rabid animal; but this is not generally used in animals perhaps because of a more variable (shorter) incubation period in animals and the expense involved. After the disease is fully developed, there is no known treatment.

RAM EPIDIDYMITIS (REO)

Ram epididymitis is an infection of the epididymis, affecting the fertility of rams. It is of major concern because of its wide geographic distribution and high incidence. Economic losses result from reduced fertility, shortened breeding life, reduced marketability, and lowered lambing rates. Additionally, an infection in the ewe causes placentitis, abortion, and in the lamb mortality about the time of birth.

SYMPTOMS OR SIGNS

This disease results in complete or partial infertility of the ram and a marked increase in dry ewes. It also causes the lambing season to be spread out over several months. In some rams, palpation of the scrotal contents reveals a lump in the epididymis.

CAUSE, PREVENTION, AND TREATMENT

The cause of the disease is the ram epididymitis organism (REO), or *Brucella ovis*. Treatment of infected rams has not been very satisfactory, but the disease can be controlled by a rigid culling and vaccination program. Based on available information, the following control program is recommended:

1. Examine all rams before the breeding season and again at shearing. Market affected rams for slaughter only.
2. Check with your veterinarian relative to the selection of the vaccine and the timing of injections.
3. Vaccinate replacement rams before putting them with older rams.

In some areas of the United States, where there have been no reports of epididymitis, a close annual inspection with rigid culling may be a preferable approach.

SALMONELLOSIS

This disease occurs worldwide and its incidence is increasing as livestock production intensifies. Lambs, kids, calves, piglets, and foals are susceptible. Mortality can reach 100% among affected animals.

SYMPTOMS OR SIGNS

The illness is acute with marked depression, fever (105 to 107°F), and death occurring in 24 to 48 hours. A watery, green diarrhea, possibly containing blood may also be noted. Stress may be a chief predisposing factor.

CAUSE, PREVENTION, AND TREATMENT

In sheep, salmonellosis is most commonly caused by *Salmonella typhimurium*, which survives in wet, warm areas for many months. The *Salmonella* gains entry into the body through the mouth from contaminated feed and water, pasture, rangeland, and possibly other items. Some animals not exhibiting any outward symptoms, following recovery, can be carriers.

Infected animals should be separated from the flock and placed in a hospital pen to lessen the chance of contracting *Salmonella*. A periodic thorough cleanup, disinfection, and an interval of vacancy are suggested. Also, major management changes should take place gradually to avoid stressful situations.

No specific treatment is effective, but sick animals can be drenched with a solution containing electrolytes and antibiotics. In order to select the most effective antibiotic, a culture and sensitivity of the organism should be conducted.

SCOURS (INFECTIOUS DIARRHEA, LAMB DYSENTERY)

Scours, which occurs in sheep throughout the world, is a highly fatal disease of young lambs, inflicting heavy financial losses in certain regions. Some sheep enterprises may lose 20 to 35% of their lambs during an outbreak.

SYMPTOMS OR SIGNS

The disease usually develops in lambs in the first few days after birth; it seldom occurs after the third week. Affected lambs are weak, depressed, and do not care to suckle. There is a profuse diarrhea (scours) that may be tinged with blood. Usually the temperature rises and the lambs may become gaunt or may bloat. Death may follow within a few hours. Many die before any signs are observed. The mortality is very high.

CAUSE, PREVENTION, AND TREATMENT

Dysentery of newborn lambs is the result of a beta toxin from *Clostridium perfringens*, Type B—a bacterium commonly residing in the soil, with cold, wet, unsanitary conditions as predisposing factors. Also, lambs born on the range are less susceptible than those born in a lambing shed.

The most successful preventive measures consist of holding the lambing ewes on dry, well-drained areas and keeping the lambs clean, warm, and dry. Moreover, the best preventive measure is the vaccination of the ewes during the last third of pregnancy—initially two vaccinations one month apart, then annually before lambing.

There is no highly effective treatment once lambs are affected. Various drugs such as sulfonamides, antibiotics and antidiarrheals may save a few.

SCRAPIE

Scrapie was first described in England in 1732. However, it was not reported in Canada until 1938 and in the United States until 1947. It is a chronic, degenerative disease of the central nervous system of sheep and goats.

Scrapie attracted much attention in the United States in the mid 1990s because of being theorized by some people, but not scientifically proven, to be the source of bovine spongiform encephalopathy (mad cow disease) in the United Kingdom.

SYMPTOMS OR SIGNS

The incubation period is measured in years. The onset is insidious, and despite intense itching, which accounts for the name *scrapie* (affected animals rub themselves against fences and other objects to relieve the itching, and in doing so scrape off wool), an uncoordinated gait and trembling are considered the most reliable symptoms. Scrapie affects the nervous system of sheep. Affected animals are restless and excitable, walk unsteadily, suffer from thirst, are weak, become paralyzed, and then die. The appetite remains good and there is no rise in temperature. Scrapie seldom appears in animals under 18 months of age, and is fatal after a course of some weeks or months. In the United States, the Suffolk breed is most often affected.

CAUSE, PREVENTION, AND TREATMENT

Scrapie is caused by an unconventional agent that can withstand heat, rendering, radiation, and formalin treatments.

The method of spreading scrapie is not fully understood, though it is usually transmitted from parents to offspring. It has been proved that scrapie is an infectious disease, and that the transmissible agent can be passed in series, indefinitely, from sheep to sheep.

No evidence of the development of antibodies has been obtained and the lesions of nerve cell degeneration do not resemble those usually associated with a virus infection.

Because the means of natural transmission are still unknown, no adequate control measures are available; nor is there any cure.

Suspected cases of scrapie should be reported by the local veterinarian to state and federal officials.

Note: The Animal, Plant, Health Inspection Service (APHIS) of the U.S. Department of Agriculture has established a voluntary scrapie flock certification program. But scrapie control programs are subject to change; so, sheep raisers should keep frequent contact with the APHIS personnel in their state for updates on the program. Unfortunately, there is no test that will detect scrapie in a live sheep or goat.

SHEATH-ROT (PIZZLE-ROT, POSTHITIS, URINE SCALD)

This condition is observed particularly in wethers on high-protein rations or good-legume pastures, though it may also occur in rams being fitted for shows or sales. It is moderately contagious and may be transmitted to ewes, causing inflammation of the vulva. Sheath-rot has also been reported in male goats.

SYMPTOMS OR SIGNS

Sheath-rot is characterized by a spreading ulceration of the skin of the prepuce and may sometimes involve the penis. Ulcers are small at first, but they become wet during urination. Gradually, they enlarge, coalesce with others, and become covered with a scab. The preputial opening may become blocked, and the sheath will become distended with foul-smelling urine and pus. Animals may be humped, may walk stiffly, and may lose condition.

CAUSE, PREVENTION, AND TREATMENT

Although the disease is associated with castration and protein-rich diets, it is caused by a bacterium indistinguishable from *Corynebacterium renale*, which acts on urea and initiates the ulceration.

For control, animals could be switched to a lower-quality ration, but this is not practical when these animals are those that the producers want to gain well.

Treatment involves the application of (1) antiseptics such as 5 to 10% copper sulfate ointment, quaternary ammonium compounds, or aluminum silicone, or (2) ointment containing penicillin or bacitracin to the affected area. Before application, any scabs, urine, pus, and debris should be removed. Changing the diet to low-quality pasture or straw and implanting testosterone, the male hormone, may hasten healing.

SHIPPING FEVER (HEMORRHAGIC SEPTICEMIA, PASTEURELLOSIS)

This is a disease or group of highly infectious diseases of sheep and goats, which most commonly affects young lambs soon after birth, older animals following shipment, and mature animals that have been exposed to bad weather, fatigue, and/or a lowered state of nutrition.

SYMPTOMS OR SIGNS

In acute outbreaks, affected sheep show depression, drooping ears, discharge from the eyes and nose, coughing, rapid respiration and pulse, and high temperature (104 to 107°F). Due to lack of appetite, they usually appear gaunt. On autopsy, hemorrhages are found throughout their bodies.

Up to 50% of the flock may be affected, but death losses rarely exceed 10%. Affected sheep may die within 1 or 2 days.

CAUSE, PREVENTION, AND TREATMENT

The actual cause of the disease is complex, but present evidence indicates that a bacterium, *Pasteurella haemolytica*, is usually involved, though other bacterial and possibly viral agents may be involved.

Prevention consists of eliminating as many as possible of the predisposing factors that lower an animal's resistance, such as overcrowding when shipping, hard driving, lack of rest, and improper shelter during inclement weather. Also, it is best to isolate newly purchased animals for 2 or 3 weeks before placing them with the rest of the flock. Vaccines against *P. haemolytica* are available, but their value is not fully established.

Some sheep producers feel that they have lowered the incidence of shipping fever in feedlot lambs by administering high levels of antibiotics or sulfas, according to manufacturers' directions, immediately upon arrival of the animals at the feedlot.

Treatment in the early stages consists of the sulfas or antibiotics. Where possible, affected sheep and goats should be well bedded, well fed, and protected from inclement weather.

SORE MOUTH (CONTAGIOUS ECTHYMA, ORF, PUSTULAR DERMATITIS, SCABBY MOUTH)

This is essentially a disease of lambs and kids. Medically, it is referred to as *contagious ecthyma*. Morbidity may reach 100%, but the mortality is low. It is transmissible to humans, causing a condition called *orf*.

Economic losses result from unthriftiness, loss of condition, slow growth, premature weaning, and some deaths. Sore mouth occurs worldwide.

SYMPTOMS OR SIGNS

The infected lambs or kids may first be noticed when they refuse feed and appear depressed. On investigation, small vesicles may be seen on the lips, gums, and tongue, causing these parts to be red and swollen. These vesicles break down and form sores that bleed easily and become encrusted with scabs. The sores may heal in 4 to 18 days if uncomplicated. The main trouble is that growth is impaired and weight is lost. Complications are common. The lesions may become infected or may spread to the teats, udders, and feet (just above the coronet) of nursing ewes. If the young go off feed for an extended time, the udders of the nursing ewes may become caked and will be predisposed to blue bag.

Fig. 8-6. Sore mouth, showing lip lesions in a ewe. This is a highly contagious disease, caused by a virus. (Courtesy, USDA)

CAUSE, PREVENTION, AND TREATMENT

The disease is caused by a poxvirus *(Parapoxvirus)*, which is extremely resistant to drying. Animals recovering from a natural infection are very resistant to reinfection.

Prevention usually entails immunization with a vaccine similar in nature and method of administration to smallpox vaccine. It is applied with a small brush to a scratched area under the tail or on the inside of the thigh. The vaccine should be used either (1) at the time of docking and castrating, or (2) at least 10 days before shipping feeder lambs in areas in which the disease is known to be very prevalent. But noninfected sheep should not be vaccinated on noninfected ground, because this will infect the premises. Scabs from soremouth lesions remain infective for several years and carry infection over from year to year, on the premises. Of course, general sanitation and isolation of infected animals should be practiced and may eliminate the necessity of vaccination when the disease is not serious in proportion.

Treatment consists of isolating, if practicable, the infected animals and preventing the invasion of screwworms and secondary invaders. Formerly, it was common practice to rub off the scabs and treat the raw surface with tincture of iodine, sheep dip, a broad spectrum antibiotic, a sulfa drug, etc. Although some persons still apply these treatments, it appears that they are of doubtful value. Providing a soft, nutritious, and relatively nonfibrous ration may benefit affected flocks.

TETANUS (LOCKJAW)

Tetanus is chiefly a wound infection disease that attacks the nervous system of horses and that of humans, though it does occur in sheep, goats, cattle, and swine. In the Southwest, it is quite common in sheep after shearing, docking, and castrating. In the central states, tetanus frequently affects lambs following castration or other wounds. It is generally referred to as *lockjaw*.

In the United States, tetanus occurs most frequently in the South, where precautions against the disease are an essential part of the routine treatment of wounds.

SYMPTOMS OR SIGNS

The incubation period of tetanus varies from 1 to 9 weeks but may be from 1 day to 2 months. It is usually associated with a wound but may not directly follow an injury The first noticeable sign of the disease is a stiffness observed about the head. The animal chews slowly and weakly and swallows awkwardly. The third eyelid is seen protruding over the forward surface of the eyeball (called *haws*). The animal then shows violent spasms or contractions of groups of muscles brought on by the slightest movement or noise. The animal usually attempts to remain standing throughout the course of the disease. If recovery occurs, it will take a month or more. In over one-half the cases, however, death ensues, usually because of paralysis of the respiratory muscles.

Fig. 8-7. Lamb with tetanus or lockjaw. In sheep this disease usually is seen following shearing, castrating, or docking. (Courtesy, Department of Veterinary Pathology and Hygiene, College of Veterinary Medicine, University of Il-

CAUSE, PREVENTION, AND TREATMENT

The disease is caused by an exceedingly powerful toxin (more than 100 times as toxic as strychnine) liberated by the tetanus organism *Clostridium tetani*. This organism is an anaerobe (lives in absence of oxygen) which forms the most hardy spores known. It may be found in certain soils, horse dung, and sometimes in human excreta. The organism usually causes trouble when it gets into a wound that rapidly heals or closes over it. In the absence of oxygen, it then grows and liberates the toxin which follows up nerve trunks. Upon reaching the spinal cord, the toxin excites the symptoms just noted.

Reducing the probability of wounds, general cleanliness, treating wounds properly, and vaccinating with tetanus toxoid in the so-called hot areas can largely prevent the disease. When an animal has received a wound from which tetanus may result, short-term immunity can be conferred immediately by the use of tetanus antitoxin, but is of little or no value after the symptoms have developed. All valuable animals should be protected with tetanus toxoid. On some infected premises, all surgery should be accompanied with tetanus antitoxin; or where tetanus is an annual problem, the entire flock should be immunized with annual boosters just before lambing.

Tetanus develops after castration or docking with rubber bands more frequently than when the surgical method is used.

All perceptible wounds should be properly treated, and the animals should be kept quiet and preferably should be placed in a dark, quiet corner free from flies. Supportive treatment is of great importance and will contribute towards a favorable course. This may entail artificial feeding. The animals should be placed under the care of a veterinarian.

TUBERCULOSIS

Tuberculosis in sheep and goats is generally regarded as a rare disease of chronic character, manifesting few symptoms and only observed on postmortem examination. Its incidence in sheep is rare, and it is of only a minor concern to the industry.

There are three kinds of tuberculosis bacilli—the human, the bovine, and the avian (bird) types. Practically every species of animal is subject to one or more of the three kinds. In sheep, most of the infections that have been typed abroad have been of the bovine type, but in the United States most cases have been of avian origin.

SYMPTOMS OR SIGNS

Since the disease is usually chronic in sheep, only emaciation and weakness are apparent. Coughing is prominent in goats, but not in sheep. Postmortem inspection reveals lesions of the lungs and lymph nodes in all cases, liver tubercles in a majority of cases, and spleen damage in about half the cases.

CAUSE, PREVENTION, AND TREATMENT

The causative agent is a rod-shaped organism belonging to the acid-fast group known as *Mycobacterium*. The disease is usually contracted when animals eat feed or drink fluids contaminated by the discharges of infected animals. Where tuberculosis is suspected in sheep, the tuberculin test may help identify and eliminate all reactors.

Tuberculosis can be transmitted to humans via goat's milk, so goats should be tested for tuberculosis.

Since spread of the disease among sheep in this country depends on contact with chickens, a campaign to eliminate tuberculosis in poultry should also be effective in controlling it in sheep and goats.

To date, medical treatment of animals has been unsatisfactory.

VIBRIONIC ABORTION (VIBRIOSIS, VIBRIO CAMPYLOBACTERIOSIS)

Vibrio Campylobacteriosis, commonly known as *vibriosis*, is the second most common cause of abortion in sheep, being exceeded only by epizootic abortion *(chlamydia)*. In affected sheep flocks, abortions may range from 5 to 70%. There are few reports of vibriosis in goats.

Transmission in sheep appears to be from the consumption of infected feed and water. Birds and rodents have been implicated as being possible carriers of the causative microorganism.

SYMPTOMS OR SIGNS

The disease is characterized by many abortions among ewes, especially during the last 4 to 6 weeks of pregnancy, and by many stillbirths and weak lambs. Ewes that abort usually raise healthy lambs in succeeding years.

CAUSE, PREVENTION, AND TREATMENT

The causative organisms are *Campylobacter jejun*, or *Campylobacter fetus* subspecies *fetus*.

Prevention consists of avoiding contact with diseased animals, and with contaminated feed, water, and other materials. Where abortion of unknown cause is encountered, one should (1) isolate aborting ewes, (2) burn or bury aborted fetuses and membranes, and (3) thoroughly clean and disinfect contaminated quarters.

To control an outbreak, the farmer or rancher should give ewes bacterin and antibiotics (penicillin or streptomycin) at the first treatment followed by antibiotics the next day. In some flocks, feeding chlortetracycline the last 7 to 8 weeks of pregnancy will reduce the incidence of abortion, but adequate sanitation is probably as effective and more practical.

Vaccination is an effective means of control if ewes are vaccinated shortly before mating, again in eight weeks, and then annually.

If a diagnosis is made early in an outbreak, the number of abortions can be reduced by antibiotics.

A vaccine for protection against vibrionic abortion is now available. It should be used according to manufacturer's directions.

Ewes that have aborted from *Vibrio fetus* are said to be immune to the disease thereafter. However, some of them may abort again 2 to 3 years later, and they can be spreaders.

PARASITES OF SHEEP AND GOATS[5]

Sheep and goats are subject to many parasites, both internal and external. From the standpoint of profitable sheep and goat production, it is paramount that parasites be controlled. In presenting information on each of the damaging parasites afflicting sheep and

[5]The use of trade names of anthelmintics (wormers) and insecticides in this section does not imply endorsement, nor is any criticism implied of similar products not named. Rather, it is recognized that sheep and goat producers, and those who counsel with them, are generally more familiar with trade names than with generic names.

goats, the author wishes to admonish the producer that (1) the responsibility of diagnosis should remain with the veterinarian or diagnostic laboratory, and (2) correct diagnosis and proper treatment are of little avail if preventive measures are not incorporated into the flock management. In order to establish effective parasite prevention and control measures, however, producers must have enlightened information relative to the life history and habits of each of the most harmful parasites with which they must contend. To this end, this section on parasites of sheep and goats is presented.

If it were not for the many parasites that plague sheep, the health problem with these animals would be a simple one. Unfortunately, sheep are seldom infected with one parasite only, but it seems that one species makes way for another and so on without end. It naturally follows, therefore, that these animals should be kept free from the most harmful parasites, thereby being able to resist others—especially when there is proper nutrition.

Only the most common and harmful parasites affecting sheep and goats are covered in the following discussion.

INTERNAL PARASITES

Though sheep and goats, like other farm animals, suffer from various infectious and noninfectious diseases, the most serious losses, especially in farm flocks or herds, are due to internal parasites. Although heavy parasite infections may result in the death of the host animal, even greater, though less startling, economic losses may be encountered in the form of unthriftiness, loss of condition, retarded growth, anemia, and other effects. Such losses take a heavy annual toll in profits, far exceeding those due to diseases. But because losses from parasites are both unspectacular and difficult to evaluate, normally they do not receive the attention they deserve.

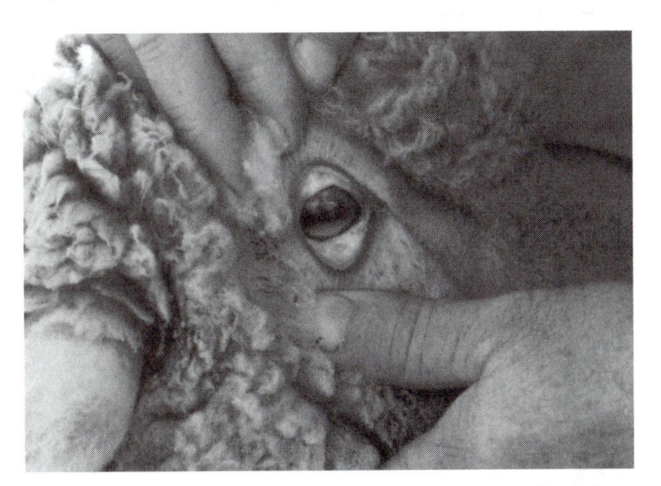

Fig. 8-8. Checking a sheep's eye for anemia—symptom of an internal parasite infection. (Courtesy, Smith-Kline Animal Health Products, Philadelphia, PA)

Fig. 8-9. Lambs heavily infected with internal parasites. Although such animals may not die, even greater, though less startling, economic losses may be encountered in the form of unthriftiness, loss of condition, retarded growth, anemia, and other effects. (Courtesy, Veterinary Research Laboratory, Montana State University)

Parasitism is most troublesome in those areas where environmental conditions are most favorable for the development of those stages which the parasites must spend outside the host and where there are relatively large numbers of sheep confined rather continuously to a small area. Such conditions are most likely to prevail among farm flocks or herds, though range bands of sheep or herds of goats are by no means free from parasites.

Although there are many different species of internal parasites, only the most common and damaging ones will be covered.

BLADDER WORMS (LARVAL TAPEWORMS)

In addition to the three most common species of tapeworms, sheep and goats are infected by the larval or immature stages of four different tapeworms found in dogs and related carnivores—the latter four species being known collectively as bladder worms. These four species are: gid tapeworms *(Coenurus cerebralis)*, sheep measles *(Cysticercus ovis)*, thin-necked bladder worms *(Cysticercus tenuicollis)*, and hydatids *(Echinococcus granulosus)*. Sheep and goats become infected with these bladder worms by ingesting the eggs voided by dogs and related carnivores on pastures or range. Each of the respective species of bladder worms will be discussed briefly in a section under the species names listed.

BROWN STOMACH WORMS OR MEDIUM WORMS (*OSTERTAGIA CIRCUMCINCTA* AND *O. TRIFURCATA*)

Although not as widespread as twisted stomach worms, brown stomach worms are more prevalent in sheep and goats in the western United States. These parasites are brown, hairlike worms about 0.5 in. long that live in the abomasum. The life history and habits of the brown stomach worm outside the host are similar to those of the common stomach worm. As infection under natural conditions is usually accompanied by infection with the common stomach worm as well as other internal parasites, it is rather difficult to attribute any specific symptoms to brown stomach worms. The prevention, control, and treatment are essentially the same as those recommended for the common stomach worm; hence, see the section headed "Stomach Worms" and Table 8-3.

COCCIDIOSIS (RED DYSENTERY)

Coccidiosis—a parasitic disease affecting sheep, goats, cattle, swine, pet stock, and poultry—is caused by microscopic protozoan organisms known as coccidia (*Eimeria* spp), which live in the cells of the intestinal lining. Each class of domestic livestock harbors its own species of coccidia; thus, there is no cross-infection between animals.

In sheep, the disease usually occurs in feeder lambs during the first three weeks after admission to the feedlots or in young lambs before weaning.

Coccidiosis usually occurs in kids 1 to 4 months of age in crowded, unsanitary pens.

Annual economic losses in lambs and kids result from weight losses, delayed marketing, and death.

DISTRIBUTION AND LOSSES

The distribution of the disease is worldwide. The chief economic loss is in lowered gains and production, but the mortality is frequently high in feedlot lambs. It is most severe in young animals.

LIFE HISTORY AND HABITS (COCCIDIA)

Infected animals or birds may eliminate daily with their droppings thousands of coccidia (in the resistant oocyst stage). Under favorable conditions of temperature and moisture, coccidia sporulate to maturity in 3 to 5 days, and each oocyst contains 8 infective sporozoites. The oocyst then gains entrance into an animal by being swallowed with contaminated feed or water. In the host's intestine, the outer membrane of the oocyst, acted on by the digestive juices, ruptures and liberates the 8 sporozoites within. Each sporozoite then attacks and penetrates an epithelial cell, ultimately destroying numerous cells through asexual and sexual reproduction. After sexual multiplication and fertilization, the parasite (oocyst) is then expelled with the feces and is again in a position to reinfect a new host.

The coccidia abound in wet, filthy surroundings; resist freezing and ordinary disinfectants; and can be carried long distances in streams.

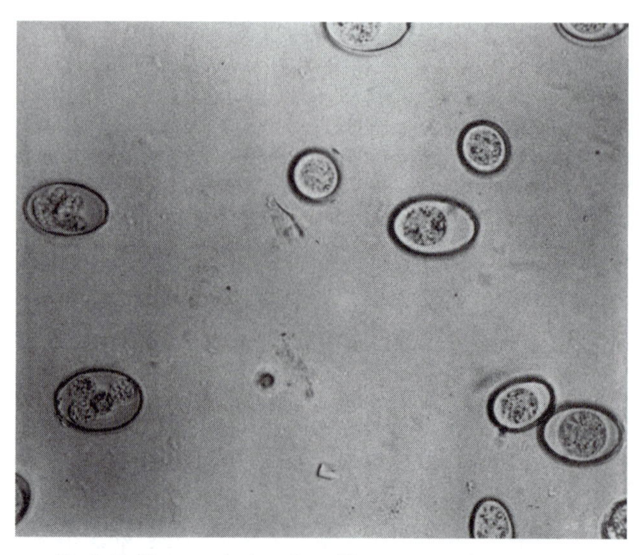

Fig. 8-10. Microscopic view of coccidia, the cause of coccidiosis. (Courtesy, Colorado State University)

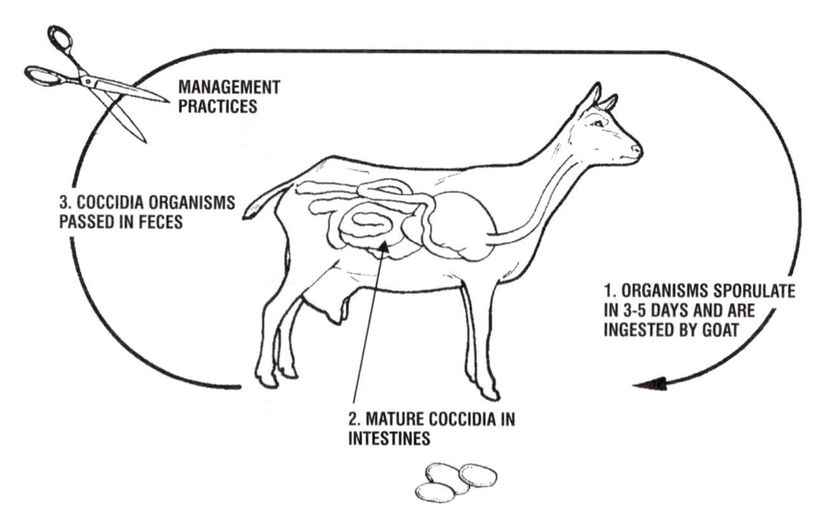

Fig. 8-11. Diagram showing the life history and habits of coccidia, which cause coccidiosis. As noted (see scissors) management practices—protecting animals from feed or water that is contaminated with the protozoa that cause the disease—are the best preventive.

DAMAGE INFLICTED; SYMPTOMS OR SIGNS OF AFFECTED ANIMALS

A severe infection with coccidia produces diarrhea, and the feces may be bloody. The bloody discharge is due to the destruction of the epithelial cells lining the intestines. Ensuing exposure and rupture of the blood vessels then produces hemorrhage into the intestinal lumen.

In addition to bloody diarrhea, affected animals usually show pronounced unthriftiness and weakness.

Fig. 8-12. Lambs with coccidiosis. Bloody diarrhea is the most prominent symptom. (Courtesy, Colorado State University)

PREVENTION, CONTROL, AND TREATMENT

In feedlot lambs, where the disease is most prevalent, good management and high natural resistance go hand in hand. This can best be achieved by (1) moving the lambs into the feedlot with as little shrink and stress as possible, (2) allowing for plenty of space so that the lambs may spread out, (3) keeping water and feed troughs free from fecal pellets, (4) maintaining drylots and bedding, (5) starting animals on grain feed carefully, making the transition from grass and milk to grain and other feeds gradually and naturally, (6) segregating affected animals if practical, and (7) feeding coccidiostats to prevent clinical disease from occurring (see Table 8-2). Where trouble is encountered, the same preventive measures and principles that apply to feedlot lambs also apply to other sheep and goats.

Although the oocyst resist freezing and certain disinfectants and may remain viable outside the body for 1 or 2 years, they are readily destroyed by direct sunlight and complete drying.

Coccidiostats such as decoquinate, lasalocid, or monensin are used effectively in feed to prevent clinical signs of coccidiosis by inhibiting the life cycle of the parasite. Coccidiostats should be mixed in feed or salt and fed to growing lambs and kids from four weeks of age until market weight.

COOPERIAS (*C. CURTICEI, C. ONCOPHORA, C. PUNCTATA, C. PECTINATA*, COOPER'S WORM)

The four species of parasites classed as Cooperias are similar to the trichostrongyles in size, and they affect both sheep and goats. Their life history and habits are not unlike those of the stomach worm; hence, the recommended measures for prevention, control, and treatment are similar (see Table 8-2).

TABLE 8-2
MAJOR ANTICOCCIDIAL DRUGS USED IN SHEEP AND GOATS

Drug[1]	Method	Use Level[2]
Amprolium	Water	50 mg/kg of body weight q24hx21d
Decoquinate	Feed	0.5 mg/kg of body weight in feed
Lasalocid	Feed	30 gm/909 kg (ton) of feed 454 gm (15% active)/23 kg (50 lb) salt
Monensin	Feed	10–30 gm/909 kg (ton) of feed
Sulfaguanidine	Feed	0.2% concentration in feed
Sulfamethazine	Feed	Up to 0.5% concentration in feed

[1]Always read label directions or check with a veterinarian for restrictions on use.

[2]To convert metric to U.S. Customary, or vice versa, see "Weights and Measures" section of the appendix.

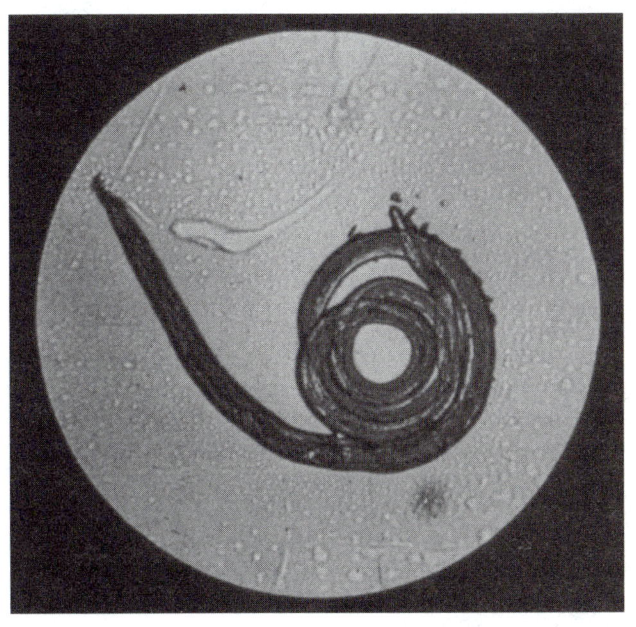

Fig. 8-13. *Cooperia* spp, small, hairlike worms less than 0.33 in. long. (Courtesy, Smith-Kline Animal Health Products, Philadelphia, PA)

GID TAPEWORMS *(COENURUS CEREBRALIS)*

The disease caused by this parasite, known as *coenurosis*, is rare in the United States.

Coenurosis, gid, is caused by *Coenurus cerebralis*, the larval stage of the tapeworm *Multiceps multiceps*, voided by dogs or other carnivorous animals. Bladder worms (cysts) develop in the brain or the spinal cord, where they cause atrophy and necrosis of nerve tissue.

The tapeworm eggs are ingested by sheep and goats grazing on infested pastures, and the embryos that emerge make their way into the bloodstream, with a few of them finally reaching the central nervous system where they occur as large cysts, or bladders, each the size of a hen's egg or larger. If there are many embryonic worms present in the early stages, the animal may die; and an autopsy will reveal a number of curving channels on the surface of the brain. If the infected animal does not die immediately, there may be no symptoms until 4 or 5 months later when *giddy* sheep will show defects in vision and disturbances in movements. The affected animals may stumble, run into objects, walk with their heads high or in circles; and there may be at least a partial paralysis of the hindquarters. There is no satisfactory treatment for animals infected with gid tapeworms. Prevention consists of the elimination of stray dogs; the examination and proper anthelmintic treatment, when necessary, of all dogs that must come in contact with sheep and goats; and the proper disposal of all carcasses or parts of carcasses coming from infected sheep and goats.

Fig. 8-14. Sheep infected with gid tapeworms. Giddy sheep show defects in vision and disturbances in movements. (Courtesy, USDA)

HOOKWORMS *(BUNOSTOMUM TRIGONOCEPHALUM)*

The hookworm, which is largely confined to the southern part of the United States, infects sheep, goats, cattle, and other animals. It is a white, thin worm varying from 0.5 to 1.0 in. in length. The life history and habits of the hookworm differ from those of the common stomach worm primarily in the method of entering the host. The hookworm larvae can infect sheep either through the mouth or through the skin. In general, the symptoms of hookworm infection are similar to those observed with stomach worm infection; namely, anemia, edema, and unthriftiness. Prevention, control, and treatment are similar to the recommendations given for stomach worms (see Table 8-3).

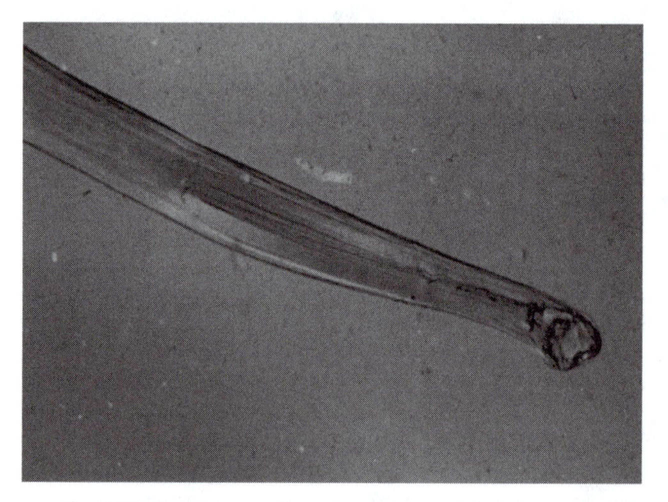

Fig. 8-15. The hookworm *Bunostomum* sp. can infect sheep and goats through the skin as well as the mouth. (Courtesy, Smith-Kline Animal Health Products, Philadelphia, PA)

HYDATIDS *(ECHINOCOCCUS GRANULOSUS)*

This species of bladder worm occurs most frequently in the liver and lungs of infected sheep and goats. It is caused by the larvae of *Echinococcus granulosus*, a tapeworm of dogs, coyotes, and wolves. Generally, infections are not detected or diagnosed in live animals but are found only on postmortem examination. Although infection is not confined to any particular area of the United States, no great numbers of sheep and goats appear to be seriously affected. The life cycle of the hydatid is similar to that already described for the other bladder worms. Elimination of tapeworms in dogs is the best approach to control.

LARGE-MOUTHED BOWEL WORM *(CHABERTIA)*

Adult large-mouthed bowel worms cause severe

damage to the mucosa of the large intestine, resulting in congestion, ulceration, and hemorrhages. The feces of infected animals are soft, contain much mucus, and may be streaked with blood. Outbreaks occur only under conditions of severe stress. Immunity develops quickly. See Table 8-3 for treatments.

LIVER FLUKES *(FASCIOLA HEPATICA)*

Four species of flukes infest various animals in different sections of the world. Two of these are found in the liver (*Fasciola hepatica* and *Fascioloides magna*); whereas the other two are found in the rumen, or paunch. The common liver fluke, *Fasciola hepatica* occurs predominantly in sheep in the western United States, and is of economic importance to producers. This parasite is a flattened, leaflike fluke, usually about 1 in. long and is found in the bile ducts. *Fascioloides magna* is lethal in sheep and goats, and is found only in areas where infected deer or elk carry the parasite. This fluke can reach 3 in. and is found only in the liver parenchyma.

Fig. 8-16. The common liver fluke *(Fasciola hepatica)*. This leaflike, brown worm—which is usually about 1 in. long—affects sheep, goats, cattle, and other animals. (Courtesy, USDA)

DISTRIBUTION AND LOSSES

The distribution of the common liver fluke is worldwide. It is found wherever low-lying, wet pastures, and intermediate snail hosts abound. In addition to causing the usual lowered meat and wool production accompanying unthriftiness, this parasite produces a large number of "fluky livers" or "rotten livers" condemned in the packing houses each year and, consequently, is of economic importance. The fluke may also carry *Clostridium novyi*, an anaerobic bacterium which causes black disease.

LIFE HISTORY AND HABITS

Flukes reproduce by means of eggs, which, after passing from the host, hatch into a free-living stage equipped with cilia that enable them to move about and seek a suitable freshwater snail. Upon encountering certain kinds of snails, they penetrate into the bodies of their intermediate hosts and eventually develop into cercariae (flukes in the larval stage), which leave the snails and become encysted on the nearby vegetation. The encysted metacercariae are then ingested by sheep and goats during grazing. Each fluke is liberated from its cyst, penetrates the intestinal wall, migrates about the abdominal cavity, and finally reaches the liver where maturity is attained approximately three months after infection.

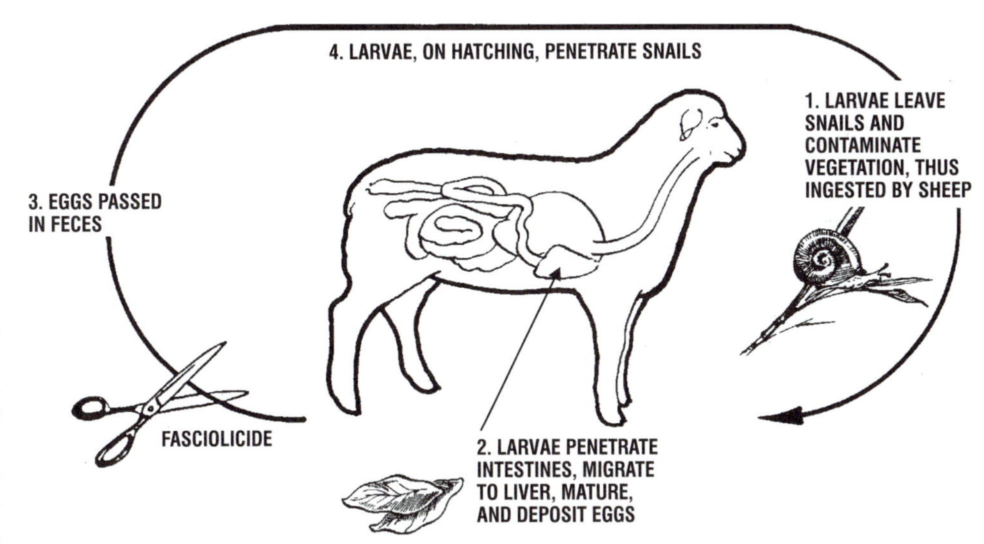

Fig. 8-17. Diagram showing the life history and habits of the common liver fluke, *Fasciola hepatica*. As noted (see scissors) the preferred treatment consists of administering a fasciolicide. In addition, wet pastures should be drained or avoided.

DAMAGE INFLICTED; SYMPTOMS OR SIGNS OF AFFECTED ANIMALS

If the flukes are present in sheep and goats in great numbers, death may occur without any definite symptoms being evident. If the condition is more chronic, the usual symptoms of parasitism are found. Sometimes there may be a distinct pot-bellied condition caused by the escape of fluids into the body cavity through damage to the liver. Flukes also may stop the flow of bile, and the ducts may be greatly enlarged. As with most parasites, positive diagnosis consists of finding eggs in the feces by microscopic examination.

If sheep or goats ingest metacercariae of the deer fluke, *Fascioloides magna*, the resulting infection is always lethal because of unrestricted migration of the flukes in the liver, and death is inevitable within six months of infection. Therefore, sheep and goats cannot be raised economically in areas where *Fascioloides magna* is present in deer or elk.

PREVENTION, CONTROL, AND TREATMENT

Prevention rests on the regular treatment of infected animals and the drainage or avoidance of wet pastures. Destruction of the snail carrier can often be accomplished in small areas by the application of 3 to 6 lb of copper sulfate to the acre. When applied in this manner, the copper sulfate should be mixed with some carrier such as sand or lime prior to application, using 1 part of copper sulfate to 4 to 8 parts of the carrier. In ponds, 1 part copper sulfate to 500,000 parts water is recommended. When used in these dilutions, copper sulfate is not injurious to grasses and will not poison farm animals. It will, however, kill such lower forms of plant life as the algae and mosses, and it may kill fish. Treatment of snail-infested areas is generally needed at least once a year. Springtime is the preferred season for treatment. Infected pastures should not be used for making hay. Other compounds that kill snails include sodium pentachlorphenate and Frescon. However, treatment of affected areas is not practical under range conditions, particularly in high rainfall areas; and it may not be permitted because of environmental toxicity.

Anthelmintics for the treatment of adult and immature flukes include: albendazole, clioxanide, clorsvion, nitroxynil, oxyclozanide, rafoxanide, and triclabendazole.

LUNGWORMS (THREAD LUNGWORMS, *DICTYOCAULUS FILARIA*, AND HAIR LUNGWORMS, *MUELLERIUS CAPILLARIS*)

In the United States, sheep and goats are frequently infected with two species of lungworms: thread lungworms, *Dictyocaulus filaria* (primarily in sheep); and hair lungworms, *Muellerius capillaris* (primarily in goats). Thread lungworms are white and may be as much as 4 in. long, whereas hair lungworms are much thinner and shorter.

DISTRIBUTION AND LOSSES

Lungworms are widely distributed in low-lying, marshy areas throughout the United States, and infections sufficiently heavy to produce death losses occasionally occur. Moreover, economic losses result from unthriftiness, retarded growth, and reduced production.

LIFE HISTORY AND HABITS

There are some differences in the life cycles of the two species, but in general, the adults lay eggs in the air passages of the lungs of sheep and goats, and there they develop into larvae. The larvae reach the upper part of the throat and may be expelled when the animal coughs, or they may be swallowed and passed out with the feces.

The larvae of the thread lungworms molt twice in the course of the next few days and reach an infective stage in about 10 days. Outside the host, the larvae of the hair lungworms appear to pass part of the life cycle in an intermediate host; namely, certain snails or slugs that may be eaten by sheep and goats. After sheep and goats have ingested the infective larvae of either the thread lungworms or the hair lungworms, the development of each species within the body of the host is very similar. The larvae penetrate the wall of the small intestine, then pass by way of the lymphatics and venous blood to the lungs. They reach the adult stage within 18 days after ingestion and produce eggs within another 8 days.

DAMAGE INFLICTED; SYMPTOMS OR SIGNS OF AFFECTED ANIMALS

The thread lungworm appears to be much more important from a disease standpoint than the hair lungworm, as no special symptoms are associated with infestations of the latter.

The thread lungworm is especially damaging to lambs and kids. The onset of symptoms is gradual. Because of the resulting irritation, the first symptom is coughing. In severe infections, the breathing is rapid and difficult because of the blocking of the air passages and the interference with the functioning of large areas of the lungs. There is unthriftiness and poor appetite. The animals may be seen with their heads lowered and their necks extended. Death may follow in 2 or 3 months, probably from suffocation or from pneumonia developing from the irritation and impaired function of the lungs.

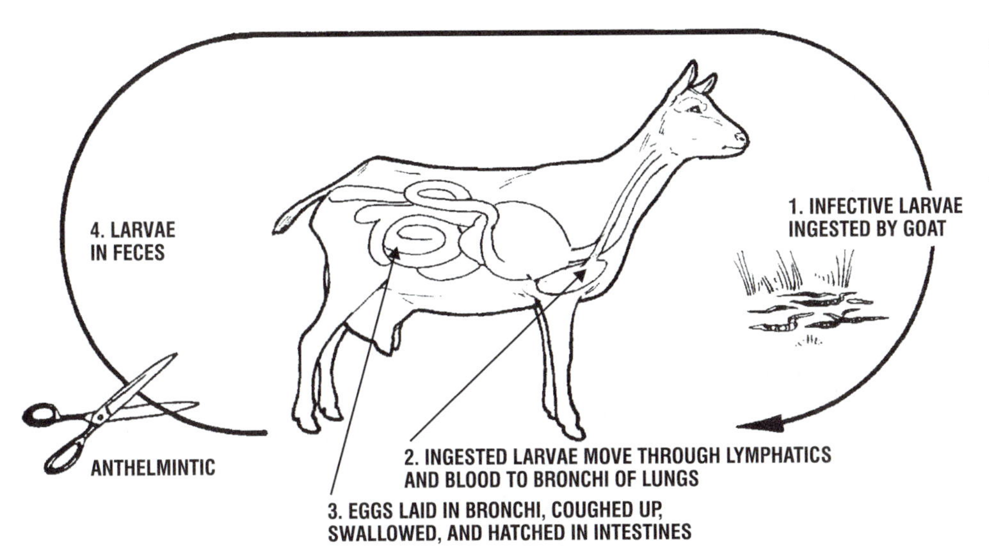

Fig. 8-18. Diagram showing the life history and habits of the lungworm *Dictyocaulus filaria*. Prevention consists of removing the animals from infected ground and placing them on dry pastures or in drylots. Moist pastures should be drained or avoided. Affected animals should be treated with the proper anthelmintic (see scissors).

PREVENTION, CONTROL, AND TREATMENT

Prevention consists of removing the animals from infected ground and placing them on dry pastures where clean water is available. Moist pastures should be drained or avoided, and the dung of infected sheep and goats should not be used to fertilize green feed for these animals. As old animals may be carriers without showing symptoms, it is recommended that lambs and kids be kept away from the mature animals whenever possible.

There are several useful drugs for treatment. Albendazole, fenbendazole, ivermectin, levamisole, and oxfendazole, administered according to manufacturer's directions, are effective against *D. filaria*. Treatment should be followed by good feeding and nursing.

NODULAR WORMS *(OESOPHAGOSTOMUM COLUMBIANUM)*

The nodular worm is a white worm about 0.6 in. long, found in the intestines of sheep and goats.

DISTRIBUTION AND LOSSES

Nodular worm infections are usually most severe in the humid areas of the central and eastern United States, with only limited damage encountered in the flocks in the West and Southwest. The presence of these parasites makes sheep unthrifty, reduces fleece and meat yields, and even causes heavy death losses. When sheep are infected with nodular worms, the infection reveals itself on postmortem examination at packing plants as nodules or knots on the small intestines. These cause all or a considerable portion of this organ to be unfit for surgical sutures (or catgut) or for casings. In the final analysis, the packer passes this loss back to the producer in the form of lower market prices.

LIFE HISTORY AND HABITS

The life history and habits are similar to those of the common stomach worm. The eggs are passed out with the feces after developing to an infective larval stage. They are ingested by grazing sheep and goats and then complete their development in the walls of the intestines. When the larvae burrow into the walls of the intestines, whitish cysts are formed, and these remain as nodules after the parasites have moved out. Hun-

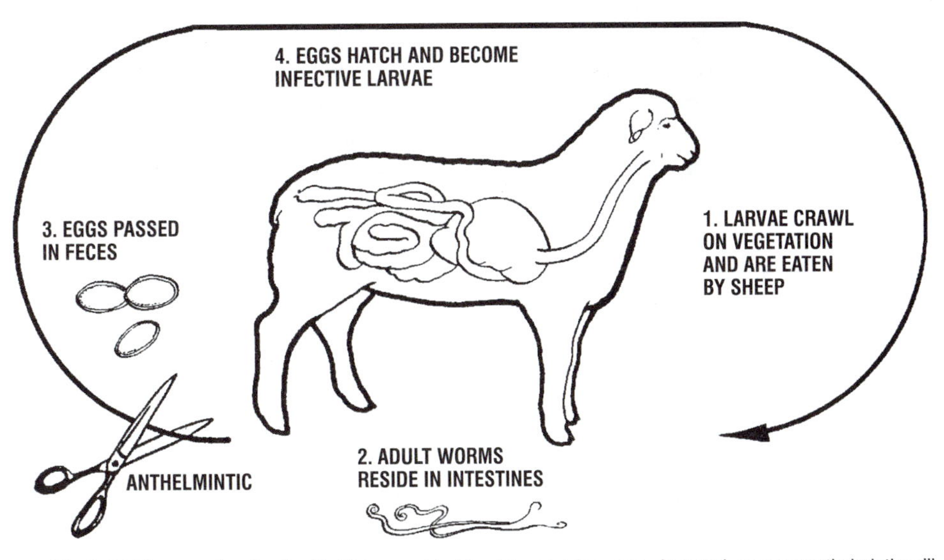

Fig. 8-19. Diagram showing the life history and habits of the nodular worm. As noted, a proper anthelmintic will remove the nodular worms from the intestines (see scissors).

dreds of these nodules may be present in an infected animal. These nodules interfere with bowel functions and may lead to impaction or telescoping of the gut. In packing-house parlance, these nodules are referred to as *knotty guts* or *pimply guts*.

DAMAGE INFLICTED; SYMPTOMS OR SIGNS OF AFFECTED ANIMALS

The symptoms accompanying severe nodular worm infection are not unlike those of general parasitism—weakness, loss of weight, diarrhea, retarded growth, and anemia.

PREVENTION, CONTROL, AND TREATMENT

The same prevention, control, and treatment recommended for stomach worms applies to nodular worms.

Anthelmintics are used for the treatment of worms (see Table 8-3). The choice of an anthelmintic depends on the worm species. Thiabendazole is still widely used, but it is less effective than other anthelmintics. Other benzimidazoles such as albendazole, fenbendazole, mebendazole, oxfendazole, and oxibendazole have been developed, and are effective against all major gastrointestinal parasites of sheep and goats. Levamisole, the pyrantel group, and ivermectin are also highly effective, safe, wide-spectrum anthelmintics. Of course, all must be used according to the manufacturer's directions.

Fig. 8-20. Sheep infected with nodular worms. (Courtesy, USDA)

SHEEP MEASLES (CYSTICERCUS OVIS)

Sheep measles is an invasion of sheep muscles by the larvae *Cysticercus ovis* of the tapeworm *Taenia ovis*. This species of bladder worm infects sheep and goats in many parts of the United States, especially in the West. The animals ingest the tapeworm eggs voided on pastures by dogs or foxes. The developing embryos penetrate through the wall of the intestine, reach the liver by way of the bloodstream, enter the veins, then travel to the heart, diaphragm, and muscles where they form mature bladder worms. No specific symptoms have been attributed to infection with these parasites, and it seems unlikely that many sheep die because of the presence of sheep measles. The chief economic loss occurs at slaughter. Infected carcasses are trimmed or condemned, according to the degree of infection. However, the parasite is not transmissible to humans, the removal or condemnation of affected carcasses merely being done because of the unattractive appearance. Tapeworms develop only when dogs or foxes eat carcasses infected with bladder worms. Sheep and goats are the intermediate hosts, and dogs and foxes are the definitive hosts.

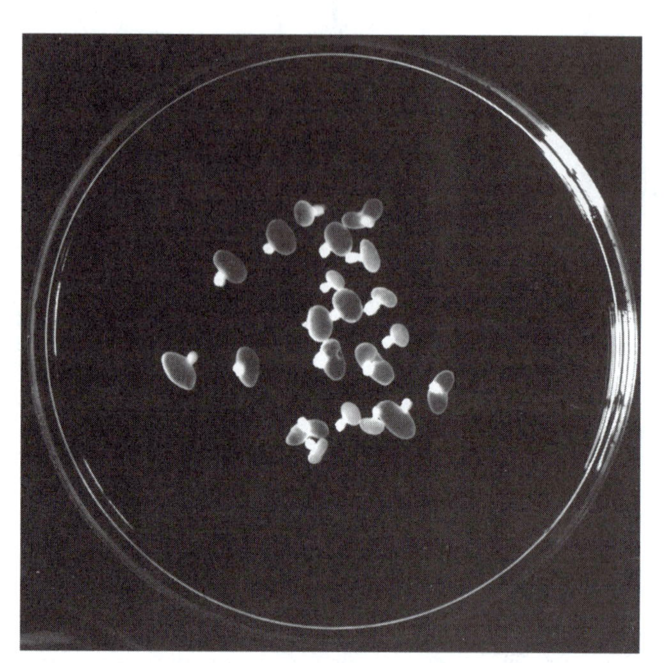

Fig. 8-21. *Cysticercus ovis*, the small, oval bladder worm responsible for sheep measles. Note that the head and neck are enclosed in a sheath about midway between the ends. (Courtesy, USDA)

There is no known anthelmintic for the removal of *Cysticercus ovis* from sheep and goats. Frequent treatment of dogs with an anthelmintic offers the best hope of control.

STOMACH WORMS (*HAEMONCHUS CONTORTUS*, TWISTED STOMACH WORMS, COMMON STOMACH WORMS, BARBER-POLE WORMS)

Haemonchus contortus, the common stomach worm, is the most common and the most destructive of the worm parasites of sheep and goats. Lambs and kids are more seriously affected than older animals. These worms are from 0.75 to 1.5 in. long and about the size of a horse hair in diameter. The live females are marked with a spiral striping, resembling a barber pole; hence, the common designation, *twisted stomach worm*. Their presence in the host is usually confined to the fourth or true stomach (the abomasum), though some of them will be found in the first part of the intestine.

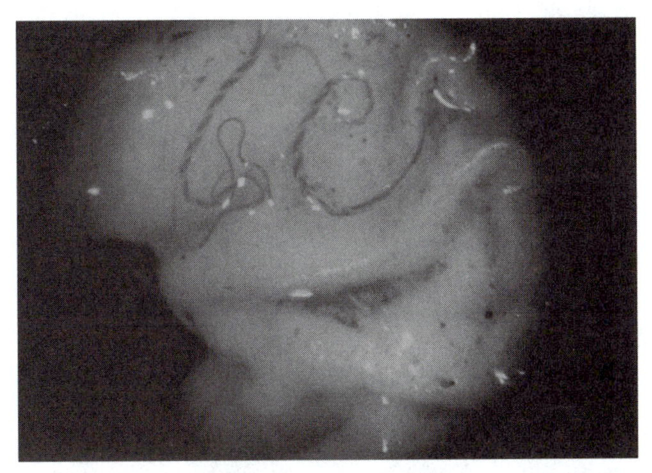

Fig. 8-22. The large stomach worm, *Haemonchus contortus*, on the stomach lining. Note the females which are striped like a barber pole. (Courtesy, Smith-Kline Animal Health Products, Philadelphia, PA)

DISTRIBUTION AND LOSSES

Most U.S. farm flocks—those sheep and goats located in the central, southern, and eastern states—are infected with this parasite. An increasing number of range bands are also acquiring these parasites, though much of the western range country is too dry for stomach worms to be much of a problem. However, irrigated pastures in the West are prime areas for common stomach worm infections. A severe infection may cause death, especially when the plane of nutrition is low, but an even greater toll is taken in the form of the setback that the whole flock suffers. A wormy flock is invariably an unprofitable flock.

LIFE HISTORY AND HABITS

The life history and habits of the stomach worm have been well established. The adult female, which normally lives in the abomasum, lays approximately 6,000 eggs daily. These microscopic-sized eggs pass out with the feces. Under conditions of favorable temperature and moisture, the eggs hatch into larvae within a few hours. Then the larvae molt twice and develop into the infective stage—all of which usually occurs within two weeks after passing from the host. When sufficient moisture is present, the young larvae then crawl up on grass blades, coming to rest with evaporation and moving onward and upward with additional moisture. At this stage the larvae are greatly resistant to changes in temperature and moisture. When clinging to a blade of grass, a larva is in a position to be swallowed by a grazing animal. Upon being ingested by the host animal, the larva travels to the fourth stomach, develops into the adult stage, and lays eggs—thus starting another cycle. The entire life cycle of the stomach worm may be completed in 21 days, conditions being favorable.

DAMAGE INFLICTED; SYMPTOMS OR SIGNS OF AFFECTED ANIMALS

Sheep and goats infected with stomach worms

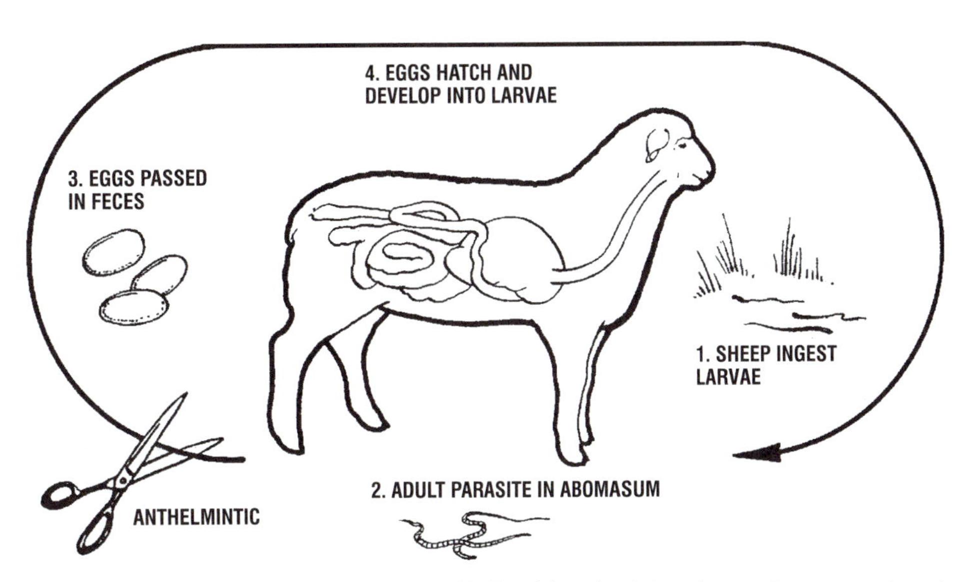

Fig. 8-23. Diagram showing the life history and habits of the twisted stomach worm (or common stomach worm). As noted, a proper anthelmintic (wormer) will remove adult stomach worms, preventing the release of more eggs (see scissors). Rotation of pastures is the best preventive.

Fig. 8-24. A goat with bottle jaw, characteristic of advanced and severe internal parasitism. (Courtesy, Christine S. F. Williams, Michigan State University)

first become unthrifty and listless and later become thin and weak. The membranes of the eyes, nose, and mouth become pale in color (anemic) from loss of blood. Diarrhea may be present. In the more severely infected animals and after a prolonged period, the wool (mohair of goats) may become loose and easy to pull out; and there may be a watery swelling under the lower jaw, which is sometimes referred to as *bottle jaw* or *poverty jaw*. There may also be swelling along the abdomen.

Unfortunately, no symptom or group of symptoms is a positive clue to the presence of stomach worms, as identical symptoms may be exhibited in other cases of infection by some other parasites. A correct diagnosis can be made only by postmortem examination of the intestinal tract.

Sacrificing a weak animal or obtaining an animal immediately after death will allow for a postmortem examination; otherwise, the worms may disintegrate and be difficult to find.

Evaluation of feces for eggs of stomach worms will provide estimates of numbers of worm eggs and will often indicate the severity of infection in the flock or herd and the numbers of animals infected. Worm-egg count can also be used to evaluate the efficacy of anthelmintics and control of parasites.

PREVENTION, CONTROL, AND TREATMENT

Practical control can be accomplished through a system of pasture rotations—changing the flock to a clean, fresh pasture at about two-week intervals. But, in the event of an outbreak of stomach worms, sheep and goats should be treated before turning them to new pastures. In a general way, this will require a minimum of three pastures and the intelligent planning of some temporary pastures to supplement the permanent pastures. Horses and hogs, which are not harmed by the common stomach worm, may be rotated with sheep and goats.

Also, pastures should never be overgrazed, because 97 to 98% of the infective larvae are found in the bottom 1 in. of grass.

■ **Treatment**—In addition to a carefully planned program of pasture management, any infection of stomach worms should be held in check by an anthelmintic (drench or injection). A number of anthelmintics are available; among them, the following: albendazole, febantel, fenbendazole, ivermectin, levamisole, pyrantel, and thiabendazole (see Table 8-3). The manufacturer's directions for use and withdrawal should always be followed. The following deworming schedule is recommended:

1. Treat four times a year. If infestation is heavy, more than four times a year may be needed in mature sheep.
2. Treat all lambs at weaning. Lambs confined to a drylot or to slotted floors may be kept practically free of stomach worms.

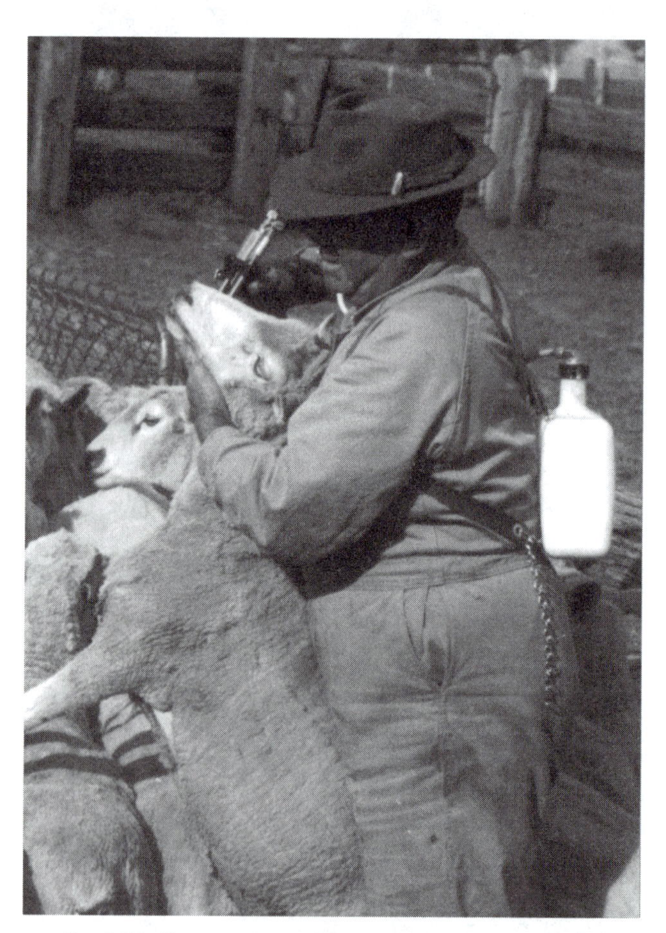

Fig. 8-25. Sheep receiving a worming drench through a calibrated gun-style dispenser. A reservoir containing the drench is worn on the operator's back. (Courtesy, Smith-Kline Animal Health Products, Philadelphia, PA)

3. Do not treat pregnant ewes within two weeks of lambing.

OTHER SPECIES OF ROUNDWORMS

Generally, sheep and goats that are heavily parasitized are infected with a number of species of roundworms. Thus, in addition to the common stomach worm, one or more of the following species may be present: (1) brown stomach worms, (2) trichostrongyles (bankrupt worms), (3) cooperias, and (4) hookworms. Each of these is discussed in a section under the names of the respective species.

TAPEWORMS (*MONIEZIA EXPANSA*, AND FRINGED TAPEWORMS, *THYSANOSOMA ACTINOIDES*)

Sheep and goats may harbor several different species of tapeworms, in both adult and larval stages. Only the two most common species will be discussed herein. The common tapeworm, *Moniezia expansa*, is a long, flat, ribbonlike worm which sometimes attains a length of several yards and a breadth of 0.75 in. Specimens 20 ft long have been found in lambs. The fringed tapeworm, *Thysanosoma actinoides*, derives its common name from the characteristic fringe which appears on the posterior of each of the segments. All species of tapeworms are commonly found in the small intestine of the host animal; but, in addition, the fringed tapeworm may occur in the ducts of the liver and the pancreas.

DISTRIBUTION AND LOSSES

Sheep and goats in all parts of the world appear to be infected with one or more species of tapeworms. The species *M. expansa* occurs mostly in the central, eastern, and southern United States, whereas the fringed tapeworm, *Thysanosoma actinoides*, appears to be confined largely to the range bands of the West. Usually, only a few members of a particular flock or band are infected at one time.

The common tapeworm is accepted as being nearly harmless. Also, there is a developed resistance to reinfection. If present in large numbers, however, these worms can block passage of food through the intestine, resulting in mild unthriftiness and digestive disturbances. On the other hand, the fringed tapeworm may cause death; and in the packing plant, livers harboring this parasite are condemned as unfit for human consumption, causing an economic loss.

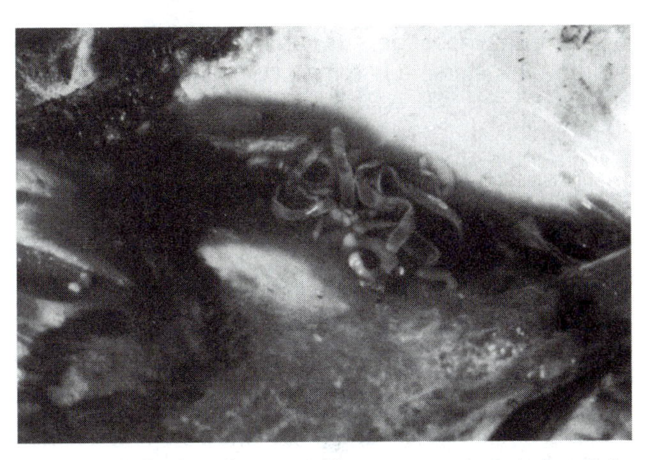

Fig. 8-26. The fringed tapeworm (*Thysanosoma* sp.) may be found in the small intestine, in bile ducts, and in pancreatic ducts. (Courtesy, Smith-Kline Animal Health Products, Philadelphia, PA)

LIFE HISTORY AND HABITS

The life history of the fringed tapeworm is not known with certainty, but it is thought to require an intermediate invertebrate host through which it must pass before it can infest sheep. The life history of the common tapeworm, *M. expansa*, appears to be about as follows: The microscopic eggs and segments con-

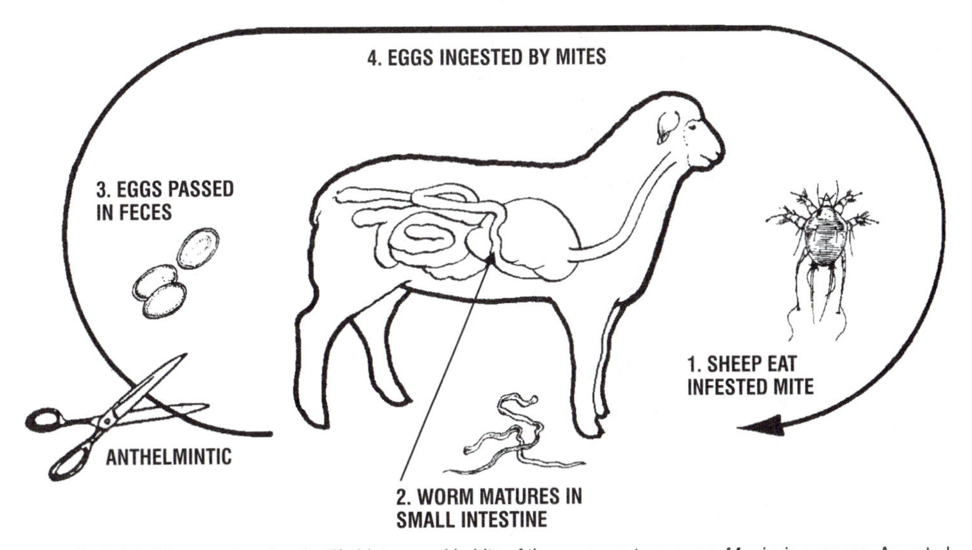

Fig. 8-27. Diagram showing the life history and habits of the common tapeworm, *Moniezia expansa*. As noted (see scissors), a suitable anthelmintic will remove the tapeworms from the intestines. The treatment should be administered according to directions, with the animals changed to clean pastures following treatment.

taining eggs pass out with the feces. On being ingested by a suitable intermediate host, oribatid mites or beetle mites, the eggs develop into an intermediate stage— larvae known as cysticercoids. Sheep and goats become infected by swallowing such larvae. The larvae travel to the small intestines of their hosts, where they develop into adult worms by the growth of segments back of the head. With reproduction, a new life cycle is started.

DAMAGE INFLICTED; SYMPTOMS OR SIGNS OF AFFECTED ANIMALS

The presence of common tapeworms does not produce any marked or specific symptoms. However, the fringed tapeworm may cause death of the host through blocking the cystic duct, the gall bladder, and the ducts of the liver and the pancreas. Infected animals usually have normal appetites.

PREVENTION, CONTROL, AND TREATMENT

Control of the common tapeworm is difficult because the infective larvae are protected by living within the hardy grass mites. These mites can be collected in abundance at all times of the year and can withstand severe cold and heat. The mites migrate up grass blades during the very early, cool, dim light of dawn, then crawl back into the protective soil as the blades become hotter. These habits provide the means for a limited control measure; namely, keeping the lambs off pasture until the sun has driven the mites from the grass blades.

The two most common treatment for control of the common tapeworm is fenbendazole (see Table 8-3).

When using the fenbendazole treatment, the producer should observe the following points: (1) the entire flock should be treated at the same time, and (2) following treatment, the animals should be moved to fresh pasture if possible.

Fenbendazole (Panacur) at 4.54 mg/lb of body weight is effective in the removal of fringed tapeworms from sheep. It is a very safe drug, and is also effective against *Moniezia, Trichuris ovis, Strongyloides*, and the other common gastrointestinal nematodes in sheep and goats.

THIN-NECKED BLADDER WORMS *(CYSTICERCUS TENUICOLLIS)*

Thin-necked bladder worms form from *Cysticercus tenuicollis*, which are the larvae of *Taenia hydatigena*, a tapeworm common to dogs. They are quite generally distributed over the United States. Their life history and habits are similar to those already described for the other bladder worms, and, consequently, the same preventive measure—the worming of dogs—applies. Thin-necked bladder worms burrow into the liver and/or thin membranes of the abdominal cavity of sheep and goats, producing tissue damage.

THREAD-NECKED INTESTINAL WORMS (*NEMATODIRUS*, TWISTED WIRE WORMS)

This parasite is endemic in some parts of the Rocky Mountain states, where it occasionally causes losses in lambs. Symptoms of heavy infections of thread-necked intestinal worms are: sudden onset, unthriftiness, profuse diarrhea, and marked dehydration, with deaths occurring 2 to 3 days after an outbreak. The parasite is commonly confined to lambs and weaners, but in low-rainfall areas older sheep may be heavily infected. See Table 8-3 for treatments.

THREADWORMS *(STRONGYLOIDES)*

This worm locates in the small intestine. Heavy infections of adult worms produce symptoms resembling trichostrongylosis. Infection is usually by skin penetration, but it can also occur via the milk. Damage to the skin between the claws, produced by the penetrating larvae, resemble early stages of footrot. See Table 8-3 for treatments.

TRICHOSTRONGYLES (*TRICHOSTRONGYLUS AXEI, T. COLUBRIFORMIS, T. VITRINUS, T. CAPRICOLA*, HAIRWORMS, BANKRUPT WORMS)

These worms—which are sometimes referred to as *hairworms, bankrupt worms* or *small stomach and intestinal worms*—are of considerable economic importance in sheep and goat husbandry. Trichostrongyles are small, hairlike worms less than 0.33 in. in length found in the fourth stomach (abomasum) and the small intestine. Because of their small size, they are often overlooked in postmortem examination. Infection is marked by severe unthriftiness, diarrhea, and dehydration. Death losses are not unusual, particularly when animals are on scant rations. The life cycle is similar to that of the common stomach worm, except that the eggs are more resistant to drying. The recommended measures for prevention, control, and treatment are similar to those outlined for the stomach worm (see Table 8-3).

WHIPWORMS *(TRICHURIS OVIS)*

Whipworms are usually found attached to the wall of the cecum, or blind gut, of sheep and goats. They are white and 1.5 to 2.0 in. long. The worms have a very slender anterior portion and a much enlarged posterior. The anterior resembles the lash of a whip and the posterior the handle; hence, the name *whipworm*.

DISTRIBUTION AND LOSSES

Although whipworms are rather widespread, seldom are infections of sheep and goats sufficiently heavy to cause great damage. However, in specific flocks damage can be severe, with numerous death losses.

LIFE HISTORY AND HABITS

The life cycle of the whipworm is simple and direct; that is, no intermediate host is necessary to complete the life cycle. Infective larvae develop in the shell, and sheep and goats, while grazing, become infected by swallowing the larvae.

DAMAGE INFLICTED; SYMPTOMS OR SIGNS OF AFFECTED ANIMALS

There are no well-defined symptoms associated with whipworm infestation in sheep and goats, though persistent bloody diarrhea and unthriftiness are usually evident.

PREVENTION, CONTROL, AND TREATMENT

Clean pastures and rotation grazing are the key to prevention and control.

Some of the same anthelmintics used for stomach worms are used for whipworms (see Table 8-3).

ANTHELMINTICS (DEWORMERS) FOR TREATMENT OF INTERNAL PARASITES OF SHEEP AND GOATS

Table 8-3 lists the anthelmintics (dewormers) commonly used for the control of internal parasites of sheep and goats. Users are admonished always to read and follow the label carefully, and consult with the veterinarian relative to approval, restrictions, and use.

EXTERNAL PARASITES

In general, sheep and goats are less harmed by external parasites than by internal ones. On the other hand, a heavy infestation of the common external parasites may result in severe irritation, restlessness, rubbing, wool loss, mohair loss, and unthriftiness.

BLOWFLIES (FLEECEWORMS, FLY STRIKE, WOOL MAGGOTS)

The flies of the blowfly group include a number of

TABLE 8-3
EFFICACY OF ANTHELMINTICS AGAINST MAJOR INTERNAL PARASITES OF SHEEP AND GOATS

		Efficacy Against Adult Parasites								
Drug[1]	Dose[2]	Brown Stomach Worms (*Ostertagia*)	Cooper's Worms (*Cooperia*)	Hairworms (*Tricho-strongylus*)	Hookworms (*Buno-stomum*)	Large-mouth Bowel Worms (*Chabertia*)	Stomach Worms (*Haemon-chus*)	Tape Worm (*Moniezia*)	Thread-necked Intestinal Worms (*Nematodirus*)	Thread-worms (*Strongy-loides*)
	(mg/kg)	*(%)*	*(%)*	*(%)*	*(%)*	*(%)*	*(%)*	*(%)*	*(%)*	*(%)*
Albendazole	10	97–100	99–100	99–100	—[3]	100	99–100	100	99–100	—
Febantel	5	95–100	—	95–100	95–100	—	95–100	—	95–100	95–100
Fenbendazole	10	95–100	95–100	95–100	95–100	80–100	95–100	85–95	95–100	—
Ivermectin	0.2 SC	95–100	95–100	95–100	—	95–100	95–100	—	95–100	95–100
Levamisole	7.5 (5 SC & IM)	95–100	95–100	95–100	95–100	95–100	95–100	—	95–100	60–85
Morantel	10	95–100	95–100	95–100	—	95–100	95–100	—	95–100	—
Oxfendazole	5	95–100	95–100	95–100	—	95–100	95–100	95–100	95–100	—
Oxibendazole	10	95–100	95–100	0–100	95–100	95–100	95–100	95–100	95–100	95–100
Pyrantel	25	—	—	0–100	—	—	95–100	—	95–100	—
Thiabendazole	44	95–100	—	95–100	60–85	95–100	95–100	—	85–100	65–85

[1]Always read the label carefully or consult a veterinarian for approvals and restrictions for use.

[2]All drugs listed are given orally unless otherwise listed (SC = subcutaneous injection; IM = intramuscular injection). To convert metric to U.S. Customary, or vice versa, see the "Weights and Measures" section in the appendix.

[3]Not effective or insufficient data to determine efficacy.

species that find their principal breeding ground in dead and putrifying flesh, although they sometimes infest wounds or unhealthy tissues of live animals and fresh or cooked meat. All the important species of blowflies except the flesh flies, which are grayish and have three dark stripes on their backs, have a more or less metallic luster.

DISTRIBUTION AND LOSSES

Although blowflies are widespread, they present the greatest problem in the Pacific Northwest and in the South and Southwest. Death losses from blowflies are not excessive, but the flies cause much discomfort to affected animals and lower production.

LIFE HISTORY AND HABITS

With the exception of the group known as gray flesh flies, which deposit tiny living maggots instead of eggs, the blowflies have a similar life cycle to the screwworm, except that the cycle is completed in about one-half the time.

DAMAGE INFLICTED; SYMPTOMS OR SIGNS OF AFFECTED ANIMALS

The blowfly causes its greatest damage by infesting wounds and the soiled fleece of sheep. Such damage, which is largely limited to the black blowfly (or woolmaggot fly), is similar to that caused by screwworms. Sheep are especially susceptible to attacks of blowflies, because their wool frequently becomes soiled or moistened by rain and accumulations of feces and urine. The maggots spread over the body, feeding on the skin surface, where they produce a severe irritation and destroy the ability of the skin to function. Infested animals rapidly become weak and fevered; and although they may recover, they may remain in an unthrifty condition for a long period.

Because blowflies infest both fresh and cooked meat, they are often a problem of major importance around packing houses and farm homes.

PREVENTION, CONTROL, AND TREATMENT

Prevention of blowfly damage consists of eliminating the pest and decreasing the susceptibility of animals to infestation. The practices of docking lambs, tagging sheep at intervals, and internal parasite control to prevent diarrhea, will materially lessen blowfly damage in this class of animals.

As blowflies breed in dead carcasses, some control is effected by promptly destroying all dead animals by burning or deep burial. The use of traps, insecticide sprays, poisoned baits, and electrified screens is also helpful in reducing trouble from blowflies. Suitable repellents, such as pine tar oil, help prevent the flies from depositing their eggs.

When the fleece of sheep and goats becomes infested with blowfly maggots, the animals should be sheared well around the affected area. In fact, complete shearing will control most outbreaks, and when possible, this is preferred to chemical treatment. When needed though, coumaphos (Co-Ral) and other insecticides (see Table 8-4) can be used as a treatment for infected animals or as a prophylactic. Insecticide use must be according to the manufacturer's recommendations.

KEDS (SHEEP TICKS, *MELOPHAGUS OVINUS*)

The so-called sheep tick is not really a tick at all, but a hairy, blood-sucking fly without wings. It is brown, has six legs, and at maturity is about 0.25 in. in length. Keds infest all sheep, though the fine-wool breeds—and sometimes Angora goats—are usually only lightly affected.

Fig. 8-28. Sheep keds or ticks *(Melophagus ovinus)*. A ked is a hairy, brown fly without wings, with six legs, and about 0.25 in. in length at maturity. (Courtesy, USDA)

DISTRIBUTION AND LOSSES

The sheep tick is widely distributed throughout the United States, but it is especially prevalent in the northern states. The chief economic losses resulting from infestation with sheep ticks are in terms of retarded growth of young animals and loss in condition of mature animals, together with damage to the fleece.

LIFE HISTORY AND HABITS

The entire life of the ked is spent on the sheep. Adult keds cannot live more than 4 or 5 days off their hosts, but the pupae may survive for a longer time, especially in warm weather. They are more common on the medium- and coarse-wooled breeds than on the fine-wooled breeds. Apparently, they are not adapted to thrive in the tight, greasy fleeces of the latter. The four stages in the life cycle of keds may be described as follows:

1. The egg is retained in the body of the mature female ked, where it develops into a larva in about 10 to 12 days. Each female deposits 10 to 15 larvae.
2. Each larva is deposited, develops a shell or puparium about itself, and remains attached to the wool fibers.
3. In 19 to 24 days from the time it is deposited, the shell of the pupa is broken open and the young ked emerges.
4. The young ked grows and reaches sexual maturity in another 3 to 4 days, depositing its first larvae within 10 to 12 days after mating.

DAMAGE INFLICTED; SYMPTOMS OR SIGNS OF AFFECTED ANIMALS

Infestation can be readily detected. If it is heavy, the host may display a marked reduction in condition and even anemia. Because of the intense irritation, the sheep may bite and scratch, thus damaging the wool and causing general unrest. Poorly housed and poorly fed animals are more likely to suffer from keds.

Note: A defect in the pelt caused by keds is known as *cockle*.

PREVENTION, CONTROL, AND TREATMENT

Prevention and control rests mainly in treating all sheep as soon as shearing cuts heal, using any one of the following insecticides, used according to the manufacturer's label: carbamates, ciodrin, Co-Ral, ivermectin, malathion, pyrethroids, toxaphene, or others (see Table 8-4).

LICE

The louse is a small, flattened, wingless insect parasite of which there are two groups—sucking lice and biting lice. Biting lice, *Bovicola (Damalinia) ovis*, are most common, but they are less harmful than sucking lice in that they live off the exfoliated epithelium and do not suck blood. The sucking species are the body lice, *Linognathus ovillus* and *L. africanus*, and the foot louse, *L. pedalis*. The sucking lice are larger and have narrower heads than the biting lice. Body lice may be found anywhere on the body where the fleece is dense, while foot lice are usually found on the legs below the knee or hock. Fortunately, lice are not so common on sheep or goats as on other domestic animals, but they do occur occasionally.

Fig. 8-29. Blue female goat louse. (Courtesy, USDA)

Most species of lice are specific for a particular class of animals. Thus, sheep harbor a number of common species of lice; but sheep lice will not remain on other farm animals, nor will lice from other animals infest sheep. Likewise, goat lice are specific for goats. Lice are always more abundant on weak, unthrifty animals and are more troublesome during the winter months than during the rest of the year.

DISTRIBUTION AND LOSSES

The presence of lice upon animals is almost universal, but the degree of infestation depends largely upon the state of animal nutrition and the extent to which the owner will tolerate parasites. The irritation caused by the presence of lice retards growth, gains, and/or production of wool, mohair, and milk.

LIFE HISTORY AND HABITS

Lice spend their entire life cycle on the host's body. They attach their eggs, or nits, to the wool, mohair, or hair near the skin, where they hatch in 9 to 18 days. Two weeks later the young females begin laying eggs, and after reproduction they die on their host. Lice do not survive more than a week when separated from their host; but, under favorable conditions, eggs clinging to detached hairs may continue to hatch for 2 or 3 weeks.

DAMAGE INFLICTED; SYMPTOMS OR SIGNS OF AFFECTED ANIMALS

Infestation shows up most commonly in winter in ill-nourished and neglected animals. There is intense irritation, restlessness, and loss of condition. As many lice are blood suckers, they devitalize their hosts. There may be severe itching, and the animals may be seen scratching, rubbing, and gnawing at the skin. The wool (mohair of goats) may be matted and thin, and may lack luster; and scabs may be evident. In some cases, the symptoms may resemble those of mange; and it must be kept in mind that the two may occur simultaneously.

With the coming of spring, when animals go to pasture, lousiness is greatly diminished.

Fig. 8-30. Goats of the same age. Growth of the smaller one was stunted, due to lice. (Courtesy, USDA)

PREVENTION, CONTROL, AND TREATMENT

Because of the close contact of domesticated animals, especially during the winter months, it is practically impossible to prevent flocks (herds of goats) from becoming slightly infested with the pests. Nevertheless, lice can be kept under control.

For control of lice, all members of the flock (herd) must be treated simultaneously at intervals.

Sheep and goats can be most effectively treated when the fleece is not too long. It is also desirable to treat the housing and the bedding.

If a power sprayer is used, the effective penetration of the wool of sheep or the mohair of goats requires about 400 psi.

Dipping vats have long been used successfully in treating sheep, goats, cattle, hogs, and sometimes horses. The chief virtue of dipping vats is that animals so treated are thoroughly covered. On the other hand, a dipping vat is rather costly to construct, lacks motility and flexibility, and results in considerable left-over dip at the finish of the operation. Dipping vats are especially effective where large numbers of animals can be assembled together.

Fig. 8-31. Dipping goats to control lice. (Courtesy, USDA)

Dusting is less effective than spraying or dipping, but may be preferable when few animals are to be treated or during the winter months.

When used in accordance with the manufacturer's directions, any of the following insecticides can be employed to treat and control lice: ciodrin, Co-Ral, ivermectin, methoxychlor, pyrethroids, and others (see Table 8-4).

MITES (SCAB, SCABIES, MANGE)

Mites produce a specific contagious disease known as mange (or scabies, scab, or itch). These small, insect-like parasites, which are just visible to the naked eye, constitute a very large group.

Each species of domesticated animals has its own peculiar species of mange mites; and, with the exception of the sarcoptic mites, the mites from one species of animals cannot live and propagate permanently on a different species.

The two chief forms of sheep mange are: sarcoptic mange, caused by mites that burrow, and psoroptic mange, caused by mites that bite the skin but do not burrow. Psoroptic or common scab is the most important form of sheep scabies in the United States. It is easily transmitted by contact from one sheep to another, and it spreads very rapidly after being introduced into a flock. The disease appears to spread most rapidly among young and poorly nourished animals. Psoroptic sheep scabies is a reportable disease.

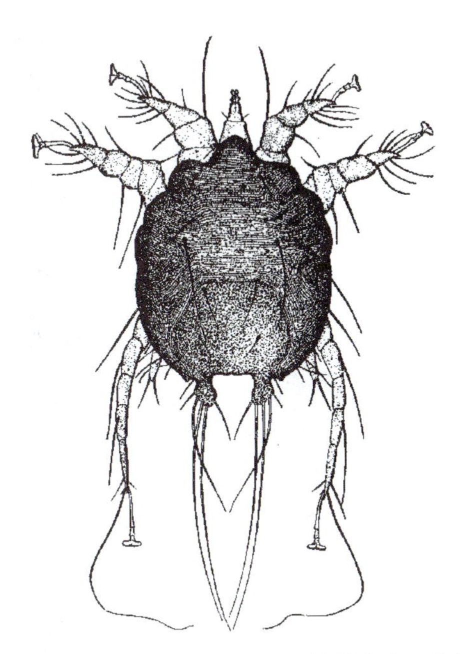

Fig. 8-32. Sheep-scab mite *(Psoroptes ovis)*. Male. Dorsal view, greatly enlarged. (After Salmon and Stiles, through courtesy of USDA)

Sarcoptic mange mites are uncommon in sheep in the United States.

Goats are afflicted by both sarcoptic and psoroptic mange mites. Additionally they are subject to demodectic mange and psoroptic ear mange.

DISTRIBUTION AND LOSSES

In a severe attack, the skins may be much less valuable for leather, and the wool of sheep and the mohair of goats may be damaged. Growth is retarded, and production of meat and milk is lowered.

LIFE HISTORY AND HABITS

The life cycle of the sheep scab mite is as follows: The eggs are deposited on the skin at the margin of the lesion; the eggs hatch in 1 to 5 days, producing six-legged larvae; the larvae feed for 2 or 3 days and then molt and become eight-legged nymphs; the nymphal stage lasts 3 or 4 days when they molt and become mature males and females which mate immediately; and the fertilized females molt once and are ready to lay eggs. The entire cycle from egg to adult is completed in 10 to 12 days. While females may live as long as 42 days on sheep, they usually die within 10 days when off their hosts. Corrals are noninfective after being vacant for 17 days.

The disease is more prevalent during the winter months, when animals are confined and in close contact with each other.

DAMAGE INFLICTED; SYMPTOMS OR SIGNS OF AFFECTED ANIMALS

When the mite pierces the skin to feed on cells and lymph, there is marked irritation, itching, and scratching. Exudate forms on the surface, and this coagulates, crusting over the surface. The crusting is often accompanied or followed by loss of wool and formation of thick crusts or scabs on the skin; hence the name scabies. Often there are secondary skin infections. The only certain method of diagnosis is to demonstrate microscopically the presence of the mites.

Fig 8-33. Sheep with mange, caused by mites. Note loss of wool. (Photo by J. C. Allen and Son, West Lafayette, IN)

PREVENTION, CONTROL, AND TREATMENT

Prevention consists of keeping animals away from diseased animals or infected premises.

Scabies is a reportable disease subject to state and federal quarantine, and control operations must be supervised by the state veterinarian's office. If scabies is suspected, a veterinarian, the county agricultural agent, and/or the state university can help in reporting.

SCREWWORMS *(COCHLIOMYIA HOMINIVORAX)*

Screwworms are the maggots (larvae) of the fly *Cochliomyia hominivorax*. They thrive on the living flesh of animals. True screwworms seldom get through the unbroken skin, but will penetrate moist pockets. They are not found in cold-blooded animals such as turtles, snakes, and lizards.

Wounds resulting from shearing, branding, castrating, dehorning, and docking afford a breeding ground for this parasite. Add to this the wounds from some types of vegetation, from fighting, and from blood-sucking insects, and ample places for propagation are provided.

DISTRIBUTION AND LOSSES

Screwworm flies, once common in the United States, are now rare because of the biological control method of releasing sterile male flies in areas where screwworms were present. However, areas of Mexico and Latin America are affected.

In infested areas, the screwworm is undoubtedly the greatest enemy of all the insect species with which the livestock owner must contend. For example, at one time in the Southwest, many ranchers reported that 50% of their normal annual livestock losses were caused by this parasite.

LIFE HISTORY AND HABITS

The primary screwworm fly is bluish green with three dark stripes on its back and red or orange below the eyes. The fly generally deposits its eggs in shingle-like masses on the edges or the dry portion of wounds. From 50 to 300 eggs are laid at one time, with a single female being capable of laying about 3,000 eggs in a lifetime. Hatching of the eggs occurs in 11 hours, and the young, whitish worms (larvae or maggots) immediately burrow into the living flesh. There they feed and grow for a period of 4 to 7 days, shedding their skin twice during this period.

When these worms have reached their full growth, they assume a pinkish color, leave the wound, and drop to the ground, where they dig beneath the surface of the soil and undergo a transformation to hard-skinned, dark brown, motionless pupae. It is during the pupae stage that a maggot changes to an adult fly. After the pupa has been in the soil from 7 to 60 days, the fly emerges from it, works its way to the surface of the ground, and crawls up on some nearby object (bush, weed, etc.) to allow its wings to unfold and otherwise to mature. Under favorable conditions, the newly emerged female fly becomes sexually mature and will lay eggs 5 days later. During warm weather, the entire life cycle is usually completed in 21 days, but under cold, unfavorable conditions, the cycle may take as many as 80 days or longer.

DAMAGE INFLICTED; SYMPTOMS OR SIGNS OF AFFECTED ANIMALS

The injury caused by this parasite is inflicted chiefly by the maggots. The early symptoms in affected animals are loss of appetite and condition, and listlessness. Unless proper treatment is administered, the great destruction of many tissues kills the host in a few days.

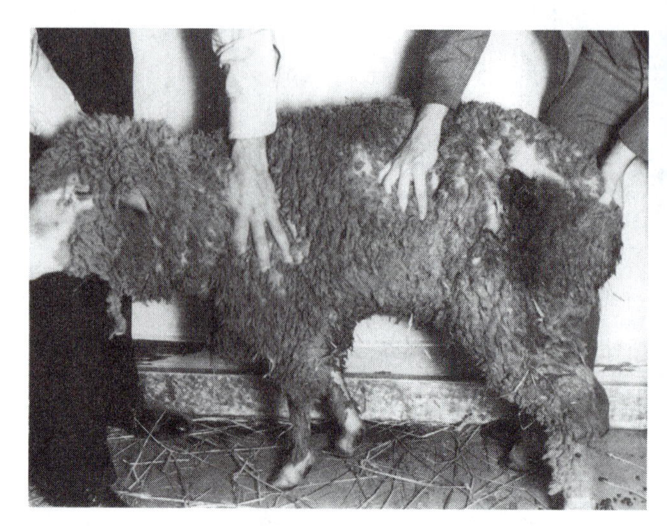

Fig. 8-34. Screwworm infestation in sheep. (Courtesy, USDA)

PREVENTION, CONTROL, AND TREATMENT

Prevention in infested areas consists mainly of keeping animal wounds to a minimum and protecting those that do materialize.

As the primary screwworm must have living warm-blooded animals in which to breed and in order to survive, it must produce a new generation during each four-month period. Therefore, the most effective control measures can be effected during the winter months. During this season, the life cycle is slowed down, and it is difficult for the fly to live and breed. Thus, the most effective control consists of preventing infestation of wounds and killing all possible maggots during the winter and spring months. Additional control is effected through timing as much as possible those farm and ranch operations that necessarily produce wounds during the winter season when the flies are least abundant and least active. Using Burdizzo pincers in castration, eradicating plants that cause injuries, breeding so that young will arrive during the seasons of least fly activity, and avoiding anything else that might produce wounds will all aid greatly in screwworm control. It is highly desirable to have a screened, flyproof area available for wounded animals. In brief, the elimination of wounds or injuries to the host constitutes effective control.

In 1958, the U.S. Department of Agriculture initiated a screwworm eradication program. This program involved the systematic and strategic release of sterile male flies into the natural fly population, thereby reducing the number of fertile matings. The eradication program has virtually eliminated all the losses caused by screwworms in the United States. Unfortunately, the states bordering on Mexico are periodically reinfested by mated female flies from Mexico. For this reason, permanent elimination of screwworms from the United States by the sterile-male technique cannot be hoped for until they are also eradicated in Mexico and Latin America.

When maggots (larvae) are found in an animal, they should be removed and sent to the proper authorities for identification, and the animal should be treated with a proper insecticide or smear.

Title 9 of the Code of Federal Regulations, Part 83, Screwworms, gives the screwworm control zone, the areas of recurring infestation, and the inspection and treatment requirements for movement from these areas.

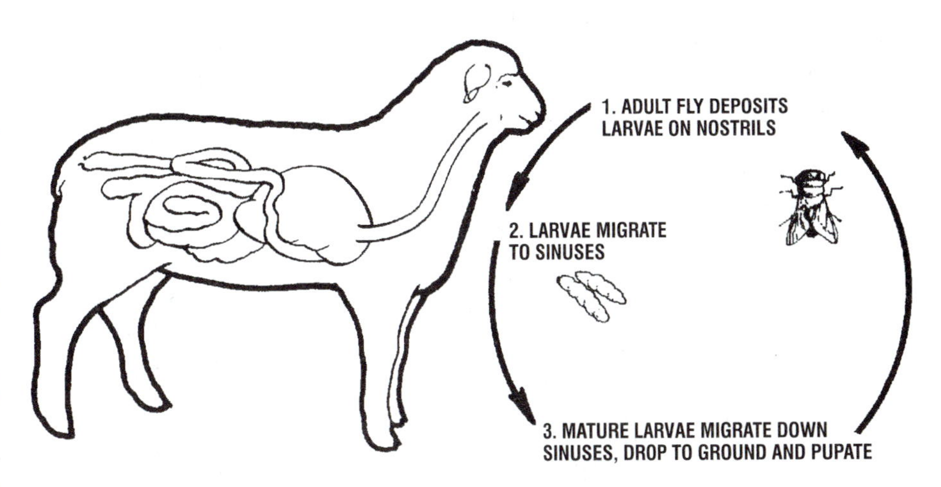

Fig. 18-36. Diagram showing the life history and habits of the sheep nasal fly.

SHEEP BOTS (*OESTRUS OVIS*; GRUB-IN-THE-HEAD; NASAL BOTFLIES)

This condition of sheep, commonly referred to as *grub-in-the-head*, is due to the larvae or grubs of the sheep nasal fly—also called the sheep gadfly and the head maggot fly. The adult nasal fly is about the size of the common horsefly. It is beelike in appearance, dark yellow or brownish, hairy, and on the wing only during warm, sunny days.

Occasionally, goats are affected by sheep bots.

DISTRIBUTION AND LOSSES

The sheep nasal fly is said to occur throughout the world. The annoyance accompanying attacks of the sheep nasal fly interferes with normal feeding and results in loss of condition. Similar unrest exists while the larvae are in the nasal passage. Death losses caused by grub-in-the-head alone are probably rare, and measurable other damage is small.

Fig. 8-35. The sheep nasal fly *(Oestrus ovis)*. It is beelike in appearance, dark yellow or brownish, hairy, and in the adult stage, about the size of the common horsefly. (Courtesy, USDA)

LIFE HISTORY AND HABITS

The sheep nasal flies deposit larvae (not eggs) around the sheep's nostrils. The larvae then crawl up the nasal passages to the frontal sinuses, where they complete their development. The mature larvae then migrate from the sinuses back to the nasal passages, drop to the ground, burrow into the soil, and pupate. In another 3 to 8 weeks, depending on temperature and moisture conditions, the flies emerge from the pupal cases, crawl to the surface, develop, and again deposit larvae. It takes from 3 to 5½ months to complete a life cycle.

DAMAGE INFLICTED; SYMPTOMS OR SIGNS OF AFFECTED ANIMALS

When the flies attempt to deposit their larvae around the nostrils of sheep, they cause great annoyance. The animals respond by ceasing to feed, becoming restless, pressing their noses against other sheep, or seeking shelter. The most generally recognized symptom of grub infestation is a nasal discharge. Because the grubs have a spiny cover, they constantly irritate the nasal passages. Breathing may be difficult, and frequent sneezing may be noted. It is unlikely that the grubs themselves cause death in many cases.

PREVENTION, CONTROL, AND TREATMENT

Sheep should be treated for bots in the fall when most larvae are small, since killing mature larvae in the sinuses may cause severe reactions in the sinus membranes. A single drench of ivermectin will control sheep nose bots.

TICKS

In addition to sheep keds, hard and soft ticks occasionally infect sheep and goats, especially in summer.

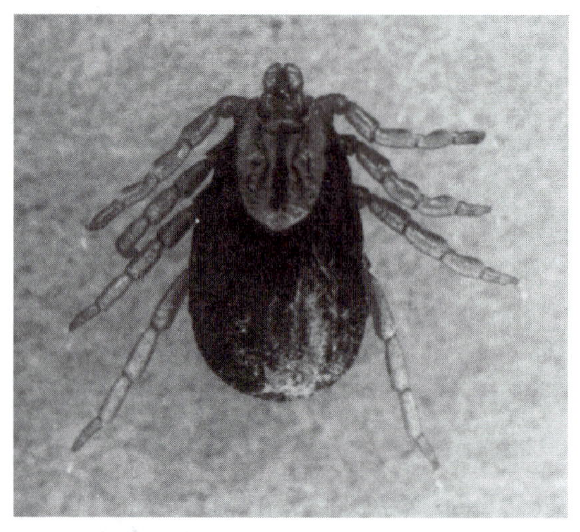

Fig. 8-37. A hard tick, *Dermacentor* spp. (Courtesy, Dr. Wm. J. Foreyt, Ph.D., Professor of Microbiology and Pathology, Washington State University.)

DISTRIBUTION AND LOSSES

Some years, ticks pose a severe problem in western range bands. Also, the same tick species may infest humans and spread tularemia and spotted fever.

LIFE HISTORY AND HABITS

There are several species of sheep ticks, depending on geographic location. Hard ticks, primarily *Dermacentor* spp, infest sheep and goats as 6-legged larva or seed tick, the 8-legged nymph, or the 8-legged adult. The latter stage is most apparent because of its size and engorging blood which is necessary for egg production. Sheep become infested by brushing against tall grass and brush on which ticks await.

The soft tick, *Otobius megnini*, infests a number of different animals including sheep and goats.

DAMAGE INFLICTED, SYMPTOMS OR SIGNS OF AFFECTED ANIMALS

Affected sheep and goats become weak due to irritation and blood loss. The hard tick can cause tick paralysis which can paralyze the animal until it dies. The soft tick locates deep down in the ear canals and results in discomfort to the animal.

PREVENTION, CONTROL, TREATMENT

There is no practical prevention. Effective treatments for hard and soft ticks are listed in Table 8-4.

PESTICIDES (INSECTICIDES) FOR TREATMENT OF EXTERNAL PARASITES OF SHEEP AND GOATS

Table 8-4 lists the pesticides (insecticides) that are commonly used for the control of external parasites of sheep and goats. Users are admonished always to read and follow the label carefully, and consult with the veterinarian relative to the approval, restrictions, and use.

TABLE 8-4
MAJOR DRUGS FOR CONTROL OF EXTERNAL PARASITES IN SHEEP AND GOATS

Drug (Trade Name)[1, 2]	Parasites Controlled	Application
Organophosphates		
Chlorfenvinphos (Supona)	Ticks, lice, flies, keds, mites	Dip, spray
Chlorpyrifos (Dursban)	Ticks, lice	Dip, spray, pour-on, dust
Coumaphos (Co-Ral)	Fleas, flies, keds, lice, ticks	Dip, spray, pour-on, dust, feed additive
Crotoxyphos (Ciodrin)	Ticks, mites, lice, flies	Dust, spray
Crufomate (Roulene)	Flies, lice, mites, ticks	Dip, spray, pour-on
Diazinon	Flies, lice, keds, ticks	Dip, dust, spray
Dichlorvos (Vapona)	Flies	Resin granules, resin strips, spray
Dimethoate (Cygon)	Flies	Spray
Dioxathion (Delnav)	Flies, keds, lice, ticks	Dips, spray
Famphur (Warbex)	Flies, lice	Pour-on, spray
Fenchlorphos (Korlan-ronnel)	Flies, lice, mites, ticks	Dust, oral, spray, pour-on
Fenthion (Tiguvon)	Sheep nose bots, lice, keds, blowflies	Spray, pour-on, spot-on
Malathion (Cythion)	Fleas, flies, lice, keds, mites, ticks	Dips, dust, spray
Phosmet (Prolate)	Flies, lice, ticks	Spray
Tetrachlorvinphos (Rabon-stirophos)	Fleas, flies, lice, ticks	Dust, ear tag, spray
Trichlorfon (Neguvon)	Flies, lice, mites	Pour-on, oral, spray
Carbamates		
Carbaryl (Sevin)	Fleas, flies, lice, mites, ticks	Dips, dust, spray
Propoxur	Flies	Spray

(Continued)

TABLE 8-4 (Continued)

Drug (Trade Name)[1, 2]	Parasites Controlled	Application
Organochlorines		
Lindane	Flies, keds, lice, mites, ticks	Dust, spray, pour-on
Methoxychlor (Marlate)	Flies, keds, lice, mites, ticks	Dust, spray
Toxaphene	Flies, keds, lice, mites, ticks	Dip, dust, spray
Pyrethroids		
Cypermethrin (P) (Curatick)	Ticks	Dip
Fenvalerate (P) (Ectrin)	Flies, ticks	Spray, ear tags
Permethrin (P) (Ectiban)	Flies, lice, mites, ticks	Dip, spray
Resmethrin (P)	Ticks	Spray
Miscellaneous [includes insect growth regulators (IGR), bacteria (B), and others]		
Ivermectin (Ivomec)	Lices, mites, nose bots	Oral, injectable (SC, IM)
Amitraz (Mitaban, Triatox, Baam)	Ticks, mites	Dip, spray
Methoprene (IGR)	Flies	Feed additive
Pyrethrum	Flies, fleas, keds, lice	Spray
Bacillus thuringiensis (B)	Aquatic insect larvae	Dried powder

[1]Always read the label direction carefully or consult a veterinarian for approvals and restrictions for use.

[2]Where trade names are used, no endorsement is intended, nor is any criticism implied of similar products not named.

IMMUNITY

No discussion of health and disease would be complete without a brief explanation of immunity. When an animal is immune to a certain disease, it simply means that it is not susceptible to that disease. There are two forms of immunity: natural and acquired.

When immunity to a disease is inherited, it is referred to as *natural immunity*. For example, when sheep are exposed to hog cholera, they never contract the disease because they have a type of natural immunity referred to as *species immunity*. Likewise, humans are naturally immune to Texas fever. Algerian sheep are said to be highly resistant to anthrax; this constitutes a type of natural immunity called *racial immunity*.

The body also has the ability, when properly stimulated by a given organism or toxin, to produce antibodies and/or antitoxins. When an animal has enough antibodies for overcoming particular disease-producing organisms, it is said to be immune to that disease. This type of immunity is referred to as *acquired*.

Acquired immunity or resistance is either active or passive. When an animal is stimulated in such a manner (by vaccination or actual disease) as to cause it to produce antibodies, it is said to have acquired active immunity. On the other hand, if an animal is injected with the antibodies (or immune bodies) produced by an actively immunized animal, it is referred to as having an acquired passive immunity. Such immunity is usually conferred by the injection of blood serum from immunized animals, the serum carrying with it the substances by which the protection is conferred. Passive immunization confers immunity upon its injection, but the immunity disappears within 3 to 6 weeks. Young mammals secure passive immunity from their mothers' colostrum the first few days following birth.

In active immunity, resistance is not developed until after 1 or 2 weeks; but it is far more lasting, for the animal apparently keeps on manufacturing antibodies. Therefore, active immunity has a great advantage. There are exceptions, however—for example, the tetanus antitoxin.

DISINFECTANTS

A disinfectant is a bactericidal or microbicidal agent that frees from infection (usually a chemical agent which destroys disease germs or other microorganisms, or inactivates viruses).

As disease-producing organisms—viruses, bacteria, fungi, and parasite eggs—accumulate in the environment, disease problems can become more severe and can be transmitted to each succeeding group of animals raised on the same premises. Under these circumstances, cleaning and disinfection become extremely important in breaking the life cycle. Also, in the case of a disease outbreak, the premises must be disinfected.

Under ordinary conditions, proper cleaning of barns removes most of the microorganisms, along with the filth, thus eliminating the necessity of disinfection.

Effective disinfection depends on five things:

1. Thorough cleaning before application.
2. The phenol coefficient of the disinfectant, which indicates the killing strength of a disinfectant as compared to phenol (carbolic acid). It is determined by a standard laboratory test in which the typhoid fever germ is used as the test organism.
3. The dilution at which the disinfectant is used.
4. The temperature; most disinfectants are much more effective if applied hot.
5. Thoroughness of application and time of exposure.

Disinfection must in all cases be preceded by a very thorough cleaning, for organic matter serves to protect disease germs and otherwise interferes with the activity of the disinfecting agent.

Sunlight possesses disinfecting properties, but it is variable and superficial in its action. Heat and some of the chemical disinfectants are more effective.

The application of heat by steam, hot water, burning, or boiling is an effective method of disinfection. In many cases, however, it may not be practical to use heat.

A good disinfectant should (1) have the power to kill disease-producing organisms, (2) remain stable in the presence of organic matter (manure, hair, soil), (3) dissolve readily in water and remain in solution, (4) be nontoxic to animals and humans, (5) penetrate organic matter rapidly, (6) remove dirt and grease, and (7) be economical to use.

The number of available disinfectants is large because the ideal universally applicable disinfectant does not exist. Table 8-5 gives a summary of the limitations, usefulness, and strength of some common disinfectants.

When using a disinfectant, *always read and follow the manufacturer's directions*.

TABLE 8-5
DISINFECTANT GUIDE[1]

Kind of Disinfectant	Usefulness	Strength	Limitations and Comments
Alcohol (ethyl-ethanol, isopropyl, methanol)	Primarily as skin disinfectant and for emergency purposes on instruments.	70% alcohol—the content usually found in rubbing alcohol.	They are too costly for general disinfection. They are ineffective against bacterial spores.
Boric acid[2]	As a wash for eyes, and other sensitive parts of the body.	1 oz in 1 pt water (about 6% solution).	It is a weak antiseptic. It may cause harm to the nervous system if absorbed into the body in large amounts. For this and other reasons, antibiotic solutions and saline solutions are replacing it.
Chlorines (sodium hypochlorate, chloramine-T)	Used for dairy equipment and as deodorants They will kill all kinds of bacteria, fungi, and viruses, providing the concentration is sufficiently high.	Generally used at about 200 ppm for dairy equipment and as a deodorant.	They are corrosive to metals and neutralized by organic materials.
Formaldehyde	Formaldehyde will kill anthrax spores, TB organisms, and animal viruses in a 1 to 2% solution. It is often used to disinfect buildings following a disease outbreak. A 1 to 2% solution may be used as a foot bath to control foot rot.	As a liquid disinfectant, it is usually used as a 1 to 2% solution. As a gaseous disinfectant (fumigant), use 1½ lb of potassium permanganate plus 3 pt of formaldehyde. Also, gas may be released by heating paraformaldehyde.	It has a disagreeable odor, destroys living tissue, and can be extremely poisonous. The bactericidal effectiveness of the gas is dependent upon having the proper relative humidity (above 75%) and temperature (above 86°F and preferably near 140°F).
Heat (by steam, hot water, burning, or boiling)	In the burning of rubbish or articles of little value, and in disposing of infected body discharges. The steam "Jenny" is effective for disinfection *if properly employed*, particularly if used in conjunction with a phenolic germicide.	10 minutes exposure to boiling water is usually sufficient.	Exposure to boiling water will destroy all ordinary disease germs, but sometimes fails to kill the spores of such diseases as anthrax and tetanus. Moist heat is preferred to dry heat, and steam under pressure is the most effective. Heat may be impractical or too expensive.
Iodine[2] (tincture)	Extensively used as skin disinfectant, for minor cuts and bruises.	Generally used as tincture of iodine, either 2% or 7%.	Never cover with a bandage. Clean skin before applying iodine. It is corrosive to metals.
Iodophors (tamed iodine)	Primarily used for dairy utensils. Effective against all bacteria (both gram-negative and gram-positive), fungi, and most viruses.	Usually used as disinfectants at concentrations of 50 to 75 ppm titratable iodine, and as sanitizers at levels of 12.5 to 25 ppm. At 12.5 ppm titratable iodine, they can be used as an antiseptic in drinking water.	They are inhibited in their activity by organic matter. They are quite expensive.

(Continued)

TABLE 8-5 (Continued)

Kind of Disinfectant	Usefulness	Strength	Limitations and Comments
Lime (quicklime, burnt lime, calcium oxide)	As a deodorant when sprinkled on manure and animal discharges; or as a disinfectant when sprinkled on the floor or used as a newly made "milk of lime" or as a whitewash.	Use as a dust; as "milk of lime"; or as a whitewash, but *use fresh*.	Not effective against anthrax or tetanus spores. Wear goggles when adding water to quicklime.
Lye (sodium hydroxide, caustic soda)	On concrete floors; in milk houses because there is no odor; against microorganisms of brucellosis and the viruses of foot-and-mouth disease. In strong solution (5%), effective against anthrax and blackleg.	Lye is usually used as either a 2% or a 5% solution. To prepare a 2% solution, add 1 can of lye to 5 gal. of water. To prepare a 5% solution, add 1 can of lye to 2 gal. of water. A 2% solution will destroy the organisms causing foot-and-mouth disease, but a 5% solution is necessary to destroy the spores of anthrax.	Damages fabrics, aluminum, and painted surfaces. Be careful, for it will burn the hands and face. Not effective against organisms of TB or Johne's disease. Lye solutions are most effective when used hot. *Diluted vinegar can be used to neutralize lye.*
Lysol (the brand name of a product of cresol plus soap)	For disinfecting surgical instruments and instruments used in dehorning, castrating, and tattooing. Useful as a skin disinfectant before surgery, and for use on the hands before castrating.	0.5 to 2.0%	Has a disagreeable odor. Does not mix well with hard water. Less costly than phenol.
Phenols (carbolic acids): 1. Phenolics—coal tar derivatives 2. Synthetic phenols	They are ideal general-purpose disinfectants. Effective and inexpensive. They are very resistant to the inhibiting effects of organic residue; hence, they are suitable for barn disinfection, and foot and wheel dip-baths.	Both phenolics (coal tar) and synthetic phenols vary widely in efficacy from one compound to another. So, note and follow manufacturers' directions. Generally used in a 5% solution.	They are corrosive, and they are toxic to animals and humans. Ineffective on fungi and viruses.
Quarternary ammonium compounds (QAC)	Very water soluble, ultra-rapid kill rate, effective deodorizing properties, and moderately priced. Good detergent characteristics and harmless to skin.	Follow manufacturers' directions.	They can corrode metal. Not very potent in combating viruses. Adversely affected by organic matter.
Sal soda	It may be used in place of lye against foot-and-mouth disease and vesicular exanthema.	10½% solution (13½ oz to 1 gal. water).	
Sal soda and soda ash (or sodium carbonate)	They may be used in place of lye against foot-and-mouth disease.	4% solution (1 lb to 3 gal. water). Most effective in hot solution.	Commonly used as cleansing agents, but have disinfectant properties, especially when used as hot solutions.
Soap	Its power to kill germs is very limited. Greatest usefulness is in cleansing and dissolving coatings from various surfaces, including the skin, prior to application of a good disinfectant.	As commercially prepared.	Although indispensable to sanitizing surfaces, soaps should not be used as disinfectants. They are not regularly effective; staphlococci and organisms which cause diarrheal disease are resistant.

[1]Where metrtic values are desired, refer to the Appendix for conversion factors.

[2]Sometimes loosely classed as a disinfectant but actually an antiseptic and practically useful only on living tissue.

POISONOUS PLANTS

Poisonous plants have been known to humans since time immemorial. Biblical literature alludes to the poisonous properties of certain plants, and history records that hemlock (made from the plant from which it takes its name) was administered by the Greeks to Socrates and other state prisoners.

No section of the United States is entirely free of poisonous plants, for there are hundreds of them. The heaviest livestock losses from poisonous plants occur on the western ranges because (1) there has been less cultivation and destruction of poisonous plants in range areas, and (2) the frequent overgrazing on some of the western ranges has resulted in the elimination of some of the more nutritious and desirable plants, and these have been replaced by increased numbers of the less desirable and poisonous species.

Some common poisonous plants are listed in Table 8-6, but the list of poisonous plants is so extensive that no attempt has been made to discuss them in this book. Instead, those who are interested in pursuing further the subject are referred to *The Stockman's Handbook*, a book by the same author and publisher as *Sheep & Goat Science*. Both the stockman and the veterinarian should have a working knowledge of the common poisonous species in the area in which they operate.

TABLE 8-6
SOME COMMON POISONOUS PLANTS AND THEIR SEASON(S)

Common Name	Scientific Name	Season(s)
Broomweed	*Gutierrezia* sp.	Spring and summer
Chokecherry	*Prunus viginiana*	Spring
Copperweed	*Oxytenia acerosa*	Summer
Death camas . . .	*Zigadenus paniculatus*	Spring
Greasewood	*Scarcobatus vermiculatus*	Fall
Halogeton	*Halogeton glomeratus*	All year
Horsebrush	*Tetradymia* sp.	Spring
Loco	*Oxytropis* sp.	Spring
Lupine	*Lupinus* sp.	Summer and fall
Milkweed	*Asclepias* sp.	Summer
Rubberweed	*Hymenoxys* sp.	Summer
Sneezeweed	*Helenium* sp.	Summer
Veratrum	*Veratrum californiacum*	Summer

DIAGNOSIS OF PLANT POISONING

The diagnosis of plant poisoning in animals is not an easy or precise procedure. Any case of sudden illness or death with no apparent cause is commonly considered to be the result of poisoning. This may not always be correct. When large numbers of animals are suddenly affected, however, a suspicion of poisoning is justified until it has been proven otherwise.

Eating poisonous plants may induce the following symptoms or signs: (1) sudden death; (2) transitory illness; (3) general body weakness; (4) disturbance of the central nervous, vascular, and endocrine systems; (5) photosensitization; (6) frequent urination; (7) diarrhea; (8) bloating; (9) chronic debilitation and death; (10) embryonic death; (11) fetal death; (12) abortion; (13) extensive liver necrosis and/or cirrhosis; (14) edema and/or abdominal dropsy; (15) tumor growths in tissues; (16) congenital deformities; (17) metabolic deficiencies; and (18) physical injury.

No general set of symptoms or signs per se irrefutably provides all the information necessary to make a diagnosis of plant poisonings. Nevertheless, a careful description of the toxic signs coupled with information pertaining to available plants provides a meaningful basis for a tentative diagnosis. Additional information essential to a poisonous plant diagnosis includes: (1) type of feed, site grazed, and availability of water; (2) identification and relative abundance of all poisonous plants available to animals; (3) amount and stage of growth of the various poisonous plants being grazed; (4) toxicity and palatability of the plants in relation to their stage of growth; (5) time from eating the plants until onset of toxic signs; (6) species, age, and sex of animals affected; (7) clinical signs of toxic reactions; (8) chemical analysis of plants; and (9) careful evaluation of all the information relative to the etiology of the disease.

PREVENTING LOSSES FROM POISONOUS PLANTS

With poisonous plants, the emphasis should be on prevention of losses rather than on treatment, no matter how successful the latter. The following are effective preventive measures:

1. **Follow good pasture or range management.** Plant poisoning is nature's sign of a sick pasture or range, usually resulting from misuse. When a sufficient supply of desirable forage is available, poisonous plants are not often eaten, for they are less palatable. On the other hand, when overgrazing reduces the available supply of the more palatable and safe vegetation, animals may, through sheer hunger, consume the toxic plants.

2. **Know the poisonous plants common to the area.** This can usually be accomplished through (a) studying drawings, photographs, and/or descriptions; (b) checking with local authorities; or (c) sending two or three fresh whole plants (if possible, include the roots, stems, leaves, and flowers) to the state agricultural college—first wrapping the plants in several thicknesses of moist paper.

By knowing the poisonous plants common to the area, it will be possible—

a. To avoid areas heavily infested with poisonous plants which, due to animal concentration and overgrazing, usually include waterholes, salt grounds, bed grounds, and trails.

b. To control and eradicate the poisonous plants effectively, by mechanical or chemical means (as recommended by local authorities), or by fencing off.

c. To recognize more surely and readily the particular kind of plant poisoning when it strikes, for time is important.

d. To know what first-aid, if any, to apply, especially when death is imminent or when a veterinarian is not readily available.

e. To graze animals with a class of livestock not harmed by the particular poisonous plant or plants, where this is possible. Many plants seriously poisonous to one kind of livestock are not poisonous to another, at least under practical conditions.

f. To shift the grazing season to a time when the plant is not dangerous, where this is possible. That is, some plants are poisonous at certain seasons of the year, but comparatively harmless at other seasons.

g. To avoid cutting poison-infested meadows for hay when it is known that the dried cured plant is poisonous. Some plants are poisonous in either green or dry form, whereas others are harmless when dry. When poisonous plants (or seeds) become mixed with hay (or grain), it is difficult for animals to separate the safe from the toxic material.

3. **Know the symptoms that generally indicate plant poisoning**, thus making for early action.

4. **Avoid turning to pasture in very early spring.** Nature has ordained many poisonous plants as early growers—earlier than the desirable forage. For this reason, as well as from the standpoint of desirable pasture management, animals should not be turned to pasture in the early spring before the usual forage has become plentiful.

5. **Provide supplemental feed during droughts or after early frost.** Otherwise, hungry animals may eat poisonous plants in an effort to survive.

6. **Avoid turning very hungry animals into areas where there are poisonous plants**, especially those that have been in corrals for shearing, branding, etc.; those that have been recently shipped or trailed long distances; or those that have been wintered on dry forage. First, feed the animals to satisfy their hunger or allow a fill on an area known to be free from poisonous plants.

7. **Avoid driving animals too fast when trailing.** On long drives, either allow them to graze along the way, or stop frequently and provide supplemental feed.

8. **Remove promptly all animals from infested areas when plant poisoning strikes.** Hopefully, this will check further losses.

9. **Treat promptly**, preferably by a veterinarian.

TREATMENT

Unfortunately, plant-poisoned animals are not generally discovered in sufficient time to prevent losses. Thus, prevention is decidedly superior to treatment.

When trouble is encountered, the owner or the caretaker should *promptly* call a veterinarian. In the meantime, the animal should be (1) placed where adequate care and treatment can be given, (2) protected from excessive heat and cold, and (3) allowed to eat only feeds known to be safe.

The veterinarian may determine the kind of poisonous plant involved (1) by observing the symptoms, and/or (2) by finding out exactly what poisonous plant was eaten through looking over the pasture and/or hay and identifying leaves or other plant parts found in the animal's digestive tract at the time of autopsy.

Knowledge is the key. Unfortunately, many poisoned animals that would have recovered had they been left undisturbed, have been killed by attempts to administer home remedies by well-meaning but untrained persons.

NATIONAL ANIMAL POISON CONTROL CENTER

Established in 1978 and maintained at the University of Illinois, Urbana-Champaign, the National Animal Poison Control Center hotline number is 1-800-548-2423. Recognizing that accidents don't wait for business hours, the Center is open 24 hours a day, every day of the week. The toxicology group is staffed to answer questions about known or suspected cases of poisoning or chemical contaminations involving any species of animal. It is not intended to replace local veterinarians or state toxicology laboratories, but to complement them.

The toxicologists at the center constantly update their files on chemicals, feed additives, human and veterinary drugs, pesticides, environmental contaminants, and plant and mold toxins. Their comprehensive file of information contains comparative species toxicity data, product ingredients, and recommended therapeutic and decontamination measures. The goal is a computer database containing 200,000 entries to facilitate quick and accurate responses to all types of poisoning/contamination incidents and inquiries.

Many times a proper treatment regime can be recommended over the telephone. When telephone consultation is inadequate or the problem is of major proportion, a team of veterinary specialists can arrive at the scene of a toxic or contamination problem within a short time.

The cost of an investigation varies according to distance traveled, personnel time, and laboratory services required. Where consultation over the telephone is adequate, there is no charge to the veterinarian or producer.

FEDERAL AND STATE REGULATIONS RELATIVE TO DISEASE CONTROL

Certain diseases are so devastating that individual owners cannot protect their herds and/or flocks against their invasion. Moreover, where human health is involved, the problem is much too important to be entrusted to individual action. In the United States, therefore, certain regulatory activities in animal-disease control are under the supervision of various federal and state organizations. Federally, this responsibility is entrusted to the following agency:

Veterinary Services
Animal and Plant Health Inspection Service
U.S. Department of Agriculture
Federal Center Building
Hyattsville, Maryland 20782

In addition to the federal interstate regulations, each of the states has requirements for the entry of livestock. Generally, these requirements include compliance with interstate regulations. States usually require a certificate of health or a permit, or both, and additional testing requirements, depending upon the class of livestock involved.

Detailed information relative to animal disease control can be obtained from federal and state animal health officials, or from accredited veterinarians in all states. Shippers are urged to obtain such information prior to making interstate shipments of livestock.

■ **Quarantine**—By quarantine many highly infectious diseases are prevented from (1) gaining a foothold in this country, or (2) spreading. *Quarantine involves (1) segregating and confining one or more animals in the smallest possible area to prevent any direct or indirect contact with animals not so restrained; and/or (2) regulating movement of animals at points of entry.* When an infectious disease outbreak occurs, drastic quarantine must be imposed to restrict movement out of an area or within areas. The type of quarantine varies from one involving a mere physical examination and movement under proper certification to one involving complete prohibition of the movement of animals, produce, vehicles, and even human beings.

■ **Indemnities**—Information relative to indemnities paid to owners by the federal government for animals disposed of as a result of outbreaks of certain diseases is given in Chapter 1, Subchapter B, Title 9, of the Code of Federal Regulations.

HARRY S. TRUMAN ANIMAL IMPORT CENTER

A federal quarantine center was authorized in Public Law 91-239, signed by the president on May 6, 1970. A 16.1-acre site for the center was selected at Fleming Key, near Key West, Florida. In 1978, the center was named the Harry S. Truman Import Center in honor of former President Truman; and in 1979, it opened under the administration of the USDA's Animal and Plant Inspection Service.

The center holds some 400 head of cattle, or other species in equivalent numbers, at one time, for a five-month quarantine period. This maximum security station enables U.S. stock producers to import breeding animals from all parts of the world, while at the same time safeguarding our domestic herds and flocks from such diseases as foot-and-mouth disease, rinderpest, piroplasmosis, and others.

QUESTIONS FOR STUDY AND DISCUSSION

1. What is your evaluation of the two primary premises on which this chapter is based; namely, (a) that sheep and goat diseases and parasites can be largely prevented through good management, proper feeding, and strict sanitation; and (b) that enlightened sheep and goat producers will more readily call a veterinarian in order that a disease may be brought under control with a minimum of loss?

2. What is the normal temperature, pulse rate, and breathing rate of sheep and of goats? How would you determine each?

3. Select a specific farm or ranch (either your own or one with which you are familiar) and outline, in logical order, a program of health, disease prevention, and parasite control for either (a) sheep or (b) goats.

4. Discuss the importance, symptoms or signs, and cause, prevention, and treatment of each of the following sheep and/or goat diseases: bluetongue, caprine arthritis encephalitis (CAE), enterotoxemia, foot rot, ram epididymitis, and scrapie.

5. Assume that a specific contagious disease (you name it) has broken out in your flock of sheep or herd of goats. What steps would you take to meet the situation? List in 1, 2, 3, order, and be specific.

6. Discuss the importance, life history and habits, and prevention, control, and treatment of each of the following sheep and/or goat internal parasites: coccidiosis, lungworms. sheep measles, and stomach worms.

7. Discuss the importance, life history and habits, and prevention, control, and treatment of each of the following sheep and/or goat external parasites: keds, lice, mites, and screwworms.

8. Why would you expect to see less of certain parasitic infestations in drylots?

9. When scabies is confirmed in a flock, why is the flock quarantined? Who does the quarantining?

10. Explain why sanitation is important to flock health, even with antibiotics, drugs, and vaccines.

11. Sheep and goat producers have been criticized for their excess use of chemicals. To lessen reasons for the criticism, what precautions should they take when using a pesticide in their health program?

12. Explain the difference between natural immunity, active immunity, and passive immunity.

13. Assume that you have, during a period of a year, encountered sheep or goat death losses from three different diseases. What kind of disinfectant would you use in each case?

14. Assume that you have encountered death losses from a certain poisonous plant (you name it). What steps would you take to meet the situation? List in 1, 2, 3, order, and be specific.

15. List and discuss effective measures for preventing losses from poisonous plants.

16. What and where is the "National Animal Poison Control Center"? Of what value is it to sheep or goat producers?

17. Why are certain regulatory activities in animal disease control under the supervision of various federal and state agencies? How can federal and state regulatory officials be of assistance to the individual sheep or goat producer?

18. What is the Harry S. Truman Animal Import Center? What is its purpose?

19. When people are ill, they call the family doctor. Isn't it just as logical that a veterinarian be called when sheep are sick? Why or why not?

SELECTED REFERENCES

Title of Publication	Author(s)	Publisher
Animal Agents and Vectors of Human Disease	E. C. Faust	Lea & Febiger, Philadelphia, PA, 1955
Animal Diseases: The Yearbook of Agriculture, 1956		U.S. Department of Agriculture, Washington, DC, 1956
Animal Health: A Layman's Guide to Disease Control	W. J. Greer J. K. Baker	Interstate Publishers, Inc., Danville, IL, 1992
Animal Health: Livestock and Pets	Staff	U.S. Department of Agriculture, 1984 Yearbook of Agriculture, Washington, DC
Animal Parasitism	C. P. Read	Prentice-Hall, Inc., Englewood Cliffs, NJ, 1972
Current Veterinary Therapy, Food Animal Practice	Ed. by J. L. Howard	W. B. Saunders Co., Philadelphia, PA, 1981
Diseases of Sheep	R. Jensen B. L. Swift	Lea & Febiger, Philadelphia PA, 1982
Diseases Transmitted from Animals to Man	Ed. by T. G. Hull	Charles C Thomas, Publisher, Springfield, IL, 1963
Disinfection, Sterilization, and Preservation	C. A. Laurence S. S. Block	Lea & Febiger, Philadelphia, PA, 1968
Farm Animal Health and Disease Control	J. K. Winkler	Lea & Febiger, Philadelphia PA, 1982
Goat Production	Ed. by C. Gull	Academic Press, Inc., New York, NY, 1981

Hagan's Infectious Diseases of Domestic Animals	D. W. Bruner J. H. Gillespie	Comstock Publishing Associates, Ithaca, NY, 1973
Handbook of Livestock Management Techniques	R. A. Battaglia V. B. Mayrose	Burgess Publishing Co., Minneapolis, MN, 1981
Losses in Agriculture, Ag. Hdbk. No. 291		U.S. Department of Agriculture, Washington, DC, 1965
Merck Veterinary Manual, The, Seventh Edition	Ed. by O. H. Siegmund	Merck & Co., Inc., Rahway, NJ, 1991
Sheep Production Handbook, The	Am. Sheep Industry Assn.	SID-INF, Englewood, CO, 1988
Stockman's Handbook, The, Seventh Edition	M. E. Ensminger	Interstate Publishers, Inc., Danville IL, 1992

In addition to the above selected references, valuable publications on different subjects pertaining to sheep diseases, parasites, and disinfectants can be obtained from the following sources:

1. Division of Publications
   Office of Information
   U.S. Department of Agriculture
   Washington, DC 20250
2. Your state agricultural college
3. Several biological, pharmaceutical, and chemical companies

CHAPTER 9

Wool—fabric of history. (Courtesy, International Wool Secretariat, London, England)

WOOL AND MOHAIR

Contents *Page*

Contents Page

Like human civilization, the story of wool began in Asia Minor during the Stone Age about 10,000 years ago. Primitive man living in the Mesopotamian Plain used sheep for three basic human needs—food, clothing, and shelter. The wool was plucked by hand. Later, primitive people learned to spin and weave.

The mobility of sheep and the warmth of wool clothing allowed people to spread civilization far beyond the warm climate of Mesopotamia.

Between 3000 and 1000 B.C., the Persians, Greeks, and Romans distributed sheep and wool throughout Europe. The Romans took sheep with them as they built their empire in what is now Spain, North Africa, and the British Isles.

The Saracens, nomadic people of the Syrian-Arabian deserts, conquered Spain in the eighth century and established a widespread wool export trade with North Africa, Greece, Egypt, and Constantinople.

During the twelfth century, weaving in Florence, Genoa, and Venice was stimulated by the Norman conquest of Greece.

In Spain, a thriving wool trade helped finance the voyages of Columbus and the Conquistadors.

Columbus took sheep to Cuba and Santo Domingo on his second voyage in 1493, and Cortez took their descendants along when he explored what is now Mexico and southwestern United States.

Despite England trying to discourage a wool industry in North America, smuggled sheep thrived. Massachusetts even passed a law requiring young people to spin and weave.

King George III of England made wool trading in the colonies a punishable offense. Despite the King's oppressive actions, the wool industry flourished in America. Both Washington and Jefferson maintained flocks of sheep. Sheep moved west, and beyond, with civilization; by the turn of the 18th century, small flocks had been established in Australia, New Zealand, and South Africa.

Through the ages, tradition and folklore grew with the sheep and wool industry, some of which has been preserved in the following vernacular:

- *Dyed in the wool* (color added to wool): Means that it's genuine.
- *Fleecing:* Refers to swindling a gullible victim.
- *Shoddy* (fabric made of reclaimed wool): Means that it's inferior.
- *Wool gathering* (wandering about collecting tufts of wool caught on brush or bushes): Said of a daydreaming person.
- *Virgin wool:* Unused.
- *Crowded like sheep:* People jammed together.
- *Sheep in people's clothing:* Sheep wearing plastic coats.
- *The land of felt:* The name given by the Chinese to an area over which the Asian nomads roamed in the fourth century.
- *Pulling the wool over one's eyes:* Hoodwinking a person.
- *Spinster* (in colonial United States, the eldest unmarried daughter in the family was responsible for spinning the wool): Hence, the term for an unmarried woman.

The discussion that follows is designed to bridge the gap between the wool producer and the manufacturer. Although each has particular problems, both are working toward a common goal. The producer breeds, feeds, and manages the flock to supply the raw material; whereas the manufacturer scours, combs, spins, and weaves the fiber into cloth.

To the end that those producing, marketing, and processing wool may better meet their competition from other fibers, it is essential that they improve their product and be fully informed of the fiber and its behavior.

WOOL IS THE NATURAL CLOTHING OF SHEEP

With all the perfection and modification in the fleece that has been wrought through centuries of domestication, breeding, selection, and improved environmental conditions, it must not be forgotten that wool is the natural hair as well as the clothing of sheep. A covering of hair or feathers performs a thermoregula-

Fig. 9-1. Rambouillet ram. (Courtesy, *Ranch and Rural Living Magazine*, San Angelo, TX)

tory function for warm-blooded animals, the original intent being to protect its growers from heat or cold. As wool fibers are poor conductors of heat, they serve to prevent any abnormal loss of heat from the body.

In the wild state, sheep carried two distinct coats. The outer or protective coat consisted of long, coarse fibers, which today are classed as hairs (known as *kemp*). The undercoat of the wild sheep was soft and curly and provided the necessary warmth for the animal. The relative development of these two coats varied with the climatic conditions under which the animals lived.

Unfortunately, up to the present time, breeding has not banished the coarse fibers entirely, for they are still evident in most breeds of sheep. This is particularly obvious in the Black-faced Highland sheep, one of the present breeds most closely related to the primitive type. However, since the days of mythology, when wool was rightfully called the Golden Fleece, sheep breeders everywhere have recognized that the elimination of black hairs, kemp, and other undesirable features must be accomplished in order to produce a fleece of the highest economic value and utility to humankind.

Domestication and proper cultivation have resulted almost exclusively in the production of true wool, with but little or none of the hairy fiber remaining. Even so, the exact character of the wool on the individual sheep varies considerably, according to its position on the body. For example, in the region of the lower britch, the wool becomes coarse and hairy; and near the feet, a short undergrowth of stiff hair is found. Too many sheep seem to produce a limited number of short, stiff fibers or undergrowth over the entire body area. These short fibers have no value as textile fibers

VIRTUES OF WOOL

Although certain other fibers may equal or even excel in one or several qualities, none can boast of the total qualities possessed by wool. The virtues of wool are as follows:

1. Wool is absorbent. It can absorb as much as 18% of its own weight in moisture without even feeling damp and up to 50% of its weight without becoming saturated. This is an important health factor in clothing because body perspiration and outer dampness are prevented from clinging to the body in heat or cold, thus removing the chill line from the body.
2. Wool generates heat in itself.
3. Wool is a superior insulator, keeping the heat of the body from escaping and the cold air from entering. Because of this quality, wool is as effective a protection from tropical heat and sun as it is against the gale-driven storms of winter.
4. Wool is light.
5. Wool is very resilient; the average fiber will stretch 30% of its normal length and still spring back into shape. Because of this resilience, wool garments resist wrinkling, stretching, or sagging during wear.
6. Wool transmits health-giving ultraviolet rays.
7. Wool takes dye beautifully, without the use of combining chemicals, and it is not likely to fade.
8. Wool is durable.
9. Wool is strong. Diameter for diameter, a wool fiber is stronger than steel.
10. Wool is flame resistant. It will stop burning almost as soon as it is taken away from a flame.
11. Wool is versatile; it weaves and knits into a wide variety of textures and weights from sheer crepes, fine gabardines, and cozy sweater knits to luxurious carpeting.

USES OF WOOL

About 15% of the wool consumed in the United States is carpet wool, and 85% is apparel wool.

The greater part of the carpet wool is used in the manufacture of floor coverings, though small quantities are used in the manufacture of press cloth, knit and felt boots, and heavy, fulled socks.

Most of the apparel wool is consumed in the spinning of woolen and worsted yarn. Some of the woolen and worsted yarn is used in the production of knit goods, including sweaters, hosiery, underwear, and gloves and mittens. The remainder is used in the weaving of fabrics, including such apparel fabrics as suitings, trouserings, dress fabrics, and coatings, and such nonapparel fabrics as blanketing, upholstery, draperies, and woven industrial felts.

MAGNITUDE OF THE U.S. WOOL AND TEXTILE INDUSTRY

Table 9-1 shows that the farmers and ranchers of the United States received from $17.9 to $92.9 million, annually, for their wool clips during the period 1975–1999. This does not include additional pulled wool, nor does it include imported wool.

Table 9-2 gives the wool production, imports and consumption, and percentage produced domestically. Annual production plus annual imports will not exactly equal annual consumption, due to stockpiling and certain other factors. However, among other things, the following noteworthy facts can be deduced from Table 9-2: (1) Virtually all of our carpet wool is imported, (2) about 24% of our apparel wool is produced domestically, and (3) about 70% of our total wool requirement (apparel and carpet) is imported and 30% is produced domestically.

In 1994, annual expenditures for clothing consumed an average of 5.9% of the income of the typical American family, but the average per capita consumption of wool in the United States that same year was only 1.4 lb.

TABLE 9-1
QUANTITY, PRICE PER POUND, AND TOTAL CASH VALUE OF WOOL PRODUCED IN THE UNITED STATES[1, 2]

Year	Shorn Wool Produced	Price per Pound	Cash Value
	(lb)	*(¢)*	*($)*
1975	119,535,000	44.8	53,505,000
1980	105,452,000	88.1	92,862,000
1985	88,055,000	63.3	55,732,000
1990	88,033,000	80.0	69,534,000
1994	68,643,000	78.0	52,419,000
1998	49,255,000	60.0	29,415,000
1999[4]	46,544,000	38.0	17,852,000

[1]*Agricultural Statistics*, USDA, 1981, p. 334, Table 493.

[2]*Agricultural Statistics*, USDA, 1995–96, p. VII-40, Table 426.

[3]Agricultural Statistics, USDA, 2001, VII-35, Table 7-56.

[4]Preliminary.

THE WOOL FIBER

Wool is the natural protective covering of sheep. It differs from other animal fibers by having a serrated surface; a crimpy, wavy appearance; an excellent degree of elasticity; and an internal structure composed of numerous minute cells. In contrast, hair has a comparatively smooth surface, lacks in crimp or waviness, and will not stretch. As a product of the skin or cuticle of vertebrate animals, wool is similar in origin and general chemical composition to the various other skin tissues found in animals—such as horns, nails, and hoofs.

CELL LAYERS

A microscope reveals that all wool fibers, from the standpoint of structure, consist of two distinct cell layers, and some fibers have a third layer. According to their positions, these layers are known as (1) the epi-

TABLE 9-2
U.S. PRODUCTION, IMPORTS, AND CONSUMPTION OF WOOL, AND PERCENTAGE PRODUCED DOMESTICALLY, CLEAN BASIS[1]

Year	Production[2]			Imports[3]			Consumption[4]			Domestic Production as Percentage of Consumption			Imports as Percentage of Consumption		
	Apparel	Carpet	Total	Apparel	Carpet	Total	Apparel	Carpet	Total	Apparel	Carpet	Total	Apparel	Carpet	Total
	- - - - - (million lb) - - - - -			- - - - - (million lb) - - - - -			- - - - - (million lb) - - - - -			- - - - - - - (%) - - - - - - -			- - - - - - - (%) - - - - - - -		
1975	67.5	—	67.5	16.6	17.0	33.6	94.1	15.9	110.0	71.7	—	61.4	17.6	104.4	30.5
1980	56.4	—	56.4	30.5	26.0	56.5	113.4	10.0	123.4	49.7	—	46.0	26.9	285.7	46.2
1985	46.5	—	46.5	50.2	29.3	79.5	106.1	10.6	116.7	40.0	—	40.0	47.3	276.4	68.1
1990	46.5	—	46.5	50.3	21.4	71.7	120.6	12.1	132.7	35.0	—	35.0	41.7	176.9	54.0
1994/5	36.2	—	36.2	64.9	26.8	91.7	129.3	12.7	152.0	24.0	—	24.0	50.0	211.0	60.0
1999	46.6	—	46.6	21.3	21.8	42.0	63.5	13.9	77.5	73.3	—	60.1	33.5	156.8	54.1

[1]*Agricultural Statistics*, USDA, 1970, p. 340, Table 497, and p. 341, Table 498; and 1981, p. 334, Table 493, p. 338, Table 498; and 1995–96, p. VII-40; and 2001, p. VII-36.

[2]Total wool production = shorn wool + pulled wool. Reported on basis of clean fiber, using conversion factor of 52.8%, and 72.9% for pulled wool production.

[3]Apparel wool includes dutiable wool. Carpet wool includes all duty-free wool.

[4]Consumption on the woolen and worsted systems.

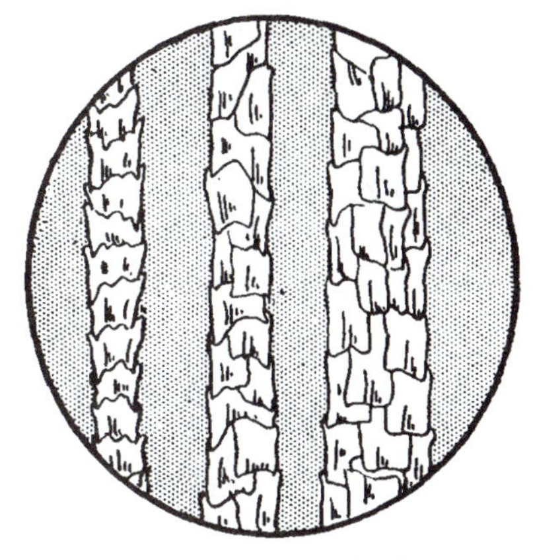

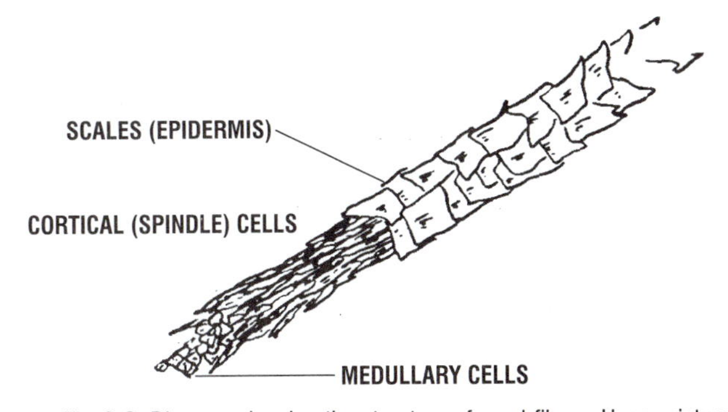

Fig. 9-2. Diagram showing the structure of wool fibers. Upper picture shows the epidermis—the outside cell layer—of fine, medium, and coarse fibers. Note that the fine wool has a greater number of serrations. Bottom picture shows the three cell layer: (1) the scales, or epidermis, (2) the cortex, and (3) the medulla. (Drawing by R. F. Johnson)

dermis or outer layer, (2) the cortex, and (3) the medulla. Although differing in characteristics, these same three cell layers are found in almost all hair. The chief characteristics of these three layers in wool fibers will be discussed briefly.

EPIDERMIS

The epidermis, which is the outside or surface layer of the fiber, is made up of flat, irregular horny cells or scales. These scales overlap, one on top of the other, much like the scales of a fish, with the free end projecting outward and upward toward the tip of the fiber in such fashion as to present a serrated appearance. Fine wool has many more serrations per inch than coarse wool. They run from 600 per inch in the low grades of wool up to 3,000 per inch in the finest of Merino wools. The epidermal cells impart spinning and felting qualities to wool.

CORTEX

This cell layer, which constitutes the principal body of the wool fiber, lies immediately below the epidermis. It is made up of long, flattened, more or less twisted, spindle-like cells. The cortical cells impart strength and elasticity to the fiber.

MEDULLA

Although usually absent in fine wools, most medium and coarse wools possess a third cell layer known as the medulla. Where present, this is the innermost layer. The cells are of various shapes, often polygonal like a honeycomb; and they may occur throughout the length of the fiber or only in certain areas. Wool fibers containing medulla are not desirable, since they lack the working properties of nonmedullated fibers. Such fibers are generally coarse, uneven in diameter, and harsh to the touch.

CHEMICAL COMPOSITION

Chemically, wool is chiefly the protein keratin, which is also the primary constituent of hair, nails, hoofs, horns, and feathers. Keratin is composed of amino acids. It is noted for its high content of the sulfur-containing amino acid cystine. A typical chemical breakdown of wool is as follows: carbon, 50%; oxygen, 22 to 25%; nitrogen, 16 to 17%; hydrogen, 7%; and sulfur, 3 to 4%.

SOME FLEECE CHARACTERISTICS

The chief fleece characteristics of interest and importance to both the producer and the manufacturer are: (1) grease, (2) length, (3) density, (4) diameter, and (5) variations of different body areas. Wide differences between breeds and individuals are recognized in such wool characters as fineness, length, density, and yield.

GREASE

In a broad sense, grease refers to all the impurities found in unscoured wool, including the yolk, suint and soluble foreign matter but not the vegetable matter. Shrinkage of fleeces varies widely, with many factors affecting it, but on the average and with all grades included, U.S. shorn grease wool shrinks about 47.2%.

The commercial value of a clip of wool is largely determined by the amount of clean wool fiber that it yields. Although a part of the impurities found in grease wool are essential for the growth and well-being of the fleece and the animal, the manufacturer is primarily concerned with securing the highest possible yield of clean wool of the finest quality. In the raw state, grease wool contains the following types of impurities:

1. **Natural impurities,** which result from the glandular secretions. This includes the yolk and the suint. The yolk, which is a mixture of a number of materials, of which the principal one is cholesterol, appears to protect the fiber against the detrimental action of the weather. When scoured out and purified, as lanolin, the yolk is used in making ointments, cosmetics, leather dressings, rope, and rust preventives. In cosmetics, in which form it retails at a very high price, few people are aware that it was derived from a sheep's back. Suint is mixed with the yolk, but it can be readily dissolved out by water. It consists primarily of potassium salts of various fatty acids and smaller amounts of sulfates, phosphates, and nitrogenous materials. Suint, which results from sweating, seems to be the source of the distinctive odor associated with sheep.
2. **Acquired impurities,** which are picked up by the animal. They include such mineral impurities as dust, sand, and dirt; vegetable materials consisting of straw, burs, twigs, and grasses; and dung.
3. **Applied impurities,** which include animal identifying substances and the residues of dips and sprays.

LENGTH

The length of the wool in a fleece is a matter of much importance to both the producer and the manufacturer. Together with quality, it constitutes the principal basis of classification and grading in buying and selling and largely determines the use to which wool will be put. The wool producer regards good length as a desirable attribute, for it gives a greater weight of wool. In judging sheep, fiber length is based on an appraisal of the annual growth, as determined by parting the fleece at three body areas—the shoulder, side, and britch. Fiber length varies anywhere from 1 to 20 in.

DENSITY

Density refers to the closeness or compactness of the fibers in a fleece and is often defined as the number of fibers per unit area of skin. Experimental studies have revealed very clearly that density differences exist between breeds, individuals, and body areas. Estimates for the number of wool fibers per animal vary from a low of about 16,000,000 for some individuals of the medium- or coarse-wool breeds to a high of 120,000,000 fibers for individuals of the Australian Merino breed. Fleece density is an attribute in determining fleece weight. By grasping the wool on the side to feel its fullness and compactness and by parting the fleece to examine the apparent closeness of the fibers, one can determine fleece density in judging. Experimental evidence shows that often this method is misleading, apparently being affected by the grease and dirt content.

DIAMETER

The fineness of wool is very important because the character of the yarns and fabrics produced is determined to a very great extent by the variations in the diameter of fiber. Wool sorting is based on fineness of fiber, and this is considered to be the soundest basis on which wool and top qualities can be classified. In the trade, the wool expert is able to estimate the fineness by visual inspection and handling. However, an ordinary sample of wool that is estimated by a wool expert as representing a certain fineness will, on close examination, usually show two or three other finenesses. That is, it is really a composite mixture. If it is placed under a microscope, the shape or contour of fibers varies greatly. As a rule, fibers are irregular and possess varying degrees of ovality or ellipticity.

In judging sheep, the number of crimps is usually accepted as an index of fineness; that this is a good indication is borne out by experiments in which more refined techniques have been used. However, average fiber diameter is most accurately measured in microns (a micron is 1/25,400 of an inch) from core samples, using one of the following four methods: (1) microprojection, (2) airflow, (3) laser (laserscan), or (4) image analysis.

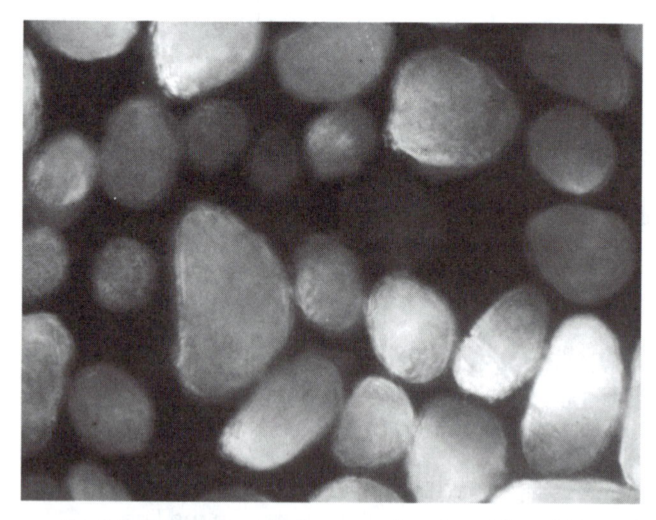

Fig. 9-3. Photomicrograph of cross-section of wool fibers. This small sample is rather typical of the variation in both fineness and shape or contour of fibers. (Photomicrograph by M. E. Ensminger)

VARIATIONS ON DIFFERENT BODY AREAS

There are wide differences between breeds and individuals in such wool characters as fineness, length, density, and yield. In extensive studies that the author conducted at the University of Massachusetts, using the Shropshire and Southdown breeds, these studies indicate that there are wide differences in wool characteristics between different body areas. On the basis of these investigations, it appears possible to reduce the wool character distribution of body areas to a common plan, with rankings from the most desirable to the least desirable character.

Fig. 9-4 illustrates a ranking of wool (1) fineness, (2) length, (3) density, and (4) clean wool yield on different areas of the sheep's body.

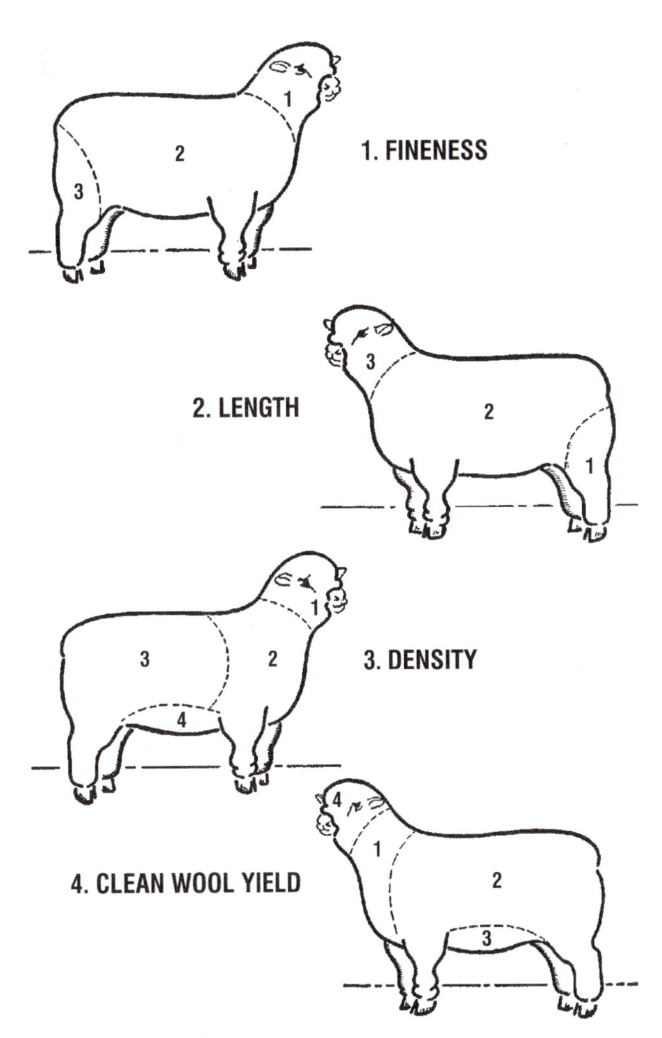

Fig. 9-4. Diagrams showing variations of different body areas in wool fineness, length, density, and clean wool yields; based on results obtained by the author in studies with animals of the Shropshire and Southdown breeds (unpublished data, University of Massachusetts). On the basis of these investigations, it appears possible to reduce the wool character distribution of body areas to a common pattern, with rankings of each character from most desirable to least desirable. Thus, as shown, the head wool is the finest, whereas the britch wool is the coarsest. The britch wool is the longest; whereas the head wool is the shortest. From the standpoint of density, the head wool ranks at the top of all body areas, but from the standpoint of yield, it is the lowest. (Drawing by R. F. Johnson)

BLACK AND MULTIPLE COLORED WOOLS

Black and multiple colored wools may be sold at a premium for hand crafting. But wool containing kemp should be discounted because it won't take a dye.

PRODUCTION AND HANDLING OF THE WOOL CLIP

If U.S. sheep producers are to survive the inroads of imports and synthetic fibers, it is imperative that they market a higher-quality product—one that does not require unnecessary processing expenditures in the textile mills.

Despite remarkable improvement in wool wrought through improved breeding, handlers, buyers, and processors of the domestic wool clip are in general agreement that the overall quality of the nation's wool clip has declined in recent years, and that the primary reason for this decline is the generally poor preparation of the clip. The most common explanations or excuses are: (1) carelessness and indifference on the part of sheep producers, and (2) lack of skilled and dedicated shearers. Whatever the cause, all are agreed that a change must be made if the domestic wool clip is to meet its increasing competition from imported wool and fabrics and from manufactured fibers.

Observance of the following wool production and handling practices will result in marketing a higher-quality product:

1. **Production.** Augment improved breeding by (a) proper nourishment and thrift of sheep throughout the year; (b) protection of on-the-back fleeces from straw, trash, and burs; and (c) removal of tags from time to time—none should be present at shearing time.
2. **Branding and identifying.** Avoid using any branding fluid if possible. Where branding or identifying is necessary for practical reasons: (a) use proven scourable products, follow the directions on the label, and do not dilute or mix with contaminating substitutes; and (b) consider using colored plastic ear tags.
3. **Management at shearing time.** Shearing is the harvesting of one of the most valuable products on a per pound basis produced on U.S. farms and ranches.

Fig. 9-5. Tender wool. Such weak spots in the fiber are generally caused by poor feeding, diseases, and/or parasites. (Courtesy, University of California-Davis)

Thus, its harvesting merits preparation commensurate with its value, including observance of the following management pointers at shearing time: (a) Remove all weeds, hay, and other vegetative wastes from holding

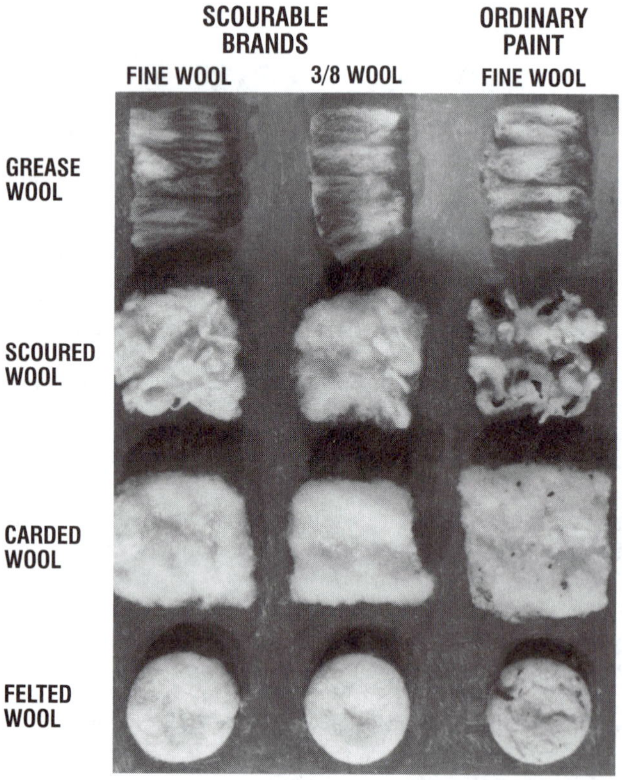

SCOURABLE BRANDS

ORDINARY PAINT

FINE WOOL 3/8 WOOL FINE WOOL

GREASE WOOL

SCOURED WOOL

CARDED WOOL

FELTED WOOL

Fig. 9-6. Wool previously marked with (1) scourable branding fluids vs. (2) insoluble paint *(right)* in various stages of manufacture. (Courtesy, USDA)

and shearing pens; (b) sprinkle shearing pens and corrals prior to bringing sheep into them, thereby alleviating dust; (c) shear only when the wool is perfectly dry; (d) keep lambs out of the shearing pens; (e) separate black sheep and shear them last, pack these fleeces separately, and identify the wool bag accordingly; and (f) shear rams separately, pack these fleeces separately, and identify the wool bag accordingly.

4. **Shearing.** In removing the natural clothing of sheep, by either machine or hand, observe the following points: (a) Shear in a clean place, and keep the shearing floor clean and dry at all times—keep brooms conveniently available, sweep the floor at intervals, place tags, dung locks, stained pieces, and other floor sweepings in a separate bag; (b) handle sheep carefully and without injury; (c) avoid second cuts—it's better to leave extra length on the sheep's back than to

Fig. 9-7. Shearing platform inside a trailer. (Courtesy, University of Wyoming)

have second cuts in the bag; (d) keep the fleece in one piece; (e) keep fleece free from straw, manure, and other extraneous matter; (f) avoid the use of burlap in shearing operations for any purpose other than packing; and (g) shear sheep with growths of 6 months, 8 months, and 12 months, and lambs separately, pack separately, and identify bags accordingly.

5. **Shearers.** Those who do the shearing should be skilled, well informed, and careful. The following points are pertinent to shearers and shearing: (a) Have a trained and well-informed person present to supervise the shearing operation; (b) use clean equipment, and provide clean oil for shearing heads; and (c) separate tags, dung locks, crutchings, face and leg wool, and stained pieces from the fleece while shearing; pack in separate bags; and label accordingly.

6. **Rolling and tying.** A well-rolled and properly

Fig. 9-8. A properly rolled and tied fleece. Note only one string of paper twine has been used. (Courtesy, USDA)

tied fleece creates a good impression with a buyer. To roll and tie correctly: (a) Put the fleece flesh side (cut side) down on the floor, fold in each of the sides to meet the center, fold in the neck about as far as the shoulder, and start at the britch end and roll toward the neck—in this manner, the finest and best wool on the entire fleece will be on the outside of the bundle when it is finished; (b) use paper twine, with the string precut to appropriate length, to tie the fleece—never use cotton, jute, sisal string, or wire; and (c) do not tie 6-month, 8-month, and lamb wool.

7. **Packing.** Proper packaging is paramount in modern marketing—in all products, and wool is no exception. To pack wool properly: (a) Use new, regulation wool bags, preferably paper lined; if a second-hand wool bag is used, make sure that it is clean, both inside and outside; avoid used grain and feed bags. (b) Pack the different kinds of wool—ewe, ram, yearling, burry and seedy, and black, also tags, crutchings, and floor sweepings—separately, and mark each bag so as to identify its contents. (c) Pack separately and identify 6-month, 8-month, 12-month, and lamb wool. (d) Pack bags firmly to facilitate handling and to reduce variations in core samples and test results. (e) Mark bags or bales with approved marking materials—never use paint. (f) Keep bags off the ground and store under shelter.

Fig. 9-9. Rejections—fleeces that are black, dead, gray, cotty, etc.—should be packed separately. (Courtesy, USDA)

8. **Shipping.** When transporting wool: (a) Clean trucks or freight cars to eliminate contamination from foreign substances and materials, and (b) keep bags dry—protect them from the elements.

Fig. 9-10. Bags of wool in a wool warehouse in Ozona, Texas. (Courtesy, Texas A&M University, College Station)

NEW SHEARING INNOVATIONS

High shearing costs and the scarcity of sheep shearers have spurred interest in finding an easier and less costly way of removing the fleece from sheep. Three experimental approaches appear promising: (1) chemical shearing, (2) laser beam shearing, and (3) computerized shears.

■ **Laser beam shearing**—A group of Australians, headed by a former sheep shearer, has patented a laser beam for shearing. These ingenious inventors reasoned that lasers, which had already been used to cut woolen cloth and steel, could be adapted to shearing sheep. The laser actually severs the wool by burning. However, the Australian developers feel that it can be governed so that it will selectively cut only wool—that it can be designed so that is will automatically switch off when the beam strikes tissue or any other material differing in density from wool.

Laser beam shearing is still experimental in Australia.

■ **Computerized shears**—This is a mechanical hand, guided by a computer (Fig. 9-11). The computer first creates a contour drawing—a computer map—of

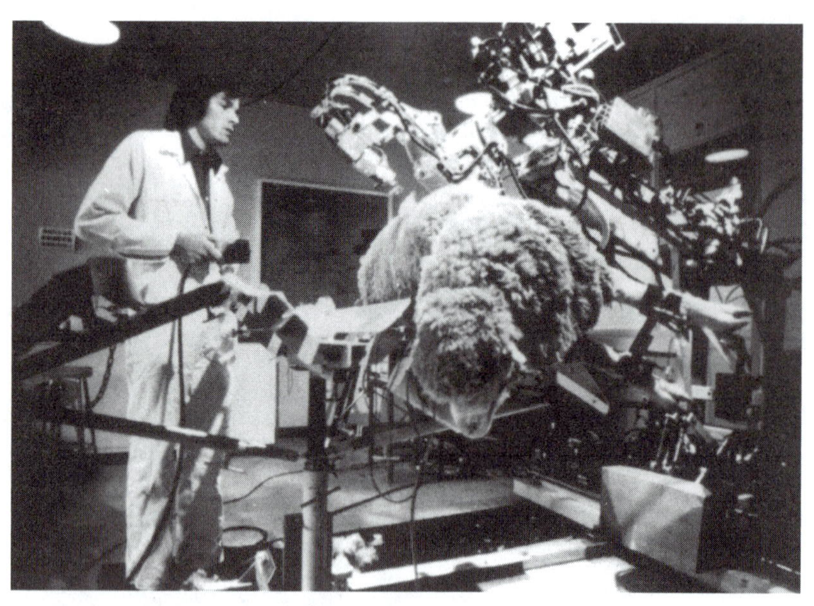

Fig. 9-11. Shearing sheep with a mechanical hand guided by a computer. The computer creates a contour drawing of the surface of the sheep and uses this to guide the shears over the animal's body. As technology improves, it may shear faster than a person. (Photo by Gary Wolinsky, through Stock, Boston, Inc., Boston, MA)

the sheep's body. This map is used by the computer to guide the mechanical hand, which holds the shears, over the animal's body. To avoid nicking, the shearer head retracts in 5/1,000 of a second when the slightest movement is detected. An experimental version at the University of Western Australia shears a sheep in three minutes, about the same as a human. Development of the computerized sheep shear was financed by the Australian Wool Corporation at a cost of $1 million.

Computerized shearing is still experimental in Australia.

REQUISITES OF WOOL

Good wool possesses character, purity, strength of fiber, good condition, cleanliness, low shrinkage, adequate but not excessive grease, uniformity, and a bright white color. These requisites apply to all wools, regardless of the class or grade.

CHARACTER

Character in wool is that inherent characteristic that makes one fleece stand out as being superior to another of the same class or grade. Character is judged largely by the crimp, which is the waviness in wool, although other factors may be considered. Fine wool of good character may have 25 to 30 crimps per inch, while a similar wool, lacking in character, would have only about 15 or fewer crimps per inch. Coarse wool may have only five or fewer crimps per inch—even if it is of good character—whereas poor wool of this type would be practically straight. Color, feel, and general makeup of the fleece also enter into character of wool. Fleeces lacking in character are usually run-out or frowsy in appearance.

PURITY

Purity in wool is a very important factor. A fleece is pure if it contains only true white fibers. Some breeds of sheep produce black or gray fibers, especially around the neck or the shins. It is impossible mechanically to separate the black fibers from the white fibers in manufacturing; hence, wool of this type cannot be used for white fabrics. Wool containing even a few black fibers is rejected from white lots, and it must be dyed a dark color.

Kemp is another type of fiber that affects the purity of wool. it is malformed, chalk white, and very brittle. It is very wasteful in manufacturing and shows up as a defect in the fabric because it does not take dyestuff. Wool with kemp must be removed in the sorting operation. Some wools are off-color in that they have a yellowish cast. They likewise are undesirable, because they do not scour out white and cannot be used for white yarns or fabrics.

STRENGTH OF FIBER

Strength of wool varies greatly from fine wool to coarse wool. Good wool has approximately the same strength throughout its length. Defective wool in this respect has one or more weak places along its length. If the weak spot is pronounced, the fiber will not stand the strain of manufacturing on the worsted system. Wool of the latter type is not worth as much and can only be used to make woolens. Some of the common causes of weak or tender wool are: illness of the sheep, radical change in feed, nutritional deficiencies, or general emotional disturbances of the sheep caused by stress.

CONDITION

Condition, when used in connection with grease wool, refers to the amount of grease, sand, dirt, and other solubles present in the wool. A wool that has a high shrinkage is said to be heavily conditioned. Wool is bought on the basis of how much clean wool it actually contains. Hence, the amount, or percentage of condition, is an important factor when determining the grease wool value.

CLEANLINESS

Grease wool is valued according to the amount of pure wool fiber it contains. Everything else that is present in grease wool must be removed in manufacturing. Sand, dirt, loam, grease, burs, chaff, grass, and seeds must all be removed. Some impurities—such as sand, dirt, and some vegetable matter—may be removed by dusting. Bur-picking removes burs. Other impurities—including suint, grease, dirt, etc.—may be removed by wool scouring. Still other impurities, as certain types of vegetable matter, may require further mechanical or chemical treatment. Grease wool requiring a minimum of effort to remove impurities brings the highest price when offered for sale. Most farmers and ranchers can do much to lessen the amount of straw, chaff, grain, and other vegetable matter found in wool. Some farmers pile their wool in the barn and allow it to collect dirt and dust for many months before they sell it. Wool that is bagged or baled at shearing time is automatically kept clean until it arrives at the market.

SHRINKAGE

Shrinkage is the weight of impurities lost in the wool scouring. The result is usually expressed as a percentage of the grease weight. An example of fleece shrinkage follows:

10.0 lb grease wool before scouring
5.3 lb clean wool after scouring
4.7 lb loss in wool scouring
4.7 lb = 47% = shrinkage
5.3 lb = 53% = yield

Yield and shrinkage percentages when added together give 100%.

Table 9-3 lists and compares the shrinkage and yield averages by breed and wool grades.

A wool buyer must possess the ability to examine grease wool and estimate its shrinkage accurately. No two fleeces of wool are exactly alike. Many factors affect shrinkage, including breed, climate, soil, feed, etc. Shrinkage consists of: (1) the natural impurities, yolk and suint; (2) acquired impurities, such as tags, sand, dirt, etc.; and (3) added impurities, such as paints and brands. Wool buyers usually estimate the amount of vegetable matter in wool and include this factor when evaluating it.

TABLE 9-3
PERCENTAGE OF WOOL SHRINKAGE AND YIELD FOR SOME BREEDS OF SHEEP AND THEIR WOOL GRADES[1]

Breed	Numerical Count Grade	Shrinkage	Yield
		(%)	(%)
Cheviot	48s–56s	25–50	50–75
Columbia	50s–60s	45–55	45–55
Corriedale	50s–60s	40–50	50–60
Debouillet	62s–80s	50–65	35–50
Delaine-Merino	64s–80s	50–65	35–50
Dorset	46s–58s	25–50	50–75
Hampshire	46s–58s	38–50	50–62
Leicester	36s–48s	20–35	65–80
Lincoln	36s–46s	20–35	65–80
Montadale	56s–58s	40–55	45–60
Oxford	46s–50s	38–50	50–62
Rambouillet	60s–80s	45–65	35–55
Romney	40s–48s	20–35	65–80
Shropshire	48s–60s	25–50	50–75
Southdown	54s–60s	45–60	40–55
Suffolk	48s–58s	38–50	50–62
Targhee	58s–64s	45–50	50–55

[1]Data from American Wool Council, a division of the American Sheep Producers Council, Inc., Wool Education Center, Denver, CO.

OIL OR GREASE IN THE FLEECE

Wool is bathed with a greasy substance during the growth of the fiber. This natural grease protects the fibers during growth and prevents adjacent fibers from becoming cotted or felted on the sheep's back. Wool grease in its natural or crude form is called *degras*, which is really a wax. It is refined and marketed as a salve and is sold under the name of *lanolin*. Wool grease is not soluble in warm or cold water. It is generally removed from wool by emulsifying it with soap, water, and alkali. In some instances, it is removed from wool by naphtha, in which it is soluble. A normal amount of wool grease, evenly distributed throughout the length of the fiber, is an asset. Too much wool grease, as is the case in some heavy-conditioned buck fleeces, is undesirable because of the high shrinkage. In addition to the grease, wool also possesses suint. The grease, suint, and other impurities are commonly referred to as the yolk.

EVENNESS OF FIBER AND UNIFORMITY

It is desirable to have wool as even and regular as possible. The grade of a fleece is a relative or comparative term, since wool of many kinds is found on one sheep. The finest wool is found on the head and shoulders, and the coarsest wool is found on the britch. Even so, a fleece that is more even and regular than another and has fewer grades is the better one. Wool sorting separates an individual fleece into its various qualities. An even fleece is easier and quicker to sort than an uneven fleece. Uniformity is the result of good breeding and good management.

COLOR

Color is an important consideration in evaluating wool for manufacturing purposes. The whiter the wool, the greater its value. Very little, if any, farm wool is true white. It usually has a yellowish tinge. The presence of pronounced yellow wool (canary-stained) is objectionable because the color does not come out in the scouring. Hence, wool of this type cannot be used for white yarn.

Black wool is in reality from dark brown to light gray in color. The very dark fleeces are easily distinguished, but it is difficult, except for a trained person, to spot a fleece that has only a few black fibers mixed intimately among the white fibers. A fleece of the latter type is discounted and sells at the same price as black wool with which it is manufactured. Some fleeces have patches of black or gray at the neck or around the edges of the shins. It is advisable for owners to examine their sheep carefully for this defect and to consider it when they are culling their flocks.

Urine- or manure-stained wool is another type of off-colored wool. Wool with this defect is usually separated out by the sorter. However, it is difficult to see all the stained pieces in grease wool, and, as a result, many mills also look over the scoured wools to remove any bits of stained wool that were overlooked by the sorter.

CLASSES OF WOOL

The wool trade recognizes two major classes of wool: *apparel wool* and *carpet wool*. As the names imply, most apparel wools are those suitable for manufacture into yarns and fabrics for human clothing, whereas most carpet wools are used in making floor covering.

Apparel wools are further classified according to use as (1) combing wool or staple wool—the long-fibered wools within the class; (2) French combing wool—the intermediate-length wools, and (3) clothing wool—the short-fibered wools. Although these three classes are based largely on length of fiber, other factors—such as supply and demand, fiber diameter, purity, condition, etc.—are important in determining the use made of wools. Thus, many wools used by the woolen industry are longer than some used in worsted manufacture; and a considerable amount of wool classed as clothing is used in the worsted industry. In general however, the manufacturer can realize the greatest profit by utilizing apparel wools according to their best adaptation as indicated by the three classes. Further carpet wool is not suited for use as apparel wool.

COMBING OR STAPLE WOOL

Combing or staple wools are usually referred to as the highest priced and best wool obtained from sheep. Both fineness and length are requisites. For example, a 64s combing wool should be 2.75 in. or more in length, with the length varying according to grade as shown in Table 9-4. By and large, combing wools are used for making worsted fabrics. They take their name from one of the main processes in worsted manufacturing, the combing operation, which separates the long fibers from the short ones. The long fibers are used to make worsted cloths, and the short fibers (called *noil*) are used to make woolen cloths. In the former, the fibers are laid parallel to each other; whereas in the latter, the shorter wool fibers that are used in making woolen cloths and felts are laid in every direction—in fact, the

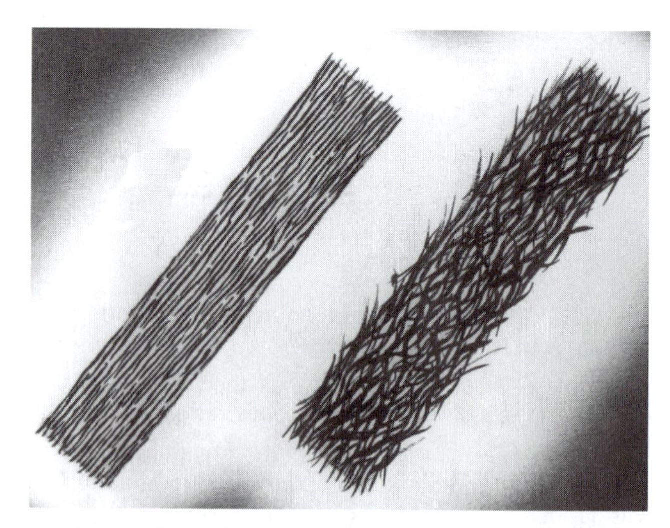

Fig. 9-12. Enlarged diagram of worsted *(left)* and woolen *(right)* yarns. The longer fibers are combed out parallel for worsteds, which explains why worsted suits hold their press better than woolen suits. (Courtesy, The Wool Bureau, Inc., New York, NY)

more mixing the better in woolens. These differences are of importance to the consumer.

In the United States, wools with sufficiently long fibers and otherwise adapted to the making of worsted cloth are commonly combed on the Noble or Bradford comb.

Prior to World War II, approximately 70% of the apparel wool used in the United States was processed on the worsted system of manufacture. However, since that time there has been a gradual decline in the use of worsteds and an increase of woolens. In 1994, about 55% of the mill consumption of wool was processed on the worsted system and 45% on the woolen system (see Table 9-7).

FRENCH COMBING WOOL

French combing wools are in between the combing wools and the clothing wools in length. These wools are manufactured on the French or Heilman comb, which is designed to use wools and still produce worsted fabrics. Thus, the French system utilizes much wool that is not long enough for manufacture on the regular worsted system known as Noble combing. This system of combing is becoming more popular.

CLOTHING WOOL

Clothing wool is the name usually given to the shortest wool. This wool is too short to be manufactured on the worsted system, but it can be used successfully on the woolen system. Although longer fibers can be used in making woolens, they are usually more expensive than short fibers and hence are reserved for making worsteds which usually sell at a slightly higher price than woolens. The term *clothing wool*, however, does not mean that the wool is suitable only for fabrics to be made into clothing. This type wool is also used to make felts.

CARPET WOOL

Carpet wools, which are usually the coarsest wools, are of low quality because they (1) contain mixtures of very coarse, hairy fibers and finer fibers, and

Fig. 9-13. A mixed band of sheep and goats owned by the Navajo Indians in northeastern Arizona. These sheep produce carpet wool. (Courtesy, Wittick Collection, Santa Fe, NM)

(2) vary markedly in fiber length. The chief requisite of carpet wool is resilience, the quality that makes it resistant to matting down and to wear under the constant scuffing of passing feet. Most of this wool comes from long-wooled sheep. Except for a small amount of carpet wool produced by the sheep kept by the Navajo Indians, most U.S. carpet wools are imported, chiefly from New Zealand.

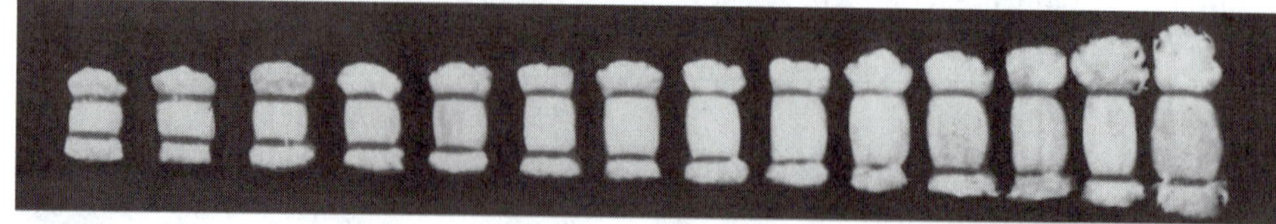

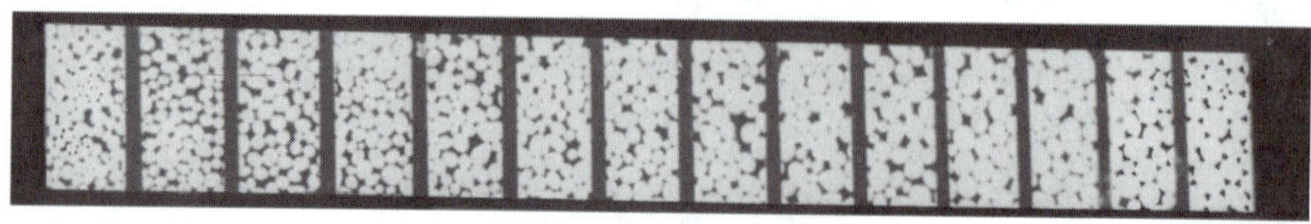

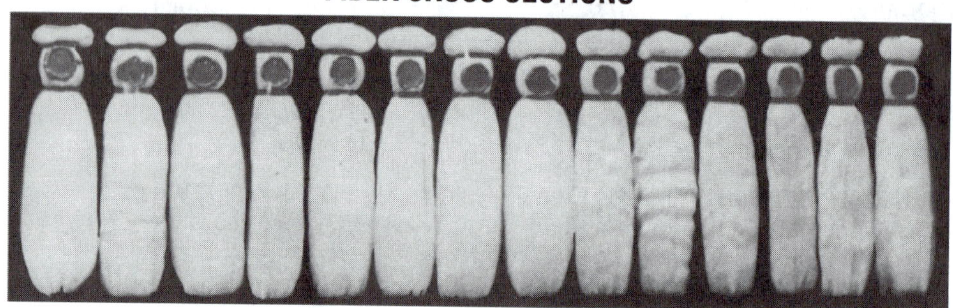

Fig. 9-15. U.S. standard grades for wool. (Courtesy, American Sheep Producers Council, Inc., Wool Education Center, Denver, CO)

WOOL GRADING

Wool grading is based primarily on fiber diameter or fineness, but consideration is also given to length. Many manufacturers desire wool of certain finenesses only. This means that the wool must be separated at the warehouse and like fleeces must be piled by themselves. This process is called *wool grading*, and it is done by a highly trained person. Grading does not infer that the wool in a pile is all of one diameter. Any single fleece of wool as it comes from the sheep may possess several different grades of wool. Thus, a 60-62s combing wool simply means that the greater part of the wool in the fleece is of that fineness and length. The manufacturer knows that some wool in these fleeces, especially on the shoulder part of the fleece, will be finer, and also that some wool, as on the britch, will be considerably coarser. Because of this, a further separation, known as *sorting*, follows. The ability to grade wool, which is acquired only with considerable experience, requires a keen sense of sight combined with the sense of touch and rare good judgment.

Fig. 9-14. Wool grading. Fleece grading consists of the grader examining the untied fleece and, on the basis of its fineness, placing it into one of the boxes or crates containing wool of comparable grade. (Courtesy, USDA)

GRADES OF WOOL

There are many factors which enter into the value of grease wool, but among the most important are diameter, length, and clean wool fiber present.

Fiber diameter is important because—

1. It contributes to fiber quality in general terms.
2. It determines use or application of wool.
3. It determines minimum lengths for combing.

4. It describes wool accurately.
5. It affects processing performance.
6. It establishes a price range for wool.

The average diameter of fiber and the limits for the variation in diameter for the various grades are shown in table 9-4. Maximum limits to the variation allowed for each grade are expressed by the statistical term—*standard deviation*. In application, if there is too much variation in fiber diameter, the wool is assigned to the next coarser grade. There are three methods of grading wool according to diameter, with several grades in each. The older method is the *blood system*; the newer methods are the *spinning count system* and the *micron system*. A comparison of these systems is contained in Table 9-4.

An experienced wool grader determines the grade of wool by the senses of sight and touch. However, for use in more objective grade determination and for arbitration purposes where there may be a dispute as to grade before final settlement, there is a scientific method of test prescribed. A copy of this method of test, which explains micro-projector equipment recommended and also sampling and testing procedures, may be obtained from the USDA, Standardization Branch, Wool Laboratory, Denver, Colorado. Testers and research workers may also use micrometers or photographic or air-flow equipment for grade determination.

The grades of wool produced vary widely between areas. Thus, about 75% of the wool produced in Texas, New Mexico, Arizona, and Nevada grades fine and one-half blood; whereas, three-eighths blood wool and one-fourth blood wool predominate in the North Atlantic, East North Central, West North Central, and South Atlantic states.

TABLE 9-4
COMPARATIVE WOOL GRADES AND CLASSES

		Standard Specifications[1]			Length Classes[2]					
			Micron System[3]							
Type of Wool	Old Blood Grade	Spinning Count Grade	Limit for Average Fiber Diameter	Variability Limit[4]	Combing Wool		French Combing Wool		Clothing Wool	
					over				under	
			(microns)[5]	(microns)[5]	(in.)	(cm)	(in.)	(cm)	(in.)	(cm)
Fine	Fine	Finer than 80s	Under 17.70	3.59	—	—	—	—	—	—
Fine	Fine	80s	17.70–19.14	4.09	2.75	*6.99*	1.25–2.75	*3.18–6.99*	1.25	*3.18*
Fine	Fine	70s	19.15–20.59	4.59	2.75	*6.99*	1.25–2.75	*3.18–6.99*	1.25	*3.18*
Fine	Fine	64s	20.60–22.04	5.19	2.75	*6.99*	1.25–2.75	*3.18–6.99*	1.25	*3.18*
Medium	1/2 blood	62s	22.05–23.49	5.89	3.0	*7.62*	1.5–3.0	*3.81–7.62*	1.5	*3.81*
Medium	1/2 blood	60s	23.50–24.94	6.49	3.0	*7.62*	1.5–3.0	*3.81–7.62*	1.5	*3.81*
Medium	3/8 blood	58s	24.95–26.39	7.09	3.25	*8.26*	2.0–3.25	*5.08–8.26*	2.0	*5.08*
Medium	3/8 blood	56s	26.40–27.84	7.59	3.25	*8.26*	2.0–3.25	*5.08–8.26*	2.0	*5.08*
Medium	1/4 blood	54s	27.85–29.29	8.19	3.5	*8.89*	2.5–3.5	*6.35–8.89*	2.5	*6.35*
Medium	1/4 blood	50s	29.30–30.99	8.69	3.5	*8.89*	2.5–3.5	*6.35–8.89*	2.5	*6.35*
Coarse	Low 1/4	48s	31.00–32.69	9.09	4.0	*10.16*	—	—	4.0	*10.16*
Coarse	Low 1/4	46s	32.70–34.39	9.59	4.0	*10.16*	—	—	4.0	*10.16*
Coarse	Common[6]	44s	34.40–36.19	10.09	5.0	*12.70*	—	—	5.0	*12.70*
Very coarse	Braid	40s	36.20–38.09	10.69	5.0	*12.70*	—	—	5.0	*12.70*
Very coarse	Braid	36s	38.10–40.20	11.19	5.0	*12.70*	—	—	—	—
Very coarse	Braid[6]	Coarser than 36s	Over 40.20	—	—	—	—	—	—	—

[1]Standards for grades of wool, as published by the USDA, August 20, 1965, *Federal Register* (7 CFR Part 31). These standards became effective January 1, 1966.

[2]There are no USDA official lengths for the different classes. The lengths given herein are in keeping with trade practices and were provided for use in this book by the Livestock Division Wool Laboratory, Standardization Branch, USDA, Denver, CO 80225.

[3]Beginning January, 1976, the unit designation terminology for wool prices changed to microns.

[4]Standard deviation maximum.

[5]A micron is 1/25,400 of an inch.

[6]Common and braid are not classified according to length because these wools are practically always of combing length. Carpet wool includes all those not suited to the three classes listed.

BLOOD SYSTEM

The blood system is the oldest system; it originated at the time of the early American colonies. It is based on the breeding (bloodlines) of the sheep. Thus, Merino and Rambouillet wool is called fine. Hence, one-half blood wool comes from sheep that are half Merino or Rambouillet and half another breed. However, the blood system cannot describe newer breeds or differences in wool from the same sheep. Nevertheless, it has persisted, but it is gradually being replaced by the spinning count system.

The blood system divides all wool, from finest to coarsest, into six market grades. These are: (1) fine, (2) one-half blood, (3) three-eighths blood, (4) one-fourth blood, (5) low one-fourth blood, and (6) common and braid. At the present time, these names indicate wool of a certain diameter only and have no connection whatsoever with the amount of Merino or Rambouillet blood in the sheep. As a matter of fact, it is possible to have three-eighths blood wool from a sheep with no Merino blood at all. The blood grades, therefore, merely identify the different grades of wool, without relationship to the breeding of the sheep, and are rapidly being replaced with the spinning count system, and the micron system.

SPINNING COUNT SYSTEM

The spinning count system divides all wool into 14 grades, and each grade is designated by a number. The numbers range from 80s for the finest wool down to 36s for the coarsest. This method gives more grades; thus, finer divisions can be made, which is more satisfactory to the wool dealers and manufacturers. Table 9-4 shows the correlation between the grade systems.

■ **Worsted spinning count**—Theoretically, the spinning count system is based on the number of hanks of yarn (each hank representing 560 yards) that can be spun from 1 lb of such wool in the form of top. Wool of 50s quality, therefore, should spin 50 × 560 yards per pound of top, if spun to the maximum on the worsted system of manufacture. Unfortunately, this is not always true; the lower grades will not spin up to their number. Moreover, in actual practice, wools are rarely spun to their maximum limit. Furthermore, spinning count is not determined by diameter alone; such factors as fiber length, moisture conditions, and the skill of the workers influence the count that may be spun. Therefore, neither the blood system nor the spinning count system denotes accurately what it is supposed to indicate according to derivation of the respective terms.

MICRON SYSTEM

The micron system is a substantially more technical and accurate measurement of the wool fiber in a lot of wool. Sixteen grades are used, and are based on the average fiber thickness as measured by a micrometer. An 80s wool, for example, averages about 18 microns, which is less than half a 36s wool that averages 39 microns. A micron is 1/25,400 of an inch. Wool too variable to fit within the limits of one grade is placed down a grade.

The micron system was largely developed at the USDA Denver Wool Laboratory. In January, 1976, the U.S. Economic Research Service began reporting the unit designation terminology for wool prices in microns. This system has become the standard for describing wools in the United States.

Table 9-4 compares the old blood system, the spinning count system, and the micron system of wool grades.

WOOL SORTING

Sorting is the operation of taking an individual fleece, untying the twine, opening the fleece, and separating it into the various grades that were grown on the different body areas. This operation is usually done in the mill, but occasionally it is done in a warehouse. The reason for this is that a mill knows exactly what qualities of wool it wishes to put into a fabric. The object of sorting is to obtain large lots of wool that are very uniform in fiber diameter, length, strength, and other characteristics. It is easy for an inexperienced person to distinguish a very fine wool from a very coarse wool, but it takes considerable training to be able to separate two consecutive grades, such as 56s and 58s. Sorting is always done on the grease wool. The dusting and scouring operations open up the fleece into small pieces and homogenize it so that sorting of scoured wool is impossible. Sorting is necessary on wool if a uniform worsted yarn with a certain spinning count is desired. The thoroughness of the sorting varies according to the type of fabric to be made from the wool.

MARKETING OF WOOL

Wool is a commodity. Its value is determined by the quantity and quality of the fibers; and clean fiber quantity and quality are affected by yield, vegetable matter, and average fiber diameter.

In the United States, average fiber diameter, diameter variability, and yield (including vegetable matter) of grease wool are the usual characteristics measured.

HARVESTING, SORTING, AND MARKETING OF WOOL

Fig. 9-16. *Left:* Yarding sheep for shearing in Australia. (Courtesy, Wool Bureau, Inc., New York, NY)

Fig. 9-17. *Below:* Shearing sheep with electric clippers just like the clippers of a barber.

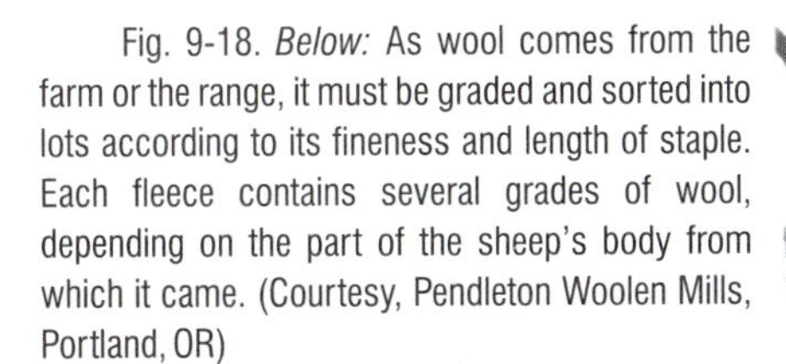

Fig. 9-18. *Below:* As wool comes from the farm or the range, it must be graded and sorted into lots according to its fineness and length of staple. Each fleece contains several grades of wool, depending on the part of the sheep's body from which it came. (Courtesy, Pendleton Woolen Mills, Portland, OR)

Fig. 9-19. *Above:* Wool auction in Wellington, New Zealand, showing buyers bidding. (Courtesy, Department of Scientific and Industrial Research, Wellington, New Zealand)

Like most industries, the wool and textile business has progressed from the status of a family enterprise. In the early days of this country—and the same pattern held true in other nations—virtually every family owned a few sheep and produced sufficient wool to meet its own needs. Under the family system, carding, spinning, and weaving were carried on by members of the household for the purpose of supplying the family with clothing. Under these conditions, there was little or no marketing. With the concentration of population in the cities, the coming of artisans, and the bringing of wool from more distant points, however, markets became a necessity.

Today, wool is one of the most important commodities of world commerce, and it might well be added that the marketing operations connected therewith are among the most intricate. In the first place, it is one of the most difficult items of commerce to classify and grade for the benefit of the trade. Second, few items have to be transported greater distances from producers to consumers. Wool production is a frontier type of industry, with the surplus-producing areas in those regions that are relatively undeveloped. On the other hand, wool consumption is greatest in the more populated regions.

U.S. sheep producers can do much to improve the marketing of their wool. First and foremost, they need to improve the production and handling of the clip, as discussed earlier in this chapter. Second, they need to do a better job of selling, including realizing that all wool is bought on a clean basis.

Fig. 9-21. Core-testing device in action in which a lot of wool is being sampled by means of a hollow tube. The core sample may be used to observe fiber length, diameter, and foreign material; and the sample may be scoured to determine shrinkage and yield.

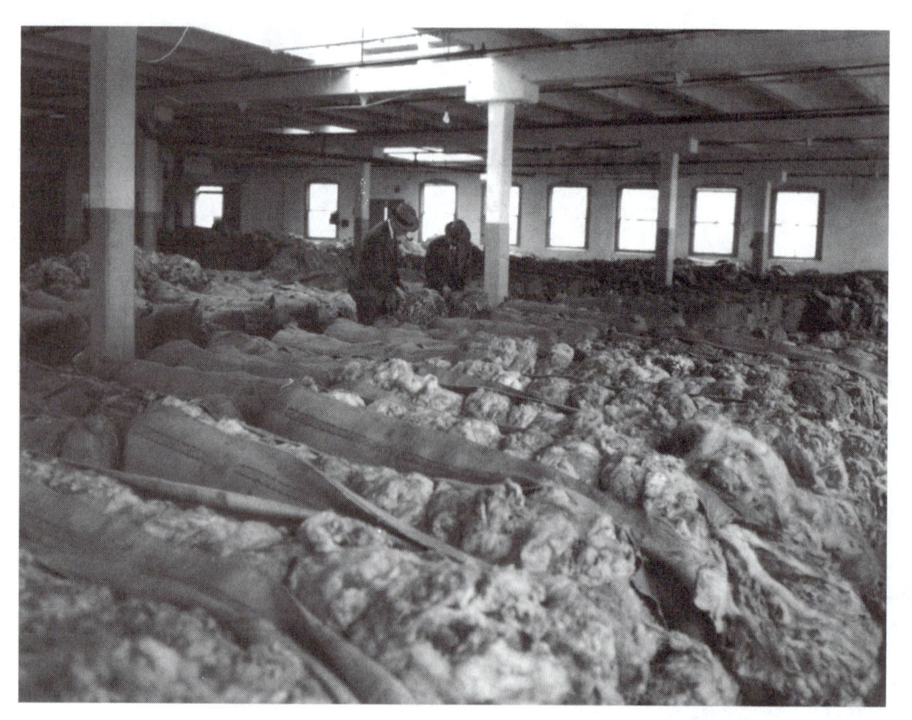

Fig. 9-20. Display of wool for prospective wool buyers. (Courtesy, USDA)

Without doubt, many buyers, in estimating shrinkage, have deliberately erred in their own favor; but carefully conducted scouring tests have conclusively demonstrated that even the so-called experts cannot accurately estimate shrinkage or clean yield. Because of this difficulty and the economic importance of accurate evaluations, numerous devices and methods for arriving at shrinkage or yield have been used. At the present time, the core method appears to be the best answer for the larger operators.

INTERNATIONAL TRADE IN WOOL

Table 9-5 lists the major exporting and importing countries. About two-thirds of the world's wool production is in the Southern Hemisphere, with the leading exporting countries consisting of Australia, New Zealand, Argentina, South Africa, and Uruguay. On the other hand, the great wool-consuming nations are in the Northern Hemisphere. They include China, Italy, United Kingdom, France, and Germany.

London has been displaced as the greatest wool-marketing center of the world by Sydney, Australia. Today, Australia's wool auctions are the standard by which all others are judged.

Boston was the leading wool market of the United States. However, since the early 1950s, the market has moved nearer the production areas, with nearly all shipments going directly to the mills.

TABLE 9-5
MAJOR WOOL EXPORTING AND WOOL IMPORTING COUNTRIES[1]

Country	Actual Weights	
	Exports	Imports
	(million lb, greasy)	(million lb, greasy)
Exporting		
Australia	1,177	
New Zealand	432	
Argentina	71	
South Africa	64	
Uruguay	15	
World	2,551	
Importing		
China		494
Italy		278
United Kingdom		225
France		205
Germany		181
World		2,513

[1]*Cotton and Wool*, USDA, ERS, November 1996, p. 61, Table 43. Data for 1995–96.

IMPORT DUTIES ON, AND SOURCES OF, WOOL AND MOHAIR

In order to protect U.S. sheep producers, there is an import duty on wool and mohair. Table 9-6 provides recent import rates for apparel wool.

TABLE 9-6
IMPORT DUTY ON APPAREL WOOL[1]

Grade	Country of Origin	
	General	Category 2[2]
	(¢/clean lb)	(¢/clean lb)
Greasy wool, including fleece-washed:		
Not finer than 40s	Free	55.1
Finer than 40s, but not finer than 44s	Free	66.1
Finer than 44s	Free	77.2
Degreased wool, not processed further:		
Not finer than 40s	Free	59.5
Finer than 40s, but not finer than 44s	Free	70.6
Finer than 44s	Free	81.6

[1]USDA (1998).
[2]Includes: Afghanistan, North Korea, Cuba.

MARKETING WOOL IN THE UNITED STATES

There are several differences between the marketing of animals on-foot and the marketing of wool. In the first place, the average producer is usually familiar with more than one of the several avenues through which live animals may be disposed; whereas, except for the larger wool growers, there is generally little knowledge concerning possible market outlets for wool. Also, practically all of the wool grown in the United States is bought and sold by private agreement, with no open- or auction-market arrangement to set values through competitive bidding. For this reason, it is difficult for growers to know what their wool is worth. On the other hand, an open market or auction would have its limitations in being informative to the grower because classes and grades of wool are so elusive and because it still is not practical to secure shrinkage determinations on small lots, even with the modern core system.

By some foreign standards, the wool producers of this country fall far short in preparing and marketing wool. In particular, U.S. wool producers might well emulate the Australians in skirting and grading fleeces at shearing time, placing uniform quality wool in labeled bales, and selling quantity lots in highly competitive auctions.

It is possible for growers to sell their wool through several agencies—including both private enterprises and cooperative associations—the chief ones of which are: (1) local dealers, (2) wool warehouses, (3) local pools, (4) commission houses, and (5) brokers or mill agents.

LOCAL DEALERS

Local or country dealers buy on a cash basis. Many such buyers merely purchase wool as a seasonal sideline, having a year-round business in such commodities as grain, livestock, produce, or junk. Local dealers may purchase wool on their own account for speculative purposes or buy on a predetermined commission basis for wool warehouses. Some local dealers have little knowledge of wool quality and buy all clips at a flat rate for the entire community. Under this system of marketing, there is little incentive for producing a quality product.

WOOL WAREHOUSES

The wool warehouses (also known as state dealers, central-market dealers, or wool merchants) buy wool on a cash basis, dealing either with local dealers or directly with the producers. They perform such important services as grading, warehousing, and assuming the risks incident to possessing large amounts of wool. Wool warehouses may sell to other wool warehouses or directly to manufacturers. Theirs is a legitimate business, but their profits are largely dependent upon being able to buy at the lowest possible price, with the chance of reselling at a sufficiently high figure to cover their services and realize a profit.

LOCAL POOLS (COOPERATIVE ASSOCIATIONS)

In recent years, the pool (or cooperative) plan of selling wool has gained in favor. This movement in the wool trade had its inception in 1918, with the organization of the Ohio Wool Growers Cooperative Association. The first cooperative was so successful that it became a permanent institution, and numerous other similar organizations scattered throughout the country have since come into existence. These groups vary widely in the manner of organization and services rendered to their membership. In general, however, they assemble and pool the wool clip of their grower-members and grade and hold it for the inspection of wool buyers. Frequently, money is advanced to the growers on the wool that is so pooled, with final settlement being made after completion of the sales and deduction of handling charges. Most cooperatives have their members execute binding signed contracts—some of which stipulate that the members must deliver their wool for sale by the agency for a period of years, whereas others carry a retainer agreement wherein growers reserve the right to sell their clips elsewhere, but with the understanding that they will pay a certain fee to the cooperative if they follow such a course. In general, cooperatives have rendered valuable service, many of them being especially helpful in the educational programs conducted among their members. Also, this method of marketing ensures growers full market value for their product, less actual marketing charges.

COMMISSION HOUSES

Wool commission houses (also known as commission merchants) perform services similar to those rendered by the commission companies on the central livestock markets. Essentially, they are agents of wool growers, operating on a percentage basis for storing and selling their clips. Usually growers are kept informed of the price offered for wool, and their sanction is obtained before the wool is sold. In some instances, however, the growers may give the commission house full responsibility for the sale. Unlike livestock commission companies, some wool commission houses also buy wool. Sometimes the latter privilege has been the subject of bitter criticism, because often the opportunities for profit are larger through purchase and speculation than through merely handling on a commission basis.

BROKERS OR MILL AGENTS

Brokers or mill agents buy for the manufacturer. Their purchases are made from any of the established marketing agencies or directly from the producer. They attempt to acquire the classes and grades that the mill can use, though usually the wool is sorted upon its arrival at the factory.

FIBER TO FABRIC

After wool is purchased by the manufacturer, it must pass through a number of intricate processes before the final product evolves. Wool fabrics are of two types—worsteds and woolens, each requiring different manufacturing and finishing processes. Woolens are made from the shorter wool fibers—carded but not combed—with the result that the little fibers in the yarn are in a criss-cross position, giving a soft, fuzzy yarn, as in broadcloth. The weave of a woolen cloth is more or less concealed by the fuzzy finish. A woolen also is usually softer and less firm than a worsted. Common worsted fabrics include serge, gabardine, rep, and coverts. Common woolen fabrics are cheviot, tweed, flannel, broadcloth, melton, kersey, cassimere, and mackinaw.

The longer fibers are used for worsteds, and they are combed in addition to being carded. The combing process places the fibers parallel, giving a yarn that is spun more tightly (as in serge) than the woolen yarns. A worsted also has a clear-cut weave and a smoother surface than a woolen.

Table 9-7 shows the amounts of wool in the United States that are used in mills on the woolen system and on the worsted system.

The steps in manufacturing both worsteds and woolens are covered in the following sections. Al-

TABLE 9-7
U.S. MILL CONSUMPTION BY GRADES ON THE WOOLEN AND WORSTED SYSTEMS, SCOURED BASIS[1]

Grade	Mill Consumption
	(million lb)
Woolen system	
60s and finer	18.4
Coarser than 60s	10.8
Worsted system	
60s and finer	27.4
Coarser than 60s	7.0
Total	63.6

[1]*Agricultural Statistics*, USDA, 2001, p. VII-36, data for 1999.

though the steps vary according to the available equipment, the kinds of wool, and the finished product desired, in general, they consist of the following: (1) sorting; (2) dusting and opening; (3) scouring; (4) drying; (5) carbonizing or bur-picking; (6) blending, oiling, and mixing; (7) carding; (8) combing; (9) spinning; (10) weaving, knitting, or felting; and (11) dyeing and finishing.

WOOL FIBER IN THE MAKING

The pictures and narrative that follow, portraying *wool fiber in the making*, were provided by Pendleton Mills, Pendleton, Oregon. Also, the fashion photos which follow, and the three fashion photos in the color section of this book, featuring both men's and women's wool clothing, were provided by Pendleton Woolen Mills.

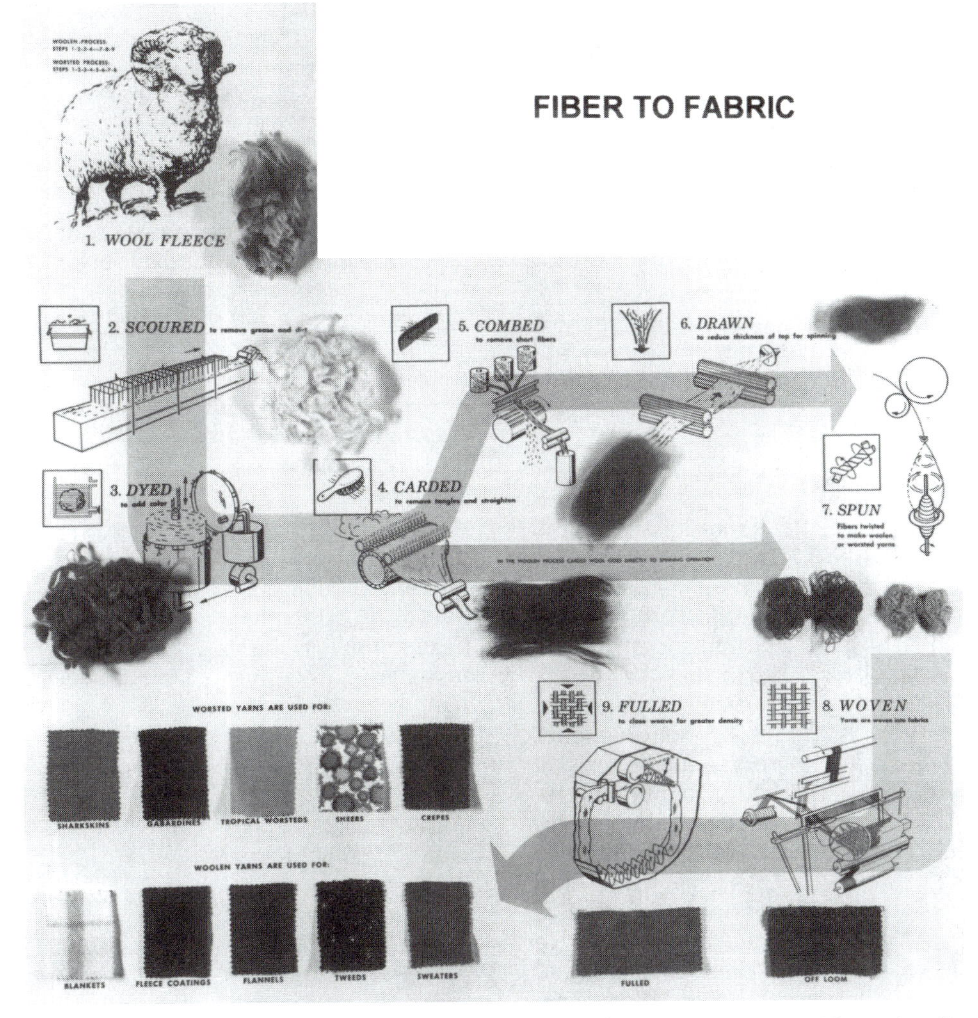

Fig. 9-22. How wool is made into fabric. Note the two processes—worsted and woolen. (Courtesy, The Wool Bureau, Inc., New York, NY)

Ewe and lamb

1. **Grading and Sorting.** The process of woolen fabric production begins with raw wool. Sheep are sheared once a year to provide this natural, renewable resource.

Measures of wool quality are fiber fineness, crimp, length, strength, and color. These attributes, which determine the grade of wool, largely depend on breed of sheep. However, climate, diet, and geographic location all influence wool quality.

Individual wool fleeces are hand graded and sorted with end use in mind. Hand sorters also remove stained wool and extraneous materials. For each wool grade, fibers from various lots are carefully blended to assure a uniform and consistent fiber mix.

2. **Scouring.** Wool is cleaned by a process referred to as *scouring*. At the beginning of this process, wool fleeces pass through a duster where they are opened and loose dirt is removed. The wool is then gently moved through a series of baths: one for soaking, two mild detergent baths, and a final clean water rinse. Between baths, the wool passes through squeeze rollers which remove water, dirt, and grease. A dryer at the end of the scouring line reduces the wool moisture content to a normal 12%. After drying, the wool is packed into 500 lb bales and shipped for further processing.

Wool scouring line

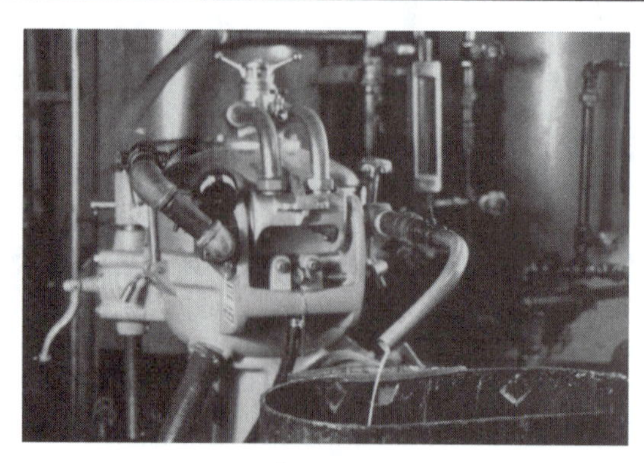

Centrifuge

3. **Lanolin recovery.** Lanolin is a natural byproduct of wool. Wool grease is separated from the discharge water of the scouring process by the use of a primary centrifuge. After passage to a holding tank, the wool grease is neutralized. The lanolin which rises to the top is skimmed off and washed in hot water. A secondary centrifuge removes water and a vacuum dryer reduces moisture content to 0.25% to produce anhydrous lanolin. The lanolin is then bleached and filtered to remove impurities. At this point, lanolin is stored in 55 gal. drums and ready for market.

Lanolin is a main ingredient in many cosmetics and pharmaceuticals, as well as a natural softener or lubricant in a variety of industries.

4. **Dyeing.** Computerization has been an important addition to the wool dye development process. A computer scanner "reads" or evaluates a sample color and generates appropriate dye formulas.

Wool can be dyed at three different stages of production: as fiber (stock dyeing), as yarn (packing dyeing), or as fabric (piece dyeing). The method selected is determined by the end fabric.

Dissolved dye is circulated by pumps during the dyeing process. Temperature and pressure are carefully controlled by electronic systems to insure rich, permanent color and exact duplication of standard colors.

To produce the subtle shading and depth of color seen in many heathered wools, an assortment of stock dyed fibers in different colors are mixed before carding.

Dye kettle

5. **Carding.** Carding is the process whereby wool fibers are combed into a fine sheet or web which is then divided into thin, continuous strands called *roving*.

Before carding, several lots of wool are blended to insure uniform color and quality. During this mixing process, a light emulsion of vegetable oil and water is applied to keep the fibers supple during the carding and spinning operations.

Carding equipment

Exact amounts of wool fiber are fed into the carding equipment in order to maintain the desired weight and subsequent size of roving. Wool stock is weighed in a hopper at the beginning of each set of cards and then passes through a series of rolls covered with fine wires which smooth out and align the wool fibers. The multi-size rolls move at different speeds and in opposite directions to produce a thin, completely uniform web of fibers. This web is divided and rubbed by rolls to form equal sized strands of roving. The roving is then wound onto large spools in preparation for the spinning frame.

6. **Spinning.** During the spinning operation, yarn is formed by the drawing out and twisting of the strands of roving. This twist adds strength to the yarn.

To spin yarn, roving spools are first mounted on the spinning frame. Roving ends are passed through two sets of rolls and then delivered to a bobbin which is mounted below. The drafting rolls stretch the roving by turning at different speeds. Twist is introduced by the turning bobbin and by a high speed, steel traveler which carries the yarn and winds it around the bobbin. After spinning, the yarn is steamed to set the twist and eliminate kinking.

Spinning frame

7. **Weaving.** Fabric is formed during the weaving operation. Through the use of a piece of machinery called a *loom*, two sets of yarns are interlaced at right angles to form cloth.

Loom

Before weaving takes place, lengthwise (warp) yarns are wound onto a warp beam. This beam is then mounted on a loom and each end of yarn is individually drawn through a wire eyelet called a heddle. Harnesses raise and lower the heddles to form a "V" shaped opening (shed) in the warp yarns. Crosswise (filling) yarn is passed through the shed by high speed grippers. During the weaving operation, the heddles change the position of the warp yarns and after each change a filling yarn is inserted. As the yarns are interlaced, the cloth is formed. The harness sequence of alternately raising and lowering the warp yarns determines the type of weave. Fabric is generally woven in 72 yd lengths.

8. **Finishing.** Finishing is the final step in woolen fabric production. The unfinished fabric requires several finishing processes before tailoring.

The finishing process, called *fulling*, is unique to the woolen industry because of the felting property of wool. During fulling, fabric is subject to controlled amounts of heat, moisture, friction, and pressure. The resulting shrinkage produces a softer, more compact fabric.

Besides fulling, wool fabric is carbonized to remove any remaining vegetable matter, washed, and dried. The cloth is sheared to produce an even surface, pressed, and sponged. When appropriate, fabrics are also napped. In the napping process, rolls covered with fine wires gently brush the fabric surface to raise individual fiber ends. This produces the soft, fuzzy nap often seen in coating or blankets.

As the finishing step is completed, fabrics are inspected, measured and rolled before the finished cloth is shipped to garment factories.

Dryer

Fig. 9-23. Fulling—a finishing treatment given to wool fabric. Fulling or controlled shrinking by felting is important in fabrics which will be napped, or which have a very fuzzy surface. (Courtesy, American Sheep Industry Assn., Inc., Englewood, CO)

2. In the manufacture of worsteds, 8.68 lb of grease wool (1.2 fleeces) furnish the following:

a. 4.34 lb of scoured wool.

b. 4.17 lb of carded sliver stock.

c. 3.62 lb of worsted top, combed fibers; plus 0.55 lb of noil, a byproduct of the combing operation.

d. 3.44 lb of roping stock ready for spinning.

e. 3.20 lb of spun-worsted yarn.

f. 3 yd and 27 in. of gray goods, ready for dyeing.

g. 3 yd and 18 in. of dyed goods, piece- or yarn-dyed; the fibers could have been stock-dyed if desired. (This operation would follow scouring and precede carding.)

h. 3 yd and 13.5 in. of finished fabric after sponging, which is enough for one worsted suit.

Fig. 9-24 shows the approximate amounts of grease wool required to make two common garments.

The average per capita consumption of wool in the United States is about 1.4 lb. Thus, the clip of one sheep cares for the needs of about five people.

The numerous operations from fleece to fabric are responsible for the often-considered elusive reasons why there is so much difference in the price of grease wool and a suit. To be more specific, it is estimated that, out of each dollar which the consumer spends for apparel and household goods made of wool, the wool producer gets only about 5¢ to 10¢.

WOOL FROM FLEECE TO FABRIC

The steps in the manufacturing process, together with the quantities of products obtained at the different stages, are clearly presented in the following summaries:

1. In the manufacturing of woolens, 6 lb of grease wool (three-fourths of one fleece) furnish the following:

a. 4.4 lb of scoured wool and noil.

b. 4.1 lb of carded roping stock.

c. 4 lb of spun-woolen yarn.

d. 3 yd and 1 in. of gray goods, ready for dyeing.

e. 3 yd of finished fabric after sponging.

f. Yardage to make one woolen coat.

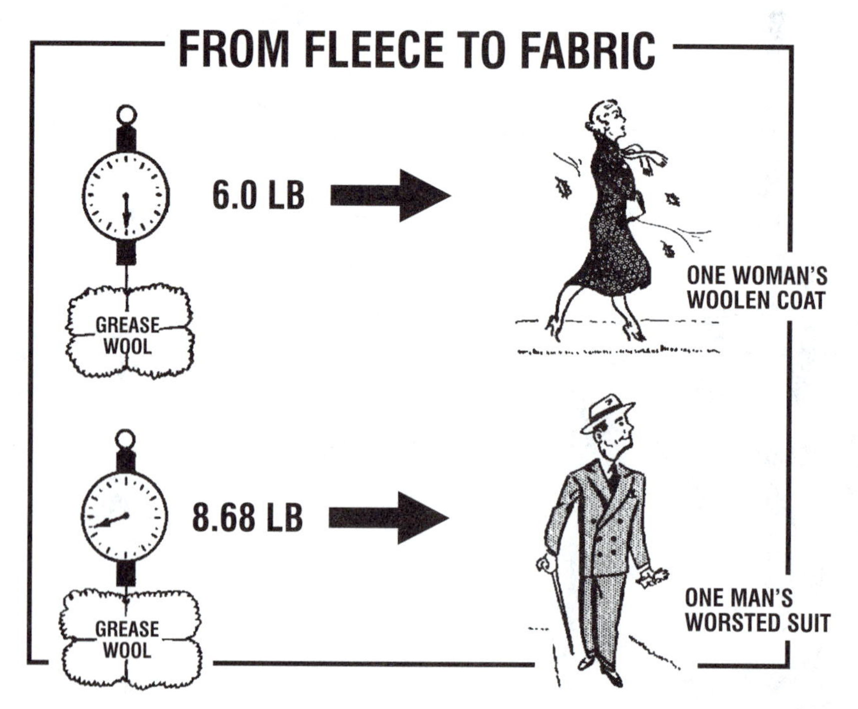

Fig. 9-24. Wool, from fleece to fabric. It requires 6.0 lb of grease wool (nearly three-fourths of an average fleece) to make one woman's woolen coat. It requires 8.68 lb of grease wool (slightly more than one fleece) to make one man's worsted suit. (Drawing by R. F. Johnson)

MADE OF WOOL

Fig. 9-25. *Left:* A fine wool plaid vest and pleated skirt for everyday dressing. (Courtesy, Pendleton Woolen Mills, Portland, OR)

Fig. 9-26. *Right:* A checked woolen jacket worn with corduroy trousers. (Courtesy, Pendleton Woolen Mills, Portland, OR)

Fig. 9-27. *Left:* Wool shag carpet and plaid wool upholstery give long wear while adding a natural look. (Courtesy, Pendleton Woolen Mills, Portland, OR)

Fig. 9-28. *Left:* Pendleton's fine gauge, multi-texture Zephyr wool sweater and classic tweed jacket reflect a return to elegance. (Courtesy, Pendleton Woolen Mills, Portland, OR)

Fig. 9-28. *Right:* A woolen suit with coordinating plaid jacket and pleated skirt. (Courtesy, Pendleton Woolen Mills, Portland, OR)

WOOL PRODUCTS LABELING ACT OF 1939

Few people, including the experts, can, on the basis of appearance, determine the composition of a fabric. Unscrupulous manufacturers and dealers formerly took advantage of this. In order to protect the consumer from such fraudulent operators, federal legislation, known as the Wool Products Labeling Act of 1939, became law on October 14, 1940. This law merely stipulates that fabrics must be labeled under one of the following categories:

1. **Wool.** The act defines wool as including the fiber from the fleece of the sheep or lamb, the hair of the Angora or the Cashmere goat, and the fibers from the camel, alpaca, llama, or vicuna—provided they are being used for the first time in the complete manufacture of a wool product.

Fig. 9-30. Labeling—guardian of wool's integrity. Read the label carefully to learn what you are buying in terms of protection, wear, and color-fastness. (Courtesy, Young American Films, Inc.)

2. **Reprocessed wool.** This includes wool fiber that has been woven or felted into a product and that, without ever having been utilized in any way by the ultimate consumer, is subsequently made into a fibrous state.
3. **Reused wool.** This consists of the resulting fiber when wool or reprocessed wool has been spun, woven, knitted, or felted into a wool product, which, after having been used in any way by the ultimate consumer subsequently has been made into a fibrous state.
4. **Virgin wool.** In addition to the three categories established by the Wool Products Labeling Act, the Federal Trade Commission, under whose direction enforcement of the law is vested, has created a fourth category by defining virgin wool or new wool as material composed wholly of new or virgin wool which has never been used or reclaimed, reworked, reprocessed, or reused from any spun, woven, knitted, felted, manufactured, or used product.

Contrary to the opinion held by some, the act does not prohibit the use of any fibers whatsoever that the manufacturer may care to use; but it does require that the contents be made known on the label, which must give the percentage of the total fiber weight represented by each kind of wool and also the percentage of other fibers present, if any.

In general, the consumer is quite correct in assuming that virgin wool is superior to reprocessed or reused wool. The very highest grades of reprocessed or reused wool, however, may be superior to low-grade virgin wool. In other words, virgin wool may be of very poor quality because of faulty breeding or feeding. Also, work quality is a major factor in determining the value of a fabric. While the Wool Products Labeling Act cannot give assurance relative to the many factors that enter into the determination of fabric quality other than the past history of the fabric, it does lend confidence to the consumer. It also benefits the wool grower, and in the final analysis, it benefits the manufacturer and the distributor.

IMPROVED WOOL PRODUCTS

Scientists have made significant advances (1) in adding to the store of knowledge of wool, and (2) in adding desired qualities to wool without losing its recognized virtues. Among these are:

1. **Yellowing.** Sometimes wool yellows. This may occur on the sheep's back, in processing, and in wearing. In the fleece, yellow stains may be caused by urine and feces, and by bacteria and fungi growing in the fleece. Alkali and light can also cause yellowing in wool products. Each kind of yellowing involves either the chemical interaction of wool protein with a colored substance or the generation in the fiber of a colored chemical group that becomes anchored to the wool protein. Whatever the cause, most such stains cannot be removed by washing, and can be removed by bleaching only with difficulty, considerable cost, and some damage.
2. **Shrinkage.** Wool shrinkage was first alleviated commercially by processes developed by the U.S. Department of Agriculture. The first development, which was widely used commercially, was known as the Wurlan process. The Wurlan treatment was then superseded by the USDA's improved treatment in which a film of polyurea instead of polyamide polymer is made on each fiber. The polyurea-treated goods exhibit excellent shrinkage resistance.
3. **Resistance to acids, alkalies, and bleaches.**

Scientists have chemically modified wool to improve its resistance to acids, alkalies, and bleaches.

4. **Durable pleats and creases.** The same principle used in setting waves in hair has been applied to the development of durable pleats and creases in wool garments. This involves the chemical rearrangement of the fiber structure.

These scientific developments, together with those yet to come, offer avenues through which wool may regain some of the markets that is has lost to synthetics and fill new uses. Further declines in per capita wool consumption are inevitable unless wool can match competing fibers in meeting consumer preferences for garments and fabrics—particularly in being easy to care for, in being lightweight, and in being fashionable, stylish, and colorful.

MOHAIR

Mohair, known as the most versatile of all fibers, is produced by the Angora goat, one of the oldest animals known to humankind. Yet, few citizens in the United States are more than casually aware of the Angora goat or its existence, even though these animals graze millions of acres of land and the hard-wearing fabrics made from their lustrous coats are used and admired from coast to coast. Mohair possesses qualities all its own, found in no other animal fiber. It has less crimp and smoother surface scales than sheep's wool. These qualities add luster, softness, and dust resistance to the other fine qualities mohair shares with wool. Mohair has remarkable resistance to wrinkles, great strength, and unequaled affinity to brilliant, deep colors that resist time, the elements, and hard wear.

Mohair is in demand for sweaters, coats, clothing, and furniture. The *Made of Mohair* pictures that follow, and the three fashion photos, featuring both men's and women's mohair clothing, in the color picture section of this book, were provided by the Mohair Council of America, San Angelo, TX.

WORLD MOHAIR PRODUCTION

At the present time, the Angora goat is bred on a commercial basis for fiber production in five countries. They are, by rank, South Africa, Turkey, the United States, Argentina, and Lesotho. Actual production figures for these countries are given in Chapter 15, Angora Goats.

There is considerable international trade associated with the raw products and the processed goods, since the processing of mohair is done largely in countries other than those in which it is produced. England, Japan, France, Italy, Spain, and Russia all import the raw product. Most U.S. mohair is exported to England.

U.S. MOHAIR PRODUCTION AND CONSUMPTION

Even though goats are rather widely distributed throughout the United States, the production of goats and mohair is of economic importance in comparatively few states only. Table 9-8 summarizes important data relative to the mohair industry of the United States.

TABLE 9-8
GOATS AND MOHAIR: NUMBER OF GOATS CLIPPED, MOHAIR PRODUCTION, AVERAGE PRICE PER POUND RECEIVED BY FARMERS, AND VALUE OF PRODUCTION, TEXAS, 1991–2000[1]

Year	Goats Clipped[2]	Average Clip per Goat	Mohair Production	Price per Pound	Value of Production[3]
	(1,000)	*(lb)*	*(1,000 lb)*	*($)*	*($1,000)*
1991	1,970	7.5	14,800	1.31	19,388
1992	2,000	7.1	14,200	0.87	12,354
1993	1,900	7.1	13,490	0.83	11,197
1994	1,600	7.3	11,680	2.62	30,602
1995	1,470	7.7	11,319	1.85	20,940
1996	1,070	7.0	7,490	1.95	14,606
1997	840	7.6	6,384	2.28	14,556
1998	620	7.5	4,650	2.59	12,044
1999	375	6.8	2,550	3.68	9,384
2000	345	6.8	2,346	4.30	10,088

[1]*Agricultural Statistics*, 2001, USDA.

[2]The number clipped is the sum of goats and kids clipped in the spring, and the kids clipped in the fall.

[3]Production multiplied by marketing year average price per pound.

Texas is by far the leading state in Angora goat numbers and mohair production. In fact, about 94% of the commercial mohair of this country is produced in the Lone Star State (see Chapter 2, Table 2-17). In 2000, the farmers and ranchers of Texas received more than $9 million from mohair; hence, mohair production is a sizable industry in Texas.

U.S. goat numbers declined in the 1960s, primarily in response to low prices for mohair, along with rising labor costs and a resurgence of the animal predator problem. Other reasons for the drastic reduction in numbers in Texas were higher revenues from cattle raising, oil rights, and hunting permits, along with the high death losses caused by severe snowfalls and freezes in 1973.

MOHAIR CHARACTERISTICS

As may be noted in Table 9-8, Angora goats, on the average, shear an annual clip of 7 to 8 lb of unscoured fleece per animal. Purebred herds often clip double this amount. Much of the domestic mohair, especially that produced in the Southwest, is taken off in two clips per year, in the spring and the fall, whereas Turkish mohair is usually allowed a full year's growth prior to shearing.

The three types of fleeces, based on the type of lock and ranked according to desirability, are: the tight or spiral lock, the flat lock, and the fluffy fleece. The tight lock hangs from the body in ringlets and is associated with the finest fibers. The flat lock is usually more wavy and coarser, but it is associated with heavy shearing weight. The fluffy fleece is objectionable because it is easily broken and is torn out by brush to a greater extent than the other types.

The length of fiber averages about 12 in. for a full year's growth and 6 in. when the animals are shorn semi-annually. Sometimes, with animals that do not have a tendency to shed and when special attention is given to tying the fleece up, mohair up to 3 ft long is produced in a period of three years.

In fineness, or diameter of fiber, mohair is somewhat coarser than wool. Length and luster are more sought than fineness. The fibers are usually very strong, high in luster, whitish in shade, fairly soft to the touch, and straight in staple appearance. Unfortunately, most mohair contains considerable kemp, which is highly undesirable from the standpoint of the manufacturer. Without doubt the amount of kemp can be lessened through breeding and selection.

Mohair shrinkage in scouring averages from 15 to 17% and does not depend on fineness, as does wool, since adult mohair shrinks as much as does kid mohair.

Chapter 23 provides more information about the mohair fiber.

PRODUCTION, HANDLING, AND MARKETING OF MOHAIR

Although mohair is usually accorded more neglect than wool, the principles involved in the economical production and advantageous marketing of a high-quality product are the same with both fibers. For practical reasons, chiefly as a means of lessening fleece losses caused by shedding or brush, more goats are shorn twice per year than is the case with sheep. Also, in the Southwest, goats are shorn twice each year because of the warm weather. Except for this difference, and because it is not recommended that the mohair fleece be tied at shearing time, the discussion already presented relative to the production, handling, and marketing of wool is equally applicable to mohair. The market channels and leading market centers for wool and mohair are identical.

It is unfortunate that a large amount of the mohair produced in this country continues to be marketed without being graded and sorted. On most shearing floors, altogether too little attention is given to keeping the fleece intact and rolling it together so that an intelligent job of grading and sorting may be done later. As long as these careless production methods are followed, mohair will neither meet the highest requirements of the manufacturer nor command a top price for the grower.

Most of the mohair is exported through local buyers, representing English and European firms. Much of it is exported in original bags, then graded and processed abroad. However, some graded mohair is exported. Also, some mohair is graded and processed into top for domestic use.

CLASSES AND GRADES OF MOHAIR

Kid hair is finest and is especially sought by mills. The fleeces from adults—especially bucks and old wethers—are the coarsest, and those from yearlings are intermediate between the other classes. These classes can be recognized by the grower and should be packed separately at shearing time. In addition, those fleeces that are extremely coarse, weak, and shorter than 6 in. or those having an excess of kemp, burs, or other foreign matter should be kept separate from clean, strong fleeces of desirable length and fineness.

The current USDA grade standards (1) for grease mohair, and (2) for mohair top are based on average fiber diameter (fineness) and fiber diameter dispersion. Grease mohair refers to the fleece as it comes from Angora goats, and before processing; mohair top is the processed fiber obtained after raw mohair has been scoured, carded, and combed.

As with wool, the grades of mohair are based primarily on the presumed spinning count obtainable on the Bradford system (or the number of 560-yard hanks to the pound). In practice, fineness is associated with softness and is recognized by the experienced touch when handled between the thumb and fingers.

Mohair has certain physical and chemical properties which are basic to its commercial value as a textile fiber. The average fiber diameter is the major consideration as this characteristic determines, to a large degree, the type of fabric or product for which the mohair may be used. Other characteristics affecting grease mohair value are its length, yield of clean mohair,

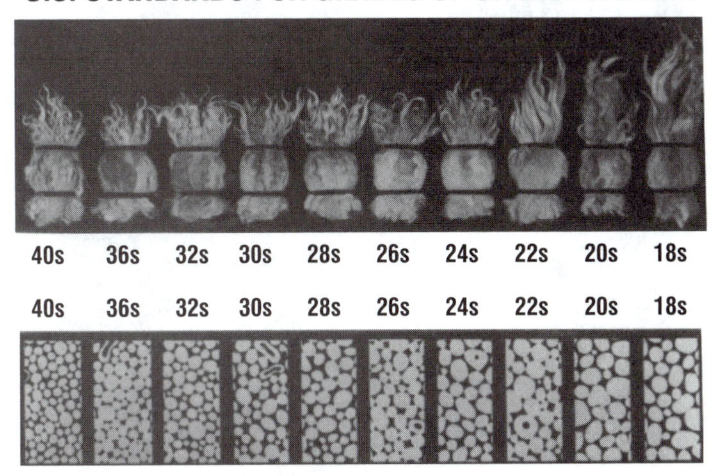

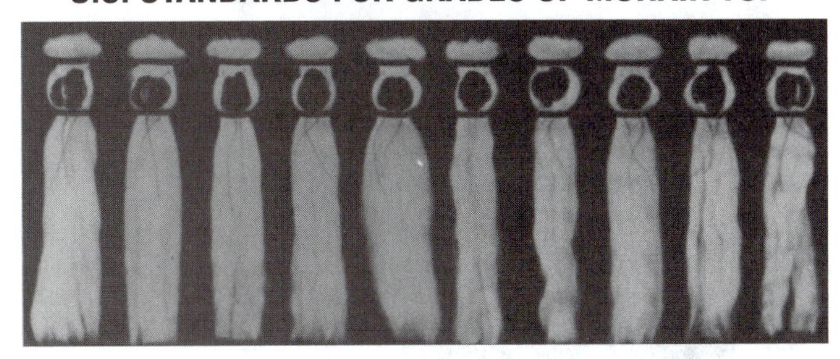

Fig. 9-31. U.S. standards for grades of mohair. (Courtesy, USDA, Livestock Division, Wool and Mohair Laboratory, Standardization Branch, Federal Center, Denver, CO)

strength, luster, color, and character. Grease mohair standards (grades) became effective August 1, 1971, and mohair top grades became effective January 1, 1973.

The official grades of grease mohair and the specifications of each are given in Table 9-9.

USES OF MOHAIR

Mohair is currently in demand for sweaters, coats, velour articles of clothing, and upholstery.

THE NATIONAL WOOL ACT

The National Wool Act, which was passed in 1954, was phased out in 1995—after 43 years. *The reason:* The dissention among the sheep and wool growers in the United States resulted in a vote, as required by law. The first vote, on February 6, 1996, was thrown out when the USDA determined that it had been flawed by misrepresentations of the voting and eligibility rules. In the second checkoff referendum held October 1, 1996,

TABLE 9-9
SPECIFICATIONS FOR THE OFFICIAL GRADES OF GREASE MOHAIR

Grade	Fiber Diameter		Approximate No. of Fiber Measurements[1]
	Limits for Average	Maximum Standard Deviation	
	(microns)	*(microns)*	
Finer than 40s . . .	Under 23.01	7.2	1,000
40s	23.01–25.00	7.6	1,000
36s	25.01–27.00	8.0	1,200
32s	27.01–29.00	8.4	1,200
30s	29.01–31.00	8.8	1,400
28s	31.01–33.00	9.2	1,400
26s	33.01–35.00	9.6	1,600
24s	35.01–37.00	10.0	1,600
22s	37.01–39.00	10.5	1,800
20s	39.01–41.00	11.0	2,200
18s	41.01–43.00	11.5	2,200
Coarser than 18s. .	Over 43.01		2,600

[1]The number of fibers to measure for each test should be the number needed to attain confidence limits of the mean within ± 0.40 micron at a probability of 95%. Measurement of the approximate number of fibers for the grades listed above may serve as a guide to meet the required confidence limits. The numbers indicated are based on mohair matchings.

MADE OF MOHAIR

Fig. 9-32. *Left:* Model in mohair sweater set. (Courtesy, Mohair Council of America, San Angelo, TX)

Fig. 9-35. *Right:* A stylish three-piece mohair outfit. (Courtesy, Mohair Council of America, San Angelo, TX)

Fig. 9-33. *Left:* Beautiful sofa, upholstered in a mohair plush, will outwear many other fabrics. (Courtesy, Mohair Council of America, San Angelo, TX)

Fig. 9-34. *Left:* Mohair sweaters and fabric are unequaled for durability. These casual outfits can be used for a variety of occasions. (Courtesy, Mohair Council of America, San Angelo, TX)

Fig. 9-36. *Right:* Two mohair suits, designed in classic styles, which means that they will be fashionable and wear beautifully through many seasons. (Courtesy, Mohair Council of America, San Angelo, TX)

the count was 47% for and 53% against. Unfortunately, fewer than 10% of the total U.S. sheep producers and feeders voted. The checkoff was expected to raise between $13 and $14 million each year, $6 to $7 million of which would be from importer contributions. So, at this point and period of time, the National Wool Act is of historic interest only.

SYNTHETIC OR MANUFACTURED FIBERS

Synthetic or manufactured fibers are of two kinds: (1) regenerated and (2) synthetic. Both are used in blends with wool.

Regenerated fibers are made by chemically altering cellulose, the large molecules (polymer) produced by plants. Two familiar regenerated fibers are rayon and acetate (cellulose acetate).

Three important groups of synthetic (noncellulose) fibers have been developed—nylons (polymides), polyesters, and acrylics. Basically, synthetic fibers are made from simple molecules, which, under certain conditions, have a coupling at each end. These molecules can be joined into long chains (called polymers) similar to box cars being hitched together to form a train. Synthetic fibers are a product of fundamental research, conducted by du Pont.

Development of the synthetic fiber industry since World War II has been a major factor in determining the demand for and the price of wool. Synthetic fibers have been relatively economical and some have desirable traits that are lacking or deficient in wool.

The impact of the synthetic fibers on the U.S. textile market is evidenced in Chapter 2, Fig. 2-12.

The future of wool and mohair appears to be brighter due to the increased cost of petroleum (energy), which is used to produce synthetic fibers. It requires about 17.8 million calories to produce a pound of synthetic fibers and only 8.3 million calories to produce a pound of wool.[1] Therefore, as the cost of petroleum rises, the price differential between wool and synthetic fibers narrows. Another factor favoring wool and mohair is the lower temperatures maintained in homes and offices during the winter. These require clothing for increased warmth, a need that can best be met by better wool and wool blends. In recent years, there has been a larger increase in blends of wool with various synthetics. The blends have the desirable attributes of both, to some degree.

[1]*The U.S. Sheep and Goat Industry: Products, Opportunities, and Limitations,* Report No. 94, Council for Agricultural Science and Technology (CAST), Ames, IA, May, 1982, p. 25, Table 6.

QUESTIONS FOR STUDY AND DISCUSSION

1. Discuss the history of wool. How many of the folklore/vernacular assertions pertaining to wool have you heard?

2. List and discuss the virtues of wool.

3. List and discuss the uses of wool.

4. Describe the wool fiber, and describe some characteristics of a good fleece.

5. List and discuss the chief fleece characteristics of interest and importance to both the producer and the manufacturer.

6. List and discuss wool production and handling practices which should be improved in order to market a higher quality product.

7. List and discuss three new shearing innovations.

8. List and discuss the requisites of good wool.

9. List and discuss the classes of wool.

10. How is wool graded and sorted?

11. May it be concluded that neither the blood system nor the spinning count system denotes accurately what it is supposed to indicate, according to derivation of the respective terms? Justify your answer.

12. How does the micron system of grading wool relate to the blood system and the spinning count system?

13. What are the import duties on wool and mohair? What is the purpose of such duties?

14. What can U.S. sheep producers do to improve the marketing of wool?

15. Outline and describe the wool manufacturing process from fiber to fabric.

16. Explain the difference between woolen and worsted.

17. What is the purpose of the "Wool Products Labeling Act of 1939"? Under what categories must fabrics be labeled?

18. What unique characteristics does mohair possess?

19. Compare the classes and grades of mohair to wool.

20. How is mohair used?

21. Discuss the purposes of the National Wool Act. Why was it phased out in 1995?

22. Why was wool unable to hold its position of prominence against the invasion of synthetic materials?

23. Discuss some factors that may increase the demand for and the price of wool in the future.

SELECTED REFERENCES

Title of Publication	Author(s)	Publisher
Cotton and Wool		USDA Economic Research Service, CWS-78, November 1994
Handbook for Woolgrowers	Ed. by G. R. Moule	Australian Wool Board, Melbourne, Australia, 1972
Textiles	M. S. Woolman E. B. McGowan	The MacMillan Company, New York, NY, 1926
Wool Handbook	W. Von Bergen	Interscience Publishers, New York, NY, 1963
Wool Handling		Eavenson & Levering Co., Camden, NJ, 1936
Wool Quality	S. G. Barker	His Majesty's Stationery Office, London, England, 1931
Wool Science	W. D. McFadden	Pruett Publishing Co., Boulder, CO, 1967

Also, valuable reference material on wool and mohair may be obtained from the following sources:

American Sheep Industry Assn., Inc
6911 South Yosemite Street
Englewood, CO 80112-1414
Phone: 303/771-3500

American Textile Manufacturers Institute
1130 Connecticut Avenue, N.W.
Suite 1200
Washington, DC 10036
Phone: 202/862-0500

Mohair Council of America
516 Norwest Bank Building
P.O. Box 5337
San Angelo, TX 76902
Phone: 915/655-3161

The Wool Bureau, Inc.
330 Madison Avenue
19th Floor
New York, NY 10017
Phone: 212/986-6222

CHAPTER 10

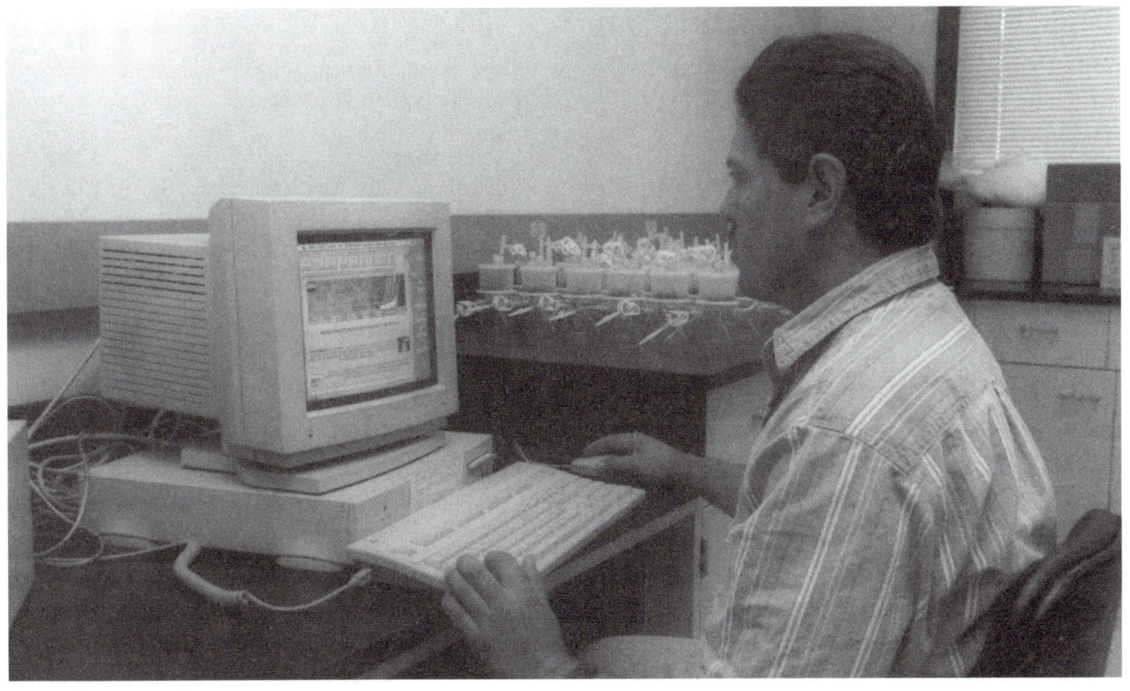

Computers are needed in modern sheep and goat production! This shows a portable computer. (Courtesy, California State University, Fresno)

BUSINESS ASPECTS OF SHEEP AND GOAT PRODUCTION

Contents *Page*

Whether one is establishing or enlarging a sheep or goat enterprise, the two most common questions are: (1) How much money will it take? and (2) How much money can I make?

Unfortunately, as important as these questions are, there is a paucity of information on which to base an answer, even though this information is needed by both sheep and goat producers and lenders. Obtaining information relative to the amount of money needed and the amount that can be made is complicated because of (1) the wide range in size of sheep and goat operations, and (2) the diverse conditions in areas where sheep and goat enterprises operate, and (3) volatile costs and returns.

Fortunately, some valuable indicators from 50- and 200-ewe farm flock enterprises, based on a recent study, are contained in a new publication issued by Washington State University.[1] The researchers make the following pertinent comments relative to the study:

[1]Warnock, W. D. and R. W. Carbner. *1995 Estimated Costs and Returns for a 50- and 200-Ewe Sheep Enterprise, Central Washington*, Washington State University, Farm Business Management Report, EB1365, 1995.

Fig. 10-1. On good pasture, 90% of the lambs are marketed off milk and grass. No other species can match this. (Courtesy, Union Pacific Railroad Co., Omaha, NE)

1. Feed represents over 80% of the total annual cash operating costs to raise sheep.
2. Sheep are converters of forage to high quality protein.
3. All labor is supplied by the owner/operator.
4. All hay is purchased at an average market price of $75 per ton.
5. All grazing is purchased at an average cost of $2.50 per ewe per month.
6. Interest on investment and operating capital is 10%.
7. Prices of all items are those prevailing in 1994.
8. A 175% lamb crop is weaned based on number of ewes wintered.
9. Twenty percent of the ewes are culled annually. Additionally, there is an annual death loss of 3% of the ewes. These ewes are replaced by ewe lambs from the flock.
10. All replacement ewe lambs are bred to lamb as yearlings.
11. Ninety percent of the lambs are marketed off milk and grass in September; and 10% are fed grain and sold in October.

TABLE 10-1
ESTIMATED ANNUAL RETURNS, COSTS, AND BREAK-EVEN LAMB PRICES FOR A 50-EWE FLOCK, CENTRAL WASHINGTON, 1995

Returns	Number	Weight	Price Per Pound	Total Value	Per Ewe
		(lb)	($)	($)	($)
Market lambs	78	110	0.60	5,148.00	102.96
Cull ewes	8	175	0.10	140.00	2.80
Rams	0.5	250	0.20	25.00	0.50
Ewe's wool	50	11	0.40	220.00	4.40
Ram's wool	2	12	0.40	9.60	0.19
Total returns				5,542.60	110.85
Less operating costs				4,050.76	81.02
Returns over operating costs				1,491.84	29.84
Break-even lamb price to cover operating costs[1]				0.43	
Less ownership costs					
Machinery and equipment				448.50	8.97
Insurance				12.00	0.24
Interest on breeding animals				110.00	2.20
Total ownership costs				570.50	11.41
Total costs				4,621.26	92.43
Break-even lamb price to cover total costs[2]				0.49	
Returns to operator's labor, management, land, and risk				921.34	18.43

[1]Operating costs, less cull sales and wool sales, divided by the pounds of lamb produced.

[2]Total costs, less cull sales and wool sales, divided by the pounds of lamb produced.

TABLE 10-2
ESTIMATED ANNUAL RETURNS, COSTS, AND BREAK-EVEN LAMB PRICES FOR A 200-EWE FLOCK, CENTRAL WASHINGTON, 1995

Returns	Number	Weight	Price Per Pound	Total Value	Per Ewe
		(lb)	($)	($)	($)
Market lambs	304	110	0.60	20,064.00	100.31
Cull ewes	40	175	0.10	700.00	3.50
Rams	1	250	0.20	50.00	0.25
Ewe's wool	200	11	0.40	880.00	4.40
Ram's wool	4	12	0.40	19.20	0.10
Total returns				21,713.20	108.57
Less operating costs				15,447.81	77.24
Returns over operating costs				6,265.39	31.33
Break-even lamb price to cover operating costs[1]				0.41	
Less ownership costs					
Machinery and equipment				1,633.50	8.17
Insurance				48.00	0.24
Interest on breeding animals				420.00	2.10
Total ownership costs				2,101.50	10.51
Total costs				17,549.31	87.75
Break-even lamb price to cover total costs[2]				0.48	
Returns to operator's labor, management, land, and risk				4,163.89	20.82

[1]Operating costs, less cull sales and wool sales, divided by the pounds of lamb produced.

[2]Total costs, less cull sales and wool sales, divided by the pounds of lamb produced.

The costs, returns, and break-even lamb prices from each (1) a 50-ewe flock, and (2) a 200-ewe flock are presented in Tables 10-1 and 10-2.

CAPITAL

Like other agricultural enterprises, sheep and goat operations require more and more capital, risks are greater, margins are smaller, greater efficiency of production is essential, computers are needed, and competent managers are in demand.

Fig. 10-2 is a balance sheet of U.S. farming; it shows (1) assets, (2) debts, and (3) equities from 1985 to 1994. Note that assets increased to $922.4 billion in 1994, debts decreased, and equity increased.

Sheep and goat producers should never invest money, either their own or borrowed, unless they are reasonably certain that it will make money. Capital will be needed for land, buildings, equipment, machinery, animals, feed, supplies, and a variety of miscellaneous items. The amount necessary will vary according to the type and size of the sheep or goat enterprise.

SOURCES OF CAPITAL (WHERE FARMERS BORROW)

Fig. 10-3 shows where farmers borrow, the amount of loans from each source, and the percentage held by each type of lender.

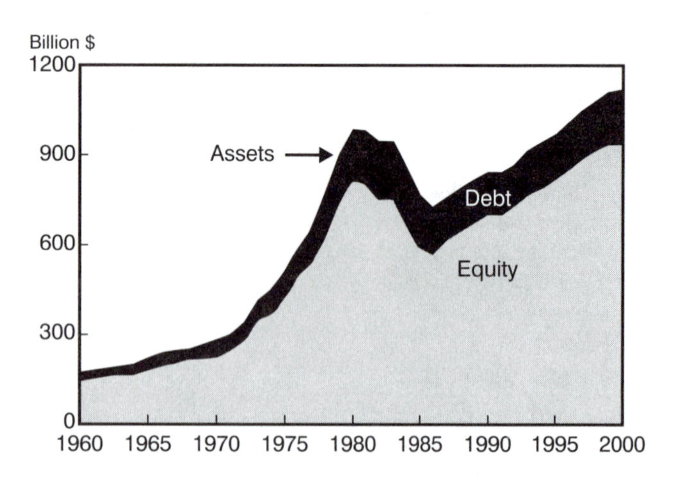

Fig. 10-2. Balance sheet of U.S. farming, showing (1) assets, (2) debts, and (3) equities. (Source: *Agricultural Income and Finance, 2001,* USDA, p. 6)

BUSINESS ASPECTS

In addition to capital, such things as credit, records and accounts, budgets, profit indicators, management, taxes, estate planning, social security, liability, and insurance require close attention. The net result is that those engaged in sheep or goat production must treat it as a business and become more sophisticated; otherwise, they will not be in business very long.

TYPES OF BUSINESS ORGANIZATIONS

The success of today's farming is very dependent on the type of business organization. No one type of organization is superior under all circumstances; rather, each situation must be considered individually. The size of the operation, the family situation, the enterprises, the objectives—all these, and more, are important in determining the best way in which to organize the farming business.

Four major types of business organizations are commonly found among farming enterprises: (1) the sole proprietorship, (2) the partnership, (3) the corporation, and (4) the cooperative.

It is not within the scope of this book to detail and compare the four major types of business organizations. Instead, the reader wishing to know more is re-

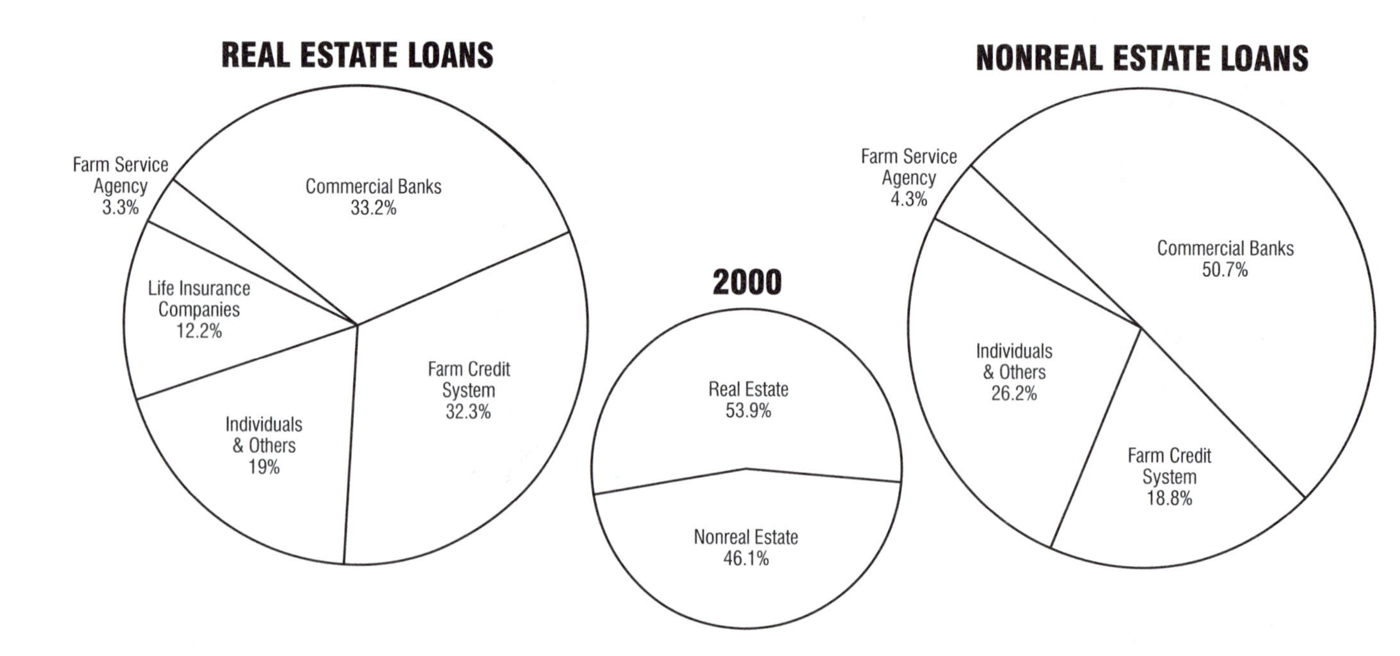

Fig. 10-3. Agricultural Income and Finance, 2001. (Source: USDA, Economic Research Service)

ferred to *The Stockman's Handbook* by the same author and publisher as this book.

CREDIT IN AGRICULTURE

The U.S. farm debt for 1994 was about $147.2 billion, while total assets were about $922.4 billion. This means that, in the aggregate, farmers have an 86% equity in their business, and 14% borrowed capital (see Fig. 10-2).

Credit is an integral part of today's sheep or goat business. Wise use of it can be profitable, but unwise use of it can be disastrous.

TYPES OF CREDIT OR LOANS

Getting the needed credit through the right kind of loan is an important part of sound financial management of a sheep or goat operation. The following three types of agricultural credit are available, based on length of life and type of collateral needed:

- **Short-term loans**—This type of loan is made for operating expenses and is usually for one year or less. It is used for the purchase of feed and for operating expenses; and it is repaid when lambs or other livestock products are sold. Security, such as a chattel mortgage, may be required by the lender.

- **Intermediate-term loans**—These loans are used to buy equipment and animals, for making land improvements, and for remodeling existing buildings. They are paid back in 1 to 7 years. Generally, they are secured by a chattel mortgage on animals and machinery.

- **Long-term loans**—These loans are secured by mortgage on real estate and are used to buy land or make major improvements to farmland and buildings or to finance construction of new buildings. They may be for as long as 40 years. Usually they are paid off in regular annual or semiannual payments. The best sources for long-term loans are insurance companies, Federal Land Banks, the Farmers Home Administration, or individuals.

CREDIT FACTORS CONSIDERED AND EVALUATED BY LENDERS

Potential money borrowers sometimes make their first big mistake by going in cold to see a lender, without adequate facts and figures—two strikes against getting a loan.

When considering and reviewing loan requests, the lender tries to arrive at the repayment ability of the potential borrower. Likewise, the borrower has no reason to obtain money unless it will make money.

Lenders need certain basic information in order to evaluate the soundness of a loan request. To this end, the following information should be submitted:

1. **Feasibility study.** Lenders are impressed with borrowers who have a written program, showing where they are now, where they are going, and how they expect to get there. In addition to spelling out the goals, this should give assurance of the necessary management skills to achieve them. Such an analysis of the present and projection into the future is imperative in big operations.

2. **The applicant, farm or ranch, and financial statement.** It is the borrowers' obligation, and in their best interest, to present the following information to the lender:

a. **Applicant**

(1) Name of applicant and spouse; age of applicant.

(2) Number of children (minors, legal age).

(3) Partners in business, if any.

(4) Years in area.

(5) References.

b. **Farm or ranch**

(1) Owner or tenant.

(2) Location—legal description and county, and direction and distance from nearest town.

(3) Type of enterprise—ewe-lamb, feedlot, goat dairy, etc.

c. **Financial statement.** This document indicates the borrowers' financial record and current financial position; their potential ahead; and their liability to others. Borrowers should always have sufficient slack so they can absorb reasonable losses due to such unforeseen happenstances as storms, droughts, diseases, and poor markets, thereby permitting the lender to stay with them in adversity and to give them a chance to recoup their losses in the future.

The financial statement should include the following:

(1) **Current assets**

(a) Livestock.

(b) Feed.

(c) Machinery.

(d) Cash. There should be reasonable cash reserves to cut interest costs and to provide a cushion against emergencies.

(e) Bonds or other investments.

(f) Cash value of life insurance.

(2) **Fixed assets**

(a) Real property, with estimated value;

(i) Farm or ranch property.

(ii) City property.

(iii) Long-term contracts.

(3) **Current liabilities**

(a) Mortgages.

(b) Contracts.

(c) Open account—to whom owed.

(d) Co-signer or guarantor on notes.

(e) Any taxes due.

(f) Current portion of real estate indebtedness due.

(4) **Fixed liabilities**—amount and nature of real estate debt

(a) Date due.

(b) Interest rate.

(c) To whom payable.

(d) Contract or mortgage.

3. **Other factors.** Shrewd lenders usually check on many things; among them:

a. **Potential borrowers.** Most lenders will tell potential borrowers that they are the most important part of the loan.

Lenders consider their—

(1) Character.

(2) Honesty and integrity.

(3) Experience and ability.

(4) Credit rating.

(5) Age and health.

(6) Family cooperation.

(7) Continuity, or line of succession.

Lenders are quick to sense the high livers—those who live beyond their means; the poor manager—the kind who would have made it except for hard luck, and to whom the hard luck happened many times; and the dishonest, lazy, and incompetent.

b. **Production records.** This refers to a good set of records showing efficiency of production. Such records should show weight and price of sheep sold, lamb-crop percentage and weaning weight, efficiency of feed utilization and rate of gain on feedlot lambs, age of ewes, ewe-lamb replacement program, weight and price of milk sold, kidding percentage, doe replacement program, depreciation schedule, average crop yield, and other pertinent information. Lenders will increasingly insist on good records.

c. **Progress with previous loans.** Has the borrower paid back previous loans plus interest? Has the borrower reduced the amount of the loan, thereby giving evidence of progress?

d. **Profit and loss (P&L) statement.** This serves as a valuable guide to the potential ahead. Preferably, this should cover the previous three years. Also, most lenders prefer that this be on an accrual basis (even if the sheep or goat producer is on a cash basis in reporting to the Internal Revenue Service).

e. **Physical plant**

(1) Is it an economical unit?

(2) Does it have adequate water, and is it well balanced in feed and livestock?

(3) Is there adequate diversification?

(4) Is the right kind of livestock being produced?

(5) Are the right crops and varieties being grown; and are approved methods of tillage and fertilizer practices being followed?

(6) Is the farmstead neat and well kept?

f. **Collateral (or security)**

(1) Adequacy to cover loan, with margin.

(2) Quality of security.

(a) Grade and age of livestock.

(b) Type and condition of machinery.

(c) If grain storage is involved, adequacy to protect from moisture and rodents.

(d) Government participation.

(3) Identification of security.

(a) Ear tags, ear marks, tattoo marks of sheep or goats.

(b) Serial numbers on machinery.

4. **Loan request.** Both sheep and goat producers are in competition for money from urban businesses. Hence, it is important that their requests for loans be well presented and supported. The potential borrower should (a) tell the purpose of the loan; (b) show how much money is needed, when it is needed, and what it is needed for; (c) present assurance of the soundness of the venture; and (d) give the repayment schedule.

CREDIT FACTORS CONSIDERED BY BORROWERS

Credit is a two-way street. It must be good for both borrowers and lenders. If borrowers meet the qualifications and are on a sound basis, more than one lender will want their business. Thus, it is usually well that borrowers shop around a bit; that they be familiar with several sources of credit and see what they have to offer. There are basic differences in length and types of loans, repayment schedules, services provided with loans, interest rates, and the ability and willingness of lenders to stick by borrowers in emergencies and times of adversity. Thus, interest rates and willingness to loan are only two of the several factors to consider. Also, if at all possible, all borrowing should be done from one source, one-source lenders will know more about borrowers' operations and be in a better position to help them.

HELPFUL HINTS FOR BUILDING AND MAINTAINING A GOOD CREDIT RATING

Sheep and goat producers who wish to build up and maintain good credit are admonished to do the following:

1. **Keep credit in one place, or in few places.** Generally, lenders frown upon split financing. Shop around for a creditor (a) who is able, willing, and interested in extending the kind and amount of credit needed and (b) who will lend at a reasonable rate of interest; then stay with the borrower.
2. **Get the right kind of credit.** Do not use short-term credit to finance long-term improvements or other capital investments.
3. **Be frank with the lender.** Be completely open and above board. Mutual confidence and esteem should prevail between borrower and lender.
4. **Keep complete and accurate records.** Complete and accurate records should be kept. By knowing the cost of doing business, decision-making can be on a sound basis.
5. **Keep an annual inventory.** Take an annual inventory for the purpose of showing progress made during the year.
6. **Repay loans when due.** Borrowers should work out a repayment schedule on each loan, and then meet payments when due. Sale proceeds should be promptly applied on loans.
7. **Plan ahead.** Analyze the next year's operation and project ahead.

INTEREST

The charge for the use of money is called interest. The basic charge is strongly influenced by the following:

1. The *basic cost* of money in the money market.
2. The *servicing costs* of making, handling, collecting, and keeping necessary records on loans.
3. The *risk* of loss.

Interest rates vary among lenders and can be quoted and applied in several different ways. The quoted rate is not always the basis for proper comparison and analysis of credit costs. Even though several lenders may quote the same interest rate, the effective or simple annual rate of interest may vary widely. The more common procedures for determining the actual annual interest rate, or the equivalent of simple interest on the unpaid balance, include: (1) simple or true annual interest on the unpaid balance, (2) installment loan with interest on the unpaid balance, (3) add-on installment loan with interest on face amount, and (4) points, each worth 1% of the face value of the loan, deducted from the total loan request with interest and repayment on the total loan request.

If interest is not stated, this formula may be used to determine the effective annual interest rate:

$$\text{Effective rate of interest} = \frac{2 \times \text{Number of payment periods in one year}^{2} \times \text{Finance charges}^{3}}{\text{Balance owed}^{4} \times \text{Number of payments in contract} + 1}$$

For example, a store advertises a refrigerator for $500. It can be purchased on the installment plan for $80 down and monthly payments of $35 for 12 months. What is the actual rate of interest if you buy on the time payment plan?

$$\text{Effective rate of interest} = \frac{2 \times 12 \times \$35}{\$420 \times (12 + 1)} = \frac{\$840}{\$5{,}460} = 15.4\%$$

RECORDS AND ACCOUNTS

The key to good business and management is records. The historian Santayana put it this way: "Those who are ignorant of the past are condemned to repeat it."

WHY KEEP RECORDS?

The chief functions of records and accounts are:

1. To provide profit and progress indicators. Production records on sheep and goat operations are profit indicators and a way to measure progress.
2. To provide information from which the sheep or goat business may be analyzed, with its strong and its weak points ascertained. From the facts thus deter-

---

[2]Regardless of the total number of payments to be made, use 12 if the payments are monthly, use 6 if payments are every other month, or use 2 if payments are semiannual.

[3]Use either the time payment price less the cash price, or the amount you pay the lender less the amount you received if negotiating for a loan.

[4]Use cash price less down payment; or, if negotiating for a loan, the amount you receive.

mined, the operator may adjust current operations and develop a more effective plan of organization.

3. To provide a net worth statement, showing financial progress during the year.

4. To furnish an accurate, but simple, net income statement for use in filing tax returns.

5. To keep production records on the sheep or goat operation and the crops.

6. To aid in making a credit statement when a loan is needed.

7. To keep a complete historical record of financial transactions for future reference.

Good records, properly analyzed and used, will increase net earnings and serve as a basis for sound management and husbandry.

KIND OF RECORD AND ACCOUNT BOOK

The record forms will differ somewhat according to the type of enterprise. For example, with dairy goats, cost per 100 lb of milk is the important thing, whereas in a sheep operation it is percentage of lambs dropped that are weaned plus growth rate. Net returns are important, but it is also necessary that records show all the items of cost and income—milk production and feed consumption of lactating goats; and mortality, feed cost, and growth rate of replacement ewes and market lambs.

Sheep and goat producers can make their own record books by ruling off the pages of bound notebooks to fit their specific needs, but the saving is negligible. Instead, it is recommended that they obtain copies of record books prepared for and adapted to their particular business. Such a book may usually be obtained at a nominal cost from the agricultural economics department of each state college of agriculture. Also, certain commercial companies distribute very acceptable record and account books at no cost.

KINDS OF RECORDS TO KEEP

Most record and account books contain simple and specific instructions relative to their use. Accordingly, it is not necessary nor is it within the realm of this book to provide such instructions. Instead, the comments made herein are restricted to the kind of records to keep.

The records should be easy to keep and should give the information desired to make a valuable analysis of the business. In general, the need for keeping records can be met by the following kinds of records:

1. **Annual inventory.** The annual inventory is the most valuable record that a sheep or goat producer can keep. It should include a list of real estate and its value, equipment, feed supplies, and all other property, including cash on hand, notes, bills receivable, and growing crops. Also, it should include a list of mortgages, notes, and bills payable. It shows sheep and goat producers what they own and what they owe—whether they are getting ahead or falling behind.

2. **Record of receipts and expenses.** Such a record is essential to any type of well-managed business. To be most useful, these entries not only should record the amount of the transaction, but should give the source of the income or the purpose of the expense, as the case may be. In other words, they should show the producer from what sources the income is derived and for what it is spent.

3. **Record of livestock and crop production.** A record of the production and sale of market lambs (or goat's milk) and of the yield of crops (if crops are grown) is most important, for the success of the operation depends upon production. Additionally, the producer should keep milk and butterfat records (for dairy goats); breeding records; herd health records, including mastitis records; and lamb or kid mortality records. Such records help in analyzing the farm business. They may be few or many, depending upon the operator's wishes.

SUMMARIZING AND ANALYZING THE RECORDS

At the end of the year, the second or closing inventory should be taken, using the same method as was followed in taking the initial inventory. The final summary should then be made, following which the records should be analyzed. In the latter connection, sheep and goat producers should remember that the purpose of the analysis is not to prove that they have or have not been prosperous. They probably know the answer to this question already. Rather, the analysis should show actual conditions on the farm and point out ways in which these conditions may be improved.

Although producers can summarize and analyze their own records, there are many advantages to having the services of a specialist for this purpose. Such a specialist is in a better position to make a cold appraisal without prejudice, and to compare enterprises with those of other similar operators. Thus, the specialist may discover that, in comparison, the sheep or goats on a given farm are requiring too much feed to produce 100 lb of lamb or milk, or that the enterprise is much less profitable than others have experienced. The local county agent can either render or recommend such specialized assistance. In some areas, it may consist of joining a cooperative farm record group or engaging the services of a consultant; in some states, such service is provided by the state agricultural college.

BUDGETS IN THE SHEEP OR GOAT BUSINESS

A budget is a projection of records and accounts and a plan for organizing and operating ahead for a specific period of time. A short-term budget is usually for one year, whereas a long-term budget is for a period of years. The principal value of a budget is that it provides a working plan through which the operation can be coordinated. Price changes, droughts, and other factors make adjustments necessary. But these adjustments are more simply and wisely made if there is a written budget to use as a reference.

HOW TO SET UP A BUDGET

It is unimportant whether a printed form (of which there are many good ones) or one made up on an ordinary ruled 8½ × 11 in. sheet placed sideways is used. The important considerations are that: (1) a budget be kept, (2) it be on a monthly basis, and (3) the operator be comfortable with whatever system or forms are to be used.

No budget is perfect. But it should be as good an estimate as can be made—despite the fact that it will be affected by droughts, diseases, markets, and many other unpredictables.

By using forms such as the three that follow (see Figs. 10-4, 10-5, and 10-6), an operator can set up a simple, easily kept, adequate budget.

ANNUAL CASH EXPENSE BUDGET

_______________________________ for _______________

(name of farm or ranch) (date)

Item	Total	Jan.	Feb.	Mar.	Apr.	May	June	July	Aug.	Sept.	Oct.	Nov.	Dec.
Labor hired													
Feed purchased													
Gas, fuel, grease													
Taxes													
Insurance													
Interest													
Utilities													
Etc.													
Total													

Fig. 10-4. The annual cash expense budget should show the monthly breakdown of various recurring items—everything except the initial loan and capital improvements. It includes labor, feed, supplies, fertilizer, taxes, interest, utilities, etc.

ANNUAL CASH INCOME BUDGET

_______________________________ for _______________

(name of farm or ranch) (date)

Item	Total	Jan.	Feb.	Mar.	Apr.	May	June	July	Aug.	Sept.	Oct.	Nov.	Dec.
Market lambs													
Cull ewes													
Oats													
Etc.													
Etc.													
Total													

Fig. 10-5. The annual cash income budget is just what the name implies—an estimated cash income budget by months. For the dairy goat operator, the form is adapted to include such items as market milk, cull does, and male kids.

ANNUAL CASH EXPENSE AND INCOME BUDGET (Cash Flow Chart)

_______________________________________ for _______________

(name of farm or ranch) (date)

Item	Total	Jan.	Feb.	Mar.	Apr.	May	June	July	Aug.	Sept.	Oct.	Nov.	Dec.
Gross income													
Gross expense													
Difference													
Surplus (+) or Deficit (–)													

Fig. 10-6. The annual cash expense and income budget is a cash flow budget, obtained from the first two forms. It is a money flow summary by months. From this, it can be ascertained when, and how much money will need to be borrowed, and the length of the loan along with a repayment schedule. It makes it possible to avoid tying up capital unnecessarily, and to avoid unnecessary interest.

HOW TO FIGURE NET INCOME

Figs. 10-4, 10-5, and 10-6 show a gross income. There are other expenses that must be taken care of before net profit can be determined; namely:

1. **Depreciation on buildings and equipment.** It is suggested that the *useful life* of buildings and equipment be as detailed in a subsequent section of this chapter headed, "Set Up Depreciation Schedule Properly."
2. **Interest on owner's money invested in farm and equipment.** This should be computed at the going rate in the area, say 10%.

Some people prefer to measure management by return on invested capital, and not by wages. This approach may be accomplished by paying management wastes first, then figuring return on investment.

ENTERPRISE ACCOUNTS

Where a sheep or goat enterprise is diversified—for example, a farm or ranch having a ewe-lamb operation, a lamb feedlot—and crops, or a goat dairy and crops—enterprise accounts should be kept. This means different accounts for each enterprise. Keeping enterprise accounts makes it possible:

1. To determine which enterprises have been most profitable, and which least profitable.
2. To compare a given enterprise with competing enterprises of like kind, from the standpoint of ascertaining comparative performance.
3. To determine the profitability of an enterprise at the margin (the last unit of production). This will give an indication as to whether to increase the size of a certain enterprise when both enterprises are profitable in total.

ANALYZING A SHEEP OR GOAT BUSINESS—IS IT PROFITABLE?

Most people are in business to make money—and sheep and goat producers are people. In some areas, particularly near cities and areas where the population is dense, land values may appreciate so as to be a very considerable profit factor. Also, a tax angle may be important. But neither of these should be counted upon. The sheep or goat operation should make a reasonable return on the investment, otherwise, the owner should not be in the business.

Owners or managers of both sheep and goat establishments need to analyze their businesses—to determine how well they are doing. With big operations, it is no longer possible to base such an analysis on the bank balance statement at the end of the year. In the first place, once a year is not frequent enough, for it is possible to go broke, without really knowing it, in that period of time. Second, a balance statement gives no basis for analyzing an operation—for ferreting out its strengths and weaknesses. In large lamb feedlots or large goat dairies, it is strongly recommended that progress be charted by means of monthly or quarterly closings of financial records.

Also, not only must sheep and goat producers compete with other sheep and goat producers down the road, but they must compete with themselves—with their records last year and the year before. They must work ceaselessly away at making progress, improving the end product, and lowering costs of production.

To analyze a sheep or goat business requires two essentials: (1) good records, and (2) yardsticks, or profit indicators, with which to measure an operation.

Profit indicators are gauges for measuring the primary factors contributing to profit. In order for sheep or goat producers to determine how well they are doing, they must be able to compare their operations with others; for example, (1) their own historical five-year average, (2) the average for the country or for their own particular area, or (3) the top 5%. The author favors the latter, for high goals have a tendency to spur superior achievement.

Like most profit indicators, the ones that follow are not perfect. But they will serve as useful guides. Also, in some establishments, there may be reason for adding or deleting some of the indicators; and this can be done. The important thing is that each sheep or goat operation have adequate profit indicators, and that these be applied as frequently as possible. In a sheep feedlot or a goat dairy this may be done monthly with some indicators.

EWE-LAMB PROFIT INDICATORS

Many factors determine the profitability of a ewe-lamb enterprise. Certainly, a favorable per animal unit capital investment in land and improvements is a first requisite. Additionally, the percentage of lamb crop weaned and weaning weight are exceedingly important, as shown in Table 10-3.

Table 10-3 will serve as a yardstick for determining (1) how you as an operator stack up with the nation's (a) average and (b) top 5% of ewe-lamb operators, and (2) where you are falling down in your ewe-lamb enterprise.

Additionally, Table 10-3 reveals that the top 5% operators have a higher investment in land, improvements, machinery, and equipment than their average counterparts. Obviously, better operators have better facilities and equipment; their savings are made in the handling of the sheep. It is not unlike selecting a ration—where it is net return, rather than cost per ton, that counts.

No claim is made that Table 10-3 is perfect. Admittedly, there are wide area differences, and no two ewe-lamb enterprises are alike. Also, there are seasonal differences; for example, a drought will materially affect the weaning weight of lambs. Nevertheless, Table 10-3 will serve as a useful guide.

TABLE 10-3
EWE-LAMB PROFIT INDICATORS[1]

Indicators	Farm Flock		Range Band	
	Average	Top 5%	Average	Top 5%
Investment/acre/animal unit (one mature ewe) in land and improvements (real estate) . . . . . ($)	500.00	600.00	400.00	450.00
Investment in machinery and equipment/ animal unit (one mature ewe). . . ($)	37.16	41.67	20.74	24.00
Labor/ewe/year . . (hr)	4.7	3.3	2.3	2
Ewes failing to lamb. . . . . . . (%)	6	3	10.0	5
Lambs born, based on no. ewes bred. . (%)	130	150	98	107
Death losses from birth to weaning . . . (%)	18	12	8.4	6
Lambs raised, based on no. ewes bred. . (%)	107	132	90	100
Average weaning wt. of lambs . . . . (lb)	80	100	70	80
Average fleece weight . . . . . (lb)	8.5	9.5	8.5	9.5
Average death losses of ewes . . . . (%)	5	3	10	6

[1]Estimates made by the author. No claim is made relative to the scientific accuracy of this data. Rather, it is presented in the hope that it will (1) serve as a useful indicator until the enterprise accumulates its own historical data, and (2) stimulate needed research along these lines.

LAMB FEEDING PROFIT INDICATORS

Lamb feedlot operators need to keep good records and make frequent analyses (at least once monthly) to determine how well they are doing. A determination of assets and liabilities at the end of the fiscal year is not good enough, primarily because it is available only once per year. Lamb feeding requires much capital; hence, records should be kept as current as possible at all times.

Operators are primarily interested in two questions: namely, (1) how well are they doing—profitwise, and (2) how do they compare with other feedlots? Table 10-4 gives some guidelines for answering these questions. It may be used in making an analysis of a specific feedlot; for determining (1) the strengths and weaknesses within the feedlot, and (2) how it stacks up with the top 5% of the nation's feedlots.

Interestingly, the top 5% operators invest more in land and equipment than their average counterparts. Obviously, they effect savings in operation rather than in physical plant.

Admittedly, profit indicators, such as those given in Table 10-4 are not perfect, simply because no two feedlots are the same. Nationally, there are wide area differences in climate, feeds, land costs, salaries and wages, and other factors. Nevertheless, indicators *per se* serve as a valuable yardstick. Through them, it is possible to measure how well a given feedlot is doing— to ascertain if it is out of line in any one category, and, if so, the extent of same.

After a few years of operation, a feedlot should evolve with its own yardstick and profit indicators, based on its own historical records and averages. Even with this, there will be year-to-year fluctuations due to seasonal differences, lamb and feed price changes, disease outbreaks, manager changes, wars, inflation, and other happenstances.

TABLE 10-4
LAMB FEEDING PROFIT INDICATORS[1]

Indicators	Average for U.S. Feedlots	Top 5% of U.S. Feedlots
Initial (new) land, feedlot and equipment cost basis/animal capacity . . . ($)	13.0	13.6
Feedlot and equipment cost charged off/animal . . . . ($)	0.7	0.5
Daily nonfeed costs/animal[2] . . . . ($)	2.0	1.8
Salaries and wages/head/day . . . ($)	0.65	0.6
Death losses . . . . (%)	3.5	2.0
Vet. fees and medicine/head. . . . (¢)	0.9	0.8
Feed/lb gain. . . . . (lb)	9.5	8.5
Daily rate of gain . . . . (lb)	0.3	0.5
Net return/head[3] . . . . ($)	3.0	3.6

[1]Estimates made by the author. No claim is made relative to the scientific accuracy of this data. Rather, it is presented in the hope that it will (1) serve as a useful indicator until the enterprise accumulates its own historical data, and (2) stimulate needed research along these lines.

[2]This embraces all costs other than lambs and feed. It includes salaries and wages; taxes, interest, insurance; utilities; gasoline, oil, grease; depreciation; repairs; veterinary, medical, consultant, and legal services; trucking; promotion; and other costs.

[3]Net return to management after deducting from gross profit all costs, including depreciation on machinery and buildings, and interest on investment.

DAIRY GOAT PROFIT INDICATORS

Many factors determine the profitability of a dairy goat enterprise. Certainly a favorable per doe capital investment in land, buildings, equipment, and does is a first requisite.

Tables 10-5 and 10-6 will serve as useful yardsticks for determining (1) how dairy producers stack up with the goals therein established, and (2) where dairy producers are falling down in their enterprises.

TABLE 10-5
LACTATING DOE PROFIT INDICATORS

Indicators	Production Goals	Achieved on Your Dairy
Investment/doe in land, buildings, equipment and does. Not more than. . . . . ($)	—[1]	______
Milk production/doe/year . . . . (lb)	2,200	______
Metabolizable energy (ME)/ cwt milk . . . . (Kcal)	56	______
Does with lactations of 305 days. More than . . . . (%)	50	______
No. cases of mastitis/month in lactating does. Not more than . . . . . (%)	8	______

[1]This should be a value which is derived from operators of similar size in the same general area.

TABLE 10-6
REPLACEMENT DOE PROFIT INDICATORS

Indicators	Production Goals[1]	Achieved by Your Dairy
Death losses from birth to first kidding. Not more than. . . . . (%)	5	______
Average age when bred . . . . (mo)	8	______
Average weight when bred . . . (lb)	70	______
Average age when joining the milking string . . . . (mo)	13	______

[1]These values are slightly above average.

COMPUTERS IN THE SHEEP OR GOAT BUSINESS

Accurate and up-to-the-minute records and controls have taken on increasing importance in all phases of agriculture, including the sheep and goat businesses, as the investment required to engage in farming and ranching has risen and profit margins have narrowed. Today's successful farmers and ranchers must have, and use, as complete records as any other businesspersons. Also, records must be kept current; it no longer suffices merely to know the bank balance at the end of the year.

With the increasing availability of small computers and the wide range of software to go with them, hand record keeping is becoming obsolete. Computerized record keeping requires less time, thereby allowing management more time for planning and decision making. Furthermore, computerized record keeping permits an all-at-once consideration of the complex interrelationships which affect the economic success of the business. This also aids planning and decision making.

Fig. 10-7. Terreton, Idaho sheep rancher Cindy Siddoway at work on a computer. (Courtesy, American Sheep Industry Assn., Englewood, CO)

Software available to sheep and goat producers, who own small home computers, offers a wide range of computerized record-keeping possibilities, which can, in turn, generate reports, thus aiding decision making and planning. With the narrowing profit margins experienced by livestock producers, the effective use of computers has become a tool of the modern producer.

There is hardly any limit to what computers can do if given the proper information and good programs. Among the difficult questions that a computer can answer for a specific operation are:

1. **How is the entire operation doing so far?** It is preferable to obtain quarterly or monthly progress reports, often making it possible to spot trouble before it is too late.
2. **What farm enterprises are making money; which ones are freeloading or losing?** By keeping records by enterprises—ewe-lamb, lamb feedlot, dairy goats, wheat, corn—it is possible to determine strengths and weaknesses, then either rectify the situation or shift labor and capital to a more profitable operation.
3. **Is each enterprise yielding maximum returns?** By having profit or performance indicators in each enterprise (see Tables 10-3 through 10-6), it is possible to compare these (a) with the historical average of the same farm or ranch, or (b) with the same indicators of other farms or ranches.
4. **How does this farm or ranch stack up with its competition?** Without revealing names, the computing center (local, state, area, or national) can determine how a given ranch compares with others— either the average, or the top (say 5%).
5. **How can you plan ahead?** By using projected prices and costs, computers can show what moves to make for the future. They can be a powerful planning tool. They can be used in determining when to plant, when to schedule farm machine use, etc.
6. **How can income taxes be cut to the legal minimum?** By keeping complete records of expenses and figuring depreciations accurately, computers make for a saving in income taxes on most farms and ranches.

There are three requisites for computerized record keeping on a farm or ranch; namely:

1. Access to a computer.
2. Computer know-how and software that fills the needs of the sheep or goat operation.
3. Accurate records.

The pioneering computer services available to farmers and ranchers were operated by universities, trade associations, and government—most of them being on an experimental basis. Subsequently, others have entered the field, including commercial data processing firms, banks, machinery companies, feed and fertilizer companies, and farm suppliers.

Modern technology has placed home computers, with large programming and storage capabilities, within the reach of most livestock producers, and there are an increasing number of companies selling software for specific types of livestock operations. Examples of some of the software available-to sheep and goat producers include:

1. Breeding records.
2. Production records.
3. Performance records.
4. Herd or flock lists.
5. Sire reports.
6. Pedigrees.
7. Lambing or kidding schedules.
8. Ewe or doe production summaries.
9. Registration form processing.
10. Sales lists.
11. General ledger.
12. Payroll.
13. Accounts receivable.
14. Accounts payable.
15. Equipment use/costs.
16. Cash flow information.
17. Net income statement.
18. Enterprise analysis.
19. Five-year histories.

In the past, the biggest deterrent to production testing on ewe-lamb operations, both purebred and commercial, was the voluminous and time-consuming record keeping involved. Keeping records *per se* does not change what an animal will transmit, but records must be used to locate and propagate the genetically superior animals if genetic improvement is to be accomplished. Currently, some universities offer computerized production-testing programs. These are described in detail in Chapter 3, Principles of Genetics. Also, production records of dairy goats are computerized through the use of the Dairy Herd Improvement (DHI) program described in the same chapter.

Another area where computers are used extensively is ration formulation. This use of computers is described in Chapter 14. Feeding Sheep.

MANAGEMENT

Four major ingredients are essential to success in the sheep or goat business: (1) good animals, (2) good feeding, (3) good records, and (4) good and aggressive management. Producers and livestock specialists rank these four factors in order of importance, putting management at the top of the list. Unfortunately, this factor is often overlooked in the present era, primarily because the accent is on scientific findings, automation, and new products.

Management gives point and purpose to everything else. The skill of the manager materially affects how well sheep, goats, and/or milk is bought and sold, the health of the animals, the results of the ration, the stress of the animals, the rate of gain and feed efficiency, the milk production of goats, the performance of labor, the public relations of the outfit, and even the expression of the genetic potential of the animals. Indeed managers must wear many hats—and they must wear each of them well.

The bigger and the more complicated the operation, the more competent the management required. This point merits emphasis because, currently, (1) bigness is a sign of the times, and (2) the most common method of attempting to bail out of an unprofitable venture is to increase its size. Although it is easier to achieve efficiency of equipment, labor, purchases, and marketing in big operations, bigness alone will not make for greater efficiency, as some owners have discovered to their sorrow, and others will experience. Management is still the key to success. When in financial trouble, owners should have no illusions on this point.

In manufacturing and commerce, the importance of and the scarcity of top managers is generally recognized and reflected in the salaries paid to persons in such positions. Unfortunately, agriculture as a whole has lagged; and altogether too many owners still subscribe to the philosophy that the way to make money out of the livestock business is to hire managers cheap, with the result that they usually get what they pay for—cheap managers.

TRAITS OF A GOOD MANAGER

There are established bases for evaluating many articles of trade, including sheep, wool, milk, hay, and grain. They are graded according to well-defined standards. Additionally, feeds are chemically analyzed and feeding trials conducted. But no such standard or system of evaluation has evolved for sheep managers, despite their acknowledged importance.

The author has prepared the Manager Check List, given in Table 10-7, which (1) owners may find useful when selecting or evaluating managers and (2) managers may apply to themselves for self-improvement purposes. No attempt has been made to assign a percentage score to each trait, because this will vary among sheep and goat establishments. Rather, it is hoped that this check list will serve as a useful guide (1) to the traits of a good manager, and (2) to what the owner wants.

TABLE 10-7
MANAGER CHECKLIST

CHARACTER—
Has absolute sincerity, honesty, integrity, and loyalty; is ethical.

INDUSTRY—
Has enthusiasm, initiative, and aggressiveness; is willing to work, work, work.

ABILITY—
Has sheep know-how and experience, business acumen—including ability to arrive at the financial aspects systematically and convert this information into sound and timely management decisions—knowledge of how to automate and cut costs, common sense, and growth potential. Is organized

PLANS—
Sets goals, prepares organization chart and job description, plans work, and works plans.

ANALYZES—
Identifies the problem, determines pros and cons, then comes to a decision.

COURAGE—
Has the courage to accept responsibility, to innovate, and to keep on keeping on.

PROMPTNESS AND DEPENDABILITY—
Is a self-starter, has "T.N.T."; which means doing it "today, not tomorrow."

LEADERSHIP—
Stimulates subordinates, and delegates responsibility.

PERSONALITY—
Is cheerful, not a complainer.

ORGANIZATION CHART AND JOB DESCRIPTION

It is important that all workers know to whom they are responsible and for what they are responsible; and the bigger and the more complex the operation, the more important this becomes. This should be written down in an organization chart and a job description designed specifically for the sheep or goat operation.

INCENTIVE BASIS FOR THE HELP

Big farms and ranches must rely on hired labor, all or in part. Good help—the kind that everyone wants—is hard to come by; it's scarce, in strong demand, and difficult to keep. And the farm work-force situation is going to become more difficult in the years ahead. There is need, therefore, for some system that will (1) get and hold top-flight help and (2) cut costs

Fig. 10-8. A good incentive basis makes hired help partners in profit.

and boost profits. An incentive basis that makes hired workers partners in profit may be the answer.

Many manufacturers have long had an incentive basis. Executives are frequently accorded stock option privileges, through which they prosper as the business prospers. Laborers may receive bonuses based on piecework or quotas (number of units, pounds produced). Also, most factory workers get overtime pay and have group insurance and retirement plans. Some industries have a true profit-sharing arrangement based on net profit as such, a specified percentage of which is divided among employees. No two systems are alike. Yet, each is designed to pay more for labor, provided labor improves production and efficiency. In this way, both owners and laborers benefit from better performance.

Family-owned and family-operated farms have a built-in incentive basis; there is pride of ownership, and all members of the family are fully cognizant that they prosper as the business prospers.

Many different incentive plans can be, and are, used. There is no best one for all operations. The various plans given in Table 10-8 are intended as guides only.

The incentive basis chosen should be tailored to fit the specific operation, with consideration given to kind and size of operation, extent of owner's supervision, present and projected productivity levels, mechanization, and other factors.

For most ewe-lamb, lamb feedlot, and goat operations, the author favors a production-sharing and prevailing-price type of incentive, examples of which are given later in this section.

HOW MUCH INCENTIVE PAY?

After (1) reaching a decision to go on an incentive basis, and (2) deciding on the kind of incentive, an operator must arrive at how much to pay. Here are some guidelines that may be helpful in determining this:

1. Pay the going base, or guaranteed, salary; then add the incentive pay above this.
2. Determine the total stipend (the base salary plus incentive) to which you are willing to go.
3. Before making any offers, always check the plan on paper to see (a) how it would have worked out in past years based on your records, and (b) how it will work out as you achieve the future projected production; for example:

> A supervisor in charge of a range band of 5,000 ewes is now producing a 90% lamb crop with average weaning weights of 75 lb that is 67.5 lb of lamb weaned per ewe bred (90% × 75 = 67.5). He/she is receiving a base salary of $800.00 per month, plus house, garden, and 300 lb of dressed meat per year. The owner prefers a production-sharing and prevailing-price type of incentive.

Step by step, here is the procedure for arriving at an incentive arrangement based on increased production:

1. By checking with local sources, it is determined that the present salary of $800 per month plus extras is the going wage; and, of course, the employee receives this regardless of what the year's lamb production or price turns out to be. It is guaranteed.
2. A study of the ewe-lamb records reveals that with a little extra care on the part of the supervisor—particularly at lambing time and in parasite control—the average weaning weight of lambs per ewe bred can be boosted enough to permit paying him/her $1,070 per month, or $270 per month more than he/she is now getting. That is $3,240 more per year. This can be fitted into the incentive plan.
3. An average increase of 5 lb of lamb weaned per ewe bred at 65¢/lb would mean $3.25 per ewe, or $16,250 on 5,000 ewes. With an 80:20 split between owner and manager, the foreman would get $3,250 or slightly over $270 per month.

REQUISITES OF AN INCENTIVE BASIS

Owners who have not previously had experience with an incentive basis are admonished not to start with any plan until they are sure of both their plan and their help. Also, it is well to start with a simple plan; then a change can be made to a more inclusive and sophisticated plan after they have acquired experience.

Regardless of the incentive plan adopted for a specific operation, it should encompass the following essential features:

1. Good owner (or manager) and good workers. No incentive basis can overcome a poor manager. Owners or managers must be good supervisors, and they must be fair to their workers. Also, in big establishments, they must prepare a written organization chart and job description so the workers know (a) to whom they are responsible, and (b) for what they are responsible. Likewise, no incentive basis can spur em-

TABLE 10-8
INCENTIVE PLANS FOR SHEEP AND GOAT ESTABLISHMENTS

Types of Incentives	Pertinent Provisions of Some Known Incentive Systems In Use	Advantages	Disadvantages	Comments
Bonuses	A flat, arbitrary bonus; at Christmas time, year-end, or quarterly or other intervals. A tenure bonus such as (1) 5 to 10% of the base wage or 2 to 4 weeks' additional salary paid at Christmas time or year-end, (2) 2 to 4 weeks' vacation with pay, depending on length and quality of service, or (3) $10.00 to $50.00 per week set aside and to be paid if employee stays on the job a specified time.	It's simple and direct.	Not very effective in increasing production and profits.	
Equity-building plan	Employee is allowed to own a certain number of animals. In ewe-lamb operations, these are usually fed without charge.	It imparts pride of ownership to the employee.	The hazard that the owner may feel that employees accord their animals preferential treatment; suspicioned if not proved.	
Production sharing	50¢ to $1/lamb weaned; $1/cwt of gain on lambs fed. 20¢ to 50¢/head on fed lambs marketed. So much per day for meeting certain levels of milk production in a goat dairy.	It's an effective way to achieve higher conception and foal crop.	Net returns may suffer. For example, a higher rate of gain than is economical may be achieved by feeding lambs more expensive feeds than are practical. This can be alleviated by (1) specifying the ration and (2) setting an upper limit on the gains to which the incentive will apply. If a high performance level already exists, further gains or improvements may be hard to come by.	Incentive payments for production above certain levels—for example, above 90 lb *(41 kg)* lamb weaned/ewe bred—are more effective than paying for all lambs produced.
Profit sharing: a. Percent of gross income in cash. b. Percent of net income	1 to 2% of the gross. 10 to 20% of the net after deducting all costs.	Net income sharing works better for managers and supervisors, than for laborers; because fewer hazards are involved to opening up the books to them. It's an effective way to get hired help to cut costs. It's a good plan for a hustler.	Percent of gross does not impart cost of production consciousness. Both (1) percent of gross income and (2) percent of net income expose the books and accounts to workers, who may not understand accounting principles. This can lead to suspicion and distrust. Controversy may arise (1) over accounting procedure; *e.g.,* from the standpoint of the owner, a fast tax write-off may be desirable on new equipment, but this reduces the net shared with the worker; and (2) because some owners are prone to overbuild and overequip, thereby decreasing net.	There must be prior agreement on what constitutes gross or net receipts, as the case may be, and how it is figured.

(Continued)

TABLE 10-8 (Continued)

Types of Incentives	Pertinent Provisions of Some Known Incentive Systems In Use	Advantages	Disadvantages	Comments
Production sharing and prevailing price	Basis may be (1) the percent of offspring weaned and (2) weaning weight or pounds of offspring weaned per female bred. For a goat dairy it can be based on measures of udder health or kidding interval. See sections that follow on "An incentive basis for ewe-lamb operators," and "An incentive basis for lamb feedlots."	It embraces the best features of both production and profit sharing, without the major disadvantages of each. It (1) encourages high productivity and likely profits, (2) is tied in with prevailing prices, (3) does not necessitate opening the books, and (4) is flexible—it can be split between owner and employee on any basis desired, and the production part can be adapted to a sliding scale or escalator arrangement—for example, the incentive basis can be higher for the 0.1 lb *(45g)* of feedlot gain made in excess of 0.5 lb *(223 g)* than for a 0.1 lb *(45 kg)* gain in excess of 0.3 lb *(136 g)*.	It is a bit more complicated than some other plans, and it requires more complete records.	When properly done, and all factors considered, this is the most satisfactory incentive basis for a sheep or goat enterprise.

ployees who are not able, interested, and/or willing. This necessitates that employees must be selected with special care when they will be on an incentive basis. Hence, the three—good owner (manager), good employees, and good incentive—go hand in hand.

2. It must be fair to both employer and employees.

3. It must be based on and must make for mutual trust and esteem.

4. It should compensate for extra performance, rather than substitute for a reasonable base salary and other considerations (house, utilities, and certain provisions).

5. It must be simple, direct, and easily understood.

6. It should compensate all members of the team; from caretakers to the manager of a ewe-lamb outfit, and from feeders and feed processors to the manager of a lamb feedlot.

7. It must be put in writing, so that there will be no misunderstanding. If some production-sharing plan is used in a lamb feedlot, it should stipulate the ration (or who is responsible for ration formulation), and the weight and grade to which finishing lambs are to be carried. On a ewe-lamb outfit, it should stipulate the ration, the culling of ewes, and other pertinent factors.

8. It is preferable, although not essential, that workers receive incentive payments (a) at rather frequent intervals, rather than annually, and (b) immediately after accomplishing the extra performance.

9. It should give the hired workers a certain degree of responsibility, from which they will benefit through the incentive arrangement.

10. It must be backed up by good records; otherwise, there is nothing on which to base incentive payments.

11. It should be a two-way street. If employees are compensated for superior performance, they should be penalized (or, under most circumstances, fired) for poor performance. It serves no useful purpose to reward the unwilling and the incompetent. For example, no overtime pay should be given to employees who must work longer because they are slow or because they are correcting mistakes of their own making. Likewise, if the reasonable break-even point on a ewe-lamb operation is an average of a 90-lb lamb weaned per ewe bred, and this production level is not reached because of obvious neglect (for example, not being on the job at lambing time), the employee(s) should be penalized (or fired).

INDIRECT INCENTIVES

Normally, incentives are thought of as monetary in nature—as direct payments or bonuses for extra production or efficiency. However, there are other ways of encouraging employees to do a better job. The latter are known as indirect incentives. Among them are:

1. Good wages.
2. Good labor relations.
3. Adequate housing plus privileges such as—
   a. Being able to use the farm truck or car.
   b. Having a horse to ride.
   c. Having electric bills paid.
   d. Being able to use a swimming pool.

e. Having hunting and fishing privileges.
f. Having meat, milk, and eggs furnished.
4. Modern buildings and equipment.
5. Vacation time with pay, time off, sick leave.
6. Group health.
7. Security.
8. Opportunity for self-improvement.
9. Right to invest in the business.
10. All-expense paid trips to short courses, shows, conventions, etc.
11. Year-end bonuses for staying all year.

These indirect incentives are being accorded the workers of more and more establishments, especially the large ones.

AN INCENTIVE BASIS FOR EWE-LAMB OPERATORS

On ewe-lamb enterprises, there is a need for some system which will encourage caretakers to be good nurses to newborn lambs, though it may mean loss of sleep, and working with cold, numb fingers. Additionally, there is a need to do all those extras which will make for the maximum percentage of lamb crop weaned at a heavy weight.

From the standpoint of the owner of a ewe-lamb enterprise, production expenses remain practically unchanged, regardless of the efficiency of the operation. Thus, the investment in land, buildings and equipment, ewes, feed, and labor differs very little with a change (up or down) in the percentage of lamb crop or the weaning weight of lambs; and income above a certain break-even point is largely net profit. Yet, owners take all the risks; hence, they should benefit the most from the profits.

In a ewe-lamb operation, the author recommends that profits beyond the break-even point (after deducting all expenses) be split on an 80:20 basis. This means that every dollar made above a certain level is split, with the owner taking 80¢ and the employees getting 20¢. Also, there is merit in an escalator arrangement, with the split changed to 70:30, for example, when a certain plateau of efficiency is reached. Moreover, the part of the profits which go to the employees should be divided on the basis of their respective contributions, all the way down the line; for example, 25% of it might go to the manager, 25% divided among the supervisors, and 50% divided among the rest of the workers.

A true profit-sharing system on a ewe-lamb outfit based on net profit has the disadvantages of (1) not benefiting employees when there are losses, as frequently happens in the livestock business and (2) requiring management to open up the books, which may lead to gossip, misinterpretation, and misunderstanding. An incentive system based on major profit factors alleviates these disadvantages.

Gross income in ewe-lamb operations, where market lamb production (rather than wool) is the main objective, is determined primarily by (1) percentage of lamb crop weaned, (2) weaning weight of lambs, and (3) price. The first two factors can easily be determined. Usually, enough lambs are sold to establish price; otherwise, the going price can be used.

The incentive basis proposed in Table 10-9 for ewe-lamb operations is simple, direct, and easily applied. As noted, it is based on average pounds of lamb weaned per ewe, which encompasses both percentage of lamb crop and weaning weight.

TABLE 10-9
A PROPOSED INCENTIVE BASIS FOR EWE-LAMB OPERATIONS

Average Pounds of Lamb Weaned/Ewe Bred	How It Works
(lb)	
70 80 81 (break-even point) 90 100 110 120 130 140 150	On this particular operation, the break-even point is assumed to be an average of 81 lb of lamb weaned/ewe bred (90% lamb crop and 90-lb weight = 81-lb lamb/ewe); and, of course, this is arrived at after including all costs of production factors. Pounds of lamb weaned/ewe bred in excess of the break-even point are sold or evaluated at the going price. If an average of 125 lb of lamb/ewe bred is weaned-marketed, and if lambs of this quality are worth 65¢ per pound, that is $28.60 more net profit/ewe. In a 1,500-ewe band, that is $42,900. With an 80:20 division, $34,320 would go to the owner, and $8,580 would be distributed among the employees. Or, if desired, and if there is an escalator arrangement, there might be a 80:20 split up to 120 lb, a 70:30 split from 120 to 130 lb, and a 65:35 split above 130 lb.

AN INCENTIVE BASIS FOR LAMB FEEDLOTS

An incentive basis for lamb feedlot workers is needed for motivation purposes, just as it is in ewe-lamb operations. It is the most effective way in which to lessen absenteeism, poor processing and mixing of feeds, irregular and careless feeding, unsanitary troughs and water, sickness, shrinkage, and other profit-sapping factors.

The incentive basis for lamb feedlots shown in Table 10-10 is simple, direct, and easily applied.

Whenever possible, the break-even points—(1) pounds feed per pound gain, and (2) daily rate of gain—should be arrived at from actual records accumulated by the specific feedlot, preferably over a period of years. Perhaps, too, they should be moving averages, based on 5 to 10 years, with older years dropped out and more recent years added from time to time, thereby reflecting improvements in efficiency and rate of gain due primarily to changing technology rather than to the efforts of the caretakers.

With a new feedlot, on which there are no historical records from which to arrive at break-even points of feed efficiency and rate of gain, the figures of other similar feedlots can be used at the outset. These can be revised as actual records on the specific feedlot become available. It is important, however, that the new feedlot start an incentive basis, even though the break-even points must be arbitrarily assumed at the time.

Because of the high correlation between feed efficiency and rate of gain, the incentive basis recommended in Table 10-10 does result in an overlapping of measures. Nevertheless, both efficiency and rate of gain are important profit indicators to lamb feeders. Because of the overlapping, however, some may prefer to choose one or the other of the measures, rather than both.

Another incentive basis followed in a few large feedlots consists of the following: A certain percentage (say 15%) of the net earnings set aside in a trust account, which is divided among the employees and applied to the account of each employee according to salary and/or length of service, and paid to employees upon retirement or after a specified period of years. The main disadvantages to this incentive basis are that there may not be any net some years, that some employees do not want to wait that long for their added compensation, and that it opens up the books of the business.

TABLE 10-10
A PROPOSED INCENTIVE BASIS FOR LAMB FEEDLOTS

Feed/Lb Gain		Daily Rate of Gain	How It Works
(lb)		*(lb)*	
12.0			On this particular lamb-finishing operation, the break-even points are assumed to be (1) 8.5 lb of feed/lb gain, and (2) 0.3 lb daily gain. Of course, for very high roughage rations, the break-even points would be different; for example, feed efficiency might be 10.0 and the daily rate of gain 0.25 lb.
11.5			Feed saved and gains made in excess of the break-even points are computed at going prices.
11.0			If the feed efficiency drops to 8.0 and the gain increases to 0.55, and if feed costs $120 per ton and lambs are worth 65¢ per pound, then these feed savings and increased gains are worth—
10.5			Feed saved . . . . 0.5 lb
9.5		0.20	Cost of feed/lb . . . . . . 6¢
9.0		0.25	Value of feed saved/lb gain . 3¢
8.5	(break-even point)	0.30	Value of feed saved on 0.55-lb gain . . . . . . 1.65¢
8.0		0.35	Gains made . . . . 0.25 lb
7.5		0.40	Per lb mkt. value of gains . 65¢
7.0		0.45	Value of incr. daily gain . . 16.25¢
6.5		0.50	**Increased profit/head/day:**
6.0		0.55	Feed saved . . . . . . 1.65¢
5.5		0.60	Gain made. . . . . . . 16.25¢
			Total . . . . 17.90¢
			On lambs fed for 60 days, that is $10.74/head. With 10,000 lambs, the total is $107,400. When divided on an 80:20 basis, that makes a total of $85,920 for the owner and $21,480 for the employees.

LIABILITY[5]

Most farmers are in such financial position that they are vulnerable to damage suits. Moreover, the number of damage suits arising each year is increasing at an almost alarming rate, and astronomical damages are being claimed. Studies reveal that about 95% of the court cases involving injury result in damages being awarded.

Comprehensive personal liability insurance protects an operator who is sued for alleged damages suffered from an accident involving his/her property or family. The kinds of situations from which a claim might arise are quite broad, including suits for personal injuries caused by animals, equipment, or personal acts.

Both workers' compensation insurance and employer's liability insurance protect farmers against claims or court awards resulting from injury to hired help. Workers' compensation usually costs slightly more than straight employer's liability insurance, but it carries more benefits to the worker. An injured employee must prove negligence by the employer before the company will pay a claim under employer's liability insurance, whereas workers' compensation benefits are established by state law, and settlements are made by the insurance company without regard to who was negligent in causing the injury. Conditions governing participation in workers' compensation insurance vary among the states.

[5]The sections on Liability, Workers' Compensation, Social Security Law, Tax Management and Reporting, and Estate Planning were authoritatively reviewed by Waymon E. Watts, CPA, Fresno, California.

WORKERS' COMPENSATION

Workers' compensation laws, now in full force in every one of the 50 states, cover on-the-job injuries and protect disabled workers regardless of whether their disabilities are temporary or permanent. Although broad differences exist among the individual states in their workers' compensation laws, principally in their benefit provisions, all statutes follow a definite pattern as to employment covered, benefits, insurance and the like.

Workers' compensation is a program designed to provide employees with assured payment for medical expenses or lost income due to injury on the job. Whenever an employment-related injury results in death, compensation benefits are generally paid to the worker's surviving dependents.

Generally all employment is covered by workers' compensation, although a few states provide exemptions for farm labor, or exempt farm employers of fewer than 10 full-time employees, for example. Farm employers in these states, however, may elect workers' compensation protection. Livestock producers in these states may wish to consider coverage as a financial protection strategy because under workers' compensation, the upper limits for settlement of lawsuits are set by state law.

This government-required employee benefit is costly for livestock producers. Costs vary among insurance companies due to dividends paid, surcharges and minimum premiums, and competitive pricing. Some companies, as a matter of policy, will not write workers' compensation in agricultural industries. Some states have a quasi-government provider of workers' compensation to assure availability of coverage for small businesses and high-risk industries.

For information, contact your area extension farm management or personnel management advisor and an insurance agent experienced in marketing workers' compensation and liability insurance.

SOCIAL SECURITY LAW

The Social Security Law covers stipulated agricultural workers, including workers on farms and ranches. Thus, owners and managers of sheep and/or goat operations should be familiar with, and follow the pertinent provisions of the Social Security Law.

The number on the social security card is very important to the farm operator as well as to the hired farm worker. It identifies the individual's social security record and is key to future benefit payments. It is important, therefore, that a person's social security number is on the social security reports for both the self-employed farmer and the agricultural worker.

Those who expect to draw social security payments later should check with the Social Security Administration every three years, especially if they change jobs frequently, to make sure that their records are in order and that their correct earnings are credited to their individual social security records.

For a social security card—either a new card or a duplicate of one that has been lost—or for more information about retirement, survivors, and disability insurance, Medicare health insurance, or Supplemental Security Income, get in touch with the nearest social security office or call Social Security's toll-free number: 1-800-772-1213.

TAX MANAGEMENT AND REPORTING

Good tax management and reporting consists in complying with the law, but in paying no more tax than is required. It is the duty of revenue agents to see that taxpayers pay the correct amount, and it is the business of taxpayers to make sure that they do not pay more than is required. From both standpoints, it is important that farmers and ranchers should familiarize themselves with as many of the tax laws and regulations as possible.

The cardinal principles of good tax management are: (1) maintenance of adequate records, and (2) conduct of business affairs to the end that the tax required is no greater than necessary. Good tax management and good farm management do not necessarily go hand in hand, and may sometimes be in conflict. When the latter condition prevails, the advantages of one must be balanced against the disadvantages of the other to the end that there shall be the greatest net return.

It is recognized that tax matters constitute a highly specialized and complex field, and each farm or ranch will need separate considerations in appropriate planning. The recent rounds of federal tax legislation have made significant changes in the procedures livestock producers must use in accounting, as well as in their approaches to financial and estate planning. More than ever, it is important that they consult competent professionals before embarking upon any business operation involving horses. It is noteworthy that, if a livestock producer's return is to be audited, under the Taxpayer Bill of Rights, the taxpayer is entitled to be represented at the audit by a CPA or other professional. Though the IRS can require the taxpayer's attendance with a special summons, this is not likely to be used at the initial meeting.

Increasingly, as local governments must make up for decreased federal support, local tax matters become more important in planning; this also makes consultation with a specialist knowledgeable in state and local tax law, crucial for effective management.

Some tax pointers of particular interest to livestock operators follow:

1. Pay estimated taxes and file your current return on time.
2. Keep adequate and accurate records and accounts.
3. Separate the farm home from the farm business if possible.
4. Keep year-to-year income as steady as possible.
5. Select the best method of accounting—cash or accrual.

CASH BASIS

Under this system, farm income includes all cash or value of merchandise or other property received during the tax year. It includes all receipts from the sale of items produced on the farm and profits from the sales of items that have been sold. It does not include proceeds from sales if the proceeds were not actually available during the tax year.

Allowable deductions include those business expenses incurred that were actually paid during the year, and depreciation on depreciable items.

ACCRUAL BASIS

This system requires the keeping of complete annual inventories. Tax is paid on all income earned during the taxable year, regardless of whether payment was actually received, and on increases of inventory values of livestock, crops, feed, produce, etc., at the end of the year as compared with the beginning of the year. All expenses incurred during the year's business are deducted from gross income regardless of whether payment is actually made, and deductions are made for any decrease in inventory values of livestock, etc., during the year.

Four methods of inventorying are available to the accrual basis farmer or rancher.

1. **Cost.** Inventory items are valued at the actual cost of producing or purchasing them.
2. **The lower of cost or market value.** The comparison is made separately for each item in the inventory, not for the entire inventory.
3. **Farm price.** Each item, raised or purchased, is valued at its market price less estimated direct cost of disposition. This method must be used for the entire inventory, except that livestock may be valued by the next method.
4. **Unit livestock price.** Animals are classified according to kind and age, and a standard unit price is used for each animal within a class. All raised livestock must be included in inventory under this method. Unit prices must reflect any costs required to be capitalized under the uniform capitalization rules. This method is usually chosen by many large operations. Producers using the unit-livestock method are permitted to elect a simplified production method for determining costs required to be capitalized.

The third and fourth methods are unique to farmers and ranchers.

DISTINGUISH CAPITAL GAINS FROM ORDINARY INCOME

There can be a difference in the maximum tax rates applied to ordinary income and capital gains. Income reported as capital gains may be taxed at a lower rate. Thus, livestock held for sale in inventory, and livestock held for breeding purposes, may produce different tax effects when sold, even if the sale prices are the same. Rates are also affected by the legal form of business operation, e.g., corporation or sole proprietorship. There continues to be developments in this area.

SET UP DEPRECIATION SCHEDULES PROPERLY

Depreciation is estimated operating expense covering wear, tear, exhaustion, and obsolescence of property used in a farm business.

Depreciation may be taken on all farm buildings (except the livestock producer's personal residence), and on everything from grain elevators to sheep clippers, including tile drains, water systems, fences, machinery and equipment.

Those who file returns on a cash basis may also take depreciation on breeding animals which were purchased, but they cannot take depreciation on livestock they raised because all costs of raising are deducted as operating expenses. On the accrual basis, depreciation may also be taken on purchased animals that are not included in inventory.

Taxpayers should list each building, and each piece of machinery on which depreciation is to be computed on the depreciation schedule. Such items as ewes and small implements may be grouped together, but such groupings should be derived from totaling of a detailed individual list kept current in a permanent farm record book.

Depreciation is not available for inventory, which would include animals held for sale to customers. After 1986, depreciable property is placed in specific classes. Because the period over which property is depreciated affects the overall tax revenues, Congress has tended to lengthen recovery periods as a means of increasing tax collections without raising taxes.

1. **Five-year property.** This includes automobiles and light-general purpose trucks, certain technological equipment and research and experimentation property.

2. **Ten-year property.** Single-purpose agricultural structures were originally recovered over 7 years, but after 1988 have a 10-year recovery period. A companion requirement limits recovery on such items to the 150% declining balance method (discussed below).

3. **Fifteen-year property.** This includes equipment used for two-way exchange of voice and data communications.

4. **27.5-year property.** This covers residential rental property.

5. **39-year property.** This covers nonresidential real property. This will include most farm buildings.

The author recommends that owners/managers seek the counsel of their tax specialist when setting up depreciation schedules on buildings and equipment.

For purchased animals, the price paid will generally determine the amount which can be depreciated. Inherited or gift animals can be depreciated. However, their value may have to be established by a qualified appraiser, if the IRS contests the taxpayer's valuation.

DO NOT OVERLOOK ADDITIONAL DEDUCTIONS

■ **Annual expensing**—The annual limitation for expensing depreciable items is $24,000 for property placed in service in 2001 and 2002, and $25,000 thereafter. However, this deduction is reduced dollar-for-dollar for qualified expenses exceeding $200,000. The amount which can be expensed is limited to taxable income derived from the trade or business. The repeal of the Investment Tax Credit and the longer recovery periods for most classes of property increases the value of this provision for the livestock producer.

■ **Soil and water conservation**—Farmers can deduct soil and water conservation expenditures only if the expenditures are consistent with a conservation plan approved by the USDA or a comparable state agency. Such expenditures include treatment or movement of earth, such as leveling, terracing or restoration of fertility, construction and protection of diversion channels, drainage ditches and earthen dams, planting of windbreaks, etc. Though land clearing expenses must be capitalized and depreciated, ordinary maintenance, including brush clearing, remains deductible. Costs of fertilizing and other conditioning of land remain deductible. The amount deducted under this election can't exceed 25% of the taxpayer's gross income from farming for the year. Part of the amount deducted may be recovered if the land is sold within 9 years of the deduction.

■ **Education expenses**—Educational expenses, such as the cost of short courses, are deductible if they are taken to maintain or improve the skills of the person in conducting the operation, or, if the person is employed by a farming operation and they are taken as a requirement of continuing that employment. However, if taken to allow the person to enter another trade or business, such expenses will not be deductible.

■ **Pay children for farm work**—The farmer/rancher must be able to show that a true employer-employee relationship exists. To do so, children should, as much as possible, be treated as are other employees. They should be assigned definite jobs at agreed-upon wages, and paid regularly.

TREAT LOSSES APPROPRIATELY

On the cash basis, no death deduction can be made for an animal that was born and raised on the farm, because the cost of raising the animal has been deducted already with operating expenses. On the accrual basis, when the value of an animal appears in the beginning-of-year inventory but not in the end-of-year inventory, the loss is automatically accounted for in the change in inventory value. Any money received from insurance or indemnity is entered as other farm income.

■ **Losses from destruction, theft, and condemnation**—Special treatment is available for certain gains or losses that are netted. The gains and losses can arise from the sale or exchange of property used in the trade or business, involuntary conversion or condemnation. If gains exceed losses, the net gain is treated as long-term capital gain. If losses exceed gains, the loss is ordinary. While this has limited significance as long as there is no tax differential between capital gain and ordinary income, the likelihood of a reintroduced capital gain preference makes the matter important to keep in mind. Gains and losses from these causes include those involving (1) sheep, regardless of age, which are held for breeding, and held for at least 24 months from the date of acquisition, and (2) other livestock, regardless of age, held for breeding, and held at least 12 months from the date of acquisition. The fact the livestock is included in inventory doesn't prevent this treatment if the animal is held for the required purposes and for the specified time.

■ **Passive activity losses**—Perhaps the most complicated addition to tax law in recent years was the passive activity loss concept, a development which will take tax lawyers years to decipher, with untold questions yet to be answered. Under this concept, all income and losses are divided between passive and nonpassive activities. A passive activity is one which involves the trade or business in which the livestock producer does *not* materially participate. Losses and credits from passive trade or business activities are disallowed to the extent they exceed aggregate passive income. Passive income does not include portfolio income (interest, dividends, or royalties). However, rental activities are (if within the definition provided in the Internal Revenue Code) always passive.

■ **Material participation**—The IRS has provided seven exclusive tests for meeting the material participation requirement as to a particular activity:

1. **The 500 hours test.** The livestock owner participates more than 500 hours in the operation during the year. Obviously, full-time livestock producers will not have significant difficulties in meeting this requirement.
2. **Substantially all test.** The producer's participation constitutes substantially all participation in the activity. Given the requirements for the care of animals, it is unlikely that this test is even necessary for producers, as they would then satisfy the first test in any case.
3. **The 100 hours test.** The individual participates for more than 100 hours and no other person participates for a greater number of hours. Again, this will not generally be relevant to livestock producers. Nevertheless, a physician who owns animals and has a full-time employee to take care of them will often fail to be an active participant under this test.
4. **The related activities test.** The livestock producer participates in a group of activities for more than 500 hours, more than 100 hours in each. This may apply where a producer has a number of operations, but only limited involvement in each.
5. **The 5 of 10 years test.** This allows a livestock producer who has materially participated in the particular activity in the past to qualify as materially participating presently, even if his/her direct involvement has fallen off somewhat.
6. **Personal service activities.** This would apply to consultants involved in the livestock industry, but not to livestock producers running their own operations.
7. **Facts and circumstances test.** This test is, according to many experts, essentially similar to the 100 hours test.

Also, certain retired livestock producers will qualify in the event of death during the year. Though these requirements will have no effect on the full-time producer, they are important factors in terms of investment planning for anyone who is considering investments in rental real estate activities and other ventures.

■ **At-risk rules**—Another provision in the Internal Revenue Code limits losses to the extent that a taxpayer is at risk with respect to a particular activity. This means generally that a taxpayer is limited to the amount of his/her personal investment and the amount as to which he/she is personally liable. This provision specifically applies to farming, which includes livestock activities. The provision was designed principally to preclude losses from tax shelters and other leveraged investments where there may be no real chance that the taxpayer will have to cover the losses. Thus, it will seldom affect livestock producers whose credit is generally limited to the amount of collateral they can provide.

AVOID OPERATING THE BUSINESS AS A HOBBY

If an activity is not engaged in for profit, deductions are generally not available for the conduct of the activity except to the extent of income from it. This requirement has often been applied when the IRS determines that a livestock operation is actually a hobby. Though the problem will generally not apply to full-time livestock producers, others who devote a smaller amount of their time to an operation may find their activity is classified by the IRS as a hobby.

The general presumption for activities is that if an activity is profitable for 3 of the 5 consecutive years ending with the year being audited, it will be presumed to be engaged in for profit. The IRS has indicated that an activity cannot be considered as engaged in for profit until there is a profit year.

In determining whether a livestock operation is a business or a hobby, the IRS will examine the following factors:

1. **The manner in which the operator carries on the activity.** The more businesslike the conduct of the activity, the more likely it is to be recognized as a business. This includes the keeping of accurate records of income and expenses. If the operation is conducted in a manner similar to other profit-making livestock operations, it is more likely to be recognized as a business. If operating methods and procedures are changed because of losses, the impression is enhanced that the operation is a business. If the operation is typical of the other operations in the vicinity, it may indicate an attempt to fit into the livestock industry.
2. **The expertise of the operator and the employees.** A study of the industry and of other success-

ful operations indicates a profit-making approach. If operators tend to ignore advice, they may have to establish that their expertise is even greater than that of their advisors.

3. **Time and effort spent in carrying out the operation.** The more time the owner devotes to the activity as a business and not as a recreational pursuit, the more likely the Service will find that the operation is a business. If the owner hires a full-time assistant to run day-to-day operations, he/she will be in a stronger position to argue that an attempt is being made to turn a profit. If the assistant is an inexperienced family member, the owner's position may, on the other hand, be weakened. Proper and rigid culling of herds will enhance the evidence for business conduct.

4. **The expectation that assets used in the activity will appreciate in value.** Even if current operations do not produce much income, the investment in land and buildings may support an argument that the owner has taken other businesslike factors into consideration. If the primary focus of the operation is breeding, it may take considerable time to get the necessary stock.

5. **Prior successes of the livestock producer.** The more experienced the producer and the more successful his/her prior livestock operations, the more he/she is likely to be seen as a serious business person. It may be important that the producer comes from a family of successful livestock producers.

6. **The operation's history of income and losses.** If losses are due to unforeseen circumstances (drought, disease, fire, theft, weather damages or other involuntary conversions, or from depressed markets), it may be possible to argue that there was nevertheless a profit motive in the operation.

7. **Occasional profits.** An occasional profit may indicate a profit motive if the investment or the losses of other years are comparatively small. The more speculative the venture, the more the livestock producer may be able to show that the losses were not due to a lack of profit intent.

8. **Financial status of the livestock producer.** The more the producer relies on the livestock operation, the more likely the producer is able to justify it as a business. If there are substantial profits from other sources, it may appear that the operation is nothing more than a private tax shelter. If this is the case, the producer may also have to worry about the effect of the limits on passive activity losses.

9. **Elements of recreation or pleasure.** Though having fun does not mean an operation is a hobby, the more the recreational element dominates the livestock producer's involvement, the more likely the livestock producer will have difficulty convincing the IRS that he/she is trying to make a profit. The presence of fishing holes, tennis courts and guest houses may indicate that the producer has a country club (a different sort of business, but not a livestock operation).

ESTATE PLANNING

Human nature being what it is, most livestock producers shy away from suggestions that someone help plan the disposition of their property and other assets after they are gone. Also, to them the subject of taxes on death seldom makes for pleasant conversation.

If a farmer has prepared a valid will, or placed the property in joint tenancy, the estate will be distributed as intended. If not, it goes to the heirs, according to the laws governing intestate (without a will) succession. The heirs are those persons whom the law appoints to succeed to the property in the event of intestacy, and are not necessarily the persons to whom the farmer would want to leave the property. These laws vary somewhat from state to state.

If no plans are made, estate taxes and settlement costs often run considerably higher than if proper estate planning is done. Today, livestock business is big business; many have well over $1 million invested in land, animals and equipment. Thus, it is not a satisfying thought to one who has worked hard to build and maintain a good establishment during their lifetime to feel that the heirs will have to sell the facilities and animals to raise enough cash to pay estate and inheritance taxes. Therefore, livestock producers should consult an estate planning specialist—for example, a lawyer or the trust department of a commercial bank specializing in this work. A limited discussion of some of the major considerations follows:

- **Valuation can be based on farming use.** Owners of farms and small businesses have been granted an estate planning advantage by means of what is called *special use valuation*. Under this concept, a farm or ranch can escape valuation for estate tax purposes at the highest and best use. Thus, a farm located in an area undergoing development may be considerably more valuable to developers than it is as a farm. Nevertheless, if the family is willing to continue the farming use for ten years, the farm can be included in the estate at its value as a farm. The aggregate reduction in fair market value cannot exceed $750,000, indexed for inflation after 1998.

In order to qualify for special use valuation, the decedent must have been a U.S. citizen or resident and the farm must be located in the U.S. The farm must have been used by the decedent or a family member at the date of the decedent's death. A lease to a nonfamily member, if not dependent on production, will not satisfy this requirement. At least 50% of the value of the decedent's estate must consist of the farm.

The property must be passed to a qualified heir, including ancestors of the decedent, the spouse and lineal descendants, lineal descendants of the spouse or parents, and the spouse of any lineal descendant. Aunts, uncles and first cousins are excluded. Legally adopted children are included.

The property must have been owned by the decedent or a family member for five of the eight years preceding the decedent's death and used as a farm in that period. The decedent or a family member must have participated in the farming operation for such a period prior to the decedent's death or disability.

■ **Electing special use valuation.** Though the procedures are clear as to how special use valuation is elected, the frequency with which mistakes are made indicates the importance of having a competent tax attorney or CPA firm prepare the estate tax return.

■ **Recapture tax.** If the farm ceases to be operated by the heir or a family member within ten years, an additional estate tax will be imposed and the advantage of the election will be substantially lost. Partition among qualified heirs will not bring about recapture. A recent change allows the surviving spouse of the decedent to lease a farm on a net cash basis to a family member without being subject to the recapture tax.

■ **Longer time to pay estate taxes.** Estates eligible for special use valuation may often be able to defer payment of estate taxes. Where more than 35% of an estate of a U.S. citizen or resident consists of a farm, the estate tax liability may be paid in up to ten annual installments beginning as late as 5 years from when the tax might otherwise be due. Thus, a portion of the estate taxes is deferred as much as 15 years. For purposes of the 35% requirement, the residential buildings and improvements on them which are on the farm are considered to be part of the farming operation.

If more than 50% of the decedent's interest in the farm is disposed of in the deferral period, then the entire unpaid portion of the estate tax liability is accelerated. The transfer of the decedent's interest in a closely held business on the death of the original heir will not cause an acceleration if the transferee is a family member of the transferor.

■ **Family-owned business deduction.** An additional deduction available to a qualifying estate after 1997 is the family-owned business deduction. Up to $675,000 in value of a qualifying family-owned business interest may be deducted. This deduction is available in addition to special use valuation. As with qualifying for special use valuation, there are several tests which must be satisfied.

■ **Use the gift tax exclusion for lifetime transfers.** The gift tax exclusion is $10,000 per donee per year. However, a husband and wife who elect gift-splitting may jointly give $20,000 per recipient per year. Gifts may be in the form of interests in the farming operation.

■ **Plan with the unlimited marital deduction.** An unlimited deduction is permitted for the value of all property included in the gross estate that passes to the decedent's surviving spouse in the specified manner. Certain *terminable* interests do not qualify for such a deduction—that is, interests as to which of the surviving spouse's interests will terminate on the happening of some event. Surviving spouses may be given *qualified terminable interests*. A common arrangement involves the surviving spouse receiving a lifetime interest in the farm, with the remainder passing on the spouse's death to others, perhaps the children of the decedent. No marital deduction is allowed if the surviving spouse is not a U.S. citizen, unless a specific trust arrangement is used.

■ **Consult a professional.** The preparation of wills, trusts, redemption agreements (if the farm is incorporated), partnership agreements, etc., requires consideration of the effects of federal and state tax law, as well as state law governing the various potential arrangements. Consequently, it is strongly advised that competent professionals be consulted in order to achieve an effective and cost-saving estate plan.

WILLS

A will is a set of instructions drawn up by or for an individual which details how the individual wishes the estate to be handled after death.

Despite the importance of a will in distributing property in keeping with the individual's wishes, many farmers and ranchers pass away without having written a will. This means that state law determines property distribution in such cases.

Every farmer/rancher should have a will. By so doing, (1) the property will be distributed in keeping with his/her wishes, (2) they can name the executor of the estate, and (3) sizable tax savings can be made by the way in which the property is distributed. Because technical and legal rules govern the preparation, validity, and execution of a will, it should be drawn up by an attorney. Wills can and should be changed and updated from time to time. This can be done either by (1) a properly drawn-up codicil (formal amendment to a will), or (2) a completely new will which revokes the old one.

It may be advisable that the same attorney prepare both the husband's and wife's wills so that a common disaster clause can be incorporated and the estate planning of each can be coordinated.

TRUSTS

A trust is a written agreement by which an owner of property (the trustor) transfers title to a trustee for the benefit of persons called beneficiaries. Both real and personal property may be placed in trust.

The trustee may be an individual(s), bank, or corporation, or a combination of two or three of these. Management skill should be considered carefully in choosing a trustee.

In general, a trust can continue for any period of time set by the owner—for a lifetime, until the youngest child reaches age 21, etc. If the trust extends beyond a lifetime, there are limitations which should be explained by an attorney.

KINDS OF TRUSTS

Basically, there are two kinds of trusts, the *living* and the *testamentary*. The living or *inter vivos* trust is in essence an agreement between the trustor and the trustee and may be revocable or irrevocable.

The *revocable trust* can be terminated or altered; under it the trustor is concerned about the here and now, rather than only the hereafter. The trustor continues to make decisions, and can call off the whole arrangement (it's revocable) if it doesn't work out as expected or if the trustor changes his or her mind about its operation. The revocable trust offers no special estate tax advantage; the assets of a revocable trust are included in the estate of the deceased creating the trust. However, it can be written in such a manner as to reduce substantially the estate taxes of the beneficiaries. Also, the revocable trust may eliminate the cost of probate—costs which may include executor's fees, attorney's fees, court costs, and appraisal fees.

The *irrevocable trust* cannot be amended, altered, revoked, or terminated. Under an irrevocable trust, the trustor must be willing to part with the trust property forever (irrevocably) and have nothing further to do with it and its administration. However, the irrevocable trust has many favorable aspects in estate planning; it will reduce estate taxes in both the estate of the trustor and the estate(s) of the life beneficiaries, and it avoids probate.

The *testamentary trust* is so-called because it is established under the provisions of the trustor's last will and testament. The testamentary trust does not become effective until after death of the trustor, followed by probate. There is no tax saving in the trustor's estate. However, the trust may be drafted to save estate taxes in the estates of the beneficiaries.

SHEEP AND GOAT INSURANCE

The ownership of any animal constitutes a risk, which means that there is a chance of financial loss. Unless owners are in such strong financial position that they alone can assume this risk, their animals should be insured.

Several good companies write livestock insurance; and, in general, the policies and rates do not differ greatly.

QUESTIONS FOR STUDY AND DISCUSSION

1. Based on Tables 10-1 and 10-2, would you recommend that a beginning farmer have a sheep enterprise? Justify your answer.

2. In the Washington study, the researchers reported that 90% of the lambs were marketed off milk and grass. Can any other species of farm animals be marketed with so little grain feeding?

3. Discuss the assets, debts, and equities of U.S. farming. Why did farm debts decrease during the period 1985-1994?

4. Why do farmers borrow from different sources for (a) real estate, and (b) non-real estate?

5. Why have the business aspects of sheep and goat production become so important in recent years?

6. List the four major types of business organizations commonly found among farming enterprises.

7. List and discuss the three types of credit available to sheep producers, and the credit factors considered by both lenders and borrowers.

8. Assume a certain kind and size of sheep or goat operation—feeding lambs, raising and finishing market lambs, or producing goat's milk—then prepare a request for credit.

9. What is a feasibility study? How would you go about making such a study or having such a study made?

10. Using Figs. 10-4, 10-5, and 10-6 as guides, develop a budget for the year ahead for your own sheep or goat operation, or for an assumed operation.

11. How would you analyze a sheep or goat operation to determine whether it is profitable? Evaluate the profit indicators given in tables 10-2, 10-3, 10-4, and 10-5. How would you change them?

12. How may computers be used on a practical basis for (a) the ewe-lamb operator, (b) the purebred sheep or goat breeder, (c) the lamb feeder, and (d) the dairy goat operator?

13. How may a student acquire the traits of a good manager?

14. Take your own sheep or goat operation, or one with which you are familiar, and develop a workable incentive basis for the help.

15. What type of incentive do you consider to be best suited for most sheep and goat operations? Justify your answer.

16. Discuss the importance of each of the following to viable sheep or goat enterprises: (a) liability, (b) workers' compensation, (c) tax management and reporting, and (e) sheep and goat insurance.

17. How can a person acquire the needed training and experience in the business aspects of sheep or goat production?

SELECTED REFERENCES

Title of Publication	Author(s)	Publisher
Beuscher's Law and the Farmer	H. H. Hannah	Springer Publishing Company, Inc., New York, NY, 1975
Contract Farming and Economic Integration	E. P. Roy	The Interstate Printers & Publishers, Inc., Danville, IL, 1972
Corporation Guide		Prentice-Hall, Inc., Englewood Cliffs, NJ, 1968
Doane's Farm Management Guide		Doane Agricultural Service, Inc., St. Louis, MO, 1965
Economics: Applications to Agriculture and Agribusiness	R. D. Little	Interstate Publishers, Inc., Danville, IL, 1997
Financial Management in Agriculture	P. J. Barry, et al.	Interstate Publishers, Inc., Danville, IL, 2000
How to Do a Private Offering—Using Venture Capital	A. A. Sommer, Jr.	Practising Law Institute, New York, NY, 1970
Lawyer's Desk Book	W. J. Casey	Institute for Business Planning, Inc., New York, NY, 1971
Livestock and Livestock Products	T. C. Vyerly	Prentice-Hall, Inc., Englewood Cliffs, NJ, 1964
Progress in Sheep and Goat Research	Ed. by A. W. Speedy	C.A.B. International, Wallingford, Oxon, U.K., 1992
Sheep Book, The	Ron Parker	Charles Scribner's Sons, New York, NY, 1983
Stockman's Handbook, The, Seventh Edition	M. E. Ensminger	Interstate Publishers, Inc., Danville, IL, 1992

PART 2
SHEEP

Chapters 11 through 19 deal specifically with the production of sheep. Nevertheless, the reader must realize that some very important aspects of sheep production are also covered in Part I; namely, genetics, reproduction, fundamentals of nutrition, pastures, behavior and environment, health and disease, wool, and business. Therefore, teachers, students, or producers interested in sheep production should combine the reading and studying of Parts I and II. The order in which they wish to do this is left to their discretion and need. However, some cross references in Chapters 11 through 19 will refer the reader to chapters in Part I.

CHAPTER 11

A pair of 1/2 East Friesian and 1/2 Rideau Arcott lambs. (Courtesy, Dr. David L. Thomas, University of Wisconsin, Madison)

TYPES AND BREEDS OF SHEEP

Contents *Page*

Contents *Page*

Contents Page

In no other class of meat animals has so many breeds evolved as in sheep. As domestic sheep were improved in various parts of the world, the producers within different geographical areas soon became convinced that the animals under their care possessed special attributes not found in more distant flocks. Out of this thinking has arisen the approximately 200 different breeds of sheep that exist today. Many of these breeds are of little importance in commercial production, with more than three-fourths of the industry of the world based on the use of not more than six breeds. Even so, breed enthusiasts are usually vociferous about the relative merits of their particular breed, no matter how small the numbers.

The fifth edition of this book listed 29 breeds and categorized them according to the type of wool produced.

But the U.S. sheep industry continued to change! The National Wool Act of 1954, which provided support prices for wool and mohair, was phased out in 1995, followed by unfavorable prices for wool in relation to the price of lamb, and the coming of hair and double-coated breeds of sheep. Also, hand-crafting of natural colors of wool became popular; and milking sheep for producing specialty cheese evolved.

Presently, the American Sheep Industry Association lists 47 breeds of sheep under the following categories: (1) dual purpose breeds, (2) fine wool breeds, (3) hair and double-coated breeds, (4) long wool breeds, (5) meat breeds, and (6) minor breeds.

In this edition of *Sheep & Goat Science*, the author has elected to list all breeds alphabetically, with the narrative of each breed giving the category of each breed, which corresponds to the classification used by the American Sheep Industry Association.

CLASSIFICATION OF THE BREEDS OF SHEEP

Breeds of sheep may be and are classified on several different bases, including (1) their degree of suitability for meat or wool production (meat or wool type), (2) color of face (white or black face), (3) presence or absence of horns (horned or polled), (4) topography of the area in which they originated (mountains, uplands, or lowlands), (5) type of wool produced, and (6) use in breeding. Each system of classification has its special merits, but, for the purposes of this book, sheep breeds will be classified according to (1) type of wool produced and (2) use in breeding.

When sheep are classified according to the type of wool produced, the following classifications are used: (1) fine-wool type, (2) medium-wool type, (3) long-wool type, (4) crossbred-wool type, (5) carpet-wool type, and (6) fur type. Table 11-1 summarizes the classification of the breeds of sheep by type of wool they produce.

In general, all of the breeds classified within each of the six wool-type categories produce wool of a similar character, especially from the standpoint of diameter and length of fibers.

Brief descriptions of each classification follow:

1. **Fine-wool breeds.** The fine-wool breeds of the United States are the American Merino, Delaine Merino, Rambouillet, and Debouillet. All of these breeds are of Spanish Merino extraction and really represent different types or ideals of Merinos brought about through selection. Because of their common ancestry, therefore, the breeds possess many characteristics in common. All of them are noted for fineness of wool and a great amount of yolk; the fleece sometimes loses over 70% in weight in scouring. In general, modern purebred animals of these breeds are of more acceptable meat conformation than formerly, although they are not equal to the meat breeds in this respect. The fine-wool breeds are hardy, gregarious, long lived, and well suited to production under range management methods throughout the world. Like the Dorset and Tunis breeds, ewes of the fine-wool breeds will breed out of season.

2. **Medium-wool breeds.** The Southdown, Shropshire, Oxford, Hampshire, and Suffolk breeds are collectively referred to as the *down* breeds, because of the

TABLE 11-1
BREEDS OF SHEEP CLASSIFIED ACCORDING TO THE TYPE OF WOOL PRODUCED

Fine-Wool Type	Medium-Wool Type	Long-Wool Type	Crossbred-Wool Type	Carpet-Wool Type	Fur Type
American Merino Debouillet Delaine-Merino Rambouillet	Cheviot Clun Forest Dorset Finnsheep Hampshire Montadale North Country Cheviot Oxford Shropshire Southdown Suffolk Tunis	Border Leicester Cotswold Leicester Longwool Lincoln Romney	Columbia Cormo Corriedale Panama Polypay Tailless, or No Tail Targhee	Navajo-Churro Scottish Highland	Karakul

nature of the country in which they were developed. This area in southern England is a country of hills or *downs*. The down breeds came into prominence during Robert Bakewell's time (late 1700s). From the beginning, they were bred primarily for mutton with special emphasis on those characteristics considered important to the nature of the particular grazing lands and climatic conditions as well as on the market demands of the regions in which they were developed. The face and leg color of all the down breeds is some shade of brown or black, and the fleece occupies a middle position between the length and coarseness of the long wools and the extreme fineness and density of the fine wools.

The medium-wool breeds have been popular in the farm-flock regions of the United States, and rams of the larger breeds have been extensively used in market lamb production on the western ranges.

3. **Long-wool breeds.** The long-wool breeds, bred chiefly for meat, are the largest of all sheep. They originated in Great Britain in an era when producers and consumers favored large, coarse, slow-maturing sheep that produced long, coarse wool, and that when liberally fed would become very fat. These conditions gave rise to such important breeds as the Cotswold, Leicester Longwool, Lincoln, and Romney. All of the long-wool breeds are large framed, have square bodies, and are somewhat rangy in build with conspicuously broad backs. As compared with the fleeces of the fine- or medium-wool breeds, those of the long-wool sheep are open, coarse, and very long.

As their size would indicate, these breeds were developed in level-lying country where feeds were abundant and could be obtained without too much travel.

At the present time, most purebred sheep of the long-wool breeds are considered too slow maturing and big to satisfy the demands of the lamb market. Moreover, their carcasses are quite coarse and overlaid with fat. They are, however, of great value in crossbreeding to improve the weight of wool in other breeds, to increase the size of little sheep, and to increase the finishing qualities of some breeds. it is claimed that the long-wool breeds, especially the Romney, will thrive in regions of excessive rainfall, as the long wool carries the water off the body and does not soak it up as a more dense fleece will do.

4. **Crossbred-wool breeds.**[1] The crossbred breeds which are descended from a long-wool X fine-wool foundation, produce medium-fine wool and are therefore often classed with the medium-wool breeds. In general, however, the crossbred-wool breeds are better adapted to the western range than are their respective parent stocks or any of the medium-wool breeds. Under range conditions, they produce better market lambs and heavier fleeces than the Rambouillet, and they are more active and have superior herding tendencies in comparison to either the long-wool or medium-wool breeds.

For many years, commercial ranchers crossed long-wool rams on grade fine-wool ewes in an effort to secure larger ewes that would yield more wool and produce heavier and superior market lambs. The results were often variable. Sometimes the meat qualities would be improved but the wool would be coarse; whereas at other times the wool yield would be greater, but the meat qualities would be disappointing.

Despite the lack of uniformity, most commercial sheep producers of the West preferred this method to the system of alternating in the use of black-faced and fine-wool rams. Topping the band with black-faced rams produced an excellent market lamb; but because of lack of the herding instinct, poor fleeces, and other deficiencies, the resulting crossbred ewe lambs were not suitable for flock replacements. In order to get desirable replacements, therefore, sheep producers had to use fine-wool rams at intervals, with the result that

[1]The listing of the crosses which produced each of the crossbred-wool breeds is given in the subsequent discussion devoted to each breed for purposes of breed history. In no sense does it imply any lack of purity of the respective breeds, or that all of them are new breeds.

the wether lambs of this breeding did not meet market demands. Thus, the need was for a type of sheep which would eliminate the ram problem that invariably plagued sheep producers in the alternate use of black-faced and fine-wool rams and which would produce lambs suitable for either market or replacement purposes. Out of this need arose the Columbia, Corriedale, Panama, and Targhee breeds of sheep—all descended from long-wool X fine-wool foundations.

5. **Carpet-wool breeds.** Wools used in the manufacture of carpets and rugs in the United States are imported from those Asian countries where the native sheep possess a coarse, wiry, tough fleece. Most American wools are too fine to be used in carpets. If so used, they would mat down and wear very rapidly. Carpet wools are quite variable, ranging from 1 to 13 in. in length and from 15 to 70 microns in diameter. In addition, carpet wools show a tremendous range of luster, strength, crimp, and resilience. Although several foreign breeds of sheep produce carpet wools, the only carpet-wool breeds known to most people in this country are the Black-faced Highland and the Karakul, and the latter breed is primarily kept for fur production (the production of carpet wool being somewhat incidental from mature breeding stock).

6. **Fur breed.** Except for the Karakul, all of the other breeds of sheep are kept primarily for lamb and wool production. Karakul sheep are bred primarily because of the suitability of the lamb pelts for fur production.

Another means of classifying sheep is based on their breeding use—whether (1) their primary use is to produce replacement ewes or breeding rams; or (2) they are dual-purpose animals, used to produce both replacement ewes and breeding rams. Within this system, the breeds are classified into (1) ewe breeds, (2) ram breeds, (3) dual-purpose breeds, and (4) other breeds. Table 11-2 summarizes the classification of sheep based on their breeding use. Details follow.

1. **Ewe breeds.** These are the white-faced breeds; they produce fine, medium, or long wool, or crosses among these types. Ewe breeds are selected for adaptability to the environment, reproductive efficiency, wool production, size, milking ability, and longevity. Replacement ewe lambs should be raised from these breed types, or from crosses among these breeds.

2. **Ram breeds.** These are the meat-type breeds. Ram breeds, or crosses of two ram breeds, are raised for the production of rams to cross on ewe breeds and selected for growth rate and carcass quality.

3. **Dual-purpose breeds.** These are the breeds that may be used to produce replacement ewes and/or breeding rams, depending on the production situation. For example, the Columbia is popular as a ewe breed in the mountain regions of the West, but it is used as a ram breed in the Midwest and South.

4. **Other breeds.** For one reason or another, these breeds fail to fit into one of the other classifications. After each breed has been studied, the reasons for its exclusion will be apparent.

5. **Milk breeds.**

BREEDS

Brief, but pertinent, information pertaining to each of the breeds of sheep follows. The breeds are listed strictly in alphabetical order, without being classified as to the type of wool they produce or their breeding use.

BARBADOS BLACKBELLY

American Sheep Industry Association category: Hair & Double-Coated Breeds

Registrations: 60 (1997)

ORIGIN AND NATIVE HOME

The Blackbelly breed originated and evolved on the island of Barbados due primarily to three major fac-

TABLE 11-2
BREEDS OF SHEEP CLASSIFIED ACCORDING TO THEIR BREEDING USE

Ewe Breeds	Ram Breeds	Dual-Purpose Breeds	Other Breeds	Milk Breeds
Corriedale Debouillet Delaine-Merino Finnsheep Polypay Rambouillet Targhee	Hampshire Oxford Shropshire Southdown Suffolk	Cheviot Clun Forest Columbia Cormo Dorset Montadale North Country Cheviot Panama Romney	Border Leicester Cotswold Karakul Leicester Longwool Lincoln Navajo-Churro Scottish Highland Tunis	1/2 East Friesian, 1/2 Rideau Arcott

tors: (1) the introduction of tropically adapted hair sheep from West Africa; (2) the introduction of some other poorly adapted, but prolific, wooled sheep; and (3) an environment of heat, humidity, and parasites, to which wooled sheep were not well adapted. Thus the Barbados Blackbelly evolved from crosses of European wooled breeds and African hair sheep.

CHARACTERISTICS

The Barbados Blackbelly varies from basic black and tan coloring to black, yellow, and variegated pinto patterns. They are black through the under parts in the basal pattern, and also up the neck, down the inside of the legs, on the plank, and the back of the thighs. The chin is black, as well as the poll and the hair inside the ears. The black contrasts with the tan to reddish coat in most other areas.

Mature rams weigh 105 to 125 lb, and have a neckpiece of hair up to six inches long extending down the neck to the brisket. Mature ewes average 100 lb. Both rams and ewes are polled or have short scurs.

Barbados Blackbelly sheep are adaptable to a variety of environments and possess high reproductive efficiency. They average two lambs per litter with a lambing interval of eight to nine months.

Fig. 11-1. Barbados blackbelly. (Courtesy, Oklahoma State University)

BLACK COTSWOLD

Registrations: 78 (1997), 95 (1996)

Black Cotswold sheep were split off from Cotswold (white) sheep. White Cotswolds from Black Cotswolds are registered as Black Cotswolds.

BLACK WELSH MOUNTAIN

American Sheep Industry Association category: Minor Breeds

Registrations: 96 (1997), 89 (1996)

The Black Welsh Mountain sheep produces black wool and premium quality meat with a favorable meat-to-bone ratio.

ORIGIN AND NATIVE HOME

The black wool and rich mutton of the Welsh Mountain sheep was prized in the Middle Ages. Breeders began to specifically select for the black fleece color during the mid-19th century, creating the Black Welsh Mountain sheep.

CHARACTERISTICS

Black Welsh Mountain sheep are small and black, with no wool on the face or on the legs below the knee and hock. Rams are usually horned and ewes polled.

Fig. 11-2. Black Welsh Mountain. (Courtesy, Oklahoma State University)

BLUEFACED LEICESTER

American Sheep Industry Association category: Minor Breeds

ORIGIN AND NATIVE HOME

Bluefaced Leicester sheep were created in Northumberland, England during the early 20th century by selecting Border Leicesters with "blue faces" (white hairs on black skin) and fine fleeces.

CHARACTERISTICS

Bluefaced Leicester sheep are medium to large in size (rams approximately 250 lb, ewes 175 lb). They are hardy, mature early, and have a lambing percentage from 220 to 250%. They are also good mothers and milkers.

Fig. 11-3. Bluefaced Leicester. (Courtesy, Oklahoma State University)

BOOROOLA MERINO

American Sheep Industry Association category: Fine Wool Breeds

ORIGIN AND NATIVE HOME

Booroola Merino sheep were originally developed on the Southern Tablelands of Australia on a private station or farm.

CHARACTERISTICS

Booroolas are similar to Merinos from which they are descended, but have a lambing average of 2.4 (with a range from one to six) and also have the ability to breed all year long.

Fig. 11-4. Booroola Merino. (Courtesy, Oklahoma State University)

BORDER LEICESTER

American Sheep Industry Association category: Long Wool Breeds

Registrations: 583 (1997), 680 (1996)

ORIGIN AND NATIVE HOME

The breed was founded in 1767 by George and Matthew Culley of Fenton, Northumberland, England by crossing some of Robert Bakewell's English stock with either Teeswater or Cheviot sheep (perhaps both). By 1850, the breed was firmly established.

CHARACTERISTICS

Border Leicesters are white, but some animals may have a bluish tinge or black spots. Both sexes are polled. The Border Leicester is slightly smaller and more stylish and active than the English Leicester. Mature rams weigh 225 to 325 lb and mature ewes weigh 175 to 275 lb. The Border type also is distinguished from its cousin in that there is no wool on any part of the head; the ears are more erect and alert; the face is cleaner cut and more refined; and the wool is shorter and denser, with a purled or twisted tip. Border Leicester ewes are considered the more prolific and heavier milkers of the two types.

Fig. 11-5. Border Leicester. (Courtesy, Oklahoma State University)

CALIFORNIA RED

American Sheep Industry Association category: Hair and Double-Coated Breeds

Registrations: 93 (1997), 83 (1996)

ORIGIN AND NATIVE HOME

California Reds are a cross between the Tunis and Barbados breeds—originating at the University of California, Davis in 1970.

CHARACTERISTICS

Lambs are born red but change to a light tan as adults, however the legs and head retain the red color.

Fig. 11-6. California Red. (Courtesy, Oklahoma State University)

Rams have a full mane growing on the chest and are polled.

CALIFORNIA VARIEGATED MUTANT

American Sheep Industry Association category: Minor Breeds

Registrations: 79 (1997), 21 (1996)

ORIGIN AND NATIVE HOME

The California Variegated Mutant (CVM) originated from all white Romeldale sheep, developed by breeding between Romneys and Rambouillets.

CHARACTERISTICS

CVMs have faces generally free of wool, although wool is sometimes found on the forehead. They are typically badger faced, with the body wool a cream color, a silver or grey back, and the belly britch and neck a darker color. Mature rams average 175 to 200 lb and ewes 120 to 150 lb. Twinning and lambing ease are emphasized, with ewes known to breed while still suckling lambs.

Fig. 11-7. California Variegated Mutant. (Courtesy, Oklahoma State University)

CHEVIOT

American Sheep Industry Association category: Meat Breeds

Registrations: 2,256 (1997), 2,160 (1996)

The Cheviot is a small and extremely hardy breed. Its meat qualities are excellent, but the fleece weight is not one of the strong points of the breed.

ORIGIN AND NATIVE HOME

The Cheviot is native to the Cheviot Hills that form about 30 mi of the border country between England and Scotland. As it has received most of its development and improvement in Scotland, however, the Cheviot is classed as a Scotch breed.

The native land of the Cheviot sheep has a harsh climate, with cold winters and heavy snows. In the formative period of the breed, the animals were given little shelter and very little feed other than what they could rustle for themselves. Under these rugged conditions, the development of a hardy breed was inevitable.

Fig. 11-8. Cheviot. (Courtesy, Oklahoma State University)

CHARACTERISTICS

The Cheviot is a beautiful, distinctive-appearing sheep. It is stylish, alert, active, short legged, and blocky in appearance. The face and legs are bare of wool and are covered with short white hairs. Conspicuous black spots often appear on the ears and occasionally on the face and legs. The nostrils, lips, and hoofs are black. The Cheviot is a small breed, with mature rams in good condition weighing 160 to 200 lb and ewes 120 to 160 lb. The fleece weight is generally light, 5- to 7-lb clips being rather common. Both sexes are hornless, although scurs are sometimes found on rams. Few breeds are so well adapted to grazing on hilly pastures as the Cheviot. The breed has not been accepted on the western ranges, primarily because it does not herd well and it produces a light fleece.

The American Cheviot Sheep Society lists any of the following characteristics as a disqualification: horns; malformed mouth (overshot or undershot jaw); split scrotum; less than two testicles in rams; and less than two teats in ewes.

CLUN FOREST

American Sheep Industry Association category: Minor Breeds

Registrations: 327 (1997), 443 (1996)

ORIGIN AND NATIVE HOME

The exact origin of the Clun Forest is obscure, but their wool suggests some Ryeland breeding. Today, the Clun Forest are the most numerous sheep of the marshes of Wales.

CHARACTERISTICS

Cluns are medium-sized sheep weighing 140 to 160 lb. The face, ears, and legs are black, and the legs and face are free of wool. Cluns produce a 6- to 8-lb fleece, grading 3/8 blood with a staple length of about 4 in. Clun ewes are said to be prolific, good mothers, good milkers, easy lambers, long-lived, and easy-to-handle animals.

Fig. 11-9. Clun Forest. (Courtesy, Oklahoma State University)

COLUMBIA

American Sheep Industry Association category: Dual Purpose Breeds

Registrations: 3,635 (1997), 5,105 (1996)

The Columbia breed is strictly a U.S. creation. In honor of its being the first breed of sheep developed in this country, it was appropriately given the name that it now bears.

ORIGIN AND NATIVE HOME

The breeding program, from which the Columbia eventually evolved, was initiated in 1912 by the Bureau of Animal Industry on the King Ranch at Laramie, Wyoming. Five years later, in 1917, the animals were transferred to the newly established U.S. Sheep Experiment Station at Dubois, Idaho. The Columbia breed is based on a crossbred foundation of Lincoln rams on Rambouillet ewes. From the very beginning, breeding animals were selected solely on the basis of utility value, with no breed fancy points to hinder progress.

The Columbia Sheep Breeders Association of America requires inspection for merit before admitting animals for registration.

CHARACTERISTICS

The Columbia breed is said to be larger than the other crossbred breeds and therefore is adapted for use on the better ranges of the West, such as those of northwestern United States. When in range condition, mature rams weigh from 225 to 300 lb and ewes from 150 to 225 lb. Under range conditions, Columbia ewes will produce a yearly clip of 10 to 16 lb, grading 1/4 to 1/2 blood. The ewes are reasonably prolific and good mothers, and the lambs are of acceptable market type. Many specimens of the breed are deficient in spring of forerib and are sloping in the rump and lacking in fullness of the rear quarters. The Columbia, however, is a relatively new breed, and these deficiencies can be improved with further selection.

The Columbia is an open-faced breed with no tendency to wool blindness. The face and legs are covered with white hair. Both sexes are polled. The breed possesses a good herding instinct.

The Columbia Sheep Breeders Association of America states that sheep showing any of the following disqualifications will not be registered: horns, scurs, or knobs; wool blindness, short staple, and uneven fleece; light fleece; wool finer than 1/2 blood or coarser than 1/4 blood (except on lower thigh); overshot or undershot jaw; colored wool or colored hair on legs; ex-

Fig. 11-10. Columbia. (Courtesy, Oklahoma State University)

cessive wrinkles or folds on neck; poor constitution; low productive capacity; malformed, weak pasterns; crooked feet; or crooked legs.

COOPWORTH

American Sheep Industry Association category: Long Wool Breeds

Registrations: 850 (1997), 950 (1996)

ORIGIN AND NATIVE HOME

Coopworth sheep were developed in New Zealand from 1956 to 1968 by I. E. Coop at Lincoln College by crossing Romneys and Border Leicesters.

CHARACTERISTICS

They are medium sized, and have been intensely selected for easy lambing, prolificacy, and good mothering ability.

Fig. 11-11. Coopworth. (Courtesy, Oklahoma State University)

CORMO

American Sheep Industry Association category: Fine Wool Breeds

Registrations: 81 (1997), 48 (1996)

This is a new breed of sheep recently imported from Australia. The United States was the third country to import the Cormo; Argentina and the People's Republic of China were the first and second.

ORIGIN AND NATIVE HOME

Cormos were developed in Tasmania, Australia, from crosses of Merino and Corriedale root stock, by Ian K. Downie. The breed was founded in 1960.

CHARACTERISTICS

Cormo ewes are noted for fertility, twinning ability, heavy milking, good mothering instincts, and strong herding instincts.

Cormos were developed to retain the finest Merino wool quality, while increasing both the amount of wool and meat. They are polled, open faced, with silky, translucent hair on their faces. Cormos give a 70% yield fleece, grading 1/2 blood to fine, with an average staple length of 4 in. Mature ewes weigh about 145 lb and rams weigh about 200 lb. Disqualifications include yellow grease wool and short staple wool.

Fig. 11-12. Cormo. (Courtesy, Oklahoma State University)

CORRIEDALE

American Sheep Industry Association category: Dual Purpose Breeds

Registrations: 3,321 (1997), 3,061 (1996)

The Corriedale is the oldest of all the so-called crossbred wool breeds. It represents an effort to combine into one breed some of the features of both the Merino and the Lincoln, neither of which in the opinion of some breeders possessed the utility qualities needed for sheep production in New Zealand and Australia.

ORIGIN AND NATIVE HOME

The Corriedale originated about 1880 in New Zealand, where sheep raising is a very important industry, and where both mutton and wool production are sought. Thus, the development of the Corriedale was an effort to develop a dual-purpose type of sheep. Lincoln and Leicester rams were crossed on Merino ewes. By inbreeding and careful selection, a uniform type was established that produced a good balance of meat and wool.

These sheep were named after the Corriedale estate of Otago, New Zealand, where the experimental crossbreeding was done.

From its long-wool ancestors, the Corriedale inherited a good meat conformation, and from its Merino parentage, it derived a dense, good-quality fleece.

CHARACTERISTICS

In general, Corriedales stand closer to the ground than Columbias, and they are smaller. Mature rams in good condition weigh 185 to 275 lb and ewes from 125 to 185 lb. On the average, animals of this breed shear 10 to 17 lb of grease wool from each year's growth. The wool usually grades 1/4 to 1/2 blood and is noted for exceptional length, brightness, softness, and a very distinct crimp. Corriedales are outstanding for their efficiency, generally producing more pounds of lamb and wool per pound of body weight than other range breeds.

Fig. 11-13. Corriedale. (Courtesy, Oklahoma State University)

The face, ears, and legs are covered with white hair, although black spots are sometimes present (brown spots are considered a defect). Wool blindness is discriminated against. Both sexes are polled, although rams sometimes have horns. The ewes are considered fair in prolificacy and milking ability. Corriedales herd well. From the standpoint of range production, many practical sheep producers consider the Corriedales too small and lacking in bone and ruggedness.

The American Corriedale Association, Inc., disqualifies for registry animals possessing any of the following faults: black or brown spots in the wool; horns; wool blindness; malformed mouth; marked lack of constitution; malformation of legs or feet; or extreme deviation from the ideal in conformation, fleece, or breed characteristics.

COTSWOLD

American Sheep Industry Association category: Long Wool Breeds

Registrations: 438 (1997), 530 (1996) American Cotswold Record Association; 524 (1997), 463 (1996) Cotswold Breeders Association

ORIGIN AND NATIVE HOME

The Cotswold breed is native to the Cotswold Hills of Gloucestershire, England. According to some historians, however, the name of the breed is derived from the two words *cote* and *wold*—the former meaning *sheep fold* and the latter meaning *open upland*. Some claim that the Cotswold is the oldest of all breeds of sheep. At least, it is very ancient, for as early as the 14th century the wool of the Cotswold region was widely and favorably known. It is also recorded that in 1462 King Edward IV permitted choice sheep from this area to be shipped to Spain. In the latter part of the 18th century, the native sheep of the Cotswold Hills were improved through the use of outside blood, principally that of Leicesters.

CHARACTERISTICS

The Cotswold is very similar to the Lincoln and Leicester breeds, both with respect to its characteristics and its use in present-day sheep production. Perhaps the most distinctive characteristic of the Cotswold breed is the natural wavy ringlets or curls in which the fleece hangs all over the body and falls in cords over the forehead. The wool is coarse and is 8 to 14 in. long and usually grades braid. The face and the legs (below the knees and hocks) are white, although grayish specks and a bluish tinge are common. The nostrils, lips, and skin about the eyes are black. Both rams and ewes are polled, although males frequently have scurs.

The Cotswold is second only to the Lincoln in size. Mature rams in good breeding condition weigh 250 to

Fig. 11-14. Cotswold. (Courtesy, Oklahoma State University)

300 lb and ewes from 175 to 225 lb. The type of the Cotswold is similar to that already described for long-wool breeds in general. The prolificacy, milking qualities, and maternal instinct of the ewes are considered superior to those of the other long-wool breeds, with the exception of the Border Leicester.

DEBOUILLET

American Sheep Industry Association category: Fine Wool Breeds

Debouillets are well-adapted for range sheep production in the southwest, where they were developed.

ORIGIN AND NATIVE HOME

The Debouillet originated on the Amos Dee Jones ranches of Roswell and Tatum, New Mexico, in 1927 to 1930. Their name signifies the origin of the breed—*De* from Delaine and *bouillet* from Rambouillet. The Debouillet Sheep Breeders Association was organized in 1954.

CHARACTERISTICS

The Debouillet is a medium-sized, white, open-faced sheep, with wool on its legs. Mature rams weigh 175 to 250 lb and mature ewes weigh 125 to 160 lb. Debouillets are hardy under arid condition, gregarious, and adaptable to unassisted pasture lambing. They produce a high-quality fine-wool fleece, weighing 10 to 18 lb, with a deep, close crimp.

Disqualification includes any of the following: overshot or undershot jaw; broken-down pasterns; undersize; brown hair on face, ears, or legs; black spots in fleece; or too light fleece.

DELAINE-MERINO

American Sheep Industry Association category: Fine Wool Breeds

Registrations: 656 (1997), 720 (1996)

The Delaine-Merino and the closely related Texas Delaine were developed from the Spanish Merino, which has an unbroken line of breeding extending back 1,200 years.

ORIGIN AND NATIVE HOME

Spanish Merinos were introduced into the United States at the beginning of the 19th century. The states of Ohio, West Virginia, and Pennsylvania pioneered the development of Merinos free from skin folds and with better meat qualities. This strain of sheep became known as Delaine-Merinos.

CHARACTERISTICS

Modern Delaine-Merinos are relatively smooth bodied and intermediate in size. Mature rams weigh 150 to 225 lb and mature ewes weigh 110 to 150 lb. They are white faced, with wool on the legs, and they are hardy, long lived, gregarious, and adapted for unassisted lambing. Delaine-Merinos produce well in extremely warm climates under relatively poor feed conditions, and breed year round. Their fleece is a high-quality fine wool, averaging 8 to 14 lb.

Disqualification from registry includes any of the following: abnormal testicles, swayback, close horns, black spots in the fleece or body, overshot or undershot jaw, and weak pasterns.

Fig. 11-15. Debouillet. (Courtesy, Oklahoma State University)

Fig. 11-16. Delaine Merino. (Courtesy, Oklahoma State University)

DORPER

American Sheep Industry Association category: Hair and Double-Coated Breeds

Registrations: 517 (1997), 220 (1996)

The Dorper is a hardy breed with good mothering abilities that requires a minimum of labor to care for and can thrive under conditions where other breeds can barely exist.

ORIGIN AND NATIVE HOME

The Dorper originated in South Africa in the 1930s by crossing Blackhead Persian ewes with Dorset Horn rams.

CHARACTERISTICS

The head and neck of the Dorper is black, with the rest of the body white, although some do have white heads (White Dorper). Its thick skin, which protects it from harsh conditions, is the most sought after sheepskin in the world.

Fig. 11-17. Dorper. (Courtesy, Oklahoma State University)

DORSET

American Sheep Industry Association category: Meat Breeds

Registrations: 11,534 (1997), 12,181 (1996)

There are horned and polled strains of Dorsets, both of which are registered by the Continental Dorset Club, Inc. Except for the absence of horns in the polled strain, horned and polled Dorsets are identical.

In the horned strain, both the rams and ewes are horned. The newer polled strain originated at North Carolina State University, from a mutation that occurred in a purebred Dorset flock, with the birth of a polled ram in November, 1953. In 1956, the Continental Dorset Club, accepted the first polled animal for registry. The inheritance of the polled trait is complex and not completely understood. For registration, polled Dorsets must show some degree of polledness, which is designated as follows: P = polled; S = scurred; and N = normal horned.

ORIGIN AND NATIVE HOME

The Dorset is native to southern England, especially to the counties of Dorset and Somerset. Although the breed is of medium-wool type, it is not one of the down breeds. The native home of the Dorset is, however, rolling to hilly.

Although the origin of the Dorset is clouded in obscurity, it is known that it developed largely through selection. There is little evidence that crossing with other breeds was of consequence. That crossing was of little importance is further attested to by the very fact that the breed is so different from the others that were available for infusion.

CHARACTERISTICS

The face, ears, and legs are white in color and practically free from wool. The nostrils, lips, and skin are pink. The hoofs are white. The breed is of medium size, with mature rams in good condition weighing 210 to 250 lb and ewes 140 to 175 lb. In type, the best representatives of the breed compare favorably with the other medium-wool breeds, though the neck and body are sometimes inclined to be too long and the fleece may lack weight. Fleece weight averages 5 to 9 lb for ewes.

The ewes will breed out of season, and they are noted for their prolificacy and heavy milk production, factors that make them ideally suited to the production of hothouse lambs. Because the greatest demand for winter lambs is in the East, most Dorsets are raised east of the Mississippi River.

Incisor teeth not meeting the dental pad, abnormal testicles, inverted eyelids, or general off-type appearance disqualifies an animal from registry.

Fig. 11-18. Dorset. (Courtesy, American Sheep Industry Association)

EAST FRIESIAN

American Sheep Industry Association category: Dual Purpose Breeds

Registrations: 1 (1997), 0 (1996)

ORIGIN AND NATIVE HOME

The East Friesian originated in northeastern Germany along the North Sea coast in the region known as Friesland.

CHARACTERISTICS

The East Friesian is considered the world's highest producing dairy sheep. With milk yield of 1,100 to 1,540 lb per lactation (6–7% milk fat). Litter size averages 2.25 lambs. Mature weight ranges from 150 to 200 lb. They are polled in both sexes with white wool. Their faces, ears, and legs are also white, but clean of wool. Their most distinctive feature is a thin and wool-free "rat tail."

East Friesian crosses with Rideau Arcott ram lambs in Wisconsin have produced a good dairy sheep.

Fig. 11-19. East Friesian. (Courtesy, Oklahoma State University)

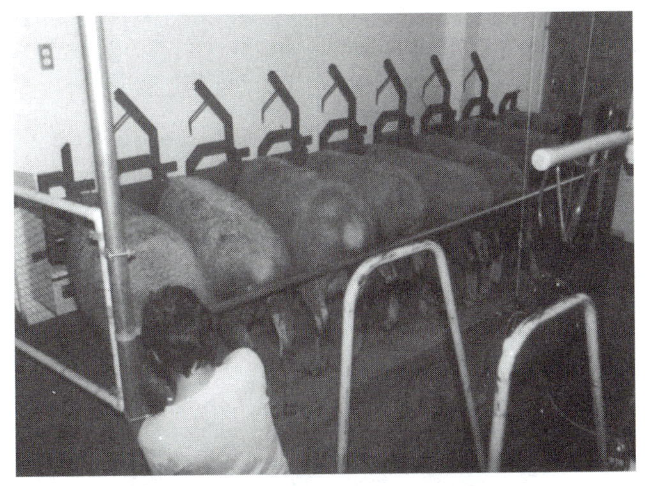

Fig. 11-20. A small dairy sheep parlor with milking pit on a farm in northwestern Wisconsin. (Courtesy, David L. Thomas, University of Wisconsin-Madison)

FINNSHEEP

American Sheep Industry Association category: Dual Purpose Breeds

Registrations: 496 (1997), 242 (1996)

Finnsheep are noted for their prolificacy. Mature ewes often produce three or more lambs per lambing and breed at 6 to 7 months of age. Finnsheep are used in many crossbreeding programs to increase productivity.

ORIGIN AND NATIVE HOME

The Finnsheep or Finnish Landrace, is, of course, native to Finland, where the breed is considered to be highly adapted to the rugged climate and high roughage feeds. In 1918, the Finnish Sheep Breeders Association was formed; and it directed the improvement of the breed.

CHARACTERISTICS

Finnsheep are adaptable to intensive management. They are highly prolific, excellent milkers, and easy lambers with excellent maternal instincts.

Mature ewes weigh 120 to 190 lb and produce a fleece that averages 4 to 8 lb and grades 1/4 to 1/2 blood. Rams weigh 150 to 200 lb at maturity. The head is white and free of wool and horns. The legs are wooled to the knees and hocks. The Finnsheep tail is undocked, since it is naturally short.

Undesirable traits include overshot or undershot jaw, horns, weak pasterns, and the production of single births.

Fig. 11-21. Finnsheep. (Courtesy, Finnsheep Breeders Assn., Inc., Zionsville, IN)

GULF COAST NATIVE

American Sheep Industry Association category: Minor Breeds

Registrations: 30 (1997), 48 (1996)

ORIGIN AND NATIVE HOME

Early importations of Spanish sheep mixed with other breeds under natural selection in the Gulf Coast states to form the Gulf Coast Native Sheep.

CHARACTERISTICS

Gulf Coast Natives have refined bone structure and small bodies. Their faces are open, their legs clean, with underlines that are white to dark brown. They are known to be resistant to internal parasites as a result of their adaptation to the hot and humid conditions of their native climate.

Fig. 11-22. Gulf Coast Native. (Courtesy, Oklahoma State University)

HAMPSHIRE

American Sheep Industry Association category: Meat Breeds

Registrations: 9,782 (1997), 10,596 (1996)

Until recently, Hampshire rams were used to a greater extent than the rams of any other breed for crossing in the United States. However, on the western range this position is being successfully challenged by the Suffolk, a breed whose crossbred get are free from wool blindness, have a smaller head and shoulders that make for less trouble at lambing time, and may have more size. Also, some range sheep producers feel that the use of the more active Suffolk rams results in a larger lamb crop than can be obtained with the use of Hampshire rams.

ORIGIN AND NATIVE HOME

The Hampshire sheep originated in the county of Hampshire in south central England. The native sheep of this area were largely of two strains: the Wiltshire and the Berkshire Knot. Both strains were described as large and coarse, with mediocre-quality meat and light fleeces. Southdowns and Cotswolds were crossed on this native stock; and out of this conglomerate breeding, followed by years of selection, the Hampshire breed of sheep was created.

CHARACTERISTICS

The face, ears, and legs of the Hampshire are a rich deep brown, approaching black; but to most observers the color is simply black. Both the ewes and rams are hornless, although small scurs may be present in the latter. The Hampshire is one of the largest of the medium-wool breeds; among the medium-wool breeds, it is exceeded in weight only by the Oxford and the Suffolk. Mature Hampshire rams in good condition weigh from 225 to 325 lb and ewes from 175 to 225 lb. The head is large, being in proportion to the rest of the skeletal makeup; and many rams possess a Roman profile. As a breed, Hampshires are of excellent meat type. In general, Hampshires do not shear a very heavy fleece; a flock average of 6 to 10 lb is common. The wool is of medium quality, but black fibers are likely to be present; and the latter are objectionable from the viewpoint of the manufacturer.

The Hampshire breed is renowned for the rapid rate of growth made by the lambs. When well cared for, the lambs usually gain a pound or more a day from the time they are born until they are marketed. The ewes are prolific and good milkers.

The American Hampshire Sheep Association lists the following as being undesirable traits: crooked legs and poor feet, inverted eyelids, abnormal sex organs, black fibers, wool blindness, broken woolcap, horns, and abnormal teeth or jaw development.

Fig. 11-23. Hampshire. (Courtesy, The American Hampshire Sheep Assn., Ashland, MO)

HOG ISLAND

Registrations: 8 (1997), 4 (1996)

ORIGIN AND NATIVE HOME

A flock of sheep of substantial Merino blood that were native to the Virginia area were established on Hog Island about 200 years ago. Introductions to the population were occasionally made over the years, the last being a Hampshire ram in 1953. These sheep remained feral until being moved to Fairfax County, Virginia in 1974.

CHARACTERISTICS

Hog Island sheep are very hardy, and fairly light, with mature rams weighing about 125 lb and ewes 90 lb. Both sexes may be horned. Most are white in color, with about 10% being black. Lambs may have spots that disappear with maturity.

Fig. 11-24. Hog Island. (Courtesy, Oklahoma State University)

ICELANDIC

American Sheep Industry Association category: Minor Breeds

Registrations: 118 (1997), 119 (1996)

ORIGIN AND NATIVE HOME

Icelandic sheep are related to such breeds as Finnsheep, Romanov, and Shetland, and are directly descended from the sheep brought to the island by Viking settlers. A few unsuccessful attempts were made to improve the breed with outside crossings, but the result was the introduction of diseases. As a consequence, producers culled all animals that were a result of these crossbreedings, and today it is illegal to import sheep into Iceland. Therefore, any improvements to Icelandic sheep have been the result of selective

Fig. 11-25. Icelandic. (Courtesy, Oklahoma State University)

breeding. Genetically, the Icelandic sheep is the same now as it was 1,100 years ago.

CHARACTERISTICS

Icelandic sheep are of medium size (ewes 150 to 160 lb, rams 200 to 220 lb) and are fine boned. They have open faces, legs and udders, and are primarily horned. The wool of Icelandic sheep is famous throughout the world, but they are raised primarily for meat production. Their flocking instinct is poor, but they are good browsers of brush and wild grasses.

JACOB

American Sheep Industry Association category: Minor Breeds

Registrations: 600 (1997), 636 (1996) Jacob Sheep Breeders Association; 69 (1997), 34 (1996) Jacob Sheep Conservancy

ORIGIN AND NATIVE HOME

The origin of the Jacob breed is unclear, but their coloring may come from Moorish sheep from Spain or

Fig. 11-26. Jacob. (Courtesy, Oklahoma State University)

Africa and their characteristic horns are similar to sheep from Scandinavia and the northern Scottish islands. They were known as Piebald sheep prior to the 20th century.

CHARACTERISTICS

Jacob sheep are slight of build, with narrow, lean carcasses. Their fleeces are white with black spots. Both rams and ewes are horned, having two, four, or even six horns. Some four-horned rams have two vertical center horns as much as two feet long, with two side horns curling along the side of the head.

Fig. 11-27. Karakul. (Courtesy, The Embassy of South Africa, Washington,

KARAKUL

American Sheep Industry Association category: Minor Breeds

Registrations: 317 (1997), 137 (1996)

Karakul lambs are eminently adapted for fur production, but the wool of the mature animal is of low market quality, grading carpet wool, and the breed is of poor meat type.

The majority of Karakul lambskins are produced in Bokhara (West Turkestan), Herat (Afghanistan), South-West Africa, Shiraz (Iran), Baghdad and Salzfelle (Iraq), and India. The best-grade lambskins come from Bokhara, the region in which the Karakul breed originated.

ORIGIN AND NATIVE HOME

The Karakul is an ancient breed of sheep from Asia, in the region of Bokhara, West Turkestan.

CHARACTERISTICS

Mature Karakuls are black or brown, with drooping ears and horns on the rams but not the ewes, and fat-tailed. Mature rams weigh 175 to 225 lb and mature ewes weigh 135 to 160 lb.

Karakul sheep are bred primarily because of the suitability of the lamb pelts for fur production. In order of their value, the pelts are classified as follows:

1. **Broadtail.** This is the most valuable, although its production is comparatively small. It is produced from prematurely born or stillborn lambs, and in some instances from those killed within a few hours after birth. The hair is undeveloped, grows in different directions, and reflects light, giving the pelt its *moiré* appearance (*moiré* means "watery design").

Such premature deaths are not forced but are the result of accidents, such as abortions.

2. **Persian lamb.** This type of fur is next in value and comes from Karakul lambs 3 to 10 days old. It has a tight, lustrous curl that must be watched carefully from the time the lamb is born; for the curl is likely to open rapidly after the fifth day, and while the value of the pelt increases with size, it is essential that the curl remain tight.

3. **Caracul.** Caracul, spelled with a "C," is a trade name given to the lustrous open type of fur that shows a wavy, *moiré* pattern free from close curls.

Caracul skins are light in weight and are best if removed when lambs are not more than 2 weeks old. But this type of pelt does not deteriorate as rapidly with growth as do those of either the broadtail or the Persian lamb types. Caracul is the least valuable of the three types of pelts.

Within each of these three groups, there are different grades of pelts, depending upon quality, tightness of curl, luster, pattern, color, and general appearance. As would be expected, the price varies with the type and quality of the fur and with the supply and demand.

As is true in making any fur garment when smaller furs are involved, large numbers of pelts are required for proper matching. For this reason, it is often difficult to sell small numbers of pelts to advantage.

Up to the present time, only a relatively few Karakul pelts have been produced in this country. For the most part, the lambs have been raised and sold to new breeders interested in establishing flocks.

Contrary to the opinion of many amateurs, the fleece of the mature Karakul is of low market quality and usually is sharply discounted in price. It grades as carpet wool. Although most lambs are black at birth, when they are about a year of age, their coats turn brownish or grayish.

New processes have resulted in the use for fur of pelts obtained from many breeds of sheep. Except for the Karakul, however, all the other breeds of sheep are still kept primarily for lamb and wool production; and any fur production is secondary.

KATAHDIN

American Sheep Industry Association category: Hair and Double-Coated Breeds

Registrations: 3,300 (1997), 2,250 (1996)

ORIGIN AND NATIVE HOME

The Katahdin breed originated at the Piel Farm in Abbot, Maine as the result of efforts to create a meat sheep that does not require shearing. Three African Hair Sheep were imported to Maine in 1957 and extensively crossed with a variety of other breeds, including Tunis, Southdown, Hampshire, Suffolk, and Cheviots. In 1985 the Katahdin Hair Sheep International was incorporated, with the first KHSI members accepted in 1987.

CHARACTERISTICS

Katahdin are hardy, adaptable, and low-maintenance, with lean, meaty carcasses. They are medium sized (rams, 180 to 250 lb; ewes, 120 to 160 lb) and lamb easily, the ewes having excellent mothering ability. They are ideal for grass and forage based systems and pasture lambing.

Fig. 11-28. Katahdin. (Courtesy, Oklahoma State University)

LEICESTER LONGWOOL

There are two strains of Leicester sheep: the Longwool (English) and the Border. They are regarded as two distinct and separate breeds. The Border Leicester is distinguished from the Leicester Longwool in that there is no wool on any part of the head; the ears are more erect and alert; the face is cleaner cut and more refined; and the wool is shorter and denser, with a purled or twisted tip.

ORIGIN AND NATIVE HOME

The native home of the Leicester breed of sheep is in the good farming country of Leicester, in central England. Although the breed is very, very old, it remained for Robert Bakewell, of Dishley, to transform the ancient, ungainly, slow-maturing Leicester—which had been bred merely for size and heavy fleece—into a compact, symmetrical, moderate-sized sheep possessing great aptitude to fatten. Bakewell's work in sheep breeding began about 1760. His methods consisted of rigid selection toward an ideal which he had clearly in mind and close breeding to fix these desired characters. Success crowned his efforts. Although Bakewell's greatest achievement was with Leicester sheep, he was also a noted improver of Shire horses and Longhorn cattle.

Fig. 11-29. Leicester Longwool. (Courtesy, Oklahoma State University)

CHARACTERISTICS

Leicester Longwools are a medium to large breed with a large, high-quality carcass. The fleece is heavy, curly and soft handling, generally weighing from 11 to 15 lb. They are currently listed by the American Livestock Breeds Conservancy as critically endangered (less than 200 annual registrations in North America, and a worldwide population under 2,000).

LINCOLN

American Sheep Industry Association category: Long Wool Breeds

Registrations: 895 (1997), 766 (1996)

Although comparatively few purebred Lincoln flocks have ever existed in the United States, the breed, through crossbreeding and establishing other breeds, has been a powerful influence in molding the sheep industry of the nation.

Lincoln sheep have made an important contribution to the world's sheep industry.

ORIGIN AND NATIVE HOME

The Lincoln originated in Lincolnshire, a fertile area on the eastern coast of England, bordering the North Sea. This breed is still the principal one in this shire, where it is known as the general-utility sheep. The Lincoln is a very old, large, coarse, slow-maturing, heavy-fleeced sheep. The breed has been produced in Lincolnshire for many years, where it has been improved by the introduction of Leicester blood and by selection.

CHARACTERISTICS

The most impressive characteristic of the Lincoln is its great size. It is reputed to be the heaviest breed of sheep in the world. Mature rams weigh 250 to 350 lb or more and ewes 225 to 250 lb. The body form is typical of the long-wool breeds. In addition to being the largest breed of sheep, the Lincoln produces the heaviest fleece of any meat breed. Although light in yolk and open, the fleece weights range from 12 to 20 lb after a year's growth, with staple 8 to 15 in. in length. The face, ears, and legs (below the knees and somewhat below the hocks) are covered with white hair, although black spots may be present. Dark nostrils, lips, and feet are preferred. Both sexes are polled. Lincoln sheep are usually sluggish, slow maturing, frequently patchy when fattened, and only average from the standpoint of prolificacy and milking ability.

Fig. 11-30. Lincoln. (Courtesy, Oklahoma State University)

MERINO

No breed of sheep has contributed so much as the Merino to the development of the worldwide sheep industry. The Merino furnished the foundation breeding so necessary for the production of fine wool and the deep-rooted flocking instinct that has made it possible for a herder to watch over the welfare of a large band.

The chief distinction between the American and Delaine-Merinos is the degree of skin folds, the more wrinkled American Merino being the A and B types and the comparatively smooth Delaine-Merino being the C type.

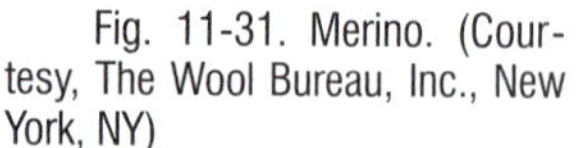

Fig. 11-31. Merino. (Courtesy, The Wool Bureau, Inc., New York, NY)

ORIGIN AND NATIVE HOME

The American and Delaine-Merino are descended entirely from the Spanish Merino; representatives of the latter were introduced into this country at the close of the 18th and the beginning of the 19th century. The word *Merino* is derived from the Spanish word for an early-day royal officer of Spain, called the *Merino*, whose duty it was to assign the various migratory flocks of the country to their respective grazing grounds. In its native land, the Merino breed of sheep had long been selected for fine wool and strong flocking instincts.

CHARACTERISTICS

For the last half century, three types of American Merinos have been recognized. Based largely upon the degree of skin folds, these are the A type, the B type, and the C type (or Delaine-Merino).

However, at the present time, the A type is practically obsolete. The Merino demand is for C type, the smooth-bodied sheep.

MONTADALE

American Sheep Industry Association category: Meat Breeds

Registrations: 3,101 (1997), 3,060 (1996)

ORIGIN AND NATIVE HOME

The Montadale breed of sheep was developed by E. H. Mattingly of St. Louis, Missouri. Although Mr. Mattingly first began crossbreeding in 1914 in an effort to produce what he visualized as a superior type of sheep, the foundation of the present Montadale breed was not laid until 1932. At that time a Columbia ram was crossed on Cheviot ewes. In subsequent years, the sire and dam of these breeds were reversed; that is, a Cheviot ram was crossed on Columbia ewes. The latter cross has been used more extensively and is recognized by the Montadale association as the foundation cross.

CHARACTERISTICS

The Montadale is of good meat type. It possesses a white face and legs that are free from wool. Both sexes are polled. Mature rams weigh 200 to 275 lb and mature ewes weigh 150 to 200 lb, producing an annual wool clip of 10 to 12 lb, averaging 3/8 blood. Montadales are hardy, easy lambers, with excellent maternal instincts, and they adapt to various climates. Disqualification includes: horns, black spots in the wool, or pink nose.

Fig. 11-32. Montadale. (Courtesy, Montadale Sheep Breeders Assn., Inc., Indianapolis, IN)

NAVAJO-CHURRO

American Sheep Industry Association category: Minor Breeds

Registrations: 200 (1997), 200 (1996)

This is the oldest sheep breed in the United States (over 400 years). The Navajo Indians use their wool to make the famous Navajo blankets and rugs.

ORIGIN AND NATIVE HOME

The Navajo-Churro was developed in the United States from the Churra, a Spanish breed.

CHARACTERISTICS

Navajo-Churro rams are horned, frequently multi-horned (four horns), but ewes are generally polled. Rams weigh about 200 lb and ewes weigh 125 to 150 lb. Ewes frequently breed out of season. Probably because of the environmental conditions under which they have been raised the past 200 to 300 years, Navajo-Churro sheep produce plenty of milk on poor range, exhibit excellent mothering instincts, and experience little lambing difficulty.

Fig. 11-33. Navajo-Churro. (Courtesy, Oklahoma State University)

Navajo-Churros yield an extremely coarse wool, which is easy to spin. The wool has a 4- to 6-in. outercoat and a 3-in. undercoat. They are white or colored, generally a single color.

NORTH COUNTRY CHEVIOT

American Sheep Industry Association category: Meat Breeds

Registrations: 585 (1997), 625 (1996)

North Country Cheviots are a hill breed that evolved on the rugged Scotch highlands, where survival was paramount.

ORIGIN AND NATIVE HOME

In 1791, Sir John Sinclair brought 500 Long Hill ewes from the Cheviot Hills near the English border to the counties of Caithness and Sutherland in the north of Scotland. He named these sheep Cheviots after the hill area from which they originated. Later on, another hill breed, the Scottish Blackface, was introduced into the ranges of central Scotland and created a definite separation between the northern counties of Caithness and Sutherland and the border region in southern Scotland. Most authorities speculate that both English and Border Leicesters may have been introduced into the North Country Cheviots in the formative period of the breed. The result was a larger sheep having a longer fleece, and maturing earlier.

Fig. 11-34. North Country Cheviot. (Courtesy, Oklahoma State University)

CHARACTERISTICS

The North Country Cheviot is a medium-sized breed. Mature ewes weigh about 180 lb and mature rams weigh around 280 lb. Ewes are excellent mothers and heavy milkers. North Countrys produce excellent meat and wool. The wool is light shrinking, white, good staple, and free from curl, hair, or kemp. Fleece weights average 6 to 10 lb, grading 1/4 to 3/8 blood. The head is covered with short, glossy white hair and has no horns. The nose may be slightly Roman.

Faults listed by the American North Country Cheviot Sheep Association include: wool between the ears; more than three grades of wool on one sheep; black spots on body where there is wool; undershot or overshot jaw; excessive folds on neck and shoulders; excess kemp; weak pasterns or splayed hoofs; and shades of tan hair on woolless parts.

OXFORD

American Sheep Industry Association category: Meat Breeds

Registrations: 1,287 (1997), 1,520 (1996)

The Oxford, the largest of the medium-wool breeds, is found scattered throughout the United States; but it is one of the less numerous breeds.

ORIGIN AND NATIVE HOME

The Oxford breed of sheep originated in south central England in the county of Oxford, near the seat of the great university that bears the same name. Beginning about 1833, several progressive Oxfordshire farmers decided to develop a new breed through crossing Hampshire rams on Cotswold ewes, with some infusion of Southdown blood. Because of the diversity of the parent stock and the ideals sought, much lack of uniformity still exists in the Oxford breed today.

CHARACTERISTICS

The most distinguishing characteristic of the Oxford is its great size. Mature rams in good condition weigh from 225 to 325 lb and ewes from 160 to 225 lb. The face, ear, and leg color is lighter than the Hampshire, varying from gray to brown. The head and ears appear quite small in comparison with the size of the body. Both sexes are polled, although the rams frequently have scurs. Many Oxfords are more upstanding than would be preferred. The frame is large, and the contour of the body is rectangular.

Perhaps because of their Cotswold heritage, Oxfords shear heavier than any other of the medium-wool breeds. They average 10 to 12 lb for unwashed fleeces. Although the fleece may be somewhat open and may lack in density, the fibers are usually 3 to 5 in. in length. This relatively long, loose fleece further accentuates the size of the breed. Although the wool extends down a little further on the face on Oxfords than on Hampshires, Oxford breeders have wisely and scrupulously avoided any tendency toward wool blindness. A top knot or tuft of wool on the forehead is characteristic, especially when animals have been groomed for show.

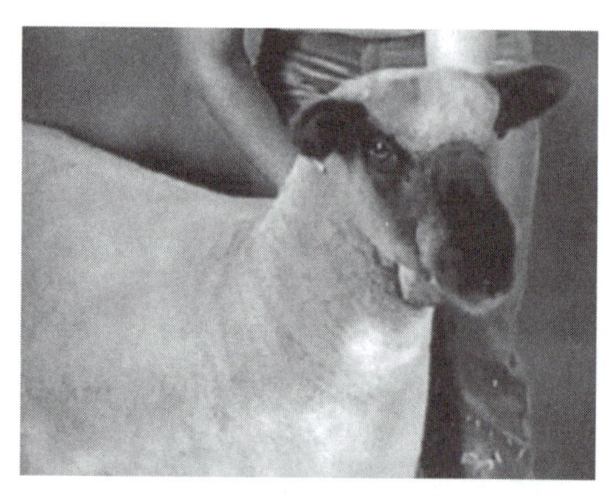

Fig. 11-35. Oxford. (Courtesy, Oklahoma State University)

Oxford ewes are prolific and good milkers with good maternal instincts. The lambs are large and vigorous at birth. Although the young make rapid growth, they do not reach market finish at as early an age as Hampshires. The Oxford is well suited to a diversified type of farming, but thus far the breed has failed to establish itself for crossbreeding purposes on the western ranges to any extent comparable with the Hampshire or Suffolk.

PANAMA

American Sheep Industry Association category:
Dual Purpose Breeds

The Panama closely resembles the Columbia, as both breeds have a similar foundation.

ORIGIN AND NATIVE HOME

The foundation of the Panama breed was started in 1912 by Laidlaw and Brockie of Muldoon, Idaho. Rambouillet rams were crossed on Lincoln ewes. These are the same two breeds that were used in developing the Columbia. The only ancestral difference between the Panama and Columbia is that in the Columbia breed the sexes were reversed; that is, Lincoln rams were crossed on Rambouillet ewes. After Laidlaw and Brockie had made the initial cross of Rambouillet rams on Lincoln ewes, the flock was closed, bred, and selected from within.

CHARACTERISTICS

Fig. 11-36. Panama. (Courtesy, American Sheep Industry Association)

As previously stated, the Panama closely resembles the Columbia. In general, it appears to have been selected a little more for meat conformation and with less stress on size. The Panama produces 3/8-blood crossbred wool, and it is best suited to the western range where ample feed is available.

PERENDALE

American Sheep Industry Association category:
Long Wool Breeds

Registrations: 47 (1997), 30 (1996)

ORIGIN AND NATIVE HOME

Perendales originated in New Zealand as a cross of Border Cheviots and Romneys.

Fig. 11-37. Perendale. (Courtesy, American Sheep Industry Association)

CHARACTERISTICS

The Perendale breed is open faced and medium in size, with bright, lofty, long-stapled, medium-wool fleeces. They are hardy and adapted to marginal forage areas.

POLYPAY

American Sheep Industry Association category:
Dual Purpose Breeds

Registrations: 2,231 (1997), 2,645 (1996)

The Polypay is a synthetic breed developed at the U.S. Sheep Experiment Station at Dubois, Idaho. Five goals were adopted at the development stages of the new breed: (1) high lifetime prolificacy, (2) good lamb crop at one year of age, (3) ability to lamb more frequently than once per year, (4) good growth rate of lambs, and (5) good carcass quality. The name Polypay was coined to suggest more than two paying crops per year—one wool and two lamb crops.

ORIGIN AND NATIVE HOME

In 1968, the U.S. Sheep Experiment Station at Dubois, Idaho, began initial trials in a series of experiments designed to improve lamb production efficiency and to increase prolificacy per lambing. Reciprocal crosses of (Dorset X Targhee) X (Finnsheep X Rambouillet) resulted in a four-breed cross from which offspring of closed inline breeding became designated as Polypay. After this initial cross, the four-breed crosses were mated *inter se* and were selected along lines for best lamb production performance. Sires were selected on the basis of the lifetime lamb production records of their dams, plus their own growth rate from birth to weaning.

The Rambouillet and Targhee were selected for the foundation because of their hardiness, size, long breeding season, herding instinct, and fleece charac-

Fig. 11-38. Polypay. (Courtesy, Oklahoma State University)

teristics. The Dorset was selected for the foundation because of its carcass quality, milking ability, long breeding season, and white fleece, and the Finnsheep because of its early puberty, postpartum fertility, and high lambing rate.

CHARACTERISTICS

Polypay ewes mature early, breed at 8-month intervals, lamb with ease, feed and raise 2 to 3 lambs, and manage easily. Mature ewes weigh 135 to 150 lb and produce about 9 lb of 3/8-blood wool.

Polypay lamb growth rate, body type, and body condition scores are comparable to or superior to those of other breeds and crosses studied at the U.S. Sheep Experiment Station.

RAMBOUILLET

American Sheep Industry Association category: Fine Wool Breeds

Registrations: 9,074 (1997), 7,286 (1996)

Like American and Delaine-Merinos, the Rambouillet is descended entirely from the Spanish Merino. For many years, grade Rambouillets have dominated the commercial range sheep industry of the western United States, though more recently the increased competition of crossbred types has become an important factor. It has been estimated that approximately 50% of all sheep in the United States carry some Rambouillet breeding.

ORIGIN AND NATIVE HOME

During the reign of Louis XVI, France was producing only a small portion of the wool that was used in its factories. In an attempt to build up the nation's flocks, Louis XVI asked, as a personal favor of the king of Spain, that he be allowed to purchase some of the famous Spanish Merinos. His request was granted, and in 1786 a total of 366 head of choice large Spanish Merinos were taken to France and put on the king's estate at Rambouillet, about 40 mi west of Paris. Other flocks were subsequently established in France and Germany, and later breeding stock from these were fused with those coming from Rambouillet. Out of these selections emerged the Rambouillet breed of sheep. From the beginning, the Rambouillet strain was selected and developed for greater size than the average Spanish Merino.

CHARACTERISTICS

Throughout the years, Rambouillet breeders have generally favored dual-purpose sheep, adapted to the production of both wool and meat. However, it is not surprising to find that, from time to time, the emphasis has shifted back and forth between the fleece and the carcass.

The B-type Rambouillets have largely disappeared, and the C-type have been greatly improved. Modern Rambouillets are large, rugged, fast-growing sheep; almost free from skin folds, though an apron across the brisket is not usually considered too objectionable; of acceptable meat conformation, though not equal to the meat breeds; good wool producers—with open faces, long staple, fair density, good uniformity, and moderate shrinkage. In brief, it is to the everlasting credit of Rambouillet breeders that they have been quick to recognize and accede to the demands of commercial sheep producers.

Fig. 11-39. Rambouillet. (Courtesy, *Ranch and Rural Living Magazine*, San Angelo, TX)

Mature rams in good condition and with full fleece weigh from 250 to 300 lb and ewes weigh from 150 to 200 lb. Most rams have large spiral horns, although polled strains exist. The face and legs are white, and the skin is pink.

The American Rambouillet Sheep Breeders Association lists any of the following as a disqualification: abnormal development of testicles or only one testicle descended in the scrotum; unsound udder or inverted teats; overshot or undershot jaw; black spots in the fleece; or black spots on other portions of the body.

RIDEAU ARCOTT

Registrations: 64 (1997), 55 (1996)

ORIGIN AND NATIVE HOME

In 1966 the Animal Research Centre in Ottawa, Canada began a program to create a breed that would reproduce rapidly. It resulted in the creation of three Arcott breeds (Canadian, Outaouais, and Rideau). In 1986, the Rideau Arcott was commercially recognized by the Canadian Sheep Breeders Association. The breed is 40% Finnsheep, 20% Suffolk, 14% East Friesian, 9% Shropshire, 8% Dorset, with the remaining 9% consisting of Border Leicester, North Country Cheviot, Romnelet, and Corriedale.

CHARACTERISTICS

The Rideau Arcott is a large, rapidly growing

Fig. 11-40. Rideau Arcott. (Courtesy, Oklahoma State University)

sheep. Adult rams weigh up to 220 lb, and ewes 200 lb. They are primarily a meat breed, but their wool is of medium quality. Rideau Arcotts are white, however the legs may show some color and the face may have some spotting. Although some rams may develop horny protuberances, they are considered a polled breed.

ROMANOV

American Sheep Industry Association category: Hair and Double-Coated Breeds

Fig. 11-41. Romanov. (Courtesy, Dr. David L. Thomas, University of Wisconsin, Madison)

Registrations: 117 (1997), 213 (1996) Canadian Livestock Record Corporation; 200 (1997), 200 (1996) North American Romanov Sheep Association

ORIGIN AND NATIVE HOME

Romanov sheep are from the Volga Valley near Moscow in Russia.

CHARACTERISTICS

Early sexual maturity (both rams and ewes are fertile by three months of age) and large litters (the North American record is seven live, healthy lambs) make the Romanov an exceptionally prolific breed. Romanovs are born black, but lighten to a silver grey as they make their fleece.

ROMNEY

American Sheep Industry Association category: Long Wool Breeds

Registrations: 2,750 (1997), 2,650 (1996)

Claims are made that the Romney breed of sheep is less susceptible than other breeds to foot rot and liver flukes, two sheep ailments common to wet areas. Although these claims have never been proved or disproved, the breed must possess these or other virtues, for Romney sheep are found in nearly every country of the world and frequently in areas of heavy rainfall.

The Romney may also be referred to as the Romney Marsh or the Kent Sheep.

ORIGIN AND NATIVE HOME

The native sheep of the Romney Marsh region in England were rangy in form, slow maturing, and coarse wooled, but they possessed a hardy constitu-

Fig. 11-42. Romney.

tion. Beginning in the early part of the 19th century, they were improved by crosses with Leicester rams and by careful selection of the offspring.

CHARACTERISTICS

In comparison with the other long-wool breeds, the Romney stands on shorter legs and is more rugged, but it is lighter in weight. Mature rams weigh from 225 to 250 lb and ewes from 160 to 200 lb. Although the Romney is classed as a longwool breed, its fleece, in comparison with the wool produced by the Lincolns, Leicesters, and Cotswold, is less open. The staple is shorter and finer, and there is less luster. Romney wool ranges in grade from braid to 1/4 blood, with average yearly clips of 8 to 12 lb expected.

In general, there is a tuft of wool on the forehead, and the cheeks are covered with short fibers. The balance of the face and legs below the knees and hocks are covered with white hair, though wool may extend below the hocks. The hoofs, nostrils, and lips are dark colored. Both sexes are polled. Romney ewes are not noted for prolificacy.

The American Romney Breeders Association lists black spots as a disqualification, but purebred colored Romneys may be registered.

SCOTTISH BLACKFACE

American Sheep Industry Association category:
Minor Breeds

Registrations: 300 (1997), 295 (1996)

ORIGIN AND NATIVE HOME

The Scottish Blackface comes from the highland country of Scotland.

Fig. 11-43. Scottish Blackface. (Courtesy, Oklahoma State University)

CHARACTERISTICS

Scottish Highland sheep are noted for their striking, stylish appearance. Their fleece consists of a long coarse outer coat and a finer inner coat. Their meat is free of superfluous fat and well known for its distinct flavor. Both sexes have horns.

SHETLAND

American Sheep Industry Association category:
Minor Breeds

Registrations: 1,179 (1997), 915 (1996)

ORIGIN AND NATIVE HOME

Shetlands were most likely brought to the Shetland Islands of Scotland by Viking settlers over a thousand years ago. They are of the Northern European short-tailed group, as are the Finnsheep, Icelandics, Romanovs, and others.

CHARACTERISTICS

The Shetland is the smallest of the British breeds, with rams usually weighing 90 to 125 lb and ewes 75 to 100 lb. Ewes are usually polled, and rams have spiral

Fig. 11-44. Shetland. (Courtesy, Oklahoma State University)

horns. Shetlands come in the widest range of colors of any sheep breed, with 11 main colors and 30 markings. They are hardy, prolific, and the ewes are good mothers producing plenty of milk.

SHROPSHIRE

American Sheep Industry Association category: Meat Breeds

Registrations: 2,458 (1997), 2,490 (1996)

In the United States, years of selection for extreme face covering, with inevitable wool blindness, resulted in the loss of some of the popularity that the breed formerly enjoyed. However, progressive breeders have rectified the situation.

ORIGIN AND NATIVE HOME

The Shropshire breed originated in central western England in the counties of Shropshire and Stafford. The early strains of sheep common to this area included the Morfe Common, Cannock Chase, and Longmynd. With the native stock as a foundation, improvement was brought about through the infusion of Southdown, Leicester, and Cotswold blood. Eventually, through breeding and selection, this mixed ancestry was molded together; and, by 1848, the type of sheep was sufficiently standardized to receive the name that it now bears, though the Royal Agricultural Society did not recognize it as a distinct breed until 11 years later.

CHARACTERISTICS

Shropshires are noted for their good crossing ability and for being hardy, early maturing, long lived, prolific, easy lambers, good milkers, and good mothers. They produce meaty lamb carcasses at light weights. When of ample size and free from wool blindness, Shropshire are considered to be excellent utility animals by farm-flock sheep raisers. Mature Shropshire rams in good breeding condition weigh from 225 to 250 lb and ewes from 145 to 175 lb.

Although they do not possess the extreme meat type of the Southdowns, the best specimens of the Shropshire breed are thick, symmetrical, smooth, and well fleshed

The Shropshire yields medium wool with a distinct crimp. It generally grades as 1/4 to 3/8 blood. Average clips of 10 lb are not uncommon.

The face, ears, and legs are dark. Both sexes are polled, though rams frequently have small scurs.

The American Shropshire Registry Association, Inc., lists the following as discriminations: black wool to any noticeable extent; excessive white on the face, throat, and legs; wool blindness; long, weak pasterns; stubs and scurs; inverted eyelids; and prolapse. Also, the Association lists any of the following as disqualification for registry: lack of breed type; horns; or any tendency toward either overshot or undershot jaw.

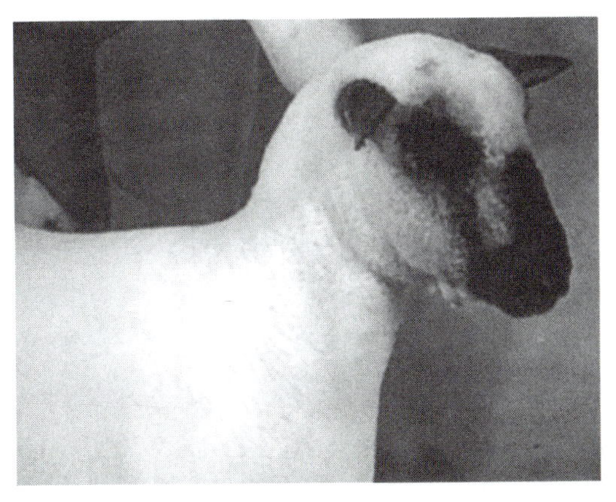

Fig. 11-45. Shropshire. (Courtesy, Oklahoma State University)

SOUTHDOWN

American Sheep Industry Association category: Meat Breeds

Registrations: 5,741 (1997), 5,711 (1996)

The Southdown breed represents the ultimate in meat type. While it is the dominant breed in only a few areas, it is known in practically all sheep-raising countries of the world.

ORIGIN AND NATIVE HOME

The breed takes its name from the chalk hills—called the South Downs—of Sussex County in extreme southeastern England, where it originated. From very early times, the native sheep of this district were famous for their fine mutton qualities. Pioneer breeders selected the best of the native stock and developed the breed known as Southdowns. As the oldest of the medium-wool sheep, they contributed to the foundation stock of all the other down breeds. Because of their early origin and almost universal selection for the best in meat type, the Southdown breed is unexcelled from the standpoint of uniformity.

CHARACTERISTICS

The body of the Southdown is compactly made, very wide and deep, full in the leg, and smooth and refined; the legs are short. The body symmetry and carcass quality are unsurpassed. The Southdown is the smallest of the down breeds, with mature rams in good

Fig. 11-46. Southdown. (Courtesy, Oklahoma State University)

Fig. 11-47. St. Croix. (Courtesy, Oklahoma State University)

breeding condition weighing 190 to 225 lb and mature ewes from 125 to 170 lb.

The fleece—which is short, close, fairly dense, and of fine quality—weighs from 5 to 8 lb unwashed. Very few animals of the breed are subject to wool blindness, although it is preferred that short wool cover the forehead, cheeks, and backs of the ears.

The color of the face, ears, and wool on the legs is gray to mousy brown.

The breed is hornless, though scurs are sometimes found on rams. The ewes are not too prolific, and they are only average as milkers.

Southdowns are early maturing. For this reason, Southdown rams are used extensively with marked success for crossbreeding, especially for the production of hothouse lambs. They also have been popular in some areas, notably Kentucky, for the production of early spring lambs.

Because of their small body size and light fleece, Southdowns have not been accepted for crossbreeding purposes on the western range, though lambs sired by Southdown rams and out of grade Rambouillet or crossbred-type ewes grade high—both on foot and dressed.

The American Southdown Breeders Association lists any of the following traits as disqualification: horns; speckled markings on face, ears, and legs; color of face and legs approaching black; only one testicle down in the scrotum (descended); incisor teeth not meeting dental pad.

ST. CROIX

American Sheep Industry Association category: Hair and Double-Coated Breeds

ORIGIN AND NATIVE HOME

St. Croix sheep are found in the Virgin Islands in the Caribbean and are probably descended from West African hair sheep.

CHARACTERISTICS

St. Croix are well adapted to their climate, resistant to parasites, fertile, and excellent foragers. Most are white with some tan, brown, or black; or white with brown or black spots. Rams have a large throat ruff, and both sexes are polled. Mature ewes average 77 lb and rams 92 lb.

SUFFOLK

American Sheep Industry Association category: Meat Breeds

Registrations: 17,329 (1997), 17,002 (1996)

Although the Suffolk is a very old breed, it did not gain prominence in the United States until recent years. Commercial range sheep operators of the West are using more and more Suffolk rams for crossing on grade Rambouillet or crossbred-type ewes for market-lamb production. The chief advantages that practical sheep producers ascribe to the use of Suffolk rams in comparison with Hampshires are: (1) less trouble at lambing time because of the smaller heads and shoulders of the progeny, (2) production of lambs that are free from wool blindness, and (3) production of lambs that are better rustlers. In addition, it is easier to secure large-type Suffolk rams.

ORIGIN AND NATIVE HOME

As a distinct breed, the Suffolk is comparatively young. It takes its name from Suffolk County, though the development occurred in Suffolk, Essex, and Norfolk counties in southeastern England, in the country bordering the English Channel. The breed was developed by crossing dark-faced Southdown rams on an old native strain of sheep known as the Norfolk. The latter animals were described as wild, hardy, active, upstanding, black faced, horned, light fleeced, and of

Fig. 11-48. Suffolk. (Courtesy, Oklahoma State University)

faulty conformation, but highly prolific and with superior texture and quality of meat.

CHARACTERISTICS

The most commanding characteristic of the Suffolk breed is its very black face, ears, and legs. The head and ears are entirely free from wool, and black hair extends to a line back of the base of the ears. There is no wool below the knees and hocks. The breed is also noted for its alertness and activity. Both rams and ewes are hornless, though the males frequently have scurs.

The size of the Suffolk approximates that of the Hampshire. Mature rams in good breeding condition weigh from 250 to 350 lb and ewes from 180 to 250 lb. Because of the lack of wool on the underline and lower legs, Suffolks are frequently mistaken as being leggy.

In general, the body conformation of the better Suffolks compares favorably with the best of the other meat breeds. The most glaring deficiency of the Suffolk breed is its light-shearing fleece, which frequently contains many black fibers. In the average flock, sheep of this breed will not produce more than 5 to 8 lb (2.3 to 3.6 kg) of grease wool in a year's time. Suffolk ewes are very prolific animals, easy lambers, and excellent milkers. Suffolk rams are highly adapted to crossbreeding with white-faced ewes for market-lamb production.

Because of its alert, active disposition, the breed is unsurpassed as a grazer and rustler.

Disqualifications listed by the National Suffolk Sheep Association include horns or scurs, excessive black fibers in the fleece, and inferior carcass conformation.

TARGHEE

American Sheep Industry Association category: Dual Purpose Breeds

Registrations: 1,794 (1997), 1,492 (1996)

The Targhee is a planned breed, developed for use on the intermediate ranges of the West.

ORIGIN AND NATIVE HOME

The Targhee was the second breed of sheep developed by the U.S. Department of Agriculture, the Columbia being the first. It evolved at the Dubois Station, in Idaho, beginning in 1926. The foundation was started by mating outstanding Rambouillet rams to ewes of Corriedale X Lincoln-Rambouillet breeding and then by interbreeding the offspring from these matings.

The breed derives its name from the Targhee National Forest, on which the U.S. Sheep Experiment Station flock grazes during the summer. Targhees were developed under range conditions, with rigid selection based on production performance.

Fig. 11-49. Targhee. (Courtesy, Oklahoma State University)

The U.S. Targhee Sheep Association was organized on September 27, 1951. Animals may be recorded in either Flock or Stud registry. All sheep eligible for registration must first be inspected and approved by a regularly qualified inspector of the Association.

CHARACTERISTICS

The Targhee is a white-faced, polled sheep of intermediate size. In range condition, mature rams weigh 200 to 300 lb and mature ewes 130 to 200 lb. Ideal animals are moderately low set, free from skin folds, and open faced. Mature ewes shear an average of 10 to 14 lb with a staple length of about 3.0 to 4.5 in.

Most fleeces grade 1/2 blood and yield about 50% clean wool. The ewes produce a high percentage of twins under range conditions, cross well to produce desirable market lambs, and herd well on the range.

Marked scurs or horns or black or brown fibers in the fleece constitute a disqualification.

TEXEL

American Sheep Industry Association category: Meat Breeds

Registrations: 877 (1997), NA (1996)

ORIGIN AND NATIVE HOME

Texel sheep originated in the Netherlands during the late 19th and early 20th centuries by crossing several breeds, including Leicester Longwools, Lincolns, and the local sheep now known as Old Texel.

Fig. 11-50. Texel. (Courtesy, David L. Thomas, University of Wisconsin, Madison)

CHARACTERISTICS

The breed is white with white legs and faces that are free of wool. The Texel is a meat type with high muscle to bone and lean to fat ratios.

TUNIS

American Sheep Industry Association category: Meat Breeds

Registrations: 1,103 (1997), 1,010 (1996)

Though the Tunis is a very old breed, it has never met with great favor among the sheep producers of this country.

ORIGIN AND NATIVE HOME

The Tunis is a fat-tailed, medium-wool sheep of Asiatic origin. It is said to have roamed the hills of Tunis and parts of Algeria, in northern Africa, prior to the Christian era. Further than this, little is known about the origin of the breed.

Fig. 11-51. Tunis. (Courtesy, Oklahoma State University)

CHARACTERISTICS

The Tunis is a medium-sized breed with red or tan face and legs, pendulous ears, and no horns. The wool is white, medium length, and of quality to grade 1/4 to 3/8 blood. The tail is distinctly broad and fat, and in it energy is stored to carry the animal over periods of famine. The Tunis is hardy, and the ewes are prolific and good mothers. Like the Dorset, Tunis ewes will mate at almost any season of the year. Modern Tunis sheep produce lambs of good quality. Newborn lambs are red or tan, but they gradually turn white.

The National Tunis Sheep Registry disqualifies for registry any animal with horns, one testicle, undershot or overshot jaw, or red or black wool.

WILTSHIRE HORN

American Sheep Industry Association category: Minor Breeds

Registrations: 11 (1997), 33 (1996)

Fig. 11-52. Wiltshire Horn. (Courtesy, Oklahoma State University)

te 1. Sheep and goats on the Edwards Plateau in Texas. urtesy, *Ranch and Rural Living Magazine,* San Angelo, TX)

te 2. Sheep in rugged Utah country. (Courtesy, American eep Industry Assn., Englewood, CO)

ate 3. Romney sheep in full fleece tend to their grazing along a eandering stream. They are being watched over by their guard g and accompanied by a Border Collie. (Courtesy, Bonnie Brae rm, Cynthiana, KY, and Francis J. Twomey, River Road Press, aston, PA)

Plate 4. Sheep crossing a stream. (Courtesy, American Sheep Industry Assn., Englewood, CO)

Plate 5. Angora nanny and kids. (Courtesy, Mohair Council of America, San Angelo, TX)

Plate 6. Meat goats on Texas rangeland. (Courtesy, *Ranch and Rural Living Magazine,* San Angelo, TX)

Plate 7. In their winter fleece, the Border Cheviots seem more interested in the snow than in the bits of dried grass remaining in their pasture. The scene makes a perfect winter postcard with the grist mill and old bake house in the background. (Courtesy, Francis J. Twomey, River Road Press, Easton, PA)

Plate 8. Romney sheep. The fleece from these handsome ewes glitters in the morning sun and is complemented by the fall colors. A spinner's paradise is found in this flock. (Courtesy, Francis J. Twomey, River Road Press, Easton, PA)

Plate 9. High on a Maine hillside, the fleeces from these sheep will become handspun yarn and other products at the Agricola Farms Country Store. (Courtesy, Francis J. Twomey, River Road Press, Easton, PA)

Plate 10. Dairy goats. (Courtesy, University of Delaware)

Plate 11. Clun Forest sheep. They originated in the border cour between England and Wales near the Forest of Clun. (Courtesy Francis J. Twomey, River Road Press, Easton, PA)

Plate 12. Considered a primitive and unimproved breed, the Icelandic is one of the oldest, as well as one of the most colorf pure breeds of sheep. The forbears of the breed were thought tc have been brought to Iceland by Celtic and Nordic people durinı its settlement in 800–900 A.D. (Courtesy, Francis J. Twomey, River Road Press, Easton, PA)

e 13. These Scottish Black Face ewes and lambs are doing t sheep do best—grazing. (Courtesy, Francis J. Twomey, er Road Press, Easton, PA)

Plate 16. North Country Cheviots. This is a "hill breed" of sheep that evolved on the rugged highlands and of necessity learned to survive on their own. They are self-reliant, resourceful, and long lived. (Courtesy, Francis J. Twomey, River Road Press, Easton, PA)

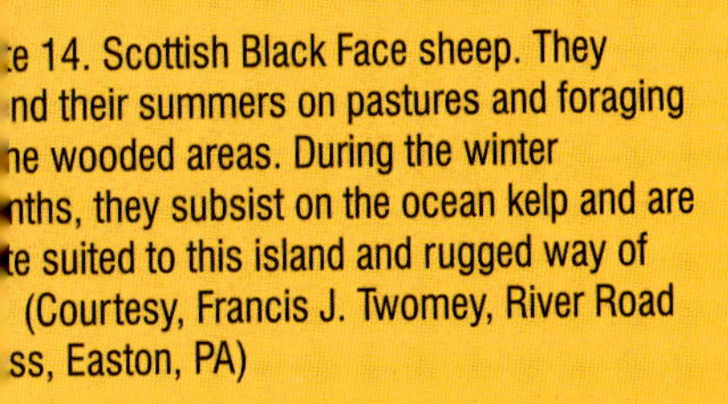

te 14. Scottish Black Face sheep. They nd their summers on pastures and foraging he wooded areas. During the winter nths, they subsist on the ocean kelp and are te suited to this island and rugged way of (Courtesy, Francis J. Twomey, River Road ss, Easton, PA)

ate 15. These hearty sheep spend their year on this private and. Summers are spent grazing on pastures and foraging in e wooded areas. (Courtesy, Francis J. Twomey, River Road ess, Easton, PA)

Plate 17. Shetland Isle sheep. The Shetland feels right at home on this island pasture. Buffeted by the wind, his luxurious fleece keeps him from feeling the cold. (Courtesy, Francis J. Twomey, River Road Press, Easton, PA)

Plate 18. Dog working sheep. (Courtesy, American Sheep Industry Assn., Englewood, CO)

Plate 21. Herding sheep in New Mexico. (Courtesy, American Sheep Industry Assn., Englewood, CO)

Plate 19. Sheep grazing in New Mexico. (Courtesy, American Sheep Industry Assn., Englewood, CO)

Plate 20. The turning of the leaves in Vermont. (Courtesy, American Sheep Industry Assn., Englewood, CO)

Plate 22. Victoria Elisabeth Uhl and James Gustav Uhl of San Antonio, Texas, are caring for a young Rambouillet friend. (Courtesy, *Ranch and Rural Living Magazine,* San Angelo, TX)

ate 23. Many types of brassicas (members of the cabbage
nily including turnips, kale, and tyfon) can provide temporary
azing during late summer and fall when pasture may be in short
pply. Here a group of ewes in Illinois graze on turnips less than
days after planting. (Courtesy, David L. Thomas, University of
isconsin–Madison)

ate 24. A modern lamb feedlot. (Courtesy, *The Sheep Producer,*
rlington, TX)

Plate 25. A headgate for self-feeding silage out of a trench silo in
Wisconsin. The headgate is suspended by chains from a beam
which lays across the top of the bunk. As the sheep consume the
silage, the beam is moved forward to allow access to more
silage. (Courtesy, Tom Cadwallader, University of
Wisconsin–Madison)

Plate 26. Lamb carcasses in a cooler. (Courtesy, American Sheep
Industry Assn., Englewood, CO)

Plate 27. Grafting stantion for
ewes in Oklahoma. Potential foster
mothers are locked in the
headgates where they can have
no visual or tactile contact with
the orphan lamb(s). After
approximately 5 days in the
stantion, over 85% of ewes will
accept the orphan lamb(s).
(Courtesy, David L. Thomas,
University of Wisconsin–Madison)

Plate 28. Lambing pens.
(Courtesy, American
Sheep Industry Assn.,
Englewood, CO)

Plate 29. Small portable creep feeder for lambs on pasture in
Oregon. The top protects the creep feed from rain. (Courtesy,
David L. Thomas, University of Wisconsin–Madison)

Plate 30. Range sheep going up the ramp into a mobile shearing
trailer. (Courtesy, Dr. Johannes E. Nel, University of Wyoming)

Plate 31. Ewes are graded on the hoof for wool quality before shearing. (See article on “Sorting ewes before shearing”) (Courtesy, Dr. Johannes E. Nel, University of Wyoming)

Plate 32. Shearing the sheep. (Courtesy, *The Sheep Producer,* Arlington, TX)

Plate 33. Scouring wool. (Courtesy, The Wool Bureau, Inc., New York, NY)

Plate 34. Roving wool. (Courtesy, The Wool Bureau, Inc., New York, NY)

Plate 35. Spinning wool. (Courtesy, The Wool Bureau, Inc., New York, NY)

Plate 36. Processing wool. (Courtesy, Pendleton Woolen Mills, Portland, OR)

Plate 37. Wool garments. (Courtesy, The Wool Bureau, Inc., New York, NY)

Plate 38. Pendleton signature plaid shirts. (Courtesy, Pendleton Woolen Mills, Portland, OR)

late 39. Saanen goats. (Courtesy, *1984 Yearbook of Agriculture* *nimal Health Livestock and Pets,* USDA)

late 40. Showing a La Mancha dairy goat. (Courtesy, University f Delaware)

late 41. The meatiness and constitution of the South African oer goat make them an important meat goat. (Courtesy, niversity of Wyoming)

late 42. Goats are browsers. (Courtesy, University of Wyoming)

Plate 43. The Powder River Resource Area in Miles City, Montana, used Angora goats to control leafy spurge in areas where chemicals could not be used, such as along the river. (Courtesy, Dalice Landers, Range Technician, Bureau of Land Management, Powder River Resource Area, Miles City, MT)

Plate 44. Angora nannies and kids. (Courtesy, Mohair Council of America, San Angelo, TX)

Plate 45. Angora goats. (Courtesy, Mohair Council of America, San Angelo, TX)

Plate 46. Cashmere goats. (Courtesy, Dr. Maurice Shelton, Texas Ag. Exp. Sta., San Angelo, TX)

Plate 47. Processed mohair. (Courtesy, Mohair Council of America, San Angelo, TX)

Plate 48. A model in a mohair garment. (Courtesy, Mohair Council of America, San Angelo, TX)

Plate 49. A model in a mohair garment. (Courtesy, Mohair Council of America, San Angelo, TX)

Plate 50. Mohair is versatile. It may be in garments or furniture. (Courtesy, Mohair Council of America, San Angelo, TX)

Plate 51. Mohair garment. (Courtesy, Mohair Council of America, San Angelo, TX)

ORIGIN AND NATIVE HOME

The Wiltshire Horn is an ancient British breed from the Chalk Downs region of England.

CHARACTERISTICS

Both ewes and rams are white, occasionally having small black spots in the undercoat. They are medium sized (rams, 250 lb; ewes, 150 lb), thrifty, and do well on good grass and hay. Wiltshire Horns also do well in both hot and cold environments, and their woollessness means no trouble with external parasites and no need to dock tails.

QUESTIONS FOR STUDY AND DISCUSSION

1. Discuss types of sheep and classify the U.S. breeds of sheep according to (a) their wool type and (b) their breeding use.

2. List those breeds of sheep common to the United States that are U.S. creations and those that are importations.

3. Classify the breeds of sheep common to the United States according to their mature size—small, medium, or large.

4. How did the polled Dorset evolve?

5. List any unique points about the following breeds: Finnsheep, Karakul, and Navajo-Churro.

6. What is Persian lamb?

7. Explain why the Merino holds a special position among the breeds of sheep, and explain the relationship of the Delaine-Merino to the Merino.

8. What need(s) prompted the development of the Columbia, Cormo, Corriedale, Panama, Polypay, and Targhee breeds of sheep?

9. What is the difference between the Cheviot and the North Country Cheviot breeds of sheep?

10. How was the Polypay breed developed?

11. Justify any preference that you may have for one particular breed of sheep.

12. Obtain breed registry association literature about your favorite breed of sheep. (See Appendix for addresses.) Evaluate the material that you receive.

SELECTED REFERENCES

Title of Publication	Author(s)	Publisher
Breeds of Livestock, The	C. W. Gay	The Macmillan Company, New York, NY, 1918
Breeds of Livestock in America	H. W. Vaughan	R. G. Adams Co., Columbus, OH, 1937
Colour and Island Sheep of the World	Ed. by G. J. Enzlin	Ram Press, Printed in the Netherlands by Transmodial Voorthuisen, 1995
Karakul Sheep, The	E. B. Bertone, *et al.*	Food and Agriculture Organization of the United Nations, Rome, Italy, 1968
Modern Breeds of Livestock	H. M. Briggs D. M. Briggs	The Macmillan Company, New York, NY, 1980
Sheep Book, The	J. McKinney	John Wiley & Sons, Inc., New York, NY, 1959
Sheep of the World in Color	K. Pouting	Blandford Press, Ltd., Poole, Dorset, England, 1980
Sheep Science	W. G. Kammlade, Sr. W. G. Kammlade, Jr.	J. B. Lippincott Co., Philadelphia, PA, 1955
Stockman's Handbook, The, Seventh Edition	M. E. Ensminger	Interstate Publishers, Inc., Danville, IL, 1992

Study of Breeds in America, The	T. Shaw	Orange Judd Co., New York, NY, 1900
Types and Breeds of Farm Animals	C. S. Plumb	Ginn and Co., Boston, MA, 1920
World Dictionary of Breeds, Types, and Varieties of Livestock, A	I. L. Mason	Commonwealth Agriculture Bureaux, Slough, Bucks, England, 1951

Also, breed literature pertaining to each breed may be secured by writing to the respective breed registry associations (see the Appendix, Table A-6, for the name and address of each association).

CHAPTER 12

Prolificacy and profits from sheep go together! This shows a Finn ewe with quads. (Courtesy, Finnsheep Breeders Assn., Zionsville, IL)

ESTABLISHING THE FLOCK; SELECTING AND JUDGING SHEEP

Contents — *Page*

Whether a large range operation or a small farm flock is being established or maintained, consideration must be given to certain factors if the venture is to be successful. In the final analysis, sheep are maintained for the production of market lambs and wool. This means that each individual in the flock should possess those characteristics making for maximum and efficient production of these two products. Furthermore, if progress is to be made in the breeding program, each succeeding generation must represent an improvement over the parent stock.

FACTORS TO CONSIDER IN ESTABLISHING THE FLOCK

The factors to consider in establishing a flock of sheep are not unlike those that the producer must face in establishing a herd of cattle or a herd of hogs. They are, however, somewhat more confusing because (1) two major products are involved—lamb and wool— instead of one; and (2) there are more breeds from which to select, and the practice of crossbreeding is more prevalent.

PUREBREDS, CROSSBREDS, OR GRADES

Generally, only the experienced breeder should undertake the production of purebreds with the intent eventually of furnishing foundation or replacement stock to other purebred breeders, or purebred rams to commercial producers. Unless prices are unusually favorable, the beginner should start with crossbred or grade ewes and a purebred ram. The vast majority of range operators, most of whom are capable sheep specialists, elect to keep bands of high-grade ewes that are mated to purebred rams.

NATIVE OR WESTERN EWES

Native ewes are those that are produced outside the western range area and that show a predominance of meat-type breeding. Western ewes are those that are produced in the range area and that either show a predominance of fine-wool breeding or represent fine-wool X long-wool crossbred types.

In starting a farm flock, the beginner must decide whether native or western ewes should be purchased. Both types of ewes are found in flocks throughout the country and may well be considered. In general, western ewes are more uniform, smaller in size, less costly, and less likely to be infested with parasites. On the other hand, native ewes are larger and produce larger lambs. If they are bought locally, a savings in price may be effected. In purchasing native ewes, however, buyers should make very certain that they are not obtaining the cull ewes (shy breeders and ewes with unsound udders or other defects) from one or several flocks. From this standpoint, the purchase of yearlings affords the best protection.

For range operations, where a large number of animals are to be handled in one or more bands, grade ewes of fine-wool breeding or ewes of the fine-wool X long-wool crossbred type are essential.

SELECTION OF THE BREED

Within certain limitations, preference is usually the deciding factor in the selection of a breed. Most commercial bands in the West, however, should carry considerable fine-wool breeding chiefly because of the herding instinct in the animals. Except for those situations, U.S. sheep producers have given little consideration to possible special-area adaptations of the different breeds. English sheep breeders, on the other hand, have long contended that some breeds are peculiarly adapted to certain conditions. This conviction stems from the fact that the several English breeds were developed in different geographical areas, each of which had different climatic, soil, and crop conditions.

According to English breeders, the small and more active breeds of sheep that were developed in the hill country of England are not adapted to the lowland region, whereas the large and more sluggish breeds developed in the fertile lowland areas will not thrive on the uplands. Moreover, the Romney breed is said to thrive better than any other breed in marsh areas similar to those of its native home in southern England.

Unfortunately, there is little experimental work either to substantiate or to refute these contentions relative to environmental adaptations of breeds. However, at the USDA Experiment Station at Dubois, Idaho, two crossbred breeds of sheep have been developed—the larger Columbias for the lush ranges of the West and the smaller Targhees for the average ranges. Also, for the production of hothouse lambs, ewes must breed out of season and milk well. Among the old established breeds, this limits the production of hothouse lambs to ewes of Dorset, Finnsheep, Merino, Rambouillet, or Tunis breeding.

In as large and variable a country as the United States, some breeds are probably better adapted to certain conditions than others. It is also quite likely that rather wide differences in environmental adaptations may exist within breeds. Perhaps there is need for experimental work to determine these differences, if any, but (1) breed preference is, and should remain, a powerful factor in the selection of a breed, and (2) it is difficult, if not impossible, to obtain a representative cross-section of a widely disseminated breed of sheep for experimental study.

In general, sheep breeds are classified into one of three groups according to their commercial use: (1) ewe breeds, (2) ram breeds, or (3) dual-purpose breeds. Ewe breeds are the white-faced sheep producing fine, medium, or long wool from which replacement ewe lambs are raised. Ram breeds are the meat-type breeds raised primarily for the purpose of crossing on the ewe breeds. Dual-purpose breeds can be used as ewe or ram breeds, depending on the production situation. Breeds falling into each of these classifications are described in Chapter 11.

SIZE OF THE FLOCK OR BAND

Sheep units vary in size all the way from small farm flocks of only a few head, which are generally part of diversified farm enterprises, to large farm flocks of 500 head or more and range bands with as many as 3,500 head (and one owner may have several bands), which constitute highly specialized operations with little or no diversification.

In determining the size of the flock, the beginner should note that the labor and equipment cost, except at lambing time, differs very little whether the flock numbers 200 or 1,500. The smaller flock will require practically the same amount of fencing to provide rotation of pastures or suitable corrals away from dogs. Furthermore, fewer sheep are much more likely to receive the neglect often accorded a minor enterprise. On the other hand, the beginner can acquire valuable practical experience with a very small flock without subjecting a large flock to the possible hazards that frequently accompany inexperience. In range operations, the size of the band varies somewhat according to the method of management, the general character of the country, and the season of the year. The number of bands run by a given operator is usually determined by the amount of range and capital available.

TIME TO START

Late summer, after the lambs have been weaned and before the ewes are bred, is usually the best time in which to start in the sheep business. At this season of the year, it is generally possible to buy some of the surplus ewes from neighboring farmers or ranchers or to obtain a wide selection of ewes from the terminal markets. Ewes can usually be purchased at reasonable prices at this period, and for the beginner there is a period of valuable training prior to lambing time. Furthermore, there is usually an abundance of meadows, grain stubble land, and other forage available on the farm or ranch that will make it possible to get the ewes in a good, thrifty condition prior to breeding time. Then, too, the purchase of a uniform flock of ewes makes possible the intelligent selection of the ram and the production of lambs with uniformity of type and quality.

UNIFORMITY

In order to produce uniform market lambs and wool, it is first desirable that the ewe flock selected be uniform—preferably from the standpoint of breeding, size, body conformation, and fleece grade and quality. Such uniformity is of decided advantage in the marketing of products at premium prices. Also, the constructive sheep producer is able to select more wisely a ram or rams for most successful mating with a uniform flock or band of ewes.

HEALTH

All ewes selected should be in a thrifty, vigorous condition. They should have every appearance of a life of usefulness ahead of them and give every evidence of being capable of producing a good fleece and raising strong, healthy lambs. Animals showing dark blue skins, paleness or lack of coloring in the lining of the nose and eyelids, listlessness or lack of vigor, and a general rundown condition should be regarded with suspicion.

AGE

The vast majority of sheep producers prefer to establish a flock by acquiring yearling ewes that are bred to lamb when approximately 24 months of age. When it can be made certain that culls are not being secured, however, older ewes are frequently the best buy. Sometimes it is to the advantage of a farmer to secure range ewes that have 1 to 2 years of usefulness left, provided that they are placed under conditions where (1) less traveling is required, (2) the pastures are more abundant, and (3) more attention is given to winter feed.

SOUNDNESS OF UDDER

In selecting ewes, sheep producers should carefully check the udders and teats. The udders should be soft and pliable and the teats normal. Ewes having hard or pendulous udders, those whose teats have been removed through careless shearing, or those with meaty or abnormally large teats should be rejected.

SIZE

There is great variation in the size of the different breeds of sheep, extending all the way from the small, refined Finnsheep to the large, ponderous Lincolns. The rank and file of commercial sheep producers in this country prefer animals with considerable size, because, based on experience, such animals are more profitable from the standpoint of the net receipts derived from the usual combination of lamb and wool production. This is because, in general, the cost of handling sheep is on a per head rather than on a per pound basis, and our markets are not sufficiently discriminating in the purchase of lamb and wool to warrant any great sacrifice in quantity in favor of quality production. On the other hand, an exception exists when a premium is paid for early-maturing, high-quality lambs, such as in hothouse lamb production. Under the latter conditions, it is preferable to use early-maturing, meat-type rams on medium-sized ewes. Also, where feed is sparse and sheep are run with a minimum of care, smaller sheep may be preferable, as they often are more efficient than larger animals. On the western range, larger-grade ewes of Rambouillet extraction have always been more popular than the smaller ewes of Merino breeding.

PRICE

The price of breeding sheep must be based upon the projected price of the products of production—lamb and wool. Although it is usually sound business to pay a premium for quality foundation stock, the ultimate objective of all sheep production is profitable market lamb and wool production; thus, the price should be governed accordingly.

SELECTION BASES

The criteria used in selecting an individual or a group of sheep or lambs will vary according to the use for which the animal or animals are intended. In general, commercial farmers or ranchers select sheep or lambs for the following purposes: (1) breeding sheep, with market lambs as the primary objective; (2) breeding sheep, with wool production as the chief objective; (3) feeder lambs for further finishing; or (4) finished lambs to send to the market. Thus, where market-lamb production is the main consideration, primary attention is given to those factors that will result in the production of a heavy finished lamb of acceptable type and grade. When sheep are maintained chiefly for wool production, the clean weight and quality of fleece are of paramount importance. Feeder lambs are selected on the basis of desirable type and quality, health and thrift, and projected gains and finish with feeding. Slaughter lambs are selected for the market, using finish, conformation, and quality as the criteria in determining value. The experienced operator can determine the degree of finish of sheep by touch, by rapidly applying the palm of one hand to each of the animals being examined, as they file through a cutting chute.

Purebred breeders may add certain breed fancy points and highly prized pedigrees to the above considerations, and market specialists find it necessary to add other considerations in arriving at the rather imposing list of market classes and grades of sheep. It is not intended, therefore, that the amateur may become proficient as a sheep judge through merely reading about the subject. There is no shortcut or substitute for long years of patient practice.

In establishing a new flock or improving an old one, however, the sheep producer has four bases of selection from which to choose; namely, (1) selection based on type or individuality, (2) selection based on pedigree, (3) selection based on show-ring winnings, and (4) selection based on production testing.

SELECTION BASED ON TYPE OR INDIVIDUALITY

A vast majority of sheep, both purebred and commercial animals, are selected on the basis of type or individuality. In limited instances, three additional criteria may be invoked effectively; namely, pedigree, show-ring, and production selection. Although production selection offers a modern approach to animal improvement and pedigree selection may be a valuable guide in certain instances, perhaps for many years to come, selection will continue to be based largely on individuality.

The judging of sheep—their selection based on individuality—differs materially from the judging of cattle, hogs, or horses because of the presence of the fleece. Not only must an additional characteristic of economic importance be considered, but the presence of the fleece makes it more difficult to determine the body type. This is especially true when the animal has been subjected to the blocking art of a clever shepherd, but the problem exists whenever there is any appreciable wool growth. Even though complete production records are not available, progressive sheep raisers can arrive at sufficient evaluation to enable them to (1) cull the dry ewes, (2) remove wool-blind ewes and rams, and (3) sort out the light-fleeced animals by the *touch method* or according to actual fleece weights. This type of selection directly affects the income, even though it may not be very important genetically.

SELECTION BASED ON PEDIGREE

Without doubt, less attention is paid to pedigree selection and family names in sheep than in any other class of farm animals. It is rare, indeed, to find commercial producers who have any concern about the ancestry of the rams that they are contemplating purchasing. Although most purebred breeders are interested in the pedigrees of stud rams, there is comparatively little interest in the ancestry of the flock ewes. In some respects, this relative lack of interest in pedigrees on the part of sheep breeders is fortunate, for breeding programs have not been subjected to the hazards of worshipping family names or selecting a breeding animal chiefly because its ancestry traces to some noted animal many generations removed. On the other hand, fancy pedigrees may have sales value from the standpoint of the purebred breeder.

SELECTION BASED ON SHOW-RING WINNINGS

The great livestock shows have been a profound influence in establishing type in the different breeds of sheep. However, when utilitarian considerations have been ignored, their influence has not always been for the good. A prime example of the latter point is the face covering that has been bred into some of the meat breeds of sheep. This has caused the affected breeds to diminish in importance and numbers, because, through sad experience, the practical operator has found that ewes subject to wool blindness produce less market lamb and wool. Utility value, therefore, should always come first, and breed fancy points second.

Winning rams and ewes and their progeny are usually in great demand, with the result that they generally bring premium prices. Provided that the type has been established wisely and is based on utilitarian considerations, the purebred breeder may find it desirable and profitable to select some animals on the basis of show-ring winnings.

SELECTION BASED ON PRODUCTION TESTING

Production testing includes (1) performance testing and (2) progeny testing. In different a
ever, the emphasis attached to each produ
will vary. For example, in the southwestern
States, where wool production is relatively more
tant than lamb production, greater stress shou
placed upon selecting animals whose progeny p
sess the maximum in fleece weight and quality. On t
other hand, in those areas where feed is more abun
dant and approximately 75 to 80% of the annual income is derived from the sale of lamb, evaluations should be based largely on the market weight, type, and finish of the offspring.

Production selection should be given greater emphasis in animal improvement, for, in most instances, it is production—of which individuality or type is merely a part—that produces the income. Also, again and again, it should be emphasized that selection based on either performance testing or progeny testing is far more accurate than any other method of selection. Without doubt, however, the progressive sheep producer will at least make increased use of production-tested rams. Despite the greater certainty and more rapid progress afforded through selection based on production, producers have traveled a long way in sheep improvement—from the rather poor meat type and average 2-lb clips of the sheep characteristic of General Washington's time to the modern sheep of today.

Production testing systems, their definitions, and relative merits are presented in Chapter 3, Principles of Genetics. Also, traits of importance in testing sheep and suggested records are given in the same chapter.

FLOCK IMPROVEMENT THROUGH SELECTION

Once the flock or band has been established, improvement can be obtained only through constant, rigid culling and careful selection of replacements. Such procedure makes the flock more profitable from the standpoint of quantity and quality lamb and wool production and affords a means of accomplishing genetic gain in the next generation.

Individual sheep identification (tag or brand) and production records (see Figs. 3-18a and 3-18b for record form) are requisite to effective culling and selection. Also, those traits that are most heritable, and that contribute most to income, should be considered. In the majority of flocks, selection and culling should be based on the following:

1. **Adaptability.** Sheep adaptable to the existing environmental conditions and management practices are necessary to establish a profitable sheep operation. This entails selecting the proper type or breed of sheep for the existing conditions. Some of the traits

(a) ability to produce, ...king instinct. Experi- ... a breed or type for

... Rapid-gaining ani- ...fficient use of feed and ...ger ages. Hence, all lambs ... weaning and/or marketing time. ...e feed is a major factor in determining ..., single lambs on good feed reach a market ... 100 lb at four months of age; and each member of twins will be only about 10 lb lighter.

3. **Fleece weight and quality.** Each fleece should be weighed at shearing time, since wool is sold by the pound. In addition to selecting for heavy-shearing sheep, the fleece should be dense, uniform, high quality, and free from dark fibers.

4. **Conformation and carcass quality.** Select meaty animals that have good size and ample bone. Also, select carcasses without excess fat, with a large loin eye, and with a maximum yield of tender lean meat.

5. **Multiple births (prolificacy).** Even though the heritability of this trait is low, it is important; and progress in increasing lambing percentage can be made through selection.

6. **Open face.** This trait is important, since an open-faced ewe will produce 11 to 12 lb more lamb than one that is wool blind. Also, this trait is highly heritable and easy to evaluate. Hence, only open-faced animals should be retained for breeding purposes.

7. **Freedom from abnormalities and defects.** Select breeding animals that are free from such defects as no teeth (gummers), undershot or overshot jaw, unsound teats and udder, one or both testicles not descended, and crooked legs or weak pasterns.

JUDGING SHEEP

The judging of sheep differs from the judging of cattle or swine in that two products of economic importance are involved, instead of one. That is, in addition to meat production, wool is a valuable product. The situation is made more trying because body conformation is often difficult to determine due to the wool under which it is hidden. The latter situation is accentuated when sheep are subjected to the art of blocking by a clever shepherd.

The essential qualifications that a good sheep judge must possess, and the recommended procedure to follow in the judging assignment, are as follows:

1. **Knowledge of parts of a sheep.** This consists of mastering the language that describes and locates the different parts of a sheep (see Fig. 12-1). In addition, it is necessary to know which of these parts is of major importance; that is, what comparative evaluation to give to the different parts.

2. **Clearly defined ideal or standard of perfection.** Successful sheep judges must know what they are looking for; that is, they must have in mind an ideal or a standard of perfection.

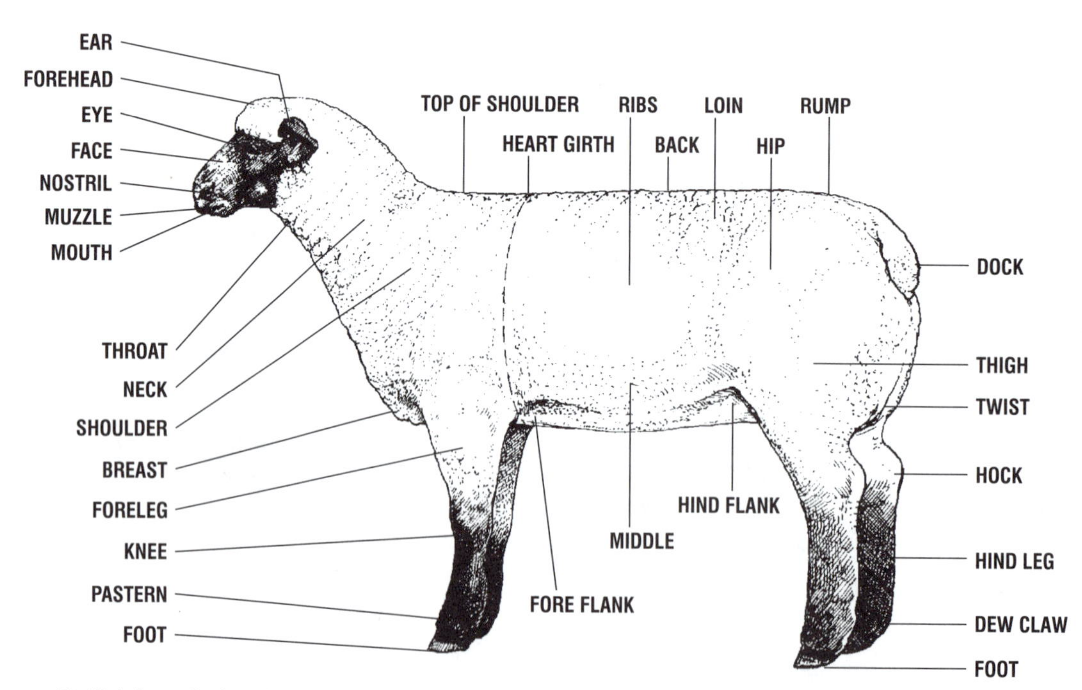

Fig 12-1. Parts of a sheep. The first step in preparation for judging sheep consists of mastering the language that describes and locates the different parts of the animal.

3. **Keen observation and sound judgment.** The good sheep judge possesses the ability to observe both good conformation and defects, and to weigh and evaluate the relative importance of the various good and bad features.

4. **Honesty and courage.** The good judge of any class of livestock must possess honesty and courage, whether it be in making a show-ring placing or in conducting a breeding and marketing program. For example, it often requires considerable courage to place a class of animals without regard to (a) placings in previous shows, (b) ownership, and (c) public applause. It may take even greater courage and honesty to discard from the flock a costly animal whose progeny has failed to measure up.

5. **Logical procedure in examining.** As in the examination of any class of livestock, the examination of sheep should be systematic and thorough. This is especially true in selecting breeding animals or in close competitive judging, such as is encountered in the show-ring, but it is neither practical nor essential in handling the large numbers of sheep in a central market.

The sheep being examined should first be looked over from a distance, so that views from the front, side, and rear may be secured. It does not make any difference which view of the animal is noted first, but it is important that the same procedure be followed each time.

From the side these observations should he made: (a) the size as indicated by the height to top of the shoulder and length from nose to dock; (b) balance and symmetry; (c) stretch, varies according to breed and age; (d) strength of back; (e) levelness of rump; (f) trimness of underline; (g) straightness of legs and strength of pasterns; (h) size of bone; (i) style; (j) breed type; and (k) freedom from wrinkles.

Observations from the rear view include: (a) uniformity of width from rear to front; (b) curve over back, loin, and round; (c) trimness of middle; (d) depth and plumpness of leg; and (e) set of hind legs.

Front view observations include: (a) shapeliness of head; (b) sex character; (c) brisket; (d) width of chest; and (e) set of front legs.

Next, the impression gained through distant inspection should be verified by handling and examining the fleece. Good judges differ as to whether one should start handling the animal from the front or rear. Perhaps as good a method as any is illustrated in Fig. 12-2 and detailed as follows:

A. Examine covering and strength of top from rump to top of shoulder.

B. Grasp neck for fullness and examine head for evidence of scurs.

C. Place hands over point of shoulders to check both width and covering.

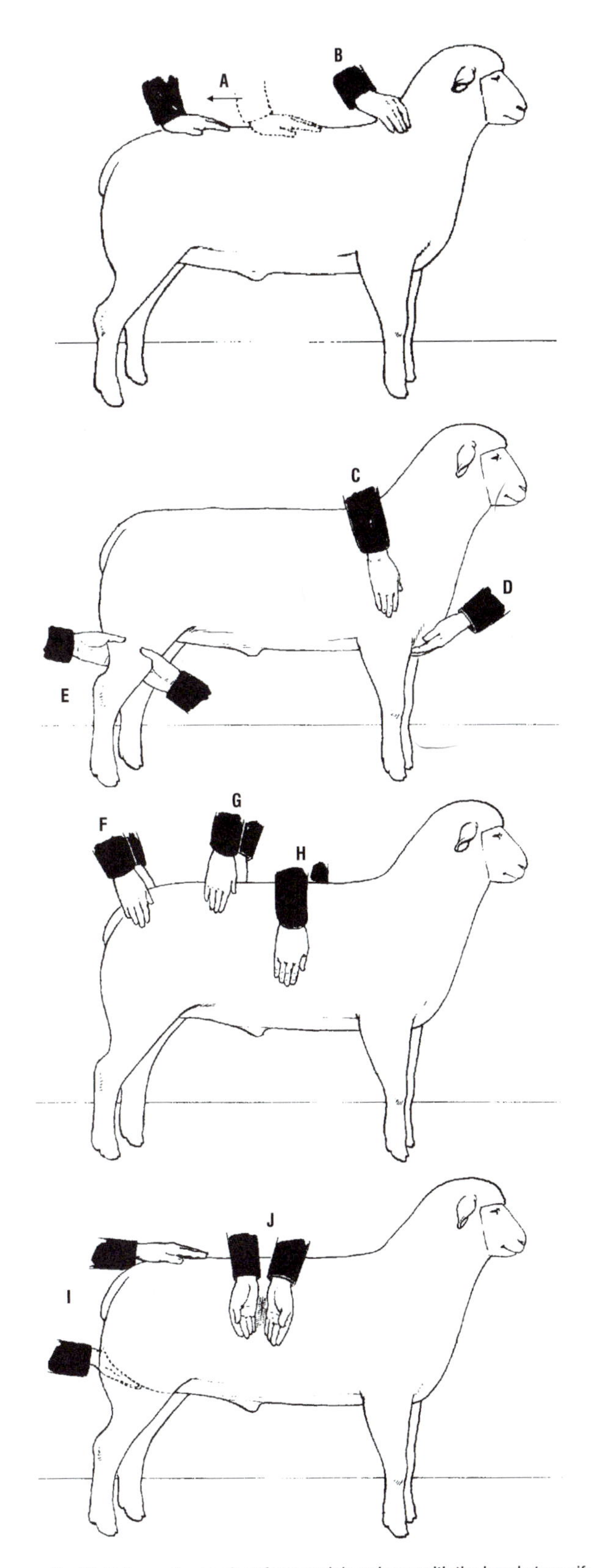

Fig 12-2. A good procedure for examining sheep with the hands to verify the visual appraisal. The hands are used with the fingers extended and joined. Feeling is accomplished with the pads of the fingers and jabbing is avoided.

D. Feel brisket's firmness; also place one hand on the top of the shoulder, the other on the floor of the chest, to determine depth of heart girth and chest.

E. With thumbs on outside of leg, fingers on inside, feel the muscling for plumpness and manner in which it carries down to the hock.

F. Check width of rump. Also, pinch the dock and measure width.

G. Check width and covering of loin edge.

H. Check spring of rib, including covering.

I. Place one hand on the top of the rump, the other in the middle of the thighs, to measure the depth of twist.

J. Check the fleece on the shoulder, side, and thigh. Part it with the back of each hand, palms held out and open to reflect light onto fleece.

With market sheep, most of the examining that is necessary in arriving at the market classes and grades is done by observation. Even so, these specialists usually like to get their hands on fat lambs; and when part of a drove of lambs is fat enough to go the slaughter route—whereas others must go as feeders—they will make the cut after handling each of them as they file through a cutting chute.

6. **Tact.** In discussing either (a) a show-ring class or (b) animals on a producer's farm or ranch, it is important that the judge be tactful. Owners are likely to resent any remarks that imply that their animals are inferior.

IDEAL MEAT TYPE AND CONFORMATION

To judge any class of livestock successfully, the judge must have clearly in mind an ideal or standard of perfection. Thus, sheep producers, regardless of whether they are producing purebred or commercial animals or whether they are farm-flock or range-band operators, should have a type or an ideal in mind and make their selections accordingly. For the meat-type breeds, and when the production of market lambs is the primary objective, this ideal means plenty of size and growthiness; heavy muscling—especially in the leg and loin; a long body; trimness and freedom from waste; straight, widely set legs; a fleece of acceptable weight and quality, and pink skin (see Fig. 12-3).

If purebred, the animals should show the characteristics of the breed represented. Rams should show boldness and masculinity and ewes should be feminine.

IDEAL FLEECE TYPE AND CONFORMATION

When the production of wool is the main source of income, weight and quality of fleece are of the utmost importance. The fibers should be long, fine, and of good crimp; the fleece should be dense, clean, and bright. Animals with fleeces that show black fibers and those with fleeces that tend to be hairy, loose, or open should be rejected. From the standpoint of body form,

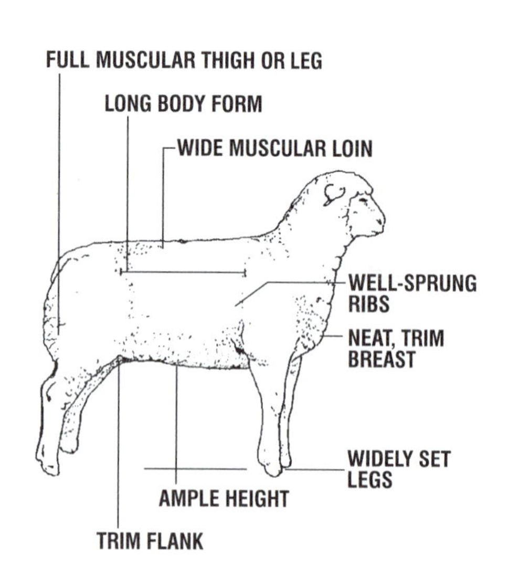

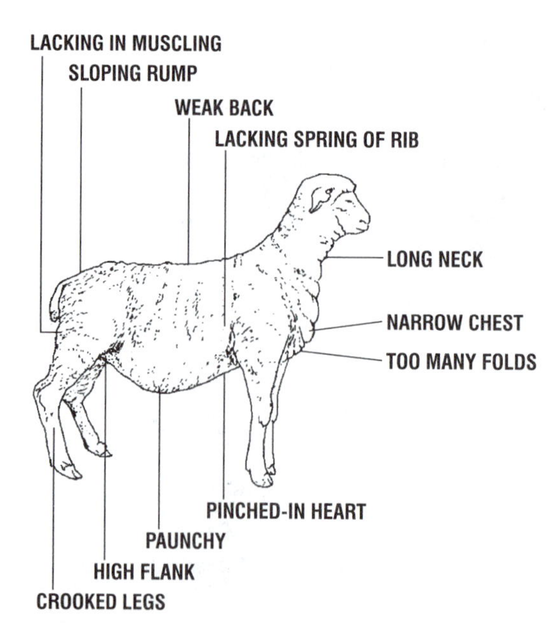

Fig 12-3. Ideal meat type *(left)* vs common faults *(right)*. Successful sheep judges must know what they are loosing for, and they must be able to recognize and appraise both the good points and the common faults.

typical wool-type sheep are quite different from those animals representing the ultimate in meat type. The former are more angular, with considerably less muscling throughout. The present-day smooth-bodied Rambouillet sheep, however, are of a much more acceptable meat type than the ideal extreme wool type of former years.

RECOGNIZING AND EVALUATING COMMON FAULTS

No animal is perfect. In judging, therefore, judges must be able to recognize and appraise common faults. Likewise, credit must be given for good points. Finally, judges weigh and evaluate the relative importance of the characteristics that they have observed, and the degree to which they are good or bad. Skill and accuracy in this art can be achieved only through patient study and long experience. Fig. 12-4 shows the ideal meat type versus some common faults.

Fig 12-4. The successful sheep judge must have clearly in mind an ideal or standard of perfection. The ideal meat type *(left)* shows plenty of size, growthiness, muscling, trimness, and freedom from waste. The ideal fleece type *(right)* is more angular and produces a dense, clean, bright fleece.

CATCHING AND HANDLING SHEEP

If sheep are to be caught and handled for any reason, they should first be confined to a small corral or shed. Sheep may best be caught around the neck, by the hind leg, or by the rear flank. Never should they be caught by the wool. Such rough treatment results in badly injured skin and tissue, which may require weeks or even months to heal. As a result of such mishandling, market sheep will exhibit damaged carcasses, and the fleece of breeding animals will lack uniformity because of the disturbance in the injured area. Sheep can be caught and held as follows:

- **Catching**—When animals are confined to a small area, move up quietly on the desired animal after working it into a position near you. With a swift, sure movement, grasp well up into the right rear flank with the right hand and while holding firmly, grasp under the lower jaw with the left hand.

- **Holding**—With the left hand firmly grasp a fold of flesh under the lower jaw. Place the right hand securely over the dock; the right hand in this position can be useful in moving the animal. As the animal quiets down, the right hand may be removed.

In handling sheep, keep the fingers together. In this way, the correct touch is obtained in the palm of the hand, and the wool is not disarranged. In observing the fleece and skin, part the wool well down on the shoulder, side, and leg. Opening the fleece on the back should be avoided, as it will allow water to run in.

DETERMINING THE AGE OF SHEEP

Mature sheep and goats have 32 teeth, of which 24 are molars and 8 are incisors. As in cattle, all incisors are in the lower jaw. The two central incisor teeth are called pinchers; the adjoining ones, first intermediates; the third pair, second intermediates; and the outer ones, corners. The temporary teeth of the lamb—milk teeth—are small and narrow, while permanent teeth are larger and broader. There are no tusks.

Until sheep are four years of age, the front teeth of the lower jaw furnish a fairly reliable guide as to their development (see Fig. 12-5). In the newborn lamb, none of the teeth may be present, though sometimes the two pinchers and also the first two intermediates are pressing through the gums or are even cut through. By three months, a full set of completely developed temporary teeth are present. Between 12 and 15 months, the temporary pinchers are replaced by the two permanent ones. By two years, the first temporary intermediates are replaced by permanent teeth, and by three years the second temporary intermediates are replaced by permanent teeth. Then by four years of age, the two temporary corner incisors are replaced by permanent teeth, giving the sheep a "full mouth."

After the sheep has a solid mouth (at four years), it is impossible to tell the exact age. With more advanced age, the teeth merely wear down and spread apart, and the degree of wearing or spreading is an indication of age. The normal number of teeth may be retained until eight or nine years, but often some are lost after about the fifth or sixth year, resulting in a broken mouth. When most of the teeth have disappeared, the animals are known as gummers.

Ewes of meat breeding usually start on the decline at about five years of age, whereas ewes of the fine-wool breeds do not begin to decline until about a year

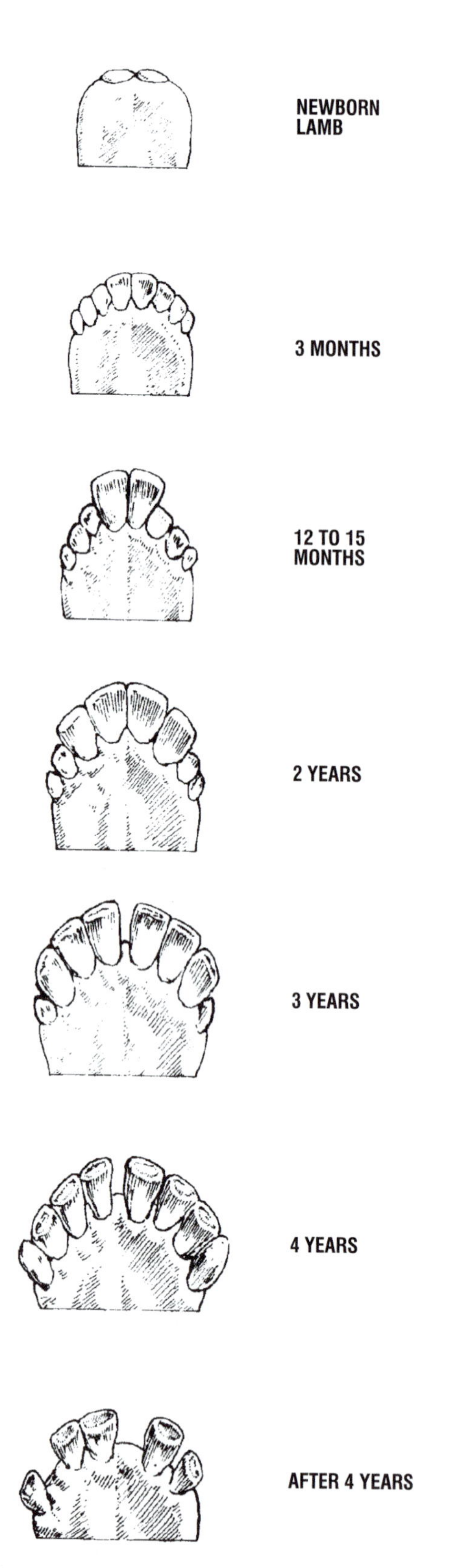

Fig 12-5. How to determine the age of sheep by the incisors in the lower jaw. The two central incisor teeth are called pinchers; the adjoining ones, first intermediates; the third pair, second intermediates; and the outer ones, corners. Temporary teeth are small and narrow.

later. Even so, outstanding producers should not be disposed of just because they have reached this age, especially in a purebred flock. Also, animals that have reached the age where it may no longer be wise to expose them to the rigors of range handling may often be retained satisfactorily and profitably in a farm flock for 1 to 2 years longer.

Teeth represent the degree of development rather than the exact age according to birth, and, therefore, they are not an infallible indication of age.

With market sheep, the break-joint or lamb-joint is also an important criterion of age. For further information on this point. the reader is referred to Chapter 18.

SCORECARD JUDGING

A scorecard is a listing of the different parts of an animal, with a numerical value assigned to each part according to its relative importance. It is a standard of excellence. The use of the scorecard involves studying each part, then assigning a score to each.

Different methods of scoring individual animals have evolved. All of them are based on visual appraisal. This point bears emphasis because producers and students often get the erroneous impression that just because some visual scoring system (scoring system based on visual appearance, in contrast to actual weights measurements, etc.) is recommended for or used in conjunction with a production testing program, it must be more accurate than all other scoring systems. This is not true. All are visual methods, and the score resulting from the use of any of them is no better than the person making it. Some method of selecting all animals by score, preferably on a systematic and written basis, is the important thing.

A breeding sheep scorecard is shown in Fig. 12-6, while a market lamb scorecard is shown in Fig. 12-7.

As noted, the scorecard gives each of several traits a value, which total 100 for a perfect score.

A scorecard is a valuable teaching aid for beginners. It systematizes judging and avoids any part of the animal being overlooked. However, a scorecard has the following limitations: (1) It is not adapted to evaluating a great number of animals, or to comparative or show-ring judging, because of the time involved in using it; (2) a nearly worthless animal may score quite high—for example, an animal that is so structurally unsound that it can hardly walk may have a rather high total score; (3) it evaluates each part of an animal, rather than the system—the skeletal system, the muscle system, etc.; (4) it is based almost entirely on consumer needs (for example, on the end product— meat, in meat animals); and (5) it accords precious little consideration as to whether, or how, an animal can be better changed to conform to human needs and desires.

		Perfect Score	ANIMAL				
			No. 1	No. 2	No. 3	No. 4	Etc.
CONFORMATION:		73					
General appearance—(25 points)							
Size and scale—big for age, roomy, heavy bone	(15)						
Type—straight lined, balanced, deep ribbed, long, stylish	(10)						
Hindquarters—(26 points)							
Leg—muscular, plump, thick, deep	(9)						
Rump—long, level, full, square dock	(7)						
Loin—wide, strong, meaty	(9)						
Twist—deep, full	(1)						
Forequarters—(22 points)							
Back—wide, straight, strong	(8)						
Ribs—bold spring, deep ribbed	(6)						
Shoulders—muscular, smooth	(4)						
Chest—deep, wide chest floor	(3)						
Neck—short, thick	(1)						
BREEDING QUALITIES:		27					
Head—clean cut, bright eyes, feminine or masculine, proper color of face, free from wool blindness	(5)						
Underpinning—strong pasterns, legs correctly and squarely placed, rugged bone	(15)						
Fleece—dense, uniform crimp, long staple, pink skin, fineness according to standard of the breed, free from black fiber	(7)						
TOTAL		100					

Fig. 12-6. Breeding sheep scorecard.

		Perfect Score	ANIMAL				
			No. 1	No. 2	No. 3	No. 4	Etc.
CONFORMATION:		55					
General appearance—(10 points)							
Straight top and underline, muscular, thick, legs set wide apart, stylish, well balanced, adequate size for age							
Hindquarters—(26 points)							
Legs—straight, set wide apart	(2)						
Twist—clean, muscular	(1)						
Leg—meaty, plump, long, deep, thick	(8)						
Rump—long, level, thickly muscled	(6)						
Loin—meaty, thick, deep loin edge, straight	(9)						
Forequarters—(13 points)							
Back—thick, straight	(6)						
Ribs—bold spring, deep forerib	(3)						
Shoulders—muscular, smooth	(2)						
Neck—short, thick	(1)						
Breast—wide, deep chest floor, trim	(0.5)						
Legs—straight, set wide apart	(0.5)						
Middle—(6 points)							
Middle—trim, free from wastiness							
FINISH:		40					
Uniformly covered with the correct amount of finish over back, ribs, loin, rump	(30)						
Covering over shoulder, dock	(5)						
Trim in flanks, cod, etc.	(5)						
QUALITY:		5					
Smooth pelt, head trim and refined, ample bone							
TOTAL		100					

Fig. 12-7. Market lamb scorecard.

QUESTIONS FOR STUDY AND DISCUSSION

1. In establishing a flock of sheep, how do the factors that should be considered differ from the factors to be considered in establishing cattle, hogs, or horses?

2. Select a certain farm or ranch (either your home farm or ranch, or one with which you are familiar). Assume that there are no sheep on this establishment at the present time. Then, outline, step by step, (a) how you would go about establishing a flock, and (b) the factors that you would consider. Justify your decisions.

3. Discuss each of the four bases of selection of sheep.

4. Cite an example when utilitarian consideration in show-ring judging of sheep was ignored, to the detriment of a certain breed of sheep.

5. List some major traits that should be considered when selecting and culling flocks, and explain their importance.

6. Suppose you wish to become an expert judge of sheep. List and discuss the essential qualifications of a good sheep judge.

7. Describe a systematic and thorough examination of sheep. Include the visual and handling approaches, noting items to be evaluated.

8. How do meat-type and fleece-type sheep differ? Have these two types moved closer together— become more alike—in recent years? Justify your answer.

9. Why is it important to catch and hold a sheep in the proper manner?

10. Why are producers more conscious about, and familiar with, the determination of the age of sheep by the teeth than they are in applying this method to cattle, hogs, or horses?

11. What is a scorecard, and what are the advantages and disadvantages of this system of judging?

SELECTED REFERENCES

Title of Publication	Author(s)	Publisher
Animal Science	M. E. Ensminger	Interstate Publishers, Inc., Danville, IL, 1991
Breeding Better Livestock	V. A. Rice F. N. Andrews E. J. Warwick	McGraw-Hill Book Company, New York, NY, 1953
Livestock Judging Guide	K. E. Gilster	Nebraska Cooperative Extension Service, University of Nebraska, Lincoln, NE, 1981
Livestock Judging, Selection, and Evaluation	R. E. Hunsley	Interstate Publishers, Inc., Danville, IL, 2001
Stockman's Handbook, The	M. E. Ensminger	Interstate Publishers, Inc., Danville, IL, 1992

CHAPTER 13

Showing sheep.

SELECTING, FITTING, AND SHOWING SHEEP

Contents Page

The winning sheep in livestock shows should reflect what the consumer wants. In the late 1980s and the early 1990s, this was not true.

A study conducted by Colorado State University showed that small-frame lambs are ready for slaughter from 104 to 120 lb, medium frame lambs should be slaughtered between 111 and 125 lb, and large frame lambs reach optimal finish between 122 and 140 lb. Yet, a Colorado packer reported that the average slaughter weight of lambs in his plant was 145 lb.[1]

Russell Cross of Texas A&M University reported that in 1989 the average lamb carcass had 19.9 lb of plate waste—defined as fat, bone, and gristle—which added up to 150 million lb.[2]

INFLUENCE OF SHEEP SHOWS

In the earlier years of agricultural fairs and expositions, animals were exhibited but not shown competitively. They provided advertising and promotion of producers' livestock programs. In 1807, Colonel Humphreys and David Livingston exhibited some of their Merino sheep at Pittsfield, Massachusetts. This is thought to be the first showing of livestock in the United States, and may have been the forerunner of American livestock shows.

Today, the show-ring is the "picture window" of the purebred sheep industry. The major sheep shows and sales provide sheep producers with an opportunity to compete.

Today, some purebred sheep breeders feel that the most important thing is to win. As a result, it is reported that there is some crossbreeding and that it is no longer possible to distinguish the breed of some animals. Although the cases are isolated, the following illegal practices should be condemned: the use of illegal drugs to enhance muscling, and the use of surgery to alter body form.

According to historical records, the shepherds of England were the first to trim sheep for the show-ring, using their art on the medium-wool breeds. Soon the practice spread to the United States, where it was widely adopted.

As most of the advertising value of the show-ring accrues to those who exhibit champions, it behooves exhibitors to prepare sheep in the most attractive manner possible. Although no amount of fitting and showing can make up for poor breeding, the expert application of the art is often the deciding factor when close placings are involved.

Fitting and showing has as its primary objective making the animals as attractive as possible through accentuating strong points and minimizing weaknesses. Accordingly, some have objected to it on the basis that it disguises animals and encourages deception. On the other hand, the experienced sheep producer or judge is able to ascertain the true merits of the individual by handling and is not misled by fancy trimming or expert skills in showing. Fitting and showing, therefore, is an accepted art designed to make the sheep appear more attractive in the show-ring or when inspected by prospective buyers. Thus, the exhibitor should know the art, which can best be mastered through long and patient practice under a skilled shepherd. The discussion that follows should not replace actual practice. but it may be of assistance to the amateur.

There can be no question that the great livestock shows throughout the land have exerted a powerful influence on U.S. agriculture in general and the livestock industry in particular. In addition to their educational value, these shows provide one of the best mediums of advertising. They also afford an opportunity for visiting livestock producers to exchange ideas, and they have been a guide in molding animal types— both good and bad. Unfortunately, in the case of sheep, utility considerations have often been pawned away in favor of such fancy points as excessive face covering, much to the detriment of the breeds involved. But this should not be construed as a reason to condemn either shows or judges. Rather, the breeders themselves must accept the blame for losing sight of the important utility considerations that are determined by both the producer and the consumer. They must also accept the blame for demanding that fancy points be substituted in place of these utility considerations. These distorted ideals have been inflicted upon judges, who either complied with the wishes of the exhibitors or failed to draw repeat assignments.

SELECTING SHOW SHEEP

The selection of the prospective show animals is the first and most important assignment. Consciously or unconsciously, in the selection of show sheep, the expert exhibitor gives consideration to the following three major factors: (1) type or individuality, (2) age of the animal as related to the show classifications, and (3) breeding as indicative of potential development or outcome.

TYPE

Today's winning-type sheep are long, tall, stylish, balanced, and well muscled, with strong, level tops, and they stand squarely on their feet and legs.

---

[1]*California Farmer*, Vol. 270, No. 7, April 1, 1989.

[2]*Ibid.*

The selection of outstanding animals is the most important single requisite in successful showing. Skillful sheep producers are able to select undeveloped animals and, within their own thinking, to determine how those animals will appear when fitted and shown. But the experienced fitters are fully aware that not all animals respond equally well to feeding. With this in mind, they start with greater numbers than will actually be shown.

The animals selected should conform rather closely to the established ideals of the breed, both from the standpoint of body conformation and fleece quality. In the breeding classes, the so-called breed fancy points must not be overlooked. For further information relative to breed characteristics and factors to consider in making selections, the reader is referred to Chapters 11 and 12.

AGE AND SHOW CLASSIFICATIONS

In most sheep shows there are two main classifications: breeding sheep and market sheep. Sometimes there is an added classification for feeder sheep.

1. **Breeding sheep.** The breeding classes are for purebreds only. Usually the number of different breeds for which classifications are indicated and the amount of premium money allotted to each vary according to the popularity of the breed within the area and the premium money support accorded by the breed associations. At most fairs, and in most breeds, a base date of September 1 is used for breeding classes, although a few use a base date of January 1. Most 4-H and FFA shows use the latter.

With the base date of September 1, lambs must have been dropped on or after September 1 of the year preceding the one in which they are shown; and yearlings must have been born on or after September 1 of the second preceding year. Aged animals include all those older than the yearling group.

The following sheep show classification has been used at most sheep shows in the past, and it is still used in many modern sheep shows:

Ram, 1 year old and under 2
Senior Ram Lamb, born after September 1
Junior Ram Lamb, born after January 1
Pair of Ram Lambs (or pen of 3)
Champion Ram
Reserve Champion Ram
Ewe, 1 year old and under 2
Senior Ewe Lamb, born after September 1
Junior Ewe Lamb, born after January 1
Pair of Yearling Ewes (or pen of 3)
Champion Ewe
Reserve Champion Ewe
Exhibitors' Flock (ram any age, 2 yearling ewes, 2 ewe lambs)

Many sheep show classifications are now designed more for spectator appeal, with their show classifications changed accordingly. For example, the following Sheep and Dairy Goat Classifications are taken from the Iowa State Fair Premium List, Aug. 7–17, 1997:

SHEEP

- **Big Ram Contest**
- **Breed Classification for:**

Cheviot
Colored Wool Breeds
Columbia
Corriedale
Hampshire
Lincoln
Long Wool
Montadale
Oxford Down
Shropshire
Southdown
Rambouillet
Suffolk

- **Age and Show Classification.** For each of the above breeds the following age and show classifications are provided:

Yearling Ram
Fall Ram Lamb
Winter Ram Lamb
Spring Ram Lamb
Pair Ram Lambs
Yearling Ewe
Yearling Ewe that has Weaned a Lamb
Pair of Yearlings
Fall Ewe Lamb
Winter Ewe Lamb
Spring Ewe Lamb
Pair Ewe Lambs
Pair of Lambs
Young Flock
Flock

- **Additionally, the following point system applies:**

Champion Ram	10 Points/Reserve 5 Points
Champion Ewe	10 Points/Reserve 5 Points
First	10 Points
Second	8 Points
Third	6 Points
Fourth	4 Points
Fifth	2 Points
Sixth	1 Point

- **4-H and FFA Lambs**
- **Sheep Shearing Contest**, with separate contests for professionals and amateurs
- **Sheep Dog Trials**

DAIRY GOATS

- **Open Senior Doe Show** of each of the following breeds:

Alpine
La Mancha
Nubian
Oberthausli
Recorded Grade
Saanen
Toggenburg

- **Age and Show Classification.** For each of the above breeds, the following age and show classifications are provided:

Under 2 years, in milk
2 years old and under 3 years old
3 years old and under 5 years old
5 years old and over
Champion Challenge

■ **Group Classes/Senior Division.** For each of the above breeds, the following group classes are provided in the Senior Division:

Mother and Daughter
Produce of Dam
Get of Sire
Breeder's Trio
Dairy Herd

■ **Championship Classes in each Breed and each Show Contest:**

Grand Champion Doe
Reserve Grand Champion Doe
Champion Challenge Doe
Best Doe of Show

■ **Dairy Goat Milking Competition**

■ **4-H and FFA Show**

2. **Wethers (market lambs).** Major fairs generally have separate classes for each of the major breeds, plus a class for grades or crossbreds. A minimum weight of 80 to 85 lb is required, depending on the breed. In addition to the class winners, fairs may have a Grand Champion Wether (Market Lamb) and a Reserve Grand Champion Wether (Market Lamb).

3. **Carcass competition.** Wethers may also be slaughtered and entered in carcass competition where carcasses are evaluated according to minimum carcass standards and carcass weight per day of age.

For complete information about show classifications, individuals can obtain premium lists of specific fairs.

BREEDING

Breeding is of importance in the selection of show animals, especially in younger animals and in the get classes. With young, undeveloped lambs, breeding may be the best indicator of future development, especially when the owner has had an opportunity to observe the development of other offspring by the same parent or parents. Also, when it is desired to feature the get of a certain sire, it may be well to select with this thought in mind.

Performance programs and genetic evaluation programs give an assist in selecting show sheep.

FEEDING AND HANDLING FOR THE SHOW

No amount of blocking or trimming can overcome poor feeding and management. This does not imply that one should minimize the importance of proper grooming from the standpoint of creating a favorable and attractive appearance; rather, experienced judges rely upon their hands to determine what is underneath an artistic job of blocking and trimming. Thus, for success in the show-ring, the animal should be fed and managed to attain maximum development in body conformation and fleece quality.

Chapters 14 and 15 contain information relative to the nutritive needs, suitable feeds, and feeding principles for sheep. In general, the feeding of show sheep differs from normal operations only in that greater effort and expense and more liberal allowances may be justified in order to produce a winner.

POINTERS IN FEEDING AND HANDLING SHOW SHEEP

The most important pointers in feeding and handling show sheep are:

1. **Keep show sheep healthy and free from parasites.** It is impossible to obtain proper growth, finish, and bloom in sheep that are unthrifty or infested with parasites. The caretaker, therefore, should scrupulously follow a program designed to assure flock health, disease prevention, and parasite control. This is especially important when fitting young lambs for show. The proper rotation of temporary pastures is the key to this type of program. Even with the best of preventive measures, generally it is necessary to spray or dip (most shepherds seem to prefer dipping to spraying, because it also cleans the fleece at the same time it destroys external parasites) prospective show sheep at least once during the year and to administer treatment for the removal of stomach worms one or more times.

2. **Provide a suitable ration.** In addition to being reasonably economical and well balanced, the ration for show sheep must be palatable. Many feed combinations meet these specifications. The ration selected is usually determined by (a) the availability and price of feed in the area, and (b) the preference and judgment of the shepherd.

Some suggested grain-fitting rations are listed in Tables 13-1 and 13-2. To each of these grain rations should be added (a) good-quality roughage—usually homegrown, and (b) salt and other minerals on a free-choice basis.

■ **Rations for lambs**—Rations in Table 13-1 are for

TABLE 13-1
RATIONS FOR SHOW LAMBS

Ingredient	Ration Number 1[1]		2		3		4		5		6		7	
	(lb)	*(kg)*	*(lb)*	*(kg)*	*(lb)*	*(kg)*	*(lb)*	*(kg)*	*(lb)*	*(kg)*	*(lb)*	*(kg)*	*(lb)*	*(kg)*
Barley	—	—	—	—	20	*9.1*	—	—	—	—	—	—	—	—
Corn	—	—	40	*18.2*	20	*9.1*	—	—	—	—	45	*20.4*	—	—
Corn, cracked	—	—	—	—	—	—	—	—	—	—	—	—	85	*38.6*
Linseed meal	17.5	*7.9*	—	—	—	—	—	—	—	—	—	—	—	—
Oats	65.0	*29.5*	40	*18.2*	30	*13.9*	70	*31.8*	80	*36.3*	45	*20.4*	—	—
Protein supplement[2]	—	—	10	*4.5*	10	*4.5*	10	*4.5*	—	—	10	*4.5*	—	—
Soybean meal	—	—	—	—	—	—	—	—	—	—	—	—	15	*6.8*
Wheat bran	17.5	*7.9*	10	*4.5*	20	*9.1*	20	*9.1*	20	*9.1*	—	—	—	—
Total	100.0	*45.4*	100	*45.4*	100	*45.4*	100	*45.4*	100	*45.4*	100	*45.4*	100	*45.4*

[1]The author has used this ration extensively. Near the end of the fitting period, the shepherd ads 50 lb *(22.7 kg)* of barley and 50 lb *(22.7 kg)* of peas to each 100 lb *(45.4 kg)* of Ration 1.

[2]Linseed, cottonseed, and/or soybean meal.

TABLE 13-2
RATIONS FOR FITTING YEARLINGS AND MATURE SHEEP

Ingredient	Ration Number 1		2		3		4	
	(lb)	*(kg)*	*(lb)*	*(kg)*	*(lb)*	*(kg)*	*(lb)*	*(kg)*
Barley	—	—	40	18.2	—	—	10	4.5
Corn	—	—	—	—	40	*18.2*	—	—
Oats	50	*22.7*	50	*22.7*	40	*18.2*	60	*27.2*
Peas (split)	40	18.2	—	—	—	—	10	*4.5*
Protein supplement[1]	—	—	—	—	10	*4.5*	10	*4.5*
Wheat bran	10	*4.5*	10	*4.5*	10	*4.5*	10	*4.5*
Total	100	*45.4*	100	*45.4*	100	*45.4*	100	*45.4*

[1]Linseed, cottonseed, and/or soybean meal.

either creep-fed or weaned lambs that are being fitted for show. Show lambs on full feed will eat about 2.5 lb of grain per head daily.

- **Rations for yearlings and mature sheep**—Rations in Table 13-2 are for fitting yearlings and mature sheep. Show yearlings will eat about 3.0 lb of grain per head daily, whereas mature sheep will eat about 3.5 lb of grain per head daily.

In general, the preceding rations are higher in protein content than rations used in commercial finishing operations, but most experienced shepherds feel that by such means they get more bloom. Exhibitors prefer to feed steamed rolled oats and barley and nutted (pea-sized) old process linseed meal.[3] Corn is usually cracked or coarsely ground, and peas are split or cracked. When pastures are not available, alfalfa is the most popular hay. But, any good legume is quite satisfactory. The lighter types of lamb rations (Table 13-1, Rations 1, 4, and 5) are usually used for summer feeding, especially when animals are being fitted for the late shows.

The rations listed in Tables 13-1 and 13-2 are intended as guides only; specific conditions may well warrant changes. For example, when wheat prices are favorable, this grain may replace part or all of the barley or corn in the ration. When either a lush pasture grass or a high-quality alfalfa is used as the roughage, it is usually advisable to reduce the protein supplement in the ration by as much as one-half. This is done for reasons of practicality and in order to avoid feeds that are too laxative.

3. **Provide suitable minerals.** All classes and ages of sheep should be fed minerals according to the requirements in Tables 14-2 and 14-3.

4. **Prepare feeds properly.** Unless grains are unusually hard, they need not be ground for sheep. The animals prefer to do their own grinding, and the feeds are no more effectively utilized when ground.

5. **Feed a suitable grain ration liberally in order to obtain proper finish.** Fine as roughages may be, they are too bulky to be used as the sole ration for show sheep. This is especially true for young lambs that require a reasonably concentrated ration because of their limited digestive capacity and because both growth and finishing are desired.

[3]Among experienced shepherds, old process linseed meal is especially popular for fine-wool sheep because of its conditioning effect on the fleece.

6. **Feed proper amounts.** No definite set of rules relative to feed allowances can be followed satisfactorily. Rather, the judgment of the skillful feeder must prevail. In general, young lambs are creep-fed and provided with all of the feed that they will clean up. Usually yearling or mature sheep are started on grain 3 to 6 months before the show season. They should be started on grain gradually. The grain ration may then be slowly increased until the sheep are receiving a full feed, which means about 2.5 lb of grain daily for lambs, and 3.5 lb for mature sheep. The skillful shepherd never overfeeds. Generally, sheep are given all the hay (or pasture) that they can consume.

Lambs will consume an amount equal to about 4% of their body weight daily, of which about two-thirds may be concentrate and one-third hay.

7. **Provide lambs with plenty of milk and grain.** For young lambs that are being fitted for show, milk is the most important feed. Therefore, the dams should be fed and handled so that there will be maximum milk production. In season, lush pastures usually will stimulate adequate milk flow. When pastures are not available, it is desirable to feed the suckling ewes 1 to 2 lb of grain daily plus 4 lb of high-quality alfalfa hay. Simple grain rations may be used for ewes, but usually two or three different feeds are included in order to obtain variety. If neither a legume hay nor a legume silage is available, a protein supplement should be included in the grain ration.

In addition to receiving plenty of milk from their dams, prospective show lambs should get a suitable grain ration through creep feeding. Young lambs are curious and usually can be taught to eat when they are 10 days to 2 weeks old. At first, only a small amount of feed should be placed in the troughs each day, any surplus being removed and given to the ewes. In this manner, the feed will be kept clean and fresh; and the lambs will not be consuming any sour or moldy feed. When 3 to 4 months of age, young lambs will consume about 0.5 lb of grain daily; the amount will vary according to the quality of the pasture or other roughage and the milk flow of the ewes.

Lambs that have been creep-fed usually do not suffer any setback at weaning time, an important consideration with prospective show animals. After the ewes are removed, the grain allowance is simply increased.

8. **Use clean, lush, temporary pastures in season or high-quality hay.** In season, clean luxuriant pastures produced on fertile soils should provide the roughage allowance for all show sheep, including both lambs and mature animals. The most desirable pastures are those that are classed as temporary and which are annuals or handled in such a manner that sheep have not grazed thereon for at least a year, for this will alleviate the hazard of parasites. Among the most common annuals used as pastures for show sheep are: rye, wheat, oats, field peas, and rape.

During hot weather, sheep will do better when kept in the barn during the heat of the day and turned to pasture in the cool of the evening or early in the morning. Also, the grain ration should be fed early and late during the summer months.

9. **Provide succulent feeds.** Sheep are very fond of such succulent feeds as cabbage, carrots, mangels (stock beets), turnips, and rutabagas (swedes). The experienced shepherd usually adds limited quantities of one of these feeds to the ration of show sheep. These succulent feeds are highly relished by sheep and appear to help their digestion and general thrift.

EQUIPMENT FOR FITTING AND SHOWING SHEEP

The essential items of equipment for fitting sheep for show are: a sharp knife or an ordinary pair of pruning shears to trim toes; shears; hone; combs; brushes; trimming stand; cards; liquid soap; and a bucket. Additional equipment that is very helpful includes: a hand-pump sprayer; and a blower-dryer.

Although these items of equipment are used in the blocking and fitting operations well in advance of the show, they are always included in the show box for use in adding the final touches prior to showing. In addition, in loading out for the show, most exhibitors take feed pans, water buckets, a fork and a broom, a saw, a hammer, a hatchet, a few nails of different sizes, some rope, a flashlight, blankets, a sleeping cot, a limited and permissible supply of mixed feed, and a sleeping bag. The show box is usually an attractive and sturdy wooden box upon which the name of the exhibitor is inscribed.

TRAINING AND GROOMING SHEEP

Training and grooming are essential in order that the judge may see the animal at its best. They are the "primping" or "beautification treatment."

GENTLING AND POSING

Show sheep should be gentle, and should pose easily and naturally.

Practice showing should be started at least six to eight weeks ahead of the fair. While standing or squatting to the left and front, the exhibitor should hold the animal by grasping the wool lightly under the chin with the left hand. The sheep should stand with the legs squarely placed, the back straight, and the head held erect.

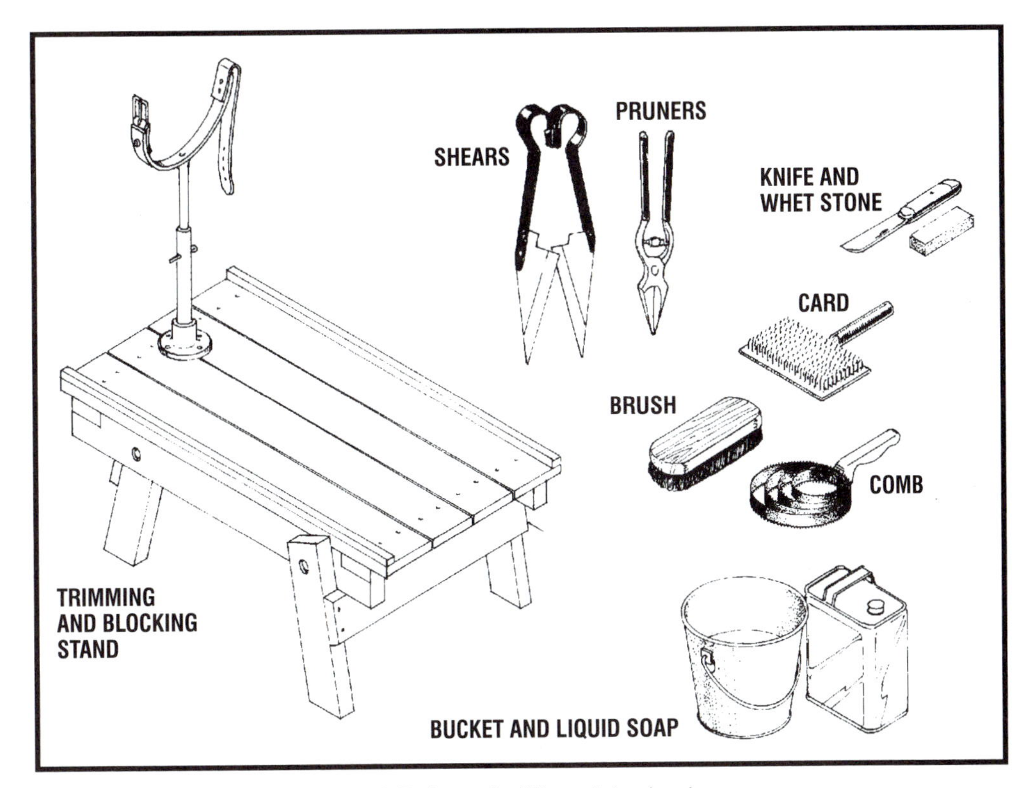

Fig. 13-1. Equipment for fitting and showing sheep.

Fig. 13-2. Correct method of gentling and posing.

TRIMMING THE FEET

The amount of foot trimming necessary will vary somewhat with soil and management conditions and with the breed (for example, the toes of a Merino grow faster than those of other breeds). The purpose of trimming the feet is to keep the sheep strong and straight on their pasterns. Usually, the feet of show sheep should be trimmed every two months, with the final trimming being done about a week in advance of the show so that there will be no soreness when the animals are exhibited. The animal is held in a sitting position (Fig. 13-3) while its hoofs are trimmed with a sharp pocket knife or an ordinary pair of pruning shears.

BLANKETING

Except during very hot weather, breeding sheep of the medium-wool breeds should be blanketed in order to keep dirt out of the fleece and to condition the fleece and keep it compact. It is not as necessary, however, to provide blankets for fat wethers. Ready-made blankets of different sizes can be purchased from the various livestock equipment companies, but some small operators or beginners make their own. Satisfactory homemade blankets may be made from heavy cotton sacks (not burlap). The seam along one side of the sack is opened up, and the opposite corner on the sewed end is cut in such a manner as to provide an

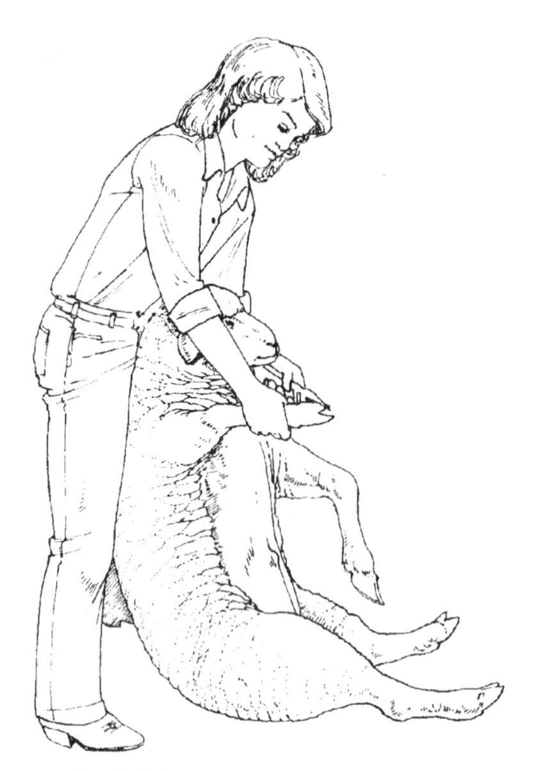

Fig. 13-3. Correct position for trimming the feet.

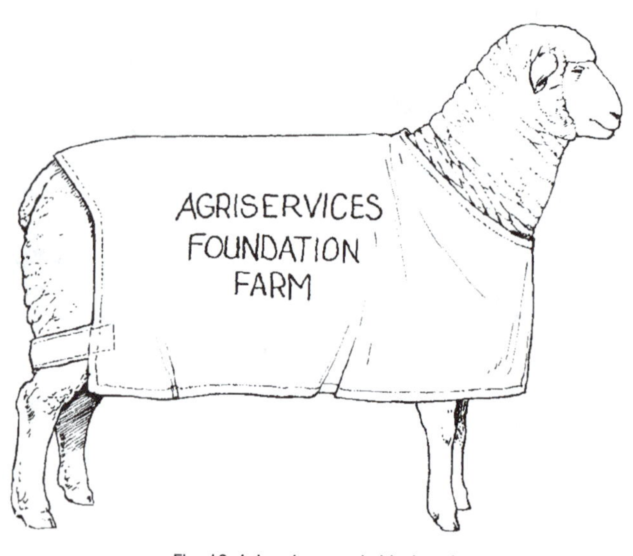

Fig. 13-4. Lamb properly blanketed.

opening for the head and neck. Cords are attached to pass around the rear legs of the sheep.

Blankets should be of such size and construction that they will not bind tightly in any place. Most important, the blanket should fit properly about the neck; otherwise, it may cause creases in the fleece that are difficult to remove.

Many shows forbid the use of blankets during certain specified hours.

PREPARING THE FLEECE

Most knowledgeable people agree that no other assignment in fitting any class of livestock requires so much skill, patience, and time as does preparing the fleece. This important operation is designed to accentuate what is considered most desirable from the standpoint of both body conformation and fleece quality.

TIME TO BEGIN TRIMMING

Most breeding sheep are trimmed several times before arriving at the fair. Most market lambs are shown in short fleece, which makes it easier to keep them clean during the summer.

The approximate time in advance of the show at which to begin trimming will depend somewhat on: (1) the shearing date; (2) the age of the sheep; (3) whether the animals are being exhibited in breeding or market classes; (4) the length, density, and condition of the fleece; and (5) the rules of the show.

Because of greater fleece growth, the trimming process on yearling or mature sheep is usually started earlier than with lambs. In mature animals—if the fleece is sufficiently long—a first rough trimming is usually administered sometime in the early summer. Then two subsequent trimmings follow before the animals are taken to the show, and a final touching up is given after arrival at the show. In no event should the first trimming of mature sheep come closer than within six weeks of the show. Earlier trimming is considered preferable.

Sometimes, because of their shorter fleeces, the trimming of lambs is delayed until 2 to 3 weeks before the show; but, if at all possible, an earlier start is desirable. Generally, the trimming of lambs starts when they are weaned and placed in the show barn, which is usually sometime in July. With lambs that are to be shown in the breeding classes, close trimming is avoided, for the added fleece length gives the appearance of greater size and scale.

The time to begin trimming also is affected by the length, density, and condition of the fleece. Prospective show sheep with long, loose fleeces or fleeces that are cotted or dirty require earlier attention than would otherwise be necessary.

Fat wethers are trimmed rather closely early in the season, for the shorter fleeces impart superior handling qualities. It is particularly desirable that wethers be trimmed very short along the topline, thus making them firmer to the touch. When wethers are crowded in hot weather, they keep cooler and feed better when their fleeces are short.

DECIDING WHAT AND HOW TO TRIM

Before beginning the trimming operations, skillful shepherds carefully study each individual animal. They arrive at a clear picture of the animal as nature made it and within their own mind visualize how it can and should appear after plying their art.

Following this preliminary study, the careful worker then begins, step by step, to bring about the desired transformation. Each bit of trimming is designed to assist in molding the final form. In this procedure, each sheep is treated as an individual, and the job of trimming is tailor-made for each animal.

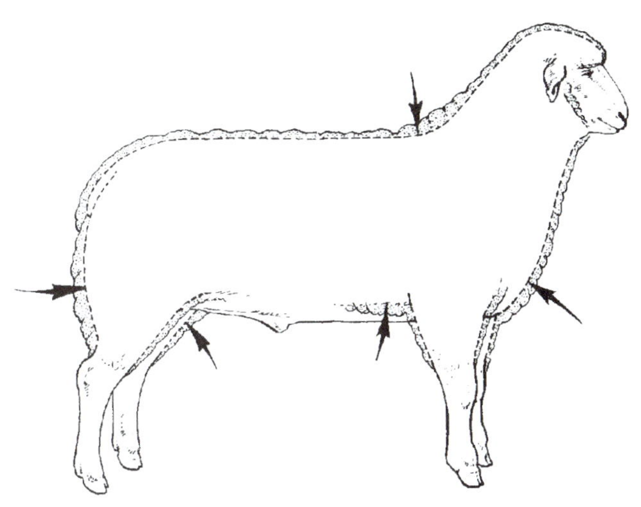

Fig. 13-5. One should first decide what to trim, in order to enhance the attractiveness of the individual through accentuating its strong points and covering up its weaknesses.

CLEANING AND WASHING FLEECES

Most experienced shepherds are opposed to washing the fleeces of breeding sheep, especially of the wool breeds, except as a last resort; for the use of soap and water removes the yolk along with dirt or other foreign matter. Yolk is necessary to keep the fleece in good condition and imparts the character so desired by experienced judges. Therefore, every effort should be made to keep show sheep clean. Secondly, before resorting to washing, the shepherd should try to clean a dirty fleece by thoroughly rubbing the outside with a flannel cloth or a burlap rag dampened in warm water.

Most showpersons wash market lambs as often, and as close to the show, as possible. This should begin about two to three weeks before the show. Lambs that are not clean-shorn at show time may need two or three washings. Special attention should be given to the belly and under the legs where dirt collects. After each washing, the lamb should be dried with towels, then carded and trimmed while still damp.

Should the latter method of cleaning fail to bring about the desired results, 2 to 3 months in advance of the show, the fleece should be washed with soap and warm water, using a specially prepared livestock soap or a clear or white-colored household liquid soap, followed by carefully rinsing all soap from the fleece to prevent discoloration. After washing, use a clean curry comb to remove as much water as possible, followed by the use of a hot-air blower-dryer. After the animal has been washed and dried, immediately blanket it and place it in a clean pen bedded with clean straw.

Many times, small dung locks must be washed from around the dock, or britch. Washing, with a solution as recommended above, is preferable to clipping these locks off with the shears, as the latter procedure may leave a depression in an area where greater fullness would have been desired. Also, it may be necessary to wash out locks on the foretop, belly, and legs; but all of this should be done with care, limiting the washing to the dirty area in order not to remove too much yolk from the balance of the fleece. Burs should be carefully cleaned from the fleece.

THE TRIMMING PROCEDURE AND METHOD

After the shepherd has cleaned the fleece, studied the individual animal, and decided what and how to trim in order to enhance the attractiveness of the individual, the trimming procedure is as follows: (1) the first rough blocking, (2) the second trimming, (3) the third and final trimming before shipment to the show, and (4) the preshow touching up.

■ **Securing the animal for trimming**—Although animals may be held fast by another person or secured by means of a rope halter or yoke, the trimming stand, which may be homemade (see Fig. 13-1), is the most satisfactory arrangement for holding sheep for trimming. The trimming stand consists of a platform on which the animal is stood, and which has an arrangement at one end for securing the head in approximate show position.

Hoofs may also be trimmed while the animal is on the trimming stand.

THE FIRST ROUGH BLOCKING

The next step in preparing the fleece consists of dampening and vigorous combing. Usually the fleece is made damp by first handspraying and then brushing with either (1) the same type of soap as was used for washing, or (2) milk oil dip (2 to 3 cupfuls per gallon of water). Where a handsprayer is not available, a brush alone will suffice. After the fleece is somewhat crusty from the dampness, it should be vigorously combed, preferably with a circular type of curry comb, although some shepherds use a Scotch comb. The combing

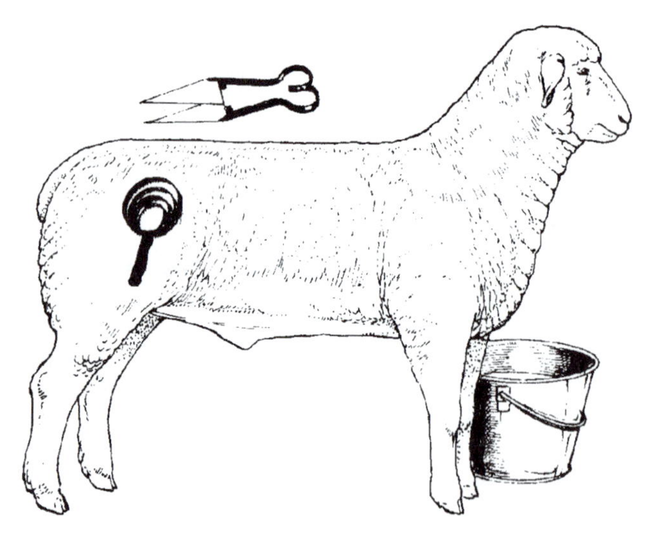

Fig. 13-6. The first rough blocking is designed to make the top level; to make the animal appear as ideal in type as possible; and to make fat wethers firm to the touch when handled over the back and loin.

process develops the desired *face* on the fleece. In preparation for the rough blocking, a comb is used rather than a card, for the simple reason that it is very difficult to card a fleece until some preliminary treatment is given.

The first rough blocking consists of flattening down the back and removing a little wool from such other parts of the body (especially the sides) as need clipping. To avoid irreparable damage, the animal is first stood squarely and correctly; and a decision is then reached on how close to trim the wool over the back, sides, and other parts of the body. The objective in flattening down the back of breeding sheep is to make the top line level. Fat wethers must be firm to the touch when handled over the back and loin.

To achieve these objectives, the shepherd usually cuts the wool on the back of breeding sheep no shorter than ¾ in., whereas wethers may be flattened down to within ¼ or ½ in. of the skin. In carrying out the back-trimming operation, the right-handed shepherd usually takes a position to the front of the left flank and faces toward the front of the animal. With sharp shears held in the right hand, trimming is started at the dock and is continued in as straight a line as possible to the top of the shoulder. This operation is repeated, starting a new swath at the dock each time, until the full width of the back has been covered.[4] The shepherd then rough trims the sides of the sheep, usually by removing some of the fleece which projects out from the center of each side. This is done to develop comparatively straight side-lines. Following this preliminary blocking, many shepherds prefer to wait for two weeks or longer before proceeding with a second and more careful trimming.

THE SECOND TRIMMING

Usually the second trimming, which is more refined and complete than the first rough blocking, is immediately preceded by dampening the fleece as previously indicated, and by a combination of brushing and carding designed to get the fibers separated and parallel to each other and to develop increased fullness in certain regions. The brushing operation consists of working over the fleece with a medium fine brush which is dipped into the water-soap solution at intervals. Brushing dampens the wool—thus making for a smoother job with the shears—and assists in straightening out the fibers. The carding operation is for the specific purpose of straightening out the wool fibers so that they can be cut off evenly, and it is also designed to give increased fullness to certain regions. Carding is accomplished by grasping the card securely in the hand, engaging it firmly in the fleece, giving it a slight forward pull in order to insert the teeth among the fibers, and then lifting up in a revolving motion so as to separate and comb out the individual fibers parallel to each other. The experienced shepherd performs the carding operation through a revolving and rotating movement of the wrist.

The regions of the forerib, rump, and thigh should be carded very carefully and thoroughly to develop the desired fullness in these areas, which frequently lack in development. The underline also should be carded.

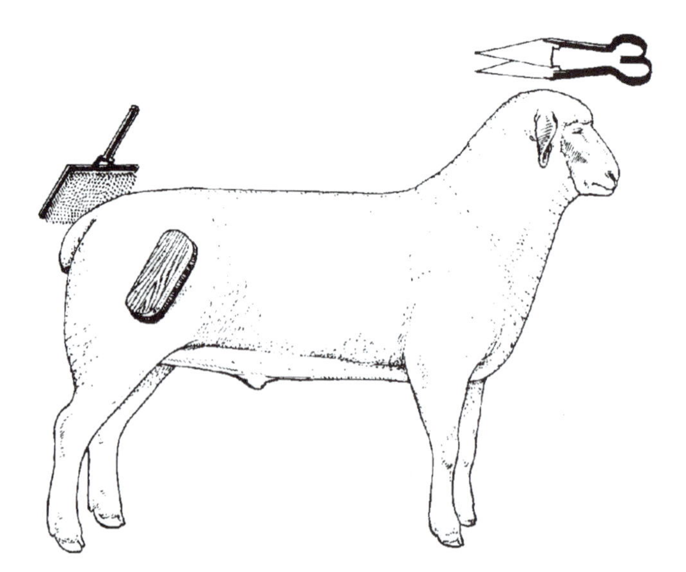

Fig. 13-7. The second trimming is designed to impart the desired appearance and to give a blending together of all body parts.

[4]In administering the back-trimming operation, other shepherds prefer to take a position to the front of the animal, working from the top of the shoulder to the dock (with the right-handed shepherd taking a position to the front of the left shoulder).

Wool cards that are filled with wool may be cleaned by pulling one card over the other or by using a bent table fork.

The next step in the trimming process is to use sharp shears to trim off the ends of the fleece fibers to an even length. Usually, it is necessary to repeat this process several times, first brushing and carding and then following with the shears. Not only will expert trimming impart the desired appearance, but it will also result in the necessary blending together of all body parts; for example, the neck will blend smoothly into the shoulder.

After the body has been trimmed, the head of the animal should be prepared to express the most desirable breed type and sex character. Consideration should be given to the amount of wool desired on the foretop, the face covering, the ears, etc. The head of wethers should be trimmed so as to accentuate a trim and clean-cut appearance. The wool on the legs of those breeds having such wool should be carded out and trimmed to impart the appearance of having lots of bone, without appearing unbalanced or abnormal.

THE THIRD AND FINAL TRIMMING BEFORE SHIPMENT TO THE SHOW

The final trimming, which is administered immediately before sheep are sent to the show and which has usually been preceded by at least two previous trimmings, should be thorough, thus making it necessary to add only the finishing touches before the animals enter the show-ring. This final trimming consists of alternate brushing, carding, and trimming with sharp shears in such manner as to enhance the natural lines of the individual to the maximum. When the trimming is completed, some shepherds pat the fleece lightly with the back of a wool card or some similar smooth-surfaced object. This packs the fleece and improves the appearance and handling qualities.

THE PRESHOW TOUCHING UP

After the sheep arrive at the show, any marks or irregularities in the fleece caused by shipment or blanketing are usually removed by going over the animals lightly with the brush, card, and shears. This also makes the animals appear fresher and gives them more bloom when led into the show-ring.

OILING OR DRESSING THE FLEECE

In general, artificial oiling of the fleece should not be resorted to unless it seems absolutely necessary, and unless it is permitted by the show(s). If the fleece is dry and harsh, as is frequently true in the long-wool breeds, a suitable dressing may be rubbed into it. But this is done well in advance of the show and before the final trimming prior to shipment.

Fig. 13-8. The preshow touching up removes any marks or irregularities in the fleece and imparts freshness and bloom. Note the sheep being trimmed, the trimming and blocking stand, and the person doing the trimming.

The most desirable dressings are those which are as nearly like the natural yolk as is possible to secure. Most exhibitors agree that wool fat, properly thinned down by heating and correctly applied through the fleece, most nearly approaches the natural condition. Wool fat is a byproduct of the wool scouring industry. Other, but less desirable, fleece dressings that are sometimes used include: olive oil, sweet oil, light mineral oil, and corn oil. The lighter oils are more easily applied than wool fat. They usually are either (1) sprayed into the fleece by a hand (atomizing) sprayer, with one person parting the fleece and another spraying the dressing around the base of the fleece, or (2) rubbed on with a brush. Wool fat may best be applied by putting a light coat on the hands and gently and carefully rubbing them through the fleece.

FITTING LONG-WOOL SHEEP

The fitting of the long-wool breeds of sheep, which produce braid wool—Lincolns, Cotswolds, and Leicesters—differs from the fitting of the medium-wool breeds in the following respect:

1. **Finish.** With the high show finish, the long-wool breeds are likely to become patchy and uneven in their fleshing. Usually this tendency can be minimized, if not eliminated, by providing more bulky rations (for example, by adding plenty of oats to the ration) and by forcing the animals to take plenty of exercise.

2. **Length of fleece.** As length of fleece is considered such an important attribute in the long-wool breeds, there is a tendency to accentuate this characteristic through stubble shearing and allowing more than 12 months of growth before showing.

3. **Foretop.** With animals of the Lincoln and Cotswold breeds intended for show, the foretop should never be removed.

4. **Protecting the fleece from the elements and foreign matter.** The fleece over the back of the long-wool breeds is likely to become harsh and cotted from exposure to rain and sunshine. Moreover, the wool along the topline is very likely to become filled with chaff and other foreign material. For the long-wool breeds, therefore, the experienced shepherd provides as much protection as is possible from the elements and from chaff and other foreign material.

5. **Adding artificial fats or oils.** Sometimes, if permitted by the show, the skillful shepherd supplements with one of the artificial fats or oils the natural yolk along the topline and sometimes the yolk along the sides of the long-wool breeds.

6. **Blankets.** Blankets must not be used immediately after dressings have been applied to long-wool fleeces, for they may cause sweating and fleece discoloration. The yolk will harden and the fleece will have more body if exposed to the air. In general, therefore, blankets are used less often on the long-wool breeds than on the medium-wool breeds.

7. **Removing certain fibers from frowsy fleeces.** In frowsy fleeces, the appearance of the fleece may often be improved by the removal of those fibers that are not a part of organized locks.

8. **No carding of long-wool fleeces and cautiously using the brush.** It is preferable that the fleece of long-wool breeds hang from the body in locks. Therefore, they are never carded, and they must be brushed with great care.

9. **Shearing only around the dock.** On the long-wool breeds, shears are used only for trimming about the dock. The ends of the locks over the body are never trimmed.

FITTING FINE-WOOL SHEEP

Although animals of the fine-wool breeds that are being fitted for show are fed and handled much like the other breeds, more attention is given to the fleece because of its greater economic importance. In preparing fine-wool sheep for show, therefore, the following factors are particularly important:

1. **Keep the fleece clean but seldom wash it.** In the feeding and management of fine-wool sheep, every effort should be made to keep the fleece free from chaff, dirt, and other foreign matter. It is preferable that the fleece of fine-wool sheep not require washing, because, unless carefully done, it will make the wool harsh and dry. Weekly rubbing of the fleece with a damp wool rag is advocated. The head, ears, and legs of fine-wool sheep are usually washed.

2. **Never block with the shears; merely smooth up the animal.** Usually, the trimming operation on fine-wool sheep merely consists of brushing the fleece with a damp brush so that the fibers will lie parallel to each other and trimming the loose and irregular ends so that a smooth appearance is obtained. The animals are never blocked, because they are raised primarily for their fleece rather than their meat qualities.

3. **Oil the outside of the fleece.** To prevent dryness or harshness and infiltrating dirt, most experienced shepherds apply, to the outside of the fleece, a light coating of a suitable dressing, such as wool fat (for best results, light oil should be used on dense, heavy fleeces, and relatively heavy oil on less dense, light fleeces). In the vernacular of a shepherd, this is known as "putting a top on the fleece." Oiling is usually done well in advance of the show, 3 to 4 weeks before. It is done early to avoid too much oil at the time of the show. However, a second light application may be given a week before the show if necessary. On fine-wool fleeces, the dressing is usually patted on the surface of the fleece and not rubbed in. *Note well:* The practice of adding artificial oils is discouraged, and it is actually outlawed in some shows.

4. **Add coloring.** Sometimes, shepherds add a coloring matter to dressings used on fine-wool sheep. Lampblack is often used, but a more natural effect is obtained by thoroughly mixing a combination of 1 tbsp of burnt amber and 2 tbsp of lampblack to each 1½ pints of dressing. Like artificial oiling, the practice of adding coloring is discouraged or outlawed in some shows.

5. **Seldom use blankets.** Blankets are always used sparingly on fine-wool sheep because their presence may cause sweating and fleece discoloration. The presence of air tends to harden the yolk and to give more body to the fleece. After the dressing or coloring matter has been applied, the blanket should be left off for a period of 3 or 4 days so that the added ingredients may dry.

6. **Stress long fleeces.** As with long-wool sheep, there is a tendency to accentuate the length of fleece through stubble shearing and allowing more than 12 months of growth before showing.

MAKING FAIR ENTRIES

Well in advance of the show, the exhibitor should request a premium list and entry blanks from the show manager or secretary. Entries generally must be filed with the show about 30 days in advance of the opening date. Most shows specify that entries be made out on printed forms and in accordance with instructions thereon. The class, age, breed, registry number, and usually the name and registry number of the sire and dam must be given. Entries must be made in all individual and group classes in which the exhibitor intends to show, but no entries are made in the championship classes, the first-place winners being eligible for the latter

PROVIDING HEALTH CERTIFICATES

Health certificates are always required for show sheep involved in interstate shipment. The health certificate must accompany the animal and must be signed by a licensed veterinarian. Some shows specify that this certificate must indicate that the veterinarian has examined each animal offered for entry and has found it free of scabies and other infectious or contagious diseases.

SHIPPING TO THE FAIR

When show sheep are shipped, it is important that the following details receive consideration:

1. Schedule the transportation so as to arrive within the limitations imposed by the show, and at least 2 to 3 days in advance of the date that the animals vie for awards.

2. Before using, thoroughly clean and disinfect all public conveyances.

3. Use clean bedding in order not to soil the fleece or introduce foreign matter into it.

4. Provide suitable and necessary partitions for separate penning of the sexes and for separating out the rams that are not accustomed to running with each other.

5. Do not overcrowd; allow sufficient space for the sheep to bed down in comfort.

6. If space is at a premium, place the feed supply, bedding, and show equipment on a deck or platform, preferably at least 5 ft above the floor in order to allow for air circulation, and place sheep under the platform.

7. When mixed feeds are used, as is usually the case in fitting rations, take along a supply adequate for the entire trip. This will reduce the hazard of animals going off feed because of feed changes.

8. Limit sheep to a half feed at the last feeding before loading out and while in transit.

9. In transit, the sheep should be handled quietly and should not be allowed to become hot nor to be in a draft.

PEN SPACE, FEEDING, AND MANAGEMENT AT THE SHOW

As soon as the show is reached, the animals should be unloaded and placed in clean pens that are freshly bedded. Most sheep pens at fairs or exhibitions will accommodate 3 to 6 lambs, 3 or 4 yearlings, or 2 or 3 older animals. Yearling or older rams should not be penned together unless they have been previously so handled.

While at the show, it is preferable that the sheep receive the same ration to which they were accustomed at home. Usually only a half ration is allowed for the first 24 hours after arrival at the show, and a normal ration is provided thereafter. So that the animals will maintain their appetite, however, it is necessary that they receive exercise while at the show. It is usually best to exercise the animals one-half hour or more in the cool of the morning. This is also a convenient time to clean out the pens while the animals are out for exercise.

It is customary for exhibitors to identify their exhibits by means of neat and attractive signs, the size of which must be within the limitations imposed by the show. Each sign usually gives the name of the breed of sheep and the name and address of the exhibitor.

After the animals have rested for 24 hours following shipment, it is usually desirable that there be a final touching up before showing. At this time, the exhibitor can remove any blanket or shipping marks or depressions in the fleece and freshen it up in general.

ETHICAL PROBLEMS OF SHEEP SHOWS

During the early 1990s, the reputation of 4-H and FFA was tarnished as champion market animals at some of the major junior livestock shows in the country were found to have been administered illegal substances or tampered with in other ways prior to, and during, the competition. Such ethical abuses resulted in some exhibitors having to forfeit premiums and ribbons; and in some instances exhibitors received a lifelong ban. It is disturbing that something as sacred as "mom and apple pie" is tarnished in this manner. However, it should be emphasized that these ethical abuses are being committed by a very small percentage of people involved in junior livestock shows.

SHOWING THE SHEEP

Some shepherds appear to excel so much in their ability to show sheep that it has often been said that "exhibitors are born and not made." Although it is reasonable to believe that some people may have more natural ability to show sheep than others, there are certain guiding principles in showing sheep that are adhered to by everyone. These are as follows:

1. During the trimming process, the sheep will have become accustomed to handling and to standing squarely on its feet with the head up and the back well supported. It should be posed in this manner in the ring.
2. The animal should be touched up with the wool card and sheep shears well in advance of show time, so that the shepherd may be prompt in parading before the judge when the class is called. Be sure that all bedding is off the sheep.
3. The exhibitor should dress neatly and in appropriate clothing for the occasion.
4. Know the birthdate of your animal.
5. Without being an eager beaver, the exhibitor should get as favorable a position as possible in the line and should keep the sheep looking its best.
6. Bring the sheep into the ring slowly. Move the sheep with one hand under the jaw and the other hand under the dock.
7. The sheep should be set up squarely, with the exhibitor stationed (either standing or squatting) to the left and front of the animal. The exhibitor's left hand should grasp the wool lightly under the chin. The right hand is kept free for use in keeping the sheep set up properly.
8. When the judge approaches, the sheep should be standing squarely, with the back strong. A properly trained animal will respond by bolstering its back and will keep its hind legs in place when the shepherd pushes against the animal's breast with the knee.
9. When the judge walks to the front, the shepherd should step aside so that a good head and front view may be obtained.
10. All movements of the exhibitor should be calm and collected, radiating confidence and experience.
11. The skillful exhibitor always keeps one eye on the judge.
12. When the judge requests that the position in line be changed, the exhibitor moves into the new position from behind the line, again sets the animal up, and continues to show it properly.
13. The experienced exhibitor is courteous to other exhibitors and the judge.
14. When the ribbons are handed out, the good exhibitor is a gracious winner or an equally good loser.
15. Some fairs prefer that showpersons not get on their knees while showing, or at least not on both knees.
16. Use halters on sheep if you prefer. Generally, those who use halters lead the sheep into the ring, but once in position they forget about the halters.

Fig. 13-9. Good exhibitors adhere to certain guiding principles when showing sheep.

AFTER THE FAIR IS OVER

Most shows have regulations requiring that all exhibits remain on the grounds until a specified time, after which signed releases must be secured from the superintendent of the sheep show. Because most exhibitors are anxious to travel once the show is over and there is considerable confusion at this time, it is usually advisable to load all equipment and left-over feed before the release of animals is secured, so that all that remains to be done is to load out the animals.

Upon returning to the farm or ranch, it is usually a good policy to isolate the show flock for about three weeks. This procedure reduces the possibility of spreading diseases or parasites to the balance of the flock.

The final assignment after the show is over consists of gradually reducing the condition of heavily fitted breeding animals. Experienced shepherds accomplish this difficult task and yet retain strong, vigorous animals through adhering to the following practices: (1) providing plenty of exercise; (2) increasing the amount of bulky feeds, such as oats, in the ration; (3) very gradually cutting down on the grain allowance; and (4) retaining the succulent feeds and increasing the pasture or hay. In caring for lambs and yearlings that are to be developed for show purposes the following year, it is important to continue to feed them an adequate, although lighter, grain ration.

When the flock is exhibited on a circuit, the shepherd must use great care in keeping the animals in show condition at all times. The peak condition should be reached at the strongest show. In order to be successful, this type of venture requires great skill on the part of the caretaker, especially from the standpoint of feeding and exercising the sheep.

4-H AND FFA SHEEP SHOWPERSON SCORECARD

The *4-H and FFA Fitting and Showing Sheep Scorecard* herewith reproduced (see Fig. 13-10) was adapted by the author from North Central Regional Extension Publication 157, sponsored by the Extension Services of Illinois, Michigan, Minnesota, Missouri, Ohio, and Wisconsin, in cooperation with the U.S. Department of Agriculture. It was prepared by C. J. Christians and W. C. Bollums.

4-H SHEEP SHOWPERSON SCORECARD
(100 Points Possible)

APPEARANCE OF LAMB — Perfect Score 40 Points

1. Cleanliness: 15
   a. Wool clean and free of foreign matter.
   b. Head, flanks, feet, and legs clean.
2. Trimming: 25
   a. Fleece of market lambs to be trimmed short.
   b. Fleece of breeding ewes trimmed to a length of 3/8 to 3/4 in., or, in the case of a wool breed, a length of which is appropriate for the standard of their breed.
   c. Back, sides, and legs of breeding sheep trimmed to emphasize merits of individual.
   d. Feet trimmed so that animal walks and stands naturally and correctly.

SHOWING LAMB — Perfect Score 40 Points

1. Posing and changing positions: 30
   a. Animal kept well posed at all times, but exhibitor not engaged in undue fussing or maneuvering.
   b. Animal shows evidence of training as indicated by responsiveness to handling.
   c. Animal shown from front when judge is viewing it from rear or left side; otherwise, animal shown from left side only.
   d. Exhibitor shows while standing or in a half-kneeling (one knee on ground) position except when moving animal to a new position.
   e. Animal led from left side with left hand under jaw and right hand at the dock.
   f. Exhibitor maintains reasonable distance from next animal at all times.
   g. Exhibitors may let animal relax, but not out of position, when judge is at other end of arena.
   h. Exhibitor may brace the animal by grasping under chin with both hands, pulling downward, and applying backward pressure to cause it to hold its back rigid and straight while being handled.
   i. Halter can be used for control.

SHOWING LAMB *(continued)* — Perfect Score 40 Points

2. Cooperation with judge: 10
   a. Awareness of position of judge maintained, but not made obvious.
   b. Body not permitted to obstruct view of judge.
   c. Animal maneuvered into position for benefit of judge's inspection before, but not during, inspection.

APPEARANCE AND MERITS OF EXHIBITOR — Perfect Score 20 Points

1. Appearance: 10
   a. Exhibitor well groomed and neat.
   b. No headgear to be worn by exhibitor.
   c. Shorts are not appropriate. Extremes in colors and fit not appropriate.
   d. White shirt (or blouse) with a 4-H T-shirt or official shirt (or blouse) provided by show management are appropriate.
2. Merits: 10
   a. Brings animal into ring promptly.
   b. Responsive to judge's and ringmaster's requests.
   c. Works quickly, but not abruptly.
   d. Minimizes faults of animal when showing.
   e. Not distracted by persons and things outside ring.
   f. Shows animal, not self.
   g. Does not leave ring until released by ring official.
   h. Displays a courteous and sportsmanlike attitude at show.
   i. Prepared to give prompt answers to questions related to the 4-H animal project.

Fig. 13-10. 4-H and FFA Fitting and Showing Sheep Scorecard.

QUESTIONS FOR STUDY AND DISCUSSION

1. Under what circumstances would you recommend that a purebred, a commercial, or a 4-H club or an FFA member (a) show, and (b) not show sheep?

2. Take and defend either the affirmative or the negative position of each of the following statements:

   a. Fitting and showing does not harm animals.

   b. Livestock shows have been a powerful force in sheep improvement.

   c. Too much money is spent on livestock shows.

   d. Unless all animals are fitted, groomed, and shown to the same degree of perfection, show-ring winnings are not indicative of the comparative quality of animals.

3. How may sheep shows be changed so that they (a) more nearly reflect consumer preference, and (b) make for greater sheep improvement?

4. Outline procedures for the feeding and handling of show sheep.

5. List and describe the major steps involved in training and grooming.

6. Detail the trimming procedures

7. Compare the fitting of long-wool sheep and fine-wool sheep to that of the meat-type sheep.

8. What show-ring advice would you give to someone showing sheep for the first time?

SELECTED REFERENCES

Title of Publication	Author(s)	Publisher
Of Sheep & Shows	J. C. P. Kroge	J. C. P. Kroge, Boulder, CO, 1972
Livestock Judging, Selection, and Evaluation	R. E. Hunsley	Interstate Publishers, Inc., Danville, IL, 2001
Sheep Production	R. V. Diggins C. E. Bundy	Prentice-Hall, Inc., Englewood Cliffs, NJ, 1958
Stockman's Handbook, The	M. E. Ensminger	Interstate Publishers, Inc., Danville, IL, 1992

CHAPTER 14

Sheep are unique among farm animals inasmuch as (1) 94% of their feed is derived from forage, primarily pastoral; and (2) lambs may be marketed for slaughter off milk and good grass. When rotated (as shown on this attractive farm) or grazed together, sheep and cattle complement each other. Sheep eat more shrubs and forbs than cattle. Cattle prefer grasses. (Courtesy, J. C. Allen and Son., Inc., West Lafayette, IN)

FEEDING SHEEP

Contents *Page*

Contents Page

Sheep consume a higher proportion of forages than any other class of livestock, it being estimated that 94% of the total feed supply of the U.S. sheep production is derived from roughages. They are naturally adapted to grazing on pastures and ranges which supply a variety of forage plants, and they thrive best on forage that is short and fine rather than high and coarse. Although sheep will eat considerable quantities of weeds and brush, they prefer choice grasses and legumes.

Except at lambing season, sheep seldom receive much grain. In the northern latitudes, farm-flock ewes are frequently given from 0.5 to 1.0 lb daily of a grain ration in addition to the roughage allowance from about six weeks before lambing to the time that they are turned to spring pasture. Higher levels of grain are fed during the suckling period than during gestation. Many of the farm flocks of the South and range bands of the Southwest, however, are kept in good thrifty condition, and the lambs are raised to the marketing stages, without the feeding of any grain. In still other areas, the ewes are fed only during periods of deep snows or extended droughts. The range bands in the colder regions of the West are normally fed alfalfa hay and grain during the period of about 3 to 4 weeks that they are confined to the lambing camp.

In general, for practical reasons, the ration of ewes should consist of as nearly year-round pastures as possible, with well-cured hay and other forages available the balance of the year, plus a limited grain allowance under certain conditions. Good quality sun-cured hay and lush pastures will not only provide most of the necessary proteins, but they are excellent sources of most of the minerals and vitamins, also.

FACTORS INVOLVED IN FORMULATING SHEEP RATIONS

Before anyone can intelligently formulate a sheep ration, it is necessary to know (1) the nutrient requirements of the particular sheep to be fed, which calls for feeding standards; (2) the availability, nutrient content, and cost of feedstuffs; (3) the acceptability and physical condition of feedstuffs: (4) the average daily consumption of the sheep to be fed; and (5) the presence of substances harmful to product quality.

NUTRITIVE NEEDS OF SHEEP

As with other classes of livestock, the nutritive needs of sheep may be classified as (1) energy, (2) protein, (3) minerals, (4) vitamins, and (5) water.

The nutritive requirements are the values considered necessary for maintenance, optimum production, and prevention of any signs of nutritional deficiency.

NATIONAL RESEARCH COUNCIL (NRC) REQUIREMENTS

The National Research Council (NRC) nutritive requirements of sheep are given in Tables 14-1 to 14-11.

The NRC requirements are adequate for average, or below average, animals. In practical rations, margins of safety should be added to provide for below-average feeds, deterioration of feeds during transportation and storage, conditions of stress (bad weather, shipment, disease, or parasitism), and above-average animals in size, stage of production, and level of production.

ENERGY

Lack of energy—hunger—is probably the most common nutritional deficiency of sheep. It may result from lack of feed or from the consumption of poor quality feed.

Inadequate amounts of feed may result from over-

Fig. 14-1. The energy needs of sheep are largely met through the consumption of hay and pasture. (Courtesy, Ralston Purina Company, St. Louis,

grazing, droughts, snow covering the feed, or from a low dry matter content of lush, washy feeds. Also, poorly digested low-quality forage leads to reduced feed intake.

The energy needs of sheep are largely met through the consumption and digestion of roughages—pasture and hay. Grains, such as corn, barley, milo, wheat, and oats, are used to raise the energy level of the ration during periods when supplementation is necessary. In general, sheep subsist on an even higher proportion of roughages to concentrates than do beef cattle, and this applies to finishing lambs. The bacterial action in the paunch of the sheep efficiently converts roughages into suitable sources of energy.

It is generally recognized that the energy requirements of sheep are affected by size, age, pregnancy, lactation, growth, and protein content of the ration. It is also affected by environment, shearing, and sex. The net energy requirements of ewes carrying different numbers of fetuses at various stages of gestation are given in Table 14-1.

■ **Symptoms of energy deficiency**—An energy deficiency is characterized by slowing and cessation of growth, loss of weight, reduced fertility or reproductive failure, lowered milk production and shortened lactation period, reduced quantity and quality of wool (including breaks in the fiber), lowered resistance to infection with internal parasites, and increased mortality.

TABLE 14-1
NE_{PREG} (NE_Y) REQUIREMENTS OF EWES CARRYING DIFFERENT NUMBERS OF FETUSES AT VARIOUS STAGES OF GESTATION[1]

Number of Fetuses Being Carried	Stage of Gestation (days)[2]					
	100	%[3]	120	%[3]	140	%[3]
	(NE_{preg} Required [kcal/day])					
1	70	100	145	100	260	100
2	125	178	265	183	440	169
3	170	243	345	238	570	219

[1]Adapted by the author from *Nutrient Requirements of Sheep*, Sixth Revised Edition, NRC-National Academy of Sciences. The (NE_Y) refers to reproductive process.

[2]For *gravid uterus* (plus contents) and mammary gland development only.

[3]As a percentage of a single fetus's requirement.

TABLE 14-2
DAILY NUTRIENT REQUIREMENTS OF SHEEP

Body Weight		Weight Change/ Day		Dry Matter per Animal[1]			Nutrients per Animal: Energy[2] TDN		DE	ME	Crude Protein		Ca	P	Vitamin A Activity	Vitamin E Activity
(kg)	(lb)	(g)	(lb)	(kg)	(lb)	(% body weight)	(kg)	(lb)	(Mcal)	(Mcal)	(g)	(lb)	(g)	(g)	(IU)	(IU)
Ewes[3]																
Maintenance																
50	110	*10*	0.02	*1.0*	2.2	2.0	*0.55*	1.2	2.4	2.0	*95*	0.21	*2.0*	*1.8*	2,350	15
60	132	*10*	0.02	*1.1*	2.4	1.8	*0.61*	1.3	2.7	2.2	*104*	0.23	*2.3*	*2.1*	2,820	16
70	154	*10*	0.02	*1.2*	2.6	1.7	*0.66*	1.5	2.9	2.4	*113*	0.25	*2.5*	*2.4*	3,290	18
80	176	*10*	0.02	*1.3*	2.9	1.6	*0.72*	1.6	3.2	2.6	*122*	0.27	*2.7*	*2.8*	3,760	20
90	198	*10*	0.02	*1.4*	3.1	1.5	*0.78*	1.7	3.4	2.8	*131*	0.29	*2.9*	*3.1*	4,230	21

(Continued)

TABLE 14-2 (Continued)

Body Weight		Weight Change/ Day		Dry Matter per Animal[1]			Nutrients per Animal: Energy[2] TDN		DE	ME	Crude Protein		Ca	P	Vitamin A Activity	Vitamin E Activity
(kg)	*(lb)*	*(g)*	*(lb)*	*(kg)*	*(lb)*	*(% body weight)*	*(kg)*	*(lb)*	*(Mcal)*	*(Mcal)*	*(g)*	*(lb)*	*(g)*	*(g)*	*(IU)*	*(IU)*
Ewes (continued)																
Flushing—2 weeks prebreeding and first 3 weeks of breeding																
50	110	100	0.22	1.6	3.5	3.2	0.94	2.1	4.1	3.4	150	0.33	5.3	2.6	2,350	24
60	132	100	0.22	1.7	3.7	2.8	1.00	2.2	4.4	3.6	157	0.34	5.5	2.9	2,820	26
70	154	100	0.22	1.8	4.0	2.6	1.06	2.3	4.7	3.8	164	0.36	5.7	3.2	3,290	27
80	176	100	0.22	1.9	4.2	2.4	1.12	2.5	4.9	4.0	171	0.38	5.9	3.6	3,760	28
90	198	100	0.22	2.0	4.4	2.2	1.18	2.6	5.1	4.2	177	0.39	6.1	3.9	4,230	30
Nonlactating—first 15 weeks gestation																
50	110	30	0.07	1.2	2.6	2.4	0.67	1.5	3.0	2.4	112	0.25	2.9	2.1	2,350	18
60	132	30	0.07	1.3	2.9	2.2	0.72	1.6	3.2	2.6	121	0.27	3.2	2.5	2,820	20
70	154	30	0.07	1.4	3.1	2.0	0.77	1.7	3.4	2.8	130	0.29	3.5	2.9	3,290	21
80	176	30	0.07	1.5	3.3	1.9	0.82	1.8	3.6	3.0	139	0.31	3.8	3.3	3,760	22
90	198	30	0.07	1.6	3.5	1.8	0.87	1.9	3.8	3.2	148	0.33	4.1	3.6	4,230	24
Last 4 weeks gestation (130–150% lambing rate expected) or last 4–6 weeks lactation suckling singles[4]																
50	110	180 (45)	0.40 (0.10)	1.6	3.5	3.2	0.94	2.1	4.1	3.4	175	0.38	5.9	4.8	4,250	24
60	132	180 (45)	0.40 (0.10)	1.7	3.7	2.8	1.00	2.2	4.4	3.6	184	0.40	6.0	5.2	5,100	26
70	154	180 (45)	0.40 (0.10)	1.8	4.0	2.6	1.06	2.3	4.7	3.8	193	0.42	6.2	5.6	5,950	27
80	176	180 (45)	0.40 (0.10)	1.9	4.2	2.4	1.12	2.4	4.9	4.0	202	0.44	6.3	6.1	6,800	28
90	198	180 (45)	0.40 (0.10)	2.0	4.4	2.2	1.18	2.5	5.1	4.2	212	0.47	6.4	6.5	7,650	30
Last 4 weeks gestation (180–225% lambing rate expected)																
50	110	225	0.50	1.7	3.7	3.4	1.10	2.4	4.8	4.0	196	0.43	6.2	3.4	4,250	26
60	132	225	0.50	1.8	4.0	3.0	1.17	2.6	5.1	4.2	205	0.45	6.9	4.0	5,100	27
70	154	225	0.50	1.9	4.2	2.7	1.24	2.8	5.4	4.4	214	0.47	7.6	4.5	5,950	28
80	176	225	0.50	2.0	4.4	2.5	1.30	2.9	5.7	4.7	223	0.49	8.3	5.1	6,800	30
90	198	225	0.50	2.1	4.6	2.3	1.37	3.0	6.0	5.0	232	0.51	8.9	5.7	7,650	32
First 6–8 weeks lactation suckling singles or last 4–6 weeks lactation suckling twins[4]																
50	110	−25 (90)	−0.06 (0.20)	2.1	4.6	4.2	1.36	3.0	6.0	4.9	304	0.67	8.9	6.1	4,250	32
60	132	−25 (90)	−0.06 (0.20)	2.3	5.1	3.8	1.50	3.3	6.6	5.4	319	0.70	9.1	6.6	5,100	34
70	154	−25 (90)	−0.06 (0.20)	2.5	5.5	3.6	1.63	3.6	7.2	5.9	334	0.73	9.3	7.0	5,950	38
80	176	−25 (90)	−0.06 (0.20)	2.6	5.7	3.2	1.69	3.7	7.4	6.1	344	0.76	9.5	7.4	6,800	39
90	198	−25 (90)	−0.06 (0.20)	2.7	5.9	3.0	1.75	3.8	7.6	6.3	353	0.78	9.6	7.8	7,650	40
First 6–8 weeks lactation suckling twins																
50	110	−60	−0.13	2.4	5.3	4.8	1.56	3.4	6.9	5.6	389	0.86	10.5	7.3	5,000	36
60	132	−60	−0.13	2.6	5.7	4.3	1.69	3.7	7.4	6.1	405	0.89	10.7	7.7	6,000	39
70	154	−60	−0.13	2.8	6.2	4.0	1.82	4.0	8.0	6.6	420	0.92	11.0	8.1	7,000	42
80	176	−60	−0.13	3.0	6.6	3.8	1.95	4.3	8.6	7.0	435	0.96	11.2	8.6	8,000	45
90	198	−60	−0.13	3.2	7.0	3.6	2.08	4.6	9.2	7.5	450	0.99	11.4	9.0	9,000	48
Ewe lambs																
Nonlactating—first 15 weeks gestation																
40	88	160	0.35	1.4	3.1	3.5	0.83	1.8	3.6	3.0	156	0.34	5.5	3.0	1,880	21
50	110	135	0.30	1.5	3.3	3.0	0.88	1.9	3.9	3.2	159	0.35	5.2	3.1	2,350	22
60	132	135	0.30	1.6	3.5	2.7	0.94	2.0	4.1	3.4	161	0.35	5.5	3.4	2,820	24
70	154	125	0.28	1.7	3.7	2.4	1.00	2.2	4.4	3.6	164	0.36	5.5	3.7	3,290	26

(Continued)

TABLE 14-2 (Continued)

Body Weight		Weight Change/ Day		Dry Matter per Animal[1]			Nutrients per Animal									
							Energy[2]									
							TDN		DE	ME	Crude Protein		Ca	P	Vitamin A Activity	Vitamin E Activity
(kg)	(lb)	(g)	(lb)	(kg)	(lb)	(% body weight)	(kg)	(lb)	(Mcal)	(Mcal)	(g)	(lb)	(g)	(g)	(IU)	(IU)
Ewe lambs (continued)																
Last 4 weeks gestation (100–120% lambing rate expected)																
40	88	*180*	0.40	*1.5*	3.3	3.8	*0.94*	2.1	4.1	3.4	*187*	0.41	*6.4*	*3.1*	3,400	22
50	110	*160*	0.35	*1.6*	3.5	3.2	*1.00*	2.2	4.4	3.6	*189*	0.42	*6.3*	*3.4*	4,250	24
60	132	*160*	0.35	*1.7*	3.7	2.8	*1.07*	2.4	4.7	3.9	*192*	0.42	*6.6*	*3.8*	5,100	26
70	154	*150*	0.33	*1.8*	4.0	2.6	*1.14*	2.5	5.0	4.1	*194*	0.43	*6.8*	*4.2*	5,950	27
Last 4 weeks gestation (130–175% lambing rate expected)																
40	88	*225*	0.50	*1.5*	3.3	3.8	*0.99*	2.2	4.4	3.6	*202*	0.44	*7.4*	*3.5*	3,400	22
50	110	*225*	0.50	*1.6*	3.5	3.2	*1.06*	2.3	4.7	3.8	*204*	0.45	*7.8*	*3.9*	4,250	24
60	132	*225*	0.50	*1.7*	3.7	2.8	*1.12*	2.5	4.9	4.0	*207*	0.46	*8.1*	*4.3*	5,100	26
70	154	*215*	0.47	*1.8*	4.0	2.6	*1.14*	2.5	5.0	4.1	*210*	0.46	*8.2*	*4.7*	5,950	27
First 6–8 weeks lactation suckling singles (wean by 8 weeks)																
40	88	*−50*	−0.11	*1.7*	3.7	4.2	*1.12*	2.5	4.9	4.0	*257*	0.56	*6.0*	*4.3*	3,400	26
50	110	*−50*	−0.11	*2.1*	4.6	4.2	*1.39*	3.1	6.1	5.0	*282*	0.62	*6.5*	*4.7*	4,250	32
60	132	*−50*	−0.11	*2.3*	5.1	3.8	*1.52*	3.4	6.7	5.5	*295*	0.65	*6.8*	*5.1*	5,100	34
70	154	*−50*	−0.11	*2.5*	5.5	3.6	*1.65*	3.6	7.3	6.0	*301*	0.68	*7.1*	*5.6*	5,450	38
First 6–8 weeks lactation suckling twins (wean by 8 weeks)																
40	88	*−100*	−0.22	*2.1*	4.6	5.2	*1.45*	3.2	6.4	5.2	*306*	0.67	*8.4*	*5.6*	4,000	32
50	110	*−100*	−0.22	*2.3*	5.1	4.6	*1.59*	3.5	7.0	5.7	*321*	0.71	*8.7*	*6.0*	5,000	34
60	132	*−100*	−0.22	*2.5*	5.5	4.2	*1.72*	3.8	7.6	6.2	*336*	0.74	*9.0*	*6.4*	6,000	38
70	154	*−100*	−0.22	*2.7*	6.0	3.9	*1.85*	4.1	8.1	6.6	*351*	0.77	*9.3*	*6.9*	7,000	40
Replacement ewe lamb[5]																
30	66	*227*	0.50	*1.2*	2.6	4.0	*0.78*	1.7	3.4	2.8	*185*	0.41	*6.4*	*2.6*	1,410	18
40	88	*182*	0.40	*1.4*	3.1	3.5	*0.91*	2.0	4.0	3.3	*176*	0.39	*5.9*	*2.6*	1,880	21
50	110	*120*	0.26	*1.5*	3.3	3.0	*0.88*	1.9	3.9	3.2	*136*	0.30	*4.8*	*2.4*	2,350	22
60	132	*100*	0.22	*1.5*	3.3	2.5	*0.88*	1.9	3.9	3.2	*134*	0.30	*4.5*	*2.5*	2,820	22
70	154	*100*	0.22	*1.5*	3.3	2.1	*0.88*	1.9	3.9	3.2	*132*	0.29	*4.6*	*2.8*	3,290	22
Replacement ram lamb[5]																
40	88	*330*	0.73	*1.8*	4.0	4.5	*1.10*	2.5	5.0	4.1	*243*	0.54	*7.8*	*3.7*	1,880	24
60	132	*320*	0.70	*2.4*	5.3	4.0	*1.50*	3.4	6.7	5.5	*263*	0.58	*8.4*	*4.2*	2,820	26
80	176	*290*	0.64	*2.8*	6.2	3.5	*1.80*	3.9	7.8	6.4	*268*	0.59	*8.5*	*4.6*	3,760	28
100	220	*250*	0.55	*3.0*	6.6	3.0	*1.90*	4.2	8.4	6.9	*264*	0.58	*8.2*	*4.8*	4,700	30
Lambs finishing—4 to 7 months old[6]																
30	66	*295*	0.65	*1.3*	2.9	4.3	*0.94*	2.1	4.1	3.4	*191*	0.42	*6.6*	*3.2*	1,410	20
40	88	*275*	0.60	*1.6*	3.5	4.0	*1.22*	2.7	5.4	4.4	*185*	0.41	*6.6*	*3.3*	1,880	24
50	110	*205*	0.45	*1.6*	3.5	3.2	*1.23*	2.7	5.4	4.4	*160*	0.35	*5.6*	*3.0*	2,350	24
Early weaned lambs—moderate growth potential[6]																
10	22	*200*	0.44	*0.5*	1.1	5.0	*0.40*	0.9	1.8	1.4	*127*	0.38	*4.0*	*1.9*	470	10
20	44	*250*	0.55	*1.0*	2.2	5.0	*0.80*	1.8	3.5	2.9	*167*	0.37	*5.4*	*2.5*	940	20
30	66	*300*	0.66	*1.3*	2.9	4.3	*1.00*	2.2	4.4	3.6	*191*	0.42	*6.7*	*3.2*	1,410	20
40	88	*345*	0.76	*1.5*	3.3	3.8	*1.16*	2.6	5.1	4.2	*202*	0.44	*7.7*	*3.9*	1,880	22
50	110	*300*	0.66	*1.5*	3.3	3.0	*1.16*	2.6	5.1	4.2	*181*	0.40	*7.0*	*3.8*	2,350	22

(Continued)

TABLE 14-2 (Continued)

Body Weight		Weight Change/Day		Dry Matter per Animal[1]			Nutrients per Animal: Energy[2]: TDN		DE	ME	Crude Protein		Ca	P	Vitamin A Activity	Vitamin E Activity
(kg)	(lb)	(g)	(lb)	(kg)	(lb)	(% body weight)	(kg)	(lb)	(Mcal)	(Mcal)	(g)	(lb)	(g)	(g)	(IU)	(IU)
Early weaned lambs—rapid growth potential[6]																
10	22	250	0.55	0.6	1.3	6.0	0.48	1.1	2.1	1.7	157	0.35	4.9	2.2	470	12
20	44	300	0.66	1.2	2.6	6.0	0.92	2.0	4.0	3.3	205	0.45	6.5	2.9	940	24
30	66	325	0.72	1.4	3.1	4.7	1.10	2.4	4.8	4.0	216	0.48	7.2	3.4	1,410	21
40	88	400	0.88	1.5	3.3	3.8	1.14	2.5	5.0	4.1	234	0.51	8.6	4.3	1,880	22
50	110	425	0.94	1.7	3.7	3.4	1.29	2.8	5.7	4.7	240	0.53	9.4	4.8	2,350	25
60	132	350	0.77	1.7	3.7	2.8	1.29	2.8	5.7	4.7	240	0.53	8.2	4.5	2,820	25

[1]To convert dry matter to an as-fed basis, divide dry matter values by the percentage of dry matter in the particular feed.

[2]One kilogram TDN (total digestible nutrients) = 4.4 Mcal DE (digestible energy); ME (metabolizable energy) = 82% of DE. Because of rounding errors, values in Tables 14-2 and 14-3 may differ.

[3]Values are applicable for ewes in moderate condition. Fat ewes should be fed according to the next lower weight category and thin ewes at the next higher weight category. Once desired or moderate weight condition is attained, use that weight category through all production stages.

[4]Values in parentheses are for ewes suckling lambs the last 4–6 weeks of lactation.

[5]Lambs intended for breeding; thus, maximum weight gains and finish are of secondary importance.

[6]Maximum weight gains expected.

TABLE 14-3
NUTRIENT CONCENTRATION IN DIETS FOR SHEEP (EXPRESSED ON 100% DRY MATTER BASIS)[1]

Body Weight		Weight Change/Day		Energy[2]: TDN[3]	DE	ME	Example Diet Proportions: Concentrate	Forage	Crude Protein	Calcium	Phosphorus	Vitamin A Activity	Vitamin E Activity
(kg)	(lb)	(g)	(lb)	(%)	(Mcal/kg)	(Mcal/kg)	(%)	(%)	(%)	(%)	(%)	(IU/kg)	(IU/kg)
Ewe[4]													
Maintenance													
70	154	10	0.02	55	2.4	2.0	0	100	9.4	0.20	0.20	2,742	15
Flushing—2 weeks prebreeding and first 3 weeks of breeding													
70	154	100	0.22	59	2.6	2.1	15	85	9.1	0.32	0.18	1,828	15
Nonlactating—First 15 weeks gestation													
70	154	30	0.07	55	2.4	2.0	0	100	9.3	0.25	0.20	2,350	15
Last 4 weeks gestation (130–150% lambing rate expected) or last 4–6 weeks lactation suckling singles[5]													
70	154	180 (0.45)	0.40 (0.10)	59	2.6	2.1	15	85	10.7	0.35	0.23	3,306	15
Last 4 weeks gestation (180–225% lambing rate expected)													
70	154	225	0.50	65	2.9	2.3	35	65	11.3	0.40	0.24	3,132	15
First 6–8 weeks lactation suckling singles or last 4–6 weeks lactation suckling twins[5]													
70	154	−25 (90)	−0.06 (0.20)	65	2.9	2.4	35	65	13.4	0.32	0.26	2,380	15
First 6–8 weeks lactation suckling twins													
70	154	−60	−0.13	65	2.9	2.4	35	65	15.0	0.39	0.29	2,500	15
Ewe lambs													
Nonlactating—first 15 weeks gestation													
55	121	135	0.30	59	2.6	2.1	15	85	10.6	0.35	0.22	1,668	15
Last 4 weeks gestation (100–120% lambing rate expected)													
55	121	160	0.35	63	2.8	2.3	30	70	11.8	0.39	0.22	2,833	15

(Continued)

TABLE 14-3 (Continued)

Body Weight		Weight Change/Day		Energy[2]			Example Diet Proportions		Crude Protein	Cal-cium	Phos-phorus	Vitamin A Activity	Vitamin E Activity
				TDN[3]	DE	ME	Concentrate	Forage					
(kg)	*(lb)*	*(g)*	*(lb)*	*(%)*	*(Mcal/kg)*	*(Mcal/kg)*	*(%)*	*(%)*	*(%)*	*(%)*	*(%)*	*(IU/kg)*	*(IU/kg)*
Ewe lambs (continued)													
Last 4 weeks gestation (130–175% lambing rate expected)													
55	121	*225*	0.50	66	2.9	2.4	40	60	12.8	0.48	0.25	2,833	15
First 6–8 weeks lactation suckling singles (wean by 8 weeks)													
55	121	*−50*	0.22	66	2.9	2.4	40	60	13.1	0.30	0.22	2,125	15
First 6–8 weeks lactation suckling twins (wean by 8 weeks)													
55	121	*−100*	−0.22	69	3.0	2.5	50	50	13.7	0.37	0.26	2,292	15
Replacement ewe lambs[6]													
30	66	*227*	0.50	65	2.9	2.4	35	65	12.8	0.53	0.22	1,175	15
40	88	*182*	0.40	65	2.9	2.4	35	65	10.2	0.42	0.18	1.343	15
50–70	110–154	*115*	0.25	59	2.6	2.1	15	85	9.1	0.31	0.17	1,567	15
Replacement ram lambs[6]													
40	88	*330*	0.73	63	2.8	2.3	30	70	13.5	0.43	0.21	1,175	15
60	132	*320*	0.70	63	2.8	2.3	30	70	11.0	0.35	0.18	1,659	15
80–100	176–220	*270*	0.60	63	2.8	2.3	30	70	9.6	0.30	0.16	1,979	15
Lambs finishing—4 to 7 months old[7]													
30	66	*295*	0.65	72	3.2	2.5	60	40	14.7	0.51	0.24	1,085	15
40	88	*275*	0.60	76	3.3	2.7	75	25	11.6	0.42	0.21	1,175	15
50	110	*205*	0.45	77	3.4	2.8	80	20	10.0	0.35	0.19	1,469	15
Early weaned lambs—moderate and rapid growth potential[7]													
10	22	*250*	0.55	80	3.5	2.9	90	10	26.2	0.82	0.38	940	20
20	44	*300*	0.66	78	3.4	2.8	85	15	16.9	0.54	0.24	940	20
30	66	*325*	0.72	78	3.3	2.7	85	15	15.1	0.51	0.24	1,085	15
40–60	88–132	*400*	0.88	78	3.3	2.7	85	15	14.5	0.55	0.28	1,253	15

[1]Values in Table 14-3 are calculated from daily requirements in Table 14-2 divided by DM intake. The exception, vitamin E daily requirements/head, are calculated from vitamin E/kg diet × DM intake.

[2]One kilogram TDN = 4.4 Mcal DE (digestible energy); ME (metabolizable energy) = 82% of DE. Because of rounding errors, values in Tables 14-2 and 14-3 may differ.

[3]TDN calculated on following basis: hay DM, 55% TDN and on as-fed basis 50% TDN; grain DM, 82% TDN and on as-fed basis 75% TDN.

[4]Values are for ewes in moderate condition. Fat ewes should be fed according to the next lower weight category and thin ewes at the next higher weight category. Once desired or moderate weight condition is attained, use that weight category through all production stages.

[5]Values in parentheses are for ewes suckling lambs the last 4–6 weeks of lactation.

[6]Lambs intended for breeding; thus, maximum weight gains and finish are of secondary importance.

[7]Maximum weight gains expected.

TABLE 14-4
NET ENERGY REQUIREMENTS FOR LAMBS OF SMALL, MEDIUM, AND LARGE MATURE WEIGHT GENOTYPES[1] (kcal/d)

Body Weight (kg)[2] NE_m Requirements[3]	**10 315**	**20 530**	**25 626**	**30 718**	**35 806**	**40 891**	**45 973**	**50 1,053**
Daily gain (g)[2] Ne_g requirements								
Small mature weight lamb[4]								
100	178	300	354	406	456	504	551	596
150	267	450	532	610	684	756	826	894
200	357	600	708	812	912	1,008	1,102	1,192
250	446	750	886	1,016	1,140	1,261	1,377	1,490
300	535	900	1,064	1,219	1,368	1,513	1,652	1,788
Medium mature weight lambs[5]								
100	155	261	309	354	397	439	480	519
150	233	392	463	531	596	658	719	778
200	310	522	618	708	794	878	960	1,038
250	388	653	771	884	993	1,097	1,199	1,297
300	466	784	926	1,062	1,191	1,316	1,438	1,557
350	543	914	1,080	1,238	1,390	1,536	1,678	1,816
400	621	1,044	1,234	1,415	1,589	1,756	1,918	2,076
Large mature weight lambs[6]								
100	132	221	262	300	337	372	407	439
150	197	332	392	450	505	558	610	660
200	263	442	524	600	674	744	813	880
250	329	553	654	750	842	930	1,016	1,099
300	394	663	785	900	1,010	1,116	1,220	1,320
350	461	775	916	1,050	1,179	1,303	1,423	1,540
400	526	885	1,046	1,200	1,347	1,489	1,626	1,760
450	592	996	1,177	1,350	1,515	1,675	1,830	1,980

[1]Approximate mature ram weights of 95 kg, 115 kg, and 135 kg, respectively.
[2]Weights and gains include fill.
[3]$NE_m = 56\ kcal \times W^{0.75} \times d^{-1}$.
[4]$NE_g = 317\ kcal \times W^{0.75} \times LWG, kg \times d^{-1}$.
[5]$NE_g = 276\ kcal \times W^{0.75} \times LWG, kg \times d^{-1}$.
[6]$NE_g = 234\ kcal \times W^{0.75} \times LWG, kg \times d^{-1}$.

TABLE 14-5
CRUDE PROTEIN REQUIREMENTS FOR LAMBS OF SMALL, MEDIUM, AND LARGE MATURE WEIGHT GENOTYPE[1] (g/d)

Body Weight (kg)[2]	10	20	25	30	35	40	45	50
Daily Gain (g)[2]								
Small mature weight lambs								
100	84	112	122	127	131	136	135	134
150	103	121	137	140	144	147	145	143
200	123	145	152	154	156	158	154	151
250	142	162	167	168	168	169	164	159
300	162	178	182	181	180	180	174	168
Medium mature weight lambs								
100	85	114	125	130	135	140	139	139
150	106	132	141	145	149	153	151	149
200	127	150	158	160	163	166	163	160
250	147	167	174	175	177	179	175	171
300	168	185	191	191	191	191	186	181
350	188	203	207	206	205	204	198	192
400	209	221	224	221	219	217	210	202
Large mature weight lambs								
100	94	128	134	139	145	144	150	156
150	115	147	152	156	160	159	164	169
200	136	166	170	173	176	174	178	182
250	157	186	188	190	192	189	192	195
300	179	205	206	207	208	204	206	208
350	200	224	224	224	224	219	220	221
400	221	243	242	241	240	234	234	234
450	242	262	260	256	256	249	248	248

[1]Approximate mature ram weights of 95 kg, 115 kg, and 135 kg, respectively.
[2]Weights and gains include fill.

TABLE 14-6
COMPOSITION OF EWE'S MILK (2.5 WEEKS POSTPARTUM)[1]

Dry matter	18.2%
Fat (5–10%)	7.1 g/100 g milk
Protein (true)	4.5 × 5.49 = 24.7% DM basis
Lactose	4.8 × 5.49 = 26.4% DM basis
Ash	0.85 g/100 g milk
Fiber	0.0 g/100 g milk
Caloric value (GE)	110 kcal/100 g × 5.49 = 6.04 Mcal/kg milk DM basis

Principal salts (g/100 g)

Na	0.040	Mg	0.016	Citrate	0.170
K	0.150	P	0.150		
Ca	0.200	Cl	0.075		

Trace minerals (mg/liter)

Fe	0.60–0.70	Mn	0.06	Zn	2.00–3.00
Cu	0.05–0.15	Al	1.70		

Vitamins (mg/liter, except where noted)

A	1,450 IU/liter	Niacin	5.0	Folacin	0.05
E (α-tocopherol)	15	B-6	0.7	B-12	0.006–0.010
Thiamin	1.0	Pantothenic acid	4.0	Ascorbic acid	40–50
Riboflavin	4.0	Biotin	0.05–0.09		

[1]Courtesy of Dr. Robert Jenness, Biochemistry Department, University of Minnesota.

PROTEIN

Sheep need protein, as do other classes of animals, for maintenance, growth, reproduction, and finishing. Additionally, sheep need protein for the production of wool—a protein product. Wool is especially rich in the sulfur-containing amino acid, cystine, but this requirement is usually amply met by the cystine of feeds or by methionine, another amino acid which is also rather widely distributed in natural sources and which is derived from rumen synthesis.

Green pastures and legume hays (alfalfa, clover, soybeans, lespedeza, etc.) are excellent and practical sources of proteins for sheep in most areas. Where the ranges are bleached and dry for an extended period, or legume hays cannot be produced for winter feeding, however, it may be desirable to provide sheep with such protein-rich supplements as soybean meal, cottonseed meal, linseed meal, canola meal, peanut meal, sunflower meal, or a commercial protein supplement, at the rate of about 0.25 to 0.33 lb per ewe per day.

The protein requirements of sheep are affected by growth, pregnancy, lactation, mature size, weight for age, body condition, rate of gain, and protein-energy ratio. Though correspondingly less because of their smaller body size and lower milk production, the protein requirements of ewes nursing lambs are much like those of lactating cows.

■ **Symptoms of protein deficiency**—A protein deficiency is characterized by reduced appetite, lowered feed intake, and poor feed efficiency. In turn, this makes for poor growth, poor muscular development, loss of weight, reduced reproductive efficiency, and reduced wool production. Under extreme conditions, there are severe digestive disturbances, nutritional anemia, and edema.

MINERALS

The sheep mineral requirements and recommended allowances are given in Tables 14-7 to 14-9.

TABLE 14-7
SHEEP MINERAL REQUIREMENTS/ALLOWANCES

Mineral	Mineral Requirements			Recommended Allowances
	Minerals/Animal/Day	Mineral Content of Ration, in % or ppm		
Major or macrominerals		*As-fed*	*M-F*	
Salt (NaCl)	*Lambs in drylot consume about 9 g of salt daily. Mature sheep in drylot may consume more.	*Salt for growing lambs: 0.38% *Na requirement of sheep: 0.08–0.16%	0.42% 0.09–0.18%	*Salt for mature sheep: 0.5% of the complete feed, or 1.0% to the concentrate portion. *Range operators commonly provide 0.5–0.75 lb *(0.25–0.33 kg)* salt/ewe/month. Mature sheep in drylot may consume more.
Calcium (Ca)	Variable, according to class, age, and weight of sheep.	*0.18–0.74%	*0.20–0.82%	Self-feed suitable mineral, or add calcium to the ration as required to bring level of total ration slightly above requirements.
Phosphorus (P)	Variable, according to class, age, and weight of sheep.	*0.14–0.34%	*0.16–0.38%	Self-feed suitable mineral, or add phosphorus to the ration as required to bring level of total ration slightly above requirements.
Magnesium (Mg)		*0.11, 0.14, and 0.16% for growing lambs, ewes in late pregnancy, and ewes in early lactation, respectively. *Where ewes in early lactation are grazing forage with high nitrogen and potassium content, the minimum level of magnesium in the ration is 0.2%.	*0.12, 0.15, and 0.18% for growing lambs, ewes in late pregnancy, and ewes in early lactation, respectively.	
Potassium (K)		*0.45% for growth of lambs. 0.63-0.72 for lactation and stress.	*0.5% for growth of lambs. 0.7-0.8 for lactation and stress.	0.7 to 1.0% of total air-dry ration.
Sulfur (S)		*Mature ewes: 0.13–0.16% *Young lambs: 0.16–0.23%	0.14–0.18% 0.18–0.26%	*It is recommended that a dietary nitrogen-sulfur ratio of 10:1 be maintained.
Trace or microminerals				
Cobalt (Co)		*0.09–0.18 ppm However, young, rapidly growing lambs may have a slightly higher requirement.	*0.1–0.2 ppm	*Feed cobalt at the rate of 1.4 g/100 lb *(2.5 g/100 kg)* of salt as cobalt chloride or cobalt sulfate.
Copper (Cu)		*6.3–20.7 ppm The Cu requirement varies with (1) the Mo content of the feed, and (2) the growth, pregnancy, lactation, and breed involved. (See the narrative section on "Copper" for details.)	*7-23 ppm	*Add copper sulfate to the salt at rate of 0.5%.

(Continued)

TABLE 14-7 (Continued)

Mineral	Mineral Requirements		Recommended Allowances
	Minerals/Animal/Day	Mineral Content of Ration, in % or ppm	
Trace or microminerals (continued)		*As-fed* *M-F*	
Fluorine (F)		*Breeding sheep should not be fed a ration containing more than 55 ppm (as-fed) or 60 ppm of fluorine on a moisture-free basis. *Finishing lambs can tolerate up to 135 ppm (as-fed) or 150 ppm of fluorine in the ration on a moisture-free basis.	
Iodine (I)		*0.09–0.72 ppm *0.1–0.8 ppm The higher levels are indicated for pregnancy and lactation. When goitrogens are present increase the iodine.	*Free access to stabilized iodized salt containing 0.0078% iodine.
Iron (Fe)		*27–45 ppm *30–50 ppm	*Intramuscular injections of iron-dextran; 2 injections, 150 mg of iron in each, given 2 to 3 weeks apart.
Manganese (Mn)		*18–36 ppm *20–40 ppm	18 ppm in as-fed ration, or 20 ppm in M-F ration.
Molybdenum (Mo)		*The minimum dietary requirement of molybdenum is not known. *The Food and Drug Administration does not recognize molybdenum as safe; hence, the law prohibits adding it to feed for sheep.	The two contrasting situations—(1) high molybdenum and copper deficiency, or (2) low molybdenum and excess copper accumulation—make it very difficult to define nutrient requirements of molybdenum and copper.
Selenium (Se)		*0.09–0.18 ppm 0.1–0.2 ppm	Selenium as either sodium selenite or sodium selenate at the rate of 0.3 ppm of complete feed. Selenium added to salt-mineral mixture fed free-choice at rate of 90 ppm. Selenium in the limit feeding (feed supplements and salt-mineral mixtures) consumption rate for sheep of 0.7 mg per head per day.
Zinc (Zn)		*Growth: 18 ppm 20 ppm *Reproduction: 30 ppm 33 ppm	

*Where preceded by an asterisk, the requirements, recommended allowances, and other facts presented herein were taken from *Nutrient Requirements of Sheep*, Sixth Revised Edition, NRC-National Academy of Sciences, Washington, DC.

NOTE: Mineral recommendations for all classes and ages of sheep, especially those fed unmixed rations or on pasture, are—

1. *When sheep are on liberal grain feeding*—Provide free access to a two-compartment mineral box, with (a) trace mineralized salt in one side, and (b) in the other side, a mixture of one-third trace mineralized salt (salt included for purposes of palatability), one-third defluorinated phosphate or steamed bone meal, and one-third ground limestone or oystershell flour.
2. *When sheep are primarily on roughage (pasture and/or hay)*—Provide free access to a two-compartment mineral box, with (a) trace mineralized salt in one side, and (b) in the other side, a mixture of one-third trace mineralized salt and two-thirds defluorinated phosphate or steamed bone meal.

Additionally, in those areas where cobalt and/or copper deficiencies exist in the soil (and plants), add cobalt and/or copper sulfate to either the salt or salt-phosphorus mixture in the proportions indicated. If desired, the mineral supplement may be incorporated in the ration in keeping with the recommended allowances given in this table.

TABLE 14-8
MACROMINERAL REQUIREMENTS OF SHEEP (PERCENTAGE OF RATION)[1]

Nutrient	Requirement	
	As-fed[2]	Moisture-free
	(%)	(%)
Sodium	0.08–0.16	0.09–0.18
Chlorine	—	—
Calcium	0.18–0.74	0.20–0.82
Phosphorus	0.14–0.34	0.16–0.38
Magnesium	0.11–0.16	0.12–0.18
Potassium	0.45–0.72	0.50–0.80
Sulfur	0.13–0.23	0.14–0.26

[1]Adapted by the author from *Nutrient Requirements of Sheep*, Sixth Revised Edition, NRC-National Academy of Sciences.

[2]As-fed was calculated using 90% dry matter (moisture-free).

TABLE 14-9
MICROMINERAL REQUIREMENTS OF SHEEP AND MAXIMUM TOLERABLE LEVELS (PPM OR MG/KG OF RATION)[1]

Nutrient	Requirement		Maximum Tolerable Level	
	As-fed[2]	Moisture-free	As-fed	Moisture-free
	(ppm or mg/kg)	(ppm or mg/kg)	(ppm or mg/kg)	(ppm or mg/kg)
Cobalt	0.09–0.18	0.1–0.2	9	10
Copper	6–10	7–11[3]	23	25[4]
Fluorine	—	—	54–135	60–150
Iodine	0.09–0.72	0.10–0.80[5]	45	50
Iron	27–45	30–50	450	500
Manganese	18–36	20–40	900	1,000
Molybdenum	0.45	0.5	9	10[4]
Selenium	0.09–0.18	0.1–0.2	1.8	2
Zinc	18–30	20–33	675	750

[1]Adapted by the author from *Nutrient Requirements of Sheep*, Sixth Revised Edition, NRC-National Academy of Sciences.

[2]As-fed was calculated using 90% dry matter (moisture-free).

[3]Requirement when dietary Mo concentrations are <1 mg/kg DM.

[4]Lower levels may be toxic under some circumstances.

[5]High level for pregnancy and lactation in rations not containing goitrogens; should be increased if rations contain goitrogens.

VITAMINS

Many phenomena of vitamin nutrition are related to solubility—vitamins are soluble in either fat or water. Consequently, it is important that both nutritionists and sheep producers be well informed about solubility differences in vitamins and make use of such differences in programs and practices. Thus, in the discussion that follows, vitamins are grouped as either (1) fat-soluble vitamins, or (2) water-soluble vitamins.

Fig. 14-2. Lamb fed a ration deficient in phosphorus. Note the knock-kneed conformation. (Courtesy, University of Idaho)

The fat-soluble vitamins are vitamin A (carotene), vitamin D, vitamin E, and vitamin K.

All of the water-soluble vitamins except vitamin C are known as B vitamins. These are *not* stored.

Table 14-10 presents sheep vitamin requirements/allowances. Table 14-11 presents the NRC-National Academy of Sciences vitamin E requirements of growing-finishing lambs.

■ **NRC-National Academy of Sciences vitamin E requirements of lambs**—Table 14-11 presents daily vitamin E requirements for lambs and the suggested amounts of alpha-tocopherol acetate to add to the rations to provide 100% of the requirements.

Fig. 14-3. Lamb with stiff-lamb disease, caused by a deficiency of vitamin E. (Courtesy, Cornell University, Ithaca, NY)

TABLE 14-10
SHEEP VITAMIN REQUIREMENTS/ALLOWANCES

Vitamin	Vitamin Requirements			Recommended Allowances
	Vitamins/Animal/Day	Vitamin Content of Ration		
Vitamin A	Variable, according to class, age, and weight of sheep, as indicated by the following:	Variable, according to class, age, and weight of sheep.		Sheep that are deficient in vitamin A and weigh 70 lb *(32 kg)* or more should receive 100,000 IU of vitamin A by injection, and their rations should be adjusted to provide recommended levels of vitamin A or carotene. Ewes deficient in vitamin A should be given vitamin A either orally or by injection prior to breeding.
		Mcg of B-carotene/day		
	Categories	**/lb body wt.**	**/kg body wt.**	
	All categories	31	*69*	
	Late gestation and in lactation	57	*125*	
	First 6–8 weeks of lactation of ewes suckling twins	67	*147*	
		IU of Vitamin A/day		
	Categories	**/lb body wt.**	**/kg body wt.**	
	All categories	21	*47*	
	Late gestation and in lactation	39	*85*	
	First 6–8 weeks of lactation of ewes suckling twins	45	*100*	
Vitamin D	*For all sheep except early-weaned lambs: 555 IU per 220 lb *(100 kg)* body weight. *For early-weaned lambs: 666 IU per 220 lb *(100 kg)* body weight.	Variable, according to class, age, and weight of sheep.		Breeding sheep, 500 to 800 IU/head/day. Feeder lambs, 500 IU/head/day.
Vitamin E				Meet the requirements given in the column headed "Vitamin Content of Ration."
		Mcg of B-carotene/day		
	Categories	**/lb**	**/kg**	
	Lambs under 44 lb *(20 kg)* liveweight	8	*18*	
	Lambs over 44 lb *(20 kg)* liveweight and pregnant ewes	6	*13.5*	
		***IU Vitamin E M-F basis**		
	Categories	**/lb**	**/kg**	
	Lambs under 44 lb *(20 kg)* liveweight	9	*20*	
	Lambs over 44 lb *(20 kg)* liveweight and pregnant ewes	7	*15*	
Vitamin K				Vitamin K_2 is normally synthesized in large amounts in the rumen; no need for dietary supplementation has been established.
Water-Soluble Vitamins				*The B vitamins are not required in the diet of sheep with functioning rumens, because the microorganisms synthesize these vitamins in adequate amounts.

*Where preceded by an asterisk, the requirements, recommended allowances, and other facts presented herein were taken from *Nutrient Requirements of Sheep*, Sixth Revised Edition, NRC-National Academy of Sciences, Washington, DC.

TABLE 14-11
VITAMIN E REQUIREMENTS OF GROWING-FINISHING LAMBS AND SUGGESTED LEVELS OF FEED FORTIFICATION TO PROVIDE 100% OF REQUIREMENTS[1]

Body Weight		Alpha-Tocopherol Acetate			Feed Intake per Lamb		Amount of Vitamin E Added to Concentrate			Amount of Vitamin E Added to Protein Supplement[2]		
(lb)	*(kg)*	*(mg/lamb/day)*[3]	*(mg/lb ration)*	*(mg/kg ration)*	*(lb)*	*(kg)*	*(mg/lb)*	*(mg/kg)*	*(mg/ton)*	*(mg/lb)*	*(mg/kg)*	*(mg/ton)*
22	*10*	5.0	44	*20*	0.50	*0.23*	9.1	*20*	18,200	133	*60*	120,000
44	*20*	10.0	44	*20*	1.00	*0.45*	9.1	*20*	18,200	60	*133*	120,000
66	*30*	15.0	33	*15*	2.10	*0.96*	6.8	*15*	13,600	45	*100*	90,000
88	*40*	20.0	33	*15*	2.86	*1.30*	6.8	*15*	13,600	45	*100*	90,000
110	*50*	25.0	33	*15*	3.50	*1.60*	6.8	*15*	13,600	45	*100*	90,000

[1]Adapted by the author from *Nutrient Requirements of Sheep*, Sixth Revised Edition, NRC-National Academy of Sciences.

[2]Assumes the concentrate diet contains 15% protein supplement.

[3]Rounded values based on approximate diet intake containing recommended vitamin E levels.

WATER

Sheep get water by drinking, and from snow, dew, and feed. The amount of water that sheep voluntarily consume is affected by temperature, rainfall, snow and dew covering, age, breed, stage of production, number of lambs carried, wool covering, respiratory rate, frequency of watering, kind and amount of feed, and exercise. On the average, mature animals consume approximately a gallon of water per day, whereas feeder lambs require about half this amount. However, sheep may go for weeks without drinking water when foraging on grasses and other feeds of high moisture content. This condition often prevails on desert ranges in the early spring and on many of the mountain ranges during the summer months.

FEED SUBSTITUTION TABLE

Successful sheep producers are keen students of values. They recognize that feeds of similar nutritive properties can and should be interchanged in the ration as price relationships warrant, thus making it possible at all times to obtain a balanced ration at the lowest cost .

Table 14-12, Feed Substitution Table for Sheep and Goats, is a summary of the comparative values of the most common U.S. feeds. In arriving at these values, the author considered two primary factors besides chemical composition and feeding value; namely, palatability and carcass quality.

In using this feed substitution table, sheep producers should recognize the following facts:

1. That individual feeds differ widely in feeding value. Barley and oats, for example, vary widely according to the hull content and the test weight per bushel, and forages vary widely according to the stage of maturity at which they are cut and how well they are cured and stored.

2. That nonlegume forages may have a higher relative value to legumes than herein indicated provided the chief need of the animals is for additional energy rather than for supplemental protein. Thus, the nonlegume forages of low value can be used to better advantage for wintering mature ewes than for wintering young lambs.

On the other hand, legumes may actually have a higher value relative to nonlegumes than herein indicated, provided the chief need is for additional protein rather than for added energy. Thus, no protein supplement is necessary for breeding ewes, provided a good-quality legume forage is fed.

3. That, based primarily on available supply and price, certain feeds—especially those of medium-protein content, such as brewers' dried grains, corn gluten feed (gluten feed), distillers' dried grains, distillers' dried solubles, and peas (dried)—are used interchangeably as (a) grains and byproducts feeds, and/or (b) protein supplements.

4. That the feeding value of certain feeds is materially affected by preparation. The values herein reported are based on proper feed preparation in each case.

Fig. 14-4. Sheep watering at the snow melter. (Courtesy, USDA, Soil Conservation Service)

For the reasons noted above, the comparative values of feeds shown in the feed substitution table (Table 14-12) are not absolute. Rather, they are reasonably accurate approximations based on average-quality feeds.

TABLE 14-12
FEED SUBSTITUTION TABLE FOR SHEEP (As-fed Basis)

Feedstuff	Relative Feeding Value (lb for lb) in Comparison with the Designated (underlined) Base Feed Which = 100	Maximum Percentage of Base Feed (or comparable feed or feeds) Which It Can Replace for Best Results	Remarks
GRAINS, BYPRODUCT FEEDS, ROOTS AND TUBERS[1] (Low- and Medium-Protein Feeds)			
Corn, No. 2	_100_	100	Grinding not necessary unless (1) for old ewes with poor teeth, (2) for lambs under 5–6 weeks, (3) for incorporation in a mixed ration.
Apple pomace, dehydrated	82–86	33.5	
Barley	90	100	It does not pay to grind barley for sheep.
Beet pulp, dried	95	33.5–50	Value of about 80% when used as the only concentrate for finishing lambs.
Beet pulp, molasses, dried	95	33.33–50	Value of about 80% when used as the only concentrate for finishing lambs.
Beet pulp, wet	25	33.33–50	
Brewers' dried grains	80–95	33.33	Not very palatable. Fed chiefly to dairy cattle.
Citrus pulp, dried	95	25–50	
Corn gluten feed (gluten feed)	85–90	50	
Distillers' dried grains	95–100	33.33–50	
Distillers' dried solubles	95–100	33.33–50	
Fat	225	5	
Hominy feed	100	100	
Molasses, beet	75	10	Actual value may be higher as an appetizer.
Molasses, cane	75	10	Actual value may be higher as an appetizer.
Molasses, citrus	75	10	
Oats	80	10–100	Lower value when used as the only grain for finishing lambs. Highest value for young lambs, for breeding animals and for starting lambs on feed. Need not be ground for sheep. Should not constitute more than one-third of finishing rations. Feeding value varies according to the test weight per bushel.
Peas, dried	100	40	
Potatoes (Irish)	25–35	85	Contrary to popular belief, potatoes can be fed successfully through the pregnancy and lactation periods.
Rice (rough rice)	55–75	100	
Rice bran	66.66–75	33.33	
Rice polishing	85–90	25	
Roots (chiefly mangels [stock beets], rutabagas [swedes], turnips, and carrots)	25–35	50	Some sheep producers believe that the feeding of high levels of roots over a long period will produce urinary calculi. Therefore, caution should be exercised in feeding them to rams and wethers (females not affected). Keep the Ca level higher than the P level. Many shepherds add roots to the ration of show sheep, for conditioning purposes.
Rye	83–87	50–100	Apparently rye is more palatable to sheep than to other classes of animals. Rye may be fed whole to sheep.

(Continued)

TABLE 14-12 (Continued)

Feedstuff	Relative Feeding Value (lb for lb) in Comparison with the Designated (underlined) Base Feed Which = 100	Maximum Percentage of Base Feed (or comparable feed or feeds) Which It Can Replace for Best Results	Remarks
GRAINS, BYPRODUCT FEEDS, ROOTS AND TUBERS[1] (Low- and Medium-Protein Feeds) (continued)			
Sorghum, milo	85–100	100	All varieties have about the same feeding value. There is no advantage in grainding sorghum for sheep.
Wheat	100–110	50	May be fed as the only grain, but it is improved by mixing with another grain. Wheat may be fed whole. Wheat-fed sheep appear to be expecially susceptible to founder.
Wheat bran	90	10–33.33	Because of its bulk and fiber, wheat bran should not constitute more than 10–15% of a finishing ration. Bran is valuable for young animals, for breeding animals, and for starting animals on feed.
Wheat mixed feed (mill-run)	90–95	10–33.33	Can be used in about the same way and in the same quantities as wheat bran for sheep.
PROTEIN SUPPLEMENTS			
Soybeam meal (41%)	_100_	_100_	
Alfalfa or clover screenings	70–75	50	Grind finely to destroy weed seeds.
Brewers' dried grains	75	100	
Copra meal (coconut meal), 21%	90–100	50	
Corn gluten feed (gluten feed)	60–70	50–100	
Corn gluten meal (gluten meal)	100	50	
Cottonseed meal (41%)	100	100	Unlike the situation with finishing cattle, cottonseed meal is about equal to linseed meal for finishing lambs.
Distillers' dried grains	90	100	Rye distillers' dried grains are about 10% lower in protein than similar products made from corn or wheat.
Distillers' dried solubles	90	100	
Linseed meal, 35%	90	100	
Peanut meal, 45%	100	100	
Peas, dried	65–75	50	
Safflower meal, with hulls, 42%	40–45	100	
Soybeans	95–100	100	It does not pay to grind soybeans for sheep.
DRY FORAGES AND SILAGES[2]			
Alfalfa hay, all analyses	_100_	_100_	
Alfalfa silage	33.33–50	50–85	When alfalfa silage replaces corn silage, more energy feed must be provided but less protein, unless grain is used as a preservative.
Barley hay	70	50	The beards may be harmful, especially to woolly faced sheep.
Beet tops, dry	70	50	
Beet tops, fresh	16–25	33.33–50	In the West, large acreages of fresh beet tops are pastured off by sheep and cattle.

(Continued)

TABLE 14-12 (Continued)

Feedstuff	Relative Feeding Value (lb for lb) in Comparison with the Designated (underlined) Base Feed Which = 100	Maximum Percentage of Base Feed (or comparable feed or feeds) Which It Can Replace for Best Results	Remarks
DRY FORAGES AND SILAGES (continued)			
Beet top silage, sugar	17–25	33.33–50	Either provide some dry forage or feed 2 oz *(57 g)* of finely ground limestone to each 100 lb *(45 kg)* of silage.
Bromegrass hay	75	100	
Clover hay, crimson	90–100	100	Crimson clover hay has a considerably lower value if not cut at an early stage.
Clover hay, red	90–100	100	If the rest of the ration is adequate in protein, clover hay will be equal to alfalfa in feeding value; otherwise, it will be lower.
Clover-timothy hay	80–90	100	
Corn fodder	75	100	Should be chopped.
Corn silage	33.33–50	50–85	Although a ration in which corn silage is the only forage is sometimes fed to sheep, most feeders prefer to limit the silage and use some hay.
Corn stover	35	50	Unsatisfactory for finishing lambs, but cut or shredded stover may be used as a part of the roughage for breeding ewes if fed along with a good legume.
Cowpea hay	95–100	100	
Grass-legume mixed hay	80–90	100	Value depends on the proportion of legumes present and the stage of maturity at which they are cut.
Grass-legume silage	32–45	50–85	Although a ration in which grass silage is the only forage is sometimes fed to sheep, most feeders prefer to limit the silage and use some hay.
Grass silage	30–45	50–85	
Johnsongrass hay	70	100	
Lespedeza hay	80–100	100	Feeding value varies considerably with stage of maturity at which it is cut.
Mint hay	80–95	75	
Oat hay	75	50	
Pea-vine hay	100–110	75	
Pea-vine silage	33.33–50	50–85	Unless grain is added as a preservative, pea-vine silage requires more energy feed, but less protein supplement than corn silage when fed to finishing lambs.
Prairie hay	65–70	100	
Reed canarygrass	70	100	
Sorghum fodder	70	100	
Sorghum silage (grain varieties)	32–47	50–85	Although a ration in which sorghum silage is the onlyu forage is sometimes fed to sheep, most feeders prefer to limit the silage and use some hay.
Sorghum silage (sweet varieties)	25–30	50–85	Nearly equal to grain varieties in value per acre because of greater yield.
Sorghum stover	35	50	Unsatisfactory for finishing lambs, but cut or shredded stover may be used as part of the roughage for breeding ewes if fed along with a good legume.
Soybean hay	85–100	100	The lower value is for finishing lambs. For other classes of sheep, it is equal to alfalfa hay.

(Continued)

TABLE 14-12 (Continued)

Feedstuff	Relative Feeding Value (lb for lb) in Comparison with the Designated (underlined) Base Feed Which = 100	Maximum Percentage of Base Feed (or comparable feed or feeds) Which It Can Replace for Best Results	Remarks
DRY FORAGES AND SILAGES (continued)			
Sudangrass hay	50–60	50	
Sweet clover hay	100	100	Value of sweet clover hay varies widely. Second-year sweet clover hay is less desirable than first-year clover hay and is more apt to cause sweet clover disease.
Timothy hay	70	50	
Vetch-oat hay	80–90	100	The higher the proportion of vetch, the higher the value.
Wheat hay	70	50	

[1]Roots and tubers are of lower value than the grain and byproduct feeds due to their higher moisture content.
[2]Silages are of lower value than dry forages due to their higher moisture content.

HOW TO BALANCE RATIONS

A balanced ration is one which provides an animal the proper proportions and amounts of all the required nutrients for a period of 24 hours.

Several suggested rations for different classes of sheep are given in subsequent sections of this chapter. Generally, these rations will suffice, but rations should vary with conditions, and many times they should be formulated to meet the conditions of a specific farm, ranch, or feedlot, or to meet the practices common to an area. Thus, where sheep are on pasture, or are receiving forage in the drylot, the added feed (generally grains, byproduct feeds, and/or protein supplements), if any, should be formulated so as to meet the nutritive requirements not already provided by the forage.

Good sheep producers know how to balance rations. They should be able to select and buy feeds with informed appraisal, to check on how well their manufacturers, dealers, or consultants are meeting their needs, and to evaluate the results.

Ration formulation consists of combining feeds that will be eaten in the amount needed to supply the daily nutrient requirements of the animal. This may be accomplished by the methods presented later in this chapter, but first the following pointers are necessary:

1. Computing rations involves more than simple arithmetic, for no set of figures can substitute for experience. Formulating rations is both an art and a science—the art comes from sheep know-how, experience, and keen observation; the science is largely founded on chemistry, physiology, and bacteriology. Both are essential for success.

2. Before attempting to balance a ration, one should consider the following major points:

a. **Availability and cost of the different feed ingredients.** Preferably, cost of ingredients should be based on delivery after processing—because delivery and processing costs are quite variable.

b. **Moisture content.** When considering costs and balancing rations, the producer should place feed on a comparable moisture basis; usually either "as-fed" or "moisture-free." This is especially important in the case of high-moisture grain or silage.

c. **Composition of feeds under consideration.** Feed composition tables ("book values"), or average analysis, should be considered only as guides, because of wide variations in the composition of feeds. For example, the protein and moisture contents of sorghum, hay, and silages are quite variable. Whenever possible, especially with large operations, it is best to take a representative sample of each major feed ingredient and have a chemical analysis made of it for the more common constituents—protein, fat, fiber, nitrogen-free extract, and moisture; and often calcium, phosphorus, and carotene. Such ingredients as oil meals and prepared supplements, which must meet specific standards, need not be analyzed so often, except as quality-control measures.

Despite the recognized value of a chemical analysis, it is not the total answer. It does not provide information on the availability of nutrients to the animal; it does not tell anything about the associated effect of feedstuffs—for example, the apparent way in which beet pulp enhances the value of ground milo; and it does not tell anything about taste, palatability, texture, or undesirable physiological effects, such as bloat and laxativeness. Nevertheless, a chemical analysis does give a sound basis on which to start the evaluation of feeds. Also, with chemical analysis at hand, and bearing in mind that it is the composition of the total feed (the finished ration) that counts, the person formulating the ration can more intelligently determine the quantity of protein to buy, and the kind and amounts of minerals and vitamins to add.

d. **Quality of feed.** Numerous factors determine the quality of feed, including—

- **Stage of harvesting**—For example, early-cut forages tend to be of higher quality than those that are mature.
- **Freedom from contamination**—Contamination from foreign substances such as dirt, sticks, and rocks can reduce feed quality, as can aflatoxins, pesticide residues, and a variety of chemicals.
- **Uniformity**—Does the feed come from one particular area or does it represent a conglomerate of several sources?
- **Length of storage**—When feed is stored for extended periods, some of its quality is lost due to its exposure to the elements. This is particularly true with forages.

e. **Degree of processing of feed.** Often, the value of feed can be either increased or decreased by processing. For example, heating some types of grains makes them more readily digestible to livestock and increases their feeding value.

f. **Soil analysis.** If the origin of a given feed ingredient is known, a soil analysis or knowledge of the soils of the area can be very helpful; for example, (1) the phosphorus content of soils affects plant composition, (2) soils high in molybdenum and selenium affect the composition of the feeds produced, (3) iodine- and cobalt-deficient areas are important in animal nutrition, and (4) other similar soil-plant-animal relationships exist.

g. **Nutrient requirements and allowances.** These should be known for the particular class of sheep for which a ration is to be formulated. Also, it must be recognized that nutrient requirements and allowances must be changed from time to time, as a result of new experimental findings.

3. In addition to providing a proper quantity of feed and to meeting the nutritive requirements, a well-balanced and satisfactory ration should be:

a. One that is palatable and digestible.

b. One that is economical. Generally, this calls for the maximum use of feeds available in the area, especially forages.

c. One that is so formulated, where ruminants are involved, as to nourish the billions of bacteria in the rumen in order that there will be satisfactory (1) digestion of forages, (2) utilization of lower quality and cheaper proteins and other nitrogenous products (thus, it is possible to use urea to constitute up to one-third of the total protein of the ration of ruminants, provided care is taken to supply enough carbohydrates and other nutrients to assure adequate nutrition for rumen bacteria), and (3) synthesis of B complex vitamins.

This means that rumen microorganisms must be supplied adequate (1) energy, including small amounts of readily available energy, such as sugars or starches; (2) nitrogen-bearing ingredients, such as proteins, urea, and ammonium salts; (3) major minerals, especially sodium, potassium, and phosphorus; (4) cobalt and possibly other trace minerals; and (5) unidentified factors found in certain natural feeds rich in protein or nonprotein nitrogenous constituents.

d. One that will enhance, rather than impair, the quality of the meat and wool produced.

4. In addition to considering changes in availability of feeds and feed prices, the feeder should alter ration formulation at stages to correspond to changes in the weight and productivity of animals.

STEPS IN RATION FORMULATION

The ideal ration is one that will maximize production at the lowest cost. A costly ration may produce phenomenal gains in livestock, but the cost per unit of production may make the ration economically infeasible. Likewise, the cheapest ration is not always the best since it may not allow for maximum production.

Therefore, the cost per unit of production is the ultimate determinant of what constitutes the best ration. Awareness of this separates successful producers from marginal or unsuccessful ones.

To formulate an economical ration, the feeder should follow these four steps in the order given:

1. **Find and list the nutrient requirements and/or allowances for the specific animal to be fed.** It should be remembered that nutrient requirements generally represent the minimum quantity of the nutrients that should be incorporated, while allowances take into consideration a margin of safety. Factors to be considered are:

a. Age.

b. Sex.

c. Body size.

d. Type of production. Are the animals being

Fig. 14-5. The final determinant of a successfully formulated ration is the cost per 100 lb of lamb produced. (Photo by J. C. Allen and Son, West Lafayette, IN)

fed for maintenance, growth, fattening, reproduction, lactation, or wool production?

e. Intensity of production. Are the growing animals gaining 0, 1, 2, or 3 lb per day? Are the lactating animals at the peak of milk production?

2. **Determine what feeds are available and list their respective nutrient compositions.** In rations for ruminants, dry matter, protein, energy, phosphorus, calcium, and vitamin A are the factors that are generally considered in ration formulation. Additional minerals are generally supplied either as a free-choice salt mix or as a premix incorporated in the ration. Animals in confinement may need some vitamin D supplementation.

3. **Determine the cost of the feed ingredients under consideration.** Not only the cost of the feed, but also the cost of mixing, transportation, and storage should be considered. Some feeds require antioxidants and/or refrigeration to prevent spoilage. Others lose nutritive value when stored for extended periods.

4. **Consider the limitations of the various feed ingredients and formulate the most economical ration.** Remember that the ultimate goal is to formulate a ration that minimizes the cost per unit of production.

ADJUSTING MOISTURE CONTENT

The majority of feed composition tables are listed on an as-fed basis, while most of the National Research Council nutrient requirement tables are on either an approximate 90% dry matter or a moisture-free basis. Since feeds contain varying amounts of dry matter, it would be much simpler, and more accurate, if both feed composition and nutrient requirement tables were on a dry basis. In order to facilitate ration formulation, the author has listed both the as-fed and the moisture-free contents of feeds in Chapter 27, Feed Composition Tables.

The significance of water content of feeds becomes obvious in the examples given in Table 14-13. When total digestible nutrients (TDN) as a measure of energy are used, the two high-moisture feeds, carrots and milk, have a higher energy value than oats on a moisture-free basis. The same principle applies to other nutrients, also.

TABLE 14-13
COMPARATIVE ENERGY VALUE OF THREE FEEDS ON AS-FED AND MOISTURE FREE BASES

Feed	Water	Dry Matter	Energy Value (TDN)	
			As-fed	Moisture-free
Oats-grain	11	89	66	74
Carrots, roots	87	13	10	77
Milk	87	13	16	123

■ **To convert as-fed rations to a moisture-free basis**—This may be done by using the following formulas:

Formula 1

When the diet is listed on an as-fed basis, and the producer wishes to compare the content of the various ingredients with the requirements on a moisture-free basis. the equation is—

$$\text{\% nutrient in dry diet (total)} = \frac{\text{\% nutrient in wet diet (total)}}{\text{\% dry matter in diet (total)}} \times 100$$

For example, a diet containing 34% dry matter and 7% protein on a wet basis becomes a 20.6% protein diet on a moisture-free basis.

Formula 2

If the dry matter content of the ingredient, the percentage of the ingredient in the wet diet, and the percentage of dry matter wanted in the diet are known, it is possible to calculate the amount of that ingredient in the diet an a moisture-free basis.

$$\text{Amount of ingredient in dry diet} = \frac{\text{\% of ingredient in wet diet}}{\text{\% dry matter wanted in diet}} \times \begin{matrix}\text{\% dry matter}\\ \text{of ingredient}\end{matrix}$$

Therefore, if a 34% dry matter diet is desired, and if an ingredient containing 25% dry matter is incorporated at a level of 30% of the wet diet, the ingredient constitutes 22% of the moisture-free weight in the diet.

Formula 3

If the producer wants to change the amounts of the ingredients from an as-fed basis to a moisture-free basis. the equation below should be used.

Parts on a wet basis = % ingredient in wet diet × % dry matter of the ingredient

This calculation should be done for each ingredient and the products added. Each product should then be divided by the sum of the products.

■ **To convert a moisture-free basis to an as-fed basis**—To convert the components of a dry diet to that of a wet diet having a given percentage of dry matter, one can use the following equation:

$$\text{Parts of ingredient in wet diet} = \frac{\text{\% ingredient in dry diet} \times \text{\% dry matter desired in diet}}{\text{\% dry matter in ingredient}}$$

The total number of parts should be summed and water added to make 100 parts

METHODS OF FORMULATING RATIONS[1]

In the sections that follow, five different methods of ration formulation are presented: (1) the square method, (2) the simultaneous equations method, (3) the 2 × 2 matrix method, (4) the trial-and-error method, and (5) the computer method. Despite the sometimes confusing mechanics of each system, if done properly, the end result of all five methods is the same—a ration that provides the desired allowance of nutrients in correct proportions economically (or at least cost), but, more important, so as to achieve the greatest net returns—for it is net profit, rather than cost, that counts. Since feed represents by far the greatest cost item in livestock production, the importance of balanced rations is evident.

An exercise in ration formulation follows for purposes of illustrating the application of each of these five methods.

SQUARE (OR PEARSON SQUARE) METHOD

The square method is simple, direct, and easy. Also, it permits quick substitution of feed ingredients in keeping with market fluctuations, without disturbing the protein content.

In a ration balanced by the square method, one specific nutrient alone receives major consideration. Correctly speaking, therefore, the square method is a method of balancing one nutrient requirement, with no consideration given to the vitamin, mineral, and other nutritive requirements.

To compute rations by the square method, or by any other method, it is first necessary to have available both feeding standards (see the Nutrient Requirement Tables 14-2 and 14-3), and feed composition tables (Chapter 27).

The following example will show how to use the square method in formulating a sheep ration:

Example. *A sheep producer wishes to feed a range supplement that is 24% protein. Corn (No. 2), containing 9.5% protein, and soybean meal, containing 41% protein, can be purchased. How many pounds of each must be mixed for each 100 lb of range supplement with 24% protein?*

Step by step, the procedure is as follows:

1. Draw a square, and place the number 24 (desired protein level) in the center thereof.

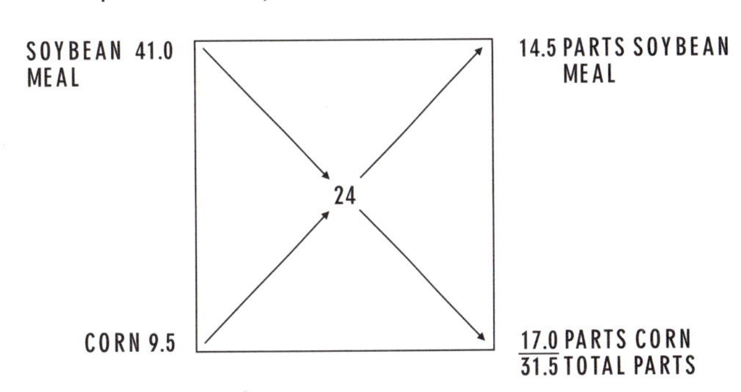

Fig. 14-6. The square method for balancing the protein content of a range supplement.

2. At the upper left-hand corner of the square, write *soybean meal* and its protein content (41); at the lower left-hand corner, write corn and its protein content (9.5).

3. Subtract diagonally across the square (the smaller number from the large number), and record the difference at the corners on the right-hand side (41 – 24 = 17; 24 – 9.5 = 14.5). The number at the upper right-hand corner gives the parts of soybean meal by weight, and the number at the lower right-hand corner gives the parts of corn by weight to make a range supplement with 24% protein.

4. To determine what percent of the ration would be corn, divide the parts of corn by the total parts: 17 ÷ 31.5 = 54% corn and the remainder, 46% would be soybean meal. In pounds, this would mean 54 lb of

---

[1]To avoid confusion, metric values are not included in the sections describing ration formulating. If metric values are desired, use conversion factors in the Appendix.

corn and 46 lb of soybean meal for each 100 lb of range supplement.

SIMULTANEOUS EQUATIONS METHOD

In addition to the square method, it is possible to formulate rations involving two sources and one nutrient quickly through the solving of simultaneous equations.

It is also possible to use simultaneous equations in the formulation of rations involving two sources and two nutrients. Since many formulations involve solving for more than one nutritional parameter, this method has many advantages.

Example. *A sheep producer is feeding 100-lb lambs. A concentrate containing 89% dry matter and 80% TDN (as-fed) and alfalfa hay containing 93% dry matter and 54% TDN (as-fed) are available. How much of each should be fed?*

From Table 14-2, Daily Nutrient Requirements of Sheep, we find that a 100-lb lamb requires 3.7 lb of dry matter and 2.6 lb of TDN daily.

Step by step, the procedure in balancing this ration is as follows:

1. Let X = amount of alfalfa hay to be fed and Y = amount of concentrate to be fed. Since the lamb requires 2.6 lb of TDN and 3.7 lb of dry matter, our equations will be:

$$0.54X + 0.80Y = 2.6 \quad \text{TDN equation}$$
$$0.93X + 0.89Y = 3.7 \quad \text{Dry matter equation}$$

2. If we compare the coefficients for alfalfa in both equations, we find that an adjustment factor of 0.58 ($0.54 \div 0.93$) is needed to balance the dry matter equation with the TDN equation in order that we may solve for Y. If we multiply the dry matter equation by 0.58, we arrive at the following:

$$[0.58\ (0.93X + 0.89Y) = 0.58\ (3.7)] = 0.54X + 0.52Y = 2.15$$

3. With the dry matter equation adjusted so that the X terms of the dry matter and TDN equations cancel out when subtracted, we can solve for Y in the following manner:

$$\begin{array}{rl} 0.54X + 0.80Y = 2.60 & \text{TDN equation} \\ -(0.54X + 0.52Y = 2.15) & \text{Dry matter equation} \\ \hline 0.00X + 0.28Y = 0.45 & \\ Y = \dfrac{0.45}{0.28} & \end{array}$$

$$Y = 1.61 \text{ lb of concentrate should be fed}$$

4. Now that we know the value of Y. we can substitute it in either original equation to solve for X.

$$\begin{aligned} 0.54X + 0.80\ (1.61) &= 2.60 \\ 0.54X + 1.29 &= 2.60 \\ 0.54X &= 1.31 \\ X &= 2.43 \text{ lb of alfalfa should be fed} \end{aligned}$$

If we substitute X and Y in our original equations, we can check our accuracy.

$$0.54\ (2.43) + 0.80\ (1.61) = 2.6 \text{ TDN equation}$$
$$0.93\ (2.43) + 0.89\ (1.61) = 3.7 \text{ DM equation}$$

TWO × TWO MATRIX METHOD

In addition to the traditional algebraic method of solving simultaneous equations, matrix algebra provides an alternative which some people find easier and quicker. The 2 × 2 matrix provides a quick and accurate way of solving for two nutritional parameters—such as energy and protein—through the use of two ingredients.

A matrix is a mathematical array which allows for the solution of unknowns through the use of a series of equations. Consider the two equations:

$$a_1 X + b_1 Y = C_1$$
$$a_2 X + b_2 Y = C_2$$

Let us assume that X represents one type of feed, and that Y represents another type. In order to solve for X and Y, we can set up a 2 × 2 matrix using their respective coefficients. C_1 and C_2 could represent two nutrient levels we want (for example, energy and protein). The 2 × 2 matrix would then be:

$$\begin{pmatrix} a_1 & b_1 \\ a_2 & b_2 \end{pmatrix}$$

The matrix would consist of two rows and two columns. In order to solve for X and Y. we must find the determinant of the matrix. The determinant is established as follows:

$$\begin{vmatrix} a_1 & b_1 \\ a_2 & b_2 \end{vmatrix} = a_1b_2 - a_2b_1$$

If the matrix is $\begin{pmatrix} 1 & 2 \\ 3 & 4 \end{pmatrix}$, the determinant would be:

$$\begin{pmatrix} 1 & 2 \\ 3 & 4 \end{pmatrix} = (1 \times 4) + (-3 \times 2) = 4 - 6 = -2$$

Note that a determinant of a square matrix is enclosed by straight vertical lines, and a square matrix is enclosed by curved lines. Through a series of derivations from the original two equations, our unknowns can be solved in the following manner:

$$X = \frac{\begin{vmatrix} c_1 & b_1 \\ c_2 & b_2 \end{vmatrix}}{\begin{vmatrix} a_1 & b_1 \\ a_2 & b_2 \end{vmatrix}} \text{ or } \frac{(c_1b_2 - c_2b_1)}{(a_1b_2 - a_2b_1)}$$

$$Y = \frac{\begin{vmatrix} a_1 & c_1 \\ a_2 & c_2 \end{vmatrix}}{\begin{vmatrix} a_1 & b_1 \\ a_2 & b_2 \end{vmatrix}} \text{ or } \frac{(a_1c_2 - a_2c_1)}{(a_1b_2 - a_2b_1)}$$

Using the same example as in the simultaneous equations section, the 2 × 2 matrix method should arrive at the same answer.

Example 1. *A sheep producer is feeding 100-lb lambs. A supplement containing 89% dry matter and 80% TDN (as-fed) and alfalfa hay containing 93% dry matter and 54% TDN (as-fed) are available. How much of each should be fed?*

From Table 14-2, Daily Nutrient Requirements of Sheep, we find that a 100-lb lamb requires 3.7 lb of dry matter and 2.6 lb of TDN daily.

1. To balance this ration by the matrix method, we will proceed as follows: Let X = amount of alfalfa hay to be fed, and Y = amount of supplement to be fed. Therefore, our equations will be:

$0.54X + 0.80Y = 2.6$ TDN equation
$0.93X + 0.89Y = 3.7$ Dry matter equation

2. From these equations, we can set up the following 2 × 2 matrix:

$$\begin{pmatrix} a_1 & b_1 \\ a_2 & b_2 \end{pmatrix} = \begin{pmatrix} 0.54 & 0.80 \\ 0.93 & 0.89 \end{pmatrix} \text{ and } \begin{pmatrix} c_1 \\ c_2 \end{pmatrix} = \begin{pmatrix} 2.6 \\ 3.7 \end{pmatrix}$$

3. Once we have set up our matrices, we can solve for X and Y by calculating the determinants, as shown below:

$$X = \frac{[2.6\ (0.89) - 3.7\ (0.80)]}{[0.54\ (0.89 - 0.93\ (0.80)]} = \frac{2.314 - 2.960}{0.481 - 0.744} = \frac{-0.646}{-0.263} = 2.46$$

We need to feed 2.46 lb of alfalfa.

$$Y = \frac{[0.54\ (3.7) - 0.93\ (2.6)]}{[0.54\ (0.89 - 0.93\ (0.80)]} = \frac{1.998 - 2.418}{0.481 - 0.744} = \frac{-0.420}{-0.263} = 1.60$$

We need to feed 1.60 lb of supplement.

These figures coincide with those calculated by the traditional form of solving simultaneous equations (X = 2.43 and Y = 1.61). The slight variation in amounts of alfalfa and supplement to be fed determined in the simultaneous equations method and the matrix method can be attributed to rounded error.

In least-cost formulations, matrix algebra is used by the computer, but the matrices that are used are much larger than 2 × 2 and are far more complicated. The 2 × 2 matrix offers a rapid means of calculating a simple ration using two feeds to fulfill two nutrients.

TRIAL-AND-ERROR METHOD

In the example that follows, the trial-and-error method is used, with consideration given to energy and protein. Also, crude protein rather than digestible protein is used because (1) this is what feed manufacturers want to know as they plan feed formulas, and (2) this is what livestock producers see on the tag when they purchase feed. In most mixed feeds, approximately 80% of the total protein is digestible.

Example. *Let us assume that we have 140-lb lactating ewes with single lambs during the first 8 weeks of lactation. The sheep producer is feeding 5 lb of good-quality grass hay per day. Corn and oats are available and the producer wishes to use these to feed a grain mixture to the ewes. What mixture of corn and oats should be used to meet the needs of the lactating ewes from the standpoint of energy and protein?*

Before we proceed further, here are some general rules and assumptions that we will follow:

1. The TDN of the complete ration of lactating ewes should be 58% or better on an as-fed basis, preferably an allowance of 62 to 64% (keeping in mind that this gives requirements, not allowances).
2. Salt, calcium, and phosphorus mineral (including trace minerals) will be added to the grain ration. Also, salt and a mineral mix will be self-fed in a two-compartment mineral box.
3. Vitamin A and vitamin D will be added to the ration at a level of 1,100 IU of vitamin A and 70 IU of vitamin D per pound of the mixed feeds, respectively.
4. The available feeds have approximately the following composition (as-fed basis):

	TDN	Crude Protein
	(%)	*(%)*
Grass hay (all analyses)	50.0	9.1
Corn (grain)	80.3	9.5
Oats (grain)	67.2	11.7

Here are the steps in balancing this ration:

1. The daily TDN and crude protein requirements of these ewes are as follows:

	TDN	Crude Protein
	(lb)	*(lb)*
2. Requirements of the ewes	3.3	0.53
3. The 5 lb of grass hay is supplying	2.5	0.45
4. The remainder to be supplied by grain mixture	0.8	0.08

5. Let us try out (that is why it is called the "trial-and-error method") a grain mix of 700 lb of corn, 280 lb of oats, 20 lb of salt, calcium, phosphorus, and trace minerals, and see how much TDN and crude protein is in 1,000 lb of the grain mix:

	TDN	Crude Protein
	(lb)	*(lb)*
Corn, 700 lb	562.0	66.5
Oats, 280 lb	188.2	32.8
Salt calcium, phosphorus, and trace minerals, 20 lb.	—	—
Total	750.2	99.3
or in percent	75.0%	9.9%

6. Divide the TDN needed from concentrate (0.8 lb) by the percentage of TDN in the mixture (75.0%). Thus, feeding 1.1 lb of the concentrate will meet the energy needs.

7. Will this level of grain mix (1.1 lb) also meet the crude protein needs? By multiplying the pounds of concentrate mixture by the percentage of crude protein (1.1 × 9.9%), we find that the proposed concentrate would supply 0.11 lb of crude protein, whereas only 0.08 lb is needed.

8. If corn is plentiful and inexpensive compared to oats, but some oats are available, the sheep producer could try increasing the corn and reducing the oats:

	Crude TDN	Protein
	(lb)	*(lb)*
Corn, 880 lb	706.6	83.6
Oats, 100 lb	67.2	11.7
Salt, calcium, phosphorus, and trace minerals, 20 lb	—	—
Total	773.8	95.3
or in percent	77.5%	9.5%

9. By referring to step No. 4, we can divide the pounds of TDN and crude protein needed from the concentrate by the percentage of TDN in the grain mix in No. 8. We find that 0.8 ÷ 0.775 = 1.0 lb needed to supply 0.8 lb of TDN. This 1.0 lb of grain mix will also adequately meet the requirement for an additional 0.08 lb of protein.

The ration for a 140-lb ewe with a single lamb during the first 8 weeks of lactation is:

	TDN	Crude Protein
	(lb)	*(lb)*
Grass hay, 5 lb	2.5	0.45
Grain mix (corn and oats), 1 lb	0.8	0.09
Total	3.3	0.54

COMPUTER METHOD

Many large sheep feedlot establishments and feed Companies now use computers in ration formulation. Also, many of the state agricultural experiment stations are offering these services to farmers within the state, on a charge basis. The county extension agent will know about these services if they are available. With the widespread availability and declining cost of home computers, their use in livestock operations for record keeping and ration formulating will continue to increase.

Despite their sophistication, there is nothing magical or mysterious about computers balancing rations. Although they can alleviate many human errors in calculations, the data which come out of a machine are no better than those which go into it. Those people back of the computer—the producer and the nutritionist who prepare the data that go into it, and who evaluate and apply the results that come out of it—become more important than ever. This is so because a computer does not know anything about (1) feed palatability; (2) bloat prevention; (3) limitations that must be imposed on certain feeds to obtain maximum utilization; (4) the goals in the feeding program—such as breeding or finishing; (5) homegrown feeds for which there may not be a suitable market; (6) feed processing and storage facilities; (7) the health, environment, and stress of the sheep; and (8) the people responsible for actual feed preparation and feeding. Additionally, a computer may even reflect, without challenge, the prejudices and whims of those who prepare the data for it.

Hand in hand with the use of computers in balancing rations, the term *least-cost ration formulation* evolved. In some respects this designation was unfortunate, for the use of least-cost rations does not necessarily assure the highest net returns—and net profit is more important than cost per ton. For example, the least-cost ration may not produce the desired daily gain.

A computer can do little more than a good mathematician. But it can do it a lot faster and can check all possible combinations. It alleviates the endless calculations and hours common to hand calculations. For example, it is estimated that there may be as many as 500 practical solutions when 6 quality specifications and 10 feedstuffs are considered for a ration.

Generally, electronic feed formulation (1) effects a greater saving when first applied to a ration than in subsequent applications, and (2) is of most use where a wide selection of feed ingredients is available and/or prices shift rather rapidly.

Setup of the Computer Formulation

A program is a systematic process for the automatic solution of a problem or problems. In least-cost formulation, all the computer does is solve a series of equations through a sophisticated system of matrix algebra. Most nutritionists and sheep producers, however, use ready-made programs for ration formulation rather than writing their own program.

Procedure for Computer Formulation

The information needed and the procedure followed in formulating rations by computer are exactly the same as in the hand method of ration formulation; namely, the (1) nutritive requirements for the particular class and kind of animal, (2) nutritive content of the feeds, and (3) ingredient costs.

Step by step, the procedure in formulating rations by computer is:

1. **List available feed ingredients, and the cost of each.** It is necessary that all of the available feeds be listed along with the unit cost (usually per ton) of each; preferably, ingredient cost should be based on market price plus delivery, storage, and processing cost.

2. **Record nutrient composition of feeds.** The more that is known about the quality of feed, the better. This is so because of the wide variation in composition and feeding value within ingredients; for example, between two samples of alfalfa hay.

Whenever possible, an actual chemical analysis of a representative sample of each ingredient under consideration should be available and used. However, the imperfections of a chemical analysis of a feedstuff should be recognized; chiefly, (a) it does not provide information on the availability of nutrients to animals, and (b) there are variations between samples.

3. **Establish ration specifications.** Set down the ration specifications—the nutrients and the levels of each that are to be met. This is exactly the same procedure as is followed in the hand method. For example, in arriving at ration specifications, nutritionists consider (a) age, weight, and grade of sheep; (b) length of feeding period; (c) probable market; (d) season of year; (e) background and stress of animals; and (f) other similar factors.

4. **Give restrictions.** Usually it is necessary to establish certain limitations on the use of ingredients; for example, (a) the maximum amount of roughage, (b) the maximum amount of urea, (c) the proportion of cottonseed hulls to alfalfa hay, (d) the proportion of one grain to another—such as 60% barley and 40% milo, (e) the upper limit of some ingredients—such as 20% hay, (f) the exact amount of the premix, and (g) the lower and upper limits of molasses, as between 5 and 10%.

The more narrow the limitations imposed on the computer, the less the choice it will have in ration formulation and the higher the cost.

5. **Stipulate feed additives.** Generally, nutritionists make rigid stipulations as to amounts of these ingredients, much as they do with added vitamins and minerals. All of them cost money, and many of them must be used in compliance with the Food and Drug Administration regulations.

6. **Obtain program.** Many programs are available from companies specializing in software and from the numerous computer companies. The nutritionist and sheep producer must pick one that meets their needs.

7. **Key in data.** The data from steps 1 through 5 (above) must be entered into the computer. Then the computer treats the data as one gigantic algebraic problem and arrives at the ration formulation in a matter of seconds. Based on available feeds, analysis, and price, the computer evolves with the mix that will meet the desired nutritive allowances at the least possible cost.

8. **Formulate as necessary.** All rations should be reviewed at frequent intervals, and reformulated when there are shifts in (a) availability of ingredients (certain ingredients may no longer be available, but new ones may have evolved), (b) price, and/or (c) chemical composition.

9. **Validate the restrictions.** That is, test or confirm them.

Information Gained Through Computer Formulation

As a consequence of the input data entered into the computer on nutrient requirements, nutrient content of the feedstuffs to be selected, ingredient restrictions and limitations, and ingredient prices, the mathematical manipulations involved in the linear program permit the development of a sizable body of useful information about the particular formula under examination, including the following:

1. Least-cost combination of feed ingredients meeting the specific nutrient requirements.

20-DEC-82 VALLEY LAMB FEEDING
ROUTE 1, BOX 1234
CENTRAL CITY, CA 91234 ( )
★FEASIBLE RATION: E389 LAMB FATTENER

(49) ID. NO. 9978

COSTS...	...$/CWT	...$/TON
BASE LP	5.048	100.97
BATCH	5.050	101.00

#	INGREDIENT	AMOUNT	BATCH	COST	LIMITS
1	GROUND BARLEY	39.674	795.00	6.00	
3	WHEAT MILLRUN	7.988	160.00	6.25	10.000U
10	ALFALFA HAY	23.501	470.00	4.50	
11	COTTONSEED HULLS	15.000	300.00	3.40	15.000U
12	ALMOND HULLS, 15 FIBER	7.500	150.00	3.75	7.500U
22	UREA, 46% N	0.486	10.00	12.00	
23	VITAMINS & TM	0.250	5.00	30.00	0.250U 0.250L
24	PLAIN SALT	0.600	12.00	4.85	0.600U 0.600L
32	CANE MOLASSES	5.000	100.00	3.15	5.000U 2.500L
	TOTALS	100.000	2002.00		

STABLE COSTS	RANGE	COST PER UNIT
5.58	40.85	0.0042I
6.09	16.11	0.0009D
6.09	33.05	0.0016I
6.60	6.56	0.0035D
2.04	24.41	0.0246I
5.75	21.18	0.0125D
	16.85	-0.0224I
5.64	13.58	0.0224D
	9.08	-0.0171I
5.46	6.40	0.0171D
		I
20.15	0.45	0.0815D
5.63	1.44	0.2437I
	0.00	-0.2437D
	1.79	-0.0078I
5.63	0.00	0.0078D
	6.10	-0.0150I
4.65	4.27	0.0150D

REQUESTED BATCH WEIGHT IS 2000.00

#	NUTRIENT	AMOUNT	LIMITS LOWER	LIMITS UPPER	COST PER UNIT DECREASE	COST PER UNIT INCREASE	EFFECTIVE RANGE DECREASE	EFFECTIVE RANGE INCREASE
1	WEIGHT	100.000	100.000	100.000	-0.056	0.056	98.8089	100.9108
7	DRY MATTER	89.842	0.000		0.025	0.061	88.9581	90.0198
8	CRUDE PROTEIN	11.000	11.000		-0.050	0.050	10.9381	11.1088
9	DIGESTIBLE PROTEIN	7.547	0.000		0.088	0.090	7.5108	7.6080
10	PROTEIN FROM NPN	1.400		1.400	0.028	-0.028	1.2876	1.4634
12	CRUDE FAT	1.950	0.000		0.207	0.050	1.9258	2.3995
13	CRUDE FIBER	17.117	0.000		0.056	0.045	16.5450	17.1884
15	CALCIUM	0.450	0.400	0.450	1.366	-1.366	0.4261	0.4500
16	PHOSPHORUS	0.257	0.250		0.691	0.249	0.2170	0.2700
33	T. D. N.	62.501	0.000		0.019	0.045	62.3333	62.6120
39	DIGESTIBLE ENERGY	1.246	1.200		0.997	2.359	1.2430	1.2484

★★ NOT USED ★★

#	INGREDIENT	COST	RELATIVE WORTH	UPPER LIMIT	COST/UNIT INCREASE	INCREASE RANGE
2	GROUND CORN	6.150	6.042		0.0011	20.640
4	BEET PULP W/MOLASSES	5.850	5.447	10.000	0.0040	7.211
5	COTTONSEED MEAL, 41.5	9.500	7.493		0.0201	0.249
6	SOYBEAN MEAL, 47.5	11.500	7.670		0.0383	0.181
20	LIMESTONE	2.000	0.000		0.4826	0.000
21	DICAL PHOSPHATE	16.500	0.000		0.4091	0.000

Fig. 14-7. A printout of a computer-formulated sheep ration. (Courtesy, Nutri-Systems, Fresno, CA)

2. Acceptance or rejection of various feedstuffs based on their cost and nutrient contribution.

3. Cost ranging, or the price range over which the current cost of a particular ingredient may vary and not cause change in the formula solution.

4. Indication of the stability of each individual ingredient within the formula, a measure of the strength of its position at the stated cost.

5. Opportunity purchasing, or indication of the price required for an ingredient to enter the least-cost solution.

6. Relative cost pressure for meeting a specific nutrient requirement and its effect on the cost of formulation.

7. Relationship of nutrient density of formula to cost of supplying equivalent nutrient intakes, and its effect on comparative economic return.

Precautions and Limitations of Computer Formulation

When utilizing the computer in the formulation of least-cost rations, the producer needs to consult an experienced nutritionist. The nutritionist can interpret the printout from the computer and make any adjustments necessary to make the ration more realistic. It must be reiterated that the computer formulates rations objectively from the information that is fed into it. What comes out of the computer may be the best solution to the mathematical problem, still it may not be completely practical or realistic.

Any nutritional specification imposed is solved exactly, regardless of the economic implications of the particular requirement, even though the method of solution proceeds in a least-cost manner. Sometimes the last unit of nutritional constraint can result in a significant cost increase because of the difficulty or "pressure" the program encounters in building the formula to the specified requirement value from the ingredients available. On the other hand, the biological values involved—either nutritional requirement figures or ingredient analysis data—in no way embody the degree of accuracy the listed specifications imply. A computer formula, therefore, should never be accepted as the final answer until it has been examined rationally.

Besides investment in equipment and in the program, one of the principal expense items involved in computer formulation arises from the need for continual review and revision of the information that must go into the system. Ingredient analysis should be monitored, the relativity of specifications and restrictions explored, and cost figures constantly updated in order to maximize the benefits of computer formulation.

The producer should be aware that radical changes in ration composition cannot be made without causing digestive disorders. Sheep sometimes need time to adapt to changes in rations, a fact which the computer does not consider.

When the computer formulates rations, it is using average values for the nutrient composition of the various feeds. Feeds can often vary in their nutrient composition, so there is a good possibility that the chemical analysis of the formulated feed will not be the same as the formulated analysis. However, in most cases, this difference is not of sufficient magnitude to create problems.

Finally, the results obtained from the computer are only as good as the person who feeds the information into the machine. If the data given to the computer are outdated or wrong, the ration that is formulated will be of little value.

HOMEGROWN VS PURCHASED FEEDS

A major factor in determining whether feeds shall be homegrown or purchased is the system of farming, three broad systems of which are practiced in the United States: (1) crop farming, (2) livestock farming, and (3) crop and livestock farming combined. Each system may be defined as follows:

1. **Crop farming.** This is that system in which crops are grown for feed, food, and/or clothing, with the income derived primarily from the sale of cash crops.
2. **Livestock farming.** This is that system in which animals are produced, with all or part of the feed bought, and with the income derived primarily from the sale of animals, meat, and/or wool.
3. **Crop and livestock farming combined.** This is that system that combines significant amounts of crop and livestock farming, with the income derived from the sale of both cash crops and animals and their byproducts.

Over a period of time, farmers and ranchers do those things which are most profitable to them, with some modification according to personal preference. Thus, profitability, with modification by personal preference, largely determines the system of farming. It follows that the choice between homegrown and purchased feeds for the strictly livestock farm or the crop and livestock farm combination is largely based on profitability.

Usually a combination of several factors determines the profits to accrue from homegrown or purchased feeds. Except for very large and intensive livestock operations, net returns are generally highest on those farms or ranches that produce most of their pasture, hay, and/or silage.

COMMERCIAL FEEDS OR MINERAL MIXTURES

Commercial feeds or mineral mixtures are just what the name implies—feeds or minerals mixed by manufacturers who specialize in the business.

The commercial feed or mineral manufacturer has the distinct advantages of (1) purchasing feeds or minerals in quantity lots, making possible price advantages; (2) using computers for purchasing and least-cost formulating; (3) having the knowledge to manufacture medicated feeds; (4) having the knowledge and the facilities to manufacture specialty feeds, such as milk replacers; (5) processing and mixing economy and control; (6) hiring scientifically trained personnel who will determine the rations; and (7) controlling quality. Most producers have neither the know-how nor the quantity of business to provide these services on their own. Because of these several advantages, commercial feeds or minerals are finding a place of increasing importance in livestock feeding.

Numerous types of commercial feeds, ranging from additives to complete rations, are on the market, with most of them designed for a specific species, age, or need. Among them are complete rations (including hay for ruminants), concentrates, pelleted or cubed forages, protein supplements (with or without reinforcements of vitamins and/or minerals), vitamin and/or mineral supplements, additives (antibiotics, hormones, etc.), milk replacers, starters, young stock rations, fitting rations, rations for different levels of production, and medicated feeds.

In summary, two good alternative sources of most feeds and rations—home mixed and commercial—exist, and the able manager will choose wisely between them. When selecting commercial feeds or minerals, the producer should be aware of (1) the reputation of the manufacturer, (2) the specific needs of the animals to be fed, and (3) the information on a properly labeled product.

BEST BUY IN FEEDS

In buying feeds, the sheep producer should check prices against values received. While there are several methods for doing this, one simple method is to compute and compare the cost per unit of nutrients, based on feed composition. Where a chemical analysis of a specific feed is not available, feed composition tables, such as those in Chapter 27 of this book, may be used as good indicators. Thus, feed composition tables may serve as a basis of feed purchasing and merchandising, as well as for ration formulation.

The use of the cost-per-unit-of-nutrients method can best be illustrated by the examples that follow:

■ **Cost per pound of protein and TDN**—If 44% protein (crude) soybean meal is selling at $9.20 per 100 pounds whereas 35% protein (crude) linseed meal is selling for $8.00 per 100 pounds, which is the better buy? Divide $9.20 by 44 to get 20.90¢ per pound of crude protein for the soybean meal. Then divide $8.00 by 35 to get 22.85¢ per pound for the linseed meal. Thus, at these prices soybean meal is the better buy—by 1.95¢ (22.85 – 20.90 = 1.95) per pound of crude protein.

When buying energy feed, one can compare the cost per pound of total digestible nutrients (TDN). For example, if corn is priced at $4.50 per 100 pounds and has a TDN of 91%, divide $4.50 by 91 and the result is 4.94¢ per pound of TDN.

If barley with 83% TDN sells for $6.00 per 100 pounds, divide $6.00 by 83, and the price is 7.22¢ per pound of TDN. Thus, corn would be the better buy by 2.28¢ (7.22 – 4.94 = 2.28) per pound of TDN.

■ **Cost per pound of phosphorus**—When buying a mineral, the sheep producer should also check prices against values received. For example, let us assume that the main need is for phosphorus, and that we wish to compare minerals, which we shall call brands "X" and "Y." Brand X contains 12% phosphorus and sells at $340 per ton, or $17 per hundredweight, whereas brand Y contains 10% phosphorus and sells at $320 per ton, or $16 per hundredweight. Which is the better buy?

COMPARATIVE VALUE OF BRANDS X AND Y
(based on phosphorus content alone)

Brand	Phosphorus	Price/Cwt	Cost/Lb Phosphorus
	(%)	($)	($)
X	12	17	1.42
Y	10	16	1.60

Brand X is the better buy, even though it costs $1 more per hundredweight, or $20 more per ton.

One other thing is important when buying minerals. Usually, the more scientifically formulated mineral mixes will have plus values in terms of trace mineral needs and balance.

Of course, many other factors affect the actual feeding value of each feed, such as (1) palatability, (2) grade of feed, (3) preparation of feed, (4) ingredients with which each feed is combined, and (5) quantities of each feed fed. It follows that, from the standpoint of the sheep producer, the most important measurement of a feed's usefulness is in terms of net returns, rather than cost per bag or cost per ton.

FEED PREPARATION

Perhaps no problem is so perplexing to the amateur feeder as the proper preparation of feeds. The type of preparation depends on the type of feed and, of course, economics.

■ **Grains**—Sheep masticate grain more thoroughly than cattle, so feed preparation for them is of less value than for cattle. If grains, like sorghum and millet, are hard, or the teeth are poor, they may be prepared by exploding, extruding, flaking, micronizing, popping, roasting, or high moisture, with cost determining the choice. Pellets are increasingly used by lamb feeders because (1) pelleted feeds are less bulky and easier to store and to handle; (2) pelleting prevents animals from selectively wasting ingredients likely to be high in certain dietary essentials; (3) pelleting alleviates wastage of relatively unpalatable feeds; and (4) pelleting reduces losses from wind blowing. Cubes or pellets are preferred for feeding sheep on pasture or range. Professional shepherds prefer flaked grain for show sheep, as the ration is lighter and there are fewer digestive disturbances.

■ **Forages**—Coarse forages, such as corn fodder and stover, should be cut or shredded for handling ease and reduction of waste. Forages should be coarsely chopped. not less than 2 in. In the West, hay is frequently chopped because (1) it is easier to handle, (2) it can be stored in a smaller area at less cost, and (3) it is fed with less waste. Also, forages may be pelleted or cubed. Many lamb feeders are using all-pelleted rations—hay and grain combined.

Breeding sheep should not be fed for extended periods of time on pellets without any long or chopped forage. A high incidence of parakeratosis—a degeneration of the rumen papilla—appears to result from feeding pellets, especially where low-forage–high-concentrate pellets are used.

ADDITIVES AND IMPLANTS

Table 14-14 summarizes the growth stimulants that are presently available and can be used. All of these products have been shown to improve gain and feed efficiency of sheep. The information presented in Table 14-14 is the most recent available. But feed additives and implants do change from time to time; new products are developed, and sometimes old products are banned by the Food and Drug Administration. So, those using additives should always confer with local authorities and read and follow manufacturers' label directions for more complete details on the use of specific drugs or combinations of drugs.

Antibiotics may improve performance when added to creep and lamb-finishing rations. Chlortetracycline and oxytetracycline are especially effective. The response to antibiotics seems to be affected by differences in management and the amount of stress to which the lambs are subjected. There is some evidence that antibiotics reduce the incidence of enterotoxemia.

In addition to the additives listed in Table 14-14, a number of feed additives are approved for treatment of specific diseases.

FEEDING BREEDING EWES

Success in the sheep business is largely measured by the percentage lamb crop raised and the pounds of lamb marketed per ewe. The most important factor affecting these criteria is the feed of the ewe. Also, the yearly feed of the ewe represents about 50%

TABLE 14-14
SHEEP FEED ADDITIVES AND IMPLANTS

Type of Additive	Method of Administering	Dosage	Effect On			Comments
			Daily Rate of Gain	Feed Efficiency	Carcass Quality	
			(% increase)	*(% increase)*		
Antibiotics (chlortetracycline and oxytetracycline)	Feeding (oral)	Aureomycin (chlortetracycline) 10 to 25 mg/lb of feed. Terramycin (oxytetracycline) 5 to 10 mg/lb of feed.	Range: 0-31 Average: 11	Range: 4-27 Average: 10	No effect to slight improvement.	Antibiotics (especially chlortetracycline and oxytetracycline) may improve performance when added to creep and lamb finishing rations. Response to antibiotics varies markedly according to differences in management and degree of stress to which lambs are subjected. There is some evidence that antibiotics reduce the incidence of enterotoxemia.
Bovatec (lasalocid)	Feeding (oral)	10-15 mg/lb complete feed, fed at rate of 15-70 mg lasalocid/day.	Range: 0-20 Average: 6-8	Range: 5-15 Average: 8-10	No effect.	Bovatec is an ionophore. In addition to increasing rate of gain and feed efficiency, Bovatec reduces rumen protein degradation and increases the amount of bypass protein. Greatest response is obtained where coccidiosis is a problem, for which purpose Bovatec was initially approved by FDA.
Ralgro (zeranol)	Implant	12 mg/head	Range: 0-25 Average: 10	Average: 6		Do not implant animals within 40 days of slaughter. Do not implant breeding animals.

Fig. 14-8. Polled Dorset ewes with a Polled Dorset ram, owned by Riverwood Farms, Powell, Ohio. (Courtesy, *Sheep Breeder and Sheepman*, Columbia, MO)

of all production costs. For purposes of convenience, the feeding of ewes will be discussed under the following headings: (1) drylot (confinement) feeding, (2) flushing ewes, (3) feeding pregnant ewes, (4) feeding at lambing time, (5) feeding lactating ewes, and (6) feeding ewes in accelerated lambing.

■ **Drylot (confinement) feeding**—The vast majority of the nation's sheep utilize pasture in season. However, some ewe-lamb producers are drylotting all or part of the year. So, now there are two alternatives, and the producer may choose between the two.

The **advantages** of drylot (confinement) production are:

1. The virtual elimination of losses from predators.
2. Freedom from the most harmful internal parasites.
3. Lowering of the energy requirement due to limited activity.
4. The opportunity to feed ewes according to their productivity and nutrient requirements rather than their appetites.
5. It results in more rapid gains, in lambs reaching market weight at an earlier age, and in improved carcass grade.

The **disadvantages** are:

1. A higher initial capital investment, especially in buildings and equipment.
2. It requires superior management.
3. All nutritive requirements must be met.
4. External parasites and contagious diseases may be increased.
5. Animal manure disposal and bedding costs will be greater.

The Table 14-20 rations are satisfactory for ewes in confinement, and the Table 14-18 and 14-19 rations are excellent for creep feeding lambs raised in confinement.

The following steps are necessary for successful drylot feeding:

1. Use a big purebred ram so that the lambs inherit rapid growth potential.
2. Have the lambs born so that they will be ready for the intended market in 4 to 4½ months.
3. After two months lactation, cut the ration of the ewes to the same amount that they were receiving one month before lambing.
4. Self-feed lambs; and wean them at 90 days of age. Do not turn lambs to pasture.
5. Market the lambs at 100 to 110 lb.
6. Vaccinate all lambs against enterotoxemia (overeating disease).

■ **Flushing ewes**—*Flushing is the practice of conditioning or having thin ewes gain in weight just prior to breeding.* Its purpose is to increase the ovulation rate and, consequently, the lambing rate.

This special feeding usually begins 2 to 3 weeks prior to breeding and continues into the breeding season. It may be accomplished by turning the ewes to a fresh luxuriant pasture 2 to 3 weeks before breeding time; or if such a pasture is not available, satisfactory results may be brought about by feeding a grain allowance of ½ to ¾ of a pound daily over a like period of time. Oats alone are excellent, or a mixture of equal parts of oats and corn is very satisfactory. Pumpkins, broken and scattered over the pasture, are also relished and are excellent for flushing purposes. Some shepherds like to feed cabbage at this season.

Although it is not likely that all of the benefits ascribed to flushing will be fully realized under all conditions, the general feeling persists that the practice will result in a 15 to 20% increase in the lamb crop, and that the ewes will breed both earlier and more nearly at the same time. Hence, it follows that the lamb crop will be earlier and more uniform in age and size.

Mature ewes appear to respond better than yearling ewes. Also, flushing may be more beneficial early and late in the breeding season than during the peak, when the ovulation rate is highest.

Fat ewes will not benefit from flushing. Instead, they should be conditioned for breeding by stepping up the exercise.

■ **Feeding pregnant ewes**—If a strong, healthy crop of lambs is to be expected, the ewes must be properly fed and cared for throughout the period of pregnancy. In general, this means the feeding of a suitable and well-balanced ration, together with the necessary minerals and vitamins as required for maintenance (and growth, if the ewe is not fully mature), growth of the

Fig. 14-9. Pregnant ewes in excellent condition. (Courtesy, American Hampshire Sheep Assn., Milo, IA)

fleece, and development of the fetus. In addition, plenty of exercise must be provided. Suitable shelter should be made available during inclement weather, and the animals should be given access to an abundance of fresh air and sunshine at all times. Ewes should gain in weight during the entire period of pregnancy, making a total gain of 20 to 30 lb for the period. They should enter the nursing period with some reserve flesh, because the lactation requirements are much more rigorous than those of the gestation period.

After the ewes are bred, they should have access to pastures as long as they are available and open. When the ground is firm, winter pasture or range, stalk, or stubble fields may be pastured to advantage. Green rye or wheat pastures furnish a very succulent feed and valuable exercise for the flock. Where winter pastures are either unavailable or inadequate, supplemental feeds must be provided. The most satisfactory forage is a good-quality legume hay—alfalfa, clover, lespedeza, or soybeans. The sheep producer, however, often seems to find it difficult to grow such roughage at satisfactory prices. Where grass hay, such as native hays or timothy, is used, every effort should be made to cut it at an early stage of maturity and to have it properly cured. Even then, a protein supplement should be provided, together with suitable minerals. Because of the known value of legumes from the standpoint of quality of proteins, minerals, and vitamins and the fact that grass hays are not recognized as too desirable for sheep, every effort should be made to supply at least a third of a good-quality legume roughage to pregnant ewes. A 150-lb ewe will eat about 4 lb of hay daily. In order to prevent waste and protect the wool from chaff and hay seeds, suitable racks should be provided.

Such succulent feeds as roots and silage are desirable in keeping the ewes healthy and doing well. Of the root crops, turnips seem to be preferable for pregnant ewes. Silage made from corn, milo, legumes, or grasses, and which is not frozen, spoiled, nor moldy, may be fed quite safely and is excellent feed. Ordinarily, the daily ration of roots or silage should not exceed 5 lb, which means that hay is usually fed in addition to the succulent feed.

During the last 4 to 5 weeks of pregnancy, the fetus develops very rapidly and the demands on the ewe are rather heavy. Also, ewes carrying twins or triplets, especially if they are a bit fat, are very prone to ketosis (lambing paralysis), which can be prevented by feeding a high-energy ration. So, during the last 4 to 5 weeks of pregnancy, ewes should be fed 0.5 to 1.0 lb of grain per head daily and gain 8 to 15 lb. Besides, ewes so fed milk better. The concentrate given to the farm flock usually consists of homegrown grains, whereas range bands are often given pelleted or cubed protein supplements.

■ **Feeding at lambing time**—As lambing time approaches, or immediately after lambing, each ewe should be placed in an individual holding or lambing pen. At this time, the grain allowance should be materially reduced, but dry roughage may be fed free choice, when it is certain that it is of good quality and palatable. Usually, some five to seven days should elapse before ewes are placed on full feed following parturition. In general, feeds of a bulky and laxative nature should be provided during the first few days. A mixture of equal parts of oats and wheat bran is excellent. Soon after lambing, the ewe should be given water with the chill removed but should not be allowed to gorge.

■ **Feeding lactating ewes**—Following lambing, the

Fig. 14-10. Polypay ewe and her newborn triplet lambs in a lambing pen. (Courtesy, American Polypay Sheep Assn., Sidney, MT)

feed allowance of the ewe should be increased according to her capacity and needs. Although there is great individual and breed variation, ewes will yield from 1 to 4 qt of milk per day. In comparison with cow's milk, ewe's milk is richer in protein and fat and higher in ash. It must also be borne in mind that, in addition to producing milk and maintaining her body, the ewe is growing wool, which is protein in character. Immature ewes are also growing. Under these circumstances, it is but natural and normal to expect ewes to lose in condition during the suckling period. The loss in weight is primarily determined by the inherent milking qualities of the individual and by the kind and amount of feed.

In general, it is considered good practice to feed lactating ewes rather liberally, for lambs make the most economical gains when suckling. It is a good plan to separate the ewes with twins from those with singles, giving the former more liberal rations or the benefit of the better pastures or ranges. In fact, some large sheep operators find this practice so advisable that they regularly separate out the twin bands.

Milk production can be greatly stimulated through the proper selection of feeds. If there is not sufficient high-quality roughage for the entire winter, the most palatable and succulent portion should be reserved for use during the suckling period. Pastures should be provided as soon as possible, but in the meantime a high-quality legume hay will take care of the roughage needs. Though varying somewhat with the size and condition of the ewe and whether there are twins or a single, an adequate ration for a lactating ewe may consist of approximately 4 lb of high-quality alfalfa hay plus 1 to 2 lb of grain daily. If a legume hay is not available, a protein supplement should be included in the grain ration.

As soon as the spring pasture season has arrived, the use of harvested feeds should be discontinued, being both uneconomical and unnecessary.

■ **Feeding ewes in accelerated lambing**—*Accelerated lambing involves ewe lambs dropping their first lambs at one year of age, and lambing at intervals of 6 to 8 months thereafter.*

Experimental studies and practical observation indicate (1) that it is feasible and profitable to have ewe lambs drop their first lambs at 1 year of age, provided they are well fed, well grown, and early dropped; and (2) that it is possible to achieve a lambing interval of 6 to 8 months, provided breeds with long breeding seasons are used (Dorsets, Rambouillets, or Merinos), there is superior nutrition, and hormones are used when a 6-month interval is planned.

Ewe lambs that are to be bred so that they lamb at 12 months of age should be liberally fed (1) from birth, using one of the creep rations given in Tables 14-18 and 14-19; and (2) during pregnancy and lactation, using one of the suggested rations in Table 14-20.

Accelerated lambing can (1) make for lower cost of production, (2) contribute to a more even supply of lamb throughout the year, and (3) allow for greater flexibility in production.

FEEDING RAMS

The rams should be fed so as to remain in vigorous, active breeding condition. In general, rams should be fed the same kind of feed as ewes but in slightly larger quantities. They need a generous allowance of relatively high-quality feed just before and during the breeding season, when pasture is not available. During the balance of the year, pasture is usually adequate when available; otherwise, the ration may be comparable to that of ewes.

Fig. 14-11. Polled Dorset ram in strong breeding condition, owned by Riverwood Farms, Powell, OK.

FEEDING RANGE SHEEP

Various geographical divisions are assumed in referring to the western range area—the native pasture area. Sometimes reference is made to the 17 range states, embracing a land area of 1.16 billion acres. At other times this area is broken down, chiefly on the basis of topography, into (1) the Great Plains area, and (2) the 11 western states. Much of the latter is federally owned.

In the early days of the range sheep industry, the animals were usually moved to lower winter ranges and expected to get their feed as best they could. There was precious little supplemental feeding. If the winter happened to be mild, and if a reasonable amount of grass was cured on the stalk, the band came through in pretty good shape. During an exceedingly cold winter, particularly when there was much snow, losses were severe and often disastrous. Today, the practical and successful range sheep producer winter feeds. The progressive rancher is also equipped to meet emergency feeding periods, of which droughts are the most common.

Ewes are normally maintained on winter grazing areas, with or without supplemental feeds, as long as possible. Usually these ranges are located at the lower altitudes and the vegetation consists of rather mature and bleached grasses or brush and browse. When the vegetation is sparse or covered by deep snow, supplemental feeds of hays, preferably alfalfa, some other legume, or concentrates are provided. Often protein supplements in the form of pellets or cubes are used, for these may be scattered about the feeding grounds, neither being blown away nor difficult for the sheep to find. Usually such expensive protein supplements are fed only when native grass hays are being utilized, high-quality alfalfa not requiring a protein supplement.

Because of the magnitude of the range sheep industry and the fact that it is a highly specialized type of operation, in the sections that follow special discussion is devoted to the feeding and management of sheep on the range.

NUTRIENT DEFICIENCIES OF RANGE FORAGE

Hunger, due to the lack of feed, is the most common deficiency on the western range. In particular, there may be a shortage of energy during droughts, late in the season, or early in the spring when grass is washy. Under such energy-deficient conditions, sheep lose weight and condition and lambs fail to grow. Also, reproduction is adversely affected.

Mature, weathered native range grass is almost always deficient in protein—being as low as 3% or less. Protein-leaching losses due to fall and winter rains may range from 37 to 73%.

Phosphorus deficiencies are rather common among range sheep, but calcium deficiency is seldom encountered.

Of the vitamins, vitamin A is most likely to be deficient in range forage, because dry, bleached range grass is very low in carotene (the precursor of vitamin A).

Fig. 14-12. Hunger, due to lack of feed, is the most common deficiency on the western ranges of the U.S. (Courtesy, Amos Dee Jones Ranches of Roswell and Tatum, NM)

RANGE SUPPLEMENTS

Four suggested range supplements, ranging from high to low protein, are given in Table 14-16.

Sheep on poor or weathered range grass should be supplemented by feeding the high protein formulation in Table 14-16, to correct the protein and phospho-

TABLE 14-15
RELATIVE RANKING OF PASTURE FORAGES FOR SHEEP

Species	Carrying Capacity	Lamb Performance	Lamb Production/ Acre	Sheep and Pasture Management Required
Alfalfa	High	High	High	Medium
Ladino clover	Low	High	Medium	Medium
Bird's-foot trefoil	Low	High	High	Medium
Bluegrass	Low	Low	Low	Low
Bromegrass	Medium	Medium	Medium	Low
Fescue	High	Low	Low	Low
Orchardgrass	High	Medium	Medium	Low
Canarygrass	High	Low	Medium	Low
Timothy	Low	Low	Low	Low
Oats	Medium	Medium	Medium	Low
Barley	Medium	Medium	Medium	Low
Sudan	High	Low	Medium	High
Rape	High	High	High	Low
Turnips	High	Low	Medium	Low

[1]Adapted by the author from *Nutrient Requirements of Sheep*, Sixth Revised Edition, NRC-National Academy of Sciences, p. 52, Table 10.

rus deficiencies. Of course, the supplements in Table 14-16 may be modified in keeping with the availability and cost of feeds, and yet meet known deficiencies. For example, if phosphorous is the only deficiency, it may be corrected by feeding a phosphorus supplement free choice.

There is no one best and most practical range supplement for any and all conditions. Many different feeds may be, and are, used; among them, (1) ranch or locally produced hay, (2) alfalfa pellets or cubes, with or without fortification, and (3) supplements of various kinds.

Also, producers can lessen the labor attendant to the daily feeding of a pasture or range supplement by (1) using protein blocks, or (2) self-feeding salt-feed mixtures.

Where salt is used for the purpose of governing consumption, the proportion of salt to feed may vary anywhere from 5 to 40% (with 30 to 33.33% salt content being most common).

Nutrient Requirements of Sheep, Sixth Revised Edition, presents (1) Table 14-15, Relative Ranking of Pasture Forages for Sheep; and (2) Table 14-16, Formulas for Range Sheep Supplements. Both tables are herewith reproduced.

TABLE 14-16
FORMULAS FOR RANGE SHEEP SUPPLEMENTS[1]

Feed[2]	Recommended Level of Protein			
	High	Medium-High	Medium-Low	Low
	(%)			
Barley, grain or corn, dent yellow, grain, grade 2 US, minimum 54 lb *24.5 kg)*/bu	5	40	75	65
Beet, sugar, molasses, or sugar cane molasses, 48% invert sugar, minimum 79.5° Brix	5	5	5	5
Cottonseed with some hulls, solvent extracted, ground, minimum 41% protein, maximum 14% fiber, minimum 0.5% fat (cottonseed meal)	66	36	—	16
Soybean, seeds, solvent extracted, ground, maximum 7% fiber, 44% protein (soybean meal)	10	10	10	10
Urea, technical, 282% protein equivalent	—	—	5	—
Alfalfa, aerial parts, dehydrated, ground, minimum 17% protein or alfalfa, hay, sun-cured, early bloom	10	5	—	—
Vitamin A (IU/lb)	—	1,818	3,636	3,636
Vitamin A (IU/kg)	—	*4,000*	*8,000*	*8,000*
Calcium phosphate, monobasic, commercial	1	1	2	1
Sodium phosphate, monobasic, technical	2	2	2	2
Salt or trace mineralized salt	1	1	1	1
Total	100	100	100	100

Composition[3]	High As-fed[4]	High M-F	Medium-High As-fed[4]	Medium-High M-F	Medium-Low As-fed[4]	Medium-Low M-F	Low As-fed[4]	Low M-F
Digestible energy (Mcal/lb)	1.4	1.5	1.4	1.5	1.4	1.5	1.3	1.4
Digestible energy (Mcal/kg)	*3.0*	*3.3*	*3.0*	*3.3*	*3.0*	*3.3*	*2.8*	*3.1*
Protein (N × 6.25) (%)	30.4	33.8	21.9	24.3	23.6	26.2	15.9	17.7
Phosphorus (%)	1.8	2.0	1.4	1.5	0.8	0.9	1.1	1.2
Carotene (mg/lb)	9.0	10.0	4.1	4.5	—	—	—	—
Carotene (mg/kg)	*19.8*	*22.0*	*9.0*	*10.0*	—	—	—	—
Vitamin A (IU/lb)	—	—	1,636.0	1,818.0	3,273.0	3,636.0	3,273.0	3,636.0
Vitamin A (IU/kg)	—	—	*3,600.0*	*4,000.0*	*7,200.0*	*8,000.0*	*7,200.0*	*8,000.0*
Rate of feeding (lb/day)	0.20–0.40	0.22–0.44	0.20–0.40	0.22–0.44	0.20–0.40[5]	0.22–0.44[5]	0.20–0.40[5]	0.22–0.44[5]
Rate of feeding (kg/day)	*0.09–0.18*	*0.1–0.2*	*0.09–0.18*	*0.1–0.2*	*0.09–0.18*[5]	*0.1–0.2*[5]	*0.09–0.18*[5]	*0.1–0.2*[5]

[1]Adapted by the author from *Nutrient Requirements of Sheep*, Sixth Revised Edition, NRC-National Academy of Sciences, p. 52, Table 11.

[2]Feeds mixed and fed in meal or pellet form.

[3]Molasses and alfalfa hay, sun-cured, early bloom not included.

[4]Estimated 90% dry matter.

[5]In emergency situations, up to 1.1 lb *(0.5 kg)* may be fed.

RATE OF SUPPLEMENTAL FEEDING

The time and rate of supplemental feeding is determined by the reason for feeding supplements. Supplements are fed for two primary purposes: (1) to balance diets by adding small quantities of a nutrient (such as protein, a mineral, or a vitamin) or a combination of nutrients; and (2) to provide nutrients during short-term emergencies. As an example of the latter, a supplement may be needed to prevent sheep from eating poisonous plants during periods when they are on the trail or when forage is covered with snow.

Supplemental feeding should be timed to start when it is needed. If phosphorus supplementation is required, it should be provided continuously, perhaps by free-choice feeding. Where energy, protein, and /or vitamin A supplementation is involved, it takes a unique skill to recognize the nutritional state of the sheep, the range condition, and the need for supplement—both in kind and amount. The successful manager develops a grazing plan that minimizes the need for supplements, yet provides the proper supplement at the proper time and in the proper amounts.

The normal range of supplementation for sheep is ¼ to ½ lb per head per day. Rates above ½ lb approach a level that will result in reduced intake of range forage. Where range vegetation is so short as to require supplementation in excess of ½ lb per head per day, consideration should be given either to moving the sheep into drylot or to moving them to a better grazing area.

Some managers divide their sheep according to age, condition, and twins vs single lambs. Of course, this is facilitated where there are several bands. By so doing, it is possible (1) to give the animals that require the highest level of nutrition the best pasture or range, and/or (2) to supplement according to need.

FEEDING GROWING-FINISHING LAMBS

The growing-finishing stage of lambs refers to that period extending from birth to weaning at 4 to 6 months of age. At no other period in the life of the sheep is growth and disease prevention so important.

Where succulent pastures are available, most practical sheep producers, including producers with both farm flocks and range bands, consider that a combination of such green forage plus the ewe's milk is ample. In fact, lambs are unique among farm animals, inasmuch as they may be marketed at top prices off grass. Although young cattle may be sold off grass without having any other feed, they will usually fail to get sufficiently fat to bring top prices.

Fig. 14-13. Growing-finishing lambs at market weight. (Courtesy, *The Dorset Journal*)

Frequently, farm-flock lambs are creep fed grain in addition to receiving their mother's milk and pastures. Usually creep feeding on the western range is too difficult. Should the range forages not be sufficiently abundant or lush to produce fat lambs, range sheep producers usually elect to sell their animals as feeders at weaning time.

Good pastures for lambs are those that are rather succulent and that are composed of plants that are palatable and nutritious. This means green, actively growing pastures in contrast to dormant or dried forages.

Hothouse lambs are born out of season, in the fall or early winter, when pastures are usually unavailable. It is necessary that these animals be crowded for slaughter at 2 to 4 months of age, when they should weigh from 40 to 60 lb. In addition to the right breeding for this specialty, therefore, hothouse lambs must be carefully fed. In the first place, the ewes should be given liberal quantities of a good succulent ration in order to stimulate the milk flow. Secondly, the lambs should be creep fed with a palatable and suitable ration from the time they are 2 weeks of age until marketing.

■ **Early weaning**—*Early weaning refers to the practice of weaning lambs earlier than usual—to weaning at 6 to 8 weeks of age or earlier.* There is much interest in early weaning because—

1. Of lambing out of season, multiple births, and more than one lamb crop per year.
2. Lactating ewes usually reach a peak in milk production 3 to 4 weeks after lambing, then decline thereafter. By 3 to 4 months after lambing, many ewes will be producing very little milk.
3. Fewer parasite problems accompany an early weaning program.
4. Increased knowledge of nutrition now makes it

possible for scientists to improve upon milk (except for colostrum), chiefly by reinforcing it with certain vitamins and minerals.

5. Young gains are cheap gains, due to (a) the higher water and lower fat content of young animals in comparison with older animals, and (b) the higher feed consumption per unit weight of young animals.

6. Following weaning, ewes can be maintained on a limited feed allowance, thereby effecting a saving in cost.

For successful early weaning, superior nutrition and management are essential; and the earlier the weaning age the more exacting these requirements.

Early weaning of lambs is, to a considerable extent, a matter of preparation, rather than the abrupt separation of lambs from their mothers. Lambs that are to be early weaned should be creep fed from the time they are old enough to eat. At weaning time, the separation should be made by removing the ewes from the lambs, rather than vice versa. By keeping the lambs in familiar surroundings, stress is minimized.

An early-weaned lamb ration should meet the following specifications: contain a minimum of 16% crude protein; be fortified with supplemental iron if the lambs are raised on slotted floors; and have a calcium:phosphorus ratio of at least 1:1 (2:1 if urinary calculi has been experienced).

Milk replacers containing approximately 30% fat and 24% protein have been used successfully in feeding lambs receiving colostrum and weaned at one day of age. Replacers with reduced lactose content (from 42 to 27% on a dry matter basis) give improved performance. The milk is fed (1) cold at 36 to 40°F, rather than warm, to reduce overeating and bacterial contamination, and (2) free choice. From the beginning, lambs are offered a very palatable solid feed in addition to the milk. The milk replacer is discontinued when the lambs are eating sufficient quantities of the dry feed, usually at 21 to 35 days of age.

■ **Creep feeding**—*The practice of supplemental feeding of nursing lambs in a separate enclosure away from their dams is known as creep feeding.* Lambs will usually consume some creep feed at 10 to 14 days of age.

Creep rations can either be hand-fed or self-fed. Many sheep producers hand-feed until the lambs begin to eat regularly, then self-feed from this point on.

The amount of creep feed consumed is inversely proportional to the ewe's milk production. For this reason, (1) twin lambs usually consume more than single lambs, and (2) significant amounts of creep feed are consumed at 6 to 8 weeks of age, at which time the ewe's milk production usually drops.

Until lambs are six weeks old, the grain should be crimped, cracked, or rolled, unless a pelleted ration is used. After this age, whole grain may be fed unless it is extremely hard (like millet).

It is important that the creep ration be very palatable. For this reason, rolled oats, wheat bran, soybean meal, and molasses are important ingredients in a creep ration. Even then, if lambs have access to lush pasture, they may prefer it to the creep feed.

Suggested creep rations are given in Tables 14-17 to 14-19, as presented in *Nutrient Requirements of Sheep*, Sixth Revised Edition.

TABLE 14-17
SUGGESTED CREEP DIETS[1]

Ingredient	Amount (As-fed Basis)		
	Diet A	**Diet B**	**Diet C**
	(%)	*(%)*	*(%)*
Simple diets (grind for lambs under 6 weeks of age; feed whole thereafter; hand- or self-feed)[2]			
Barley, grain	38.5	—	—
Corn, dent yellow, grain, ground, grade 2 US, min 54 lb/bu	40.0	60.0	88.5
Oats, grain	—	28.5	—
Wheat, bran, dry milled	10.0	—	—
Linseed meal, soybean meal, or sunflower meal	10.0	10.0	10.0
Limestone, ground, min. 33% calcium	1.0	1.0	1.0
Trace mineralized salt with selenium	0.5	0.5	0.5
Total	100.0	100.0	100.0
Alfalfa hay, sun-cured, early bloom should be fed free choice in conjunction with any of the above diets.			
Commercially mixed diets (hand-or self-fed as meal, but usually as pellets)[3]			
Alfalfa, sun-cured, early bloom or dehydrated alfalfa	—	10.0	20.0
Barley, grain	20.0	—	—
Corn, dent yellow, grain, grade 2 US, min 54 lb/bu	54.5	34.5	44.5
Oats, grain	—	30.0	10.0
Linseed, soybean, or sunflower meal	10.0	10.0	10.0
Bran, wheat	10.0	10.0	10.0
Beet or cane molasses	4.0	4.0	4.0
Limestone, ground, min 33% calcium	1.0	1.0	1.0
Trace mineralized salt with selenium	0.5	0.5	0.5
Total	100.0	100.0	100.0
Chlortetracycline or oxytetracycline (mg/kg)	15–25	15–25	15–25
Vitamin A (IU/kg)	500.0	500.0	500.0
Vitamin D (IU/kg)	50.0	50.0	50.0
Vitamin E (IU/kg)	20.0	20.0	20.0

[1]Adapted by the author from *Nutrient Requirements of Sheep*, Sixth Revised Edition, NRC-National Academy of Sciences, p. 53, Table 12.

[2]Limestone will separate from whole grain, so a combination of protein supplement with 10% limestone may be top dressed on the whole grain. Equal parts of trace mineralized salt and limestone is an additional way to maintain adequate calcium intake and prevent urinary calculi.

[3]The addition of 0.25 to 0.50% ammonium chloride will minimize urinary calculi. Corn may be substituted for all the barley and oats. Weight gains are depressed when barley or oats exceed 40% of the ration.

TABLE 14-18
SOME EXCELLENT CREEP RATIONS (AS-FED BASIS)[1]

	Unpelleted		Pelleted	
	First 2 Months	2 Months to Market	First 2 Months	2 Months to Market
	(%)	(%)	(%)	(%)
Ground corn	80	60	40	50
Ground oats		20	15	—
Soybean meal	20	10	20	10
Alfalfa hay	—	—	10	35
Bran		10	10	10
Molasses	—	—	5	5
Trace mineral salt	0.5	0.5	0.5	0.5
Limestone	1.0	1.0	1.0	1.0
Antibiotic . . . (mg/lb)[2]	50	20	50	15
Vitamin A . . . (IU/lb)	1,000	1,000	1,000	1,000
Vitamin D . . . (IU/lb)	200	200	200	200
Vitamin E . . . (mg/lb)	20	20	20	20

[1]The addition of 0.25 to 0.50% ammonium chloride will minimize urinary calculi.

[2]Chlortetracycline (Aureomycin) or oxytetracycline (Terramycin).

Feeding Directions:

1. Lambs should be started on creep feed about 10 days after birth. Although they will not consume significant amounts of feed until 3–4 weeks of age, the small amounts consumed at earlier ages are critical for establishing both rumen function and the habit of eating.

2. Feed high quality legume hay in a separate rack. Feed hay and creep ration twice daily to keep them fresh.

3. The amount of creep feed consumed by lambs 2 to 6 weeks of age is affected by the palatability of the ration (ration consumption and ration form) and the location and environment of the creep area. A well-bedded, well-lighted area located close to where the ewes congregate is preferred.

TABLE 14-19
SOME SIMPLE CREEP RATIONS (AS-FED BASIS)[1]

	Unpelleted		Pelleted	
	First 2 Months	2 Months to Market	First 2 Months	2 Months to Market
	(%)	(%)	(%)	(%)
Ground corn	49	89	64	59
Crushed oats	30	—	—	—
Soybean meal	20	10	20	10
Limestone	1.0	1.0	1.0	1.0
Trace mineral salt	0.5	0.5	0.5	0.5
Alfalfa	—	—	10	25
Molasses	—	—	5	5

[1]The addition of 0.25 to 0.50% ammonium chloride will minimize urinary calculi.

Feeding Directions:

Same as presented with Table 14-18, so, see the latter.

FEEDING ORPHAN ("BUMMER") LAMBS

Sheep producers estimate that about 10% of their lamb crop dies from starvation during the first week after birth. Some starvation results from newborn lambs sucking the scrotum and/or navel of another lamb. But most starved lambs are orphans (bummers) resulting from (1) the mother dying at lambing, (2) rejection by the mother, (3) the mother not being able to suckle the lamb because of mastitis or some similar problem, or (4) multiple births beyond the ewe's nursing capacity. Whatever the cause, the most satisfactory arrangement for the orphan is to provide a foster mother—to transfer (graft) the lamb to another ewe. The alternate to a foster mother arrangement is artificial rearing.

Fig. 14-14. Feeding orphan lambs. (Courtesy, *California Wool Grower*)

Observance of the following principles and practices will increase the chances of raising orphan lambs artificially:

1. **Give colostrum.** Colostrum makes for a good start in life. A newborn lamb needs 3.2 oz of colostrum per pound body weight during the first 18 hours after birth, according to the Modern Research Institute in Scotland. Colostrum contains antibodies which impart immunity to infections for the first few weeks of life. This is important because the lamb's own immune system does not develop until it is 3 to 4 weeks old. Colostrum may be stored for this purpose. If a ewe either drops a stillborn lamb or loses her lamb within a day of birth, she may be milked and the colostrum frozen. Then it can be warmed to 100°F and fed as needed. If ewe colostrum is not available, colostrum from a cow or a goat may be used, although it will not impart immunity to certain infections that are specific to sheep.

2. **Inject orphans.** When orphan lambs are placed in the nursery, inject them with (1) vitamins A, D, and E, (2) iron-dextran, and (3) selenium in selenium-deficient areas.

Also, enterotoxemia should be prevented. If the ewes were vaccinated with Type D toxoid prior to lambing, orphan lambs will receive colostral protection for 2 to 3 weeks; then, they should be vaccinated at 4 to 6 weeks of age. If the ewes were not vaccinated, the lambs should be vaccinated with the toxoid at 3 weeks of age and again at 5 weeks of age.

3. **Use milk replacer.** A number of commercially prepared milk replacers are on the market. Best results will be obtained by using a replacer containing 25 to 30% fat, 20 to 25% protein provided by spray-dried milk products, and not to exceed 30 to 35% lactose; with the milk replacer diluted, mixed, and fed according to the manufacturer's directions.

Where several orphans are being fed, up to 12 lambs of similar size and age may be grouped together in a small pen and self-fed from a multiple nipple container, allowing one nipple for each 2 to 4 lambs. When self-feeding, cool milk (50° to 60°F) should be fed because (1) it does not sour as quickly, and (2) the lambs are not apt to engorge on it. However, hand-fed lambs can safely be given warm milk fed twice daily.

Do not use cow's milk or calf milk replacer for lambs. They contain too much lactose (milk sugar) for lambs and will cause scours.

Because milk replacer is expensive, the liquid-feeding period should be as short as possible. Lambs can be successfully weaned from milk replacer at 18 to 28 days of age.

4. **Provide a good starter feed and water from day one.** From day one, orphan lambs should be provided access to a palatable dry ration to accustom them to eating dry feed and stimulate rumen development. This ration should be (1) palatable, (2) high in energy, (3) high in protein (22 to 24% crude protein on as-fed basis), (4) reinforced with minerals and vitamins, and (5) fed in finely ground (mash) form. A good starter ration follows:

Fig. 14-15. Orphan lambs being self-fed milk replacer from a multiple-nipple container. (Courtesy, Ralston Purina Co., St. Louis, MO)

Lamb Starter Rations:

Ingredients	(%)
Soybean meal (49% CP)	40.0
Ground corn	27.0
Alfalfa meal	15.0
Dextrose (corn sugar)	10.0
Fat (e.g. vegetable oil)	5.0
Limestone	2.0
Trace mineral salt	0.7
Vitamin premix	0.3
Total	100.0

Once the lambs have fully adjusted to the starter ration, they can be slowly switched onto the regular creep or grower ration. (See Tables 14-18 and 14-19.)

5. **Maximize sanitation, observation, and TLC.** The successful artificial rearing of orphan lambs necessitates that the caretaker maximize sanitation, observation, and TLC—tender loving care.

FEEDER LAMBS[2]

The primary objective of the sheep producer is that of producing milk-fat lambs suitable for slaughter at weaning time. Only when pasture is inadequate are lambs sold via the feeder route. Almost all feeder lambs come from the range area. Some range areas produce only a small percentage of lambs which are classed as feeders, whereas in other areas almost all the lambs must be sold as feeders because the vegetation is not sufficient to promote rapid growth and finishing. It is estimated that, for the range area as a whole, an average of at least 50% of all lambs produced in one year receive additional feed after they are removed from the range prior to slaughter.

■ **Field finishing**—This method of finishing lambs is somewhat comparable to the pasture finishing of cattle, except that a greater variety of feeds is used by lamb feeders. Field feeding requires relatively little labor and equipment, and the manure is dropped back on the land where it will do the most good. Death losses run higher than in drylot feeding because the feed consumption cannot be controlled.

The kind of field feeding varies from area to area and even between farms within the same locality. Most of the feeder lambs are shipped to these feeding areas in August and September at the time the lambs are nor-

---

[2]Many helpful suggestions for this section were received from the following authority: Dr. Clair Acord, Ph.D., Utah Wool Growers Assn., Salt Lake City, UT.

Fig. 14-16. "Sheeping down" corn on an Iowa farm. First the crabgrass, weeds, and lower corn leaves are eaten. (Photo by A. M. Wettach, Mt. Pleasant, IA)

Fig. 14-17. Lambs finishing on winter wheat in western Kansas. (Courtesy, Rufus F. Cox, Kansas State University, Manhattan)

mally weaned from range ewes. Usually these field-fed lambs are ready for market in November and early December, though it is not uncommon for a small percentage of thin lambs to be held back for additional drylot finishing of 30 to 60 days.

Throughout the Corn Belt, feeder lambs are usually used as scavengers during the early part of the field feeding process. Frequently, the lambs are pastured in the stubble fields or on the meadow until all these feeds are consumed, after which they are turned into corn fields.

In Kansas, Oklahoma, Nebraska, and Texas, thousands of lambs are finished primarily by fall pasturing of the wheat fields. In the Pacific Northwest, a limited number of lambs are finished by gleaning pea stubble.

■ **Drylot feeding**—Drylot feeding is, as the name indicates, feeding under restricted conditions. This may be either (1) shelter or barn feeding, or (2) open-yard feeding.

Large quantities of alfalfa and sugar beet byproducts are utilized in drylot lamb feeding. As it is often most economical to use a maximum of roughages, the feeding period may extend for as long as 4 to 5 months. Also, because of the desire to use a large proportion of roughage to grain, the lambs in these western feedlots are usually hand-fed twice daily rather than self-fed. In order to save labor, however, the practice of self-feeding is increasing. Lambs may be self-fed successfully, but it is recommended that the following precautions be taken in order to lessen the incidence of overeating disease: (1) good management, (2) vaccination against enterotoxemia, and (3) more roughage and less concentrate be used.

■ **Basic considerations in finishing lambs**—Although no rules of success are applicable to any and

Fig. 14-18. Drylot feeding of lambs.

all conditions, the following basic considerations in finishing lambs are worth noting:

1. In lamb-feeding operations, the purchase price of the lambs represents 60-70% of all costs. This indicates the importance of keeping death losses to a minimum.

2. Experienced feeders normally expect to lose about 1 to 2% of lambs on feed. This is about twice the loss that occurs in commercial cattle-feeding operations.

3. Lamb feeding is seasonal in nature, usually extending from August to about the following May. This seasonal condition is due to the fact that (a) suitable feeder lambs are not available until the late summer and fall months, and (b) following the growing and harvesting seasons, the feeders have available quantities of marketable and unmarketable feeds which may be utilized by lambs.

4. As in cattle feeding, feedlot gains are expensive, usually costing more per pound than the selling price on the market. Thus, a reasonable margin or difference between the cost and selling price per hundredweight is necessary.

5. Feed accounts for approximately 66% of the cost of finishing feedlot lambs, exclusive of the initial purchase price of the feeder lambs.

6. Though the situation varies according to the kind of feed and the age of animals, it requires about the following amounts of feed per pound gain:

Age	Lb Feed/Lb Gain
Preweaned lambs.	2.0 to 2.5
Early weaned lambs.	2.5 to 4.0
Late weaned lambs	6.0 to 8.0

7. In a 250-mile shipment, lambs will shrink about 5%. If properly fed, watered, and cared for en route, lambs may be shipped 1,500 to 2,000 miles without much greater shrinkage than this.

8. Most feeder lambs weigh between 60-80 lb when placed on feed and from 105-125 lb following a 90- to 120-day feeding period.

9. Wool is of importance in selecting feeder lambs because it has a bearing on their market value, the pelt being the most valuable slaughter byproduct.

10. Range feeder lambs are more plentiful than native feeders, thus allowing for greater selection; and usually they are more uniform and are less heavily infected with parasites.

11. Lambs are frequently fed on a contract basis, with many and varied agreements being used.

12. Wether lambs appear to make slightly more rapid gains than ewe lambs, but they do not finish quite so early as ewe lambs.

13. Where western lambs have undergone a long shipment immediately after being taken from their mothers, special care is necessary in starting them on feed. After rest following shipment, lambs are usually started on grain by feeding about ¼ lb per head daily. Gradually this allowance is increased so that the lambs are getting a full feed of about 2 lb of grain per head daily and about the same amount of hay when on full feed 4 weeks later.

14. A great variety of feeds can be used in lamb feeding. In general, the successful feeder balances out the ration by selecting those feeds which are most readily available at the lowest possible price.

15. Unless such extremely hard seeds as millet are included in the ration, it does not pay to grind feeds for finishing lambs.

FEED ALLOWANCE AND SUGGESTED RATIONS

Sheep rations vary with the section of the country, depending chiefly on available local feeds. Fortunately, many feeds of similar nutritive properties can be interchanged in the ration as price relationships warrant. This makes it possible at all times to obtain a balanced ration at the lowest cost.

Except at lambing time or when emergencies occur as a result of drought or inclement weather, western bands receive little supplemental feed. Even with farm flocks, a minimum of grain is fed to breeding animals. Grain feeding usually is limited to the latter part of gestation and to the lactation period prior to turning to pasture.

Tables 14-20 and 14-21 contain some rations that have been used by successful sheep operators in various sections of the country.

The rations in Table 14-21 are nutritionally adequate and balanced with respect to Ca:P and N:S ratios. The mineral and vitamin mixture given in Table 14-22 may replace the trace mineral salt in all the Table 14-21 rations.

TABLE 14-20
DAILY RATIONS FOR BREEDING EWES AT VARIOUS STAGES OF PRODUCTION

Ration Number	Moisture Basis[1] A-F (as-fed) M-F (moisture-free)	Hay[2]		Corn Silage		Haylage		Corn Straw		Stover (Stalks)		Grain[3]		Protein Supplement[4]	
		(lb)	(kg)	(lb)	(kg)	(lb)	(kg)	(lb)	(kg)	(lb)	(kg)	(lb)	(kg)	(lb)	(kg)
Maintenance															
1	**A-F**	**3.0**	***1.4***												
	M-F	3.3	*1.5*												
2	**A-F**			**6.0**	***2.7***									**0.20**	***0.09***
	M-F			6.7	*3.0*									0.22	*0.10*
3	**A-F**					**6.0**	***2.7***								
	M-F					6.7	*3.0*								
4	**A-F**							**3.0**	***1.4***					**0.40**	***0.18***
	M-F							3.3	*1.5*					0.44	*0.20*
Gestation, early (first 15 weeks)															
1	**A-F**	**3.5**	***1.6***												
	M-F	3.9	*1.8*												
2	**A-F**	**2.0**	***0.9***									**1.0**	***0.45***		
	M-F	2.2	*1.1*									1.1	*0.49*		
3	**A-F**	**1.8**	***0.8***									**0.6**	***0.27***	**0.20**	***0.09***
	M-F	2.0	*0.9*									0.7	*0.31*	0.22	*0.10*
4	**A-F**			**8.0**	***3.6***									**0.20**	***0.09***
	M-F			8.9	*4.0*									0.22	*0.10*
5	**A-F**					**7.0**	***3.2***					**0.20**	***0.09***		
	M-F					7.8	*3.5*					0.22	*0.10*		
6	**A-F**	**2.0**	***0.9***							**2.0**	***0.9***	**0.5**	***0.23***		
	M-F	2.2	*1.0*							2.2	*1.0*	0.6	*0.27*		
7	**A-F**	**1.0**	***0.45***							**2.0**	***0.9***	**0.5**	***0.23***	**0.30**	***0.14***
	M-F	1.1	*0.49*							2.2	*1.0*	0.6	*0.27*	0.33	*0.15*
Gestation, late (last 4 weeks): Add 0.5–1.0 lb *(0.23–0.45 kg)* grain per ewe daily to any of the above rations.															
Lactation															
1	**A-F**	**4.0**	***1.8***									**2.0–3.0**	***0.9–1.4***	—	—
	M-F	4.4	*2.0*									2.2–3.3	*1.1–1.5*	—	—
2	**A-F**			**10.0**	***4.5***							**1.5**	***0.7***	**0.25**	***0.11***
	M-F			11.1	*5.0*							1.7	*0.8*	0.28	*0.13*
3	**A-F**	**1.0**	***0.45***	**8.0**	***3.6***							**1.5**	***0.7***	**0.20**	***0.09***
	M-F	1.1	*0.49*	8.9	*4.0*							1.7	*0.8*	0.22	*0.10*
4	**A-F**					**8.9**	***3.6***					**2.0–3.0**	***0.9–1.4***		
	M-F					8.9	*4.0*					2.2–3.3	*1.1–1.5*		

[1]As-fed was calculated using an average figure of 90% dry matter. When using silages, roots, and other wet feeds, these feeds should be converted to a moisture-free basis and the ration calculated using the moisture-free data.

[2]Alfalfa hay, midbloom, preferred.

[3]Grain may consist of corn, barley, wheat, oats, and/or grain sorghum.

[4]Protein supplement may consist of soybean, cottonseed, linseed, sunflower, safflower, or rapeseed meal.

Feeding Directions:

1. These rations are formulated to meet the requirements of a 154 lb *(70 kg)* ewe in average condition, and are designed for hand-feeding. The daily feed allowance can be increased or decreased, depending on the actual size of the ewe and the body condition.
2. Some of these rations are deficient in calcium and/or phosphorus; therefore, a supplement containing 50% trace mineral salt (for sheep) and 50% dicalcium phosphate should be fed free choice. The consumption of 0.05 lb *(0.02 kg)* per sheep per day of this mixture will provide the amounts of calcium and phosphorus needed for maintenance and the first 15 weeks of gestation; and 0.10 lb *(0.05 kg)*/day will provide the needed Ca and P for late gestation and lactation. Vitamins A and E should be added to the salt-mineral mix when sheep are fed the wheat straw and corn stover rations.
3. Ewes should gain 15 to 25 lb *(6.8 to 11.4 kg)* during gestation. During early gestation (first 15 weeks), they should gain 0.05 lb *(0.02 kg)*/day. During late gestation (last 4 weeks), they should gain 0.5 lb *(0.23 kg)*/day. If, during the last 4 weeks of pregnancy, ewes are fed 0.5 to 1.0 lb *(0.23 to 0.45 kg)* grain per head daily and gain 8 to 15 lb *(3.6 to 6.8 kg)*, ketosis (lambing paralysis) can be prevented almost entirely.
4. During maintenance and early gestation, each ewe should have 14 in. *(36 cm)* of bunk feed space. In late gestation and during lactation, bunk space should be increased to 15 to 18 in. *(38 to 46 cm)*.

TABLE 14-21
GROWING—FINISHING RATIONS FOR LAMBS[1]

Ingredient	Moisture Basis[2] A-F (as-fed) M-F (moisture-free)	Rations Using Corn/Alfalfa Hay/Soybean Meal					Rations Using Milo/Cottonseed Hulls/Cottonseed Meal				
		1	2	3	4	5	1	2	3	4	5
		(%)	(%)	(%)	(%)	(%)	(%)	(%)	(%)	(%)	(%)
Corn grain (dent yellow)		31.0	41.5	51.7	63.0	73.3	—	—	—	—	—
Sorghum grain (milo)		—	—	—	—	—	19.5	32.7	46.2	60.7	73.7
Alfalfa hay (mature)		55.0	45.0	35.0	25.0	15.0	15.0	15.0	15.0	15.0	15.0
Cottonseed hulls		—	—	—	—	—	40.0	30.0	20.0	10.0	—
Soybean meal (solvent 44% CP)		7.0	6.5	6.0	5.5	5.0	—	—	—	—	—
Cottonseed meal (solvent 41% CP)		—	—	—	—	—	17.5	14.0	10.5	7.0	4.0
Molasses (cane)		6.0	6.0	6.0	5.0	5.0	6.0	6.0	6.0	5.0	5.0
Calcium carbonate		—	—	0.3	0.5	0.7	1.0	1.3	1.3	1.3	1.3
Trace mineral salt (sheep)		0.5	0.5	0.5	0.5	0.5	0.5	0.5	0.5	0.5	0.5
Ammonium chloride		0.5	0.5	0.5	0.5	0.5	0.5	0.5	0.5	0.5	0.5
Nutritional Content											
Dry matter (%)	**A-F**	**87.5**	**87.5**	**87.5**	**87.6**	**87.6**	**89.0**	**88.9**	**88.7**	**88.7**	**88.5**
	M-F	100	100	100	100	100	100	100	100	100	100
TDN (%)	**A-F**	**75.3**	**80.1**	**84.6**	**89.2**	**93.1**	**67.9**	**72.3**	**77.2**	**82.0**	**86.8**
	M-F	65.9	70.1	74.0	78.1	82.0	60.4	64.3	68.5	72.7	76.8
Net energy maintenance (Mcal/lb)	**A-F**	**0.80**	**0.86**	**0.91**	**0.98**	**1.04**	**0.71**	**0.76**	**0.82**	**0.88**	**0.94**
	M-F	0.70	0.75	0.80	0.86	0.91	0.63	0.68	0.73	0.78	0.83
Net energy maintenance (Mcal/kg)	***A-F***	***0.36***	***0.39***	***0.41***	***0.44***	***0.47***	***0.32***	***0.35***	***0.37***	***0.40***	***0.43***
	M-F	*0.32*	*0.34*	*0.36*	*0.39*	*0.41*	*0.29*	*0.32*	*0.33*	*0.35*	*0.38*
Net energy for gain (Mcal/lb)	**A-F**	**0.40**	**0.47**	**0.53**	**0.59**	**0.66**	**0.33**	**0.38**	**0.45**	**0.52**	**0.59**
	M-F	0.35	0.41	0.46	0.52	0.58	0.29	0.34	0.40	0.46	0.52
Net energy for gain (Mcal/kg)	***A-F***	***0.18***	***0.21***	***0.24***	***0.27***	***0.30***	***0.15***	***0.17***	***0.20***	***0.24***	***0.27***
	M-F	*0.16*	*0.19*	*0.21*	*0.24*	*0.26*	*0.13*	*0.15*	*0.18*	*0.21*	*0.24*
Crude protein (%)	**A-F**	**17.1**	**16.6**	**16.0**	**15.4**	**14.8**	**17.0**	**16.3**	**15.8**	**15.2**	**14.8**
	M-F	15.0	14.5	14.0	13.5	13.0	15.1	14.5	14.0	13.5	13.1
Protein bypass (%)	**A-F**	**42.3**	**44.1**	**45.8**	**47.9**	**49.7**	**44.6**	**47.6**	**51.0**	**54.7**	**57.9**
	M-F	37.0	38.6	40.1	42.0	43.5	39.7	42.3	45.2	48.5	51.2
Calcium (%)	**A-F**	**0.86**	**0.71**	**0.70**	**0.67**	**0.66**	**0.85**	**0.96**	**0.94**	**0.91**	**0.89**
	M-F	0.75	0.62	0.61	0.59	0.58	0.76	0.85	0.83	0.81	0.79
Phosphorus (%)	**A-F**	**0.29**	**0.30**	**0.31**	**0.31**	**0.32**	**0.37**	**0.37**	**0.36**	**0.36**	**0.36**
	M-F	0.25	0.26	0.27	0.27	0.28	0.33	0.33	0.32	0.32	0.32

[1]Adapted by the author from *The Sheepman's Production Handbook*, published by the Sheep Industry Development Program, Inc., Denver, CO, p. N-44, Table 13.

[2]As-fed was calculated using an average figure of 90% dry matter. When using silages, roots, and other wet feeds, these feeds should be converted to a moisture-free basis and the ration calculated using the moisture-free data.

Feeding Directions:

1. These rations can be fed once daily in troughs or bunks if there is capacity for a day's feed. They can also be self-fed if the feeders are designed to handle such feed without bridging.
2. Offering lambs a good quality hay for 1-3 days along with rations 1 or 2 (provided free choice) can be used to start lambs on feed.
3. About 3 in. *(7.6 cm)* of self-feeder or trough space must be provided per lamb for self-feeding and about 12 in. *(30.5 cm)* if hand-fed.
4. Gradually adapt the lambs to the higher energy rations by allowing 4-7 days on a ration before switching to the ration with the next higher energy level.
5. Complete mixing to prepare a uniform ration is important.
6. Lambs must not be allowed to be without feed even for a short period of time.
7. The mineral and vitamin mixture given in Table 14-22 may replace the trace mineral salt in all Table 14-21 rations.

The rations in Table 14-21 are nutritionally adequate and balanced with respect to Ca:P and N:S ratios. The mineral and vitamin mixture given in Table 14-22 may replace the trace mineral salt in all the Table 14-21 rations.

TABLE 14-22
MINERAL AND VITAMIN MIXTURE FOR LAMB RATIONS[1]

Ingredient	Lb/Ton	Kg/Ton	Contribution to Complete Ration[2]
Salt, plain fine mixing	1,729.613	*785.244*	0.43 salt
Sulfur, elemental[3]	200.00	*90.80*	0.05% S
Cobalt carbonate ($CaCO_3$)	0.087	*0.039*	0.1 ppm Co
Ethylenediamine dihydro-iodidie (EDDI)	0.10	*0.045*	0.2 ppm I
Manganese oxide (MnO)	10.30	*4.676*	20 ppm Mn
Zinc oxide (ZnO)	10.30	*4.676*	20 ppm Zn
Vitamin A[4]	17.60	*8.0*	600 IU/lb
Vitamin E[5]	32.00	*14.5*	10 IU/lb

[1]Adapted by the author from *The Sheepman's Production Handbook*, published by the Sheep Industry Development Program, Inc., Denver, CO, p. N-45, Table 14.

[2]Contribution to the complete ration when 10 lb *(4.5 kg)* of the mineral and vitamin mixture is added to 1 ton of complete lamb ration.

[3]In complete rations containing ammonium sulfate instead of ammonium chloride for prevention of urinary calculi, sulfur should not be added to the mineral and vitamin premix (0.5% NH_4SO_4 contributes 0.12% S to the ration).

[4]Contains 13,607,700 IU of vitamin A per pound.

[5]Contains 125,000 IU of vitamin E per pound.

FITTING RATIONS

In addition to being reasonably economical (mostly homegrown) and well-balanced, the ration for show and sale sheep must be palatable. Many feed combinations meet these specifications. The ration selected is usually determined by (1) the availability and price of feed in the area, and (2) the preference and judgment of the feeder.

Some suggested grain fitting rations are given in Tables 14-23 and 14-24. To each of these grain rations, good quality roughage should be added.

Fig. 14-19. A beautifully fitted lamb, Champion in the Omaha, Nebraska 4-H and FFA show. (Courtesy, American Hampshire Sheep Assn., Ashland, MO)

TABLE 14-23
FITTING CONCENTRATE MIX FOR SHOW LAMBS

Ingredient	Percent
Cracked corn	50
Whole or rolled oats	35
Soybean meal	10
Molasses	4
Mineral (limestone/sheep salt-mineral mix)	1
Total	100

TABLE 14-24
RATIONS FOR FITTING YEARLING AND MATURE SHEEP

Ingredient	Ration Number 1		2		3		4	
	(lb)	*(kg)*	*(lb)*	*(kg)*	*(lb)*	*(kg)*	*(lb)*	*(kg)*
Barley, rolled	—	—	40	*18.2*	—	—	10	*4.5*
Corn, cracked	—	—	—	—	40	*18.2*	—	—
Oats, rolled	50	*22.7*	40	*18.2*	40	*18.2*	60	*27.2*
Peas (split)	40	*18.2*	—	—	—	—	10	*4.5*
Protein supplement[1]	—	—	10	*4.5*	10	*4.5*	10	*4.5*
Wheat bran	10	*4.5*	10	*4.5*	10	*4.5*	10	*4.5*
Total	100	*45.4*	100	*45.4*	100	*45.4*	100	*45.4*

[1]Cottonseed, linseed, rapeseed (canola), soybean, and/or sunflower meal.

Lambs (either creep fed or weaned lambs that are being fitted for show) will eat about 2.5 lb of Table 14-23 fitting concentrate, per head daily when on full feed.

Yearlings will eat about 3 lb per head daily of one of the rations shown in Table 14-24 when on full feed, whereas mature sheep will eat about 3.5 lb of one of these rations.

Exhibitors prefer to feed steam rolled oats and barley and nutted (pea-sized) old process, linseed meal.[3] Corn is usually cracked or coarsely ground, and peas are split or cracked. When pastures are not available, alfalfa is the most popular hay. But any good legume is quite satisfactory.

In fitting animals for show or sale, most successful shepherds feed a limited quantity of sliced carrots, cabbage, mangels (stock beets), rutabagas (swedes), or turnips. These succulent feeds are highly relished by sheep and appear to help their digestion and general thrift.

The following points also are pertinent in feeding sheep:

1. All classes and ages of sheep should be al-

[3]Among experienced shepherds, old-process linseed meal is especially popular for fine-wool sheep because of its conditioning effect on the fleece.

lowed free access to a double compartment mineral box, with loose salt in one compartment and a mixture of one-third salt and two-thirds steamed bone meal, or other suitable mineral, in the other.

2. Unless grains are unusually hard, they need not be ground for sheep. The animals prefer to do their own grinding, and the feeds are no more effectively utilized when ground.

FEEDING AS AN ART

The feed requirements of animals do not necessarily remain the same from day to day or from period to period. The age and size of the individual animals; the kind and degree of activity; the climatic conditions; the kind, quality, and amount of feed; the system of management; and the health, condition, and temperament of the individual animals are all continually exerting a powerful influence in determining the nutritive needs. How well the feeder understands, anticipates, interprets, and meets these requirements usually determines the success or failure of the ration and the results obtained. Although certain principles are usually followed by all good feeders, no book knowledge or set of instructions can substitute for experience and born livestock intuition. Skill and good judgment are essential to the feedlot. Indeed, there is much truth in the adage that "the eye of the master fattens the animals."

Several points, however, deserve emphasis:

1. **Amount to feed.** Prolonged underfeeding is to be avoided, since there is a loss in body weight and condition. Overfeeding is also undesirable, being wasteful of feeds and creating a health hazard. When overfeeding exists, there is usually considerable leftover feed and wastage, and there is a high incidence of bloat, lactic acidosis, scours, and even death.

2. **Frequency, regularity, and order of feeding.** In general, finishing lambs are fed twice daily. With animals that are being fitted for show, where maximum consumption is important, it is not uncommon to find three or even four feedings daily. When self-fed, animals eat at more frequent intervals, though they consume smaller amounts each time.

Sheep learn to anticipate their feed. Accordingly, they should be fed *with great regularity*, as determined by a timepiece. During warm weather, they will eat better if the feeding hours are early and late, in the cool of the day.

Unless a complete, mixed ration (hay and grain combined) is fed, the grain ration is usually fed first,

TABLE
NUTRITION-RELATED

Disorder	Cause	Symptoms and Signs (or age group most affected)	Distribution and Losses Caused By
Bloat (Tympany)	Most common on lush legume pastures. Incidence on wheat pasture has been increasing in recent years, Pasture bloat is a frothy bloat caused by interaction of several factors—plant, animal, and microbial, Soluble plant proteins play a prominent role in permitting stable froth formation. Genetic tendency or physiological abnormality.	First observed as distention of paunch on left side in front of hipbone, This is followed by distention of right side, protrusion of anus, respiratory distress, cyanosis of tongue, struggling, and death if not treated.	Widespread, although some areas appear to have more bloat than others. Often results in death.
Ketosis (Pregnancy disease)	A metabolic disorder, thought to be a disturbance in the carbohydrate metabolism. May involve adrenal insufficiency.	In ewes and goats, ketosis or pregnancy disease generally strikes during last 2 weeks of pregnancy. Usually affected ewes are carrying twins or triplets. Symptoms include grinding of teeth, dullness, weakness, frequent urination, and trembling when exercised—with the final stage being complete collapse, followed by death in 90% of the cases.	Worldwide. Ketosis or pregnancy disease of sheep affects farm flocks more than range bands, the losses in the former sometimes being as high as 25%.

with the roughage following. In this manner, the animals eat the bulky roughages more leisurely.

3. **Abrupt feed changes.** Sudden changes in diet are to be avoided, especially when changing from a less-concentrated ration to a more-concentrated one. When this rule of feeding is ignored, digestive disturbances result, and sheep go off feed. In either adding or omitting one or more ingredients, the feeder should make the change gradually. Likewise, caution should be exercised when sheep are turned out to pasture or when they are transferred to more lush grazing. If it is not convenient to accustom them to the new pasture gradually, they should at least be well filled with hay (or with the former pasture) before they are turned out.

4. **Selection of feeds.** In general, the successful feeder balances out the ration through selecting those feeds which are most readily available at the lowest possible cost. In addition, consideration is given to supplying quality proteins, the proper minerals, and the necessary vitamins. Attention is also given to the laxative or constipating qualities of feeds and the palatability of the ration. Furthermore, the relation of the feeds to the quality of product produced should not be overlooked.

5. **Attention to details.** The successful sheep producer pays great attention to details. In addition to maintaining the health and comfort of the animals and filling their feed troughs, the producer should consider their likes.

It is important to avoid exercising the animal excessively, which results in loss of energy through unnecessary muscular activity. Rough treatment, excitement, and noise usually result in nervousness and inefficient use of feed. Finishing lambs should not be required to exercise any more than is deemed necessary for the maintenance of health.

NUTRITION-RELATED DISORDERS

Besides specific nutrient deficiencies and toxic levels of some microminerals, which are discussed in Chapter 5, there are some other feed and management aspects of great importance in sheep production. Generally, these may be classed as nutrition-related disorders. Foot rot (foul foot) and other contagious diseases, along with (1) parasites of sheep and goats, and (2) immunity, disinfectants, poisonous plants, National Animal Poison Control Center, federal and state regulations relative to disease control, and Harry S. Truman Animal Import Center are covered in Chapter 8, Sheep and Goat Health, Disease Prevention, and Parasite Control, of this book.

14-25
DISORDERS

Treatment	Control and Eradication	Prevention	Remarks
Time permitting, severe cases of bloat should be treated by a veterinarian. Puncturing of the paunch should be a last resort. Mild cases may be home treated by (1) keeping the animals on their feet and moving, and (2) drenching either with (a) vegetable oils, or (b) poloxalene (Therabloat®).	When there is high incidence of bloat, it may be desirable to change the feed. Where legume bloat is encountered, use poloxalene (Bloat Guard®) according to manufacturer's directions.	The incidence is lessened by (1) keeping animals off straight legume pastures, (2) feeding dry forage along with pasture, (3) keeping animals from getting a rapid fill from an empty start, (4) keeping animals continuously on pasture after they are once turned out, (5) keeping salt and water conveniently accessible at all times, and (6) keeping animals off frosted pastures. Use poloxalene (Bloat Guard®), a nonionic surfactant, according to manufacturer's directions for the control of legume bloat.	Legume pastures, alfalfa hay, and barley appear to be associated with a higher incidence of bloat than many other feeds. Legume pastures are particularly hazardous when moist, after a light rain or dew.
4 oz *(118 ml)* of propylene glycol, given orally twice daily.	Avoid feeding animals that are in early pregnancy so liberally that they become obese. Feed rather liberally in the last 6 weeks of pregnancy.	Ensure adequate feed intake in late pregnancy. Avoid any changes in feeding or management that might reduce the plane of nutrition or result in stress.	The clinical findings are similar in the case of affected sheep and cattle, but it usually strikes ewes just before lambing, whereas cows are usually affected within the first 1-6 weeks after calving.

(Continued)

TABLE 14-25

Disorder	Cause	Symptoms and Signs (or age group most affected)	Distribution and Losses Caused By
Lactic acidosis (Founder)	Engorgement on such feeds as barley, wheat, corn, rye, green whole corn, sugar beets, mangoes, potatoes and fruits by animals not accustomed to these feeds. Rapid change from a high-roughage diet to a high-concentrate diet.	Within 1–3 days after engorgement, the affected sheep develop depressed appetite, depression and weakness, rumen immobilization, and dehydration. As condition progresses, diarrhea, incoordination, acute laminitis, coma, and death.	Occurs in all breeds, sexes, and ages. More common in farm flocks than range or pasture bands. Occurs in all sheep-producing countries in temperate climates. High incidence in western, midwestern, and southwestern states where lamb-fattening industry has been developed.
Lambing sickness (Hypocalcemia, milk fever, transport tetany)	Decline of calcium levels in the blood from 8–12 mg/100 mi to 3–6 mg/100 mi. Drop in calcium levels possibly caused by (1) calcium-deficient diet, (2) sudden change from dry to lush pastures, (3) fasting 2 to 6 days plus transporting, (4) exercise such as driving, (5) late pregnancy and early lactation, and (6) other stressful situations.	Stiff and uncoordinated, spraddle stances, muscular tremors, weakness, apprehension, and rapid breathing. Eventually, ewes go down, develop paralysis, pass into coma, and die.	Occurs in all breeds and sexes beyond weaning age, though primarily in fat ewes during last 6 weeks of pregnancy and first week of lactation. Highest incidence in Australia, New Zealand, the United States, Great Britain, and South Africa.
Mycotoxins (Toxin-producing molds; e.g., *Aspergillus flavus, Penicillium cyclopium, P. islandicum*, and *P. palitans*)	Aflatoxin (most studied of the group) associated with peanuts, Brazil nuts, silage, corn and most other cereals, hay, and grasses. The mold can produce toxic compounds on virtually any food (even synthetic) that will support growth. While aflatoxin appears to cause most of the problem, it is not the only mycotoxin to be feared. Other mycotoxins are being studied.	Molds affect sheep in a variety of ways, from decreased production to sudden death. Usually, the first sign is toss of appetite and weight. A few animals will abort.	Widely distributed throughout the world. Generally, sheep and other ruminants appear to tolerate higher levels of mycotoxins and longer periods of intake than simple-stomached animals. In addition to the effect of mycotoxins on the animals' health, milk is contaminated by the residues or mycotoxins, or their metabolic products.
Nitrate poisoning (Oat hay poisoning, cornstalk poisoning)	Forages (vegetative part) of most grain crops (oats, wheat, barley, rye, corn, sorghum), Sudangrass, and numerous weeds, especially (1) when plants are under stress such as drought, insufficient sunlight, or after spraying with weed killer (herbicide); or (2) following heavy nitrate fertilization of soils (commercial, green manure crop, barnyard manure). Some nitrate may be formed after forage is stacked. Inorganic nitrate or nitrite salts, or fertilizer left where animals have access to them, or where they may be mistaken for salt. Pond or shallow well into which surface runoff from barnyard or well-fertilized soil might drain.	Accelerated respiration and pulse rate; diarrhea; frequent urination; loss of appetite; general weakness; trembling and staggering gait; frothing from the mouth; lowered milk production; abortion; blue color of the mucous membrane, muzzle, and udder due to lack of oxygen in blood; death within 4½–9 hours after nitrate consumption. A rapid and accurate diagnosis of nitrate poisoning may be made by examining blood. Normal blood is red and becomes brighter when exposed to air, whereas blood from cows toxic with nitrates is brown due to formation of the methemoglobin. Nitrates oxidize ferrous hemoglobin (oxyhemoglobin) to ferric hemoglobin (methemoglobin), which is not an efficient oxygen transporter. Animals essentially suffocate from lack of oxygen in tissues, When three-fourths of the oxyhemoglobin is converted to methemoglobin, the animals will die.	Excessive nitrate content of feeds is an increasingly important cause of poisoning in farm animals, due primarily to more and more high nitrogen fertilization. But nitrate toxicity is not new, having been reported as early as 1850, and having occurred in semi-arid regions of this and other countries for years.

NUTRITION-RELATED DISORDERS

Treatment	Control and Eradication	Prevention	Remarks
Effective only during early stages. Administer (1) 0.3-1.0 g of tetracycline or 500,000 units of penicillin, (2) mineral oil and antiferment into rumen, and (3) a solution of bicarbonate into rumen or blood to restore acid-base balance.	Keep animals away from grain and other highly fermentable feeds. Gradual stepwise change in ration of roughage to high concentrate over 10 days or more.	Appropriate management (see Control and Eradication).	May be confused with urea poisoning, urinary calculi, bloat, or foot rot.
Prompt response to intravenous or subcutaneous administration of 50-100 ml of 20% calcium gluconate.	Avoid letting animals get into calcium-depressing situations described under Cause. Supplement calcium-deficient diets with good sources of calcium.	Same as Control and Eradication.	Despite years of research, lambing sickness is not completely understood. May be confused with grass tetany (hypomagnesemia), ketosis, or enterotoxemia.
Remove the source of the mold. Animals suffering from molds frequently respond to vitamin B injections. Iron therapy may be helpful, since hemorrhaging is a frequent problem.	The prime cause of aflatoxin is moisture; hence, proper harvesting, drying, and storage are important factors in lessening contamination and toxin production. Propionic and acetic acids, and sodium propionate, will inhibit mold growth; hence, their use in preserving high-moisture grains is encouraged.	Same as Control and Eradication.	Certain molds produce toxins, or mycotoxins. Ultraviolet irradiation and anhydrous ammonia under pressure will reduce the toxicity of aflatoxins and, if continued long enough, will deactivate them entirely. Not all toxins are harmful. For example, zeranol is being commercially produced as a growth-promotant hormone for sheep and cattle.
A 2% solution of methylene blue (in a 5% glucose or a 1.8% sodium sulfate solution) administered by a veterinarian intravenously at the rate of 100 ml/1,000-1b *(22 ml/100-kg)* liveweight. The Missouri Station recently reported that yeast cultures were effective in preventing nitrate poisoning. The recommendation: 1.6 oz *(45.4 g)* per day for lambs.	(See Prevention.)	More than 0.9% nitrate (dry basis) may be considered as potentially toxic. Analyze feed when in doubt, by using a simple test to detect presence of nitrates (qualitative); if present, follow with a quantitative test to determine how much is present. Nitrate poisoning may be reduced by (1) feeding high levels of grains and other high-energy grains and other high-energy feeds (molasses) and vitamin A, (2) limiting the amount of high-nitrate feeds, (3) ensiling forages which are high in nitrates, (fermentation reduces some nitrates to gas, but care must be taken to avoid nitric oxide and nitrogen dioxide released in early stages of fermentation) and to avoid feeding until they have been in storage for 3-4 weeks.	Nitrate form of nitrogen does not appear to cause the actual toxicity. During digestion, the nitrate is reduced to nitrite, a far more toxic form (10-15 times more toxic than nitrates). In sheep and cows this conversion takes place in the rumen. Lethal dose varies with (1) nutritional state, size, and type of animal; and (2) the consumption of feed other than nitrate-containing material. Methods of reporting nitrates (dry basis) in rations in relation to death losses follow: Potentially Lethal Levels (%) (ppm) Nitrate (NO_3) . . . . . over 0.9 — 9,000 Nitrate nitrogen (NO_3N) . . . . over 0.21 — 2,100 Potassium nitrate (KNO_3) . . . . over 1.5 — 15,000

(Continued)

TABLE 14-25

Disorder	Cause	Symptoms and Signs (or age group most affected)	Distribution and Losses Caused By
Osteomalacia	Inadequate phosphorus (some times inadequate calcium). Lack of vitamin D in confined animals. Incorrect ratio of calcium to phosphorus.	Phosphorus-deficiency symptoms are depraved appetite (gnawing on bones, wood, or other objects; or eating dirt); lack of appetite, stiffness of joints, failure to breed regularly, decreased milk production, and an emaciated appearance. Calcium-deficiency symptoms are fragile bones, reproductive failures, and lowered lactations. Mature animals most affected. Most of the acute cases occur during pregnancy and lactation.	Southwestern U.S. is classed as a phosphorus-deficient area, where as calcium-deficient areas have been reported in parts of Florida, Louisiana, Nebraska, Virginia, and West Virginia.
Oxalate poisoning	Ingestion of halogeton and greasewood, which contain toxic levels of soluble oxalate, up to 30%. Oxalate interferes with energy metabolism and causes formation of crystals which damage the rumen and kidneys.	Dullness, loss of appetite, lowering of head, reluctance to follow the flock, and irregular gait are early signs. Advanced signs include drooling and frothing at the mouth, nasal discharge, progressive weakening, rapid and shallow breathing, coma, and finally death.	Plants grow in the arid and semi-arid saline regions of the West. Losses may be high if sheep are abnormally thirsty and receive water in a halogeton area because of indiscriminate grazing.
Rickets	Lack of calcium, phosphorus, or vitamin D; or an incorrect ratio of the two minerals.	Enlargement of the knee and hock joints, and the animal may exhibit great pain when moving about. Irregular bulges (beaded ribs) at juncture of ribs with breastbone, and bowed legs. Rickets is a disease of lambs and kids.	Worldwide. It is seldom fatal.
Salt sick (Cobalt deficiency)	Cobalt deficiency, associated with copper and perhaps iron deficiencies.	Loss of appetite, emaciation, depraved appetite, scaly skin, listlessness, and lack of thrift.	Australia, Western Canada, Florida, New Hampshire, Michigan, Wisconsin, New York, and North Carolina. On sandy soils.
Urinary calculi (Gravel, stones, water belly, urolithiasis)	Unknown, but it does seem to be nutritional. Experiments and experiences have shown a higher incidence of urinary calculi when there is (1) a high-potassium in take, (2) a high-phosphorus–low calcium ratio (from the standpoint of preventing urinary calculi, the Ca:P ratio should be about 2:1), (3) a high-silica content in the ration, or a high proportion of high-silica grains and forages, such as native grasses, wheat straw, sugar beet leaves or pulp, sorghums, and cottonseed meal. A deficiency of vitamin A may be a contributing factor.	Frequent attempts to urinate, dribbling or stoppage of the urine, pain, and renal colic. Usually only males affected, the females being able to pass the concretions. Bladder may rupture, with death following. Otherwise, uremic poisoning may set in.	Worldwide. Affected animals seldom recover completely.

QUESTIONS FOR STUDY AND DISCUSSION

1. In what ways may a sheep producer take practical advantage of the fact that sheep consume a higher proportion of forages than any other class of livestock?

2. What is the economic importance of feed for sheep?

3. Why should margins of safety be provided over and above the NRC requirements?

4. Explain the difference between nutrient requirements and nutrient allowances.

NUTRITION-RELATED DISORDERS

Treatment	Control and Eradication	Prevention	Remarks
Select natural feeds that contain sufficient quantities of calcium and phosphorus. Feed a special mineral supplement or supplements. If this disease is far advanced, treatment will not be successful.	(See Treatment.)	Feed balanced rations, and allow animals free access to a suitable phosphorus and calcium supplement. Increase the calcium and phosphorus content of feed through fertilizing the soils.	Calcium deficiencies are much more rare than phosphorus deficiencies in sheep.
None effective.	Careful range management and livestock management practices.	Sheep should have both water and good feed before they enter a halogeton area. Supplement with grain, alfalfa pellets, or other pellets during stress periods of trailing or trucking. Careful range management practices can reduce halogeton since it competes poorly with other vegetation.	Fasted animals are more apt to consume a toxic dose, and the toxic dose is much smaller for a fasted animal. After sheep have been grazing halogeton for 2-3 days, the lethal dose is increased by nearly 50%. Poisoning of sheep from the oxalate in greasewood is not as common.
If the disease has not advanced too far, treatment may be successful by supplying adequate amounts of vitamin D, calcium, and phosphorus, and/or adjusting the ratio of calcium to phosphorus.	(See Prevention.)	Provide (1) sufficient calcium, phosphorus, and vitamin D, and (2) a correct ratio of the two minerals.	Rickets is characterized by a failure of growing bone to ossify, or harden, properly.
Provide 0.2–0.5 oz cobalt salt/100 lb *(12.5–31.3 g/100 kg)* of salt—or feed a suitable trace mineral supplement.	(See Prevention.)	Mix 0.2–0.5 oz of cobalt chloride, cobalt sulfate, or cobalt carbonate/100 lb *(12.5–31.3 g/100 kg)* of either (1) salt, or (2) whatever mineral mix is being used.	Cobalt is needed especially for rumen microbial synthesis of vitamin B-12.
When calculi develop, it may be advisable to dispose of the affected animals, since treatments have limited success. Treatment: (1) incorporate 20% alfalfa in the ration, (2) administer muscle relaxants to help the passage of calculi from the bladder or (3) surgically remove the calculi; however, males will become nonbreeders after such an operation.	If severe outbreaks of urinary calculi occur in finishing iambs, it is usually well to dispose of them if they are carrying acceptable finish. For feedlot wether lambs, add to the ration 0.5% ammonium chloride (1 oz/head/day *[30 ml/head/day]*) or 0.9% ammonium sulfate.	Good feed and management appear to lessen the incidence. Feed salt at level equivalent to 4% of total diet in order to induce more water consumption. Feed ammonium chloride at level of 7 g per head per day to reduce the alkalinity of the urine. Keep high-phosphorus–low-calcium intake in the animals to a minimum; hence, increase calcium supplement as required. Provide adequate vitamin A, salt, and water.	Calculi are stonelike concretions in the urinary tract which almost always originate in the kidneys. These stones block the passage of urine, resulting in the condition commonly referred to as "water belly." The mineral deposits may be of variable sizes, shapes, and composition. Ammonium chloride (see Control and Eradication) appears to be the product of choice, However, ammonium sulfate may be used, at the rate of 1.7–2.0 oz/head/day *(50–59 ml/head/day)*. Add it to the ration when an outbreak occurs.

5. Of what value are feed substitution tables? What factors are usually considered in arriving at the comparative values of feeds listed in a feed substitution table?

6. List some factors which the sheep producer must consider before attempting to balance a ration by any method.

7. List the five methods of balancing rations, and give the advantages and disadvantages of each.

8. Select a specific class of sheep and formulate a balanced ration using those feeds that are available at the lowest cost. List the ration in terms of both as-fed and moisture-free bases. Justify your answer.

9. Will the least-cost ration always make for the greatest net returns? If not, why not?

10. What precautions should the sheep producer take when using computer-formulated rations?

11. Under what circumstances would you (a) buy a commercial sheep feed, or (b) home mix a feed for sheep?

12. What type of feed preparation would you recommend when barley is used in the ration?

13. Discuss the place of feed additives and implants for sheep.

14. What is the purpose of flushing? How would you flush a range band of ewes?

15. What are the most common nutrient deficiencies of range forage for sheep?

16. Step by step, how would you go about determining the kind of range supplement needed for sheep?

17. Range sheep operators generally agree that where range vegetation is so short as to require supplementation in excess of 0.5 lb per head per day, consideration should be given to moving the sheep either into a drylot or to a better grazing area. What prompts producers to set an upper limit of 0.5 lb of supplement per head per day?

18. Why is early weaning of lambs likely to increase?

19. Under what circumstances would you (a) recommend creep feeding lambs, or (b) recommend against creep feeding lambs?

20. Lambs may be self-fed successfully, provided (a) they are vaccinated against enterotoxemia (overeating disease), and (b) considerable roughage and/or other bulky feeds are incorporated into the ration. What advantages does self-feeding have over hand feeding?

21. Why is lamb feeding so seasonal in nature; more so than cattle feeding?

22. It requires 6 to 8 lb of feed to produce 1 lb of on-foot feeder lamb. How does this efficiency of feed utilization compare with finishing cattle and growing-finishing hogs?

23. Explain why feeding sheep is as much an art as a science.

24. Describe management practices that will help prevent (a) bloat, (b) ketosis, (c) lactic acidosis, (d) lambing sickness, (e) mycotoxins, (f) nitrate poisoning, (g) oxalate poisoning, and (h) urinary calculi.

SELECTED REFERENCES

Title of Publication	Author(s)	Publisher
Animal Science, Ninth Edition	M. E. Ensminger	Interstate Publishers, Inc., Danville, IL, 1992
Effect of Environment on Nutrient Requirements of Domestic Animals	National Research Council	National Academy Press, Washington, DC, 1981
Fundamentals of Nutrition	E. W. Crampton L. E. Lloyd	W. H. Freeman & Co., San Francisco, CA, 1978
Nutrient Requirements of Sheep	National Academy of Sciences, National Research Council, R. M. Jordan, University of Minnesota, Chairman of Subcommittee on Sheep Nutrition	National Academy of Sciences, Washington, DC, 1985
Sheepman's Production Handbook, The	American Sheep Industry Association	Sheep Industry Development Program (SID), Denver, CO, 1988
Stockman's Handbook, The, Seventh Edition	M. E. Ensminger	Interstate Publishers, Inc., Danville, IL, 1992

In addition to the above selected references, valuable publications on feeding and feeds for sheep can be obtained from:

1. Division of Publications
   Office of Information
   U.S. Department of Agriculture
   Washington, DC 20250
2. Your state agricultural college
3. Feed manufacturers and pharmaceutical houses

CHAPTER 15

Ewe with newborn lambs in a lambing cubicle. Lambing cubicles prevent lambs from becoming separated from the ewe shortly after birth and prevent "lamb stealing" by other ewes because the birth ewe only has a two-foot opening to defend against other ewes. (Courtesy, Harold Gonyou, University of Illinois, Urbana-Champaign)

SHEEP MANAGEMENT

Contents — *Page*

Although it is not possible to arrive at any overall certain formula for success, those sheep producers who have made money have paid close attention to the details of management. *Management is the art of caring for, handling, and controlling sheep.* In a general sort of way, good management principles are much the same. While no attempt has been made to cover all management practices, some sheep production management systems, and some facts relative to and methods of accomplishing certain sheep management practices, follow.

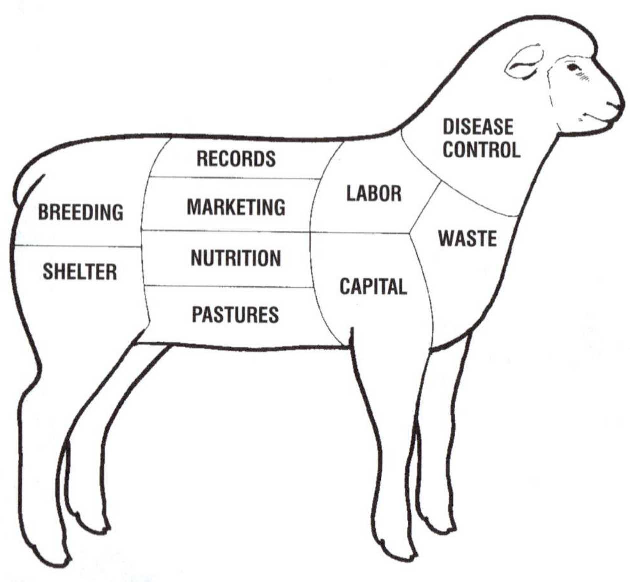

Fig. 15-1. Good management has many components. All are important— especially when missing or poorly performed.

SYSTEMS OF SHEEP PRODUCTION

It is logical that different systems of sheep production should exist in a country as large and diverse as the United States. This is fortunate, both from the standpoint of (1) the most efficient utilization of production factors, especially feeds, and (2) the meeting of consumer needs. For purposes of convenience and discussion, the different systems of sheep production are herein grouped into the following categories: (1) the farm-flock method, (2) the purebred-flock method, (3) the rangeband method, and (4) lamb feeding. Although the system of sheep production and the size of the enterprise may introduce new problems, the fundamental nature of sheep remains the same. For this reason, there is neither as much difference, nor as many secrets to success in different systems of sheep production, as some would have us believe. Essentially, success in any area or system depends upon (1) the maintenance of a healthy and highly productive flock that is economically managed, and (2) the advantageous marketing of the lamb and wool crop.

FARM-FLOCK METHOD

The farm-flock method is the common system of sheep production in the humid farming areas of the central, southern, and eastern United States. In general, it accompanies a diversified and intensive system of farming. Since it stresses market-lamb production, with wool production of secondary importance, most of the sheep in the farm-flock states are the meat type. Farm flocks range in size from a few head to several hundred head. The commercial farm flock is not under the care of a special shepherd or herder, but its handling is entrusted to a farm worker who has other responsibilities. For this reason, it is apt to be neglected as a minor enterprise, especially when the flock is very small and the caretaker has other assignments which are considered more important and remunerative.

During the grazing season, farm flocks usually compete with beef and dairy cattle for the use of permanent or seeded pastures. On many farms, sheep are considered scavengers, and as such they are given the assignment of keeping down the weeds and grass of fence rows, lanes, and draws. For best results, especially from the standpoint of greater efficiency of production and control of parasites, however, the farm flock should be given better pastures, and a system of pasture rotation should be followed. Also, confinement or partial-confinement production should be considered.

Fig. 15-2. The farm-flock method. Ewes and lambs are shown utilizing land not suitable for cultivation.

PUREBRED-FLOCK METHOD

For the most part, purebred flocks are comparatively small, and the vast majority of purebred breeders are located in the farm-flock states. There are, however, a sizable number of purebred flocks in the range states, some of which are quite large and are handled by range-band methods.

In general, the objectives of the purebred sheep breeder are the sale of rams to commercial sheep producers and the sale of both rams and ewes to established or new purebred breeders. Over a period of years, the most successful purebred breeders are keen students of commercial sheep production, (1) keeping ever in mind that the ultimate objective in all sheep production is the sale of market lambs and wool. and (2) gauging their type or ideal accordingly.

Most purebred flocks are given much closer attention than the average commercial flock. Frequently, a full-time caretaker is assigned. Also, in keeping with the general requirements for successful purebred production—regardless of the class or breed of farm animals—more attention is given to (1) the location of the farm, (2) individual records, (3) the careful study and selection of individual sheep rather than of flocks or bands, (4) matings that will produce animals with great inherent possibilities, (5) the maximum development of animals through feeding, and (6) different methods of advertising, including showing. In brief, the production of purebred sheep is a highly specialized business, and only a few sheep producers should attempt this system of production.

RANGE-BAND METHOD

Nearly 70% of the sheep of the United States are located on the western range, with the vast majority of these animals handled according to the range-band method. Each band is under the care of an experienced herder who moves the animals over a comparatively large area of unenclosed land. However, in the southwestern United States—in Texas, Oklahoma, and much of northern New Mexico—nearly all of the range is fenced. Under the latter conditions, the management of the sheep is very similar to that existing in Australia and New Zealand.

The relative emphasis on lamb and wool production in the range area varies according to the rainfall and vegetation. In the arid and semi-arid regions of the Southwest, where feed is not sufficient for satisfactory grass-finished lamb production, the production of wool is of greater relative importance than in those areas where the vegetation is more abundant. Even so, at the present time, lambs constitute the major source of income in these areas. In general, however, the sheep of this area are smaller and produce finer-quality wool, and practically all the lambs go the feeder route for finishing in more distant areas where feeds are more abundant.

Fig. 15-3. The range-band method, showing a band of ewes on a western range. Note herder and dogs.

In the semi-arid and sub-humid areas of the West and Northwest, where the ranges are more lush, many of the lambs go the slaughter route at weaning time, being finished entirely on milk and grass. Most operators in this area produce feeder lambs only when the range has been poorer than normal. Their income, therefore, is derived mainly from the sale of lambs, with the income from wool being of secondary importance. The sheep of this area are larger—large Rambouillets or long-wool X fine-wool crossbreds predominating—produce coarser wool, and yield lambs of more desirable shape and meat quality.

LAMB FEEDING

Lamb feeding is a highly specialized system of sheep production. In general, where pasture or range conditions are sufficiently good, the sheep producer strives to produce grass-finished lambs at weaning time. When the vegetation is not good or if for other reasons the lambs lack finish at weaning time, they are usually sold for further finishing. Seldom do producers attempt to finish out their own feeders. Most lamb feeders are large, specialized operators whose feedlots are located in close proximity to adequate and economical feed supplies, such as irrigated valleys, mill centers, or areas where winter wheat fields may be grazed. A full discussion of lamb-feeding methods and problems is contained in Chapter 14.

Fig. 15-4. Crossbred range lambs on feed in North Dakota.

SYSTEMS OF LAMB RAISING

In general, four systems of lamb raising are practiced in the United States—namely, (1) hothouse lambs, (2) Easter lambs, (3) spring lambs, and (4) lambs raised on grass.

HOTHOUSE LAMBS

These are lambs produced out of season—principally for the Boston and New York markets—and sold from December to April. The lambs are dropped in the fall or early winter months and are ready for market in 6 to 12 weeks, weighing 30 to 60 lb liveweight. As the ewes must be bred out of season for this type of market-lamb production, those individuals carrying a predominance of Dorset, Tunis, Rambouillet, or Merino breeding are usually used in the production of hothouse lambs. Generally, these ewes are mated to rams of the meat breeds.

This system represents a highly specialized business, and the person entering it must, first of all, have assurance of a market. Then, in addition, heavy milking ewes of the proper breeding must be secured. Ewes and lambs must be properly housed and skillfully fed in order to produce the fancy carcasses in demand. Hothouse lambs should be castrated; but peculiarly enough, on certain markets, notably the Boston market, the consumer demand is for lambs that have not been docked. The usual explanation of the demand for undocked hothouse lambs is that the city consumer associates the tailless condition with older sheep only, thus having considerable question about the age of a hothouse lamb should it be docked. Because of the high-moisture content of the young carcasses, the pelts are usually left on hothouse lambs until the carcasses are ready for delivery.

EASTER LAMBS

The Easter lamb trade has been a variable one, the demand being within a rather wide range both as to weight and quality of lambs. During the past several years, there has been a considerable Easter demand for light lambs, 20 to 30 lb in weight and not carrying much finish. At the same time, there is always a rather constant demand for heavier lambs of higher quality for which good prices are paid.

SPRING LAMBS

Young lambs marketed in the spring of the year, and prior to the first of July, are referred to as *spring lambs*. This class is not to be confused with lambs born the previous year but which are marketed the following spring, perhaps following a period in the feedlot. The first spring lambs usually come from the southwestern states and California.

After the first of July, the market classification of spring lambs no longer exists, animals of similar birth simply being known as lambs. Likewise, after this date those previously designated as lambs are yearlings (wethers) or yearling ewes.

LAMBS RAISED ON GRASS

The fourth general system of market lamb production is that of having the lambs arrive in the spring and producing them to marketable weight entirely on milk and grass. This is the most common system of lamb production throughout the United States and the nearly universal method on the range. Lambs handled in this manner must have the benefit of lush pastures, otherwise it will be necessary to market them via the feeder route. They also are generally more subject to parasites than earlier lambs, and they sell on a somewhat lower market; but the cost of production is relatively low.

PASTURE VERSUS CONFINEMENT SHEEP PRODUCTION

Sheep and grass have long been associated together. But altogether too often this relationship has lessened the efficiency of the sheep industry. It has resulted in (1) the relegation of sheep to the least productive areas of the farm, (2) small and inefficient flocks, (3) poor management, (4) too many parasites, and (5) unprofitable operations. Some sheep producers feel that these ills could be alleviated by raising sheep under intensive conditions and in confinement.

Without a doubt, parasite control ranks as the number one reason why people are interested in raising sheep in confinement. Also, interest in confinement sheep production has been accentuated by (1) the extent and success of confinement production in poultry, swine, and cattle; (2) scarcity of good pastures; (3) rise in land prices; (4) the growth of larger and more specialized farming operations; (5) the use of slotted floors and environmental control; and (6) the increase in multiple births and number of lambings per year.

This does not mean that the use of pasture for sheep production is antiquated. Rather, there now exist two alternatives for sheep producers, instead of just one; and the able manager will choose wisely between the two, or combine them.

Although more cost studies are needed, it appears that, on a per sheep basis, the total cost of confinement and pasture sheep production does not differ greatly. Despite this fact, a number of motivating forces have caused, and will continue to cause, more and more confinement sheep production. Among the factors **favorable** to confinement sheep production are:

1. It practically eliminates the problem of internal parasites.
2. It permits substituting automation for labor, especially in feeding and watering. Also, the added shelter and equipment costs of confinement production may be less than the cost of maintaining pasture fences and extensive water systems, plus the labor entailed in combating parasites, dogs, and bloat.
3. It frees land for more remunerative uses; for example, most Corn Belt farmers can make more money from growing corn and soybeans than from sheep pastures.
4. It makes for higher per acre yields of such crops as silage and dry forage than farmers can obtain by pasturing them with sheep.
5. It allows sheep producers to intensify their operations—to carry more sheep—without enlarging their acreage.
6. It results in more rapid gains and in lambs reaching market weight at an earlier age, thereby decreasing labor, risk, and production expense. Also, it results in improved carcass grade.
7. It requires less time in docking and castrating, vaccinating, controlling parasites, and loading for the market.
8. It makes for less distance to transport feed.
9. It facilitates sorting and lotting groups for size, uniformity, and single vs multiple births.
10. It facilitates environmental control.
11. It makes it possible to employ superior caretakers, because such operations are usually sufficiently large to warrant same.
12. It enhances the application of the latest in research, and the best in breeding, feeding, management, marketing, and business.
13. It lessens the chance of sheep being accorded the neglect of a minor enterprise, because it is a sizable business, and treated as such.
14. It generally makes for more favorable feed prices, due to purchase of quantity lots.
15. It makes it possible to feed ewes according to their requirements, rather than according to their appetites. This is important for, in general, pastures that will finish lambs provide more feed than ewes need.

On the other hand, the following factors are **favorable** to conventional sheep production on pasture, all or in part, over confinement production:

1. Lower building and equipment costs.
2. The use of pasture often makes for the most desirable land use, crop rotation, and soil conservation.
3. Pasture sheep operations are more flexible than confinement programs—an important consideration where renters are involved.
4. Pastures are especially valuable for the breeding flock—providing a combination of needed exercise, forage, and nutrients.
5. Fewer cases of wool eating, primarily caused by the stress and boredom of confinement, occur among sheep on pasture.
6. Sheep production on pasture does not require as high levels of skill and management as are necessary to make confinement production work

Also, the monetary value of pastures varies rather widely, being affected by (1) the quality and quantity of the forage; (2) the length of the pasture season; (3) the relative price of protein supplements, grains, and dry forages; and (4) the class and age of the sheep.

Without doubt, many sheep producers can advantageously combine pasture and confinement production—using pastures for the breeding flock and confinement production for young lambs. Therefore, sheep confinement systems will not put good pasture systems out of business; rather, confinement production, all or in part, will increase and replace more and more of the conventional practice of grazing all sheep throughout the pasture season.

SYSTEMS OF PASTURE UTILIZATION

For those sheep producers who wish neither (1) to continue the conventional practice of grazing ewes and lambs together until the lambs are marketed, nor

(2) to switch strictly to confinement raising, the following alternative pasture systems are being evaluated, both experimentally and by practical operators:

1. Keep ewes and lambs together in drylot at night, but place them on separate pastures during the day.
2. Keep ewes and lambs together in drylot during the day, but place them on separate pastures at night.
3. Keep ewes and lambs together in drylot at night, place ewes on pasture during the day, and keep lambs in the barn on creep feeders.
4. Keep ewes and lambs together in drylot during the day, place ewes on pasture at night, and keep lambs in the barn on creep feeders.
5. Keep lambs in confinement until early weaned (8 to 12 weeks of age), then either (a) place lambs on separate (apart from the ewes), clean pastures until ready for market, or (b) finish lambs in drylot.

Where pastures are to be utilized for sheep, the particular system decided upon will depend upon the lambing season, temperature, quantity and quality of available pasture, and parasite situation.

Fig. 15-5. Early weaned lambs. (Photo by J. C. Allen and Son, West Lafayette, IN)

GRAZING SHEEP AND CATTLE TOGETHER

Limited experiments, along with observations, indicate that it is more efficient to graze sheep and cattle together than to graze either specie alone. Joint grazing results in (1) the production of more total pounds of meat, or greater carrying capacity, per acre, and (2) more complete and uniform grazing than pasturing by either specie alone. This is attributed to the difference in the grazing habits of the two classes of livestock. Cattle tend to leave patches of forage almost untouched especially areas around urine spots and manure droppings. Also, cattle take larger bites and are less selective in their eating habits than sheep. Sheep tend to be selective of plant parts and will strip the leaves of plants. Also, sheep will graze many common weeds, even when good-quality grasses and legumes are abundant.

EARLY WEANING

Early weaning refers to the practice of weaning lambs earlier than usual and rearing them to market weight away from their mothers. Usually, it implies weaning at 8 to 12 weeks of age, or earlier, rather than the normal 5 months of age.

The age at which lambs are weaned is influenced by (1) when they are born (early or late in the year), (2) whether there is an accelerated lamb program (ewes lambing every eight months), (3) percentage of multiple births, (4) creep feeding, (5) availability of harvested feed and pasture, (6) parasite problems, (7) type of sheep raised, and (8) market prices and price outlook.

There is much interest in early weaning. Without doubt, the practice will increase in the future.

For successful early weaning, superior nutrition and management are essential; and the earlier the weaning age the more exacting these requirements.

The early weaning of lambs is covered more extensively in Chapter 14, Feeding Sheep.

SOME SHEEP MANAGEMENT PRACTICES

The almost innumerable sheep management practices vary widely between both areas and individual operators. In a general sort of way, the principles of good management of farm flocks and range bands are much alike. The main differences arise from the sheer size of the range enterprise, which means that things must be done in a big way. Without attempting to cover all management practices, the author presents some facts relative to—and methods of accomplishing—the following important sheep management practices.

DOCKING AND CASTRATING LAMBS

All lambs should be docked before they are 14 days of age. If ram lambs are to be castrated, it should be done when they are between 2 and 4 weeks of age. With strong, vigorous lambs, both operations may be performed at the same time, making for greater convenience and a saving in labor. When both operations are performed at the same time, castration should be done first. It is best to do this work early in the morning when the weather is not too cold and when the sun is shining brightly. Should weather conditions change to rain or snow soon after, the lambs should not be allowed to be out. With late lambs, a clean pasture is the ideal place to turn them after docking and castrating. With early lambs and when the work is done inside, the bedding should be clean and dry. Care should be taken not to run or otherwise excite the lambs prior to or immediately after docking and castrating.

There are a number of methods in use for docking and castrating, all of which are quite satisfactory if properly performed. Castrating may be done with a knife, a Burdizzo, or an elastrator. When cutting instruments are used, the hands should first be thoroughly washed with soap and water and rinsed in a good disinfectant. Likewise, all instruments should be thoroughly disinfected prior to the work and between operations. The lamb is usually held with its back to the assistant, who grasps the hind and front legs of the same side in one hand. When the operation is performed with a knife, the lower third of the scrotum is removed, exposing the testicles, which are then drawn out. Some sheep producers prefer to draw the testicles with the teeth, thus securing greater speed and keeping the danger of infection to a minimum. Whether the teeth or the thumb and forefinger are used, each testicle must be pulled from the scrotum slowly until the cord breaks. Then a 7% iodine solution should be applied to the cut edges of the scrotum.

In recent years, the elastrator, which is a bloodless method, has been used in docking and castrating lambs. It consists of placing a small, strong, rubber ring around the tail or scrotum when the lambs are only a few days old (see Fig. 15-7). The rubber ring stops the blood flow to the testicles and scrotum and they shrink and slough off in 2 to 3 weeks. As no wound is inflicted, infection seldom follows this method. The elastrator is the invention of A. O. Hammond, a sheep farmer of Blenheim, New Zealand.

Elastrators may also be used to make *short-*

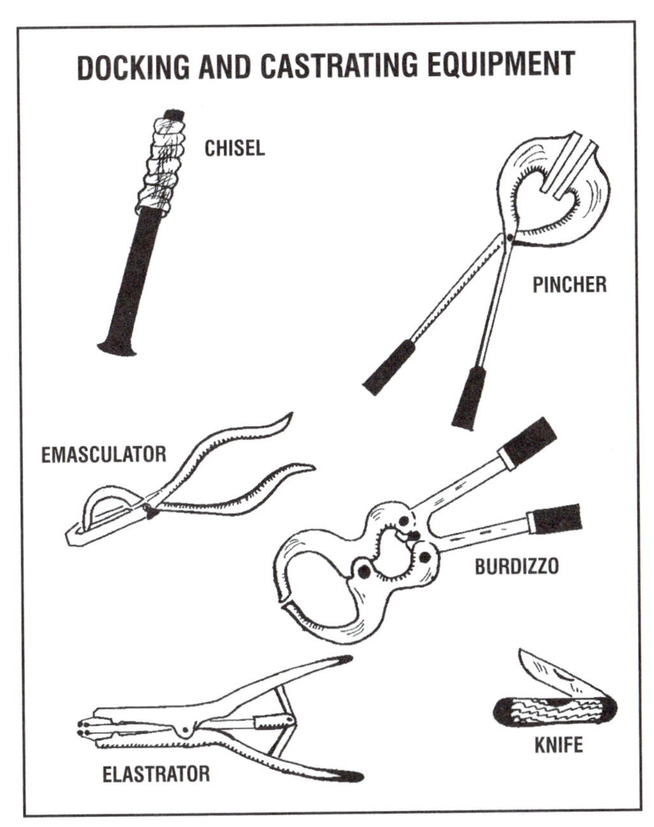

Fig. 15-6. Common instruments used for docking and castrating lambs. (Drawings by R. F. Johnson)

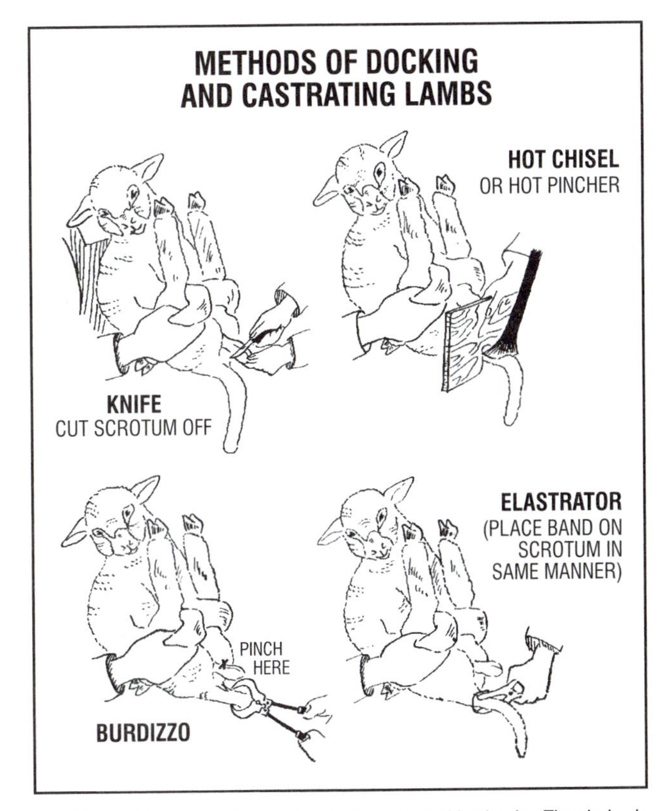

Fig. 15-7. Some methods of castrating and docking lambs. The choice is generally one of personal preference.

scrotum lambs. To do this, the testicles are pushed into the body cavity and the elastrator band is placed at the top of the scrotum, holding the testicles in the body cavity. After about two weeks the scrotum drops off but the testicles remain in the body cavity where the temperature is too high for sperm production but the male hormone testosterone is still produced. Short-scrotum lambs, like ram lambs, gain faster and yield a higher percentage of lean meat than wether lambs.

The Burdizzo, or emasculatome, can be used on larger or older lambs, where castration with a knife may be dangerous and may cause too much trauma. This method crushes the spermatic cord preventing the transport of sperm and reduces the blood supply to the testicles causing them to shrink. The scrotum remains intact.

Deciding whether or not to castrate is something that producers will have to determine. When lambs are to be marketed before reaching puberty, or around five months of age, a good case can be made against castration, including:

1. Ram lambs grow faster than wethers.
2. Ram lambs show a greater feed efficiency.
3. Ram lambs produce a carcass with less fat and more lean than do wethers.

On the other hand, depending on the operation, there may be some advantages to castrating, including:

1. Wether lambs are quieter than ram lambs; and castrating alleviates the hazard of ram lambs breeding some of the ewes.
2. Wether lambs carry more finish and have a higher dressing percentage than ram lambs.
3. Wether lambs generally bring a higher price than ram lambs after about five months of age, chiefly because ram lambs are discriminated against.
4. Wether lambs that carry insufficient finish for slaughter may go the feeder route, whereas feeders do not want ram lambs at any price.

As in castrating, there are a number of methods for docking lambs. The important thing, however, is to use one of them and to use it at the proper time. It may be done with a knife or a pair of shears, a hot iron (pincer or chisel), an emasculator, a Burdizzo pincer, or an elastrator. There is less loss of blood, and the danger of infection is not as great, with the use of the hot docking iron or chisel. However, the wound heals more slowly than when the operation is done with a knife. For most rapid healing when a heated instrument is used, the instrument should not be heated beyond a very dull red, and the tail should be severed rather quickly, avoiding any more burning than is necessary to prevent bleeding.

The cut is usually made about 1 in. from the body as measured on the underside of the tail, though the operator who is producing fancy show sheep may want to get it closer.

When the work is done with a heated instrument a 1-in. thick notched board over the tail next to the body protects the lamb from the hot irons (see Fig. 15-7). When a pair of shears or a knife is used, the skin should first be pressed toward the body before the tail is removed, leaving loose skin above the cut which will close over the wound. After docking, a 7% iodine solution should be applied to the dock. Excessive bleeding can be controlled by (1) pinching the dock for a few minutes, or (2) tying a string tourniquet above the dock for an hour.

It is important that a disinfectant be used with unheated instruments.

With small flocks, some sheep producers make a practice of tying a string or placing a rubber band around the tail prior to docking, preventing a loss of blood in this manner. If this is done, the string or band should be removed 3 or 4 hours later. For understandable reasons, however, this extra trouble is not practical with a range band.

Lambs can also be docked with an elastrator. This method is bloodless and the tail sloughs off in 1 to 2 weeks. However, the dead tissue in the tail may become a source of tetanus, and for this reason, some shepherds will cut the tails off after 24 hours.

The chief reasons for docking lambs are:

1. Docked lambs are cleaner.
2. Long tails interfere with breeding ewes.
3. Docked lambs and sheep present a more uniform and blocky appearance.
4. Feeders try to avoid long-tailed lambs. (This fact is of great importance to the western producer.)
5. The dressed carcass has a more attractive appearance.

CREEP FEEDING LAMBS

The practice of self-feeding concentrates to young animals in a separate enclosure away from their dams is known as creep feeding.

With farm flocks, creep feeding may be advantageous for the following reasons:

1. **To reach an early market and bring a higher price.** April to June lamb prices are usually highest, following which prices decline very sharply. This means that lambs born in February have only about 100 days in which to reach the top market price of May and June. The following data point up this situation:

Lambing Date	Days to June 1
Jan. 1	151
Jan. 15	136
Feb. 1	120
Feb. 15	105
Mar. 1	92
Mar. 15	77
Apr. 1	61

2. **To take advantage of extremely good feed conversion during the early life of the lambs.** Two to three pounds of added creep feed will produce 1 lb of gain during this early period of growth. Therefore, the added feed cost will increase the returns, depending on the market prices of top spring lambs.

3. **To maintain the finish of nursing lambs.** Unless the pasture is good, top lambs will not be produced without creep feeding; that is, the lambs will not acquire sufficient finish and marketing will be delayed.

4. **To avoid added handling.** Creep-fed lambs will go to market in 100 to 120 days, thus eliminating the need for worming, tagging, and handling lambs later in the season. Also, ram lambs may not need to be castrated.

Creep-fed lambs need to be started on feed early and carefully and be provided with a good ration. The subject of creep rations as such is fully covered in Chapter 14.

Fig. 15-8. A properly tagged (crutched) ewe. The wool has been shorn from around the udder, flank, dock, and eyes. Tagging is a valuable aid to successful lambing, nursing, and breeding. (Courtesy, California Wool Growers Association, Sacramento, CA)

TAGGING[1]

Tagging is the practice of cutting the dung locks off sheep from around the udders and docks. Usually this is done immediately prior to lambing and shearing, and again prior to the breeding season.

Usually the wool around the eyes of animals subject to wool blindness is clipped at the same time.

FOOT TRIMMING

Most ewes keep their feet worn off sufficiently. However, some need to have their feet trimmed to avoid lameness. The rams' feet should always be trimmed prior to the breeding season.

Fig. 15-9 shows a convenient way to hold a sheep—seated on the ground, with the head held between the legs of the handler—for feet trimming. Then the bottom of the exterior hoof wall which has overgrown the sole should be trimmed until it is level with the fleshy center portion of the toe. This should be done to the outer and inner portions of all toes, with the trimmer being careful not to cut too deeply. Any sheep showing signs of foot rot should be treated at this time. Long hoofs may require successive trimmings.

Fig. 15-9. Trimming sheep's feet, using pruning shears. (Photo by J. C. Allen and Son, West Lafayette, IN)

SHEARING

The production and handling of the wool crop is more fully covered in Chapter 9, but a discussion of

[1]In Australia. this practice is known as *crutching*.

certain items of special importance from the standpoint of good management will follow.

Generally, shearing is an annual event, though some flocks and bands in the South and Southwest are shorn twice each year. The shearing season is almost entirely determined by climate, starting early in the spring in the South and ending in June or July in the North. Where the animals are turned to pasture immediately after shearing and no shelter is available—the usual situation on the western range—cold rains or stormy weather at this time may result in excessive losses from pneumonia.

In addition to the problem of inclement weather in determining the time of shearing, consideration must also be given to the lambing season. Generally, the wool is removed before lambing, but there are good reasons for delaying under certain conditions. For example, in the Northwest where the production of grass-fat lambs is the primary objective and warm weather comes rather late, few shed-lambed ewes are shorn before lambing.

The main **advantages** of shearing before lambing include:

1. The lamb can nurse more easily without the obstruction of the fleece.
2. The wool is cleaner, with fewer taggy locks and a higher percentage of clean fleece weight.
3. Less wool is lost on the brush or other objects.
4. Usually with range bands, shearing may be done at a more convenient location, nearer to headquarters.

On the other hand, there are certain **disadvantages** to shearing before lambing, chief of which include:

1. The possible loss from exposure should inclement weather follow early shearing.
2. Possible abortions due to rough handling of the ewes.
3. Possible lower fleece weights.

Some operators have permanent-type shearing plants, whereas others rely on portable outfits. The permanent-type plant generally consists of an unpretentious shelter with a shearing floor, holding pens, counting pens, a larger *sweat pen*, facilities for handling the wool, and adjacent corrals of ample size to handle one band. Usually, only one band is corralled at a time in order to save feeding. The portable-type plant consists of the very minimum equipment, with canvas providing such cover as seems necessary. In general, in those areas where the production of grass-fat lambs is the primary objective, portable outfits are preferred, chiefly because the lambs suffer less setback when shearing is done right on the range. By contrast, with the permanent central type, it is generally necessary to trail the band a considerable distance for shearing, and, in altogether too many instances, the plant is located in an area where feed is scant.

When there is danger of either inclement weather or blistering from the sun, some operators take the added precaution of insisting that the sheep be shorn with thick combs or by hand shears, thus leaving a small amount of wool for protection. Most shearing is done by unionized professional shearers who travel from band to band, charging on a per head basis. Expert machine shearers will shear up to 200 head per day. In addition to the experience of the shearer, the speed of shearing is affected by the size of the sheep, the degree of wrinkles, and the density and condition of the fleece. On the whole, the professional shearers do excellent work, but occasionally they may have to be cautioned against too many ugly cuts or unnecessary rough handling.

Fig. 15-10. Shearing in a permanent-type shearing plant. (Courtesy, University of Wyoming)

DIPPING OR SPRAYING

Soon after the shear cuts heal, the entire flock or band, including lambs, should be sprayed or dipped to kill external parasites and to check skin diseases. The operation should take place on a bright, sunny day. After dipping, animals should be held on a drain platform for at least five minutes and allowed to drip. A discussion of spraying and dipping operations, together with the proper solutions to use, is found in Chapter 8.

DRENCHING

Drenching is a management procedure used to control and treat internal parasites. There are a variety of chemicals used for drenching and these are covered in Chapter 8. Equipment for drenching can be a 2- or 4-oz dose syringe with a 6-in. metal nozzle. Also, there are a number of automatic drench guns, each of which has a reservoir attached to the gun via a tube. The gun can be adjusted for various dosages and the reservoir speeds refilling the gun between animals.

In small flocks, sheep can be confined to small pens, which limits movement. Then, each sheep is caught, drenched, and marked. Where a large number of animals are to be drenched, sheep should be confined to a holding pen which funnels sheep into a running chute. Then as each animal comes through the chutes it is caught and drenched.

The actual drenching process proceeds as follows:

1. Do not excite or frighten the sheep.
2. Straddle the animal and hold securely.
3. Raise the sheep's head so that the nose is level.
4. Insert the nozzle of the syringe or gun back between the molars (grinders) and extend it back over the base of the tongue. If an automatic drenching gun is used, the longer curved tube allows placement further down the throat.
5. Discharge the drench, allowing the sheep to swallow. Care should be exercised to prevent forcing the drench into the lungs.
6. Keep the sheep relaxed.
7. Withdraw the nozzle of the syringe or gun.
8. Mark the drenched sheep with wool-marking chalk if sheep are confined to a pen.

GRAFTING LAMBS

A lamb may be orphaned when its mother dies or because its mother is not able to suckle it. The most satisfactory arrangement for the orphan is to provide a foster mother. The good shepherd will try to have every ewe raise a lamb. There may be a ewe that has just lost her lamb or a strong, healthy ewe with just one lamb. When a lamb dies at birth and it is desired to transfer, or graft, another lamb on the ewe, any one of four procedures is possible: (1) slime grafting, (2) wet grafting (3) skin grafting, or (4) fostering pens.

1. **Slime grafting.** The lamb to be transferred must be thoroughly saturated with the placental fluids and membranes from the new mother before she has completely identified her own new offspring. This means the grafting must be done immediately after birth.

2. **Wet grafting.** Wet grafting can be used when it is too late to use slime grafting. With this method, the ewe's own lamb and the extra lamb needing milk are completely immersed in a salt-water solution. Then the lambs are thoroughly rubbed together and the extra, or new, lamb may be rubbed with any placental membranes available. Since lambs may be rejected after this procedure, the new family is placed in a lambing pen and observed until the ewe establishes a firm mother-offspring bond with both lambs.

3. **Skin grafting.** When a ewe loses her own lamb, the dead lamb can be skinned and the pelt tied over the lamb to be adopted. The success of skin grafts varies but tying a ewe in a lambing pen for several days will help the adoption process. After the lamb has been adopted, the skin may be removed gradually, a piece at a time.

4. **Fostering pens.** Often a ewe with only one lamb but adequate milk for two lambs is not identified soon enough to use slime, wet, or skin grafting. When this is the case, a fostering pen may be used.

Fostering pens are similar in size to lambing pens, but each is provided with a stanchion to restrain the ewe. The stanchions are designed so the ewes can lie down or stand up as they desire, but their heads are held so they cannot turn to see or smell the lambs. However, the lambs are free to walk in front of the ewes.

Constant association of two lambs in the same environment and drinking the same milk will probably result in both lambs smelling alike. Thus, with dimmed visual memory and forced adoption to two nursing lambs that have the same smell, the ewe is not able to distinguish one lamb from the other. So she accepts both.

Establishing a good graft usually requires 4 or 5 days. The younger the lambs and the more recently the ewe has lambed, the less time required. Relatively old lambs may require 5 to 10 days. The ewes and lambs should be observed closely for acceptance behavior and thriftiness. As soon as a good relationship exists, the ewe and lambs should be placed in a trial jug or the ewe should be turned loose in the fostering pen where she is free to observe and suckle her lambs. The new family should be left in this trial situation for 1 or 2 days. If acceptance appears good, the ewe and lambs can be turned into a small mixing pen with two or three other ewes and their lambs. If no problems develop within a day or so, the new family can be managed the same as all other sheep. With good management, one can expect a success rate of about 85%.

When an orphan lamb cannot be transferred, it should be fed as directed in the section entitled "Orphan Lambs," in Chapter 14.

MARKING OR IDENTIFYING SHEEP

Proper identification is necessary to maintain good records on a flock.

The common methods of marking or identifying sheep are:

1. **Paint brands.** The indiscriminate use of insoluble paint on sheep should be avoided, for it will lower the value of the wool. Ordinary lead-base or tar paints will not come out in the normal wool-scouring operation. For this reason, the wool sorter must separate out and clip off these brands with shears. Small particles of paint that escape the sorter go through into the cloth and cause defects.

Despite these facts, practical sheep operators often find it necessary to brand sheep for one or more of the following reasons:

a. Identification of western sheep on ranges, especially public land, where the brands of different owners may get mixed.

b. Identification with a *buck brand* at breeding time.

c. Identification of ewes and lambs at lambing time.

The solution to the problem lies in using a brand which will satisfactorily serve for identification purposes, but scour out.

There are now on the market commercial wool-branding paints which will remain on the sheep for a year, but can be removed from the wool by normal scouring methods.

Likely, satisfactory scourable branding fluids will cost more than ordinary paint. However, it is costly to remove insoluble brands. Some of this saving will be passed back to the wool producer, or the manufacturer may pay a premium for wool upon which scourable branding fluids have been used.

2. **Ear marks.** Identification for sheep can be provided by ear marks, made with either a sharp knife or a regular ear notcher. Such marks are permanent and easily recognized, but unattractive.

Where individual identity is desired, as is necessary in a purebred flock, a definite value is assigned to each area location. With a commercial band the same kind of mark is administered to all animals. Few purebred breeders, however, use ear marks since it detracts from the appearance of sheep that are exhibited.

3. **Metal or plastic ear tags.** Most purebred sheep are provided with metal or plastic ear tags, and sometimes each animal has two. Where two tags are used, generally one of them is the individual or flock number assigned by the owner, whereas the other is the individual number assigned by the breed registry association.

Ear tags are easily attached, but some are easily lost. Also, they frequently rub or scratch the skin, thus making openings for infestation.

4. **Ear tattoos.** This method of marking consists of piercing the skin with instruments equipped with needle points which form letters or numbers. This operation is followed by rubbing indelible ink into the freshly pierced area. On dark-skinned animals, tattoos are difficult to read.

Table 15-1, Marking or Identifying Guide for Registered Sheep, summarizes the pertinent regulations of the sheep associations relative to marking or identifying.

5. **Marking the ram.** This subject is fully covered in Chapter 4.

TABLE 15-1
MARKING OR IDENTIFYING GUIDE FOR REGISTERED SHEEP

Breed	Association Rules Relative to Marking
Cheviot	All sheep for which application is made for registry must be marked with either (1) a metallic label in the ear, or (2) tattoo numbers. The metal label or tattoo must bear the breeder's or applicant's name or initials thereon. Numbers shall not be duplicated. Should any sheep be labeled with another breeder's label, it shall not be changed. If a label is lost, it must be promptly replaced with a duplicate of the original. Association ear labels will be furnished by the secretary as part of the registration charges, but there is a specific charge for duplicate labels.
Clun Forest	All sheep belonging to members of the society must bear identification tattoos. Individual sheep must be tattooed in the producer's flock letters in the sheep's right ear, and the sheep's flock number and year letter in the left ear.
Columbia	All flock numbers should be placed in the left ear. Leave the right ear blank for association tattoo and tag. Flock numbers are individual identification numbers. Each sheep should have a different flock number.
Corriedale	All sheep should be ear tagged with the owner's individual number of the sheep and the name or initials of the owner. The use of duplicate ear tags, one for each ear, or of tattoo is recommended.
Cotswold	Breeder must use private ear labels with name and number, which must be given when application is being made for registry. The association provides a metal ear tag for the registry number.
Debouillet	All sheep must be ear tagged by the inspector, preferably in the right ear. Ear tag identifies the flock number (all breeders have their own flock number), the year of birth, whether permanent registry, and the individual number.

(Continued)

TABLE 15-1 (Continued)

Breed	Association Rules Relative to Marking
Delaine-Merino	**American and Delaine-Merino Association:** Each registered animal will have two ear tags—one with the flock owner's number and one furnished by the association upon registration. **Black-Top Delaine-Merino Sheep Breeders' Association:** Each animal admitted to record shall have a number assigned to it, which shall correspond with the number of the ear tag furnished by the secretary.
Dorset	Sheep offered for registration must bear (in either ear) the owner's private flock record tag with number, and this number must be given on the application form. When an animal is approved for registration, an association tag is issued with certificate of registry. The latter tag must be inserted in the other ear. Tattoo marking is accepted, but the association tag must be used, also.
Finnsheep	Breeder's tag, carrying the name and initials of the owner of the dam when lamb was dropped, plus the lamb's individual number and tattoo, placed in left ear. Association number and ear tags will be furnished by association and shall be placed with tattoo in right ear.
Hampshire	When an animal is approved for registration, an individual ear tag (with association numbers) is forwarded to the owner, along with a certificate of registry. The association ear tag shall be inserted in the animal's ear in addition to the owner's private identification.
Karakul	There are no rules for marking individual sheep although they are registered individually. Each certificate of registration contains descriptions of the sheep with respect to ear type and size, tail type and size, and coloring.
Lincoln	A tattoo or metal ear tag may be used for the private or flock number. The association provides a loop-style metal ear tag for the registry number.
Montadale	The Montadale Association provides a standard ear tag for each animal accepted for registry. Each breeder is required to purchase flock tags with name, address, and consecutive numbers.
Oxford	Breeder must use private ear number, which must be given on the pedigree when making application for registry. Association ear tag number is furnished when animal is registered.
Panama	At the time the sheep are inspected for registration, the inspector must tattoo in the right ear all sheep that are accepted. The tattoo shall consist of the breed insignia followed by the flock number of the respective flock. Also, at the time of inspection, a small V-shaped notch is cut into the extreme tip of the right ear of all animals accepted for registration.
Polled Dorset	Each sheep offered for registration must bear tag or tattoos with number, which must be given on application. This refers to the owner's private ear tag flock number.
Polypay	All sheep must be continuously identified by ear tag or tattoo. A registration number will be issued for association record purposes, but the individual flock tags will identify the animals in the flock.
Rambouillet	The ear label record is a numerical listing, by ear tag number, of all sheep registered by each breeder, with the association number assigned to each ear tag. Spaces are skipped for the numbers not registered, so that they will still be in numerical order if registered later.
Romney	Each breeder should use a private ear tag or tattoo with a unique number on it for each animal. This must be done prior to or at weaning. Breeders must also choose a farm name for their respective flocks, or use some form of their own name and initials, along with the farm tag number, when registering sheep.
Shropshire	Prior to recording, each sheep must wear a permanent identification in the ear. This may be either a tag or a tattoo, and it must show the number and the applicant's name or initials thereon. No number shall be duplicated; hence, no two sheep can be registered with the same number. The association assigns a registry number and furnishes an ear tag bearing numbers corresponding to those shown on the certificate of registry. This tag must be inserted in the ear of the sheep, promptly. Thus, all registered Shropshires should be wearing two ear tags (one in each ear). Should either tag get lost, it should be replaced with a duplicate of the original.
Southdown	Sheep to have a permanent ear tag with the same name and number as the flock number on the registration paper. Double ear tags or a permanent tattoo is encouraged.
Suffolk	**American Suffolk Sheep Society:** Each registered sheep and any eligible for registration must be positively identified at all times by means of a metal ear tag. If an animal loses its private flock owner's ear tag, the Suffolk may be retagged, but it cannot be renumbered in the NSSA's records. Once a private flock name and number has been officially recorded in the NSSA records, it remains with the animal's record permanently regardless of any change in ownership of the animal.
Targhee	*Left ear:* Belongs to the breeder. Breeder's flock tag is minimum identification. Information required on the registration application: flock number for ear tag; birthdate; type of birth; sex; sire identification; dam identification. *Right ear:* Belongs to the association. Sheep passing inspection have association individual number, corresponding to ear tag number, tattooed in right ear. Association serially numbered ear tag in right ear. When ear tags are lost, replacement tags should have same numbers.
Tunis	When an animal is approved for registration, an individual ear tag (with association numbers) is forwarded to the owner. This tag must be inserted either in place of the private tag or in addition thereto.

SHEEP COATS

The ancient Greek shepherds spread skins over their animals to protect their fleeces from inclement weather; and at the height of the glory of Rome, sheep were blanketed so that a luster and gloss might be imparted to the wool. In recent years, researchers in Australia, New Zealand, and the United States have studied various types of blankets and coats as a means of protecting fleeces from damage by weathering, dirt, burs, and seeds.

In general, there is agreement that coating sheep will result in lower shrinkage, whiter wool, heavier clean-fleece weight, longer staple, and fewer death losses of animals exposed to inclement weather immediately following shearing.

Scientists in Australia project that if newly shorn sheep were provided with plastic coats, death losses of one million sheep from exposure after shearing would be alleviated. Their studies showed that plastic coats raised the fleece surface temperature of shorn sheep by 16°F if the sheep were dry when coated, and by 11°F if they were wet when covered. Throwaway water-repellent paper coats for sheep may also be a possibility.

BEDDING SHEEP

Bedding or litter is used primarily for the purposes of keeping animals clean and comfortable. But bedding has the following added values from the standpoint of the manure:

1. It soaks up the urine which contains about one-half the total plant food of manure (see Fig. 15-13).
2. It makes manure easier to handle.
3. It absorbs plant nutrients, fixing both ammonia and potash in relatively insoluble forms that protect them against losses by leaching.

Currently, there is a trend to eliminate bedding, especially in some of the larger confinement-type operations, primarily because it reduces costs and makes cleaning easier. This is being accomplished by the use of slotted floors.

KIND AND AMOUNT OF BEDDING

The kind of bedding material selected should be determined primarily by (1) availability and price, (2) absorptive capacity, (3) cleanness (this excludes dirt or dust which might cause odors or stain wool), (4) ease of handling, (5) ease of cleanup and disposal, (6) nonirritability from dust or components causing allergies, (7) texture or size (for example, material that will not get into the wool of sheep), and (8) fertility value or plant nutrient content. In addition, desirable bedding should not be excessively coarse, and should remain well in place and not be too readily kicked aside.

Table 15-2 lists some common bedding materials and gives the average water absorptive capacity of each.

Naturally, the availability and price per ton of various bedding materials vary from area to area, and from year to year. Thus, straws are more plentiful in the central and western states.

Table 15-2 shows that bedding materials differ considerably in their relative capacities to absorb liquid. Other facts of importance relative to certain bedding materials and bedding uses are:

1. **Cut straw.** Cut straw is not suited for bedding sheep because the fine particles get into the fleece.
2. **Fertility value.** From the standpoint of the value of plant food nutrients per ton of air-dry material, peat moss is the most valuable bedding, and wood products the least valuable. The suspicion that sawdust or shav-

TABLE 15-2
WATER ABSORPTION OF BEDDING MATERIALS

Material	Water Absorbed by Air-Dry Bedding
	(lb/100 lb or kg/100 kg)
Barley straw	210
Cocoa shells	270
Corn stover (shredded)	250
Corncobs (crushed or ground)	210
Cottonseed hulls	250
Flax straw	260
Hay (mature, chopped)	300
Leaves (broadleaf)	200
(pine needles)	100
Oat hulls	200
Oat straw (long)	280
(chopped)	375
Peanut hulls	250
Peat moss	1,000
Rye straw	210
Sand	25
Sawdust (top-quality pine)	250
(run-of-the-mill hardwood)	150
Sugarcane bagasse	220
Tree bark (dry, fine)	250
(from tanneries)	400
Vermiculite[1]	350
Wheat straw (long)	220
(chopped)	295
Wood chips (top-quality pine)	300
(run-of-the-mill hardwood)	150
Wood shavings (top-quality pine)	200
(run-of-the-mill hardwood)	150

[1]This is a micalike mineral mined chiefly in South Carolina and Montana.

ings will hurt the land is rather widespread, but unfounded. These products decompose slowly, but this process can be expedited by the addition of nitrogen fertilizers. Also, when plowed under, they increase soil acidity, but the change is both small and temporary.

The minimum desirable amount of bedding to use is the amount necessary to absorb completely the liquids in manure. Some helpful guides to this end are as follows:

1. Per 24-hour confinement, the minimum daily bedding requirement, based on uncut wheat or oats straw, of sheep is 1 lb. With other bedding materials, these quantities will vary according to their respective absorptive capacities (see Table 15-2). Also, more than these minimum quantities of bedding may be desirable where cleanliness and comfort of the animal are important. Lying down makes animals more comfortable and helps them utilize a higher proportion of the energy of the feed for productive purposes (sheep require 9% less energy when lying down than when standing up).
2. Under average conditions, and for all farm animals, about 500 lb of bedding are used for each ton of excrement; less than this is required for sheep because of their dry feces.

REDUCING BEDDING NEEDS

In most areas, bedding materials are becoming scarcer and higher in price, primarily because (1) geneticists are breeding plants with shorter straws and stalks, (2) there are more competitive and remunerative uses for some of the materials, and (3) the current trend toward more confinement rearing of livestock requires more bedding.

Sheep producers may reduce bedding needs and costs as follows:

1. **Use deep-bedding system.** For wintering sheep, and under certain other conditions, a deep-litter system—letting the bedding build up beneath, and adding a light sprinkling of fresh bedding on top at intervals—will keep the animals warm and dry, and save in bedding.
2. **Ventilate quarters properly.** Proper ventilation lowers the humidity and keeps the bedding dry.
3. **Feed and water away from sleeping quarters.** Animals should be fed and watered in areas removed from their sleeping quarters. With this type of arrangement, they defecate less in the sleeping area.
4. **Provide exercise area.** Where possible and practical, provide for winter exercise in well-drained, dry pastures or corrals, without confining animals to or near their sleeping quarters.
5. **Consider slotted floors.** More and more slotted or wire floors are being used for different classes of livestock, especially poultry, hogs, and sheep.

MANURE

The term manure refers to a mixture of animal excrements (consisting of undigested feeds plus certain body wastes) and bedding.

No doubt, the manure pollution problem will persist. The collection, transport, storage, and use of manure must meet sanitary and pollution control regulations.

Modern sheep buildings and equipment should be designed to handle the manure produced by the animals that they serve; and this should be done efficiently, with a minimum of labor and pollution, so as to retrieve the maximum value of the manure, and make for maximum animal sanitation and comfort.

Common sheep manure management systems include (1) pastures or ranges, (2) open lots with shelters on solid floors, and (3) confinement on slotted floors. Manure from each of the management systems ultimately will be disposed of on the land. Pastures or ranges, of course, do not require extra handling for land disposal. Manure from open lots with shelters and confinement on slotted floors requires planning for storage and equipment for removal and disposal.

AMOUNT, COMPOSITION, AND VALUE

The quantity, composition, and value of manure produced vary according to specie, weight, kind and amount of feed, and kind and amount of bedding. Fig. 15-11 compares the yearly manure production of sheep with that of the other livestock species.

Sheep manure, free of bedding, is produced in the approximate quantities shown in Table 15-3.

Many factors affect the composition of manure; for example, the farm, the ration, the method of collection and storage, the bedding, the added water, and the time and method of application. In general terms, about 75% of the nitrogen (N), 80% of the phosphorus (P), and 90% of the potassium (K) contained in sheep

TABLE 15-3
APPROXIMATE DAILY MANURE PRODUCTION EXCLUSIVE OF BEDDING AND SPILLED WATER[1]

Animal Unit	Size		Manure Production per Day			
			Weight		Volume	
	(lb)	*(kg)*	*(lb)*	*(kg)*	*(cu ft)*	*(m^3)*
Rams	180–300	*82–136*	10	*4.5*	0.15	*0.004*
Dry ewes	150–200	*68–91*	6	*2.7*	0.10	*0.003*
Feeder lambs	30–110	*14–50*	4	*1.8*	0.065	*0.002*
Ewes with lambs	—	—	7	*3.2*	0.12	*0.003*

[1]Adapted by the author from *Sheep Housing and Equipment Handbook*, Midwest Plan Service, Ames, IA.

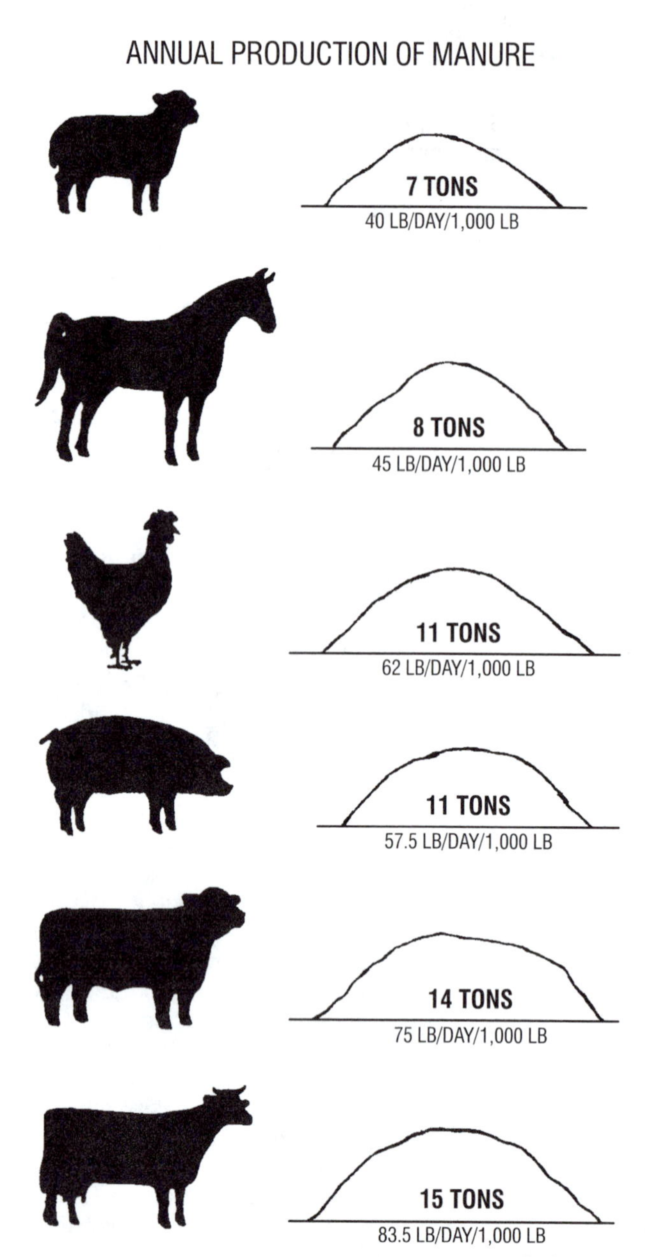

Fig. 15-11. On the average, each class of livestock produces per year per 1,000 lb the tonnages of manure (urine and feces) shown above. Actual values vary widely due to differences in ration, animal age, and management practices.

feeds are returned as manure, as shown in Fig. 15-12. In addition, about 40% of the organic matter in feeds is excreted as manure. As a rule of thumb, it is commonly estimated that 80% of the total nutrients in feeds are excreted by animals as manure.

The urine comprises 40% of the total weight of the excrement of swine and 20% of that of horses. These figures represent the two extremes in farm animals. Yet, the urine contains nearly 50% of the nitrogen, 4% of the phosphorus, and 55% of the potassium of average manure—roughly one-half of the total nutrient content of manure (see Fig. 15-13). Also, the plant nutrients in the liquid portion of manure are more readily available to plants than those in the solid portion. Conservation of urine is important.

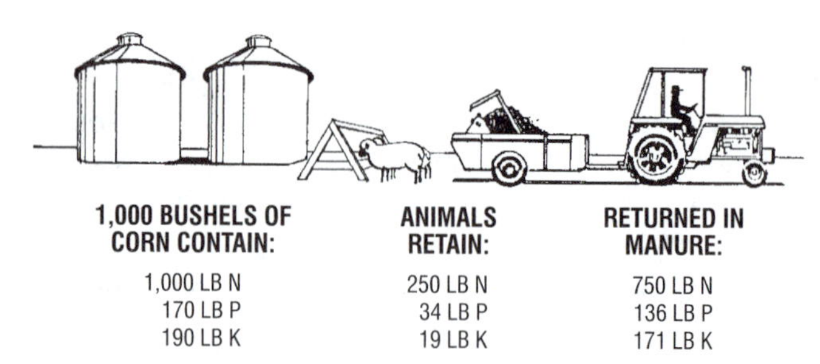

Fig. 15-12. Sheep retain about 20% of the nutrients in feed. The rest is excreted in manure.

The actual monetary value of manure can and should be based on (1) increased crop yields, and (2) equivalent cost of a like amount of commercial fertilizer. Numerous experiments and practical observations have shown the measurable monetary value of manure in increased crop yields.

Each year, all livestock combined produce about 1.5 billion tons of manure, with a potential value of about \$6 billion. Guidelines relative to the fertilizer value of sheep manure are given in Table 15-4. The nitrogen (N), phosphorus (P_2O_5), and potassium (K_2O) content of sheep manure is greater than that of livestock in general due to the higher percentage of dry matter.

Assuming that nitrogen (N) retails at 29¢ per pound, phosphorus (P_2O_5) at 30¢ per pound, and potassium (K_2O) at 12¢ per pound, then a ton of sheep manure is worth about \$9.76, using the values in Table 15-4 for manure without bedding.

Of course, the value of manure cannot be measured alone in terms of increased crop yields and equivalent cost of a like amount of commercial fertilizer. Manure has additional value for the organic matter which it contains, which almost all soils need, and which farmers and ranchers cannot buy in a sack or a tank.

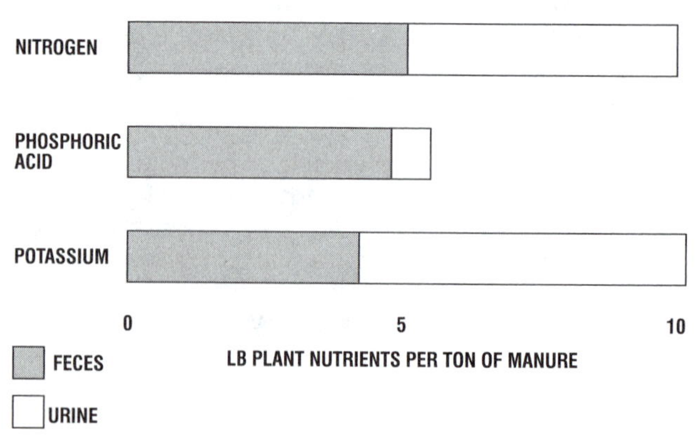

Fig. 15-13. Distribution of plant nutrients between liquid and solid portions of a ton of average farm manure. As noted, the urine contains about half the fertility value of manure.

TABLE 15-4
APPROXIMATE DRY MATTER AND NUTRIENT COMPOSITION OF RAW SHEEP MANURE[1]

Waste System	Dry Matter	Nitrogen (N)[2]				Phosphorus (P_2O_5)[3]		Potassium (K_2O)[4]	
		Available		Total					
	(%)	*(lb/ton)*	*(kg/metric ton)*	*(lb/ton)*	*(kg/metric ton)*	*(lb/ton)*	*(kg/metric ton)*	*(lb/ton)*	*(kg/metric ton)*
With bedding.	28	5	2.5	18	9.0	11	5.5	26	13.0
Without bedding	28	5	2.5	14	7.0	9	4.5	25	12.5

1Adapted by the author from *Sheep Housing and Equipment Handbook*, Midwest Plan Service, Ames, IA.

2Available nitrogen (N) is primarily ammonium-N, which is readily available to plants during the growing season, while total N includes ammonium-N plus organic N, which is released slowly.

3To convert to elemental P, multiply by 0.44.

4To convert to elemental K, multiply by 0.83.

Manure produces rather lasting benefits due to the slower availability of its nitrogen and its contribution to the soil humus, which may continue for many years. Approximately one-half the plant nutrients in manure are available to and effective upon the crops in the immediate cycle of the rotation to which the application is made. Of the unused remainder, about one-half, in turn, is taken up by the crops in the second cycle of the rotation; one-half of the remainder in the third cycle, etc. Likewise, the continuous use of manure through several rounds of a rotation builds up a backlog which brings additional benefits, and a measurable climb in yield levels.

Sheep producers sometimes fail to recognize the value of this barnyard crop because (1) it is produced whether or not it is wanted, and (2) it is available without cost.

WAYS OF HANDLING

Modern handling of manure involves maximum automation and a minimum loss of nutrients. There are a number of methods; among them, the use of scrapers, power loaders, conveyors, industrial-type vacuums, slotted floors with the manure stored underneath or emptying into irrigation systems, storage vats, spreaders (including those designed to handle liquids alone or liquids and solids together), dehydrators, and lagoons. Actually, there is no one best manure management system for all situations; rather, it is a matter of designing and using that system which will be most practical for a particular set of conditions.

When considering manure storage in a settling basin, a holding pond, or a pit under slotted floors, one can compute the storage capacity as follows: Storage capacity = number of animals × daily manure production (see Table 15-3) × desired storage in days + extra water.

If bedding is used, the storage capacity will have to be increased by about one-half the volume of bedding used.

Generally, several months' storage is desirable.

MANURE AS A FERTILIZER—PRECAUTIONS FOR USE

Historically, manure has been used as a fertilizer, and it is expected that it will continue to be used primarily as a fertilizer for many years to come.

With today's heavy animal concentrations in restricted locations, the question is being asked: How many tons of manure can be applied per acre without depressing crop yield, creating salt problems in the soil, creating nitrate problems in feed, contributing excess nitrate to groundwater or surface streams, and/or violating state regulations?

Based on earlier studies in the Midwest, before the rise of commercial fertilizers, it would appear that one can apply from 5 to 20 tons of manure per acre year after year with benefit.

Heavier than 20-ton applications can be made, but probably should not be repeated every year. With higher rates annually, there may be excess salt and nitrate buildup. Excess nitrate from manure can pollute streams and groundwater and result in toxic levels of nitrate in crops. Without doubt the maximum rate at which manure can be applied to the land will vary widely according to soil type, rainfall, and temperature. Moreover, state regulations may control or limit manure application to the soil.

When farmers have sufficient land, they should use rates of manure which supply only the nutrients needed by their crops rather than the maximum possible amounts suggested for pollution control.

■ **Precautions**—The following precautions should be observed when manure is used as a fertilizer:

1. Avoid applying waste closer than 100 ft to waterways, streams, lakes, wells, springs, or ponds.
2. Do not apply where percolation of water down through the soil is not good, or where irrigation water is very salty or inadequate to move salts down.
3. Do not spread on frozen ground.
4. Distribute the waste as uniformly as possible on the area to be covered.

5. Incorporate (preferably by plowing or discing) manure into the soil as quickly as possible after application. This will maximize nutrient conservation, reduce odors, and minimize runoff pollution.

6. Minimize odor problems by—

a. Spreading raw manure frequently, especially during the summer.

b. Spreading early in the day as the air is warming up, rather than late in the day when the air is cooling.

c. Spreading only on days when the wind is not blowing toward populated areas.

7. In irrigated areas, (a) irrigate thoroughly to leach excess salts below the root zone, and (b) allow about a month after irrigation before planting, to enable soil microorganisms to begin decomposition of manure.

QUESTIONS FOR STUDY AND DISCUSSION

1. Define *management*. What are the components of sheep management?

2. Compare the various systems of sheep production.

3. Compare the various systems of raising lambs.

4. For your sheep enterprise, or one with which you are familiar, would you use confinement or pasture production, or a combination of both? Justify your decision.

5. How may pastures be utilized to best advantage in sheep production?

6. Why does the grazing of sheep and cattle together result in the production of more pounds of meat, or higher carrying capacity, per acre?

7. Define *early weaning*. What factors influence the age at which lambs are weaned?

8. By what method would you dock and castrate lambs in your area? At what age should they be docked and castrated? Give reasons for your answer.

9. Should male lambs be castrated? Justify your answer.

10. Under what circumstances would you recommend creep feeding of lambs? Under what specific conditions would you not recommend creep feeding?

11. What is the importance of tagging sheep?

12. Why is foot trimming an important management practice?

13. In your area, should farm-flock sheep producers do their own shearing or hire professional shearers?

14. Define *drenching*.

15. What is meant by "grafting lambs"? Name and describe four procedures for grafting.

16. What method of marking or identifying sheep would you select? Justify your selection and tell how you would apply this method.

17. What type of sheep bedding is commonly used on your farm or ranch (or a farm or ranch with which you are familiar)? Would some other type of bedding be more practical? If so, why?

18. For your farm or ranch (or one with which you are familiar) is it preferable and practical to apply manure to the land, or should commercial fertilizers be used instead? Justify your answer.

19. Suppose you maintained a farm flock of 200 ewes, how much manure would they produce in one year, and what would be the approximate value of this manure as a fertilizer? Be certain to consider the lambs produced during the year and four rams. Assume a lamb crop of 125%.

20. What type of training would you recommend for a young person who wishes to become the manager of a sheep operation?

SELECTED REFERENCES

Title of Publication	Author(s)	Publisher
Handbook of Livestock Management Techniques	R. A. Battaglia V. B. Mayrose	Burgess Publishing Co., Minneapolis, MN, 1981
Sheep Housing and Equipment Handbook		Midwest Plan Service, Iowa State University, Ames, IA, 1982
Sheepman's Production Handbook, The	American Sheep Industry Assn.	Sheep Industry Development Program (SID), Denver, CO, 1998
Stockman's Handbook, The, Seventh Edition	M. E. Ensminger	Interstate Publishers, Inc., Danville, IL, 1992

CHAPTER 16

This is from an original painting by the noted artist, Tom Phillips. It portrays the artist's conception of *Home on the Range*, immortalized in song by Brewster Highley in 1873. The setting is a valley in Jackson Hole, Wyoming, with the majestic Tetons in the distance.

RANGE SHEEP MANAGEMENT[1]

Contents | Page

[1]Mr. Eric Luse, Range Conservationist, Bureau of Land Management, United States Department of the Interior, Washington, DC, provided much of the data and information for updating Chapter 16.

Sheep management on the western range differs from sheep management in the rest of the United States primarily in the size of the respective enterprises. Regardless of area, however, the fundamental nature of sheep remains the same. For this reason, there are neither as many differences nor as many secrets to successful management in various areas of the United States as some sheep producers think.

THE WESTERN RANGE

Various geographical divisions are assumed in referring to the western range area—the native pasture area. Sometimes reference is made to the 17 range states, embracing a land area of approximately 1.16 billion acres. At other times, this larger division is broken down, chiefly on the basis of topography, into (1) the Great Plains area (the six states of North Dakota, South Dakota, Nebraska, Kansas, Oklahoma, and Texas); and (2) the 11 western states (Arizona, California, Colorado, Idaho, Montana, Nevada, New Mexico, Oregon, Utah, Washington, and Wyoming). In addition to these major and commonly referred to geographical divisions, numerous other groupings exist. These are of importance to the sheep producer in that they affect the type of management and, to some extent, the kind of animals kept.

Fig. 16-1. Range ewes, on summer range near Browning, Montana. (Courtesy, *Western Livestock Reporter*)

Almost half (47.9%) of the land area in the 11 western states is federally owned. Domestic livestock graze on 73% of this area.[2]

It is estimated that 2,000,000 cattle and 1,700,000 sheep, plus some wild or domestic horses and donkeys, are using the 165,000,000 acres of public rangelands. In 1996, $14,488,800 in grazing fees were collected.

In the fiscal year 1995, the estimated number of game animals on public lands were as follows:

Antelope	373,245
Barbary Sheep	531
Bear	15,885
Bighorn Sheep	19,347
Buffalo	803
Caribou	940,000
Deer	1,275,775
Elk	201,904
Wild Boar	15,320
Moose	30,395
Mountain Goat	759
Turkey	30,010

Because of the magnitude of the range livestock industry and because it is a highly specialized type of operation, considerable discussion will be devoted to the range area and the care and management of sheep in the range method.

[2]Data in this paragraph provided by Eric K. Luse, Range Management Specialist, United States Department of the Interior, Bureau of Land Management, Washington, DC.

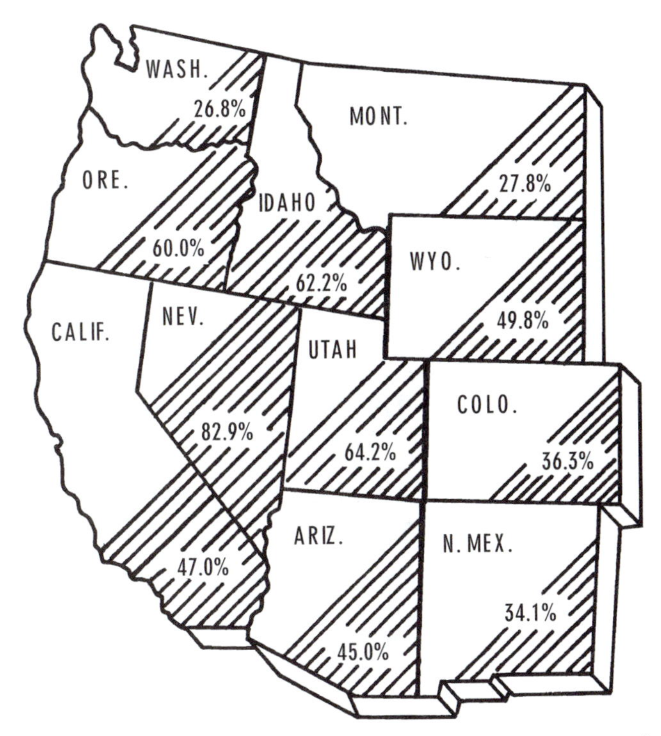

Fig. 16-2. A map showing the 11 western states and the percentage of land in each of these states that is owned by the U.S. government. (Source: Data provided in a letter to the author, by Eric K. Luse, United States Department of the Interior, Bureau of Land Management, Washington, DC)

The carrying capacity of much of the western range is low, and little of it provides year-long grazing. Moreover, variation in vegetative types, climate, and topography in the range country is accompanied by great diversity in the seasonal use made of it. As a result, rangelands are usually grazed during different parts of the year, and the herds and flocks migrate with the season, moving to the mountains and higher elevations in summer and returning to the lower ranges in winter.

From the standpoint of vegetation and utilization by livestock, ranges differ from cultivated pastures as follows:

1. **They are less productive.** Generally, their productive capacity is lower. This is as one would expect, for they are largely made up of the residue remaining after the usable agricultural lands have been taken up. Also, plant growth on rangelands frequently is limited by low and undependable rainfall (even drought), short growing seasons, shallow or rocky soil, alkali or salt accumulations, steep topography, etc. Under such conditions, forage plants are usually less resistant to grazing damage than those growing under a more favorable environment.

2. **They are more likely to progress to less palatable plants.** Range vegetation consists of a mixture of native and introduced plants, varying greatly in palatability, nutritive value, and productive ability. Grazing animals select the most palatable plants first. Thus, unless careful management is practiced, the best plants are crowded out through a combination of grazing injury and competition from the ungrazed low-value plants. Continued poor management can result in good forage plants being almost completely replaced by low-value annual, weedy, or shrubby vegetation, or left denuded and subject to severe erosion.

3. **They are more difficult to restore when depleted.** Once a range becomes depleted, it is a slow process to rebuild it. Plowing and drilling are impractical on most rangelands; thus, very often the only feasible way of restoring a range to good condition is to stock it conservatively and manage it well.

4. **They often serve multiple uses.** Rangelands often have other uses in addition to grazing values. Among such uses are: water production, timber production, mineral production, wildlife production, and recreation (camping, hiking, picnicking, etc.).

Thus, many people, in addition to the livestock producer, have an interest in the grazing management practiced on ranges. This is part of the justification given for federal government ownership of large tracts of rangeland.

HISTORY OF THE RANGE SHEEP INDUSTRY

Buffalo, deer, elk, mountain sheep, antelope, and other forms of wildlife, large and small, were the first users of the range. Beginning about 1840, stockmen began pushing westward, with the big boom in the range-cattle industry occurring between 1880 and 1885. Somewhat later, and extending over a longer period of time—between 1880 and 1910—range-sheep

Fig. 16-3. A herd of buffalo. These animals—along with deer, elk, mountain sheep, antelope, and other forms of wildlife—were the first users of the range. (Courtesy, N. H. Rose Collection of Old Time Photographs, San Antonio, TX)

numbers rose dramatically. It is noteworthy that in 1850 only 1,500,000 sheep, or 7.6% of the total number in the United States, were found west of the Mississippi River; whereas by 1910 approximately 37,500,000 head, or 71.8% of all sheep in the United States, were raised west of the Mississippi.

These huge sheep numbers, appearing almost without warning, aroused deep resentment on the part of cattle owners. There was much overgrazing and strife over water holes and previous occupation. The cattle-sheep feuds waxed hot. Deadlines were set up and enforced by the side that could muster the greatest strength. Herds and flocks were scattered and even stampeded over precipices. Firearms were used, bringing death and destruction to both owners and animals.

Some 15 to 20 years of this unrestricted and uncontrolled competition resulted in the depletion of many ranges.

Fig. 16-4. Early-day cattle raisers warning sheep herders away from their range. Frequently the cattle-sheep feuds waxed hot. (Courtesy, The Bettmann Archive, Inc., New York, NY)

In the period from 1910 to 1915, wool prices declined and market-lamb prices became remunerative. This brought about a reduction in wethers kept solely for wool production and a shift to combined lamb and wool production. Following closely in period of time, there was a greater desire to own at least a part of the land used for grazing, and public land control measures became a stabilizing factor. But these changes were not alone. The coming of the homesteaders and the railroads and the creation of the national forests had far-reaching effects upon the sheep industry. The less efficient operators were driven out of business. Scientific management replaced much of the frontier spirit and the romantic, adventurous life of the shepherd.

THE RANGE AREA ADAPTED TO SHEEP RAISING

Despite uncontrolled and abusive use of many western ranges in the past, most agriculturists agree that the vast majority of the range area will continue to be utilized chiefly for sheep and cattle production. Except for the irrigated areas, rangeland is not suited to the production of grains.

Certainly, the western range area must be well adapted to sheep raising; for the conquering spirit of the shepherd could not alone have withstood all the vicissitudes of time, the privations of western expansion, the competition of hordes of wild animals and herds of cattle for grass and eventual range depletion, the cattle-sheep feuds, a slow and sometimes confused public-land policy, and the labor and price dilemma during and since World War II. Although many of these problems have caused serious fluctuations in sheep numbers, the industry appears to be as much of the West as the mountains.

Fig. 16-5. Much of the range area is rough and otherwise unsuited to the production of grains. Sheep and cattle share in the utilization of such areas. (Courtesy, Union Pacific Railroad Company, Omaha, NE)

TYPES OF RANGE VEGETATION

Sheep producers and students alike—whether they reside in the East, West, North, or South—should be well informed concerning range grasslands, the very foundation of the range livestock industry. This is so because this vast area is one of the greatest sheep countries in the world and a potential competitor of every U.S. sheep producer. Since sheep and cattle compete successfully with each other in utilizing most range forages, both classes of animals necessarily will be mentioned in the discussion which follows relative to types and uses of range vegetation.

Chiefly because of climate, topography, and soil, the character and composition of native range vegetation is quite variable. Ten broad types of vegetation native to the western ranges of the United States are discussed in this chapter.

TALL-GRASS TYPE

It is estimated that about 20 million acres of tall-grass range remain, most of which is in the eastern Great Plains region, in a rainfall area varying from 20 to 40 in. Although differing with the soil, topography, and rainfall, the dominant native tall-grass species include the bluestem (big and little), Indiangrass, switchgrass, side oats grama, and sloughgrass. Such famous grazing areas as the Flint Hills of Kansas, the Osage country of Oklahoma, and the Sand Hills of Nebraska belong to the tall-grass type of vegetation. For the most part, this type of range is utilized by cattle, although sheep do graze some of it. Each fall, thousands of slaughter cattle are marketed after being finished on tall-grass vegetation without a grain supplement. The carrying capacity of these ranges is very high.

SHORT-GRASS TYPE

The short-grass range represents the largest and most important grassland type in the United States, embracing an area of approximately 280 million acres. This extends from the Texas Panhandle to the Canadian border and from the foothills of the Rocky Mountains eastward midway into the Dakotas. The common grasses of this area include the grama grasses, buffalograss, and western wheatgrass, all of which are well adapted to making their growth during the time of favorable moisture conditions in the late spring and early summer. Although they become bleached and cured on the stalk, because of the small amount of leaching in the fall and winter months, these plants retain sufficient nutrients to furnish valuable winter grazing. Because the forages in the short-grass area are dry during much of the year and droughts are rather frequent, the smaller fine-wool breeds of sheep are most numerous in the area, and most of the lambs go the feeder route. Also, cow-and-calf operations predominate in the cattle industry of the area; and most of the calves and older steers are finished in Corn Belt feedlots or in irrigated valleys prior to slaughter.

SEMIDESERT-GRASS TYPE

The semidesert-grass type—which predominates in an area characterized by low rainfal!, frequent droughts, and mild winters—embraces about 93 million acres of grasslands in central and southwestern Texas, Arizona, and New Mexico. It provides year-round grazing. Because of great differences in climate and soil, the vegetation is quite variable. The most common grasses are grama, curly-mesquite, and black grama. Scattered among the more or less sparse grasses are many scraggly shrubs, dwarf trees, yuccas, and cacti. Some of these—especially saltbush, mesquite, ratany, and scrub oak—are rather palatable and are browsed effectively by goats. For the most part, the semidesert-grass area is utilized by commercial cattle as a cow-and-calf proposition. But bands of breeding sheep are found throughout the area. Both sheep and goats are common in southwestern Texas. The sheep of this area are kept primarily for wool production, and production of feeder lambs is secondary.

PACIFIC BUNCHGRASS TYPE

The Pacific bunchgrass area embraces about 60 million acres in western Montana, eastern Washington and Oregon, northern and southwestern Idaho, and central California. Much of the original bunchgrass area, including the famous Palouse area of eastern Washington and northern Idaho, is now devoted to the production of wheat and peas. Though well adapted to the dry summers and moist winters of the area, the native tall bunchgrasses or tuft-growing grasses of this area—bluebunch wheatgrass, Idaho fescue, Sandberg bluegrass, and California needlegrass— did not withstand overgrazing and have largely been replaced by such annuals as alfileria, bur-clover, and cheatgrass (in the Northwest), and wild oatgrasses (in California). These ranges furnish excellent grazing in the spring and fall months but are too dry for summer use. The Pacific bunchgrass area is best adapted for spring, fall, and winter grazing by sheep and cattle.

SAGEBRUSH-GRASS TAPE

The sagebrush-grass area, which is the third largest of all range types, embraces between 90 and 100 million acres extending from northern New Mexico and Arizona northwestward into Montana and to the east slope of the Cascades in the Pacific Northwest. This type of vegetation is characteristic of low rainfall areas where most of the meager precipitation occurs during the winter and spring seasons. Interspersed among

Fig. 16-6. Sheep on sagebrush-grass type of vegetation in southeastern Idaho. (Courtesy, U.S. Forest Service)

Fig. 16-7. Sheep grazing on winter range (salt-desert shrub type) in Millard County, Utah. (Courtesy, U.S. Forest Service)

the ever-present sagebrush, of which there are several kinds, are many species of native grasses among which are bluebunch and western wheatgrasses, needle-and-thread, Indian ricegrass, Sandberg bluegrass, and numerous species of weeds. The sagebrush, which varies from 2 to 7 ft in height, provides little forage except when winter snows blanket the grasses. For the most part, the sagebrush type of vegetation is used for early spring and late fall grazing for sheep and cattle. It furnishes interim pasture until more distant summer and winter grazing areas may be used. In recent years, studies have shown that the carrying capacity of sagebrush areas may be increased by destruction of the sagebrush, which encourages greater growth of the grasses.

SALT-DESERT SHRUB TYPE

About 40 million acres in central Nevada, Utah, southwestern Wyoming, western Colorado, and southern Idaho are covered with a mixture of low shrubs and scattered grasses. The common browse species of the area are shadscale, saltbush, black sagebrush, winterfat, rabbitbrush, greasewood, spiny hopsage, and horsebrush; and the rather sparse grass species include blue grama, sand dropseed, galleta, and Indian ricegrass. Because there is not any dependable source of water and because of the high temperature and dryness during the summer months, the use of much of this area by sheep and cattle is restricted to the winter months when there is snow. Other areas cannot be grazed because of the high alkali content of the soil.

SOUTHERN-DESERT SHRUB TYPE

Approximately 50 million acres, located chiefly in southeastern California, southern Nevada, and southwestern Arizona, are classed as southern-desert shrub vegetation. The common shrubs are the creosotebush and different kinds of cacti. Normally, the scant rainfall, extremely high temperatures, and sparse vegetation of this area make it rather poor grazing for sheep and cattle. However, when moisture conditions are favorable, there is growth of such annuals as alfileria, Indian wheat, bur-clover, black grama, tobosa, dropseed, and six-week fescue. When forage and water are available, nearby ranchers make use of the southern-desert shrub area, primarily for winter grazing, although it is used for spring and fall grazing and in a few cases throughout the year.

PIÑYON-JUNIPER TYPE

The piñon-juniper type of vegetation forms the transition zone from the shrub and grass areas of the lower elevations to the forests of the mountains. The 76 million acres in this general type area extend all the way from southwest Texas to south central Oregon. As the name would indicate, piñon and juniper trees are common to the area. These scattered trees range in height from 20 to 40 ft and interspersed among them are such low-growing shrubs as sagebrush, bitterbrush, mountain-mahogany and cliffrose, and grasses like the gramas, bluebunch and bluestem wheatgrass, and galleta. For the most part, this area is used for spring and fall grazing by sheep and cattle; but in the Southwest, where forage cures on the ground and retains much of its nutritive value through the winter, year-long grazing is prevalent.

Fig. 16-8. Ewes and lambs on spring and fall range. (Courtesy, University of Wyoming)

WOODLAND-CHAPARRAL TYPE

This type of vegetation is characteristic of parts of California and Arizona. It varies all the way from an open forest of park-like oak and other hardwood trees with an undergrowth of herbaceous plants and shrubs to dense *chaparral* thickets of no value to animals. Alfileria, slender oatgrass, and bur-clover have been introduced in the more open areas. Though somewhat restricted, woodland-chaparral is used for fall, winter, and spring grazing by sheep and cattle.

OPEN-FOREST TYPE

The 130 million acres of open forests, found scattered in practically all the mountain ranges, constitute the second largest range type of vegetation. This is the summer range of the West, which provides grazing for large numbers of sheep, cattle, and big game. Many grass-fat lambs and cattle are sent to market directly off these cool, lush, high-altitude ranges. For the most part, the tree growth common to the area consists of pine, fir, and spruce; and the grasses include blue grama, fescues, bluestem, wheatgrasses, timothy, bluegrasses, sedges, and many others. More than half of these mountain ranges are federally owned as national forestlands. In addition to serving as valuable grazing areas, the open forests are important for lumbering and recreational purposes.

Fig. 16-9. Sheep grazing on open-forest type range in Colorado (Courtesy, U.S. Forest Service)

GRAZING PUBLICLY OWNED LANDS

The ownership of U.S. land is summarized in Table 16-1.

TABLE 16-1
OWNERSHIP OF U.S. LAND (50 States)[1]

Ownership	Acreage		Percentage of Total
	(million acres)	*(million ha)*	(%)
Private ownership	1,329	*538.0*	42.2
Indian land	51	*20.4*	1.6
Public ownership	885	*358.3*	28.1
Federal	730	*295.5*	23.2
State and local governments	155	*62.7*	4.9

[1]*Statistical Abstracts of the United States*, 1982, p. 182, Table 318.

About 42.3% of the public lands are in Alaska. Because of its remoteness and northern location, land development has been slow in this state. As a result, the federal government still owns over 66.4% of all the lands in Alaska.

Today, in the 11 western public land states, the federal government owns and administers approximately 262 million acres on which grazing is allowed. At one time or another during the year, domestic cattle and sheep graze on about half of these public lands. More of the public lands are used for this purpose than for any other economic activity.

AGENCIES ADMINISTERING PUBLIC LANDS

Livestock grazing was regulated first in 1897 on the Forest Reserves; then, in 1934 on the rest of the public rangelands.

Because much of the grazing land that ranchers

rely upon to maintain their cattle and sheep enterprises is built up into operating units by leasing or by obtaining use permits from several federal and state agencies, private corporations, and individuals, it is imperative that the owner have a working knowledge of the most important of these agencies. Some range operators are placed in the position of using range rented from as many as six landlords—private, state, and/or federal.

The bulk of federal land is administered by the following six agencies: the Bureau of Land Management, the Forest Service, the Bureau of Indian Affairs, the Department of Defense, the National Park Service, and the Bureau of Reclamation. The largest land area from the standpoint of grazing permits and utilization of grazing areas by animals is administered by the first three of these agencies.

1. **Bureau of Land Management.** The Bureau of Land Management of the Department of the Interior administers about 264 million acres of federally owned lands, most of it in the 12 western states, but about one-third of the land it manages is in Alaska. The remainder is almost entirely in the 11 western states. Its lands are used for a variety of activities, such as grazing, mining, hunting, fishing, and camping.

From the standpoint of the livestock producer, the most important function of the Bureau of Land Management is its administration of the 54 grazing districts established under the Taylor Grazing Act of 1934 and of the unreserved public land situated outside these districts which are subject to grazing lease under Section 15 of the Act. This federal act and its amendments authorize the withdrawal of public domain from homestead entry and its organization into grazing districts administered by the Department of the Interior. Also, this legislation, as amended, allows the Bureau of Land Management to administer state and privately owned lands under a cooperative arrangement.

The Bureau of Land Management operates its federal lands under units called *grazing districts*. Each grazing district is administered by a range manager, who is a technically trained employee of the Bureau of Land Management. Managers are responsible to their state bureau offices for the proper use, management, and welfare of the range resources of their districts. In turn, the state offices are responsible to the director's office in Washington, DC.

Grazing privileges are allocated to individual operators, associations, and corporations on the basis of (a) priority of use, (b) ownership or control of base property dependent on grazing district land for forage during certain seasons of the year, or control of permanent water needed to graze district land, (c) proximity of home ranch to the grazing district, and (d) adequate property to supply the feed needed, along with grazing privileges, to maintain throughout the year the livestock permitted on public range. All of these lands are subject to classification and disposal under Sections 7 and 14 of the Taylor Grazing Act and the Classification and Multiple-Use Act, for any higher use or other appropriate purpose. Grazing privileges may, therefore, be cancelled whenever such lands are determined to be more suitable for other purposes.

A unit month is the amount of forage needed to sustain one cow (including her calf), bull, yearling bovine, horse or donkey, and five sheep or goats that are six months of age or older. In 1997, the grazing fee for western public lands administered by the Bureau of Land Management was $1.35 per animal unit month (AUM). The formula used for calculating the fee, which was established by Congress in the 1978 Public Rangelands Improvement Act, has continued under a Presidential Executive Order issued in 1986. The annually adjusted grazing fee is computed by using a 1966 base value of $1.23 per AUM for grazing on public lands in the western states. The figure is then adjusted to three factors: (a) current private grazing land lease rates, (b) beef cattle prices, and (c) the cost of livestock production.

The Taylor Grazing Act has been responsible for many changes, not all of which have been popular. Some livestock producers complain about the loss of their ranges; others tell of increased costs; and there are those who resent government controls, and, above all, there is the confusion which results from dealing with several agencies. Without doubt, many of these criticisms are justified, and some errors in administration should be rectified; but those who would be fair agree that the ranges as a whole have improved under the supervision of the Bureau of Land Management and that further improvements are in the offing.

2. **Forest Service.** The Forest Service, an agency of the U.S. Department of Agriculture, manages 191 million acres of federal lands for multiple use, including grazing, in 44 states, Puerto Rico, and the Virgin Islands. The national forests are used for grazing under a system of permits issued to local farmers and ranchers by the Forest Service.

The Forest Service issues term grazing permits and annual permits. Among other things, a permit prescribes the boundaries of the range which may be used, the maximum number of animals allowed, the season when grazing is permitted, and the expiration date of term permits.

Temporary permits may be waived back to the government when the permitees sell livestock or base property. Then, the purchaser of the permitted livestock or base property may apply for and be issued a permit if qualified.

The requisites in order to qualify for a permit are:

a. **U.S. citizenship.**

b. **Ownership.** The ownership of both the livestock and commensurate ranch property.

A grazing preference is not a property right. Rather, it is a privilege granted to a qualified livestock operator to graze a definite number of livestock on a specified area of national forest range for a definite period of time. Permits may be revoked in whole or in part for a clearly established violation of the terms of the permit, the regulations upon which it is based, or the instructions of forest officers issued thereunder.

A ranger administers the grazing use on each National Forest Ranger District. Several districts (usually 3 to 6 or more) comprise a national forest. A forest supervisor and staff administer the national forest. Several national forests, under the direction of a regional forester and staff, comprise a forest region. The chief administers the Forest Service from Washington, DC, under the supervision of the Secretary of Agriculture.

As is true in the administration of Taylor grazing districts, local farmers and ranchers act in an advisory capacity in the details of administration of the national forests in reviewing allotment management plans and the use of betterment funds.

Forest Service grazing fees are based on a formula which takes into account livestock prices over the past 10 years, the quality of forage on the allotment, and the cost of ranch operation. In 1997, the U.S. Forest Service charged a grazing fee of $1.35 per AUM on national forests in Arizona, California, Colorado, Idaho, Kansas, Montana, Nebraska, Nevada, New Mexico, North Dakota, Oklahoma, Oregon, South Dakota, Texas, Utah, Washington, and Wyoming.

Although shortcomings exist in the management of the national forests, these ranges have been vastly improved under the administration of the Forest Service. Many of them now approach the quality that existed in their virgin state. Perhaps the most heated arguments between livestock producers and the Forest Service arise over the relative importance attached to the multiple use of big game and other wildlife, recreation, etc.

3. **Bureau of Indian Affairs.** Indian lands, comprising over 51 million acres, are really not public lands. Rather, most of these lands are held in trust for the benefit or use of the Indians and are merely administered by the Bureau of Indian Affairs of the Department of the Interior.

Many of the Indian lands have suffered serious vegetative depletion, but a concerted effort is now being made to control livestock numbers in keeping with available feed supplies and to improve the management of the range resources as well as the quality of animals produced. However, overstocking continues to be a difficult problem on the Navajo, Hopi, and Papago reservations.

4. **State and local government-owned lands.** A total of 134 million acres are owned by state and local governments. For the most part, the management of these areas is diverse and confused, each state and local government having established different regulations relative to the lands under its ownership. In general, however, such lands are operated on a stipulated lease arrangement. On many such areas, range depletion has been severe.

5. **Railroad-owned lands.** Recognizing that the main deterrent to rapid settlement and development of the West was the lack of adequate transportation facilities, the federal government very early encouraged the construction and westward extension of the railroads by means of large grants of land. It was intended that the railroads should sell or otherwise utilize these lands in financing their costs of construction. These initial grants, totaling 94,149,866 acres, consisted of alternate sections extending in a checkerboard fashion for a distance of from 10 to 40 mi on each side of the right of way. Today, less than 20 million acres of these lands are held by railroads. Many of these holdings are leased to livestock producers; but because of inconvenience, past abuses, or other reasons, some of these lands are considered worthless for grazing. In general, railroad lease agreements do not restrict the number of stock to be grazed or the season during which the land may be so used.

KINDS OF SHEEP FOR THE RANGE

As most western sheep are handled in large bands and kept out in the open much of the year, a gregarious (or flocking) instinct and a rugged constitution are prime essentials. Because these two requisites are best embodied in sheep of fine-wool breeding, western sheep carry considerable Rambouillet breeding. The

Fig. 16-10. Grade Rambouillet ewes and lambs sired by black-faced rams on summer range. Sheep owned by John Redd. Ewes showing a predominance of Rambouillet breeding are most common on the western range. (Courtesy, *The Record Stockman*, Denver, CO)

Rambouillet is of Spanish Merino extraction. In Spain, hardy migratory flocks were developed many centuries ago.

As has already been noted, the western range embraces a large and variable area, thus making for differences in adaptation within its boundaries. These variations are primarily due to vegetative differences that result from differences in moisture, soil, and altitude. Normally, therefore, most of the southwestern arid and semi-arid range (which includes Texas, New Mexico, Arizona, and southern Colorado) does not produce sufficiently lush vegetation for grass-finished lamb production. Wool is of comparatively greater importance than in the more lush range areas to the north, and smaller-grade Rambouillet sheep predominate. On the other hand, in the semi-arid and sub-humid areas of the West and Northwest, where grasses are more lush and abundant and the objective is the production of grass-finished lambs, large Rambouillet ewes or longwool X Rambouillet crossbred types predominate. Moreover, except where replacement ewe lambs are being raised, a considerable number of black-faced rams are used in the latter area.

FENCED RANGE

Many of the range flocks of Australia and New Zealand are kept on large tracts of enclosed land, without the assistance of herders. Fenced ranges are also common to many of the ranges of the southwestern United States. For understandable reasons, fenced ranges must be limited to privately owned lands; and as much of the rangeland in the West is publicly owned and likely to continue so, fences will not be built. Sheep cannot be easily switched from pasturing on fenced range to herding, for, under such circumstances, they are untrained and unmanageable. It is imperative, therefore, that year-round fenced ranges be available if one is going to use fenced holdings without herders.

Fig. 16-11. Grazing sheep inside a fenced range, protected from dingos (the wild dogs of Australia). (Australia official photo by W. Brindle)

Among the claims made in favor of fenced ranges in comparison with the herding system are: (1) less labor is required, and it is said that over a period of years, the fences cost less than herders; (2) the carrying capacity of the range is increased; (3) the fences afford some protection against predatory animals; (4) the sheep thrive better, primarily because they scatter out more in grazing and they are not required to walk so far in bedding down, securing water, and keeping from other bands. Where both the enclosure and the herding methods have been tried, sheep owners prefer the enclosure system.

HERDING METHOD

As the designation implies, the herding method consists of entrusting the welfare of a band of sheep to a caretaker, known as the herder. This is the most common method of management employed in the West and is the only practical system for handling sheep on unfenced and publicly owned range. The herding method, in addition, is well adapted to a migratory type of flock husbandry in which seasonal use is made of the range, a practice which was first followed by the early-day Spanish flockmasters who trailed their flocks to the North in the summer and returned them to the South in the winter.

SIZE OF BAND

In general, the size of a band is determined by the number of sheep that can be efficiently handled by one person. This will vary according to (1) the season of the year, (2) the topography of the country and type of vegetation, and (3) the condition of the ewes, whether they are dry, lambing, or suckling. Bands are larger, therefore, when on the winter range than at other seasons; larger when in a lush-plains–type country than in a broken or mountainous country or where the vegetation is sparse; and larger when the band is composed of dry ewes than when lambing-out or suckling.

Because herders frequently are scarce and their salaries plus keep represent a heavy item of expense, there is a strong tendency to give herders more animals than they can properly care for. In the end, this practice is unsound and uneconomical. After all, the care given by the herder greatly affects (1) the quantity and quality of lambs and wool produced; (2) the losses suffered from accidents, predatory animals, and disease; and (3) to some extent, the status of the range vegetation. Within itself, a band of sheep represents a very sizable investment.

Range bands are not and should not be the same size. In fact, they vary from 800 to 3,500.

HERDER AND EQUIPMENT

The profession of sheep herding is very, very old. Also, people of various nationalities and repute herd sheep. As in all professions, there are wide differences in the capabilities of the people. Some are shiftless and unreliable, whereas others revere the profession and take great pride in their heritage as members of families who have been honorable herders for many generations.

The standard equipment of the herder consists of merely a couple of dogs and the minimum articles with which to make camp. Over rough terrain and where one-night bedding (known as *bedding-out*) is followed, the herders live in tents, and the supplies are moved about on the backs of patient burros or horses. The pack animals graze along with the sheep. But, to make sure that the camp supplies will be on hand at night, the burros or horses are given a little inducement in the form of a few handfuls of grain once or twice each day.

In most range areas and almost without exception on desert type of winter ranges, herders use camp wagons which contain all the essential equipment for living a great distance from headquarters. These covered wagons are generally equipped with a stove, a bed, one or two chairs, cupboards, and a few dishes and cooking utensils. Most modern wagons are mounted on rubber tires and are moved from place to place hitched to trucks, whereas, in more rugged country the camp wagons are pulled by horses. These wagons are moved at frequent intervals as the band migrates to new feeding and bedding grounds. In some areas, the herders do much of their work on horseback.

Fig. 16-13. A herder's camp wagon. Home on the winter range. (Courtesy, Utah State University, Logan, UT)

Although most herders insist on having a couple of dogs, the minimum use of such assistance is recommended. By nature, sheep are timid, thus making it desirable to handle the band quietly and to guard it from alarm. The caretakers who follow approved *open* herding, rather than *close* herding and who consistently secure the greatest production of lamb and wool confine the use of dogs to safeguarding the band against predatory animals and to locating strays. Above all, dogs should not be allowed to chase ewes during the lambing season or ewes and lambs on the summer range. Although sheep dogs vary in breeding and appearance, to qualify as good ones they must work as directed, have sheep sense, and have sufficient stamina to endure the rugged existence of the range.

Fig. 16-12. This sheep herder sleeps in the "teepee" at the left, while caring for a band on the rolling ranges of Wyoming. (Courtesy, Charles J. Belden, Pitchfork, WY)

Fig. 16-14. In some areas, herders do much of their work on horseback. Dogs are standard equipment. Note the herder's camp wagon and more horses to the left. (Courtesy, U.S. Forest Service)

DUTIES AND RESPONSIBILITIES OF THE HERDER

Herders are busy if they are conscientious and practice the most approved methods of handling the bands under their supervision. Their day begins as soon as the sheep leave the bed ground about daybreak. Their work at this hour consists of keeping a sharp lookout for predatory animals and guiding their bands in the general direction where they are to feed and rest for the day. After grazing for 3 to 4 hours in the morning, the bands may *shade up* during the heat of the day, during which time herders can take care of their cooking and chores. Usually, two of the herder's three meals are eaten while the band is shaded up—one soon after the sheep have come to rest and the other in the middle of the afternoon just before the bands again start on the move.

The animals resume foraging about mid-afternoon and will graze until almost dark. Skillful herders (1) protect their bands from enemies; (2) direct the course of travel as quietly as possible; (3) prevent their charges from mixing with other bands and from straying or trespassing; (4) encourage open rather than close herding; (5) use dogs a very minimum; and (6) change bedding and feeding grounds frequently. Such grazing produces fatter lambs and more wool and constitutes better range management.

At some convenient time during the day, herders select the bedding ground for the night. The most desirable spot is of sufficient size (3 to 5 acres or more for a band of sheep), relatively open, and fairly high. Although some trees are permissible, the area should be free from brush, fallen timber, or other obstructions, because predatory animals are more hazardous in such areas and there is more danger of injury. On the open range, a high spur or ridge is preferable to a canyon or swale, for sheep prefer high areas and will rest more contentedly under such conditions. In no instance should the bedding ground be near cliffs or ravines, for the animals might be injured if they wandered into the darkness or became frightened.

After the herders have selected the bed ground for the night, the course of travel is directed so that the place is reached about the time the sheep will be ready to bed down. As the animals begin to settle down, herders determine if their bands are intact. This conclusion is based upon the general contentment and appearance of the flocks and the presence of certain easily recognized individuals—black, belled, and odd-looking individuals known as *markers, counters*, or *spotters*. Normally, there is one marker to each 40 to 75 sheep in the band. At night, if herders find that one of these markers is missing, they may well conclude that several sheep have strayed. It is then their duty to make a diligent search for these strays and, if at all possible, to bring them back into the fold before they are permanently scattered or killed by predatory animals. If the herders fail to locate the lost sheep, the camp tenders take over the assignment and either find them or pursue the search until all hope is lost. When all members of the band have bedded down for the night, the day is ended, but the herders must watch over their flocks by night.

In addition to the multiple responsibilities already mentioned, good herders keep to a minimum the losses from poisonous plants. This is accomplished by (1) recognizing poisonous plants and avoiding grazing areas heavily infested with them, (2) providing ample feed and salt, (3) avoiding long drives followed by rapid filling in areas where poisonous plants are present, and (4) exercising special care along sheep driveways.

Herders must also know the location of water and see that their bands drink at such intervals as necessary. Like other animals, sheep prefer clean, fresh, running water; but if such a supply is not available and they are thirsty, they will drink any kind. The frequency of watering is determined chiefly by the weather and the character of the forage. In cool weather and when sheep are grazing succulent forage, watering once in 3 or 4 days may be sufficient. On the other hand, when the weather is hot and the forage is dry, bands will thrive best if watered daily. In experiments conducted on the forest in central Utah, it was found that sheep can be successfully grazed for 100 days without any source of water, provided that the vegetation consists of succulent broad-leaved herbs and there are heavy dews.[3] Skillful herders sense the water needs of their bands and allow them to spread out and graze to and from the source of water at the proper intervals, avoiding both thirst and long drives.

Herders must also provide their bands with an adequate supply of salt. When the feed is green and succulent, more salt is required than when the forage is dry. Little or no salt is required, however, in areas where the soils are alkali. Usually herders take care of the salt needs by distributing about 100 lb of crystal rock salt, or unrefined salt, on the bedding ground at 4- to 5-day intervals, for each 1,200 ewes. Others prefer giving a more limited salt supply each evening, with 20 to 30 lb provided daily for each band of about 1,200 head. Sometimes the salt is placed in portable troughs of wood or canvas, but in most instances it is simply distributed on a grassy turf.

The welfare of the bands and the profit derived therefrom are so dependent upon the skill and diligence with which the herders exercise their duties

---

[3]Jardine, J. T., "Grazing Sheep on Range Without Water," *The National Wool Grower,* V, No. 9, 7-10.

and responsibilities that many sheep producers believe it would pay owners to place their herders on an incentive basis, giving them a bonus for lamb and wool production above and over a pre-estimated average.

CAMP TENDING

In plains areas, the position of the camp tender generally entails more responsibility than that of the herder, and usually more pay. In such areas, where a large operator has several bands, the camp tender may look after as many as five flocks, sometimes with the aid of an assistant camp tender and usually under the direction of a foreman. Their duties consist of selecting the new camp site and moving camp, keeping the camp provided with food and other necessities, bringing in feed for the horses or burros and salt for the sheep, keeping the herder in ammunition, and occasionally bringing in mail. Camp tenders also help with the cooking, check the count on the sheep, and assist in locating strays. In addition to locating the camp site where the feed is good, camp tenders must be thoroughly familiar with the territory, making certain that the herders under their supervision keep their respective bands on an area covered by ownership, leases, or permits. When on winter range, camp tenders may on occasion have to provide supplemental feed for the sheep, and it is their responsibility to melt snow for camp use and to water the saddle and pack stock.

Where several flocks are being looked after, camp tenders may have under their supervision a larger investment than many businesspersons in cities, and for successful operation even greater ability and ingenuity may be required. In the northwestern United States, where fat lambs are produced on summer mountain ranges, one camp tender usually accompanies each herder, with the latter serving as the boss.

ONE-NIGHT CAMP SYSTEM VS USE OF SAME CAMP SITE FOR SEVERAL SUCCESSIVE NIGHTS

The one-night camp system is sometimes referred to as *bedding-out*, or the blanket or burro system. Although this method has much to commend it, the more common practice consists of using the same camp site for several nights, usually until the surrounding forage is exhausted, and then moving to a new camp site where the process is repeated.

In Biblical days, and in the early days on the western ranges of this country, it was common practice for the herders to camp together at night, separating their flocks as best they could the next morning. In those days, this nightly banding of herds and herders was a necessary means of protection. With range depletion, the ending of the cattle-sheep feuds, the only remaining virtue of this system was the sociability afforded herders. Thus, this banding together gave way to the practice of keeping the bands separated while using the same camp site for several successive nights.

Although it presents somewhat more difficulties for the camp tender and the herder, the most modern and desirable method of handling the range band is the one-night camp system. In studies made by the U.S. Forest Service[4]—in which the one-night camp system was compared to bedding the band down in the same camp site many nights in succession—it was found that with the one-night system two-thirds as much range is required, the sheep and lambs make larger gains, and the lambs are ready for market at an earlier date. Contrary to the opinion held by some operators, no greater problem with predatory animals has been encountered in the one-night system.

CALENDAR OF OPERATIONS

Because of the magnitude of and the area covered by the range sheep industry, and because the handling of a large band is a highly specialized type of enterprise, certain phases of management peculiar to this method will be discussed at length. In presenting a calendar of operations, however, the author is fully cognizant of the considerable differences which should and do exist between areas with respect to both timing and management. Much of this diversity arises from differences in severity of winter, in fenced vs open ranges, in availability of seasonal ranges, and in the relative emphasis on lamb and wool production. Despite variations in management and timing caused by differences in environment and objectives, however, sheep remain the same physiologically. This means that the basic principles of breeding, feeding, health, disease prevention, parasite control, and marketing of lamb and wool remain the same regardless of the system of production.

FALL OPERATIONS

The chief fall operations in handling the range band are (1) return to the fall range, and (2) breeding.

---

[4]Jardine, J. T., *The Pasturage System of Handling Range Sheep*, U.S. Forest Service Circ. 178.

FALL RANGE

In those areas where sheep utilize summer ranges in the mountains, they are moved back to the fall range before the onset of inclement weather, especially snow. This means that by late September or in October most of the sheep are brought out of the mountains, even if feed is abundant; for the weather is usually hazardous thereafter. Generally, the grass-milk–fat lambs and cull ewes are marketed at the time the band is taken from the lush summer ranges, and the ewes and the balance of the lambs are shipped or driven to the fall range on the foothills or plains. These movements vary in distance anywhere from 50 mi to as much as 350 mi.

Fig. 16-15. Sheep watering on a fall range in Colorado. (Courtesy, Bureau of Land Management, U.S. Department of the Interior)

BREEDING SHEEP ON THE RANGE

The general principles of breeding sheep as discussed in Chapter 4 are applicable to the farm flock and range band alike. For this reason, only those differences peculiar to the range will be discussed herein.

The primary requisites of range rams (commonly referred to as *bucks* by western sheep producers) are that they be large and that they have utility value (freedom from wool blindness, etc.). In addition, the range operator insists that rams be of acceptable meat conformation, possess good bone and ruggedness, have a strong constitution, and stand on sturdy feet and legs. Most range operators prefer purebred rams, although they do not look upon such ancestry as either sacred or indispensable; and they are quite willing to accept either grade or crossbred rams if they meet their approval from the standpoint of type.

Except during the breeding season, range rams are usually kept separate from the rest of the sheep. In some areas, during the nonbreeding season and especially in the summer months, several operators combine their buck flocks, making up what is known as a *buck band*. In general, range rams are not fed supplemental feed, with the possible exception of the period immediately preceding and during the breeding season. At breeding time, the general practice is to run 3 rams to each 100 ewes, although cases may be found where the number of ewes per ram is as few as 20 or as many as 70. In determining the proper number of rams to use, practical operators give consideration to the age, vigor, breed, and method of handling. Usually the rams are transported from headquarters to the range where the ewes happen to be at the time of the desired breeding season.

The breeding season on the western range is as variable as that of the farm-flock states, with the breeding of individual bands extending anywhere from July to January and lambing-out from November until June. On many ranges in California and the Southwest, the ewes are bred for November and December lambing, the objective being to market hothouse lambs. On the other hand, throughout the area where pasture lambing is practiced, the breeding season is very late, ranging from November until the first of February. On the shed lambing ranges of the Northwest—in Washington, Oregon, and Idaho—and in certain other parts of the range, the breeding season is intermediate between the above two extremes, for the most part extending from August to October. Regardless of the season, once a decision has been reached, it is important that the entire flock be bred in the shortest possible period of time, preferably with the bulk of lambing bunched within three weeks and not extending for more than six weeks. This is advisable because (1) extra labor and feed must be provided for the lambing season, and (2) lambs of a uniform age can be more easily managed until marketed.

WINTER OPERATIONS

After the animals' arrival on the winter range, either directly from the mountains or after spending a little time on a fall range in the foothills, the chief concern is to make certain that water is available and to provide such supplemental feed as may be necessary, especially when the range is covered with snow. Many winter ranges have so little water that they cannot be grazed unless there is snow or unless water is hauled to the band. On the colder northern ranges, some operators provide shelter for stormy weather, but for the most part shelter is not essential except with early lambing in the North and Northwest.

WINTER RANGE

The most desirable winter ranges are found on the plains or desert areas, which comprise millions of acres of flat or gently rolling lands throughout the West. In general, these areas are characterized by low-growing shrubs interspersed with sparse-growing grasses. Although the vegetation is too limited for fattening, it apparently is palatable and serves well for winter maintenance of sheep. The animals travel great distances for feed, with from 3 to 20 acres required to maintain one sheep for the winter months. The rainfall in most of the winter grazing area ranges from 5 to 15 in. with the growth of vegetation and the carrying capacity being in proportion to the moisture conditions.

Fig. 16-16. Sheep on winter range in Utah. (Courtesy, U.S. Forest Service)

WINTER FEED AND SHELTER

The common supplemental feeds used on the winter range when the grasses and shrubs are sparse or covered with snow include hay; cottonseed, linseed or soybean cake; and corn. Progressive sheep producers have found that proper winter feeding lessens mortality and results in increased lamb and wool production. The severity of the winter and the kind and location of the winter range determine the duration and amount of supplemental feeding. In New Mexico and Arizona, March is the only month likely to require supplemental feeding each day, whereas some ranges to the north may require almost daily feeding for 3 to 5 months.

Most range flocks are not provided with shelter during the winter months, an exception being made during the lambing season where shed lambing is practiced. In general, range sheep secure ample protection through such natural windbreaks as are afforded by the topography and shrubs.

SPRING OPERATIONS

The spring operations are the most numerous and important of any season, including such tasks as (1) tagging, (2) lambing, (3) docking and castrating, (4) migrating to the spring range, (5) shearing, (6) dipping or spraying, (7) utilizing the spring range, and (8) moving to the summer range.

LAMBING SEASON

The lambing season on the range is the most important of all operations, largely determining the profit or loss derived from the enterprise. For success, there must be a large, healthy lamb crop, with ownership by the mothers. Although many variations of each exist, the two common systems of lambing-out are (1) the pasture or open-range lambing system, and (2) the shed lambing system.

PASTURE OR OPEN-RANGE LAMBING SYSTEM

The pasture or open-range lambing system is limited to those areas having a mild climate or to the spring months after the passing of inclement weather. At the present time, this system, which is a relic of the old days of free range, is largely limited to the southwestern United States, but it still has its adherents throughout the range area. Sometimes in pasture lambing the ewes and lambs are placed in corrals for a few days, and these enclosures may even be subdivided into individual pens and larger group pens. At other times, the pasture or open-range method is strictly what the name implies—lambing on the open range. Under certain conditions, it is referred to as the *broadcast method*. But the pasture method need not imply or necessitate neglect, for, at its best, one or more herders are with the band by day and night. Usually those ewes that are heavy with lamb are kept together in what is known as the *drop band*, whereas those which have lambed-out are placed together in a *lamb band*. On some ranches, a *drop wagon* or *drop sled*, partitioned off with compartments for handling several individual ewes and their lambs, is used to collect the ewes with newborn lambs. When a load has been gathered, they are taken to the lamb band, which may be elsewhere on the range or temporarily stationed in a corral.

SHED LAMBING SYSTEM

Most modern ranches now make use of some sort

of lambing shed or tent. Some of these are permanent structures; others are quite cheap and temporary in construction. Some have arrangements for heat (usually by a stove), and others have no artificial heat except that which is purely incidental from the light of the lanterns. Although there is hardly any limit to the number of different designs, the essential features of most lambing sheds consist of a narrow central alley with lambing pens (commonly called *jugs* or *cells* in the West) on either side, sometimes one or more large holding pens under cover, and adjacent corrals of various sizes and shapes. With the larger units, usually there is a feed room, several stacks of hay, and a cook house. The better operators design the lambing quarters so as to get the maximum of efficiency and sanitation.

Where careful attention is given, each ewe and her newborn lamb or lambs are given identical brands or marks soon after birth, thus making it possible to identify them should they become separated. They are kept in the individual pens until the lambs have nursed and they are thoroughly acquainted with each other. Unless there has been trouble in lambing or there is a problem of ownership, usually 8 to 12 ewes with single lambs or 4 to 6 ewes with twins are placed in a single pen the first day. In another two days, they may be placed with a like number of about the same age. In still another two days, the group may be redoubled. When lambs are coming fast and weather and other conditions are favorable, subsequent combinations continue at about the same interval of time, finally culminating in a lamb band. Some operators make it a practice to keep ewes with twin lambs separate from those with singles, giving the former special feed and attention while in the lambing camp until there is no doubt of ownership by the mothers or of their ability to raise two lambs. Others always keep ewes with twins separate from those with singles, eventually assigning the twin lamb band to the choicest range.

Although the operations in a well-managed lambing camp usually pay dividends in terms of a larger percentage of lamb crop saved and owned by their mothers, the feed and labor items are very expensive. It is always hoped, therefore, that the breeding of the ewes may have been such that the lambing season is bunched as much as possible and that all goes well in order that the ewes and lambs may be turned to the nearby range within the shortest possible time. In order to save on feed costs during the daytime, many operators regularly take the drop bands out on the range near the lambing sheds.

Although the best operations in a lambing camp represent mass production at its best, the principles in caring for the ewes and newborn lambs are identical whether there be few or thousands.

SPRING RANGE

In those areas where seasonal ranges are available, it is common practice to migrate the band to the spring range soon after lambing. Spring ranges are usually located in the foothills or on the plains either adjacent to or in close proximity to the lambing camp or area. Compared with the usual winter range vegetation, the growth on the spring ranges is more lush and succulent, thus stimulating the milk flow and giving the lambs a good start in life.

MIGRATE TO THE SUMMER RANGE

Spring ranges are usually rather limited in area, thus not providing grazing for a very long period of time. Moreover, more desirable summer ranges in the mountains are usually ready from about the middle of June to the first of July. At the proper time, therefore, the band should be moved to the summer range. Most range bands are trailed to summer ranges, but where great distances are involved, they may be shipped by rail or truck.

SUMMER OPERATIONS

Once the band has arrived at the summer range in the mountains, there are usually no further major operations until the lamb crop is marketed at the end of the summer grazing season.

SUMMER RANGE

There are three essentials in the production of grass-fat lambs: (1) proper breeding, (2) adequate feed, and (3) good herding. Suboptimum conditions in any one of these categories will result in lambs for which the market outlet is limited to the feeder route at

Fig. 16-17. Sheep on southern Utah summer range. (Courtesy, Utah State University)

weaning time. For example, on the drier plains areas, such as prevail in much of the Southwest, the forage is a limiting factor and grass-fat lambs are seldom produced, no matter how superior the breeding or herding.

Where available, there is no finer summer range than that afforded by the open forests of the mountains. The cool temperature of the high altitude, the abundant forage of great variety, and plenty of clear, fresh water provide ideal summer range for sheep. With proper herding, the great bulk of lambs from bands grazed on mountain ranges are fat and ready for slaughter by the end of the summer grazing season.

RANGE MANAGEMENT CONSIDERATIONS

Good range management may be achieved if an inventory or analysis is made of the forage resources and all contributing factors, followed by a sound plan of management based upon the analysis. Consideration should be given to such factors as proper stocking rate, and safe degree of use, season of use, condition and trend of forage, stability of soil, system of use, improvements needed, etc.

STOCKING RATE

The key to the successful long-term operation of rangeland lies in making (1) a reliable determination of the land that is suitable or adaptable to grazing use over a long period of time; (2) a realistic estimate of grazing capacity for this land; and (3) a flexible stocking rate, even within a single season, followed by (a) application of proper stocking intensity, and (b) frequent observations to determine the effect of the stocking rate upon changes in condition of the forage cover. Too light stocking wastes forage, while too heavy stocking results in a change of forage plant cover from an abundance of valuable forage plants to an abundance of worthless plants.

Of course, the stocking rate for any given unit may vary widely from year to year, and within a given season, depending on the forage production as affected by weather and other factors. For this reason, stocking should either be adjusted to forage yield each year, and within season, or be set at a constant rate that will assure a sustained yield of most valuable forage plants (constant stocking at about 25% below average capacity will usually achieve the latter).

Animals do not graze uniformly over a range unit. Certain areas are more attractive to them. Consequently, some areas produce most of the grazed forage, while others may go practically unused. Sheep prefer hillsides. If herded, sheep can be moved rather uniformly over a range. If not herded, they tend to congregate in the portion of the pasture in the direction of the prevailing winds, and to spot graze more than cattle. For the purpose of determining grazing capacity, the key areas—those rather extensive parts of the range which are most heavily grazed—must be given greatest consideration. If preferred or key areas are maintained in good condition, the whole unit will generally remain in good condition. Conversely, if key areas are allowed to deteriorate, the grazing capacity of the whole unit will be endangered.

Grazing capacity determinations are relatively complex and require careful study over a period of several years. They are arrived at most simply and accurately by observing soil stability conditions and changes in plant cover. If the best plants are being destroyed and soil movement is observed, the numbers of animals or season of use should be reduced; conversely, if excessive forage remains at the end of the grazing season, the numbers should be slowly increased until a balance is struck.

In arriving at grazing capacity, it is generally wise to seek assistance from qualified range technicians, who need to know:

1. The potential of the particular range.
2. The present state of the vegetation as it relates to potential on each site.
3. The alternative methods of changing present conditions to meet management objectives, including such things as flexible stocking, seeding, and brush control, fences, watering, and trails.

SEASON OF USE

A prime requisite of successful management for sheep is that there be as nearly year-round grazing as possible and that both the animals and the range thrive.

Because a range band of sheep can be moved and herded on unenclosed areas with greater ease than a herd of cattle and because investigations in range livestock management have been conducted more extensively with sheep, greater seasonal use of ranges is made with sheep.

Some pertinent points in determining the proper season of use of the range follow:

1. **Elevation.** Generally speaking, vegetative development is delayed 10 to 15 days by each 1,000-ft increase in elevation. Also, severe storms occur later in the spring and earlier in the fall at higher altitudes than at low desert locations.
2. **Availability of water.** Certain desert areas are so poorly watered that only the occurrence of winter snows makes their use practical.

3. **Early forage washy.** Early spring forage is extremely washy, and may be incapable of supporting stock. Spring grazing should be delayed until the plants are developed enough to meet the nutritive needs of animals.

4. **Soil tramping.** Soil tramping may be serious in early spring. In order to avoid plant damage and soil compaction, grazing should be delayed until the soil is firm.

5. **Poisonous plants.** Most poisonous plants are very early growers and cause their greatest damage when animals are turned out too early. Poisoning losses are usually negligible if sheep are detained until the best forage plants have made suitable growth.

6. **Winter range should be saved.** If stock are allowed to remain on winter ranges too long after spring growth begins, the next winter's feed will be reduced, because the forage produced on these ranges grows mainly during the spring and early summer.

RANGE GRAZING SYSTEMS

Consciously or unconsciously, ranchers follow one or various combinations of the following grazing systems: (1) continuous grazing, (2) rotation grazing, (3) rotation-deferred grazing, (4) rest-rotation grazing, (5) Savory grazing method.

Currently, most progressive ranchers of the northwestern United States use combinations of the five grazing systems, adjusting the systems to fit their range needs. In the Great Plains and other regions of good summer rainfall, continuous grazing is most widely used.

CONTINUOUS GRAZING

Perhaps most ranges of the West are grazed more or less continuously, although some rest is given them through the use of seasonal ranges (such as in migrating to summer ranges in the mountains). Where continuous grazing is moderate, it is perhaps more suitable and practical than rotation-deferred or rest-rotation grazing under the following conditions: (1) when the construction of fences or barriers is very costly, (2) when the important forage species are not dependent upon reseeding for reproduction, and (3) when seasonal ranges are available and used.

Until and unless more research studies reveal that rotation grazing is superior, from the standpoint of both stock and vegetation, continuous grazing will be followed most extensively.

ROTATION GRAZING

Rotation grazing is that system in which the grazing of areas is alternated at internals throughout the season. A heavy concentration of animals is placed on a given area for a few weeks, after which all the sheep are moved on to another area or areas and are finally returned to the first field when the growth is sufficient to withstand another period of grazing. This system is best adapted to the utilization of cultivated pastures in the irrigated valleys of the West or to the humid regions of the United States. However, if a high-intensity–low-frequency system is followed, it will work on the arid ranges of the West.

Many sheep allotments on national forests are managed on a rotation basis, on a once-over system. Other sheep allotments are used every other year, particularly in high, rough country.

ROTATION-DEFERRED GRAZING

In rotation-deferred grazing, the range usually is divided into 3 to 5 or more units. The grazing on at least one unit is deferred each year until after the seed crop has matured or through a complete growing season or period. The next year a second area is deferred and the grazing on the first area is delayed as late as possible to afford opportunity for the young seedlings to become established. By so treating a new unit each year, the entire area is rested, allowed to reseed itself, and grazed in rotation.

Sometimes rotation-deferred grazing is used as a part of rotation grazing and continuous grazing to improve plant vigor, to ensure natural seed production and establishment, or in conjunction with practical range reseeding and brush control.

REST-ROTATION GRAZING

This is a relatively new range grazing system. It consists of resting one subunit (range or area), while grazing the others. The length of the rest period and the intensity of grazing varies depending upon the system used.

SAVORY GRAZING METHOD

The Savory method is a short-duration rest-rotation grazing system. Sheep are concentrated into substantial herds wherever possible for the desired herd effect of trampling, dunging, and urinating as they move around the paddock. Then animals are held in each paddock for a very short time during the range's growing season. Ideally, these short periods range from 1 to 5 days. The short grazing periods are interspersed with short rest periods of about 30 to 60 days. The originator of the system, Allen Savory, subscribes to the philosophy that most ranges are "overgrazed and understocked." This is a highly sophisticated method of range management and should not be applied to a ranch on which the general management is not up to handling it, as failure will likely result.

MULTIPLE USE/CONSERVATION OF LAND

Fig. 16-18. Buffalo, once almost extinct, at home on the range. (Courtesy, USDA, Soil Conservation Service)

The multiple use of publicly-owned lands evolved in response to public pressure. The multiple use of privately-owned lands followed, in response to economic pressure—the need to increase net returns. Soil and water conservation evolved on both public and private lands in recognition that they are national resources that should be preserved for posterity.

The sections that follow pertain to the multiple use/conservation aspects of both public and private lands—to both introduced (seeded) pastures and native (range) pastures.

Fig. 16-19. Recreation on the range—trail rides are popular and a source of income for ranchers. (Courtesy, H. Dietz, USDA, Soil Conservation Service, Fort Worth, TX)

MULTIPLE-USE CONCEPT

Multiple-use of land is the management of all the various resources of lands, both public and private, so that they are utilized in combination. With federal lands, multiple use is based on their most profitable use, and management decisions are made privately by owners/managers. Important multiple uses include livestock grazing, mining, national heritage preservation, occupancy, recreation, water, wildlife, and wood/timber production.

The multiple use concept developed as a compromise relative to the use of public lands; it evolved as a result of attempting to placate individuals and groups who wish to have the land used for purposes which they consider desirable or to prevent others from using the land for purposes which they consider to be undesirable.

MULTISPECIES GRAZING

Grazing two or more species of livestock together or separately on the same land unit in a single growing season is known as multispecies grazing. Research indicates that multispecies grazing contributes to better and more uniform forage use and higher economic returns from livestock.

Multispecies grazing evolved in regions with diverse vegetation types and suitable climates. Grazing by a mix of domestic and wild animals can often result in more efficient use of forage and browse, more total animal gains, and a more vigorous plant community. While multispecies grazing is a common management practice on rangelands of the West, it is much less commonly practiced on the pasture lands of eastern United States.

Western rangelands are characterized by vast diversity in elevation, precipitation, temperature, and other climatic factors. These differences make for a multitude of range sites dispersed among several major vegetation or habitat types. It follows that great potential exists on these lands for multispecies grazing by livestock and wildlife to maintain forage production and species diversity.

Where multispecies grazing is practiced, cattle and sheep dominate. In the Southwest, goats are sometimes a component. Goats are without a peer in rough, unimproved areas and as browsers. Sheep prefer steeper terrain and eat more shrubs and forbs than cattle. Cattle stick to the more gentle slopes and prefer grasses. So, multispecies grazing can result in more complete and uniform utilization of multiplant species pastures and greater animal production. However, predators and labor problems have caused decreased sheep and goat numbers. In turn, this has resulted in lower income on many ranges.

In the past, wildlife has generally been incidental.

Now, and in the future, economic pressures dictate that wildlife will be an integral part of multiple land use.

WILDLIFE

Wild animals and birds are becoming more valuable to today's landowner. Higher livestock production costs and demand for outdoor recreation have prompted practical landowners to seek means of increasing income by providing game for hunters. In some areas, wildlife income exceeds livestock income. This has caused landowners to include wildlife in farm and ranch planning.

There is a close association between kinds of plants and animals present in the habitat. Also, livestock numbers and grazing patterns can be manipulated to enhance wildlife habitat. Through proper habitat management, the farmer and rancher can maintain healthy, abundant wildlife populations. To accomplish this, land managers must place wildlife high on their priority list and consider wildlife in overall farm/ranch planning.

■ **Kinds of wildlife**—Many kinds of wild animals and birds live on pastures and ranges. Identifying the kinds is necessary because management will vary for different species. Deer, for example, need browse, forbs, and grasses for feed, and timber and brushy areas for cover. Quail feed on weed seeds, nuts, and seeds of certain grasses and shrubs; and they prefer a mixture of wooded and open areas with small plots of low shrubs and vines for cover. Normally, management will involve meeting the needs of several different kinds of animals and birds.

■ **Wildlife management**—Wildlife can exist in harmony with livestock operations provided (1) wildlife needs and species are inventoried and included in the management plan, and (2) the following management aspects prevail:

1. **Grazing system.** For proper use of vegetation, it is important that the grazing system allow livestock and wildlife to harvest the forage without overgrazing. This protects the quality of the forages, provides wildlife cover, and prevents erosion. Grazing systems where livestock are concentrated and rotated between pastures reduce some of the competition with wildlife.
2. **Revegetation.** Some grasslands do not have adequate plants to meet the needs of livestock, wildlife, and erosion control. Such areas may be in need of reseeding. When reseeding, consideration should be given to including plants that have special value for wildlife. Many native forbs and shrubs have been selected and released and are now available for this purpose.
3. **Water.** Water is as important to wildlife as it is to domestic animals. Reliable and well distributed supplies of water should be provided and maintained.
4. **Brush control.** Controlling brush can improve grasslands for wildlife and domestic animals. But it must be done properly in harmony with other conservation practices. If poorly planned and not followed with good grassland management, it can harm the habitat for wildlife. One method of brush control is prescribed burning. If this method is used, a burning plan should be developed to meet the objectives of the owner/manager. Patterned brush control, or leaving strips or mottes of brush in pastures, increases the edge effect and enhances wildlife habitat for many species.

■ **Elk Ranching**[5]—People are captivated by the Wapiti, as the Shawnee Indians called the elk of the rugged outdoors and western wilderness.

Both sexes provide annual income, with the cows producing calves, and the bulls producing velvet antlers, coveted in the Far East for their alleged medical qualities. Fuzz-covered antlers averaged $66.00 per pound in 1996. The velvet harvest occurs in May, when the antlers are about 80 days, or about halfway through their growth cycle. After applying a tourniquet around the antler base, the harvester saws off the antlers about 2 in. above the pedicule—the point where they join the skull. The nubs are treated to stop blood flow and repel flies. The green antlers are frozen until they can be processed.

The largest markets are in Korea and China, where velvet antlers have been used for medical purposes for 5,000 years.

The leading elk farming states of the United States, by rank, are: Colorado, Minnesota, Wisconsin, Iowa, and Indiana. Texas and Oklahoma also have significant numbers.

Elk can endure the Southwest's hot summers provided they have wooded areas for shade and cover, and ponds for drinking and swimming.

The elk was indigenous to North America. At one time, it ranged from the Atlantic to the Pacific, and from Mexico to Alberta, Canada.

In 1997, elk cow prices ranged from $3,000 to $30,000 per head; and bulls ranged up to $400,000 each.

SOIL EROSION CONTROL

Soil erosion control is any management plan to reduce soil and water losses.

Soil erosion is a natural occurrence. However, it may be increased by activities that disturb the natural

---

[5]This section is based on an article that appeared in *The Cattleman*, August 1997.

Fig. 16-20. An Iowa farmer planting corn between terraces in the stubble of the previous year's crop, using a 16-row no-till planter. Conservation tillage and terraces help protect the soil from erosion. (Courtesy, USDA, Soil Conservation Service)

Fig. 16-21. Range furrowing in low rainfall area to stop water runoff and increase water penetration. (Courtesy, USDA, Soil Conservation Service)

balance of the pasture or range ecosystem. Poor grazing management is a major cause of erosion.

A raindrop that hits bare soil has a different effect from one that falls on a plant or litter. A racing raindrop smashes against bare soil with great force, splashing water and soil particles and packing the surface soil together; it seals pores, with the result that little water goes into the soil and runoff occurs. By contrast, when a raindrop hits a plant or litter, its force is broken and it trickles into the soil. Grasses are very effective in catching and holding moisture.

When plant cover is reduced by poor management and the distance between plants allows wind to reach the soil, wind erosion may result.

The amount of plant cover on the soil surface at the time of a rain or wind storm is the primary factor in preventing erosion. Both the bulk of cover and the distribution over the surface are important in reducing erosion.

The primary methods of controlling erosion on grasslands include brush control, deferred grazing, reseeding, and mechanical land treatments. Among the latter are contour furrowing, pitting, small dams, and diversions.

The combination of practices used for erosion control will gradually result in better production of grass, improved condition of the range, and a better water supply for domestic animals and wild game. Also, and most important, it will lessen the sedimentation due to soil erosion, which (1) reduces channel capacity and reservoir storage, and (2) results in increasing flooding and reduced water supply.

WATER

Water is often a limiting factor in pasture productivity, affecting forage production and/or drinking water for grazing animals.

The water cycle is the never-ending movement of water from clouds to soil, through plants, and back to clouds again. The cycle begins when precipitation strikes the land and ends when the water leaves the land either through runoff or evaporation. During the intervening time, a sheep producer should store as much water as possible, in the soil and in reservoirs. The shortage of water over much of the West is particularly important. In addition to limiting livestock production, lack of water may limit stream flow for fish, cultivated crops, and industries.

There are various types of stock water developments. These include natural water supplies such as lakes, ponds, streams, springs, and seeps, and made developments such as wells, reservoirs, dugouts, sand tanks, and catchment basins. A combination of two or more types of water development is often more advantageous than one type only.

QUESTIONS FOR STUDY AND DISCUSSION

1. Why is so much of the range area of the west publicly owned and unenclosed? Is it good or bad to have so much public domain? Justify your answer.

2. Do you concur in the policy which permits a sizable number of wild animals to feed on privately owned land? Justify your answer.

3. List and describe four of the most common types of range vegetation.

4. What agencies administer public lands of most importance to sheep producers?

5. What similarities and differences characterize the Bureau of Land Management and the U.S. Forest Service?

6. The U.S. Forest Service issues term grazing permits and annual permits. Why does such a permit prescribe the boundaries of the range, the maximum number of animals allowed, and the season when grazing is permitted?

7. What characteristics are essential in range sheep? What breed or breeds meet these requisites best?

8. Where and why are fenced sheep ranges used instead of herders?

9. Discuss each of the following as they pertain to the herding method: (a) the size of band, (b) the herder and equipment, (c) the duties and responsibilities of the herder, (d) the camp tending, and (e) the one-night camp system vs the use of the same camp site for several successive nights.

10. Discuss the seasonal use of western ranges.

11. Discuss open-range lambing vs shed lambing.

12. Describe the Savory grazing method.

13. What is meant by the multiple use concept of the Western range?

14. What is meant by multispecies grazing? Where multiple species grazing is practiced, what species most commonly dominate?

15. Discuss wildlife on the Western range.

16. Discuss soil erosion on the Western range.

17. Discuss the importance of water on the Western range.

SELECTED REFERENCES

Title of Publication	Author(s)	Publisher
Agricultural Biotechnology: A Public Conversation About Risk	Ed. by J. F. Macdonald	National Agricultural Biotechnology Council, Ithaca, NY, 1993
Alternative Agriculture		National Academy Press, National Research Council, Washington, DC
Feeding a Billion	Wittwer, Sylvan, *et al.*	Michigan University Press, East Lansing, MI, 1987
Forages: Resources of the Future	J. E. Oldfield, Plainsman Task Force	Council of Agricultural Science and Technology, Ames, IA, 1986
Grazing on Public Lands	W. A. Laycock, Chair	CAST, Ames, IA, 1996
Holistic Resource Management	Alan Savory	Island Press, Washington, DC, 1988
Intensive Grazing Management: Forage, Animals, Men, Profits	B. Smith P. S. Leung G. Love	The Graziers Hui, Kanuela, HI, 1986
Rangeland Health		National Academy Press, National Research Council, Washington, DC, 1994
Range Management	A. W. Sampson	John Wiley & Sons, Inc., New York, NY, 1952
Range Management	L. A. Stoddart A. D. Smith	McGraw-Hill Book Co., New York, NY, 1955
Sheep and Man—An American Saga	C. A. Kilker C. R. Koch	American Sheep Producers Council, Inc., Denver, CO, 1978
Stockman's Handbook, The, Seventh Edition	M. E. Ensminger	Interstate Publishers, Inc., Danville, IL, 1992
Western Range Livestock Industry, The	L. A. Stoddart A. D. Smith	McGraw-Hill Book Co., New York, NY, 1950

CHAPTER 17

Some sheep producers, especially those in the southwestern United States, successfully operate the enterprise with little or no shelter. Here is a band of sheep that have sought a tree shelter during a blizzard on the range. (Courtesy, Charles J. Belden, Pitchfork, WY)

BUILDINGS AND EQUIPMENT FOR SHEEP

Contents — *Page*

Contents | Page

Sheep buildings and equipment are changing with the shift to more intensive production methods. This is especially true in the farm-flock states, where the primary emphasis is on lamb production. There is particular interest in confinement production, environmental control, and slotted floors.

Naturally, no standard set of buildings and equipment can be expected to be well-adapted to such diverse conditions and systems of sheep production as exist in the United States. In the discussion and illustrations which follow, it is intended that they be considered as guides only. Detailed plans and specifications for buildings and equipment can usually be obtained (1) through the local county agricultural agent or the FFA instructor, (2) by writing to the state college of agriculture, or (3) from some of the references listed at the end of this chapter.

CONFINEMENT AND ENVIRONMENTAL CONTROL FOR SHEEP

Increasing land costs, restricted grazing, and predator problems have caused more and more producers to consider some type of confinement—all or part of the year. Confinement offers certain advantages like:

1. Increased lamb production from highly productive lands by the use of harvested forages.
2. Reduced losses from predators.
3. Automation that will lessen labor costs.
4. Easier control of internal parasites.
5. Lower feed requirements because of limited activity.
6. Being able to pen and feed ewes according to productivity.

Even with the advantages, however, many successful operations encompass both confinement and pasture production systems. Moreover, individuals considering confinement production should be aware that superior managerial ability is required to maintain the optimal environment.

Limited basic research has shown that animals are more efficient—that they produce and perform better, and require less feed—if raised under ideal conditions of temperature, humidity, and ventilation. The primary reason for having sheep buildings, therefore, is to modify the environment. Properly designed barns and other shelters, shades, insulation, and ventilation can be used to approach the optimal environment. Naturally, the investment in environmental control facilities must be balanced against the expected increased returns; and there is a point beyond which further expenditures for environmental control will not increase returns sufficiently to justify the added cost. This point of diminishing returns will differ between sections of the country, between sheep of different ages (higher expenditures for environmental control can be justified for newborn lambs than for older sheep), and between operators. Labor and feed costs will enter into the picture, also.

Environmental control is of particular importance when lambs are born out of season and for a short period after sheep have been sheared in cold climates.

HEAT PRODUCTION BY SHEEP

The heat produced by sheep varies according to body weight, rate of feeding, environmental conditions, and degree of activity. Nevertheless, Table 17-1 can be used as a guide.

As noted, Table 17-1 gives both *total heat produc-*

TABLE 17-1
HEAT PRODUCTION OF SHEEP[1]

Heat Source	Heat Production, Btu/hr[2]			Heat Production, Kcal/hr[2]		
	Temperature	Total	Sensible	Temperature	Total	Sensible
Sheep, 100 lb *(45.4 kg)*	0.039-in. fleece length:			*0.1-cm fleece length:*		
	45	560	500	*7*	*141.1*	*126.0*
	70	320	245	*21*	*80.6*	*61.7*
	3.937-in. fleece length:			*10.0-cm fleece length:*		
	45	245	185	*7*	*61.7*	*46.6*
	70	260	125	*21*	*65.5*	*31.5*

[1]Adapted by the author from *Agricultural Engineers Yearbook*, St. Joseph, MI, 1979, ASAE Data D270.4, p. 400.

[2]One Btu (British thermal unit) is the amount of heat required to raise the temperature of 1 lb of water 1°F, while one Kcal (Kilocalorie) is the amount of heat required to raise the temperature of 1 kg of water 1°C.

tion and sensible heat production. Total heat production includes both sensible heat and latent heat combined. Latent heat refers to the energy involved in a change of state, which cannot be measured with a thermometer; evaporation of water and respired moisture from the lungs are examples. *Sensible heat is that portion of the total heat, measurable with a thermometer,* that can be used for warming air, compensating for building losses, etc.

VAPOR PRODUCTION BY SHEEP

Sheep give off moisture during normal respiration; and the higher the temperature, the greater the amount of moisture. This moisture should be removed from buildings through the ventilation system. Most building designers govern the amount of winter ventilation by the need for moisture removal. Also, moisture removal in the winter is lower than in the summer; hence, less air is needed. However, lack of heat makes moisture removal more difficult in the wintertime. Table 17-2 gives the information necessary for determining the approximate amount of moisture to be removed.

Since ventilation also involves a transfer of heat, it is important to conserve heat in the building to maintain desired temperatures and reduce the need for supplemental heat. In a well-insulated building, mature animals may produce sufficient heat to provide a desirable balance between heat and moisture; but young lambs will usually require supplemental heat.

RECOMMENDED ENVIRONMENTAL CONDITIONS FOR SHEEP

The comfort of animals (or humans) is a function of temperature, humidity, and air movement. Likewise, the heat loss from animals is a function of these three items.

Temperature, humidity, and ventilation recommendations for sheep are given in Table 17-3. This table will be helpful in obtaining a satisfactory environment in confinement sheep buildings, which require careful planning and design. Confinement of sheep seldom implies a totally enclosed warm building. Rather, it is generally an open building, with one or more sides open, or a cold building that is enclosed but the inside temperature fluctuates with the outside temperature.

When choosing and designing buildings, operators should consider numerous factors; among them, the following:

- **Effects of temperature and humidity**—Rate of

TABLE 17-2
VAPOR PRODUCTION OF SHEEP[1]

Vapor Source	Temperature		Vapor Production		Vapor Production	
	(°F)	*(°C)*	(lb/hr)	(Btu/hr)	*(kg/hr)*	*(kcal/hr)*
			0.039-in. fleece length:		*0.1-cm fleece length:*	
	45	*7*	0.06	60	*0.03*	*15.1*
Sheep, 100 lb *(45.4 kg)*	70	*21*	0.07	75	*0.03*	*18.9*
			3.937-in. fleece length:		*10.0-cm fleece length:*	
	45	*7*	0.06	60	*0.03*	*15.1*
	70	*21*	0.13	135	*0.06*	*34.0*

[1]Adapted by the author from *Agricultural Engineers Yearbook*, St. Joseph, MI, 1979, ASAE Data D270.4, p. 400.

TABLE 17-3
RECOMMENDED ENVIRONMENTAL CONDITIONS FOR SHEEP

Class	Temperature				Acceptible Humidity	Commonly Used Ventilation Rates[1]				Drinking Water			
	Comfort Zone		Optimum			Winter[2]		Summer		Winter		Summer	
	(°F)	*(°C)*	(°F)	*(°C)*	(%)	(cfm)	*(m³/min.)*	(cfm)	*(m³/min.)*	(°F)	*(°C)*	(°F)	*(°C)*
Ewe	45–75	*7–24*	55	*13*	50–75	20–25	*0.6–0.7*	40–50	*1.1–1.4*	35–37	*1.7–2.8*	60–75	*15–24*
Feeder lamb	40–70	*5–21*	50–60	*10–15*	50–75	15	*0.3*	30	*0.65*	35–37	*1.7–2.8*	60–75	*15–24*
Newborn lamb	75–80	*24–27*	—	—	—	—	—	—	—	—	—	—	—

[1]Generally two different ventilating systems are provided; one for winter and an additional one for summer. Hence, as shown in this table, the winter ventilating system in a sheep barn should be designed to provide 20–25 cfm (cu ft/minute) for each ewe. Then, the summer system should be designed to provide an additional 20–25 cfm, thereby providing a total of 40–50 cfm for summer ventilation.

In practice, cold buildings are ventilated by ridge openings and adjustable wall openings.

[2]Provide approximately one-fourth the winter rate continuously for moisture removal.

Fig. 17-1. Lambs under a heat lamp. Heat lamps can be used in lambing pens. (Photo by J. C. Allen and Son, West Lafayette, IN)

gain and feed efficiency are lowered when sheep (lambs) endure temperatures appreciably below or above the optimal (Table 17-3) of 50 to 60°F shown in Table 17-4. Therefore, for promoting maximum efficiency, it is desirable to provide buildings, shelters, shades, and windbreaks, which modify extreme changes in environmental temperature. Also, the influences of humidity are tied closely to temperature. However, compared to swine, cattle, and poultry, sheep are more tolerable of environmental extremes, because of the insulatory value of the fleece.

■ **Insulation**—This refers to materials that have a high resistance to the flow of heat. The measured resistance of a material is its R-value (insulating value). Good insulating materials have high R-values. Polystyrene, polyurethane, and mineral wool have high R-values, while wood and concrete have very low R-values. To save the most money, operators should insulate at the time of construction. Cold buildings and

TABLE 17-4
AVERAGE DAILY GAIN AND FEED EFFICIENCY OF SHORN LAMBS GROWN AT DIFFERENT AMBIENT TEMPERATURES AND FED *AD LIBITUM*[1]

Temperature		Average Daily Gain		Feed Efficiency
(°F)	*(°C)*	*(lb)*	*(kg)*	*(gain/feed)*[2]
23	–5	0.161	0.073	0.04
32	0	0.287	0.130	0.08
41	5	0.374	0.170	0.11
50	10	0.423	0.192	0.15
59	15	0.434	0.197	0.14
68	20	0.406	0.184	0.13
86	30	0.236	0.107	0.08
95	35	0.090	0.041	0.04

[1]Adapted by the author from Ames, D. R., and D. R. Brink, "Effect of Temperature on Lamb Performance and Protein Efficiency Ratio," *Journal of Animal Science*, Vol. 44, 1977, p. 137, Table 2.

[2]Lb gain/lb feed (or kg gain/kg feed).

warm buildings use insulation. The amount of insulation depends on insulation type, barn (building) type and size, number and size of animals, supplemental heat requirements, and climate. In mild and moderate climates, the walls can be about R-14 and the ceiling about R-22 to R-25. In cold climates the walls should be around R-20 and the ceilings around R-33.

■ **Vapor barrier**—Moisture in enclosed sheep buildings can build up. When the amount of water vapor in the house is greater than in the outside air, the vapor will tend to move from inside to outside. The moisture enters the wall and moves outward, condensing when it reaches a cold enough area. Condensed water in the wall greatly reduces the value of the insulation and may damage the wall. Since warm air holds more water vapor than cold air, the movement of vapor is most pronounced during the winter months. The effective way to combat this problem in a sheep building is to use a vapor barrier with the insulation. It should be placed on the warm side of or inside the house. Common vapor barriers are 4-mil plastic film and some of the asphalt-impregnated building papers.

■ **Ventilation**—Ventilation refers to the changing of air. Its purpose is (1) the replacement of foul air with fresh air, (2) the removal of moisture, (3) the removal of odors, and (4) the removal of excess heat in hot weather. The prime function of the winter ventilation system is to control moisture, whereas the summer ventilation system is primarily to control temperature. If air in livestock barns is supplied at a rate sufficient to control moisture—that is, to keep the inside relative humidity in winter below 75%—then this will usually provide the needed fresh air, help suppress odors, and prevent an ammonia buildup.

Ventilation of sheep buildings is accomplished by

natural ventilation—openings in the building—or by fans through inlets and outlets, or by a combination of these.

■ **Supplemental heat**—Some warm barns may require heat in addition to that supplied by heat lamps in lambing pens. The amount of heat needed depends on (1) animal density, (2) whether or not sheep are shorn, (3) amount of insulation and other construction features, (4) ventilation, (5) manure management, and (6) climate. For ewes and lambs, 100 to 200 Btu per 100 lb approximates the need.

■ **Automation**—*Automation is a coined word meaning the mechanical handling of materials.* Sheep producers automate to lessen labor and cut costs. Confinement buildings lend themselves to automating, because of their compactness. Obviously, the feed and water facilities of confinement buildings should be automated. The extent and type of automation depends largely on the production system.

SHEEP BUILDINGS[1]

Sheep do not require expensive or elaborate buildings and equipment, but this statement should not be construed to mean that the facilities for the sheep enterprise should not be carefully planned. On the contrary, it pays well to plan and construct sheep buildings and equipment that will promote sheep health and conserve feed and labor.

The shelter should be of such a nature as to protect the flock from becoming soaked with rain or wet snow. Dry snow or bitter cold has no harmful effect, and up until lambing time, a shelter open to the south on well-drained ground may be entirely satisfactory.

Sheep buildings can be grouped into three general types:

■ **Open housing**—One or more sides of the building are left open. Sheep are allowed to move in and out of the building at all times. The inside temperature is about the same as the outside temperature.

■ **Cold housing**—The building is enclosed and the inside temperature fluctuates according to the outside temperature. The inside temperature can be slightly warmer. Natural ventilation is required to remove excess moisture from the building. Normally, sheep are not penned inside the building.

[1]The perspectives of several sheep buildings pictured in this section are reproduced with permission from Midwest Plan Service, Iowa State University, Ames. Detailed plans and specifications for these buildings may be obtained from Midwest Plan Service.

Fig. 17-2. Fenceline feeding. (Courtesy, *The Sheepman's Production Handbook*, Denver, CO)

■ **Warm housing**—The building is enclosed and the temperature is maintained between 45 and 65°F. Ventilation is required to remove excess moisture from the building. Insulation is recommended to minimize the required heat needed to maintain the desired temperature.

Fig. 17-3. Many sheep buildings are open—generally to the south. One or more sides of the building are left open and sheep are allowed to move in and out at all times. The temperature in the building is about the same as the outside temperature, though there is protection from wet and prevailing winds. (Photo by J. C. Allen and Son, West Lafayette, IN)

SHEEP-FINISHING FACILITIES

Many lambs are fed in drylots. These are areas where feed is fed under restricted conditions. Drylot feeding may be either open-yard feeding or shelter or barn feeding

OPEN-YARD FEEDING

Open-yard feeding is the common method of finishing lambs in the irrigated areas of the West, though a few eastern lamb-feeding operations are in open yards. In this system, equipment costs are kept to a minimum—the facilities merely consisting of an enclosed and well-drained yard which may or may not have a natural or constructed windbreak, and the necessary feed bunks. Open-yard feeding is often used by large operators who feed thousands of lambs.

SHELTER OR BARN FEEDING

Because of inclement weather in the fall and early winter, many of the lamb-feeding operations in the central and eastern states are in drylots which afford shelter. In some instances, the lambs are kept under cover without an exercising lot. These barns may consist of anything from an open shed to more costly and elaborate structures, including slotted floors.

CONFINEMENT OF SHEEP ON SLOTTED FLOORS

Slotted floors are floors with slots through which the feces and urine pass to a storage area immediately below.

Today, there is much interest in confinement sheep production and slotted floors, among both ewe-lamb producers and lamb feeders. For both groups, it offers new hope for eliminating internal parasites, lessening labor, improving environmental control, lessening space requirements, increasing gains and saving feed, eliminating bedding, and lessening mud, odor, and fly problems. Additionally, for the ewe-lamb producer, it fits in with early weaning and multiple lambing; and for the lamb feeder, it provides a way in which to achieve greater animal comfort during hot weather.

Without doubt, the interest of sheep producers in slotted floors has been accentuated by the extent and success of slotted floors in poultry and swine production.

The shelter above the slotted floor may range all the way from a mere shade over the top of the floor to a completely enclosed, environmentally controlled building; or it may be somewhere between those two extremes—for example, a shed open to one side.

DESIGN REQUIREMENTS

Unfortunately, there is a paucity of experimental work on which to base design recommendations for slotted floors for sheep. Nevertheless, the recommendations in Table 17-5 serve as useful guides.

In addition to the recommendations in Table 17-5, there should be no more than 100 pregnant ewes, or 50 ewes and lambs, or 500 feeder lambs in each group. Also, slat material and width and spacing (slot width) are governed by animal comfort and cleaning efficiency. Many types of slotted floor material have been used. However, some slotted material has been found to cause injury to the feet. Unflattened ¾ in. No. 9 expanded metal provides a good slotted floor for sheep. Slotted floors are supported underneath.

With slotted floors, hay should always be chopped or ground. Long hay may pile up on the slotted floors and prevent manure from dropping through, thereby providing a place for internal parasite development and causing the sheep to befoul themselves.

TABLE 17-5
SPACE REQUIREMENTS FOR VARIOUS AGES AND WEIGHTS OF SHEEP ON SLOTTED FLOORS[1]

Class, Age, and Size of Animal	Slotted Floor Space per Head		Feeder Space per Head[2]				Waterer		
			Limit Fed		Self Fed		Bowl or Nipple	Tank	
	(sq ft)	*(m²)*	*(in.)*	*(cm)*	*(in.)*	*(cm)*	*(head/bowl or nipple)*	*(head/ft)*	*(head/m)*
Ewes with 5- top 30-lb *(2.3- to 13.6-kg)* lambs	10–12[3]	0.9–1.1	16–20	41–51	6–8	15–20	40–50	15–20	49–66
Feeder lambs, 30 to 110 lb *(13.6 to 49.9 kg)*	4–5	0.4–0.5	9–12	23–30	1–2	3–5	50–75	25–40	82–131
Dry ewes, 150 to 200 lb *(68.0 to 90.7 kg)*	8–10	0.7–0.9	16–20	41–51	4–6	10–15	40–50	15–25	49–66
Rams, 180 to 300 lb *(81.6 to 136.1 kg)*	14–20	1.3–1.9	12	30	6	15	10	2	7

[1]Adapted by the author from *Sheep Housing and Equipment Handbook*, Midwest Plan Service, Iowa State University, Ames.

[2]Feeder space per animal depends on: animal size, shorn vs unshorn, breed, pregnancy stage, number of times fed per day, and feed quality.

[3]For lambing rates above 170%, increase floor space 5 sq ft *(0.5 m²)* per head.

Fig. 17-4. Ewes and lambs on slotted floor of ¾ in. expanded metal.

Access to the pit through the slotted floor should be provided by (1) removing floor sections, (2) hinging and tilting up the floor sections, or (3) raising the entire slotted floor with a winch.

MANURE PRODUCTION AND STORAGE

Sheep manure may be handled either as a liquid or as a solid. Because sheep feces are rather dry, in comparison with the feces of cattle and hogs, the manure lends itself to handling as a solid.[2] On the other hand, handling sheep manure as a liquid may, in some operations (*e.g.*, early weaned lambs on a liquid diet), offer some advantages from the standpoint of automation and labor savings. In either case, storage area beneath the slotted floor is required.

The quantity of manure produced varies according to species, kind and amount of feed, and amount of bedding. Table 17-6 shows the daily excretion (free of bedding) of sheep.

When sheep manure is handled as a liquid (rather than as a solid), extra water will need to be added to liquefy the wastes. From one-fifth to three-fifths of the storage volume may be needed for extra water if the manure is to be pumped. For irrigation, there should be about 95% water and 5% manure.

TABLE 17-6
RAW MANURE (FECES AND URINE) PRODUCTION OF SHEEP

Class of Sheep	Daily Volume		Daily Weight		Water
	(cu ft)[1]	*(m^3)*	(lb)	*(kg)*	(%)
Feeder lamb . .	0.065	*0.002*	4	*1.8*	75
Ewe	0.100	*0.003*	6	*2.7*	75
Ram	0.150	*0.004*	10	*4.5*	75

[1]There are about 34 cu ft *(0.96 m^3)* in 1 ton *(907 kg)* of manure.

[2]Fresh sheep manure is about 14% lower in moisture than cow manure.

Of course, the total manure storage capacity will depend on the frequency of cleaning. Here is how to determine how much manure will need to be stored:

Storage capacity = No. of sheep × daily manure production × desired storage time (days) + extra water if handled as a liquid

Pits range up to 10 ft deep. With finishing lambs, a 5-ft pit, cleaned every 100 days, will suffice.

Where manure is to be handled as a liquid, storage tank dimensions and proportions should follow the recommendations of the manure agitator manufacturer. Access to the pit is usually provided from the slotted floor, via either a steel grid or removable slats.

Where manure is handled as a solid, the height of the storage area is determined by two factors: (1) frequency of cleaning, and (2) method of cleaning. The manure may be removed by a tractor-mounted loader or a scraper. Less working height is required where the building is arranged so floor sections may be removed. But the removing of floor sections does require labor. So, consideration should be given to having greater floor height, with access to the pit via doors or removable panels opening from the ends or sides.

Regardless of whether sheep manure is handled as a liquid or as a solid, the area from the floor to the ground must be completely enclosed to prevent drafts.

■ **Manure gases**—When stored inside a building, gases from liquid wastes create a hazard and undesirable odors. Most (95% or more) of the gas produced by manure decomposition is methane, ammonia, hydrogen sulfide, and carbon dioxide. Several have undesirable odors or possible animal toxicity, and some promote corrosion of equipment.

Animals and people can be killed (asphyxiated) because methane and carbon dioxide displace oxygen.

Most gas problems occur when manure is agitated or when ventilation fans fail.

No one should enter a storage tank, unless (1) the space over the wastes is first ventilated with a fan, (2) another person is standing by to give assistance if needed, and (3) that person is wearing self-contained breathing equipment.

Maximum building ventilation must be provided when wastes are being agitated or pumped from a pit. Also, an alarm system (loud bell) to warn of power failures in tightly enclosed buildings is important, because there can be a rapid buildup of gases when forced ventilation ceases.

SHEEP EQUIPMENT[3]

Except for its much smaller size, sheep equipment closely resembles that used for cattle, and its functions and requisites are similar.

HAY RACKS

The essentials of a satisfactory hay rack for sheep are: (1) adequate capacity, (2) ready availability of feed to animals, (3) minimum waste of feed, and (4) protection of the fleece from chaff and from loss of wool by pulling or rubbing. These requisites may be met satisfactorily by hay racks of several different designs.

Frequently, the feed receptacle is designed to handle the feeding of both hay and grain. Combination receptacles of this type are usually superior to separate hay racks from the standpoint of preventing waste of hay containing a high percentage of loose leaves. The loose leaves are saved by the same arrangement that holds grain. In general, these combination racks are of two types: (1) straight-sided, flat-bottomed bunks, and (2) sloping, solid- or slat-sided racks with grain troughs below.

Fig. 17-5 shows a straight-sided, flat-bottomed rack. In this type of rack, the hay and grain are placed on the same bottom, but usually the grain is consumed before the hay is fed. This kind of rack is one of the easiest to build and to keep clean. It is also rather well suited to all kinds of feeds, especially when hand feeding is practiced; though there may be some waste of long hay. These racks may be made any desired length, and the sides may have one long, continuous opening or divided by upright slats into a series of 8-in. openings. A width of about 24 in., outside measurement, is common.

Fig. 17-6 shows a sloping, solid-sided rack with the grain trough below.

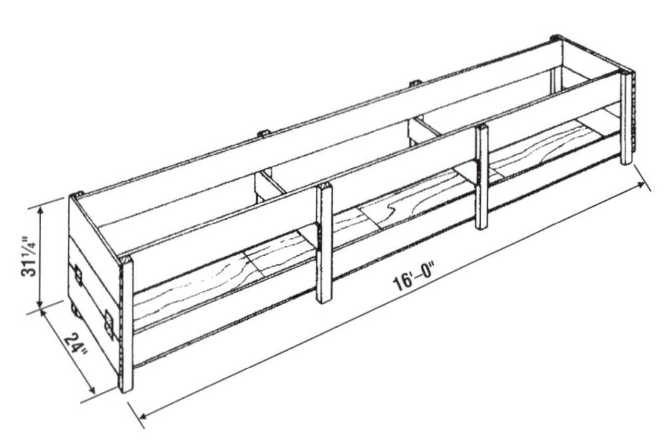

Fig. 17-5. Hay and grain bunk, skid mounted. (From *Sheep Housing and Equipment Handbook*, Midwest Plan Service, Iowa State University, Ames)

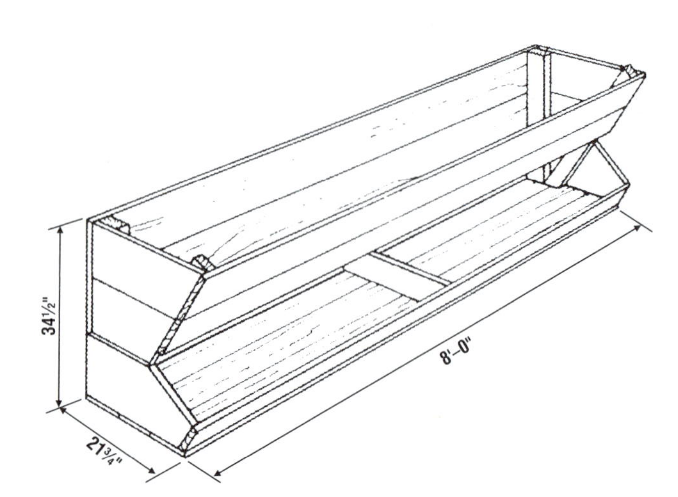

Fig. 17-6. A combination hay and grain bunk with sloping sides. (From *Sheep Housing and Equipment Handbook*, Midwest Plan Service, Iowa State University, Ames)

Fig. 17-7 illustrates a flexible hay feeder whose 12-ft sides can be joined together for occasional use or used separately for portable feeding panels.

Where large, round bales are used, a feeding fence can be constructed, or square or round tubing can be welded to construct a self-feeder such as the one shown in Fig. 17-8.

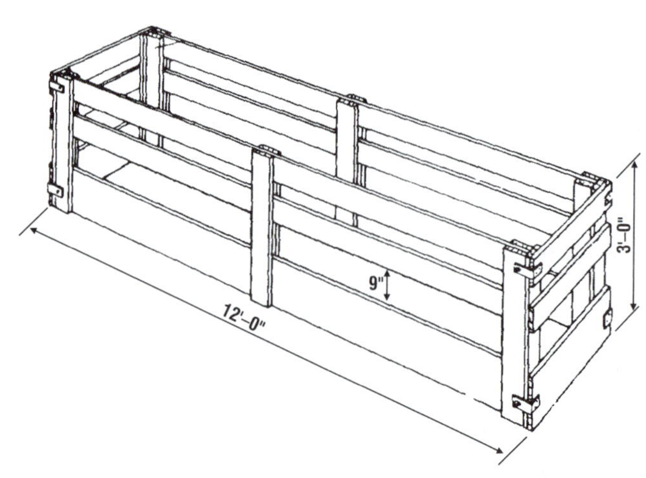

Fig. 17-7. Hay feeder. This feeder can be joined with hasps for frequent use or wired together for occasional use. Also, the sides can be used separately for portable feeding panels. (From *Sheep Housing and Equipment Handbook*, Midwest Plan Service, Iowa State University, Ames)

[3]The perspectives of several pieces of sheep equipment pictured in this section are reproduced with permission from Midwest Plan Service, Iowa State University, Ames. Detailed plans and specifications for these pieces of equipment may be obtained from Midwest Plan Service.

Fig. 17-8. Large bale self-feeder. The sides collapse as the bale gets smaller, and the bottom panel saves leaves and stems. (Courtesy, Shalom Valley Sheep Equipment, Inc., Gary, SD)

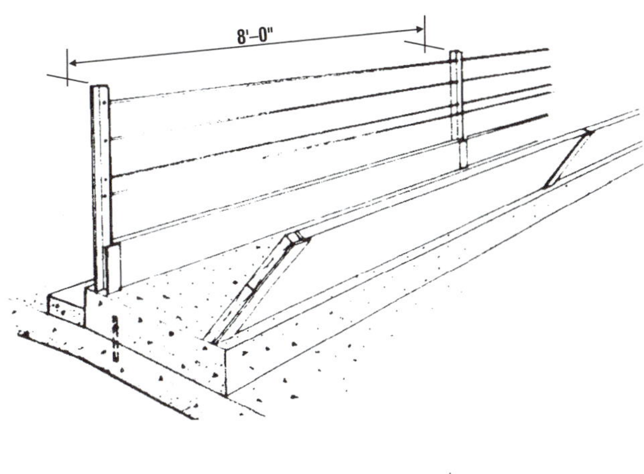

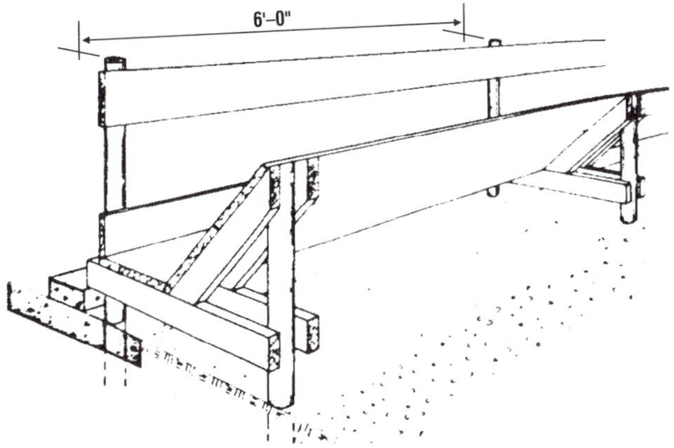

Fig. 17-10. Fenceline bunks, two examples of construction. (From *Sheep Housing and Equipment Handbook*, Midwest Plan Service, Iowa State University, Ames)

GRAIN TROUGHS AND FENCELINE BUNKS

There are many types of troughs for feeding grain, silage, and roots. When hay is fed in separate racks, feed troughs are usually 10 to 12 in. wide, 3 to 4 in. deep, with a throat height 10 to 15 in. from the ground. They may be of any desired length. They should be constructed so that they cannot be easily pushed over, and there should be a guard rail along the top to strengthen them and to keep lambs from getting into them. Troughs should be easily cleaned. Fig. 17-9 shows a desirable type of sheep grain trough.

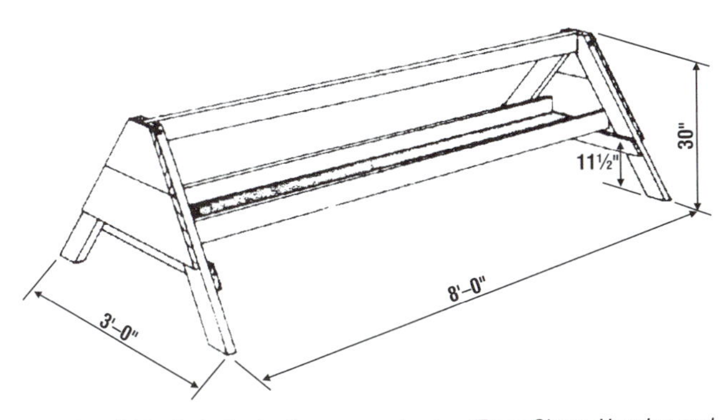

Fig. 17-9. Grain feeder for ewes or lambs. (From *Sheep Housing and Equipment Handbook*, Midwest Plan Service, Iowa State University, Ames)

In some situations, fenceline bunks are used for feeding grain, silage, or roots to sheep. There are many variations of fenceline bunks. Two examples are shown in Fig. 17-10.

SELF-FEEDERS

Self-feeders are frequently used by lamb feeders. In comparison with hand feeding, less labor is required; and equal or superior gains may be secured without any greater death losses from overeating, provided that the grain is mixed with chopped roughage in the proportions of approximately 45% grain and 55% roughage. For greatest efficiency of labor, self-feeders must be relatively large so that frequent refilling is avoided. In hand feeding, about 12 in. of space is needed for each lamb; whereas in self-feeding this same amount of space will accommodate three lambs. Self-feeders may need attention several times a day to keep the feed poked down so that it is always available. How well self-feeders feed-down depends on the nature of the feed, the weather, and the kind of feeder.

Sometimes, chopped hay is placed in large self-feeders with an overhanging type of roof. Feeders of this type are convenient for wintering breeding sheep in fenced pastures. They also are practical near a cen-

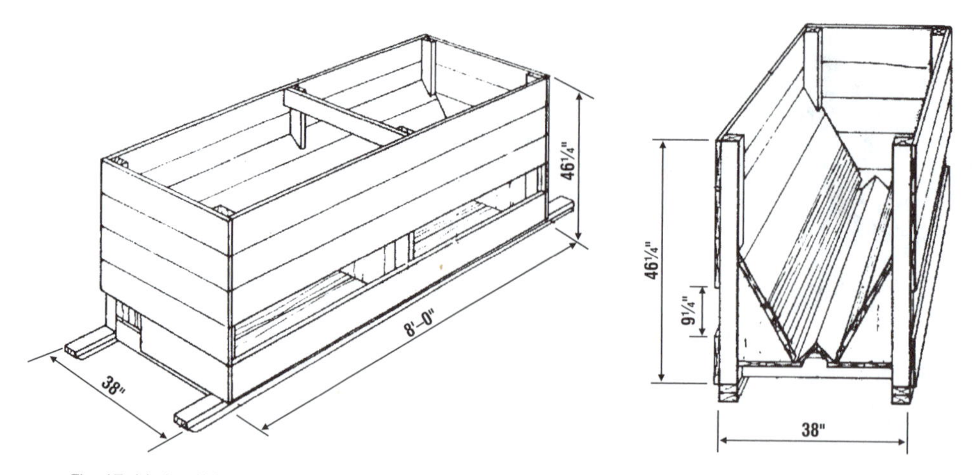

Fig. 17-11. A self-feeder for grain or pellets for feeder lambs. Based on the dimensions shown, it holds 40 bushels of feed. The diagram on the right is a cross-section of the feeder. It can also be covered with a roof. (From *Sheep Housing and Equipment Handbook*, Midwest Plan Service, Iowa State University, Ames)

Fig. 17-12. A headgate for self-feeding silage out of a plastic silo bag in Oregon. As the ewes eat the silage, they push the headgate forward on the wooden track on either side of the bag. When the headgate reaches the end of the track, the track is moved forward. (Courtesy, David L. Thomas, University of Wisconsin, Madison)

Fig. 17-13. A headgate for self-feeding silage out of a bunk silo in Wisconsin. The headgate is suspended by chains from a beam which lays across the top of the bunk. As the ewes consume the silage, the beam is moved forward to allow access to more silage. Note the lack of wasted silage in the bottom of the bunk. (Courtesy, Tom Cadwallader, University of Wisconsin, Madison)

tral lambing shed where ewes are held until the lambs are old enough to go on the range.

MINERAL FEEDERS

When minerals (salt and other needed minerals) or salt mixtures are provided on pasture, the container should be constructed to protect the materials from rain. Fig. 17-14 shows several suitable mineral feeders.

WATERING FACILITIES

As indicated in Table 17-7 sheep will consume from 1 to 3 gal. of water per head daily, with variations according to size of animal, season, type of feed, and temperature of water. When on pastures or ranges, sheep frequently drink water from reservoirs, springs, lakes, and streams; or they may even get water by consuming winter snow. For sheep in confinement, water may be piped to tanks, tubs, troughs, barrels, or other equipment. The main essentials of water containers are: (1) adequate size or number for the flock or band, (2) protection or heating devices to prevent freezing in cold weather, (3) ease of cleaning, and (4) convenience to animals to encourage frequent watering.

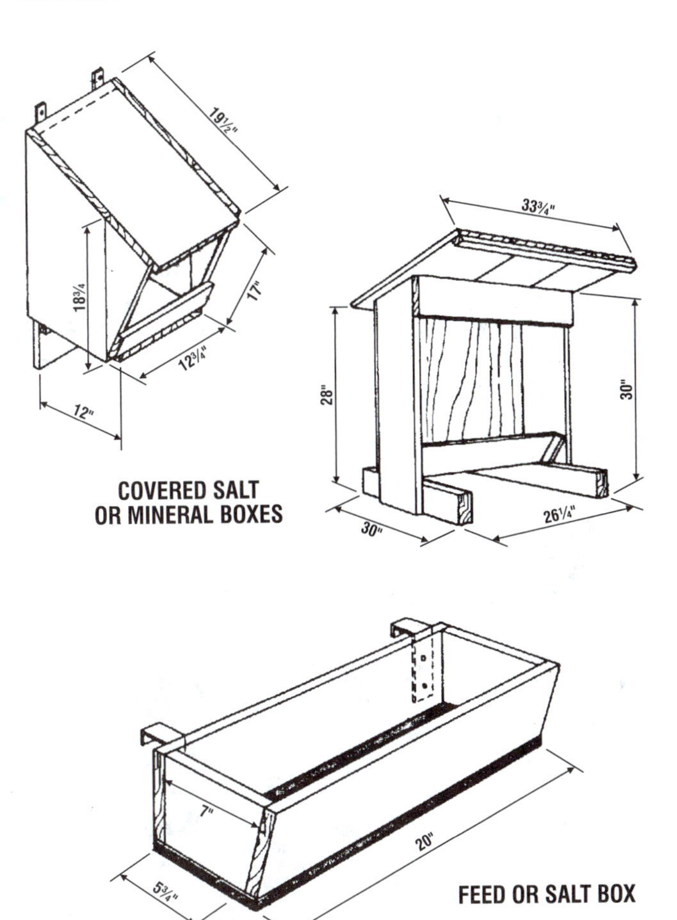

Fig. 17-14. Three designs for mineral or salt feeders. (From *Sheep Housing and Equipment Handbook*, Midwest Plan Service, Iowa State University, Ames)

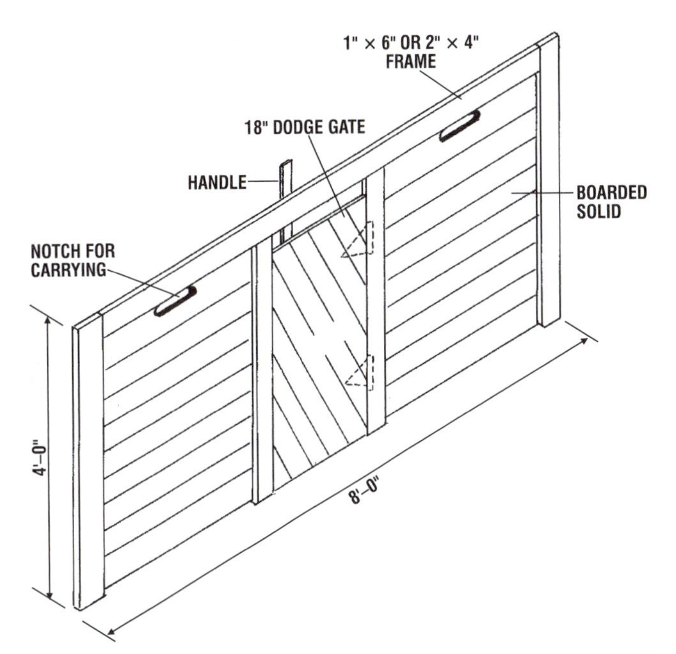

Fig. 17-15. Sheep cutting panel and dodge gate. The sheep cutting panel may be used singly or in pairs for more than a two-way cut. (Courtesy, Washington State University)

Though convenience is important, it is preferable that water containers be some distance from feeders, otherwise sheep will carry feed in their mouths and foul the water.

CUTTING CHUTE

A cutting chute is easily made, inexpensive, and extremely useful for sorting sheep. By means of one dodge gate, the flock or band may be sorted into two lots. With additional gates, more sorts may be made; or more than two sorts may be made with a single dodge gate arrangement simply by reworking one or both lots following the first sort. The specifications of a desirable cutting panel and dodge gate are given in Fig. 17-15. By flaring the top out, both ewes and lambs may be sorted in the same chute. Dodge gates should swing from one side of the chute to the other to permit cutting into separate pens. In addition to making for greater convenience and reducing labor, a cutting chute saves much disturbance and running of animals in working over the flock or band.

SQUEEZE

A sheep squeeze restrains an animal for examination or treatment. Many squeezes tilt sheep to a supine position (lying on the back) permitting regular inspection and trimming of the feet.

Fig. 17-16. A tilting sheep squeeze. The sheep can be tipped on its side or back for examination or treatment. (Courtesy, Sheepman Supply Co., Barboursville, VA)

DIPPING OR SPRAYING EQUIPMENT

For the control of external parasites (ticks, lice, and scab mites), most flocks or bands need to be treated with an insecticide at least once a year, preferably soon after shearing. This operation may be satisfactorily done by either dipping or spraying. Rectangular dipping vats are of variable length, but the rest of the dimensions are rather standard as follows: 4 ft deep, 10 to 12 in. inside width at the bottom, and 20 to 24 in. wide at the top. The better vats are constructed with suitable entrances, an inclined exit, and a drain platform. Some operators prefer the round dipping vat for sheep and goats; such a vat is 5 ft deep and 5 ft in diameter, with a catch pen and drain pen, each 12 ft × 12 ft.

A number of insecticides are registered for control of ticks and lice as a spray or a dip. If an insecticide is applied as a spray, the equipment should be able to provide sufficient pressure to soak the animals thoroughly.

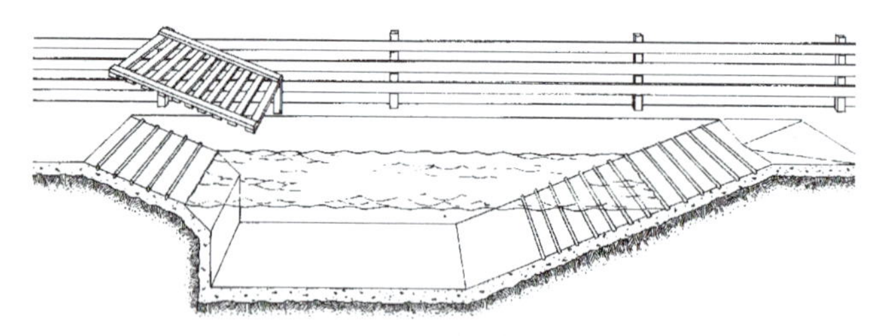

Fig. 17-17. A cross-section of a permanent concrete dipping vat. The entry and exit ramps are created and on a 20- to 25-degree angle. The steeper portion of the entry is hidden under water and is at a 45-degree angle. The low roof over the entry keeps animals from jumping into the vat and splashing.

LAMBING CUBICLE

Lambing cubicles prevent lambs from becoming separated from the ewe shortly after birth and prevent "lamb stealing" by other ewes because the birth ewe only has a two-foot opening to defend against other ewes.

Lambing cubicles should be provided for ewes and newborn lambs. Fig. 17-18 shows a satisfactory type. These pens are usually 4 ft square and are made by placing together two hinged hurdles, which are then set against the walls of the sheep barn.

Fig. 17-18. Ewe with newborn lambs in a lambing cubicle. (Photo from Harold Gonyov, University of Illinois, Urbana-Champaign. Courtesy, Dr. David L. Thomas, University of Wisconsin, Madison)

TABLE
SPACE REQUIREMENTS OF BUILDINGS

Age, Class, and Size of Animal	Barn or Shed			Feedlot		Shades	
	Floor Area per Animal[3]	Height of Ceiling	Window Space (not including open sheds)	Area if Dirt	Area if Paved	Area per Animal	Height
	(sq ft)	*(ft)*	*(sq ft)*	*(sq ft)*	*(sq ft)*	*(sq ft)*	*(ft)*
Dry ewes (150–200 lb)	16	8½–10	1 sq ft window space per 35 sq ft floor space	25–40	16	10–12	8–10
Ewes with lambs (5–30 lb)	20[5]	8½–10	"	30–50	20	14	8–10
Feeder lambs (30–110 lb)	8–10	8½–10	"	25–30	10	6–8	8–10
Stud rams (180–300 lb)	20–30	8½–10	"	30–60	25	15	8–10

[1]For conversion to metric, refer to the Appendix. Wide barn doors are needed to prevent crowding and possible injury to pregnant ewes. Doors at least 8 ft wide are preferable. Provide a paved area of at least 5 ft around waterers, feed bunks, roughage racks, and entrances to sheds.

[2]Approximate water requirements are as follows: 2.0 gal./day for dry ewes; 3.3 gal./day for ewes with lambs; 1.5 gal./day for feeder lambs; and 3.0 gal./day for rams. Water requirements vary widely with time of year and ration. In the summer maintain water temperature below 75°F and in the winter above 35°F.

[3]For specifications for slotted floors, see section entitled "Confinement of Sheep on Slotted Floors."

GRAFTING PENS

Potential foster mothers are locked in the headgates of a grafting pen, where they can have no visual or tactile contact with the orphan lamb(s). After approximately five days in the stantion, over 85% of ewes will accept the orphan lamb(s).

LAMB CREEP

A lamb creep is an enclosure with openings large enough for lambs to enter, but small enough to keep out older sheep. A lamb creep makes it possible to feed concentrates to lambs in a separate enclosure away from their dams.

Fig. 17-19. Grafting stantion for ewes in Oklahoma. (Courtesy, Dr. David L. Thomas, University of Wisconsin, Madison)

Fig. 17-20. A small portable lamb creep for lambs on pasture in Oregon. (Courtesy, Dr. David L. Thomas, University of Wisconsin, Madison)

SHADES

Providing adequate shade to protect sheep from the hot sun is among the more important and widely used devices for improving the environment of sheep in hot climates.

The most satisfactory shades (1) provide 4 to 6 sq ft of shade per animal, (2) are 4 to 8 ft high, (3) are located with a north-south placement because they are drier—the sun can get underneath them to dry out the manure and urine, and (4) are open all around, so as to permit maximum air movement. Portable shades may be used.

SPACE REQUIREMENTS OF BUILDINGS AND EQUIPMENT FOR SHEEP

Average and conservative space requirements of buildings and equipment for sheep are presented in Table 17-7.

17-7
AND EQUIPMENT FOR SHEEP[1]

Hay or Silage Manager or Rack (hand feeding)				Feed Trough (for grain or roots; hand feeding)				Self-feeder (for concentrate or roughage)	Watering Equipment[2]	
Length per Animal	Width if Feeds from One Side	Width if Feeds from Both Sides	Height at Throat	Length per Animal	Width if Feeds from One Side	Width if Feeds from Both Sides	Height at Throat	Trough Length if Feeder is Kept Filled	Bowl or Nipple	Tank
(in.)	*(in.)*	*(in.)*	*(in.)*	*(in.)*	*(in.)*	*(in.)*	*(in.)*	*(in.)*	*(head/unit)*	*(head/ft)*
12	14–16	20–24	12–15	12	14–16	20–24	10–15	4–6[4]	40–50	15–25
12	14–16	20–24	12–15	12	14–16	20–24	10–15	6–8[6]	40–50	15–25
12	12–14	18–22	10–12	12	14–16	12	8–12	1–3	50–75	25–40
12	14–16	20–24	12–15	20–24	14–16	20–24	10–15	6	10	2

[4]When slat is used as a governor, 3 in. will suffice.

[5]For lambing rates above 170%, increase floor space 5 sq ft/head. Also, lambs require 1.5 to 2.0 sq ft of creep space/lamb.

[6]Lambs need 2 in. of space at creep feeder.

RECOMMENDED MINIMUM WIDTH OF SERVICE PASSAGES

In general, the requirements for service passages are similar, regardless of the kind of animal. Accordingly, the suggestions contained in Table 17-8 are equally applicable to sheep, cattle, swine, and horse barns.

TABLE 17-8
RECOMMENDED MINIMUM WIDTHS FOR SERVICE PASSAGES

Kind of Passage	Use	Minimum Width	
		(ft)	(m)
Feed alley . . .	For feed cart	4	1.2
Driveway. . . .	For wagon, spreader, or truck	9	2.7
Doors and gate .	Drive-through	9	2.7
Doors and gate .	To small pens	4	1.2

STORAGE SPACE REQUIREMENTS FOR FEED AND BEDDING

The space requirements for feed storage for the livestock enterprise—whether it be for sheep, cattle, hogs, or horses, or as is more frequently the case, a combination of these—vary so widely that it is difficult to provide a suggested method of calculating space requirements applicable to such diverse conditions. The amount of feed to be stored depends primarily upon: (1) length of pasture season, (2) method of feeding and management, (3) kind of feed, (4) climate, and (5) proportion of feeds produced on the farm or ranch in comparison to the proportion purchased. Normally, the storage capacity should be sufficient to handle all feed grain and silage grown on the farm and to hold purchased supplies. Forage and bedding may or may not be stored under cover. In those areas where weather conditions permit, hay and straw are frequently stacked in the fields or near the barns in loose, baled, or chopped form. Sometimes poled framed sheds or a cheap cover of plastic or wild grass is used for protection. Other forms of low-cost storage include temporary upright silos, trench silos, temporary grain bins, and open-wall buildings for hay.

Table 17-9 gives the storage space requirements for feed and bedding. This information may be helpful to the individual operator who desires to compute the barn space required for a specific livestock enterprise. This table also provides a convenient means of estimating the amount of feed or bedding in storage.

TABLE 17-9
STORAGE SPACE REQUIREMENTS FOR FEED AND BEDDING[1]

Kind of Feed or Bedding	Pounds per Cubic Foot	Cubic Feet per Ton	Pounds per Bushel of Grain	Cubic Feet per Bushel
Hay–straw				
1. Loose				
Alfalfa	4.4–4.0	450–500		
Nonlegume	4.4–3.3	450–600	—	—
Straw	3.0–2.0	670–1,000		
2. Baled				
Alfalfa	10.0–6.0	200–300		
Nonlegume	8.0–6.0	250–330	—	—
Straw	5.0–4.0	400–500		
3. Chopped				
Alfalfa	7.0–5.5	285–360		
Nonlegume	6.7–5.0	300–400	—	—
Straw	8.0–5.7	250–350		
Silage				
Corn or sorghum in tower silos.	40	50		
Corn or sorghum in trench silos	35	57		
Grain				
Corn, shelled[2]	45	45	56	1.25
Corn, ear	28	72	70	2.50
Corn, shelled, ground . .	38	—	48	—
Barley.	39	51	48	1.25
Barley, ground	28	—	37	—
Oats	26	77	32	1.25
Oats, ground	18	106	23	—
Rye	45	44	56	1.25
Rye, ground	38	—	48	—
Sorghum	45	44	56	1.25
Wheat	48	42	60	1.25
Wheat, ground	43	46	50	—
Mill feed				
Bran	13	154		
Middlings	25	80		
Linseed meal.	23	88		
Cottonseed meal	38	53		
Soybean meal	42	43		
Alfalfa meal	15	134		
Miscellaneous				
Soybeans	48	56		
Salt, fine	50	40		
Pellets, mixed feed . . .	37	58		
Shavings, baled.	20	100		

[1]For metric conversions, refer to the Appendix.

[2]These values are for corn with 15% moisture. Corn that is 30% moisture weights 51 lb/cu ft shelled and 36 lb/cu ft ground or 68 lb/bu shelled and 90 lb/bu ground.

FENCES FOR SHEEP

Good fences (1) maintain farm boundaries, (2) make livestock operations possible, (3) reduce losses to both animals and crops, (4) increase land values, (5) promote better relationships between neighbors, (6) lessen accidents resulting from animals getting on roads, and (7) add to the attractiveness and distinctiveness of the premises.

The discussion which follows will be limited primarily to wire fencing, although it is recognized that such materials as rails, poles, boards, stone, hedge, pipe, and concrete have a place and are used under certain circumstances. Also, where there is a heavy concentration of animals, such as in corrals and feed yards, there is need for a more rigid type of fencing material than wire. Moreover, certain fencing materials have more artistic appeal than others; and this is an especially important consideration on the purebred establishment.

The kind of wire to purchase should be determined primarily by the class of animals to be confined.

The following additional points are pertinent in the selection of wire:

1. **Styles of woven wire.** The standard styles of woven wire fences are designated by numbers such as 958, 1155, 849, 1047, 741, 939, 832, and 726. The first 1 or 2 digits represent the number of line (horizontal wires): the last 2 the height in inches.
2. **Mesh.** Generally, a close-spaced fence with stay or vertical wires 6 in. apart (6-in. mesh) will give better service than a wide-spaced fence.
3. **Weight of wire.** A fence made of heavier-weight wires will usually last longer and prove cheaper than one made of lighter wires. Heavier or larger-size wire is designated by a smaller-gauge number.
4. **Standard-size rolls or spools.** Woven wire comes in 20- and 40-rod rolls; barbed wire in 80-rod spools.
5. **Wire coating.** The kind and amount of coating on wire definitely affect its lasting qualities. Galvanized coating is most commonly used to protect wire from corrosion. Coatings are specified as Class I, Class II, and Class III. The higher the class number, the greater the coating thickness.

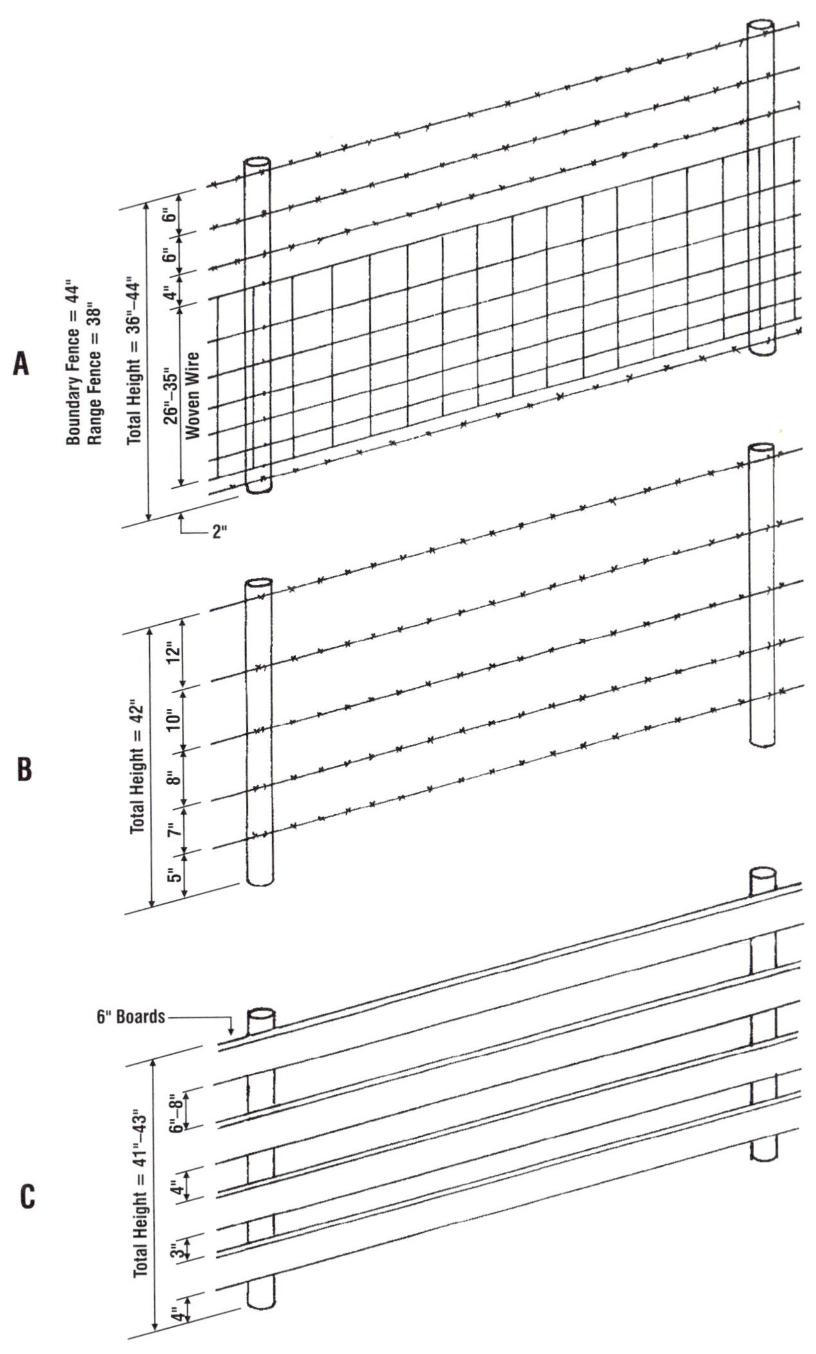

Fig. 17-21. Fences for sheep: (A) a woven wire plus barbed wire fence, which can be made to discourage dogs from digging under by adding a strand of barbed wire at ground level; (B) a barbed wire fence for pasture; and (C) a plank fence. (Fence types and most wire spacings are from recommendations in *Sheep Housing and Equipment Handbook*, Midwest Plan Service, Iowa State University, Ames)

Three kinds of material are commonly used for fence posts: wood, metal, and concrete. The selection of the particular kind of posts should be determined by (1) the availability and cost of each, (2) the length of service desired (posts should last as long as the fencing material attached to it, or the maintenance cost may be too high), (3) the kind and number of livestock to be confined, and (4) the cost of installation.

1. **Wood posts.** Osage orange, black locust, chestnut, red cedar, black walnut, mulberry, and catalpa—each with an average life of 15 to 30 years without treatment—are the most durable wood posts, but they are not available in all sections. Untreated posts of the other and less durable woods will last 3 to 8 years only, but they are satisfactory if properly butt treated (to 6 to 8 in. above the ground line) with a good wood preservative.

■ **Treating posts**—The less durable types of fence posts will last about five times longer when treated than when untreated. This effects yearly savings in two ways: (1) in the cost of posts, and (2) in the labor involved in fence construction.

Although the relative durability of posts does not materially affect initial fencing costs, the length of life of the posts is the greatest single factor in determining the cost of a fence on an annual basis.

Farmers and ranchers should check with their local County Extension Agent, or other local authority, relative to approved wood preservatives; and preservatives should always be used in keeping with the directions of their manufacturers.

Pressure treating is preferable because it forces the preservative to the center of the post—leaving none of the wood untreated.

2. **Metal posts.** Metal posts last longer, require less storage space when not in use, and require less labor in setting than wood posts.

Metal line posts are made in different styles and cross-sections. Heavier studded "T" or "Y" section posts are most popular for livestock, though lighter channel posts may be used for temporary and movable fences.

3. **Concrete posts.** When properly made, concrete posts give excellent service over many years. In general, however, they are expensive.

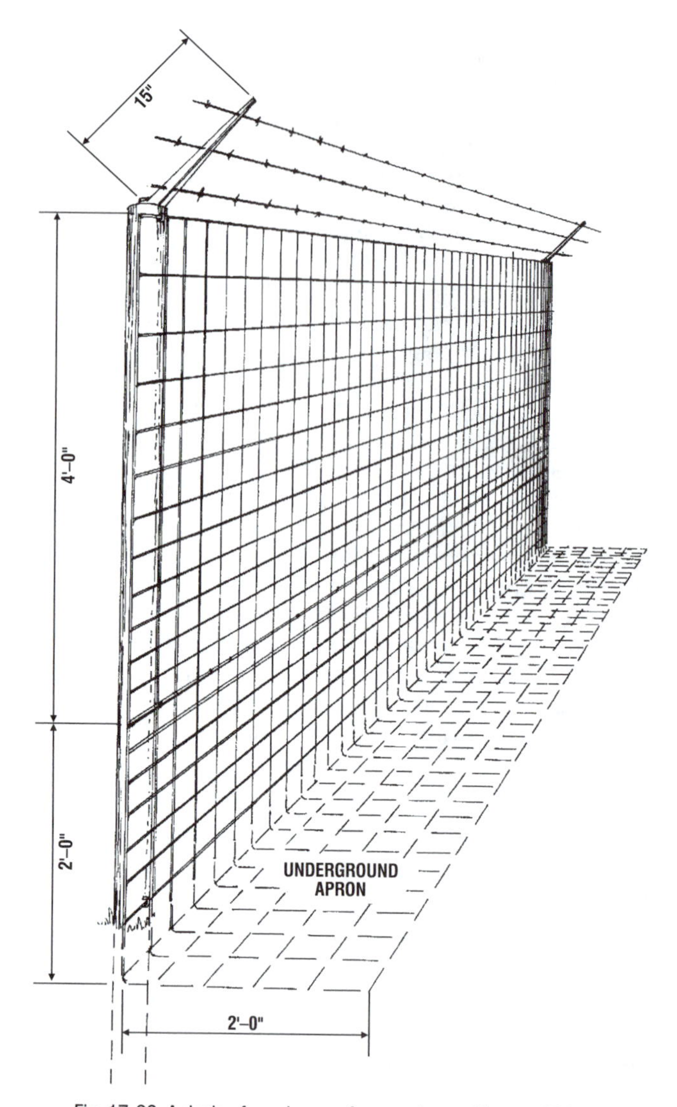

Fig. 17-22. A design for a dogproof or coyoteproof fence which prevents digging under or jumping over. Stays on metal posts should be added between wood posts 30 ft or more apart. The two rolls of woven wire can be joined with hog rings.

DOGPROOF FENCES

Sheep corrals should be fenced with dogproof fencing. Although such fencing is difficult to construct and expensive, it will pay dividends in the protection afforded.

A satisfactory dogproof fence should be a minimum of 6 ft in height (preferably 7 ft). Two rolls of woven wire are used as shown in Fig. 17-22. To keep dogs from digging under the fence, the wide-spaced horizontal wires are buried to form an apron. To prevent dogs from jumping the fence, three strands of barbed wire run across the top on an arm as shown in Fig. 17-22.

ELECTRIC FENCES

Where a temporary enclosure is desired or where existing fences need bolstering from roguish or breachy animals, it may be desirable to install an electric fence, which can be done at minimum cost.

The following points are pertinent in the construction and use of an electric fence:

1. **Safety.** When farmers or ranchers plan to install and use electric fences, they should first check into the regulations of their individual states relative to the installation and use of electric fences and they should then take the necessary safety precautions

against accidents to both persons and animals. *An electric fence can be dangerous!*

2. **Charger.** The charger should be safe and effective—one made by a reputable manufacturer.

3. **Wire height.** As a rule of thumb, the correct wire height for an electric fence is about three-fourths the height of the animal, with two wires provided for sheep (see Fig. 17-23).

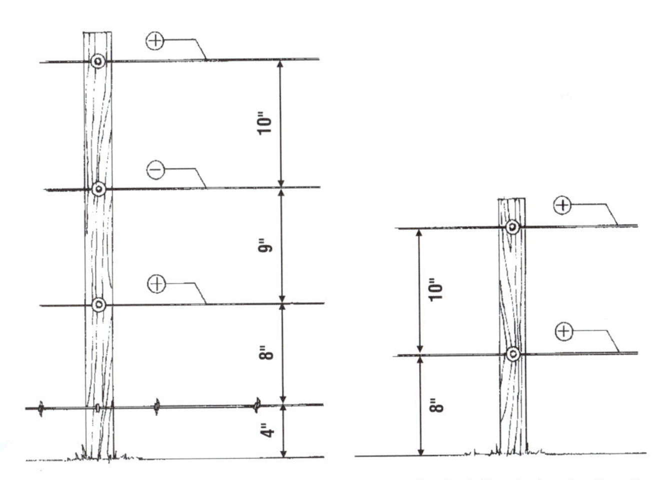

Fig. 17-23. Two designs for electric fences. On the left a design for dry climates with a barbed strand that can be added when lambs are present. On the right a double-circuit electric fence. The shock between two wires is twice the shock between either wire and the ground. (Wire spacings and arrangement are from recommendations in *Sheep Housing and Equipment Handbook*, Midwest Plan Service, Iowa State University, Ames)

4. **Posts.** Wood, steel, plastic, or fiberglass posts may be used for electric fencing. Corner posts should be as firmly set and well braced as required for any nonelectric fence so as to stand the pull necessary to stretch the wire tight. Line posts (a) need only be heavy enough to support the wire and withstand the elements, and (b) may be spaced 25 to 40 ft apart for sheep.

5. **Wire.** The most popular galvanized and aluminized wire ranges from 12½ gauge high-tensil wire for permanent fences used in rotational grazing systems. In those states where barbed wire is legal for electric fences, 12½ gauge hog wire is preferred. Never use rusty wire, because rust is an insulator.

6. **Insulators.** Wire should be fastened to the posts by insulators and should not come into direct contact with posts, weeds, or the ground. Inexpensive solid glass, porcelain, or plastic insulators should be used, rather than old rubber or necks of bottles.

7. **Grounding.** Inadequate grounding is the root of power-surge problems. Always check with the manufacturer or distributor to determine proper grounding methods. *An electric fence should never be grounded to a water pipe, because it could carry lightning directly to connecting buildings. A lightning arrestor should be installed on the ground wire.* If electric fences are installed in dry areas with little soil moisture, ground wires need to be installed on the fence along with hot wires to ensure that animals will receive an electrical shock when touching the hot wire.

8. **Conditioning the sheep.** Wool (especially dry wool) is a poor conductor of electricity, so the body areas without wool (face and legs) should receive the initial electrical shock. For the first exposure, the flock should be moved slowly into the fenced area. If they are frightened, they may crowd through the fence and perhaps break the electric fence wires. Special attempts to attract the attention of sheep to the electric wires have proved worthwhile. One method of accomplishing this is to attach small aluminum pans to the wires to attract the sheep to the wires, so that they will receive the initial shock.

TWENTY-FIRST CENTURY BUILDINGS AND EQUIPMENT FOR SHEEP

The section which follows is from the *Midwest Plan Service*, Fourth Edition, 1994, Iowa State University, Ames. The author has selected and reproduced from this publication perspectives, tables, and other pertinent information which he feels should be considered by progressive and modern sheep producers. Detailed plans and specifications may be obtained from the Midwest Plan Service.

REMODELING EXISTING BUILDINGS AND EQUIPMENT

Use existing buildings and equipment only if their size, location, condition, and arrangements fit the overall plan. If remodeling costs will exceed 60% of the cost of new buildings and equipment, consider new buildings and equipment.

BARN STYLES

Barn styles are determined by the roof shape; the most common are shed, gable, and offset gable.

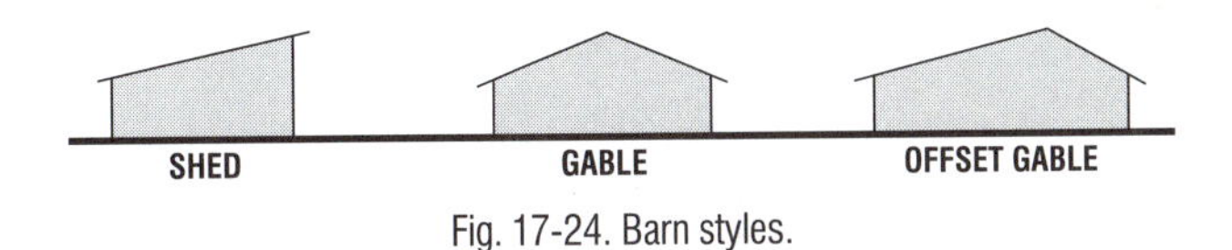

Fig. 17-24. Barn styles.

ESTIMATED DAILY FEED STORAGE NEEDS FOR SHEEP

Use Table 17-10 to estimate feed storage, not to estimate daily rations.

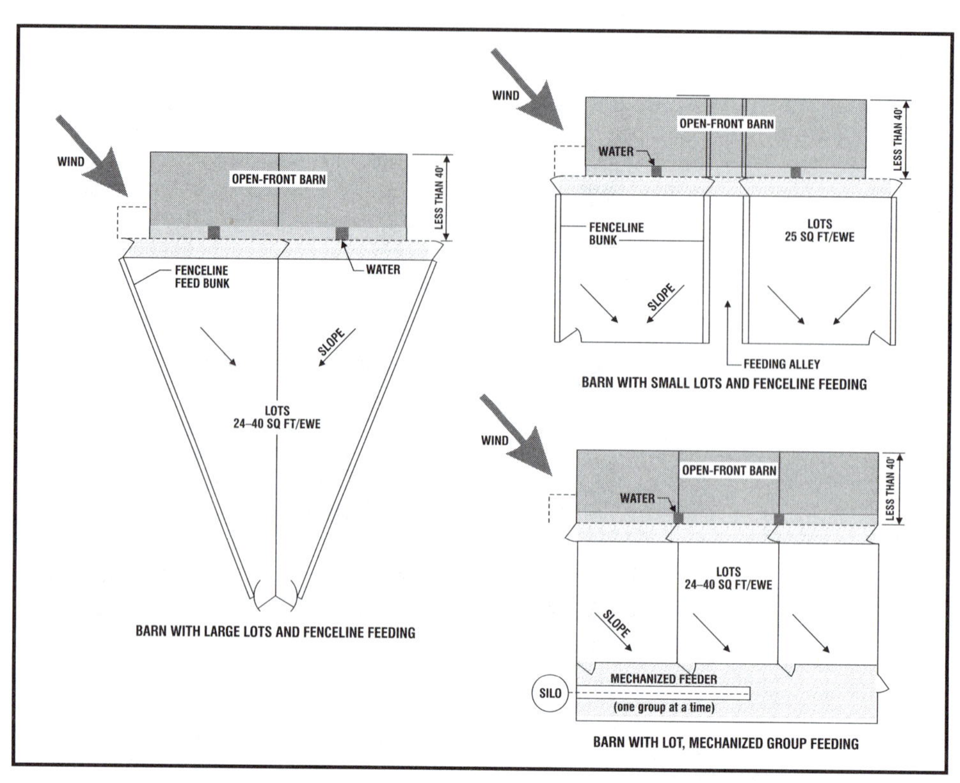

Fig. 17-25. Ewe feeding layout. Use 10 to 12 sq ft/ewe barn space; 16 to 20 in./ewe feeding space; and 40 sq ft/ewe lot space with 25 in. or more annual rain or 25 sq ft/ewe with good lot drainage and less than 25 in. annual rainfall.

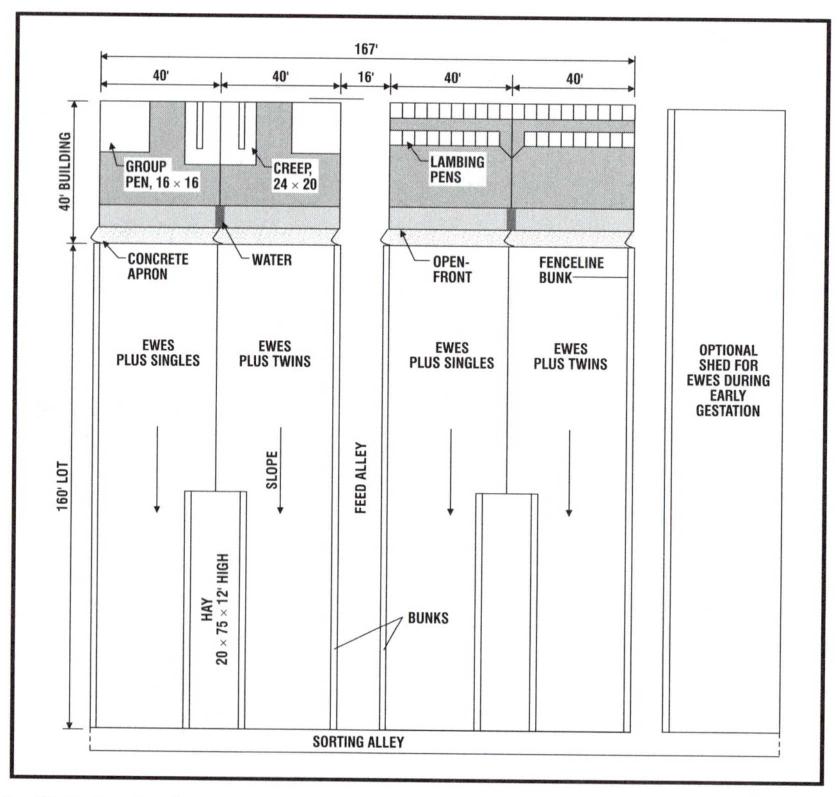

Fig. 17-26. Four hundred ewes, outside feeding, separation of twins and singles. Once-a-year lambing.

TABLE 17-10
ESTIMATED DAILY FEED STORAGE NEEDS FOR SHEEP

	Roughage[1]				
	Hay	Haylage	Corn Silage	Grain[2]	Protein Supplement
	(lb/animal)				
Ewe					
Maintenance	2.5–4	6–7	7–8		0.125–0.25
Breeding	2.5–4	7–8	8–9	0.5–1.0	
Early gestation	2.5–4	7–8	8–9		0.125–0.25
Late gestation	4–5[3]	9–10[3]	10–11[4]	0.5–1.0	0.125–0.25
Lactation	5–7[3]	10–12[3]	11–13[4]	1.0–2.5	0.25–0.50
Ram	4–7	8–10	11–15	0.5–2.5	0–0.25
Replacements	2–4	6–7		1.0–2.0	0.25–0.50
Feeder lambs (30–110 lb)	0.5–2	2–4	4–6	1.0–3.5	0.25–0.50

[1]Reduce amounts fed when more than one roughage fed at one time.
[2]Fed in addition to the roughage.
[3]Protein supplement not required.
[4]Mineral supplement required.

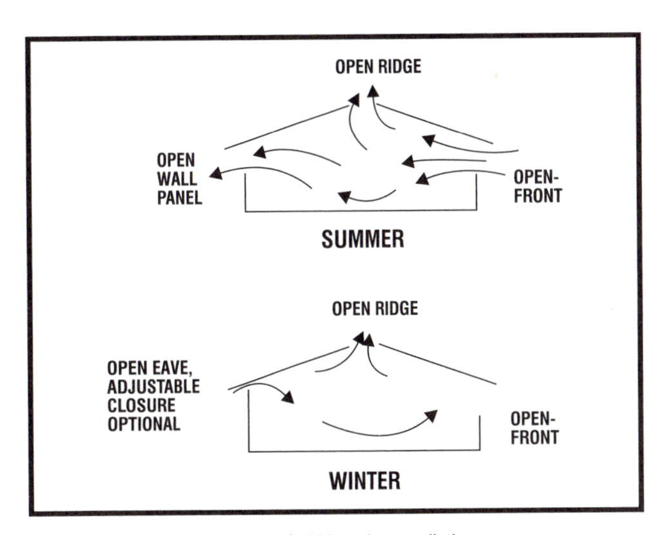

Fig. 17-27. Cold housing ventilation.

COLD HOUSING VENTILATION

Inside temperatures are usually a few degrees warmer in cold weather, and a few degrees cooler in hot weather, than outside temperature extremes. Buildings must have well-planned and located inlets (fresh air) and outlets (moist air) for adequate air movement throughout the barn. Cross ventilation and roof shade in hot weather help reduce inside temperatures. Protect lambing area with temporary walls or drop curtains during cold weather lambing.

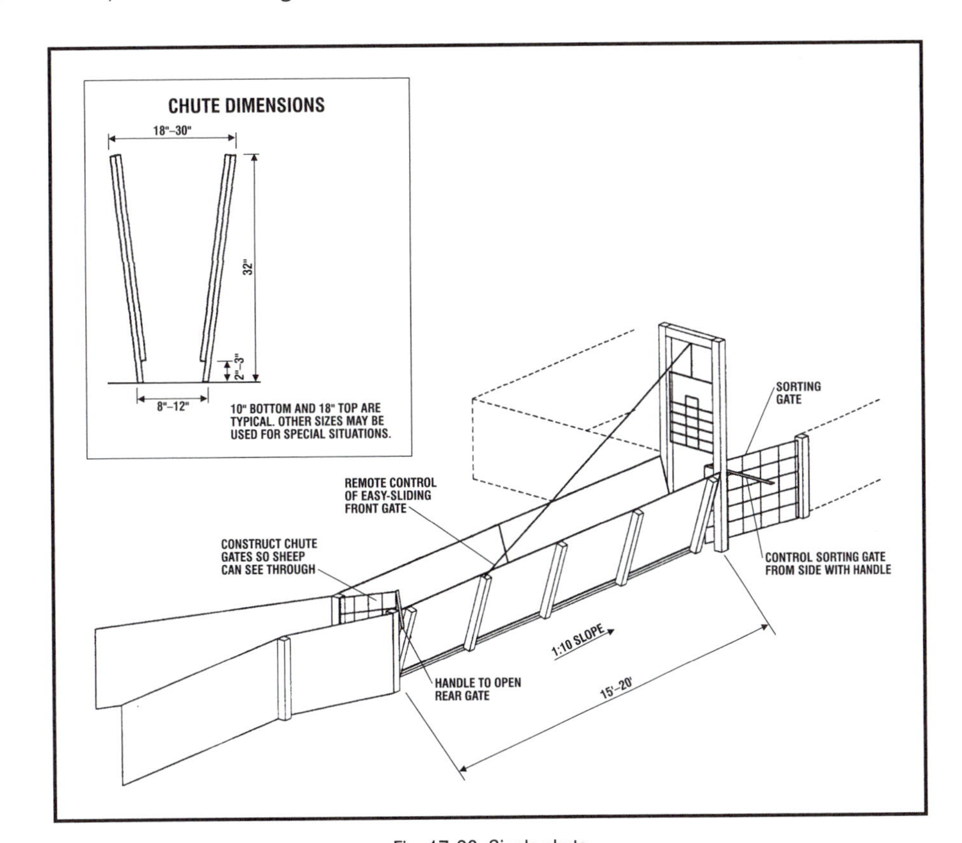

Fig. 17-28. Single chute.

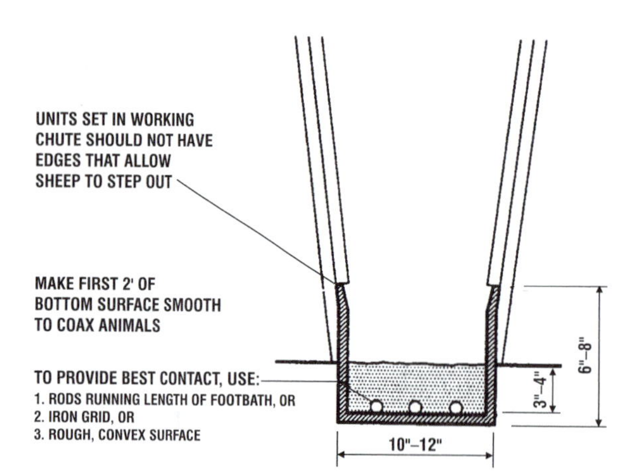

Fig. 17-29. Footbath.

QUESTIONS FOR STUDY AND DISCUSSION

1. Why is there increasing interest in some type of confinement for sheep?

2. When confinement is considered, why is it necessary to know how much heat the bodies of sheep produce?

3. In buildings for sheep, the major requirement of winter ventilation is moisture removal, whereas the major requirement of summer ventilation is temperature control. Why the difference?

4. List and discuss some environmental factors that are important for an operator to consider when choosing and designing buildings for sheep.

5. Describe the three general types of sheep buildings.

6. Describe (a) open-yard feeding and (b) shelter facilities for feeding lambs.

7. Describe the requirements for sheep on slotted floors.

8. Calculate the manure storage capacity necessary for 500 feeder lambs on slotted floors.

9. What factors must be considered when a sheep raiser is selecting feeders? Consider hay, grain, and mineral feeders.

10. Why are cutting chutes important and necessary equipment for handling sheep?

11. List the essentials of a well-designed dipping vat, lambing pen, lamb creep, and shade.

12. One of the first and frequently one of the most difficult problems confronting the sheep producer who wishes to construct a building or a piece of equipment is that of arriving at the proper size or dimensions. In planning to construct new buildings and equipment for sheep, what factors and measurements for buildings and equipment should sheep producers consider?

13. Make a critical study of your own sheep barn or shed, or one with which you are familiar, and determine its (a) desirable and (b) undesirable features.

14. Diagram and describe (a) a dogproof fence and (b) an electric fence for sheep in your area.

SELECTED REFERENCES

Title of Publication	Author(s)	Publisher
Advances in the Study of Behavior, Vol. 26		Academic Press, a division of Harcourt Brace & Co., San Diego, CA, 1997
Farm Builder's Handbook	R. J. Lytle	Structures Publishing Co., Farmington, MI, 1973
Farm Buildings: From Planning to Completion	R. E. Phillips	Doane-Western, Inc., St. Louis, MO, 1981
Sheep Housing and Equipment Handbook		Midwest Plan Service, Iowa State University, Ames, IA, 1994
Sheepman's Production Handbook, The	American Sheep Industry Foundation	Sheep Industry Development Program (SID), Denver, CO, 1998
Stockman's Handbook, The, Seventh Edition	M. E. Ensminger	Interstate Publishers, Inc., Danville, IL, 1992
Structures and Environment Handbook		Midwest Plan Service, Iowa State University, Ames, IA

CHAPTER 18

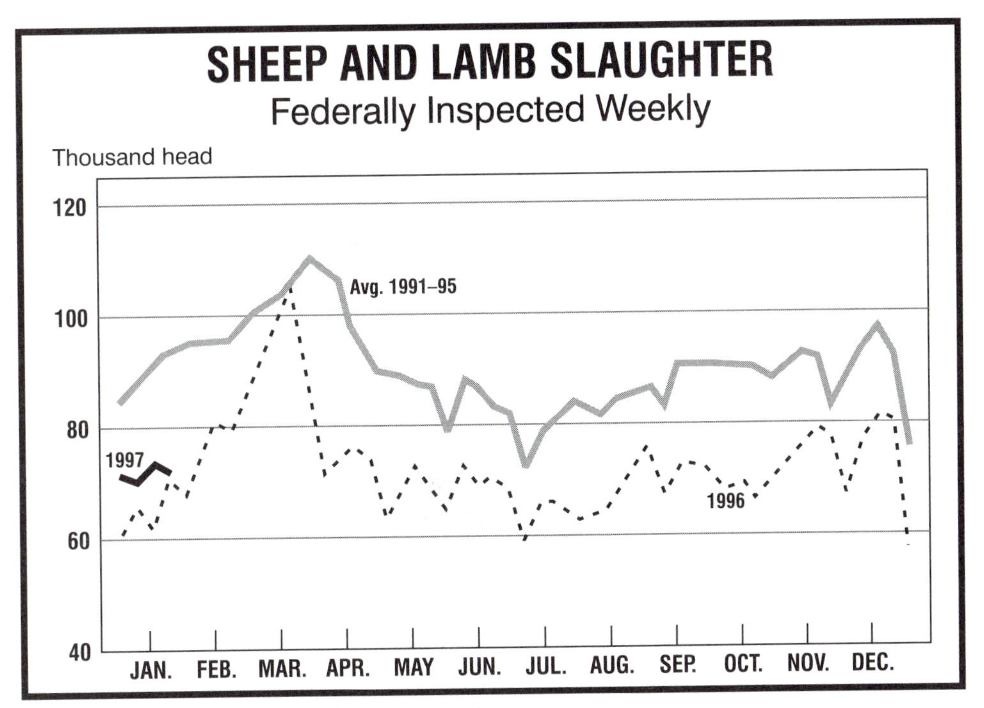

Sheep and lamb slaughter. (Courtesy, Livestock Marketing Information Center, Lakewood, CO)

MARKETING AND HARVESTING SHEEP AND LAMBS[1]

Contents *Page*

[1]Helpful review suggestions for this chapter were received from the following, who have great expertise in the marketing and harvesting of sheep and lambs: Terence R. Dockerty, Ph.D., Director of Meat Science Programs, and H. Kenneth Johnson, Vice President, both staff members of the National Livestock and Meat Board, 444 North Michigan Avenue, Chicago, Illinois; and Jim Wise, Ph.D., USDA, Standardization Branch—AMS, 14th and Independence SW, Washington, DC.

Contents Page

Contents Page

Livestock marketing embraces those operations beginning with loading animals out on the farm and extending until they are sold to go into processing channels.

In the early history of the United States, sheep were produced almost exclusively for wool. As other fibers became more common, lamb became more important. Today, lamb rather than wool is the major source of income to sheep producers. About 75 to 80% of the sheep producer's income comes from the sale of sheep and lambs and only 20 to 25% from wool.

Marketing sheep and lambs is the job of pricing, assembling, sorting, transporting, and processing them, and of distributing their subsequent products. Marketing sheep and lambs is a complex and costly job, complicated by two situations: (1) The animals are widely dispersed over farms and ranches located in every state; and (2) the decline in slaughter sheep numbers and quantity of retail product puts lamb at a disadvantage relative to other meats.

The trend toward fewer markets handling live sheep and lambs, and fewer, larger, and more specialized plants killing and processing them, makes the decision of where and how to market them a major consideration.

The significance of sheep marketing is further attested to by the following facts:

1. In recent years, an average of 6,942,000 sheep and lambs were marketed in the United States. [2]

2. Livestock markets establish values of all animals, including those down on the farm or ranch. In recent years, an average of 8,756,500 head of sheep and lambs were on U.S. farms and ranches with an average aggregate value of $709,670,000 or $81.78 per head.[3]

Also, the marketing of sheep and lambs differs from the marketing of cattle and hogs as follows:

1. In general, there is a greater transportation problem, because a preponderance of the lambs are raised west of the Mississippi while most of the lamb is consumed east of the Mississippi.

2. Age is a greater factor in determining relative market values of sheep than it is in determining those of cattle and hogs; thus, 95% of the sheep slaughter is from lambs and yearlings.

3. More sheep than cattle are sold on the farm.

4. Because of the pelt, the byproducts of sheep are more valuable than are the byproducts obtained in cattle or hog slaughter.

5. Sheep producers take more breeders back to the country than do cattle or hog producers.

MARKET CHANNELS FOR SHEEP

Sheep producers are confronted with the perplexing problem of determining where and how to market their animals due to diminishing markets in recent years. Usually there is a choice of outlets, and the one selected often varies with different classes and grades of sheep and lambs and among sections of the country. Thus, the methods of marketing also differ between slaughter and feeder lambs, and both of these differ from the marketing of purebreds.

Most slaughter sheep and lambs are sold directly to packers. Auction markets and terminal markets rank second and third, respectively, as market channels for

TABLE 18-1
MARKET CHANNELS FOR SLAUGHTER SHEEP AND LAMBS[1]

Market Channel	Packer Purchases Through Different Outlets	
	1972	1994
Direct, country dealers, etc. .	74.6	83.6
Auction markets, public markets[2]	25.4	16.4

[1]*Packers and Stockyards Resume*, USDA, Vol. 7, No. 13, 1969, Table 2, and Vol. 19, No. 5, 1982, Table 10.

[2]Includes terminals and auctions.

[2]*Agricultural Statistics*, USDA, 1998, p. VII-34, Table 7-50.

[3]*Ibid.*, p. VII-28. Table 7-42.

slaughter sheep and lambs (see Table 18-1). A discussion of these market channels follows, in order of their importance.

DIRECT METHOD (COUNTRY DEALERS)

Direct selling refers to producers' sales of livestock directly to packers or local dealers or farmers, without the support of commission firms, selling or buying agents, or brokers. Direct selling does not involve a recognized market. The selling usually takes place at the farm or feedlot or some other nonmarket buying station or collection yard. Some country buyers purchase livestock at fixed establishments similar to packer-owned country buying points. Direct selling is similar to terminal market selling with respect to price determination; both are by private treaty and negotiation. But it permits producers to observe and exercise some control over selling while it takes place, whereas consignment to distant terminal markets usually represents an irreversible commitment to sell. Larger and more specialized livestock farmers feel competent to sell their livestock directly.

Prior to the advent of terminal public markets in 1865, country selling accounted for virtually all sales of livestock. Sales of livestock in the country declined with the growth of terminal markets until the latter method reached its peak of selling at the time of World War I. But country buying was accelerated by the large nationwide packers following World War I in order to meet the increased buying competition of the small interior packers.

Improved highways and trucking facilitated the growth of country selling. Farmers were no longer tied to outlets located at important railroad terminals or river crossings. Livestock could move in any direction. Improved communications, such as the radio and telephone, and an expanded market information service, also aided in the development of country selling of livestock, especially in sales directly to packers.

■ **Producer or seller's direct selling cost**—The out-of-pocket cost to the producer for direct selling is zero. Time is the only selling expense which, of course, is not a direct or out-of-pocket cost to the producer.

AUCTION MARKET METHOD

Auction markets (also referred to as sales barns, livestock auction agencies, community sales, and community auctions) are trading centers where animals are sold by public bidding to the buyers who offer the highest price per hundredweight or per head. Auctions may be owned by individuals, partnerships, corporations, or cooperative associations. Auctions handle about 13% of the public market sales of sheep and lambs.

This method of selling livestock in this country is very old, probably being copied from Great Britain where auction sales date back many centuries.

Apparently the auction method of selling was used in many of the colonies as a means of disposing of property, imported goods, second-hand household furnishings, farm utensils, and animals.

According to available records, the first public livestock auction sale was held in Ohio in 1836 by the Ohio Company, whose business was importing English cattle.

Although there are some records of occasional livestock auction sales during the 19th century, there is no indication of any auction market that continued operation throughout the period of the greatest development of terminal public stockyards markets. Livestock auction markets had their greatest growth after 1930, both in numbers established and in the extensiveness of the area in which they operate. About 200 auctions were operating by 1930; by 1937, this number had increased to 1,345. The peak in numbers was reached in 1949, when over 2,500 different livestock auctions were holding sales; but by 1980, the total number declined to about 1,832.

Several factors contributed to the phenomenal growth in auction markets during the 1930s, chief of which were the following:

1. The decentralization of markets, and the improvement and extension of hard-surfaced roads accompanied by the increased use of trucks as a means of transporting livestock.
2. The development of more uniform class and grade designations for livestock.
3. Improvements made by the federal government in providing more extensive collection and dissemination of market news.
4. The greater convenience afforded in disposing of small lots of livestock and in purchasing stockers, feeders, and breeding animals.
5. The recognized educational value of these nearby markets, which enabled producers to keep currently informed of local market conditions and livestock prices.
6. The depression of 1930–33 which created the need to lower the cost of marketing and transportation.
7. The abnormal feed distribution caused by the droughts of 1934 and 1936 in the western Corn Belt and the range states.
8. The desire to sell near home.

Prior to the advent of community livestock auctions, small livestock operators had two main market outlets for their animals: (1) shipping them to the nearest terminal public market, or (2) selling them to buyers who came to their farms or ranches. Sometimes, the first method was too expensive because of the transportation distance involved and the greater expense in shipping small lots. The second method pretty much put producers at the mercy of the buyers, because they had no good alternative to taking the price offered, and often they did not know the value of their animals. By contrast, big operators are not particularly concerned about these things. Because of their large scale, usually they can take advantage of any of several terminal markets, and they know enough about values that they can deal satisfactorily with buyers who come to their farms or ranches. Thus, community livestock auctions are really of greatest importance to the small operators.

The auction market method of selling is similar to the terminal market in that both markets (1) are an assembly or collection point for livestock being offered for sale, (2) furnish or provide all necessary services associated with the selling activity, (3) are supervised by the federal government in accordance with the provisions of the National Packers and Stockyards Act, and (4) are characterized by buyers purchasing their animals on the basis of visual inspection.

But there are several important differences between terminals and auctions; among them, auction markets (1) are not always terminal with respect to livestock destination; (2) are generally smaller; (3) are usually single-firm operations; (4) sell by bid, rather than by offer and counteroffer; and (5) are completely open to the public with respect to bidding, and all buyers present have an equal opportunity to bid on all livestock offered for sale, whereas the terminal method is by private treaty (the negotiation is private).

■ **Charges to the producer or seller**—The cost to the producer of using the auction market is the combined cost of selling, yardage, feed, and services. Auction market charges tend to be somewhat higher than terminal charges. Estimated marketing costs of sheep and lambs at auctions average $1.80 per head.

TERMINAL OR CENTRAL MARKET METHOD

Terminal markets (also referred to as terminals, central markets, public stockyards, and public markets) are livestock trading centers, which generally have several commission firms and an independent stockyards company. Formerly, terminal markets were synonymous with private treaty selling. Today, however, many terminal markets operate their own sale ring and all, or almost all, of their livestock are sold by auction.

The first terminal market was established at Chicago in 1865, and most of the larger terminal markets were established in the latter half of the 19th century.

Up through World War I, the majority of slaughter livestock in the United States was sold through terminal markets by farmers and local buyers shipping to them. Since then, the importance of these markets has declined in relation to other outlets. In 1923, slaughterers with federally inspected operations purchased 90% of their sheep and lambs at terminal markets.[4] By 1968, packers purchased 18.6% of their sheep and lambs on terminal markets, and by 1980, this declined further to 7.5% (see Table 18-1).

The terminal or central market method entails the following distinct steps:

1. Producers must decide when to sell their animals.
2. Producers must deliver their animals to the terminal.
3. Producers must consign their animals to a commission firm.
4. The commission firm must accept the animals upon their arrival at the terminal.
5. The commission firm must pen, water, feed, sort, sell, and attend to all the necessary tasks from the time the animals arrive until they are sold.
6. The commission firm collects from the buyers and pays the sellers.
7. The buyers take title to the animals when they are weighed.

The stockyards company performs a number of useful services concerning the physical operation of the market. But, the stockyards company is a service company. Generally, it takes no active part in either buying or selling animals.

■ **Markets for sheep and lambs have declined**—Nevertheless, they provide an established outlet, indi-

TABLE 18-2
TOTAL RECEIPTS OF 8 LEADING TERMINAL SHEEP AND LAMB MARKETS, BY RANK[1]

Market	Receipts
	(No.)
Sioux Falls	185,798
South St. Paul	84,486
West Fargo	24,758
Sioux City	17,832
National Stock Yards	16,145
Omaha	13,426
South St. Joseph	8,566
Oklahoma City	4,148
U.S. total, 28 markets in 1993	1,284,700

[1]Animal Products Branch, Economic Research Service, USDA, 1301 New York Avenue, NW, Washington, DC.

[4]*Livestock Marketing*, ERS-322, USDA, October, 1966, p. 90.

cate yearly variations, and show where sheep and lambs are being raised and marketed.

In 1996, 4,184,100 head of sheep and lambs were slaughtered at an average live weight of 128 lb. *Note:* About 60% of the sheep and lambs marketed are slaughtered. Many feeder lambs are marketed, but taken to feedlots for finishing before slaughter, and some market ewes are returned to farms and ranches for breeding.

The following market prices for sheep and lambs are noteworthy:

Market Class	Market Price 1996	Market Price 1997
	($)	($)
Sheep	34.50	49.24
Lambs	85.68	89.26

On January 1, 1998, there were 7,616,000 sheep and lambs in the United States, 1,500,000 of which were in Texas—the nation's leading sheep state.

■ **Charges to the producers or sellers**—The direct or out-of-pocket cost to the producers is the combined cost of yardage, sales commission, feed, and service. These costs vary somewhat from market to market. Commission fees are the largest terminal market cost. They usually account for one-half the total charges, with yardage plus feed and bedding accounting for the other half. The cost of feed is based on the actual cost plus a specified margin.

ELECTRONIC MARKETING

Traditional marketing procedures require the physical presence of buyers, sellers, and animals. With fewer slaughtering plants, lowered sheep numbers, and increasing costs for transportation, traditional marketing has become one of the biggest problems in the sheep industry. In some areas of the country, producers (sellers) have been limited to one buyer who could virtually pay any price. Electronic marketing overcomes some of the problems plaguing sheep marketing.

Electronic marketing is based on selling by description and does not require the physical presence of the buyers, sellers, or animals. It includes telephones and computerized systems. By linking buyers and sellers with these electronic systems, it can bring together more buyers and sellers from a larger geographical area with a rapid exchange of information between them. Furthermore, electronic marketing decreases the transportation time and cost for the producers and buyers, and reduces the handling of livestock. Some initial electronic marketing of lambs demonstrated a reduction in the marketing bill and increased prices to producers.

When electronic marketing is conducted as a telephone auction, it operates like a conference call conducted by an auctioneer. The auctioneer describes the lambs offered for sale and then invites bids from the buyers participating in the conference call. Buyers place bids using a preassigned number so they remain anonymous to each other. After the final bid is accepted, arrangements are made with the buyer, by phone, to take delivery of the lambs at a central point. Sellers (consignors) are responsible for delivery of their lambs to the pick-up point.

With the increased availability of small computers and portable computer terminals, these can be tied to a central computer via the telephone, thereby, computerizing electronic marketing.

Where computers are used, several programs form the marketing system. The sellers enter the number, location, estimated grade, weight, and other pertinent information about the lambs being sold. Buyers access a program that generates a description of the lambs being offered for sale. Then at an appointed time, buyers, each at a computer terminal tied to the main computer containing the sales information, via telephone, access an auction program. As the bidding proceeds, all buyers can see the high bid on their terminal screens, but only the buyer who holds the high bid knows this. Bidders place their bids, in set increments, by pressing the proper key on the terminal keyboard. When the bidding ends, the successful bidder is informed of the purchase, and all other bidders only know of the sale price. The identity of the successful buyer is never revealed to the other buyers. At the end of all bidding, all buyers receive a summary of their purchases and any special instructions on contracts or pick-up arrangements.

Electronic marketing procedures, when properly planned and conducted, offer a viable marketing alternative which enhances competition and efficiency. Electronic marketing may well be the marketing method of the future.

CARCASS GRADE AND WEIGHT BASIS

Carcass grade and weight (yield) selling is that type of transaction in which the packer pays according to the quality of the carcass as well as its weight.

Carcass grade and weight selling tends to focus the marketing transaction on the actual end product—the amount and quality of meat contained in the animal. It is increasing as a method of marketing. From 1971 to 1994, the percentage of sheep and lambs marketed by grade and weight increased from 7.4 to 46.6%.[5]

Many barriers have retarded the use of grade and weight marketing. Chief among them is the sheep producers' lack of familiarity with the mechanics of the operation and the necessary calculations required to compare prices and net returns. Sheep producers also tend to lack confidence in grade and weight selling. Many producers are suspicious of a marketing system where livestock quality and returns to the producer are determined by an unseen and unknown individual employed by the buyer after the sale is made.

There are, however, some factors **favorable** to selling on the basis of carcass grade and weight; namely:

1. It encourages the breeding and feeding of quality animals.
2. It provides the most unassailable evaluation of the product.
3. It eliminates wasteful filling on the market.
4. It makes it possible to trace losses from condemnations, and bruises to the producer responsible for them.
5. It is the most effective approach to animal improvement.

But some factors are **unfavorable** to selling on the basis of carcass grade and weight; namely:

1. The procedure is more time-consuming than the conventional basis of buying.
2. There is less flexibility in the operations.
3. The physical difficulty of handling the vast U.S. volume of animals on this basis is great.

The USDA guidelines, wherein meat packers buy livestock on the basis of carcass grade, carcass weight, or a combination of the two, are:

1. Make known to the seller, before the sale, the significant details of the purchase contract.
2. Maintain the identity of each seller's livestock and carcasses.
3. Maintain sufficient records to substantiate settlement for each purchase transaction.
4. Use hooks, rollers, gambrels, and similar equipment of uniform weight in weighing carcasses from the same species of livestock in each packing plant; and include only the weight of this equipment in the actual weight of the container.
5. Make payment on the basis of final USDA carcass grades or furnish the seller with detailed written specifications for any other grades used in determining final payment.
6. Grade carcasses by the close of the second business day following the day of slaughter.

The need for a system of marketing which favors payment for a high cut-out value of primal cuts and a quality product is recognized. Selling on the basis of carcass grade and weight fulfills these needs. This type of selling benefits those who produce superior animals, but the producers of lower-quality animals may feel that this method unjustly discriminates against them.

CHOICE OF MARKET OUTLETS

Marketing is dynamic; thus changes are inevitable in types of market outlets, market structures, and market services. Some outlets have gained in importance; others have declined.

The choice of a market outlet represents the seller's evaluation of the most favorable market among the number of alternatives available. No simple and brief statement of criteria can be given as a guide to the choice of the most favorable market channel. Rather, an evaluation is required of the contributions made by alternative markets in terms of available services offered, selling costs, the competitive nature of the pricing process, and ultimately the producer's net return. Thus, an accurate appraisal is not simple.

From time to time, producers can be expected to shift from one type of market outlet to another. Because price changes at different market outlets do not take place simultaneously, nor in the same amount, nor even in the same direction, one market may be the most advantageous outlet for a particular class and grade of livestock at one time, but another may be more advantageous at some other time. The situation may differ for different classes and kinds of livestock and from one area to another.

Regardless of the channel through which producers market their livestock, in one way or another they pay, or bear, either in the price received from the livestock or otherwise, the entire cost of marketing. Because of this, they should never choose a market because of convenience or habit, or because of personal acquaintance with the market and its operator. Rather, the choice should be determined strictly by the net returns from the sale of livestock; effective selling and net returns are more important than selling costs.

SELLING PUREBRED SHEEP

Selling purebred animals is a highly specialized and scientific business. Purebred sheep are usually sold at private treaty directly to other purebred breeders or commercial flocks. Only the elite rams are

[5] *Packers and Stockyards Statistical Report*, USDA, 1996, p. 41, Table 11.

retained with the hope of effecting further breed improvement in purebred flocks. On the other hand, the sale of purebred ewes is fairly well restricted to meeting the requirements for replacement purposes in existing purebred flocks or for establishing new purebred flocks.

Most consignment sales are sponsored by a breed association, either local, statewide, or national in character. Purebred auction sales are conducted by highly specialized auctioneers. In addition to being good salespeople, such auctioneers must have a keen knowledge of values and be familiar with the breeding of the sheep.

MARKETING CONSIDERATIONS PECULIAR TO SHEEP AND LAMBS

Although sheep and lambs are marketed through many of the same channels as cattle and hogs, the following differences exist:

■ **Marketing on carcass weight and grade**—It is expected that the trend toward increased carcass yield grades will continue unabated. When marketing lambs on a carcass weight basis, the USDA guidelines call for weights to be taken and payments to be made on the basis of hot carcass weights without pencil shrink.

■ **Pencil shrink**—In direct sales, it is common to deal on the basis of a pencil shrink: 3 to 4% for feeder lambs, and 4% for slaughter lambs. Producers should know how to convert or compare *asking* or *offer* prices for various shrinkage levels. For example, an offer of $60.00 with 4% pencil shrink is equivalent to a price of $57.60 without this reduction.

■ **Spot (cash) sales**—The term *spot sales* appears to be more of the lexicon of sheep marketing than of cattle or hog marketing. Simply stated, it refers to cash sales. Of course, cash sales are also common in the marketing of cattle and hogs. In the marketing of sheep and lambs, however, the term *spot sales* is used primarily for the purpose of differentiating cash sales from contractual sales. Spot sales occur in the traditional market channels—direct, terminals, auctions, dealers, etc.— both on a live and a carcass basis. Among the advantages of spot sales are the following:

1. The price is determined quickly.
2. The title is transferred promptly.
3. The marketing costs, if any, are known and paid for soon after the transaction.
4. Producers receive their money immediately, or within a few days.

■ **Ownership and control of lambs**—Most breeding sheep are owned by producers. However, an increasing number of lambs on feed are either (1) owned by packers, or (2) controlled through packers by some form of contractual or custom feeding arrangement. The vast majority of this ownership control is centered in about a dozen firms. Their stated reasons for involvement in lamb feeding are (1) it increases plant efficiency, and (2) it improves merchandising programs.

Also, contracting for future delivery of either feeder lambs or slaughter lambs is common in the sheep industry. For example, a lamb feeder may contract for feeder lambs in July for delivery in September, or a packer may contract for slaughter lambs to be delivered the following week or later.

As a result of this ownership and control of lambs being centered in few packers, and of fewer lambs moving through traditional price-determining and price-reporting channels (like terminals and auctions), many knowledgeable sheep producers feel that a suitable market channel price barometer no longer exists.

■ **Higher marketing costs**—Although the commission and yardage charges on a per head basis for lambs at central markets are about one-fourth that of cattle, on a per hundredweight basis sheep and lamb costs are about three times higher than for cattle. This is attributed to low volume and problems in handling, as compared to other classes of livestock.

■ **Discount plans for heavyweight live lambs**—Western packers and dealers use the following four plans to discount live lambs at such times as wholesale markets discount heavyweight lamb carcasses:

1. **Sliding scale.** Under the sliding scale plan, the price decreases a specified amount per hundredweight for each pound that a shipment averages over a stipulated weight. For example, a 50¢ per hundredweight discount might be made for each pound that a shipment exceeds 110 lb. Thus, if the contract price is $60.00, and if lambs average 115 lb, the settlement or paying price would be $57.50.
2. **Stop weight.** Under this plan, the packer or dealer specifies an upper weight limit for the average liveweight of a lot, with no payment made for any weight in excess of this limit. For example, if the stop weight is 110 lb, and if the lot averaged 115 lb, no payment is made for the 5 lb overweight.
3. **Guaranteed yield.** Under this plan, the buyer specifies that the shipment shall yield a certain dressing percentage; for example, 52%. If the lambs yield better than the guarantee, the payment is based on the actual at-plant liveweight. However, if the lambs fail to yield the specified percentage, the packer determines the pay weight by dividing the shrunk carcass weight by the guaranteed yield. Generally, packers weigh the carcasses while hot and use an arbitrary, calculated cooler shrink (usually 2.5 to 3%).

The following example will show how the guaranteed yield plan works: Lambs averaging 110 lb were bought on a 52% guaranteed yield basis, equivalent to an average weight of 57.2 lb carcass weight shrunk. If the carcasses only average 56 lb after the pencil cooler shrink is applied to the hot weight, the packer would pay the seller for lambs averaging 107.69 lb liveweight (56 ÷ 0.52 = 107.69).

4. **Double dressing.** The double-dressed weight plan is a variation of the guaranteed yield. The shrunk dressed weight of a lot is doubled. This is the live pay weight, regardless of the actual liveweight. The double-dressed method assumes a 50% carcass yield. Thus, for lambs whose carcasses average 54 lb after the cooler shrink adjustment, payment is made on the basis of an average weight of 108 lb for the live lambs.

Although heavyweight lambs may be overfat and wasty, and, therefore, worth less per hundredweight than lightweight lambs, carefully selected and properly finished heavyweight lambs may actually have higher cutability carcasses and be more valuable per hundredweight than many lightweight lambs. This is so because:

a. The carcasses may yield a higher percentage of retail cuts.

b. More pounds of retail cuts may be processed from heavyweight lambs for each work minute. It takes no longer to prepare a larger, higher cutability carcass for the meat case than for a light one; yet, the merchandiser has more meat to sell.

c. The cuts may be a more desirable size for processing and merchandising and more preferred by the consumer.

LIVESTOCK MARKET NEWS SERVICES

Accurate market news is essential to the efficient marketing of livestock, both from the standpoint of the buyer and the seller. In the days of trailing, the meager market reports available were largely conveyed by word of mouth. Moreover, the time required to move livestock from the farm or ranch to market was so great that detailed market information would have been of little benefit even if it had been available. With the speed in transportation afforded by railroads and trucks, late information on market conditions became important.

The Federal Market News Service was initiated by the U.S. Department of Agriculture beginning in 1916. This service was established for the purpose of providing unbiased and uniformly interpretable market information. It depends on voluntary cooperation in gathering information. There is no legal compulsion for buyers and sellers to divulge purchase and sale information. Reports on direct sales are obtained largely by telephone and teletypewriter, augmented by interviews made at feedlots, packing plants, and farms. Then, for disseminating market reports, the Federal Market News Service relies upon local and privately owned newspapers, radio stations, and TV stations—merely supplying them with the information. Because at least a part of the readers or listeners are interested in this type of information, the local papers and radio stations are usually glad to serve as media for releasing these reports.

Other important sources of market information include farm and trade magazines. Also, many market agencies—such as commission firms, auction markets and related organizations—prepare and distribute market information. By means of weekly market newsletters or cards, they commonly emphasize the price and market conditions of the particular market they serve.

PREPARING AND SHIPPING SHEEP AND LAMBS

Improper handling of sheep immediately prior to and during shipment may result in excessive shrinkage; high death, bruise, and crippling losses; disappointing sales; and dissatisfied buyers. Unfortunately, many sheep producers who do a superb job of producing animals dissipate all the good things that have gone before by doing a poor job of preparing and shipping to market. Generally, such omissions are due to lack of know-how, rather than any deliberate attempt to take advantage of anyone. Even if a sale is consummated prior to delivery, negligence at shipping time will make for a dissatisfied customer. Buyers soon learn what to expect from various producers and place their bids accordingly.

In preparing sheep for shipment and in transporting them to market, sheep producers should consider the following:

1. **Select the best suited method of transportation.** Today, truck shipments account for the majority of receipts at markets.

All major operators clean, disinfect, and bed facilities prior to loading, but it is always well that shippers make their own inspection to make sure that these matters have been handled to their satisfaction. Generally sheep are bedded in keeping with the recommendations given in Table 18-4. If shippers desire a type of bedding other than that normally supplied by the carrier, however, they may provide it at their own expense.

2. **Consider health certificates and permits in interstate shipments.** When sheep are to be shipped into another state, the shipper should check into and comply with the state regulations relative to health cer-

tificates and permits. Usually, a local veterinarian will have this information. Knowledge of and compliance with such regulations well in advance of shipment will avoid frustrations and costly delays.

3. **Avoid shipping during extremes in weather.** Whenever possible, avoid shipping when the weather is either very hot or very cold. During such times, shrinkage and death losses are higher than normal. During warm weather, avoid transporting animals during the heat of the day; travel at night or in the evening or early morning.

4. **Feed and water properly prior to loading.** Never ship sheep on excess fill. Instead, feed and water lightly. Withhold grain feeding 12 hours before loading (omit one feed), and do not allow access to water within 2 to 3 hours of shipment. Sheep may be allowed free access to dry, well-cured grass hay up to loading time, but more laxative-type hays, such as alfalfa and clover, should not be fed within 12 hours of shipment even if the animals were accustomed to them previously. Likewise, sheep on green succulent feed should be conditioned to dry feeds prior to shipment.

Sheep that are too full of concentrated feeds or that have eaten succulent feeds will scour and urinate excessively. As a result, the floors become dirty and slippery and the animals befoul themselves. Such animals undergo a heavy shrink and present an unattractive appearance when unloaded.

5. **Keep sheep quiet.** Prior to and during shipment, sheep should be handled carefully. Hot, excited animals are subject to more shrinkage and injury, and higher mortality.

If sheep are trailed on-foot to the shipping point, they should be moved slowly and allowed to rest and to drink moderately prior to loading.

Although loading may be exasperating at times, take it easy; never lose your temper. Avoid hurrying, loud hollering, and striking. Never beat an animal with such objects as pipes, sticks, canes, or forks; instead, use either (a) a flat, wide canvas slapper with a handle, or (b) a broom. In addition to avoiding excitement of the sheep, loading them slowly will prevent their crowding against sharp corners.

6. **Use partitions when necessary.** When mixed loads (consisting of sheep, hogs, and/or cattle) are placed in the same truck, partition each class off. Also, all crippled and weak animals should be properly partitioned. In sheep shipments, separate rams.

PREVENTING BRUISES, CRIPPLING, AND DEATH LOSSES

Losses from bruises, crippling, and death that occur during the marketing process represent a part of the cost of marketing livestock; and, indirectly, the producer foots most of the bill!

The following precautions are suggested (in addition to those already covered under the main heading, "Preparing and Shipping Livestock") as a means of reducing sheep marketing losses from bruises, crippling, and death:

1. Remove projecting nails, splinters, and broken boards from feed containers and fences.
2. Keep feedlots free from old machinery, trash, and any obstacles that may bruise.
3. Do not feed grain heavily just prior to loading.
4. Remove protruding nails, bolts, or any sharp objects in truck.
5. Bed properly (see Table 18-4).
6. Use good loading chutes; not too steep.
7. With two or more decks, have upper deck(s) high enough to prevent back bruises on animals below.
8. Use partitions in trucks that are not fully loaded, to keep animals closer together; and in very long trucks, to keep animals from crowding from one location to another.
9. Drive trucks carefully. Slow down on sharp turns and avoid sudden stops.
10. Inspect load en route to prevent trampling of animals that may be down. If an animal goes down, get it back on its feet immediately.
11. Back truck slowly and squarely against unloading dock.
12. Unload slowly. Do not drop animals from upper to lower deck; use cleated inclines.
13. Never lift sheep by the wool.

All of these precautions are simple to apply; yet all are violated every day of the year.

NUMBER OF SHEEP AND LAMBS IN A TRUCK

Overcrowding of market animals causes heavy losses. Sometimes a truck is overloaded in an attempt to effect a saving in hauling charges. More frequently, however, it is simply the result of not knowing space requirements.

The suggested number of animals, Table 18-4, should be tempered by such factors as haul distance, livestock class, weather conditions, and road conditions.

Table 18-3 shows the number of sheep for safe transportation by truck.

TABLE 18-3
NUMBER OF SHEEP PER DECK FOR SAFE LOADING IN A TRUCK[1]

Floor Length		Average Weight of Sheep[2]			
		60 lb (27 kg)	80 lb (36 kg)	100 lb (45 kg)	120 lb (55 kg)
(ft)	(m)	(no.)			
8	2.4	28	23	20	18
10	3.0	35	29	26	23
12	3.7	43	35	31	28
15	4.6	54	45	40	36
18	5.5	65	54	48	43
20	6.1	73	60	54	48
24	7.3	88	73	65	58
28	8.5	103	85	76	68
30	9.1	110	92	81	73
32	9.8	118	98	87	78
36	11.0	133	110	98	88
42	12.8	145	128	115	103

[1]Authoritative recommendations of Livestock Conservation, Inc., Chicago, IL.

[2]Always partition mixed loads into separate classes.

TABLE 18-4
GUIDE RELATIVE TO BEDDING AND FOOTING MATERIAL WHEN TRANSPORTING LIVESTOCK[1,2,3]

Class of Livestock	Kind of Bedding for Moderate or Warm Weather; Above 50°F (10°C)	Kind of Bedding for Cool or Cold Weather; Below 50°F (10°C)
Sheep and goats	Straw	Straw
Cattle.	Sand, 2 in. (5.1 cm)	Sand; for calves use sand covered with straw
Swine.	Sand, 0.5 to 2 in. (1.3 to 5.1 cm)[4]	Sand covered with straw
Horses and mules.	Sand	Sand

[1]Straw or other suitable bedding (covered over sand) should be used for protection and cushioning breeding stock that are loaded lightly enough to permit their lying down in the car or truck.

[2]Sand should be clean and medium-fine, and free from brick, stones, coarse gravel, dirt, or dust.

[3]Fine cinders may be used as footing for cattle, horses, and mules, but not for sheep and hogs. They are picked up by and damage the wool of sheep, and they damage hog casings.

[4]In hot weather, wet sand down before loading and while en route. Drench hogs when necessary, but never apply water to the backs of hot hogs—it may kill them.

KIND OF BEDDING TO USE FOR SHEEP IN TRANSIT

Among the several factors affecting sheep losses, perhaps none is more important than proper bedding and footing in transit.

Footing, such as sand, is required at all times of the year, to prevent the car or truck floor from becoming wet and slick, thus predisposing animals to injury by slipping or falling. Bedding, such as straw, is recommended for warmth in the shipment of sheep during extremely cold weather, and as cushioning for breeding stock or other animals loaded lightly enough to permit their lying down. Recommended kinds and amounts of bedding and footing materials are given in Table 18-4.

SHRINKAGE IN MARKETING LAMBS

The shrinkage (or drift) refers to the weight loss encountered from the time animals leave the feedlot until they are weighed over the scales at the market. Thus, if a lamb weighed 100 lb at the feedlot and had a market weight of 94 lb the shrinkage would be 6 lb or 6.0%. Shrink is usually expressed in terms of percentage. Most of this weight loss is due to excretion in the form of feces and urine and the moisture in the expired air. On the other hand, there is some tissue shrinkage, which results from metabolic or breakdown changes.

The most important factors affecting shrinkage are:

1. **The fill.** Naturally, the larger the fill animals take upon their arrival at the market, the smaller the shrinkage.
2. **Time and distance in transit.** The longer the animals are in transit and the greater the distance, the higher the total shrinkage. Also, the shrink takes place at a rapid rate during the first part of the haul, and then decreases as time in transit progresses.
3. **Season.** Extremes in temperature, either very hot or very cold weather, result in higher shrinkage. Shrink is at a minimum between 20 and 60°F.
4. **Age and weight.** Young animals of all species shrink proportionally more than older animals because of their lower carcass yield caused by less body fat and greater amount of fill in proportion to liveweight.
5. **Overloading and underloading.** Overloading always results in abnormally high shrinkage. Unless animals are partitioned off properly, underloading will also result in excess shrinkage.
6. **Rough ride, abnormal feeding, and mixed loads.** Each of these factors will increase shrinkage.

On the average, sheep shrink from 6 to 10%, the highest of all livestock.

AIR AND SEA TRANSPORTATION

With modern communication and transportation, the world is becoming smaller and smaller. Along with this, there is an increasing need and desire to move animals efficiently between countries separated by great distances. Ancestors of the present-day U.S. sheep endured long sea voyages, which were often a hardship on humans. For many years, ships were considered the only economical method of moving large numbers of animals across the seas, but animals never adapted well to the pitching and tossing of sea travel. In some cases, losses of up to 50% were not uncommon. Then, during the late 1960s, the concept of air transportation of large numbers of animals was born. Whole planes were adapted for livestock comfort and maximum capacity—literally flying corrals. With this modern concept, animals can arrive at any destination in the world within hours, thereby minimizing stress and virtually eliminating death losses. Numerous cattle, calves, sheep, pigs, horses, chicks, and exotic animals have been successfully airlifted worldwide. The number of animals per load depends on the size of the individual animals and the size of the aircraft, but some of the larger aircraft can carry 1,500 sheep, 950 calves, 900 pigs, 196 dairy cattle, or 50 horses.

While modern air transportation of sheep is fast, modern sea transportation of sheep is massive. Recently, some oil tankers have been converted into large sheep ships capable of accommodating as many as 125,000 sheep. From time to time, sea transportation is used, as in the shipment of sheep from Australia to the Persian Gulf, where Muslim religious tradition dictates specific methods of slaughter.[6] Aboard ship, conveyors and piping systems feed and water the animals automatically. Shipping live animals to the growing Middle East market is a boon to Australia's sheep industry.

Fig. 18-1. Sheep being loaded onto a "flying corral." (Courtesy, Transamerica Airlines, Oakland, CA)

Fig. 18-2. World's largest "floating corral" leaving port for its first voyage. The Kuwaiti *Al-Shuwaikh*, a former Norwegian oil tanker, transports sheep from Australia to the Persian Gulf. Its 14 stable decks can accommodate 125,000 sheep. (Photo from United Press International, New York, NY)

MARKET CLASSES AND GRADES OF SHEEP

While no official grading of live sheep and lambs is done by the U.S. Department of Agriculture, grades do form a basis for uniform reporting of livestock marketing. Also, grades of live animals are intended to be directly related to the grades of the carcasses they produce.

The market classes and grades of sheep (see Table 18-5) follow closely the pattern for the classes and grades of cattle and swine. One notable difference is that a sizable number of sheep are sold as breeders. For the most part, this class is made up of mature western ewes that are sold to country buyers for the purpose of producing one or two more crops of lambs before again being returned to the market. Usually such ewes can be acquired at a lower cost than ewe lambs. In certain sections of the country, many flock rams are

[6]According to Islamic (Muslim) law, no animal food, except fish and locusts, is considered lawful unless it has been slaughtered following the proper ritual. At the instant of slaughter, the person killing the animal must repeat, "In the name of God, God is great."

acquired on the market, though no market quotation is given for such breeders and the practice is not considered too sound. Another difference between sheep and other species is that one feeder class, namely the shearers, is based on wool value as well as adaptability for further feeding.

FACTORS DETERMINING MARKET CLASSES OF SHEEP

The disposition or use to be made of sheep is determined by: (1) whether they are sheep or lambs, (2) use selection, (3) sex, (4) age, and (5) weight.

SHEEP AND LAMBS

The first major market subdivision of the ovine species separates the sheep from the lambs. Lambs include those animals that are approximately one year old and under. When there is any question as to whether animals should be classified as sheep or lambs, a final decision is usually based upon an examination of the teeth. If the first pair of larger, broader per-

TABLE 18-5
MARKET CLASSES AND GRADES OF SHEEP

Sheep or Lambs	Use Selection	Sex Classes	Age	Weight Division	(Pounds)	*(Kilograms)*	Commonly Used Grades
Sheep	Slaughter sheep	Ewes	Yearlings	Light Medium Heavy	90 down 90–100 100 up	*40.9 down* *40.9–45.4* *45.4 up*	Prime, Choice, Good, Utility[1]
			Mature (2-year-olds or older)	Light Medium Heavy	120 down 120–140 140 up	*54.5 down* *54.5–63.6* *63.6 up*	Choice, Good, Utility, Cull[1]
		Wethers	Yearlings	Light Medium Heavy	100 down 100–110 110 up	*45.4 down* *45.4–49.9* *49.9 up*	Prime, Choice, Good, Utility1
			Mature (2-year-olds or older)	Light Medium Heavy	115 down 115–130 130 up	*52.2 down* *52.2–59.0* *59.0 up*	Choice, Good, Utility, Cull[1]
		Rams	Yearlings	All weights			Prime, Choice, Good, Utility[1]
			Mature (2-year-olds or older)	All weights			Choice, Good, Utility, Cull[1]
	Feeder sheep	Ewes and wethers	Yearlings	All weights			Fancy, Choice, Good, Medium, Cull
		Ewes	Mature (2-year-olds or older)	All weights			Choice, Good, Medium, Cull
	Breeding sheep	Ewes (rams occasionally purchased as breeders, but not listed in market reports)	Yearlings, 2-, 3-, or 4-year-olds and older	All weights			Fancy, Choice, Good, Medium, Cull
Lambs	Slaughter lambs	Ewes, wethers, and rams	Hothouse lambs	60 down			Prime, Choice, Good, Utility[1]
		Ewes, wethers, and rams	Spring lambs	Light Medium Heavy	70 down 70–90 90 up	*31.8 down* *31.8–40.9* *40.9 up*	Prime, Choice, Good, Utility[1]
		Ewes, wethers, and rams	Lambs	Light Medium Heavy	75 down 75–95 95 up	*34.0 down* *34.0–43.1* *43.1 up*	Prime, Choice, Good, Utility[1]
	Feeder lambs	Ewes and wethers	All ages	All weights			Fancy, Choice, Good, Medium, Cull
	Shearer lambs	Ewes and wethers	All ages	All weights			Choice, Good, Medium

[1]In addition to the above quality grades, there are five yield grades applicable to all lamb and mutton carcasses, denoted by numbers 1 through 5, with the Yield Grade 1 representing the highest degree of cutability. Thus, slaughter sheep and lambs are graded for both quality and yield grades.

manent teeth is about fully developed, the animal is classified as a yearling; for this change in the teeth takes place at about 12 months of age.

At the present time, lambs and yearlings make up approximately 95% of the total sheep slaughter.

USE SELECTION OF SHEEP AND LAMBS

Sheep and lambs are divided into six market groups based on the uses to be made of them or the purposes for which they are best suited: slaughter sheep, feeder sheep, breeding sheep, slaughter lambs, feeder lambs, and shearer lambs. A brief description of each of these classes follows:

- **Slaughter sheep**—Yearlings or older animals intended for immediate slaughter.
- **Feeder sheep**—Yearlings or older animals best suited for further finishing.
- **Breeding sheep**—Largely mature western ewes that are returned to the country for further reproduction. In addition to market grade in considering the suitability of ewes for breeding purposes, it is important that attention be given to the condition of the teeth and to the breed, health, and general potentialities as a breeder.
- **Slaughter lambs**—Young animals under one year of age that are sufficiently finished for immediate slaughter.
- **Feeder lambs**—Young animals under one year of age that carry insufficient finish for slaughter purposes but which show indications of making good gains if placed on feed.
- **Shearer lambs**—Those intended for shearing and further finishing prior to slaughtering. This classification is of importance on certain markets during the late winter and early spring months. Typical shearer lambs carry nearly a full year's growth of wool, but they are not finished enough to be market-topping slaughter lambs. Such lambs are usually shorn out and finished before returning to the market. The term *shorn lamb* is used to designate those lambs that have had their fleece removed within 60 days prior to marketing. Those fed lambs that have not been shorn are usually differentiated from shorn lambs by adding the term *wooled* to the class name.

SEX CLASSES

The sex classes for sheep and lambs are: ewes, wethers, and rams. At the lamb and yearling stage, ewes and wethers are equally suitable for slaughter purposes. Ram lambs are usually somewhat discounted in price, and they are almost never used for feeder purposes. A definition of each sex class follows:

- **Ewe**—A female sheep or lamb.
- **Wether**—A male ovine animal that was castrated at an early age, before reaching sexual maturity and before developing the physical characteristics peculiar to rams.
- **Ram**—An uncastrated male ovine animal of any age. The term *buck* is sometimes applied to animals of this sex class.

AGE GROUPS

Each of the major age divisions is further separated into more exacting age groups. Thus, mature sheep may be designated as yearlings, two-year-olds, three-year-olds, four-year-olds, or mature sheep. Yearlings are much more acceptable for slaughter purposes than mature sheep.

In a general way, age groups in lambs are indicated by the terms *hothouse lambs, spring lambs*, and *lambs*. Each of these age groups may be described briefly as follows:

- **Hothouse lambs**—These lambs are considered by Epicureans as being the most delectable of all the lamb age groups. *Hothouse lambs are very young lambs—usually less than three months of age at slaughter— which are born and marketed out of season.* Such milk-finished lambs are usually marketed during the period from Christmas to the Easter holidays at weights ranging from 30 to 60 lb. Hothouse lambs may consist of ewe, wether, or ram lambs. Although sex class is unimportant, these lambs should be undocked if sold on the Boston market.

Fig. 18-3. Hothouse lambs "hog dressed" (meaning head and pelt on, but front feet and viscera removed). This method holds shrinkage to a minimum and maintains a pink carcass color in young lambs. The pluck is left in. In this case, the pluck consists of the liver, heart, lungs, gullet, and windpipe. (Courtesy, The Pennsylvania State University)

■ **Spring lambs**—New-crop lambs arriving at the market in the spring of the year. Usually these are lambs born in the late fall or early winter and marketed prior to July 1. After July 1, animals of like birth are simply designated as lambs on the market. The most desirable spring lambs range from 3 to 7 months in age and weigh from 70 to 100 lb. Because of the young age, sex class is unimportant in spring lambs. They may be ewe, wether, or ram lambs.

■ **Lambs**—All young ovine animals that do not classify as either hothouse or spring lambs. This is by far the most numerous class of market sheep under one year of age. These animals are usually born in the late winter or early spring of the year and are marketed at 7 to 12 months of age. In general, they subsist on milk and grass. Lambs fed grain prior to marketing are designated as *fed lambs* in order to differentiate them from lambs.

■ **Yearlings**—Young sheep between approximately one and two years of age and which have cut their first pair of permanent incisor teeth but not the second pair.

■ **Two-year-olds**—Sheep that are between 24 and 36 months old and which have cut their second pair of permanent incisor teeth.

■ **Three-year-olds**—Sheep that are between 36 and 48 months old and which have cut their third pair of permanent incisor teeth.

■ **Four-year-olds**—Sheep that are between 48 and 60 months old and which have a full set of permanent incisors.

■ **Mature sheep**—Usually animals that are two years old or over. With further age and the loss of teeth, they are referred to as *broken-mouthed*. If all the incisors are missing or worn down to the gums, they are known as *gummers*. If the teeth are long and spread apart at the surface, the ewes are called *spreaders*. All gummers and broken-mouthed ewes should be rejected when breeding sheep are being purchased, for such animals are not likely to hold up in flesh on winter feeds.

WEIGHT DIVISIONS

Weight is an especially important price factor in the case of lambs. As a rule, heavy lambs (110 lb and up) are in less demand on the market than lighter-weight animals of similar grade. This is largely due to the consumer preference for small- or medium-sized cuts of lamb. Thus, weight is an important factor in both feeder and slaughter lambs. Heavy, fat ewes usually sell at a discount; they are not considered desirable for either slaughter or breeding purposes.

MARKET GRADES OF SHEEP AND LAMBS

Grades for live animals—slaughter lambs, yearlings, and sheep—are directly related to the quality and yield grades of the carcasses they will produce. Before significant progress can be made in improving lamb production and marketing, it is necessary for producers and those engaged in marketing to have a clear understanding of the meaning and relative importance of the various live and carcass grades.

Sheep and lambs are graded for both (1) quality grades, and (2) yield grades. Quality grades relate to the eating characteristics of the meat, whereas yield grades describe the volume or total proportion of trimmed retail cuts a carcass contains.

Fig. 18-4. A plant employee rolls a carcass with a quality grade. (Courtesy, American Sheep Industry Assn., Englewood, CO)

The quality grades are designed to identify differences in (1) palatability characteristics, such as tenderness, juiciness, and flavor; and (2) carcass shape or conformation. Evaluation of lamb carcass quality is based on the quantities of fat in the flank area in relation to the maturity of the animal, with carcass conformation also being a grading factor. Lamb is produced from animals less than a year old. Meat from older sheep is called yearling or mutton; where graded, these words must be stamped on the meat along with the grade. Grades for yearling and mutton are the same as for lamb, except that there is no Prime grade for mutton, but there is a Cull grade. About 80% of all lambs slaughtered commercially are quality graded. Of this number, over 98% grade Choice or higher.

The quality grades of slaughter sheep and lambs are: Prime, Choice, Good, Utility, and Cull. The corresponding grades of feeder sheep and lambs are: Fancy, Choice, Good, Medium, and Cull.

When both quality and yield grades are used, two separate evaluations may be obtained: (1) the palatability-indicating characteristics of the lean and conformation; and (2) the estimated percentage of closely trimmed, boneless major retail cuts (leg, loin, hotel rack, and shoulder) to be derived from the carcass.

More information about the carcass grades of sheep and lambs is presented in Chapter 19, Sheep Meat, Milk, and Byproducts.

OTHER SHEEP MARKET TERMS AND FACTORS

In addition to the rather general terms used in designating the different market classes and grades of sheep, the following terms and factors are frequently of importance.

Fig. 18-5. Market grades of slaughter lambs. (Courtesy, USDA)

NATIVE AND WESTERN SHEEP

Native sheep are predominantly of meat breeding and are produced in the central, eastern, and southern states, in the mixed farming areas. *Westerns* are predominantly of grade Rambouillet breeding or Rambouillet X meat breed crossbreds that come from the western ranges. In the case of western sheep, their state of origin is frequently used as a designation rather than the broad classification as westerns.

WHITE-FACED AND BLACK-FACED SHEEP

White-faced and *black-faced* are sometimes used to designate the general type and breeding of sheep and lambs. In general, white-faced sheep and lambs are predominantly of fine-wool breeding and do not possess quite the blockiness and general excellent meat conformation of the black-faced sheep. Black-faced sheep are either predominantly of meat breeding or are sired by a black-faced ram of one of the meat breeds.

DISCOUNTING LONG-TAILED AND RAM LAMBS

Long-tailed and buck lambs are frequently docked up to $2 and $5 per hundredweight on a discriminating market. Thin lambs, which should go the feeder route, are discounted more than slaughter animals. Despite the dock in price of ram lambs that are subsequently culled out and marketed for slaughter, purebred breeders can often afford to retain all male lambs as rams until weaning age, at which time more intelligent selections may be made. Of course, this can only be justified when some of the ram lambs are valuable for breeding purposes.

VALUE OF THE PELT IN MARKET SHEEP

When wool prices are good, the degree of desirability of the pelt is a factor of importance in determining the price of market sheep and lambs.

The value of the wool, or fleece, will depend on (1) whether the animal has recently been shorn or is carrying a full fleece, and (2) the price of wool. Pelt prices are difficult to obtain, but they are generally reported in major industry publications.

SOME SHEEP MARKETING CONSIDERATIONS

Enlightened and shrewd marketing practices generally characterize the successful sheep enterprise. Sheep and lamb prices tend to vary cyclically over the years and seasonally within a year. A knowledge of these price variations may be useful in planning production and marketing so that animals are available at the most advantageous time. But cyclical and seasonal price information is based on history and should be used only as a rough guide.

CYCLICAL TRENDS

The price cycle is that period of time during which the price for a certain kind of livestock advances from a low point to a high point and then declines to a low point again. From another standpoint, a cycle is the change in animal numbers which livestock producers make in response to prices; it is a production cycle. Although there is considerable variation in the length of the cycle within any given class of stock, over a long period of time it has been observed that the price cycle of the different classes of animals is about as follows: sheep, 9 to 10 years; hogs, 3 to 5 years; and cattle, 10 to 12 years.

The species' cycles are a direct reflection of the rapidity with which the numbers of each class of farm animals can be shifted under practical conditions to meet consumer meat demands. Litter-bearing and early-producing swine can be increased in numbers much more rapidly than either sheep or cattle. When market sheep and lamb production is sufficiently profitable to induce increased production, breeding stock to increase production must come from either (1) a reduc-

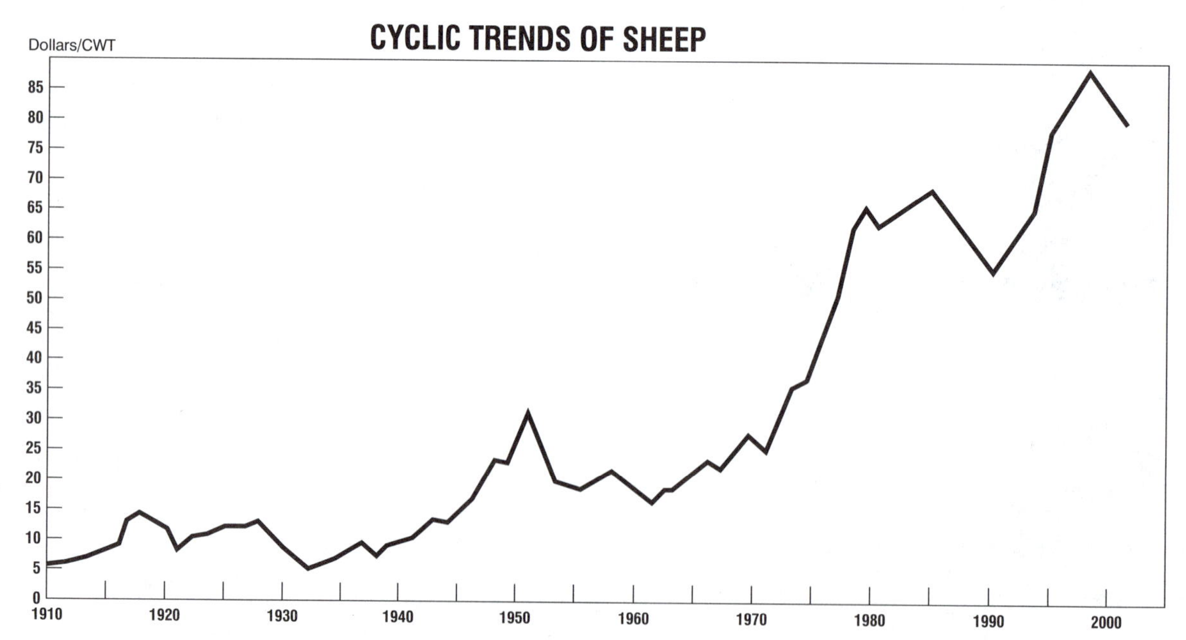

Fig. 18-6. Average price for lambs received by U.S. sheep producers, 1910–2000. In general, the price cycle for sheep is 9 to 10 years. (From *Agricultural Statistics*, 2001, USDA)

tion of the rate of ewe slaughter, or (2) a holding back of ewe lambs to build breeding flocks. In either case, slaughter sheep and lamb supply is further reduced and even higher prices occur in the short run. Conversely, when prices are sufficiently low to cause producers to sell off breeding stock and/or keep fewer replacement lambs, this further adds to the sheep and lamb supply and results in lower prices.

Normal cycles are disturbed by droughts, wars, general periods of depression or inflation, and federal controls.

The normal price cycles of lambs can be made less acute—that is, prices may be made more stable year after year—by (1) applying technological advances to get away from seasonable breeding and lambing, (2) informing producers of buildups and shortages, and (3) contracting for delivery at specified times.

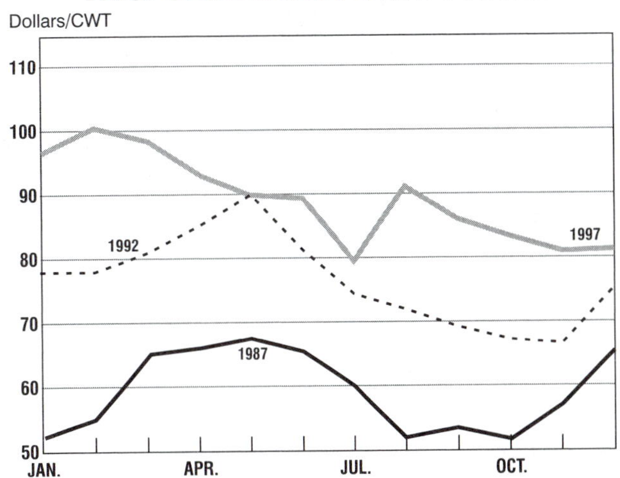

Fig. 18-7. Slaughter lamb prices vary seasonally, largely because of the supply. This shows slaughter lamb prices at four markets for the years 1987, 1992, and 1997. (From Steve Meyer, Livestock Marketing Information Center, Lakewood, CO)

SEASONAL VARIATIONS

Seasonal prices are the usual price changes from season to season during the year. Prices vary seasonally due to the variations in market receipts. Progressive sheep producers study seasonal changes as a valuable guide (1) when making management decisions, and (2) when planning their buying and selling operations.

As would be expected, (1) seasons of high market prices are generally associated with light marketings, and seasons of low market prices with heavy marketings, and (2) seasons of high and low market prices vary with different classes of livestock. However, the normal seasons of high and low market prices may be changed by such factors as (1) federal farm programs and controls, (2) business conditions and general price levels, (3) feed supplies and weather conditions, (4) wars, etc.

In recent years, seasonal patterns have not been as reliable as they used to be. Year-round finishing of lambs in large commercial feedlots has created more uniform marketings throughout the year, thereby lessening seasonality. Thus, when arriving at livestock forecasts and marketing advice, sheep producers should exercise proper reservation in considering seasonal patterns. Anyway, it is not always wise to plan production to hit the highest market, for sometimes that would push up production costs more than enough to offset the gains from higher prices. Nevertheless, a careful study of normal seasonal prices will help in deciding the best time to buy and sell sheep in order to reap the greatest profits.

Figs. 18-7 and 18-8 show the seasonal price variations for slaughter lambs and feeder sheep for three different years.

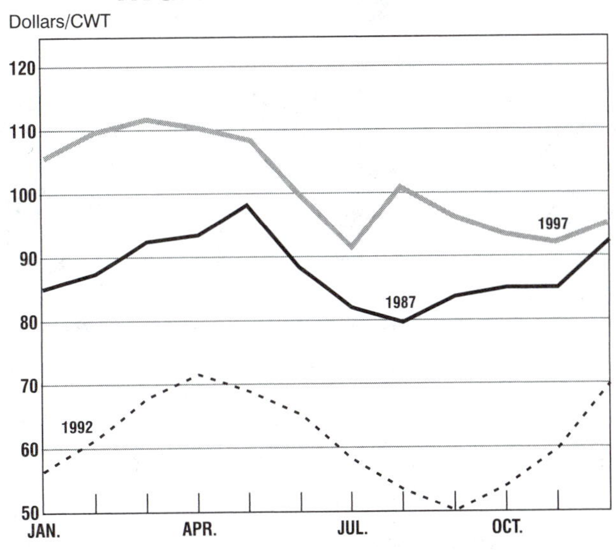

Fig. 18-8. Feeder lamb prices may vary seasonally, largely because of the supply. This shows choice feeder lamb prices at four markets for the years 1987, 1992, and 1997. (From Steve Meyer, Livestock Marketing Information Center, Lakewood, CO)

PACKER SLAUGHTERING AND DRESSING OF SHEEP AND LAMBS

Table 18-6 shows the proportion of sheep and lambs dressed commercially—slaughtered in federally inspected and other wholesale and retail establishments. The total figure refers to the number dressed in

TABLE 18-6
PROPORTION OF SHEEP AND LAMBS
HARVESTED COMMERCIALLY[1]

Year	Sheep and Lambs		
	Total Number Harvested (commercially and noncommercially)	Number Harvested Commercially	Percentage Harvested Commercially
	(1,000 head)	(1,000 head)	(%)
1996	4,249	4,184	98.4
1997	3,969	3,907	98.4
1998	3,861	3,804	98.5
1999	3,766	3,701	98.2
2000	3,527	3,460	98.1
5-yr. avg.	3,874	3,811	98.3

[1]*Agricultural Statistics 2001*, p. VII-33, Table 7-52.

Fig. 18-9. In order to avoid undue excitement and make for ease in handling, gregarious sheep and lambs are usually led to the packers' slaughtering pens by an old goat, commonly referred to as *Judas*. A reasonable fast and quiet prior to slaughtering are especially important for sheep. (Courtesy, Monfort of Colorado, Greeley, CO)

all establishments and on farms. Additionally, in 1980, 93.4% of the lamb and yearling slaughter was under federal inspection.

Whether sheep are harvested in large, federally inspected packing plants, in small, local packing houses, or on farms, the harvesting procedure is much the same. Because most sheep are harvested commercially, rather than on farms, only the large, packer harvesting procedure will be discussed.

FEDERAL MEAT INSPECTION

The federal government requires supervision of establishments which harvest, pack, render, and prepare meats and meat products for interstate shipment and foreign export; it is the responsibility of the respective states to have and enforce legislation governing the harvesting, packaging, and handling of meats shipped intrastate, but state standards cannot be lower than federal levels. The meat inspection laws do not apply to farm harvest for home consumption, although all states require inspection if the meat is sold.

The meat inspection service of the U.S. Department of Agriculture was inaugurated, and is maintained under, the Meat Inspection Act of June 30, 1906. This act was updated and strengthened by the Wholesome Meat Act of December 15, 1967. The latter statute (1) requires that state standards be at least to the levels applied to meat sent across state lines; and (2) assures consumers that all meat sold in the United States is inspected either by the federal government or by an equal state program. The Animal and Plant Health Inspection Service of the U.S. Department of Agriculture is charged with the responsibility of meat inspection.

The purposes of meat inspection are (1) to safeguard the public by eliminating diseased or otherwise unwholesome meat from the food supply, (2) to enforce the sanitary preparation of meat and meat products, (3) to guard against the use of harmful ingredients, and (4) to prevent the use of false or misleading names or statements on labels. Personnel for carrying out the provisions of the act are of two types: professional or veterinary inspectors who are graduates of accredited veterinary colleges, and nonprofessional food inspectors who are required to pass a Civil Service examination. In brief, the inspections consist of the following two types:

1. **Antemortem (before death)** inspection is made in the pens or as the animals move from the scales after being weighed. The inspection is performed to detect evidence of disease or any abnormal condition that would indicate a disease. Suspects are provided with a metal ear tag bearing the notation "U.S. Suspect No...." and are given special postmortem scrutiny. If in the antemortem examination there is definite and conclusive evidence that the animal is not fit for human consumption, it is condemned, and no further postmortem examination is necessary.
2. **Postmortem (after death)** inspection is made at the time of slaughter and includes a careful examination of the carcass and the viscera (internal organs). All good carcasses (no evidence of disease) are stamped "U.S. Inspected and Passed," whereas the inedible carcasses are stamped "U.S. Inspected and Condemned." The latter are sent to the rendering tanks, the products of which are not used for human food.

In addition to the antemortem and postmortem inspections referred to, the government meat inspectors have the power to refuse the application of the mark of inspection to meat products produced in a plant that is not sanitary. All parts of the plant and its equipment must be maintained in a sanitary condition at all times. In addition, plant employees must wear clean, washable garments, and suitable lavatory facilities must be provided for hand washing.

Meat inspection regulations require the condemnation of all or affected portions of carcasses of animals with various disease conditions, including pneumonia, peritonitis, abscesses and pyemia, uremia, tetanus, rabies, anthrax, tuberculosis, various neoplasms (cancer), arthritis, actinobacillosis, and many others.

Most of the larger meat packers are under federal inspection: hence, they are allowed to ship interstate.

STEPS IN HARVESTING AND DRESSING SHEEP AND LAMBS

Modern sheep and lamb harvesting establishments use highly mechanized equipment, including moving rails. The steps in the packing house harvest procedure are as follows:

1. **Rendering insensible, shackling and hoisting.** The sheep are rendered insensible,[7] a shackle is placed around the hind leg just above the foot; and the animals are delivered to an overhead rail by means of a wheel hoist.
2. **Bleeding.** A double-edged knife is inserted into the neck just below the ear so that it severs the large blood vessel in the neck.
3. **Removal of front feet.** After bleeding, the front feet are removed. Lambs and most yearlings will break at the break-joint or lamb-joint, a temporary joint characteristic of young sheep which is located immediately above the ankle. In the dressing of mature sheep, the front feet are removed at the ankle, leaving a round point on the end of the shank bone.
4. **Removal of pelt.** Next, the pelt is removed. Caution is taken to prevent damage to the *fell*—the thin, tough membrane covering the carcass immediately under the pelt.
5. **Removal of hind feet and head.** Next, the hind feet and head are removed.
6. **Opening of carcass and eviscerating.** The carcass is opened down the median line; the internal organs, windpipe, and gullet are removed; and the breast bone is split. The kidneys are left intact.
7. **Shaping.** The forelegs are folded at the knees and are held in place by a skewer. A spread-stick is inserted in the belly to allow for proper chilling and to give shape to the carcass.
8. **Washing.** Finally the carcass is washed, wiped, and promptly sent to the cooler. Some of the better-grade carcasses are wrapped in special coverings for marketing.

[7]By federal law (known as the Humane Slaughter Act) passed in 1958, effective June 30, 1960, and amended in 1978, unless a packer uses humane slaughter methods, federal inspection is suspended; thus no slaughter is allowed until the problem is resolved. The law lists the following methods as humane: by rendering insensible to pain by a single blow or gunshot or an electrical, chemical, or other means that is rapid and effective, before being shackled, hoisted, thrown, cast, or cut. These methods pertain to all livestock, except those being slaughtered ritually.

KOSHER HARVEST[8]

Meat for the Jewish trade—known as kosher meat—is harvested, washed, and salted according to ancient Biblical laws, called *Kashruth*, dating back to the days of Moses, more than 3,000 years ago. The Hebrew religion holds that God issued these instructions directly to Moses, who, in turn, transmitted them to the Jewish people while they were wandering in the wilderness near Mount Sinai.

The Hebrew word *kosher* means "fit" or "proper," and this is the guiding principle in the handling of meats for the Jewish trade. Also, only those classes of animals considered clean—those that both chew the cud and have cloven hoofs—are used. Thus, sheep goats, and cattle—but not hogs—are koshered (Deuteronomy 14:4-5 and Leviticus 11:1-8).

The killing is performed by a rabbi of the Jewish church or a specially trained representative—a person called the *shohet* or *shochet*, meaning "slaughterer."

In kosher harvest, an animal is hoisted without stunning and is cut across the throat with a special razor-sharp knife, known as a *chalaf*. With one quick, clean stroke the throat is cut through the jugular vein and other large vessels, together with the gullet and windpipe. Two reasons are given for using this method of killing instead of the more conventional method of stunning and sticking; namely (1) it produces instant death with less pain, and (2) it results in more rapid and complete bleeding, which Orthodox Hebrews consider essential for sanitary reasons.

The shohet also makes an inspection of the lungs, stomach, and other organs during the dressing. If the carcass is acceptable, it is marked on the brisket with a cross inside a circle. The mark also gives the date of slaughter and the name of the inspector.

Both forequarters and hindquarters of kosher-slaughtered cattle, sheep, and goats may be used by

[8]Authoritative information relative to kosher harvest was obtained from the *Foods & Nutrition Encyclopedia*, Second Edition, by A. H. Ensminger, M. E. Ensminger, J. E. Konlande, and J. R. K. Robson, CRC Press, 1994.

Orthodox Jews. However, the Jewish trade usually confines itself to the forequarters. The hindquarters are generally sold as nonkosher for the following reasons:

1. The Sinew of Jacob ("the sinew that shrank," now known as the sciatic nerve), which is found in the hindquarters only, must be removed by reason of the Biblical story of Jacob's struggle with the Angel, in the course of which Jacob's thigh was injured and he was made to limp.

Actually, the sciatic nerve consists of two nerves; an inner long one located near the hip bone which spreads throughout the thigh, and an outer short one which lies near the flesh. Removal of the sinew (sciatic nerve) is very difficult.

The Biblical law of the sciatic nerve applies to sheep, goats, and cattle, but it does not apply to birds because they have no spoon-shaped hip (no hollow thigh).

2. The very considerable quantity of forbidden fat *(Heleb)* found in the hindquarters, especially around the loins, flanks, and kidneys, must be removed; and this is difficult and costly. Forbidden fat refers to fat (tallow) (a) that is not intermingled (marbled) and the flesh of the animal, but forms a separate solid layer; and (b) that is encrusted by a membrane which can be easily peeled off.

The Biblical law of the forbidden fat applies to sheep, goats, and cattle, but not birds and nondomesticated animals.

3. The blood vessels must be removed, because the consumption of blood is forbidden; and such removal is especially difficult in the hindquarters.

Forbidden fat and blood (the blood vessels) must be removed from both the forequarters and the hindquarters, but such removal is more difficult in the hindquarters than in the forequarters. However, the Sinew of Jacob (the sciatic nerve), which must also be removed, is found in the hindquarters only.

Because of the difficulties in processing, meat from the hindquarters is not eaten by Orthodox Jews in many countries, including England. However, the consumption of the hindquarters is permitted by the Rabbinic authorities where there is a special hardship involved in obtaining alternative supplies of meat; thus, in Israel the sinews are removed and the hindquarters are eaten.

Because the forequarters do not contain such choice cuts as the hinds, the kosher trade attempts to secure the best possible fores; thus, this trade is for high-grade slaughter animals.

Kosher meat must be sold by the packers or the retailers within 72 hours after slaughter, or it must be washed (a treatment known as *begiss*, meaning "to wash") and reinstated by a representative of the synagogue every subsequent 72 hours. At the expiration of 216 hours after the time of harvest (after begissing three times), however, it is declared *trafeh*, meaning "forbidden food," and is automatically rejected for the kosher trade. It is then sold in the regular meat channels. Because of these regulations, kosher meat is moved out very soon after slaughter.

Kosher sausage and prepared meats are made from kosher meats which are soaked in water ½ hour, sprinkled with salt, allowed to stand for an hour, and washed thoroughly. This makes them kosher indefinitely.

The Jewish law also provides that before kosher meat is cooked, it must be soaked in water for ½ hour. After soaking, the meat is placed on a perforated board in order to drain off the excess moisture. It is then sprinkled liberally with salt. One hour later, it is thoroughly washed. Such meat is then considered to remain kosher as long as it is fresh and wholesome.

Since neither packers nor retailers can hold kosher meat longer than 216 hours (and even then it must be washed at 72-hour intervals), rapid handling is imperative. This fact, plus the heavy concentration of Jewish people in the eastern cities, results in large numbers of live animals being shipped from the markets farther west to be slaughtered in or near the eastern consuming areas.

BREAK-JOINT OR LAMB-JOINT

In general, the packer classes as lamb all carcasses in which the forefeet are removed at the breakpoint or lamb-joint. This joint can be severed on all lambs, most yearling wethers, and some yearling ewes (ewes mature earlier than wethers or males). It is a temporary cartilage located just above the ankle. In lambs,

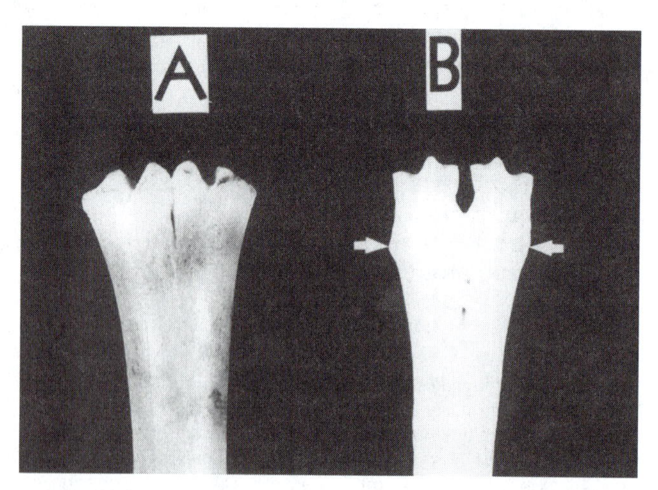

Fig. 18-10. The two types of joints of the foreleg of a sheep: A, the breakpoint or lamb-joint, and B, the round-joint or spool-joint. Arrow indicates the location of the breakpoint or ossification. All carcasses possessing the round-joint are sold as mutton rather than lamb. (Courtesy, Washington State University)

Fig. 18-11. The breaking line where lamb carcasses are cut into smaller and different cuts. (Courtesy, American Sheep Industry Assn., Englewood, CO)

the break-joint has four well-defined ridges that are smooth, moist, and red. In yearlings, the breakpoint is more porous and dry. In mature sheep, the cartilage is knit or ossified and will no longer break, thus making it necessary to take the foot off at the ankle instead. This makes a round-joint, commonly called spool-joint. All carcasses possessing the round-joint are sold as mutton rather than lamb.

DRESSING PERCENTAGE

In order to yield a high percentage of carcass, sheep must be (1) light in pelt, (2) well finished and heavily muscled, and (3) free from paunchiness. Because of the higher value of the offal of sheep, especially the high value of the pelt, a high dressing percentage in sheep is not as important as in cattle or hogs. In fact, the wool is usually worth more per pound than the carcass of mutton or lamb. Wool yield, therefore, is usually an important item in slaughter return and dressing return, and dressing percentage lowered by heavier wool yield may actually mean a greater total return. For this reason, the fleece should be as heavy as is consistent with the production of mutton or lamb of high quality.

Table 18-7 gives the dressing percentages that may be expected from the different grades of sheep and lambs. As would be expected, the highest dressing percentage is obtained when animals are slaughtered following shearing. Lambs of the mutton breeds yield a somewhat higher percentage of carcass than those of the so-called wool breeds. The offal and byproducts from the slaughter of sheep and lambs consist of the blood, pelt, feet, head, and viscera.

The average liveweight of sheep and lambs dressed by federally inspected meat-packing plants, and their percentage yield in meat for the year 1996. was as shown in Table 18-8.

TABLE 18-7
DRESSING PERCENTAGE OF LAMBS AND SHEEP (MUTTON) BY GRADE[1]

Lambs (wooled)			Sheep (excluding yearlings)		
Grade	Range	Average	Grade	Range	Average
Prime	49–55	52	Choice	49–54	52
Choice	47–52	50	Good	47–52	49
Good	45–49	47	Utility	44–48	46
Utility	43–47	45	Cull	40–46	43
Cull	40–45	42			

[1]From USDA sources.

TABLE 18-8
AVERAGE LIVEWEIGHT, CARCASS YIELD, AND DRESSING PERCENTAGES OF ALL SHEEP AND LAMBS COMMERCIALLY HARVESTED IN THE U.S.[1]

	Average Liveweight	Average Dressed Weight	Average Dressing Percentage
	(lb)	*(lb)*	*(%)*
Sheep and lambs	137	68	49.6

[1]*Agricultural Statistics*, USDA, 2001, p. VII-4, Table 7-53.

WITHDRAWAL PERIOD PRIOR TO HARVEST

When certain sprays and dips, drenches, feed additives, or implants are used, a waiting period must be observed from last use to harvest. Individuals using such products should always read and heed the label directions and contact the suppliers or manufacturers' agents for more complete details on the use of specific products or combinations.

QUESTIONS FOR STUDY AND DISCUSSION

1. Define *livestock marketing*.

2. Present facts to substantiate the importance of sheep marketing.

3. List and discuss each of the market channels for sheep.

4. Since World War I, terminal public markets have declined in importance, while livestock auction markets and direct selling have increased. Why has this happened?

5. In what ways does the selling of purebred sheep differ from the selling of commercial sheep?

6. What method of marketing (what market channel) do you consider most advantageous for the sheep sold out of your area, and why?

7. List and describe several ways in which the marketing of sheep and lambs is peculiar in comparison to cattle and hogs.

8. How can a sheep producer keep informed relative to prevailing market prices?

9. Outline, step by step, how you would prepare and ship sheep to market.

10. List the precautions that a sheep producer should take in order to lessen marketing losses from bruises, crippling, and death.

11. How should sheep be bedded for transit?

12. Discuss the practical ways and means of lessening shrinkage in market sheep.

13. Discuss the place and importance of air and sea transportation of sheep at the present time.

14. List the on-foot market (a) classes and (b) grades of sheep and lambs, and tell of their value.

15. Why is it important that a sheep producer know the market classes and grades of sheep and lambs and what each implies?

16. What are "quality grades" and what are "yield grades"? Why is each important?

17. Discuss practical ways through which the sheep producer can minimize (a) cyclical trends, and (b) seasonal changes.

18. Define and give the special significance of the following sheep and lamb market terms: (a) breeding sheep; (b) shearer lambs; (c) hothouse lambs; (d) spring lambs vs lambs; (e) native and western sheep; (f) breakpoint; (g) fell.

19. Describe the federal meat inspection of sheep.

20. Outline the sheep slaughtering process, step by step.

21. What is the Humane Slaughter Act, and how does it affect sheep slaughtering?

22. How does kosher slaughter differ from regular slaughter?

SELECTED REFERENCES

Title of Publication	Author(s)	Publisher
Animal Science, Ninth Edition	M. E. Ensminger	Interstate Publishers, Inc., Danville, IL, 1991
Livestock and Meat Marketing	J. H. McCoy	Avi Publishing Co., Westport, CT, 1972
Livestock Marketing	A. A. Dowell K. Bjorka	McGraw-Hill Book Co., New York, NY, 1941
Meat We Eat, The, Fourteenth Edition	J. R. Romans, et al.	Interstate Publishers, Inc., Danville, IL, 2001
Sheepman's Production Handbook, The	Ed. by G. E. Scott	Sheep Industry Development Program (SID), Denver, CO, 1975
Stockman's Handbook, The, Seventh Edition	M. E. Ensminger	Interstate Publishers, Inc., Danville, IL, 1992

CHAPTER 19

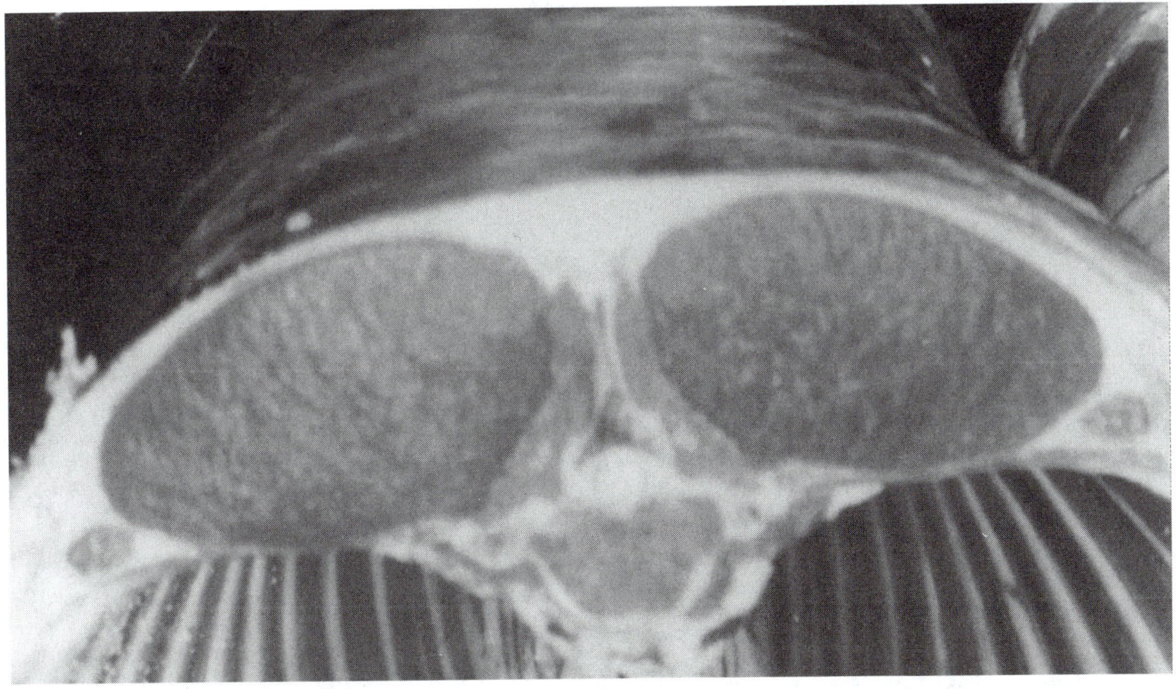

Cross section of the loin of lamb with heavy muscling. (Courtesy, Texas Agricultural Experiment Station, San Angelo, TX)

SHEEP MEAT, MILK, AND BYPRODUCTS[1]

Contents *Page*

[1]Helpful review suggestions for this chapter were received from the following who have great expertise relative to lamb, mutton, and byproducts of sheep slaughter: Terence R. Dockerty, Ph.D.; H. Kenneth Johnson, Executive Director, Value Based Meat Systems, National Cattlemen's Beef Association, 444 North Michigan Avenue, Chicago, IL 60611; and Jim Wise, Ph.D., USDA, Standardization Branch—AMS, 14th and Independence SW, Washington, DC.

Contents *Page*

Contents Page

Lamb is the edible flesh of young sheep. Mutton is the edible flesh of mature sheep. Byproducts include all products, edible and inedible, other than carcass meat.

Although this chapter is devoted primarily to the final product—meat—the top grades of lamb represent the culmination of years of progressive breeding, the best nutrition, vigilant sanitation and disease prevention, superior care and management, and modern marketing, slaughtering, processing, and distribution.

LAMB OVER THE COUNTER—THE ULTIMATE OBJECTIVE

The end product of all breeding, feeding, care, management, marketing, and processing is lamb over the counter. It is imperative, therefore, that the progressive sheep producer, the student, the sheep scientist, and all those engaged in marketing and distribution have a working knowledge of lamb and mutton and of the byproducts from sheep slaughter. Such knowledge will be of value in selecting animals and in determining policies relative to their handling.

Of course, the type of animals best adapted to the production of lamb has changed in a changing world. Thus, in the early history of this country, the very survival of animals was often dependent upon their speed, hardiness, and ability to fend for themselves. Moreover, long legs and plenty of bone were important attributes when it came time for animals to trail as drovers took them to market. The multi-colored sheep of colonial times were adapted to these conditions.

With the advent of rail transportation and improved care and feeding methods, the ability of animals to travel and survive under rugged conditions diminished in importance. It was then possible, through selection and breeding, to produce meat animals better suited to the needs of more critical consumers. With the development of large cities, artisans and their successors in industry required fewer calories than those who were engaged in the more arduous tasks of logging, building railroads, etc. Simultaneously, the American family decreased in size. The demand shifted, therefore, to smaller and less fatty cuts of meats; and, with greater prosperity, high-quality lamb legs and chops were in demand. To meet the needs of the consumer, the producer gradually shifted to the breeding and marketing of younger animals with maximum cutout value of the

Fig. 19-1. Lamb counter. (Courtesy, American Sheep Industry Assn., Englewood, CO)

primal cuts. Gradually, these younger animals have become bigger and meatier—a 65-lb dressed lamb in 4½ months.

Thus, through the years, consumer demand has influenced the type of lambs produced. To be sure, it is necessary that such production factors as prolificacy, economy of feed utilization, rapidity of gains, etc., receive due consideration along with consumer demands. But once these production factors have received due weight, sheep producers—whether they be purebred or commercial operators—must remember that lamb over the counter is the ultimate objective.

Now, and in the future, sheep producers need to select and feed so as to obtain increased red meat without excess fat. Production testing programs need to be increased and to be augmented with more and more carcass work; and livestock shows need to be oriented to give greater emphasis to consumer demands.

QUALITIES IN LAMB DESIRED BY THE CONSUMER

Because consumer preference is such an important item in the production of lamb, it is well that the sheep producer, the packer, and the retailer be familiar with these qualities, which are summarized as follows:

1. **Palatability.** First and foremost, people eat lamb because they like it. Palatability is influenced by the tenderness, juiciness, and flavor of the lean. In the United States, this calls for lamb—not mutton.

2. **Attractiveness.** The general attractiveness is an important factor in selling lamb to the consumer. The color of the lean, the degree of fatness, and the marbling are leading factors in determining buyer appeal. Most consumers prefer a pinkish-white fat and light red in the lean. The lean of mutton is much darker than lamb.

3. **Maximum muscling; minimum fat.** Maximum thickness of muscling influences materially the acceptability by the consumer. Also, consumer resistance to fat on all meats has been very marked in recent years.

4. **Smaller cuts.** Most purchasers prefer to buy cuts of meat that are of a proper size to meet the needs of their respective families. Because the American family has decreased in size, this has meant smaller cuts. This, in turn, has had a profound influence on the type of animals and on market age and weight.

5. **Ease of preparation.** In general, consumers prefer to select lamb that will give them the greatest amount of leisure time; thus they select those cuts that can be prepared with the greatest ease and the least time. Often this means chops and legs.

6. **Tenderness.** Lamb may be aged, but it is not necessary since the meat is tender.

The texture of lamb meat is fine, whereas that of yearlings and mutton is coarser.

7. **Repeatability.** Consumers want to be able to secure a standardized product—meat of the same eating qualities as their previous purchase.

Fig. 19-2. Shoulder lamb chops. (Courtesy, National Live Stock and Meat Board, Chicago, IL)

If the preceding seven qualities are not met by lamb, other products will meet them. Recognition of this fact is important, for competition is keen for space on the shelves of a modern retail food market.

To this end, more experimental studies on consumer preference and demand need to be conducted. Then, this information needs to be accurately reflected in price differentials at the market place, and, in turn, in production.

SHIFTS IN RETAIL LAMB PRICES

Traditionally, lamb has been sold in whole or half carcasses to retail outlets. Retail stores have often found it difficult to sell the less desirable cuts—breasts, necks, flanks, and shanks. They have attempted to derive enough revenue from the desirable cuts—chops and leg of lamb—to pay for the cost of the unsalable parts. The increased price of the desirable cuts depressed the demand for lamb.

Today, with packing plants boxing vacuum-packed primal cuts which have extended shelf life, the retail outlets need only to buy the cuts they can sell. Furthermore, technology such as mechanical deboning and restructuring offers the potential of creating desirable retail cuts sold at a profit.

CONSUMER-PREFERRED LAMB CARCASS

To ensure success, by meeting consumer preference, U.S. poultry producers agreed on the "chicken of tomorrow," and U.S. swine producers agreed on the "standard of excellence in hog type." Then, producers set out to produce products that met these goals.

While the U.S. sheep industry has not defined the consumer-preferred lamb to the extent of the poultry and swine industries, it is generally agreed that consumers prefer a meatier, leaner type of lamb. Moreover, lambs should be growthy and fast gaining, and should reach market finish at a weight compatible with their genetic background and frame size. Further, lamb should grade Choice or better, have a high cutability, and carry enough fat cover (at least 0.15 in.) to avoid excess carcass shrink loss in shipment.

Consumer-preferred lamb carcasses that meet the above specifications can be produced through the following procedure:

1. **Selecting breeding stock that will produce meat-type lambs.** This calls for more production testing and carcass evaluation. Rams and ewes that produce superior meat-type lambs need to be identified and used more extensively.
2. **Feeding properly.** Lambs will produce Choice grade carcasses (a) on suitable pasture, (b) in drylot on a high-concentrate diet, or (c) in a wide variety of intermediate situations. The level of feeding, rather than the kind of feed, is the most important feed factor affecting lamb-carcass quality. It influences condition, or degree of finish, which is important in determining (a) carcass cut-out value, and (b) ultimate consumer acceptance of the cuts.
3. **Marketing at desirable weights.** Consumers do not want fat lambs.

FEDERAL GRADES OF LAMB AND MUTTON

The grade of lamb and mutton may be defined as a measure of the degree of excellence based on quality or eating characteristics of the meat and the yield or total proportion of primal cuts. It is intended that the specifications for each grade shall be sufficiently definite to make for uniform grades throughout the country and from season to season, and that on-hook grades shall be correlated with on-foot grades.

Fig. 19-3 shows that the proportion of lamb federally graded exceeds beef.

Both producers and consumers should know the federal grades of meats and should have a reasonably clear understanding of the specifications of each grade.

PROPORTION OF U.S. COMMERCIAL MEAT SLAUGHTER FEDERALLY GRADED

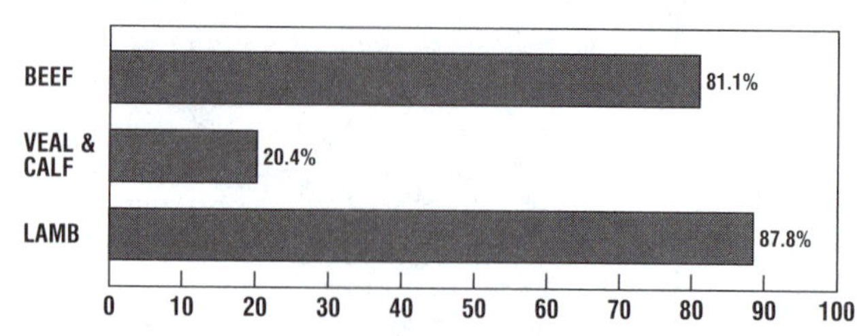

Fig. 19-3. Proportion of U.S. commercial beef, veal & calf, and lamb federally graded in 1986. (Data provided by Jim Wise, Ph.D., USDA, Standardization Branch—AMS, Washington, DC)

From the standpoint of consumers, this is important because there is less opportunity for consumers to secure the counsel and advice of the meat cutter when they are making purchases, and the average consumer is not the best judge of the quality of the various kinds of meats on display in the meat counter.

If lamb is graded, it must be both quality and yield graded. On a voluntary basis, lamb may be graded (1) for quality alone, (2) for yield grade alone, or (3) for both quality and yield grades.

The present quality carcass grades are: Prime, Choice, Good, Utility, and Cull. These grades primarily reflect differences in palatability characteristics, such as tenderness, juiciness, and flavor. However, differences in conformation are also a factor in determining quality grade.

In quality grading, one can best evaluate the lean flesh by considering its texture, firmness, and marbling, as observed in a cut surface. However, in the grading of carcasses, direct observation of these characteristics is not possible. Therefore, one can evaluate the quality of the lean indirectly by considering the amount of streaking of fat within and upon the inside flank muscles, with a minimum of firmness specified for each grade.

Fig. 19-4 illustrates the relationship between flank fat streakings, maturity, and quality, even though young lamb, older lamb, yearling mutton, and mutton are no longer graded for maturity.

In quality grading, superior conformation (1) implies a high proportion of edible meat to bone and a high proportion of the weight of the carcass in the more demanded cuts; and (2) is reflected in carcasses which are very wide and thick in relation to their length and which have a very plump, full, and well-rounded appearance. Inferior conformation (1) implies a low proportion of edible meat to bone and a low proportion of the weight of the carcass in the more demanded cuts; and (2) is reflected in carcasses which are very narrow in relation to their length and which have a very angular, thin, and sunken appearance.

In addition to the quality grades, yield grades were

introduced in 1969; and in 1992, yield grades were made mandatory if carcasses were to be quality graded (*i.e.*, if a packer chooses to grade lamb carcasses, they must be both quality and yield graded. The yield grade is determined by a single measurement of the external fat at the 12th rib.

In lamb grading, two separate evaluations are obtained: (1) palatability-indicating characteristics of the lean, incorporating conformation (quality grade), and (2) the estimated percentage of closely trimmed, boneless major retail cuts from the leg, loin, rib, and shoulder (yield grade).

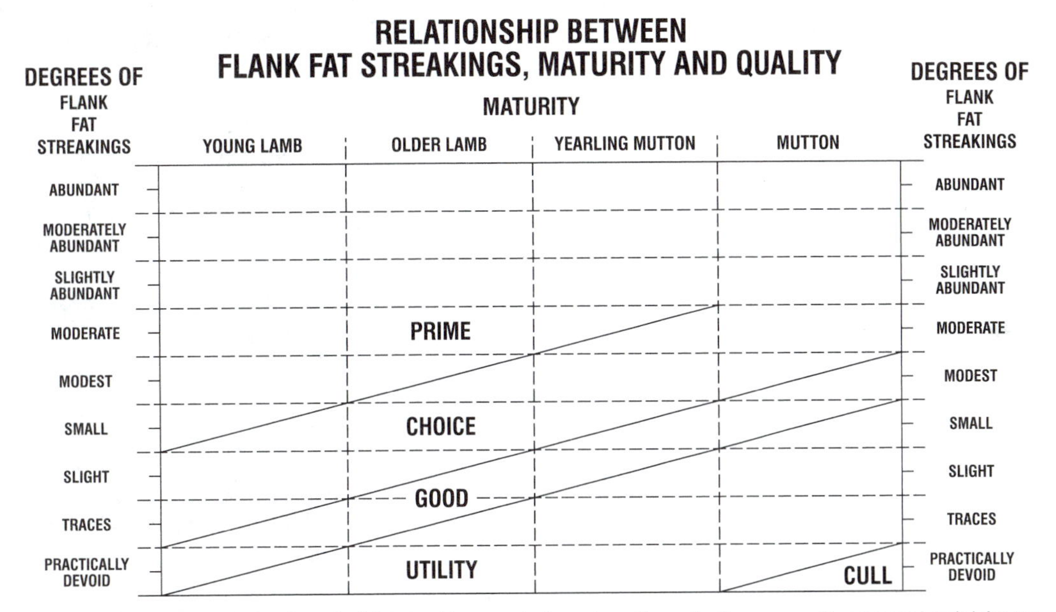

Fig. 19-4. Relationship between flank fat streakings, maturity, and quality grade. Carcasses with only one break-joint are classified as lamb if other maturity characteristics are typical of lamb. (Courtesy, USDA)

A study conducted by Texas A&M University showed a $15.86 per hundredweight difference in carcass value between a Yield Grade 1 lamb and a Yield Grade 5 lamb. More astounding yet, nearly 30% of the weight of a Yield Grade 5 lamb was waste fat.

The quality and yield grade descriptions are defined primarily in terms of carcasses. However, the grading standards also are applicable to the grading of sides.

Unlike meat inspection, government grading is purely voluntary, on a charge basis. Official graders are subject to the call of anyone who wishes their services (packer, wholesaler, or retailer), for a per hour charge.

AGING LAMB AND MUTTON

Only the ribs and loins of high-quality lamb and mutton are aged, and then only rarely. Typically, lamb carcasses are not aged.

THE LAMB CARCASS AND ITS WHOLESALE CUTS

Whether a lamb carcass is cut up in the home or by an expert, it should always be cut across the grain of the muscle tissue, and the thick cuts should be separated from the thin cuts and the tender cuts from the less tender cuts.

The lamb carcass may be divided into the hindsaddle and the foresaddle, between the twelfth and thir-

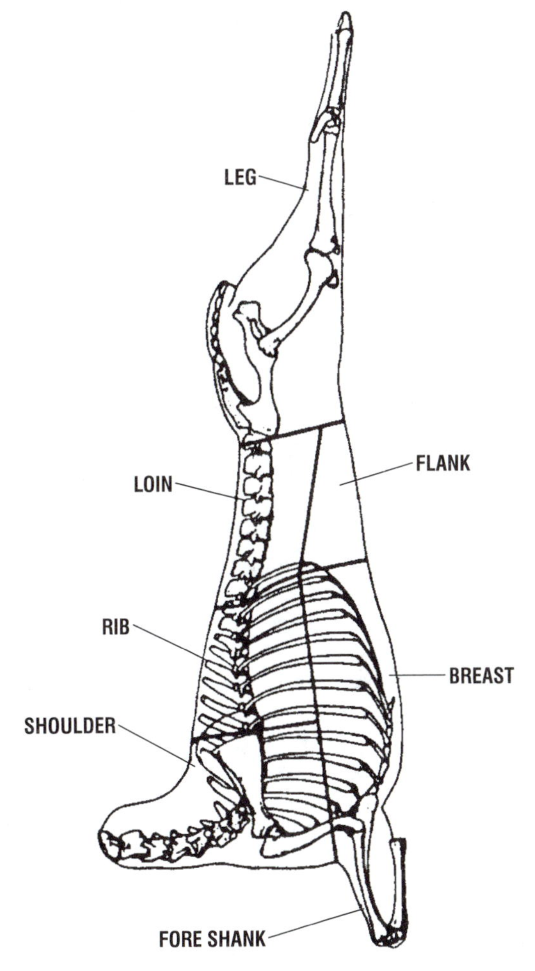

Fig. 19-5. Wholesale cuts of lamb and their bone structure.

teenth ribs. Each of these two cuts represents about 50% of the carcass weight.

The approximate yields of the wholesale cuts are listed in Table 19-1.

TABLE 19-1
APPROXIMATE YIELDS FROM WHOLESALE CUTS OF LAMB

Wholesale Cut	Yield
	(%)
Foresaddle	50
Shoulder (five ribs)	26
Rib	9
Shank	5
Breast	10
Handsaddle[1]	50
Leg (sirloin on)	39
Loin, trimmed	7
Flank	2

[1]Kidneys and suet comprise 2%.

DISPOSITION OF THE LAMB AND MUTTON CARCASS AND CUTS

The lamb industry is turning to boxed lamb, much as the beef and pork industries have done. Through breaking, trimming, and boxing, slaughterers and retailers can save millions of dollars in transportation costs alone by not shipping fat and bones. In addition, where wholesale cuts are sealed in a plastic skintight container, the shrinkage is reduced from the normal 6 to 7% to less than 1%. Storage life is also increased, reducing perishability. To gain optimum economies, packaging, as well as cutting, must be moved outside the retail store. Also, block-ready and case-ready cuts can help solve butcher resistance to lamb at the store level. The meat department is more apt to display lamb, to put a better product in the meat case, and to buy only the cuts needed.

About two-thirds of all U.S. meats are sold as fresh meats, a highly perishable commodity. The proportion of lamb sold fresh is even higher. For efficiency reasons, the long-time, but gradual, trend is toward prepackaged frozen meats.

The retail cuts of lamb are similar to those of other species. The leg is used largely as roasts, the shoulder as roasts and chops, and the rib and loin as chops.

Carcasses or wholesale cuts that are not considered desirable for the trade, or for which there is not sufficient demand, are processed at the packing plant. Eventually, they are sold as prepared meats and meat food products.

In recent years, the following marketing methods and innovations have given a new look to the sheep industry.

■ **Cuts and names**—The wholesale (primal) cuts of lamb are the leg, loin, rib, shoulder, and breast (including the fore shank, breast, and flank areas).

In the case of lamb, names may be a disadvantage to merchandising. Meat derived from lamb is called lamb, whereas meat from cattle becomes beef, meat from calf becomes veal, and meat from hogs becomes pork. Not only that, but many lamb cuts are the actual names of live animal parts—leg, shoulder, breast, flank, etc. By contrast, the corresponding parts of other species generally bear alluring cut names which are not associated with the live animal part, such as chuck for shoulder of beef.

■ **Boneless lamb and convenient forms**—Boneless cuts, such as boneless leg and boneless shoulder, may be cooked in smaller containers, and they are easier to carve. In certain areas, these boneless cuts have been merchandised in the form of netted roasts. The netting reduces the labor required for tying roasts and holds them in very uniform shape.

■ **Frozen cuts**—Frozen lamb cuts are being promoted in some quarters, as a means of expanding the market for lamb and making it available to consumers throughout the year.

■ **Specialty forms**—As a specialty product, with a high margin, some processors are interested in such items as frozen shish kabobs, ground lamb combinations with other meats, and other similar items.

Also, processors are moving steadily in the direction of completely prepared frozen products that require a minimum of preparation; many of these products are combined with sauces, gravies, or other foods. Likely, greater quantities of precooked and frozen meats, including lamb, will be available in the future.

■ **Retail display**—Retail stores normally allocate meat case display space proportionate to the sales volume of the product. Because lamb is a low-volume item, it usually has the smallest display allocation of any kind of meat. Thus, block-ready and case-ready cuts can help solve meat cutter resistance to lamb at the store level.

LAMB RETAIL CUTS AND HOW TO COOK THEM

Fig. 19-6 illustrates the retail cuts of lamb, and gives the recommended method or methods of cooking each.

This informative figure may be used as a guide to wise buying, in dividing the lamb carcass into the greatest number of desirable cuts, in becoming familiar with the types of cuts, and in preparing each type of cut by the proper method of cookery.

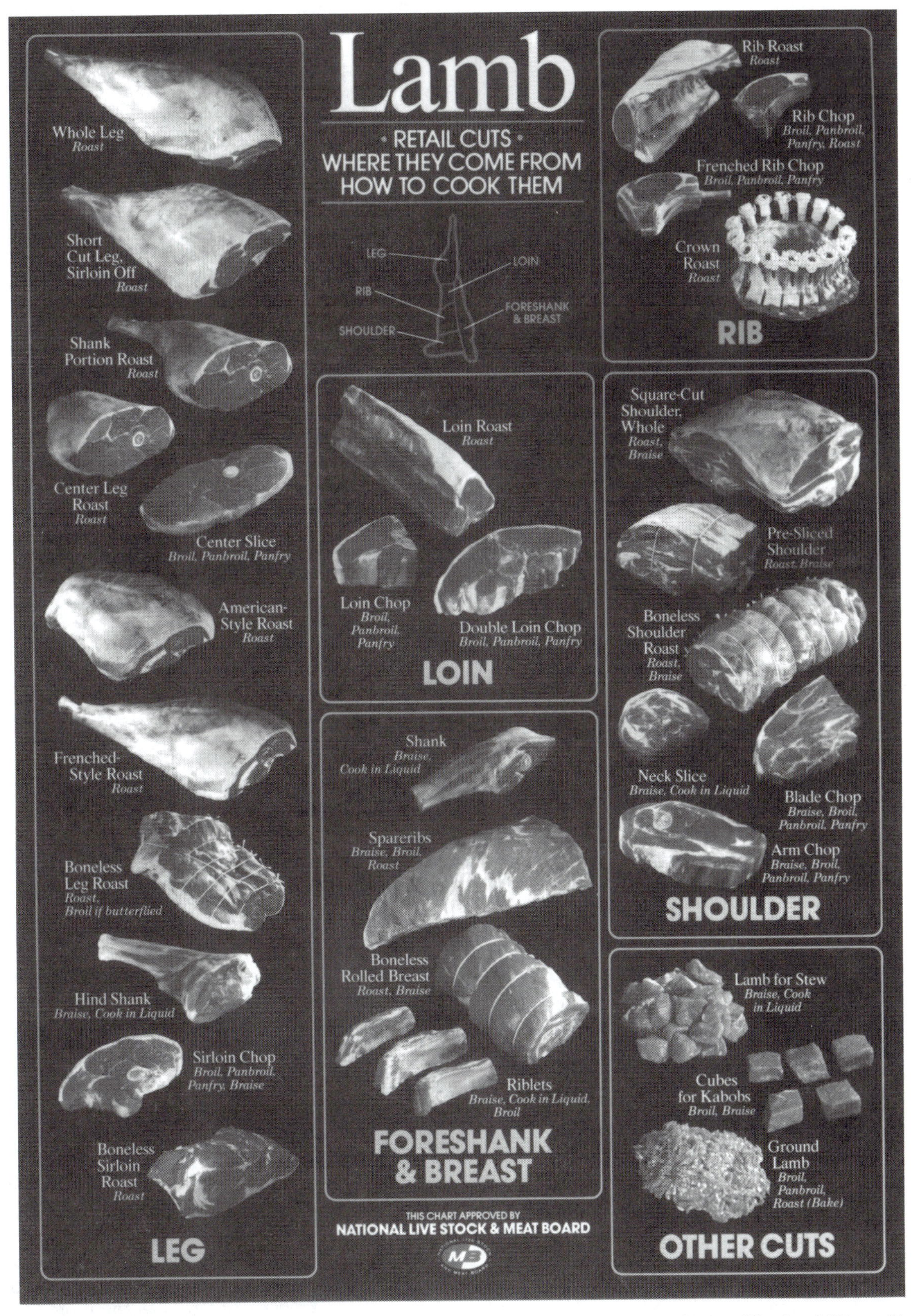

Fig. 19-6. Retail cuts of lamb—where they come from, and how to cook them. (Courtesy, National Live Stock and Meat Board, Chicago, IL)

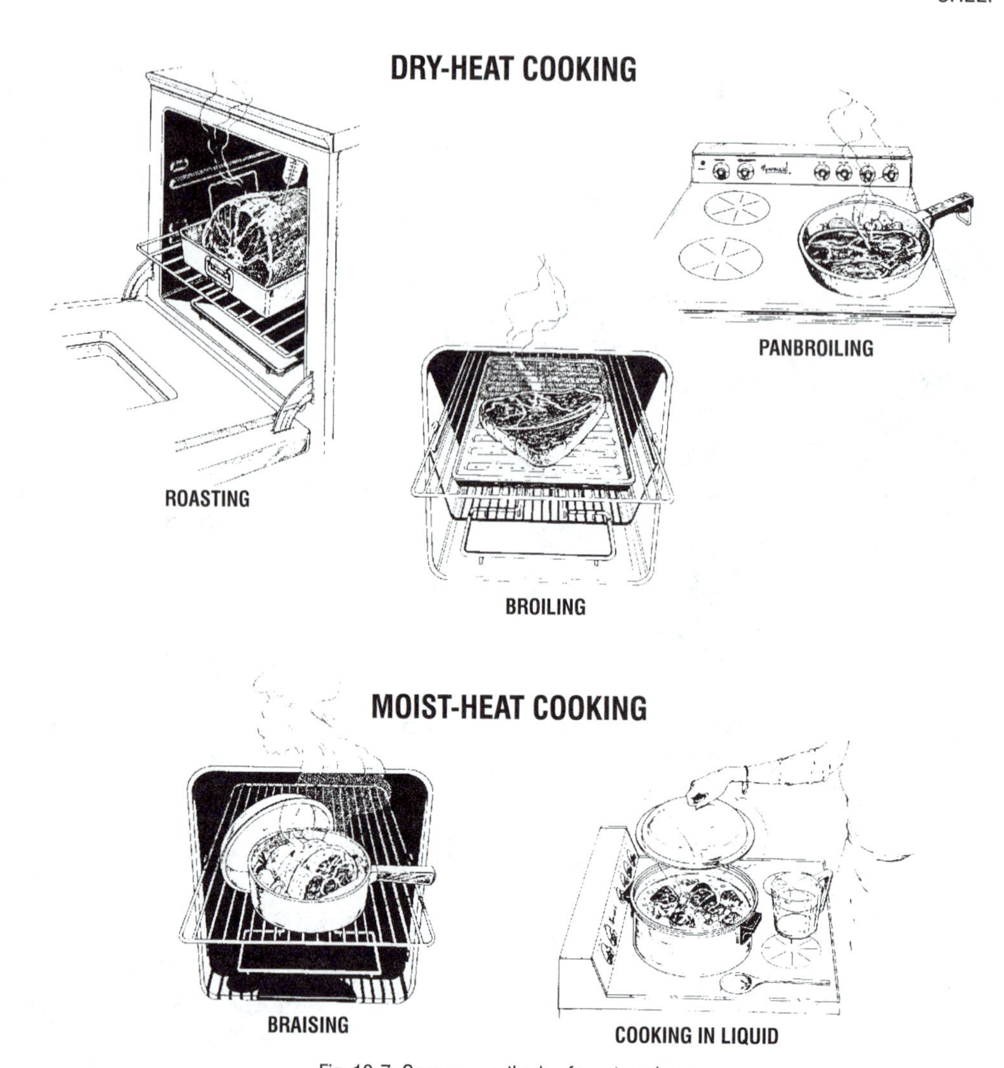

Fig. 19-7. Common methods of meat cookery.

Lamb should be cooked to a minimum finished temperature of 145°F, as recommended by the U.S. Department of Agriculture for safety reasons.[2]

The methods used in meat cookery depend on the nature of the cuts. In general, the types of meat cookery may be summarized as follows (see Fig. 19-7):

1. **Dry-heat cooking.** Dry-heat cooking is used in preparing the more tender cuts, those that contain little connective tissue. This method of cooking consists of surrounding the meat by dry air in the oven or under the broiler. The common methods of cooking by dry heat are (a) roasting, (b) broiling, and (c) panbroiling.

a. **How to roast:**

(1) Season with salt and pepper, if desired.

(2) Place fat side up on rack in an open roasting pan.

(3) Insert a meat thermometer.

(4) Roast in a preheated oven to a minimum finished temperature of 145°F.

(5) Do not add water, or cover, or baste.

(6) Roast until the meat thermometer registers rare, medium, or well-done, as desired.

For best results, a meat thermometer should be used to test the doneness of roasts (and thick steaks and chops). It takes the guesswork out of meat cooking. Allowing a certain number of minutes to the pound is not always accurate. For example, rolled roasts take longer to cook than roasts with bones.

The thermometer is inserted into the cut of meat so that the end reaches the center of the largest muscle, and so that it is not in contact with fat or bone. Naturally, frozen roasts need to be partially thawed before the thermometer is in-

[2]*Facts About Lamb*, USDA special report.

Fig. 19-8. Broiled shoulder lamb chops. Parsleyed potatoes fill the center of this lamb chop platter. Round bone shoulder lamb chops, shown here, are equally as flavorful and often less expensive than the loin chops. Add a salad of sliced tomatoes and cucumbers, warm yeast rolls, and a rhubarb-pineapple pie to the meal featuring broiled chops. (Courtesy, American Meat Institute, Washington, DC)

serted, or a metal skewer or ice pick will have to be employed in order to make a hole in frozen meat.

As the oven heat penetrates, the temperature at the center of the meat gradually rises and is registered on the thermometer. Most meat can be cooked as desired—rare, medium, or well-done.

b. **How to broil:**

(1) Set the oven regulator for broiling.

(2) Place the meat on the rack of the broiler pan and cook 2 to 5 in. from the heat.

(3) Broil until the top of the meat is brown.

(4) Season the top side with salt and pepper, if desired.

(5) Turn the meat and brown the other side.

(6) Season, if desired, and serve at once.

c. **How to panbroil:**

(1) Place the meat in a heavy frying pan or on a griddle.

(2) Do not add fat or water.

(3) Cook slowly, turning occasionally.

(4) If fat accumulates, pour it off.

(5) Brown the meat on both sides.

(6) Do not overcook. Season, if desired, and serve at once.

2. **Moist-heat cooking.** Moist-heat cooking is generally used in preparing the less tender cuts, those containing more connective tissues that require moist heat to soften them and make them tender. In this type of cooking the meat is surrounded by hot liquid or steam. The common methods of moist-heat cooking are: (a) braising, and (b) stewing or cooking in water.

a. **How to braise:**

(1) Brown the meat on all sides in a heavy utensil.

(2) Season with salt and pepper, if desired.

(3) Add a small amount of liquid, if necessary.

(4) Cover tightly.

(5) Cook at simmering temperature, without boiling, until tender.

(6) Make sauce or gravy from the liquid in the pan, if desired.

b. **How to cook large cuts in water:**

(1) Brown meat on all sides, if desired.

(2) Cover the meat with water or stock.

(3) Season with salt, pepper, herbs, spices and/or vegetables, if desired.

(4) Cover kettle and simmer (do not boil) until tender.

(5) If the meat is to be served cold, let it cool and then chill in the stock in which it was cooked.

(6) When vegetables are to be cooked with the meat, as in "boiled" dinners, add them whole or in pieces, just long enough before the meat is tender to be cooked.

c. **How to cook stews:**

(1) Cut meat in uniform pieces, usually 1- to 2-in. cubes.

(2) If a brown stew is desired, brown meat cubes on all sides.

(3) Add just enough water, vegetable juices, or other liquid to cover the meat.

(4) Season with salt, pepper, herbs and/or spices, if desired.

(5) Cover kettle and simmer (do not boil) until meat is tender.

(6) Add vegetables to the meat just long enough before serving to be cooked.

(7) When done, put meat and vegetables in a pan or casserole or on a platter and keep hot.

(8) If desired, thicken the cooking liquid with flour for gravy.

(9) Serve the hot gravy (or thickened liquid) over the meat and vegetable or serve separately in a sauce boat.

(10) Meat pies may be made from the stew; a meat pie is merely a stew with a top on it. (The top may be made of pastry, biscuits or biscuit dough, mashed potatoes, or cereal.)

3. **Frying.** When a small amount of fat is added before cooking, or allowed to accumulate during cooking, the method is called panfrying. This is suitable for

preparing comparatively thin pieces of tender meat, or those pieces made tender by pounding, scoring, cubing, or grinding, or for preparing left-over meat. When meat is cooked by being immersed in fat, it is called deep-fat frying. This method of cooking, sometimes used for preparing brains, liver, and left-over meat. Usually the meat is coated with eggs and crumbs or a batter, or dredged with flour or cornmeal.

a. **How to panfry:**

(1) Brown meat on both sides in a small amount of fat.

(2) Season with salt and pepper, if desired.

(3) Do not cover the meat.

(4) Cook at moderate temperature, turning occasionally, until done.

(5) Remove from pan and serve at once.

b. **How to deep-fat fry:**

(1) Use a deep kettle and a wire frying basket.

(2) Heat fat to frying temperature.

(3) Place meat in frying basket.

(4) Brown meat and cook it through.

(5) When done, drain fat from meat into kettle and remove meat from basket.

(6) Strain fat through cloth; then cool the meat.

4. **Microwave cooking.** In electronic or microwave cooking, food is cooked by the heat generated in the food itself.

The primary advantage of microwave cooking is speed.

There are still some problems with microwave cooking, especially with regard to the cookery of basic meat cuts. For example, electronically cooked roasts have a higher shrinkage; microwaves frequently do not penetrate roasts uniformly, with the result that there may be variations in doneness of meat; the amount of food electronically cooked greatly affects cooking time; and foods cooked solely by microwave do not brown. The microwave is, however, very acceptable for reheating meat.

NUTRITIVE QUALITIES OF LAMB

Perhaps most people eat lamb simply because they like it. They derive a rich enjoyment and satisfaction therefrom.

But lamb is far more than just a very tempting and delicious food. From a nutritional standpoint, it contains certain essentials of an adequate diet: high-quality protein, calories, minerals, and vitamins. This is important, for how we live and how long we live are determined in large part by our diet.

Fig. 19-9 illustrates the contribution of lamb toward fulfilling the Recommended Daily Dietary Allowances (RDA) for calories (energy), protein, and certain minerals and vitamins.

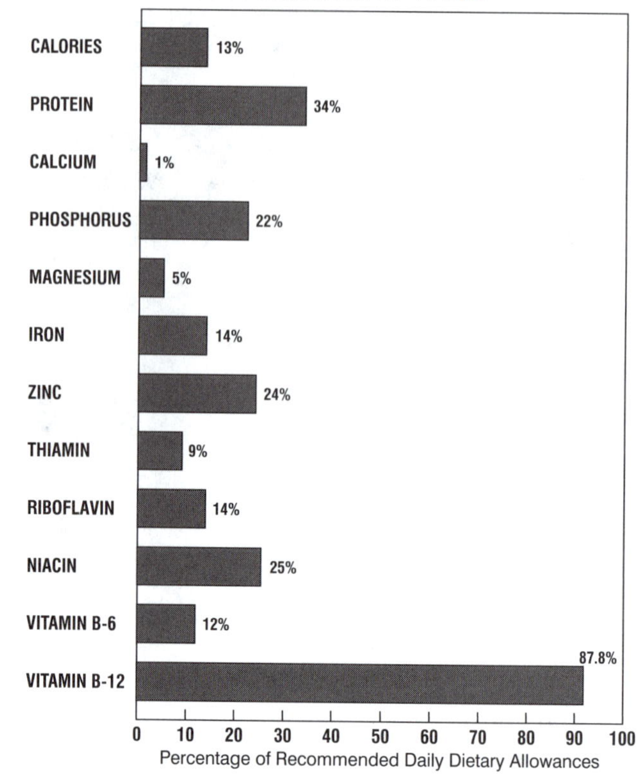

Fig. 19-9. Nutritional value of a 3-oz serving of leg of lamb. (Based on the average values for cooked lean leg of lamb and the National Research Council's 1989 Recommended Daily Dietary Allowances for a 25- to 50-year-old

Effective lamb promotion necessitates full knowledge of the nutritive qualities of meats, the pertinent facts of which follow:

1. **Proteins.** The word *protein* is derived from the Greek word *proteios*, meaning *in first place*. Protein is recognized as a most important body builder. Fortunately, meat contains the proper quantity and quality of protein for the building and repair of body tissues. All meats provide more protein than formerly. As trimmed and eaten today, lamb leg and chops contain 26 to 28% protein. Also, meat protein contains all of the amino acids, or building blocks, which are necessary for the making of new tissue. Lamb is a high-quality protein—a complete protein—because it contains all the essential amino acids.

2. **Minerals.** Minerals are necessary in order to build and maintain the body skeleton and tissues and to regulate body functions. Lamb is a rich source of several minerals, but is especially good as a source of zinc, phosphorus, and iron.

Zinc is needed for normal skin, bones, and hair, and it is a component of many enzymes of energy metabolism and protein synthesis. Meat is generally a better source of dietary zinc than plants because the phytate present in many plant sources complexes the zinc and makes it unavailable.

Phosphorus combines with calcium in building bones and teeth. Phosphorus also enters into the structure of every body cell, helps to maintain the alkalinity of the blood, is involved in the output of nervous energy, and has other important functions.

Iron is necessary for the formation of blood, and its presence protects against nutritional anemia. It is a constituent of the hemoglobin or red pigment of the red blood cells. Thus, it helps to carry the life-giving oxygen to every part of the body. Iron from meat, such as lamb, is absorbed most readily by the body, and it also increases the absorption of iron from vegetable sources.

3. **Vitamins.** As early as 1500 B.C., the Egyptians and Chinese discovered that eating livers would improve one's vision in dim light. It is now known that livers furnish vitamin A, a very important factor for night vision. In fact, medical authorities recognize that night blindness, glare blindness, and poor vision in dim light are all common signs that the individuals so affected are not getting enough vitamin A in their diet.

Meat is one of the richest sources of the important B group of vitamins, especially thiamin, riboflavin, niacin, and vitamin B-12 (see Table 19-2)

These B vitamins are indispensable in our daily diet. They are necessary for energy metabolism, synthesis of new tissue (growth), nervous function, and many other functions. A marked deficiency of thiamin causes beriberi. Niacin prevents and cures pellagra. The only dietary source of vitamin B-12 is foods of animal origin. Indeed, one of the reasons for the rapid decline in B-vitamin deficiencies in the United States may well be the increased amount of meat in the daily diet.

4. **Digestibility.** Finally, in considering the nutritive qualities of meats, it should be noted that they are highly digestible. About 97% of meat proteins are digested.

TABLE 19-2
VITAMIN CONTENT OF LAMB[1]

Lamb Cut[2]	Thiamin	Riboflavin	Niacin	Vitamin B-12
	- - - - - - - (mg/100 g)[3] - - - - - - -			(mcg/100 g)[3]
Leg	0.14	0.25	5.2	2.0
Loin (chop)	0.12	0.23	5.0	—
Rib.	0.12	0.21	4.6	—
Shoulder	0.13	0.23	4.7	2.0

[1]Ensminger, A. E., M. E. Ensminger, J. E. Konlande, and J. R. K. Robson, *Foods & Nutrition Encyclopedia*, CRC Press, Boca Raton, FL, 1994.

[2]Prime or Choice, cooked, total edible.

[3]One hundred grams is approximately equal to 3½ oz.

CONSUMER HEALTH CONCERNS

Much has been spoken and written linking the consumption of meat to certain health-related problems in humans, including heart disease, cancer, high blood pressure (hypertension), and harmful residues. Typical of many well-publicized issues, some myths have arisen. Four of these myths along with the facts follow:

■ **Myth: Meat fats cause coronary heart disease**—Coronary heart disease (CHD) is the leading cause of death in the United States, accounting for one-third of all deaths. The major form of CHD results from atherosclerosis, a condition characterized by fatty deposits in the coronary arteries. These deposits are rich in cholesterol, a complex fatlike substance. Also, in general, serum cholesterol levels are relatively high among individuals with atherosclerosis. This led some persons to hypothesize that high intakes of cholesterol and of saturated fats cause CHD. Because foods from animals are major dietary sources of cholesterol and saturated fats, it was deduced that they caused heart disease.

Fact: Both cholesterol and saturated fat are synthesized in the human body; and cholesterol is an essential constituent of all body cells.

Opinions of physicians are divided as to whether manipulating the dietary intake of cholesterol, saturated fats, and polyunsaturated fats has been or will be effective in combating heart disease.

Epidemiological data (studies of populations) showing the relationship between death rates from coronary heart disease and per capita meat consumption for selected countries fail to indicate a relationship between meat consumption and coronary heart disease. Moreover, deaths from heart disease in the United States have declined since 1950, though consumption of meat and poultry has greatly increased during this same period (from 169 lb in 1950 to 242 lb in 1979).

■ **Myth: Meat causes bowel cancer**—This question has been prompted by the following reports: (1) that the age-adjusted incidence of colon cancer has been found to increase with the per capita consumption of meat in countries; (2) that, in a study done in Hawaii, the incidence of colon cancer in persons of Japanese ancestry was found to be greater among those who ate Western-style meals, especially those who ate beef; and (3) that an examination of (a) international food consumption patterns, and (b) food consumption survey data from the United States showed that a higher incidence of colon cancer occurred in areas with greater beef consumption.

Fact: A direct cause-effect relationship between diet and cancer has not been established. Such studies as the three cited provide valuable leads for researchers who are trying to determine the cause of a certain disease such as colon cancer, but they do not establish the cause. The reason is that the factor measured and found associated with the incidence of colon cancer or other condition is not the only difference among the population groups studied, and the factor measured in the study may be only associated in some way with the real cause.

■ **Myth: Meat causes high blood pressure (hypertension)**—Epidemiological studies have implicated meat as a cause of high blood pressure.

Fact: There is no evidence that meat per se has any major effect on high blood pressure. However, consumption of cured meat containing large amounts of salt should be minimized as should the amount of salt used as a condiment on meat and other foods.

■ **Myth: Meat contains harmful residues**—Some reports have suggested meats contain harmful toxic metals, pesticides, animal drugs, and additives.

Fact: If one pushed the argument of how safe is safe far enough, nothing would be safe to eat.

The following are the facts relative to toxic metals, pesticides, animal drugs, and additives in meats:

1. All metals are present in at least trace amounts in soil and water. It follows that they are also present in small amounts in plants and animals. But samplings of meat have not shown any toxic levels of these metals; hence, the normal intake of these metals in meats does not present any known hazard.
2. By proper application of pesticides to control pests, food and fiber losses can be, and are being, reduced substantially—perhaps by as much as 30 to 50%. Only when improperly used, have pesticide residues been found in meat.
3. More than 1,000 drugs and additives are approved by the Food and Drug Administration (FDA) for use by livestock producers. The FDA requires withdrawal times on some of them, in order to protect consumers from residues. Additionally, federal agencies (USDA and/or FDA), as well as certain state and local regulatory groups, conduct continuous surveillance, sampling programs, and analyses of meats and other food products on their content of drugs and additives.

LAMB PRICES AND SPREAD

During those periods when lamb is high in price, especially the choicest cuts, there is a tendency on the part of the consumer to blame either or all of the following: (1) the farmer, (2) the packer, (3) the meat retailer, (4) the government; and these four may blame each other. Vent to such feelings is sometimes manifested in political campaign propaganda, consumer boycotts, and sensational news stories.

Also, some people are prone to compare what the packer is paying for lambs on-foot to what they are paying for a pound of lamb chops over the counter. Then they scream, "Why $2.30 lamb chops from 60¢ lambs?"

Is there any real justification for this often vicious criticism? Who or what is to blame for high meat prices and for the spread? If good public relations are to be maintained, it is imperative that each member of the meat team—the producer, the packer, and the retailer—be fully armed with documented facts and figures with which to answer such questions and to refute such criticisms. Also, the consumer should be informed about the situation. The sections which follow are designed to ferret out the facts relative to lamb prices and spread.

WHAT DETERMINES LAMB PRICES?

Lamb prices are determined by the laws of supply and demand: The price of meat is largely dependent upon what consumers as a group are able and willing to pay for the available supply.

AVAILABLE SUPPLY OF LAMB

When one considers the available supply of lamb, two factors make for differences that are not encountered to the same extent in beef and pork. First, certain ranges are not adapted to use by other meat animals, which means that they will be used by sheep, if indeed they are used at all. Second, lamb production is, biologically, more seasonal than beef or pork production. Over and above these differences, the available supply of lamb is affected by the same factors as are other meats.

Because lamb is a perishable product, the supply of this food is very much dependent upon the number and weight of lambs available for slaughter. In turn, the number of market animals is largely governed by the relative profitability of the sheep enterprise in comparison with other agricultural pursuits. Farmers—like other good businesspersons—generally do those things that are most profitable to them. Thus, a short supply of market animals usually reflects the unfavorable and unprofitable production factors that existed some months earlier and which caused curtailment of breeding and feeding operations.

Historically, when short sheep supplies exist, lamb prices rise, and the market price on slaughter lambs usually advances, making sheep production more profitable. But, unfortunately, sheep breeding and feeding operations cannot be turned on and off like a spigot.

History also shows that if sheep prices remain high and food abundant, producers will step up their breeding and feeding operations as fast as they can within the limitations imposed by nature, only to discover when market time arrives that too many other producers have done likewise. Overproduction, disappointingly low prices, and curtailment in breeding and feeding operations are the result.

Nevertheless, the operations of livestock farmers do respond to market prices, producing so-called cycles. Thus, the intervals of high production, or cycles in sheep, occur every 9 to 10 years.

DEMAND FOR LAMB

The demand for lamb is primarily determined by buying power and competition from other products. Stated in simple terms, demand is determined by the spending money available and the competitive bidding of millions of consumers who are the chief home purchasers of meats. On a nationwide basis, high buying power and great demand for meats exist when most people are employed and wages are high.

Often, in boom periods—periods of high personal income—meat purchases are affected in three ways: (1) More total meat is desired; (2) there is a greater demand for the choicest cuts; and (3) because of the increased money available and shorter working hours, there is a desire for more leisure time, which in turn increases the demand for those meat cuts or products that require a minimum of time in preparation (such as lamb chops). In other words, during periods of high buying power, not only do people want more meats, but they also compete for the choicer and more easily prepared cuts of meats.

Because of the operation of the old law of supply and demand, when the choicer and more easily prepared cuts of lamb are in increased demand, they advance proportionately more in price than the cheaper cuts. This results in a great spread in prices, with some lamb cuts very much higher than others. Thus, while lamb chops may be selling for four or five times the cost per pound of the live animal, less demanded cuts may be priced at less than half the cost of the more popular cuts. This is so because a market must be secured for all the cuts.

Thus, when the national income is exceedingly high, there is a demand for the choicest but limited cuts of lamb from the very top grades. This is certain to make for high prices, for the supply of such cuts is limited, but the demand is great. Under these conditions, if prices did not move up to balance the supply with demand, there would be a marked shortage of the desired cuts at the retail counter.

WHERE THE CONSUMER'S FOOD DOLLAR GOES

On the surface, the above heading appears to pose a very controversial subject. Actually, much of it is due to misunderstanding. Most people know too little about the other person's business. The following sections are designed to ferret out the facts relative to lamb prices and profits.

LAMB PRICES AND PRODUCERS' PROFITS

It is preposterous to think that sheep producers could control prices, for producers are well-known individualists and the competition between them is too great.

Unfortunately, all too many lamb consumers fail to realize that a ewe lamb ordinarily is not bred until she is past one year of age, that the pregnancy period requires another 148 days, and, finally, that the young are usually grown and finished five months before marketing. Thus, under the most favorable conditions, this process, which is under biological control and cannot be speeded up, requires about 18 months in which to produce a new generation of market lambs. Most of these critics also fail to realize that, in addition to breeding stock and feed costs, there are shipping charges, interest on borrowed money, death losses, marketing charges, taxes, and numerous other costs, before the lambs finally reach the packers.

For the lambs, and all the expenses and services that go into their production, the producer gets only 36% of the consumer's dollar (see Fig. 19-10). And these figures do not tell the whole story! They do not relate the bedeviling effect of well-meant, imposed, or threatened legislation, or the shivering hours spent in the lambing quarters as numb fingers attempt to bring life into a newborn lamb; they do not tell of disease problems, or of the sweat of a long work week which begins at 40 hours.

Obviously, farmers are not receiving excessive profits in lamb production. With knowledge of these facts, and considering the long-time risks involved, it is doubtful if very many people will object to the too few good years when producers have a chance to recoup the losses they suffer during the lean years.

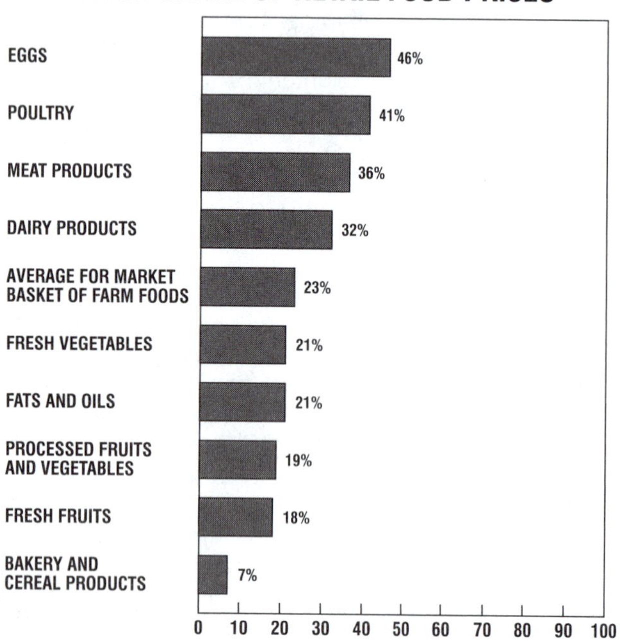

Fig. 19-10. Farm share of retail food prices. Note that the farmer receives 36% of the retail price for meat products, including lamb. For products requiring considerable processing, such as processed fruits and vegetables and bakery and cereal products, the farmer's share is less. (Courtesy, USDA)

LAMB PRICES AND MEAT-PACKERS' PROFITS

Of course, meat-packing companies, like other businesses, are owned and operated by people; and all people want to be paid for their work. Therefore, they are entitled to a fair and reasonable profit; otherwise, they would not stay in business. The only questions are—do they control prices and are they making too lush profits? Here are the pertinent facts on which to base an answer to this question:

1. Packers do not and cannot control prices because there is too intensive competition between them; there are more than 5,000 meat packers and meat processors in the United States.

2. Meat-packers' net earnings amount to less than 1¢ per dollar of sales—not enough (see Fig. 19-11). By comparison, the net earnings per dollar of sales of other companies in the United States are 10 times greater. However, meat packers deal in higher volume sales than many other industries.

3. During 1980 and 1981, the net profit of packing companies per 100 lb of live animal averaged 65¢; and for 100 lb of dressed meat, it averaged $1.01. Thus, on the average, for each 115 lb of lamb purchased, the packer netted about $1.16[3] in profit. Certainly this is a reasonable and legitimate earning. Of course, the volume of sales (the number of animals processed in a year), the efficiency of operations, and the utilization of byproducts make it possible for the industry to operate on these comparatively small margins.

Actually, the packers' profits are so small that, were they eliminated entirely, they would have practically no effect on the ultimate selling price of the retail cuts. The meat packer is in the middle of an impossible situation; on the one hand, the nation's lamb producers want high prices for all the animals that they can sell, while, on the other hand, more than 235 million consumers desire to buy as much meat as they wish at low prices.

LAMB PRICES AND MEAT-RETAILERS' PROFITS

Finally, let us consider meat retailers. Like producers and packers, they, too, are in business to make money; thus, they buy lamb carcasses at as low a price as possible and they sell the retail cuts at as high a price as consumers will pay. But do meat retailers control meat prices, and are they making excessive profits? Here are the facts:

[3] *Annual Financial Review of the Meat Packing Industry*, American Meat Institute, Washington, DC, 1982, p. 3, Table 1-1.

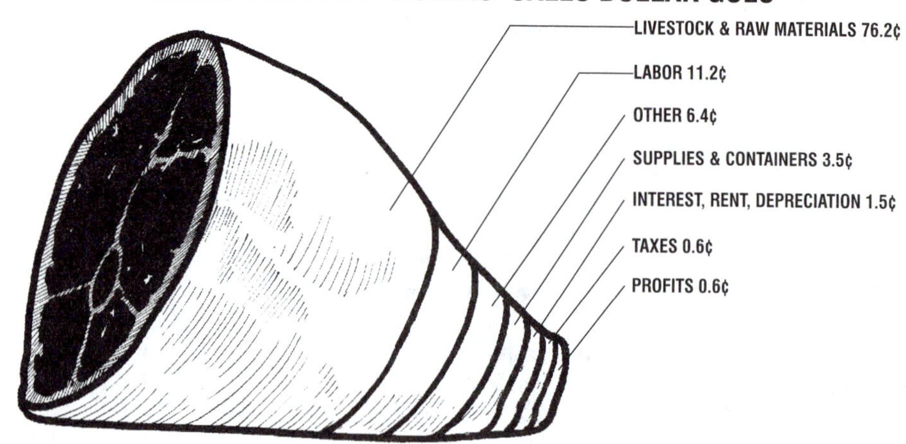

Fig. 19-11. For each dollar of sales, meat packers earn an average of 0.6¢. (Based on data from the American Meat Institute, Washington, DC)

1. There are about 186,000 grocery and other food stores in the United States, most of which handle meat, plus about 286,000 establishments serving meals or snacks which may also be retail meat outlets.[4] There is much competition among them. As a result, no meat retailer can keep prices out of line for very long; otherwise, consumers will just quit patronizing the market.

2. On the average, year after year, the meat retailer nets less than 1.5¢ after taxes on each $1.00 of sales.

3. There is an old axiom in the meat business which says, "You sell it or you smell it." This simply means that as a perishable product, it must be moved promptly into consumption channels and sold for whatever it will bring. If the retailer attempted to get a higher price than the market afforded, the amount of the product demanded would diminish, the meat would not be sold, and it would soon spoil.

CONSUMER CONTROL OF LAMB PRICES

The preceding facts should make it obvious that lamb prices are not controlled by the producer, the packer, or the retailer. Unknowingly, consumers really dictate the price of lambs on-foot, the price of dressed lambs, and the price of retail cuts. Consumers actually put the price tag on retail cuts. For purposes of illustrating their impact on lamb and sheep prices, let us assume the following: The producer sells the packer a Choice lamb at 76¢ per pound, the packer sells the retailer the on-hook carcass at $2.00 per pound, and the retailer prices the lamb chops at $7.50 per pound.

When a consumer walks along the counter, that person, alone, determines what he/she will buy. If the consumer feels that $7.50 a pound is too much to pay for lamb chops, he/she may buy a leg of lamb or lamb stew, or perhaps move down the counter a few feet and buy some broilers or some fish. He/she does not say anything to the retailer about the high price of the lamb chops. There is no organized boycott on the part of the consumers, but it does not take retailers long to discover a price on lamb chops that is out of line.

If lamb chops will not sell at $7.50, the retailer may lower the price to $7.25 a pound, and likely try to raise the prices on some other lamb cuts to take care of the loss on chops. If the retailer is unable to make up the deficit, there is only one thing to do—tell the packer that he/she cannot afford to pay $2.00 a pound for lamb carcasses. The retailer may tell the packer that he/she can pay $1.90 a pound, and in all probability the packer will refuse this offer; so the retailer will likely buy less lamb and double the order for beef and poultry. Then, in 2 or 3 days, as lamb carcasses begin to back up in the coolers, the packer will probably call the retailer and make a deal. They may compromise on $1.95 a pound, following which the retailer starts buying the normal quantity of lamb.

If the packer cannot get $2.00 a pound for lamb carcasses, it does not take very long to tell the sheep buyer that he/she cannot afford to pay 76¢ a pound for live lambs and that the buyer will have to cut the price to 72¢. If the run is heavy, the packer may be able to buy all the needed lambs at 72¢. But if the run is light, he/she may be forced to pay 74¢, realizing that less money will be lost by paying 74¢ for lambs than by not having enough work to keep employees busy. Packers face a dilemma: They must keep enough livestock coming into their plants to keep them in operation, but they must also buy livestock at low enough prices so that the dressed meat can be sold at a profit.

As in the case of all other commodities on a free market, meat prices are determined by supply and demand—by what consumers as a group are able and willing to pay for the available supply. In essence, what all America eats tonight will determine tomorrow's lamb prices.

WHAT DETERMINES THE SPREAD BETWEEN ON-FOOT LAMB PRICES AND RETAIL LAMB PRICES?

When a sheep producer receives a check for $91.00 for a 120-lb lamb—76¢ per pound—and on the way home stops at a retail meat market and buys lamb chops at $7.50 per pound, that person must be prone to think that being on the other side of the counter as a meat packer or a meat retailer would be more profitable.

Why is there so much spread between the price of a lamb on-foot and the price of a pound of lamb chops? This is a straightforward question which deserves a straightforward answer. Here are the facts.

Lambs are not all meat, and the carcasses are not all chops. It is important, therefore, that those who produce and slaughter animals and those who purchase wholesale and/or retail cuts know the approximate (1) percentage yield of chilled carcass in relation to the weight of the animal on-foot, and (2) yield of different retail cuts. For example, a lamb weighing 120 lb on-foot will only yield 58.8 lb of salable retail cuts (the balance consists of internal organs, etc.). Thus, only about 49% of a live lamb can be sold as retail cuts of lamb. In other words, the price of lamb at retail would have to be more than double the live cost even if there were no processing and marketing charges at all. Second, the higher priced cuts make up only a small

[4] *Statistical Abstracts of the United States*, U.S. Bureau of the Census, 1998, p. 766.

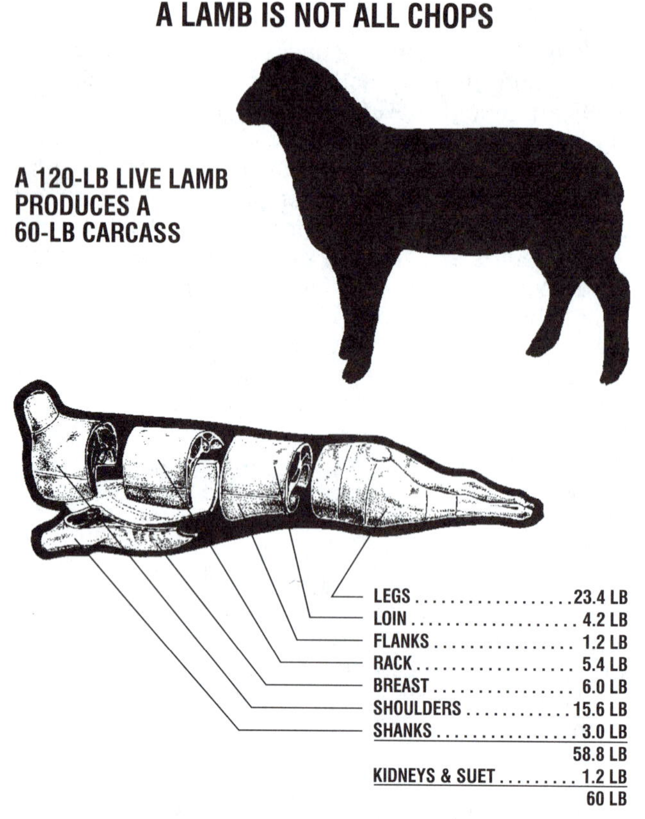

Fig. 19-12. Lambs are not all chops! Only about 58.8 lb of salable retail cuts can be obtained from 120 lb of live lamb. And the loin comprises only 7% of the carcass.

part of the carcass. Thus, this 58.8 lb will cut out only about 4.2 lb of loin chops. The other cuts retail at lower prices than do these choice cuts; also, there are bones, fat, and cutting losses which must be considered.

INCREASED SERVICES AND ATTRACTIVENESS

Since more than half of the nation's women are now in the labor force and many people have more leisure time, it is understandable why consumers want more convenience—more processing and packaging. They desire that food purchases be largely prepared for immediate cooking, for their kitchen and meal time is limited. Without realizing it, U.S. consumers have 1.7 million people working for them in food processing and distributing alone. These mysterious persons do not do any work in the kitchen; they are the people who work on the food from the time it leaves the farms and ranches until it reaches the nation's kitchens. They are the people who make it possible for the consumer to choose between quick-frozen, dry-frozen, quick-cooking, ready-heat, ready-eat, and many other convenient foods. All this is fine, but these 1.7 million individuals engaged in processing and distributing foods must be paid, for they must eat too. So, all of these services have increased the farm-consumer spread, as illustrated in Fig. 19-13.

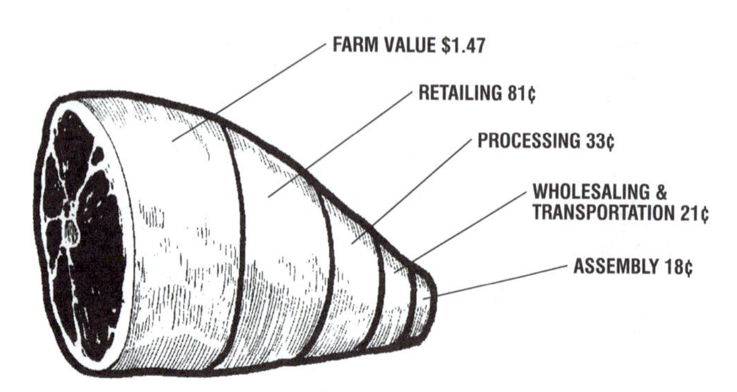

Fig. 19-13. The ultimate objective—lamb over the counter—requires that each business be paid for goods and services.

MARKETING AND PROCESSING CHARGES AND PROFITS

Everyone and everything connected with the meat industry influences the spread between on-foot and retail prices. Investment capital is not free; it must yield returns comparable with other industries which compete for the use of capital. In addition, there are costs for labor, rent, supplies, transportation, and equipment. Over and above these costs, there should be a reasonable and fair profit.

What about decreasing margins by reducing the profits of the marketing agencies? The two major organizations involved are the meat packing industry and the retail stores. As previously stated, the average net profit of each—the meat packer and the meat retailer—amounting to less than 1 and 2%, respectively, on each $1 of sales, is very small. Were the profits from both businesses eliminated entirely—and if producers received all of this additional amount— lamb prices would be raised only 0.5¢ a pound.

Producers and consumers also need to recognize that when the demand or the supply changes for lamb, there is of necessity much more change in the price of live lambs than there is in the price of lamb over the counter. This is so because of the tendency for marketing margins to be more fixed—that is, the costs of labor, rent, supplies, transportation, and equipment do not fluctuate rapidly.

OTHER FACTORS AFFECTING LAMB PRICES AND SPREAD

These factors include those forces other than supply and demand, per se, which affect lamb prices and may help to explain why lamb chops may cost four times the price of lamb on-foot.

CONSUMERS WITH HIGH INCOMES DEMAND CHOICE CUTS AND TOP GRADES

In periods of prosperity—when incomes are rising—consumers place a premium on the preferred kind, cut, and quality of meat, rather than show any marked increase in total meat consumption. Also, people on higher incomes eat more lamb, beef, and veal, and less pork, and they eat more of the expensive cuts of lamb, such as chops and legs, and fewer of the cheaper cuts.

Also, due to the increased money available and shorter working hours, there is a desire for more leisure time, which, in turn, increases the demand for those meat cuts or products which require the minimum time in preparation. In many respects, these two factors operate together; in other words, with high buying power, people hunt the choicer and more easily prepared cuts of meat—chops and legs.

All this suggests that producers, processors, and distributors have much to gain by taking steps to provide the desired kind and quality of products—by breeding for increased carcass quality, by feeding to enhance grade, and by processing and preparing a higher-quality and more attractive product.

NATURE MAKES FEW CHOICE CUTS

The novice may wonder why these choice cuts are so scarce, even though people are able and willing to pay a premium for them. The answer is simple; nature does not make many choice cuts, regardless of price. Lambs are born with only two hind legs; and only so many chops can be obtained from each carcass upon slaughter. The balance of the cuts are equally wholesome and nutritious, but they are better adapted for use as stews, etc.

Thus, when the national income is exceedingly high, there is a demand for the choicest but limited cuts from the very top but limited grades of meats. This is bound to make for extremely high prices for such cuts—for the supply is limited, but the demand is great.

CONSUMER TREND TO LESS LAMB AND MORE BEEF AND OTHER MEATS

Another underlying trend which depresses the price of lamb is the shift in consumer preference to beef. From 1945 to 1980, per capita lamb consumption dropped from 7.3 to 1.6 lb, whereas during this same period of time, beef consumption increased from 59.4 to 103.4 lb.

COMPETITION WITH OTHER FOOD PRODUCTS AND SERVICES

Meat must compete with other food products for the consumer's dollar. Thus, in addition to preference, relative price is an important factor. Perhaps poultry is one of the biggest competitors for the consumer's dollar. The poultry industry is aggressive and innovative; the best evidence of this is the steady increase in the per capita consumption of poultry.

Also, meat must compete with certain nonfood items, for there are people who would go hungry in order to spend their money for other purposes. On the average, consumers spend about 4% of their disposable income, or about 20 to 25% of their food budget, for meat.

Finally, sometimes consumers are prone to blame their budget troubles on food, and meat in particular, simply because they forget that they are spending part of their income on things which they did not have before—including such things as VCRs or computer hardware and software.

PACKING HOUSE BYPRODUCTS FROM SHEEP SLAUGHTER

The meat or flesh of animals is the primary object of slaughtering. The numerous other products are obtained incidentally. Thus, all products other than the carcass meat are designated as byproducts, even though many of them are wholesome and highly nutritious items in the human diet. Upon slaughter, sheep yield an average of 51% of products other than carcass meat. When meat packers buy lamb, they buy far more than the cuts of meat that will eventually be obtained from the carcasses, since only about 49% of a meat animal is meat.

In the early days of the meat-packing industry, the only salvaged byproducts were wool, hides, tallow, and tongue. The remainder of the offal was usually carted away and dumped into the river or burned or buried. In some instances, packers even paid for having the offal taken away. In due time, factories for the manufacture of glue, fertilizer, soap, buttons, and numerous other byproducts sprang up in the vicinity of the packing plants.

Some factories were company owned; others were independent industries. Soon much of the former waste was being converted into materials of value.

Naturally, the relative value of carcass meat and byproducts varies according to the class of livestock and from year to year. The long-time trend for byproducts has been downward relative to the value of the live animal, due largely to technological progress in competitive products derived from nonanimal sources.

Although many of the byproducts from sheep, cattle, and hogs are utilized in a like manner, there are a few special products which are peculiar to the class of animals (*e.g.*, wool and catgut from sheep).

The complete utilization of byproducts is one of the chief reasons why large packers are able to compete so successfully with local butchers. Were it not for this conversion of waste material into salable form, the price of meat would be much higher than under existing conditions.

Obviously, all of the 100 or more byproducts obtained from animal slaughter cannot be described in this book. Rather, only a few of the more important ones obtained from sheep slaughter will be listed and briefly discussed (see Fig. 19-14).

1. **Sheep pelts.** The sheep pelt is by far the most valuable byproduct of sheep slaughtering. Pelts are handled in one of two ways: either (a) fresh, or (b) salted, with the manner of handling determined by the distance from the plant where the wool is processed.

Sheep skins with short wool, 0.75 in. or less in

Fig. 19-14. Some of the more important items from which byproducts obtained from slaughter are used—items which contribute to the convenience, enjoyment, and health of people in all walks of life. (Courtesy, American Meat Institute, Washington, DC)

1. Bone for bone china.
2. Horn and bone handles for carving sets.
3. Hides and skins for leather goods.
4. Rennet for cheese making.
5. Gelatin for marshmallows, photographic film, printers' rollers.
6. Stearin for making chewing gum and candies.
7. Glycerin for explosives used in mining and blasting.
8. Lanolin for cosmetics.
9. Chemicals for tires that run cooler.
10. Binders for asphalt paving.
11. Medicines such as various hormones and glandular extracts, insulin, pepsin, epinephrine, ACTH, cortisone; and surgical sutures.
12. Drumheads and violin strings.
13. Animal fats for soap.
14. Wool for clothing.
15. Camel's hair (actually from cattle ears) for artists' brushes.
16. Cutting oils and other special industrial lubricants.
17. Bone charcoal for high-grade steel, such as ball bearings.
18. Special glues for marine plywoods, paper, matches, window shades.
19. Curled hair for upholstery. Leather for covering fine furniture.
20. High-protein livestock feeds.

length, are usually tanned with the wool on and are used for coats, robes, rugs, pelts, slippers, and other articles. Pelts with longer wool are sent to the pullery. The pulling process consists of applying a depilatory solution (made of sodium sulfide, slaked lime, and water) to the skin side of the pelt and then pulling the wool loose from the skin after the chemical action has loosened the hold on the fibers.

Pelts are sorted into different grades, based primarily on length and quality of wool, as determined by breeding, nutrition, climatic conditions, and whether previously clipped and, if so, the lapse of time since the last clipping. The quality of the skin is of secondary importance, except for torn or broken skins. The grades and groups of pelts are:

a. **Shearlings and fall clips.** As the designation implies, these are pelts from sheep and lambs that have been shorn and the fleece has not grown out sufficiently to be classed as wool pelts. Fleece measurement is taken at the shortest place on the pelt; and pelts are divided into the following groups:

(1) No. 1 shearlings, wool measuring over 0.5 in.

(2) No. 2 shearlings, wool measuring from 0.25 to 0.5 in.

(3) No. 3 shearlings, wool measuring from 0.125 to 0.25 in.

(4) No. 4 shearlings, wool measuring from bare to 0.125 in.

(5) All shearlings with wool count under 56s. These are the coarse hairy, open pelts.

All cut, torn, scabby, black and mottled pelts must be kept separate, as they are not acceptable for making garment leather or for making articles of clothing used by the military.

b. **Wool pelts.** These come from either (1) sheep that have been shorn in the spring and a new coat of wool has grown out again, or (2) spring lambs that have never been shorn. They are divided into five groups as follows:

(1) Wool sheep, wool 1.5 in. and up.

(2) Genuine spring lambs, not shorn.

(3) Fall clip lambs, carrying wool from 1.0 to 1.5 in.

(4) California genuine spring lambs, not shorn.

(5) Yearling lambs, carrying wool measuring 1.5 in. and up.

c. **Miscellaneous pelts.** These include the 219 groups shown in each of the two preceding grades—shearlings and fall clips, and wool pelts. But they are burry or seedy.

d. **Mexican hairy skins.** These pelts contain hair, commonly called kemp, mixed in the wool.

2. **Hides.** Hides are particularly valuable as a byproduct of cattle slaughter, though skins from sheep and other classes of animals are processed and used. The leather from sheep hides is used for diplomas and certain other leather goods.

3. **Fats.** Next to hides and pelts, fats are the most valuable byproduct derived from slaughtering. Products rendered from them are used in the manufacture of soaps, animal feeds, lubricants, leather dressings, candles, fertilizers, shaving creams, salves, and chemicals.

Oleomargarine (margarine), which is perhaps the best known of the products in which rendered animal fat is incorporated, can be a mixture of vegetable oils and select animal fats. Oleo oil, one of the chief animal fats of this product, is obtained from beef and mutton or lamb.[5]

4. **Variety meats.** The heart, liver, brains, kidneys, tongue, cheek meat, sweetbreads (thymus and pancreatic glands), and tripe (pickled rumen of sheep and cattle) are sold over the counter as variety meats or fancy meats.

5. **Blood.** The blood is used in the refining of sugar; in making blood sausage and stock feeds; in making buttons; in making shoe polish; and in medicine.

6. **Meat scraps and muscle tissue.** After the grease is removed from meat scraps and muscle tissue, they are made into meat meal, or tankage.

7. **Bones.** The bones and cartilage are converted into stock feed, fertilizers, glue, crochet needles, dice, knife handles, buttons, toothbrush handles, and numerous other articles.

8. **Intestines and bladders.** Intestines and bladders are used as sausage, lard, cheese, and snuff and putty containers. Lamb casings are used in making surgical sutures, strings for various musical instruments (violins, cellos, harps, and ukuleles), and strings for tennis rackets.

9. **Glands.** Various glands are used in the manufacture of numerous pharmaceutical preparations. These include the pituitary, thyroid, ovary, pancreas, adrenals, and other glands. People afflicted with diabetes, arthritis, and many other diseases are treated with insulin, and many other medicines made in whole or in part from the glands of lambs.

Proper preparation of glands requires quick chilling and skillful handling. Moreover, a very large number of glands must be collected in order to obtain any appreciable amount of most of these pharmaceutical products. For example, the adrenals from more

[5]Oleomargarine was first perfected in 1869 by the Frenchman Mege-Mouries, who won a prize offered by Napoleon III for a palatable table fat which would be cheaper than butter, keep better, and be less subject to rancidity. Mouries used oleo oil, salt, milk, and annatto, but today most fat used in margarine is entirely of plant origin.

than 100,000 lambs are necessary to produce 1 lb of epinephrine (adrenaline), a powerful heart stimulant.

10. **Collagen.** The collagen of the connective tissues—sinews, lips, head, knuckles, feet, and bones—is made into glue and gelatin. The most important uses for glue are in the woodworking industry; as coated abrasives, in the match industry; and as a binder for ignition chemicals. Gelatin is used in canning large cuts of meat, baking, ice cream making, capsules for medicine, coating for pills, photography, culture media for bacteria, etc.

11. **Contents of the stomach.** Contents of the stomach are used in making fertilizer.

Thus, in a modern packing plant, there is no waste; literally speaking, "everything but the baaa" is saved. These byproducts benefit the human race in many ways. Moreover, their utilization makes it possible to slaughter and process lamb at a much lower cost. But this is not the end of accomplishment! Scientists are continually striving to find new and better uses for packing house byproducts in an effort to increase their value.

LAMB PROMOTION

Promotion will bring lamb to the attention of the consumer. Unless it is a quality product, however, the consumer may buy it once, but never again. It takes repeat sales—consumers wanting the product again and again—to make for increased business. The latter can be achieved only with a quality product; and this calls for lambs with more red meat (especially in the high-priced cuts) and less fat.

Effective lamb promotion—which should be conceived in a broad sense and embrace research, and educational and sales approaches—also necessitates full knowledge of the nutritive qualities of the product. To this end, we need to recognize that (1) lamb contains 20 to 30% high-quality protein, when trimmed and eaten; (2) lamb is a rich source of energy, the energy value being dependent largely upon the amount of fat it contains; (3) lamb is a rich source of several minerals, but it is especially good as a source of phosphorus and iron; (4) lamb is one of the richest sources of the important B group of vitamins, especially thiamin, riboflavin, niacin, and vitamin B-12; and (5) lamb is highly digestible, about 97% of meat proteins and 96% of meat fats being digested. Thus, lamb is one of the best foods with which to alleviate human malnutrition—a most important consideration. Recently, however, meat and animal products in general have received "bad press" for a variety of reasons, including health, world hunger, and animal welfare. Thus, there is a place and a need for increased lamb promotion; thereby increasing lamb consumption and price. The promotion of lamb should include:

1. Making lamb more generally and continuously available everywhere.
2. Improving lamb merchandising.
3. Maintaining a consistently high-quality product.
4. Making lamb more competitively priced due to greater efficiency of production, processing, and distribution.

The National Wool Act, which was passed in 1954, was phased out in 1995—following 43 years. The reason: The dissention among the sheep and wool growers in the United States resulted in a vote, as required by law. The first vote, on February 6, 1996, was thrown out when the USDA determined that it had been flawed by misrepresentations of the voting and eligibility rules. In the second checkoff referendum held October 1, 1996, the count was 47% for and 53% against. Unfortunately, fewer than 10% of the total U.S. sheep producers and feeders voted. The checkoff was expected to raise between $13 and 14 million each year, $6 to 7 million of which would be from importer contributions. So, at this point and period of time, the National Wool Act is of historic interest only.

SHEEP MILK[6]

In many countries of the world, particularly those surrounding the Mediterranean, sheep dairying has been an important agricultural enterprise for many years. There is a long tradition of milking sheep and

Fig. 19-15. Milk yield of crossbred ewes being measured at the University of Minnesota. (Courtesy, Bill Boylan, University of Minnesota, St. Paul; and David L. Thomas, University of Wisconsin, Madison)

[6]In the preparation of this section, the author drew heavily from *Sheep Dairying in Idaho*, published by the University of Idaho in 1993.

Fig. 19-16. A large commercial dairy sheep operation in southern Europe. (Courtesy, David L. Thomas, University of Wisconsin, Madison)

manufacturing cheese in those countries; for example, the dairy sheep industry of France using the Lacaune breed and the production of the world famous Roquefort cheese. This cheese is made in France from the milk of about one million dairy ewes. About 16,000 tons of Roquefort are made annually, of which less than 10% is exported, the rest being eaten by the French themselves.

There are over 50 cheese varieties made from sheep's milk, along with such non-cheese products as yogurt, ice cream, and butter.

It is noteworthy that, in 1996, there were 1,320,081 cattle and 1,047,720 sheep in the world.[7]

All over the Middle East and North Africa, ewes are milked; 40 million of them in Turkey and 16 million in Morocco. Holland, Norway, Sweden, and the United States are seeing an upsurge.

In Greece, 560,000 tons of ewes' milk are produced each year, along with 4,000 tons of sheep butter, many kinds of cheeses—including the famous Feta—a white, crumbly cheese, preserved in a mixture of brine and whey, usually served on top of salads.

Italy has a long history of milking ewes. Cheeses such as the hard Pecorino, with its excellent keeping qualities, and the light fresh Ricotta, which is a byproduct, have been known and appreciated since early times.

Information relative to the U.S. sheep dairy industry is limited. Dr. W. J. Boylan, Animal Science Department, University of Minnesota, started milking sheep in 1984 in his genetics research.

It is noteworthy that the North American Dairy Sheep Association (NADSA), which was organized in 1988 to promote sheep dairying and dairy sheep milk products, grew from four members in 1988 to 121 in 1992.

[7]*FAO Yearbook*, Food and Agriculture Organization of the United Nations, Rome, Italy.

QUESTIONS FOR STUDY AND DISCUSSION

1. Why should sheep producers have a reasonable working knowledge of the end product, lamb?

2. Consumer studies indicate the following transition in preferences relative to lamb:

   a. Preference for more red meat and less fat.

   b. Preference for more highly processed meat; that is, meat that is boned-out, trimmed, etc., prior to purchase.

   c. Preference for lamb chops.

   Discuss the impact of each of these trends from the standpoint of the producer, the processor, and the consumer.

3. What qualities do you desire in lamb? Are these qualities reflected adequately in the top federal grades of lamb?

4. Do you feel that we should have more federal grading of lamb? Justify your answer

5. How and why does the disposition of the lamb carcass differ from that of beef and pork?

6. Discuss the relationship of choice of retail lamb cuts and method of cookery.

7. Discuss the nutritional value of lamb.

8. Is eating lamb a health hazard? Support your answer.

9. What determines lamb prices?

10. Choose and then debate either the affirmative or the negative for each of the following statements:

    a. Lamb prices are controlled by (1) the producers, (2) the packers, or (3) the retailers.

    b. Excessive profits are made by (1) the producers, (2) the packers, or (3) the retailers.

11. Discuss the factors that determine the spread between on-foot lamb prices and retail lamb prices. Where does the consumer's dollar go?

12. Explain the significance of the statement "Lambs are not all meat, and the carcasses are not all chops."

13. What factors contributed to the drastic decline in the U.S. per capita consumption of lamb?

14. List and describe some of the byproducts from lamb slaughter.

15. How is lamb promoted?

16. Why have dairy sheep been of major importance in many parts of the world, but of minor importance in the United States?

17. Do you foresee any great future for dairy sheep in the United States? Justify your answer.

SELECTED REFERENCES

Title of Publication	Author(s)	Publisher
Animal Science, Ninth Edition	M. E. Ensminger	Interstate Publishers, Inc., Danville, IL 1991
Developments in Meat Science—2	Ed. by R. Lawrie	Applied Science Publishers, Inc., Englewood Cliffs, NJ, 1981
Foods & Nutrition Encyclopedia, Second Edition	A. H. Ensminger M. E. Ensminger J. E. Konlande J. R. K. Robson	CRC Press, Boca Raton, FL, 1994
Lessons on Meat		National Live Stock and Meat Board, Chicago, IL, 1976
Livestock and Meat Marketing	J. H. McCoy	Avi Publishing Co., Westport, CT, 1979
Meat Board Meat Book, The	B. Bloch	McGraw-Hill Book Co., New York, NY, 1977
Meat Handbook	A. Levie	Avi Publishing Co., Westport, CT, 1979
Meat, Poultry, and Seafood Technology	R. L. Henrickson	Prentice-Hall, Inc., Englewood Cliffs, NJ, 1978
Meat We Eat, The, Fourteenth Edition	J. R. Romans, et al.	Interstate Publishers, Inc., Danville, IL, 2001
Practical Meat Cutting and Merchandising: Pork, Lamb, and Veal, Vol. 2	T. Fabbricante W. J. Sultan	Avi Publishing Co., Westport, CT, 1975
Science of Meat and Meat Products, The	Ed. by J. F. Price B. S. Schweigert	W. H. Freeman & Co., San Francisco, CA, 1971
Sheepman's Production Handbook, The	Ed. by G. E. Scott	Sheep Industry Development Program (SID), Denver, CO, 1975
Stockman's Handbook, The, Seventh Edition	M. E. Ensminger	Interstate Publishers, Inc., Danville, IL, 1992

Also, literature on lamb may be secured by writing to meat packers and processors and trade organizations; in particular, the following two trade organizations:

American Sheep Industry Assn.
6711 S. Yosemite
Englewood, CO 80112-1414

National Cattlemen's Beef Assn.
444 N. Michigan Avenue
Chicago, IL 60611

PART III
GOATS

Chapters 20 through 25 deal specifically with goat production, including dairy goats, Angora and Cashmere goats, meat (Spanish) goats, and pygmy goats, with dairy goats receiving the most coverage. Nevertheless, the reader must realize that some very important aspects of goat production are also covered in Part I; namely, genetics, reproduction, fundamentals of nutrition, pastures, behavior and environment, health and disease, cashmere, mohair, and business. Therefore, teachers, students, or producers interested in goat production must combine the reading and studying of Parts I and III. The order in which they wish to do this is left to their discretion and need. However, some cross references in Chapters 20 through 25 will refer the reader to chapters in Part I.

CHAPTER 20

Angora goats. (Courtesy, *Ranch & Rural Living*, San Angelo, TX)

TYPES AND BREEDS OF GOATS

Contents *Page*

Worldwide, there are numerous types and breeds of goats. In many areas, goats tend to be multiple-purpose animals—used for milk, meat, fiber, and land clearing. Thus, each area or country tends to have unique breeds of goats. Because of the many types and breeds of goats, no attempt will be made to discuss each of them in this chapter. Rather, the important types and breeds in the United States will be discussed.

TYPES AND BREEDS OF GOATS

There are five major types of goats in the world: (1) Angora (mohair-bearing) goats, (2) Cashmere goats, (3) dairy goats, (4) meat (Spanish) goats, and (5) pygmy goats. In total, there are about 609,488,000 goats of all types and breeds in the world, 61.2% of which are located in Asia.[1]

ANGORA GOATS

The Angora breed derives the name from Angora, a province in Turkey, in which land it originated. The Angora goat was first introduced into the United States in 1849.

The leading mohair-producing countries of the world, ranked in descending order, are: South Africa, United States, Turkey, Argentina, Australia, and New Zealand (see Table 2-15). Texas has a commanding lead in number of Angora goats in the United States (see Table 2-16).

The vast majority of the goats in the United States belong to the mohair-bearing Angora breed. Although there are about one million head of these strange-looking, heavy-coated creatures in this country, few people outside the Angora goat districts know what they look like.

ORIGIN AND NATIVE HOME

Angora, Turkey (now Ankara), is a high plateau, lying from 1,000 to 4,000 ft above sea level. In 1881, the Sultan of Turkey passed an edict prohibiting the exportation of Angoras, expecting thereby to confine the mohair industry to Asia Minor and forever hold a monopoly on the mohair trade. Thirty years later, in 1910, the Union of South Africa followed suit, passing a law for the same purpose. Subsequent events proved that both Turkey and the Union of South Africa were too late in their efforts to hold a monopoly on the mohair trade, for some of the choicest Angora blood had already been brought to the United States.

EARLY AMERICAN IMPORTATIONS

The Angora goat was first introduced into this country by Dr. James B. Davis, of Columbia, South Carolina. Dr. Davis had been sent to Turkey by President Polk in answer to a request made by the sultan for someone to experiment in the production of cotton in that country. Upon returning to the United States in 1849, 32 years before the sultan's edict, Dr. Davis brought with him nine choice Angora goats, including seven does and two bucks. These and subsequent importations founded in the United States the Angora industry which continued to thrive in this new land despite later restrictions imposed by Turkey.

CHARACTERISTICS

At the present time, the Angora goat of this country is considerably larger and more rugged than its Turkish ancestor. This transformation in type has been accomplished through long-continued selective breeding and some infusion of common or so-called Mexican goats. In range condition, mature bucks weigh from 125 to 175 lb and does from 80 to 90 lb. The breeders of this country prefer considerable size as long as it is not necessary to sacrifice fleece quality, because the added size gives a larger surface area upon which to produce a heavy fleece.

Angoras are almost always pure white, though a black one appears occasionally. Red kids will shed

Fig. 20-1. Angora buck. Winner of B-type yearling in the Texas Angora Goat Raisers' Association Show and Sale. Bred and shown by Nancy Haby, Leckey, Texas; purchased by Jack Richardson, Uvalde, Texas. (Courtesy, Nancy Haby)

[1]*FAO Production Yearbook*, Vol. 48, 1994, p. 192, Table 90, data from 1994.

Fig. 20-2. Angora goats—the source of mohair—are managed primarily under range conditions. (Courtesy, George F. W. Haenlein, University of Delaware)

their hair and produce white mohair later, but it is recommended that these animals be culled out and sent to slaughter.

The outer coat of the animal is made up of long locks or strands of hair, known commercially as mohair, which covers the animal's body. The fleece should cover all parts of the body except the face and should be characterized by fine quality, close curl, and high luster. It should be as free from kemp as possible.

Under range conditions, does and kids shear 3.5 to 4.5 lb and wethers 4.0 to 5.0 lb of mohair annually, which is usually removed in two clips. The best fleeces possess ample quantities of natural oil, and the best strains of Angoras show no tendency to shed.

The body conformation should be symmetrical and denote a good constitution. Both sexes are usually horned, but polled individuals occur. The rather thin, long, pendulous ears droop out of the hair.

Fig. 20-3. The Powder River Resource Area in Miles City, Montana, used Angora goats to control leafy spurge in areas where chemicals could not be used, such as along the river. Angora goats were herded along the Powder River floodplain on public domain in Powder River County. The goats did a very good job controlling leafy spurge, but finding goats and a herder has become a difficult task. Now, in some areas, grazing sheep are used instead. (Photo by Terry Wilson, Rangeland Management Specialist for the Bureau of Land Management, Powder River Resource Area, Miles City District Office, Miles City, MT)

Chapter 23, Angora Goats, details the characteristics of Angora goats that should be considered when one is selecting breeding animals.

REGISTRATION

The first U.S. breed registry association for Angora goats, known as the American Angora Goat Breeders Association, was established in 1900. From that time until 1924, all animals registered by this association were either the original inspected stock and their progeny or goats subsequently imported.

As in sheep production, the breeding of registered Angoras is considered a specialty business, and few large range goat operators engage therein. Yet, most practical commercial goat producers prefer to use registered bucks and patronize the purebreed breeders in order to secure bucks of a type that they hope will improve their herds.

CASHMERE GOATS

The Cashmere goat is a double-fibered goat that produces a fine undercoat known as *down*. It embraces several varieties rather than a distinct breed. Cashmere goats are native to the Central Asian mountains in Kashmir. It was named *Cashmere* after the old spelling of the country.

Cashmere is the finest animal fiber used in commercial trade. Clothing made from cashmere is exceptionally light, soft, warm, and in the luxury class. On a comparable weight basis, cashmere is reported to have three times the insulating value of wool. The most common uses for cashmere are for sweaters, ladies' dress goods, shawls, and coatings.

Fig. 20-4. Cashmere buck kids, 9 or 10 months old. (Courtesy, Tom and Ann Dooling, Professional Cashmere Marketers' Assn., Inc., Dillon, MT 59725)

Fig 20-5. Cashmere goats are gaining in popularity in many parts of the country. (Courtesy, Dr. Johannes E. Nel, University of Wyoming)

Cashmere goats have been introduced into the United States in recent years, primarily from Australia. There appears to be a bright future in the United States for Cashmere goats and cashmere fiber. However, U.S. producers face two major problems in developing a cashmere industry: (1) the small amount of fiber produced by a Cashmere goat—about 0.5 lb annually; and (2) combing (shearing) is a tedious, time-consuming job, and labor costs in this country may be too high for it to become practical.

DAIRY GOATS

Dairy goats produce 2.0% of the total world milk supply.

Worldwide, on a pounds per capita basis, the species production of milk in descending order is: cow milk, 179.6 lb; buffalo milk, 18.9 lb; goat milk, 4.1 lb; and sheep milk, 3.1 lb.[2]

The goat has long been a popular milk animal in the Old World, where it is often referred to as the poor peoples' cow.

There are about 150,000 dairy goats in the United States. They are generally kept near large cities to provide milk for people who cannot afford to buy, or are allergic to, cow's milk. The leading states in dairy goat numbers are: California, Texas, Ohio, Wisconsin, New York, Tennessee, Missouri, Washington, Oregon, and Michigan.

The goat has long been a popular milk animal in the Old World, where it is often referred to as the cow of the poor. When traveling or vacationing, Asiatics frequently take their goats with them in order to be assured of a supply of milk. Mahatma Ghandi of India took two milk goats with him on his last visit to England.

Dairy goats were introduced to America in early times—the first settlers in the Virginia colonies bringing their milk goats from their homeland. However, improved strains were not imported until many years later, for the first purebreds said to have been brought to the United States were four Toggenburgs imported to Ohio in 1893. Today, the milk goat industry is growing, and these small animals are supplying nature's finest food—milk—to many children who would otherwise be undernourished. They are especially well adapted for furnishing a milk supply for low-income families in small towns and the suburbs of large cities where there is not enough feed available for cows. A doe can often secure much of her feed from lawn clippings or garden and kitchen waste or by grazing in vacant lots.

A good milking doe will average 3 qt of milk per day over a lactation period of 10 months, with superior animals producing as much as 4 to 5 qt. The highest official milk production on record in the United States was made by a Toggenburg doe in 1960; she produced 5,750 lb of milk in 305 days. The butterfat record is held by an Alpine that produced 249 lb in 305 days.[3]

In addition to the specific breed disqualifications given under the respective breeds, the American Dairy Goat Association lists the following as disqualifications in any breed: total blindness; permanent lameness or difficulty in walking; blind or nonfunctioning half of udder; blind teat; double teats; extra teats that interfere with milking; hermaphrodism; navel hernia; crooked face in bucks; and extra teats, teats cut off, or double orifice in bucks.

The American Dairy Goat Association recognizes six breeds of dairy goats: the Alpine, the La Mancha, the Nubian, the Oberhasli, the Saanen, and the Toggenburg.

ALPINE

Alpines are hardy, adaptable animals that thrive in most climates, while maintaining good health and excellent production. This breed is composed of several varieties, including the British, Rock, and Swiss Alpines. The French Alpine, however, is by far the most numerous of the Alpines registered in the United States.

ORIGIN AND NATIVE HOME

French Alpines are found throughout the goat-producing districts of France. They developed from Swiss foundation stock. The British Alpine is the result of Alpine goats imported from Switzerland in 1903 to

[2]*FAO Production Yearbook*, Vol. 48, 1994.

[3]*American Dairy Goat Association Yearbook,* Spindale, NC, Vol. 27, 1981, pp. 3 and 12.

Fig. 20-6. French Alpine doe, *GCH Quesa Mess Jefe de Elegance EM*, owned by Country Ridge. Her color is described as sungau. (Courtesy, Alpines International, Ft. Pierce, FL)

Britain to upgrade British goats. Rock Alpine goats were developed in the United States by Mary Edna Rock. Swiss Alpines are now called Oberhaslis—a name change approved in 1978 by the American Dairy Goat Association. They originated near Bern, Switzerland. Swiss Alpines, or Oberhaslis, are discussed as a separate breed in a subsequent section.

Fig. 20-7. French Alpine doe, *Green Oaks Top Girl*. (Courtesy, International Dairy Goat Registry, Inc., Dublin, TX)

EARLY AMERICAN IMPORTATIONS

The French Alpine is a relatively new breed of dairy goat in the United States. All registered animals of the breed in this country trace to an importation of 3 bucks and 18 does made by Dr. Charles DeLangle of California in 1922. Dr. DeLangle was a member of the French Academy and a personal friend of Joseph Crepin—a leading French authority on goat raising at that time. In the fall of 1922, these two men selected these goats and took them to Paris for shipment via steamer to Cuba for quarantine before arriving in New Orleans and finally California.

CHARACTERISTICS

French Alpines are medium to large, and rugged in build, and the does are good milkers. Some have horns at birth and are disbudded; others are hornless. Their hair is medium to short. They have erect ears and straight noses. There are a number of color patterns typical of French Alpines. These are listed in Table 20-1.

A Roman nose, pendulous ears, Toggenburg color and markings, or Saanen color are discriminated against.

TABLE 20-1
FRENCH ALPINE COLOR PATTERNS

Pattern Name	Description
Cou blanc	White front quarters and black hindquarters with black or gray markings on head.
Cou clair	Front quarters tan, saffron, off-white, or shading to gray with black hindquarters.
Cou noir	Black front quarters and white hindquarters.
Chamoisee.	Brown or bay color. Characteristic markings are black face, dorsal stripe, feet and legs, and sometimes a martingale running over the withers and down to the chest.
Broken chamoisee	A solid chamoisee broken with another color being banded or splashed, etc.
Two-tone chamoisee . . .	Light front quarters with brown or gray hindquarters. Not a cou blanc or a cou clair, as these terms are reserved for animals with black hindquarters.
Pied	Spotted or mottled.
Sungau	Black with white markings such as under body, facial stripes, etc.

Fig. 20-8. French Alpine buck, *+B GCH Seneca Valley's Malcolm X*, pictured at five years of age. *Malcolm X* was classified as excellent by the American Dairy Goat Association. (Courtesy, Dorothea Custer, President, Oberhasli Breeders of America, Harvard, IL)

LA MANCHA

The La Mancha is a fairly new breed of dairy goat. La Manchas were accepted as a breed for registry in January, 1958.

ORIGIN AND NATIVE HOME

The exact background of La Manchas remains a mystery, but references were made to short-eared goats as far back as ancient Persia. Spanish missionaries colonizing California brought with them a short-eared breed of goat. During the 1930s, Eula F. Frey of Oregon bred some of these goats lacking external ears to some top Swiss and Nubian sires, producing a new dairy breed. It retains the vestigial ear characteristic which is probably controlled by a single gene. The name *La Mancha* is likely derived from the La Mancha region of Spain.

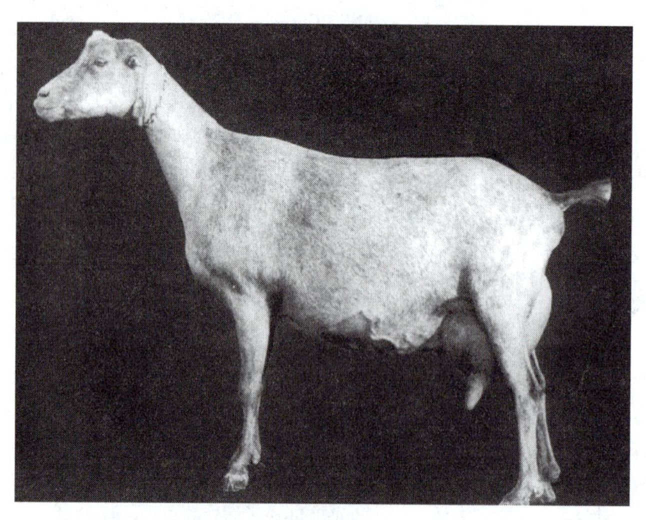

Fig. 20-9. La Mancha doe, *GCH Ananda Hill's Rashana 2*M L5663*. She produced 3,200 lb of milk and 125 lb of butterfat. (Courtesy, Barbara Backus, Quixote La Manchas, St. Helena, CA)

CHARACTERISTICS

The ears, or lack of ears, are the distinctive breed characteristic. La Manchas can be any color or color combinations. Their faces should be straight or slightly dished, and their hair short, fine, and glossy. Mature bucks should weigh about 150 lb, and does should weigh about 130 lb.

Fig. 20-10. Showing La Mancha goat, male. (Courtesy, Prof. G. Haenlein, University of Delaware)

There are two types of La Mancha ears. In does, one type of ear has no advantage over the other. The *gopher ear* has a maximum length of approximately 1 in. But, preferably, the ear should be nonexistent and have very little or no cartilage. The end of the ear must turn up or down. Only bucks with gopher ears are eligible for registration. The *elf ear* has a maximum length of approximately 2 in., and the end of the ear must turn up or down; cartilage shaping the ear is allowed.

Overall, La Manchas are of excellent dairy temperament, sturdy, productive, and adaptable.

NUBIAN

The Nubian is the most popular breed of registered dairy goat in the United States today. Nubians are noted for their premium-quality milk with a butterfat content higher than that of all the other dairy breeds.

ORIGIN AND NATIVE HOME

Nubia is the region which includes northern Sudan and southern Egypt. Likely, the name *Nubian* is derived from the name of this region, but it is difficult to find Nubian goats in Nubia. The Nubian of the United States is entirely derived from the Anglo-Nubian of England. This breed was in turn formed by crossing the native English stock with lop-eared goats of Egypt and India.

EARLY AMERICAN IMPORTATIONS

Three Anglo-Nubian goats—one buck and two does—were imported to this country in 1909. They became the foundation of the breed in the United States.

CHARACTERISTICS

The Nubian is a large, distinctive-looking goat with long, drooping ears and a turned-up tail. Some animals are born with horns and disbudded; others are hornless. The does are beardless. There is a characteristic Roman nose and prominent forehead.

The head is the distinctive breed characteristic, the facial profile between the eyes and the muzzle being strongly convex. The ears are long—extending at least 1 in. beyond the muzzle when held flat along the face—wide, and pendulous. They lie close to the head at the temple and flare slightly out and well defined at the rounded tin forming a bell shape. The ears are rather thin, with the cartilage well defined. The hair is short, fine, and glossy.

Any color or colors, solid or patterned, are acceptable. Black and white, tan and white, and red and white are common colors, but Nubians may be any of these colors without white markings. A few are spotted or piebald.

In general, Nubian does are not as heavy milkers as some of the other breeds, but the milk is very rich, the breed often being referred to as the "Jersey of milk goats."

Fig. 20-12. Hay-feeding Nubian goat. (Courtesy, Prof. G. Haenlein, University of Delaware)

OBERHASLI

The Oberhasli has also been known as the Oberhasli-Brienzer, the Graubunden, the Chamoise, the Brown Alpine, and the Rehbraun in Europe. When first imported to the United States, it was known as the Swiss Alpine.

Fig. 20-11. Nubian doe, *Shadow Hills Charis 2*M*. (Courtesy, Bill Scheufele, Public Relations Chairman, International Nubian Breeders Association, Stenwood, WA)

Fig. 20-13. Oberhasli doe, *Seneca Valley's Little Gypsy 2*M*. (Courtesy, Dorothea Custer, President, Oberhasli Breeders of America, Harvard, IL)

ORIGIN AND NATIVE HOME

The Oberhasli (Oberhasli-Brienzer) was developed near Bern, Switzerland, where it is still found today.

EARLY AMERICAN IMPORTATIONS

Some Oberhaslis were imported in 1906 by Fred Stucker, followed by a second importation in 1920 by August Bonjean, but the descendants of these first importations were not maintained as purebreds. All the purebred Oberhaslis of today trace their ancestry back to one buck and four does imported in 1936 by Dr. H. O. Pence of Kansas City, Missouri. When introduced into this country, the Oberhasli-Brienzer goats were misnamed Swiss Alpine. In 1978, the American Dairy Goat Association granted approval of a name change from Swiss Alpine to Oberhasli and approved the breed standard. Then, in 1979, the American Dairy Goat Association voted to give the Oberhasli breed its own herdbooks. Furthermore, in 1980, the American Dairy Goat Association decided to retrieve all American and part-Oberhasli animals from other herdbooks and to reregister them in the appropriate Oberhasli herdbooks.

CHARACTERISTICS

The Oberhasli is of medium size, tending to be long bodied and fine boned, vigorous, and alert in appearance. The face is straight, and a Roman nose is discriminated against. Its most striking characteristic is the rich red coat with black trim. This color is chamoisee—ranging from a light to a deep red bay, with the latter being most desirable. The black trim should be (1) two black stripes down the face from above each eve to a black muzzle; (2) forehead nearly all black, black stripes from the base of each ear coming to a point just back of the poll and continuing along the neck and back as a dorsal stripe to the tail; (3) black belly and udder; (4) black legs below the knees and hocks; and (5) ears black inside and bay outside. Bucks often have more black on their heads than does, black whiskers, and black hair along the shoulders and lower chest, with a mantle of black along the back. Bucks frequently have more white hairs through the coat than does.

Milk production by Oberhasli goats is moderate with persistent lactation and a fairly high butterfat.

SAANEN

The Saanen is the most widely distributed of the improved breeds. It has been exported from Switzerland to many countries of the world, and in many countries of Europe local goats have been graded up to the Saanen, and new breeds formed.

ORIGIN AND NATIVE HOME

Saanens originated in the Saane Valley in the Bernese Oberland—the western and northwestern part of Switzerland. They are considered the best milkers in Switzerland and the largest breed.

Fig. 20-14. Oberhasli buck, *Seneca Valley's Maxwell*, pictured at 11½ years of age. (Courtesy, Dorothea Custer, President, Oberhasli Breeders of America, Harvard, IL)

Fig. 20-15. Saanen doe, *Pandy Line Sno Bunny S343452p*, a two-year-old milker. (Courtesy, Dennis Hoeter, President, National Saanen Breeders Association, Lafayette, IN)

EARLY AMERICAN IMPORTATIONS

Once the advanced stage of breed development was developed in Switzerland, 10 Saanen goats were imported to the United States in 1904. Other importations followed.

CHARACTERISTICS

Saanens are medium to large in size with rugged bone and plenty of vigor. Does should be feminine, however, and not coarse. Saanens are white or light cream, with white preferred. Spots on the skin are not discriminated against. Small spots of color on the hair are allowable, but not desirable. The hair should be short and fine, though a fringe over the spine and thighs is often present. Ears should be erect and alertly carried, preferably pointing forward. The face should be straight or dished. A tendency toward a Roman nose is discriminated against.

Fig. 20-16. Saanen buck, *CH T & J Roxies Motivation S333145*, photographed at three years of age. (Courtesy, Dennis Hoefer, President, National Saanen Breeders Association, Lafayette, IN)

SABLE

Occasionally, two pure white Saanen parents produce a dark- or sable-colored kid. Such offspring do not meet the Saanen breed standard. This occurs because the parents are heterozygous for the white. They possess a dominant white gene and a recessive black gene. Thus, on the average, the matings between heterozygous Saanens will produce about 25% Sable offspring. These Sable goats, when mated, always produce Sable (black) offspring. Except for color, the Sable possesses characteristics similar to the Saanen.

Currently, the Sable Breeders Association is maintaining a herd registry program. The American Dairy Goat Association is considering accepting the Sable as a separate breed.

Fig. 20-17. Sable doe, *West's Rosy GE 416306*. (Courtesy, Donald West, Director, Sable Breeders Association, Rillito, AZ)

TOGGENBURG

Toggenburgs along with the Saanens were the first two breeds of dairy goats established in the United States. Like Saanens they have been exported from their native Switzerland to many parts of the world, though less extensively.

ORIGIN AND NATIVE HOME

The Toggenburg originated in Obertoggenburg and Werdenberg in St. Gallen, a canton in northwestern Switzerland. In 1884, Toggenburgs were imported to England where they became the first separate breed recognized.

EARLY AMERICAN IMPORTATIONS

In 1893, four Toggenburgs were imported from England by William A. Shafer of Ohio; and, in 1906, 16 were brought directly from Switzerland by F. S. Peer. Then, between 1906 and the 1930s, over 200 more were imported. All of the U.S. purebreds were derived from these importations.

CHARACTERISTICS

Toggenburgs are of medium size, sturdy, vigorous, and alert in appearance. The hair is short or medium in

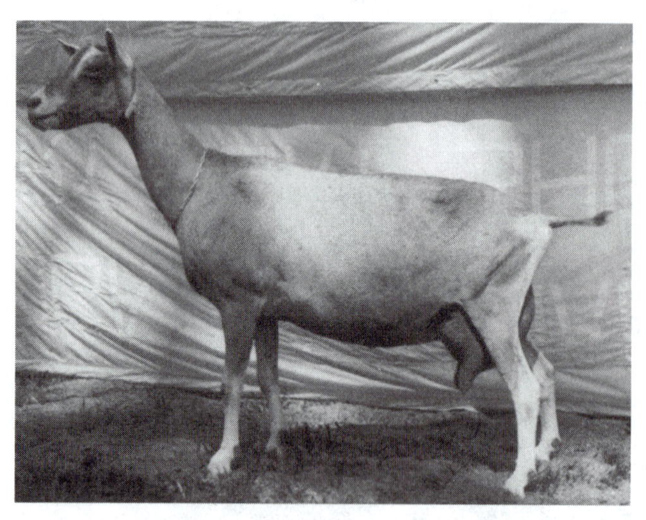

Fig. 20-18. Toggenburg doe, a 1982 grand champion. (Courtesy, American Dairy Goat Association, Spindale, NC)

length, soft, fine, and lying flat. Their color is solid, varying from light fawn to dark chocolate, with no preference for any shade. Distinct white markings are as follows: (1) white ears with dark spot in middle; (2) two white stripes down the face from above each eye to the muzzle; (3) hind legs white from hocks to hoofs; (4) forelegs white from knees downward with dark lien (clock) below knee acceptable; (5) a white triangle on either side of the tail; and (6) white spots may be present at root of wattles or in that area if no wattles are present. Varying degrees of cream markings instead of pure white are acceptable, but not desirable. The ears are erect and carried forward. Facial lines may be dished or straight, never Roman.

It is characteristic of Toggenburgs to hold up production the last half of their lactation. They are persistent milkers.

Fig. 20-19. Toggenburg goat, male. (Courtesy, Prof. G. Haenlein, University of Delaware)

POPULARITY AND PRODUCTION OF BREEDS

The number of goats registered by breeds each year are not available from the American Dairy Goat Association, which registers about 55,000 goats each year. However, a fairly recent summary of 10 years of lactation records by the U.S. Department of Agriculture can be used to indicate the relative popularity of the breeds. After 107,345 lactation records were screened for certain requirements, 51,531 lactation records were studied. Nubians and Alpines represented over 50% of the records studied, as shown in Table 20-2.

In terms of purebred registrations, the records of the American Goat Society, Inc., also indicate that Nubians are the leading breed, accounting for almost half of the registrations, as shown in Table 20-3.

TABLE 20-2
GOAT LACTATIONS BY BREED[1]

Breed	Number of Lactations	Percentage of Total
Nubian	14,005	27.2
Alpine	12,712	24.7
Toggenburg	7,534	14.6
Unknown	7,401	14.4
Saanen	5,718	11.1
La Mancha	3,094	6.0
Experimental	780	1.5
Mixed	287	0.6
Total	51,531	100

[1]Grossman, M., and G. R. Wiggans, "Dairy Goat Lactation Records and Potential for Buck Evaluation," *Journal of Dairy Science*, Vol. 63, No. 11.

TABLE 20-3
PUREBRED GOATS REGISTERED BY THE AMERICAN GOAT SOCIETY, INC.

Breed	Number[1]
Nubian	50,292
Saanen	18,233
French Alpine	17,475
Toggenburg	13,630
Pygmy	2,814
La Mancha	1,029
Swiss Alpine	236
Total	103,709

[1]As of the end of 1982.

The same USDA summary cited in Table 20-2 also provides information relative to the productivity of each breed, as indicated in Table 20-4.

TABLE 20-4
DAIRY GOAT MILK AND FAT PRODUCTION, BY BREED, FOR DOES IN MILK 305 DAYS[1]

Breed	Production				
	Milk		Fat		
	(lb)[2]	*(kg)*	*(lb)*	*(kg)*	*(%)*
Saanen	2,121.3	962.2	74.5	33.8	3.5
Alpine	2,099.4	952.3	73.2	33.2	3.5
Toggenburg	2,030.6	921.1	67.0	30.4	3.3
Unknown	1,928.4	874.7	71.6	32.5	3.7
Experimental	1,880.5	853.0	68.8	31.2	3.7
Mixed	1,862.4	844.8	69.9	31.7	3.8
La Mancha	1,800.3	816.6	69.0	31.3	3.8
Nubian	1,776.2	805.7	81.3	36.9	4.6
Average[3]	1,981.5	898.8	73.2	33.2	3.7

[1]Grossman, M., and G. R. Wiggans, "Dairy Goat Lactation Records and Potential for Buck Evaluation," *Journal of Dairy Science*, Vol. 63, No. 11.

[2]A gallon *(3.8 liter)* of milk weighs 8.64 lb *(3.92 kg)* at 60°F *(15.6°C)*.

[3]This number represents the average of 12,446 lactation records.

MEAT (SPANISH) GOATS

The term *Spanish goat* is used in the United States to refer to goats of mixed-breed origin. Since they are kept largely for meat production, the term *meat goat* is also used. These terms are employed to distinguish these goats from the Angora and dairy breeds. Most Spanish goats are of the same origin as the Mexican Criollo—a breed derived from the Granada, Murcia, and Malaga breeds of Spain.

There are about 500,000 meat (Spanish) goats in the United States, most of which are located in Texas.

Meat goats are adaptable and unsurpassed in their ability to exist largely upon brush and still yield acceptable quantities of edible meat.

Meat goats are highly variable in appearance and performance. Some show traces of Nubian and Toggenburg breeding. Others lack external ears or have very small ears. As would be expected, the colors and markings of Spanish goats vary widely—there being no ideal or accepted color or marking. Colors range from solid black, brown, and white to fawn and brown with black points and a black stripe down the back. There are also many combinations of spotting—black and white, brown and white, black and brown, and some blue-gray. Most of these animals, both males and females, are horned. The horns of the males grow much larger and heavier than those of the females. A few Spanish goats are polled.

BOER GOATS

Boer goats originated in South Africa. They are a noted meat breed, with a docile disposition.

Fig. 20-20. Young Boer rams on the range. (Courtesy, Dr. Johannes E. Nel, University of Wyoming)

PYGMY GOATS

These miniature goats have been known by a variety of names—West African Dwarfs, African Pygmies, and Cameroons. They have proven to be very adaptable animals and valuable research animals. Pygmy goats are hardy, alert, animated, good-natured, gregarious, and docile. They are responsive pets.

ORIGIN AND NATIVE HOME

These small goats originated in the western part of Africa, particularly in the Cameroons. Until recently their name was African Pygmies.

EARLY AMERICAN IMPORTATIONS

The original Pygmy goats reached zoos in the United States via imports from zoos in Sweden and West Germany sometime during the 1950s. In 1961, Dr. James Metcalfe established a colony at the University of Oregon Medical School at Portland by purchasing goats from two zoos. In 10 years, the colony produced over 400 offspring.

CHARACTERISTICS

The Pygmy goat is small, cobby, and compact. It is full-barreled and well-muscled, and the body circumference in relation to height and weight is proportionately greater than that of other breeds. Also, the head

Fig. 20-21. Pygmy goat. These goats, originally from Africa, measure only 16 to 23 in. at the withers. (Couresty, Lydia Hale, National Pygmy Goat Association, Sherborn, MA)

and legs are short relative to body length. Hornlessness is considered a disqualifying fault.

Preferred colors range from white to gray and black in a predominantly grizzled (agouti) pattern. Muzzle, forehead, eyes, and ears are accented in lighter tones. Front and rear hoofs and cannons (socks) are black, as are the crown and dorsal stripe, or martingale. Coat length and density vary with climates.

Pygmy goats are precocious breeders, bearing 1 to 4 young every 9 to 12 months after a 5-month gestation period. Normally 1 or 2 kids are produced, weighing an average of 4 lb each. Does are usually bred for the first time when they are about 9 months of age, though they may conceive as early as 3 months if care is not taken to separate them early from the bucks. Mature females may produce 4 or more lb of milk of 6 to 9% butterfat at the peak of a 4- to 6-month lactation period.

The National Pygmy Goat Association unites breeders of these animals.

QUESTIONS FOR STUDY AND DISCUSSION

1. List the place of origin and the distinguishing characteristics of Angora goats and of each of the dairy goat breeds. Discuss the importance of each breed.

2. Justify any preference that you may have for one particular breed of dairy goat.

3. Tables 20-2 and 20-3 indicate the relative popularity of the dairy goat breeds. How do you account for the popularity of the Nubian?

4. How many pounds of milk, gallons of milk, and pounds of butterfat could you expect from a herd of 100 well-managed dairy goats? Express this on a daily and a yearly basis.

5. How important are breed characteristics? Can dairy goat breed differences be detected in the milk? Justify your answer.

6. Why are Pygmy goats suitable for biomedical studies?

7. Obtain breed registry information and breed association literature about your favorite breed of goat. Evaluate the soundness and value of the material that you receive. (See Tables A-7 and A-8 in the Appendix for addresses.)

SELECTED REFERENCES

Title of Publication	Author(s)	Publisher
Colour Island Sheep of the World	Ed. by G. J. Enzlin	Ram Press, printed in the Netherlands by Transmodial Voothuiser, 1995
Goat Production: Breeding & Management	Ed. by C. Gall	Academic Press, Inc., New York, NY, 1981
Modern Milk Goats	I. Richards	J. B. Lippincott Co., Philadelphia, PA, 1921
Texas Angora Goat Production, Bull. B-926	J. A. Gray J. L. Groff	Texas A&M University, College Station, TX, 1970

Also, breed literature pertaining to each breed of goats may be secured by writing to the respective breed association or club. The name and address for each of these is listed in the Appendix, Tables A-7 and A-8.

CHAPTER 21

Angora goats on pasture. (Courtesy, Mohair Council of America, San Angelo, TX)

FEEDING GOATS

Contents *Page*

Contents Page

Contents Page

There are more than 450 million goats in the world, of which about one-third are in Africa. They contribute 1.4% of the world meat supply and 1.5% of the world milk supply.

Goats provide nearly one-third of the total meat produced in India and from 7 to 16% of the total meat produced in Turkey, Morocco, Indonesia, Nigeria, and Cyprus. In a number of countries, goat meat is preferred to other meats.

The dairy goat has long been a popular milk animal in the Old World, where it is often referred to as "the cow of the poor." In some countries, goat milk accounts for up to 50% of the total milk production. Southeast Asia, Africa, and the Near East lead in the production of goat milk.

The goats of the world also produce 36 million lb of mohair and cashmere and 33 million skins, annually. The three leading mohair-producing countries of the world, by rank, are South Africa, Turkey, and the United States.

There are approximately 2,950,000 goats in the United States, consisting of about 1,600,000 Angoras, 850,000 dairy goats, and 500,000 Spanish goats. Ninety-five percent of the Angoras are located in Texas. About 85% of the gross income from Angoras is from mohair and 15% from meat. Most of the Spanish goats are also in Texas. California is the most important dairy goat state.

Most Angora and Spanish goat herds are large and produced under extensive range conditions, whereas most dairy goats are found in small herds and on small farms or farms operated on a part-time basis.

Nutritional deficiencies and diseases in goats are of special concern because they have such widely-differing uncontrolled diets as a consequence of the great variety of conditions under which they are produced. Moreover, their nutritional and management practices differ according to their primary end products. Thus, Angoras which produce mohair, and Spanish goats which produce meat, are raised primarily on rangelands, often without supplemental feed. But dairy goats, which produce milk, require well-balanced rations high in energy and protein, similar to those of lactating dairy cows.

So, the goat industry of America can be divided into three distinct types of production: (1) dairy, (2) mohair, and (3) meat. Also, mention should be made of the pygmy goat.

The dairy goat, which is kept primarily for milk production, is gaining in popularity in the United States. Presently, it consists of the following important breeds: Alpine, American La Mancha, Nubian, Oberhasli (Swiss Alpine), Saanen, and Toggenburg.

The most numerous goat breed in America is the mohair-bearing Angora, the heavy-coated creatures kept for fiber production and brush control. Yet, few people outside the Angora district, characterized by rugged grazing lands, know what they look like. Angora goats are used to produce (1) a beautiful, long, lustrous fiber known as mohair, (2) meat, and (3) to augment other brush-control methods.

The Spanish goat is kept for meat production and brush control. It is of uncertain origin, but in all probability its ancestors were brought from Spain by early

Fig. 21-1. Three distinct types and uses of goats in America. *Upper left:* Dairy goats for milk. *Lower center:* Angora goats for mohair. *Upper right:* Spanish goats for meat. (Photos: Left, courtesy University of New Hampshire, Penacook; center and right, Texas A&M University, College Station)

explorers. Subsequently, dairy goat breeds have been infused. Colors vary from solid black, brown, or white to striped, to spotted. In Mexico, the meat from young milk-fed Spanish goat kids, known as *cabrito* (Spanish for little goat), is considered a delicacy. In the United States, Spanish goats are usually slaughtered when a little older at which time the meat is known as *chevon*. Spanish goats are usually left to survive on range forages with little or no feed supplementation.

Also, small numbers of pygmy goats are found in the United States. The American pygmies are descended mostly from West African dwarf goats found in Nigeria, Ghana, and the Cameroons. They are small, adaptable animals, used for meat, milk, and research. The pygmy is smaller than the other recognized types and breeds in the United States; full grown bucks stand about 20 in. high at the withers, and does are even smaller. Pygmies may be fed the same rations as their larger counterparts, but in smaller quantities.

Because the goat industry is so diverse, the feeding methods vary accordingly. For this reason this chapter, devoted to feeding goats, is presented in three parts: Part I—Nutritive Needs and Feeds, covering the principles and practices of feeding that are applicable to all goats, regardless of type; Part II—Feeding Dairy Goats; Part III—Feeding Angora and Spanish Goats, with Angora and Spanish goats discussed together because both are produced under similar conditions.

PART I—NUTRITIVE NEEDS AND FEEDS

ECONOMIC IMPORTANCE OF FEED FOR GOATS

Due to their smaller numbers, along with producers being prone to let the animals fend for themselves, in terms of tonnage of purchased feed ingredients and feed costs, goats are unimportant in comparison to other domestic animals.

Angora and Spanish goats utilize rough, brushy range areas that are not suited to other species—many of these ranges would not otherwise be utilized and would revert to brush and wilderness. On such ranges, Angora goats are supplemental fed to a limited degree only, whereas most Spanish goats live entirely off the land and are rarely supplemented. This does not mean, however, that Angora and Spanish goats would not benefit from, and increase production, with supplemental feeding, especially during the critical periods—just before breeding (for flushing), just before and after kidding, and when feed is short.

Modern dairy goat producers generally feed well-balanced rations that are high in energy and protein and contain adequate minerals and vitamins. Many of them use commercially prepared feeds during lactation and for the young kids.

NUTRITIVE NEEDS OF GOATS

In the past, efforts to set nutritional requirements for goats have relied heavily on the extrapolation of values derived from cattle and sheep studies. Despite their similarities as ruminants, goats exhibit significant differences from cattle and sheep in grazing habits, feed selection, water requirements, physical activities, milk composition, carcass composition, metabolic disorders, and parasites. So, the nutrient requirements of goats should be treated separately from those of other ruminants.

The hearty appetite of goats makes for a significant species difference. Lactating and growing goats will consume from 3.5 to 5.0% of their body weight (moisture-free basis) in one day, while cattle and sheep normally eat only 2.5 to 3.0%. It follows that their large feed capacity in relation to body weight makes it possible for them to consume large quantities of low quality materials. This characteristic, along with their ability to select the high-quality parts of plants, makes it possible to maintain goats successfully on poor pastures.

Since the nutritional requirements of the goat are distinctly different for milk, mohair, and meat production, specific requirements and allowances are discussed in separate feeding sections. Despite these distinctly different quantitative needs, the basic nutritional physiology of all goats is similar; hence, certain fundamentals relative to their nutritive needs—energy, protein, minerals, vitamins, and water—apply to all goats regardless of the purpose for which they are kept.

NATIONAL RESEARCH COUNCIL (NRC) REQUIREMENTS

The most up-to-date feeding standards for goats in the United States are those published by the National Research Council (NRC) of the National Academy of Sciences. Through the use of these standards, rations can be formulated for the different classes and categories of goats by proper use of available feedstuffs.

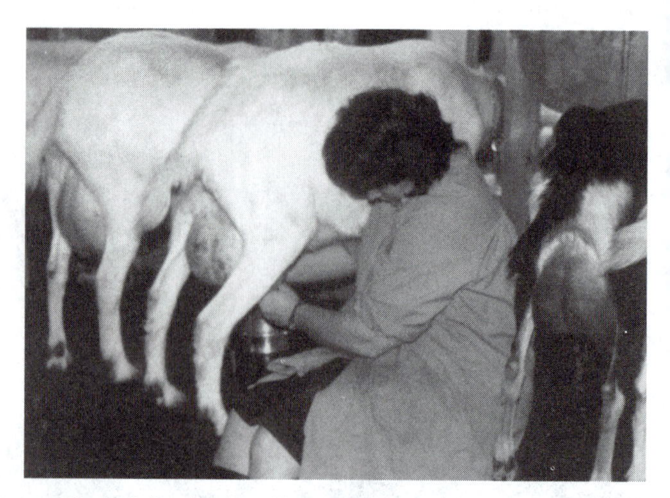

Fig. 21-2. Despite a hearty appetite and a large feed capacity, it is virtually impossible for a high-producing lactating doe to consume enough feed to meet the demands of body maintenance plus milk production during peak lactation; so, she must draw upon body reserves. (Courtesy, University of New Hampshire, Penacook)

The nutritive requirements of goats for maintenance, various levels of activity, late pregnancy, and growth, are given in Table 21-1. Additional nutrient requirements for milk production at different fat percentages are given in Table 21-2; and additional nutrient requirements for mohair production are given in Table 21-3.

ENERGY

Efficient utilization of nutrients depends on an adequate supply of energy, which is of paramount importance in determining the productivity of goats. Energy deficiency retards kid growth, delays puberty, reduces fertility, and depresses milk production. With continued deficiency the animals show a concurrent reduction in resistance to infectious diseases and parasites. The problem may be further complicated by deficiencies of proteins, minerals, and vitamins.

Energy limitations may result from inadequate feed intake or from a low-quality ration. Low energy intake that results from either feed restriction or low ration component digestibility prevents goats from meeting their requirements and from attaining their genetic potential. High water content of forages may also become a limiting factor.

Energy requirements are affected by age, body size, growth, pregnancy, and lactation. Energy requirements are also affected by the environment, hair growth, muscular activity, and relationships with other nutrients in the ration, which, for best results, need to be supplied in adequate amounts. Temperature, humidity, sunshine, and wind velocity may increase or decrease energy needs, depending upon the region. Stress of any kind may increase energy requirements.

TABLE 21-1
DAILY NUTRIENT REQUIREMENTS OF GOATS[1]

Body Weight (BW)		Dry Matter per Animal[2]: 0.9 Mcal ME/lb (2 Mcal ME/kg) Total		% of BW	Dry Matter per Animal[2]: 1.09 Mcal ME/lb (2.4 Mcal ME/kg) Total		% of BW	Feed Energy: TDN		DE	ME	NE	Protein: TP	DP	Minerals: Ca	P	Vitamin A	Vitamin D
(lb)	(kg)	(lb)	(kg)	(%)	(lb)	(kg)	(%)	(lb)	(kg)	(Mcal)	(Mcal)	(Mcal)	(g)	(g)	(g)	(g)	(1,000 IU)	(IU)
Maintenance only (includes goats under stable-fed conditions, minimal activity, and early pregnancy)																		
22	*10*	0.6	*0.28*	2.8	0.5	*0.24*	2.4	0.4	*0.16*	0.70	0.57	0.32	22	15	1	0.7	0.4	84
44	*20*	1.1	*0.48*	2.4	0.9	*0.40*	2.0	0.6	*0.27*	1.18	0.96	0.54	38	26	1	0.7	0.7	144
66	*30*	1.4	*0.65*	2.2	1.2	*0.54*	1.8	0.8	*0.36*	1.59	1.30	0.73	51	35	2	1.4	0.9	195
88	*40*	1.8	*0.81*	2.0	1.5	*0.67*	1.7	1.0	*0.45*	1.98	1.61	0.91	63	43	2	1.4	1.2	243
110	*50*	2.1	*0.95*	1.9	1.7	*0.79*	1.6	1.2	*0.53*	2.34	1.91	1.08	75	51	3	2.1	1.4	285
132	*60*	2.4	*1.09*	1.8	2.0	*0.91*	1.5	1.3	*0.61*	2.68	2.19	1.23	86	59	3	2.1	1.6	327
154	*70*	2.7	*1.23*	1.8	2.2	*1.02*	1.5	1.5	*0.68*	3.01	2.45	1.38	96	66	4	2.8	1.8	369
176	*80*	3.0	*1.36*	1.7	2.5	*1.13*	1.4	1.7	*0.75*	3.32	2.71	1.53	106	73	4	2.8	2.0	408
198	*90*	3.3	*1.48*	1.6	2.7	*1.23*	1.4	1.8	*0.82*	3.63	2.96	1.67	116	80	4	2.8	2.2	444
220	*100*	3.5	*1.60*	1.6	3.0	*1.34*	1.3	2.0	*0.89*	3.93	3.21	1.81	126	86	5	3.5	2.4	480
Maintenance plus low activity (basic plus 25% increment, includes goats under intensive management, tropical range, and early pregnancy)																		
22	*10*	0.8	*0.36*	3.6	0.7	*0.30*	3.0	0.4	*0.20*	0.87	0.71	0.40	27	19	1	0.7	0.5	108
44	*20*	1.3	*0.60*	3.0	1.1	*0.50*	2.5	0.7	*0.33*	1.47	1.20	0.68	46	32	2	1.4	0.9	180
66	*30*	1.8	*0.81*	2.7	1.5	*0.67*	2.2	1.0	*0.45*	1.99	1.62	0.92	62	43	2	1.4	1.2	243
88	*40*	2.2	*1.01*	2.5	1.9	*0.84*	2.1	1.2	*0.56*	2.47	2.02	1.14	77	54	3	2.1	1.5	303
110	*50*	2.6	*1.19*	2.4	2.2	*0.99*	2.0	1.5	*0.66*	2.92	2.38	1.34	91	63	4	2.8	1.8	357
132	*60*	3.0	*1.36*	2.3	2.5	*1.14*	1.9	1.7	*0.76*	3.35	2.73	1.54	105	73	4	2.8	2.0	408
154	*70*	3.4	*1.54*	2.2	2.8	*1.28*	1.8	1.9	*0.85*	3.76	3.07	1.73	118	82	5	3.5	2.3	462
176	*80*	3.7	*1.70*	2.1	3.1	*1.41*	1.8	2.1	*0.94*	4.16	3.39	1.91	130	90	5	3.5	2.6	510
198	*90*	4.1	*1.85*	2.1	3.4	*1.54*	1.7	2.3	*1.03*	4.54	3.70	2.09	142	99	6	4.2	2.8	555
220	*100*	4.4	*2.00*	2.0	3.7	*1.67*	1.7	2.4	*1.11*	4.91	4.01	2.26	153	107	6	4.2	3.0	600
Maintenance plus medium activity (basic plus 50% increment, includes goats on semiarid rangeland, slightly hilly pastures, and early pregnancy)																		
22	*10*	0.9	*0.43*	4.3	0.8	*0.36*	3.6	0.5	*0.24*	1.05	0.86	0.48	33	23	1	0.7	0.6	129
44	*20*	1.6	*0.72*	3.6	1.3	*0.60*	3.0	0.9	*0.40*	1.77	1.44	0.81	55	38	2	1.4	1.1	216
66	*30*	2.2	*0.98*	3.3	1.8	*0.81*	2.7	1.2	*0.54*	2.38	1.95	1.10	74	52	3	2.1	1.5	294
88	*40*	2.7	*1.21*	3.0	2.2	*1.01*	2.5	1.5	*0.67*	2.97	2.42	1.36	93	64	4	2.8	1.8	363
110	*50*	3.2	*1.43*	2.9	2.6	*1.19*	2.4	1.8	*0.80*	3.51	2.86	1.62	110	76	4	2.8	2.1	429
132	*60*	3.6	*1.64*	2.7	3.0	*1.37*	2.3	2.0	*0.91*	4.02	3.28	1.84	126	87	5	3.5	2.5	492
154	*70*	4.1	*1.84*	2.6	3.4	*1.53*	2.2	2.2	*1.02*	4.52	3.68	2.07	141	98	6	4.2	2.8	552
176	*80*	4.5	*2.03*	2.5	3.7	*1.69*	2.1	2.5	*1.13*	4.98	4.06	2.30	156	108	6	4.2	3.0	609
198	*90*	4.9	*2.22*	2.5	4.1	*1.85*	2.0	2.7	*1.24*	5.44	4.44	2.50	170	118	7	4.9	3.3	666
220	*100*	5.3	*2.41*	2.4	4.4	*2.01*	2.0	3.0	*1.34*	5.90	4.82	2.72	184	128	7	4.9	3.6	723
Maintenance plus high activity (basic plus 75% increment, includes goats on arid rangeland, sparse vegetation, mountainous pastures, & early pregnancy)																		
22	*10*	1.1	*0.50*	5.0	0.9	*0.42*	4.2	0.6	*0.28*	1.22	1.00	0.56	38	26	2	1.4	0.8	150
44	*20*	1.9	*0.84*	4.2	1.5	*0.70*	3.5	1.0	*0.47*	2.06	1.68	0.94	64	45	2	1.4	1.3	252
66	*30*	2.5	*1.14*	3.8	2.1	*0.95*	3.2	1.4	*0.63*	2.78	2.28	1.28	87	60	3	2.1	1.7	342
88	*40*	3.1	*1.41*	3.5	2.6	*1.18*	3.0	1.7	*0.78*	3.46	2.82	1.59	108	75	4	2.8	2.1	423
110	*50*	3.7	*1.67*	3.3	3.1	*1.39*	2.7	2.1	*0.93*	4.10	3.34	1.89	128	89	5	3.5	2.5	501
132	*60*	4.2	*1.92*	3.2	3.5	*1.60*	2.7	2.3	*1.06*	4.69	3.83	2.15	146	102	6	4.2	2.9	576
154	*70*	4.7	*2.14*	3.0	3.9	*1.79*	2.6	2.6	*1.19*	5.27	4.29	2.42	165	114	6	4.2	3.2	642
176	*80*	5.2	*2.37*	3.0	4.4	*1.98*	2.5	2.9	*1.32*	5.81	4.74	2.68	182	126	7	4.9	3.6	711
198	*90*	5.7	*2.59*	2.9	4.8	*2.16*	2.4	3.2	*1.44*	6.35	5.18	2.92	198	138	8	5.6	3.9	777
220	*100*	6.2	*2.81*	2.8	5.2	*2.34*	2.3	3.4	*1.56*	6.88	5.62	3.17	215	150	8	5.6	4.2	843

(Continued)

TABLE 21-1 (Continued)

Body Weight (BW)		Dry Matter per Animal[2] 0.9 Mcal ME/lb (2 Mcal ME/kg) Total		% of BW	1.09 Mcal ME/lb (2.4 Mcal ME/kg) Total		% of BW	Feed Energy TDN		DE	ME	NE	Protein TP	DP	Minerals Ca	P	Vitamin A	Vitamin D
(lb)	(kg)	(lb)	(kg)	(%)	(lb)	(kg)	(%)	(lb)	(kg)	(Mcal)	(Mcal)	(Mcal)	(g)	(g)	(g)	(g)	(1,000 IU)	(IU)
Additional requirements for late pregnancy (for all goat sizes)																		
		1.6	0.71		1.3	0.59		0.9	0.40	1.74	1.42	0.80	82	57	2	1.4	1.1	213
Additional requirements for growth—weight gain at 1.75 oz (50 g) per day (for all goat sizes)																		
		0.4	0.18		0.3	0.15		0.2	0.10	0.44	0.36	0.20	14	10	1	0.7	0.3	54
Additional requirements for growth—weight gain at 3.5 oz (100 g) per day (for all goat sizes)																		
		0.8	0.36		0.7	0.30		0.4	0.20	0.88	0.72	0.40	28	20	1	0.7	0.5	108
Additional requirements for growth—weight gain at 5.3 oz (150 g) per day (for all goat sizes)																		
		1.2	0.54		1.0	0.45		0.7	0.30	1.32	1.08	0.60	42	30	2	1.4	0.8	162

[1]Adapted by the author from *Nutrient Requirements of Goats*, No. 15, NRC-National Academy of Sciences.

[2]Good-quality roughages furnish about 0.9 Mcal ME/lb *(2 Mcal ME/kg)* of dry matter. Roughage-concentrate mixed rations are sometimes necessary to increase the energy content of the diet to 1.09 Mcal/lb *(2.5 or 3.0 Mcal ME/kg)* of dry matter when early weaned kids or high-producing dairy goats are being fed.

TABLE 21-2
ADDITIONAL NUTRIENT REQUIREMENTS FOR MILK PRODUCTION PER POUND AT DIFFERENT FAT PERCENTAGES[1]

Fat	Feed Energy TDN		DE	ME	NE	Protein TP	DP	Minerals Ca	P	Vitamins A	D
(%)	(lb)	(kg)	(Mcal)	(Mcal)	(Mcal)	(g)	(g)	(g)	(g)	(1,000 IU)	(IU)
2.5	0.33	0.151	0.67	0.54	0.31	26.7	19.1	0.9	0.6	1.7	345
3.0	0.34	0.153	0.68	0.55	0.31	29.0	20.4	0.9	0.6	1.7	345
3.5	0.34	0.155	0.68	0.56	0.31	30.8	21.8	0.9	0.6	1.7	345
4.0	0.35	0.157	0.69	0.57	0.32	32.7	23.1	1.4	1.0	1.7	345
4.5	0.35	0.159	0.70	0.57	0.32	34.9	24.5	1.4	1.0	1.7	345
5.0	0.36	0.161	0.71	0.58	0.33	37.2	25.9	1.4	1.0	1.7	345
5.5	0.36	0.163	0.72	0.59	0.33	39.0	27.2	1.4	1.0	1.7	345
6.0	0.37	0.166	0.73	0.59	0.34	40.8	28.6	1.4	1.0	1.7	345

[1]Adapted by the author from *Nutrient Requirements of Goats*, No. 15, NRC-National Academy of Sciences. These requirements are in addition to those listed in Table 21-1. They include requirements for nursing single, twin, or triplet kids at the respective milk production level. To convert to requirements for milk production per kg, multiply by 2.205.

TABLE 21-3
ADDITIONAL NUTRIENT REQUIREMENTS FOR MOHAIR PRODUCTION BY ANGORA GOATS AT DIFFERENT FLEECE PRODUCTION LEVELS[1]

Annual Fleece Yield		Feed Energy TDN		DE	ME	NE	Protein TP	DP
(lb)	(kg)	(lb)	(kg)	(Mcal)	(Mcal)	(Mcal)	(g)	(g)
4.4	2	0.035	0.016	0.07	0.06	0.03	9	6
8.8	4	0.075	0.034	0.15	0.12	0.07	17	12
13.2	6	0.110	0.050	0.22	0.18	0.10	26	18
17.6	8	0.146	0.066	0.29	0.24	0.14	34	24

[1]Adapted by the author from *Nutrient Requirements of Goats*, No. 15, NRC-National Academy of Sciences. These requirements are in addition to those listed in Table 21-1.

Shearing mohair from Angora goats and pashmina from Cashmere goats decreases insulation and results in increased energy needs, especially during cold weather. Goats are more active and travel greater distances than sheep, which increases energy requirements. Maintenance requirements of goats on pasture, browse, and range, especially on mountainous and/or seasonal grazing, are considerably higher than those of stable-fed animals. The magnitude of this increase is presented in Table 21-1 at three levels of activity, which results from the availability of feed and water, and from the topography, elevation, and distance traveled in grazing.

Good-quality roughages furnish about 0.9 Mcal

Fig. 21-3. Roughages furnish most of the energy required by goats. However, concentrates must be added to meet the high energy needs of rapidly growing kids and high-lactating dairy goats. (Courtesy, Rocking M Ranch, Hilmar, CA)

metabolizable energy (ME) per pound dry matter (DM). However, roughage-concentrate mixed rations are sometimes necessary to increase the energy content of the ration to 1.09 to 1.40 Mcal ME/lb DM when feeding early weaned kids or high producing dairy goats. So, in Table 21-1, under the heading "Dry Matter Per Animal," provision is made for two different energy levels of feeds: (1) good-quality roughage alone (0.9 Mcal ME/lb), and (2) a roughage-concentrate mix (1.09 Mcal ME/lb). The efficiency with which energy is utilized for weight gain, pregnancy, and lactation usually increases with increasing levels of ME concentration in the ration.

The energy requirement for goats as herein reported are the amounts needed (1) for maintenance, pregnancy, and growth (Table 21-1); (2) for milk production (Table 21-2); and (3) for mohair production (Table 21-3). Note that the energy requirements for the various categories are expressed as total digestible nutrients (TDN), digestible energy (DE), metabolizable energy (ME), and net energy (NE).

In Table 21-1, in addition to the energy requirement for maintenance only, energy values are given for three different levels of activity: (1) light, (2) medium, and (3) high. Also, additional energy allowances are made for: (1) late pregnancy (the last 60 days); and (2) three different levels of growth—50, 100, 150 g per day.

In Table 21-2, energy requirements in addition to those listed in Table 21-1 are given for milk production per pound at different fat percentages.

In Table 21-3, energy requirements in addition to those listed in Table 21-1 are given for mohair production of Angora goats of different fleece weights.

Carbohydrates and fats supply virtually all of the energy needs of the body, though a small portion may be derived from protein catabolism. Rumen microorganisms serve essential roles in the digestion of many of the complex carbohydrates consumed by goats, especially range feeds. Carbohydrates and fats, accounting for 60 to 70% of the total energy derived from the feed, are converted to volatile fatty acids (primarily acetate, propionate, and butyrate) in the rumen which are, in turn, absorbed through the rumen wall and used to supply energy.

Although fats serve as carriers of the fat-soluble vitamins and other fat-soluble substances, they are used primarily as a concentrated source of energy. A general rule of thumb is that fats supply 2.25 times the energy of carbohydrates. However, excessive amounts of fat in the ration usually reduce palatability and make the ration more susceptible to oxidation and subsequent spoilage.

PROTEIN

Proteins are the principal constituents of the animal body and are continuously needed in the feed for growth and cell repair. The transformation of feed protein into body protein is an important process of nutrition and metabolism. Proteins consist of amino acids, which are the building blocks of all body cells. Secretions such as enzymes, hormones, mucin, and milk make for additional amino acid requirements. Proteins are, therefore, vital for animal maintenance, growth, reproduction, and milk production. However, in goats as in other ruminants, nonprotein nitrogen (NPN) can substitute for parts of the required proteins for these functions.

In Table 21-1, protein requirements are given for maintenance, activity, late pregnancy, and growth, along with the energy requirements.

In Table 21-2, protein requirements in addition to those listed in Table 21-1 are given for milk production per pound at different fat percentages.

In Table 21-3, protein requirements in addition to those listed in Table 21-1 are given for mohair production of different fleece weights of Angora goats.

Total protein (TP) is considered to be the most accurate guide for converting proteins from feed composition tables to the quantities required, but digestible protein (DP) values are also used.

Ruminal microorganisms can utilize either protein or nonprotein nitrogen to synthesize microbial protein. The microbial protein, along with the undigested feed protein, passes from the rumen-reticulum through the omasum, then to the small intestine where it is subjected to digestive processes similar to those of the nonruminant and broken down to amino acids which are absorbed and utilized by goats. Thus, goats can utilize protein and nonprotein nitrogen (such as urea) in their rations. So, the protein available for digestion in the small intestine of goats consists of microbial protein and feed protein that has escaped microbial breakdown in the rumen. But it has been shown that the protein produced by ruminal synthesis does not supply all of the amino acids in the quality or quantity needed for maximum growth of kids or milk production of does. Thus, quality and degradability of protein fed to goats is more important than formerly thought. Moreover, it has been found that protein efficiency can be increased by protecting protein from the degradation of the microbes in the rumen and increasing the escape of protein from the rumen to the intestines where it is digested and absorbed. This technology of manipulating the quantity of dietary protein rumen fermentation, thereby increasing the supply of protein (amino acids) in the small intestine, is known as *protein bypass*.

The most commonly used protein supplements are brewers' dried grains, cottonseed meal, linseed meal, and soybean meal. One of the most economical sources of protein is good-quality alfalfa hay—fed as long hay, chopped hay, range cubes, or pellets. It can either be fed separately or mixed with the concentrate portion of the ration in a complete feed. Dehydrated alfalfa is an excellent source of protein, but it is more expensive than sun-cured hay.

Protein deficiencies in the ration deplete stores in

TABLE
GOAT MINERAL

Mineral Which May Be Deficient Under Normal Conditions	Conditions Usually Prevalent Where Deficiencies Are Reported	Function of Mineral	Deficiency Symptoms/Toxicity
Major or Macrominerals:			
Salt (NaCl)	Negligence, for salt is inexpensive. Lactating does may require additional salt as milk contains high amounts of sodium.	Sodium chloride helps maintain osmotic pressure in body cells, upon which depends the transfer of nutrients to the cells, the removal of waste materials, and the maintenance of water balance among the tissues. Also, sodium is important in making bile, which aids in the digestion of fats and carbohydrates; and chlorine is required for the formation of hydrochloric acid in the gastric juice so vital to protein digestion. It is noteworthy that when salt is omitted, sodium expresses its deficiency first.	**Deficiency symptoms**—Loss of appetite, depraved appetite and consumption of soil and debris, emaciation, decline in milk production, a general rough appearance with poor coat and lusterless eyes. Acute deficiency symptoms include shivering, weakness, cardiac disturbances, and ultimately death. **Toxicity**—The maximum tolerable level of salt for sheep is 9.0%. For goats, a similar level of salt will likely be toxic.
Calcium (Ca)	Goats in heavy lactation. Lack of vitamin D. Calcium-deficient areas (where pasture and range forages are deficient in calcium) are FL, LA, NE, VA, and WV. Feeds that contain primarily cereal grains.	Essential for the development and maintenance of good strong bones and teeth; maintains the contractability, rhythm, and tonicity of the heart muscles; antagonizes the action of the sodium and potassium on the heart; is required for normal coagulation of the blood; is necessary for proper nerve irritability; and appears to be essential for selective cellular permeability.	**Deficiency symptoms**—In young kids, retarded growth and abnormal bone development. Also, a deficiency of calcium may cause rickets in young animals and osteomalacia in adults. In lactating does, depressed milk yields and fragile bones. Milk fever can occur when calcium levels in the blood drop. **Toxicity**—If there is adequate phosphorus, sheep can tolerate a calcium:phosphorus ratio of 7:1 and as much as 2% calcium in the diet. It is postulated that goats can tolerate a similar level of calcium.
Phosphorus (P)	When goats subsist on pastures in phosphorus-deficient areas. When goats subsist for long periods on mature, dry forages. Lack of vitamin D.	Essential for sound bones and teeth, and for the assimilation of carbohydrate and fats. A vital ingredient of the proteins in all body cells. Necessary for enzyme activation. Acts as a buffer in blood and tissue. Occupies a key position in biologic oxidation and reactions requiring energy.	**Deficiency symptoms**—Slowed growth, depraved appetite (chewing bones, wood, hair), unthrifty appearance, rickets in young animals, osteomalacia in mature animals, and depressed milk yields in lactating does. **Toxicity**—There is no known phosphorus toxicity in goats. However, excess phosphorus consumption may decrease the absorption of calcium. Also, when phosphorus is high in relation to calcium, urinary calculi may be formed.

the blood, liver, and muscles, and predispose animals to a variety of serious and even fatal ailments. Below a minimum level of 6% crude protein (CP) in the ration, feed intake will be reduced, which leads to a combined deficiency of energy and protein. This deficiency further reduces rumen function and lowers the efficiency of feed utilization. Long-term protein deficiencies retard fetal development, lead to low birth weights, affect kid growth, and depress milk production.

MINERALS

If goats are fed a good concentrate, along with a good-quality hay produced on land that has been properly fertilized, few problems arising from mineral deficiencies occur.

The mineral requirements of goats are given in Tables 21-1, 21-2, and 21-4.

GOAT MINERAL CHART

Table 21-4, Goat Mineral Chart, gives a summary of the different factors involved with mineral nutrition in the goat. Further elucidation of certain minerals is contained in the accompanying narrative. Fluorine is discussed because of its toxicity.

21-4
CHART

Nutrient Requirements[2]	Recommended Allowances[2]	Practical Sources	Comments
	Salt should be provided free-choice or as a component of the ration. In a complete feed, 0.5 to 1.0% salt is recommended, with proportionately higher levels in supplements.	Iodized salt in iodine-deficient areas. Can be offered free-choice or incorporated into the ration. In alkaline areas, water may contain enough salt to meet the requirements.	In range areas, salt may be added to feed to limit feed intake. If self-feeders are located near water, the level of salt in the ration should be high (25-40%). If self-feeders are some distance from water, the level of salt in the ration should be reduced. In arid regions, the salt content of some water sources can reduce intake of water and feed.
Variable according to age, sex, and class (see Tables 21-1, and 21-2).	Because milk is high in calcium, lactating does need rations with high calcium levels. In % of ration: 0.78 M-F 0.70 A-F	Ground limestone, steamed bone meal, dicalcium phosphate, and oyster shell.	The recommended ratio of calcium to phosphorus ranges from 2:1 to 4:1. If the ratio falls below 2:1, urinary calculi may develop in males. Under grazing conditions, calcium is seldom a problem with either Angora or meat-type goats.
Variable according to age, sex, and class (see Tables 21-1 and 21-2).	Can be offered free-choice or incorporated into the ration. In % of ration: 0.45 M-F 0.40 A-F	Cereal grains. Defluorinated phosphate, dicalcium phosphate, steamed bone meal, monosodium phosphate.	Phosphorus is the mineral most likely to be deficient in range forages. It is, therefore, recommended that it be supplied in range supplements. *The calcium-to-phosphorus ratio should not drop below 1.2:1.

(Continued)

TABLE 21-4

Mineral Which May Be Deficient Under Normal Conditions	Conditions Usually Prevalent Where Deficiencies Are Reported	Function of Mineral	Deficiency Symptoms/Toxicity
Major or Macrominerals (Continued)			
Magnesium (Mg)	Animals grazing lush green grass or winter cereal pastures fertilized with nitrogen and potassium.	Required for many enzyme systems and for proper functioning of the nervous system. Also, closely associated with the metabolism of calcium and phosphorus.	**Deficiency symptoms**—Loss of appetite, excitability, and calcification of soft tissues. The most noted problem associated with low magnesium is grass tetany. **Toxicity**—Magnesium toxicity of goats has not been reported under practical conditions.
Potassium (K)	When goats are grazing mature range forage during winter or drought periods. High concentrate rations.	It (1) affects osmotic pressure and acid-base balance within the cells, and (2) aids in activating several enzyme systems involved in energy transfer and utilization, protein synthesis, and carbohydrate metabolism.	**Deficiency symptoms**—Marginal deficiencies result in reduced feed intake, retarded growth, and reduced milk production. Severe deficiencies cause emaciation and poor muscular tone. **Toxicity**—The maximum tolerable level of potassium for sheep is about 3% of the ration DM. It is postulated that the toxicity level of goats is similar.
Sulfur (S)	Possibly with liberal intake of tannic acid-containing plants. This is of concern with range goats, which liberally graze and browse such plants.	Essential for synthesis of the sulfur amino acids (cystine and methionine). Sulfur is particularly high in goat hair.	**Deficiency symptoms**—Depressed appetite, loss of weight, poor growth, excessive salivation, tearing, loss of mohair, depressed milk yields. **Toxicity**—Elemental sulfur is practically devoid of toxicity.
Trace or Microminerals:			
Cobalt (Co)	In cobalt deficient areas when the cobalt level in the feed drops to 0.04 to 0.07 ppm or lower.	The only function of cobalt is that of being an integral part of vitamin B-12.	**Deficiency symptoms**—The deficiency symptoms are actually vitamin B-12 deficiencies. They are: loss of appetite, emaciation, weakness, anemia, and decreased production. **Toxicity**—In sheep, about 204.5 mg cobalt/100 lb live weight is toxic. Likely, the same applies to goats.
Copper (Cu) and Molybdenum (Mo)	Copper and molybdenum are interrelated in animal metabolism; hence, they should be considered together. The most common problem occurs when a normal or low level of copper is accompanied by a high level of molybdenum, resulting in copper being excreted and producing a copper deficiency. This condition can be corrected by adding copper.	Copper and iron are mutually involved in the formation of hemoglobin—the red pigment which carries oxygen.	Few studies on copper and molybdenum have been conducted with goats. It appears that sheep are sensitive to copper toxicity and resistant to molybdenosis, but it is not known whether this is also the case with goats.
Fluorine (F)		Necessary for sound bones and teeth.	**Deficiency symptoms**—Fluorine deficiency appears to be rare. Rather, the hazard is fluorine toxicity. **Toxicity**—With sheep, fluorine toxicity occurs at levels above 200 ppm. So, it is postulated that the toxicity level for goats is similar.
Iodine (I)	Iodine-deficient areas or soils (in northwestern U.S., and in the Great Lakes and Rocky Mountain Regions), unless iodized salt is fed.	Formation of thyroxin, a hormone of the thyroid gland.	**Deficiency symptoms**—Enlarged thyroid gland, a condition called goiter. Kids born weak or dead. **Toxicity**—The maximum tolerable level for sheep is 45 ppm A-F or 50 ppm M-F. It is postulated that the toxicity level for goats is similar.

(Continued)

Nutrient Requirements[2]	Recommended Allowances[2]	Practical Sources	Comments
	In % of ration: 0.25 M-F 0.22 A-F	Plant protein supplements and plant by-product feeds are excellent sources of magnesium. The common magnesium supplements are magnesium carbonate, magnesium oxide, and magnesium sulfate.	*Goats have a marginal ability to compensate for low dietary magnesium by reducing the rate of excretion.
*In growing sheep, the potassium requirement is 0.5% of the ration. In lactating dairy cattle, the requirement is 0.8% of the complete ration. These levels are also postulated as the requirements of growing and lactating goats, respectively.	In % of ration: 1.0 M-F 0.9 A-F	Roughage-based rations. Common potassium supplements are potassium chloride, potassium bicarbonate, and potassium sulfate.	
	In % of ration: 0.20 M-F 0.18 A-F A sulfur-to-nitrogen ratio of 1:10 is recommended.	Sulfates, such as sodium sulfate and ammonium sulfate, are the most available forms of sulfur for ration formulation.	Because of mohair production, Angora goats may have an elevated sulfur requirement.
*A level of 0.1 ppm in the M-F ration.	In % of ration: 0.1 to 0.2 ppm M-F 0.09 to 0.18 ppm A-F	*Cobalt sulfate or cobalt chloride added at the rate of 5.45 g per 100 lb *(12 g per 100 kg)* of salt.	
	Add copper sulfate to the salt at the rate of 0.5% Copper in total ration: 5.0 ppm M-F 4.5 ppm A-F	Salt containing 0.5% copper sulfate.	
Iodine in the ration: A-F, 0.09–0.72 ppm; M-F, 0.1–0.8 ppm. The higher levels are indicated for pregnancy and lactation.	Free access to stabilized iodized salt containing 0.0078% iodine. In total ration: 0.5 ppm M-F 0.45 ppm A-F	Iodized salt.	Iodized salt should not be used as a feed-limiter because it could lead to excessive intakes of iodine.

(Continued)

TABLE 21-4

Mineral Which May Be Deficient Under Normal Conditions	Conditions Usually Prevalent Where Deficiencies Are Reported	Function of Mineral	Deficiency Symptoms/Toxicity
Trace or Microminerals (Continued)			
Iron (Fe)	Iron deficiency may occur in young goat kids because of their minimal body stores at birth and the low iron content of milk.	As a component of blood hemoglobin required for oxygen transport. Iron is also required for some enzyme systems.	**Deficiency symptoms**—Anemia, poor growth, lethargy, increased respiration rate, decreased resistance to infection, and in severe cases high mortality. **Toxicity**—Free iron ions are very toxic, causing loss of appetite, diarrhea, below normal temperature, shock, acidosis, and death.
Manganese (Mn)	High calcium and iron may increase manganese requirements.	Skeletal development and reproduction.	**Deficiency symptoms**—Reluctance to walk, deformity of the forelegs, delayed estrus, more inseminations per conception, more abortions, and 20% reduction in birth weights. **Toxicity**—1,000 ppm appears to be the maximum tolerance level for sheep; so, it is postulated that the toxicity level for goats is similar.
Selenium (Se)			**Deficiency symptoms**—White muscle disease in young kids from birth to a few months of age, which may take one of two forms: (1) sudden unexplained death, or (2) muscular paralysis, particularly of the hind limbs, or stiffness and inability to rise. **Toxicity**—All livestock species, including goats, are susceptible to selenium toxicity. Selenium toxicity in sheep occurs from prolonged consumption of plants containing over 3 ppm Se. It is postulated that the toxicity level for goats is about the same as for sheep.
Zinc (Zn)	Rations excessively high in calcium adversely affect zinc utilization.	Needed for normal skin, bones, and hair. A component of several enzyme systems involved in digestion and respiration.	**Deficiency symptoms**—Reduced feed intake, weight loss, parakeratosis, stiffness of joints, excessive salivation, swelling of the feet and horny overgrowth, small testicles, and low libido. **Toxicity**—Levels of 1,000 ppm may be toxic.

[1]Where preceded by an asterisk, the requirements, recommended allowances, and other facts presented herein were adapted from *Nutrient Requirements of Goats*, No. 15, NRC-National Academy of Sciences.

[2]As used herein, the distinction between "nutrient requirements" and "recommended allowances" is as follows: In nutrient requirements, no margins of safety are included intentionally, whereas in recommended allowances, margins of safety are provided in order to compensate for variations in feed composition, environment, and possible losses during storage or processing.

MAJOR OR MACROMINERALS

Seven major or macrominerals are considered dietary essentials for livestock, including goats. These are: salt (sodium and chlorine), calcium, magnesium, phosphorus, potassium, and sulfur.

SALT (NaCl)

Of the various macrominerals demonstrated to be required by goats, salt is probably the most likely to be deficient in goat rations and is one of the easiest to supply. Goats require both sodium and chlorine, but sodium is the mineral element most likely to be lacking in common feeding practices.

Goats should have access to salt free-choice in loose form at all times, whether on the range, on seeded pasture, or in a corral. Where iodine is lacking, iodized salt is recommended.

Salt may be offered free-choice in a mineral mix or provided as a feed intake governor in concentrate mixes that are offered free-choice. In operations where salt is not used to control feed intake, a trace mineralized salt is usually mixed at a ratio of 1:1 with either

(Continued)

Nutrient Requirements[2]	Recommended Allowances[2]	Practical Sources	Comments
*0.03% ferrous iron in the ration.	*Iron-dextran (150 mg) may be injected in kids at 2 to 3 week intervals if iron deficiencies are observed. In total ration: 50 ppm M-F 45 ppm A-F	Iron-dextran is recommended as an injection; and ferrous sulfate and ferric citrate are recommended for incorporating in rations.	Iron deficiency seldom occurs in mature grazing goats.
	In total ration: 40 ppm M-F 36 ppm A-F	Manganese gluconate.	
	In total ration: 0.15 ppm M-F 0.13 ppm A-F		
*Direct and indirect evidence indicates minimum requirements of 10 ppm in the ration.	In total ration: 50 ppm M-F 45 ppm A-F	Zinc carbonate. Zinc sulfate.	

steamed bone meal or dicalcium phosphate and offered free-choice.

If goats are deprived of salt for an extended period and are then given salt free-choice, they may consume too much initially and become sick; hence, a period of adjustment is required. Where goats have been salt-starved, they should be hand-fed salt for a period of time, with the amount increased daily until salt is left in the box from the previous day. At this stage, it is safe to provide all they will consume.

CALCIUM (Ca) AND PHOSPHORUS (P)

Under grazing conditions, calcium is seldom a problem for either Angora or Spanish goats, but it can be very important for high-producing dairy goats, because goat's milk is rich in calcium. A phosphorus deficiency in grazing goats is likely when they are consuming phosphorus-deficient range forages.

Production increases the demands for calcium and phosphorus. Both growth (bone development) and lactation (deposition of minerals in milk) require substantial quantities of these minerals. If there is a severe imbalance of these minerals during pregnancy and early lactation, milk fever may occur. In males, an imbalance of calcium to phosphorus often leads to the development of urinary calculi.

Most forages are good sources of dietary calcium, but are rather low in phosphorus. Legumes are excellent calcium sources, while the grasses and silages

Fig. 21-4. Alfalfa hay is an excellent source of calcium for goats. (Courtesy, University of Delaware, Newark)

tend to be substantially lower in calcium content. Conversely, the calcium:phosphorus ratio is important. Ideally, the goat ration should have a calcium:phosphorus ratio of 1.4:1 to 4:1, but ruminants have been observed to tolerate ratios of up to 7:1.

The source and quantity of mineral supplementation depends on the mineral composition of the total ration. Where calcium is needed, ground limestone is generally the mineral of choice. Where phosphorus is the primary need, it is usually provided in the form of monosodium phosphate, disodium phosphate, sodium tripolyphosphate, or feed-grade phosphoric acid. Where both calcium and phosphorus are needed, the most frequently used supplements are dicalcium phosphate, defluorinated phosphate, or steamed bone meal. The mineral supplement(s) can either be incorporated in the ration or offered free-choice. Quite often, steamed bone meal or dicalcium phosphate is mixed with equal amounts of salt and made available to goats free-choice.

MAGNESIUM (Mg)

Magnesium is a constituent of bone. Also, it is necessary for many enzyme systems and for proper functioning of the nervous system. A deficiency of magnesium may result in grass tetany, characterized by stiff legs and head retraction. Where does graze forage with high nitrogen and potassium content, the minimum level of magnesium in the ration should be 0.22% as-fed basis.

POTASSIUM (K)

Although potassium is required in relatively large amounts, it is usually sufficient in roughage-based rations. For growing sheep, the potassium requirement is 0.5% of the ration, whereas for lactating dairy cattle, the requirement is 0.8% of the ration. These levels are also postulated as meeting the requirements of growing and lactating goats, respectively. Ration values below these levels are infrequently encountered, and are usually caused by high-concentrate rations or severely weathered range forage.

Marginal potassium deficiencies result in reduced feed intake, retarded growth, and reduced milk production. Severe deficiencies cause emaciation and poor muscular tone.

SULFUR (S)

Since sulfur is a key constituent in two important amino acids (methionine and cystine), it should be added to goat rations when nonprotein nitrogen sources are used. This additional sulfur can then be incorporated into amino acids by the microorganisms in the rumen. Either elemental sulfur or sulfur in sulfate form can be used effectively. When urea or other NPN sources are used, the ratio of sulfur-to-nitrogen should be 1:10. More sulfur is required if tannic acid is high in the forage.

TRACE OR MICROMINERALS

Nine trace or microminerals are required by goats. These are: cobalt, copper and molybdenum, fluorine, iodine, iron, manganese, selenium, and zinc.

Goat rations are seldom deficient in trace minerals. However, for does in heavy lactation or for goats grazing on sandy soils, it may be advisable to supply a limited amount of a broad-based trace mineral mixture to guard against possible deficiencies. Care should be taken when using a trace mineral mixture because of the many interrelationships among the minerals which affect their availability for absorption.

COBALT (Co)

Cobalt is a component of vitamin B-12, for the synthesis of which it is essential. In sheep, a cobalt intake of 0.1 ppm is considered adequate. It is postulated that the same level is adequate for goats.

Cobalt should be ingested frequently, preferably daily. This is best accomplished by adding cobalt to the salt at a level of 5.45 g/100 lb of salt, fed free-choice.

Deficiency signs include loss of appetite, emaciation, weakness, anemia, and decreased production.

COPPER (Cu) AND MOLYBDENUM (Mo)

Copper and molybdenum are interrelated in animal metabolism; hence, herein they are considered together. The most common problem occurs when a normal or low level of copper is accompanied by a high level of molybdenum. In this case, copper is excreted and a deficiency occurs. This condition can be corrected by adding copper.

Few studies on copper and molybdenum have been conducted with goats. It appears that sheep are sensitive to copper toxicity and resistant to molybdenosis, but it is not known whether this is also the case with goats.

FLUORINE (F)

In small amounts, fluorine helps develop strong bones and teeth, but in excessive amounts bones become porous and soft and teeth become mottled and easily worn down. Fluorine deficiency is rare; rather, the hazard is fluorine toxicity, which may be caused by high fluorine levels in ground water or in crude mineral supplements (raw rock phosphate).

IODINE (I)

Iodine is necessary for the formation of thyroxin, a hormone of the thyroid gland. A deficiency of iodine results in an enlargement of the thyroid gland, a condition called goiter. Also, kids may be born weak or dead. Iodine-deficient areas are widespread throughout the world, including parts of the United States (in northwestern United States, and in the Great Lakes and Rocky Mountain regions). Deficiencies are readily corrected by feeding iodized salt. Iodized salt should not be used for the purpose of limiting feed consumption because it could lead to excessive intakes of iodine.

IRON (Fe)

Iron is a component of blood hemoglobin that is required for oxygen transport. It is also required for some enzyme systems. Although iron deficiency seldom occurs in mature grazing animals, it may occur in young goat kids because of their minimal body stores of iron at birth and the low iron content of milk.

If an iron deficiency is observed in young kids on a milk diet, injection of iron-dextran (150 mg) at 2 to 3 week intervals is recommended. Ferrous sulfate and ferric citrate are recommended for incorporation in rations. A level of 45 ppm of iron in an as-fed ration appears to be adequate.

MANGANESE (Mn)

Manganese is an essential mineral in the ration of goats, required for skeletal development and reproductive efficiency.

Deficiency symptoms are: reluctance to walk, deformity of the forelegs, delayed estrus, more inseminations per conception, more abortions, and 20% reduction in birth weights.

In an experiment involving two groups of female goats for the first year of life—with one group on low manganese, 20 ppm for the first year of life and 6 ppm during the following year; and the controls receiving 100 ppm—the low manganese ration did not affect growth or bone structure, but it did affect reproduction adversely.

SELENIUM (Se)

Selenium is essential, but only in minute amounts. It is a component of glutathione peroxidase, the metabolic role of which is to protect against oxidation of polyunsaturated fatty acids and resultant tissue damage. Also, selenium is interrelated with vitamin E—they spare each other, and with the sulfur-containing amino acids.

All livestock species, including goats, are susceptible to selenium toxicity. Selenium toxicity in sheep occurs from prolonged consumption of plants containing over 3 ppm selenium. It is postulated that the toxicity level for goats is about the same as for sheep.

ZINC (Zn)

Zinc is essential for goats. Deficiency symptoms include reduced feed intake, weight loss, parakeratosis, stiffness of joints, excessive salivation, swelling of the feet and horny overgrowth, small testicles, and low libido. Zinc must be supplied continuously because little is stored in the body in readily available form. Direct and indirect evidence indicates minimum ration requirements of 10 ppm. Levels of 1,000 ppm may be toxic.

VITAMINS

Typical range or pasture diets of goats usually contain adequate levels of vitamins or vitamin precursors to maintain the normal health of goats. However, young kids, goats kept in confinement, and high-producing dairy goats may need supplemental vitamins.

GOAT VITAMIN CHART

Table 21-5, Goat Vitamin Chart, gives, in summary form, the following pertinent information relative to each vitamin listed: (1) conditions usually prevailing where deficiencies are reported, (2) function, (3) deficiency symptoms, (4) nutrient requirements, (5) recommended allowances, and (6) practical sources.

TABLE
GOAT VITAMIN

Vitamin Which May Be Deficient Under Normal Conditions	Conditions Usually Prevalent Where Deficiencies Are Reported	Function of Vitamin	Deficiency Symptoms
Fat-Soluble Vitamins:			
A	During extended dry periods when the supply of green forage is limited.	Required for normal vision. Aids in reproduction and lactation. Needed for maintaining normal epithelial tissue. Aids in resistance to infection.	Keratinization of the epithelia of the respiratory, alimentary, reproductive, and urinary tracts, and of the eye. Multiple infections, poor bone development, birth of abnormal offspring, and vision impairment, including night blindness.
D	Young goats kept in confinement where they have little or no access to sunlight.	Absorption of calcium and phosphorus.	Bone abnormalities, including rickets. Depressed growth.
E	Abnormally high levels of nitrates may produce vitamin E deficiencies. Where soils are very low in selenium.	Serves as a physiological antioxidant. In dairy goats, the vitamin E transferred to the milk is important because of the antioxidant properties that aid in milk storage.	Evidence of spontaneous vitamin E deficiency signs in goats is lacking. The probability of lowered productivity in goats as a result of a vitamin E deficiency is remote.
K	Vitamin K deficiency may occur when the dicoumarol content of hay is excessively high, as when moldy sweet clover hay is fed.	Vitamin K or K_2 is necessary in the blood clotting mechanism.	
Water-Soluble Vitamins:			
B vitamins Vitamin C	B vitamin deficiencies may be evident in poorly fed and unhealthy animals. B-12 may be deficient if cobalt is absent or at extremely low levels, as cobalt is required for the synthesis of vitamin B-12.	B-1 participates as a coenzyme in the utilization of carbohydrates.	

[1]As used herein, the distinction between "nutrient requirements" and "recommended allowances" is as follows: In nutrient requirements, no margins of safety are included intentionally; whereas in recommended allowances, margins of safety are provided to compensate for variations in feed composition, environment, and possible losses during storage or processing.

FAT-SOLUBLE VITAMINS

A discussion of the fat-soluble vitamins of goats follows.

VITAMIN A

Vitamin A is not contained in forages, but its precursors, the carotenes, are common in plants. Beta-carotene is the standard form of provitamin A. One milligram of beta-carotene is equivalent to approximately 400 IU of vitamin A.

Vitamin A deficiencies are likely to occur during extended dry periods when the supply of green forage is limited, or when poor-quality hays are fed. Hays that are badly weathered or have been stored for long periods generally have lost most of their carotene; hence, vitamin A should be supplemented.

Goats that are deficient in vitamin A exhibit night blindness, poor reproductive performance, a keratinization of the epithelial cells throughout the body, and bone deformities. Vitamin A supplements can be administered two ways: (1) as an additive to feed, or (2) as an intramuscular injectable in a slow-release form.

2-5
CHART

Nutrient Requirements[1]	Recommended Allowances	Practical Sources	Comments
Variable according to size, sex, age, and class (see Tables 21-1 and 21-2).	The recommended allowances should provide margins of safety over and above the requirements. So, add 10 to 20% to the requirements given in Tables 21-1 and 21-2.	Synthetic vitamin A. Injectable vitamin A. Yellow corn. Green forages.	Young animals, which have not built up vitamin A reserves, are more susceptible to a vitamin A deficiency than are mature animals. Goats that have had access to green feed can store sufficient vitamin A in the liver and fat to last for 3 months on a low carotene ration without showing signs of vitamin A deficiency.
Variable according to size, sex, age, and class (see Tables 21-1 and 21-2).	Add 10 to 20% to the requirements given in Tables 21-1 and 21-2 to provide a margin of safety.	Sunlight action on ergosterol, a plant sterol, and on 7-dehydrocholesterol, a sterol of animal origin. Sun-cured hays. Irradiated yeast. Vitamin D_2 or vitamin D_3, which goats use equally well.	Vitamin D should be of little concern when goats are maintained on pasture or range.
		Alpha-tocopherol, added to the diet or injected intramuscularly. Grains are generally high in vitamin E.	Most goat rations contain adequate vitamin E. Hence, there is little need for vitamin E supplementation.
		Green leafy materials of any kind, fresh or dry are good sources of K_1. Vitamin K_2 is normally synthesized in large amounts in the rumen; no need for dietary supplementation has been established.	
		Only vitamin B-12 (cobalamin) is likely to be deficient in goats with functioning rumens, because the microorganisms synthesize these vitamins in adequate amounts. Adequate vitamin C is synthesized in body tissues to satisfy requirements.	The B vitamins should be included in the diets of very young kids, animals with poorly functioning rumens, sick animals, and those with radically changed diets.

VITAMIN D

Since vitamin D is abundant in sun-cured forages and can be synthesized in the body through exposure to sunlight, there is little need for dietary supplementation. It is noteworthy, however, that the physiological requirements for vitamin D increase when there is an imbalance of calcium and phosphorus. Young kids that are housed without adequate exposure to sunlight should be given supplemental vitamin D.

VITAMIN E

Vitamin E is normally found in large quantities in goat rations, and supplementation is not necessary. In dairy goats, the vitamin E transferred to the milk is important because of its antioxidant properties that aid in milk storage.

VITAMIN K

In adult goats, the microorganisms of the functioning rumen synthesize vitamin K.

WATER-SOLUBLE VITAMINS

Unlike monogastric animals, goats do have the

ability to synthesize a number of vitamins, due primarily to the action of the ruminal microflora. In adult goats, the microorganisms of the functioning rumen synthesize the B complex vitamins. Vitamin C is synthesized in the tissues. However, when the newborn kid starts to eat, the rumen is not well developed and the microflora of the rumen are not of sufficient magnitude to synthesize adequate amounts of the B vitamins; hence, the B complex vitamins are supplied through the milk or milk replacer.

WATER

Water, an essential for all metabolic processes, it important for goats. The amount of water required depends on that needed for the maintenance of normal water balance and to provide for satisfactory levels of production. The body water content of the goat varies with age and amount of fat in the body, but it may be expected to exceed 60% of the body weight.

The water requirement may be met by water consumption (drinking), but other important sources include water contained in the feed ingested and metabolic water resulting from oxidation of feed energy sources. The major avenues of water losses are those from urine, lactation, evaporation, and perspiration.

Factors affecting the water intake of goats are lactation level, environmental temperature, water content of the forage consumed, amount of exercise, and the salt and other minerals in the ration.

Goats are among the most efficient domestic animals in the use of water, approaching the camel in the low rate of water turnover per unit of body weight. They appear to be less subject to high temperature stress than wooled sheep or many breeds of cattle and require less water evaporation to control body temperature. They also have the ability to conserve water by reducing losses in urine and feces.

Lactating goats should always have a fresh, clean supply of water readily available. Nonlactating goats can get by with only a small amount of drinking water when given access to good, succulent range or pasture. However, lactating goats require from 1 to 4 gal. of water per day plus about 2½ qt of water for every quart of milk produced. If there is insufficient water during lactation, milk yields will be depressed. Some water can be obtained from succulent feeds and from dew on the vegetation early in the morning.

Normally, goats consume 1.4 to 1.7 lb of water per 1 lb of dry matter, whereas cattle consume 2.1 lb of water per 1 lb of dry matter.

Wherever possible, water should be within ½ mile of the grazing area. A running stream is best. If a tank or trough is used, precautions should be taken to prevent the kids from getting into it and being unable to get out. Water troughs, bowls, tanks, or containers should be cleaned frequently to avoid a reduction in water consumption, as goats are more particular about water quality than some other animals.

Water should be no higher in salts than would be acceptable to the taste of the caretaker. Soluble salt content should be less than 3,000 ppm, but animals can become accustomed to water with salt levels as high as 6,000 ppm. Diarrhea can occur when goats are initially exposed to water with a high salt content, but they usually adjust to the water if the salt content is not extremely high. Goats tend to be more sensitive to certain salts (especially magnesium sulfate) than other animals.

NONNUTRITIVE FACTORS

Nonnutritive factors are substances that cannot be classified as metabolic nutrients but can aid in the utilization of the nutrients in the feed, such as bulk and feed additives.

BULK

Forages and browse-type feeds, as well as coarse textured concentrates, contain considerable bulk. In many cases, animals can utilize bulky feed more efficiently than if the feed were finely ground because finely ground feeds pass through the digestive tract more rapidly. Finely ground feeds are not exposed to microbial fermentation and enzymatic degradation for periods sufficient to maximize utilization. Hence, they tend to be digested somewhat inefficiently.

ANTIBIOTICS AND OTHER FEED ADDITIVES

As in the case with most other farm animals, antibiotics and other compounds are often added to the rations of goats in order to improve production performance and health. While the effects of these drugs can be beneficial to the overall production scheme, certain precautions must be taken in their use. At the present time, there are over 1,000 drugs approved by the Food and Drug Administration for use in livestock production, and more than one-half of these drugs require preslaughter withdrawal from treated feed or milk-discard periods to prevent problems of contamination in the respective products.

The withdrawal and milk-discard periods of the various drugs are constantly being reviewed and revised; hence, the producer must always read and heed the label of the drug that is being used. *The label is the ultimate guide to the producer as to the proper use of the drug.* Unless it is followed, costly condemnations or seizures can result.

FEEDS FOR GOATS

Goats can effectively use the same kinds of feeds as are consumed by other ruminants—grasses and legumes; hays and other dry roughages; silages, haylages, and root crops; concentrates; milk and milk replacers; and commercial feeds. Additionally, goats have a unique preference for, and succeed in feeding on, a wide assortment of browse and forbs on which other species fail.

BROWSE, FORBS, AND GRASSES/LEGUMES

Browse refers to the edible parts of woody vegetation, such as leaves, stems, and twigs from bushes.

Forbs refers to nongrasslike range herbs which animals eat (forbs are commonly called weeds by western ranchers).

For using browse and forbs, goats are without a peer. Mohair and meat-type goats are used extensively to graze unimproved pastures and range areas where vegetation is generally of low quality. Since goats are good browsers, they can be used effectively to control brush and undergrowth. As a result, they have been exploited as "mobile pruning weapons" against encroaching browse and forbs in range areas.

Numerous types of shrubs and woody plants can be utilized as feed for goats with varying degrees of success. Table 21-6, Types of Brush Utilized by Goats, shows the relative feeding values of several types of brush that are commonly found in range areas.

TABLE 21-6
TYPES OF BRUSH UTILIZED BY GOATS[1]

Common Name	Scientific Name	Efficiency of Utilization
Black persimmon	*Discaria* spp	–
Catclaw	*Acacia gregii*	++
Cedar	*Juniper ashei; J. penchoti*	+
Coral bean	*Erythrina corrallodendron*	+
Elm	*Ulmus* spp	++
Guajillo	*Acacia berlandieri*	+++
Ill-scented sumac	*Rhus* spp	++
Mesquite	*Prosopis*	–
Oak, live	*Quercus virginicus*	+++
Post oak	*Quercus stellata*	++
Shin oak	*Quercus havardii*	++
Small-leaved sumac (red)	*Rhus glabra*	++
White brush	*Lippia liguestrina*	–
Wild plum	*Prunus* spp	+
Yaupon	*Ilex vomitoria*	++

[1]Adapted by the author from *Texas Angora Production*, Texas Ag. Exp. Sta. Bull., B-926.

[2]Excellent utilization = +++; Good = ++; Fair = +; Poor = –.

While goats can utilize a number of types of browse and forbs that other livestock refuse, poisonous plants must be avoided. Also, goat producers should be aware that many palatable browse species are limited in value because of one or more inhibitors that bind or otherwise prevent utilization of the nutrients contained in plants. Among such inhibitors are high levels of (1) lignin in woody twigs, which is practically indigestible; (2) essential oils (terpene-based organic compounds), which inhibit growth of rumen bacteria; and (3) tannins, which depress digestion by binding and/or inhibiting enzyme activity.

Forages can provide the vast bulk of the nutrients required for maintenance. Thus, the goat rancher should have a good knowledge of the feeding value of the forages available and supplement the forages when necessary. Generally, range forages are very low in phosphorus and salt, and often marginal in levels of vitamin A, calcium, and trace minerals. As forages mature, their nutrient value and digestibility decline, as shown in Table 21-7, which lists several range forages.

Good-quality pasture and a supply of minerals are all that are required to feed goats at maintenance levels. For the lactating doe, pasture can replace up to one-half of the concentrate in the ration. When pastures are short or when winter limits the availability of good, fresh grass, it is advisable to provide some supplemental feed. This may consist of whole corn, range cubes, or a salt-feed mixture to limit feed intake.

Improved pasture is a necessity for lactating dairy does. In order to prevent overgrazing, grass should be allowed to get 3 to 4 in. high before animals are allowed to graze. A good management practice with goats is to divide the pasture into lots and rotate the animals from the various lots every 10 to 14 days. An

Fig. 21-5. Goats are superb browsers. (Courtesy, J. C. Allen and Son)

TABLE 21-7
ENERGY, PROTEIN, AND PHOSPHORUS COMPOSITION OF VARIOUS RANGE PLANTS AS INFLUENCED BY STAGE OF MATURITY[1]

Plant and Stage of Growth	Energy			Protein			Phosphorus
	TDN	DE[2]		Crude	Digestibility	Digestible	
	(%)	(kcal/lb)	(kcal/kg)	(%)	(%)	(%)	(%)
Curly mesquite and buffalograss:							
Green growth	60	1,200	*2,646*	12.0	65	7.8	0.11
Partly cured forage	54	1,000	*2,205*	8.8	55	4.8	0.09
Cured forage	47	900	*1,984*	5.3	28	1.5	0.06
Gramas (blue, black, and side-oats):							
New growth	56	1,100	*2,426*	11.5	76	8.7	0.14
Fruiting	52	1,000	*2,205*	7.7	49	3.8	0.12
Mostly mature	49	1,000	*2,205*	6.4	39	2.5	0.09
Mature and weathered	40	800	*1,764*	3.5	0	0.0	0.05
Bluestems:							
Very young	68	1,400	*3,087*	14.5	66	9.5	0.14
Green growth	57	1,100	*2,426*	10.4	61	6.3	0.08
Past maturity	44	900	*1,984*	3.7	0	0.0	0.04
Three-awn (purple):							
Past maturity	49	900	*1,984*	5.5	31	1.7	0.06
Mixed cured and green	48	900	*1,984*	6.1	38	2.3	0.08
Rescue grass:							
Foliage, preheading	70	1,400	*3,087*	17.0	74	12.6	0.28
Plants in head	60	1,200	*2,646*	11.7	65	7.6	0.14
Texas wintergrass:							
Luxuriant green growth	65	1,300	*2,866*	14.4	70	10.1	0.10
Green growth	56	1,100	*2,426*	10.0	59	5.9	0.12
Green growth, mature plants	50	1,000	*2,205*	7.2	46	3.3	0.08
Live oak leaves:							
New foliage	55	1,100	*2,426*	17.7	31	5.4	0.26
Foliage, mostly mature	51	1,000	*2,205*	8.9	30	2.7	0.12
Foliage	50	1,000	*2,205*	9.6	30	2.9	0.10
Shin oak leaves	50	1,000	*2,205*	9.4	31	2.9	0.15
Various forms:							
Winter and spring	65	1,300	*2,866*	16.0	76	12.2	0.20
Summer and fall	45	900	*1,984*	18.0	79	14.2	0.15

[1]Adapted by the author from *Nutritional Requirements of the Angora Goat* by J. E. Huston, M. Shelton, and W. C. Ellis, Texas Ag. Exp. Sta. Bull. B-1105.
[2]Digestible energy (DE) in kilocalories per pound of dry matter was calculated on the basis that 1 lb of TDN is approximately equal to 2,000 kcal of DE.

electric fence provides an easy way of setting up pasture lots. Not only does this practice prevent overgrazing, it helps to break up the life cycle of internal parasites which can create health problems as well as reduce production. Some of the grasses and legumes that can be effectively used in pasture management for goats are alfalfa, alfalfa-brome mix, clover, clover and grass, timothy, and bluegrass. Since goats are ruminants, care should be exercised when fresh, lush legume pastures are first used, as bloat problems can result.

Temporary pasture can provide excellent forage for lactating dairy goats. Rye, wheat, and barley are excellent for early spring or late fall grazing. Sudangrass and millet are excellent summer pastures. Rape (canola) (or a combination of rape and oats) has been used with considerable success in cooler areas.

When on pasture, high-producing dairy goats should be properly supplemented with concentrates and minerals.

Green chop (fresh herbage cut and chopped in the field), instead of pasture, is sometimes fed to lactating

Fig. 21-6. All animals enjoy good pastures and these Angora nannies and kids are no exception. (Courtesy, Mohair Council of America, San Angelo, TX)

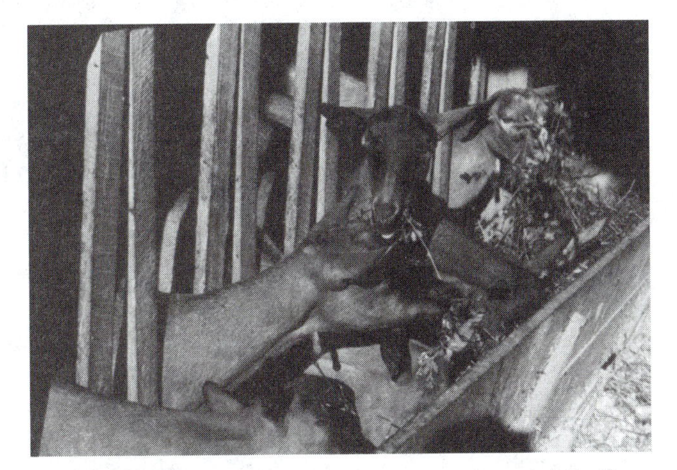

Fig. 21-7. French Alpine does consuming green-chopped alfalfa. (Courtesy, Dr. C. Lu, The American Dairy Goat Research Institute, Langston University, Langston, OK)

goats in confinement. However, this type of forage is labor intensive because the feed must be harvested daily. (Also see Chapter 7 for additional information on pasture management.)

HAYS AND OTHER DRY ROUGHAGES

With the exception of pasture and range feeds, hay and dry roughages are usually the most economical feeds for goats. A good-quality legume hay or a mixed legume and grass hay provide an excellent source of highly digestible nutrients. Mixed hays should be at least 50% legume, especially if hay is to be the primary source of feed. Grass hays require supplementation with concentrate. Except for dairy goats in lactation, it is not necessary to provide large amounts of concentrates to goats, especially if they are on maintenance or low production levels.

Hays with the highest nutritive values are those that have tender stems and are leafy. For this reason, hay from second cuttings are generally better utilized than first cuttings. Palatability of coarse, stemmy hay can be improved by crushing. The stage at which hay is cut has a direct influence on its feeding value. As the grass or legume matures, there is a steady decrease in crude protein content along with an increase in crude fiber content.

The following kinds of hays are most commonly fed to goats: alfalfa, alsike clover, red clover, ladino clover, soybean, vetch, birdsfoot trefoil, and mixed legume and grass.

In the South, cottonseed hulls are also a popular dry roughage for goats. They are bulky, containing about 43% fiber and 40% TDN.

Fig. 21-8. High-producing dairy goats need high-quality hay, preferably alfalfa or clover. (Courtesy, University of New Hampshire, Penacook)

SILAGES, HAYLAGES, AND ROOT CROPS

Silages and haylages have never been used extensively for feeding goats because of the following practical reasons: (1) their high water content makes it impractical to feed them at great distances, thereby alleviating their use on most Angora and Spanish (meat) goat operations; and (2) the small number of animals in most dairy goat operations makes it impossible to feed the top 3-4 in. of silage or haylage that must be removed from the silo daily to prevent spoilage.

Dairy goats are sometimes fed silage or haylage (1) where the herd is associated with a cattle operation, thereby making it practical to share silage with the goats; or (2) where a large commercial dairy goat herd is involved. About 3 lb of silage or 2 lb of haylage may replace 1 lb of hay.

Goats are quite fond of root crops and garden products; and these types of feeds can be effectively incorporated in the ration for a change of routine. Carrots, beets, turnips, and cabbage are especially relished by goats. These types of feeds are high in moisture and should be fed in the same manner as silage. Roots should be chopped in order to lessen choking.

In order to prevent off-flavors in milk, it is recommended that silage, haylage, and root crops such as turnips be fed either after milking or in amounts that will be consumed 3 to 4 hours prior to milking.

Fig. 21-9. Haylage feeding at Irish Hills Goat Farm, Tipton, Michigan, owned by Gary and Nancy Abner. (Courtesy, Dr. Christine S. D. Williams, Michigan State University, East Lansing)

CONCENTRATES

The concentrates used in goat rations can be classified as either energy feeds or protein supplements.

ENERGY FEEDS

The common energy ingredients are corn, oats, barley, milo, and wheat, along with their byproducts. The amount of energy feed to include in the ration should be determined by the production demands. A dry doe requires little or no supplementation, while a doe at the peak of lactation requires substantial amounts of energy.

Molasses, an excellent energy source, is commonly used to reduce the dustiness of feed and to increase palatability. If too much molasses is included in the ration, the feed becomes sticky and lumpy; so, it is usually limited to 5 to 8% of the mixture.

Fig. 21-10. Dairy goat eating a concentrate while on a stand ready for milking. (Courtesy, University of New Hampshire, Penacook)

PROTEIN FEEDS

A wide variety of protein supplements can be used in rations for goats. As is the case with most other species of livestock, the oil meals are used extensively. Cottonseed meal and soybean meal are probably the most widely used sources of protein for goats, but other meals can be and are used, depending on their respective prices and availability. Among the alternative sources are copra meal, peanut meal, sunflower meal, safflower meal, rapeseed (canola) meal, corn gluten feed, brewers' dried grains, and distillers' dried grains.

Urea and other nonprotein nitrogen (NPN) sources are often used in rations for goats, but several precautions should be taken when they are incorporated in the ration. Urea can constitute up to 1% by weight of the total concentrate mix, or supply one-third of the protein equivalent in the total ration. Rations containing NPN should be introduced very gradually in the feeding scheme of goats, as a period of adaptation is required by the microorganisms of the rumen. In addition to urea, other NPN sources are ammoniated cottonseed meal, ammoniated rice hulls, ammoniated citrus pulp, and ammoniated beet pulp.

MILK AND MILK REPLACERS

Unless extenuating circumstances prevail, newborn kids are seldom taken from Angora or Spanish goat mothers. However, with dairy goats, normally kids are allowed to nurse for 2 to 4 days only, or not at all. It

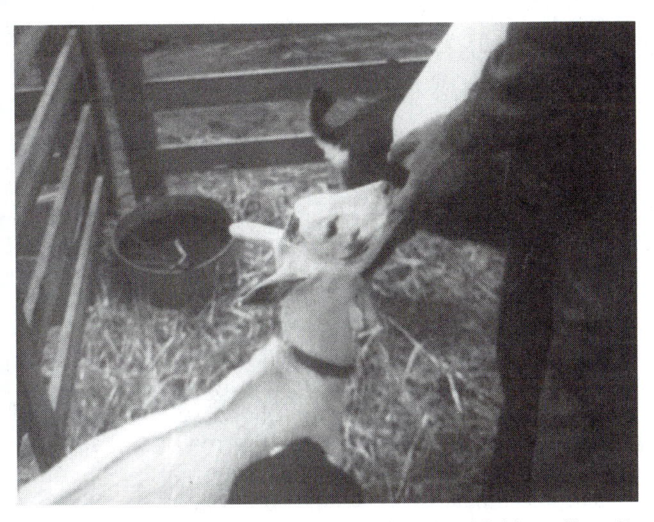

Fig. 21-11. Kid being bottle fed. (Courtesy, University of Delaware, Newark)

is usually easier to train kids to nipple- or pan-feeding if they have never nursed their mothers.

Recent research indicates that disease organisms, especially caprine arthritis-encephalitis (CAE), may pass from doe to kid through the milk, and that transmission may be avoided through the use of frozen colostrum from does tested and shown to be free of CAE. Colostrum from does that are known CAE carriers should be pasteurized to prevent transmission to newborn kids. (See section headed Caprine Arthritis-Encephalitis [CAE].)

When newborn kids are to be raised separately from their dams, they may be fed cow's milk or a milk replacer. The most desirable milk replacers for kids contain a minimum of 20% fat and 24% crude protein; are skimmed milk-based, rather than grain-based; and are fortified with minerals and vitamins. Because of varying formulations, care should be taken to follow the manufacturer's directions.

COMMERCIAL FEEDS

Commercial feeds, containing a variety of ingredients, including minerals and vitamins, are used by many goat producers. Special ingredients such as molasses and/or fat may be added to reduce dustiness and increase palatability.

Cottonseed hulls may be added to improve texture and provide fiber. In some cases, the commercial feeds are pelleted.

Commercial feeds may be available as (1) complete feeds (roughage and concentrate combined), (2) concentrates, or (3) protein supplements.

FEED PREPARATION

Grain can best be utilized by goats when it is processed to a limited extent. Cracking, rolling, crimping, flaking, or coarse grinding all aid in making grains more digestible. Grinding feed to a fine powder form is not desirable, especially in low-roughage rations, because it generally results in lowered palatability and digestibility. Large grains, such as corn, can be fed whole on the ground in range areas and effectively utilized with little wastage. When grains are to be mixed with hay and other ingredients or to be used on the range, pelleting is advisable.

In rangeland areas, feed supplements are sometimes offered free-choice, using salt as a governor to limit feed consumption. The proportion of salt to feed may vary from 5 to 40%. By varying the proportion of salt in the mixture, it is possible to hold the consumption of feed supplement to any level desired. The amount of salt should vary according to the level of feed consumption desired. A commonly used mixture consists of 1 part salt, 1 part cottonseed meal, and 3 parts grain. Iodized salt should never be used as a feed intake inhibitor because toxicity problems can result with the high intake of iodine. When high-salt mixes are fed, water should always be readily available.

Pelleting and cubing are used when roughages and grains are mixed together. Unless these feeds are pelleted or cubed, the various ingredients can separate out, resulting in waste and inefficient utilization of feed. Hay is sometimes chopped to reduce waste. It should be emphasized that any processing of feed creates additional feed costs; hence, feed should not be processed if the additional treatment does not increase feed efficiency or reduce waste sufficiently to offset the added cost.

When urea or other nonprotein nitrogen is to be included in the ration, the ingredients must be mixed thoroughly. It may be unwise to mix such rations in farm mixers unless (1) the urea is premixed with other ingredients to ensure its uniform distribution throughout the mix, and (2) some means is employed to prevent the urea from settling out with other fines. A molasses-urea mixture is often used to provide for these controls in mixers equipped to handle liquid molasses.

FEED SUBSTITUTION TABLE

Sometimes, goat producers have an opportunity to obtain a feed ingredient at a favorable price, but they may not know the relative feeding value of the product with respect to the feed currently in use. In order to assist producers in making these managerial decisions, a feed substitution table giving the comparative value of feeds is needed.

Feed substitutions for goats and sheep are similar.

NUTRITION RELATED DISORDERS

Goats are subject to several nutrition-related disorders, most of which occur in sheep, and many of which occur in other species. Abortion, caprine arthritis-encephalitis (CAE), enterotoxemia, and posthitis of goats are discussed in the sections that follow.

Fig. 21-12. Signs of good health, evidenced by Angora kids playing on pasture. (Courtesy, Mohair Council of America, San Angelo, TX)

ABORTION

Infectious diseases such as brucellosis are capable of causing abortions in goats. However, herein reference is made to a particular type of abortion caused by a metabolic disturbance of the functional corpus luteum, to which Angora goats are predisposed. Under normal production conditions, this malady commonly causes a low level of abortion, but catastrophic losses sometimes occur. Most abortions occur in response to stress between 90 and 110 days of gestation. Undernutrition during the critical stage of rapid fetal development and competition for nutrients between fetal and maternal organisms appear to be one explanation. It is noteworthy that the incidence of abortion is reduced in herds in which replacement does are fed for proper size and development prior to the first breeding season and during gestation.

CAPRINE ARTHRITIS-ENCEPHALITIS (CAE)

It is estimated that more than 80% of the dairy goats of the United States are infected by the retrovirus that causes CAE. In kids, it causes paralysis; in adults, it causes arthritis, which, in the late stages, is similar to rheumatoid arthritis in humans.

In most herds, the expression of the disease ranges from 0 to 25%. In those animals that do show clinical symptoms, the rate of progression and the severity of the disease varies markedly. In kids, the disease may vary from a barely noticeable unsteadiness of gait to a rapid fatal paralysis. In mature goats, the joints (front knees, hocks, and stifle joints) become swollen and disfigured, accompanied by a loss in body weight and a drop in production. The severity of the arthritis varies from years of intermittent lameness or stiffness to complete debilitation.

CAE is caused by a retrovirus, a slow acting virus that is latent, with the result that many animals may not show signs of the disease until after a long incubation period. The virus is transmitted from the doe to the kid(s) through the colostrum and milk. Does not showing symptoms of CAE may carry the virus and transmit it.

The rate of the infection in newborn goats can be reduced by more than 90% by (1) removing kids from infected does at the time they pass from the birth canal; (2) providing them colostrum that is from does identified as negative with the agar-gel immunodiffusion (AGID) test or that has been heated to 132°F for 1 hour; (3) raising them on pasteurized goat's milk or a milk replacer; and (4) keeping them in isolation from infected goats. The AGID test can be used to monitor infection.

Fig. 21-13. Several small housing units for dairy kids, placed side by side. Kids should be fed and cared for separate and apart from the milking herd in order to lessen exposure to disease. (Courtesy, Rocking M Ranch, Hilmar, CA)

ENTEROTOXEMIA (OVEREATING DISEASE)

Enterotoxemia, also known as *overeating disease* or *toxic indigestion*, may be the most insidious, most often undiagnosed, and, in many herds, the most important of all goat diseases. It is a toxic reaction to *Clostridium perfringens*, Types C and D, characterized by diarrhea, depression, lack of coordination, digestive upsets, coma, and death.

In baby kids, excess feeding or sudden access to palatable feed, changes in feed, or feeding following an unusual period of starvation, may cause acute enterotoxemia and death. Causative factors for enterotoxemia in mature goats include sudden changes in concentrate feed, excessive feeding of concentrates to animals not accustomed to such feeds, sudden access to highly palatable forage, sudden change to lush pasture, sudden access to feed, and overeating by hungry goats.

Prevention consists in proper feeding along with a vaccination program, using *Clostridium perfringens* toxoid, Types C and D. Initially, all goats should be given two separate doses of the toxoid, at 2-4 week intervals. Then, all does should be given an annual booster toxoid about 1 month before kidding; all kids should be vaccinated at 3-4 weeks of age, followed by a second dose of toxoid 2 weeks later. All goats in the herd, including bucks, should receive at least two doses of toxoid annually; one when does are in late pregnancy, and another when the kids are 4 to 5 months old.

POSTHITIS (SHEATH-ROT, PIZZLE-ROT, URINE SCALD)

Posthitis, a moderately contagious disease primarily of male goats and sheep, is characterized by ulcerative lesions with scab formation on the prepuce of males and less frequently on the vulval lips of females. It is caused by a high-protein ration in combination with the organism, *Corynebacterium renale*. The problem appears to be aggravated by confinement to areas where irritation or infection are more likely to occur. Posthitis may be serious in mature Angora wethers kept for mohair production; and it may also occur in individual breeding bucks kept in confinement. Reducing the protein intake tends to reduce the incidence and severity of the disease.

PART II—FEEDING DAIRY GOATS

A good dairy doe will average 5 lb of milk, or more, per day over a lactation period of 10 months, whereas superior animals will average 10 lb or more. Based on 200,000 official lactation records since 1968, average milk production per goat per lactation is 1,643 lb with 3.8% fat. The highest official milk production on record in the United States was made by a Saanen doe that

Fig. 21-14. With confinement housing, dairy goats need adequate feeding space, dry bedding, and good ventilation. (Courtesy, University of New Hampshire, Penacook)

Fig. 21-15. All-time, all-breed world record milk production holder. This Saanen doe produced 6,850 lb milk and 296 lb fat, in 305 days, in 1984. Bred by Gary and Sharon Swanson, Renton, Washington; owned by Gary Lee Cox, Eagle Point, Oregon. (Courtesy, T. H. Teh, Ph.D., Prairie View A&M University Research Center, Prairie View, TX)

produced 6,850 lb of milk in 305 days. The butterfat record is held by a Nubian doe that produced 384 lb in 305 days.

In comparison with cow's milk, goat's milk has a higher percentage of small fat globules, and a sweeter flavor. Goat's milk forms a fine, soft curd during digestion, thus making it more easily digested than cow's milk for some children and for older people who cannot tolerate cow's milk. If milked in clean quarters and away from the bucks, goat's milk will be free from any unpleasant flavor or odor, but when a buck is around the milking quarters, his strong odor is quickly absorbed by warm milk. The dairy goat produces milk that is very similar in composition to that of the dairy cow (see Table 21-8).

TABLE 21-8
COMPARISON OF THE COMPOSITION OF GOAT'S MILK AND COW'S MILK

Source	Water	Protein	Fat	Sugar	Ash
	(%)	(%)	(%)	(%)	(%)
Goat	86.0	4.4	4.6	4.2	0.8
Cow	87.7	3.3	3.6	4.6	0.7

FEEDING GUIDE

Table 21-9, Dairy Goat Feeding Guide, provides a good basic guide for the nutritional program of the dairy goat.

TABLE 21-9
DAIRY GOAT FEEDING GUIDE[1]

Age	Feed	Amount Each Day
Birth to 3 days	Colostrum.	All the kid wants 3 to 4 times daily.
3 days to 3 weeks	Whole milk (cow or goat, or milk replacer). Water, minerals.	2 to 3 pt *(0.9 to 1.4 liter)*, feed 3 times daily. All the kid wants.
3 weeks to 4 months	Whole milk. Creep/starter feed (1). Alfalfa hay (2). Water, minerals.	2 to 3 pt *(0.9 to 1.4 liter)*, up to 8 weeks, feed 2 times daily. All the kid will eat, up to 1 lb *(0.45 kg)* per day. All the kid will eat. All the kid wants.
4 months to freshening	Grain mixture/grower feed (3b). Alfalfa hay or pasture (2). Water, minerals.	Up to 1 lb *(0.45 kg)* of high-protein feed. All the doe will eat. All the doe wants.
Milking doe	Grain mixture (3a). Alfalfa hay (2). Water, minerals.	Minimum of 1 lb *(0.45 kg)* up to 2 qt *(1.89 liter)* of milk per day. Add 1 lb *(0.45 kg)* grain mixture for each additional 2 qt *(1.89 liter)* of milk. All the doe will eat. All the doe wants.
Dry pregnant	Grain mixture (3b). Hay/silage or pasture (2).	Up to 1 to 2 lb *(0.45 to 0.91 kg)* mix for a dry animal. All the doe will eat.
Bucks	Roughages. Grain. Water, minerals.	All the buck wants. 1 to 2 lb *(0.45 to 0.91 kg)* during breeding season if needed. Free-choice.

(1) Creep or starter feed (see Table 21-13 for starter ration).
(2) Alfalfa hay of extremely high quality, fine stemmed, leafy, and green.
(3) Suggested grain mixtures:
 (a) For a lactating doe (or select a grain mix from Table 21-11)—
  55 lb *(25.0 kg)* barley or oats
  15 lb *(6.8 kg)* beet pulp
  20 lb *(9.1 kg)* wheat, mixed feed, or mill run
  10 lb *(4.5 kg)* linseed, cottonseed, or soybean oil meal
 (b) For a growing or a dry doe (or select growing ration from Table 21-13)—
  15 lb *(6.8 kg)* beet pulp
  50 lb *(22.7 kg)* barley or oats
  15 lb *(6.8 kg)* wheat, mixed feed, or mill run
  20 lb *(9.1 kg)* linseed, cottonseed, or soybean oil meal
If you use commercial dairy cow or dairy goat feed, use it according to your goat's stage of growth—growing, drying, or lactating.

[1]Adapted by the author from *Your Dairy Goat*, Western Regional Extension Publication, WREP 47, University of California, Davis.

The dairy goat producer must combine the available feeds so as to achieve the most profitable production. At its best, developing a dairy goat feeding program involves combining the art and the science of feeding. Also, there should be a ration for each specific need—for lactating does, for dry does, for kids, for yearlings and replacements, and for bucks.

FEEDING LACTATING DAIRY DOES

Following parturition, the feed intake should be increased gradually.

The nutritional demands upon lactating does are tremendous. It is essentially impossible for the doe to consume enough to meet the demands for body maintenance and milk production during the first few months of lactation; so, she must draw upon her body reserves to augment the nutrients consumed. Based on the National Research Council recommendations (see Tables 21-1 and 21-2), Table 21-10 shows the combined requirements for maintenance and milk production at various levels for dairy goats of three different sizes producing 4% milk. Note that provision is made for (1) five different levels of milk production, ranging from 2.5 to 20 lb daily; (2) three different body weights—130, 160, and 190 lb; (3) TDN and net energy, (4) crude protein, and (5) calcium and phosphorus. As shown in Table 21-10, a dairy goat weighing 160 lb and producing 10 lb of 4% milk daily would need 5.5 lb of TDN daily (5 Mcal of net energy), 1.0 lb of crude protein, 20 g of calcium, and 14 g of phosphorus.

Fig. 21-16. Alpine does at feed at The Coach Dairy Goat Farm, Pine Plains, New York. (Courtesy, The Coach Dairy Goat Farm, Pine Plains, NY)

In order to formulate a ration to meet the requirements given in Table 21-10, available feeds and feed compositions must be known (see Chapter 27, Feed Composition Tables). Also, consideration must be given to dry matter intake (DMI)—the ability of the animal to consume. Total DMI becomes a critical factor in balancing a ration because the total combination of daily forage and concentrate consumption must (1) come within this range, and (2) meet nutrient requirements. So, the lower the quality of the roughage, the higher the nutrient density of the concentrate to accomplish this objective. Experiments and experiences have shown that the daily DMI of lactating does usually ranges from 4 to 5% of their body weight. The main goal is to meet the nutrient needs at minimal cost and yet maximize production.

TABLE 21-10
COMBINED REQUIREMENTS FOR MAINTENANCE AND MILK PRODUCTION AT VARIOUS LEVELS FOR DAIRY GOATS AT THREE DIFFERENT SIZES PRODUCING 4% MILK FAT

Daily Milk		Body Weight		TDN		Net Energy	Crude Protein		Ca	P
(lb)	*(kg)*	*(lb)*	*(kg)*	*(lb)*	*(kg)*	*(Mcal)*	*(lb)*	*(kg)*	*(g)*	*(g)*
2.5	*1.1*	130	*59.0*	2.8	*1.3*	2.4	0.42	*0.19*	8	5
		160	*72.6*	3.0	*1.4*	2.7	0.45	*0.20*	9	6
		190	*86.3*	3.5	*1.6*	2.9	0.50	*0.23*	10	6
5	*2.3*	130	*59.0*	3.4	*1.5*	3.2	0.60	*0.27*	12	8
		160	*72.6*	3.8	*1.7*	3.4	0.65	*0.30*	13	9
		190	*86.3*	4.0	*1.8*	3.7	0.70	*0.32*	14	10
10	*4.5*	130	*59.0*	5.2	*2.4*	4.8	1.95	*0.89*	19	13
		160	*72.6*	5.5	*2.5*	5.0	1.00	*0.45*	20	14
		190	*86.3*	5.8	*2.6*	5.3	1.04	*0.47*	21	15
15	*6.8*	130	*59.0*	6.9	*3.1*	6.5	1.31	*0.59*	26	18
		160	*72.6*	7.2	*3.3*	7.0	1.35	*0.61*	27	19
		190	*86.3*	7.5	*3.4*	7.4	1.40	*0.64*	28	20
20	*9.1*	130	*59.0*	8.6	*3.9*	7.9	1.67	*0.76*	33	23
		160	*72.6*	8.9	*4.0*	8.2	1.71	*0.78*	34	24
		190	*86.3*	9.2	*4.2*	8.5	1.76	*0.80*	35	25

Some suggested rations for lactating dairy goats are given in Table 21-11. These rations meet the requirements set forth in Table 21-10. Note that when a legume hay (alfalfa or clover) is fed, a 12–14% crude protein grain mix will suffice. But, when a grass hay is fed, a 16–18% crude protein grain mix should be fed.

FEEDING DRY DAIRY DOES

It is important to dry off dairy goats about 6 to 8 weeks prior to kidding. This gives the doe a brief rest from the heavy demands of lactation, enables her to meet the nutrient needs of the rapidly growing fetus, and allows her to build body reserves with which to meet the rigorous requirements of lactation which follow.

Depending on the kind and quality of the forage and the size and condition of the doe, 1 to 2 lb of 12 to 16% protein concentrate ration should be fed during the dry period. Trace-mineralized salt and water should also be available.

In the last 6 weeks of gestation, the fetus gains 70% of its birth weight, so the nutrition of the doe during this period is critical.

Some suggested rations for dry does are given in Table 21-12.

TABLE 21-11
RATIONS FOR LACTATING DAIRY GOATS (160-LB BODY WEIGHT)[1]

Feedstuffs	Daily Milk Production (lb)							
	2.5		5.0		10.0		15.0	
	Daily Feed							
	(lb)	(kg)	(lb)	(kg)	(lb)	(kg)	(lb)	(kg)
Ration #1:								
Alfalfa clover hay (16% CP)	2.0	*0.9*	3.0	*1.4*	3.5	*1.6*	4.5	*2.0*
Grain mix to be selected from 4 mixes listed below (14–16% CP, 70% TDN)	3.0	*1.4*	4.0	*1.8*	6.0	*2.7*	8.0	*3.6*
Ration #2:								
Grass hay (7% CP)	2.5	*1.1*	2.5	*1.1*	3.0	*1.4*	4.0	*1.8*
Grain mix to be selected from 4 mixes listed below 16–18% CP, 65% TDN)	3.4	*1.5*	4.6	*2.1*	7.0	*3.2*	9.0	*4.1*
Grain mixes:								
Ration	#1		#2		#3		#4	
Level of crude protein (CP) in grain mix	14%		16%		18%		20%	
Lb or kg per ton	(lb)	(kg)	(lb)	(kg)	(lb)	(kg)	(lb)	(kg)
Ingredients								
Rolled corn	900	*409.0*	800	*363.6*	720	*327.3*	656	*298.2*
Crimped oats	421	*191.0*	300	*136.0*	240	*109.1*	200	*91.0*
Beet-citrus pulp	200	*91.0*	200	*91.0*	200	*91.0*	200	*91.0*
Dried brewers' grains	—	—	150	*68.0*	200	*91.0*	200	*91.0*
40% protein supplement	300	*136.0*	—	—	516	*234.5*	—	—
Soybean meal	—	—	356	*161.8*	—	—	600	*272.7*
Molasses	150	*68.0*	150	*68.0*	100	*45.5*	100	*45.5*
Trace mineralized salt	10	*4.5*	20	*9.1*	10	*4.5*	20	*9.1*
Dicalcium phosphate	—	—	10	*4.5*	10	*4.5*	20	*9.1*
Monosodium phosphate	15	*7.0*	10	*4.5*	—	—	—	—
Magnesium oxide	4	*1.8*	4	*1.8*	4	*1.8*	4	*1.8*
Vitamins[2]								

[1]Adapted by the author from *Extension Goat Handbook*, "Feeding," by R. S. Adams, Pennsylvania State University; B. Harris, University of Florida; M. F. Hutjens, University of Illinois; and E. T. Oleskie and F. A. Wright, Rutgers University.

[2]During winter, all mixtures should be supplemented with 6 million IU of vitamin A and 3 million IU of vitamin D per ton of grain mix.

TABLE 21-12
RATIONS FOR PREGNANT OR DRY DOES

Ration	Ingredients	Amount	
		(lb)	*(kg)*
1	Pasture plus good mixed grass/legume hay	*ad libitum*	
	16% protein supplement.	1.1	*0.5*
2	Mixed hay	1.1	*0.5*
	Silage	1.1	*0.5*
	Beet pulp	0.7	*0.3*
	16% protein supplement.	1.1	*0.5*
3	Alfalfa hay	1.1	*0.5*
	Beets	2.2	*1.0*
	Beet pulp	1.1	*0.5*
	16% protein supplement.	1.1	*0.5*

FEEDING DAIRY KIDS

It is important to get 2 to 4 oz of colostrum in a kid as quickly as possible after birth.[1] Colostrum contains higher levels of total protein, milk solids, globulins, fat, and vitamin A than normal milk. It is also a laxative. Most important, it contains antibodies against disease to which the doe has immunity. Young kids are able to absorb this antibody protection effectively at birth, but by the time they are three days old, this ability almost disappears.

Fig. 21-17. Dairy kids at a nipple bar, an efficient method of feeding. (Courtesy, University of New Hampshire, Penacook)

[1]Because of the hazard of caprine arthritis-encephalitis (CAE) virus, scientists recommend that kids be fed (a) colostrum only from does negative to the CAE test, (b) heat-treated colostrum (132°F for 1 hour), or (c) no colostrum at all. (See section headed Caprine Arthritis-Encephalitis [CAE].)

During the first two days of life, kids should receive at least three colostrum feedings per day. A kid will consume 1½ to 2 pt daily.

If the kid is to be raised without its mother's milk, it should be allowed to nurse for 2 or 3 days only or not at all. Most commercial dairy goat producers favor putting the newborn kid directly on bottle- or pan-feeding without allowing it to nurse its mother; after a kid nurses its mother a couple of times, it is very difficult to get it to accept a bottle or a pan. Initially, the nipple bottle is generally preferred to pan feeding as it tends to be more natural, thereby resulting in less ingested air. However, as soon as kids are strong enough and can drink milk easily, experienced dairy goat producers prefer to train them to bucket or pan feeding, which is faster and allows for easier cleaning and maintenance of utensils.

After the kid is removed from colostrum feeding, it can be fed cow's milk or a milk replacer, along with a starter ration, with the change made gradually. The following points are pertinent to feeding and managing kids:

1. **Be clean and sanitary.** Wash and sanitize any feeding utensils which will be used for feeding.
2. **Prepare the milk or milk replacer properly.** The milk or milk replacer should be heated to about 103°F for feeding. A young kid will consume about 2 to 3 pt of milk daily. During the first 3 days, feed should be offered 3 to 4 times daily; thereafter, twice a day feeding is adequate.
3. **Think economy.** If cow's milk is cheaper than goat's milk, use it. It is also possible to prepare a suitable kid starter on the farm by mixing 30 lb each of cornmeal, ground oats, and wheat bran, plus 10 lb of an oil meal.
4. **Get kids on dry starter as soon as possible.** At about one week of age, the starter can be made available to kids. Also, fine-quality hay should be fed in order to enhance the development of the rumen.
5. **Guard against overfeeding or underfeeding.** Feeding too much or too little should be avoided.
6. **Weaning.** Kids can be weaned from milk as early as 5 to 6 weeks of age, although most goat breeders delay it until 3 to 4 months of age. As young kids approach weaning age, gradually add warm water to their milk diet. This will provide them with the necessary fluids for rumen development and ease the stress of weaning them. After the kids are weaned from the milk, feed them all the bright green forage they will eat, plus ¾ to 1 lb of any good grower ration.

Suggested starter and grower rations for kids are given in Table 21-13. *Note well:* Kids should receive colostrum for the first three days (see Table 21-9).

TABLE 21-13
STARTER AND GROWING RATIONS FOR DAIRY KIDS[1]

	Kid Starter[2]	Growing Ration[3]
	(%)	(%)
Corn	27.6	12.9
Crimped oats	37.9	10.0
Soybean meal (44%)	10.0	8.6
Alfalfa leaf meal	18.0	10.0
Cane molasses	5.0	5.0
Cottonseed hulls	—	51.9
Trace mineralized salt	1.0	1.0
Limestone	0.3	0.4
Vitamins A, D, and E (premix)	0.2	0.2

[1]From: *Raising Goat Kids*, by T. H. Teh *et al.*, Texas A&M, International Dairy Goat Research Center, Prairie View A&M Research Center, Prairie View, Texas, Vol. A1, No. 1.

[2]On a dry matter basis, the kid starter should contain a minimum of 80% TDN, 16% protein, 0.6% calcium, and 0.4% phosphorus. Also see Table 21-9.

[3]The grower ration may be fed free-choice after four months.

FEEDING DAIRY YEARLINGS AND REPLACEMENTS

In feeding young animals, the object is to provide enough nourishment for body maintenance and growth. But too much feed, especially concentrates, causes animals to fatten, which may lead to difficulties in breeding. Beginning at 4 to 6 months of age, animals should have a good pasture (if available), high-quality hay, and a place to exercise. If the forage is good, ½ lb of grain per day should lead to ample growth. If the forage is poor, animals may require 1 to 1½ lb of grain daily. Yearlings can be fed the same grain mix that is given to the lactating does (Table 21-11). Low-quality forages should be supplemented with a 14–16% protein grain mixture. A mineral mixture of equal parts of trace mineralized salt and dicalcium phosphate is suitable for free-choice feeding.

A suggested grower ration for weaned kids and yearlings is given in Table 21-13.

FEEDING DAIRY BUCKS

Since the buck is larger than the doe, he will require more forage. Additionally, he should receive 1 to 2 lb of grain daily. Overweight bucks make inefficient breeders, so it is important to have the buck in good physical condition; but he should not be fat or have excessive fleshing. As breeding season approaches, the amount of grain should be increased to help the buck prepare for this heavy production period. Exercise is necessary to keep the herd buck healthy and sexually active. A 1:1 mixture of salt and dicalcium phosphate

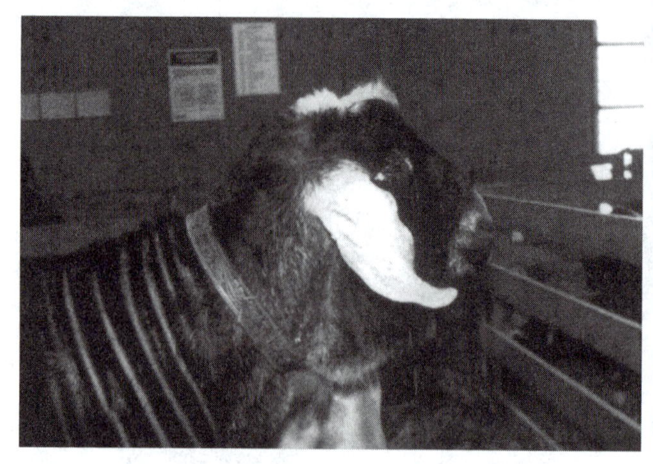

Fig. 21-18. Nubian buck. (Courtesy, University of Delaware, Newark)

should be available free-choice, as well as an abundant supply of water.

CAUTION: In bucks, the incidence of urinary calculi is increased by a high concentrate ration in which there is a mineral imbalance involving excessive phosphorus intake.

A suggested ration for bucks is given in Table 21-14.

TABLE 21-14
SUGGESTED RATIONS FOR BREEDING BUCKS

	Ingredients	Amount	
		(lb)	(kg)
Bucks (breeding)			
Out of breeding season	Good hays and pasture	*ad libitum*	
In breeding season	Good hays and pasture plus	*ad libitum*	
	14% protein supplement plus	1.1–2.2	*0.5–1.0*
	mineral supplements and salt		

FEEDING FOR SHOW

For success in the show-ring, an animal should be fed and managed to attain maximum development in body conformation. In general, the feeding of dairy show goats differs from normal operations only in that greater effort and expense and more liberal feed allowance may be justified in order to produce a winner. In addition to being well balanced, the ration for show goats must be palatable. No definite set of rules relative to feed allowances can be followed satisfactorily. Rather, the judgment of the skillful feeder must prevail.

Fig. 21-19. Dairy goat properly fitted and shown by a 4-H Club member. (Courtesy, University of New Hampshire, Penacook)

PART III—FEEDING ANGORA AND MEAT (SPANISH) GOATS

The basic nutrient requirements established by the National Research Council (NRC) are suited for both Angora and meat goats (see Table 21-1). These requirements are based upon the weight of the animals being fed, the level of activity, the stage of pregnancy, and the growth rate. The type of range generally governs the activity of goats. For example, goats that range on sparsely vegetated grassland and on seasonal mountainous ranges must travel long distances daily for grazing and watering; whereas, slightly less activity is required by goats that graze on semiarid rangeland or on slightly hilly land. For mohair production, nutrient requirements in excess of those in Table 21-1 are needed. These are provided in Table 21-3 according to the level of mohair production. Furthermore, the nutrient requirements of lactating does as given in Table 21-2 must also be considered for goats nursing kid(s). Therefore, based on Tables 21-1 and 21-3, a 44-lb Angora goat kid gaining 0.22 lb per day, having medium activity, and producing 4.4 lb of mohair per year has the following requirements for energy and protein:

	Energy	Total Protein
	(Mcal DE/day)	*(g/day)*
Maintenance (medium activity)	1.77	55
Growth	0.88	28
Mohair	0.07	9
Total	2.72	92

The feed recommendations listed herein should be considered to be minimum levels. Thus, goat producers should adapt the recommendations to fit the particular operation; and, in many cases, it is advisable to use higher levels than are listed.

Since Angora and Spanish goats characteristically travel across large areas, eating the available browse (shrubs and woody plants), they have been routinely used to control brush on ranges. Brush, such as live oak, can be cut to leave a stump about 3 ft high; and the sprouts will provide additional feed. In order to maximize the production potential of range areas, goats are often placed in areas where other types of livestock are being grazed. Goats preferentially consume a certain amount of browse. Cattle prefer grasses, whereas sheep readily consume forbs (weeds) in rangelands. By combining several types of

Fig. 21-20. Angoras on typical range on a west Texas ranch. (Courtesy, *Sheep & Goat Raiser*)

livestock, the producer can utilize these three types of plants and increase the gain per unit of land as well as manage the vegetative makeup of the land.

Available browse and forages will satisfy many of the nutritive needs of goats that are raised on ranges, but, for maximum performance, it is advisable to provide supplemental feed when range conditions become adverse. Twenty percent protein range cubes are a popular supplement. Also, shelled corn can be fed on the ground to goats, with very little waste. Usually, about ¼ to 1 lb of supplement per head per day is adequate in the winter or during dry periods when green feed is scarce. Some examples of concentrate supplements for range goats are found in Table 21-15.

TABLE 21-15
CONCENTRATE SUPPLEMENTS FOR RANGE GOATS[1]

Ingredients	Supplement A	Supplement B	Supplement C
	20% Protein	30% Protein	40% Protein
	(%)	(%)	(%)
Corn or sorghum	82	58	25
Cottonseed meal or soybean meal	14	37	70
Urea	2	3	3
Dicalcium phosphate	2	2	2
Vitamin A supplement	—[2]	—[2]	—[2]
Approximate nutrient composition			
Energy:			
TDN (%)	75	72	70
DE (Mcal/lb)	1.50	1.45	1.40
Protein:			
Crude (%)	20	30	40
Digestible (%)	16	24	32
Phosphorus (%)	0.55	0.65	0.77

[1]Adapted by the author from *Nutrient Requirements of the Angora Goat* by J. E. Huston, M. Shelton, and W. C. Ellis, Texas Ag. Exp. Sta. Bull. B-1105.

[2]Sufficient supplement to provide 2,500 IU of vitamin A per lb of feed (5 million IU/ton of total concentrate supplement mix).

A summary of the recommended schedule for providing supplemental feed to Angora goats on average range is given in Table 21-16. If the range is extremely poor, the producer should refer to Table 21-17.

Note well: The same rations and feeding practices presented herein for Angora goats are suited for meat goats, also. However, supplemental feeding of meat goats is not the norm. Yet, meat goats do respond well to supplemental feeding during critical periods—just before breeding, just before and following kidding, and when range feed is short.

TABLE 21-16
SCHEDULE FOR SUPPLYING SUPPLEMENTAL FEED TO ANGORA GOATS ON AVERAGE RANGE[1]

Class & Weight	Period	Type of Supplement[2]	Lb/Day/ 100 Head
Wethers & dry does:			
50–80 lb	July 15 – Nov. 15	A	20
	Nov. 15 – Mar. 15	B	40
Above 80 lb	Nov. 15 – Mar. 15	C	20
Breeding does:			
50–80 lb	July 15 – Nov. 15	A	20
	Nov. 15 – Mar. 15	B	32–50
Above 80 lb	July 15 – Nov. 15	A	5–10
	Nov. 15 – Mar. 15	C	20–30
Growing kids and yearlings:			
Below 40 lb	July 15 – Nov. 15	A	35
	Nov. 15 – Mar. 15	B	40
Above 40 lb	July 15 – Nov. 15	A	20
	Nov. 15 – Mar. 15	B	40
Developing billies:			
80–120 lb	July 15 – Nov. 15	A	40
	Nov. 15 – Mar. 15	B	50

[1]Adapted by the author from *Nutrient Requirements of the Angora Goat* by J. E. Huston, M. Shelton, and W. C. Ellis, Texas Ag. Exp. Sta. Bull. B-1105.

[2]Compositions of the recommended supplements are listed in Table 21-15.

Fig. 21-21. Typical Boer (meat) goats. (Courtesy, R. M. Robinson, Kerrville, TX)

TABLE 21-17
FEEDING REGIMEN FOR ANGORA GOATS ON EXTREMELY POOR RANGE[1]

Class	Hay Plus Concentrate			Concentrates Only			
	Hay	Concentrate[2]		Supplement[3]		Grain	
	(% body wt.)	(lb/head/day)	(kg/head/day)	(lb/head/day)	(kg/head/day)	(lb/head/day)	(kg/head/day)
Wethers and dry does	1	0.75	0.34	0.3	0.14	1.0	0.45
Breeding does:							
Dry	1	0.75	0.34	0.3	0.14	1.0	0.45
Pregnant	1	1.50	0.68	0.3	0.14	1.6	0.73
Lactating	1	1.50	0.68	0.3	0.14	2.0	0.91
Growing kids and yearlings	1	1.25	0.57	0.3	0.14	1.0	0.45
Developing billies	1	1.25	0.79	0.2	0.09	1.5	0.68

[1]Adapted by the author from *Nutrient Requirements of the Angora Goat* by J. E. Huston, M. Shelton, and W. C. Ellis, Texas Ag. Exp. Sta. Bull. B-1105.

[2]Type of concentrate depends on hay fed. If alfalfa (or other quality legume hay) is fed, corn is adequate. If grass hay is used, a supplement similar to Supplement A (Table 21-15) should be fed.

[3]Supplement should be similar to Supplement C (Table 21-15).

FEEDING DOES

On good range in spring, mature dry does will consume enough feed to satisfy all their nutrient demands except salt and phosphorus. During lactation, they may need ½ to ¾ lb of a supplement of type A as given in Table 21-15. (The total ration, supplement plus range grass, will usually run about 11 to 12% protein.)

During the summer and early fall, the quality of range feed is reduced and a higher protein supplement, such as B (Table 21-15), should be provided at the rate of 1 lb up to 10 does. It may also be advisable to include some trace mineralized salt. Immature does (yearlings and those not fully developed) should be provided 1 lb of supplement for each 5 does. If the range is exceptionally poor, double the amount of supplement.

In late fall and winter, ranges tend to be at their lowest nutritive value. Poor ranges require supplemental feeding—supplement C, Table 21-15—at levels of 1 lb for each 3 to 5 mature does and 1 lb for each 1 to 3 yearling or underdeveloped does. These supplements should take care of the needs of late pregnancy and of lactation. For ranges with new growth, supplements should follow the recommendations given previously for good spring range.

Does with more than two kids should be given 25 to 50% more of the supplement during lactation than is recommended for does with singles or twins. Does which have a history of giving birth to more than one kid should be fed at a high rate of supplementation at least three weeks before expected parturition (kidding).

Additional daily supplementation of 0.25 to 0.33 lb of grain or range cubes should be fed to does 1 to 2 weeks prior to turning the bucks in for the breeding season. This added feed (called flushing) improves conception rate by having does in a positive nutrient balance during breeding. When the practice of flushing is to be used, feed should be increased gradually. Likewise, at the end of the breeding season, the feed allowance should be reduced gradually, so as to avoid upsetting the appetite and the digestive tract.

■ **Supplemental feeding with salt as governor**—In large, rough, and sometimes brushy pastures, hand feeding is impractical, so grain and meal mixtures with salt as an inhibitor are fed in self-feeders. The proportion of concentrates to salt varies with the amount of feed the goats should consume; usually it is 3:1, 4:1, or 5:1, with the lower amounts of salt allowing the higher feed intake.

■ **Supplemental feeding of Spanish does**—Although Spanish goats are hardy and will survive on poor brushy range with little or no supplement, the better operators provide a supplemental feed to does of 0.25 to 0.5 lb of cottonseed cake or 0.33 to 0.67 lb of yellow corn per head per day during the following critical periods: (1) beginning three weeks prior to breeding (for flushing), (2) beginning three weeks prior to kidding, (3) when kids are very young, and (4) when feed is short.

FEEDING KIDS (NURSING AND WEANED)

As long as kids are receiving adequate amounts of milk from their mothers, they do very well on good range. Additional supplementation, however, makes for more rapid growth and better prepares them for market or breeding. One pound of supplement for each 2½ to 3 kids should be provided. Older and larger kids may have their supplement reduced to 1 lb daily for each five

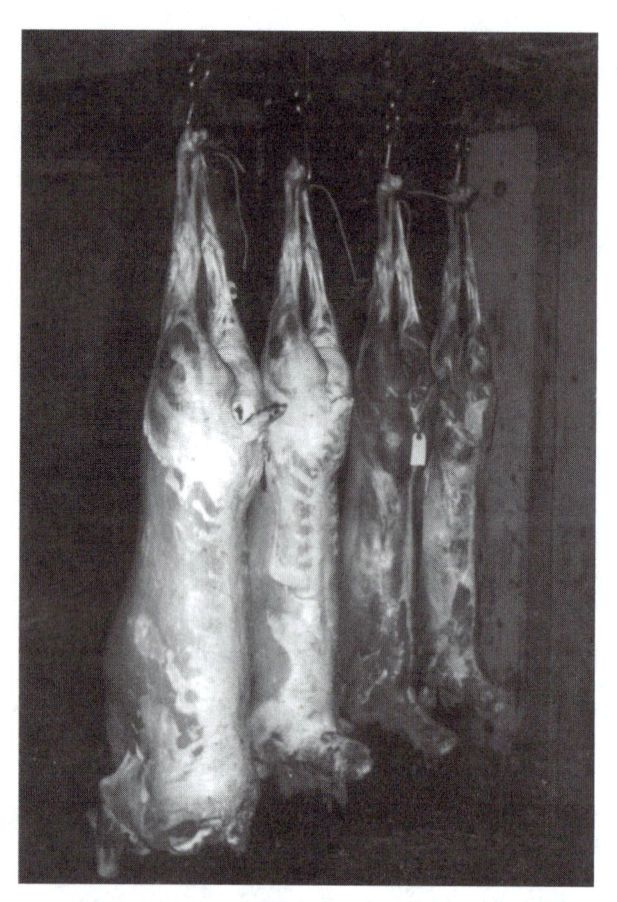

Fig. 21-22. Two lamb carcasses *(left)* and two Spanish goat carcasses (known as chevon) in a meat packing plant in San Angelo, Texas. Note that the goat carcasses have less external fat. Supplemental feeding of kids makes for more rapid growth and better prepares for market. (Courtesy, Texas A&M University, College Station)

kids if the range conditions are good. When range is poor, the grain and supplement should be increased to provide 1.0 to 1.33 lb of grain daily per kid. In addition, kids should have access to good-quality hay.

A suitable ration for kids and yearlings is given in Table 21-18.

TABLE 21-18
RATION FOR ANGORA AND SPANISH KIDS AND YEARLINGS

Age	Ingredients	Amount
		(%)
Kids/yearlings	Alfalfa hay	32
	Cottonseed hulls	28
	Sorghum grain	18
	Barley grain	8
	Molasses	6
	Cottonseed oilmeal	6
	Salt/mineral mix	2
	Total	100

FEEDING YEARLINGS AND REPLACEMENTS

Yearling range goats fit into one of two categories: (1) those which are being retained for the breeding herd, or (2) those which are being prepared for market. Replacement does and bucks should be fed rations that allow for growth. Those which are being prepared for market should be fed so as to put on more flesh, because goats tend to be leaner than most meat animals.

To keep a replacement animal growing, it is advisable to provide good-quality forage with a minimum amount of grains in the supplement. They should be given adequate protein, minerals, and vitamin A. Playfulness is a sign of good goat health and vigor. When fed properly, they will be large enough to breed as yearlings, without interfering with their continued growth to mature size.

FEEDING WETHERS AND DRY DOES

Under present conditions, Angora wethers are kept for several years for mohair production. Hence, they are rarely fed for meat production. For those goats being prepared for market, supplemental grain should be increased gradually. If kids are to be marketed at weaning (4 to 5 months of age), it is unnecessary to castrate them. If they are to be held over until yearlings, they can remain uncastrated until about 6 to 9 months of age.

Both wethers (or muttons, as they are sometimes called) and dry does do well without supplemental feed unless the range is poor. Angora does and wethers are generally not allowed to run together. Wethers can graze higher on brush and more vigorously than does, thereby reducing the available browse for does that have a more critical need for nutrients. When supplementation is needed, dry does and wethers are fed about the same amount of grain or supplement—1 lb for each five animals during the summer and fall months and twice this amount during the winter. On poor range, the amount may be increased to 0.75 lb per head daily when hay is also fed and to 1.33 lb per head daily when no hay is fed.

FEEDING BUCKS

Young bucks being raised for breeding purposes should be grown in much the same manner as replacement does, except that they will require more feed because their growth is more rapid. They should be separated from the does soon after weaning. If young bucks are permitted to remain with the does, early sexual maturity in the bucks may result in their breeding young females not intended to be bred. Bucks should be growthy, but they should not be permitted to put on extra fat which may impede their muscular, skeletal, and sexual development. They may be used lightly as yearlings when properly grown out; but they should not be used if they have been on sparse feed and are not prepared for the stresses of heavy breeding.

Whether the bucks are young or mature, it is advisable to feed them 0.33 to 0.5 lb of grain or supplement daily per head for a week or so before turning them in with the does, with the supplementation continued throughout the breeding season. If the bucks get too fat or become inactive in their mating habits, grain can be withdrawn as one means of improving their effectiveness in breeding.

Bucks that are not in active service do not need supplemental grain or concentrates unless the range feed becomes too sparse or too mature. In the latter case, they may be given about 1 lb of supplement for each 2 to 2½ animals daily. This type of supplementation is especially important for young bucks that are still growing. Some of the older bucks can get along well without supplementation unless the range is extremely poor.

Fig. 21-23. An outstanding Angora buck. (Courtesy, Texas A&M University, College Station)

QUESTIONS FOR STUDY AND DISCUSSION

1. Can experimental work and instruction relative to goats be justified on the basis of their contributions in terms of animal numbers, meat, milk, and fiber; (a) worldwide, and (b) in the United States?

2. Name and describe the three major types of goats, and tell how they are similar and how they are different from the standpoint of nutrition/feeding.

3. Discuss the economic importance of feed for goats.

4. Angora goats are supplementally fed on a limited basis whereas meat goats are seldom supplemented at all. Does this mean that it would not pay to supplement meat goats during critical periods—just before breeding, before and after kidding, and during feed scarcity?

5. Since cattle, sheep, and goats are all ruminants, why go to the trouble and expense of evolving with separate nutritional requirements for goats?

6. Why did the National Research Council (NRC) evolve with different nutrient requirements of goats for maintenance, various levels of activity, late pregnancy, growth, milk production, and mohair?

7. What are the signs of energy deficiency?

8. Discuss microbial synthesis in goats, and explain how it makes it possible for goats to utilize both protein and nonprotein nitrogen.

9. Discuss what happens to goats when they are fed protein-deficient rations.

10. Give the functions, deficiency symptoms, and practical sources of each of the following minerals for goats: salt, calcium, phosphorus, cobalt, iodine, and selenium.

11. Give the functions, deficiency symptoms, and practical sources of each of the following vitamins for goats: vitamin A, vitamin D, and vitamin E.

12. What are the important ways of meeting the water requirements of goats, and what are the major avenues of water losses from the body?

13. Define (a) browse, and (b) forbs. If goats can effectively use grasses and legumes, why provide them with browse and forbs?

14. What are the advantages to dividing goat pastures into lots and rotating their use every 10 to 14 days?

15. Discuss the effects of stage of maturity and cutting on hay for goats.

16. Why haven't silages and haylages been used extensively for goats?

17. What precautions should be taken when feeding silage and root crops to dairy goats?

18. How much molasses can safely be incorporated into a goat ration? Why is it not advisable to incorporate more?

19. What is the maximum recommended amount of nonprotein nitrogen that can be added to goat feeds?

20. What's a milk replacer? How and why may a milk replacer be used for dairy kids?

21. What determines the concentration of salt in rations that are to be offered free-choice to goats?

22. Discuss the cause and control/prevention of each of the following nutrition related disorders: abortion, caprine arthritis-encephalitis, enterotoxemia, and posthitis.

23. Outline a dairy goat feeding guide.

24. Discuss the nutritional demands upon lactating dairy does, and tell how they may be met.

25. What are the reasons for drying off dairy does six weeks before kidding?

26. List and discuss briefly pertinent points relative to feeding dairy kids.

27. How much supplement should be given to dairy goat bucks, and when should it be given?

28. Discuss the feeding of Angora and Spanish does.

SELECTED REFERENCES

Title of Publication	Author(s)	Publisher
Dairy Goat, The, Bull. No. 1160	W. F. Brannon	Cornell University Extension, Ithaca, NY, 1967
Dairy Goat Management, Bull. No. 334	G. W. VanderNoot D. M. Kniffen	Rutgers University Extension, New Brunswick, NJ
Dairy Goats, Pub. No. 439	B. E. Colby, *et al.*	University of Massachusetts Cooperative Extension, Amherst, MA
"Evolution of Range Ecology Practices and Policy"	W. B. Kessler	*Rangelands*, 15(3), June, 1993
Extension Goat Handbook	G. F. W. Haenlein D. L. Ace	Extension Service, USDA, Washington, DC, 1984
Forest Service Program for Forest and Rangeland Resources, The		USDA-Forest Service, Washington, DC, 1995
Goat Production	C. Gall	Academic Press, New York, NY, 1981
Goat Production in the Tropics	C. Devendra M. Burns	Unwin Brothers Limited, Old Woking, Surrey, England, 1983
Grazing on Public Lands	CAST	Iowa State University, Ames, Dec. 1996
Grazing Statistical Summary 1995		USDA-Forest Service, Washington, DC, 1995
Managing Interior Northwest Rangelands		USDA-Forest Service, Washington, DC, 1989
Nutrient Requirements of Goats, No. 15	National Research Council	National Academy Press, Washington, DC
Nutritional Requirements of the Angora Goat, Bull. B-1105	J. E. Huston M. Shelton W. C. Ellis	Texas Agricultural Experiment Station, San Angelo, TX
Observations on the Goat	M. H. French	FAO of the United Nations, Rome, Italy
Sharing Common Ground on Western Rangelands		USDA-Forest Service, Washington, DC, 1996
Sheep and Goat Handbook	The International Stockmen's School staff	Agriservices Foundation, Clovis, CA
Texas Angora Goat Production, Bull. B-296		Texas Agricultural Extension Service, San Angelo, TX
U.S. Sheep and Goat Industry, The	C. S. Menzies, Chairman, Task Force	Council for Agricultural Science and Technology (CAST), Report No. 94, Ames, IA

CHAPTER 22

Angora goats controlled by two sheep dogs. (Courtesy, *Ranch and Rural Living*, San Angelo, TX)

ANGORA AND CASHMERE GOATS

Contents *Page*

Angora goats are native to the Turkish province of Ankara, formerly Angora. Even today, Turkey is one of the leading Angora-raising countries.

Unlike the wool from sheep, the mohair from Angora goats represents a large proportion of the income derived from the animals. Wool provides only about 22% of the gross income from sheep, but mohair represents about 85% of the income from Angora goats. Fiber therefore, is the major economic product, and mohair prices have a controlling influence on the Angora goat industry. In recent years, the mohair produced in the United States has been worth an average of $2.82 per pound, or about $36.4 million each year.[1] Moreover, the United States is a leading producer of mohair, exporting mohair to the United Kingdom Italy, and Spain. Most of the Angora goats in the United States are in Texas, but other states have established herds.

Fig. 22-1. Angora goats in Turkey. It is reported that goats have declined in quality in their native land, with black and multi-colored animals in evidence in practically all the herds. Note the sparse vegetation.

DISTRIBUTION OF ANGORA PRODUCTION

On a worldwide basis, Table 22-1 lists the five leading mohair-producing countries. Interestingly, the processing of mohair is done largely in countries other than those in which it is produced. Thus, there is considerable international trade associated with the raw product and the processed goods. England, Japan, France, Italy, Spain, and Russia all import the raw product.

Table 22-2 lists the leading Angora goat-producing states in the United States based on the most recent count available. More recent data for Texas alone indicates that 1,270,000 goats were clipped, producing 9.7 million lb of mohair.[2]

TABLE 22-1
LEADING COUNTRIES, BY RANK, IN MOHAIR PRODUCTION[1]

Country	Production	
	(million lb)	*(million kg)*
Republic of South Africa . .	13.4	*6.1*
Turkey	9.9	*4.5*
United States (Texas) . . .	9.0	*4.1*
Argentina	2.2	*1.0*
Lesotho	1.3	*0.6*
Total	35.8	*16.3*

[1]Van Der Westhuysen, J. M., "Mohair as a Textile Fibre," *Proceedings of the Third International Conference on Goat Production and Disease*, Dairy Goat Journal Publishing Co., Scottsdale, AZ, p. 265, Table 2.

TABLE 22-2
U.S. ANGORA GOAT INVENTORY[1]

State	Number of Angora Goats	Number of Farms	Number per Farm
Texas.	1,496,037	2,791	536
Arizona	73,184	74	989
New Mexico	62,548	126	496
Oklahoma	46,798	252	186
Missouri.	18,161	90	84
North Dakota.	12,479	96	189
Kansas	9,486	108	88
Arkansas	7,342	75	98
United States.	1,702,168	6,150	278

[1]1992 Data from USDA.

Texas raises about 88% of the Angora goats and produces about 85.7% of the mohair. Arizona, however has the largest average herd per farm, and produces the most mohair per farm. The huge goat population of the Lone Star State is due principally to the large area of rugged grazing land which is well-adapted to utilization by this species. The center of goat raising in Texas is in the south central part of the state, a region generally known as the Edwards Plateau. The area is characterized by rolling hills, somewhat rough and broken, covered by grasses, cedar and oak trees, and a considerable amount of brush. The elevation is between 1,500 and 3,000 ft and the rainfall varies from 15 to 25 in. per year.

Generally, the densest Angora goat populations in this country are found in those areas in which the grazing is too scanty, rough, or brushy for cattle or sheep. On the other hand, a very large percentage of Texas Angora goats are handled on the same ranges that support cattle and sheep. The goats utilize browse

[1]*Agricultural Statistics*, USDA, 1997, Table 7-65.

[2]*Ibid.*

Fig. 22-2. Angora goats. Few people are familiar with these strange-looking, heavy-coated creatures. (Courtesy, Gary Cutrer, *Ranch & Rural Living*, San Angelo, TX)

which would be of little or no value to sheep and cattle and keep the brush from crowding out the natural grasses. On cut-over lands of the Pacific Coast states and in the Ozarks of the central states, Angora goats are frequently used for clearing land of brush; but if this end is to be accomplished, the area must be heavily stocked and closely grazed.

Recently, Angoras have moved north, some as far north as Michigan's upper peninsula and parts of Canada. These goats are managed much differently from those raised under the extensive range conditions of Texas.

Fig. 22-3. Angora nannies and kids. (Courtesy, Mohair Council of America, San Angelo, TX)

ESTABLISHING THE HERD

Angora herds vary from 25 to 30 head in farm herds to several hundred in range herds. Before establishing a herd, a rancher must consider several items, including the selection of animals to maintain and improve it.

FACTORS TO CONSIDER IN ESTABLISHING THE HERD

Doe and kid operations consist of a herd of healthy animals of productive age, 3 to 6 years of age. Operators should raise replacement animals to make improvements in the herd.

Commercial flocks usually are produced in areas with suitable range, mainly for the mohair and the wether goats that go to market for meat, or for controlling brush.

Using goats to control sprouts on cut-over, bulldozed, or chained brush country is popular in certain sections. Wether goats are more popular for controlling sprouts because they usually are larger and stronger, and can withstand more cold after shearing than kids and does. Very often, ranchers who are interested in controlling brush are not interested in breeding goats.

Some herds of wethers are maintained for mohair production. The trend toward finer-quality mohair makes it important to keep the herd young and the quality of clip fine. Thus, most goats are disposed of following the fourth shearing, or the operator stands the chance of being penalized on the clip price.

Commercial farm herds of Angora goats are also useful in controlling brush and undergrowth along draws and fence rows, in rough pastures and around the edges of fields. Goats are easily handled under farm-herd conditions and do not require much attention.

The production of registered Angora goats is not recommended unless the rancher has had experience in commercial production. Breeding registered Angora goats is detailed business and requires accurate records in breeding and kidding dates, sires, and dams. Individuals entering the registered goat business should have a good knowledge of genetics and nutrition, as well as know-how in advertising and selling. Registered herds usually are smaller than commercial herds; they supply bucks to commercial producers.

SELECTION BASES

To establish a herd or to pick replacement animals, a rancher should use the criteria which stress primarily the animals' ability to produce quality and quantity mohair. In Angora goats, this means, selecting animals on the basis of type or individuality, and possibly production testing. Selection on the basis of type or individuality requires judgment since some traits are not obvious and some are interrelated. There are, however, some traits which are obvious and can easily be selected for or against.

HERD IMPROVEMENT THROUGH SELECTION

Since most of the income from Angora goats is derived from the mohair, too much emphasis may be placed on selecting for mohair quality, resulting in goats that lose size and thriftiness. On the other hand, the large, open-faced rugged goats are better able to fend for themselves and produce kids and mohair. These larger, more rugged animals do, however, produce a coarser type of mohair. Thus, goat selection and mohair marketing are in direct competition. Nevertheless, a system of selection and breeding is necessary for improving Angora goat production. When selecting animals, ranchers should consider the following items—with about 50% allotted to body size and conformation and 50% to mohair quality:

- **Size and weight for age**—A yearling buck should weigh at least 80 lb and a yearling doe, at least 60 lb. Size at maturity varies greatly, depending on the amount of feed available during growth. Small goats usually are not desirable breeding animals.

- **Constitution and vigor**—It is impossible to determine the constitution and vigor of a goat by visual appraisal only, but an animal with a wide, deep chest, full heart girth, and full spring of ribs is often vigorous and possesses a strong constitution.

- **Conformation**—Goats should have medium length of leg with good width and depth of body. They should have fairly good length of body and not be short and dumpy. The back should be straight and strong with adequate width across the back and loin. A broad loin is essential to the development of strong Angora goats.

- **Amount of bone**—When bred for mohair quality, Angora goats may lack size of body and bone. Bone size, indicated by bone development below the knees and hocks, usually indicates an animal's ruggedness. Bones should be clean and proportionate to the animal's size. Straight legs should be placed squarely under the animal.

- **Angora breed type**—Angora breed type is indicated by head, horns, ears, and color markings. White hair on the face and lower parts of the legs should be free from colored fibers. Freckles or brown spots in the skin around the nose are not objectionable. Horns are widely set and on a buck should spiral outward and backward. Black horns disqualify an animal from registration.

- **Kemp or colored fibers**—An important point in considering the fleece of Angora goats is absence of kemp. Kemp fibers are large, chalky, white hairs. They are commonly found at the base of the neck, along the backbone, around the tail and often on lower parts of the thighs or the britch. Kemp is highly undesirable to manufacturers because the fibers are brittle, and they take dye and reflect light differently from true mohair. Kemp should be eliminated from the herd through a strict selective breeding program. Also, animals showing any colored fibers should be culled from the herd.

- **Uniformity and completeness of covering**—Angora goats should have a bright fleece of white mohair that is uniform in fineness and length from front to rear. The animals should be covered uniformly except for the face below the eyes. This deserves maximum selection pressure.

- **Fineness of the fleece**—Fine mohair generally is more desirable. The coarsest mohair usually is on the underside of the neck, so this is an important place to check for uniformity of fineness. Fleeces coarsen with age until animals are eight years old, at which time they tend to get finer again, because of loss of thriftiness in the animals.

  Mohair should feel soft rather than wiry or harsh to the touch.

- **Luster of the fleece**—This is the fiber's brightness or shininess, or the way it reflects light. In good-quality mohair, luster is highly developed and is one of the most desirable characteristics of mohair as a textile fiber.

- **Oil in the fleece**—Natural oil in the fleece is desirable to the condition, feel, and appearance of the fleece. It should contain enough oil to protect from weathering and to preserve the natural quality of the fleece.

- **Density of the fleece**—The fleece of the Angora goat should be dense, as determined by the number of fibers per unit of area. It is difficult to measure density, but it may be estimated in two ways: (1) The more skin that is exposed when the fleece is parted, the less the density; and (2) when a handful of mohair is grasped on each side of the animal and lifted, fuller handfuls and heavier weight indicate more mohair—more density. Goats with light, fluffy fleeces should be eliminated from the breeding herd because this affects total fleece weight and mohair quality.

- **Character of the fleece**—This refers largely to the type of lock, which is more important to registered breeders than to commercial producers. Lock types, ranked according to desirability, are: the tight or spiral lock, the flat lock, and the fluffy fleece. The tight lock hangs from the body in ringlets and is associated with the finest fibers. The flat lock is usually more wavy and coarser, but it is associated with heavy shearing weight. The fluffy fleece is objectionable because it is easily broken and is torn out by brush to a greater extent than the other types. Regardless of the lock type, the fleece should be uniform over the entire body.

Physical disqualifications for breed registration include: all blue or black horns or hoofs, a deformed mouth, deteriorated pasterns, deformed feet, crooked legs, divided scrotum or abnormalities of the testicles, closely set distorted horns, and a sway back. Fleece disqualifications for breed registration include: excessive kemp, colored hair, "sheepy" fleece, and straight beard-type hair in the foretop or on the back.

Animals can be selected before the fall or spring shearing, but fall is preferred since credit can be given to the does that raised kids. It is difficult to make accurate selections on kid goats before their first shearing because of differences in age. So, another selection should be made before the second shearing when all animals have had an equal chance to develop and produce mohair.

MOHAIR PRODUCTION

When expressed as rate of fiber growth, amount produced per unit of body weight maintained, or efficiency of production, the Angora is generally recognized as the world's best fiber producer. While the biology of mohair production is similar to other types of animal fiber, it is a special type since the production from the secondary follicles has been developed to a higher degree. Apart from the inherited traits, mohair production is markedly affected by the age of the goat.

THE MOHAIR FIBER—BIOLOGY OF GROWTH

Mohair fiber consists of keratinized protein similar to other tissues such as horns, nails, and hair. Fibers are produced in structures known as *follicles*. These follicles are arranged in groups or bundles, and the fiber density is then affected by the number of follicle bundles and the number of fibers in a bundle. Within a bundle, the follicles are of two types: primary and secondary. The primary follicles give rise to long, coarse guard hairs known as the outercoat, while the secondary follicles produce a large number of fine, short fibers known as the undercoat, or down. In the Angora goat, the secondary fiber production has been selected for to the extent that it masks the primary fibers. However, the Angora goat is born with a primary coat consisting mainly of long kemp fibers. Then, during the first six months of the goat's life, there is a change in the activity of hair-producing follicles, resulting in a coat of fibers from the secondary follicles.

The fiber is actually produced by a group of rapidly dividing cells located in the bulb at the base of the follicles. These cells are some of the most rapidly dividing body cells as evidenced by the rate of fiber production.

Mohair differs from wool with respect to the absence of crimp, in having fewer scales that do not protrude as distinctly. These characteristics give mohair its smooth feel and high luster for which the fiber is renowned. Also, since mohair takes dye readily and since it is mostly white, it can be dyed brilliant colors.

EXPECTED PRODUCTION

In the United States, the average yearly clip of mohair per goat is about 7 lb. A good Angora goat should produce 1 in. of mohair per month, producing a 6-in. staple for each 6-month clip.

Mohair production is, however, markedly influenced by age. Mohair production in terms of amount per unit of body weight peaks when the animals are about two years of age and then gradually declines. Fiber diameter increases as the animals mature, from 25 micrometers in six-month-old goats to as much as 40 micrometers in adult goats. Fiber diameter increases fairly rapidly when the animals are between 6 months and 3 to 4 years of age, and then it plateaus. The rate of growth remains very constant over the life of the animals.

MANAGEMENT OF ANGORA GOATS

The proper care of goats differs from that which should be accorded sheep under similar conditions. Like sheep, Angora goats pay dividends for good management. Unfortunately, there is a widespread and common belief that goats will thrive despite neglect. This popular conception is not true. Successful goat raisers apply the same care and management to goat raising as is given to any other profitable livestock enterprise. Success with goats can be achieved in no other way. In the discussion which follows, particular attention is given to principles or systems of management wherein goats differ markedly from sheep. Needless repetition on such matters as parasite control, watering, etc., is omitted.

HERDING

A large number of the Angora goats in the United States are maintained under range conditions, where they are grazed in herds much like range bands of sheep. These herds vary in numbers from a few hundred to over two thousand head, with an average herd numbering about 1,200. On rough range or in thick brush, a larger number than 1,200 can seldom be properly controlled by a single herder.

The principles and practices of good herding with goats are almost identical to those with sheep, as discussed in Chapter 16. There is one distinct difference: Rarely do sheep herders work ahead of the band; whereas, it is common practice for goat herders to work in front, turning the lead goats back to avoid unnecessary travel.

At the present time, most of the goats of Texas and other southwestern states are loose grazed (unherded) in wolfproof fenced ranges, in the same manner that sheep are handled in this area.

FEEDING

Unfortunately, there is a widely prevailing belief that goats will eat and do well on anything from newspapers to rusty tin cans. This is erroneous. Like other animals that are hungry or suffering from mineral or vitamin deficiencies, they will develop depraved appetites and chew on many things; but they prefer good-quality, wholesome feeds, and they will pay dividends when so fed. Moreover, high-producing Angora goats have high-protein requirements.

Goats are naturally browse animals; but a good goat range, in addition to furnishing abundant palatable evergreen brush (not cedar or other coniferous vegetation), should provide a mixture of grasses and broadleaved herbs. On the most desirable goat ranges, this feed combination is available the year round, though browse is usually the principal winter feed. Some common brushes that are utilized by goats include: guajillo, live oak, briar, catclaw, elm, ill-scented sumac, post oak, shin oak, small-leaved sumac, Spanish oak, and yaupon.

Complete details of feeding Angora goats plus the nutrient needs of Angora goats are given in Chapter 21, Feeding Goats.

BREEDING

Angora goats are seasonal in their breeding habits. Does come in season in the fall and kid in the spring. This characteristic is so well-established in Angoras that many ranchers leave bucks with does year round. Others turn bucks in with does during October and November, and the does kid in March and April. Some ranchers prefer earlier kidding. The gestation period varies from 147 to 155 days, an estimated length of 5 months.

Does per buck vary with size, roughness, and brushiness of pastures. Three to four bucks per 100 is the usual rate for good breeding. Three to four weeks before placement with the does, bucks should be fed supplements, such as whole oats, to condition them. Buck kids are not sexually mature and should not be used for breeding purposes.

Does can be flushed by putting them on a rested pasture or by supplementing their grazing with 0.25 to 0.33 lb per head daily of corn or range cubes.

Angora does abort easier than most other animals. Therefore, they must receive adequate nutrition, and stressful situations must be eliminated.

KIDDING TIME

Goats require much more care than sheep at the time the newborns arrive. Young kids are much more delicate than young lambs—neither being able to endure as much cold or damp weather nor being able to follow their mothers to the range as early in life. It is not surprising, therefore, to discover that an 80% kid crop is considered excellent for range herds. The average is 40 to 50%, but ranges from 0 to 100%.

If heavy losses of young kids are to be averted, a safe system of kidding must be followed. The two most common systems of kidding followed by progressive and successful goat raisers of today are: (1) the toggle or staking system, and (2) the pen or corral system.

TOGGLE OR STAKING SYSTEM

Under the toggle system, each young kid is staked with about 15 in. of rope attached to a swivel on the stake end, with a loop on the other end being attached to the fetlock joint of the kid. This loop should be changed to another leg once daily. Each kid is provided with a box, usually a small A-shaped structure made of 12-in. boards with one end open, which furnishes protection from the elements. The camp should have sufficient stakes and boxes to allow for as many kids as may be dropped within any 10-day period. Ten feet between stakes is considered about the right distance, providing sufficient space to avoid confusion, affording room for handling, and minimizing quarreling of the does. Whenever a kid is staked, it and its mother are branded with corresponding marks, thus avoiding any possibility of confusion in identity. Some ranchers clip the does' tails and place the numbers there, whereas others place brand marks on similar body areas of the does and kids.

The ideal toggle camp should have a good slope for drainage. With such a slope, the older does will spend the night on the higher side; so their kids should

Fig. 22-4. A toggle camp in which the young kids are staked near small A-shaped boxes. (Courtesy, *Sheep and Goat Raiser*)

be staked there, with the kids of the younger does being placed in the lower part of the camp. Disturbance will be avoided if the kids or does that fight or habitually overturn the kid boxes are placed at one end or corner of the camp. On cold, rainy nights, the does should be confined or otherwise kept out of the toggle camp; for each time a doe passes through the yard to her kid, other kids will come out of their boxes and suffer possible exposure.

A doe that disowns her kid should be staked with it. Kids may usually be grafted if the foster mother and orphan kid are staked or penned together. The presence of dogies, as orphan kids are known, is evidence that such details have not been given adequate attention around a kidding camp.

The careful herder passes through the toggle camp each morning and evening, ascertaining whether each kid has nursed. If a kid is gaunt or restless, its mother should be brought to the stake; if a doe has a distended udder, she should be taken to the kid or kids that carry the corresponding number or mark. Does with extra large teats may have to be hand milked until the kids have learned to nurse the abnormally large teats.

Generally, kids should not be staked in the toggle camp for more than 10 days. At that time, they should be herded with their mothers. Some of the large operators, especially those in the southwestern United States, release the kids directly from the toggle camp to the range with the does. These operators insist that kids handled in this manner develop into better foragers and possess more muscle and bone. Others prefer to make the change from the toggle camp to the range more gradually. Usually, the latter operators first transfer the does and their kids in small numbers from the toggle camp to a small field. The size of the field and herd is then gradually increased until the kid band may be turned on the range when the young average about six weeks of age. A modification of the latter system consists of confining the kids to a corral for a few weeks while the does depart via a *jump board* (a structure about 18 in. high) to forage on the range in the daytime.

PEN OR CORRAL SYSTEM

The pen or corral system is gaining in popularity, because less work is involved in the handling of the animals and the results are as good as those obtained in the toggle system. For a herd of 1,200 does, the pen system usually involves having about 8 corrals of ample size to enclose 50 does and their kids, 12 small pens each large enough to hold an individual doe and her kid, and 1 or 2 larger corrals that may be adequate for 400 to 500 does and their kids.

Even the toggle system should be supplemented with a few individual kidding pens, which may be used as *bum pens* to force does to accept disowned kids that are their own or orphan kids that are to be grafted. In general, in the pen or corral system about 50 does and their kids of about the same age group are kept in one pen or corral for the first 2 or 3 weeks; the kids are confined to the corral, whereas the does are permitted to go out to the range in the daytime. As the kids get older, groups are combined and turned to larger areas, eventually to travel to the range in a kid herd consisting of 1,000 to 1,200 does and their kids.

Kidding on the stake or in the pen usually is practiced by registered breeders because they need to keep records of the sire and dam for registration. Typically, though, does are kidded on the range without any attempt to provide supervision.

ORPHANS

Orphaned kids should receive colostrum by being allowed to nurse does that have just kidded. Then they can be bottle raised on cow's milk or milk replacer. Feedings should consist of small amounts at least four times daily for two weeks. After which, feedings can be reduced to three times daily. Kids also can receive grain at about two weeks of age.

SHEARING

Angora goats usually are shorn twice each year in Texas. Spring shearing starts in February in the southern counties and lasts through April in the northern counties. Fall shearing begins about mid-July and lasts until the first of October.

Prior to shearing, kids and older goats are separated. Then kid mohair and the adult mohair are packed separately and labeled—spring kid, fall kid, yearling, adult. Also, in many herds, the finer-haired adult goats are sheared separately from the coarser-haired goats.

Shearing requires a good, clean location with either a concrete or a wooden shearing floor. Where shearing floors are not available, portable shearing boards may be used.

Individual fleeces should be kept together as much as possible, and bags should be packed firmly and uniformly. Burlap bags will weigh 280 to 320 lb, depending on the fineness of the mohair.

If the goats are wet or damp, they must not be sheared. Moreover, bags of mohair must be stored in a dry, clean place.

Cold, driving rains for as long as six weeks after shearing cause the heaviest death losses in Angora goats. This hazard is the motive for retaining a cape on goats at shearing time. The cape is a strip of mohair 3 to 4 in. wide down a goat's back. It protects the goat from the weather after shearing, but may not be sufficient in extremely severe weather. Most capes are left in the spring, since the weather usually is more severe during that season.

Capes present some problems. If capes are left until fall shearing, a percentage of long mohair is produced in the clip. If they are sheared off after six weeks, a percentage of short mohair is produced. Buyers prefer the capes being sheared off because short mohair is more desirable than 12 months of mohair growth.

Some ranchers provide raised combs for shearers to use on goats. Raised combs have runners built up so they leave 0.25 to 0.50 in. of stubble on the goats. This gives about as much protection as the cape and eliminates undesirable staple length.

Shelter may also be used to protect newly sheared goats from cold, wet weather. Goat sheds are generally low structures, covered with metal roofs and boarded up on one or two sides. When well fed and not newly sheared, goats seldom succumb to cold and, therefore, do not need shelter

MARKING AND CASTRATING

The time of marking and castrating varies more among Angora goats than other classes of livestock. Some ranchers mark and castrate when most of the kids have been born, or when the oldest kids are 1 to 1½ months old. Others mark the kids in the spring and castrate in November, December, or January. This lets the kids develop faster and grow more horn. When castrated at 6 to 8 months of age, the wether has horns more like those of a buck, which facilitates separating muttons and does when run together. Ranchers running only wether goats usually prefer goats castrated at 6 to 8 months of age.

Marking is strictly for identification. Ear tags are unpopular because they are pulled out in the brush. Most registered breeders use a system of earnotching and tattooing. The approved earnotching system is shown in Fig. 22-5. The same number notched in a goat's ear should be tattooed inside the ear. To be eligible for registration, Angora goats must be identified by any two of the following three methods: (1) metal or plastic ear tag, (2) number tattooed or burned in one ear, or (3) earnotched according to Fig. 22-5.

Other ear marks include swallow fork, underbit, overbit, overslope, underslope, crop, and various combinations of these marks. Marks should be registered in the county of the ranch's location.

Some ranchers use a fire brand on the nose or cheek. Others paint the base of the horns, but paint may deteriorate the horns and cause them to break easily.

Fig. 22-5. Approved system of marking registered Angora goats with earnotches.

RECORDS

Records are a part of good management. Those records necessary to measure progress include: (1) average fleece weight for spring and fall clips, (2) length of staple, (3) shorn body weight, and (4) percentage of kid crop.

Registered breeders need to keep more accurate and complete records than commercial producers. Individual records on does and bucks are useful in building up the herd through selection. Some registered breeders include comments on the type of lock produced at each shearing and find that the type of lock may change from one shearing to the next.

PASTURE MANAGEMENT

One of the greatest drawbacks to mohair production is vegetable defect, in the form of needle and spear grass, grass burs, cockle burs, horehound, and other plant contaminants. Much of this can be avoided by careful pasture management. For example, pastures containing one or more contaminants can be grazed during the growing season and before seeds mature. Then goats can be moved to pastures or fields containing few contaminating plants.

Early kidding, accompanied by early shearing of kids, (1) often keeps contaminants out of fleeces, (2) helps kids develop, and (3) prevents spear and needle damage. In some cases, changes in shearing dates of adult goats may cut down on vegetable contamination. Also, pasture improvement programs should include the control of these noxious plants.

MARKETING ANGORA GOATS AND MOHAIR

Young Angora wethers usually are in demand by ranchers interested in controlling brush sprouts on chained, rootplowed, or bulldozed land. Also, these animals are used in controlling the spread of shinnery.

When mohair production begins to decline on the aged wethers, ranchers usually ship them to auction sales specializing in goat handling. Some auction sales specialize in handling goats for stocker and packer purposes. Other auction sales handle goats but do not specialize in them.

Most goats sold on the San Antonio market are processed locally. The meat finds a ready market among the Mexican population. At San Antonio, goat meat is canned and prepared for shipment to foreign countries.

Cabrito is meat from 5- to 6-month-old kids weighing 30 to 40 lb alive. It is regarded as a delicacy and highly prized for barbecuing in the Southwest.

Chapter 24 contains more information about meat from goats.

Warehouses handle wool and mohair. Some handle these products on consignment only, while others buy on their own account and handle consignments. Handling charges vary, but competition keeps them in line.

Some warehouses provide grading services so mohair may be sold on a graded basis. Producers with good-quality clips profit by selling mohair on a graded basis. Grading also gives producers valuable information about their clips, which they can use to improve their breeding programs. Many warehouses handle livestock supplies, feed, salt, and minerals, or serve as lending agencies and give advances on unsheared wool and mohair.

Chapter 9, Wool and Mohair, furnishes a more complete discussion on the marketing of mohair.

CASHMERE

Cashmere is goat down. This fine undercoat of goat hair is produced by the secondary follicles. The term *cashmere* is apparently derived from the geographic region of Kashmir in Asia. *Pashmina* is a term used somewhat synonymously with cashmere, especially in India.

Superior grades of the cashmere fiber range from 13 to 16 micrometers in diameter and seldom more than 2.4 in. long. Combing in the late spring just as the animals start to shed removes the cashmere from the goat. Or, it may be sheared from the goat. After coarse hairs are removed, cashmere is spun and woven, or knitted on conventional woolen and worsted textile machinery.

For centuries, cashmere has been the byproduct of goats raised primarily for milk and meat.

FUTURE OF THE ANGORA GOAT INDUSTRY

Angora goats will, in some areas, continue to be important for controlling brush regrowth after initial clearing, thereby reducing fire hazards and providing income from the sale of meat and mohair. Use of goats for this purpose can save millions of dollars, and prove to be a valuable asset on rangelands that otherwise have little income-producing potential.

QUESTIONS FOR STUDY AND DISCUSSION

1. Where is most of the mohair produced in (1) the United States, and (2) the world? What accounts for these areas of mohair production?

2. Supposing you wished to start a herd of purebred Angoras, what things should you consider?

3. Outline the items you would consider when selecting Angora goats for breeding purposes with the intention of improving production. Indicate the amount of emphasis you would place on each item.

4. What is mohair. How does it differ from wool?

5. What factors affect mohair production?

6. Discuss each of the following management aspects of Angora goats: herding, feeding, breeding, kidding, shearing, marking, castrating, and records.

7. Compare the toggle or staking system of kidding to the pen or corral system.

8. Subjecting recently shorn Angora goats to cold, wet weather can result in heavy death losses. How can this problem be overcome?

9. What is meant by "vegetable defect" in mohair production?

10. Through what channels may Angora goats and mohair be marketed?

11. What is cashmere, and where is it produced?

12. What future do you see for the Angora goat industry?

SELECTED REFERENCES

Title of Publication	Author(s)	Publisher
Goat Production: Breeding & Management	Ed. by C. Gall	Academic Press, Inc., New York, NY, 1981
Proceedings of the Third International Conference on Goat Production and Disease		Dairy Goat Journal Publishing Co., Scottsdale, AZ, 1982
Sharing Common Ground on Western Rangelands		USDA Forest Service, Washington, DC, 1996
Sheep Research Journal, Special Issue: 1994	Ed. by Maurice Shelton	American Sheep Industry Assn., Englewood, CO, 1994

CHAPTER 23

Mama, don't move!
(Courtesy, Jodi Frediani, Santa Cruz, CA)

DAIRY GOATS

Contents *Page*

Contents *Page*

Dairy goats, "the cows of the poor," accompanied explorers and immigrants. Christopher Columbus and Captain James Cook took goats with them. Immigrants to the New World brought dairy goats from their native England, France, Germany, Greece, Italy, Spain, and Switzerland. Early merchant ships left goats behind in U.S. ports. Additionally, there were some intentional importations of valuable breeding stock, though these numbers were small.

From early beginnings, the dairy goat industry of the United States has survived and grown.

DISTRIBUTION OF DAIRY GOATS

Table 23-1 summarizes, by areas of the world, the amount of goat's milk produced, and the goat population. As noted, Asia leads in number of goats, whereas Europe leads in milk production per goat. About 80% of the goat's milk of Asia is produced in Bangladesh, China, India, Iran, Pakistan, and Turkey. In Africa, the countries of Algeria, Ethiopia, Mali, Niger, Somalia, and Sudan produce about 74% of the goat's milk. Mexico and Brazil produce most of the goat's milk in North and South America, respectively. In Europe, France, Greece, Russia, and Spain produce about 70% of the goat's milk.[1]

TABLE 23-1
WORLDWIDE GOAT MILK PRODUCTION AND GOAT POPULATION[1]

Area	Number of Goats	Goat Milk Production	Production per Goat[2]	
	(1,000 hd)	*(1,000 metric ton)*	*(lb)*	*(kg)*
Asia	435,729	5,577	28,200	*12,700*
Africa	176,401	2,078	26,000	*11,800*
South America	23,308	184	17,367	*7,890*
Europe	18,908	2,156	250,857	*114,025*
North America	15,596	148	20,877	*9,490*
Russia	6,685	327	107,614	*48,915*
World	674,139	10,144	33,103	*15,047*

[1]*FAO Production Yearbook*, Food and Agriculture Organization of the United Nations, Rome, Italy, Vol. 50, 1996, Tables 90 and 100. Data for 1996.

[2]This was derived by dividing goat milk production (column 3) by the number of goats (column 2). While it is not an accurate representation of yearly production of goat's milk, the value does give some idea of the importance of goat's milk in the area.

Table 23-2 lists the 10 leading dairy goat states based on the analysis of their lactation records for 10 years.

In 1992, the U.S. Department of Agriculture computed state summaries of number of herds, does per herd, and total number of does, and national summaries of dairy goat herd averages for production. Some results of these computations for the 10 leading dairy goat states are given in Table 23-3. Most dairy goat herds are small—10 or fewer does per herd. The most common herd size is 5 to 10 does, but about 2% of the herds have over 50 does.

TABLE 23-2
TOP 10 DAIRY GOAT STATES[1]

State	Number	Percentage of Top Ten
California	18,592	28
Texas	11,727	18
Wisconsin	7,677	12
New York	5,746	8.6
Ohio	4,953	7.4
Michigan	4,184	6.2
Washington	3,731	5.5
Oklahoma	3,393	5.0
Indiana	3,247	4.8
Tennessee	3,154	4.5
Total for 10 states	66,404	100

[1]*1992 Census of Agriculture*, U.S. Bureau of Census.

TABLE 23-3
LEADING DAIRY GOAT STATES BY THE NUMBER OF HERDS[1]

State	Number of Herds	Does per Herd	Total Number of Does
Texas	902	16.5	14,883
Ohio	834	9.9	8,257
California	694	20.2	14,007
Missouri	608	9.2	5,579
Pennsylvania	496	8.2	4,068
Indiana	435	8.9	3,857
Oregon	413	11.5	4,755
Michigan	404	12.7	5,136
Tennessee	383	9.0	3,445
Washington	382	13.3	5,080
Wisconsin	381	21.2	8,077
Kentucky	341	6.5	2,224
Colorado	339	8.9	3,032
United States	11,559	12.0	138,708

[1]Courtesy, U.S. Department of Agriculture.

ESTABLISHING THE HERD

Whether a large dairy goat operation or a small dairy goat herd is being established or maintained,

[1]*FAO Production Yearbook*, Food and Agriculture Organization of the United Nations, Rome, Italy, Vol. 50, 1996, pp. 219–220, Table 100.

consideration must be given to certain factors if the venture is to be successful. In the final analysis, dairy goats are maintained for the production of milk. This means that each individual herd should possess those characteristics creating maximum and efficient production of milk. Furthermore, if progress is to be made in the breeding programs, each succeeding generation must represent an improvement over the parent stock.

FACTORS TO CONSIDER IN ESTABLISHING THE HERD

Small goat herds are established as a hobby and/or for personal use, and large herds are established for commercial production. Whatever the reason for establishing a herd, there are a variety of factors to keep in mind, such as selection of breed, number of animals, uniformity, health, age, soundness of udder, and price.

SELECTION OF BREED

According to USDA research at Beltsville, Maryland, the differences between families and individuals within each goat breed appear to be greater than differences between breeds, and crossbreds and grades may be more profitable in a dairy than purebreds. Therefore, several factors should be considered in the selection of a breed of dairy goat: personal preference, purpose of dairy goat enterprise, and popularity of a breed in a locality. The latter makes it much easier to secure replacement stock and to have good bucks available for breeding purposes. The quality of the individual animals available must also be kept in mind. It is preferable to buy good animals of a less-preferred breed to inferior animals of another breed. Also, for those planning to show, the breed selected should be one that will have enough entries for good competition.

NUMBER OF ANIMALS

The beginner can acquire valuable practical experience with a few goats without subjecting a large herd to the possible hazards that frequently accompany inexperience. But, two animals are better than one, since a single goat will be noisy and lonesome. Ultimately, the size of the herd is determined by facilities and capital.

UNIFORMITY

A person wishing to raise goats for breeding purposes should strive for uniformity of color, type, and other characteristics of the breed, as well as those points which apply to all breeds. These points include: straight top, wide chest, large heart girth, plenty of feed capacity, rugged muzzle, straight legs, standing solidly on the hoofs, good bone, loose and pliable coat, proper udder and teat placement, shape and texture in females, and masculinity and ruggedness in bucks.

HEALTH

All does selected should be in a thrifty, vigorous condition. They should have every appearance of a life of usefulness ahead of them and give every evidence of being good milkers and being able to raise strong, healthy kids. Animals showing dark blue skins, paleness or lack of coloring in the lining of the nose and eyelids, listlessness or a lack of vigor, and a general rundown condition, should be regarded with suspicion. If the health of an animal is questionable, it is prudent to have a veterinarian examine and certify that the animal is free from disease and parasites.

AGE

Age of the animals is a matter of preference and possibly economics. Some individuals may find it advantageous to buy aged, proven does past their prime, then use them as foundation breeders. Others wishing a small initial investment may buy doe kids. Although raising kids to maturity and production means a considerable investment in time and feed, invaluable experience will be gained. Those wishing to invest the maximum and leave less to chance may wish to buy does entering their second to fifth lactation, about the time they go dry.

SOUNDNESS OF UDDER

In selecting does, one should give particular attention to the udders and teats. Each udder should show plenty of capacity and be well held up to the body by the suspension ligament so it will not become injured by hitting stones or other objects in the pasture or around the barn. The low-slung udder is called pendulous and is very undesirable. To the touch, the udder should be pliable and soft, not hard and meaty. Hard lumps in the udder or teats should be discounted in judging or selection. The teats on the udder should be large enough to be easily milked. The milk goat udder should be balanced in shape, with teats hanging the same length. The teats should be uniformly placed on the udder and slightly tilted forward. After a milking, the udder should be collapsed and pliable like a soft leather glove (see Fig. 23-1).

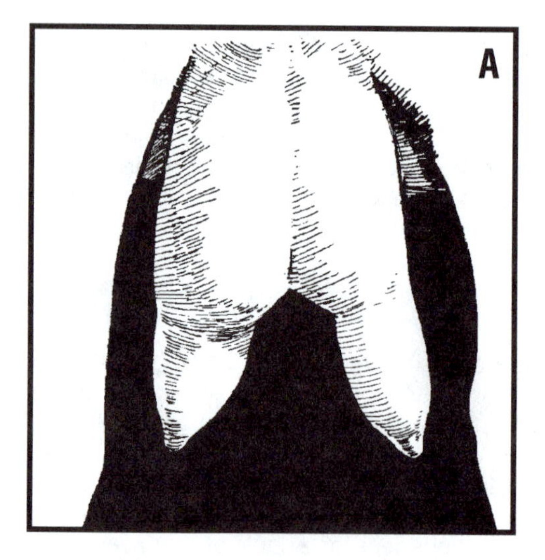

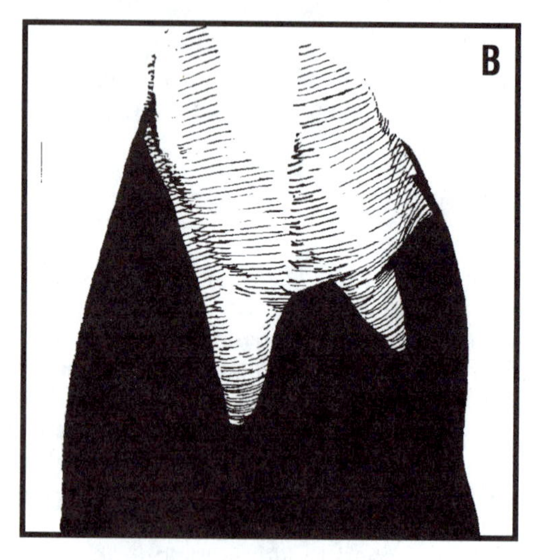

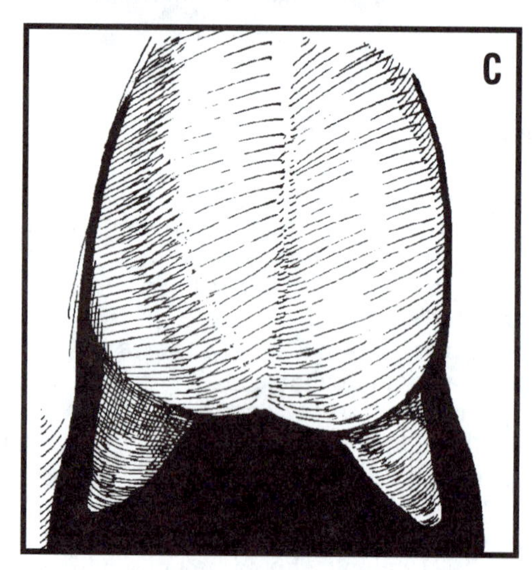

Fig. 23-1. Evaluation of the udder and teats is important. Viewed from behind, **A** shows an undesirable udder with teats that are too large; **B** shows an udder that is not balanced and teats that are poorly formed; and **C** shows a good udder with balance and teats of the same length tilted slightly forward.

PRICE

Generally, grades are less expensive than purebreds, but there are good and poor grades just as there are good and poor purebreds. Kids, whether purebreds or grades, cost less than milkers. Purebred milkers and purebred doelings may be the most expensive, depending upon their genetic capability.

SELECTION BASES

The criteria used in selecting an individual or group of dairy goats are those that influence the production of milk. Purebred breeders may add certain fancy breed points and highly prized pedigrees to the criteria to be considered. With all the criteria to be considered, it is not intended that the amateur become a proficient dairy goat judge through merely reading about the subject. There is no short-cut or substitute for long years of experience.

Establishing a new herd or improving an old one, however, involves four bases of selection: (1) selection based on type or individuality, (2) selection based on pedigree, (3) selection based on show-ring winnings, and (4) selection based on production records.

1. **Selection based on type or individuality.** Another name for this kind of selection is judging. A large number of goats are selected on this basis. In selecting animals on this basis, the producer must judge how well a particular animal conforms to the description of an ideal type of dairy goat. Moreover, animals with obvious defects can be removed from the herd or prevented from entering the herd.

2. **Selection based on pedigree.** Pedigree is a record of an individual's heredity or inheritance. If the ancestry is good, it lends confidence in projecting how well young animals may perform.

However, mere names and registration numbers are meaningless. A pedigree may be considered as desirable only when the ancestors close up in the lineage—the parents and grandparents—were superior individuals and outstanding producers. Too often, purebred goat breeders are prone to play up one or two outstanding animals back in the third or fourth generation—a type of ancestor worship. If pedigree selection is to be of any help, one must be familiar with the individual animals listed therein. Regardless of whether animals are purebreds, grades, or crossbreds, good ancestry is important since it gives more assurance of the production of high-quality kids that are uniform and true to type.

3. **Selection based on show-ring winnings.** There can be no question that livestock shows of the land have had a profound influence in establishing type in livestock. When utilitarian considerations have been ignored, however, the influence of shows has not

always been for the good. Shows may emphasize some fancy point that ultimately interferes with efficient production. Therefore, some scrutiny should be given to the type of animals winning in the show, especially to ascertain whether such animals are of the type that are efficient from the standpoint of the producer.

4. **Selection based on production records.** No criterion that can be used in selecting an animal is as accurate or as important as performance. Individuals maintaining breeding herds should keep official production records—pounds of milk and butterfat per lactation. Also, production records should be maintained in any commercial herd. By enrolling in the local Dairy Herd Improvement Association, a producer can maintain these records. At the very least, daily milk weights should be recorded for each milking doe once each month to secure a yearly milk record. High production records as close as the first or second generation are of great value in the selection of an animal. When an individual is buying a mature goat, the animal's milk production records and the production of its offspring are valuable.

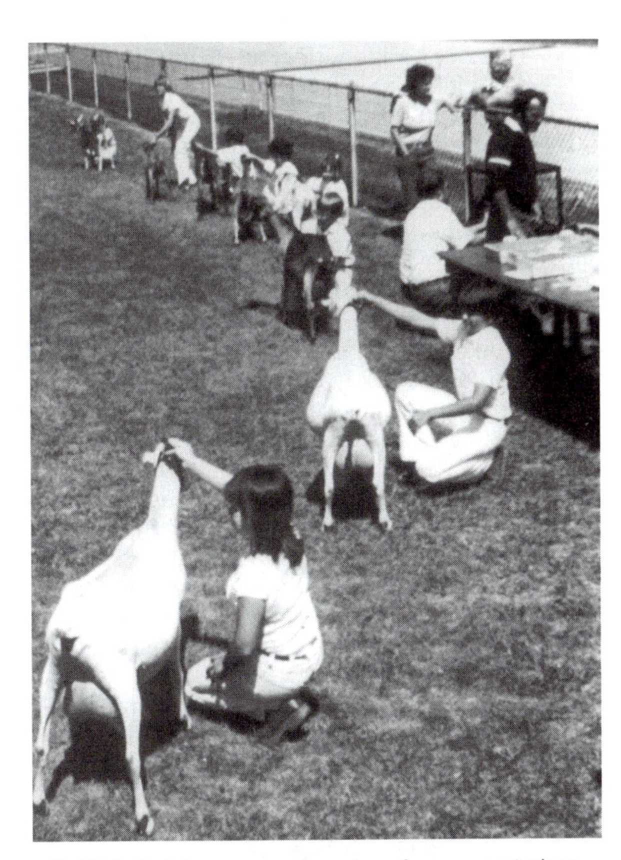

Fig. 23-2. Goat shows are good experience for young goat raisers.

HERD IMPROVEMENT THROUGH SELECTION

Once the herd has been established, improvement can be obtained only through constant, rigid culling and careful selection of replacements—both does and kids. Such procedure makes the herd more profitable from the standpoint of quantity and quality of milk production and affords a means of accomplishing a genetic gain in the next generation.

Individual goat identification and production records (see Figs. 23-13 and 23-14 for record forms) are requisite to effective culling and selection. Also, those traits that are most heritable, and that contribute most to income, should be considered. In the majority of herds, selection and culling should be based on the following: (1) milk, butterfat, and protein yield per lactation, (2) milking time on milking machine, (3) fertility and prolificacy, (4) conformation and vigor, and (5) freedom from abnormalities and defects.

JUDGING DAIRY GOATS

Often, the only method other than pedigree available to evaluate dairy goats is by what is commonly known as judging.

Judging—as practiced in shows, in contests, or on the farm—is an attempt to place or rank animals in the order of their excellence in body type. Scoring, or type classifying, an animal accomplishes the same thing, in that the individual being scored is classified and compared to an animal that is theoretically perfect, and a rating is assigned on this basis.

Admittedly, there is considerable question as to the degree of correlation between type and production, for appearance is not always indicative of a goat's productive ability. Yet, it is generally recognized that desirable type (1) is indicative of a goat's wearing ability, and (2) does not negate functional value. Moreover, it is generally recognized that attractiveness, and desirable type, enhances the market value of purebred animals. Also, well-attached udders are less subject to injury and mastitis infection, and strong legs hold up longer than weak legs and feet.

Good and successful owners and managers are generally good judges of dairy goats.

QUALIFICATIONS OF A GOOD DAIRY GOAT JUDGE

The essential qualifications which a good dairy goat judge must possess are:

1. **Knowledge of the parts of an animal.** This consists of mastering the language that describes and locates the different parts of a dairy goat.

2. **A clearly defined ideal or standard of perfection.** Successful dairy goat judges must know the type for which they are looking.

3. **Keen observation and sound judgment.** Good judges possess the ability to observe both good conformation and defects, and to weigh and evaluate the relative importance of the various good and bad features.

4. **Honesty and courage.** Good judges of any class of livestock must possess honesty and courage, whether it be in making a show-ring placing or in conducting a breeding and marketing program. For example, it often requires considerable courage to place a class of animals without regard to (a) placings in previous shows, (b) ownership, and (c) public applause. It may even take greater courage and honesty to discard from the herd a costly animal whose progeny has failed to measure up.

5. **Logical procedure in evaluating.** There is always great danger of the beginner making too close an inspection. Good judging procedure consists of the following three separate steps: (a) observing at a distance (20 to 30 ft) and securing a panoramic view where several animals are involved, (b) using close inspection and handling, and (c) moving the animal in order to observe action. Also, it is important that a logical method be used in viewing an animal from all directions, as for example (a) side view, (b) rear view, and (c) front view; thus avoiding overlooking anything and making it easier to retain the observations that are made.

6. **Tact.** In discussing either (a) a show-ring class or (b) animals on an owner's farm or ranch, the judge must be tactful. Owners are likely to resent any remarks which indicate that their animals are inferior.

Having acquired this knowledge, judges must spend long hours in patient study and practice in comparing animals. Even this will not make expert and proficient judges in all instances, for there may be a grain of truth in the statement that "the best judges are born and not made." Nevertheless, training in judging and selecting animals is effective when directed by a competent instructor or an experienced producer.

PARTS OF A GOAT

One of the characteristics of good judges is that they possess a thorough knowledge of animals. In speaking of the characteristics of dairy goats, they usually refer to parts rather than to the individual as a whole. It is important, therefore, to become familiar with the names of the parts. Fig. 23-3 shows an animal and identifies by name the various parts of the body. This figure should be studied until each part of the animal can be easily and quickly identified by location and name. Nothing so quickly sets a real dairy goat producer apart from a novice as a thorough knowledge of the parts and the language commonly used in describing them.

IDEAL TYPE AND CONFORMATION

A major requisite in judging or selection is to have clearly in mind a standard or an ideal. After becoming familiar with the dairy type, a person is in a much better position to select an animal and to know whether it is fair, good, very good, or excellent.

A dairy goat should be angular, not round; the hip bones should be prominent; the thighs should be thin; and the animal should possess considerable length of neck and a long body. Any tendency to be short and thick of body, short of neck, thick in the thighs, or in any way fat and meaty, indicates lack of dairy type. Meatiness is the opposite of dairyness.

The good dairy goat will be sleek and alert, not fat and sluggish. The doe should be as straight as possible on top and especially strong in the chine and loin areas. From the hip bones back to the pin bones (bones on each side of the tail), there will be some slope on nearly every animal, but the object should be to get this line as straight as possible.

The shoulder should be refined and not coarse. It should blend into the middle smoothly. The withers or top of the shoulder should be sharp and refined, and not rounded as in a meat-type animal.

The middle should be long and the ribs well sprung, making adequate room for roughage, plus two or more kids. The ribs should be long and far enough apart to slide one finger between the ribs. This openness of rib denotes dairy temperament in the goat. There should be some width in the floor of the chest so the front legs are not too close together. Width plus depth of body denotes lung capacity and constitution and is associated with strength and ruggedness.

The legs should be straight, with adequate bone for strength but not so wide as to appear coarse. The animal should walk easily and freely so it can forage on pasture. The hoofs should be well-trimmed so the feet do not become deformed. Long pasterns make the legs look crooked; they should have some angle but not be so long that the dewclaws touch the ground. Breeding bucks, particularly, will be heavily discounted in the show-ring if they are weak in the pasterns.

The skin should be smooth, thin, and pliable. The hair should be reasonably fine to denote quality, but this varies considerably with the breed.

The udder should show plenty of capacity and be well held up to the body by the suspensory ligament so it will not be injured by banging on stones or other objects in the pasture or around the barn. A low-slung, pendulous udder is undesirable. The udder should be

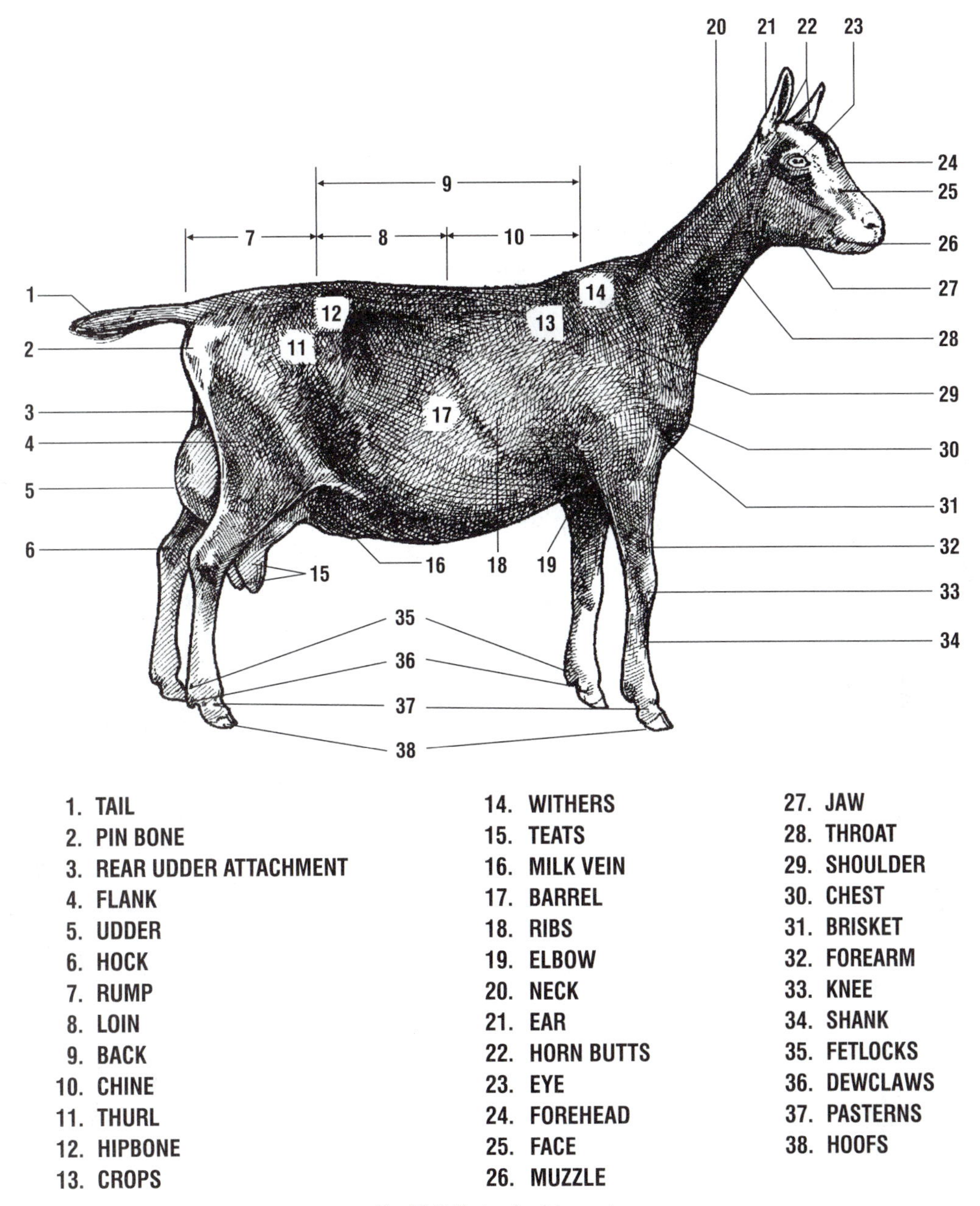

Fig. 23-3. Parts of a dairy goat.

pliable and soft, not hard and meaty. Hard bunches in the udder or teats will be discounted in judging or selection. The udder should be balanced in shape, with teats hanging the same and slightly tilted forward. The teats should be large enough to be easily milked. After a milking, the udder should be collapsed and pliable like a soft leather glove.

The head should have an alert, intelligent appearance, and the ears and head shape should conform to the particular breed. (See Fig. 23-4 and Fig. 23-5.)

RECOGNIZING AND EVALUATING COMMON FAULTS

No animal is perfect. In judging, therefore, one must be able to recognize and appraise common faults. Likewise, credit must be given to good points. Some faults of dairy goats include pinched muzzle, coarse neck, shallow body, pendulous udder, roach back, and thick, meaty shoulders.

Fig. 23-4. There are differences! The top goat is fat and nonproductive; the middle goat lacks depth of body; and the bottom goat displays lots of depth of body and body capacity. (Courtesy, Christine S. F. Williams, Michigan State University)

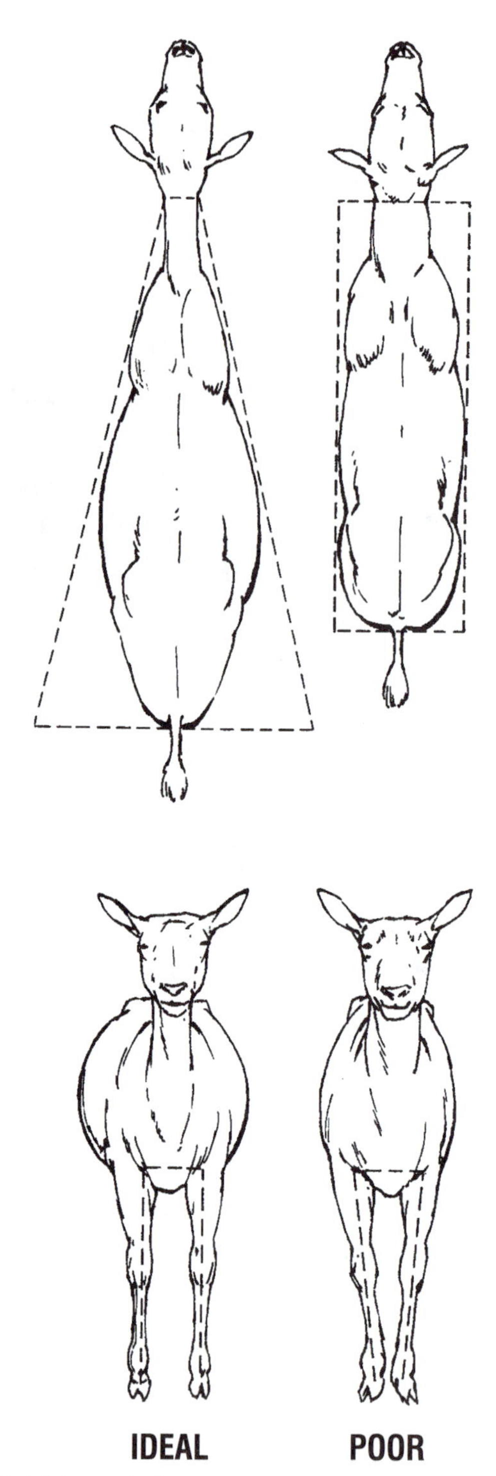

Fig. 23-5. Dairy goat body types, viewed from above and from the front.

SCORECARD JUDGING

A scorecard is a listing of the different parts of an animal, with a numerical value assigned to each part according to its relative importance. It is a standard of excellence. The use of the scorecard involves studying each part, then assigning a score to it.

Show-ring judging or actual selection on the farm is not accompanied by a scorecard. But a scorecard is a valuable teaching aid in acquainting students and beginners with the various parts of an animal and the relative importance of each. It systematizes judging and avoids any part of the animal being overlooked.

The American Dairy Goat Association (ADGA) provides its judges with a scorecard to evaluate dairy goats. The scorecard for does assigns 30% to general appearance, 30% to the mammary system, 20% to dairy character, and 20% to body capacity. For bucks, the scorecard assigns 40% to general appearance, 25% to dairy character, 25% to body capacity, and 10% to the mammary and reproductive systems. In addition, the judge must see that each animal meets the breed standard.

Figs. 23-6 and 23-7 show the ADGA Dairy Goat Scorecard for bucks and does.

ADGA DAIRY GOAT SCORECARD FOR BUCKS

	Points	
1. GENERAL APPEARANCE AND BREED CHARACTERISTICS		
Attractive individuality revealing vigor, masculinity with a harmonious blending and correlation of parts; impressive style and majestic carriage; graceful and powerful walk.		
Head – medium in length, clean cut; broad muzzle with large, open nostrils; lean, strong jaw; full, bright eyes; forehead broad between the eyes; ears medium size, alertly carried (except Nubians and La Manchas).	10	
Color – appropriate for breed.	5	
Shoulder blades – set smoothly against the chest wall and withers, forming neat junction with the body. **Back** – strong and appearing straight with vertebrae well defined. **Loin** – broad, strong, and nearly level. **Rump** – long, wide, and nearly level. **Hips** – wide, level with back. **Thurls** – wide apart. **Pin bones** – wide apart, lower than hips, well defined. **Tail head** – slightly above and neatly set between pin bones. **Tail** – symmetrical with body.	10	
Legs – wide apart, squarely set, clean-cut and strong with forelegs straight. **Hind legs** – nearly perpendicular from hock to pastern. When viewed from behind, legs wide apart and nearly straight. Bone strong, flat and flinty; tendons well defined. Pasterns of medium length, strong and springy. Hocks cleanly moulded. **Feet** – short and straight, with deep heel and level sole.	15	
SUBTOTAL 40		
2. DAIRY CHARACTER		
Animation, angularity, general openness, and freedom from excess tissue. **Neck** – medium length, strong and blending smoothly into shoulders and brisket. **Withers** – well defined and wedge-shaped with the dorsal process of the vertebrae rising slightly above the shoulder blades. **Ribs** – wide apart; rib bone wide, flat, and long. **Flank** – deep, arched, and refined. **Thighs** – incurving to flat from the side; apart when viewed from the rear. **Skin** – fine textured, loose, and pliable. Hair fine.	25	
SUBTOTAL 25		
3. BODY CAPACITY		
Relatively large in proportion to the size of the animal, providing ample digestive capacity, strength, and vigor. **Barrel** – deep, strongly supported; ribs wide apart and well sprung; depth and width tending to increase toward rear of barrel.	13	
Heart girth – large, resulting from long, well-sprung foreribs; wide chest floor between the front legs, and fullness at the point of elbow.	12	
SUBTOTAL 25		
4. MAMMARY AND REPRODUCTIVE SYSTEM		
Mammary – two rudimentary teats of uniform size and showing no evidence of extra orifices, extra teats, spur teats or teats that have been removed. Teats should be squarely placed below a wide arched escutcheon.	5	
Reproductive – two testicles of appropriate size for age of animal both showing evidence of being in a viable, healthy, breeding condition. All visible parts of reproductive system showing no evidence of disease or disability.	5	
SUBTOTAL 10		
TOTAL 100		

Fig. 23-6. ADGA dairy goat scorecard for bucks. The user of this card must consider ideals of type and breed characteristics. Items on the card are listed in order of observation.

ADGA DAIRY GOAT SCORECARD FOR DOES

Item	Points	Score
1. GENERAL APPEARANCE AND BREED CHARACTERISTICS		
Attractive individuality revealing vigor, femininity with a harmonious blending and correlation of parts; impressive style and attractive carriage; graceful walk.		
Head – medium in length, clean cut; broad muzzle with large, open nostrils; lean, strong jaw; full, bright eyes; forehead broad between the eyes; ears medium size, alertly carried (except Nubians).	10	
Shoulder blades – set smoothly against the chest wall and withers, forming neat junction with the body.		
Back – strong and appearing straight with vertebrae well defined.		
Loin – broad, strong, and nearly level.		
Rump – long, wide, and nearly level.		
Hips – wide, level with back.		
Thurls – wide apart.		
Pin bones – wide apart, lower than hips, well defined.		
Tail head – slightly above and neatly set between pin bones.		
Tail – symmetrical with body.	8	
Legs – wide apart, squarely set, clean-cut and strong with forelegs straight.		
Hind legs – nearly perpendicular from hock to pastern. When viewed from behind, legs wide apart and nearly straight. Bone flat and flinty; tendons well defined. Pasterns of medium length, strong and springy. Hocks cleanly moulded.		
Feet – short and straight, with deep heel and level sole.	12	
SUBTOTAL 30		
2. DAIRY CHARACTER		
Animation, angularity, general openness, and freedom from excess tissue, giving due regard to period of lactation.		
Neck – long and lean, blending smoothly into shoulders and brisket, clean-cut throat.		
Withers – well defined and wedge-shaped with the dorsal process of the vertebrae rising slightly above the shoulder blades.		
Ribs – wide apart; rib bone wide, flat, and long.		
Flank – deep, arched, and refined.		
Thighs – incurving to flat from the side; apart when viewed from the rear, providing sufficient room for the udder and its attachments.		
Skin – fine textured, loose, and pliable. Hair fine.	20	
SUBTOTAL 20		
3. BODY CAPACITY		
Relatively large in proportion to the size of the animal, providing ample digestive capacity, strength, and vigor.		
Barrel – deep, strongly supported; ribs wide apart and well sprung; depth and width tending to increase toward rear of barrel.	12	
Heart girth – large, resulting from long, well-sprung foreribs; wide chest floor between the front legs, and fullness at the point of elbow.	8	
SUBTOTAL 20		
4. MAMMARY AND REPRODUCTIVE SYSTEM		
A capacious, strongly-attached, well-carried udder of good quality, indicating heavy production and a long period of usefulness.		
Udder		
Capacity and Shape – long, wide, and capacious; extended well forward; strongly attached.	10	
Rear attachment – high and wide. Halves evenly balanced and symmetrical.	5	
Fore attachment – carried well forward, tightly attached without pocket, blending smoothly into body.	6	
Texture – soft, pliable, and elastic; free of scar tissue; well collapsed after milking.	5	
Teats – uniform, of convenient length and size, cylindrical in shape, free from obstructions, well apart, squarely and properly placed, easy to milk.	4	
SUBTOTAL 30		
TOTAL 100		

Fig. 23-7. ADGA dairy goat scorecard for does. The user of this card must consider ideals of type and breed characteristics. Items on the card are listed in order of observation.

DETERMINING THE AGE OF GOATS

Mature goats have 32 teeth of which 24 are molars and 8 are incisors. All incisors are in the lower jaw. The temporary teeth of the kid—milk teeth—are small and narrow, while permanent teeth are larger and broader.

These incisors in the lower jaw may be used as rough guides to determining age. The incisors are small and sharp in animals less than one year of age. At about one year, the center pair of teeth will drop out and are replaced by two large permanent teeth. At about the twenty-fourth month, two more large front teeth appear, one on each side of the first two yearling teeth. The 3- to 4-year-old has six permanent teeth, two more than the 2-year-old and these come in, one on each side of the 2-year-old teeth. The 4- to 5-year-olds have a complete set of eight permanent teeth in front. After this age, the approximate age can be told by the amount of wear in the front teeth. As the animal gets older, the teeth spread apart and finally become loose and some drop out. At this age the animal begins to lose its usefulness as a grazing animal. If the animal is still valuable for breeding purposes, however, it may be fed specially prepared feeds. (See Fig. 23-8.)

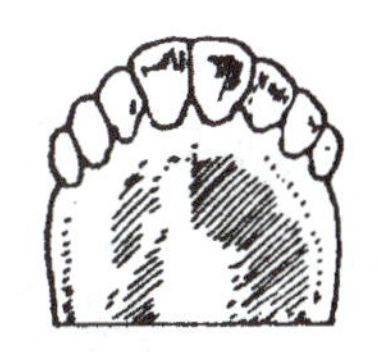

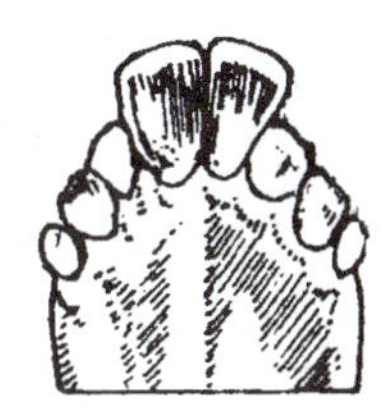

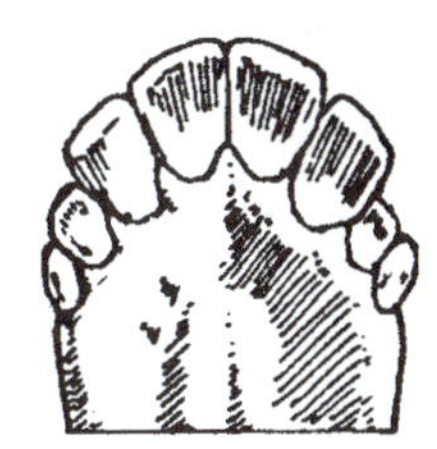

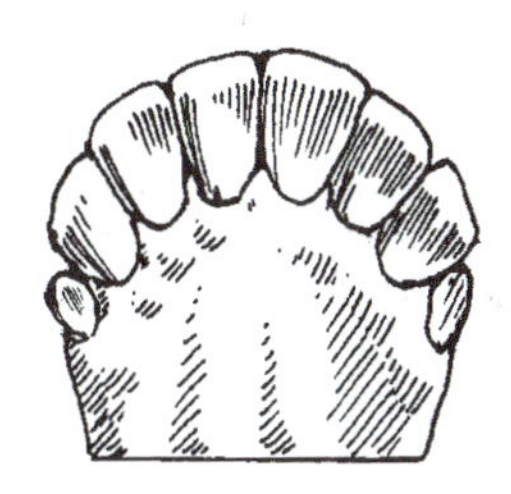

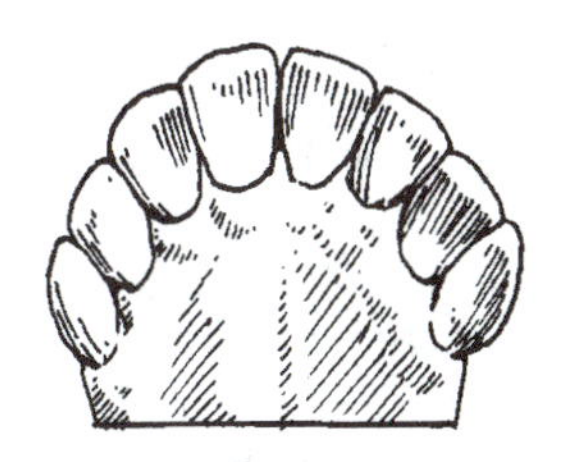

Fig. 23-8. Incisors in the lower jaw of a goat. Temporary teeth are small and narrow, while permanent teeth are larger and broader.

DAIRY GOAT SHOWS

The primary purpose of dairy goat shows is to select animals which come the closest to the ideal of a sound, productive type of dairy goat. Awards and publicity accruing from shows encourage breeders to adhere to the ideal for continued improvement of the dairy goat. Shows put quality dairy goats in the public view to improve their public image. Also, shows give dairy goat breeders recognition for long months of work in the barn, plus an opportunity to talk with other breeders.

FITTING AND SHOWING

Dairy goats need special feeding beginning 6 to 8 weeks prior to show day, depending on the condition of each animal. The animals should not be fattened, but should be fed enough extra grain to add bloom to the coat. Additional bedding is needed to keep the animals clean and to avoid stains. Regular leading and posing several weeks prior to the show day will prove very helpful in contests. To look their best, animals should have their hoofs trimmed and polished. Daily grooming with a stiff-bristled, not densely bristled brush, keeps the goats clean, stain-free, and slick-coated. Show animals should be carefully clipped, particularly the long hair along the backbone and flank. All of this handwork on the hair and skin improves the quality of the hide and hair. Even adding saddle soap to the leather show straps will add to the general appearance of a well-prepared show animal.

After all this work on the goat, the exhibitor should not be forgotten. For show day, the exhibitor should be immaculate and wear the prescribed white clothes.

Goat keepers must take feed and supplies with them to shows. These include a water bucket, possibly milk utensils, a brush, polish and cloth, and a goat blanket.

All goat shows should meet state health requirements. Upon arrival at a show, animals should be inspected before being unloaded. If signs of illness are found, the goats should not be unloaded. After the show, show animals should be isolated and watched closely for at least 10 days.

DIVISIONS OF THE SHOW

Shows are first divided by breed—Alpines, La Manchas, Nubians, Oberhaslis, Saanens, and Toggenburgs. Not all breed classes are provided at all shows. Some shows combine two or more breeds into a division called *All Other Purebreds* when few entries are expected in those breeds. Another breed division found at many shows is *Recorded Grade*, for does whose ancestry makes them ineligible to compete in the purebred classes.

Within each breed, animals are further divided by sex, and then into age classes. Thus, each animal competes against others of its own breed and sex and near its own age.

First-place winners from each class compete for champion. There may be junior and senior champions, which then compete for grand champion. A reserve champion is chosen from the remaining first-place winners plus the animal which placed second in its class behind the animal chosen champion.

In addition to age classes, many shows have group classes, in which animals are sorted by criteria other than age. Group classes commonly seen are: (1) get of sire, (2) produce of dam, (3) dam and daughter, (4) dairy herd, (5) breeder's trio, and (6) best udder. The final class of the day is usually Best in Show, from which the judge selects a winner from the grand champions of each breed.

Some shows have competitions in show techniques, in which the judges consider the appearance of the animals and the exhibitors, as well as the actual showing, looking for the exhibitors who show their animals to best advantage without undue fussing and maneuvering.

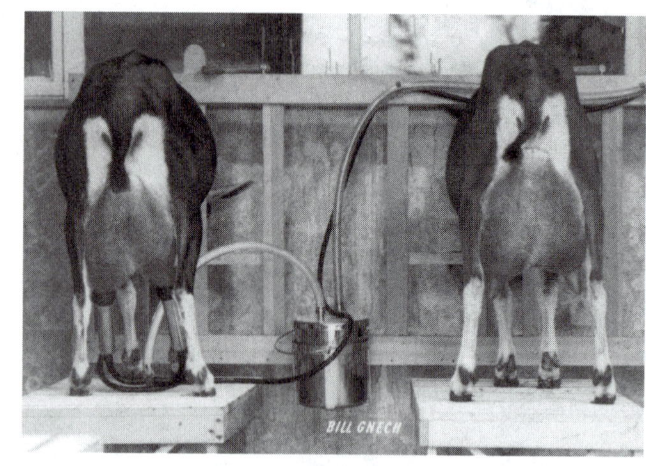

Fig. 23-9. Two grand champion Alpine does in milking parlor, GCH Seneca Valley's Malcolmite X *M *(left)* and GCH Seneca Valley's Gretchen *(right)*. (Courtesy, Dorothea Custer, President, Oberhasli Breeders of America, Harvard, IL)

MILK PRODUCTION BY DAIRY GOATS

Like dairy cows, goats vary greatly in their producing ability, but excellent producers are those producing 3,000 to 4,500 lb of milk, or 347 to 521 gal. per lactation. When produced by healthy animals in sanitary surroundings, goat's milk is a highly nutritious and healthful human food. It is usually pure white, like cow's milk. The primary difference between goat's milk and cow's milk is the relative size of the fat globules and the consistency of the curd. Goat's milk has smaller fat globules and a softer curd.

Milk and butterfat production leaders of the different breeds are given in Table 23-4.

LACTATION

Lactation is the secretion or formation of milk, or the time during which a female produces milk. It is governed by a variety of factors, including hormones from the anterior and posterior pituitaries, the thyroid, the pancreas, the adrenals, the placenta, and the ovaries.

The dairy goat producers should have some understanding of milk formation, ejection, and withdrawal, and what to expect from a lactation period.

TABLE 23-4
MILK AND BUTTERFAT RECORD HOLDERS, BY BREEDS[1]

Breed	Name and Number of Animal	Production		Year
		(lb)	*(kg)*	
Milk Production				
Alpine	Donnie's Pride Lois A177455P 3*M	6,416	2,910	1982
La Mancha . . .	Tyler Mt. May's Priscilla AL618876 2*M	5,400	2,449	1991
Nubian.	Skyhill's Elisha N904515 7*M	5,940	2,694	1996
Oberhasli. . . .	Catoico Summer Storn B0935588 2*M	4,665	2.116	1997
Saanen	JC-Reed's Cloverhoof Haley AS0894085 2*M	6,571	2,980	1997
Toggenburg. . .	Western-Acres Zephyr Rosemary AT0926741 4*M	7,965	3,612	1997
Butterfat Production				
Alpine	Donnie's Pride Lois A177455P 3*M	309	139.7	1982
La Mancha . . .	Pansy DK Mardi Gras L931457 2*M	180	81.6	1997
Nubian.	Pacem Faun's Folly N324844P 2*M	384	174.1	1984
Oberhasli. . . .	Catoico Summer Storn B0935588 2*M	135	61.2	1997
Saanen	Quality Crest JR Sugar Storm AS346611 4*M	225	102.0	1984
Toggenburg. . .	Western-Acres Zephyr Rosemary AT0926741 4*M	240	108.8	1997

[1]Courtesy, American Dairy Goat Association, www.adga.org.

MILK FORMATION, EJECTION, AND WITHDRAWAL

The basic milk-producing unit of the udder is a very small bulb-shaped structure with a hollow center called the *alveolus*. Alveoli are lined with a single layer of epithelial cells which are responsible for secreting milk. Their functions are threefold: (1) remove nutrients from the blood, (2) transform these nutrients into milk, and (3) discharge the milk into the lumen. Each alveolus is surrounded by a network of capillaries from which nutrients are extracted and by a specialized type of muscle cell, called the myoepithelial cell, which is sensitive to the effects of the hormone oxytocin from the posterior pituitary. When oxytocin is secreted into the blood, it stimulates contraction of these muscle cells, thereupon initiating milk ejection.

Groups of alveoli empty into a duct thereupon forming a functional unit called a *lobule*. Several lobules empty into another duct system forming a larger unit called a *lobe*. The ducts of lobes empty into what is referred to as a *galactophore*, which, in turn, empties into the gland cistern.

The alveolus is, in effect, a milk factory. It has the ability to take nutrients from the blood and transform them into one of nature's most perfect foods. *Galactopoiesis is the term used to describe the biosynthesis of milk.*

The ducts of the udder provide a storage area for milk and a means of transporting it to the outside. No milk secretion, per se, occurs within the ducts. The cells lining the cisterns and duct systems consist of two layers of epithelium. Myoepithelial cells are arranged in a longitudinal organization allowing the ducts to shorten and increase the diameter to facilitate the flow of milk.

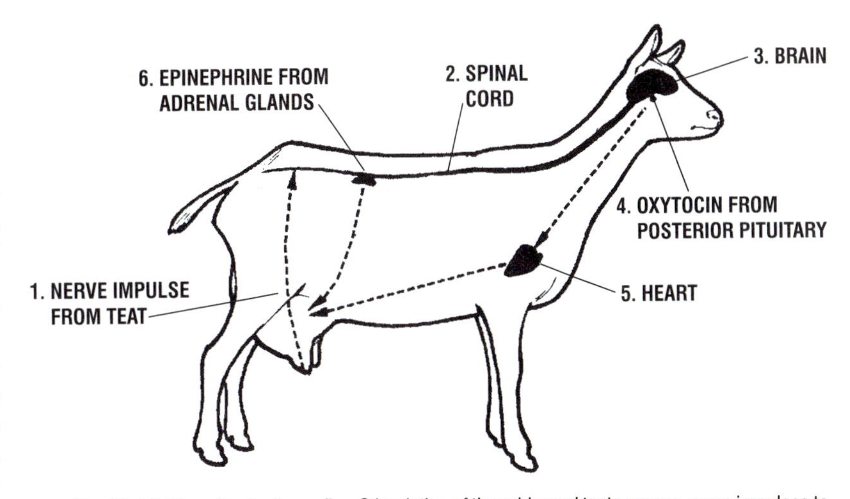

Fig. 23-10. The milk ejection reflex. Stimulation of the udder and teats causes nerve impulses to travel to the posterior pituitary at the base of the brain where the hormone oxytocin is released into the bloodstream. Oxytocin travels via the bloodstream to the udder where it causes milk to be ejected from the alveoli. Release of the hormone epinephrine can inhibit the action of oxytocin.

Milk secretion is regulated primarily through hormonal mechanisms. However, milk let-down is initiated largely through neural mechanisms. In the udder, there is a network of afferent (sensory) and efferent (motor) nerves. Receptors in the udder are sensitive to touch, temperature, and pain. During the preparation for milking, the washing and cleaning of the udder stimulates these receptors, and the process of milk ejection is initiated. Motor nerves transmit impulses from the brain and regulate blood flow and smooth muscle activity around the ducts and in the teat sphincter.

When the goat is startled or subjected to pain, the hormone epinephrine is released, and the sympathetic nervous system is stimulated. Blood vessels constrict so that blood can be shunted to other parts of the body (*e.g.*, skeletal muscle) and milk ejection is slowed and production depressed.

Before milk is available to the kid or milker, it has to be forced from the alveoli into the larger ducts and cisterns. This process is known as the *let-down* of the milk. Here is how it works (see Fig. 23-10): When the udder (especially the teats) is stimulated by a kid of a milker, (1) impulses are conducted along the nerves to the posterior pituitary at the base of the goat's brain; (2) the posterior pituitary stores and releases the hormone oxytocin into the bloodstream; (3) the blood transports oxytocin back to the udder; and (4) the oxytocin causes the smooth, musclelike cells surrounding each alveolus to contract, thereby forcing the milk out of them into the large ducts and cisterns of the udder.

Besides stimulation of the udder and teats, other stimuli, such as familiar sounds, smells, or sights, can cause the release of oxytocin.

Discharge of the hormone epinephrine into the bloodstream results in a poor milk ejection reflex. Epinephrine is released in response to aggression, fright, or a sudden change in the environment. This should be avoided since the ejection reflex must be strong in order to discharge as much oxytocin as possible to eject the alveolar milk and stimulate the mammary gland, enhancing its yield.

Milking is the act of removing milk from the udder. It is routinely carried out through one of three methods: (1) kid suckling, (2) hand milking, or (3) machine milking.

LENGTH OF LACTATION

Depending on the breed and the environmental conditions, the length of lactation varies from 200 to 350 days. A normal termination of lactation is considered to be 305 days.

The lactation curve—the amount of milk produced each day over the 305-day period—resembles that of the cow. The curve tends to rise in production, peaking at weeks 2 to 10. After peaking, milk production declines steadily at about 10% monthly. High-producing goats reach a higher peak production than low-producing goats, but following this, they decline at a similar rate.

FACTORS AFFECTING MILK YIELD

A number of factors affect the amount of milk, among them, the following:

1. **Body size and weight.** There is a positive relationship between body weight and milk yield. Larger does produce more milk, but research indicates that only about 10% of the variation in milk yield can be accounted for by body weight.

2. **Age.** Age affects milk yield, but it is closely tied to body weight. Age accounts for much of the increase in body weight. Peak milk yield is reached when the doe is between 4 and 8 years of age.

3. **Udder size and shape.** A weak udder attachment is considered a major defect, but udder volume is highly correlated with milk yield. The larger the udder, the greater the yield.

4. **Growth.** There is a tendency for late-maturing animals, with a flat growth curve, to be more productive.

5. **Litter size.** The results of a number of studies indicate that mammary growth is regulated by the number of kids born—the more kids, the greater the mammary growth during pregnancy. This seems to be reasonable since more milk would be needed.

6. **Season of kidding.** To some extent, the season of kidding influences milk yield. Yields from lactation beginning early in the year (January, February, and March) are lower than those beginning later in the year (October, November, and December). A French study showed this difference to be about 441 lb.

7. **Nutrition.** The mammary gland needs glucose to form lactose, which in turn largely controls the movement of water into milk. A reduction in feed intake quickly lowers the milk yield, since there is very little glucose stored in the body.

8. **Temperature.** Exposure of lactating goats to cold reduces milk secretion. For example, one study showed that the milk yield at 31°F was about 30% below that obtained from goats in an environmental temperature of 68°F.

9. **Multiple relationships.** Many body relationships are interrelated. Therefore, the comparison of a variety of body measurements—size, udder volume, abdominal volume, and volume of the muscles—should result in a better understanding of type-production relationships.

10. **Disease.** Disease lowers milk production, with the degree of the effect determined by the kind and severity of the disease.

PRODUCTION OF HIGH-QUALITY MILK

High-quality milk is safe, pure, stable, uniform—and legal. The primary requisites for the production of high-quality goat's milk are cleanliness and low temperature.

CLEANLINESS—SANITATION

There are three components to cleanliness and sanitation: (1) the animals, (2) the area, and (3) the equipment.

- **Milking animals**—The flanks, udder, and belly should be free from dirt. This means that goats should be clipped closely on the udder and flank areas and brushed or washed frequently. Before milking, the udder and teats should be cleaned and treated with a sanitizing solution and then dried just before milking. It is best to dry each with a different cloth or paper towel. Frequently used chemical sanitizers include chlorine, iodine, and quaternary ammonia. These must be employed at the proper concentrations to be effective.

- **Milking area**—Milking should be performed in an area apart from the living and feeding areas of the goat herd. The room should be clean; well-lighted; well-ventilated; free of dust, odors, and flies; and easily washed down after each milking.

- **Milking equipment**—The production of high-quality milk requires strict cleaning and sanitizing procedures for all equipment which comes into contact with the milk. Because of the complex nature of milk, its residues pose difficulty in cleaning. To reduce this difficulty to a minimum, all multi-use equipment and utensils should be made of smooth, nonabsorbent, noncorrosive, nontoxic materials; for example, glass and stainless steel. All surfaces contacting milk should be cleaned after each use. Four basic steps applied to cleaning and sanitizing include:

1. Rinse with warm water, about 104°F.

2. Wash with an alkaline cleaning solution. Preparation of the solution and the temperature vary according to the nature and amount of residue to be removed.

Fig. 23-11. Milking parlor of a commercial dairy in Michigan. The worker's "backwards apron" of heavy denim protects his clothing from soiling when he sits on the platform to strip out does. (Courtesy, Judy Kapture and Jim Vandergriff, Caprine Supply, Shawnee, KS)

Washing can be achieved by hand brushing or by a high-velocity turbulent flow.

3. Rinse in hot water; this removes milk and chemical residues, presenting a clean surface that can be effectively sanitized. Use a mildly acid rinse at regular intervals to prevent the buildup of milkstone—a film from the reaction between milk residues and cleaning chemicals.

4. Sanitize the equipment surface. Either heat 171°F for at least five minutes or use chemicals as sanitizers. In most situations, chemical sanitization is less costly and more convenient. There are several approved dairy sanitizers.

LOW TEMPERATURE

Warm milk deteriorates rapidly. It absorbs odors from the surroundings readily, contains enzymes which can produce undesirable chemical changes, and is an ideal medium for the growth of microorganisms. Upon removal, milk should be moved from the milking area to minimize the exposure to odors, and cooled as rapidly as possible to approximately 32°F. Depending on the type of operation, equipment to cool the milk will vary. If commercial equipment is not available, milk will cool much more rapidly if it is first placed in circulating ice water, then in a refrigerator. Cooling slows the growth of bacteria, slows the action of enzymes, and limits the absorption of odors.

PREVENTION OF OFF-FLAVORS

Most seasoned dairy goat producers recommend that bucks should not be kept in the same buildings where milking does are kept, as they impart an off-flavor to the milk. There are other factors influencing milk flavor, and most consumers base the quality of any food product on its flavor. They want milk that tastes good. The flavors most often found in milk and their causes and prevention are:

1. **Feed and weed.** Many weed and woody plants—for example, ragweed, wild onion, goldenrod, honeysuckle, elderberry, grape leaves, and many others—produce off-flavors in goat's milk. The surest way to prevent feed odors is to keep the animals in a drylot and feed hay or silage of known content. No feed should be given within two hours of milking, especially green chop forages and silage. Usually, concentrate feeds can be fed while does are being milked without affecting the flavor of the milk.

2. **Oxidized.** These are sometimes described as cardboard flavors. Some causes of oxidized flavors are (a) metallic contamination from copper and iron, which may be alleviated by using stainless steel; (b) exposure to sunlight or just daylight; (c) foaming; and (d) drylot feeding. Feeding vitamin E to milking herds will reduce or eliminate oxidized flavors.

3. **Rancid.** This flavor is caused by a breakdown of the butterfat which releases strong-flavored acids. This action is caused by the enzyme lipase, which is present in all milk. The primary causes of rancid milk are (a) goats well-advanced in lactation; (b) excessive agitation of milk; and (c) slow cooling with foaming.

4. **Barney.** This flavor(s) is caused by dirty stables, poor ventilation, unclean milking, and unclean goats—all of which can be alleviated.

5. **Salty.** This flavor, which masks the slightly sweet flavor of milk, is caused by mastitis, or certain individual goats. Milk from goats that have mastitis should not be marketed.

6. **Malty.** Malty flavor is primarily due to a high bacteria count. The remedy is to keep bacteria out of milk as much as possible, and to prevent growth of those that do get into it. Clean and cold milk will practically eliminate malty flavor. Also, milk handlers should pick up all the milk and not leave any of it in the farm bulk tank.

7. **High-acid sour milk.** This is due to a very high bacterial count. In these days of mechanical refrigeration, there is no excuse for sour milk; it should simply be cooled as rapidly as possible from the 90°F temperature of the milk pail to near 32°F.

8. **Unnatural or foreign.** This refers to flavors that come from medicinal agents and disinfectants. The control of such off-flavors consists of (a) handling medicines and disinfectants so that the flavors or odors from them will not get into the milk, and (b) using chemical sanitizers only in the concentrations indicated by the directions. Milk from drug-treated goats should not be marketed for at least 72 hours after the last treatment, or longer if so prescribed on the drug label or by the veterinarian.

For good-tasting milk, the dairy producer should keep it clean; keep it from contact with air and light; keep it cold; feed silage after a milking; use good-quality feed; and not include milk from problem goats. Properly produced and handled, good goat's milk does not have any stronger flavor than good cow's milk.

MILK COMPOSITION

Goat's milk is a healthful and nutritious food. It is whiter than cow's milk because it contains no carotene which causes fat to have various degrees of yellow coloring. Thus, butter made from the cream of goat's milk is white. The composition of goat's milk is similar to cow's milk, and varies both within and between goat breeds. Additionally, the composition varies depending on milk yield, stage of lactation, and level and quality of feeding. Table 23-5 compares the composition of goat's milk to that of cows, sheep, and humans. There are some noticeable differences between the species.

TABLE 23-5
AVERAGE COMPOSITION OF MILK FROM GOATS, SHEEP, COWS, AND HUMANS[1]

Species	Weight	Water	Food Energy	Protein	Fats	Carbo-hydrates	Calcium	Phos-phorus	Iron	Vitamin A	Thiamin	Ribo-flavin	Niacin	Vitamin B-12
	(g)	(%)	(kcal)	(g)	(g)	(g)	(mg)	(mg)	(mg)	(IU)	(mg)	(mg)	(mg)	(mcg)
Goat	100	87.5	67.0	3.3	4.0	4.6	129.0	106.0	0.05	185.0	0.04	0.14	0.30	0.07
Sheep	100	80.7	108.0	6.0	7.0	5.4	193.0	158.0	0.10	147.0	0.07	0.36	0.42	0.71
Cow	100	87.2	66.0	3.3	3.7	4.7	117.0	151.0	0.05	138.0	0.03	0.17	0.08	0.36
Human	100	88.3	69.1	1.0	4.4	6.9	33.0	14.0	0.02	240.0	0.01	0.04	0.20	0.04

[1]Data from Ensminger, A. H., M. E. Ensminger, J. E. Konlande, and J. R. K. Robson, *Foods & Nutrition Encyclopedia*, CRC Press, Boca Raton, FL, 1994, Table F-36.

Fat is the most variable component of milk. It is composed primarily of triglycerides, and it forms globules which are suspended in milk as an emulsion. These fat globules have diameters similar to those found in milk from cows, but goat's milk has a higher percentage of small fat globules—a characteristic that seems to be partly responsible for the better digestibility of goat's milk. Also, because of the small fat globules, the cream rises very slowly and never as thoroughly as in cow's milk.

MILKING RECORDS

Records are an essential part of a profitable dairy goat operation. Records tell if individual does are paying for their feed and keep with their milk production. The type and extent of the records kept will vary with the individual operation.

METHODS AND EXTENT OF RECORDING

The novice will probably begin with a chart on the wall near scales which weigh in tenths of pounds. Then, each milking is weighed and entered on the chart.

Fig. 23-12 shows a form that can be used to record the milk production of 10 does for 1 week. Such records help to cull poor producers and detect disease. Also, the total milk production compared to the feed bill indicates the financial status of the operation.

More elaborate record keeping can involve individual lifetime records. These records can include such items as birth date, pedigree, health treatments, notes on production, breeding dates, freshening dates, lactation time, and total milk and fat produced during the lactation. Figs. 23-13 and 23-14 show a suggested record form that can be printed on the front and back of a 5 × 8 in. card.

DAIRY HERD IMPROVEMENT PROGRAMS

The Dairy Herd Improvement Association (DHIA) programs of the U.S. Department of Agriculture are conducted through the various state agricultural experiment stations. DHIA programs are available almost everywhere, and they are under the direction of the local county extension agents. These programs provide the services of an approved tester who regularly visits herds, weighs the milk, and performs butterfat tests. Dairy producers receive the official record of milk and butterfat production on each doe in their herds.

The testing program can be either Dairy Herd Improvement (DHI) or Dairy Herd Improvement Registry (DHIA).

- **DHI**—Regulation of this testing plan is done by a national coordinating group which has both goats and cattle under its supervision. DHI provides official production records but does not give the registry associations any of the data collected.

- **DHIA**—This testing plan provides official records recognized by the registry associations. It requires that the individual dairy producer (1) apply to the registry association, (2) pay the state and/or county DHI fees and follow its rules and regulations, and (3) pay additional fees to the registry association and follow its special rules and regulations. The testing may be either (1) Standard or (2) Group. Group testing is usually less expensive than Standard, but a number of rules must be known and followed.

Information on DHIA can be obtained from the state extension dairy producer.

MANAGEMENT OF DAIRY GOATS

Certain principles of dairy goat management are similar to those of meat goat, Angora goat, and sheep management. These have been discussed elsewhere. Also, certain chapters of this book are devoted to some important management principles such as feeding, reproduction, genetics, behavior, and business aspects. So, these topics will be mentioned only briefly in this section. Nevertheless, some aspects of management are unique to dairy goats and require some essential skills.

MILK RECORD

From ____________________
To ____________________ Year __________

Name or Number																					
		Lb	10ths	Lb	10ths	Lb	10ths	Lb	10ths	Lb	10ths	Lb	10ths	Lb	10ths	Lb	10ths	Lb	10ths	Lb	10ths
Sunday	a.m.																				
	p.m.																				
Monday	a.m.																				
	p.m.																				
Tuesday	a.m.																				
	p.m.																				
Wednesday	a.m.																				
	p.m.																				
Thursday	a.m.																				
	p.m.																				
Friday	a.m.																				
	p.m.																				
Saturday	a.m.																				
	p.m.																				
Total for Week																					

Fig. 23-12. A form for recording the dairy milk production and weekly totals for 10 does.

INDIVIDUAL GOAT LIFETIME RECORD

Name ____________________ **Sex** __________ **Birth Date** __________

Registration No. ____________________ **Tattoo: RE** __________
LE __________

Pedigree on Dam:

__________ (Dam) — __________ (Dam's Sire)
__________ (Dam's Dam)

Pedigree on Sire:

__________ (Sire) — __________ (Sire's Sire)
__________ (Sire's Dam)

Treatments:

Date	Type	Date	Type	Date	Type
Date	Type	Date	Type	Date	Type
Date	Type	Date	Type	Date	Type

Notes ____________________

Date and Reason for Removal from Herd ____________________

Fig. 23-13. A suggested individual lifetime record, front side.

BREEDING RECORD AND MILK PRODUCTION

Lactation	Date Bred	Date Due	Date Fresh	Kid Disposal	Days Milked	Milk (lb)	Fat Test	Fat (lb)
1st								
2nd								
3rd								
4th								
5th								
6th								
7th								
8th								
9th								
10th								

Fig. 23-14. Back side of individual lifetime record shown in Fig. 23-13.

FEEDING

Even the best bred dairy goat will not produce satisfactorily unless fed properly. Knowledge of the nutritional requirements of the dairy goat is necessary, but these requirements have not been as accurately determined as they have for other types of livestock. The National Research Council (NRC) recommendations on the nutrient requirements of goats and other aspects of feeding are covered in Chapters 5 and 21. Feed allowances should take into account the age, mature size, physiological state (pregnant, nonpregnant, and/or lactating), activity level, and milk production. With the application of suitable feeding standards, rations and feeding programs can be improved.

BREEDING

Goats are seasonal breeders. The breeding season usually is from late August through mid-March. Does come in heat at intervals of 17 to 21 days, and usually remain in heat from 1 to 2 days. Signs of heat are usually easily detected and include uneasiness, riding other animals, standing for riding, unusual amounts of tail wagging, frequent urination, abnormal amounts of bleating, reddish and perhaps swollen vulvas, and moisture under their tails. Conception is highest from the middle to the latter part of the heat period. For this reason, if signs of heat are first noticed in the morning, it is recommended that goats be bred late in the afternoon; if heat is detected in the afternoon, goats should be bred late the following morning. Good records of all heat periods and breeding dates are a component of good management.

When keeping a buck cannot be justified or when there is need to improve a herd through the use of proven bucks, artificial insemination (AI) or embryo transfer (ET) can be employed. This requires heat detection and records of the heat periods.

Does may be bred when they reach 65 to 70 lb. If does are well fed, they should reach this weight by 8 to 9 months of age. Thin does are usually in production by 13 to 14 months of age.

The gestation period is 148 to 154 days. The expected kidding date may simply be calculated by counting forward 5 months (150 days) from the date of breeding. Goats produce 1 to 3 kids.

Does should be bred to freshen once each year, with a dry period of 6 weeks to 2 months. The dry period allows the mammary system time to repair and regenerate mammary tissue and to gain new stimulation for lactation following kidding. To dry off a doe, milking is ceased whenever the dry period is to begin, but high-producing does (more than 5 lb per day) require a gradual slackening in milking until milk production drops below 5 lb daily. Then, after 5 to 7 days during which the doe is not milked, the udder should be hand stripped. During the dry period, the doe should be well fed, as this is the time the unborn kid(s) gains 70% of its weight.

Chapter 3, Genetics and Selection, and Chapter 4, Reproduction in Sheep and Goats, provide more complete information relative to breeding goats.

CARE AT KIDDING

Shortly before the expected kidding date, the doe's udder, hindquarters, and tail should be clipped for greater cleanliness during kidding. A few days before she is due, a laxative feed such as bran or beet pulp should replace some grain feeding.

Signs of approaching kidding include: rising of the tail bone; loose to the touch around tail; sharp hollows on either side of the tail and flank; rapid cud chewing; restlessness; pawing at bedding; low, plaintive bleating; rapidly filling udder; and a mucous discharge from the vulva. Normal presentation of the kid is front feet first, with the heal resting on the feet. Heavy labor for some time without the expulsion of the kid indicates trouble, and a veterinarian or experienced producer should be summoned before the doe becomes exhausted.

Quiet, draft-free quarters with clean bedding should be provided. As soon as the doe has completed kidding, she should be offered warm water if the weather is cold and cool water if the weather is hot, her pen should be cleaned and her bedding freshened. Also, at this time, each kid's umbilical cord should be treated with tincture of iodine. After the kid(s) have nursed, the doe should be checked, and, if necessary, relieved of excess milk.

Following kidding, the first milk is called *colostrum*. It is not normally used for human consumption, but it is essential to the well-being of the kid(s).[2]

Fig. 23-15. Does give birth to 1 to 3 kids after a gestation of about 150 days. At kidding they should be provided with clean, warm, draft-free Quarters. (Courtesy, *Dairy Goat Guide*, Waterloo, WI)

[2]Because of the presence of the Caprine Arthritis Encephalitis (CAE) virus, scientists recommend that kids be fed (a) Nostrum only from does identified as negative with the Agar Gel Immunodiffusion Test (AGID) for antibody against CAE, (b) heat-treated colostrum (135°F for 1 hour), or (c) no colostrum at all.

RAISING AND FEEDING KIDS

It is important that the kid receive colostrum as soon as possible after birth, and for at least two days following birth. Colostrum provides antibodies which give the kid resistance to disease and, in addition, acts as a mild laxative which aids in cleaning the digestive residue from the newborn kid. Colostrum also is high in nutrient value, especially vitamin A, B-vitamins, proteins, and minerals. But overfeeding colostrum or other milk can cause scours. Extra colostrum may be saved, refrigerated, or frozen and at some later date warmed to body temperature and fed. After the first few days on colostrum, the kid may be left on goat's milk or changed to cow's milk or a commercial milk replacer. Further details about feeding kids are given in Chapter 21.

Fig. 23-16. Milk feeders can save time when kids are being raised once the kids are trained. About 10 kids can be fed at one time. (Courtesy, Judy Kapture and Jim Vandergriff, Caprine Supply, Shawnee, KS)

In dairy goat operations, the does do not generally raise the kids. They are allowed to clean and dry the kids, and possibly give them their first nursing. Then the kids are removed from the does. When the kids are taken from their mothers, they must have a warm, dry place to sleep. Deep wooden boxes with slanted floors that are raised off the ground to provide drainage make good beds for new kids. The boxes should be well-bedded and draft-free.

The kids also must have plenty of exercise and as much sunshine as possible. Older kids should have something on which to climb and jump. Boxes or barrels can be provided for this purpose. Buck kids should be separated from the does at about two months of age to avoid premature breeding.

MILKING

Goats should be milked twice a day on a regular schedule—preferably every 12 hours. They should be milked from the side, on a raised platform, either by hand or machine. Their udders and teats should be washed and then dried thoroughly with clean towels. Milking should begin within 1 to 2 minutes after the udders have been washed. The first stream of milk from each teat should be discarded into a strip cup to check for signs of mastitis. After the milking, it is advisable to dip each teat in a dairy teat dip or a solution of 4 parts Clorox and 1 part water, since teat dips have proven successful in the reduction and prevention of mastitis. The milk should be filtered and cooled immediately (see earlier section, "Production of High-Quality Milk").

HAND MILKING AND MACHINE MILKING

Despite the many technological advances, most of the goat's milk in the world is obtained via hand milking. It is widely used in the lesser developed countries where labor is cheaper than automation, and in many small dairies that cannot justify milking machines.

In hand milking, the teat is grasped between the thumb and forefinger; then, by applying pressure with the other fingers, the milker forces the milk from the teat cistern through the streak canal. Through this method, more milk can be obtained than by the use of a milking machine.

A milking machine consists of three major parts: (1) vacuum supply, (2) pulsation, and (3) milking unit. Each milking unit has two individual teat cups attached by hoses to either a suspended or a floor-type machine that houses a unit pulsator; or in cases where a master pulsator is used, a slave pulsator.

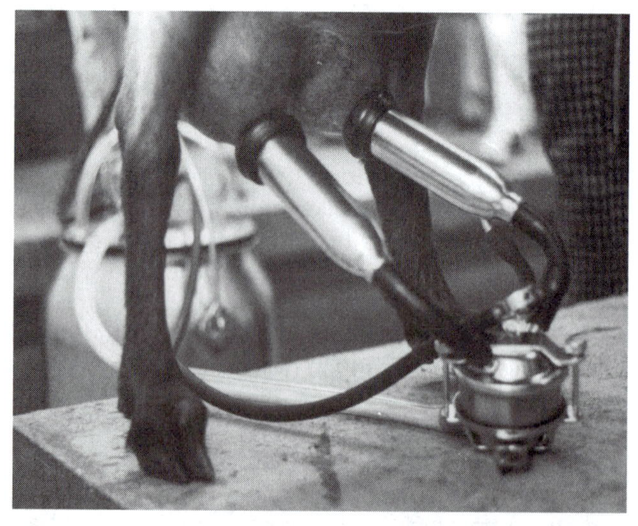

Fig. 23-17. Any standard milking machine used for cows can be adapted to goats. The major adaptation is, of course, from four teats to two. The Universal claw shown above is especially designed for goat milking. (Courtesy, Judy Kapture and Jim Vandergriff, Caprine Supply, Shawnee, KS)

Each teat cup contains an inflation tube called a teat cup liner which is surrounded by a metal outer shell, generally of stainless steel. When teat liners lose their elasticity, teats become very susceptible to injury because the liners produce a sharp slapping action.

In the first phase of milking, called the *expansion phase*, pressure decreases in the space between the outer shell and the liner. This action dilates the teat canal and promotes milk flow. The pressure inside the teat cistern is greater than the pressure outside the teat, and milk is forced through the streak canal. The *massage phase,* or nonmilking phase, is initiated when air is pumped into the space between the liner and the outer shell. The teat cup liner then collapses around the teat. This massaging action promotes circulation in the teat and allows the teat a brief moment of rest. If the massage phase is too short and the expansion phase too long, circulation is impaired, and the teat becomes injured due to congestion.

The decision to purchase a milking machine must be an economic one. Very little time can be saved by machine milking instead of hand milking small- or medium-size herds. An investment in mechanization may, however, be justified because of labor shortage and cost, as well as for acceptable working conditions of milkers.

MILKING SYSTEMS

The choice of a milking system is determined by a variety of factors, including the type of milking plant—bucket, milker, pipeline—the kind and number of milking platforms, stalls, and units; the number of animals; and the means of handling the animals. Many of the systems devised have been inspired by concepts of dairy cow milking, and other systems have resulted from some practical observations of dairy goat operators. In general, there are three types of milking systems: (1) milking in the barn, (2) milking parlors with various platform arrangements, and (3) rotary or turntable milkers.

ESSENTIAL MANAGEMENT SKILLS

Good dairy goat operators must possess an endless list of skills, not all of which can be discussed. However, some very basic, and often performed, skills follow.

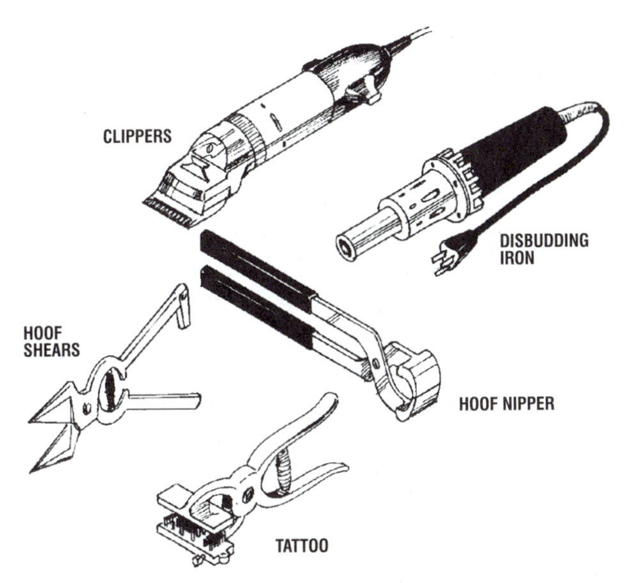

Fig. 23-18. Some of the tools used to perform some essential management practices.

DISBUDDING

Some kids are born hornless, but as a practical management procedure, every kid—male and female—should be examined for horn buds. Horns are a danger to the operator and to the herd. *Disbudding is the practice of removing horn buds from very young kids.* The best method for disbudding is the use of a disbudding iron which heats a circular tip to a red-hot temperature.

It is best to disbud kids when they are a few days old and the horn buds can be clearly felt. To disbud, preheat the iron. Then, follow these steps:

1. With a finger, locate the horn buds.
2. Clip as much of the hair as possible from the area of the horn buds.
3. Center the iron on each horn bud, applying it with a circular motion and light pressure for 5 to 10 seconds, depending on the size and development of the horn bud.
4. If necessary, reheat the iron between horn buds.
5. If a horn bud is large, a second burn after a few minutes' rest is advisable.

A proper burn leaves two copper-colored circles.

If kids have visible horns, the horn tips can be removed first and then burned around the base.

Following the disbudding, kids should be checked every 2 to 3 months for scur growth. Scurs should be immediately removed with the disbudding iron.

While kids can be held between the knees or thighs for the disbudding procedure, a holding box makes the task easier for one person (see Chapter 25, Buildings and Equipment for Goats).

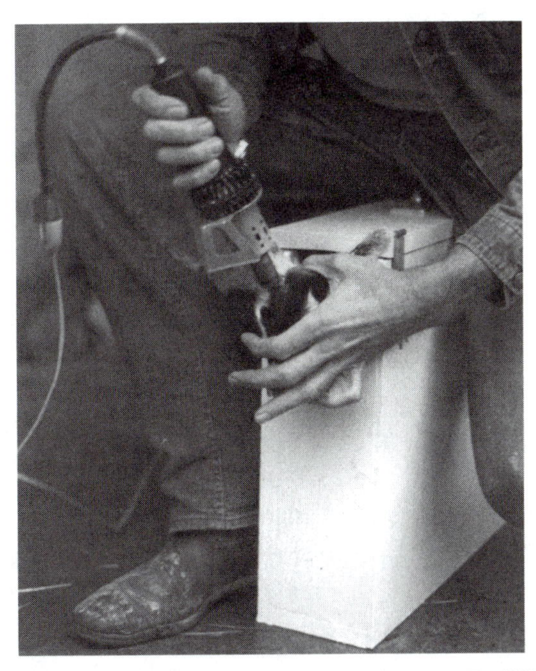

Fig. 23-19. Disbudding a kid with a red-hot electric iron. The kid is in a specially designed holding box which is used for disbudding and ear tattooing. (Courtesy, Judy Kapture and Jim Vandergriff, Caprine Supply, Shawnee, KS)

Caustic sticks (potassium or sodium hydroxide) can be used to disbud, but they require extreme care. They may injure the person making the application, other goats, or the goats being disbudded.

DEODORIZING

Bucks are noted for bad smelling. They label their territory and everything in it with their buck goat aroma by rubbing their heads against everything claimed. The buck's shed, posts, and feed buckets; the operators boots and clothing, and even the does will acquire the buck aroma.

The scent or musk glands responsible for this odor are located immediately behind and along the inside edge of each horn base. In a polled kid they are in the same position.

Making two additional burns at the time of disbudding will easily deodorize goats as kids. The red hot disbudding iron should be applied about 0.5 in. behind and toward the center from the disbudded horn buds for about 10 seconds. The hair should be clipped from these areas, also. This results in two sets of overlapping circles as shown in Fig. 23-20.

Older animals may also be deodorized. They require more restraint, though. After the hair over the area has been clipped or shaved, the scent glands can be identified as areas of skin that are shiny and darker than the surrounding skin. These areas are burned for 10 to 15 seconds with a red-hot disbudding iron.

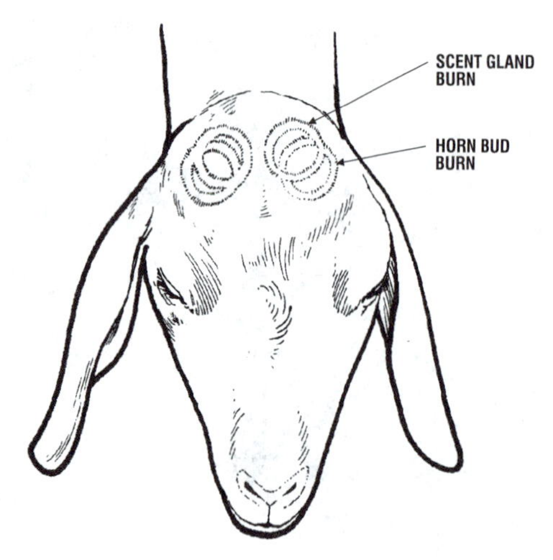

Fig. 23-20. Disbudding and deodorizing with a red-hot iron leaves two sets of overlapping circles.

CASTRATION

Most goat owners should castrate every buck kid, since there is little to be gained genetically in small herds. Castration is performed when the kid is between 1 and 14 days old. Castrating may be done with a knife, an elastrator, or an emasculatome.

A knife is the surest and least expensive method of castration. When a knife is used, the hands should be thoroughly washed with soap and water and rinsed in a good disinfectant. Likewise, the knife or scalpel should be thoroughly disinfected prior to beginning the work and between operations. The kid is usually held with his back to the assistant, who grasps the hind and front legs of the same side in each hand (see Fig. 23-21). The scrotum should also be rinsed with a mild disinfectant before the following procedure:

Fig. 23-21. Method of holding a buck kid for castration.

1. Cut off the lower one-third of the scrotum.
2. Draw each testicle out with the attached cords. Use a firm, steady downward pull.
3. In young animals, draw testicle out until the cord breaks.
4. In animals more than 4 or 5 weeks old, scrape the cord with the knife blade until the cord abrades through.
5. Treat wound with disinfectant.
6. Return animal to dry, clean, well-bedded quarters.

Castration with a knife should be done before fly season, and care should be taken not to excite kids before or immediately after the castration.

The emasculatome and elastrator may also be used for castrating. An emasculatome crushes each cord leading to the testicles, destroying the blood supply. It is a bloodless operation, but special care is necessary to be certain the cords are properly crushed. An elastrator is a forcep that is used to apply a heavy rubberband to the top of the scrotum, cutting off the blood supply to the testicles and scrotum. Care must be exercised to be certain that both testicles are in the scrotum before the rubberband is applied.

TATTOOING

A variety of methods are available for identifying goats, but tattoos are permanent—when they are done properly.

A tattooing outfit consists of a pair of tongs and numbers or letters made of sharp-pointed, needlelike projections that fit into the jaws of the tongs (see Fig. 23-18). The letters or numbers arranged in the proper sequence in the jaws of the tongs pierce the skin. Then, a tattoo ink forced into the punctures remains visible after the punctures heal. Goats are usually tattooed on the ears, with the exception of La Manchas which are tattooed on the tail webs.

After the necessary equipment—tattoo tongs, digits, ink, disinfectant, and cotton swabs—has been assembled, the tattoo is applied as follows:

1. Before piercing the ear, test the tattoo on a piece of cardboard to be certain the digits are set up properly.
2. Keep the tattoo tongs and unused digits in a disinfectant before and between usage.
3. Catch and restrain the animal, making certain the head can be held motionless during tattooing.
4. Examine the ear and find the widest space between the cartilage ribs.
5. Clean front and back of ear with disinfectant.
6. Position the tattoo tongs over the edge of the ear and puncture the skin quickly, but smoothly and forcefully, by squeezing the tong handle.
7. Release the tongs carefully and place in disinfectant.
8. Rub tattoo ink or paste thoroughly into the needle punctures.

It is best to wait until goats are 5 or 6 months of age before tattooing. If kids are tattooed very young, the growth of the ear will cause the tattoo to spread, making it difficult to read.

For registration of an animal with the American Dairy Goat Association, letter designations are tattooed in the left ear. The right ear should have the owner's initials or other letters of identification. Tattoo letters for recent and future years are listed in Table 23-6.

Most goat owners number kids consecutively as they are born. For example, a kid with the tattoo L-13 would indicate the thirteenth kid born in 1998.

All dairy goats must be tattooed before they can be accepted for registry or recordation in the herd books of the American Dairy Goat Association. It is strongly recommended that all breeds be tattooed in the ears, except the La Mancha which should be tattooed in the tail web. A maximum of four letters and/or numerals are allowed in each of two locations.

The International Dairy Goat Registry requires a tattoo, in either ear, and a color description for registration. For tattoos, the same letter designation as listed in Table 23-6 is followed.

TABLE 23-6
AMERICAN DAIRY GOAT ASSOCIATION
TATTOO LETTER DESIGNATIONS TO IDENTIFY YEAR OF BIRTH

Year	Letter	Year	Letter
1995	H	2003	S
1996	J	2004	T
1997	K	2005	U
1998	L	2006	V
1999	M	2007	W
2000	N	2008	Y
2001	P	2009	Z
2002	R		

DEHORNING

If for some reason, beyond the control of the dairy goat operator, a mature animal sports a set of horns, they should be removed. It is far better, however, to be certain kids are disbudded very young.

To dehorn a mature animal is quite a task and may be best if done by a veterinarian who can use a general or local anesthetic. Nevertheless, dehorning can be done using a hacksaw or wire saw to remove the horns along with 0.25 to 0.5 in. of the skin growing up from the base. Before the horns are removed, the hair at their base should be clipped so the skin line can easily be seen, and the horn area washed and painted with iodine. When a wire saw is used, very little bleeding is experienced. After the horns are removed, sulfa powder and a bandage should be applied. In some cases injections of tetanus antitoxin and antibiotics may be recommended.

HOOF TRIMMING

Hoofs must be trimmed often. Unless they are kept properly trimmed, feet and leg problems develop. This shortens an animal's productive life, besides detracting from the appearance of the animal. So, hoofs should be checked once a month.

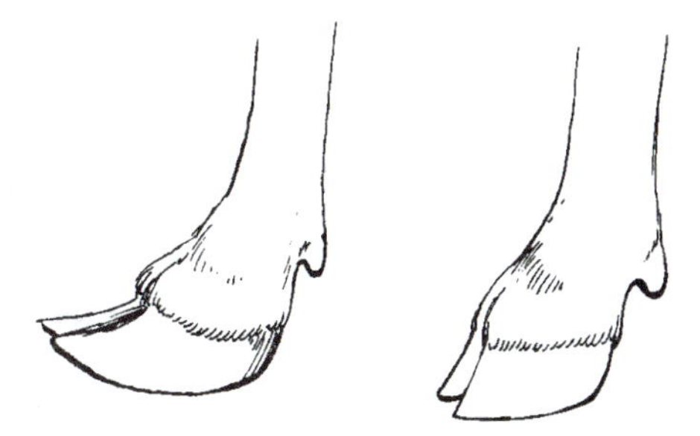

Fig. 23-22. A hoof needing trimming *(left)* and a perfectly trimmed hoof *(right)*. The walking surface should be parallel to the hairline at the top of the hoof.

A number of tools to trim hoofs are available, some of which are shown in Fig. 23-18. Sharp shrubbery pruning shears are handy for cutting toes down, and a good knife is also a valuable tool. A ferrier's knife, which does not have a sharp point, is ideal.

Several positions can be used for hoof trimming: (1) placing the animal's head in a stanchion on a raised platform, (2) tying the animal's head to a rail or board fence, and (3) setting the animal on its backside in a well-bedded pen—the *shear position*.

The goal of hoof trimming is a hoof that rests parallel and flat on the ground. To do this, follow these steps for each hoof:

1. Clean out the underneath of the hoof with the point of a knife or trimming shears. Before attempting to clean them, wipe muddy feet with a rag.
2. Trim any portion of the rim folded over the hoof wall and cut the sidewalls down even with the pad. Work one toe at a time. Cut from the heel toward the toe.
3. Level the bearing surface by alternately cutting thin slices from the heels and toes. Do not trim any growth from the center area until the heels and toes are level with it. The color of the bottom of the hoof will become pink as trimming nears the blood supply.
4. Trim the toe if it is too long.

MANAGEMENT CLIPS

From the standpoint of cleanliness, removing long and superfluous hair is essential. In warm climates, the entire body can be clipped, if desired. Regardless of the climate, it is important to give does a milking clip to produce quality milk. Fig. 23-23 indicates the area on a doe that is included in a milking clip. Briefly, a milking clip is (1) from the tailhead diagonally forward and downward into the flank area to a point about 5 in. in front of the udder; (2) everything to the rear of this path, including the sides and rear of the udder and the entire rear legs; and (3) a swath about as wide as the udder extending 6 to 8 in. forward.

Bucks also should be clipped to help control the buck odor. Fig. 23-23 indicates the areas on the buck included in a buck clip—the belly, brisket, front legs, and beard.

Management clips can be repeated as often as necessary.

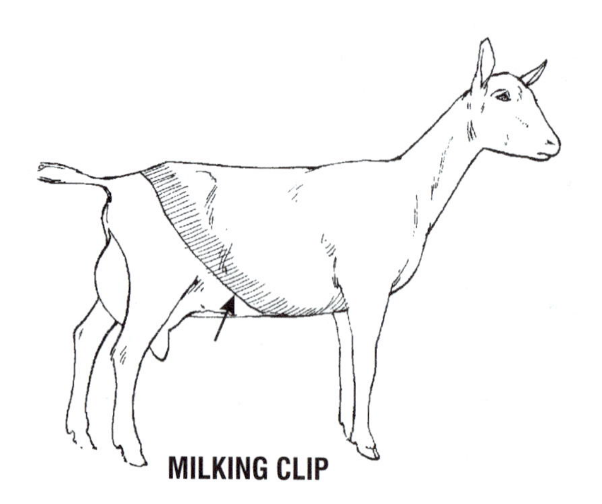

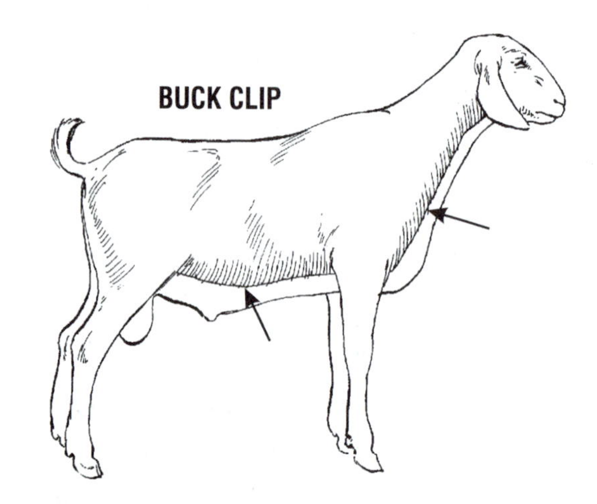

Fig. 23-23. Management clips ensure clean milk and control buck odor.

WATTLE REMOVAL

Wattles are fleshy appendages on the sides of the necks of goats. They occur on males and females, and in all breeds. They are a source of injury, and a possible site of infection. Wattles should be removed whenever they are large enough.

Before the wattles are removed, the skin and areas surrounding the site should be disinfected. Then, while the animal is restrained, each wattle is held out from the neck and snipped off, at its point of attachment, with a pair of disinfected scissors. The procedure is nontraumatic and simple.

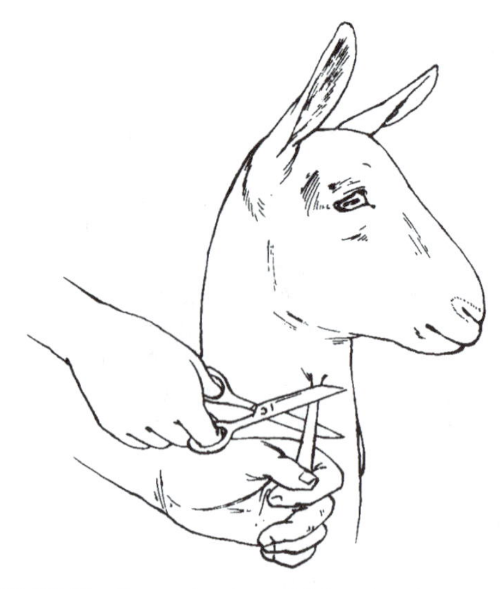

Fig. 23-24. After the area has been disinfected, wattles can be snipped from the neck with a pair of scissors. The procedure is so simple that it should be combined with some other management procedures.

MARKETING DAIRY GOAT PRODUCTS

Much of the goat's milk produced is used to provide milk for family consumption; some is sold or given to neighbors and friends as raw milk. Goat's milk is, however, a marketable item, and it can be made into a variety of marketable products like cow's milk. Although there are organized goat milk markets—local, statewide, and regional—in various locations, likely less than 20% of the dairy goats in the United States are milked on commercial dairy farms. Most of these commercial goat dairies, that have ventured into retail sales, have concentrated primarily on health food stores and some cheese specialty stores.

As for a market, the market for goat's milk and goat's milk products can only be defined in vague terms. Reliable figures reporting the consumption of dairy goat products in the United States do not exist. A few small surveys in specific areas have identified potential markets, but no systematic merchandising efforts have been attempted on a broad scale.

Offering a good product is essential to establishing a viable market, but overall, the dairy goat's image to most U.S. consumers is negative. Before goat's milk and products can be successfully marketed, a positive image must be established, from effective promotion and packaging. Also, other important considerations to developing a viable market include: (1) choice of distribution channels, (2) choice of packaging to meet a specific market and entice customers, (3) choice of label to satisfy government agencies and to attract consumers, and (4) a price with a reasonable profit margin and yet acceptable to consumers.

While there are a few commercial goat dairies that cater to specific markets, like certain ethnic groups, and health food enthusiasts, the marketing of dairy goat milk and products is still very much in its infancy.

GOAT'S MILK AND PRODUCTS

Goat's milk obtained from properly fed and managed goats can be utilized for essentially the same purposes as milk from cows.

■ **Pasteurized milk**—*Pasteurization is the heating of raw milk to a sufficiently high temperature for a specified length of time to destroy disease-causing bacteria.* Raw milk, even when handled under ideal conditions, may contain some disease-causing bacteria. The requirements for producing Grade A pasteurized milks are quite uniform from state to state. Most states have adopted the Grade A Pasteurized Milk Ordinance to cover the inspection and licensing of dairies. To design a large plant efficiently to pasteurize, process, and package goat milk requires large amounts of capital where the advanced methods now used in the United States are employed. Since the commercial-scale pasteurization is generally quite complex, operators should consult their local and/or state public health agencies for detailed requirements.

For the individual who milks goats for family use or the individual who buys raw goat's milk, the milk can be pasteurized at home on the stove. Small electric pasteurizers are also available for home use. The latter operate automatically and guarantee better temperature control than stove-top pasteurization.

Should fluid milk marketing reach a large scale, the employment of separation, clarification, and homogenization processes is useful and practical because they ensure a uniform product.

Goat's milk can be manufactured into a variety of products, including: butter, cheeses, ice cream, yogurt, condensed milk, dry milk, cottage cheese, and cream cheese. Consideration should be given to the manufacture of these products in a marketing program.

ECONOMICS OF MILK PRODUCTION

The high cost of labor per unit of goat's milk produced means the price received for goat's milk must be 1.5 to 2.5 times higher than that received by the dairy cow producer. To be profitable, the producer must pay close attention to measures of efficiency such as the amount of milk produced per hour of labor and the milk yield per doe. However, some marketing experiences have shown that goat's milk and products are relatively price insensitive compared to cow's milk and products. Should goat's milk and products penetrate into the supermarket channel, prices would have to be more competitive.

Some feasibility studies have reported that a minimum of 500 milking does with year-round even production is essential for a profitable, viable, full-time operation. Moreover, a combination of fluid milk and processing to handle surpluses and returns is necessary.

The contribution of goat's milk to the U.S. economy is presently small, but much remains unknown and untried in the economics of goat milk production.

FUTURE OF THE DAIRY GOAT INDUSTRY

The U.S. dairy goat industry holds the potential for growth. Over the past 15 to 20 years there has been a steady increase in the interest in dairy goats. This is manifested by increased numbers of subscribers to dairy goat magazines, increased dairy goat book sales, increased numbers of dairy goat shows, increased purebred dairy goat registrations, and increased enrollments in official milk testing programs.[3] Also, dairy goats have attracted the attention of universities and private industry for research purposes and development. But the dairy goat industry has miles to go!

There are many challenges ahead. One of the easiest challenges to meet is that of assembling a unified national collection of complete and reliable data on all aspects of the U.S. dairy goat industry. Another challenge is that of developing a national program to help identify superior, proven bucks that will make rapid genetic improvements in the commercial herds. This sire-proving system then could be combined with frozen semen and artificial insemination to make superior sires more readily available.[4] Another challenging area is the seasonality of goat milk production. By altering seasonality in reproduction, a consistent annual flow of milk could be maintained.

The development of a professional marketing and promotional system is part of the challenge. But because of the small size of the industry, supplies are limited and products cannot be made available regularly and in consistent quality. It does little good, therefore, to promote a product to create a demand in a new market if a constant supply of products cannot be provided. Along with this, research is needed in the area of milk product development.

Another aspect of promotion is the need to dispel existing prejudices toward goats resulting from unreliable information and/or poor-quality control of dairy

---

[3]Leach, K., "Trends in Dairy Goats," *Journal of Dairy Science*, Vol. 63, No. 10, p. 1601.

[4]The first step in a national program for genetic improvement was the first National Dairy Goat Buck Evaluation for Milk and Fat released by the USDA.

goat products. Many people regard goats as smelly garbage eaters.

Research and development by universities and industries needs to concentrate on a number of areas to fill some gaps in the knowledge about dairy goats. Equipment and facilities that are specifically for goats and that will reduce labor, increase health and sanitation, and reduce wastage and spoilage need to be developed. The nutritional requirements and feeding of dairy goats needs more research. Numerous products that have been developed and registered for other animals are efficacious for goats, but have not been cleared for goats because of the high cost of registration relative to the limited market. Productivity could be increased if goats were allowed to benefit from available technology and continued innovative research programs.

Worldwide, goats offer hope to feed hungry populations. Goats have been considered a number one priority in foreign aid programs. Goats supply meat and milk, increasing the availability of animal protein to the rural communities and people who cannot afford these products or who cannot afford to raise larger ruminants. At least for a time, the role of dairy goats in the United States will be similar—primarily to supplement the family food supply.

QUESTIONS FOR STUDY AND DISCUSSION

1. What countries lead in the production of goat's milk?
2. What state leads in the production of goat's milk?
3. Suppose you wish to select some goats to start a dairy operation. How would you begin? List some of the factors you would consider.
4. Suppose you have an established herd of dairy goats that you wish to improve? Describe how you would go about doing this.
5. What are the essential qualifications of a good dairy goat judge?
6. Describe the ideal doe of each of the following breeds: Alpine, La Mancha, Nubian, Oberhasli, Saanen, and Toggenburg.
7. What is scorecard judging? What are its advantages; and what are its disadvantages?
8. Why may a goat's age be estimated by its teeth?
9. List some reasons for fitting and showing dairy goats.
10. Outline the process of milk formation and ejection.
11. List the factors which affect milk yield.
12. Discuss the importance of (a) cleanliness and (b) low temperature in the production of high quality goat's milk.
13. List some factors that may cause off-flavors in goat's milk, and describe how you would prevent their occurrence.
14. Compare goat's milk to cow's milk.
15. What kinds of records should a dairy goat operator keep? What is their value to the operation?
16. What is unique to dairy goats of each of the following: feeding, breeding, care at kidding, raising and feeding kids, and milking?
17. List and describe four essential management skills, and give the reason or reasons for performing each of them.
18. Under what conditions is the production of goat's milk economically feasible?
19. What does the future hold for the dairy goat industry? Justify your answer.

SELECTED REFERENCES

Title of Publication	Author(s)	Publisher
Livestock Waste Facilities Handbook	Staff	Midwest Plan Service, Ames, IA, 1985
Management and Diseases of Dairy Goats	S. B. Guss	Dairy Goat Journal Publishing Co., Scottsdale, AZ, 1977
Meat & Poultry Facts, 1997	H. Kenneth Johnson, Executive Director	National Cattleman's Beef Assn., Chicago, IL, 1997
Sheep Book, The	Ron Parker	Charles Scribner's Sons, New York, NY, 1983
Sheep Production Handbook (SID)	Charles F. Parker, Director	American Sheep Industry Assn., Englewood, CO, 1998
Sheep Research Journal	Ed. by Maurice Shelton	American Sheep Industry Assn., Englewood, CO, 1994

CHAPTER 24

Meat goats (commonly known as Spanish goats) on the Walter Pope Ranch near Sonora, Texas. (Courtesy, *Ranch and Rural Living*, San Angelo, TX)

MEAT GOATS

Contents *Page*

The importance of goats as meat-producing animals varies worldwide. In general, the production of goat meat in less industrialized countries greatly exceeds that in industrialized countries; and in industrialized countries, the demand for goat meat is primarily the result of the preferred diets of ethnic groups, who prepare specialty dishes. Moreover, the significance of the production of goat meat is complicated by (1) the limited number of reports available on the production of goat meat, and (2) the numerous breeds of goats kept under widely different conditions. Therefore, knowledge of the distribution of goat meat production and of the sources of goats for meat production are requisite to a discussion of (1) establishing the herd, (2) herd improvement, (3) meat production, (4) management of meat goats, and (5) marketing goat meat.

DISTRIBUTION OF GOAT MEAT PRODUCTION

Few accurate statistics are available relative to the distribution of goat meat production. In countries where goat meat is important in the diet, it is produced almost entirely for local consumption. Table 24-1 gives the number of goats slaughtered, the average carcass weight, and the total goat meat production of some major areas of the world. Note that little goat meat is produced in North America. Note, too, that Mexico and Haiti account for most of the production in North America, and that very little is produced in the United States.

The relatively minor importance of goat meat is evidenced by the following statistics: Worldwide production of goat meat equals 34% of the total world production of lamb mutton, only 4% of the beef and veal, and only 4% of the pork.[1]

TABLE 24-1
WORLDWIDE GOAT MEAT PRODUCTION, BY MAJOR REGIONS[1]

	No. Slaughtered	Carcass Weight	Total Meat Production
		(hg/an)	*(metric tons)*
Asia	217,150,500	124	2,696,449
Africa	57,522,020	119	682,290
Europe	11,468,200	98	112,480
So. America	6,347,132	111	70,388
Oceania	625,280	206	12,857
World	296,446,200	122	3,620,766

[1]FAO Database. Data for 1996.

[1]*FAO Production Yearbook*, Food and Agriculture Organization of the United Nations, Rome, Italy, Vol. 50.

SOURCES OF GOATS FOR MEAT PRODUCTION

Goat meat is derived from four types of goats: (1) Angora goats, (2) dairy goats, (3) meat goats, and (4) feral goats. While Angora goats are kept primarily for mohair production, old goats and culls are slaughtered for their meat. Male kids, slaughtered at a very young age, and culled older animals are the meat sources of the dairy goat industry. Thus, meat is a secondary product of the Angora and dairy goat industries. Meat goats are, of course, kept primarily for the production of meat. These animals generally vary in size, shape, and color. In the United States, meat goats are commonly called Spanish goats or brush goats—mongrel animals descended from most of the major breeds of milk goats. Of all breeds, however, the South African Boer goat is the breed that most exhibits pronounced meat production qualities. Feral goats—escaped descendants of imported domestic goats—provide a source of meat for the export trade in Australia and New Zealand.

While much of this chapter relates to all sources of meat goats, the discussion pertains primarily to Spanish goats.

ESTABLISHING THE HERD

Spanish goats are able to withstand the rigors of Texas rangeland where they receive little care and attention. Most of them are produced on the Edwards Plateau of Texas, where they utilize and control browse species.

Spanish goats vary widely in their genetic background. They come in a variety of colors, ranging from solid black, brown, and white to fawn and brown with black points and a black stripe down the back. Some are blue-gray. Also, there are many combinations of spotting—black and white, brown and white, and black and brown. Most of these animals, both males and females, are horned, but the horns of the males grow much larger and heavier than those of the females. A few Spanish goats are polled.

A few ranchers have tried to establish herds of uniform color. Apparently black is fairly easy to establish. Also, white is popular and fairly easy to establish. However, most herds are multicolored. When these goats are to be run with fine-wool sheep, white or light-colored animals are preferred. Black and dark-colored hair from the goats may shed and contaminate the fleeces of the sheep.

HERD IMPROVEMENT

A great potential exists for the improvement of meat production in goats, but few attempts have been made to improve Spanish goats through selective breeding, or through the introduction of new genetic material. However, some attempts have been made to infuse dairy breeds to increase the size and milk production of the native does. Using meat goats from other parts of the world to introduce new genetic material offers the potential for improvement, but this is hampered by the inability to import animals due to health restrictions. If restrictions were not the problem, such breeds as the Boer from South Africa, the Jamnapari from India, and the Beetal from Pakistan and India, as well as the Nubian and Saanen, could be used to improve body type and milk production.

A selective breeding program among a herd of Spanish goats is similar to that of other livestock: Mate the best does to the best bucks. The bucks should have good conformation and large size, and be muscular. They should grow rapidly from birth to weaning. Selected bucks should be changed about every two years to prevent inbreeding and loss of vigor in the herd.

Some important points to be considered in a selective breeding program include: large size, multiple births, good conformation (good muscling), twice-a-year kidding, rapid growth, and straight legs with good bone.

Some points that should receive less consideration may include: color, horned or polled, type of ears, long or short coat.

Considerations for culling include: weakness of conformation, bad mouth, weak feet and legs, or failure to reproduce.

MEAT PRODUCTION

Compared to the other four-footed farm animals, goats are not efficient meat producers. This is because (1) they are raised under extensive (range) conditions, which include lack of feed and management, and (2) their adaptation to normal production conditions relates more to survival than to production.

As with any meat-producing animal, the producer is concerned with growth rate, feed conversion, carcass characteristics, nutritive value of goat meat, meat and byproducts, and acceptability of goat meat.

GROWTH RATE

Despite the high prolificacy of meat goats, their growth rate is disappointingly low, considerably lower than that of sheep. For example, goats grow 0.33 to 0.50 lb per day, while sheep grow 0.66 to 0.90 lb per day on a similar ration. Nevertheless, through genetic improvement and proper feeding, increased growth rates may be obtained.

FEED CONVERSION

Poor feed conversions are often due to low growth rates, which, in turn, may be due to poor nutritional conditions as well as to limited genetic potential. However, when growth rate increases, feed conversion also improves; in kids, this may be similar to lambs.

CARCASS CHARACTERISTICS

The goal of raising meat goats is edible, salable carcasses; and, depending on the area, these may be different. For instance, in South Africa, the meat of a young Boer goat is an alternative to lamb, while the meat of mature goats is sought in other countries. The meat of the young kid is referred to as *cabrito* and is regarded as a delicacy for the barbecue trade in the southwestern United States. Whatever the age, goat and kid carcasses are typically thin, shallow carcasses, becoming thicker and more compact as the carcass weight increases. Furthermore, a common feature of goat carcasses is their thin fat cover. Compared to sheep, there is a lack of subcutaneous fat cover on the carcasses, especially over the loins. The body fat of goats is concentrated in the visceral and intermuscular areas.

Dressing percentage is influenced by nutritional status and diet. Reports show dressing percentages ranging from 37 to 55%, which is lower than those of sheep.

NUTRITIVE VALUE OF GOAT MEAT

Table 24-2 shows the nutritive value of goat meat compared to lamb and beef. In general, goat meat is similar to beef and lamb, but it contains less fat. Although not shown in Table 24-2, goat meat is also a good source of phosphorus, zinc, niacin, pantothenic acid, riboflavin, thiamin, vitamin B-6, and vitamin B-12. Meat and animal products are the only major food sources of vitamin B-12.

MEAT AND BYPRODUCTS

Meat-packing plants wholesale goat meat in deboned form or as whole carcasses. Carcasses to be deboned are first chilled, then the meat is put into 110- to 120-lb boxes and frozen until sold. Most all of the boned meat from mature goats finds a ready market as sausage or chili meat in the southwestern United States. Carcass goat meat may be sold over the counter, just like beef, pork, and lamb. Most of this is sold

TABLE 24-2
NUTRITIVE VALUE OF GOAT MEAT COMPARED TO LAMB AND BEEF[1]

Meat	Weight	Moisture	Food Energy	Protein	Fat	Calcium	Iron
	(g)	(%)	(kcal)	(g)	(g)	(mg)	(mg)
Goat	100	71	165.0	18.7	9.4	11.0	2.2
Lamb[2]	100	61	263.0	16.5	21.3	10.0	1.2
Beef[3]	100	48	340.0	23.6	27.3	10.0	3.1

[1]Ensminger, A. H., M. E. Ensminger, J. E. Konlande, and J. R. K. Robson, *Foods & Nutrition Encyclopedia*, CRC Press, Boca Raton, FL, 1994, pp. 914, 920, and 922, Table F-36.
[2]Choice composite of cuts.
[3]Choice rump roast.

near the border of Mexico. The cabrito, or weaned kid, is a delicacy that is becoming more popular with the general public. Most cabrito carcasses, along with the heart, liver, and lungs, are barbecued or baked in ovens. Cabritos are purchased on the range or at markets and butcher shops.

Byproducts of goats include: bones, offal, liver, lungs, heart, tripe, tongue, cheeks, hide, and hair. The brains, hoofs, offal, and horns are sold to rendering plants. The tripe, cheeks, and tongue are edible. Goat pelts are used in all types of leather products. They are much stronger than sheep pelts, and Spanish goat pelts are sturdier than Angora pelts.

Some new and innovative developments in processing could make goat meat more profitable and more desirable. Mechanical deboning increases the supply of meat for further processing, reduces waste, conserves energy, and reduces labor. Goat meat could be restructured—pieces of muscles held together by muscle exudate—into boneless roasts and chops of uniform size, shape, and composition. These restructured, boneless products are attractive for use in the retail trade and the food service industry. Also, hot boning combined with electrical stimulation of the carcasses is a potential energy saver. Furthermore, electrical stimulation of goat carcasses is reported to increase tenderness, brighten muscle color, enhance flavor, and reduce the need for aging. While the technology for each of these processes exists, there is, currently, little demand to put it to use.

ACCEPTABILITY OF GOAT MEAT

Most people have never eaten goat meat due to its lack of availability in the United States and other industrialized countries, but the acceptability of any meat is greatly influenced by local custom and preference. Only a few studies have been conducted to determine if there is something unacceptable about goat meat. In general, a wide range of studies confirm that goat is less tender than lamb, and that, in terms of flavor, goat meat cannot be distinguished from lamb provided the animals are quite young. The lack of tenderness in goat meat may be due to muscle-cold shortening of the poorly covered goat carcasses. Electrical stimulation of goat carcasses could increase the tenderness.

MANAGEMENT OF MEAT GOATS

Most meat goats are managed under extensive conditions. Because of this, there is limited practical or financial reason to modify management procedures to implement programs necessary to increase production. Nevertheless, management changes which would increase kidding rates, reduce kid losses, increase kidding frequency, and increase weaning weight could be made.

GRAZING AND FEEDING OF MEAT GOATS

Angora and Spanish goats are produced under extensive conditions, on diets consisting primarily of range plants. Spanish goats are used to control sprouts of certain species of brush in cut-over and bulldozed areas, and at the same time to convert forage into meat. The stocking rate for goats should be calculated at five does to the animal unit. These animals can be used in the mixed grazing of livestock, provided they are stocked according to the recommendations for the area. They complement sheep and cattle and help provide the most efficient use of rangelands.

Rotation grazing should be practiced to improve the ranges and to help control internal parasites.

Range improvement practices consistent with the economics of the area and of the ranch should be practiced.

Supplemental grazing, such as stubble fields, small grain, Sudan, and irrigated pastures, should be used to supplement native pastures when available.

Spanish goats thrive and reproduce often without any supplemental feeding. Those ranchers who do provide supplements, however, report kid crops that are higher and herds that are easier to handle. During winter months or prolonged dry periods, 0.25 to 0.5 lb of cottonseed cake or 0.33 to 0.75 lb of corn may be fed daily to each animal. In some rough or brushy pastures, self-feeding, with salt as a governor, may be practiced. For this, a popular mixture is 3 parts ground sorghum (milo), 1 part cottonseed meal, and 1 part salt.

Nutritive requirements of Angora and meat goats, plus more complete feeding details, are covered in Chapter 21, Feeding Goats.

BREEDING MEAT GOATS

Spanish goats differ considerably from Angoras in their breeding habits. They are not as seasonal; some will breed almost every month of the year. Also, many Spanish does breed back while they are nursing kids. It is a common practice to leave the bucks with the does the year round. Some producers prefer to have the kids come at certain seasons of the year. This is especially true in areas where eagle predation is common. In such cases, the bucks should be removed and run separately from the does until breeding time.

The gestation period varies from 147 to 155 days, with an average of 5 months (150 days).

Three to four bucks per 100 does are recommended, depending on the size, roughness, and brushiness of the pastures. It is a good practice to condition the bucks with supplemental feed about two weeks before turning them with the does. One-half to one lb of grain or stock cubes will do a good job of conditioning.

Feeding does 0.25 to 0.33 lb of grain or range cubes per head daily, or moving them to fresh, rested pasture about two weeks before turning the bucks out, will flush does.

There are several systems of mating. Some producers prefer to leave the bucks with the does all the time. In such situations, doe kids may be bred before reaching a desirable size. A better method may consist of putting the bucks out during February and March, removing them, and putting them back during September and October. This system will allow for better management of the doe kids.

Replacement does should be weaned at 4 to 5 months of age, and selected from early-born kids and from does that kid twice each year. They should be weaned in a drylot and taught to eat supplemental feed. When replacement does reach adequate size, or are about one year of age, they can return to the breeding herd.

CARE AT KIDDING

No special attention is given to the does at kidding time. The best policy is to leave them alone and stay out of the pastures as much as possible during the kidding season. Pastures that have been idled are good for kidding. Also, small-grain pastures are good for kidding, since they provide excellent feed for milk production.

ESSENTIAL MANAGEMENT SKILLS

Since most kids are marketed at live weights under 50 lb, or at about 4 to 5 months of age, it is not necessary to mark or castrate the kids.

If sore mouth is present, the kids should be vaccinated at 2 to 4 weeks of age.

Both internal and external parasites should be controlled through proper management practices involving the use of drenches and sprays. Internal and external parasites affecting goats are discussed in Chapter 8, Sheep and Goat Health, Disease Prevention, and Parasite Control.

Records are, of course, necessary for income tax purposes and for planning a more efficient operation. They are a mark of good management. Accurate records of the percentage of kid crop and the weight of kids at market time are important to the meat goat raiser.

MARKETING GOAT MEAT

Kids are marketed at 4 to 5 months of age, at or before weaning. Some buyers will pick the kids up at the ranch.

Markets handling a number of Spanish goats are located in San Antonio, Uvalde, Lampasas, Golthwaite, and Junction, Texas. San Antonio and Los Angeles are major markets for the cabritos (weaned kids)—prized for barbecue. In areas where it is favored, the demand for cabrito exceeds the supply.

While it is not a market, many ranchers keep a few goats for their employees as a meat source.

Castrated male Angoras are maintained for fiber production until advancing age reduces hair production and quality, at which time they are marketed for meat.

Culled young in excess in large goat dairies are often sold when they are only a few days old to buyers for their own use as meat.

Marketing of goats at both the live and the carcass stages is characterized by a number of problems, including (1) seasonal production, (2) relatively high cost per pound for slaughtering and processing, (3) difficulty of selling less desirable cuts at retail, (4) high condemnation rate of culled breeding stock, (5) increased cost of marketing resulting from decreases in the number of plants that slaughter goats, and (6) consumer resistance to the products as a result of myths and misinformation.

■ **Chevon**—In 1922, a widely publicized contest was held by the Sheep and Goat Raisers Association of Texas to select a trade name for goat meat comparable to beef, pork, mutton, and lamb. The name *chevon* was adopted by the Association in August, 1922. It was also accepted by the U.S. Department of Agriculture and by *Webster's*. Apparently, the name *chevon* was contrived from the French word *chevre* for goat and the *on* from the word *mutton*. Although campaigns were conducted to get chevon into common usage, it is not widely used by people in the meat industry. Nevertheless, the name is occasionally used, and one should be aware that it refers to goat meat.

FUTURE OF THE MEAT GOAT

In many countries, primarily developing countries, goats are, and will continue to be, an important source of food. They harvest forage unavailable for or underutilized by other animals. Their carcasses are much smaller than those of cattle, and tend to be smaller than hogs, making them suitable for developing countries where refrigeration (storage) is limited. The small carcasses can easily be consumed before spoiling.

Opportunities exist for increasing the production of meat goats in the United States. With their unique adaptation for using extensive rangeland efficiently, these opportunities include the presence of underutilized land and feed resources in some areas of the country and the economic stimulus for better land use. Since goat production systems require relatively low energy input to control unwanted plants, they are energy efficient. Also, there is the potential for applying new scientific and technological developments and approaches for efficient marketing.

(1) Using the best available technology in range livestock, (2) using grazing areas not now used for sheep and goats, and (3) combining or alternating the grazing of sheep and goats with the grazing of cattle could increase goat production in the major range areas by at least 50%.[2]

Goats are considered to be a minor livestock enterprise in U.S. agriculture, and this smallness is a limiting factor. Small size means a weak marketing and distribution system and limited markets, slaughter facilities, and retail outlets. The production of a limited amount of meat lessens the need of promoting and creating a demand. Also, producers in isolated production areas have difficulty obtaining supplies, materials, and skilled labor.

[2] *The U.S. Sheep and Goat Industry: Products, Opportunities, and Limitations,* Council for Agricultural Science and Technology (CAST), Report No. 94, p. 19.

QUESTIONS FOR STUDY AND DISCUSSION

1. Worldwide, where is most of the goat meat consumed? Why do these areas (countries) consume relatively large amounts of goat meat?
2. In the United States, what are the sources of goat meat?
3. Describe the type of goats raised solely for meat.
4. Through selection and breeding, how might meat goats be improved?
5. Compare the production of meat from goats to that of meat from cattle, pigs, and sheep.
6. Compare goat meat to beef, pork, and lamb.
7. Outline the management of Spanish goats under extensive range conditions.
8. What is chevon?
9. In your opinion, what does the future hold for the meat goat? Describe some of the hopes and some of the limitations.

SELECTED REFERENCES

Title of Publication	Author(s)	Publisher
Goat Production: Breeding & Management	Ed. by C. Gall	Academic Press, Inc., New York, NY
Proceedings of the Third International Conference on Goat Production and Disease		Dairy Goat Journal Publishing Co., Scottsdale, AZ

CHAPTER 25

Automated feeding of green chop at a 1,600 dairy goat operation.

BUILDINGS, EQUIPMENT, AND MANAGEMENT FOR GOATS

Contents — *Page*

A discussion of buildings and equipment for goats is complicated by two factors: (1) the scarcity of information on buildings and equipment designed specifically for goats, and (2) the wide range of conditions under which goats are produced. So, the purpose of this chapter is to make some recommendations, provide some general details, and furnish some general ideas. Ultimately, goat raisers should study their own particular needs, evaluate other operations, and plan buildings and equipment.

Except at kidding time, housing and equipment are seldom provided for Angora goats and meat goats. Where and when needed, their building and equipment needs and designs are very similar to those of sheep. (See Chapter 17, Buildings and Equipment for Sheep.) So, Chapter 25, Buildings and Equipment for Goats, pertains primarily to dairy goats.

HOUSING AND MILKING SYSTEMS FOR DAIRY GOATS

Housing for goats ranges from the very simple to the elaborate, though most types of housing tend to be somewhere in between. The essential requirement of goat housing is a building that is dry, well ventilated, and draft free. Herein, a few general types of housing and their requirements are discussed.

Fig. 25-1. Overall view of a large housing and milking system for dairy goats. Each corral holds about 100 does.

When choosing and designing buildings, one should consider numerous factors, but some of the important ones include:

■ **Heat production by goats**—The heat produced by goats varies according to body weight, rate of feeding, environmental conditions, and degree of activity. As a guide though, each 100 lb of goats produces a total of about 560 Btu/hr at 45°F, which means that the sensible heat production is about 500 Btu/hr.[1] Sensible heat is that portion of the total heat production, measurable with a thermometer, that warms the air.

■ **Vapor production by goats**—During normal respiration, goats give off moisture—the higher the temperature the greater the amount of moisture. For every 100 lb of goats, vapor production amounts to about 0.06 lb/hr, at 45°F. This moisture should be removed through the ventilation system. Most building designers govern the amount of winter ventilation by the need for moisture removal.

■ **Ventilation**—Ventilation of goat buildings is accomplished by natural ventilation—openings in the building—or by fans through inlets and outlets, or by a combination of these. When fans are used in warm buildings, fresh air intakes and exhaust fans should be designed so that 25 to 200 cu ft of air per minute per 1,000 lb of goats can be moved.

■ **Insulation; vapor barrier**—The insulation and vapor barrier requisites of goat housing are the same as for sheep.

■ **Supplement heat**—The comfort zone for dairy goats is between 55 and 70°F. In some areas, supplemental heat may be required to maintain buildings at 40°F or above for the comfort of the goats and the operators. The amount of heat needed depends on (1) animal density, (2) amount of insulation and other construction features, (3) ventilation, (4) manure management, and (5) climate.

■ **Manure production**—The design of any facilities for goats may give some consideration to the amount of manure that will be produced so that it can be efficiently handled. A successful manure-handling system efficiently uses an increasingly valuable fertilizer while preventing water pollution and offending neighbors. The quantity of manure produced varies according to the kind and amount of feed, the number of animals, and the amount of bedding. Table 25-1 can be used as a guide for the daily excretion (free of bedding) of individual coats of various weights.

■ **Laborsavings**—All goat facilities must be designed with laborsaving features in mind to cut cost. Some may consider automation, mechanical handling of materials—of at least feed and water facilities. The extent and type of automation depends largely on the operation.

On a very broad basis, all housing can be classi-

[1]One Btu (British thermal unit) is the amount of heat required to raise the temperature of 1 lb of water 1°F, while one Kcal is the amount of heat required to raise the temperature of 1 kg of water 1°C.

TABLE 25-1
DAILY FECES AND URINE PRODUCTION BY INDIVIDUAL GOATS

Weight of Goat		Daily Production[1]			
(lb)	*(kg)*	*(lb)*	*(kg)*	*(cu in.)*	*(cm³)*
7	*3.2*	0.3	*0.1*	8	*131*
27	*12.2*	1.1	*0.5*	31	*508*
47	*21.3*	1.9	*0.9*	53	*869*
67	*30.4*	2.7	*1.2*	76	*1,245*
87	*39.5*	3.5	*1.6*	98	*1,606*
107	*48.5*	4.3	*1.9*	121	*1,983*
127	*57.6*	5.1	*2.3*	143	*2,343*
147	*66.7*	5.9	*2.7*	166	*2,720*
167	*75.8*	6.7	*3.0*	188	*3,081*
187	*84.8*	7.5	*3.4*	211	*3,458*

[1]One cu ft equals 1.728 cu in. One ton *(907 kg)* of manure (feces and urine) occupies about 34 cu ft *(0.96 m³)*.

fied as cold housing or warm housing, and each (or both) of these types has a place in some goat operations.

■ **Cold housing**—Cold buildings are enclosed shells that keep the rain and snow off the goats and protect them from the wind. The inside temperatures are usually a few degrees warmer in cold weather and a few degrees cooler in hot weather than outside temperature extremes. Cold buildings use natural air movement through adjustable and fixed openings which must be well planned and well located for adequate air movement throughout the building.

Warming a cold building by tightly closing it during cold weather causes a severe moisture buildup and condensation on the walls and roof, with only a small temperature rise. It is much more important in goat housing to maintain a dry and draft-free building than to raise the temperature a few degrees and cause a moisture buildup.

■ **Warm housing**—Warm buildings are enclosed and the temperature is maintained between 45 and 65°F. Supplemental heat may be needed to maintain desired temperature. These buildings are designed for fan ventilation in winter, but natural ventilation is often used in summer through adjustable windows and wall panels.

LOOSE HOUSING

Loose housing is that system in which the herd is handled on a group basis except at milking time. It may be open or enclosed.

There is considerable interest in this system, primarily because it requires less labor than the stall system.

The following functional areas are involved in most loose housing systems:

1. **Resting or loafing area.** There are two rather distinct types of resting or loafing arrangements:

   a. **Group housing.** In this system, the goats rest in a common, bedded area on a manure pack which is at least 15 to 18 in. deep. About 20 sq ft of bedded area should be provided per goat. The bedding material should be deep to begin with; then each day the droppings should be removed and fresh bedding added.

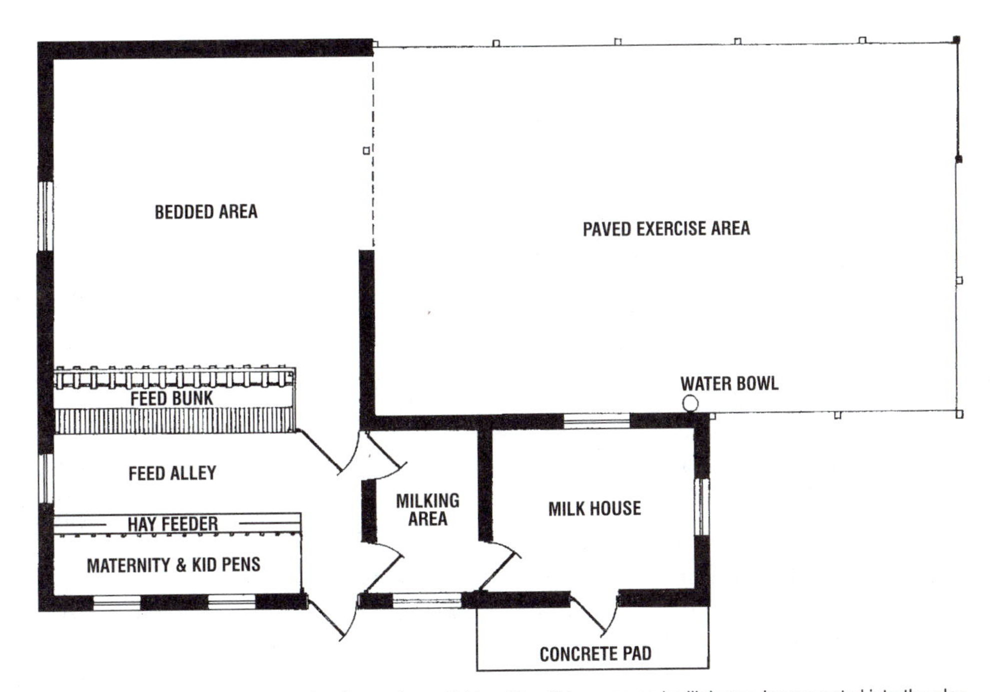

Fig. 25-2. Floor plan of loose housing for goats and kids with milking area and milk house incorporated into the plan.

The advantage of group housing over individual-stall housing is that the cost per goat is less.

b. **Free-stall housing.** This is a modification of the group housing system. It consists of individual open stalls, which are bedded. The stalls should be 18 to 22 in. wide and about 36 in. long, with the rear curbs of the stalls at least 8 in. above the floor. Partitions between stalls can be expanded metal, plywood, or nonclimbable wire, and the front of the stalls should be plywood. The bottom of the free-stall should be built up level to about 1 in. from the top of the rear curb, using tamped clay. Horizontal bars, placed about 1 ft back from the front of the stall and about 30 in. from the stall floor, help prevent does from standing too far forward.

Free-stalls save bedding and labor and increase sanitation by alleyways behind goat stalls which can be scraped and/or washed down regularly. Also, less space per goat is necessary.

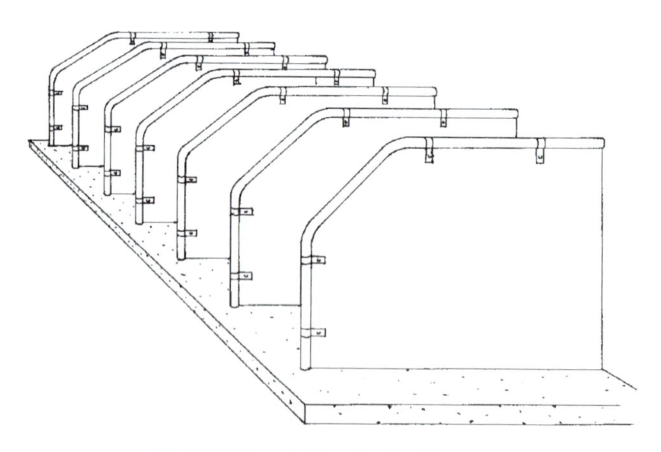

Fig. 25-3. Free-stall housing for dairy goats.

2. **Feeding area.** Separate feeding and bedded areas are preferred. Fenceline feeders, or keyhole feeders, mineral feeders, grain bunks, and/or hay racks should be located in the drylot some distance from the manure pack.

3. **Exercise yard.** This is usually an area which serves as a place for exercise and feeding. Approximately 25 to 40 sq ft per goat for dirt lots and 16 sq ft for paved lots should be allowed. It should be designed so that it can be cleaned by scraping.

4. **Milking area and milkroom.** These rooms are often located adjacent to, or incorporated with, the other areas allotted to loose housing. The milking area and milkroom should, however, be rooms separated from the stable areas.

STALL HOUSING OR INDIVIDUAL CONFINEMENT

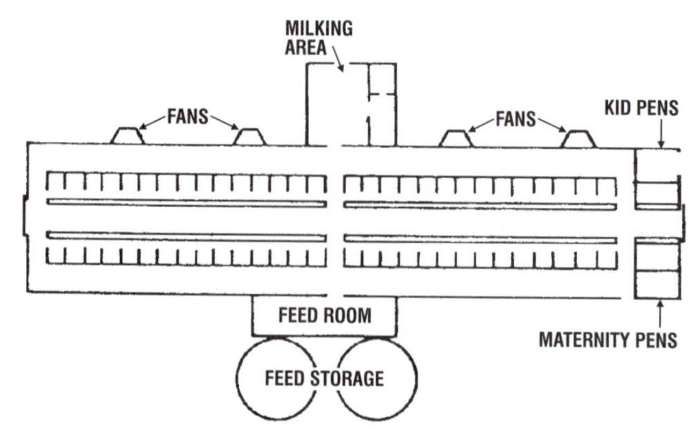

Fig. 25-4. Floor plan of a tie-stall dairy with 60 stalls.

Animals tied or otherwise restrained in individual stalls can be individually fed and cared for, with dominant animals prevented from bothering timid animals. Stall housing requires mechanical ventilation and supplementary heat. The labor and housing costs are much greater than for loose housing.

The stall barn consists of one or two rows of goats that are usually confined to stanchions, though tie stalls may be used. In tie stalls, each goat is individually tied with a strap or a chain. This offers goats considerably more freedom than stanchion stalls, but requires more labor.

The floors are usually concrete, though a raised, slatted platform provides easy cleaning and comfort. Sometimes dairy goat stalls are covered with rubber mats. The concrete floor slopes toward a gutter, and the feces and urine may be flushed into a storage tank. Individual stalls should be 18 to 20 in. wide and 40 in. in length. The gutter is 12 in. wide and 6 in. deep.

Animals are released for milking and exercise. The milkroom is located on the side of the stall building opposite the feed room. Fig. 25-4 shows a simple floor plan of a stall building, feed storage, and milkroom.

SHEDS

Sheds are versatile and widely used shelters throughout the United States. They are particularly suitable for housing dry goats and young stock.

Sheds usually open to the south or east, preferably opposite the direction of the prevailing winds and toward the sun. They are enclosed on the ends and sides. Sometimes the front is partially closed, and in severe weather drop-doors may be used. The latter arrangement is especially desirable when the ceiling height is sufficient to accommodate a power manure loader.

So that the bedding will be kept reasonably dry, it is important that sheds be located on high, well-drained ground; that eave troughs and downspouts drain into suitable tile lines, or surface drains; and that the structures have sufficient width to prevent rain and snow from blowing to the back end. Sheds should be a minimum of 24 ft in depth, front to back. A height of 8.5 ft is necessary to accommodate some power-operated manure loaders. When this type of equipment is to be used in the shed, a minimum ceiling height of 10 ft is recommended. The extra 1.5 ft allows for the accumulation of manure. Lower ceiling heights are satisfactory when a blade is to be used in cleaning the building.

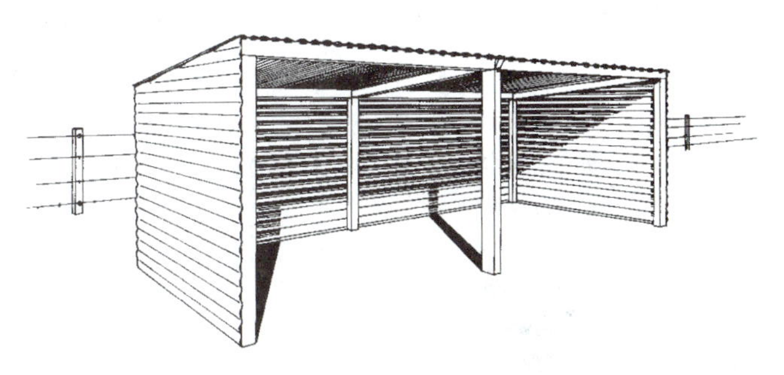

Fig. 25-5. A three-sided shed which faces the sun will give adequate protection in the Southwest.

The length of the shed can be varied according to needed capacity. Likewise, the shed may be a single long shed, or it may be L- or T-shaped. The long arrangement permits more corral space. When an open shed is contemplated, thought should be given to feed storage and feeding problems.

Sometimes hay racks are built along the back wall of sheds, or next to an alley, if the shed is very wide or if there is some hay storage overhead. Most generally, however, hay racks, feed bunks, and watering troughs are placed outside the structure.

MILKING AREA

The milking area is part of the housing plan, but it should be separate from the stable area. For the small operator, the milking area may be as simple as a small room with one milking stand. A room 5 × 8 ft with a concrete floor and a drain is adequate.

The best planned milking barns have a room for milking and a room for cooling milk, for storing and handling milk, and for washing utensils.

■ **Milkroom**—This is an area where milk is cooled and stored, and where milking equipment can be cleaned, stored, sanitized, and maintained. Storage of milk may be in bulk tanks or cans depending on the size of the dairy. Cooling equipment may consist of refrigeration coils in the tank and/or in line tube or plate coolers, or for cans, a water bath-type cooler. Milkroom equipment includes at least: (1) a double sink, (2) hot and cold water, (3) a small table, (4) a rack for drying and storing equipment, and (5) refrigeration for cooling the milk.

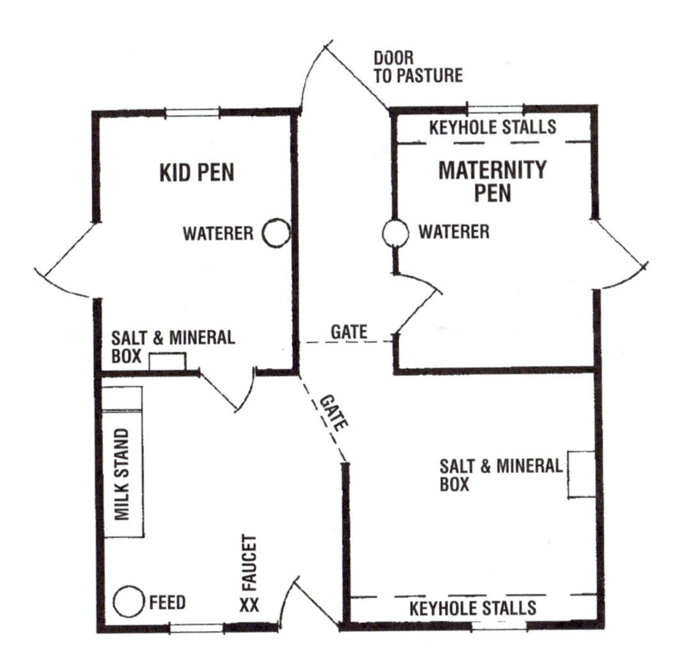

Fig. 25-6. Floor plan of a small building for a few goats. The building provides shelter for kids, pregnant does, and milking does and a small area for milking.

While a room with a single milking stand may qualify as a milking parlor, most people think of a milking parlor as being larger. For the sake of efficiency, most dairy farms would want to accommodate 6 to 10 does at a time.

Fig. 25-7. Milking parlor with herringbone arrangement and keyhole feeders. (Courtesy, Christine S. F. Williams, Michigan State University)

The purpose of the milking parlor is improved labor efficiency, working conditions, and sanitation surrounding the milking operation. Parlors are of various designs. But all of them have many features in common, including:

1. An all-weather drive leading to the facility for bulk milk access.

2. A facility that (a) is well insulated, (b) can be heated to 50 to 60°F in the wintertime, (c) is well ventilated, (d) has sufficient size to permit herd expansion, and (e) has walls constructed of an easy-to-clean material.

Although there is a wide range of milking parlor arrangements, the four most common ones are:

1. **Side-by-side.** Goats stand parallel to each other on a platform and secured in a stanchion. The goats are milked from behind.

2. **Tandem.** The goats stand "Indian file" in line and broadside to the operator's pit. The goats are milked from the side.

3. **Herringbone.** The goats stand in groups at an angle to the operator's pit. This parlor, which was developed in New Zealand, takes its name from the arrangement of the goats in the parlor.

4. **Rotary (or carousel).** The parlor, which may incorporate either the tandem or herringbone principle, rotates around the operator.

Fig. 25-8. Milking parlor with side-by-side stanchion arrangement using Surge Jersey bucket milkers. (Courtesy, Judy Kapture and Jim Vandergriff, Caprine Supply, Shawnee, KS)

Fig. 25-9 shows diagrams of different milking parlor arrangements.

In general, if milk is to be used by the family, the degree of sanitation and cleanliness, and the size and type of building constructed, can be based on the owner's judgment, pride, and finances. On the other hand, if milk is to be sold to the public, or to a processor, a total plan for the dairy must be submitted to state health officials and approved before construction is begun. Most states, however, have no regulations for goat dairies per se, but they extend dairy regulations to cover goat dairies. This leads to some requirements which are unnecessary for goats but appropriate for cows. Such problems should be reconciled with state officials as they arise.

Fig. 25-10 shows a schematic diagram of a large goat dairy.

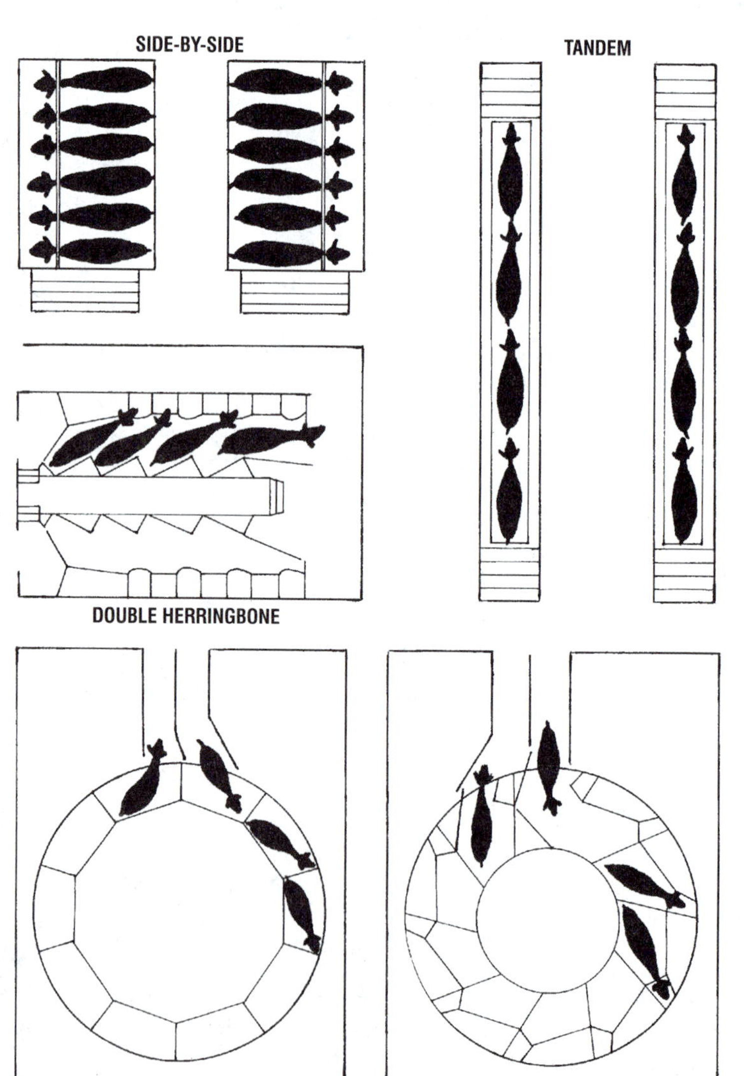

Fig. 25-9. Milking parlor arrangements.

KID HOUSING; BUCK HOUSING

As indicated in several previous drawings, housing for kids and other young stock may be included in the housing plan for the milking herd, but the kid area must be separate from the milking herd to reduce exposure to diseases. Often

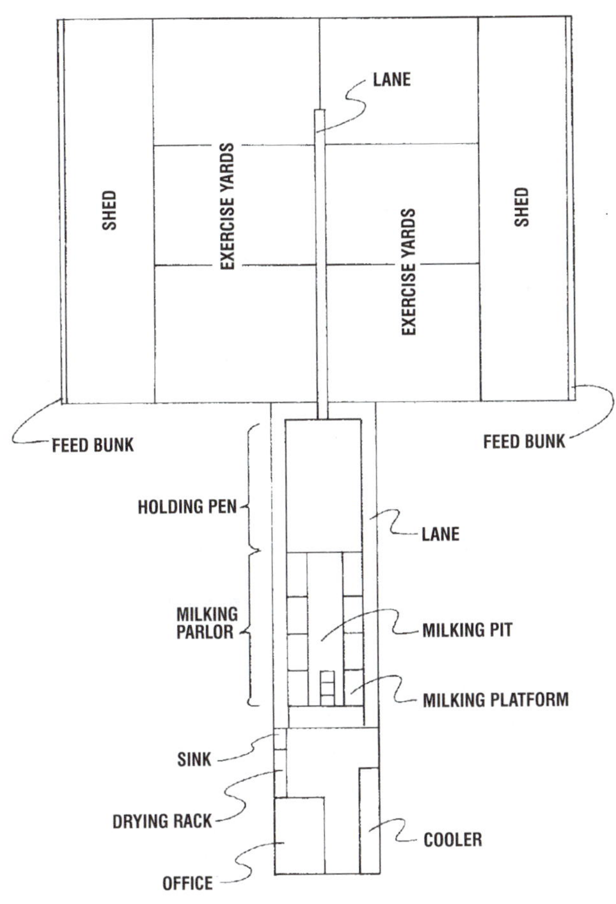

Fig. 25-10. Schematic diagram of a large goat dairy.

kids are kept in 4-ft square box stalls with at least one side slatted to permit air movement. It may be advantageous to supply newborn and sick kids with a heat lamp. Older kids can be housed together in cold housing similar to that used for calves, as shown in Figs. 25-11 and 25-12. For large operations, several of these units can be placed side by side.

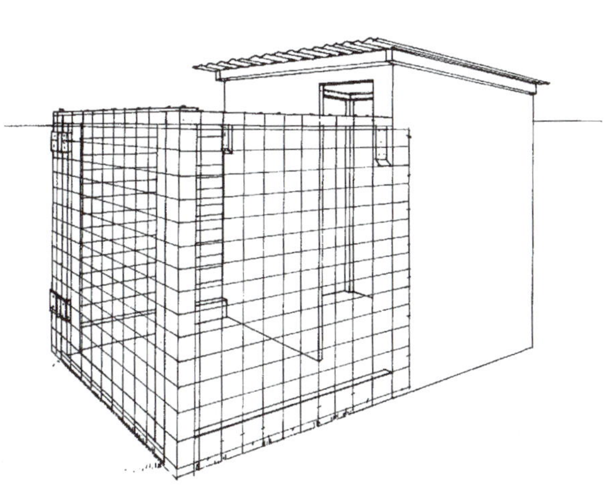

Fig. 25-11. Housing for small groups of kids.

Fig. 25-12. Several housing units for young kids, placed side-by-side. (Courtesy, Rocking M Ranch, Hilmar, CA)

Buck housing has one important requirement—complete separation from the does, and preferably downwind. The housing does not have to be elaborate. An open shed 6 × 8 ft, facing south and opening into an exercise lot, is adequate. Several bucks can share housing if adequate space is provided in the shed and exercise lot. It may be advantageous to construct a mound of earth 5 to 6 ft tall and 5 × 5 ft at its top in the center of the exercise lot for a single buck. This provides exercise and prevents boredom. The mound must be at least 8 ft from the fence.

Fig. 25-13. A well-constructed buck pen with access to the doe herd. Does in heat stay around the pen. (Courtesy, Christine S. F. Williams, Michigan State University)

SPACE REQUIREMENTS OF BUILDINGS AND EQUIPMENT

One of the first and frequently one of the most difficult problems confronting the individual who wishes to construct or remodel a building or an item of equipment is that of arriving at the proper size or dimensions. Table 25-2 contains some conservative average figures which may prove helpful. In general, too little space may jeopardize the health and well-being of the

TABLE 25-2
SPACE REQUIREMENTS OF BUILDING AND EQUIPMENT FOR GOATS

Class of Animal	Building Floor Space		Lot Space				Feeder Space				Waterer Space		
			Dirt		Paved		Limit Fed		Self Fed		Bowl or Nipple	Tank	
	(sq ft/hd)	(m^2/hd)	(sq ft/hd)	(m^2/hd)	(sq ft/hd)	(m^2/hd)	(in./hd)	(cm/hd)	(in./hd)	(cm/hd)	(hd/bowl or nipple)	(hd/ft)	(hd/m)
Does	12–18	*1.1–1.7*	25–40	*2.3–3.7*	16	*1.5*	16–20	*41–50*	4–6	*10–15*	40–50	15–20	*49–66*
Young kids	8–10	*0.7–0.9*	20–30	*1.9–2.9*	10	*0.9*	9–12	*23–30*	1–2	*3–5*	50–75	25–40	*82–131*
Bucks	30–40	*2.8–3.7*	100	*9.3*	—	—	12	*30*	6	*15*	10	2	*7*

animals, whereas too much space may make the buildings and equipment more expensive than is necessary.

RECOMMENDED MINIMUM WIDTH OF SERVICE PASSAGES

The requirements for service passages given in Table 17-8 of Chapter 17, Buildings and Equipment for Sheep, will be applicable to most dairy facilities.

STORAGE SPACE REQUIREMENTS FOR FEED AND BEDDING

The space requirements for feed storage for the dairy enterprise vary so widely that it is difficult to provide a suggested method of calculating space requirements applicable to such diverse conditions. The amount of feed to be stored depends primarily upon (1) length of pasture season, (2) method of feeding and management, (3) kind of feed, (4) climate, and (5) the proportion of feeds produced on the farm in comparison with purchased feeds. Normally, the storage capacity should be sufficient to handle all feed grain and silage grown on the farm and to hold purchased supplies. Forage and bedding may or may not be stored under cover. In those areas where weather conditions permit, hay and straw are frequently stacked in the fields or near the barns in loose, baled, or chopped form. Sometimes poled framed sheds or cheap covers of plastic or wild grass are used for protection. Other forms of low-cost storage include temporary upright silos, trench silos, temporary grain bins, and open-wall buildings for hay.

Table 17-9 in Chapter 17, Buildings and Equipment for Sheep, lists the storage space requirements for feed and bedding. This information may be helpful to the individual operator who desires to compute the barn space required for a specific dairy enterprise. This table also provides a convenient means of estimating the amount of feed or bedding in storage.

GOAT EQUIPMENT

It is not proposed that all of the numerous types of dairy goat equipment be described herein. Only some common articles will be discussed. Individuals searching for ideas for goat equipment can possibly glean some ideas for adaptation from a number of books describing equipment for dairy cattle and sheep. Some of these books are listed at the end of this chapter. Also, Chapter 17, Buildings and Equipment for Sheep, may be helpful. Suitable equipment saves feed and labor and increases production.

HAY RACKS

Whatever the design of the hay rack, it should prevent goats from wasting feed. One popular design, used in many variations, is the keyhole feeder shown in Figs. 25-14 and 25-15. The round head openings are about 8 in. in diameter and the neck slots are about 4 in. wide. The feeder itself should be 14 to 18 in. wide for adult goats.

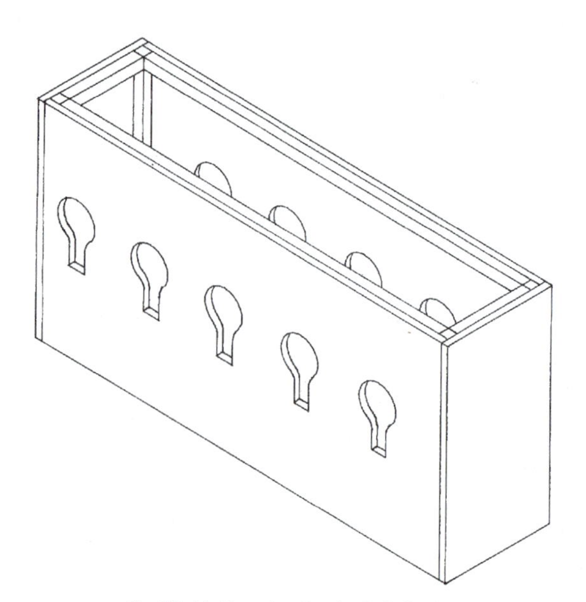

Fig. 25-14. Free-standing keyhole feeder.

Fig. 25-15. Keyhole-style mangers in barn. Also, note the octagonal keyhole feeder in the background. (Courtesy, Judy Kapture and Jim Vandergriff, Caprine Supply, Shawnee, KS)

MINERAL FEEDERS

Mineral feeders for supplying salt and other minerals should be located in a building or protected from the weather. A wooden trough about 8 to 10 in. wide, 4 to 6 in. deep, and 4 to 8 ft long makes an adequate mineral feeder. It should stand about 26 to 30 in. high, as shown in Fig. 25-16.

Fig. 25-16. Mineral feeder used at Natura Goat Dairy, Puerto Rico. (Courtesy, Alpines International, Ft. Pierce, FL)

WATERING FACILITIES

Goats will drink from water troughs, bowls, and even nipples. The primary requisite for watering facilities is that the water be kept clean. Goats are often more sensitive and reluctant than other species to drink from foul-tasting water sources. Also, in cold weather, goats relish warm water.

If bowls or troughs are used, they should be elevated to prevent fouling from dirt, goat droppings, and urine. Bowls should be installed so that the lips are 30 in. above the floors. Using nipples ensures a clean water supply, but goats may be somewhat sloppy and drip considerable water under the nipples.

Fig. 25-17. Automatic water bowls, raised about 30 in. off the ground, provide a clean supply of water for goats, and they can be heated. (Courtesy, George F. W. Haenlein, University of Delaware)

MILKING UTENSILS

For the individual who has a few goats and wishes to milk them by hand, the basic utensils necessary include: a hooded pail, a strip cup, a strainer, and a tote

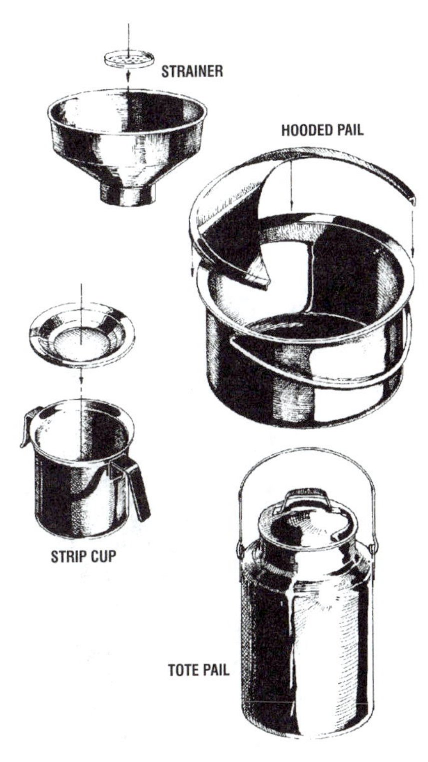

Fig. 25-18. Some basic milking utensils.

pail. These items should be made of seamless, stainless steel, for ease and thoroughness of cleaning. The hooded pail protects the milk from falling dirt. A strip cup is used to detect mastitis. Fresh milk should be strained as it is poured into the tote (container for carrying) and before it is cooled.

MILKING STAND

For the sake of cleanliness and the comfort of the milker, goats should be milked on a stand. The stand is about 18 in. high and 3 ft long, with a stanchion and feed tray at the front, a seat for the milker on one side, and railing along the other side. Steps may lead up the rear, or goats may be trained to jump onto the stand. Milking stands can be constructed from wood (Fig. 25-19) or metal (Fig. 25-20). Metal may be preferred for ease of cleaning.

Fig. 25-19. Milking stand constructed of wood.

Fig. 25-20. Metal milking stands. (Courtesy, Linda S. Campbell, Khimaira Kaprine Kreations, Luray, VA)

KID-HOLDING BOX

A kid-holding box (Fig. 25-21) is designed to be used for newborn kids and kids several weeks old. The holding box makes disbudding or tattooing an easier one-person task. Most boxes are 5 in. wide, 24 in. deep, and 16 in. high.

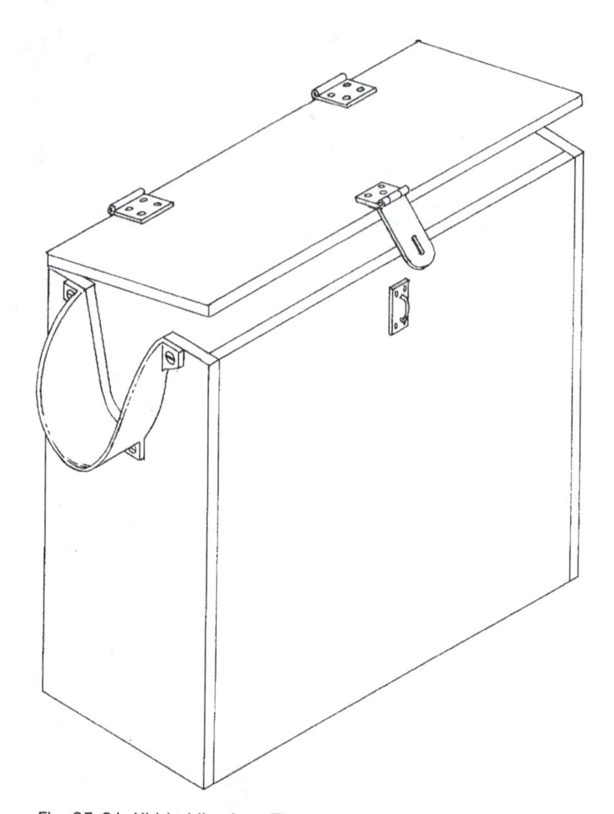

Fig. 25-21. Kid-holding box. The top, bottom, and ends can be made of 1-in. lumber. The sides can be made of plywood or masonite.

SHADES

Providing adequate shade to protect goats from the sun is an important and widely used means for improving the environment of goats in hot climates.

The most satisfactory shades (1) provide about 15 sq ft of shade per adult goat, (2) are located with north-south placement, and (3) are open all around, thus permitting maximum air movement. The north-south placement allows the sun to get underneath the shade, keeping it dry.

FENCES FOR GOATS

Of all the four-legged livestock species, goats perhaps provide the biggest fencing challenge. They are apt to jump or climb a fence that would keep other live-

Fig. 25-22. Shade at the Natura Goat Dairy. (Courtesy, Alpines International, Ft. Pierce, FL)

stock contained. Whatever the type of fencing, it should be a minimum of 48 in. tall; and 54 in. tall is better. A buck fence, however, must be 5.0 to 5.5 ft tall.

Barbed wire fencing is not the first choice for goats. It will catch or tear their hides, since they will run into it, lean against it, or stick their heads through it.

Rail, pipe, or board fences can be used, though they will become very expensive if they are 48 to 54 in. high and kidproof—spaced 6 to 8 in. apart.

In pastures, electric fencing can be used for cross-fencing. The wires should be placed 12, 27, and 42 in. above the ground. Pieces of rag or tape serving as flags along the fence help goats see the fence and learn about it. Also, if goats are unaccustomed to an electric fence, a training pen can be used to teach them respect for the fence. For goats that are accustomed to an electric fence, two wires, 12 and 28 in. from the ground, may be sufficient in some cases.

Fig. 25-23. Unless woven wire fences are properly constructed, goats will stand on them, eventually causing them to sag. (Courtesy, Christine S. F. Williams, Michigan State University)

Woven wire fencing can be used, but goats will stand on it and rub against it causing it to sag. Also, a woven wire fence requires corner braces which goats will use as a bridge to climb out. Running a strand or two of electric wire on the goats' side of the fence, at 12 and 24 in. above the ground and about 12 in. out from the fence, will help prevent goats from climbing on it.

When choosing woven wire for fencing goats, one should select the heaviest gauge.

A chain-link fence makes a good fence for goats. It can be purchased without the metal posts and metal top rail, thereby reducing the cost. Then, regular wooden or regular metal posts can be used. Also, another good, sturdy fence for goats is the stock or hog panels. These come in 52-in. heights, 16 ft long with horizontal stay rods 3 in. apart at the bottom and 6 in. apart at the top, and vertical rods 6 in. apart.

More pertinent information regarding fencing can be found at the end of Chapter 17, Buildings and Equipment for Sheep.

TETHERING GOATS

In some unfenced areas, goats can be tethered to salvage the feed. To do this, collars, rings, swivels, several lengths of chain, and stakes are needed. Each stake should have a freely revolving ring at its top. The

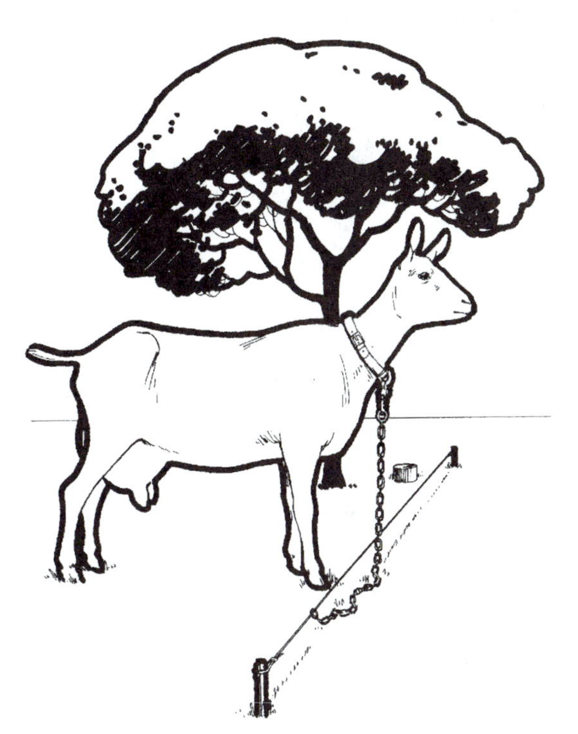

Fig. 25-24. Where a few goats are involved and good fences are not available, goats can be tethered. Water and shade should be readily available.

swivel attached to the chain is attached to the ring on the goat's collar and the other end of the chain is attached to the ring on the stake, which is driven into the ground. As the goat grazes, the chain can be lengthened.

Another form of tethering is known as a *running tether* (see Fig. 25-24). A stout wire is stretched between two stakes 10 to 100 ft apart. Then a short chain, 2 to 3 ft long, attached to the goat's collar with a swivel, is clipped to the stretched wire. This allows the goat to graze parallel to the wire as the chain slides along it.

Goats should never be tethered without shade, water, or protection from dogs; and they are usually not left tethered overnight.

QUESTIONS FOR STUDY AND DISCUSSION

1. Discuss the importance of each of the following as it relates to the construction of buildings for goats: (a) heat production by goats, (b) vapor production by goats, (c) ventilation, (d) insulation, (e) vapor barrier, (f) supplemental heat, (g) manure production, and (h) laborsavings.

2. Suppose you plan to house 50 does. Assuming the does all kid, what amount of manure must you expect to handle in a year's time?

3. Compare cold housing to warm housing.

4. Compare group housing to free-stall housing as two alternatives of loose housing.

5. Compare loose housing to stall housing (tie-stalls). Which would you recommend and why?

6. Describe some uses for sheds.

7. Define and give the essential components of a (a) milking area, (b) milkroom, and (c) milking parlor.

8. Explain the differences between the following milking parlor arrangements: (a) side by side, (b) tandem, (c) herringbone, and (d) rotary (or carousel).

9. In all goat herds, some allowances must be made for kid housing and buck housing. What would you recommend?

10. Suppose you are going to build a loose housing system (group housing) for 20 does. How much floor space should you allow in the bedded shed area and how much in the dirt exercise lot? Also, how much feeder space will be necessary and how many waterers will be needed?

11. Assume you are feeding 100 lb of grain each month. How much storage space will be needed for a two-month supply?

12. Why is the keyhole-style feeder so popular?

13. Sketch some plans for a mineral feeder. (Refer to Chapter 17 for additional ideas.)

14. What is the most important consideration of watering facilities for goats?

15. Suppose you have two goats to milk for household milk. What utensils do you need to ensure the collection of clean, sanitary milk?

16. Why are goats difficult to fence? What type of fencing would you recommend and why?

17. Under what conditions and how would you tether a goat?

SELECTED REFERENCES

Title of Publication	Author(s)	Publisher
Barns and Buildings for Dairy Goats	A. L. Klingbiel	American Supply House, Columbia, MO
Dairy Cattle Science	M. E. Ensminger	Interstate Publishers, Inc., Danville, IL, 1993
Dairy Housing & Equipment Handbook		Midwest Plan Service, Iowa State University, Ames, IA
Sheep Housing and Equipment Handbook		Midwest Plan Service, Iowa State University, Ames, IA
Stockman's Handbook, The	M. E. Ensminger	Interstate Publishers, Inc., Danville, IL, 1992

PART IV
GLOSSARY/
FEED COMPOSITION TABLES

CHAPTER 26

Left: Two-year-old Finn ewe. (Courtesy, Finnsheep Breeders Assn., Inc., Zionsville, IN) *Right:* Goats are great browsers. (Courtesy, *1984 Yearbook of Agriculture Animal Health Livestock and Pets*, USDA)

GLOSSARY OF SHEEP AND GOAT TERMS

The mark of distinction of good sheep and goat producers is that they are familiar with the common terms used to describe the production, marketing, processing, and distribution of sheep and goats and their byproducts. They "speak the language." Moreover, modern producers must also be able to communicate effectively with professionals. The following glossary of sheep and goat terms is presented in order that the amateur and the established producer may be able to acquire and maintain a mastery of the correct vocabulary.

Many terms that are defined or explained elsewhere in this book are not repeated in this chapter. Thus, if a particular term is not listed herein, the reader should look in the Index or in the particular chapter and section where it is discussed.

A

ABATTOIR. A slaughterhouse.

ABOMASUM. The fourth compartment of the ruminant stomach where enzymatic digestion occurs. It is often called the true stomach.

ABORTION. The expulsion or loss of the fetus before the completion of pregnancy, and before it is able to survive.

ABSCESS. A localized collection of pus.

ACCELERATED LAMBING. Breeding a ewe flock to average more than one lamb crop each year, usually three lamb crops in two years.

ACCLIMATIZATION. The process of becoming accustomed to a new climate or other environmental conditions.

ACUTE. Having a rapid onset, a short course, and pronounced signs, as of some diseases.

ADAPTATION. The adjustment of an organism to a new or changing environment.

ADDITIVE. An ingredient or a substance added to a basic feed mix, usually in small quantities, for the purpose of fortifying it with certain nutrients, stimulants, and/or medicines.

ADIPOSE TISSUE. Fatty tissue.

AD LIBITUM. Free to eat at will. Same as *free-choice.*

AEROBE. A microorganism that requires oxygen to live and reproduce.

AFTERBIRTH. The placenta and allied membranes associated with the fetus that are expelled from the uterus at lambing.

AI. Abbreviation for artificial insemination.

AIR DRY (approximately 90% dry matter). Referring to feed that is dried by means of natural air movement, usually in the open. It may be either an actual or an assumed dry matter content; the latter is approximately 90%. Most feeds are fed in the air dry state.

ALKALI. Usually a soluble salt or a mixture of soluble salts present in some soils of arid or semiarid regions in a quantity detrimental to ordinary agriculture.

ALLELOMIMETIC. Doing the same thing.

AMBIENT TEMPERATURE. The prevailing or surrounding temperature.

AMERICAN FEED MANUFACTURERS ASSOCIATION, INC. (AFMA). The American Feed Manufacturers Association is a nationwide organization of feed manufacturers banded together (1) to improve the quality and promote the use of commercial feeds, (2) to encourage high standards on the part of its members, and (3) to protect the best interests of the feed manufacturer and the stock producer in legislative programs. The address is 1701 N. Fort Myer Drive, Arlington, VA 22209.

AMINO ACIDS. Nitrogen-containing compounds that constitute the building blocks or units from which proteins are formed. They contain both an amino (NH_2) group and a carboxyl (COOH) group.

ANABOLISM. The conversion of simple substances into more complex substances by living cells (constructive metabolism).

ANAEROBE. A microorganism that normally does not require air or free oxygen to live and reproduce.

ANESTRUS.

- The nonbreeding season.
- Females not in heat (estrus).

ANOREXIA. A lack or loss of appetite for food.

ANOXIA. A lack of oxygen in the blood or tissues. This condition may result from various types of anemia, reduction in the flow of blood to tissues, or lack of oxygen in the air at high altitudes.

ANTHELMINTIC. A drug that kills or expels intestinal worms.

ANTHRAX. A highly dangerous, infectious disease caused by *Bacillus anthracis*, commonly affecting animals and sometimes humans. In the human, a form of this infection, commonly referred to as *wool sorter's disease*, may be contracted, most likely through a skin abrasion from the handling of fleeces from infected animals. The infection, however, may localize in the lungs or in the intestines. Workers handling carpet wools and other fleeces imported from some tropical countries, where the disease is more prevalent than in this country, are more constantly exposed to anthrax than handlers of other wools.

ANTIBODY. A protein substance (modified type of blood-serum globulin) developed or synthesized by lymphoid tissue of the body in response to an antigenic stimulus. Each antigen elicits production of a specific antibody. In disease defense, the animal must have an encounter with the pathogen (antigen) before a specific antibody can be developed in its blood.

ANTIGEN. A high molecular-weight substance (usually protein) which, when foreign to the bloodstream of an animal, stimulates formation of a specific antibody and reacts specifically *in vivo* or *in vitro* with its homologous antibody.

ANTIOXIDANT. A compound that prevents oxidative rancidity of polyunsaturated fats. Antioxidants are used to prevent rancidity in feeds and foods.

ANTISEPTIC. A chemical substance that prevents the growth and development of microorganisms.

APPAREL WOOLS. All wools that are manufactured into cloth for use as clothing.

APPETITE. The immediate desire to eat when food is present. Loss of appetite in an animal is usually caused by illness or stress.

ARTIFICIAL INSEMINATION. The introduction of semen into the female reproductive tract by a technician, using a pipette and/or other instruments.

AS-FED (A-F). Referring to the way feed is normally fed to animals. It may range from 0 to 100% dry matter.

ASH. The mineral matter of a feed. The residue that remains after complete incineration of the organic matter.

ASSAY. Determination of (1) the purity or potency of a substance, or (2) the amount of any particular constituent of a mixture.

ASSIMILATION. The physiological processes by

which the nutrients in feed are made available to and used by the body; the processes include digestion, absorption, distribution, and metabolism.

ATROPHY. A wasting away of a part of the body, usually muscular, induced by injury or disease.

AUTOSOMES. All chromosomes except the sex chromosomes.

AVERAGE DAILY GAIN (ADG). The average daily liveweight increase of an animal.

AVOIRDUPOIS WEIGHTS AND MEASURES. The old English system of weights and measures, or U.S. customary weights and measures, to differentiate it from the metric system. *Avoirdupois* is a French word meaning *to weigh*.

B

BABY COMBING WOOL. Short, fine wool which is usually manufactured on the French system of worsted manufacture. This term is synonymous with *French combing wool*.

BACKCROSS. The mating of a crossbred (F_1) animal to one of the parental breeds.

BACTERIA. Microscopic, single-cell plants, found in most environments, often referred to as microbes; some are beneficial, others are capable of causing disease.

BACTERICIDE. A product that destroys bacteria.

BACTERIN. A suspension of killed bacteria (vaccine) used to increase disease resistance.

BAG. The udder.

BALANCED RATION. A ration which provides an animal with the proper amounts and proportions of all the required nutrients.

BALE. A way of handling wool in some countries. These bundles differ in size; thus, Australian and New Zealand wools weigh about 330 lb per bale, whereas South American bales weigh approximately 1,000 lb each.

BAND. The total number of sheep in one group. Often used to designate large numbers on the range, while *flock* is used in connection with small numbers on the farm.

BASAL METABOLIC RATE (BMR). The heat produced by an animal during complete rest (but not sleeping) following fasting, when using just enough energy to maintain vital cellular activity, respiration, and circulation, the measured value of which is called the basal metabolic rate (BMR). Basal conditions include thermoneutral environment, resting, postabsorptive state (digestive processes are quiescent), consciousness, quiescence, and sexual repose.

BELLY WOOL. The wool that grows on the belly of the sheep and occasionally extends up the side in irregular patches. It is usually uneven, different in grade (generally grades one grade or more finer) from the body of the fleece. It is shorter and less desirable because of its poor lock formation, and it usually lacks the character of the body of the fleece. Some sheep with poor belly fleece are inclined toward bare belliness, excessive crimp, and low wool production.

BIOLOGICAL CONTROL. The destruction or suppression of undesirable insects or other animals or plants by the introduction or propagation, encouragement, artificial increase, and dissemination of their natural enemies.

BIOLOGICAL VALUE OF A PROTEIN. The percentage of the protein of a feed or feed mixture which is usable as a protein by the animal. Thus, the biological value of a protein is a reflection of the kinds and amounts of amino acids available to the animal after digestion. A protein which has a high biological value is said to be of *good quality*.

BIOSYNTHESIS. The production of a new material in living cells or tissues.

BLACK-TOP WOOL. Wool containing a large amount of wool grease combined at the tip of the wool staples with dirt, usually obtained from Merino sheep. This wool is usually fine in quality, of good character, and desirable in type, but the shrinkage is high.

BLACK WOOL. Any wool containing black fibers. A fleece having only a few black fibers is rejected by a grader and goes into black wool because there is no way of separating the few black fibers in the manufacturing processes. Black wool is usually run in lots that are to be dyed black.

BLEAT. The sound or cry made by sheep and goats.

BLOAT. A digestive disorder of ruminants, usually characterized by an abnormal accumulation of gas in the rumen.

BLOCKY. Deep, wide, and often low-set, referring to an animal.

BLOOD. The degree of fineness of wool, measured as low 1/4, 1/4, 3/8, and 1/2 blood.

BOLUS.

- Regurgitated food that has been chewed and is ready to be swallowed.
- A large pill for animals.

BOMB CALORIMETER. An instrument used to measure the gross energy content of any material, in which the feed (or another substance) tested is placed and burned in the presence of oxygen.

BOTANY WOOLS. All fine Australian wools. This designation is extensively used in Great Britain and in the United States. Also, it is a general term for all classes of fine wool.

BRADFORD SYSTEM. The system of combing and spinning wool commonly used in the United States. The longer wools are used for this system of manufacture. Bradford system yarns are used to make worsted fabrics. Same as English system.

BRAID WOOL. The coarsest of the U.S. grades of wool, according to the blood system of classification. It is a very coarse and lustrous wool.

BRAND. In sheep, the markings made on wool with a special branding fluid for identification purposes.

BRAND NAME. Any word, name, symbol, or device, or any combination of these, often registered as a trademark or name, which identifies a product and distinguishes it from others.

BREAK. Weak at a certain point, but strong above and below the weak spot, as opposed to tender, which signifies a generally weak fiber.

BRED.

- Referring to an animal that is pregnant.
- Sometimes used synonymously with the term *mated*.

BREECH.

- The buttocks.
- A birth in which the rear portion of the fetus is presented first.

BREED.

- Animals that are genetically pure enough to have similar external characteristics of color and conformation, and when mated together produce offspring with the same characteristics.
- The mating of animals.

BREED TYPE. The combination of characteristics that makes an animal better suited for a specific purpose.

BRIGHT WOOL. Light-colored farm wool, as compared to semi-bright wool, which is darker due to soil conditions. Both types scour out to a good white.

BRITCH WOOL. Wool from the thigh and twist region of the sheep. It is the coarsest and poorest wool on the entire fleece. It is usually manure encrusted and urine stained.

BRITISH THERMAL UNIT (Btu). The amount of energy required to raise 1 lb of water 1°F; equivalent to 252 calories.

BROADTAIL. A type of skin obtained from lambs of the Karakul breed.

BROKEN MOUTH. A condition of sheep that have lost some, but not all, of their teeth.

BROWSE. Woody or bushy plants.

BUCK. A male goat; but may also refer to a male sheep—a ram.

BUCK FLEECE. A fleece from a ram. The wool usually has a heavy shrinkage due to excessive wool grease; thus, wool of this type is not worth as much in the grease as similar wool from ewes or wethers.

BUFFER. A substance in a solution that makes the degree of acidity (hydrogen-ion concentration) resistant to change when an acid or a base is added.

BURRY WOOL. Wool heavy in vegetable matter—including burs, leaves, seeds, and twigs, which requires special and expensive processing in removal.

BUSHEL. A unit of capacity equal to 2,150.42 cu. in. (approximately 1.25 cu ft).

BYPRODUCT FEEDS. The innumerable roughages and concentrates obtained as secondary products from plant and animal processing, and from industrial manufacturing.

C

CAKE (presscake). The mass resulting from the pressing of seeds, meat, or fish in order to remove oils, fats, or other liquids.

CALCIFICATION. The process by which organic tissue becomes hardened by a deposit of calcium salts.

CALORIC. Pertaining to heat or energy.

CALORIE (cal). A small calorie is the amount of heat required to raise the temperature of 1 g of water from 14.5 to 15.5°C. This is equivalent to 4.185 joules.

CALORIMETER. An instrument for measuring the amount of energy.

CANARY-STAINED WOOL. A yellowish coloration in the wool which cannot be removed by ordinary scouring methods. Such stain has been observed in the wools obtained from New Zealand in Corriedale and Romney sheep, West Australian and Queensland Merino sheep, and in the Cape Merino wools.

CAPE WOOL OR CAPES. Wool from South Africa.

CAPRICULTURE. Goat husbandry or goat science.

CAPRINE. Pertaining to or derived from a goat.

CARACUL. A type of pelt produced by young lambs of Karakul breeding.

CARBONIZING. The process of treating wool with chemicals, usually acids, to destroy and remove the burs without seriously damaging the wool. Wool so treated is known as *carbonized wool*.

CARCASS. The dressed body of a meat animal, the usual items of offal having been removed.

CARCASS WEIGHT. Weight of the carcass of an animal following harvest, as it hangs on the rail, expressed either as warm (hot) or chilled (cold) carcass weight.

CARCASS YIELD. The carcass weight as a percentage of the liveweight.

CARD. A machine used to separate the wool fibers by opening the locks or tufts of wool. The machine contains multiple rolls with teeth. Hand cards are used chiefly in the fitting of show sheep.

CARDING. An operation which converts loose, clean, scoured wool into continuous, untwisted strands.

CARDING WOOLS. Wools that are too short to be manufactured by either the Bradford or the French system, and so must be manufactured by the woolen system. Synonymous with *clothing wools*.

CARPET WOOL. Coarse, harsh, strong wool that is more suitable for carpets than for fabrics. Very little wool of this type is produced in the United States. Some of the choicer carpet wools are used to make tweeds or other rough sport clothing.

CARRIER.

- A disease-carrying animal.
- A heterozygote for any trait.
- An edible material to which ingredients are added to facilitate their uniform incorporation into feeds.

CARRYING CAPACITY. The number of animal units a property or an area will carry on a year-round basis. This includes the land grazed plus the land necessary to produce the winter feed.

CASTRATE.

- To remove the testicles or ovaries.
- An animal that has had its testicles or ovaries removed.

CASTRATION. The process of removing the testicles or crushing the cords of the male animal.

CATABOLISM. The conversion or breaking down of complex substances into more simple compounds by living cells (destructive metabolism).

CENTIGRADE (C). A means of expressing temperature. To convert to Fahrenheit, multiply by 9/5 and add 32.

CEREAL. A plant in the grass family (Gramineae), the seeds of which are used for human and animal food; *e.g.*, corn (maize) or wheat.

CHAFFY WOOL. Wool containing a considerable amount of chaff—finely chopped straw.

CHARACTER. The evenness, distinctiveness, and uniformity of crimp characteristic of respective wool classes.

CHEVON. The trade name for goat meat.

CHOLESTEROL. A white, fat-soluble substance found in animal fats and oils, bile, blood, brain tissue, nervous tissue, the liver, kidneys, and adrenal glands. It is important in metabolism and is a precursor of certain hormones. Some have implicated cholesterol as a factor in arteriosclerosis, but direct experimental proof is lacking.

CLEAN CONTENT. The amount of clean scoured wool remaining after removal of all vegetable and other foreign material, and containing 12% by weight of moisture and 1.5% by weight of ingredients removable by pretreating with alcohol.

CLEAN WOOL. Usually refers to scoured wool but occasionally it describes grease wool that has a minimum amount of vegetable matter.

CLINICAL. Having to do with direct observation.

CLIP. Usually the total annual production from any given flock, state, or country.

CLOSEBREEDING. A form of inbreeding, such as brother to sister or sire to daughter.

CLOTHING WOOLS. Wools under 1.5 in. in length and distinguished from combing wools by their shorter length. Principal properties include softness, crimpiness, and felting ability; also called *carding wools*.

CLOUDY WOOL. Wool that is off-color. It may be due to wool becoming wet while in a pile.

COARSE WOOL. Wool that has a Blood Grade of 1/4 blood or common, or a Numerical Count Grade of 44s, 46s, or 48s.

COEFFICIENT OF DIGESTIBILITY. The percentage value of a food nutrient that is absorbed. For example, if a food contains 10 g of nitrogen and it is found that 9.5 g are absorbed, the digestibility is 95%.

COLD TEST. Cold shrinkage. After wool scouring, the hot wool is allowed to stand in a cool room until it has taken on a normal amount of moisture from the air. The wool is then weighed, and the shrinkage figured to give the cold shrinkage, or cold test.

COLLAGEN. A white, papery transparent type of connective tissue which is of protein composition. It forms gelatin when heated with water.

COLOR DEFECT. Any color that is not removable in wool scouring, due to urine stain, dung stain, canary yellow stain, or black fibers.

COLOSTRUM. The milk secreted by mammalian females for the first few days following parturition, which is laxative and high in antibodies.

COMBING. An operation in the worsted system of removal of the short fibers (noil) and foreign impurities. The long fibers are straightened out and laid parallel.

COMBING WOOLS. Wools having sufficient length and strength to comb. According to the standards, the length of fibers for strictly fine combing must be over 2.75 in., with an increase in length as the wool becomes coarser.

COMBUSTION. The combination of substances with oxygen accompanied by the liberation of heat.

COMMERCIAL FEEDS. Feeds mixed by manufacturers who specialize in the feed business.

COMMON. One of the U.S. grades of wool. It is next to the coarsest grade, and derives its name because it presumably comes from sheep of common ancestry.

COMPLETE RATION. All feedstuffs (forages and grains) combined in one feed. A complete ration fits well into mechanized feeding and the use of computers to formulate least-cost rations.

CONCENTRATES. A broad classification of feedstuffs which are high in energy and low in crude fiber (under 18%). For convenience, concentrates are often broken down into (1) carbonaceous feeds, and (2) nitrogenous feeds.

CONDITION.

- In grease wool, the amount of yolk and foreign impurities it contains.
- The state of health, as evidenced by the fleece and general appearance.
- The amount of flesh or finish (fat covering).

CONFORMATION. The shape and design of an animal.

CONTAGIOUS. Transmissible by direct or indirect contact.

CONTENTMENT. A stress-free condition exhibited by healthy animals.

COOKED. Heated to alter chemical or physical characteristics or to sterilize.

CORDOVA. Long, coarse wool from Argentina, largely used for carpets.

COSSET. A lamb raised without the help of its mother.

COTTY WOOL. Wool that has matted or felted on the sheep's back. Caused by insufficient wool grease being produced by the sheep, usually due to breeding, injury, or sickness. This type of defective wool is more common in the medium and coarse wools. The fibers cannot be separated without excessive breakage in manufacturing.

COUNT. The fineness to which yarn may be spun. Generally referred to as *spinning count*.

COWTAIL. A coarse, hairylike fleece or a very low-quality britch wool.

CRACKED. Reduced by combined breaking and crushing action to particle size.

CREEP. An enclosure or a feeder used for supplemental feeding of nursing young, which excludes their dams.

CRIMP. The wave effect in the wool fiber. Usually the finer wools show the most crimp. Uniformity of desired crimp generally indicates superior wool.

CROSSBRED WOOL. In the United States, wool obtained from sheep of long-wool X fine-wool breeding. Usually this wool grades 3/8 or 1/2 blood.

CROSSBREEDING. Mating animals of different breeds.

CRUDE FAT. Material extracted from moisture-free feeds by ether. It consists largely of fats and oils with small amounts of waxes, resins, and coloring matter. In the energy value calculation of a feed, the fat is considered to have 2.25 times as much energy as either nitrogen-free extract or protein.

CRUTCHING. An Australian term. It refers to the practice of clipping the wool from the rear end of ewes before turning them to green pasture in the spring and prior to lambing. Usually the wool around the eyes of animals subject to wool blindness is clipped at the same time.

CRYPTORCHID. A male lamb or kid, which is often sterile because one or both testicles are retained in the abdominal cavity.

CUD. A bolus of regurgitated food, common only to ruminants.

CULL. An animal taken out of a flock (herd, in goats) because it is below flock (herd) standards.

CUTTING.

- Removing the testicles.
- Separating one or more animals from a flock (herd, in goats).

CUTTING CHUTE. A narrow chute allowing animals to pass through in single file and be sorted into specified pens by opening and closing gates in the chute.

DAIRY HERD IMPROVEMENT. Known generally as DHI or DHIA (Dairy Herd Improvement Association); this is a system for testing a herd for milk production under standard rules.

DAM. The female parent.

DAMP WOOL. Wool that has become damp or wet before or after bagging and may mildew. This weakens the fibers and seriously affects the spinning properties.

DEAD WOOL. Wool that is pulled from dead (not slaughtered) sheep. Wool recovered from sheep that have been dead for some time is sometimes referred to as *merrin*.

DEFECATION. The evacuation of fecal material from the rectum.

DEFECTIVE WOOL. Wool that contains excessive vegetable matter, such as burs, seeds, and straw, or which is kempy, cotty, tender, or otherwise faulty. These defects lower the value from a manufacturer's viewpoint.

DEFICIENCY DISEASE. A disease caused by a lack of one or more basic nutrients, such as a vitamin, a mineral, or an amino acid.

DEGRAS. The commercial designation given to crude wool grease.

DEGREASED WOOL. Wool that has been cleansed by the naphtha process.

DEHYDRATE. To remove most or all moisture from a substance for the purpose of preservation, primarily through artificial drying.

DELAINE WOOL. Fine, strictly combing wool, usually obtained from the state of Ohio. Delaine wool does not necessarily have to come from the Delaine-Merino; however, that breed is noted for this class of wool.

DEMI-LUSTER WOOL. Wool that has some luster but not enough to be classed as luster wool. Wool of this type is produced by the Romney and similar breeds.

DENSITY. An index of the number of wool fibers per unit area of a sheep's body. Fine-wool breeds show greater fleece density than the coarser wool breeds.

DENTAL PADS. The very firm gums in the upper front jaw of sheep, goats, and cattle.

DEPILATORY. A solution or a paste—usually consisting of sodium sulfide, sulfuric acid, and ground oystershells—applied to the flesh side of pelts in order to loosen the fibers from the skin in preparation for the wool-pulling process.

DERMATITIS. Inflammation of the skin.

DESICCATE. To dry completely.

DIET. Feed ingredient or mixture of ingredients, including water, which is consumed by animals.

DIGESTIBLE NUTRIENT. The part of each feed nutrient that is digested or absorbed by the animal.

DIGESTION COEFFICIENT (coefficient of digestibility). The difference between the nutrients consumed and the nutrients excreted, expressed as a percentage.

DINGY. Wool that is dark or grayish in color and generally heavy in shrinkage.

DISBUDDING. Removal of horns from a kid by a hot iron or a chemical.

DISINFECTANT. A chemical capable of destroying disease-causing microorganisms or parasites.

DISPOSITION. The temperament or spirit of an animal.

DIURESIS. An increased excretion of urine.

DOCK.

- To cut off the tail.
- To reduce the rate of pay.
- The remaining portion of the tail of a sheep that has been docked.

DOE. An adult female goat (also, a rabbit or a deer).

DOELING. Generally a female goat from its first to second birthday, or from its first birthday until its first freshening.

DOGGY WOOLS. Wools that have no character and show the results of lack of breeding. These wools are usually short, coarse, and lacking in feel.

DOMESTIC WOOLS. (1) It includes all wools grown in the United States as against foreign wools, and (2) it includes wools produced on farms east of the intermountain and range regions of the United States, which are also known in the trade as fleece wools. Texas wools are not included in the latter group.

DOMINANT. Describes a gene which, when paired with its allele, covers up the phenotypic expression of that gene.

DOWN WOOL. Wool of medium fineness produced by such down breeds as the Southdown and the Shropshire. These wools are lofty and well suited for knitting yarn. Much of the down wool runs 1/4 to 3/8 blood in quality.

DRENCH. Liquid medicine given to a sheep or goat by mouth.

DRESSING PERCENTAGE. The percentage of the live animal that becomes the carcass at harvest.

DRIED. Describes materials from which water or other liquids have been removed.

DRUGS. Substances of mineral, vegetable, or animal origin used in relieving pain or for curing disease.

DRY. Nonlactating. The dry period is the time between lactations (when a female is not secreting milk).

DRYLOT. A relatively small enclosure without vegetation, either (1) with shelter, or (2) in an open yard, in which animals may be confined.

DRY MATTER BASIS. A method of expressing the level of a nutrient contained in a feed on the basis that the material contains no moisture.

DUNG. The feces or excrement of animals.

DUNG LOCKS. Britch wool locks that are encrusted in hardened dung.

E

EARLY LAMBS. Lambs born early enough to go to market as spring lambs.

EARLY-MATURING. Completing sexual development at an early age.

EARLY WEANING. Practice of weaning young animals earlier than usual; weaning lambs 8 to 12 weeks of age or earlier than 5 months.

EARNOTCHING. Making slits or perforations in an animal's ears for identification purposes.

EASTERN PULLED WOOL. Wool from sheep and

lambs harvested in the East. The wool is pulled from the skins after it has been loosened, usually by a depilatory. Pulled wool should not be confused with dead wool.

EASY KEEPER. An animal that grows or fattens rapidly on limited feed.

EDEMA. Swelling of a part or all of the body due to the accumulation of excess water.

EFFICIENCY OF FEED CONVERSION. The ratio expressing the number of units of feed required for one unit of production by an animal. The value is commonly expressed as pounds of feed eaten per pound gain in body weight. It may also refer to the production of meat or milk per unit of feed consumed.

ELASTICITY. The ability of wool to return to its original length after having been stretched. The elasticity varies greatly with the character of the wool. Wool that is sound and has good character has considerable elasticity.

ELASTRATOR. A mechanical device used to apply an elastic band to the tail or scrotum of a lamb, or the scrotum of a kid, so that this tissue will atrophy and drop off.

EMACIATED. Having excessive loss of flesh.

ENDOCRINE. Pertaining to glands and their secretions that pass directly into the blood or lymph instead of into a duct (secreting internally). Hormones are secreted by endocrine glands.

ENDOGENOUS. Originating within the body; *e.g.*, hormones and enzymes.

ENERGY.

- Vigor or power in action.
- The capacity to perform work.

ENERGY FEEDS. Feeds that are high in energy and low in fiber (under 18%), and that generally contain less than 20% protein.

ENTERITIS. Inflammation of the intestines.

ENVIRONMENT. The sum total of all external conditions that affect the life and performance of sheep and goats.

ERGOSTEROL. A plant sterol which, when activated by ultraviolet rays, becomes vitamin D_2. It is also called provitamin D_2 and ergosterin.

ERGOT. A fungus disease of plants.

ERUCTION. The elimination of gas by belching.

ESSENTIAL AMINO ACIDS. Those amino acids which cannot be made in the body from other substances or which cannot be made in sufficient quantity to supply the animal's needs.

ESSENTIAL FATTY ACID. A fatty acid that cannot be synthesized in the body or that cannot be made in sufficient quantities for the body's needs.

ESTROUS CYCLE. The time and the physiological events from one estrus (heat) period to the next, about 17 days in ewes, and 20 days in does.

ESTRUS. The period when the ewe will accept service by the ram and the doe will accept service by the buck.

ETHER EXTRACT (EE). Fatty substances of feeds and foods that are soluble in ether.

EVAPORATED. Reduced to a denser form; concentrated as by evaporation or distillation.

EWE. A female sheep of any age.

EXCRETA. The products of excretion—primarily feces and urine.

EXPERIMENT. A procedure used to discover or to demonstrate a fact or a general truth. The word *experiment* is derived from the Latin *experimentum*, meaning *proof from experience*.

F

FALL WOOL OR FALL-SHORN WOOL. Wool shorn in the fall (usually in Texas or California) following 4 to 6 months of growth. Only a small percentage of the total U.S. wool production is the product of twice-a-year shearing.

FAT. Frequently used in a general sense to include both fats and oils, or a mixture of the two. Both fats and oils have the same general structure and chemical properties, but they have different physical characteristics. The melting points of most fats are such that they are solid at ordinary room temperatures, while oils have lower melting points and are liquids at these temperatures.

FATTENING. The deposition of energy in the form of fat within the body tissues.

FECES. The excreta discharged from the digestive tract through the anus.

FECUNDITY. Ability to produce many offspring.

FEED (or feedstuff). Any naturally occurring ingredient, or material, fed to animals for the purpose of sustaining them.

FEED ADDITIVE. An ingredient or a substance added to a feed to improve the rate and/or efficiency of gain of animals, to prevent certain diseases, and to preserve the feed.

FEED EFFICIENCY. The ratio giving the number of units of feed required for one unit of production (body weight, meat, or milk) by an animal. In lambs, this value is commonly expressed as pounds of feed eaten per pound of gain in body weight.

FEEDER LAMBS. Young animals under one year of age that carry insufficient finish for slaughter purposes but will make good gains if placed on feed.

FEEDER'S MARGIN. The difference between the cost per hundredweight of feeder animals and the selling price per hundredweight of the same animals when finished.

FEED GRAIN. Any of several grains most commonly used for livestock or poultry feed, such as corn, sorghum, oats, or barley.

FEEDLOT. A lot or plot of land on which animals are fed or finished for market.

FEED OUT. To feed an animal until it reaches market weight.

FEEDSTUFF. Any product, of natural or artificial origin, that has nutritional value in the diet when properly prepared.

FELLMONGER. A dealer in pelts who pulls the wool from the skins, scours the wool, and tans or pickles the skins.

FELT. Woven or pressed and heavily fulled goods as distinct from woolens and worsteds.

FELTING. The interlocking of wool fibers when rubbed together under conditions of heat, moisture, and pressure. No other fiber can compare with wool in felting properties.

FERMENTATION. Chemical changes brought about by enzymes produced by various microorganisms.

FETUS. A young organism in the uterus from the time the organ systems develop until the animal is born.

FIBER CONTENT OF A FEED. The amount of hard-to-digest carbohydrates. Most fiber is made up of cellulose and lignin.

FILL.

- The fullness of the digestive tract of an animal.
- The amount of feed and water consumed by market animals upon their arrival at the market and prior to being sold.

FILLING. Commonly termed *weft*. The crosswise threads or yarns used in a fabric to fill the warp threads which run lengthwise. Filling yarn is usually made from shorter wools and does not need to be as strong as the warp yarn.

FINE WOOL. The finest grade of wool—64s or finer, according to the numerical count grade. Also, the wool from any of the Merino breeds of sheep.

FINISH. To fatten a slaughter animal. Also, the degree of fatness of such an animal.

FITTING. The conditioning of an animal for show or sale, which usually involves a combination of special feeding plus exercise and grooming.

FLEECE. The wool from one sheep, or the mohair from one goat, either as it comes from the animal, or after it is rolled into a bundle and tied.

FLEECE WOOLS. Wools produced on farms in areas east of the Rockies, inclusive of those produced east of the Mississippi River.

FLOCK. A small group of sheep.

FLORA. The plant life present. In nutrition, it generally refers to the bacteria present in the digestive tract.

FLUSHING. The practice of feeding females more generously 2 to 3 weeks before breeding. The beneficial effects attributed to this practice are (1) more eggs (ova) are shed, and this results in more offspring; (2) the females come in heat more promptly; and (3) conception is more certain.

FOOD AND DRUG ADMINISTRATION (FDA). The federal agency in the Department of Health and Human Services that is charged with the responsibility of safeguarding American consumers against injury, unsanitary food, and fraud. It protects industry against unscrupulous competition, and it inspects and analyzes samples and conducts independent research on such things as toxicity (using laboratory animals), disappearance curves for pesticides, and long-range effects of drugs.

FORAGE. Vegetable material in a fresh (pasture), dried (hay), or ensiled (silage) state which is fed to livestock.

FORTIFY. Nutritionally, to add one or more feeds or feedstuffs.

FREE-CHOICE. Free to eat a feed or feeds at will.

FREE WOOL. Usually means wool that is free from defects, such as vegetable matter.

FRENCH COMBING WOOLS. Wools that are intermediate in length between strictly combing and clothing. French combs can handle fine wools from 1.25 to 2.5 in. in length. The yarn is softer and loftier than Bradford (worsted) yarn.

FRESHENING. Giving birth or kidding.

FRIBS. Short second cuts resulting from faulty shearing; also small-sized dirty or dungy locks.

FROWSY WOOL. A wasty, lifeless-appearing, dry, harsh wool, lacking in character. In direct contrast to lofty.

FULL FEED. Indicating that animals are being provided as much feed as they will consume safely without going off feed.

FULLING. The operation of shrinking and felting a woolen fabric to make it thicker and denser. The individual yarns cannot be distinguished on a fulled fabric.

FUMIGANT. A liquid or solid substance that forms vapors that destroy pathogens, insects, and rodents.

FUNGI. Plants, such as molds, mushrooms, toadstools, and yeasts, that contain no chlorophyll, flowers, or leaves. They get their nourishment from either dead or living organic matter.

G

GASTROINTESTINAL. Pertaining to the stomach and intestines.

GESTATION (pregnancy). Time between breeding and lambing, about 148 days for ewes and 150 days for does.

GET. The offspring of a male animal—his progeny.

GIGOT. A leg of mutton (venison or veal) that is trimmed and ready for consumption.

GLUCOSE. A hexose monosaccharide obtained upon the hydrolysis of starch and certain other carbohydrates. Also called dextrose.

GOITROGENIC. Producing or tending to produce goiter.

GRADE.

- Offspring of parents that cannot be registered by a breed association.
- A measure of how well an animal or a product fulfills the requirements for the class; for example, the federal grades of lambs and their carcasses are a specific indication of the degree of excellence.
- A means of designating wool or mohair, primarily according to the fineness and length of fiber.

GRADING. Classification of the unopened or untied fleeces according to fineness, staple length, character, soundness, etc.

GRADING UP. The continued use of purebred sires of the same breed in a grade flock (herd).

GRAIN. Seed from cereal plants.

GRAY WOOL. Fleeces with a few dark fibers, a rather common occurrence in the medium wools produced by down or black-faced breeds.

GRAZE. To consume standing vegetation.

GREASE WOOL. Wool in its natural condition as it comes from the sheep, either shorn or pulled.

GREGARIOUS. Having the flocking instinct—the tendency of sheep to bunch together.

GRIND. To reduce to small segments by impact, shearing, or attrition (as in a mill).

GROATS. Grain from which the hulls have been removed.

GROUP-FED. Feeding system where all animals in a group are fed at one time (compare *self-fed*).

GROW OUT. To feed animals so that they attain a certain desired amount of growth with little or no fattening.

GROWTH. The increase in size of the muscles, bones, internal organs, and other parts of the body.

GROWTHY. Describing an animal that is large and well developed for its age.

GRUEL. A feed prepared from ground ingredients mixed with hot or cold water.

GUMMER. A sheep with all the incisor teeth missing—a sign of age.

GUMMY WOOL. Grease wool that has an excessive amount of yolk; or scoured wool that still has some yolk in it.

H

HALF-BLOOD WOOL. Designation of a grade classifying immediately below the fine grade.

HAND FEEDING. Providing a certain amount of a ration at regular intervals.

HANDLE. The hand of the skilled grader or sorter aids the eye in classification as to fineness, softness, length, and strength of the wool.

HAND MATING. Controlled breeding with confined rams or bucks rather than allowing the males to run loose with groups of unbred females.

HAND-WASHED WOOL. Wool washed before it is shorn from the sheep. This practice is not recommended.

HANK. A standard length of yarn from a reel. A hank of worsted yarn contains 560 yd.

HARD KEEPER. An animal that is unthrifty and grows or fattens slowly regardless of the quantity or quality of feed.

HARVEST. The process of killing the animal; formerly called slaughter.

HAY. Dried forage.

HAYLAGE. Ensiled forage crop about 50% dry matter.

HEAT (estrus). The period when the female will accept service by the male.

HEAT INCREMENT (HI). The increase in heat production following consumption of feed when the animal is in a thermoneutral environment.

HEAT LABILE. Unstable to heat.

HEAVY WITH LAMB. Female sheep approaching parturition as indicated by their full sides and bellies and looseness around genital parts.

HEAVY WOOL. Wool that has considerable grease or dirt and will have a high shrinkage in scouring.

HEMOGLOBIN. The oxygen-carrying, red-pigmented protein of the red corpuscles.

HERDER. A person who takes care of goats; the counterpart of a shepherd for sheep. However, the caretaker of a range band of sheep is commonly referred to as a herder.

HERNIA. The protrusion of some of the intestine through an opening in the body wall—commonly called a *rupture*.

HETEROSIS (hybrid vigor). Amount the F_1 (filial) generation exceeds the P_1 (parent) generation for a given trait, or the amount the crossbreds exceed the average of the two purebreds that are crossed to produce the crossbreds.

HOGGET. A sheep from weaning until its first shearing.

HOMOGENIZED. Describes particles broken down into evenly distributed globules small enough to remain emulsified for long periods of time.

HOMOLOGOUS CHROMOSOMES. Chromosomes having the same size and shape, and containing the genes affecting the same characteristics.

HORMONE. A body-regulating chemical secreted by an endocrine gland into the bloodstream, thence transported to another region within the animal where it elicits a physiological response.

HOTHOUSE LAMBS. Lambs born in the fall or early winter and marketed when they are from 9 to 16 weeks of age or from Christmas to May to a special trade.

HOT TEST. A shrinkage test made on wool as it comes hot from the wool drier after scouring, and before the wool has had time to take on moisture from the air. This gives a quick result, but it is not as accurate as a cold test.

HULLS. Outer coverings of grains or other seeds, especially when dry.

HYPERVITAMINOSIS. An abnormal condition resulting from the intake of an excess of one or more vitamins.

HYPOCALCEMIA. Below normal concentration of ionic calcium in blood, resulting in convulsions, as in tetany or parturient paresis (milk fever).

HYPOGLYCEMIA. A reduction in concentration of blood glucose which is below normal.

HYPOMAGNESEMIA. An abnormally low level of magnesium in the blood.

HYPOTHALAMUS. A portion of the brain found in the floor of the third ventricle. It regulates body temperature, appetite, hormone release, and other functions.

I

IMMUNITY. The ability of an animal to resist or overcome an infection to which most members of its species are susceptible.

IMMUNOGLOBULINS. A family of proteins found in body fluids which have the property of combining with antigens; and, when the antigens are pathogenic, sometimes inactivating them and producing a state of immunity. Also called *antibodies*.

INBREEDING. The mating of individuals that are more closely related than average individuals in a population. It increases homozygosity.

INFLAMMATION. The reaction of tissue to injury, characterized by redness, swelling, pain, and heat.

INGEST. To eat or take in through the mouth.

INGESTA. Food or drink taken into the stomach.

INGESTION. The taking in of food and drink.

INGREDIENT. A constituent feed material.

INSULIN. A hormone secreted by the pancreas into the blood, which regulates sugar (glucose) metabolism.

INTRADERMAL. Into, or between, the layers of the skin.

INTRAMUSCULAR. Within the muscle or muscles.

INTRAPERITONEAL. Within the peritoneal cavity.

INTRAVENOUS. Within the vein or veins.

IN VITRO. Occurring in an artificial environment, as in a test tube.

IN VIVO. Occurring in the living body.

INVOLUTION. Return of an organ to its normal size and condition after enlargement, as of the uterus after lambing or kidding.

IRRADIATED YEAST. Yeast that has been irradiated. Yeast contains considerable ergosterol, which, when exposed to ultraviolet light, produces vitamin D.

IRRADIATION. Exposure to ultraviolet light.

IS BY. Indicates the male parent in animal breeding.

IU (international unit). A standard unit of potency of a biologic agent (*e.g.*, a vitamin, a hormone, an antibiotic, an antitoxin) as defined by the International Conference for Unification of Formulae. Potency is based on bioassay that produces a particular effect agreed on internationally. Also called a USP unit.

J

JAIL. A small pen only large enough to hold one ewe and her offspring.

JOULE. A proposed international unit (4,184 j = 1 calorie) for expressing mechanical, chemical, or electrical energy, as well as heat. In the future, energy requirements and feed values will likely be expressed by this unit.

JUG. A penned area where the lamb(s) is kept with the ewe for a few days after birth. It is also called a *lambing pen*.

K

KED. An external parasite that affects sheep.

KEMP. A white, straight, opaque, coarse, nonfelting, inelastic fiber having a thick central medulla with hollow interspaces. It will not take a dye; hence, its presence in wool is most objectionable.

KERNEL. The whole grain of a cereal. The meat of nuts and drupes (single-stoned fruits).

KID. A young goat up to the first year of age.

KIDDING. The act of a female goat giving birth—parturition.

KJELDAHL. A method of determining the amount of nitrogen in an organic compound. The quantity of nitrogen measured is then multiplied by 6.25 to calculate the protein content of the feed or compound analyzed. The method was developed by the Danish chemist J. G. C. Kjeldahl in 1883.

L

LABILE. Unstable. Easily destroyed.

LACTATION. The period in which an animal is producing milk.

LACTOSE (milk sugar). A disaccharide found in milk having the formula $C_{12}H_{22}O_{11}$. It hydrolyzes to glucose and galactose. Commonly known as milk sugar.

LAMB.

- A sheep less than one year old.
- To give birth to a lamb.

LAMB HOG. A male lamb from weaning time until it is shorn.

LAMBING PEN OR JUG. A small pen in which a ewe is placed for lambing, and until the ewe and lamb(s) are able to run with the flock.

LAMB'S WOOL. Wool shorn from lambs, usually when they are less than 7 or 8 months old. It is soft and has spinning qualities superior to fleeces of similar quality produced by older sheep.

LANOLIN. Purified wool grease, chiefly a mixture of cholesterol esters. It is used in salves, cosmetics, grease paints, and ointments.

LARVA. The immature form of insects and other small animals.

LAXATIVE. A feed or drug that will induce bowel movements and relieve constipation.

LIBIDO. Sexual desire.

LIMITED FEEDING. Feeding animals less than they would like to eat. Giving sufficient feed to maintain weight and growth, but not enough for their potential production or finishing.

LINEBREEDING. A form of inbreeding which attempts to concentrate the inheritance of some ancestor in the pedigree.

LINECROSS. A cross of two inbred lines.

LINE FLEECE. A fleece of wool midway between two grades in quality and length, which can be thrown into either grade.

LITTER.

- The young produced by a female animal which bears more than one or two at one time.
- Materials used for bedding farm animals.

LIVER ABSCESSES. Single or multiple abscesses on the liver, observed at harvest. Usually an abscess consists of a central mass of necrotic liver surrounded by pus and a wall of connective tissue. At harvest, those livers affected with abscesses are condemned for human food.

LIVEWEIGHT. Weight of an animal on foot.

LOCK. A tuft or group of wool fibers that cling naturally together in the fleece.

LOFTY WOOL. Wool that is open, springy, and bulky in comparison to its weight. This type of wool is desirable.

LONG WOOL. Wool from such English breeds as the Lincoln, Leicester, and Cotswold. It is large in diameter and up to 12 or 15 in. in length.

LOW-SET. Designating a short-legged animal

LOW WOOL. Wool of low 1/4 blood or lower in quality. Same as *coarse wool*.

LUMEN. The cavity inside a tubular organ—the lumen of the intestine.

LUSTER. The natural gloss or sheen characteristic of the fleeces of long-wool breeds of sheep and Angora goats.

LYMPH. The slightly yellow transparent fluid occupying the lymphatic channels of the body.

M

MACROMINERALS. The major minerals—calcium, phosphorus, sodium, chlorine, potassium, magnesium, and sulfur.

MAINTENANCE REQUIREMENT. A ration which is adequate to prevent any loss or gain of tissue in the body when there is no production.

MALNUTRITION. Any disorder of nutrition. Commonly used to indicate a state of inadequate nutrition.

MALTOSE. A disaccharide, also known as malt sugar, having the formula $C_{12}H_{22}O_{11}$. Obtained from the partial hydrolysis of starch. It hydrolyzes to glucose.

MANURE. A mixture of animal excrements (consisting of undigested feeds plus certain body wastes) and bedding.

MARGIN (spread). The difference between the purchase price and the selling price.

MARKET CLASS. The grouping of animals according to the use to which they will be put, such as slaughter or feeder.

MARKET GRADE. The grouping of animals within a market class according to their value.

MASTICATION. The chewing of feed.

MASTITIS. Inflammation of the mammary glands.

MATCHING. Grouping or throwing together of corresponding qualities from many fleeces.

MEAL.

- A feed ingredient having a particle size somewhat larger than flour.
- A mixture of concentrate feeds, usually in which all of the ingredients are ground.

MEATS.

- Animal tissues used as food.
- The edible parts of nuts and fruits.

MECONIUM. Excrement accumulated in the bowels during fetal development.

MEDICATED FEED. Any feed which contains drug ingredients intended or represented for the cure, mitigation, treatment, or prevention of diseases of animals (other than humans).

MEDIUM WOOLS. Usually 1/4 blood, 3/8 blood, and 1/2 blood wools, or wools grading 50s to 62s.

METABOLISM. Process describing all the changes that take place in the nutrients after they are absorbed from the digestive tract, including (1) the building-up processes in which the absorbed nutrients are used in the formation or repair of body tissues, and (2) the breaking-down processes in which nutrients are oxidized for the production of heat and work.

MICROBE. Same as *microorganism*.

MICROFLORA. Microbial life characteristic of a region, such as the bacteria and protozoa populating the rumen.

MICROINGREDIENT. Any ration component, such as a mineral, a vitamin, an antibiotic, or a drug, normally measured in milligrams or micrograms per kilogram, or in parts per million.

MICROORGANISM. Any organism of microscopic size, applied especially to bacteria and protozoa.

MILK EJECTION OR LET-DOWN. The process, controlled by the hormone oxytocin, in which milk is forced from the alveoli, where it is stored, into the larger ducts and cisterns, where it is available to sucklings or the milker.

MILL BYPRODUCT. A secondary product obtained in addition to the principal product in milling practice.

MINERALS (ash). The inorganic elements of animals and plants, determined by burning off the organic matter and weighing the residue, which is called *ash*.

MINERAL SUPPLEMENT. A rich source of one or more of the inorganic elements needed to perform certain essential body functions.

MOHAIR. The long, lustrous fleece covering of the Angora goat.

MOISTURE. The amount of water that is contained in feeds—expressed as a percentage.

MOISTURE-FREE (M-F, oven-dry, 100% dry matter). Referring to any substance that has been dried in an oven at 221°F until all the moisture has been removed.

MOLDS (fungi). Fungi which are distinguished by the formation of mycelium (a network of filaments or threads), or by spore masses.

MORBIDITY. A state of sickness or rate of sickness.

MORTALITY. Death or death rate.

MOUTON. A modern fur made by a chemical treatment and processing of sheep pelts carrying designated lengths and qualities of wool on the skin.

MUMMIFIED FETUS. A shriveled or dried fetus that remains in the uterus instead of being expelled or aborted after dying.

MUNGO. Wool fibers recovered from old and new hard worsteds and woolens of firm structure. The fibers are less than 0.5 in. in length, and owing to their reduced spinning and felting qualities, they are restricted largely to use in the cheaper woolen blends. Mungo fibers are usually shorter than shoddy fibers.

MUSHY WOOL. Wool that is dry and wasty in manufacturing.

MUTTON. The flesh of a mature sheep.

MYCOTOXINS. Toxic metabolites produced by molds during growth. Sometimes present in feed materials.

N

NANO. A prefix meaning one-billionth (10^{-9}).

NATIONAL RESEARCH COUNCIL (NRC). A division of the National Academy of Sciences established in 1916 to promote the effective utilization of scientific and technical resources. Periodically, this private nonprofit organization of scientists publishes bulletins giving nutrient requirements and allowances of domestic animals, copies of which are available on a charge basis through the National Academy of Sciences, National Research Council, 2101 Constitution Avenue, N.W., Washington, DC 20418.

NATIVE WOOL. Farm wool produced east of the Mississippi River.

NECROPSY. An examination of the internal organs of a dead animal to determine the apparent cause of death—an autopsy or a postmortem.

NECROSIS. Death of tissue.

NEGATIVE PRESSURE VENTILATING SYSTEM. Mechanical ventilating system where fans exhaust air out of building and air enters through inlets. Air distribution is controlled by the inlets.

NEONATE. A newborn.

NEPHRITIS. Inflammation of the nephrons of the kidneys.

NICK. The result of a certain mating which produces an animal of high order of excellence, sometimes from mediocre parents.

NITROGEN. A chemical element essential to life. Animals get it from protein feeds; plants get it from the soil; and some bacteria get it directly from the air.

NITROGEN BALANCE. The nitrogen in the feed intake minus the nitrogen in the feces, minus the nitrogen in the urine.

NITROGEN FIXATION. Conversion of free nitrogen of the atmosphere to organic nitrogen compounds by symbiotic or nonsymbiotic microbial activity.

NITROGEN-FREE EXTRACT (NFE). It consists principally of sugars, starches, pentoses, and nonnitrogenous organic acids. The percentage is determined by subtracting the sum of the percentages of moisture, crude protein, crude fat, crude fiber, and ash from 100.

NOIL. The short fibers that are removed from the staple wool in the combing or top-making process. Noil is satisfactory for the manufacture of felts and woolens.

NONPROTEIN NITROGEN (NPN). Nitrogen which comes from a source other than protein but which may be used by a ruminant in the building of protein. NPN sources include compounds like urea and anhydrous ammonia, which are used in feed formulations for ruminants only.

NORTH CENTRAL REGION. The states of Illinois, Indiana, Iowa, Kansas, Michigan, Minnesota, Missouri, Nebraska, North Dakota, Ohio, South Dakota, and Wisconsin.

NUMDAH RUG. A woolen, felted rug imported from India. This washable, practical bedroom rug is embroidered with wool threads to add to the effect.

NUMERICAL COUNT SYSTEM. A grading system. It divides all wools into 14 grades, and each grade is designated by a number.

NUTRIENT ALLOWANCES. Nutrient recommendations that allow for variations in feed composition; possible losses during storage and processing; day-to-day and period-to-period differences in needs of animals; age and size of animal; stage of gestation and lactation; kind and degree of activity; amount of stress; system of management; health, condition, and temperament of the animal; and kind, quality, and amount of feed—all of which exert a powerful influence in determining nutritive needs.

NUTRIENT REQUIREMENTS. Nutrients which meet the animal's minimum needs, without margins of safety, for maintenance, growth, fitting, reproduction, lactation, and work. To meet these nutrient requirements, the different classes of animals must receive sufficient feed to furnish the necessary quantity of energy (carbohydrates and fats), proteins, minerals, and vitamins.

NUTRIENTS. The necessary chemical substances found in feed materials that are used for the maintenance, production, and health of animals. The chief classes of nutrients are carbohydrates, fats, proteins, minerals, vitamins, and water.

NUTRITION. The sum total of processes that provide nutrients to the component cells of an animal.

NUTRITIVE RATIO (NR). The ratio of digestible protein to other digestible nutrients in a feedstuff or a ration. (The NR of shelled corn is about 1:10.)

OFFAL. All organs or tissues removed from the carcass in slaughtering.

OFF-SORTS. The byproducts of sorting—shorts; britch wool, kemp, gray wool, stained wool, etc.

OIL. One of several kinds of fatty or greasy liquids that are lighter than water, burn easily, are not soluble in water, and are composed principally, if not exclusively, of carbon and hydrogen. The melting points of oils are such that they are liquid at ordinary room temperatures.

OIL CROPS. Crops grown primarily for oil, including soybeans, cottonseed, peanuts, flaxseed, sunflowers, safflowers, and castor beans.

OMASUM. The third compartment of the ruminant stomach. It is often called the *manyplies* due to its structure.

OPEN-FACED. Having little or no wool on the face, especially around the eyes.

OPEN WOOL. Wool that is not dense on the sheep and shows a distinct part down the ridge or middle of the back. Usually found in the coarser wool breeds.

ORPHAN LAMB. A lamb without a mother.

OSSIFICATION. The process of bone formation; the calcification of bone with advancing maturity.

OSTEITIS. Inflammation of a bone.

OSTEOMALACIA. A bone disease of adult animals caused by lack of vitamin D, inadequate intake of calcium or phosphorus, or an incorrect dietary ratio of calcium and phosphorus.

OSTEOPOROSIS. Abnormal porosity and fragility of bone as the result of (1) a calcium, phosphorus, and/or vitamin D deficiency; or (2) an incorrect dietary ratio of calcium and phosphorus.

OUTCROSS. The introduction of genetic material from some outside and unrelated source, but of the same breed, into a flock or herd which is more or less related.

OUT OF. *Mothered by*, in animal breeding.

OVERFEEDING. Excess feeding.

OVERFINISHING. Excess finishing or fatness—a wasteful practice.

OXIDATION. Any chemical change which involves the addition of oxygen or its chemical equivalent. The animal combines carbon from feedstuffs with inhaled oxygen to produce carbon dioxide, energy (as ATP), water, and heat.

OXYTOCIN. The hormone that controls milk letdown.

P

PALATABILITY. The result of the following factors sensed by the animal in locating and consuming feed: appearance, odor, taste, texture, and temperature. These factors are affected by the physical and chemical nature of the feed.

PANTOTHENIC ACID. One of the B vitamins. It is a constituent of coenzyme A, which plays an essential role in fat and cholesterol synthesis.

PARROT MOUTH. A peculiar condition in the shape of the mouth resulting from one jaw crossing over the other.

PARTS PER BILLION (ppb). It equals micrograms per kilogram or microliters per liter.

PARTS PER MILLION (ppm). It equals milligrams per kilogram or milliliters per liter.

PARTURITION. The process of giving birth—lambing or kidding.

PATHOGENIC. Disease-causing.

PEARLED GRAINS. Dehulled grains reduced into smaller and smoother particles by machine brushing, or abrasion.

PEDIGREE. A written statement giving the record of an animal's ancestry.

PELT. The skin from a harvested sheep before the wool on it has been pulled.

PERFORMANCE TEST. The evaluation of an animal by its own performance.

PER ORAL. Administration through the mouth.

PER OS. Oral administration (by the mouth).

PERSIAN LAMB. A type of skin obtained from lambs of the Karakul breed.

PERSIAN RUGS. Rugs made from wool in Iran, which are near perfection from the standpoint of loom technique and actual weaving. Floral patterns, mosaic designs, and splendid backgrounds are characteristics of these highly sought rugs.

pH. A measure of the acidity or alkalinity of a solution. Values range from 0 (most acid) to 14 (most alkaline), with neutrality at pH 7.

PHASE FEEDING. Changing the animal's diet (1) to adjust for age and stage or production, (2) to adjust for season of the year and for temperature and climatic changes, (3) to account for differences in body weight and nutrient requirements of different strains of animals, and/or (4) to adjust one or more nutrients as other nutrients are changed for economic or availability reasons.

PHENOTYPE. The characteristics of an animal that can he seen and/or measured.

PHOTOPERIOD. Length of daylight or artificial light.

PHOTOSYNTHESIS. The process whereby green plants utilize the energy of the sun to build up complex organic molecules containing energy.

PHYSIOLOGICAL. Pertaining to the science which deals with the functions of living organisms or their parts.

PHYSIOLOGICAL FUEL VALUES. Units, expressed in calories, used in the United States to measure food energy in human nutrition. Similar to metabolizable energy.

PHYSIOLOGICAL SALINE. A salt solution (0.9% NaCl) having the same osmotic pressure as blood plasma.

PIEBALD. An animal that is spotted in two colors, especially white and black or another dark color.

PIECES. The skirtings and other pieces of wool removed from the main part of the fleece and packed separately in the Australian system.

PINNING. Collection of dung around the anal opening of very young lambs that has dried to the point of interfering with normal bowel movements.

PIZZLE. The penis of a bull, ram, or goat.

PLANT PROTEINS. This group includes the common oilseed byproducts—soybean meal, cottonseed meal, linseed meal, peanut meal, safflower meal, sunflower meal, rapeseed meal.

PNEUMONIA. Inflammation of the lungs.

POLLED. Naturally hornless.

POLYNEURITIS. Neuritis of several peripheral nerves at the same time, caused by metallic and other poisons, infectious disease, or vitamin deficiency. In humans, alcoholism is also a major cause of polyneuritis.

POLYUNSATURATED FATTY ACIDS. Fatty acids having more than one double bond. Linoleic acid, which contains two double bonds, is the primary dietary essential fatty acid of humans.

POOL. The assembling of several clips for sale at one time.

POSITIVE PRESSURE VENTILATING SYSTEM. Mechanical ventilating system where fans blow air into the structure creating a positive pressure.

POSTNATAL. Occurring after birth, when referring to the offspring.

POSTPARTUM. Occurring after the birth of the offspring, when referring to the ewe or the doe.

POT-BELLIED. Designating an abnormally large abdomen in an animal or a human.

PRECURSOR. A compound that can be used by the body to form another compound; for example, carotene is a precursor of vitamin A.

PREHENSION. The seizing (grasping) and conveying of feed to the mouth.

PREMIX. A uniform mixture of one or more microingredients and a carrier, used in the introduction of microingredients into a larger mixture.

PRENATAL. Before birth, when referring to the offspring.

PREPOTENCY. The ability of an individual to transmit its own qualities to its offspring.

PRESERVATIVES. Materials available to incorporate into feeds, with claims made that they will improve the preservation of nutrients, the nutritive value, and/or the palatability of the feed.

PRODUCE. A female's offspring. The produce-of-dam commonly refers to two offspring of one dam.

PROGENY. The offspring of animals.

PROGENY TESTING. An evaluation of an animal on the basis of the performance of its offspring.

PROLAPSE. Abnormal protrusion of a part or an organ of the body.

PROLIFIC. Capable of producing numerous offspring.

PROSTAGLANDINS. A large group of chemically related 20-carbon hydroxy fatty acids with variable physiological effects in the body. The most familiar to stock producers is prostaglandin $F_2\propto$, which is available commercially.

PROTEIN. From the Greek, meaning *of first rank, importance*. Complex organic compounds made up chiefly of amino acids are present in characteristic proportions for each specific protein. At least 24 amino acids have been identified and may occur in combinations to form an almost limitless number of proteins. Protein always contains carbon, hydrogen, oxygen, and nitrogen; and, in addition, it usually contains sulfur and frequently phosphorus. Crude protein is determined by finding the nitrogen content and multiplying the result by 6.25. The nitrogen content of proteins averages about 16% (100 ÷ 16 = 6.25). Proteins are essential in all plant and animal life as components of the active protoplasm of each living cell.

PROTEIN SUPPLEMENTS. Products that contain more than 20% protein or protein equivalent.

PROUD FLESH. Excess flesh growing around a wound.

PROVITAMIN. The material from which an animal may produce vitamins; *e.g.*, carotene (provitamin A) in plants is converted to vitamin A in animals.

PROVITAMIN A. Carotene.

PROXIMATE ANALYSIS. A chemical scheme for evaluating feedstuffs, in which a feedstuff is partitioned into the six fractions: (1) moisture (water) or dry matter (DM); (2) total (crude) protein (CP or TP – N × 6.25); (3) ether extract (EE) or fat; (4) ash (mineral salts); (5) crude fiber (CF)—the incompletely digested carbohydrates; and (6) nitrogen-free extract (NFE)—the more readily digested carbohydrates (calculated rather than measured chemically).

PUBERTY. The age at which the reproductive organs become functionally operative—sexual maturity.

PULLED WOOL. Wool pulled from skins of harvested sheep. The wool is pulled from the skins after treatment of the fleshy side of skins with a depilatory. Pulled wool should not be confused with dead wool.

PUREBRED. An animal of pure breeding, registered or eligible for registration in the herd book of the breed to which it belongs.

PURIFIED DIET. A mixture of the known essential dietary nutrients in a pure form that is fed to experimental (test) animals in nutrition studies.

PURITY. Freedom from dark fibers, kemp, etc., in unshorn fleeces.

PURULENT. Consisting of or forming pus.

PUS. A liquid inflammatory product consisting of leukocytes (white blood cells), lymph, bacteria, dead tissue cells, and the fluid derived from their disintegration.

QUALITATIVE TRAITS. Traits in which there is a sharp distinction between phenotypes, usually involving only one or two pairs of genes.

QUALITY. The desirability and/or acceptance of an animal or feed product.

QUALITY OF PROTEIN. The amino acid balance of

protein. A protein is said to be of good quality when it contains all the essential amino acids in proper proportions and amounts needed by a specific animal; and it is said to be poor quality when it is deficient in either content or balance of essential amino acids.

QUANTITATIVE TRAITS. Traits in which there is no sharp distinction between phenotypes, usually involving several pairs of genes and the environment. These include such economic traits as gestation length, birth weight, weaning weight, rate and efficiency of gain, and carcass quality.

QUARANTINE.

- Compulsory segregation of exposed susceptible animals for a period of time equal to the longest usual incubation period of the disease to which they have been exposed.
- An enforced regulation for the exclusion or isolation of an animal to prevent the spread of an infectious disease.

QUARTER-BLOOD WOOL. One of the grades in the standards for wool.

R

RADIOACTIVE. Giving off atomic energy in the form of alpha, beta, or gamma rays.

RAM. A male sheep of any age. Also called a buck.

RANCID. Describing fats that have undergone partial decomposition—spoiled.

RANGE WOOL. Wool produced under range conditions in the West and Southwest. With the exception of Texas and California wools, it is usually classified as *territory wool*.

RANGY. Designating an animal that is long, lean, leggy, and not very muscular.

RATE OF PASSAGE. The time taken by undigested residues from a given meal to reach the feces. (A stained undigestible material is commonly used to estimate rate of passage.)

RATIO. The performance of an individual in relation to the average of all animals of the same group. It is calculated as follows:

$$\frac{\text{Individual record}}{\text{Average of animals in group}} \times 100$$

It is a record or an index of individual deviation from the group average, expressed in percentage. A ratio of 100 is average for a particular group. Thus, ratios above 100 indicate animals above average, whereas ratios below 100 indicate animals below average.

RATION(S). The amount of feed supplied to an animal for a definite period, usually for 24 hours. However, by practical usage, the word *ration* implies the feed fed to an animal without limitation to the time in which it is consumed.

RAW WOOL. Wool in the grease. Same as *grease wool*.

RECLAIMED WOOL. Wool that is reclaimed from new or old fabrics.

RED MEAT. Meat that is red when raw, due to the red coloration of myoglobin, the pigment of muscle. Red meats include beef, veal, pork, mutton, and lamb muscle tissue with attendant fat and bone.

REGAIN. Referring to an increase in weight due to water absorbed by bone-dry scoured wool under standard atmospheric conditions.

REGISTERED. Designating purebred animals whose pedigrees are recorded in the breed registry.

REGURGITATION. The casting up (backward flow) of undigested food from the stomach to the mouth, as by ruminants.

REJECTS OR REJECTIONS. Off-grades thrown to one side by the wool grader; fleeces with excessive black fibers, kemp, dead fibers, vegetable matter, etc.

REPLACEMENT. An animal selected to be kept for the breeding flock or herd.

REPROCESSED WOOL. Scraps and clips of woven and felted fabrics made of previously unused wool. These remnants are *garnetted*; that is, shredded back into a fibrous state and used in the manufacture of woolens.

RETAINED PLACENTA. A placenta not expelled at parturition or shortly thereafter.

RETICULUM. The second compartment of the ruminant stomach. It has a honeycomb-texture lining; so, it is often called the *honeycomb*.

REUSED WOOL. Old wool which has actually been worn or used, including the rags and miscellaneous old clothing collected by rag dealers. These are cleaned and shredded into fibers again, and then blended to make utility fabrics. The consumer has no way of telling how much the original desirable qualities of wool have been impaired by this previous use. Also called *shoddy*.

REWORKED WOOL. Wool that has been previously used. Also called *shoddy* and *mungo*.

RIDGELING (rig). Any male animal whose testicles fail to descend into the scrotum—a cryptorchid.

RIG. The line down through the middle of the back on coarse wools where the wool separates, some falling down one side, the rest falling down the other side.

RIGOR MORTIS. The stiffness of body muscles that is observed shortly after death.

ROUGHAGE. Feed consisting of bulky and coarse plants or plant parts, containing high-fiber content and low total digestible nutrients, arbitrarily defined as feed with over 18% crude fiber. Roughage may be either dry or green.

RUMEN. The first stomach compartment of a ruminant—the paunch. It is a large nutrient-producing vat.

RUMEN FLORA. The microorganisms of the rumen.

RUN. A standard of length for woolen yarn. One run equals 1,600 yd of woolen yarn weighing 1 lb.

RUN-ON PASTURE. To graze or pasture.

RUN-OUT FLEECE. A fleece that varies greatly in quality, lacks character, and carries a large percentage of britch wool and possibly kemp.

SACCHARIDES. Sugars. The prefixes *mono-, di-, tri,* and *poly-* denote the number of sugars contained in the saccharide.

SALIVA. A clear, somewhat viscid solution secreted by glands within the mouth. It may contain the enzymes salivary amylase and salivary maltase.

SALMONELLA. A pathogenic, diarrhea-producing organism, of which there are over 100 known strains, sometimes present in contaminated feeds.

SATIETY. Full satisfaction of desire; may refer to satisfaction of appetite.

SATURATED FAT. A completely hydrogenated fat—each carbon atom is associated with the maximum number of hydrogens; there are no double bonds.

SCOURING.

- Persistent diarrhea in young animals, possibly due to feeding practices, management practices, environment, or disease.
- A process of washing or cleansing wool of grease soil, and suint in a water-soap-alkali solution.

SCREENED FEEDSTUFF. A feedstuff that has been separated into various sized particles by passing over or through screens.

SCURS. Small, rounded portions of horn tissue attached to the skin at the horn pits of polled animals. They may also be called *buttons*.

SECONDARY INFECTION. Infection following an infection already established by other pathogens.

SECOND CUTS. Fribs, or short lengths of wool resulting from cutting wool fibers twice in careless shearing. An excessive number of second cuts decreases the average fiber length and, hence, depreciates spinning quality.

SEEDY WOOL. Wool containing numerous seeds or an appreciable amount of another type of vegetable matter.

SELECTION. The determination of which animals in a population will produce the next generation. Humans practice artificial selection while nature practices natural selection.

SELENIUM. An element that functions with glutathione peroxidase, an enzyme which enables the tripeptide glutathione to perform its role as a biological antioxidant in the body. This explains why deficiencies of selenium and vitamin E result in similar signs—loss of appetite and slow growth.

SELF-FED. Providing a part or all of the ration on a continuous basis, thereby permitting animals to eat at will.

SELF-FEEDER. A feed container by means of which animals can eat at will. (See *AD LIBITUM*.)

SEMEN. The fluid containing the sperm that is ejaculated by the male.

SEMI-BRIGHT WOOL. Grease wool that lacks brightness due to the environment under which it is produced, though it is white after scouring.

SERRATIONS. The outer or epidermal scaly edges on the wool fiber which can be seen under a microscope. Usually the finer the wool the greater the number of serrations. Serrations assist in felting by interlocking.

SERUM. The colorless fluid portion of blood remaining after clotting and removal of corpuscles. It differs from plasma in that the fibrinogen has been removed.

SERVICE. The mating of a female by a male.

SETTLED. Indicating that an animal has become pregnant.

SHAFTY WOOL. Wool of extra good length, sound, and well grown.

SHEARING. The process of removing the fleece of wool from the sheep by means of hand shears or machine clippers. This operation must be carefully done or the value of the wool will be materially reduced.

SHEARLINGS. Pelts of harvested sheep carrying 0.25 to 1.0 in. growth of wool. Also, it refers to short wool pulled from sheep recently shorn and to yearling sheep in England.

SHEEPSKIN.

- The parchment for college degrees.
- The wool still on the pelt or skin.

SHINES. Small particles of vegetable matter other than burs present in wool.

SHODDY. Wool fibers recovered from either new or used woven or felted cloth and which must be designated as reprocessed or reused. Wool fibers included in this classification usually run 0.5 in. or more in length.

SHORTS. Short pieces or locks of wool that are dropped out while fibers are being sorted.

SHOW BOX (tackbox). A container in which to keep all show equipment and paraphernalia.

SHRINKAGE.

- The amount of loss in body weight when animals are exposed to adverse conditions, such as being transported, being out in severe weather, or having a shortage of feed.
- The loss in carcass weight during the aging process.
- The loss of weight in wool resulting from the removal of the yolk and other foreign matter in scouring or carbonizing.

SIB. A brother or a sister.

SIB TESTING. A method of selection in which an animal is selected on the basis of the performance of its brothers or sisters.

SIRE.

- The male parent.
- To father or beget.

SISAL. A vegetable fiber that is made into strong, coarse twine. It is used for binder twine, but should not be used to tie wool fleeces.

SKIRTING. A practice, common in Australia, of removing from the edges of the whole fleece, at shearing time, all stained and inferior parts.

SLAUGHTER. The term now known as harvest; harvesting.

SLOTTED FLOORS. Floors with slots through which the feces and urine pass to a storage area below or nearby.

SLOUGHING. The process whereby a mass of dead tissue separates from a surface.

SOLUTION. A uniform liquid mixture of two or more substances molecularly dispersed within one another.

SORTING. The process (usually done in a mill) of separating a fleece into its various qualities according to diameter, length, color, strength, and other factors. This is usually the first operation after the grease wool arrives at the mill.

SOUND WOOL. Wool that has a strong staple. Wool buyers or graders test the soundness of the wool by holding a staple at either end and snapping their fingers across the middle of it.

SPECIFIC DYNAMIC ACTION (SDA). The increased production of heat by the body as a result of a stimulus to metabolic activity caused by ingesting food.

SPINNING COUNT. The fineness to which a yarn may be spun. The number of hanks of 560 yd each in length in 1 lb of top. Thus, 1 lb of fine tops that will spin 64 hanks is called 64s.

SPOOK. To scare or frighten animals.

SPRING WOOL. The 6 to 7 months of wool produced by sheep shorn in the spring (usually in Texas and California) following fall shearing.

STABILIZED. Made more resistant to chemical change by the addition of a particular substance.

STAG. A male that was castrated after the secondary sexual characteristics developed sufficiently to give the appearance of a mature male.

STAINED WOOL. Wool that has become discolored by urine, dung, or whatever, which will not scour out white. Badly stained pieces should be removed at shearing pens before the fleeces are packed. Wool from which stains cannot be removed must be used in fabrics to be dyed a dark color.

STANCE. Position or posture adopted when an animal is stationary.

STANCHION. A device for holding a goat for feeding or for milking.

STAPLE. A cluster or group of wool fibers naturally clinging together in the fleece by binder fibers which commonly transverse one or more of these clusters. Also, fine strictly combing wools produced in the western half of the United States.

STENCIL. Paint branding numbers or letters on the backs or sides of sheep.

STERILE. Incapable of reproducing.

STILLBORN. Born lifeless; dead at birth.

STONER. The dried stalks and leaves, but not the grain portion, of corn or milo.

STRAGGLER. An animal that strays or wanders from a flock or herd.

STRAW. The plant residue remaining after separation of the seeds in threshing.

STRESS. Any physical or emotional factor to which an animal fails to make a satisfactory adaptation. Stress may be caused by excitement, temperament, fatigue, shipping, disease, heat or cold, nervous strain, number of animals together, previous nutrition, breed, age, and/or management. The greater the stress, the more exacting the nutritive requirements.

STRIPPING. The removal of the last of the milk from the udder.

STUBBLE SHEARING. The practice of shearing or cutting a portion of the wool at varying lengths, from sheep used for show purposes.

SUBCUTANEOUS. Situated or occurring beneath the skin.

SUCKLE. To nurse at the breasts or mammary glands.

SUGAR. A sweet, crystallizable substance that consists essentially of sucrose, and that occurs naturally in the most readily available amounts in sugarcane, sugar beets, sugar maple, sorghum, and sugar palm.

SUINT. Generally referred to as the perspiration of sheep, which consists largely of potassium salts of various fatty acids and small quantities of sulfate, phosphate, and nitrogenous substances. It is soluble in water.

SUPPLEMENT. A feed or feed mixture used to improve the nutritional value of basal feeds (*e.g.*, protein supplement—soybean meal). Supplements are usually rich in proteins, minerals, vitamins, antibiotics, or a combination of part or all of these; and they are usually combined with basal feeds to produce a complete feed.

SUPPURATION. Formation of pus.

SWEATING PROCESS. The practice of putting sheep skins in a warm, moist room to loosen the wool in preparation for pulling.

SWEAT SHED OR PEN. A properly ventilated shed or enclosure used to hold the sheep so that their natural body heat causes them to sweat. Such a practice immediately preceding the shearing operation softens the yolk and makes shearing easier.

SYMMETRY. A balanced development of all parts.

SYNTHESIS. The bringing together of two or more substances to form a new material.

SYNTHETICS. Artificially produced products that may be similar to natural products.

T

TAGGING. The practice of cutting the dung locks off sheep. Usually this operation is done immediately prior to shearing, and it may be done prior to lambing.

TAGS. Large locks of britch wool clotted with dung and dirt.

TALLOW. The extracted fat of cattle and sheep.

TARE. The weight of the bag or container pack which packages the wool. The buyer pays for only net weight of the wool.

TATTOO. Permanent identification of animals produced by placing indelible ink under the skin; generally put in the ears of young animals.

TDN. (See TOTAL DIGESTIBLE NUTRIENT.)

TEART. Molybdenosis of farm animals caused by feeding on vegetation grown on soil that contains high levels of molybdenum.

TEASER RAM. A ram that is incapable of impregnating ewes, but which is used to find ewes in heat. The ram may be vasectomized or may wear an apron to prevent copulation.

TEG. A sheep two years of age.

TEND. To care for a flock of sheep or a herd of goats.

TENDER. Weak at one or more places along its length. Such wool will not withstand the tension put on it in worsted manufacturing and is rejected and used to manufacture woolens.

TERRITORY WOOL. A designation originally given to wools originating in regions west of the Missouri River. Now applies to western range wools, not including Texas and California. Territory wools seldom equal similar domestic wools.

TETANY. A condition in an animal in which there are localized spasmodic muscular contractions.

TETHER. To tie an animal with a rope or a chain to allow feeding but to prevent straying.

THERMAL. Pertaining to heat.

THERMOGENESIS. The chemical production of heat in the body.

THERMONEUTRALITY. The state of thermal (heat) balance between an animal and its environment. The thermoneutral time is referred to as the comfort zone.

THRIFTY. Healthy and vigorous in appearance.

TIPPY WOOL. Staples which are encrusted with wool grease and dirt at the weather end.

TOCOPHEROL. Any of four different forms of an alcohol; also known as vitamin E.

TONIC. A drug, medicine, or feed designed to stimulate the appetite.

TOP. A continuous untwisted strand of wool fibers of predetermined length from which the short fibers (noil) have been removed in the combing process. Tops represent an intermediate stage in the manufacture of worsted yarn.

TOTAL DIGESTIBLE NUTRIENT (TDN). The energy value of a feedstuff. It is computed by the following formula:

$$\%\ \text{TDN} = \frac{\text{DCP} + \text{DCF} + \text{DNFE} + (\text{DEE} \times 2.25)}{\text{Feed consumed}} \times 100$$

Where DCP = digestible crude protein; DCF = digestible crude fiber; DNFE = digestible nitrogen-free extract, and DEE = digestible ether extract. One pound of TDN = 2,000 kcal of digestible energy.

TOXIC. Of a poisonous nature.

TRACE ELEMENT. A chemical element used in minute amounts by organisms and held essential to their physiology. The essential trace elements are cobalt, copper, iodine, iron, manganese, selenium, and zinc.

TRACE MINERAL. A mineral nutrient required by animals in micro amounts only (measurable in milligrams per pound or smaller units).

TUBER. A short, thickened, fleshy stem or terminal portion of a stem or rhizome that is usually formed underground, bears minute scale leaves, each with a bud capable under suitable conditions of developing into a new plant, and constitutes the resting stage of various plants such as the potato or the Jerusalem artichoke.

TUP. A ram.

TURKISH RUGS. Expensive woolen rugs of various designs.

TYPE.

- The physical conformation of an animal.
- The combination of all those physical attributes that contribute to the value of an animal for a specific purpose.

U

UDDER. The encased group of mammary glands with each gland provided with a nipple or teat.

UNDERFEEDING. Not providing sufficient energy. The degree of lowered production therefrom is related to the extent of underfeeding and the length of time it exists.

UNIDENTIFIED FACTORS. Factors that have not yet been isolated or synthesized in the laboratory. Nevertheless, rich sources of these factors and their effects have been well established. There is evidence that the growth factors exist in dried whey, marine and packing house byproducts, distillers' solubles, antibiotic fermentation residues, alfalfa meal, and certain green forages. Most unidentified or unknown factor sources are added to the diet at a level of 1 to 3%.

UNMERCHANTABLE WOOL. Wool containing excessive vegetable matter, hence, necessitating carbonizing. Also, wool damaged by fire, or weathering on sheep's back before shearing. A term now more or less obsolete.

UNSATURATED FAT. A fat having one or more double bonds; not completely hydrogenated.

UNSATURATED FATTY ACID. Any one of several fatty acids containing one or more double bonds, such as oleic, linoleic, linolenic, and arachidonic acids.

UNTHRIFTINESS. Lack of vigor, poor growth or development; the quality or state of being unthrifty.

UNWASHED WOOL. Wool in its original condition as it comes from a sheep.

USP (United States Pharmacopoeia). A unit of measurement or potency of biologicals that usually coincides with an international unit. (Also see IU.)

V

VACCINATION (shot). An injection of vaccine, bacterin, antiserum, or antitoxin to produce immunity or tolerance to disease.

VACCINE. A suspension of attenuated or killed microorganisms (bacteria, viruses, or rickettsiae) administered for the prevention, improvement, or treatment of infectious diseases.

VARIETY MEATS. Liver, brain, heart, kidney—all excellent sources of many essential nutrients.

VASECTOMY. The surgical removal of a section of the vas deferens, thereby rendering a male sterile without affecting his libido. Sperm cannot be transported from the testicles at the time of ejaculation.

VECTORS. Living organisms which carry pathogens.

VEGETABLE MATTER. Any kind of bur, seed, chaff, grass, or other vegetable matter found in grease wool. Some types are removable by mechanical dusters or bur-pickers. Some are removable in carding or combing. Waste wool from carding or combing is materially reduced in value by the presence of this vegetable matter. Wools containing excessive vegetable matter must be carbonized. This is an added expense, and wools of this type are heavily discounted by the wool buyer.

VERMIFUGE (vermicide). Any chemical substance given to animals to kill internal parasitic worms.

VIRGIN WOOL. Wool that has been clipped from a live sheep and that has not previously been advanced in manufacturing to the stage where it contains twist. Noil is merely separated from long fibers in combing and is considered virgin wool.

VIRUS. One of a group of submicroscopic infectious agents. It lacks independent metabolism and can only multiply within living host cells.

VISCERA. Internal organs of the body, particularly in the chest and the abdominal cavity.

VITAMINS. Complex organic compounds that function as parts of enzyme systems essential for the transformation of energy and the regulation of metabolism of the body, and required in minute amounts by one or more animal species for normal growth, production, reproduction, and/or health. All vitamins must be present in the ration for normal functioning, except for B vitamin in ruminants (cattle, sheep, and goats) and vitamin C.

VITAMIN SUPPLEMENTS. Rich synthetic or natural feed sources of one or more of the complex organic compounds, called vitamins, that are required in minute amounts by animals for normal growth, production, reproduction, and/or health.

VOID. To evacuate feces and/or urine.

VOMITING. The forcible expulsion of the contents of the stomach through the mouth.

W

WARP. The yarns running lengthwise in a fabric. They are usually stronger than the filling yarn in order to withstand the strain of weaving.

WASHED WOOL. Wool washed in cold water while on the sheep's back before shearing. This practice has now practically disappeared in this country.

WASTY.

- Describes a carcass with too much fat, requiring excessive trimming.
- Describes paunchy live animals.

WASTY WOOL. Wool that is short, weak, and tangled, which often carries a high percentage of dirt or sand and which is wasty in processing.

WATTLES. Small, fleshy appendages usually hanging from the neck of a goat. They are hereditary, but do not occur on all goats. They have no purpose.

WEANING. The stopping of young animals from suckling their mothers.

WETHER. A male sheep or goat castrated before sexual maturity.

WIRY WOOL. Wool that is inelastic and has poor spinning capacity. It is usually straight and is the result of poor breeding.

WOOL. The crimpy, serrated fiber that grows out of the skin of the sheep.

WOOL BLIND. Obstruction of vision (often complete) due to heavy cover on face and around eyes.

WOOLEN. Yarn made from shorter cleaned fibers, usually by mixing, oiling, picking, carding, and spinning.

WOOL IN THE GREASE. Wool in its natural condition as it is shorn from a sheep.

WOOL SORTER'S DISEASE. A misnomer. (See ANTHRAX.)

WORSTED. Yarn made from top under the Bradford and French systems.

Y

YEARLING. A sheep, goat, beef cow, or dairy cow that is 12 to 18 months of age.

YEAN. To give birth, especially by sheep and goats.

YIELD. The amount of scoured wool obtained from a definite amount of grease wool. Usually the result is expressed as a percentage.

YOLK. The natural grease and suint covering on the wool fibers of the unscoured fleece, and excreted from glands in the sheep's skin. Usually the finer the wool, the more abundant the yolk. Yolk serves to prevent entanglement of the wool fibers and mechanical injury during growth of the fleece.

CHAPTER 27

Feeds. (Courtesy, American Sheep Industry Assn., Englewood, CO)

FEED COMPOSITION TABLES

Contents *Page*

Nutritionists, sheep producers, and goat producers should have access to accurate and up-to-date composition of feedstuffs in order to formulate rations for maximum production and net returns. The ultimate goal of feedstuff analysis, and the reason for feed composition tables, is to be able to predict the productive response of animals when they are fed rations of a given composition. In recognition of this need and its importance, the author spared no time or expense in compiling the feed composition tables presented in this section. At the outset, a survey of the industry was made in order to determine what kind of feed composition tables would be most useful, in both format and content. Second, it was decided to utilize, to the extent available, the monumental work of L. E. Harris,[1] along with the feed composition tables of the National Academy of Sciences. Additional feeds and compositions were provided by the authors with compositions obtained from experimental reports, industries, and other reliable sources.

FEED NAMES

Ideally, a feed name should conjure up the same meaning to all those who use it, and it should provide helpful information. This was the guiding philosophy of the author when he chose the names given in the feed composition tables. Genus and species—Latin names—are also included. To facilitate worldwide usage, the International Feed Number of each feed is given. To the extent possible, consideration was also given to source (or parent material), variety or kind, stage of maturity, processing, part eaten, and grade.

MOISTURE CONTENT OF FEEDS

It is necessary to know the moisture content of feeds in ration formulation and buying. Usually, the composition of a feed is expressed according to one or more of the following bases:

1. **As-fed; A-F (wet, fresh).** This refers to feed as normally fed to animals. It may range from 0% to 100% dry matter.
2. **Air-dry (approximately 90% dry matter).** This refers to feed that is dried by means of natural air movement, usually in the open. It may either be an actual or an assumed dry matter content; the latter is approximately 90%. Most feeds are fed in an air-dry state.

Fig. 27-1. Alfalfa hay is one of the basic feedstuffs to overwinter sheep. (Courtesy, Dr. Johannes E. Nel, University of Wyoming)

3. **Moisture-free; M-F (oven-dry, 100% dry matter).** This refers to a sample of feed that has been dried in an oven at 221°F until all the moisture has been removed.

Where available, feed compositions are presented on both As-fed (A-F) and Moisture-free (M-F) bases.

PERTINENT INFORMATION ABOUT DATA

The information which follows is pertinent to the feed composition tables presented in this section.

- **Variations in composition**—Feeds vary in their composition. Thus, actual analysis of a feedstuff should be obtained and used wherever possible, especially where a large lot of feed from one source is involved. Many times, however, it is impossible to determine actual compositions or there is insufficient time to obtain such analysis. Under such circumstances, tabulated data may be the only information available.

- **Feed compositions change**—Feed compositions change over a period of time, primarily due to (1) the introduction of new varieties, and (2) modifications in the manufacturing process from which byproducts evolve.

- **Biological value**—The response of animals when fed a feed is termed the biological value, which is a function of its chemical composition and the ability of the animal to derive useful nutrient value from the feed. The latter relates to the digestibility, or availability, of the nutrients in the feed. Thus, soft coal and shelled corn may have the same gross energy value in a bomb calorimeter but markedly different useful energy val-

[1]Professor Emeritus of Nutrition; Department of Animal, Dairy, and Veterinary Sciences; International Feedstuffs Institute; Utah State University; Logan.

ues (TDN, digestible energy, metabolizable energy, and net energy) when consumed by an animal. Biological tests of feeds are more laborious and costly to determine than chemical analyses, but they are much more accurate in predicting the response of animals to a feed.

■ **Where information is not available**—Where information is not available or reasonable estimates could not be made, no values are shown. Hopefully, such information will become available in the future.

■ **Calculated on a dry matter (DM) basis**—All data were calculated on a 100% dry matter basis (moisture-free), then converted to an as-fed basis by multiplying the decimal equivalent of the DM content times the compositional value shown in the table.

■ **Fiber**—Four values for fiber are given in the feed composition tables—crude fiber, cell walls or NDF, acid detergent fiber, and lignin.

Crude fiber, methods for the determination of which were developed more than 100 years ago, is declining as a measure of low digestible material in the more fibrous feeds. The newer method of forage analysis, developed by Van Soest and associates of the U.S. Department of Agriculture, separates feed dry matter into two fractions—one of high digestibility (cell contents) and the other of low digestibility (cell walls)—by boiling a 0.5- to 1.0-g sample of the feed in a neutral detergent solution (3% sodium lauryl sulfate buffered to a pH of 7.0) for 1 hour, then filtering. Also, the amount of lignin in the cell walls is determined.

Fig. 27-2. Cull onions being self-fed to ewes on an Oregon farm. (Courtesy, Dr. David L. Thomas, University of Wisconsin, Madison)

To date, relatively few investigators have examined the value of the Van Soest system for predicting the energy value of feed ingredients for sheep and goats. Likely, this system of analysis will become more widely used for sheep and goat feed ingredients in the future.

1. **Crude fiber (CF).** This is the residue that remains after boiling a feed in a weak acid, and then in a weak alkali, in an attempt to imitate the process that occurs in the digestive tract. This procedure is based on the supposition that carbohydrates which are readily dissolved also will be readily digested by animals, and that those not soluble under such conditions are not readily digested. Unfortunately, the treatment dissolves much of the lignin, a nondigestible component. Hence, crude fiber is only an approximation of the indigestible material in feedstuffs. Nevertheless, it is a rough indicator of the energy value of feeds. Also, the crude fiber value is needed for the computation of TDN.

2. **Cell walls (CW) or neutral detergent fiber (NDF).** This is the insoluble fraction resulting from boiling a feed sample in a neutral detergent solution. It contains cellulose, hemicellulose, silica, some protein, and lignin. Cell wall, or NDF, components are of low digestibility and entirely dependent on the microorganisms of the digestive tract for any digestion that they undergo; hence, they are essentially undigested by nonruminants. This fraction of a forage affects the volume it will occupy in the digestive tract, a principal factor limiting the amount of feed consumed. Animals fed such forages are often unable to consume enough feed to produce weight gains or milk economically.

The soluble fraction—the cell contents—consists of sugars, starches, fructosans, pectin, proteins, nonprotein nitrogen, lipids, water, soluble minerals, and vitamins. This portion is highly digestible (about 98%) by both ruminants and nonruminants.

3. **Acid detergent fiber (ADF).** This involves boiling a 1.0-g sample of air-dry material in a specially prepared acid detergent solution for one hour, then filtering. The insolubles, or residue, make up what is known as acid detergent fiber (ADF) and consists primarily of cellulose, lignin, and variable amounts of silica.

Acid detergent fiber differs from neutral detergent fiber in that NDF contains most of the feed hemicellulose and a limited amount of protein, not present in ADF.

ADF is the best predictor of forage digestible dry matter and digestible energy.

Fig. 27-3. Feeding Angora goats near Sonora, Texas. (Courtesy, Dr. Gary C. Smith, Department of Animal Science, Texas A&M University, College Station, TX)

4. **Lignin.** This fraction is essentially indigestible by all animals and is the substance that limits the availability of cellulose carbohydrates in the plant cell wall to rumen bacteria.

The acid detergent fiber procedure is used as a preparatory step in determining the lignin of a forage sample. Hemicellulose is solubilized during this procedure, while the lignocellulose fraction of the feed remains insoluble. Cellulose is then separated from lignin by the addition of sulfuric acid. Only lignin and acidinsoluble ash remain upon completion of this step. This residue is then ashed, and the difference of the weights before and after ashing yields the amount of lignin present in the feed.

■ **Nitrogen-free extract**—The nitrogen-free extract was calculated with mean data as: mean nitrogen-free extract (%) = 100 – % ash – % crude fiber – % ether extract – % protein.

■ **Protein values**—Both crude protein and digestible protein values are given. Crude protein represents Kjeldahl nitrogen value times 100/16, or 6.25, since protein contains 16% nitrogen on the average. The ruminant values represent a pooling of cattle, sheep, and goat data.

■ **Energy**—Many of the energy values given in the feed composition tables were derived from complex formulas developed by L. E. Harris and other animal scientists.

The following four measures of energy are shown:

1. **TDN.** This value is given because there are more of them, and because it has been the standard method of expressing the energy value of feeds for many years. However, the following disadvantages are inherent in the TDN system: (1) Only digestive losses are considered—it does not take into account other important losses, such as those in the urine, gases, and increased heat production; (2) there is a poor relationship between crude fiber and NFE digestibility in certain feeds; and (3) it overestimates roughages in relation to concentrates when animals are fed for high rates of production, due to the higher heat loss per pound of TDN in high-fiber feeds.

2. **Digestible energy (DE).** Digestible energy is that portion of the gross energy in a feed that is not excreted in the feces. It is roughly comparable to TDN.

For most animals, digestible energy is relatively easy to determine. With poultry, however, true digestibility is very difficult to measure because undigested residues and urinary wastes are excreted together.

3. **Metabolizable energy (ME).** Metabolizable energy represents that portion of the gross energy that is not lost in feces, urine, and gas (mainly methane). It does not take into account the energy lost as heat, commonly called heat increment. As a result, it overevaluates roughages compared with concentrates, as do TDN and DE.

Fig. 27-4. Feeding sheep. (Courtesy, American Sheep Industry Assn., Englewood, CO)

4. **Net energy (NE).** Net energy represents the energy fraction in a feed that is left after the fecal, urinary, gas, and heat losses are deducted from the GE. Because of its greater accuracy, net energy is being used increasingly in ration formulations, especially in computerized formulations for large operations. However, net energy is difficult to determine.

Two systems of net energy evaluation are presently used: (1) net energy for maintenance (NE_m) and net energy for gain (NE_g), and (2) net energy for lactation (NE_{lc}).

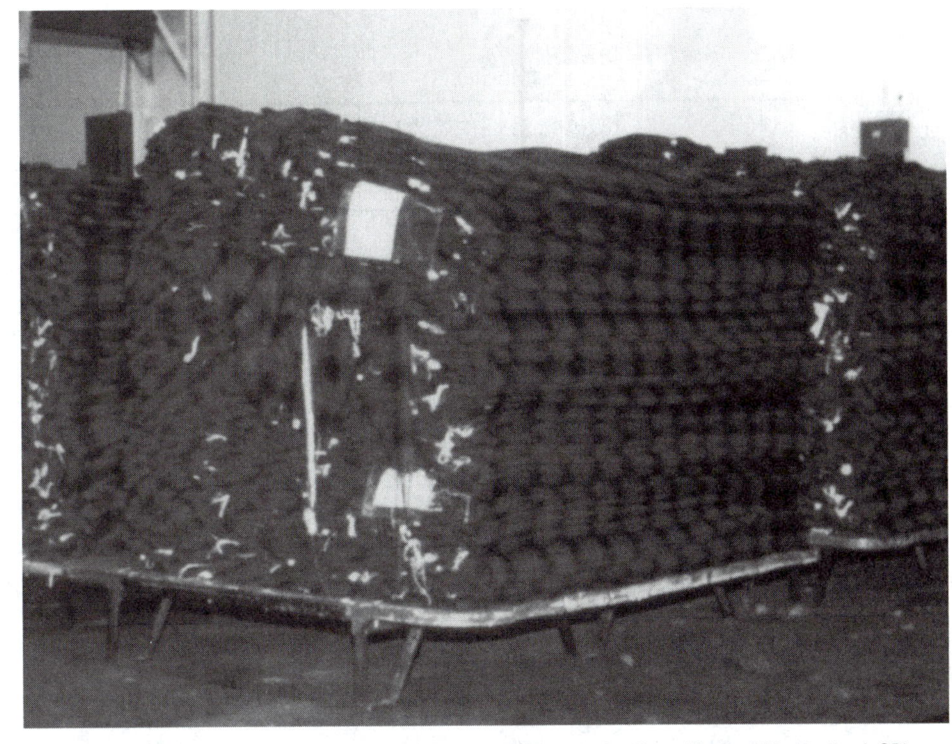

Well-fed sheep produce plenty of warm wool blankets! (Courtesy, Pendleton Woolen Mills, Portland, OR)

■ **Minerals**—The level of minerals in forages is largely determined by the mineral content of the soil on which the feeds are grown. Calcium, phosphorus, iodine, and selenium are well-known examples of soil nutrient–plant nutrient relationships.

■ **Carotene**—Where carotene has been converted to vitamin A, the conversion rate of the rat has been used as the standard value, with 1 mg of β carotene equal to 1,667 IU of vitamin A. Generally, it is unwise to rely on harvested feeds as a source of carotene (vitamin A value), unless the forage being fed is fresh (pasture or green chop) or of a good green color and not over a year old.

TABLE 27-1, COMPOSITION OF FEEDS, DATA EXPRESSED AS-FED AND MOISTURE-FREE

In this table, the commonly used sheep and goat feeds are listed on both an as-fed and a moisture-free basis; and their chemical analysis, TDN, digestible energy, metabolizable energy, net energy, mineral composition, and vitamin composition are given. Note that four pages (two double-page spreads) are devoted to each feed, with presentations as follows:

First page (left-hand): chemical analysis—dry matter, ash, crude fiber, cell walls or NFD, acid detergent fiber, lignin, ether extract (fat), N-free extract, and crude protein.

Second page (right-hand): digestible protein, TDN, digestible energy, metabolizable energy, and net energy.

Third page (left-hand): macrominerals and microminerals.

Fourth page (right-hand): fat-soluble and water-soluble vitamins.

TABLE 27-2, COMPOSITION OF MINERAL SUPPLEMENTS, DATA EXPRESSED AS-FED AND MOISTURE-FREE

This table includes rich natural and synthetic sources of macrominerals and microminerals. Also where applicable, the ash, ether extract, nitrogen-free extract, and protein content are given for the mineral sources.

TABLE

COMPOSITION OF FEEDS, DATA

Entry Number	Feed Name Description	International Feed Number[1]	Moisture Basis: A-F (as-fed) or M-F (moisture-free)	Chemical Analysis: Dry Matter	Ash	Crude Fiber	Cell Walls or NDF	Acid Detergent Fiber	Lignin	Ether Extract (Fat)	N-Free Extract	Crude Protein
				%	%	%	%	%	%	%	%	%
	ALFALFA (LUCERNE) *Medicago sativa*											
1	-ALL ANALYSES	2-00-196	A-F	24	2.2	6.4	—	—	—	0.8	10.1	4.7
			M-F	100	9.2	26.5	—	—	—	3.2	41.7	19.4
2	-PREBLOOM	2-00-181	A-F	21	2.1	5.3	—	—	—	0.6	8.8	4.5
			M-F	100	9.9	24.9	—	—	—	2.8	41.2	21.2
3	-EARLY BLOOM	2-00-184	A-F	24	2.3	6.7	—	—	—	0.7	9.5	4.6
			M-F	100	9.5	28.0	—	—	—	3.1	40.0	19.4
4	-MIDBLOOM	2-00-185	A-F	24	2.1	7.3	—	—	—	0.6	9.5	4.7
			M-F	100	8.7	30.1	—	—	—	2.5	39.3	19.4
5	-FULL BLOOM	2-00-188	A-F	25	2.1	8.2	—	—	—	0.7	9.8	4.1
			M-F	100	8.5	33.0	—	—	—	2.8	39.4	16.3
6	-HAY, SUN-CURED, ALL ANALYSES	1-00-078	A-F	90	8.1	27.2	42.9	32.6	6.3	2.3	36.9	15.9
			M-F	100	9.0	30.1	47.5	36.1	7.0	2.6	40.8	17.6
7	-HAY, SUN-CURED, PREBLOOM	1-00-054	A-F	86	7.8	21.1	36.1	28.1	5.3	3.6	35.1	18.8
			M-F	100	9.0	24.4	41.8	32.6	6.1	4.1	40.7	21.7
8	-HAY, SUN-CURED, EARLY BLOOM	1-00-059	A-F	89	8.0	24.6	39.0	29.4	5.8	2.4	36.7	17.7
			M-F	100	9.0	27.5	43.5	32.9	6.4	2.7	41.0	19.8
9	-HAY, SUN-CURED, MIDBLOOM	1-00-063	A-F	90	7.8	26.0	44.4	33.9	6.5	2.9	36.6	16.9
			M-F	100	8.6	28.8	49.2	37.6	7.2	3.2	40.6	18.8
10	-HAY, SUN-CURED, FULL BLOOM	1-00-068	A-F	89	8.0	30.6	44.5	34.9	6.7	1.6	35.8	13.0
			M-F	100	9.0	34.4	50.0	39.1	7.6	1.8	40.2	14.6
11	-HAY, SUN-CURED, MATURE	1-00-071	A-F	91	6.9	34.7	—	—	—	1.1	36.8	12.0
			M-F	100	7.5	37.9	—	—	—	1.2	40.3	13.1
12	-HAY, SUN-CURED, STEMMY	1-00-093	A-F	91	6.7	36.2	—	—	—	1.0	34.8	12.5
			M-F	100	7.4	39.7	—	—	—	1.1	38.2	13.6
13	-HAY, SUN-CURED, BROWN	1-00-103	A-F	89	9.2	25.3	—	—	—	1.5	36.7	16.1
			M-F	100	10.4	28.4	—	—	—	1.7	41.3	18.2
14	-HAY, SUN-CURED, RAINED ON	1-00-130	A-F	86	6.3	31.6	—	—	—	1.1	32.1	14.7
			M-F	100	7.3	36.8	—	—	—	1.3	37.4	17.2
15	-LEAVES, SUN-CURED, GROUND	1-00-246	A-F	92	11.3	13.9	—	—	—	2.9	38.8	20.5
			M-F	100	12.2	20.4	—	—	—	3.1	42.0	22.2
16	-LEAVES, MEAL, DEHY	1-00-137	A-F	92	11.2	18.1	—	—	—	3.1	39.7	20.0
			M-F	100	12.2	19.6	—	—	—	3.4	43.1	21.7
17	-MEAL, DEHY, 15% PROTEIN	1-00-022	A-F	91	8.8	25.9	—	33.2	—	2.3	39.0	15.4
			M-F	100	9.6	28.3	—	36.4	—	2.5	42.7	16.9
18	-SILAGE, ALL ANALYSES	3-00-212	A-F	30	2.6	9.8	15.5	12.8	—	1.1	11.0	5.1
			M-F	100	8.9	33.1	52.3	43.2	—	3.7	37.1	17.3
19	**ALFALFA-BROMEGRASS, SMOOTH** *Medicago sativa, Bromus inermis*	2-00-262	A-F	21	2.0	5.2	—	—	—	0.7	8.7	4.1
			M-F	100	9.8	25.3	—	—	—	3.6	41.7	19.6
20	-SILAGE, 30–50% DRY MATTER	3-08-147	A-F	46	—	15.3	—	—	—	—	—	7.2
			M-F	100	—	33.0	—	—	—	—	—	15.5
	ALFALFA-ORCHARDGRASS *Medicago sativa, Dactylis*											
21	-SILAGE, 30–50% DRY MATTER	3-08-144	A-F	40	—	12.6	—	—	—	—	—	6.9
			M-F	100	—	31.6	—	—	—	—	—	17.2
	ALKALI SACATON *Sporobolus airoides*											
22	-MIDBLOOM	2-05-601	A-F	40	3.7	14.5	—	—	—	0.7	18.6	2.5
			M-F	100	9.3	36.3	—	—	—	1.8	46.5	6.3
23	**ALSIKE CLOVER** *Trifolium hybridum*	2-01-316	A-F	22	2.3	5.3	—	—	—	0.7	9.9	4.1
			M-F	100	10.2	23.7	—	—	—	3.3	44.6	18.3

Footnote on last page of table.

C O M P O S I T I O N   O F   F E E D S

27-1

EXPRESSED **AS-FED AND MOISTURE-FREE**

Entry Number	Digestible Protein	TDN	Digestible Energy		Metabolizable Energy		Net Energy			
	Ruminant	Ruminant	Ruminant		Ruminant		Ruminant NE_m		Ruminant NE_g	
	%	%	Mcal		Mcal		Mcal		Mcal	
			lb	kg	lb	kg	lb	kg	lb	kg
1	3.5	14	0.28	0.62	0.23	0.51	0.14	0.30	0.07	0.14
	14.5	58	1.16	2.56	0.95	2.10	0.56	1.24	0.27	0.59
2	3.4	13	0.25	0.56	0.21	0.46	0.12	0.27	0.06	0.13
	15.9	59	1.18	2.61	0.97	2.14	0.57	1.27	0.29	0.63
3	3.6	14	0.28	0.62	0.23	0.51	0.14	0.30	0.07	0.15
	15.0	59	1.19	2.62	0.97	2.15	0.58	1.27	0.29	0.64
4	3.5	15	0.29	0.64	0.24	0.52	0.14	0.31	0.07	0.16
	14.3	60	1.20	2.65	0.98	2.17	0.58	1.29	0.30	0.66
5	2.9	13	0.27	0.59	0.22	0.48	0.13	0.28	0.05	0.10
	11.5	54	1.07	2.36	0.88	1.94	0.52	1.14	0.19	0.42
6	11.7	51	1.13	2.48	0.96	2.12	0.59	1.30	0.33	0.73
	12.9	56	1.25	2.75	1.07	2.35	0.65	1.44	0.37	0.82
7	14.8	51	1.22	2.69	1.00	2.21	0.63	1.38	0.39	0.85
	17.2	59	1.42	3.12	1.16	2.56	0.73	1.60	0.45	0.99
8	13.4	51	1.10	2.43	0.95	2.10	0.58	1.28	0.33	0.72
	14.9	57	1.23	2.72	1.06	2.35	0.65	1.44	0.37	0.81
9	12.6	51	1.12	2.46	0.91	2.02	0.55	1.22	0.29	0.64
	14.0	56	1.24	2.73	1.01	2.24	0.61	1.36	0.32	0.71
10	9.1	48	1.03	2.28	0.85	1.87	0.51	1.12	0.23	0.52
	10.2	54	1.16	2.56	0.95	2.10	0.57	1.26	0.26	0.58
11	8.4	50	1.01	2.22	0.82	1.82	0.49	1.08	0.19	0.43
	9.1	55	1.10	2.43	0.90	1.99	0.54	1.19	0.21	0.47
12	8.5	49	0.99	2.18	0.78	1.73	—	—	—	—
	9.3	54	1.09	2.40	0.86	1.90	—	—	—	—
13	9.8	49	0.98	2.17	0.81	1.78	0.48	1.06	0.20	0.43
	11.0	55	1.11	2.44	0.91	2.00	0.54	1.19	0.22	0.49
14	9.9	46	0.91	2.01	0.75	1.65	0.44	0.98	0.16	0.34
	11.5	53	1.06	2.34	0.87	1.92	0.52	1.14	0.18	0.40
15	—	55	1.10	2.42	0.90	1.98	0.54	1.19	0.26	0.58
	—	59	1.19	2.62	0.98	2.15	0.59	1.30	0.29	0.63
16	14.0	58	1.17	2.57	0.96	2.11	0.58	1.28	0.32	0.70
	15.2	63	1.27	2.79	1.04	2.29	0.63	1.40	0.35	0.76
17	—	52	1.05	2.31	0.86	1.90	0.51	1.13	0.23	0.51
	—	57	1.15	2.53	0.94	2.08	0.56	1.24	0.25	0.56
18	3.6	17	0.33	0.72	0.28	0.62	0.17	0.36	0.08	0.17
	12.1	58	1.10	2.42	0.94	2.08	0.56	1.23	0.26	0.57
19	—	14	0.27	0.60	0.22	0.49	0.14	0.31	0.08	0.18
	—	66	1.32	2.90	1.08	2.38	0.66	1.46	0.38	0.84
20	4.3	25	0.48	1.06	0.42	0.87	0.23	0.51	0.08	0.18
	9.0	53	1.04	2.30	0.91	1.90	0.50	1.10	0.18	0.40
21	3.8	22	0.44	0.96	0.36	0.80	0.22	0.48	0.07	0.16
	9.4	55	1.09	2.40	0.91	2.00	0.54	1.20	0.18	0.40
22	—	22	0.44	0.96	0.36	0.79	0.21	0.47	0.08	0.18
	—	55	1.09	2.40	0.89	1.97	0.53	1.17	0.21	0.45
23	—	15	0.31	0.68	0.25	0.56	0.16	0.34	0.09	0.21
	—	69	1.39	3.06	1.14	2.51	0.71	1.56	0.43	0.95

(Continued)

TABLE 27-1

Entry Number	Feed Name Description	Moisture Basis: A-F (as-fed) or M-F (moisture-free)	Dry Matter	Macrominerals							Microminerals						
				Calcium (Ca)	Phosphorus (P)	Sodium (Na)	Chlorine (Cl)	Magnesium (Mg)	Potassium (K)	Sulfur (S)	Cobalt (Co)	Copper (Cu)	Iodine (I)	Iron (Fe)	Manganese (Mn)	Selenium (Se)	Zinc (Zn)
			%	%	%	%	%	%	%	%	ppm or mg/kg	ppm or mg/kg	ppm or mg/kg	%	ppm or mg/kg	ppm or mg/kg	ppm or mg/kg
	ALFALFA (LUCERNE) *Medicago sativa*																
1	-ALL ANALYSES	A-F	24	0.46	0.06	0.05	0.12	0.06	0.52	0.10	0.024	2.6	—	0.011	15.2	—	4.3
		M-F	100	1.91	0.26	0.19	0.50	0.24	2.13	0.41	0.099	10.9	—	0.046	62.9	—	17.6
2	-PREBLOOM	A-F	21	0.48	0.08	0.04	0.08	0.05	0.50	0.13	—	—	—	—	5.9	—	—
		M-F	100	2.26	0.35	0.20	0.35	0.25	2.36	0.60	—	—	—	—	27.7	—	—
3	-EARLY BLOOM	A-F	24	0.55	0.07	—	—	0.01	0.46	—	—	—	—	—	—	—	—
		M-F	100	2.33	0.31	—	—	0.03	1.92	—	—	—	—	—	—	—	—
4	-MIDBLOOM	A-F	24	0.49	0.07	0.04	0.11	0.06	0.50	0.07	—	—	—	—	—	—	—
		M-F	100	2.01	0.28	0.16	0.45	0.26	2.06	0.29	—	—	—	—	—	—	—
5	-FULL BLOOM	A-F	25	0.38	0.07	0.04	0.11	0.07	0.53	0.08	—	—	—	0.011	38.7	—	—
		M-F	100	1.53	0.27	0.15	0.43	0.27	2.13	0.31	—	—	—	0.043	155.2	—	—
6	-HAY, SUN-CURED, ALL ANALYSES	A-F	90	1.38	0.20	0.14	0.31	0.30	1.98	0.25	0.109	12.6	—	0.018	27.3	—	15.1
		M-F	100	1.53	0.22	0.15	0.34	0.33	2.19	0.28	0.120	13.9	—	0.020	30.2	—	16.7
7	-HAY, SUN-CURED, PREBLOOM	A-F	86	2.12	0.26	0.19	0.30	0.22	2.37	0.25	—	—	—	0.022	29.6	—	—
		M-F	100	2.45	0.30	0.22	0.34	0.25	2.75	0.29	—	—	—	0.025	34.3	—	—
8	-HAY, SUN-CURED, EARLY BLOOM	A-F	89	1.28	0.23	0.13	—	0.33	2.27	0.25	0.081	19.4	—	0.018	27.6	—	15.4
		M-F	100	1.44	0.26	0.15	—	0.37	2.54	0.28	0.090	21.7	—	0.020	30.9	—	17.2
9	-HAY, SUN-CURED, MIDBLOOM	A-F	90	1.32	0.21	0.14	0.34	0.30	1.56	0.25	0.082	13.0	—	0.015	24.2	—	15.5
		M-F	100	1.47	0.24	0.15	0.38	0.33	1.73	0.27	0.090	14.4	—	0.017	26.8	—	17.2
10	-HAY, SUN-CURED, FULL BLOOM	A-F	89	1.03	0.19	—	—	0.29	1.61	0.23	—	12.0	—	0.013	24.9	—	—
		M-F	100	1.16	0.22	—	—	0.32	1.81	0.26	—	13.4	—	0.014	28.0	—	—
11	-HAY, SUN-CURED, MATURE	A-F	91	1.07	0.16	0.14	—	0.32	1.80	0.18	0.083	12.3	—	0.018	30.2	—	15.7
		M-F	100	1.17	0.17	0.15	—	0.35	1.97	0.20	0.090	13.4	—	0.020	33.1	—	17.2
12	-HAY, SUN-CURED, STEMMY	A-F	91	0.89	0.20	—	—	0.29	1.80	—	—	—	—	0.014	—	—	—
		M-F	100	0.98	0.21	—	—	0.32	1.98	—	—	—	—	0.015	—	—	—
13	-HAY, SUN-CURED, BROWN	A-F	89	1.47	0.28	—	—	—	—	—	—	—	—	—	—	—	—
		M-F	100	1.66	0.32	—	—	—	—	—	—	—	—	—	—	—	—
14	-HAY, SUN-CURED, RAINED ON	A-F	86	—	—	—	—	—	—	—	—	—	—	—	—	—	—
		M-F	100	—	—	—	—	—	—	—	—	—	—	—	—	—	—
15	-LEAVES, SUN-CURED, GROUND	A-F	92	1.69	0.24	0.10	0.46	0.35	2.08	—	0.199	10.6	—	0.036	41.3	—	—
		M-F	100	1.83	0.27	0.11	0.50	0.38	2.25	—	0.216	11.5	—	0.039	44.8	—	—
16	-LEAVES, MEAL, DEHY	A-F	92	1.64	0.23	0.06	0.31	0.35	2.07	0.60	0.199	10.1	—	0.036	35.9	—	15.8
		M-F	100	1.79	0.26	0.07	0.34	0.38	2.25	0.65	0.216	11.0	—	0.039	39.0	—	17.2
17	-MEAL, DEHY, 15% PROTEIN	A-F	91	1.27	0.22	0.07	0.44	0.28	2.25	0.17	—	9.5	0.118	0.027	28.1	0.284	19.6
		M-F	100	1.39	0.24	0.08	0.48	0.30	2.47	0.18	—	10.5	0.129	0.030	30.8	0.311	21.4
18	-SILAGE, ALL ANALYSES	A-F	30	0.51	0.08	0.05	0.13	0.11	0.66	0.10	0.027	3.5	—	0.007	12.5	—	5.1
		M-F	100	1.72	0.27	0.17	0.45	0.36	2.21	0.35	0.090	11.9	—	0.024	42.3	—	17.2
19	**ALFALFA-BROMEGRASS, SMOOTH** *Medicago sativa, Bromus inermis*	A-F	21	—	—	—	—	—	—	—	—	—	—	—	—	—	—
		M-F	100	—	—	—	—	—	—	—	—	—	—	—	—	—	—
20	-SILAGE, 30–50% DRY MATTER	A-F	46	—	—	—	—	—	—	—	—	—	—	—	—	—	—
		M-F	100	—	—	—	—	—	—	—	—	—	—	—	—	—	—
	ALFALFA-ORCHARDGRASS *Medicago sativa, Dactylis*																
21	-SILAGE, 30–50% DRY MATTER	A-F	40	—	—	—	—	—	—	—	—	—	—	—	—	—	—
		M-F	100	—	—	—	—	—	—	—	—	—	—	—	—	—	—
	ALKALI SACATON *Sporobolus airoides*																
22	-MIDBLOOM	A-F	40	—	—	—	—	—	—	—	—	—	—	—	—	—	—
		M-F	100	—	—	—	—	—	—	—	—	—	—	—	—	—	—
23	**ALSIKE CLOVER** *Trifolium hybridum*	A-F	22	0.29	0.07	0.10	0.17	0.07	0.56	0.04	—	1.3	—	0.010	26.0	—	—
		M-F	100	1.30	0.31	0.46	0.77	0.32	2.54	0.17	—	5.8	—	0.046	116.9	—	—

Continued)

Entry Number	Fat-soluble Vitamins					Water-soluble Vitamins									
	A (1 mg Carotene = 1667 IU Vit A)	Carotene (Provitamin A)	D	E (α-tocopherol)	K	B_{12}	Biotin	Choline	Folic Acid (Folacin)	Niacin (Nicotinic Acid)	Pantothenic Acid	Pyridoxine (B_6)	Riboflavin (B_2)	Thiamin (B_1)
	IU/g	ppm or mg/kg	IU/kg	ppm or mg/kg	ppm or mg/kg	ppb or mcg/kg	ppm or mg/kg	ppm or mg/kg	ppm or mg/kg	ppm or mg/kg	ppm or mg/kg	ppm or mg/kg	ppm or mg/kg	ppm or mg/kg
1	78.6 324.1	47.1 194.4	39 161	147.5 608.5	— —	— —	0.12 0.49	377 1556	0.61 2.51	12 49	8.8 36.3	1.62 6.68	3.2 13.4	1.4 5.9
2	— —	— —	— —	— —	— —	— —	— —	— —	— —	— —	— —	— —	— —	— —
3	69.4 291.1	41.6 174.6	— —	— —	— —	— —	— —	— —	— —	— —	— —	— —	— —	— —
4	57.1 236.4	34.3 141.8	— —	— —	— —	— —	— —	— —	— —	— —	— —	— —	— —	— —
5	— —	— —	— —	— —	— —	— —	— —	— —	— —	— —	7.8 31.3	— —	— —	— —
6	68.4 75.8	41.1 45.4	1447 1602	83.0 91.9	15.89 17.58	2.0 2.2	0.20 0.22	— —	3.07 3.40	38 42	28.6 31.6	5.73 6.35	12.0 13.3	2.7 3.0
7	— —	— —	— —	— —	— —	— —	— —	— —	— —	— —	— —	— —	— —	— —
8	204.2 228.2	122.5 136.9	1786 1996	— —	— —	— —	— —	— —	— —	— —	62.3 69.7	— —	— —	— —
9	94.8 105.1	56.9 63.1	1392 1544	— —	— —	— —	— —	— —	— —	— —	62.8 69.7	— —	9.5 10.6	— —
10	17.8 20.0	10.7 12.0	— —	— —	— —	— —	— —	— —	— —	— —	— —	— —	— —	— —
11	6.5 7.2	3.9 4.3	1290 1411	— —	— —	— —	— —	— —	— —	— —	— —	— —	— —	— —
12	— —	— —	— —	— —	— —	— —	— —	— —	— —	— —	— —	— —	— —	— —
13	6.5 7.4	3.9 4.4	— —	— —	— —	— —	— —	— —	— —	— —	— —	— —	— —	— —
14	4.1 4.8	2.5 2.9	811 946	— —	— —	— —	— —	— —	— —	— —	— —	— —	— —	— —
15	202.1 219.1	121.2 131.4	344 373	— —	— —	— —	0.28 0.31	1097 1189	5.98 6.48	48 53	29.9 32.4	— —	16.7 18.1	4.8 5.2
16	229.4 249.4	137.6 149.6	356 387	— —	— —	— —	0.33 0.36	— —	— —	39 43	33.5 36.4	— —	15.1 16.4	5.5 5.9
17	125.3 137.3	75.2 82.3	— —	82.7 90.6	9.68 10.61	— —	0.26 0.28	1588 1739	1.58 1.73	42 46	20.9 22.9	6.33 6.94	10.7 11.7	3.0 3.3
18	42.5 143.3	25.5 86.0	86 289	— —	— —	— —	— —	— —	— —	— —	— —	— —	— —	— —
19	35.1 169.1	21.0 101.4	— —	— —	— —	— —	— —	— —	— —	— —	— —	— —	— —	— —
20	— —	— —	— —	— —	— —	— —	— —	— —	— —	— —	— —	— —	— —	— —
21	— —	— —	— —	— —	— —	— —	— —	— —	— —	— —	— —	— —	— —	— —
22	— —	— —	— —	— —	— —	— —	— —	— —	— —	— —	— —	— —	— —	— —
23	— —	— —	— —	— —	— —	— —	— —	— —	— —	— —	— —	— —	4.4 19.6	2.0 8.8

(Continued)

TABLE 27-1

Entry Number	Feed Name Description	International Feed Number[1]	Moisture Basis: A-F (as-fed) or M-F (moisture-free)	Chemical Analysis								
				Dry Matter	Ash	Crude Fiber	Cell Walls or NDF	Acid Detergent Fiber	Lignin	Ether Extract (Fat)	N-Free Extract	Crude Protein
				%	%	%	%	%	%	%	%	%
24	-HAY, SUN-CURED	1-01-313	A-F	88	7.6	25.8	—	—	—	2.6	38.9	13.1
			M-F	100	8.6	29.3	—	—	—	3.0	44.2	14.8
	ANIMAL											
25	-MEAT WITH BLOOD, MEAL, TANKAGE RENDERED	5-00-386	A-F	92	21.8	2.2	—	—	—	9.0	−0.3	59.5
			M-F	100	23.6	2.4	—	—	—	9.7	−0.4	64.7
26	-MEAT WITH BONE, MEAL, RENDERED	5-00-388	A-F	93	28.4	2.0	—	—	—	9.9	2.2	50.5
			M-F	100	30.5	2.2	—	—	—	10.6	2.4	54.3
	ANIMAL—POULTRY											
27	-FAT	4-00-409	A-F	99	—	—	—	—	—	98.9	—	—
			M-F	100	—	—	—	—	—	99.9	—	—
	BARLEY *Hordeum vulgare*											
28	-HAY, SUN-CURED	1-00-495	A-F	87	6.6	23.2	—	—	—	1.8	48.0	7.4
			M-F	100	7.6	26.7	—	—	—	2.0	55.2	8.5
29	-STRAW	1-00-498	A-F	91	6.1	38.0	—	—	—	1.7	41.9	3.6
			M-F	100	6.7	41.6	—	—	—	1.9	45.9	3.9
	BARLEY *Hordeum vulgare*											
30	-GRAIN, ALL ANALYSES	4-00-549	A-F	88	2.3	5.0	—	—	—	1.9	66.6	12.2
			M-F	100	2.6	5.6	—	—	—	2.2	75.7	13.9
31	-GRAIN, PACIFIC COAST	4-07-939	A-F	90	2.8	6.3	—	—	—	1.7	69.6	9.6
			M-F	100	3.1	7.0	—	—	—	1.9	77.3	10.7
32	-GRAIN SCREENINGS	4-00-542	A-F	89	3.1	8.7	—	—	—	2.3	63.2	11.7
			M-F	100	3.4	9.8	—	—	—	2.6	71.1	13.1
	BEET, MANGEL *Beta vulgaris macrorhiza*											
33	-TOPS WITH CROWNS	2-00-632	A-F	13	2.4	1.4	—	—	—	0.5	6.1	2.1
			M-F	100	19.2	11.4	—	—	—	4.2	48.2	17.0
	BEET, SUGAR *Beta vulgaris saccharifera*											
34	-TOPS WITH CROWNS	2-00-649	A-F	18	3.6	1.9	—	—	—	0.4	9.2	2.6
			M-F	100	20.4	10.7	—	—	—	2.1	51.9	15.0
35	-MOLASSES, MORE THAN 48% INVERT SUGAR, MORE THAN 79.5 DEGREES BRIX	4-00-668	A-F	78	8.7	—	—	—	—	0.1	62.2	6.0
			M-F	100	11.2	—	—	—	—	0.2	79.7	7.7
36	-PULP, WET	4-00-671	A-F	12	0.6	3.7	—	—	—	0.3	6.1	1.4
			M-F	100	4.7	30.6	—	—	—	2.2	51.1	11.3
37	-PULP, DEHY	4-00-669	A-F	90	4.8	17.9	—	—	—	0.5	58.1	8.7
			M-F	100	5.3	19.9	—	—	—	0.5	64.5	9.7
38	-PULP WITH MOLASSES, DEHY	4-00-672	A-F	92	5.6	15.4	—	—	—	0.6	61.0	9.4
			M-F	100	6.1	16.7	—	—	—	0.7	66.3	10.2
	BERMUDAGRASS *Cynodon dactylon*											
39	-HAY, SUN-CURED	1-00-703	A-F	91	7.2	25.8	—	—	—	1.8	47.4	9.0
			M-F	100	7.9	28.2	—	—	—	2.0	52.0	9.8
	BERMUDAGRASS, COASTAL *Cynodon dactylon*											
40	-HAY, SUN-CURED	1-00-716	A-F	91	5.7	27.6	68.7	33.0	4.6	2.1	44.3	11.1
			M-F	100	6.3	30.4	75.6	36.3	5.1	2.3	48.8	12.3
	BIRDSFOOT TREFOIL *Lotus corniculatus*											
41	-HAY, SUN-CURED	1-05-044	A-F	92	6.5	26.6	43.4	33.4	8.1	2.4	40.3	16.3
			M-F	100	7.1	28.9	47.1	36.3	8.8	2.6	43.8	17.6

Footnote on last page of table.

(Continued)

Entry Number	Digestible Protein	TDN	Digestible Energy		Metabolizable Energy		Net Energy			
	Ruminant	Ruminant	Ruminant		Ruminant		Ruminant NE_m		Ruminant NE_g	
	%	%	Mcal		Mcal		Mcal		Mcal	
			lb	kg	lb	kg	lb	kg	lb	kg
24	8.7	51	1.02	2.24	0.83	1.84	0.50	1.10	0.23	0.50
	9.9	58	1.15	2.54	0.95	2.08	0.57	1.25	0.26	0.57
25	—	67	1.34	2.95	1.10	2.42	0.68	1.50	0.44	0.97
	—	73	1.46	3.21	1.19	2.63	0.74	1.63	0.48	1.05
26	46.0	66	1.32	2.90	1.08	2.38	0.68	1.49	0.42	0.92
	49.4	71	1.42	3.12	1.16	2.56	0.73	1.61	0.45	0.99
27	—	183	3.66	8.08	3.00	6.62	2.21	4.87	1.71	3.78
	—	185	3.70	8.16	3.03	6.69	2.23	4.92	1.73	3.82
28	4.1	50	1.00	2.20	0.82	1.81	0.49	1.08	0.22	0.49
	4.7	57	1.15	2.53	0.94	2.08	0.56	1.24	0.25	0.56
29	0.6	44	0.87	1.93	0.72	1.58	0.43	0.94	0.08	0.17
	0.7	48	0.96	2.11	0.79	1.73	0.47	1.03	0.09	0.19
30	9.6	73	1.45	3.21	1.19	2.63	0.78	1.72	0.52	1.15
	10.9	82	1.65	3.64	1.36	2.99	0.89	1.96	0.60	1.31
31	7.2	76	1.53	3.37	1.25	2.76	0.83	1.82	0.56	1.23
	8.0	85	1.70	3.74	1.39	3.07	0.92	2.03	0.62	1.37
32	9.0	72	1.43	3.15	1.17	2.59	0.76	1.66	0.50	1.11
	10.1	80	1.61	3.54	1.32	2.91	0.85	1.87	0.57	1.25
33	1.8	8	0.16	0.36	0.13	0.30	0.09	0.19	0.05	0.11
	13.9	65	1.30	2.87	1.07	2.36	0.66	1.45	0.37	0.82
34	2.0	11	0.23	0.50	0.19	0.41	0.12	0.26	0.07	0.14
	11.3	65	1.29	2.85	1.06	2.34	0.65	1.43	0.36	0.80
35	3.3	59	1.19	2.62	0.97	2.15	0.61	1.35	0.40	0.89
	4.2	76	1.52	3.36	1.25	2.75	0.79	1.73	0.52	1.14
36	0.7	9	0.18	0.40	0.15	0.33	0.10	0.21	0.06	0.14
	6.2	76	1.52	3.35	1.25	2.75	0.80	1.76	0.52	1.14
37	4.3	66	1.32	2.90	1.08	2.38	0.67	1.47	0.43	0.96
	4.8	73	1.46	3.22	1.20	2.64	0.74	1.64	0.48	1.06
38	6.2	71	1.42	3.12	1.16	2.56	0.75	1.64	0.49	1.07
	6.7	77	1.54	3.39	1.26	2.78	0.81	1.79	0.53	1.16
39	4.6	45	0.89	1.96	0.73	1.61	0.43	0.95	0.09	0.21
	5.0	49	0.98	2.15	0.80	1.77	0.48	1.05	0.10	0.23
40	6.9	50	1.03	2.26	0.84	1.86	0.50	1.11	0.22	0.48
	7.6	55	1.13	2.49	0.93	2.04	0.55	1.22	0.24	0.53
41	11.2	54	0.93	2.05	0.76	1.68	0.45	0.99	0.12	0.27
	12.2	59	1.01	2.22	0.83	1.82	0.49	1.08	0.13	0.29

(Continued)

COMPOSITION OF FEEDS

TABLE 27-1

Entry Number	Feed Name Description	Moisture Basis: A-F (as-fed) or M-F (moisture-free)	Dry Matter	Macrominerals							Microminerals						
				Calcium (Ca)	Phosphorus (P)	Sodium (Na)	Chlorine (Cl)	Magnesium (Mg)	Potassium (K)	Sulfur (S)	Cobalt (Co)	Copper (Cu)	Iodine (I)	Iron (Fe)	Manganese (Mn)	Selenium (Se)	Zinc (Zn)
			%	%	%	%	%	%	%	%	ppm or mg/kg	ppm or mg/kg	ppm or mg/kg	%	ppm or mg/kg	ppm or mg/kg	ppm or mg/kg
24	-HAY, SUN-CURED	A-F	88	1.16	0.26	—	—	0.36	2.17	—	—	5.3	—	0.023	60.7	—	—
		M-F	100	1.32	0.29	—	—	0.41	2.46	—	—	6.0	—	0.026	69.0	—	—
	ANIMAL																
25	-MEAT WITH BLOOD, MEAL, TANKAGE RENDERED	A-F	92	5.80	2.99	1.68	1.73	0.34	0.57	0.70	0.154	38.7	—	0.210	19.1	—	—
		M-F	100	6.31	3.25	1.82	1.88	0.36	0.62	0.76	0.167	42.1	—	0.228	20.8	—	—
26	-MEAT WITH BONE, MEAL, RENDERED	A-F	93	10.16	4.89	0.73	0.74	1.13	1.28	0.26	0.180	1.5	1.313	0.050	13.3	0.262	95.3
		M-F	100	10.92	5.26	0.78	0.80	1.22	1.38	0.28	0.193	1.6	1.412	0.054	14.3	0.282	102.5
	ANIMAL—POULTRY																
27	-FAT	A-F	99	—	—	—	—	—	—	—	—	—	—	—	—	—	—
		M-F	100	—	—	—	—	—	—	—	—	—	—	—	—	—	—
	BARLEY *Hordeum vulgare*																
28	-HAY, SUN-CURED	A-F	87	0.18	0.27	0.12	—	0.17	1.29	0.15	0.056	3.8	—	0.026	34.3	—	—
		M-F	100	0.21	0.31	0.14	—	0.19	1.49	0.17	0.064	4.4	—	0.030	39.4	—	—
29	-STRAW	A-F	91	0.22	0.05	0.13	0.62	0.14	1.84	0.16	0.060	9.3	—	0.029	15.3	—	—
		M-F	100	0.24	0.05	0.14	0.68	0.15	2.01	0.17	0.066	10.1	—	0.032	16.8	—	—
	BARLEY *Hordeum vulgare*																
30	-GRAIN, ALL ANALYSES	A-F	88	0.04	0.33	0.03	0.18	0.14	0.40	0.15	0.099	8.0	0.044	0.008	16.1	0.176	45.2
		M-F	100	0.05	0.37	0.03	0.20	0.15	0.45	0.18	0.113	9.1	0.050	0.009	18.3	0.199	51.4
31	-GRAIN, PACIFIC COAST	A-F	90	0.05	0.34	0.02	0.15	0.12	0.53	0.15	0.088	8.2	—	0.010	16.2	0.102	15.4
		M-F	100	0.05	0.38	0.02	0.17	0.13	0.58	0.17	0.098	9.1	—	0.012	18.0	0.114	17.1
32	-GRAIN SCREENINGS	A-F	89	0.23	0.29	—	—	—	1.23	—	—	—	—	—	—	—	—
		M-F	100	0.26	0.33	—	—	—	1.38	—	—	—	—	—	—	—	—
	BEET, MANGEL *Beta vulgaris macrorhiza*																
33	-TOPS WITH CROWNS	A-F	13	—	—	—	—	—	—	—	—	—	—	—	—	—	—
		M-F	100	—	—	—	—	—	—	—	—	—	—	—	—	—	—
	BEET, SUGAR *Beta vulgaris saccharifera*																
34	-TOPS WITH CROWNS	A-F	18	0.18	0.04	0.10	0.08	0.19	1.02	0.10	—	2.4	—	0.003	9.6	—	—
		M-F	100	1.01	0.22	0.55	0.44	1.07	5.79	0.57	—	13.6	—	0.017	54.5	—	—
35	-MOLASSES, MORE THAN 48% INVERT SUGAR, MORE THAN 79.5 DEGREES BRIX	A-F	78	0.12	0.02	1.10	1.50	0.19	4.63	0.47	0.374	17.1	—	0.007	4.5	—	—
		M-F	100	0.15	0.03	1.41	1.92	0.24	5.94	0.60	0.480	22.0	—	0.009	5.7	—	—
36	-PULP, WET	A-F	12	0.10	0.01	—	—	—	0.02	—	—	—	—	—	—	—	—
		M-F	100	0.86	0.10	—	—	—	0.19	—	—	—	—	—	—	—	—
37	-PULP, DEHY	A-F	90	0.65	0.09	0.21	0.04	0.27	0.17	0.20	0.065	12.4	—	0.030	34.4	—	0.7
		M-F	100	0.72	0.10	0.23	0.04	0.30	0.19	0.22	0.072	13.8	—	0.033	38.3	—	0.8
38	-PULP WITH MOLASSES, DEHY	A-F	92	0.56	0.08	0.54	—	0.14	1.63	0.39	0.209	14.7	—	0.019	24.4	—	1.5
		M-F	100	0.61	0.09	0.59	—	0.16	1.78	0.42	0.227	16.0	—	0.021	26.5	—	1.6
	BERMUDAGRASS *Cynodon dactylon*																
39	-HAY, SUN-CURED	A-F	91	0.40	0.19	—	—	0.16	1.39	—	—	—	0.105	0.026	—	—	—
		M-F	100	0.43	0.21	—	—	0.17	1.52	—	—	—	0.115	0.029	—	—	—
	BERMUDAGRASS, COASTAL *Cynodon dactylon*																
40	-HAY, SUN-CURED	A-F	91	0.35	0.17	—	—	0.15	—	—	—	—	—	—	—	—	0.9
		M-F	100	0.38	0.19	—	—	0.17	—	—	—	—	—	—	—	—	1.0
	BIRDSFOOT TREFOIL *Lotus corniculatus*																
41	-HAY, SUN-CURED	A-F	92	1.54	0.25	0.81	—	0.47	1.66	—	0.102	8.5	—	0.021	13.8	—	71.1
		M-F	100	1.67	0.27	0.88	—	0.51	1.80	—	0.110	9.3	—	0.023	15.0	—	77.2

(Continued)

Entry Number	Fat-soluble Vitamins					Water-soluble Vitamins								
	A (1 mg Carotene = 1667 IU Vit A)	Carotene (Provitamin A)	D	E (α-tocopherol)	K	B_{12}	Biotin	Choline	Folic Acid (Folacin)	Niacin (Nicotinic Acid)	Pantothenic Acid	Pyridoxine (B_6)	Riboflavin (B_2)	Thiamin (B_1)
	IU/g	ppm or mg/kg	IU/kg	ppm or mg/kg	ppm or mg/kg	ppb or mcg/kg	ppm or mg/kg	ppm or mg/kg	ppm or mg/kg	ppm or mg/kg	ppm or mg/kg	ppm or mg/kg	ppm or mg/kg	ppm or mg/kg
24	326.2	195.7	—	—	—	—	—	—	—	—	—	—	16.7	4.5
	370.5	222.2	—	—	—	—	—	—	—	—	—	—	19.0	5.2
25	—	—	—	—	—	236.6	—	1704	1.54	37	2.4	—	2.3	0.4
	—	—	—	—	—	257.2	—	1852	1.67	40	2.6	—	2.5	0.4
26	—	—	—	0.9	—	109.1	0.10	2010	0.37	49	4.2	8.73	4.5	0.6
	—	—	—	1.0	—	117.3	0.11	2162	0.40	53	4.5	9.39	4.9	0.7
27	—	—	—	7.8	—	—	—	—	—	—	—	—	—	—
	—	—	—	7.9	—	—	—	—	—	—	—	—	—	—
28	—	—	959	—	—	—	—	—	—	—	—	—	—	—
	—	—	1103	—	—	—	—	—	—	—	—	—	—	—
29	3.5	2.1	604	—	—	—	—	—	—	—	—	—	—	—
	3.9	2.3	662	—	—	—	—	—	—	—	—	—	—	—
30	4.7	2.8	—	15.8	—	—	0.14	903	0.55	85	8.2	6.57	1.6	4.4
	5.4	3.2	—	18.0	—	—	0.16	1026	0.62	96	9.3	7.46	1.8	5.0
31	—	—	—	21.1	—	—	0.15	1003	0.51	48	7.1	2.93	1.6	4.3
	—	—	—	23.5	—	—	0.17	1114	0.56	53	7.9	3.26	1.7	4.7
32	—	—	—	—	—	—	—	—	—	—	7.9	—	1.4	—
	—	—	—	—	—	—	—	—	—	—	8.8	—	1.6	—
33	—	—	—	—	—	—	—	—	—	—	—	—	—	—
	—	—	—	—	—	—	—	—	—	—	—	—	—	—
34	10.4	6.2	—	—	—	—	—	—	—	—	—	—	1.2	—
	58.6	35.2	—	—	—	—	—	—	—	—	—	—	6.6	—
35	—	—	—	—	—	—	—	829	—	41	4.5	—	2.3	—
	—	—	—	—	—	—	—	1063	—	53	5.7	—	2.9	—
36	—	—	—	—	—	—	—	—	—	—	—	—	—	—
	—	—	—	—	—	—	—	—	—	—	—	—	—	—
37	0.4	0.2	573	—	—	—	—	810	—	17	1.4	—	0.7	0.4
	0.4	0.2	637	—	—	—	—	900	—	19	1.5	—	0.8	0.4
38	0.4	0.2	—	—	—	—	—	—	—	—	—	—	—	—
	0.4	0.2	—	—	—	—	—	—	—	—	—	—	—	—
39	87.6	52.5	—	—	—	—	—	—	—	—	—	—	—	—
	96.0	57.6	—	—	—	—	—	—	—	—	—	—	—	—
40	123.8	74.3	—	—	—	—	—	—	—	—	—	—	—	—
	136.2	81.7	—	—	—	—	—	—	—	—	—	—	—	—
41	331.2	198.7	1422	—	—	—	—	—	—	—	—	—	14.8	6.3
	359.5	215.7	1544	—	—	—	—	—	—	—	—	—	16.1	6.8

(Continued)

TABLE 27-1

Entry Number	Feed Name Description	International Feed Number[1]	Moisture Basis: A-F (as-fed) or M-F (moisture-free)	Chemical Analysis: Dry Matter	Ash	Crude Fiber	Cell Walls or NDF	Acid Detergent Fiber	Lignin	Ether Extract (Fat)	N-Free Extract	Crude Protein
				%	%	%	%	%	%	%	%	%
	BLUEGRASS, CANADA *Poa compressa*											
42	-ALL ANALYSES	2-00-764	A-F	31	2.6	8.5	—	—	—	1.1	13.9	4.6
			M-F	100	8.6	27.6	—	—	—	3.7	45.2	14.9
43	-IMMATURE	2-00-763	A-F	26	2.4	6.6	—	—	—	1.0	11.1	4.8
			M-F	100	9.1	25.5	—	—	—	3.7	43.0	18.7
44	-HAY, SUN-CURED	1-00-762	A-F	92	6.9	26.9	—	—	—	2.5	44.9	10.8
			M-F	100	7.5	29.2	—	—	—	2.7	48.8	11.7
	BLUEGRASS, KENTUCKY *Poa pratensis*											
45	-IMMATURE	2-00-777	A-F	31	3.0	8.0	—	—	—	1.1	13.9	5.5
			M-F	100	9.4	25.4	—	—	—	3.5	44.2	17.5
46	-EARLY BLOOM	2-00-779	A-F	34	2.4	9.4	—	—	—	1.3	15.4	5.7
			M-F	100	7.1	27.4	—	—	—	3.9	44.9	16.6
47	-MILK STAGE	2-00-782	A-F	42	3.1	12.7	—	—	—	1.5	19.8	4.9
			M-F	100	7.3	30.3	—	—	—	3.6	47.2	11.6
	BLUESTEM *Andropogon* spp											
48	-IMMATURE	2-00-821	A-F	27	2.4	6.7	—	—	—	0.7	13.6	3.4
			M-F	100	8.9	24.9	—	—	—	2.8	50.6	12.8
49	-MATURE	2-00-825	A-F	59	3.3	20.2	—	—	—	1.4	30.6	3.4
			M-F	100	5.6	34.2	—	—	—	2.4	51.9	5.8
	BREWERS' GRAINS											
50	-WET	5-02-142	A-F	22	1.1	3.4	—	—	—	2.3	10.0	5.3
			M-F	100	4.8	15.3	—	—	—	10.5	45.5	24.0
51	-DEHY	5-02-141	A-F	92	3.8	14.4	—	—	—	6.9	42.0	25.0
			M-F	100	4.2	15.6	—	—	—	7.4	45.6	27.2
	BROMEGRASS *Bromus* spp											
52	-HAY, SUN-CURED, ALL ANALYSES	1-00-890	A-F	92	6.7	31.2	56.0	34.8	—	2.0	43.5	8.1
			M-F	100	7.3	34.1	61.2	38.0	—	2.1	47.6	8.8
53	-IMMATURE	2-00-892	A-F	35	3.7	7.2	—	—	—	1.6	17.2	5.1
			M-F	100	10.7	20.8	—	—	—	4.5	49.4	14.6
54	-MATURE	2-00-898	A-F	56	—	18.5	—	—	—	—	—	3.6
			M-F	100	—	33.0	—	—	—	—	—	6.4
55	**BUFFALOGRASS** *Buchloe dactyloides*	2-01-010	A-F	55	6.9	15.8	—	—	—	0.9	26.8	5.0
			M-F	100	12.5	28.6	—	—	—	1.7	48.4	9.0
56	**CACTUS, PRICKLY PEAR** *Opuntia* spp	2-01-061	A-F	18	3.8	2.3	—	—	—	0.4	10.6	0.9
			M-F	100	21.2	12.7	—	—	—	2.0	59.3	4.7
57	**CANARYGRASS, REED** *Phalaris arundinacea*	2-01-113	A-F	27	2.2	7.8	—	—	—	0.9	12.6	3.2
			M-F	100	8.4	29.0	—	—	—	3.5	47.1	12.0
58	-HAY, SUN-CURED	1-01-104	A-F	91	7.3	30.3	57.5	31.4	3.2	2.9	39.7	11.3
			M-F	100	8.0	33.1	62.9	34.3	3.5	3.1	43.4	12.3
	CLOVER, ALSIKE *Trifolium hybridum*											
59	-HAY, SUN-CURED	1-01-313	A-F	88	7.6	25.8	—	—	—	2.6	38.9	13.1
			M-F	100	8.6	29.3	—	—	—	3.0	44.2	14.8
	CLOVER, CRIMSON *Trifolium incarnatum*											
60	-HAY, SUN-CURED	1-01-328	A-F	87	9.3	26.7	—	—	—	2.1	33.0	16.0
			M-F	100	10.7	30.7	—	—	—	2.4	37.9	18.4

Footnote on last page of table.

(Continued)

Entry Number	Digestible Protein	TDN	Digestible Energy		Metabolizable Energy		Net Energy			
	Ruminant	Ruminant	Ruminant		Ruminant		Ruminant NE_m		Ruminant NE_g	
	%	%	Mcal		Mcal		Mcal		Mcal	
			lb	kg	lb	kg	lb	kg	lb	kg
42	—	18	0.37	0.81	0.30	0.67	0.18	0.41	0.09	0.20
	—	60	1.20	2.64	0.98	2.17	0.59	1.31	0.29	0.65
43	—	17	0.35	0.76	0.28	0.63	0.18	0.39	0.10	0.23
	—	67	1.34	2.95	1.10	2.42	0.68	1.49	0.40	0.88
44	4.6	55	1.10	2.42	0.90	1.99	0.54	1.20	0.27	0.59
	5.0	60	1.19	2.63	0.98	2.16	0.59	1.30	0.29	0.64
45	—	21	0.42	0.92	0.34	0.75	0.21	0.46	0.12	0.27
	—	66	1.33	2.92	1.09	2.40	0.67	1.48	0.39	0.86
46	4.1	24	0.48	1.05	0.39	0.86	0.24	0.53	0.15	0.32
	12.0	69	1.39	3.06	1.14	2.51	0.71	1.56	0.43	0.95
47	—	26	0.53	1.18	0.44	0.97	0.27	0.59	0.15	0.32
	—	63	1.27	2.80	1.04	2.30	0.64	1.40	0.35	0.77
48	—	18	0.36	0.80	0.30	0.66	0.19	0.41	0.11	0.24
	—	68	1.35	2.99	1.11	2.45	0.69	1.52	0.41	0.90
49	—	31	0.63	1.38	0.51	1.13	0.31	0.67	0.11	0.24
	—	53	1.06	2.34	0.87	1.92	0.52	1.14	0.18	0.40
50	3.9	15	0.29	0.65	0.24	0.53	0.15	0.33	0.09	0.19
	17.5	67	1.34	2.95	1.10	2.42	0.68	1.50	0.40	0.88
51	18.3	62	1.24	2.73	1.01	2.24	0.61	1.35	0.37	0.82
	19.9	67	1.34	2.96	1.10	2.43	0.67	1.47	0.41	0.89
52	4.1	50	1.05	2.32	0.86	1.90	0.52	1.14	0.23	0.51
	4.5	54	1.15	2.53	0.94	2.07	0.56	1.24	0.25	0.56
53	4.2	28	0.56	1.23	0.46	1.01	0.30	0.66	0.20	0.44
	12.0	80	1.61	3.54	1.32	2.90	0.86	1.89	0.57	1.25
54	1.8	33	0.70	1.50	0.50	1.20	0.40	0.80	0.10	0.50
	3.2	60	1.20	2.70	1.00	2.20	0.60	1.40	0.40	0.80
55	2.7	31	0.62	1.37	0.51	1.12	0.30	0.66	0.13	0.28
	4.8	56	1.12	2.46	0.92	2.02	0.55	1.21	0.23	0.50
56	0.4	10	0.21	0.46	0.17	0.38	0.10	0.23	0.05	0.10
	2.1	58	1.16	2.55	0.95	2.09	0.57	1.25	0.26	0.57
57	—	15	0.30	0.66	0.25	0.54	0.15	0.33	0.06	0.14
	—	56	1.12	2.47	0.92	2.02	0.55	1.21	0.23	0.51
58	7.2	52	1.12	2.47	0.92	2.03	0.55	1.22	0.29	0.64
	7.9	57	1.23	2.71	1.01	2.22	0.61	1.34	0.32	0.70
59	8.7	51	1.02	2.24	0.83	1.84	0.50	1.10	0.23	0.50
	9.9	58	1.15	2.54	0.95	2.08	0.57	1.25	0.26	0.57
60	11.0	49	0.99	2.17	0.81	1.78	0.48	1.06	0.21	0.46
	12.7	57	1.13	2.50	0.93	2.05	0.56	1.22	0.24	0.53

(Continued)

TABLE 27-1

Entry Number	Feed Name Description	Moisture Basis: A-F (as-fed) or M-F (moisture-free)	Dry Matter	Macrominerals							Microminerals						
				Calcium (Ca)	Phosphorus (P)	Sodium (Na)	Chlorine (Cl)	Magnesium (Mg)	Potassium (K)	Sulfur (S)	Cobalt (Co)	Copper (Cu)	Iodine (I)	Iron (Fe)	Manganese (Mn)	Selenium (Se)	Zinc (Zn)
			%	%	%	%	%	%	%	%	ppm or mg/kg	ppm or mg/kg	ppm or mg/kg	%	ppm or mg/kg	ppm or mg/kg	ppm or mg/kg
	BLUEGRASS, CANADA *Poa compressa*																
42	-ALL ANALYSES	A-F	31	0.12	0.11	—	—	0.05	0.56	—	—	—	—	—	24.3	—	—
		M-F	100	0.41	0.36	—	—	0.16	1.82	—	—	—	—	—	79.1	—	—
43	-IMMATURE	A-F	26	—	—	—	—	—	—	—	—	—	—	—	—	—	—
		M-F	100	—	—	—	—	—	—	—	—	—	—	—	—	—	—
44	-HAY, SUN-CURED	A-F	92	0.28	0.24	—	—	0.30	1.73	—	—	—	—	—	85.2	—	—
		M-F	100	0.30	0.26	—	—	0.33	1.88	—	—	—	—	—	92.6	—	—
	BLUEGRASS, KENTUCKY *Poa pratensis*																
45	-IMMATURE	A-F	31	0.14	0.13	—	—	0.05	0.71	—	—	—	—	—	—	—	—
		M-F	100	0.43	0.41	—	—	0.17	2.26	—	—	—	—	—	—	—	—
46	-EARLY BLOOM	A-F	34	0.16	0.13	—	—	0.04	0.69	—	—	—	—	—	—	—	—
		M-F	100	0.46	0.39	—	—	0.11	2.01	—	—	—	—	—	—	—	—
47	-MILK STAGE	A-F	42	—	—	—	—	—	—	—	—	—	—	—	—	—	—
		M-F	100	—	—	—	—	—	—	—	—	—	—	—	—	—	—
	BLUESTEM *Andropogon* spp																
48	-IMMATURE	A-F	27	0.17	0.05	—	—	—	0.46	—	—	12.6	—	0.024	28.5	—	—
		M-F	100	0.63	0.20	—	—	—	1.72	—	—	47.0	—	0.090	106.3	—	—
49	-MATURE	A-F	59	0.23	0.07	—	—	0.04	0.30	—	—	15.6	—	0.063	35.9	—	—
		M-F	100	0.40	0.12	—	—	0.06	0.51	—	—	26.5	—	0.108	60.9	—	—
	BREWERS' GRAINS																
50	-WET	A-F	22	0.07	0.11	—	—	—	0.02	—	—	—	—	—	—	—	—
		M-F	100	0.30	0.51	—	—	—	0.08	—	—	—	—	—	—	—	—
51	-DEHY	A-F	92	0.03	0.53	0.21	0.12	0.16	0.09	0.30	0.085	21.6	0.065	0.025	37.8	—	27.3
		M-F	100	0.32	0.58	0.22	0.13	0.17	0.09	0.33	0.093	23.5	0.071	0.027	41.1	—	29.6
	BROMEGRASS *Bromus* spp																
52	-HAY, SUN-CURED, ALL ANALYSES	A-F	92	0.30	0.14	0.02	—	0.08	2.01	—	—	—	—	—	—	—	—
		M-F	100	0.33	0.16	0.02	—	0.09	2.19	—	—	—	—	—	—	—	—
53	-IMMATURE	A-F	35	—	—	—	—	—	—	—	—	—	—	—	—	—	—
		M-F	100	—	—	—	—	—	—	—	—	—	—	—	—	—	—
54	-MATURE	A-F	56	0.17	0.14	—	—	—	0.70	—	—	—	—	—	—	—	—
		M-F	100	0.30	0.26	—	—	—	1.25	—	—	—	—	—	—	—	—
55	**BUFFALOGRASS** *Buchloe dactyloides*	A-F	55	0.29	0.10	—	—	0.07	0.30	—	—	—	—	—	—	—	—
		M-F	100	0.53	0.18	—	—	0.13	0.54	—	—	—	—	—	—	—	—
56	**CACTUS, PRICKLY PEAR** *Opuntia* spp	A-F	18	1.73	0.02	0.05	0.04	0.25	0.40	0.04	—	—	—	—	—	—	—
		M-F	100	9.61	0.12	0.30	0.21	1.38	2.21	0.23	—	—	—	—	—	—	—
57	**CANARYGRASS, REED** *Phalaris arundinacea*	A-F	27	0.11	0.09	—	—	—	0.97	—	—	—	—	—	—	—	—
		M-F	100	0.42	0.35	—	—	—	3.64	—	—	—	—	—	—	—	—
58	-HAY, SUN-CURED	A-F	91	0.34	0.22	0.36	—	0.28	1.70	0.37	0.020	8.8	—	0.014	97.4	—	—
		M-F	100	0.37	0.24	0.39	—	0.31	1.86	0.41	0.022	9.6	—	0.015	106.5	—	—
	CLOVER, ALSIKE *Trifolium hybridum*																
59	-HAY, SUN-CURED	A-F	88	1.16	0.26	—	—	0.36	2.17	—	—	5.3	—	0.023	60.7	—	—
		M-F	100	1.32	0.29	—	—	0.41	2.46	—	—	6.0	—	0.026	69.0	—	—
	CLOVER, CRIMSON *Trifolium incarnatum*																
60	-HAY, SUN-CURED	A-F	87	1.21	0.21	0.34	0.55	0.25	2.46	0.24	—	—	0.058	0.061	181.8	—	—
		M-F	100	1.39	0.24	0.39	0.63	0.29	2.82	0.28	—	—	0.066	0.070	208.7	—	—

(Continued)

Entry Number	Fat-soluble Vitamins					Water-soluble Vitamins								
	A (1 mg Carotene = 1667 IU Vit A)	Carotene (Provitamin A)	D	E (α-tocopherol)	K	B_{12}	Biotin	Choline	Folic Acid (Folacin)	Niacin (Nicotinic Acid)	Pantothenic Acid	Pyridoxine (B_6)	Riboflavin (B_2)	Thiamin (B_1)
	IU/g	ppm or mg/kg	IU/kg	ppm or mg/kg	ppm or mg/kg	ppb or mcg/kg	ppm or mg/kg	ppm or mg/kg	ppm or mg/kg	ppm or mg/kg	ppm or mg/kg	ppm or mg/kg	ppm or mg/kg	ppm or mg/kg
42	—	—	—	—	—	—	—	—	—	—	—	—	—	—
	—	—	—	—	—	—	—	—	—	—	—	—	—	—
43	—	—	—	—	—	—	—	—	—	—	—	—	—	—
	—	—	—	—	—	—	—	—	—	—	—	—	—	—
44	—	—	—	—	—	—	—	—	—	—	—	—	—	—
	—	—	—	—	—	—	—	—	—	—	—	—	—	—
45	252.3	151.3	—	—	—	—	—	—	—	—	—	—	—	—
	803.4	481.9	—	—	—	—	—	—	—	—	—	—	—	—
46	—	—	—	—	—	—	—	—	—	—	—	—	—	—
	—	—	—	—	—	—	—	—	—	—	—	—	—	—
47	—	—	—	—	—	—	—	—	—	—	—	—	—	—
	—	—	—	—	—	—	—	—	—	—	—	—	—	—
48	97.9	58.7	—	—	—	—	—	—	—	—	—	—	—	—
	365.3	219.1	—	—	—	—	—	—	—	—	—	—	—	—
49	—	—	—	—	—	—	—	—	—	—	—	—	—	—
	—	—	—	—	—	—	—	—	—	—	—	—	—	—
50	—	—	—	—	—	—	—	—	—	—	—	—	—	—
	—	—	—	—	—	—	—	—	—	—	—	—	—	—
51	—	—	—	25.9	—	—	0.96	1670	7.11	43	8.1	0.66	1.3	0.6
	—	—	—	28.1	—	—	1.05	1815	7.73	47	8.8	0.72	1.4	0.6
52	46.0	27.6	942	—	—	—	—	—	—	—	—	—	—	—
	50.3	30.1	1029	—	—	—	—	—	—	—	—	—	—	—
53	266.9	160.1	—	—	—	—	—	—	—	—	—	—	—	—
	765.9	459.4	—	—	—	—	—	—	—	—	—	—	—	—
54	—	—	—	—	—	—	—	—	—	—	—	—	—	—
	—	—	—	—	—	—	—	—	—	—	—	—	—	—
55	70.1	42.0	—	—	—	—	—	—	—	—	—	—	—	—
	126.4	75.8	—	—	—	—	—	—	—	—	—	—	—	—
56	1.8	1.1	—	—	—	—	—	—	—	—	—	—	—	—
	9.9	6.0	—	—	—	—	—	—	—	—	—	—	—	—
57	—	—	—	—	—	—	—	—	—	—	—	—	—	—
	—	—	—	—	—	—	—	—	—	—	—	—	—	—
58	77.4	46.4	—	—	—	—	—	—	—	—	—	—	8.3	3.5
	84.7	50.8	—	—	—	—	—	—	—	—	—	—	9.1	3.8
59	326.2	195.7	—	—	—	—	—	—	—	—	—	—	16.7	4.5
	370.5	222.2	—	—	—	—	—	—	—	—	—	—	19.0	5.2
60	—	—	—	—	—	—	—	—	—	—	—	—	—	—
	—	—	—	—	—	—	—	—	—	—	—	—	—	—

(Continued)

TABLE 27-1

Entry Number	Feed Name Description	International Feed Number[1]	Moisture Basis: A-F (as-fed) or M-F (moisture-free)	Chemical Analysis: Dry Matter	Ash	Crude Fiber	Cell Walls or NDF	Acid Detergent Fiber	Lignin	Ether Extract (Fat)	N-Free Extract	Crude Protein
				%	%	%	%	%	%	%	%	%
	CLOVER, LADINO *Trifolium repens*											
61	-HAY, SUN-CURED	1-01-378	A-F	90	9.0	19.4	—	—	—	2.3	40.0	19.1
			M-F	100	10.0	21.6	—	—	—	2.6	44.5	21.3
	CLOVER, RED *Trifolium pratense*											
62	-EARLY BLOOM	2-01-428	A-F	21	2.3	5.5	—	—	—	1.1	7.9	4.1
			M-F	100	11.0	26.4	—	—	—	5.3	38.0	19.4
63	-FULL BLOOM	2-01-429	A-F	26	2.0	6.7	—	—	—	0.7	12.5	3.7
			M-F	100	7.8	26.1	—	—	—	2.9	48.8	14.5
64	-HAY, SUN-CURED, ALL ANALYSES	1-01-415	A-F	89	7.5	25.7	—	—	—	2.4	39.0	14.1
			M-F	100	8.5	29.0	—	—	—	2.8	44.0	15.8
	CORN, DENT YELLOW *Zea mays indentata*											
65	-GRAIN, ALL ANALYSES	4-02-935	A-F	88	1.3	2.1	—	—	—	3.9	71.1	9.6
			M-F	100	1.4	2.4	—	—	—	4.5	80.8	10.9
66	-GRAIN, GRADE 2, 54 LB/BUSHEL OR 695 G/LITER	4-02-931	A-F	89	1.3	2.0	—	—	—	3.9	72.7	8.7
			M-F	100	1.5	2.2	—	—	—	4.4	81.7	9.8
67	-EARS, GROUND (CORN-AND-COB MEAL)	4-02-849	A-F	86	1.6	8.3	—	—	—	3.2	65.0	7.8
			M-F	100	1.9	9.6	—	—	—	3.7	75.6	9.1
	CORN *Zea mays*											
68	-SILAGE, ALL ANALYSES	3-02-822	A-F	26	1.4	6.6	—	—	—	0.7	15.4	1.8
			M-F	100	5.3	25.5	—	—	—	2.8	59.3	7.0
69	-SILAGE, IMMATURE	3-02-817	A-F	23	1.7	5.6	—	—	—	0.7	12.6	2.2
			M-F	100	7.3	24.6	—	—	—	3.1	55.3	9.7
70	-EARS WITH HUSKS, SILAGE	3-02-839	A-F	50	1.4	5.7	—	—	—	1.9	36.4	4.5
			M-F	100	2.8	11.5	—	—	—	3.8	72.9	9.0
71	-FODDER WITH EARS, SUN-CURED	1-02-775	A-F	71	4.9	19.8	—	—	—	1.8	38.6	6.1
			M-F	100	6.9	27.8	—	—	—	2.5	54.2	8.6
72	-STOVER WITHOUT EARS, WITHOUT HUSKS, SUN-CURED	1-02-776	A-F	85	6.1	28.8	—	—	—	1.1	43.0	5.7
			M-F	100	7.2	34.0	—	—	—	1.3	50.8	6.7
73	-COBS, GROUND	1-02-782	A-F	90	1.5	32.9	—	—	—	0.7	52.3	2.9
			M-F	100	1.7	36.4	—	—	—	0.8	57.9	3.2
	CORN, SWEET *Zea mays saccharata*											
74	-CANNERY RESIDUE	2-02-975	A-F	47	2.0	12.2	—	—	—	1.0	27.5	3.9
			M-F	100	4.2	26.2	—	—	—	2.2	59.0	8.4
	COTTON *Gossypium* spp											
75	-SEEDS, MEAL, MECH EXTD, 41% PROTEIN	5-01-617	A-F	93	6.2	11.3	—	—	—	4.6	30.0	40.9
			M-F	100	6.6	12.1	—	—	—	5.0	32.3	44.0
76	-SEEDS, MEAL, PRE-PRESSED, SOLV EXTD, 48% PROTEIN	5-07-874	A-F	92	7.2	8.5	—	—	—	1.1	26.7	50.0
			M-F	100	7.8	9.2	—	—	—	1.2	28.9	54.0
77	-SEEDS, MEAL, SOLV EXTD, LOW GOSSYPOL	5-01-633	A-F	93	5.8	12.7	—	—	—	1.2	31.6	41.6
			M-F	100	6.3	13.7	—	—	—	1.3	34.0	44.8
78	-BOLLS, SUN-CURED	1-01-596	A-F	92	7.1	29.2	—	—	—	2.4	42.7	10.3
			M-F	100	7.7	31.8	—	—	—	2.7	46.6	11.3
79	-HULLS	1-01-599	A-F	91	2.5	42.8	—	—	—	1.6	40.0	3.8
			M-F	100	2.8	47.2	—	—	—	1.7	44.1	4.2
	CRESTED WHEATGRASS *Agropyron* spp											
80	-HAY, SUN-CURED	1-05-418	A-F	93	6.8	31.9	—	—	—	1.8	43.8	8.9
			M-F	100	7.3	34.2	—	—	—	2.0	47.0	9.5

Footnote on last page of table.

(Continued)

Entry Number	Digestible Protein	TDN	Digestible Energy		Metabolizable Energy		Net Energy			
	Ruminant	Ruminant	Ruminant		Ruminant		Ruminant NE_m		Ruminant NE_g	
	%	%	Mcal		Mcal		Mcal		Mcal	
			lb	kg	lb	kg	lb	kg	lb	kg
61	14.5	59	1.18	2.60	0.97	2.13	0.60	1.31	0.34	0.75
	16.2	66	1.31	2.89	1.08	2.37	0.66	1.46	0.38	0.84
62	2.9	14	0.29	0.63	0.24	0.52	0.15	0.32	0.09	0.20
	13.9	69	1.37	3.03	1.13	2.48	0.70	1.54	0.42	0.93
63	—	17	0.34	0.75	0.28	0.61	0.17	0.38	0.10	0.22
	—	66	1.32	2.92	1.09	2.39	0.67	1.47	0.39	0.85
64	8.7	53	0.93	2.06	0.77	1.69	0.46	1.01	0.15	0.34
	9.8	60	1.05	2.32	0.86	1.90	0.51	1.13	0.17	0.38
65	7.1	82	1.63	3.60	1.34	2.96	0.91	2.00	0.62	1.37
	8.0	93	1.86	4.10	1.52	3.36	1.03	2.28	0.71	1.56
66	6.7	82	1.66	3.64	1.34	2.94	0.93	2.05	0.61	1.34
	7.5	92	1.86	4.10	1.50	3.30	1.04	2.30	0.68	1.50
67	4.1	71	1.43	3.14	1.17	2.58	0.77	1.69	0.51	1.13
	4.7	83	1.66	3.66	1.36	3.00	0.89	1.97	0.60	1.32
68	0.9	18	0.35	0.78	0.29	0.64	0.18	0.39	0.11	0.24
	3.6	68	1.37	3.01	1.12	2.47	0.68	1.50	0.42	0.92
69	—	15	0.29	0.65	0.24	0.53	0.14	0.32	0.08	0.19
	—	65	1.29	2.85	1.06	2.34	0.63	1.40	0.37	0.81
70	2.4	37	0.74	1.63	0.61	1.34	0.39	0.85	0.25	0.55
	4.9	74	1.49	3.28	1.22	2.69	0.77	1.71	0.49	1.09
71	2.9	46	0.92	2.04	0.76	1.67	0.46	1.02	0.26	0.58
	4.1	65	1.30	2.86	1.06	2.34	0.65	1.44	0.37	0.81
72	2.4	50	1.01	2.23	0.83	1.83	0.50	1.10	0.25	0.54
	2.8	60	1.19	2.63	0.98	2.15	0.59	1.30	0.29	0.64
73	−0.3	46	0.91	2.01	0.29	0.65	0.18	0.40	0.11	0.24
	−0.4	51	1.01	2.23	1.13	2.49	0.70	1.55	0.42	0.93
74	—	31	0.62	1.38	0.51	1.13	0.32	0.70	0.19	0.41
	—	67	1.34	2.95	1.10	2.42	0.68	1.49	0.40	0.87
75	—	70	1.39	3.07	1.14	2.52	0.73	1.60	0.47	1.03
	—	75	1.50	3.30	1.23	2.71	0.78	1.73	0.50	1.11
76	40.4	69	1.38	3.04	1.12	2.48	0.71	1.56	0.46	1.01
	43.7	75	1.50	3.30	1.22	2.70	0.77	1.70	0.50	1.10
77	—	66	1.32	2.90	1.08	2.38	0.68	1.49	0.42	0.92
	—	71	1.41	3.12	1.16	2.56	0.73	1.60	0.45	0.99
78	2.7	42	0.84	1.85	0.69	1.52	0.41	0.91	0.04	0.08
	2.9	46	0.91	2.02	0.75	1.65	0.45	0.99	0.04	0.09
79	−0.3	43	0.87	1.91	0.71	1.57	0.42	0.94	0.08	0.17
	−0.3	48	0.96	2.11	0.78	1.73	0.47	1.03	0.08	0.18
80	5.6	56	1.18	2.61	0.97	2.14	0.59	1.30	0.32	0.72
	6.1	60	1.27	2.80	1.04	2.30	0.64	1.40	0.35	0.77

(Continued)

TABLE 27-1

Entry Number	Feed Name Description	Moisture Basis: A-F (as-fed) or M-F (moisture-free)	Dry Matter	Macrominerals							Microminerals						
				Cal-cium (Ca)	Phos-phorus (P)	Sodium (Na)	Chlo-rine (Cl)	Mag-nesium (Mg)	Potas-sium (K)	Sulfur (S)	Cobalt (Co)	Copper (Cu)	Iodine (I)	Iron (Fe)	Man-ganese (Mn)	Sele-nium (Se)	Zinc (Zn)
			%	%	%	%	%	%	%	%	ppm or mg/kg	ppm or mg/kg	ppm or mg/kg	%	ppm or mg/kg	ppm or mg/kg	ppm or mg/kg
	CLOVER, LADINO *Trifolium repens*																
61	-HAY, SUN-CURED	A-F	90	1.19	0.28	0.12	0.27	0.41	2.32	0.19	0.147	8.6	0.269	0.028	74.5	—	15.3
		M-F	100	1.32	0.31	0.13	0.30	0.45	2.58	0.22	0.163	9.5	0.300	0.031	82.9	—	17.0
	CLOVER, RED *Trifolium pratense*																
62	-EARLY BLOOM	A-F	21	0.47	0.08	—	—	—	0.52	—	—	—	—	—	—	—	—
		M-F	100	2.26	0.38	—	—	—	2.49	—	—	—	—	—	—	—	—
63	-FULL BLOOM	A-F	26	0.26	0.07	—	—	0.13	0.50	—	—	—	—	—	—	—	—
		M-F	100	1.01	0.27	—	—	0.51	1.96	—	—	—	—	—	—	—	—
64	-HAY, SUN-CURED, ALL ANALYSES	A-F	89	1.32	0.22	0.16	0.28	0.38	1.48	0.15	0.139	9.7	0.217	0.019	65.0	—	15.3
		M-F	100	1.49	0.25	0.18	0.32	0.43	1.66	0.17	0.156	10.9	0.244	0.021	73.3	—	17.2
	CORN, DENT YELLOW *Zea mays indentata*																
65	-GRAIN, ALL ANALYSES	A-F	88	0.03	0.27	0.01	0.05	0.12	0.31	0.12	0.033	3.2	—	0.002	4.9	0.069	18.3
		M-F	100	0.04	0.30	0.01	0.05	0.13	0.36	0.14	0.037	3.6	—	0.003	5.6	0.079	20.8
66	-GRAIN, GRADE 2, 54 LB/BUSHEL OR *695 G/LITER*	A-F	89	0.02	0.30	0.01	0.04	—	0.28	—	0.018	—	—	—	5.0	—	10.0
		M-F	100	0.02	0.34	0.01	0.04	—	0.31	—	0.020	—	—	—	5.6	—	11.2
67	-EARS, GROUND (CORN-AND-COB MEAL)	A-F	86	0.07	0.23	0.04	—	0.12	0.45	0.19	0.228	6.6	0.022	0.008	24.1	0.074	15.5
		M-F	100	0.08	0.27	0.05	—	0.13	0.52	0.22	0.265	7.7	0.026	0.010	28.0	0.086	18.0
	CORN *Zea mays*																
68	-SILAGE, ALL ANALYSES	A-F	26	0.09	0.07	0.01	0.05	0.06	0.30	0.03	0.019	2.2	—	0.005	12.9	—	6.6
		M-F	100	0.35	0.28	0.03	0.18	0.23	1.15	0.12	0.075	8.7	—	0.019	49.8	—	25.3
69	-SILAGE, IMMATURE	A-F	23	0.12	0.07	—	—	0.07	0.37	—	—	—	—	0.011	—	—	—
		M-F	100	0.53	0.33	—	—	0.31	1.64	—	—	—	—	0.049	—	—	—
70	-EARS WITH HUSKS, SILAGE	A-F	50	0.07	0.15	—	—	0.04	0.24	—	—	—	—	—	—	—	—
		M-F	100	0.14	0.30	—	—	0.09	0.48	—	—	—	—	—	—	—	—
71	-FODDER WITH EARS, SUN-CURED	A-F	71	0.31	0.16	0.02	0.14	0.21	0.68	0.10	—	5.5	—	0.007	48.7	—	—
		M-F	100	0.43	0.23	0.03	0.19	0.29	0.95	0.14	—	7.7	—	0.010	68.2	—	—
72	-STOVER WITHOUT EARS, WITHOUT HUSKS, SUN-CURED	A-F	85	0.51	0.08	0.06	—	0.38	1.38	0.14	—	4.3	—	0.019	115.3	—	—
		M-F	100	0.60	0.10	0.07	—	0.45	1.63	0.17	—	5.1	—	0.022	136.0	—	—
73	-COBS, GROUND	A-F	90	0.11	0.04	—	—	0.06	0.82	0.42	—	—	—	—	—	—	—
		M-F	100	0.12	0.04	—	—	0.07	0.91	0.47	—	—	—	—	—	—	—
	CORN, SWEET *Zea mays saccharata*																
74	-CANNERY RESIDUE	A-F	47	—	0.29	—	—	—	—	0.07	—	3.3	—	—	—	—	—
		M-F	100	—	0.63	—	—	—	—	0.15	—	7.0	—	—	—	—	—
	COTTON *Gossypium spp*																
75	-SEEDS, MEAL, MECH EXTD, 41% PROTEIN	A-F	93	0.20	1.01	0.06	0.04	0.53	1.28	0.40	0.167	18.2	—	0.013	23.0	—	—
		M-F	100	0.22	1.09	0.06	0.04	0.57	1.38	0.43	0.179	19.5	—	0.014	24.8	—	—
76	-SEEDS, MEAL, PRE-PRESSED, SOLV EXTD, 48% PROTEIN	A-F	92	0.16	1.01	0.05	—	0.46	1.26	—	0.093	17.9	—	0.011	22.8	—	73.3
		M-F	100	0.17	1.09	0.05	—	0.50	1.36	—	0.100	19.4	—	0.012	24.6	—	79.2
77	-SEEDS, MEAL, SOLV EXTD, LOW GOSSYPOL	A-F	93	—	—	—	—	—	—	—	—	—	—	—	—	—	—
		M-F	100	—	—	—	—	—	—	—	—	—	—	—	—	—	—
78	-BOLLS, SUN-CURED	A-F	92	0.62	0.11	—	—	—	2.34	—	—	—	—	—	—	—	—
		M-F	100	0.67	0.11	—	—	—	2.55	—	—	—	—	—	—	—	—
79	-HULLS	A-F	91	0.14	0.08	0.02	0.02	0.13	0.79	—	0.047	1.6	—	0.014	107.3	—	14.8
		M-F	100	0.15	0.08	0.02	0.02	0.14	0.87	—	0.052	2.8	—	0.015	118.3	—	16.3
	CRESTED WHEATGRASS *Agropyron spp*																
80	-HAY, SUN-CURED	A-F	93	0.25	0.14	—	—	—	—	—	0.222	—	—	—	—	—	—
		M-F	100	0.27	0.16	—	—	—	—	—	0.238	—	—	—	—	—	—

COMPOSITION OF FEEDS

(Continued)

Entry Number	Fat-soluble Vitamins					Water-soluble Vitamins								
	A (1 mg Carotene = 1667 IU Vit A)	Carotene (Provitamin A)	D	E (α-tocopherol)	K	B_{12}	Biotin	Choline	Folic Acid (Folacin)	Niacin (Nicotinic Acid)	Pantothenic Acid	Pyridoxine (B_6)	Riboflavin (B_2)	Thiamin (B_1)
	IU/g	ppm or mg/kg	IU/kg	ppm or mg/kg	ppm or mg/kg	ppb or mcg/kg	ppm or mg/kg	ppm or mg/kg	ppm or mg/kg	ppm or mg/kg	ppm or mg/kg	ppm or mg/kg	ppm or mg/kg	ppm or mg/kg
61	136.4	81.8	—	—	—	—	—	—	—	10	31.8	—	15.3	3.8
	151.9	91.1	—	—	—	—	—	—	—	11	35.4	—	17.0	4.2
62	—	—	—	—	—	—	—	—	—	—	—	—	—	—
	—	—	—	—	—	—	—	—	—	—	—	—	—	—
63	—	—	—	—	—	—	—	—	—	—	—	—	—	—
	—	—	—	—	—	—	—	—	—	—	—	—	—	—
64	48.6	29.2	1699	—	—	—	0.09	—	—	38	27.3	—	15.8	2.0
	54.7	32.8	1914	—	—	—	0.11	—	—	43	30.7	—	17.8	2.2
65	3.7	2.2	—	22.6	—	—	0.06	536	0.30	30	6.6	5.18	1.3	2.1
	4.2	2.5	—	25.7	—	—	0.07	609	0.34	34	7.5	5.89	1.5	2.3
66	2.9	1.8	—	22.0	—	—	0.06	620	0.36	24	4.8	7.00	1.3	3.5
	3.3	2.0	—	24.7	—	—	0.07	697	0.40	27	5.4	7.87	1.5	4.0
67	1.3	0.8	—	—	—	—	0.04	357	—	17	4.1	5.54	0.9	2.9
	1.6	0.9	—	—	—	—	0.05	415	—	19	4.7	6.44	1.0	3.4
68	19.1	11.5	114	—	—	—	—	—	—	12	—	—	—	—
	73.8	44.3	439	—	—	—	—	—	—	47	—	—	—	—
69	42.1	25.2	—	—	—	—	—	—	—	—	—	—	—	—
	184.5	110.7	—	—	—	—	—	—	—	—	—	—	—	—
70	6.4	3.8	—	—	—	—	—	—	—	—	—	—	—	—
	12.9	7.7	—	—	—	—	—	—	—	—	—	—	—	—
71	6.5	3.9	943	—	—	—	—	—	—	—	—	—	—	—
	9.1	5.5	1323	—	—	—	—	—	—	—	—	—	—	—
72	—	—	935	—	—	—	—	—	—	—	—	—	—	—
	—	—	1103	—	—	—	—	—	—	—	—	—	—	—
73	—	—	—	—	—	—	—	—	—	—	—	—	—	—
	—	—	—	—	—	—	—	—	—	—	—	—	—	—
74	10.5	6.3	—	—	—	—	—	—	—	—	—	—	—	—
	22.5	13.5	—	—	—	—	—	—	—	—	—	—	—	—
75	0.4	0.2	—	32.5	—	—	0.77	2787	1.80	32	9.9	5.39	4.8	6.6
	0.4	0.2	—	34.9	—	—	0.83	2997	1.94	35	10.6	5.79	5.2	7.2
76	—	—	—	—	—	—	—	—	—	—	—	—	—	—
	—	—	—	—	—	—	—	—	—	—	—	—	—	—
77	—	—	—	—	—	—	—	—	—	—	—	—	—	20.3
	—	—	—	—	—	—	—	—	—	—	—	—	—	21.9
78	—	—	—	—	—	—	—	—	—	—	—	—	—	—
	—	—	—	—	—	—	—	—	—	—	—	—	—	—
79	—	—	—	—	—	—	—	—	—	—	—	—	3.8	—
	—	—	—	—	—	—	—	—	—	—	—	—	4.1	—
80	120.9	72.5	—	—	—	—	—	—	—	—	—	—	—	—
	129.8	77.9	—	—	—	—	—	—	—	—	—	—	—	—

(Continued)

TABLE 27-1

Entry Number	Feed Name Description	International Feed Number[1]	Moisture Basis: A-F (as-fed) or M-F (moisture-free)	Chemical Analysis								
				Dry Matter	Ash	Crude Fiber	Cell Walls or NDF	Acid Detergent Fiber	Lignin	Ether Extract (Fat)	N-Free Extract	Crude Protein
				%	%	%	%	%	%	%	%	%
	DESERT MOLLY (SUMMER CYPRESS) *Kochia vestita*											
81	-DORMANT	2-08-843	A-F	75	—	16.5	—	—	—	1.9	36.0	6.8
			M-F	100	—	22.0	—	—	—	2.5	48.0	9.0
	DROPSEED, SAND *Sporobolus cryptandrus*											
82	-DORMANT	2-05-596	A-F	86	—	—	—	—	—	—	—	—
			M-F	100	—	—	—	—	—	—	—	—
	FESCUE, MEADOW *Festuca elatior*											
83	-HAY, SUN-CURED	1-01-912	A-F	88	7.5	29.5	—	—	—	2.2	41.0	8.1
			M-F	100	8.5	33.4	—	—	—	2.5	46.5	9.1
	GALLETA *Hilaria jamesii*											
84	-DORMANT	2-05-594	A-F	55	6.9	17.6	—	—	—	0.8	27.5	0.2
			M-F	100	12.6	32.0	—	—	—	1.4	50.0	4.0
	GRAMA *Bouteloua* spp											
85	-IMMATURE	2-02-163	A-F	41	4.6	11.2	—	—	—	0.8	19.0	5.4
			M-F	100	11.3	27.2	—	—	—	2.0	46.4	13.1
86	-MATURE	2-02-166	A-F	63	7.2	20.7	—	—	—	1.1	30.2	4.1
			M-F	100	11.4	32.7	—	—	—	1.7	47.7	6.5
	GRASS-LEGUME											
87	-SILAGE, ALL ANALYSES	3-02-303	A-F	28	2.2	9.3	—	—	—	1.0	12.5	3.4
			M-F	100	7.7	32.8	—	—	—	3.6	44.0	11.9
	LESPEDEZA, COMMON *Lespedeza striata*											
88	-HAY, SUN-CURED, ALL ANALYSES	1-08-591	A-F	91	5.8	27.5	—	—	—	2.7	41.4	13.3
			M-F	100	6.4	30.3	—	—	—	3.0	45.6	14.7
89	-HAY, SUN-CURED, PREBLOOM	1-20-881	A-F	89	6.4	22.7	—	—	—	2.7	43.0	14.3
			M-F	100	7.2	25.5	—	—	—	3.0	48.3	16.0
90	-HAY, SUN-CURED, MIDBLOOM	1-02-554	A-F	94	—	—	—	—	—	—	—	—
			M-F	100	—	—	—	—	—	—	—	—
	MEADOW, INTERMOUNTAIN											
91	-HAY, SUN-CURED	1-03-181	A-F	92	7.7	27.0	—	—	—	2.5	45.6	8.8
			M-F	100	8.4	29.5	—	—	—	2.8	49.7	9.6
	MILK											
92	-FRESH (COW'S)	5-01-168	A-F	13	0.8	—	—	—	—	3.7	4.9	3.6
			M-F	100	5.9	—	—	—	—	28.8	37.9	27.4
	MOLASSES AND SYRUP											
93	-BEET, SUGAR, MOLASSES, MORE THAN 48% INVERT SUGARS, MORE THAN 79.5 DEGREES BRIX	4-00-668	A-F	78	8.7	—	—	—	—	0.1	62.2	6.0
			M-F	100	11.2	—	—	—	—	0.2	79.7	7.7
94	-CORN, MOLASSES, MORE THAN 43% DEXTROSE EQUIVALENT, MORE THAN 50% TOTAL DEXTROSE, MORE THAN 78 DEGREES BRIX	4-02-888	A-F	73	8.1	—	—	—	—	—	64.6	0.3
			M-F	100	11.0	—	—	—	—	—	88.6	0.4
95	-SUGARCANE, MOLASSES (BLACKSTRAP), MORE THAN 46% INVERT SUGARS, MORE THAN 79.5 DEGREES BRIX	4-04-696	A-F	75	7.7	—	—	—	—	0.1	64.2	3.9
			M-F	100	10.3	—	—	—	—	0.1	85.7	5.2

Footnote on last page of table.

(Continued)

Entry Number	Digestible Protein	TDN	Digestible Energy		Metabolizable Energy		Net Energy			
	Ruminant	Ruminant	Ruminant		Ruminant		Ruminant NE_m		Ruminant NE_g	
	%	%	Mcal		Mcal		Mcal		Mcal	
			lb	*kg*	lb	*kg*	lb	*kg*	lb	*kg*
81	—	43	0.86	*1.90*	0.70	*1.55*	0.42	*0.93*	0.19	*0.42*
	—	57	1.15	*2.53*	0.94	*2.07*	0.56	*1.24*	0.25	*0.56*
82	4.3	50.7	1.03	*2.27*	0.86	*1.89*	0.43	*0.95*	0.26	*0.57*
	5.0	59.0	1.20	*2.64*	1.00	*2.20*	0.50	*1.10*	0.30	*0.66*
83	4.3	51	1.01	*2.24*	0.83	*1.83*	0.50	*1.10*	0.23	*0.50*
	4.9	58	1.15	*2.54*	0.94	*2.08*	0.57	*1.25*	0.26	*0.56*
84	—	30	0.59	*1.31*	0.48	*1.07*	0.29	*0.64*	0.10	*0.24*
	—	54	1.08	*2.38*	0.88	*1.95*	0.53	*1.16*	0.19	*0.43*
85	—	25	0.49	*1.09*	0.41	*0.89*	0.24	*0.54*	0.12	*0.27*
	—	60	1.21	*2.65*	0.99	*2.18*	0.60	*1.32*	0.30	*0.66*
86	—	35	0.69	*1.53*	0.57	*1.25*	0.34	*0.74*	0.13	*0.29*
	—	55	1.09	*2.41*	0.89	*1.97*	0.53	*1.18*	0.21	*0.46*
87	1.7	16	0.33	*0.72*	0.27	*0.59*	0.16	*0.35*	0.07	*0.16*
	6.1	58	1.15	*2.54*	0.95	*2.09*	0.56	*1.23*	0.26	*0.58*
88	6.1	47	0.95	*2.09*	0.78	*1.71*	0.46	*1.02*	0.15	*0.33*
	6.7	52	1.04	*2.30*	0.86	*1.89*	0.51	*1.12*	0.17	*0.36*
89	9.9	50	1.01	*2.25*	—	—	—	—	—	—
	11.1	56	1.13	*2.50*	—	—	—	—	—	—
90	—	—	—	—	—	—	—	—	—	—
	—	—	—	—	—	—	—	—	—	—
91	5.3	63	1.27	*2.80*	1.04	*2.29*	0.65	*1.43*	0.39	*0.87*
	5.7	69	1.38	*3.05*	1.14	*2.50*	0.71	*1.56*	0.43	*0.95*
92	3.4	17	0.33	*0.73*	0.27	*0.60*	0.20	*0.43*	0.14	*0.31*
	26.1	128	2.56	*5.64*	2.10	*4.62*	1.51	*3.33*	1.08	*2.38*
93	3.3	59	1.19	*2.62*	0.97	*2.15*	0.61	*1.35*	0.40	*0.89*
	4.2	76	1.52	*3.36*	1.25	*2.75*	0.79	*1.73*	0.52	*1.14*
94	—	—	—	—	—	—	—	—	—	—
	—	—	—	—	—	—	—	—	—	—
95	1.1	55	1.10	*2.42*	0.90	*1.98*	0.56	*1.23*	0.36	*0.80*
	1.5	73	1.46	*3.22*	1.20	*2.64*	0.74	*1.64*	0.48	*1.06*

(Continued)

COMPOSITION OF FEEDS

TABLE 27-1

Entry Number	Feed Name Description	Moisture Basis: A-F (as-fed) or M-F (moisture-free)	Dry Matter	Macrominerals							Microminerals						
				Calcium (Ca)	Phosphorus (P)	Sodium (Na)	Chlorine (Cl)	Magnesium (Mg)	Potassium (K)	Sulfur (S)	Cobalt (Co)	Copper (Cu)	Iodine (I)	Iron (Fe)	Manganese (Mn)	Selenium (Se)	Zinc (Zn)
			%	%	%	%	%	%	%	%	ppm or mg/kg	ppm or mg/kg	ppm or mg/kg	%	ppm or mg/kg	ppm or mg/kg	ppm or mg/kg
	DESERT MOLLY (SUMMER CYPRESS) *Kochia vestita*																
81	-DORMANT	A-F	75	—	—	—	—	—	—	—	—	—	—	—	—	—	—
		M-F	100	—	—	—	—	—	—	—	—	—	—	—	—	—	—
	DROPSEED, SAND *Sporobolus cryptandrus*																
82	-DORMANT	A-F	86	0.49	0.05	—	—	—	—	—	—	—	—	—	—	—	—
		M-F	100	0.57	0.06	—	—	—	—	—	—	—	—	—	—	—	—
	FESCUE, MEADOW *Festuca elatior*																
83	-HAY, SUN-CURED	A-F	88	0.51	0.32	—	—	0.52	1.53	—	—	—	—	—	21.6	—	—
		M-F	100	0.57	0.37	—	—	0.59	1.74	—	—	—	—	—	24.5	—	—
	GALLETA *Hilaria jamesii*																
84	-DORMANT	A-F	55	0.19	0.04	—	—	—	—	—	—	—	—	—	—	—	—
		M-F	100	0.34	0.08	—	—	—	—	—	—	—	—	—	—	—	—
	GRAMA *Bouteloua spp*																
85	-IMMATURE	A-F	41	0.22	0.08	—	—	—	—	—	—	2.3	—	—	18.2	—	—
		M-F	100	0.53	0.19	—	—	—	—	—	—	5.5	—	—	44.3	—	—
86	-MATURE	A-F	63	0.22	0.08	—	—	—	0.22	—	0.115	8.1	—	0.082	30.1	—	—
		M-F	100	0.34	0.12	—	—	—	0.35	—	0.181	12.8	—	0.130	47.4	—	—
	GRASS-LEGUME																
87	-SILAGE, ALL ANALYSES	A-F	28	0.24	0.08	—	0.30	—	0.47	0.20	0.036	—	—	—	15.6	—	—
		M-F	100	0.85	0.27	—	1.06	—	1.67	0.72	0.126	—	—	—	55.1	—	—
	LESPEDEZA, COMMON *Lespedeza striata*																
88	-HAY, SUN-CURED, ALL ANALYSES	A-F	91	1.08	0.23	0.14	0.68	0.23	1.04	0.32	0.213	9.6	—	0.031	146.1	0.075	24.3
		M-F	100	1.19	0.26	0.15	0.76	0.26	1.15	0.35	0.235	10.5	—	0.034	161.1	0.083	26.8
89	-HAY, SUN-CURED, PREBLOOM	A-F	89	1.03	0.20	—	—	0.21	1.07	—	—	—	—	0.300	159.0	—	—
		M-F	100	1.16	0.22	—	—	0.24	1.20	—	—	—	—	0.340	178.4	—	—
90	-HAY, SUN-CURED, MIDBLOOM	A-F	94	1.16	0.16	—	—	0.25	0.96	—	—	—	—	0.032	294.3	—	—
		M-F	100	1.23	0.17	—	—	0.27	1.02	—	—	—	—	0.034	312.4	—	—
	MEADOW, INTERMOUNTAIN																
91	-HAY, SUN-CURED	A-F	92	0.62	0.18	—	—	—	—	—	—	—	—	—	—	—	—
		M-F	100	0.68	0.20	—	—	—	—	—	—	—	—	—	—	—	—
	MILK																
92	-FRESH (COW'S)	A-F	13	0.12	0.10	0.05	0.20	—	0.14	—	—	0.0	—	—	—	—	—
		M-F	100	0.93	0.75	0.39	1.56	—	1.11	—	—	0.3	—	—	—	—	—
	MOLASSES AND SYRUP																
93	-BEET, SUGAR, MOLASSES, MORE THAN 48% INVERT SUGARS, MORE THAN 79.5 DEGREES BRIX	A-F	78	0.12	0.02	1.10	1.50	0.19	4.63	0.47	0.374	17.1	—	0.007	4.5	—	—
		M-F	100	0.15	0.03	1.41	1.92	0.24	5.94	0.60	0.480	22.0	—	0.009	5.7	—	—
94	-CORN, MOLASSES, MORE THAN 43% DEXTROSE EQUIVALENT, MORE THAN 50% TOTAL DEXTROSE, MORE THAN 78 DEGREES BRIX	A-F	73	—	—	—	—	—	—	—	—	—	—	—	—	—	—
		M-F	100	—	—	—	—	—	—	—	—	—	—	—	—	—	—
95	-SUGARCANE, MOLASSES (BLACKSTRAP), MORE THAN 46% INVERT SUGARS, MORE THAN 79.5 DEGREES BRIX	A-F	75	0.78	0.09	0.17	2.78	0.35	2.85	0.35	0.908	60.4	1.577	0.019	42.9	—	22.0
		M-F	100	1.05	0.11	0.22	3.71	0.47	3.80	0.46	1.210	80.5	2.103	0.026	57.1	—	30.0

(Continued)

Entry Number	A (1 mg Carotene = 1667 IU Vit A)	Carotene (Provitamin A)	D	E (α-tocoph-erol)	K	B_{12}	Biotin	Choline	Folic Acid (Folacin)	Niacin (Nicotinic Acid)	Panto-thenic Acid	Pyri-doxine (B_6)	Ribo-flavin (B_2)	Thiamin (B_1)
	Fat-soluble Vitamins					Water-soluble Vitamins								
	IU/g	ppm or mg/kg	IU/kg	ppm or mg/kg	ppm or mg/kg	ppb or mcg/kg	ppm or mg/kg	ppm or mg/kg	ppm or mg/kg	ppm or mg/kg	ppm or mg/kg	ppm or mg/kg	ppm or mg/kg	ppm or mg/kg
81	— —	— —	— —	— —	— —	— —	— —	— —	— —	— —	— —	— —	— —	— —
82	— —	— —	— —	— —	— —	— —	— —	— —	— —	— —	— —	— —	— —	— —
83	— —	— —	— —	119.6 135.6	— —	— —	— —	— —	— —	— —	— —	— —	— —	— —
84	6.2 11.2	3.7 6.7	— —	— —	— —	— —	— —	— —	— —	— —	— —	— —	— —	— —
85	— —	— —	— —	— —	— —	— —	— —	— —	— —	— —	— —	— —	— —	— —
86	32.2 50.7	19.3 30.4	— —	— —	— —	— —	— —	— —	— —	— —	— —	— —	— —	— —
87	97.0 341.8	58.2 205.0	82 289	— —	— —	— —	— —	— —	— —	13 46	— —	— —	— —	— —
88	79.6 87.8	47.8 52.7	— —	230.7 254.4	— —	— —	0.32 0.35	1357 1496	3.51 3.87	40 44	29.1 32.1	— —	15.5 17.1	2.8 3.1
89	— —	— —	— —	— —	— —	— —	— —	— —	— —	— —	— —	— —	— —	— —
90	— —	— —	— —	— —	— —	— —	— —	— —	— —	— —	— —	— —	— —	— —
91	65.3 71.3	39.2 42.8	— —	— —	— —	— —	— —	— —	— —	— —	— —	— —	— —	— —
92	1.5 11.5	0.9 6.9	— —	— —	— —	— —	— —	— —	— —	1 10	2.9 22.4	— —	1.8 13.5	0.4 2.7
93	— —	— —	— —	— —	— —	— —	— —	829 1063	— —	41 53	4.5 5.7	— —	2.3 2.9	— —
94	— —	— —	— —	— —	— —	— —	— —	— —	— —	— —	— —	— —	— —	— —
95	— —	— —	— —	5.0 6.7	— —	— —	0.71 0.94	744 992	0.11 0.15	41 54	39.2 52.2	6.50 8.67	2.9 3.8	0.9 1.2

(Continued)

TABLE 27-1

Entry Number	Feed Name Description	International Feed Number[1]	Moisture Basis: A-F (as-fed) or M-F (moisture-free)	Chemical Analysis								
				Dry Matter	Ash	Crude Fiber	Cell Walls or NDF	Acid Detergent Fiber	Lignin	Ether Extract (Fat)	N-Free Extract	Crude Protein
				%	%	%	%	%	%	%	%	%
	NAPIERGRASS *Pennisetum purpureum*											
96	-PREBLOOM	2-03-158	A-F	26	2.2	8.8	—	—	—	0.8	12.2	1.6
			M-F	100	8.6	34.5	—	—	—	3.0	47.5	6.4
97	-LATE BLOOM	2-03-162	A-F	23	1.2	9.0	—	—	—	0.3	10.8	1.8
			M-F	100	5.3	39.0	—	—	—	1.1	46.8	7.8
	NEEDLE-AND-THREAD *Stipa comata*											
98	-DORMANT	2-07-989	A-F	—	—	—	—	—	—	—	—	—
			M-F	100	10.8	33.8	—	—	—	2.1	48.4	4.9
	OATS *Avena sativa*											
99	-GRAIN, ALL ANALYSES	4-03-309	A-F	89	3.0	10.9	—	—	—	4.9	58.0	12.1
			M-F	100	3.4	12.2	—	—	—	5.6	65.2	13.6
100	-GRAIN, PACIFIC COAST	4-07-999	A-F	91	3.8	11.3	—	—	—	4.9	61.8	9.2
			M-F	100	4.2	12.4	—	—	—	5.4	67.9	10.1
101	-HAY, SUN-CURED, ALL ANALYSES	1-03-280	A-F	90	6.8	29.1	55.6	25.5	3.5	2.3	43.5	8.0
			M-F	100	7.6	32.4	62.0	28.5	3.9	2.6	48.5	8.9
102	-STRAW	1-03-283	A-F	92	6.9	37.0	—	—	—	2.1	42.5	4.0
			M-F	100	7.4	40.0	—	—	—	2.2	46.0	4.3
	ORCHARDGRASS *Dactylis glomerata*											
103	-HAY, SUN-CURED, ALL ANALYSES	1-03-438	A-F	91	7.3	32.5	59.6	34.5	3.8	2.7	39.5	9.2
			M-F	100	8.0	35.6	65.4	37.9	4.2	3.0	43.4	10.1
	PEA *Pisum spp*											
104	-VINES WITHOUT SEEDS, SILAGE	3-03-596	A-F	25	2.2	7.3	—	—	—	0.8	11.0	3.2
			M-F	100	9.0	29.8	—	—	—	3.3	44.9	13.1
	POTATO *Solanum tuberosum*											
105	-TUBERS, FRESH	4-03-787	A-F	23	1.1	0.6	—	—	—	0.1	19.0	2.2
			M-F	100	4.8	2.5	—	—	—	0.4	82.8	9.5
	PRAIRIE, MIDWEST (PRAIRIE HAY)											
106	-HAY, SUN-CURED, IMMATURE	1-03-183	A-F	90	8.0	28.3	—	—	—	2.4	41.6	9.7
			M-F	100	8.9	31.4	—	—	—	2.7	46.2	10.8
107	-HAY, SUN-CURED, MATURE	1-03-187	A-F	91	6.0	32.2	—	—	—	2.3	45.7	4.6
			M-F	100	6.6	35.4	—	—	—	2.6	50.3	5.1
	REDTOP *Agrostis alba*											
108	-FULL BLOOM	2-03-891	A-F	26	1.8	6.6	—	—	—	0.9	14.8	2.1
			M-F	100	7.0	25.1	—	—	—	3.5	56.3	8.1
	RICE *Oryza sativa*											
109	-HULLS, AMMONIATED	1-05-698	A-F	92	—	44.7	—	—	—	0.9	16.9	10.4
			M-F	100	—	48.6	—	—	—	1.0	18.4	11.3
	RUSSIAN THISTLE (TUMBLEWEED) *Salsola kali tenuifolia*											
110	-HAY, SUN-CURED	1-03-988	A-F	88	13.5	25.1	—	—	—	1.6	38.2	10.1
			M-F	100	15.3	28.4	—	—	—	1.8	43.2	11.4
	RYE *Secale cereale*											
111	-GRAIN, ALL ANALYSES	4-04-047	A-F	87	1.6	2.2	—	—	—	1.5	69.7	12.0
			M-F	100	1.9	2.5	—	—	—	1.7	80.1	13.8
112	-STRAW	1-04-007	A-F	90	4.5	38.5	—	—	—	1.5	42.2	3.6
			M-F	100	5.0	42.6	—	—	—	1.7	46.7	4.0

Footnote on last page of table.

(Continued)

Entry Number	Digestible Protein	TDN	Digestible Energy		Metabolizable Energy		Net Energy			
	Ruminant	Ruminant	Ruminant		Ruminant		Ruminant NE_m		Ruminant NE_g	
	%	%	Mcal		Mcal		Mcal		Mcal	
			lb	kg	lb	kg	lb	kg	lb	kg
96	1.0	14	0.27	0.61	0.23	0.50	0.14	0.30	0.05	0.11
	3.7	54	1.07	2.36	0.88	1.94	0.52	1.15	0.19	0.42
97	0.8	11	0.23	0.50	0.19	0.41	0.11	0.24	0.02	0.05
	3.6	49	0.98	2.16	0.81	1.78	0.48	1.05	0.11	0.24
98	—	—	—	—	—	—	—	—	—	—
	1.5	58	1.17	2.60	1.00	2.20	—	—	—	—
99	9.6	68	1.32	2.91	1.08	2.39	0.67	1.48	0.44	0.97
	10.8	77	1.48	3.27	1.22	2.68	0.76	1.67	0.49	1.09
100	6.9	71	1.42	3.13	1.17	2.57	0.74	1.63	0.49	1.08
	7.6	78	1.56	3.44	1.28	2.82	0.81	1.79	0.54	1.19
101	4.6	51	1.12	2.47	0.92	2.02	0.56	1.23	0.30	0.66
	5.2	57	1.25	2.75	1.02	2.26	0.62	1.37	0.33	0.73
102	0.8	46	1.12	2.46	0.92	2.02	0.55	1.22	0.28	0.61
	0.9	50	1.21	2.67	0.99	2.19	0.60	1.32	0.30	0.67
103	5.3	53	1.23	2.71	1.01	2.22	0.62	1.37	0.37	0.81
	5.8	58	1.35	2.97	1.11	2.44	0.68	1.51	0.40	0.89
104	1.9	14	0.28	0.62	0.23	0.51	0.14	0.30	0.06	0.14
	7.7	57	1.14	2.52	0.94	2.07	0.55	1.22	0.25	0.56
105	1.4	19	0.37	0.82	0.31	0.67	0.20	0.44	0.13	0.29
	5.9	81	1.62	3.56	1.33	2.93	0.87	1.91	0.58	1.27
106	—	57	1.13	2.49	0.93	2.05	0.56	1.24	0.30	0.67
	—	63	1.26	2.77	1.03	2.27	0.63	1.38	0.34	0.74
107	—	44	0.88	1.93	0.72	1.59	0.43	0.95	0.08	0.19
	—	48	0.97	2.13	0.79	1.75	0.47	1.04	0.09	0.21
108	—	19	0.37	0.82	0.31	0.67	0.19	0.42	0.12	0.26
	—	71	1.42	3.13	1.16	2.56	0.73	1.61	0.45	1.00
109	—	—	—	—	—	—	—	—	—	—
	—	—	—	—	—	—	—	—	—	—
110	6.5	40	0.80	1.76	0.65	1.44	0.39	0.86	0.02	0.05
	7.4	45	0.90	1.98	0.74	1.63	0.44	0.98	0.03	0.06
111	8.7	73	1.46	3.22	1.20	2.64	0.79	1.74	0.53	1.17
	10.0	84	1.68	3.70	1.38	3.03	0.91	2.00	0.61	1.34
112	−1.8	41	0.82	1.82	0.67	1.48	0.40	0.89	0.04	0.08
	−2.0	46	0.91	2.01	0.75	1.65	0.45	0.99	0.04	0.09

(Continued)

TABLE 27-1

Entry Number	Feed Name Description	Moisture Basis: A-F (as-fed) or M-F (moisture-free)	Dry Matter	Macrominerals							Microminerals						
				Calcium (Ca)	Phosphorus (P)	Sodium (Na)	Chlorine (Cl)	Magnesium (Mg)	Potassium (K)	Sulfur (S)	Cobalt (Co)	Copper (Cu)	Iodine (I)	Iron (Fe)	Manganese (Mn)	Selenium (Se)	Zinc (Zn)
			%	%	%	%	%	%	%	%	ppm or mg/kg	ppm or mg/kg	ppm or mg/kg	%	ppm or mg/kg	ppm or mg/kg	ppm or mg/kg
	NAPIERGRASS *Pennisetum purpureum*																
96	-PREBLOOM	A-F	26	—	—	—	—	—	—	—	—	—	—	—	—	—	—
		M-F	100	—	—	—	—	—	—	—	—	—	—	—	—	—	—
97	-LATE BLOOM	A-F	23	—	—	—	—	—	—	—	—	—	—	—	—	—	—
		M-F	100	—	—	—	—	—	—	—	—	—	—	—	—	—	—
	NEEDLE-AND-THREAD *Stipa comata*																
98	-DORMANT	A-F	—	—	—	—	—	—	—	—	—	—	—	—	—	—	—
		M-F	100	—	—	—	—	—	—	—	—	—	—	—	—	—	—
	OATS *Avena sativa*																
99	-GRAIN, ALL ANALYSES	A-F	89	0.06	0.33	0.16	0.11	0.12	0.39	0.21	0.057	5.8	0.008	0.008	37.0	0.210	36.6
		M-F	100	0.07	0.37	0.18	0.12	0.14	0.44	0.23	0.064	6.5	0.099	0.009	41.6	0.236	41.1
100	-GRAIN, PACIFIC COAST	A-F	91	0.10	0.31	—	—	—	—	0.21	—	—	—	—	—	0.076	—
		M-F	100	0.11	0.34	—	—	—	—	0.23	—	—	—	—	—	0.083	—
101	-HAY, SUN-CURED, ALL ANALYSES	A-F	90	0.26	0.24	0.15	0.47	0.67	1.10	0.27	0.064	3.9	—	0.035	108.0	—	—
		M-F	100	0.30	0.26	0.17	0.52	0.75	1.23	0.30	0.071	4.4	—	0.039	120.4	—	—
102	-STRAW	A-F	92	0.23	0.06	0.39	0.72	0.17	2.19	0.21	—	9.6	—	0.018	34.5	—	—
		M-F	100	0.25	0.07	0.42	0.78	0.19	2.37	0.23	—	10.4	—	0.020	37.3	—	—
	ORCHARDGRASS *Dactylis glomerata*																
103	-HAY, SUN-CURED, ALL ANALYSES	A-F	91	0.32	0.28	—	0.37	0.18	2.74	0.24	0.019	12.5	—	0.009	227.3	—	16.5
		M-F	100	0.35	0.31	—	0.41	0.20	3.01	0.26	0.021	13.7	—	0.010	249.4	—	18.1
	PEA *Pisum spp*																
104	-VINES WITHOUT SEEDS, SILAGE	A-F	25	0.32	0.06	—	—	—	—	—	—	—	—	—	—	—	—
		M-F	100	1.31	0.24	—	—	—	—	—	—	—	—	—	—	—	—
	POTATO *Solanum tuberosum*																
105	-TUBERS, FRESH	A-F	23	0.01	0.06	0.02	0.07	0.03	0.50	0.02	—	6.5	—	0.002	9.6	—	—
		M-F	100	0.04	0.24	0.09	0.28	0.14	2.18	0.09	—	28.3	—	0.008	41.7	—	—
	PRAIRIE, MIDWEST (PRAIRIE HAY)																
106	-HAY, SUN-CURED, IMMATURE	A-F	90	0.44	0.21	—	—	0.22	0.97	—	—	—	—	0.008	—	—	—
		M-F	100	0.49	0.23	—	—	0.24	1.08	—	—	—	—	0.009	—	—	—
107	-HAY, SUN-CURED, MATURE	A-F	91	0.34	0.14	—	—	0.26	—	—	—	21.1	—	0.010	44.9	—	—
		M-F	100	0.38	0.16	—	—	0.29	—	—	—	23.3	—	0.011	49.5	—	—
	REDTOP *Agrostis alba*																
108	-FULL BLOOM	A-F	26	—	—	—	—	—	—	—	—	—	—	—	—	—	—
		M-F	100	—	—	—	—	—	—	—	—	—	—	—	—	—	—
	RICE *Oryza sativa*																
109	-HULLS, AMMONIATED	A-F	92	0.15	0.19	—	—	—	—	—	—	—	—	—	—	—	—
		M-F	100	0.16	0.21	—	—	—	—	—	—	—	—	—	—	—	—
	RUSSIAN THISTLE (TUMBLEWEED) *Salsola kali tenuifolia*																
110	-HAY, SUN-CURED	A-F	88	1.45	0.19	—	—	0.79	6.06	0.37	—	—	—	—	—	—	—
		M-F	100	1.64	0.22	—	—	0.89	6.85	0.42	—	—	—	—	—	—	—
	RYE *Secale cereale*																
111	-GRAIN, ALL ANALYSES	A-F	87	0.06	0.32	0.02	0.03	0.12	0.45	0.15	—	6.7	—	0.006	54.4	0.382	31.4
		M-F	100	0.07	0.36	0.03	0.03	0.14	0.52	0.17	—	7.7	—	0.007	62.5	0.439	36.1
112	-STRAW	A-F	90	0.22	0.08	0.12	0.22	0.07	0.88	0.10	—	3.6	—	—	6.0	—	—
		M-F	100	0.24	0.09	0.13	0.24	0.08	0.97	0.11	—	4.0	—	—	6.6	—	—

(Continued)

Entry Number	Fat-soluble Vitamins					Water-soluble Vitamins								
	A (1 mg Carotene = 1667 IU Vit A)	Carotene (Provitamin A)	D	E (α-tocopherol)	K	B_{12}	Biotin	Choline	Folic Acid (Folacin)	Niacin (Nicotinic Acid)	Pantothenic Acid	Pyridoxine (B_6)	Riboflavin (B_2)	Thiamin (B_1)
	IU/g	ppm or mg/kg	IU/kg	ppm or mg/kg	ppm or mg/kg	ppb or mcg/kg	ppm or mg/kg	ppm or mg/kg	ppm or mg/kg	ppm or mg/kg	ppm or mg/kg	ppm or mg/kg	ppm or mg/kg	ppm or mg/kg
96	— —	— —	— —	— —	— —	— —	— —	— —	— —	— —	— —	— —	— —	— —
97	— —	— —	— —	— —	— —	— —	— —	— —	— —	— —	— —	— —	— —	— —
98	— —	— —	— —	— —	— —	— —	— —	— —	— —	— —	— —	— —	— —	— —
99	0.2 0.2	0.1 0.1	— —	12.9 14.5	— —	— —	0.24 0.27	1013 1138	0.34 0.39	14 15	7.1 8.0	2.50 2.81	1.5 1.5	6.4 7.2
100	— —	— —	— —	20.2 22.2	— —	— —	— —	918 1009	— —	14 16	11.7 12.8	— —	1.2 1.3	— —
101	98.2 109.5	58.9 65.7	1385 1544	— —	— —	— —	6.34 7.07	— —	— —	— —	— —	— —	4.7 5.3	3.0 3.3
102	5.9 6.3	3.5 3.8	611 662	— —	— —	— —	— —	216 234	— —	— —	— —	— —	— —	— —
103	66.6 73.0	39.9 43.8	— —	174.2 191.1	— —	— —	— —	— —	— —	— —	— —	— —	6.2 6.8	2.6 2.9
104	77.2 315.0	46.3 188.9	— —	— —	— —	— —	— —	— —	— —	— —	— —	— —	— —	— —
105	— —	— —	— —	— —	— —	— —	— —	— —	— —	17 74	— —	— —	0.5 2.0	1.1 5.0
106	— —	— —	— —	— —	— —	— —	— —	— —	— —	— —	— —	— —	— —	— —
107	14.5 16.0	8.7 9.6	— —	— —	— —	— —	— —	— —	— —	— —	— —	— —	— —	— —
108	66.9 254.4	40.1 152.6	— —	— —	— —	— —	— —	— —	— —	— —	— —	— —	— —	— —
109	— —	— —	— —	— —	— —	— —	— —	— —	— —	— —	— —	— —	— —	— —
110	6.8 7.7	4.1 4.6	— —	— —	— —	— —	— —	— —	— —	— —	— —	— —	— —	— —
111	16.9 19.4	10.1 11.6	— —	14.9 17.2	— —	— —	0.32 0.37	— —	0.61 0.70	20 23	8.8 10.2	2.55 2.94	1.6 1.8	3.0 3.4
112	— —	— —	— —	— —	— —	— —	— —	— —	— —	— —	— —	— —	— —	— —

(Continued)

COMPOSITION OF FEEDS

TABLE 27-1

Entry Number	Feed Name Description	International Feed Number[1]	Moisture Basis: A-F (as-fed) or M-F (moisture-free)	Chemical Analysis								
				Dry Matter	Ash	Crude Fiber	Cell Walls or NDF	Acid Detergent Fiber	Lignin	Ether Extract (Fat)	N-Free Extract	Crude Protein
				%	%	%	%	%	%	%	%	%
	RYEGRASS, ITALIAN *Lolium multiflorum*											
113	-ALL ANALYSES	2-04-073	A-F	26	3.6	6.2	—	—	—	0.8	11.6	3.8
			M-F	100	13.8	24.0	—	—	—	3.0	44.6	14.6
	SAGE, BLACK *Salvia mellifera*											
114	-BROWSE, DORMANT	2-05-564	A-F	38	2.1	—	—	—	—	4.1	—	3.2
			M-F	100	5.5	—	—	—	—	10.8	—	8.5
	SAGEBRUSH, BIG *Artemisia tridentata*											
115	-BROWSE, DORMANT	2-07-992	A-F	38	3.6	—	—	—	—	3.1	—	3.4
			M-F	100	9.6	—	—	—	—	8.2	—	9.0
	SAGEBRUSH, BUD *Artemisia spinescens*											
116	-BROWSE, IMMATURE	2-07-991	A-F	25	—	—	—	—	—	—	—	4.3
			M-F	100	—	—	—	—	—	—	—	17.3
117	-BROWSE, PREBLOOM	2-04-125	A-F	27	5.8	6.1	—	—	—	0.7	9.6	4.7
			M-F	100	21.6	22.7	—	—	—	2.5	35.7	17.5
	SAGEBRUSH, FRINGED *Artemisia frigida*											
118	-BROWSE, MIDBLOOM	2-04-129	A-F	32	2.1	10.6	—	—	—	0.6	15.6	3.0
			M-F	100	6.5	33.2	—	—	—	2.0	48.9	9.4
119	-BROWSE, MATURE	2-04-130	A-F	38	6.5	12.1	—	—	—	1.3	16.1	2.1
			M-F	100	17.1	31.8	—	—	—	3.4	42.3	5.4
	SALTBUSH, NUTTALL *Atriplex nuttallii*											
120	-BROWSE, DORMANT	2-07-993	A-F	75	—	—	—	—	—	—	—	5.4
			M-F	100	—	—	—	—	—	—	—	7.2
	SALTBUSH, SHADSCALE *Atriplex confertifolia*											
121	-BROWSE, DORMANT	2-05-565	A-F	45	11.4	—	—	—	—	1.2	—	3.2
			M-F	100	25.4	—	—	—	—	2.6	—	7.1
	SALTGRASS *Distichlis* spp											
122	-ALL ANALYSES	2-04-170	A-F	74	5.4	26.0	—	—	—	1.9	37.9	3.1
			M-F	100	7.3	34.9	—	—	—	2.6	51.0	4.2
123	-OVERRIPE	2-04-169	A-F	74	5.4	26.0	—	—	—	1.9	37.9	3.1
			M-F	100	7.3	34.9	—	—	—	2.6	51.0	4.2
124	-HAY, SUN-CURED	1-01-168	A-F	89	11.4	28.3	—	—	—	1.8	40.0	8.0
			M-F	100	12.7	31.6	—	—	—	2.1	44.7	8.9
125	**SALTGRASS, DESERT** *Distichlis stricta*	2-04-171	A-F	29	2.0	8.6	—	—	—	0.5	16.2	1.7
			M-F	100	6.8	29.7	—	—	—	1.7	55.9	5.9
	SEDGE *Carex* spp											
126	-HAY, SUN-CURED	1-04-193	A-F	89	6.4	28.0	—	—	—	2.1	44.4	8.4
			M-F	100	7.2	31.3	—	—	—	2.4	49.7	9.4
	SORGHUM *Sorghum vulgare*											
127	-GRAIN, ALL ANALYSES	4-04-383	A-F	90	1.7	2.4	—	—	—	2.8	71.7	11.4
			M-F	100	1.9	2.7	—	—	—	3.1	79.6	12.6
128	-FODDER, SUN-CURED	1-07-960	A-F	87	7.5	24.0	—	—	—	2.0	45.6	7.5
			M-F	100	8.7	27.7	—	—	—	2.3	52.6	8.6
	SORGHUM, SORGO *Sorghum vulgare saccharatum*											
129	-SILAGE	3-04-468	A-F	27	1.6	7.8	—	—	—	0.7	15.5	1.7
			M-F	100	6.0	28.5	—	—	—	2.6	56.7	6.2

Footnote on last page of table.

(Continued)

Entry Number	Digestible Protein	TDN	Digestible Energy		Metabolizable Energy		Net Energy			
	Ruminant	Ruminant	Ruminant		Ruminant		Ruminant NE_m		Ruminant NE_g	
	%	%	Mcal		Mcal		Mcal		Mcal	
			lb	kg	lb	kg	lb	kg	lb	kg
113	1.6	15	0.30	*0.67*	0.25	*0.55*	0.15	*0.33*	0.07	*0.16*
	6.3	59	1.17	*2.59*	0.96	*2.12*	0.58	*1.27*	0.27	*0.60*
114	1.7	19	0.36	*0.80*	0.18	*0.40*	0.13	*0.30*	0.00	*0.00*
	4.5	49	0.96	*2.12*	0.47	*1.04*	0.35	*0.78*	0.00	*0.00*
115	1.4	16	0.33	*0.74*	0.19	*0.43*	0.14	*0.30*	0.00	*0.00*
	3.8	43	0.88	*1.95*	0.51	*1.13*	0.36	*0.79*	0.00	*0.00*
116	3.4	13	0.26	*0.58*	0.36	*0.80*	0.14	*0.30*	0.06	*0.12*
	13.7	52	1.04	*2.30*	0.86	*1.90*	0.54	*1.20*	0.23	*0.50*
117	—	14	0.28	*0.62*	0.23	*0.51*	0.14	*0.30*	0.05	*0.10*
	—	52	1.04	*2.30*	0.86	*1.89*	0.51	*1.12*	0.17	*0.36*
118	—	19	0.37	*0.81*	0.30	*0.67*	0.18	*0.40*	0.08	*0.18*
	—	58	1.15	*2.54*	0.95	*2.08*	0.57	*1.25*	0.26	*0.57*
119	—	19	0.38	*0.84*	0.31	*0.69*	0.19	*0.65*	0.05	*0.11*
	—	50	1.00	*2.21*	0.82	*1.81*	0.49	*1.07*	0.13	*0.28*
120	2.5	27	0.55	*1.20*	0.44	*0.98*	—	—	—	—
	3.4	36	0.73	*1.60*	0.59	*1.30*	—	—	—	—
121	1.6	15	0.28	*0.61*	0.17	*0.38*	0.16	*0.34*	0.00	*0.00*
	3.6	33	0.62	*1.36*	0.38	*0.85*	0.35	*0.76*	0.00	*0.00*
122	—	40	0.79	*1.75*	0.65	*1.44*	0.39	*0.85*	0.14	*0.30*
	—	53	1.07	*2.35*	0.88	*1.93*	0.52	*1.15*	0.19	*0.41*
123	—	40	0.79	*1.75*	0.65	*1.44*	0.39	*0.85*	0.14	*0.30*
	—	53	1.07	*2.35*	0.88	*1.93*	0.52	*1.15*	0.19	*0.41*
124	—	45	0.90	*1.99*	0.74	*1.64*	0.44	*0.97*	0.12	*0.27*
	—	51	1.01	*2.23*	0.83	*1.83*	0.49	*1.09*	0.14	*0.30*
125	—	17	0.34	*0.75*	0.28	*0.62*	0.17	*0.47*	0.08	*0.18*
	—	59	1.18	*2.60*	0.97	*2.13*	0.58	*1.28*	0.28	*0.62*
126	4.2	47	0.93	*2.06*	0.77	*1.69*	0.45	*1.00*	0.15	*0.33*
	4.7	52	1.05	*2.31*	0.86	*1.89*	0.51	*1.12*	0.17	*0.37*
127	7.9	80	1.60	*3.53*	1.31	*2.89*	0.88	*1.94*	0.60	*1.32*
	8.8	89	1.78	*3.92*	1.46	*3.21*	0.98	*2.15*	0.67	*1.47*
128	3.0	51	1.01	*2.24*	0.83	*1.83*	0.50	*1.11*	0.24	*0.53*
	3.5	59	1.17	*2.59*	0.96	*2.12*	0.58	*1.27*	0.27	*0.60*
129	0.4	16	0.31	*0.67*	0.25	*0.55*	0.15	*0.33*	0.06	*0.14*
	1.4	58	1.11	*2.46*	0.91	*2.01*	0.54	*1.19*	0.23	*0.50*

(Continued)

TABLE 27-1

Entry Number	Feed Name Description	Moisture Basis: A-F (as-fed) or M-F (moisture-free)	Dry Matter	Macrominerals: Calcium (Ca)	Phosphorus (P)	Sodium (Na)	Chlorine (Cl)	Magnesium (Mg)	Potassium (K)	Sulfur (S)	Microminerals: Cobalt (Co)	Copper (Cu)	Iodine (I)	Iron (Fe)	Manganese (Mn)	Selenium (Se)	Zinc (Zn)
			%	%	%	%	%	%	%	%	ppm or mg/kg	ppm or mg/kg	ppm or mg/kg	%	ppm or mg/kg	ppm or mg/kg	ppm or mg/kg
	RYEGRASS, ITALIAN *Lolium multiflorum*																
113	-ALL ANALYSES	A-F	26	0.17	0.11	—	—	0.09	0.52	—	—	—	—	—	—	—	—
		M-F	100	0.65	0.41	—	—	0.35	2.00	—	—	—	—	—	—	—	—
	SAGE, BLACK *Salvia mellifera*																
114	-BROWSE, DORMANT	A-F	38	0.31	0.06	—	—	—	—	—	—	—	—	—	—	—	—
		M-F	100	0.81	0.17	—	—	—	—	—	—	—	—	—	—	—	—
	SAGEBRUSH, BIG *Artemisia tridentata*																
115	-BROWSE, DORMANT	A-F	38	0.37	0.08	—	—	—	—	—	—	—	—	—	—	—	—
		M-F	100	0.98	0.21	—	—	—	—	—	—	—	—	—	—	—	—
	SAGEBRUSH, BUD *Artemisia spinescens*																
116	-BROWSE, IMMATURE	A-F	25	0.24	0.08	—	—	—	—	—	—	—	—	—	—	—	—
		M-F	100	0.97	0.33	—	—	—	—	—	—	—	—	—	—	—	—
117	-BROWSE, PREBLOOM	A-F	27	0.16	0.11	—	—	0.13	—	—	—	—	—	—	—	—	—
		M-F	100	0.60	0.42	—	—	0.49	—	—	—	—	—	—	—	—	—
	SAGEBRUSH, FRINGED *Artemisia frigida*																
118	-BROWSE, MIDBLOOM	A-F	32	—	—	—	—	—	—	—	—	—	—	—	—	—	—
		M-F	100	—	—	—	—	—	—	—	—	—	—	—	—	—	—
119	-BROWSE, MATURE	A-F	38	—	—	—	—	—	—	—	—	—	—	—	—	—	—
		M-F	100	—	—	—	—	—	—	—	—	—	—	—	—	—	—
	SALTBUSH, NUTTALL *Atriplex nuttallii*																
120	-BROWSE, DORMANT	A-F	75	1.66	0.16	—	—	—	—	—	—	—	—	—	—	—	—
		M-F	100	2.21	0.21	—	—	—	—	—	—	—	—	—	—	—	—
	SALTBUSH, SHADSCALE *Atriplex confertifolia*																
121	-BROWSE, DORMANT	A-F	45	1.08	0.04	—	—	—	—	—	—	—	—	—	—	—	—
		M-F	100	2.41	0.08	—	—	—	—	—	—	—	—	—	—	—	—
	SALTGRASS *Distichlis spp*																
122	-ALL ANALYSES	A-F	74	0.16	0.06	—	—	0.22	0.18	—	—	—	—	0.014	115.2	—	—
		M-F	100	0.22	0.08	—	—	0.30	0.24	—	—	—	—	0.019	154.8	—	—
123	-OVERRIPE	A-F	74	0.17	0.05	—	—	0.22	—	—	—	—	—	—	—	—	—
		M-F	100	0.23	0.07	—	—	0.30	—	—	—	—	—	—	—	—	—
124	-HAY, SUN-CURED	A-F	89	—	—	—	—	—	—	—	—	—	—	—	—	—	—
		M-F	100	—	—	—	—	—	—	—	—	—	—	—	—	—	—
125	**SALTGRASS, DESERT** *Distichlis stricta*	A-F	29	0.05	0.03	—	—	—	—	—	—	—	—	—	—	—	—
		M-F	100	0.16	0.09	—	—	—	—	—	—	—	—	—	—	—	—
	SEDGE *Carex spp*																
126	-HAY, SUN-CURED	A-F	89	—	—	—	—	—	—	—	—	—	—	—	—	—	—
		M-F	100	—	—	—	—	—	—	—	—	—	—	—	—	—	—
	SORGHUM *Sorghum vulgare*																
127	-GRAIN, ALL ANALYSES	A-F	90	0.03	0.30	0.03	0.09	0.18	0.35	0.15	0.264	9.8	0.022	0.005	15.5	0.805	14.5
		M-F	100	0.03	0.33	0.03	0.10	0.20	0.39	0.16	0.293	10.8	0.025	0.005	17.3	0.894	16.1
128	-FODDER, SUN-CURED	A-F	87	0.63	0.18	0.02	—	0.26	1.21	—	—	—	—	—	—	—	—
		M-F	100	0.72	0.21	0.02	—	0.30	1.39	—	—	—	—	—	—	—	—
	SORGHUM, SORGO *Sorghum vulgare saccharatum*																
129	-SILAGE	A-F	27	0.08	0.04	0.04	0.02	0.07	0.32	—	—	8.5	—	0.006	16.7	—	—
		M-F	100	0.29	0.16	0.15	0.06	0.27	1.15	—	—	31.1	—	0.020	61.0	—	—

(Continued)

Entry Number	Fat-soluble Vitamins					Water-soluble Vitamins								
	A (1 mg Carotene = 1667 IU Vit A)	Carotene (Provitamin A)	D	E (α-tocopherol)	K	B_{12}	Biotin	Choline	Folic Acid (Folacin)	Niacin (Nicotinic Acid)	Pantothenic Acid	Pyridoxine (B_6)	Riboflavin (B_2)	Thiamin (B_1)
	IU/g	ppm or mg/kg	IU/kg	ppm or mg/kg	ppm or mg/kg	ppb or mcg/kg	ppm or mg/kg	ppm or mg/kg	ppm or mg/kg	ppm or mg/kg	ppm or mg/kg	ppm or mg/kg	ppm or mg/kg	ppm or mg/kg
113	173.3 668.5	104.0 401.0	— —	— —	— —	— —	— —	— —	— —	— —	— —	— —	— —	— —
114	— —	— —	— —	— —	— —	— —	— —	— —	— —	— —	— —	— —	— —	— —
115	— —	— —	— —	— —	— —	— —	— —	— —	— —	— —	— —	— —	— —	— —
116	9.8 39.7	5.9 23.8	— —	— —	— —	— —	— —	— —	— —	— —	— —	— —	— —	— —
117	— —	— —	— —	— —	— —	— —	— —	— —	— —	— —	— —	— —	— —	— —
118	— —	— —	— —	— —	— —	— —	— —	— —	— —	— —	— —	— —	— —	— —
119	— —	— —	— —	— —	— —	— —	— —	— —	— —	— —	— —	— —	— —	— —
120	— —	— —	— —	— —	— —	— —	— —	— —	— —	— —	— —	— —	— —	— —
121	— —	— —	— —	— —	— —	— —	— —	— —	— —	— —	— —	— —	— —	— —
122	— —	— —	— —	— —	— —	— —	— —	— —	— —	— —	— —	— —	— —	— —
123	— —	— —	— —	— —	— —	— —	— —	— —	— —	— —	— —	— —	— —	— —
124	— —	— —	— —	— —	— —	— —	— —	— —	— —	— —	— —	— —	— —	— —
125	— —	— —	— —	— —	— —	— —	— —	— —	— —	— —	— —	— —	— —	— —
126	— —	— —	— —	— —	— —	— —	— —	— —	— —	— —	— —	— —	— —	— —
127	1.9 2.1	1.2 1.3	—	10.9 12.1	—	—	0.25 0.28	629 699	0.21 0.24	41 46	11.7 13.0	4.68 5.19	1.3 1.4	4.2 4.7
128	40.5 46.9	24.3 28.1	— —	— —	— —	— —	— —	— —	— —	— —	— —	— —	— —	— —
129	13.5 49.4	8.1 29.7	— —	— —	— —	— —	— —	— —	— —	— —	— —	— —	— —	— —

(Continued)

TABLE 27-1

Entry Number	Feed Name Description	International Feed Number[1]	Moisture Basis: A-F (as-fed) or M-F (moisture-free)	Chemical Analysis: Dry Matter	Ash	Crude Fiber	Cell Walls or NDF	Acid Detergent Fiber	Lignin	Ether Extract (Fat)	N-Free Extract	Crude Protein
				%	%	%	%	%	%	%	%	%
	SOYBEAN *Glycine max*											
130	-SILAGE	3-04-581	A-F	27	2.6	7.6	—	—	—	0.7	11.4	4.8
			M-F	100	9.7	28.0	—	—	—	2.6	42.0	17.7
131	-HAY, SUN-CURED	1-04-558	A-F	90	7.5	29.9	—	—	—	2.4	35.5	14.4
			M-F	100	8.3	33.3	—	—	—	2.7	39.6	16.1
132	-HULLS (SOYBEAN FLAKES)	1-04-560	A-F	92	4.1	37.3	—	—	—	2.2	37.0	11.0
			M-F	100	4.4	40.7	—	—	—	2.4	40.5	12.0
133	-STRAW	1-04-567	A-F	88	5.6	38.9	—	—	—	1.3	37.4	4.6
			M-F	100	6.4	44.3	—	—	—	1.4	42.7	5.2
134	-SEEDS, MEAL, MECH EXTD, 41% PROTEIN	5-04-600	A-F	90	6.0	6.0	—	—	—	4.7	36.1	43.8
			M-F	100	6.7	6.7	—	—	—	5.2	34.6	48.7
	SQUIRRELTAIL *Sitanion* spp											
135	-DORMANT	2-05-566	A-F	50	8.5	—	—	—	—	1.1	—	1.5
			M-F	100	17.0	—	—	—	—	2.2	—	3.1
	SWEET CLOVER, YELLOW *Melilotus officinalis*											
136	-HAY, SUN-CURED	1-04-754	A-F	89	7.5	32.1	—	—	—	1.5	35.6	12.5
			M-F	100	8.4	36.0	—	—	—	1.7	39.8	14.0
	TIMOTHY *Phleum pratense*											
137	-HAY, SUN-CURED, ALL ANALYSES	1-04-893	A-F	90	5.2	30.0	61.4	33.6	4.5	2.4	44.9	7.5
			M-F	100	5.7	33.3	68.2	37.3	5.0	2.7	49.9	8.4
138	-MIDBLOOM	2-04-905	A-F	30	2.0	10.0	—	—	—	0.9	14.3	2.7
			M-F	100	6.6	33.5	—	—	—	3.0	47.9	9.1
	TREFOIL, BIRDSFOOT *Lotus corniculatus*											
139	-HAY, SUN-CURED	1-05-044	A-F	92	6.5	26.6	43.4	33.4	8.1	2.4	40.3	16.3
			M-F	100	7.1	28.9	47.1	36.3	8.8	2.6	43.8	17.6
140	**TUMBLEWEED (RUSSIAN THISTLE)** *Salsola kali tenuifolia*	2-03-990	A-F	48	7.7	14.0	—	—	—	1.0	19.3	6.4
			M-F	100	15.8	29.0	—	—	—	2.0	40.0	13.1
	VETCH *Vicia* spp											
141	-HAY, SUN-CURED	1-05-106	A-F	91	8.6	28.1	52.1	39.5	8.7	2.3	35.6	16.5
			M-F	100	9.5	30.9	57.2	43.3	9.5	2.5	39.1	18.1
	WHEAT *Triticum aestivum*											
142	-IMMATURE	2-05-176	A-F	22	3.0	3.9	—	—	—	1.0	8.1	6.3
			M-F	100	13.3	17.4	—	—	—	4.4	36.3	28.6
143	-HAY, SUN-CURED	1-05-172	A-F	88	5.9	25.7	—	—	—	1.7	47.0	7.5
			M-F	100	6.7	29.3	—	—	—	1.9	53.6	8.5
144	-GRAIN, ALL ANALYSES	4-05-211	A-F	88	1.6	2.5	—	—	—	1.8	67.3	14.9
			M-F	100	1.8	2.8	—	—	—	2.0	76.4	16.9
145	-BRAN	4-05-190	A-F	89	6.2	10.4	—	—	—	3.9	53.4	15.2
			M-F	100	6.9	11.6	—	—	—	4.4	60.0	17.1
146	-GRAIN SCREENINGS	4-05-216	A-F	89	6.0	7.5	—	—	—	3.4	57.9	14.2
			M-F	100	6.7	8.4	—	—	—	3.8	65.1	16.0
147	-CHAFF	1-05-192	A-F	93	16.7	30.1	—	—	—	1.5	38.7	5.7
			M-F	100	18.0	32.5	—	—	—	1.7	41.7	6.2
148	-STRAW	1-05-175	A-F	90	7.0	36.9	—	—	—	1.4	40.8	3.8
			M-F	100	7.8	41.1	—	—	—	1.6	45.4	4.2

Footnote on last page of table.

(Continued)

Entry Number	Digestible Protein	TDN	Digestible Energy		Metabolizable Energy		Net Energy			
	Ruminant	Ruminant	Ruminant		Ruminant		Ruminant NE_m		Ruminant NE_g	
	%	%	Mcal		Mcal		Mcal		Mcal	
			lb	*kg*	lb	*kg*	lb	*kg*	lb	*kg*
130	3.1	15	0.30	*0.67*	0.25	*0.55*	0.15	*0.32*	0.06	*0.14*
	11.5	56	1.11	*2.45*	0.91	*2.01*	0.54	*1.19*	0.23	*0.50*
131	9.7	50	1.02	*2.25*	0.84	*1.84*	0.50	*1.11*	0.22	*0.49*
	10.8	56	1.14	*2.51*	0.93	*2.06*	0.56	*1.23*	0.25	*0.54*
132	5.7	71	1.43	*3.15*	1.17	*2.58*	0.76	*1.67*	0.50	*1.10*
	6.3	78	1.56	*3.44*	1.28	*2.82*	0.82	*1.82*	0.54	*1.19*
133	1.3	38	0.76	*1.67*	0.62	*1.37*	0.38	*0.83*	0.00	*0.00*
	1.5	43	0.87	*1.91*	0.71	*1.56*	0.43	*0.94*	0.00	*0.00*
134	38.3	76	1.51	*3.33*	1.22	*2.70*	0.86	*1.89*	0.58	*1.26*
	42.6	84	1.68	*3.70*	1.36	*3.00*	0.95	*2.10*	0.64	*1.40*
135	0.0	25	0.46	*1.01*	0.38	*0.85*	0.23	*0.51*	0.04	*0.08*
	0.0	50	0.92	*2.02*	0.77	*1.70*	0.46	*1.02*	0.07	*0.15*
136	9.1	48	1.12	*2.47*	0.92	*2.02*	0.56	*1.23*	0.30	*0.66*
	10.2	54	1.25	*2.76*	1.03	*2.27*	0.63	*1.38*	0.34	*0.74*
137	4.0	51	1.01	*2.22*	0.83	*1.82*	0.49	*1.09*	0.21	*0.46*
	4.5	57	1.12	*2.47*	0.92	*2.02*	0.55	*1.21*	0.23	*0.51*
138	1.5	18	0.37	*0.81*	0.30	*0.66*	0.18	*0.40*	0.10	*0.21*
	4.9	62	1.23	*2.72*	1.01	*2.23*	0.61	*1.35*	0.32	*0.71*
139	11.2	54	0.93	*2.05*	0.76	*1.68*	0.45	*0.99*	0.12	*0.27*
	12.2	59	1.01	*2.22*	0.83	*1.82*	0.49	*1.08*	0.13	*0.29*
140	—	21	0.42	*0.92*	0.34	*0.76*	0.21	*0.45*	0.00	*0.00*
	—	43	0.87	*1.91*	0.71	*1.57*	0.43	*0.95*	0.00	*0.00*
141	10.6	51	1.02	*2.24*	0.83	*1.84*	0.50	*1.10*	0.21	*0.46*
	11.6	56	1.12	*2.46*	0.92	*2.02*	0.55	*1.20*	0.23	*0.50*
142	—	—	—	—	—	—	—	—	—	—
	—	—	—	—	—	—	—	—	—	—
143	3.9	49	0.97	*2.14*	0.80	*1.76*	0.48	*1.05*	0.19	*0.43*
	4.5	55	1.11	*2.44*	0.91	*2.00*	0.54	*1.19*	0.22	*0.48*
144	11.9	78	1.56	*3.45*	1.28	*2.83*	0.86	*1.89*	0.58	*1.29*
	13.5	89	1.78	*3.92*	1.46	*3.21*	0.97	*2.15*	0.66	*1.47*
145	11.7	63	1.26	*2.77*	1.03	*2.27*	0.64	*1.42*	0.40	*0.88*
	13.2	71	1.41	*3.11*	1.16	*2.55*	0.72	*1.60*	0.45	*0.98*
146	10.2	67	1.34	*2.96*	1.10	*2.43*	0.70	*1.55*	0.45	*1.00*
	11.5	75	1.51	*3.33*	1.24	*2.73*	0.79	*1.74*	0.51	*1.12*
147	2.6	35	0.69	*1.52*	0.57	*1.25*	0.36	*0.79*	0.00	*0.00*
	2.8	37	0.74	*1.64*	0.61	*1.35*	0.39	*0.85*	0.00	*0.00*
148	0.1	38	0.76	*1.68*	0.63	*1.38*	0.38	*0.84*	0.00	*0.00*
	0.1	42	0.85	*1.87*	0.70	*1.54*	0.42	*0.93*	0.00	*0.00*

(Continued)

TABLE 27-1

Entry Number	Feed Name Description	Moisture Basis: A-F (as-fed) or M-F (moisture-free)	Dry Matter	Macrominerals							Microminerals						
				Calcium (Ca)	Phosphorus (P)	Sodium (Na)	Chlorine (Cl)	Magnesium (Mg)	Potassium (K)	Sulfur (S)	Cobalt (Co)	Copper (Cu)	Iodine (I)	Iron (Fe)	Manganese (Mn)	Selenium (Se)	Zinc (Zn)
			%	%	%	%	%	%	%	%	ppm or mg/kg	ppm or mg/kg	ppm or mg/kg	%	ppm or mg/kg	ppm or mg/kg	ppm or mg/kg
	SOYBEAN *Glycine max*																
130	-SILAGE	A-F	27	0.38	0.12	—	—	—	0.25	—	—	—	—	—	—	—	—
		M-F	100	1.39	0.46	—	—	—	0.93	—	—	—	—	—	—	—	—
131	-HAY, SUN-CURED	A-F	90	1.10	0.24	0.10	0.50	0.53	1.17	0.30	0.160	9.4	0.217	0.032	64.3	0.074	22.8
		M-F	100	1.23	0.26	0.12	0.55	0.59	1.30	0.34	0.178	10.4	0.242	0.036	71.8	0.083	25.5
132	-HULLS (SOYBEAN FLAKES)	A-F	92	0.42	0.13	0.05	—	—	0.94	—	0.110	16.3	—	0.030	29.4	—	22.0
		M-F	100	0.45	0.15	0.05	—	—	1.03	—	0.120	17.8	—	0.032	32.1	—	24.1
133	-STRAW	A-F	88	1.40	0.05	—	—	0.81	0.49	—	—	—	—	—	44.9	—	—
		M-F	100	1.59	0.06	—	—	0.92	0.56	—	—	—	—	—	51.1	—	—
134	-SEEDS, MEAL, MECH EXTD, 41% PROTEIN	A-F	90	0.27	0.63	0.24	0.07	0.25	1.71	0.33	0.180	18.0	—	0.016	32.3	—	—
		M-F	100	0.30	0.70	0.27	0.08	0.28	1.90	0.37	0.200	20.0	—	0.018	35.9	—	—
	SQUIRRELTAIL *Sitanion* spp																
135	-DORMANT	A-F	50	0.18	0.03	—	—	—	—	—	—	—	—	—	—	—	—
		M-F	100	0.37	0.06	—	—	—	—	—	—	—	—	—	—	—	—
	SWEET CLOVER, YELLOW *Melilotus officinalis*																
136	-HAY, SUN-CURED	A-F	89	1.14	0.19	0.08	0.33	0.44	1.30	0.40	—	8.9	—	0.014	96.6	—	—
		M-F	100	1.27	0.22	0.09	0.37	0.49	1.46	0.45	—	10.0	—	0.015	108.2	—	—
	TIMOTHY *Phleum pratense*																
137	-HAY, SUN-CURED, ALL ANALYSES	A-F	90	0.34	0.19	0.09	0.46	0.14	1.35	0.16	0.076	4.5	0.033	0.017	55.5	—	15.3
		M-F	100	0.38	0.21	0.10	0.51	0.15	1.50	0.18	0.084	5.0	0.037	0.018	61.7	—	17.0
138	-MIDBLOOM	A-F	30	0.08	0.08	0.06	0.19	0.04	0.51	0.04	—	3.3	—	0.005	57.4	—	—
		M-F	100	0.25	0.25	0.19	0.63	0.13	1.71	0.13	—	11.2	—	0.016	192.5	—	—
	TREFOIL, BIRDSFOOT *Lotus corniculatus*																
139	-HAY, SUN-CURED	A-F	92	1.54	0.25	0.81	—	0.47	1.66	—	0.102	8.5	—	0.021	13.8	—	71.1
		M-F	100	1.67	0.27	0.88	—	0.51	1.80	—	0.110	9.3	—	0.023	15.0	—	77.2
140	**TUMBLEWEED (RUSSIAN THISTLE)** *Salsola kali tenuifolia*	A-F	48	1.24	0.08	—	—	0.41	2.51	0.08	0.082	9.3	—	—	16.1	—	—
		M-F	100	2.56	0.17	—	—	0.85	5.18	0.17	0.170	19.2	—	—	33.3	—	—
	VETCH *Vicia* spp																
141	-HAY, SUN-CURED	A-F	91	1.11	0.25	0.22	0.69	0.26	2.16	0.31	0.242	9.6	0.448	0.037	49.6	0.075	24.4
		M-F	100	1.22	0.27	0.24	0.76	0.29	2.37	0.34	0.265	10.5	0.492	0.041	54.4	0.083	26.8
	WHEAT *Triticum aestivum*																
142	-IMMATURE	A-F	22	0.09	0.09	—	—	0.05	0.78	—	—	—	—	—	—	—	—
		M-F	100	0.42	0.40	—	—	0.21	3.50	—	—	—	—	—	—	—	—
143	-HAY, SUN-CURED	A-F	88	0.13	0.17	0.25	—	—	0.88	0.21	—	—	—	—	—	—	—
		M-F	100	0.15	0.19	0.28	—	—	1.00	0.24	—	—	—	—	—	—	—
144	-GRAIN, ALL ANALYSES	A-F	88	0.03	0.38	0.03	0.07	0.15	0.36	0.16	0.118	5.7	0.087	0.006	36.7	0.222	44.7
		M-F	100	0.04	0.43	0.03	0.08	0.17	0.41	0.18	0.134	6.5	0.098	0.006	41.7	0.253	50.8
145	-BRAN	A-F	89	0.11	1.26	0.03	0.05	0.52	1.46	0.22	0.101	12.7	0.065	0.011	109.9	0.375	103.9
		M-F	100	0.12	1.42	0.04	0.06	0.59	1.64	0.25	0.113	14.3	0.073	0.012	123.4	0.422	116.7
146	-GRAIN SCREENINGS	A-F	89	0.15	0.36	—	—	—	—	—	—	—	—	—	14.4	—	—
		M-F	100	0.17	0.40	—	—	—	—	—	—	—	—	—	16.2	—	—
147	-CHAFF	A-F	93	0.19	0.09	—	—	—	0.52	—	—	—	—	—	—	—	—
		M-F	100	0.21	0.09	—	—	—	0.56	—	—	—	—	—	—	—	—
148	-STRAW	A-F	90	0.19	0.07	0.13	0.29	0.11	1.05	0.17	0.041	2.9	—	0.016	40.4	—	—
		M-F	100	0.21	0.07	0.14	0.32	0.12	1.16	0.19	0.045	3.2	—	0.017	45.0	—	—

(Continued)

Entry Number	Fat-soluble Vitamins					Water-soluble Vitamins									
	A (1 mg Carotene = 1667 IU Vit A)	Carotene (Provitamin A)	D	E (α-tocopherol)	K	B_{12}	Biotin	Choline	Folic Acid (Folacin)	Niacin (Nicotinic Acid)	Pantothenic Acid	Pyridoxine (B_6)	Riboflavin (B_2)	Thiamin (B_1)
	IU/g	ppm or mg/kg	IU/kg	ppm or mg/kg	ppm or mg/kg	ppb or mcg/kg	ppm or mg/kg	ppm or mg/kg	ppm or mg/kg	ppm or mg/kg	ppm or mg/kg	ppm or mg/kg	ppm or mg/kg	ppm or mg/kg
130	**50.7** 186.9	**30.4** 112.1	— —	— —	— —	— —	— —	— —	— —	— —	— —	— —	— —	— —
131	**49.0** 54.6	**29.4** 32.8	**949** 1059	**160.9** 179.5	— —	— —	**0.32** 0.35	**1341** 1496	**3.47** 3.87	**40** 44	**28.8** 32.1	— —	**16.7** 18.7	**2.8** 3.1
132	— —	— —	— —	— —	— —	— —	— —	**632** 690	— —	— —	— —	— —	— —	**1.6** 1.8
133	— —	— —	— —	— —	— —	— —	— —	— —	— —	— —	— —	— —	— —	— —
134	**0.3** 0.3	**0.2** 0.2	— —	**6.6** 7.3	— —	— —	**0.30** 0.33	**2673** 2940	**6.60** 7.73	**30** 34	**14.9** 16.6	— —	**3.5** 3.9	**4.0** 4.9
135	— —	— —	— —	— —	— —	— —	— —	— —	— —	— —	— —	— —	— —	— —
136	**68.2** 76.4	**40.9** 45.8	**1673** 1874	— —	— —	— —	— —	— —	— —	— —	— —	— —	— —	— —
137	**34.5** 38.3	**20.7** 23.0	**1925** 2138	**56.8** 63.1	— —	— —	**0.06** 0.07	**730** 811	**2.06** 2.29	**26** 29	**7.1** 7.9	— —	**11.2** 12.4	**1.5** 1.7
138	— —	— —	— —	— —	— —	— —	— —	— —	— —	— —	— —	— —	— —	— —
139	**331.2** 359.5	**198.7** 215.7	**1422** 1544	— —	— —	— —	— —	— —	— —	— —	— —	— —	**14.8** 16.1	**6.3** 6.6
140	**54.7** 113.0	**32.8** 67.8	— —	— —	— —	— —	— —	— —	— —	— —	— —	— —	— —	— —
141	**662.0** 726.4	**397.1** 435.8	— —	**231.8** 254.4	— —	— —	**0.32** 0.35	**1363** 1496	**3.53** 3.87	**40** 44	**29.3** 32.1	— —	**17.0** 18.7	**2.9** 3.1
142	**192.5** 866.9	**115.5** 520.1	— —	— —	— —	— —	— —	— —	— —	**13** 57	**4.7** 21.2	— —	**6.1** 27.6	— —
143	**125.0** 42.3	**75.0** 85.4	**1356** 1544	— —	— —	— —	— —	— —	— —	— —	— —	— —	**14.9** 17.0	— —
144	**16.9** 19.3	**10.2** 11.6	— —	**13.7** 15.6	— —	**0.9** 1.0	**0.10** 0.11	**1005** 1142	**0.40** 0.45	**57** 65	**9.7** 11.6	**5.00** 5.68	**1.4** 1.6	**4.2** 4.8
145	**4.4** 4.9	**2.6** 2.9	— —	**20.9** 23.4	— —	— —	**0.57** 0.64	**1880** 2113	**1.23** 1.39	**274** 308	**31.7** 35.7	**10.18** 11.44	**4.8** 5.4	**6.5** 7.3
146	— —	— —	— —	— —	— —	— —	— —	— —	— —	— —	— —	— —	— —	**6.4** 7.2
147	— —	— —	— —	— —	— —	— —	— —	— —	— —	— —	— —	— —	— —	— —
148	**3.3** 3.7	**2.0** 2.2	**595** 662	— —	— —	— —	— —	— —	— —	— —	— —	— —	**2.2** 2.4	— —

(Continued)

COMPOSITION OF FEEDS

TABLE 27-1

Entry Number	Feed Name Description	International Feed Number[1]	Moisture Basis: A-F (as-fed) or M-F (moisture-free)	Chemical Analysis								
				Dry Matter	Ash	Crude Fiber	Cell Walls or NDF	Acid Detergent Fiber	Lignin	Ether Extract (Fat)	N-Free Extract	Crude Protein
				%	%	%	%	%	%	%	%	%
	WHEATGRASS, CRESTED *Agropyron cristatum*											
149	-EARLY BLOOM	2-05-422	A-F	37	2.7	12.2	—	—	—	0.6	18.7	2.7
			M-F	100	7.3	33.1	—	—	—	1.6	50.7	7.3
150	-FULL BLOOM	2-05-424	A-F	45	4.2	13.6	—	—	—	1.6	21.2	4.4
			M-F	100	9.3	30.3	—	—	—	3.6	47.0	9.8
151	-OVERRIPE	2-05-428	A-F	80	3.3	32.2	—	—	—	1.0	41.0	2.5
			M-F	100	4.1	40.3	—	—	—	1.2	51.3	3.1
152	-HAY, SUN-CURED	1-05-418	A-F	93	6.8	31.9	—	—	—	1.8	43.8	8.9
			M-F	100	7.3	34.2	—	—	—	2.0	47.0	9.5
	YELLOW BRUSH (RABBIT BRUSH) *Chrysothamnus stenophyllus*											
153	-BROWSE, DORMANT	2-07-997	A-F	70	—	—	—	—	—	—	—	4.6
			M-F	100	—	—	—	—	—	—	—	6.6

(Continued)

Entry Number	Digestible Protein	TDN	Digestible Energy		Metabolizable Energy		Net Energy			
	Ruminant	Ruminant	Ruminant		Ruminant		Ruminant NE_m		Ruminant NE_g	
	%	%	Mcal		Mcal		Mcal		Mcal	
			lb	*kg*	lb	*kg*	lb	*kg*	lb	*kg*
149	—	21	0.42	*0.93*	0.35	*0.76*	0.21	*0.46*	0.09	*0.20*
	—	57	1.14	*2.52*	0.94	*2.07*	0.56	*1.24*	0.25	*0.55*
150	—	27	0.55	*1.21*	0.45	*0.99*	0.27	*0.60*	0.14	*0.31*
	—	61	1.22	*2.69*	1.00	*2.20*	0.60	*1.33*	0.31	*0.69*
151	—	39	0.78	*1.72*	0.64	*1.41*	0.38	*0.84*	0.08	*0.18*
	—	49	0.97	*2.15*	0.80	*1.76*	0.47	*1.05*	0.10	*0.22*
152	5.6	56	1.18	*2.61*	0.97	*2.14*	0.59	*1.30*	0.32	*0.72*
	6.1	60	1.27	*2.80*	1.04	*2.30*	0.64	*1.40*	0.35	*0.77*
153	2.2	35	0.70	*1.54*	0.54	*1.19*	0.31	*0.70*	0.03	*0.07*
	3.1	50	1.00	*2.20*	0.77	*1.70*	0.45	*1.00*	0.05	*0.10*

TABLE 27-1

Entry Number	Feed Name Description	Moisture Basis: A-F (as-fed) or M-F (moisture-free)	Dry Matter	Macrominerals							Microminerals						
				Calcium (Ca)	Phosphorus (P)	Sodium (Na)	Chlorine (Cl)	Magnesium (Mg)	Potassium (K)	Sulfur (S)	Cobalt (Co)	Copper (Cu)	Iodine (I)	Iron (Fe)	Manganese (Mn)	Selenium (Se)	Zinc (Zn)
			%	%	%	%	%	%	%	%	ppm or mg/kg	ppm or mg/kg	ppm or mg/kg	%	ppm or mg/kg	ppm or mg/kg	ppm or mg/kg
	WHEATGRASS, CRESTED *Agropyron cristatum*																
149	-EARLY BLOOM	**A-F**	**37**	—	—	—	—	—	—	—	—	—	—	—	—	—	—
		M-F	100	—	—	—	—	—	—	—	—	—	—	—	—	—	—
150	-FULL BLOOM	**A-F**	**45**	**0.18**	**0.13**	—	—	—	—	—	—	—	—	—	—	—	—
		M-F	100	0.39	0.28	—	—	—	—	—	—	—	—	—	—	—	—
151	-OVERRIPE	**A-F**	**80**	**0.22**	**0.06**	—	—	—	—	—	**0.199**	**6.7**	—	—	**42.3**	—	—
		M-F	100	0.27	0.07	—	—	—	—	—	0.249	8.4	—	—	52.9	—	—
152	-HAY, SUN-CURED	**A-F**	**93**	**0.25**	**0.14**	—	—	—	—	—	**0.222**	—	—	—	—	—	—
		M-F	100	0.27	0.16	—	—	—	—	—	0.238	—	—	—	—	—	—
	YELLOW BRUSH (RABBIT BRUSH) *Chrysothamnus stenophyllus*																
153	-BROWSE, DORMANT	**A-F**	**70**	**1.33**	**0.07**	—	—	—	—	—	—	—	—	—	—	—	—
		M-F	100	1.90	0.10	—	—	—	—	—	—	—	—	—	—	—	—

[1]The first digit is the feed class, coded as follows: (1) dry forages and roughages; (2) pasture, range plants, and forages fed green; (3) silages; (4) energy feeds; and (5) protein supplements.

(Continued)

Entry Number	Fat-soluble Vitamins					Water-soluble Vitamins								
	A (1 mg Carotene = 1667 IU Vit A)	Carotene (Provitamin A)	D	E (α-tocoph-erol)	K	B_{12}	Biotin	Choline	Folic Acid (Folacin)	Niacin (Nicotinic Acid)	Panto-thenic Acid	Pyri-doxine (B_6)	Ribo-flavin (B_2)	Thiamin (B_1)
	IU/g	ppm or mg/kg	IU/kg	ppm or mg/kg	ppm or mg/kg	ppb or mcg/kg	ppm or mg/kg	ppm or mg/kg	ppm or mg/kg	ppm or mg/kg	ppm or mg/kg	ppm or mg/kg	ppm or mg/kg	ppm or mg/kg
149	—	—	—	—	—	—	—	—	—	—	—	—	—	—
	—	—	—	—	—	—	—	—	—	—	—	—	—	—
150	—	—	—	—	—	—	—	—	—	—	—	—	—	—
	256.9	154.1	—	—	—	—	—	—	—	—	—	—	—	—
151	0.3	0.2	—	—	—	—	—	—	—	—	—	—	—	—
	0.4	0.2	—	—	—	—	—	—	—	—	—	—	—	—
152	120.9	72.5	—	—	—	—	—	—	—	—	—	—	—	—
	129.8	77.9	—	—	—	—	—	—	—	—	—	—	—	—
153	—	—	—	—	—	—	—	—	—	—	—	—	—	—
	—	—	—	—	—	—	—	—	—	—	—	—	—	—

COMPOSITION OF MINERAL SUPPLEMENTS,

Entry Number	Feed Name Description	Interna-tional Feed Number	Moisture Basis: A-F (as-fed) or M-F (moisture-free)	Chemical Analysis						Digestible Protein		
				Dry Matter	Ash	Crude Fiber	Ether Extract (Fat)	N-Free Extract	Crude Protein (6.25 × N)	Ruminant	Non-ruminant	Horse
				(%)	(%)	(%)	(%)	(%)	(%)	(%)	(%)	(%)
1	AMMONIUM CHLORIDE	6-08-814	A-F	—	—	—	—	—	—	—	—	—
			M-F	100	—	—	—	—	160.0	—	—	—
2	AMMONIUM PHOSPHATE, MONOBASIC	6-09-338	A-F	97	34.5	—	—	—	68.8	—	—	—
			M-F	100	35.6	—	—	—	70.9	—	—	—
3	AMMONIUM PHOSPHATE, DIBASIC	6-00-370	A-F	97	34.5	—	—	—	112.4	—	—	—
			M-F	100	35.6	—	—	—	115.9	—	—	—
4	AMMONIUM-POLYPHOSPHATE SOLUTION FROM DEFLUORINATED PHOSPHORIC ACID	6-08-042	A-F	60	—	—	—	—	62.5	—	—	—
			M-F	100	—	—	—	—	104.2	—	—	—
5	AMMONIUM POLYPHOSPHATE SOLUTION FROM FURNACE PHOSPHORIC ACID	6-26-401	A-F	—	—	—	—	—	68.7	—	—	—
			M-F	100	—	—	—	—	—	—	—	—
6	BONE, BLACK, SPENT	6-00-404	A-F	90	—	—	—	—	8.5	—	—	—
			M-F	100	—	—	—	—	9.4	—	—	—
7	BONE, CHARCOAL	6-00-402	A-F	90	—	—	—	—	8.5	—	—	—
			M-F	100	—	—	—	—	9.4	—	—	—
8	BONE MEAL	6-00-397	A-F	95	—	—	9.6	—	17.8	—	—	—
			M-F	100	—	—	10.2	—	18.8	—	—	—
9	BONE MEAL, STEAMED	6-00-400	A-F	97	77.0	1.4	11.3	0.0	12.8	8.7	—	—
			M-F	100	79.3	1.4	11.6	0.0	13.2	9.0	—	—
10	CALCIUM CARBONATE	6-01-069	A-F	99	—	—	—	—	—	—	—	—
			M-F	100	—	—	—	—	—	—	—	—
11	CALCIUM PHOSPHATE, MONOBASIC, FROM DEFLUORINATED PHOSPHORIC ACID	6-01-082	A-F	97	—	—	—	—	—	—	—	—
			M-F	100	—	—	—	—	—	—	—	—
12	CALCIUM PHOSPHATE, MONOBASIC, FROM FURNACE PHOSPHORIC ACID	6-26-334	A-F	—	—	—	—	—	—	—	—	—
			M-F	100	—	—	—	—	—	—	—	—
13	CALCIUM PHOSPHATE, DIBASIC, FROM DEFLUORINATED PHOSPHORIC ACID	6-01-080	A-F	97	91.0	—	—	—	—	—	—	—
			M-F	100	93.8	—	—	—	—	—	—	—
14	CALCIUM PHOSPHATE, DIBASIC, FROM FURNACE PHOSPHORIC ACID	6-26-335	A-F	97	—	—	—	—	—	—	—	—
			M-F	100	93.8	—	—	—	—	—	—	—
15	CALCIUM PHOSPHATE, TRIBASIC, FROM FURNACE PHOSPHORIC ACID	6-01-084	A-F	—	—	—	—	—	—	—	—	—
			M-F	100	—	—	—	—	—	—	—	—
16	CALCIUM SULFATE ANHYDROUS (GYPSUM)	6-01-087	A-F	85	—	—	—	—	—	—	—	—
			M-F	100	—	—	—	—	—	—	—	—
17	COBALT CARBONATE	6-01-566	A-F	81	—	—	—	—	—	—	—	—
			M-F	100	—	—	—	—	—	—	—	—
18	COBALT OXIDE	6-01-560	A-F	—	—	—	—	—	—	—	—	—
			M-F	100	—	—	—	—	—	—	—	—
19	COBALT SULFATE	6-01-564	A-F	55	—	—	—	—	—	—	—	—
			M-F	100	—	—	—	—	—	—	—	—
20	COLLOIDAL CLAY (SOFT ROCK PHOSPHATE)	6-03-947	A-F	—	—	—	—	—	—	—	—	—
			M-F	100	—	—	—	—	—	—	—	—
21	COPPER (CUPRIC) CHLORIDE	6-01-705	A-F	79	—	—	—	—	—	—	—	—
			M-F	100	—	—	—	—	—	—	—	—
22	COPPER (CUPRIC) OXIDE	6-01-711	A-F	—	—	—	—	—	—	—	—	—
			M-F	100	—	—	—	—	—	—	—	—
23	COPPER (CUPRIC) SULFATE	6-01-720	A-F	64	—	—	—	—	—	—	—	—
			M-F	100	—	—	—	—	—	—	—	—
24	CURACAO PHOSPHATE	6-05-586	A-F	—	—	—	—	—	—	—	—	—
			M-F	100	100.0	—	—	—	—	—	—	—
25	DEFLUORINATED PHOSPHATE FROM PHOSPHORIC ACID	6-26-336	A-F	—	—	—	—	—	—	—	—	—
			M-F	100	—	—	—	—	—	—	—	—
26	DIAMMONIUM PHOSPHATE	6-00-370	A-F	97	34.5	—	—	—	112.4	—	—	—
			M-F	100	35.6	—	—	—	115.9	—	—	—
27	DICALCIUM PHOSPHATE (CALCIUM PHOSPHATE, DIBASIC)	6-00-080	A-F	96	90.0	—	—	—	—	—	—	—
			M-F	100	93.8	—	—	—	—	—	—	—

27-2

DATA EXPRESSED **AS-FED** AND **MOISTURE-FREE**

Entry Number	Macrominerals							Microminerals						
	Calcium (Ca)	Phosphorus (P)	Sodium (Na)	Chlorine (Cl)	Magnesium (Mg)	Potassium (K)	Sulfur (S)	Cobalt (Co)	Copper (Cu)	Iodine (I)	Iron (Fe)	Manganese (Mn)	Selenium (Se)	Zinc (Zn)
	(%)	(%)	(%)	(%)	(%)	(%)	(%)	(ppm or mg/kg)	(ppm or mg/kg)	(ppm or mg/kg)	(%)	(ppm or mg/kg)	(ppm or mg/kg)	(ppm or mg/kg)
1	—	—	—	—	—	—	—	—	—	—	—	—	—	—
	—	—	—	66.28	—	—	—	—	—	—	—	—	—	—
2	0.50	24.00	0.06	—	0.45	—	0.70	—	80	—	1.200	400	—	300
	0.52	24.74	0.06	—	0.46	—	0.70	—	82	—	1.237	412	—	309
3	0.57	20.00	0.04	—	0.45	0.01	2.50	—	91	—	1.200	400	—	342
	0.59	20.60	0.04	—	0.46	0.01	2.60	—	94	—	1.237	412	—	353
4	0.10	14.50	—	—	—	—	—	—	—	—	—	—	—	—
	0.17	24.20	—	—	—	—	—	—	—	—	—	—	—	—
5	—	16.00	—	—	—	—	—	—	—	—	—	—	—	—
	—	—	—	—	—	—	—	—	—	—	—	—	—	—
6	27.10	12.73	—	—	0.53	0.14	—	—	—	—	—	—	—	—
	30.11	14.14	—	—	0.59	0.16	—	—	—	—	—	—	—	—
7	27.10	12.73	—	—	0.53	0.14	—	—	—	—	—	—	—	—
	30.11	14.14	—	—	0.59	0.16	—	—	—	—	—	—	—	—
8	25.95	12.42	—	—	—	—	—	—	—	—	—	—	—	—
	27.32	13.07	—	—	—	—	—	—	—	—	—	—	—	—
9	29.82	12.49	5.53	—	0.32	0.18	2.44	—	11	33,196	0.085	22	—	126
	30.71	12.86	5.69	—	0.33	0.19	2.51	—	11	34,188	0.088	23	—	130
10	37.62	0.04	0.02	0.04	0.50	0.06	0.09	—	24	—	0.034	277	—	—
	38.00	0.04	0.02	0.04	0.50	0.06	0.09	—	24	—	0.034	280	—	—
11	15.91	20.95	—	—	—	—	—	—	—	—	0.002	—	—	—
	16.40	21.60	—	—	—	—	—	—	—	—	0.002	—	—	—
12	—	—	—	—	—	—	—	—	—	—	—	—	—	—
	22.00	23.00	—	0.00	—	—	—	—	80	—	0.002	—	—	220
13	21.30	18.70	—	—	0.60	0.07	—	—	—	—	—	304	—	—
	22.00	19.30	—	—	0.62	0.07	—	—	—	—	—	313	—	—
14	26.30	18.70	—	—	0.62	0.07	—	—	—	—	—	304	—	—
	27.10	19.30	—	—	0.62	0.07	—	—	—	—	—	313	—	—
15	38.00	19.50	—	—	—	—	—	—	—	—	—	—	—	—
	—	—	—	—	—	—	—	—	—	—	—	—	—	—
16	22.02	0.01	—	—	2.21	—	20.01	—	—	—	0.171	—	—	—
	25.90	0.01	—	—	2.61	—	23.54	—	—	—	0.201	—	—	—
17	—	—	—	—	—	—	—	460,000	—	—	—	—	—	—
	—	—	—	—	—	—	—	570,000	—	—	—	—	—	—
18	—	—	—	—	—	—	—	—	—	—	—	—	—	—
	—	—	—	—	—	—	0.20	720,000	—	—	0.050	—	—	—
19	—	—	—	0.00	0.04	—	11.40	210,000	—	—	0.001	20	—	—
	—	—	—	0.00	0.07	—	20.65	380,000	—	—	0.002	36	—	—
20	17.00	9.00	—	—	—	—	—	—	—	—	—	—	—	—
	—	—	—	—	—	—	—	—	—	—	—	—	—	—
21	—	—	—	41.64	—	—	0.04	—	373,000	—	0.005	—	—	—
	—	—	—	52.71	—	—	0.04	—	472,000	—	0.006	—	—	—
22	—	—	—	—	—	—	—	—	—	—	—	—	—	—
	—	—	—	0.01	—	—	0.13	—	799,000	—	—	—	—	—
23	—	—	—	0.00	—	—	12.84	—	254,000	—	0.003	—	—	—
	—	—	—	0.00	—	—	20.06	—	396,000	—	0.005	—	—	—
24	—	—	—	—	—	—	—	—	—	—	—	—	—	—
	34.00	15.00	—	—	—	—	—	—	—	—	—	—	—	—
25	32.00	18.00	—	—	—	—	—	—	—	—	—	—	—	—
	—	—	—	—	—	—	—	—	—	—	—	—	—	—
26	0.57	20.00	0.04	—	0.45	0.01	2.50	—	91	—	1.200	400	—	342
	0.59	20.60	0.04	—	0.46	0.01	2.60	—	94	—	1.237	412	—	353
27	26.00	18.50	—	—	0.60	0.07	—	—	—	—	—	300	—	—
	27.10	19.30	—	—	0.62	0.07	—	—	—	—	—	313	—	—

(Continued)

TABLE 27-2

Entry Number	Feed Name Description	International Feed Number	Moisture Basis: A-F (as-fed) or M-F (moisture-free)	Chemical Analysis: Dry Matter	Ash	Crude Fiber	Ether Extract (Fat)	N-Free Extract	Crude Protein (6.25 × N)	Digestible Protein: Ruminant	Non-ruminant	Horse
				(%)	(%)	(%)	(%)	(%)	(%)	(%)	(%)	(%)
28	DISODIUM PHOSPHATE	6-04-286	A-F	—	—	—	—	—	—	—	—	—
			M-F	100	—	—	—	—	—	—	—	—
29	DOLOMITE LIMESTONE (LIMESTONE, MAGNESIUM)	6-02-633	A-F	99	—	—	—	—	—	—	—	—
			M-F	100	—	—	—	—	—	—	—	—
30	IRON (FERRIC) OXIDE	6-02-431	A-F	—	—	—	—	—	—	—	—	—
			M-F	100	—	—	—	—	—	—	—	—
31	IRON (FERROUS) CHLORIDE	6-01-865	A-F	64	—	—	—	—	—	—	—	—
			M-F	100	—	—	—	—	—	—	—	—
32	IRON (FERROUS) SULFATE	6-01-869	A-F	55	—	—	—	—	—	—	—	—
			M-F	100	—	—	—	—	—	—	—	—
33	LIMESTONE, GROUND	6-02-632	A-F	99	95.9	—	—	—	—	—	—	—
			M-F	100	96.9	—	—	—	—	—	—	—
34	LIMESTONE, MAGNESIUM (DOLOMITE)	6-02-633	A-F	99	—	—	—	—	—	—	—	—
			M-F	100	—	—	—	—	—	—	—	—
35	MAGNESIUM CARBONATE	6-02-754	A-F	81	—	—	—	—	—	—	—	—
			M-F	100	—	—	—	—	—	—	—	—
36	MAGNESIUM OXIDE	6-02-756	A-F	—	—	—	—	—	—	—	—	—
			M-F	100	—	—	—	—	—	—	—	—
37	MAGNESIUM SULFATE (EPSOM SALTS)	6-02-758	A-F	49	—	—	—	—	—	—	—	—
			M-F	100	—	—	—	—	—	—	—	—
38	MANGANESE CARBONATE	6-03-036	A-F	—	—	—	—	—	—	—	—	—
			M-F	100	—	—	—	—	—	—	—	—
39	MANGANESE CHLORIDE	6-03-038	A-F	64	—	—	—	—	—	—	—	—
			M-F	100	—	—	—	—	—	—	—	—
40	MANGANESE DIOXIDE	6-03-042	A-F	—	—	—	—	—	—	—	—	—
			M-F	100	—	—	—	—	—	—	—	—
41	MONOAMMONIUM PHOSPHATE	6-09-338	A-F	97	34.5	—	—	—	68.8	—	—	—
			M-F	100	35.6	—	—	—	70.9	—	—	—
	MONOCALCIUM PHOSPHATE (SEE CALCIUM PHOSPHATE, MONOBASIC)											
42	MONOSODIUM PHOSPHATE, ANHYDROUS	6-04-288	A-F	87	—	—	—	—	—	—	—	—
			M-F	100	—	—	—	—	—	—	—	—
43	ORGANIC IODIDE (ETHYLENEDIAMINE DIHYDROIODIDE)	6-01-842	A-F	—	—	—	—	—	—	—	—	—
			M-F	100	—	—	—	—	—	—	—	—
44	OYSTERSHELL, GROUND (FLOUR)	6-03-481	A-F	99	89.7	—	—	—	1.0	—	—	—
			M-F	100	90.6	—	—	—	1.0	—	—	—
45	PHOSPHATE, ROCK, GROUND (RAW)	6-03-945	A-F	—	—	—	—	—	—	—	—	—
			M-F	100	—	—	—	—	—	—	—	—
46	PHOSPHATE ROCK, DEFLUORINATED	6-01-780	A-F	—	—	—	—	—	—	—	—	—
			M-F	100	100.0	—	—	—	—	—	—	—
47	PHOSPHATE ROCK, LOW FLUORINE	6-03-946	A-F	—	—	—	—	—	—	—	—	—
			M-F	100	—	—	—	—	—	—	—	—
48	PHOSPHATE SOFT ROCK (COLLOIDAL CLAY)	6-03-947	A-F	—	—	—	—	—	—	—	—	—
			M-F	100	—	—	—	—	—	—	—	—
49	PHOSPHORIC ACID, DEFLUORINATED, WET PROCESS	6-20-534	A-F	—	—	—	—	—	—	—	—	—
			M-F	100	—	—	—	—	—	—	—	—
50	PHOSPHORIC ACID, FEED GRADE (ORTHO)	6-03-707	A-F	75	—	—	—	—	—	—	—	—
			M-F	100	—	—	—	—	—	—	—	—
51	POTASSIUM CHLORIDE	6-03-755	A-F	—	—	—	—	—	—	—	—	—
			M-F	100	—	—	—	—	—	—	—	—
52	POTASSIUM IODIDE	6-03-759	A-F	92	—	—	—	—	—	—	—	—
			M-F	100	—	—	—	—	—	—	—	—
53	POTASSIUM SULFATE	6-08-098	A-F	—	—	—	—	—	—	—	—	—
			M-F	100	—	—	—	—	—	—	—	—

(Continued)

Entry Number	Macrominerals							Microminerals						
	Calcium (Ca)	Phosphorus (P)	Sodium (Na)	Chlorine (Cl)	Magnesium (Mg)	Potassium (K)	Sulfur (S)	Cobalt (Co)	Copper (Cu)	Iodine (I)	Iron (Fe)	Manganese (Mn)	Selenium (Se)	Zinc (Zn)
	(%)	(%)	(%)	(%)	(%)	(%)	(%)	(ppm or mg/kg)	(ppm or mg/kg)	(ppm or mg/kg)	(%)	(ppm or mg/kg)	(ppm or mg/kg)	(ppm or mg/kg)
28	— —	21.50 —	32.00 —	0.00 —	— —	— —	— —	— —	— —	— —	0.001 —	— —	— —	— —
29	22.08 22.30	0.04 0.04	— —	0.12 0.12	9.87 9.99	0.36 0.36	— —	— —	— —	— —	0.076 0.077	— —	— —	— —
30	— —	— —	— —	— —	— —	— —	— 0.13	— —	— —	— —	— 69.940	— —	— —	— —
31	— —	— —	— —	35.78 55.91	— —	— —	0.03 0.05	— —	— —	— —	28.090 44.628	— —	— —	— —
32	— —	— —	— —	0.00 0.00	0.05 0.09	— —	11.60 21.10	— —	— —	— —	20.080 36.709	— —	— —	— —
33	33.66 34.00	0.02 0.02	0.06 0.06	0.03 0.03	2.04 2.06	0.11 0.12	0.04 0.04	— —	— —	— —	0.347 0.350	269 270	— —	— —
34	22.08 22.30	0.04 0.04	— —	0.12 0.12	9.89 9.99	0.36 0.36	— —	— —	— —	— —	0.076 0.077	— —	— —	— —
35	0.02 0.02	— —	— —	0.00 0.00	24.96 30.81	— —	— —	— —	— —	— —	0.002 0.002	— —	— —	— —
36	— 0.05	— —	— 0.50	— 0.01	— 60.31	— 0.01	— —	— —	— —	— —	— 0.010	— —	— —	— —
37	0.02 0.04	— —	0.00 0.01	— —	9.89 20.18	0.00 0.01	13.02 26.58	— —	— —	— —	0.000 0.001	— —	— —	— —
38	— —	— —	— —	— 0.02	— —	— —	— 0.07	— —	— —	— —	— 0.002	— 43	— —	— —
39	— —	— —	— —	36.04 56.32	— —	— —	0.07 0.11	— —	— —	— —	0.001 0.001	28 44	— —	— —
40	— —	— —	— —	— 0.01	— —	— —	— 0.02	— —	— —	— —	— 0.050	— 632,000	— —	— —
41	0.50 0.52	24.00 24.74	0.06 0.06	— —	0.45 0.46	— —	0.70 0.70	— —	80 82	— —	1.200 1.237	400 412	— —	300 309
42	— —	22.18 25.50	16.53 19.00	— —	— —	— —	— —	— —	— —	— —	0.001 0.001	— —	— —	— —
43	— —	— —	— —	— —	— —	— —	— —	— —	— —	800,000 —	— —	— —	— —	— —
44	37.62 38.00	0.07 0.07	0.21 0.21	0.01 0.01	0.30 0.30	0.10 0.10	— —	— —	— —	— —	0.284 0.287	133 134	— —	— —
45	35.00 —	13.00 —	— —	— —	— —	— —	— —	— —	— —	— —	— —	— —	— —	— —
46	— 32.00	— 16.25	— 4.00	— —	— —	— 0.09	— —	— —	— 22	— —	— 0.920	— 220	— —	— 44
47	36.00 —	14.00 —	— —	— —	— —	— —	— —	— —	— —	— —	— —	— —	— —	— —
48	17.00 —	9.00 —	— —	— —	— —	— —	— —	— —	— —	— —	— —	— —	— —	— —
49	0.20 —	23.70 —	— —	— —	— —	— —	— —	— —	— —	— —	— —	— —	— —	— —
50	— —	23.70 31.60	0.01 0.03	— —	— —	0.01 0.01	0.05 0.07	— —	— —	— —	0.002 0.003	— —	— —	— —
51	— —	— —	0.01 —	47.30 —	— —	50.50 —	— —	— —	— —	— —	0.000 —	— —	— —	— —
52	— —	— —	0.01 0.01	0.01 0.01	— —	21.67 23.56	— —	— —	— —	699,200 760,000	— —	— —	— —	— —
53	0.15 —	— —	0.09 —	1.52 —	0.60 —	43.10 —	17.70 —	— —	3 —	— —	0.070 —	9 —	— —	4 —

(Continued)

TABLE 27-2

Entry Number	Feed Name Description	International Feed Number	Moisture Basis: A-F (as-fed) or M-F (moisture-free)	Chemical Analysis						Digestible Protein		
				Dry Matter	Ash	Crude Fiber	Ether Extract (Fat)	N-Free Extract	Crude Protein (6.25 × N)	Ruminant	Non-ruminant	Horse
				(%)	(%)	(%)	(%)	(%)	(%)	(%)	(%)	(%)
54	RAW ROCK PHOSPHATE	6-03-945	A-F	—	—	—	—	—	—	—	—	—
			M-F	100	—	—	—	—	—	—	—	—
55	ROCK PHOSPHATE (PHOSPHATE ROCK)		A-F	—	—	—	—	—	—	—	—	—
			M-F	100	—	—	—	—	—	—	—	—
56	SODIUM BICARBONATE	6-04-273	A-F	—	—	—	—	—	—	—	—	—
			M-F	100	—	—	—	—	—	—	—	—
57	SODIUM CHLORIDE	6-04-152	A-F	—	—	—	—	—	—	—	—	—
			M-F	100	—	—	—	—	—	—	—	—
58	SODIUM IODIDE	6-04-279	A-F	—	—	—	—	—	—	—	—	—
			M-F	100	—	—	—	—	—	—	—	—
59	SODIUM PHOSPHATE, MONOBASIC, FROM FURNACE PHOSPHORIC ACID, ANHYDROUS	6-04-287	A-F	87	—	—	—	—	—	—	—	—
			M-F	100	—	—	—	—	—	—	—	—
60	SODIUM PHOSPHATE, DIBASIC, FROM FURNACE PHOSPHORIC ACID	6-04-286	A-F	—	—	—	—	—	—	—	—	—
			M-F	100	—	—	—	—	—	—	—	—
61	SODIUM TRIPOLYPHOSPHATE	6-08-076	A-F	96	—	—	—	—	—	—	—	—
			M-F	100	—	—	—	—	—	—	—	—
62	SODIUM SULFATE	6-04-292	A-F	44	—	—	—	—	—	—	—	—
			M-F	100	—	—	—	—	—	—	—	—
63	ZINC CARBONATE	6-05-549	A-F	—	—	—	—	—	—	—	—	—
			M-F	100	—	—	—	—	—	—	—	—
64	ZINC CHLORIDE	6-05-551	A-F	—	—	—	—	—	—	—	—	—
			M-F	100	—	—	—	—	—	—	—	—
65	ZINC OXIDE	6-05-553	A-F	—	—	—	—	—	—	—	—	—
			M-F	100	—	—	—	—	—	—	—	—
66	ZINC SULFATE	6-05-555	A-F	56	—	—	—	—	—	—	—	—
			M-F	100	—	—	—	—	—	—	—	—

(Continued)

	Macrominerals							Microminerals						
Entry Number	Calcium (Ca)	Phosphorus (P)	Sodium (Na)	Chlorine (Cl)	Magnesium (Mg)	Potassium (K)	Sulfur (S)	Cobalt (Co)	Copper (Cu)	Iodine (I)	Iron (Fe)	Manganese (Mn)	Selenium (Se)	Zinc (Zn)
	(%)	(%)	(%)	(%)	(%)	(%)	(%)	(ppm or mg/kg)	(ppm or mg/kg)	(ppm or mg/kg)	(%)	(ppm or mg/kg)	(ppm or mg/kg)	(ppm or mg/kg)
54	35.00	13.00	—	—	—	—	—	—	—	—	—	—	—	—
	—	—	—	—	—	—	—	—	—	—	—	—	—	—
55	—	—	—	—	—	—	—	—	—	—	—	—	—	—
	—	—	—	—	—	—	—	—	—	—	—	—	—	—
56	—	—	—	—	—	—	—	—	—	—	—	—	—	—
	—	—	27.36	—	—	0.01	—	—	—	—	0.001	—	—	—
57	—	—	39.34	60.66	—	—	—	—	—	—	—	—	—	—
	—	—	—	—	—	—	—	—	—	—	—	—	—	—
58	—	—	—	—	—	—	—	—	—	—	—	—	—	—
	—	—	15.33	0.01	—	—	—	—	—	847,000	0.001	—	—	—
59	—	22.18	16.53	—	—	—	—	—	—	—	0.001	—	—	—
	—	25.50	19.00	—	—	—	—	—	—	—	0.001	—	—	—
60	—	—	—	—	—	—	—	—	—	—	—	—	—	—
	—	21.50	32.00	0.00	—	—	—	—	—	—	0.001	—	—	—
61	—	24.00	28.80	—	—	—	—	—	—	—	0.004	—	—	—
	—	25.00	30.00	—	—	—	—	—	—	—	0.004	—	—	—
62	—	—	14.27	0.00	—	—	9.96	—	—	—	0.000	—	—	—
	—	—	32.36	0.00	—	—	22.59	—	—	—	0.001	—	—	—
63	—	—	—	—	—	—	—	—	—	—	—	—	—	—
	—	—	—	0.00	—	—	0.14	—	—	—	0.002	—	—	560,000
64	—	—	—	—	—	—	—	—	—	—	—	—	—	—
	—	—	—	52.03	—	—	0.07	—	—	—	0.001	—	—	480,000
65	—	—	—	—	—	—	—	—	—	—	—	—	—	—
	—	—	—	—	0.00	—	0.03	—	—	—	0.001	—	—	730,000
66	—	—	—	—	0.00	—	11.15	—	—	—	0.001	—	—	227,000
	—	—	—	—	0.00	—	19.84	—	—	—	0.001	—	—	404,000

MINERAL SUPPLEMENTS

Fig. 27-6. A flock of sheep grazing in Washington state. (Courtesy, Bill Reuter, Kent, WA)

(Photo by David Cornwell; courtesy, American Sheep Industry Assn., Englewood, CO)

APPENDIX

Contents *Page*

This Appendix is essential to the completeness of *Sheep & Goat Science.* It provides useful supplemental information relative to (1) animal units, (2) weights and measures, (3) some uses of weights and measures, (4) breed registries and associations, (5) sheep and goat magazines, (6) colleges and universities, and (7) poisoning.

ANIMAL UNITS

An animal unit is a common animal denominator, based on feed consumption. It is assumed that one mature cow represents an animal unit. Then, the comparative (to a mature cow) feed consumption of other age groups or classes of animals determines the proportion of an animal unit which they represent. For example, it is generally estimated that the ration of one mature cow will feed seven ewes. Table A-1 gives the animal units for different classes and ages of livestock.

TABLE A-1
ANIMAL UNITS

Type of Livestock	Animal Units
Cattle:	
Cow, with or without unweaned calf at side, or heifer 2 yrs. old or older	1.0
Bull, 2 yrs. old or older	1.3
Young cattle, 1 to 2 yrs. old	0.8
Weaned calves to yearlings	0.6
Horses:	
Horse, mature	1.3
Horse, yearling	1.0
Weanling colt or filly	0.75
Sheep:	
7 mature ewes, with or without unweaned lambs at side	1.0
7 rams, 2 yrs. old or older	1.3
7 yearlings	0.8
7 weaned lambs to yearlings	0.6
Swine:	
Sow	0.4
Boar	0.5
Pigs to 200 lb *(91 kg)*	0.2
Chickens:	
75 layers or breeders	1.0
325 replacement pullets to 6 mo. of age	1.0
650 8-week-old broilers	1.0
Turkeys:	
35 breeders	1.0
40 turkeys raised to maturity	1.0
75 turkeys to 6 mo. of age	1.0

WEIGHTS AND MEASURES

Weights and measures are standards employed in arriving at weights, quantities, and volumes. Even among primitive people, such standards are necessary; and with the growing complexity of life, they become of greater and greater importance.

Weights and measures form one of the most important parts of modern agriculture. This section contains pertinent information relative to the most common standards used by U.S. sheep and goat producers.

METRIC SYSTEM[1]

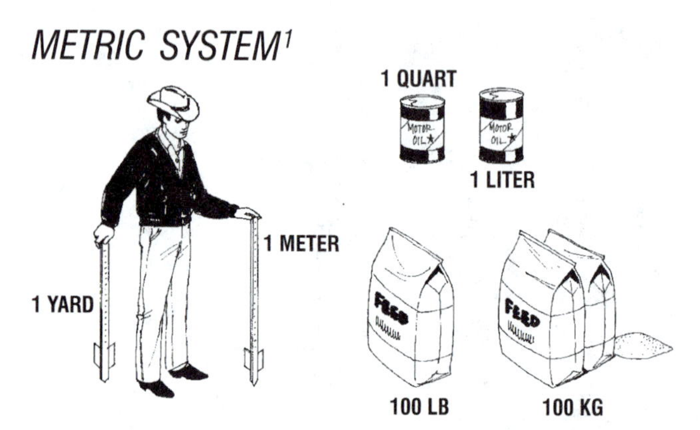

Fig. A-1. Metric vs U.S. customary—length, weight, and volume comparisons.

The United States and a few other countries use standards that belong to the *customary,* or English, system of measurement. This system evolved in England from older measurement standards, beginning about the year 1200. All other countries—including England—now use a system of measurements called the *metric system,* which was created in France in the 1790s. Increasingly, the metric system is being used in the United States. Hence, everyone should have a working knowledge of it.

The basic metric units are the *meter* (length/ distance), the *gram* (weight), and the *liter* (capacity). The units are then expanded in multiples of 10 or made smaller by 1/10. The prefixes, which are used in the same way with all basic metric units, follow:

"milli-"	=	1/1000
"centi-"	=	1/100
"deci-"	=	1/10
"deca-"	=	10
"hecto-"	=	100
"kilo-"	=	1,000

[1]For additional conversion factors, or for greater accuracy, see *Misc. Pub. 223,* the National Bureau of Standards.

The following tables will facilitate conversion from metric units to U.S. customary, and vice versa:

- Table A-2, Weight-Unit Conversion Factors
- Table A-3, Weight Equivalents
- Table A-4, Weights and Measures per Unit
- Table A-5, Weights and Measures
  - Length
  - Surface or Area
  - Volume
  - Weight

TABLE A-2
WEIGHT-UNIT CONVERSION FACTORS

Units Given	Units Wanted	For Conversion Multiply By
lb	*g*	453.6
lb	*kg*	0.4536
oz	*g*	28.35
kg	lb	2.2046
kg	*mg*	1,000,000
kg	*g*	1,000
g	*mg*	1,000
g	μg	1,000,000
mg	μg	1,000
mg/g	*mg*/lb	453.6
mg/kg	*mg*/lb	0.4536
μ*g/kg*	μ*g*/lb	0.4536
Mcal	*kcal*	1,000
kcal/kg	*kcal*/lb	0.4536
kcal/lb	*kcal/kg*	2.2046
ppm	μ*g/g*	1
ppm	*mg/kg*	1
ppm	*mg*/lb	0.4536
mg/kg	%	0.0001
ppm	%	0.0001
mg/g	%	0.1
g/kg	%	0.1

TABLE A-3
WEIGHT EQUIVALENTS

1 lb	=	*453.6 g*	=	*0.4536 kg*	=	16 oz
1 oz	=	*28.35 g*				
1 kg	=	*1,000 g*	=	2.2046 lb		
1 g	=	*1,000 mg*				
1 mg	=	*1,000* μ*g*	=	*0.001 g*		
1 μ*g*	=	*0.001 mg*	=	*0.000001 g*		

1 μ*g* per *g* or *1 mg* per *kg* is the same as ppm

TABLE A-4
WEIGHTS AND MEASURES PER UNIT

Unit	Is Equal To
Volume per Unit Area	
1 liter/hectare	0.107 gal/acre
1 gallon/acre	9.354 liter/ha
Weight per Unit Area	
1 kilogram/cm²	14.22 lb/in.²
1 kilogram/hectare	0.892 lb/acre
1 pound/square inch	0.0703 kg/cm²
1 pound/acre	1.121 kg/ha
Area per Unit Weight	
1 square centimeter/kilogram	0.0703 in.²/lb
1 square inch/pound	14.22 cm²/kg

TEMPERATURE

One centigrade (C) degree is 1/100 the difference between the temperature of melting ice and that of water boiling at standard atmospheric pressure. *One centigrade degree equals 1.8°F.*

One Fahrenheit (F) degree is 1/180 the difference between the temperature of melting ice and that of water boiling at standard atmospheric pressure. *One Fahrenheit degree equals 0.556°C.*

To Change	To	Do This
Degrees centigrade . .	Degrees Fahrenheit . . .	Multiply by 9/5 and add 32
Degrees Fahrenheit . .	Degrees centigrade . . .	Subtract 32, then multiply by 5/9

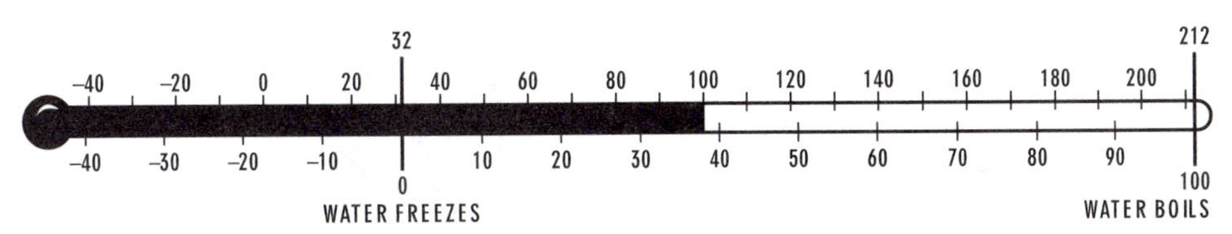

Fig. A-2. Fahrenheit-centigrade (Celsius) scale for direct conversion and reading.

TABLE A-5
WEIGHTS AND MEASURES (METRIC AND U.S. CUSTOMARY)

LENGTH

Unit	Is Equal To	
Metric System		**(U.S. Customary)**
1 millimicron (mμ)	0.000000001 m	0.000000039 in.
1 micron (μ)	0.000001 m	0.000039 in.
1 millimeter (mm)	0.001 m	0.0394 in.
1 centimeter (cm)	0.01 m	0.3937 in.
1 decimeter (dm)	0.1 m	3.937 in.
1 meter (m)	1 m	39.37 in.; 3.281 ft; 1.094 yd
1 hectometer (hm)	100 m	328 ft, 1 in.; 19.8338 rd
1 kilometer (km)	1,000 m	3,280 ft, 10 in.; 0.621 mi
U.S. Customary		**(Metric)**
1 inch (in.)		25 mm; 2.54 cm
1 hand*	4 in.	
1 foot (ft)	12 in.	30.48 cm; 0.305 m
1 yard (yd)	3 ft	0.914 m
1 fathom** (fath)	6.08 ft	1.829 m
1 rod (rd), pole, or perch	16.5 ft; 5.5 yd	5.029 m
1 furlong (fur.)	220 yd; 40 rd	201.168 m
1 mile (mi)	5,280 ft; 1,760 yd; 320 rd; 8 fur.	1,609.35 m; 1.609 km
1 knot or nautical mile	6,080 ft; 1.15 land mi	
1 league (land)	3 mi (land)	
1 league (nautical)	3 mi (nautical)	

*Used in measuring height of horses.

**Used in measuring depth at sea.

CONVERSIONS

To Change	To	Multiply By
inches	centimeters	2.54
feet	meters	0.305
meters	inches	39.73
miles	kilometers	1.609
kilometers	miles	0.621

(To make opposite conversion, divide by the number given instead of multiplying)

(Continued)

TABLE A-5 (Continued)

SURFACE OR AREA

Unit	Is Equal To	
Metric System		**(U.S. Customary)**
1 square millimeter (mm^2)	0.000001 m^2	0.00155 $in.^2$
1 square centimeter (cm^2)	0.0001 m^2	0.155 $in.^2$
1 square decimeter (dm^2)	0.01 m^2	15.50 $in.^2$
1 square meter (m^2)	1 centare (ca)	1,550 $in.^2$; 10.76 ft^2; 1.196 yd^2
1 are (a)	100 m^2	119.6 yd^2
1 hectare (ha)	10,000 m^2	2.47 acres
1 square kilometer (km^2)	1,000,000 m^2	247.1 acres; 0.386 mi^2
U.S. Customary		**(Metric)**
1 square inch ($in.^2$)	1 in. × 1 in.	6.452 cm^2
1 square foot (ft^2)	144 $in.^2$	0.093 m^2
1 square yard (yd^2)	1,296 $in.^2$; 9 ft^2	0.836 m^2
1 square rod (rd^2)	272.25 ft^2; 30.25 yd^2	25.29 m^2
1 rood	40 rd^2	10.117 a
1 acre	43,560 ft^2; 4,840 yd^2; 160 rd^2; 4 roods	4,046.87 m^2; 0.405 ha
1 square mile (mi^2)	640 acres	2.59 km^2; 259 ha
1 township	36 sections; 6 mi^2	

CONVERSIONS

To Change	To	Multiply By
square inches	square centimeters	6.452
square centimeters	square inches	0.155
square yards	square meters	0.836
square meters	square yards	1.196

(To make opposite conversion, divide by the number given instead of multiplying.)

VOLUME

Unit	Is Equal To		
Metric System Liquid and Dry		**(U.S. Customary)**	
		(Liquid)	**(Dry)**
1 milliliter (ml)	0.001 liter	0.271 dram (fl)	0.061 $in.^3$
1 centiliter (cl)	0.01 liter	0.338 oz (fl)	0.610 $in.^3$
1 deciliter (dl)	0.1 liter	3.38 oz (fl)	
1 liter (l)	1,000 cc	1.057 qt; 0.2642 gal (fl)	0.908 qt
1 hectoliter (hl)	100 liter	26.418 gal	2.838 bu
1 kiloliter (kl)	1,000 liter	264.18 gal	1,308 yd^3

(Continued)

TABLE A-5 (Continued)

VOLUME (Continued)

Unit	Is Equal To			
U.S. Customary *Liquid*		**(Ounces)**	**(Cubic Inches)**	**(Metric)**
1 teaspoon (t)	60 drops	0.1666		5 ml
1 dessert spoon	2 t			
1 tablespoon (T)	3 t	0.5		15 ml
1 fl oz		1	1.805	29.57 ml
1 gill (gi)	0.5 c	4	7.22	118.29 ml
1 cup (c)	16 T	8	14.44	236.58 ml; 0.24 l
1 pint (pt)	2 c	16	28.88	0.47 l
1 quart (qt)	2 pt	32	57.75	0.95 l
1 gallon (gal)	4 qt	8.34 lb	231	3.79 l
1 barrel (bbl)	31.5 gal			
1 hogshead (hhd)	2 bbl			
Dry		**(Ounces)**	**(Cubic Inches)**	**(Metric)**
1 pint (pt)	0.5 qt		33.6	0.55 l
1 quart (qt)	2 pt		67.20	1.10 l
1 peck (pk)	8 qt		537.61	8.81 l
1 bushel (bu)	4 pk		2,150.42	35.24 l
Solid **Metric System**		**(Metric)**	**(U.S. Customary)**	
1 cubic millimeter (mm^3)		0.001 cc		
1 cubic centimeter (cc)		1,000 mm^3	0.061 $in.^3$	
1 cubic decimeter (dm^3)		1,000 cc	61.023 $in.^3$	
1 cubic meter (m^3)		1,000 dm^3	35.315 ft^3; 1.308 yd^3	
U.S. Customary				**(Metric)**
1 cubic inch ($in.^3$)				16.387 cc
1 board foot (fbm)		144 $in.^3$		2,359.8 cc
1 cubic foot (ft^3)		1,728 $in.^3$		0.028 m^3
1 cubic yard (yd^3)		27 ft^3		0.765 m^3
1 cord		128 ft^3		3.625 m^3

CONVERSIONS

To Change	To	Multiply By
ounces (fluid)	cubic centimeters	29.57
cubic centimeters	ounces (fluid)	0.034
quarts	liters	0.946
liters	quarts	1.057
cubic inches	cubic centimeters	16.387
cubic centimeters	cubic inches	0.061
cubic yards	cubic meters	0.765
cubic meters	cubic yards	1.308

(To make opposite conversion, divide by the number given instead of multiplying.)

(Continued)

TABLE A-5 (Continued)

WEIGHT

Unit	Is Equal To	
Metric System		**(U.S. Customary)**
1 microgram (mcg)	0.001 mg	
1 milligram (mg)	0.001 g	0.015432356 grain
1 centigram (cg)	0.01 g	0.15432356 grain
1 decigram (dg)	0.1 g	1.5432 grains
1 gram (g)	1,000 mg	0.03527396 oz
1 decagram (dkg)	10 g	5.643833 dr
1 hectogram (hg)	100 g	3.527396 oz
1 kilogram (kg)	1,000 g	35.274 oz; 2.2046223 lb
1 ton	1,000 kg	2,204.6 lb; 1.102 tons (short); 0.984 ton (long)
U.S. Customary		**(Metric)**
1 grain	0.037 dr	64.798918 mg; 0.064798918 g
1 dram (dr)	0.063 oz	1.771845 g
1 ounce (oz)	16 dr	28.349527 g
1 pound (lb)	16 oz	453.5924 g; 0.4536 kg
1 hundredweight (cwt)	100 lb	
1 ton (short)	2,000 lb	907.18486 kg; 0.907 (metric) ton
1 ton (long)	2,200 lb	1,016.05 kg; 1.016 (metric) ton
1 part per million (ppm)	1 microgram/gram; 1 mg/liter; 1 mg/kg; 0.0001%; 0.00013 oz/gal	0.4535924 mg/lb; 0.907 g/ton
1 percent (%) (1 part in 100 parts)	10,000 ppm; 10 g/liter; 1.28 oz/gal; 8.34 lb/100 gal	

CONVERSIONS

To Change	To	Multiply By
grains	milligrams	64.799
ounces (dry)	grams	28.35
pounds (dry)	kilograms	0.4535924
kilograms	pounds	2.2046223
milligrams/pound	parts/million	2.2046223
parts/million	grams/ton	0.90718486
grams/ton	parts/million	1.1
milligrams/pound	grams/ton	2
grams/ton	milligrams/pound	0.5
grams/pound	grams/ton	2,000
grams/ton	grams/pound	0.0005
grams/ton	pounds/ton	0.0022
pounds/ton	grams/ton	453.5924
grams/ton	percent	0.00011
percent	grams/ton	9,072
parts/million	percent	move decimal four places to left

(To make opposite conversion, divide by the number given instead of multiplying.)

WEIGHTS AND MEASURES OF COMMON FEEDS

In calculating rations and mixing concentrates, the farmer or rancher usually needs to use weights rather than measures. However, in practical feeding operations it is often more convenient for the farmer or rancher to measure the concentrates. Table A-6 will serve as a guide in feeding by measure.

TABLE A-6
WEIGHTS AND MEASURES OF COMMON FEEDS

Feed	Approximate Weight[1]	
	Lb per Quart	Lb per Bushel
Alfalfa meal	0.6	19
Barley	1.5	48
Beet pulp (dried)	0.6	19
Brewers' grain (dried)	0.6	19
Buckwheat	1.6	50
Buckwheat bran	1.0	29
Corn, husked ear	—	70
Corn, cracked	1.6	50
Corn, shelled	1.8	56
Corn meal	1.6	50
Corn-and-cob meal	1.4	45
Cottonseed meal	1.5	48
Cowpeas	1.9	60
Distillers' grain (dried)	0.6	19
Fish meal	1.0	35
Gluten feed	1.3	42
Linseed meal (old process)	1.1	35
Linseed meal (new process)	0.9	29
Meat scrap	1.3	42
Milo (grain sorghum)	1.7	56
Molasses feed	0.8	26
Oats	1.0	32
Oats, ground	0.7	22
Oat middlings	1.5	48
Peanut meal	1.0	32
Rice bran	0.8	26
Rye	1.7	56
Sorghum (grain)	1.7	56
Soybeans	1.7	60
Tankage	1.6	51
Velvet beans, shelled	1.8	60
Wheat	1.9	60
Wheat bran	0.5	16
Wheat middlings, standard	0.8	26
Wheat screenings	1.0	32

[1]To convert to metric, refer to Table A-5.

ESTIMATING WEIGHT OF SHEEP AND GOATS

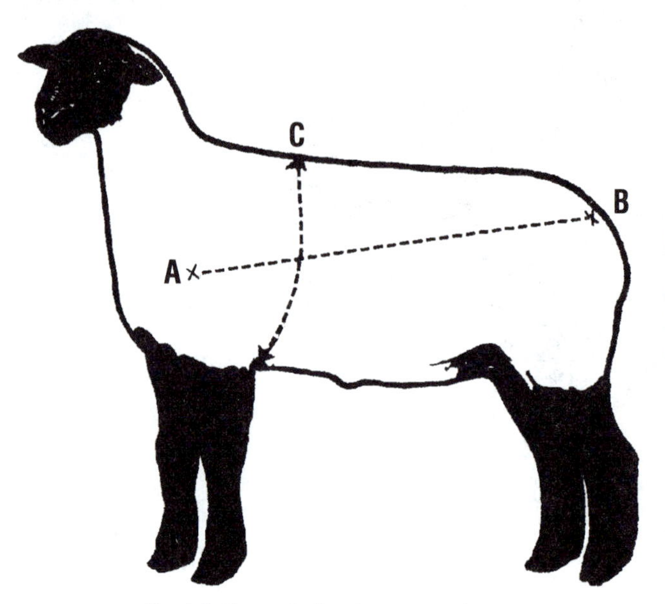

Fig. A-3. How and where to measure shape.

Feeders who finish large numbers of animals have scales in their feedyards for use in determining in-weights, out-weights, and interim weight gains of animals while they are on feed. Likewise, both purebred and commercial breeders usually have scales. However, those with only one animal, or a few head—such as 4-H Club and FFA members, and part-time farmers—may not have scales. As a result, rations cannot be accurately evaluated, rate of gain cannot be calculated, and an animal's "weight readiness" for a livestock show or for market cannot be determined. Under such circumstances, a simple but reasonably accurate method of estimating body weight is very useful. Fortunately, animal weights may be determined with reasonable accuracy by taking two body measurements (length and heart girth), then applying a certain formula.

Here is how to do it:

Step 1—Measure the circumference (heart girth), from a point slightly behind the shoulder blade, thence down over the foreribs and under the body, behind the elbow (distance C in Fig. A-3).

Step 2—Measure the length of body, from the point of the shoulder to the point of the rump (pinbone), in inches (distance A-B in Fig. A-3).

Step 3—Take the values obtained in Steps 1 and 2 and apply the following formula to calculate body weight:

Heart girth × heart girth × body length ÷ 300 = weight in pounds

For example, assume that the heart girth of a ram measures 39 in. and the body length, 33 in. How much does the animal weigh?

39 × 39 = 1,521
1,521 × 33 = 50,193
50,193 ÷ 300 = 167 lb

ESTIMATING WEIGHT OF GRAIN IN A BIN

Sometimes producers need to estimate the weight of grain in storage. Such estimates are difficult to make because of differences in moisture content, depth of material stored, and other factors. However, the following procedure will enable one to figure feed quantities fairly closely.

1. **Corn (shelled) or small grain in rectangular cribs or bins.** Multiply the width by the length by the average depth (all in feet) and multiply by 0.8 to get the number of bushels (multiplying by 0.8 is the same as dividing by 1.25, the number of cubic feet in a bushel).
2. **Ear corn in rectangular cribs or bins.** Multiply the width by the length by the average depth (all in feet) and multiply by 0.4 to get the number of bushels (multiplying by 0.4 is the same as dividing by 2.5, the number of cubic feet in a bushel of ear corn).
3. **Round bins or cribs.** To find the cubic feet in a cylindrical bin, multiply the squared radius by 3.1416 by the depth.

Thus, the volume of a round bin 20 ft in diameter and 10 ft deep is determined as follows:

a. The radius is half the diameter, or 10 ft
b. 10 × 10 = 100
c. 100 × 3.1416 = 314.16
d. 314.16 × 10 = 3,141.6 cubic ft
e. Where shelled corn or small grain is involved, multiply 3,141.6 × 0.8, which equals 2,513.28 bushels of grain that it would hold if full.
f. Where ear corn is involved, multiply 3,141.6 × 0.4 which equals 1,256.64 bushels of ear corn that it would hold if full.

BREED REGISTRY ASSOCIATIONS

A breed registry association consists of a group of breeders banded together for the purposes of. (1) recording the lineage of their animals, (2) protecting the purity of the breed, (3) encouraging further improvement of the breed, and (4) promoting the interest of the breed. The sheep registry associations are given in Table A-7 and the goat ones are listed in Table A-8.

TABLE A-7
SHEEP REGISTRY ASSOCIATIONS

Breed	Association and Address
General	American Sheep Industry Association 6911 S. Yosemite Street Englewood, CO 80112-1414
	Sheep & Goat Raisers' Association P. O. Box 2290 San Angelo, TX 76902
	Texas Sheep And Goat Raisers Association P. O. Box 2678 San Angelo, TX 76902
Border Leicester	American Border Leicester Association 1039 State Route 168 Darlington, PA 16115
California Red	California Red Sheep Registry 1850 E. Reilly Road Merced, CA 95340
Cheviot	American Cheviot Sheep Society Rt. 1, Box 100 Clarks Hill, IN 47930
Clun Forest	North American Clun Forest Association RR 1, Box 4173 Houston, MN 55943
Columbia	Columbia Sheep Breeders Assn. of America Box 272 Upper Sandusky, OH 43351
Cormo	American Cormo Sheep Association RR 59 Broadus, MT 59317
Corriedale	American Corriedale Association, Inc. P. O. Box 391 Clay City, IL 62824
Cotswold	American Cotswold Record Association 18 Elm Street P. O. Box 59 Plympton, MA 02367
	Cotswold Breeders, Inc. 11803 N. Cotswold Lane Coolidge, AZ 85228
Debouillet	American Sheep Industry Association 6911 S. Yosemite, Ste. 200 Englewood, CO 80112
Delaine-Merino	American & Delaine-Merino Record Assn. 1026 Co. Rd 1175, Rt. 3 Ashland, OH 44805
	Black Top & National Delaine-Merino Sheep Association 290 Beach Street Muse, PA 15350
	Texas Delaine Sheep Association Route 1 Burnet, TX 78611
Dorper	American Dorper Sheep Breeders' Society 18202 120th Street Westgate, IA 50681

(Continued)

TABLE A-7 (Continued)

Breed	Association and Address
Dorset	Continental Dorset Club P. O. Box 506 WN Hudson, IA 50643
Finnsheep	National Finnsheep Breeders Association P. O. Box 260 Dousman, WI 53118
Hampshire	American Hampshire Sheep Association 1557 173rd Avenue Milo, IA 50166
Icelandic	Jager Farm Icelandic Sheep 75 Mountain Street Haydenville, MA 01039
Jacob	Jacob Sheep Breeders Association 6350 County Road 56 Fort Collins, CO 80524
Karakul	American Karakul Fur Registry Rt. 1, Box 179 Rice, WA 99167
Katahdin	Katahdin Hair Sheep Association P. O. Box 115 Fairview, KS 66425
Lincoln	National Lincoln Sheep Breeders Assn. 1557 173rd Avenue Milo, IA 50166
Montadale	Montadale Sheep Breeders Assn., Inc. P. O. Box 603 Plainfield, IN 46168
National Colored Wool . .	Natural Colored Wool Growers Association P. O. Box 487 Willits, CA 95490
Navajo-Churro	Navajo-Churro Sheep Association P. O. Box 94 Ojo Caliente, NM 87549
North Country Cheviot . . .	American North Country Cheviot Sheep Association 8708 So. County Road 500 W. Reelsville, IN 46171
Oxford Down	American Oxford Sheep Association 1960 E. 2100 North Rd. Stonington, IL 62567
Panama	American Panama Registry Association HC 71, Box 2020 Cascade, ID 83611
Polypay	American Polypay Sheep Association 609 S. Central, Ste. 9 Sidney, MT 59270
Rambouillet	American Rambouillet Sheep Breeders Assn. 2709 Sherwood Way San Angelo, TX 76901
Romanov	North American Romanov Sheep Assn. P. O. Box 1126 Pataskala, OH 43062

(Continued)

TABLE A-7 (Continued)

Breed	Association and Address
Romney	American Romney Breeders Association 29515 NE Weslinn Drive Corvallis, OR 97333
Shropshire	American Shropshire Registry Association P. O. Box 635 Harvard, IL 60035-0635
Southdown	American Southdown Breeders Association HCR 13, Box 220 Fredonia, TX 76842
Suffolk	American Suffolk Sheep Society P. O. Box 256 Newton, UT 84327
	National Suffolk Sheep Association 3316 Ponderosa Street Columbia, MO 65201
Targhee	U.S. Targhee Sheep Association P. O. Box 462 Jordan, MT 59337
Texel	North American Texel Sheep Association Rt. 1, Box 927 740 Lower Myrick Road Laurel, MS 39440
Tunis	National Tunis Sheep Registry 819 Lyons Street Ludlow, MA 01056

TABLE A-8
GOAT REGISTRY ASSOCIATIONS

Breed	Association and Address
General	American Dairy GoaAssociBox 865 Spindale, NC 28160
	American Goat Society, Inc. RR 1, Box 56 Esperance, NY 12066-9704
	American Livestock Breeds Conservancy P. O. Box 477 Pittsboro, NC 27312
	American Meat Goat Association P. O. Box 333 Junction, TX 76849
	International Dairy Goat Registry, Inc. P. O. Box 309 Chickamauga, GA 30707
	International Goat Association 1015 Louisiana Street Little Rock, AR 72202
	Sheep & Goat Raisers' Association P. O. Box 2290 San Angelo, TX 76902
	Texas Sheep And Goat Raisers Association P. O. Box 2678 San Angelo, TX 76902

(Continued)

TABLE A-8 (Continued)

Breed	Association and Address
Alpines	Alpines International 3746 Rt. 96 Shortsville, NY 14548
Angora	American Angora Goat Breeders Assn. P. O. Box 195 Rocksprings, TX 78880
Boer	American Boer Goat Association 232 W. Beauregard, Ste. 104 San Angelo, TX 76903
	International Boer Goat Association Rt. 3, Box 111 Bonham, TX 75418
	U.S. Boer Goat Association P. O. Box 830 Jamestown, TN 38556
Cashmere	Professional Cashmere Marketers' Assn. 3299 Anderson Lane Dillon, MT 59725
Kiko	American Kiko Goat Association, Inc. P. O. Box 11293 Robinson, TX 76716
Kinder	Kinder Goat Breeders Association P. O. Box 1575 Snohomish, WA 98291-1575
La Mancha	American La Mancha Club 1924 Mt. Pleasant Rd Port Angeles, WA 98362
Nubian	International Nubian Breeders Association RR 2, Box 216 Inman, KS 97546-8931
Oberhasli	Oberhasli Breeders of America 11620 Sunset Ct. Montague, CA 96064
Pack Goat	American Pack Goat Registry 1918 N. 3400 W Malad, ID 83252
Pygmy	National Pygmy Goat Association 11047 E. Miller Rd. Gaines, MI 48436
Pygora	Pygora Breeders Association P. O. Box 51 Clackamas, OR 97015
Saanen	National Saanen Breeders Association 8555 Sypes Canyon Rd. Bozeman, MT 59715
Sable	Sable Breeders Association Rt. 2, Box 104-B Potlach, ID 83855
Toggenburg	National Toggenburg Club P. O. Box 373 Enumclaw, WA 98022

SHEEP AND GOAT MAGAZINES

Many magazines publish news items and informative articles of special interest to sheep and goat producers. However, in the compilation of the list herewith presented, the authors did not attempt to list the numerous general livestock magazines. Only those magazines which are devoted exclusively to sheep are included in Table A-9, and those devoted to goats are found in Table A-10.

TABLE A-9
SHEEP MAGAZINES

Publication	Address
American Southdown, The	1125 Danielson Pike North Scituate, RI 02857
The Banner Sheep Magazine	P. O. Box 500 Cuba, IL 61427
The Corriedale Extra (semi-annual)	P. O. Box 391 Clay City, IL 62824
Hampshire Heartbeat	1557 173rd Ave. Milo, IA 50166
The Marker	SR, Box 48 Brooks, CA 95606
Montadale Mover	P. O. Box 603 Plainfield, IN 46168
Montana Woolgrower	Box 1693 Helena, MT 59601
National Lamb & Wool Grower, The	6911 S. Yosemite Street Englewood, CO 80112-1414
Ranch & Rural Living	P. O. Box 2678 301 W. First St. San Angelo, TX 76902-2678
Sheep Breeder And Sheepman	P. O. Box 796 Columbia, MO 65205
Sheep	W. 2997 Market Road Helenville, WI 53137
Sheep Country (annual)	c/o American Sheep Industry Assn. 6911 S. Yosemite St. Englewood, CO 80112-1414
The Shepherd	5696 Johnston Road New Washington, OH 44854
The Sheep Producer	Rt. 2, Box 131-A Arlington, KY 42021
Shropshire Voice	P. O. Box 635 Harvard, IL 60033-0635
Speaking of Columbias	Box 272 Upper Sandusky, OH 43351
Suffolk News	3316 Ponderosa Street Columbia, MO 65201
The Wool Sack	P. O. Box 328 Brookings, SD 57006

TABLE A-10
GOAT MAGAZINES

Publication	Address
Countryside & Small Stock Journal	N 2601 Winter Sports Rd. Withee, WI 54498-9317
Dairy Goat Journal	128 E. Lake St. P. O. Box 10 Lake Mills, WI 53551-1605
Goat Farmer, The	Fibre News Ltd. P. O. Box 641 Whangare, New Zealand
Meat Goat News	P. O. Box 2678 San Angelo, TX 76902
Pygmy Goat World, The	4728 W. Taylor Rd. Cheney, WA 99004
Ranch & Rural Living	P. O. Box 2678 301 W. First St. San Angelo, TX 76902-2678
United Caprine News	P. O. Drawer 365 Granbury, TX 76048-0365

U.S. STATE COLLEGES OF AGRICULTURE AND CANADIAN PROVINCIAL UNIVERSITIES

U.S. livestock producers can obtain a list of available bulletins and circulars, and other information regarding livestock, by writing to (1) their state agricultural college (land-grant institution), or (2) the U.S. Superintendent of Documents, Washington, DC; or by going to the local County Extension Office (farm advisor) of the county in which they reside. Canadian producers may write to the department of agriculture of their province or to their provincial university. A list of U.S. Land-Grant Institutions and Canadian Provincial Universities follows in Table A-11.

TABLE A-11
U.S. LAND-GRANT INSTITUTIONS AND CANADIAN PROVINCIAL UNIVERSITIES

State	Address
Alabama	School of Agriculture, Auburn University, Auburn, AL 36830
Alaska	Department of Agriculture, University of Alaska, Fairbanks, AK 99701
Arizona	College of Agriculture, The University of Arizona, Tucson, AZ 85721
Arkansas	Division of Agricutlure, University of Arkansas, Fayetteville, AR 72701
California	College of Agriculture and Environmental Sciences, University of California, Davis, CA 95616
Colorado	College of Agricultural Sciences, Colorado State University, Fort Collins, CO 80521
Connecticut	College of Agriculture and Natural Resources, University of Connecticut, Storrs, CT 06268
Delaware	College of Agricultural Sciences, University of Delaware, Newark, DE 19711
Florida	College of Agriculture, University of Florida, Gainesville, FL 32611
Georgia	College of Agriculture, University of Georgia, Athens, GA 30602
Hawaii	College of Tropical Agriculture, University of Hawaii, Honolulu, HI 96822
Idaho	College of Agriculture, University of Idaho, Moscow, ID 83843
Illinois	College of Agriculture, University of Illinois, Urbana–Champaign, IL 61801
Indiana	School of Agriculture, Purdue University, West Lafayette, IN 47907
Iowa	College of Agriculture, Iowa State University, Ames, IA 50010
Kansas	College of Agriculture, Kansas State University, Manhattan, KS 66506
Kentucky	College of Agriculture, University of Kentucky, Lexington, KY 40506
Louisiana	College of Agriculture, Louisiana State University and A&M College, University Station, Baton Rouge, LA 70803
Maine	College of Life Sciences and Agriculture, University of Maine, Orono, ME 04473
Maryland	College of Agriculture, University of Maryland, College Park, MD 20742
Massachusetts	College of Food and Natural Resources, University of Massachusetts, Amherst, MA 01002
Michigan	College of Agriculture and Natural Resources, Michigan State University, East Lansing, MI 48823

(Continued)

TABLE A-11 (Continued)

State	Address
Minnesota	College of Agriculture, University of Minnesota, St. Paul, MN 55101
Mississippi	College of Agriculture, Mississippi State University, Mississippi State, MS 39762
Missouri.	College of Agriculture, University of Missouri, Columbia, MO 65201
Montana	College of Agriculture, Montana State University, Bozeman, MT 59715
Nebraska	College of Agriculture, University of Nebraska, Lincoln, NE 68503
Nevada	The Max C. Fleischmann College of Agriculture, University of Nevada, Reno, NV 89507
New Hampshire	College of Life Sciences and Agriculture, University of New Hampshire, Durham, NH 03824
New Jersey	College of Agriculture and Environmental Science, Rutgers University, New Brunswick, NJ 08903
New Mexico	College of Agriculture and Home Economics, New Mexico State University, Las Cruces, NM 88003
New York	New York State College of Agriculture, Cornell University, Ithaca, NY 14850
North Carolina	School of Agriculture, North Carolina State University, Raleigh, NC 27607
North Dakota.	College of Agriculture, North Dakota State University, State University Station, Fargo, ND 58102
Ohio	College of Agriculture and Home Economics, The Ohio State University, Columbus, OH 43210
Oklahoma	College of Agriculture and Applied Science, Oklahoma State University, Stillwater, OK 74074
Oregon	School of Agriculture, Oregon State University, Corvallis, OR 97331
Pennsylvania	College of Agriculture, The Pennsylvania State University, University Park, PA 16802
Puerto Rico	College of Agricultural Sciences, University of Puerto Rico, Mayagüez, PR 00708
Rhode Island.	College of Resource Development, University of Rhode Island, Kingston, RI 02881
South Carolina	College of Agricultural Sciences, Clemson University, Clemson, SC 29631
South Dakota.	College of Agriculture and Biological Sciences, South Dakota State University, Brookings, SD 57006
Tennessee	College of Agriculture, University of Tennessee, P.O. Box 1071, Knoxville, TN 37901
Texas	College of Agriculture, Texas A&M University, College Station, TX 77843
Utah	College of Agriculture, Utah State University, Logan, UT 84321
Vermont.	College of Agriculture, University of Vermont, Burlington, VT 05401
Virginia	College of Agriculture, Viriginia Polytechnic Institute and State University, Blacksburg, VA 24061
Washington	College of Agriculture, Washington State University, Pullman, WA 99163
West Virginia.	College of Agriculture and Forestry, West Virginia University, Morgantown, WV 26506
Wisconsin	College of Agricultural and Life Sciences, University of Wisconsin, Madison, WI 53706
Wyoming	College of Agriculture, University of Wyoming, University Station, P.O. Box 3354, Laramie, WY 82070
Canada	**Address**
Alberta	University of Alberta, Edmonton, Alberta T6H 3K6
British Columbia	University of British Columbia, Vancouver, British Columbia V6T 1W5
Manitoba	University of Manitoba, Winnipeg, Manitoba R3T 2N2
New Brunswick.	University of New Brunswick, Federicton, New Brunswick E3B 4N7
Ontario	University of Guelph, Guelph, Ontario N1G 2W1
Québéc	Faculty d'Agriculture, L'Université Laval, Québéc City, Québéc G1K 7D4; and Macdonald College of McGill University, Ste. Anne de Bellevue, Québéc H9X 1C0
Saskatchewan	University of Saskatchewan, Saskatoon, Saskatchewan S7N 0W0

POISON INFORMATION CENTERS

With the large number of chemical sprays, dusts, and gases now on the market for use in agriculture, accidents may arise because operators are careless in their use. Also, there is always the hazard that a child may eat or drink something that may be harmful. Centers have been established in various parts of the country where doctors can obtain prompt and up-to-date information on treatment of such cases, if desired.

Local physicians have information relative to the Poison Information Centers in their area, along with some of the names of their directors, their telephone numbers, and their street addresses. When calling any of these centers, one should ask for the "Poison Information Center." If this information cannot be obtained locally, the U.S. Public Health Service at Atlanta, Georgia, or Wenatchee, Washington, should be contacted.

Also, the *National Poison Control Center* is located at the University of Illinois, Urbana-Champaign. It is open 24 hours a day, every day of the week. The *hot line* number is: 1-800-548-2423. The toxicology group is staffed to answer questions about known or suspected cases of poisoning or chemical contaminations involving any species of animal. It is not intended to replace local veterinarians or state toxicology laboratories, but to complement them. Where consultation over the telephone is adequate, there is no charge to the veterinarian or producer. Where telephone consultation is inadequate or the problem is of major proportion, a team of veterinary specialists can arrive at the scene of a toxic or contamination problem within a short time. The cost of a personal visitation varies according to the distance traveled, personnel time, and laboratory services required.

Index

A

B

C

D

E

F

G

H

I

J

K

L

Q

R